The Scientific Foundation of Space Weather

Daniel Baker • André Balogh • Tamas Gombosi •
Hannu E. J. Koskinen • Astrid Veronig •
Rudolf von Steiger
Editors

The Scientific Foundation of Space Weather

Previously published in *Space Science Reviews* in the Topical Collection "The Scientific Foundation of Space Weather"

Springer

Editors
Daniel Baker
Laboratory for Atmospheric & Space Physics
University of Colorado at Boulder
Boulder, CO, USA

Hannu E. J. Koskinen
Department of Physics
University of Helsinki
Helsinki, Finland

André Balogh
Imperial College London
Hertfordshire, UK

Astrid Veronig
Institute of Physics
University of Graz
Graz, Steiermark, Austria

Tamas Gombosi
College of Engineering
University of Michigan
Ann Arbor, MI, USA

Rudolf von Steiger
International Space Science Institute
Bern, Switzerland

ISSN 1385-7525 Space Sciences Series of ISSI
ISBN 978-94-024-1587-2

Library of Congress Control Number: 2018960247

Cover Image: Two images representing the origin of space weather at the Sun and one of its spectacular consequences at the Earth. Left: hot coronal loops on the Sun observed with TRACE (source: NASA/GSFC). Right: Aurora Borealis observed in Kiruna (source: www.kirunaaurora.com)

Printed on acid-free paper

This Springer imprint is published by the registered company Springer Nature B.V.
The registered company address is: Van Godewijckstraat 30, 3311 GX Dordrecht, The Netherlands

Contents

Space Sci Rev (2017) 212:1137–1157
DOI 10.1007/s11214-017-0390-4

Achievements and Challenges in the Science of Space Weather

Hannu E.J. Koskinen[1,2] **· Daniel N. Baker**[3] **·**
André Balogh[4] **· Tamas Gombosi**[5] **· Astrid Veronig**[6] **·**
Rudolf von Steiger[7]

Received: 29 March 2017 / Accepted: 7 June 2017 / Published online: 1 August 2017

Abstract In June 2016 a group of 40 space weather scientists attended the workshop on Scientific Foundations of Space Weather at the International Space Science Institute in Bern. In this lead article to the volume based on the talks and discussions during the workshop we review some of main past achievements in the field and outline some of the challenges that the science of space weather is facing today and in the future.

Keywords Space weather · Coronal mass ejections · Solar flares · Solar wind · Shocks · Solar energetic particles · Magnetosphere · Magnetospheric storms · Substorms · Ionosphere

1 Introduction

The concept of space weather started to penetrate through the academic solar-terrestrial physics (STP) community in the early 1990s. Science behind, and technological consequences of, space weather have, of course, much longer history (see Lanzerotti's history review, this issue). We do not need to go as far as Galileo's observations of sunspots or the

The Scientific Foundation of Space Weather
Edited by Rudolf von Steiger, Daniel Baker, André Balogh, Tamás Gombosi, Hannu Koskinen and Astrid Veronig

✉ H.E.J. Koskinen
Hannu.E.Koskinen@helsinki.fi

1 Department of Physics, University of Helsinki, Helsinki, Finland

2 Finnish Meteorological Institute, Helsinki, Finland

3 University of Colorado Boulder, Boulder, USA

4 Imperial College London, London, UK

5 University of Michigan, Ann Arbor, USA

6 University of Vienna, Vienna, Austria

7 International Space Science Institute, Bern, Switzerland

investigations of Anders Celcius and Olof Hjorter showing that the magnetic needle reacts on bright auroral displays, both of which still today are among the most readily accessible space weather phenomena. One of the most important early events was the large flare on the Sun on September 1, 1859, observed by Carrington and Hodgson. Only 17 hours later an exceptionally strong magnetic storm, perhaps the strongest ever recorded using scientific instruments, commenced on the Earth. However, it took a long time to understand what really had happened and how the effects of solar activity could be carried to the Earth.

Today we have good reasons to assume that the agent must have been a large and very fast coronal mass ejection (CME) released in association with the flare, but CMEs were not identified until 1972. Finding these giant magnetic and plasma ejecta was a major step forward in understanding Sun–Earth connections. CMEs not only turned out to be major drivers of strong magnetospheric storms but also to play a critical role in the acceleration of solar energetic particles to tens of MeV energies.

We have today a highly sophisticated understanding of many aspects of solar and solar-terrestrial physics, and the study of space weather phenomena form a subset of the field. Thus why are we concerned about the scientific foundations of space weather when we already have STP? We can try to answer the question from the viewpoint of the practical rationale behind space weather. Space weather emphasizes applications and societal relevance much more explicitly than STP, which can be seen as "pure" science. We can take magnetic reconnection as an example. In basic space plasma physics the question, what breaks the frozen-in condition for electrons, is highly relevant to ask whereas in space weather it is much more important to find out when and where the reconnection really takes place, and also to be able to forecast the event and its consequences in advance.

Space weather is science with applications. In the absence of direct applications, nobody would have ever introduced the concept of space weather. In space weather the applications pose the most important and hardest scientific questions. Particularly, the progress needed to improve both short- and long-term forecasting challenges our understanding of the scientific foundations. Sometimes the path from basic research to applications may be short but more often it is long. In this field the easy questions have already been answered and the difficult ones will be with us for a long time to come.

This is the lead-article to papers based on the one-week workshop on Scientific Foundations of Space Weather at the International Space Science Institute in Bern, Switzerland, in June 2016. During the workshop the participants looked for answers to questions:

- What kind of progress have we made in the understanding of the scientific foundations of space weather during the past few decades?
- What are the holes in our understanding that can be filled in the near future?
- What is needed for long-term progress?

This article aims at setting the tone for the rest of the volume. We do not want to pre-empt the content of the other articles. Instead we wish to highlight some of the key achievements in the past and challenges for the future.

2 Space Weather Phenomena from the Sun to the Earth

In science we often tend to be reductionists, studying our pet corner of the universe, whether it is a certain decay channel of the Higgs boson or the very first microseconds after the Big Bang. The need to understand space weather motivates the STP community to form a more integrated view of the entire chain of processes from the solar activity to near-Earth space and the surface of our planet. Progress on this more basic level, beyond the phenomenological descriptions of space weather, has only begun. That is why workshops like this are needed.

2.1 Space Weather Chains Connecting the Sun to the Earth

There are three main routes along which the Sun affects the Earth's space environment: the electromagnetic connection, the plasma chain mediated by the solar wind, and the energetic particle chain consisting of high-energy charged particles accelerated in the solar corona and along their way to the Earth.

The electromagnetic solar radiation does much more than transferring information of solar activity and eruptions. The solar EUV radiation ionizes the upper layers of the atmosphere forming the partially ionized ionosphere. Strong X-ray flares can also cause observable ionization events in the atmosphere. In fact, a 110-nT perturbation in the Greenwich magnetogram was observed at the time of the Carrington flare (e.g., Cliver and Svalgaard 2004), much before anybody knew of X-rays and how these can lead to generation of electric currents in the space environment that give rise to disturbances in the terrestrial magnetic field. In rare cases strong radio bursts associated with solar flares can also cause significant problems by directly introducing noise in the wireless communication systems (Lanzerotti et al. 1999). The most extreme case thus far observed at GPS frequencies on 6 December, 2006, will be discussed further in Sect. 2.9.

The understanding of the decisive role of the plasma chain developed gradually with the progress of solar-terrestrial physics. Already before the Carrington flare, Schwabe (1844) had found the quasi-periodic approximately 11-year cycle in the number of sunspots. Hale (1908) was the first to observe strong magnetic fields in the sunspots. By that time magnetic variations both on the Sun and on the Earth were connected to phenomena that we today understand as elements of space weather. But how sufficient amounts of energy could be transported from the Sun to the Earth to generate the global-scale disturbances, remained a mystery.

During the early 20th century some scientists (e.g., Chapman 1918; Lindeman 1919; Chapman and Ferraro 1931a,b) proposed that ejections from the Sun of quasi-neutral gas streams, or plasma, could affect the geomagnetic environment when hitting the Earth's magnetic field. They were on the right track but they did not yet understand how the particles could escape from the gravitational pull of the Sun. The escape velocity from the solar surface is 618 $\mathrm{km\,s^{-1}}$, which is much larger than the thermal velocity of the gas on the solar surface at the black-body temperature of 5778 K. With improving spectroscopy in the 1940s the solar corona was found to be much hotter, up to a couple of million kelvin, the corresponding thermal speed of a proton thus being sufficient for escape from the Sun. How the corona is heated remains, despite much observational, theoretical and modelling progress, a largely unsolved problem. The reason why this question is still unanswered lies in the complexity of processes in the solar atmosphere that prevents the identification of unique and simple heating mechanisms (for a recent review, see Klimchuk 2015).

In the 1950s Biermann (1951, 1957) demonstrated the existence of a continuous interplanetary corpuscular flow from the Sun by a careful study of cometary tails. In addition to the slowly escaping dust tail lagging behind the comet, he recognized a much faster, almost anti-sunward, outflow from the comet, which must be swept by a fast flow originating from the Sun. Alfvén (1957) pointed out that this flow had to be in the form of magnetized plasma that was picking up charged particles from the cometary environment. In 1958 Parker published the first physics-based model for plasma escape from the Sun (Parker

1958). Soon thereafter the first spacecraft moving outside of the Earth's magnetosphere directly proved the existence of continuously flowing magnetized solar wind (Neugebauer and Snyder 1962). The existence and basic properties of the plasma chain connecting the Sun and the Earth were firmly established.

However, it still remained unclear how the solar wind can shake the magnetosphere as strongly as the most severe magnetic storms indicated. There was also the enigma of solar energetic particles (SEPs). There was a correlation between solar flares and SEPs but the correlation was far from perfect and the strong SEP fluxes often lasted long after the flare activity on the Sun had subsided. That the magnetosphere was occasionally hit by interplanetary shock waves was known and these were understood to be potential particle accelerators, but the main drivers of the shocks were unknown. Another early observation was that the flux of galactic cosmic rays transiently decreased during strong solar magnetic activity (Forbush 1938); these became later known as Forbush decreases. Somehow the enhanced solar magnetism seemed to provide a strengthened magnetic shield against galactic cosmic ray particles.

The solution to these questions came with the first observations of coronal mass ejections in 1972 (Tousey 1973). In addition to the slow and fast solar wind, the Sun was found, from time to time, to be able to expel huge plasma and magnetic clouds to the interplanetary medium. A large fraction of these propagate with super-Alfvénic speed with respect to the background solar wind, producing shock waves that are efficient particle accelerators. A fast and large CME, in particular when its magnetic field has a strong southward component (B_S), is the most efficient driver of magnetospheric activity and such a structure lies behind all strongest magnetospheric storms.

2.2 Solar Drivers of Space Weather

The ultimate driver of space weather phenomena is the variable plasma flow in the convective zone of the Sun that shapes the structure and dynamics of the solar atmosphere. The variations in the plasma flow take place on many spatial and temporal scales leading to magnetically active regions and sunspots on the solar surface. On the largest, coronal, scales the solar magnetic field forms regions of open and closed magnetic flux. The fast solar wind is understood to escape from the open-flux coronal holes but the origin of slow solar wind is much more unclear.

On smaller scales the magnetic flux emerges as magnetic arcades above the solar surface in the form of flux ropes with both ends tied to the surface and below. The line-tying to the solar surface is strong and when the footpoints of a flux rope move, the rope becomes twisted. If the twisting continues long enough, a macroscopic plasma instability sets in and the flux rope may erupt and escape from the Sun as a CME. This is by no means the only scenario to explain the birth of a CME, but fundamentally it is an event where a flux rope is destabilized through some plasma physical process. It is worth noting that the eruptions take place in the corona without any significant changes on the surface.

The CMEs are often, but not always, associated with solar flares. On the other hand, there is much flaring activity on a very wide range of energy scales without any signs of CMEs. Large enough flares are observable at all electromagnetic wavelengths from radio waves to hard X-rays, even to γ-rays, and they accelerate electrons to relativistic energies and protons up to the MeV–GeV range. Until the discovery of CMEs solar flares were thought to be the main drivers of magnetospheric storms. This "solar flare myth" actually survived in parts of the magnetospheric research community until the early 1990s, when a landmark paper by Gosling (1993) shifted the emphasis from flares to CMEs. It is, however, important to recognize that solar eruptive phenomena are dynamically very complex and interrelated. Thus it would be misleading to discard the role of flares in solar phenomena affecting the Earth's environment (Hudson et al. 1995). The origins of solar eruptions are discussed further by Green et al. (this issue).

While separate phases of the eruptions can be studied with sophisticated numerical models, here lies a serious modelling challenge. The plasma parameters from the subsurface convection zone to the hot and tenuous corona vary so largely that different approaches must be applied to describe the different regimes. Even with the present pace of computing power evolution and development of ingenious numerical schemes this state of matters will remain work-in-progress for the foreseeable future. The causal chains of physical processes that shape, from the photosphere to the outer corona, the structures and their dynamics are at the heart of the predictability of solar phenomena vital for quantitative space weather forecasting. Without significant progress in understanding these causal chains, advancing the science of space weather will remain a challenge.

Some immediate needs for modelling improvements were identified at this workshop. More parametric studies with simple and idealistic models are required to understand the basic physics of eruptions, for example, torus instability and flare reconnection. As the heat flux is an essential element in the solar environment, more efforts to develop thermodynamic MHD approach are called for.

2.3 Propagation of Solar Perturbations in the Inner Heliosphere

The solar wind is traditionally divided into the slow (around 300 to 550 $\mathrm{km\,s^{-1}}$) and the fast (around 650 to 900 $\mathrm{km\,s^{-1}}$) wind. At 1 AU the electron number density of the slow wind is typically about 10 electrons per cm^3. The positive charges are almost all protons ($\sim$95%) with small fractions of helium nuclei ($\sim$4%) and heavier elements ($<$1%). Electron temperature of the slow wind is typically 1000 K and proton temperature 3×10^4 K, whereas the fast wind is more tenuous, some 3 electrons per cm^3, and much hotter, electron temperature being typically 10^5 K, proton temperature 2×10^5 K.

There are large variations in these numbers, and it has recently become questioned whether the solar wind distribution really is bimodal or is there some kind of underlying dynamic continuum of states. This question is related to origins of the different types of solar wind (see Cranmer et al., this issue). It is fairly clear that the fast wind emerges from the coronal holes but there is no consensus about the coronal sources of the slow wind.

The Ulysses mission clearly demonstrated the out-of-ecliptic bimodality of the solar wind. Around solar minimum, the uniformly high-speed solar wind originated in the large solar polar coronal holes, while the slower, much more variable solar wind streams were associated with the more complex, magnetically closed corona (McComas et al. 1998). It was found that the charge state composition of the solar wind was an objective measure of the origin of individual streams. In particular, the ratio of O^{7+} to O^{6+} indicated the freezing-in temperature of the stream (where the plasma becomes collisionless). The temperature distribution was found to be strongly bipolar and is strongly anti-correlated to the solar wind speed (Geiss et al. 1995).

The bimodality is not as evident the in-ecliptic observations, and thus in the near-Earth space. Although bimodality can be observed, it is less clear-cut and more variable. Even if solar wind streams may be well differentiated at their coronal origin, the dynamic effects of the interactions in the non-uniform solar wind modify and mix the kinetic signatures, so that bimodality based on speed becomes obscured. Consequently, the kinetic parameters (speed, density and temperature) at 1 AU are not an unambiguous guide to the origin of the solar wind streams. Simultaneous measurements of ion charge composition would be necessary—unfortunately these are rarely available.

Both the fast wind and the stream interaction regions (SIR), often called corotating interaction regions (CIR), between the fast and slow wind are important drivers of weak and moderate magnetospheric storms. While fast and slow winds escape radially from adjacent regions, the solar rotation forming the Parker spiral leads to structures where either the fast wind pushes against the slower wind causing a fast forward compressional wave, or the slow wind lags behind the fast wind and an in the solar wind frame reverse rarefaction wave emerges. The fast forward wave gradually evolves toward a shock. In most cases the wave does not steepen to a fully developed shock until beyond the Earth's orbit, but it can already be sufficiently strongly compressed to drive a moderate storm in the magnetosphere, in particular if the compression has led to sufficiently enhanced southward component of the magnetic field.

But where do the interplanetary propagating CMEs, often called ICMEs, belong in this scheme of fast and slow wind? At and after the eruption, CMEs are accelerated to very different speeds (e.g., Yurchyshyn et al. 2005). Sometimes the CME is slower than the solar wind in which it is embedded. In such cases the ICME is gradually dragged close to the speed of the ambient wind. The process can be described in terms of anomalous viscosity, i.e., wave–particle interactions. Similar interaction decelerates an ICME that is just slightly faster than the ambient wind. The relative velocity between these ICMEs and the ambient wind is too small to drive a bow shock ahead of the ICME. Nevertheless, the magnetic field may have fairly large southward component when the structure hits the magnetosphere and even a strong storm can result.

Fast CMEs causing the strongest storms can erupt with very high speeds. The acceleration must come from a rapid release of magnetic tension in the erupting structure and can be expressed in terms of the Ampère force ($\mathbf{J} \times \mathbf{B}$). The famous extreme event on July 23, 2012, left the Sun with the speed of 3050 ± 260 km s^{-1} and had the speed of 2246 ± 110 km s^{-1} at 1 AU (Liu et al. 2014). It led to an extreme magnetic compression, the peak magnetic field exceeding 100 nT. This event also illustrates another important feature of the interplanetary propagation of CMEs. Sometimes two or more CMEs are expelled from the same region after each other. It appears that the first CME ploughs the road for the following CME, which can catch the leading CME and the two structures merge, sometimes called CME cannibalism. The combined, complicated structure can lead to very strong events, as was the case in this particular event.

Also ICMEs compressed between the heliospheric plasma sheet, typically a slow wind domain, and fast wind can lead to large magnetic fields and become very geoeffective, as was the case in the March 17, 2015, event (Kataoka et al. 2015). For further review of the large variety of the interactions of ICMEs with their surroundings and each other, including the interactions between ICME shocks, we refer to Manchester et al. (this issue).

For space weather in the magnetosphere the southward component B_S of the interplanetary magnetic field (IMF) is a key parameter. It is important to keep in mind that there are different causes for strong ICME-related B_S (e.g., Kilpua et al., this issue). The flux ropes with strong helical magnetic field form in the corona, whereas the enhanced sheath region B_S accumulates during the propagation. A major difficulty in understanding evolution of the ICMEs during the propagation from the Sun is that there are no detailed observations between the corona and 1 AU. Observing and tracing radio emissions associated with the propagating CME and its leading shock wave has been used to reconstruct some aspects of propagation characteristics (Reiner et al. 2007; Corona-Romero et al. 2015). The heliospheric imagers such as onboard STEREO help but, for example, the essential parameter, the strength of the magnetic field can be measured only very near the Sun and thereafter not until the ICME has propagated to 1 AU, unless there happens to be an interplanetary spacecraft with magnetic field instrument in the right place at the right time.

2.4 Shocks and Solar Energetic Particles

Solar energetic particle (SEP) events are traditionally divided into impulsive and gradual SEP events (e.g. Cane et al. 1986). The impulsive events are relatively short-lived, the energetic particles being observable for a few hours following the onset. These events are electron-rich and the ^{3}He/^{4}He ratio is of the order of one, sometimes even larger than 10. The impulsive events are typically associated with impulsive hard and soft X-ray flares but not necessarily with any CMEs. The gradual events are proton-rich, they are observable for days, and ^{3}He/^{4}He is much smaller than in the impulsive cases. The associated X-ray flares are typically soft. Almost all gradual events, some 96% of them, have been found to be associated with CMEs (Kahler et al. 1984).

Particles in impulsive events are thought to be accelerated close to the Sun by the rapid energy release in the impulsive phase of a solar flare and by the consequent strong wave activity. The high ^{3}He content is a curious fact because only some 0.05% of all helium in the solar atmosphere is ^{3}He. This indicates that the acceleration process must be highly selective to ^{3}He, more precisely ^{3}He^{++} whose gyro frequency is $2/3$ of the proton gyro frequency. Thus interaction with Alfvén ion cyclotron waves at frequencies below the proton gyro frequency is a plausible candidate. The same acceleration mechanism that preferentially accelerates ^{3}He appears to act, in a similar way, to enhance the fraction of partially ionized heavier atoms (Mason et al. 2016).

In the pre-CME times it was much harder to explain the gradual events, as practically the only solar activity they could be correlated with were flares. It is instructive to refer to the early studies as they shed an unbiased (if partly ignorant) light on the properties of the observed energetic particle populations, then termed Energetic Storm Particle (ESP) events (see, e.g. Bryant et al. 1962; Rao et al. 1967). It is true that large gradual events are associated with big flares, but today we know that the protons and other ions in gradual events can be accelerated to high energies at the shocks driven by fast ICMEs. Once the CMEs entered to the scene, the picture with a clear separation between impulsive and gradual events developed.

However, analogous to the problems with a bimodal picture of fast and slow solar wind, also this bimodality has problems with statistics, as discussed by Klein and Dalla (this issue). When correlating SEPs with flares, positive correlations are found. Flares and CMEs also correlate, in particular in the case of large flares. Even though the causal relationship between CMEs and flares is not straightforward, it could well be indirect with complex intermediary processes. Kahler (1982) introduced the concept of "big flare syndrome" to describe the danger of associating SEPs with flares based on the simultaneous appearance of big flares, CMEs and SEPs. The likely cause of the correlations is the dramatic reconfiguration of coronal structures, involving widespread magnetic reconnection and clearly releasing large amounts of stored magnetic energy. It is these underlying processes that are likely to trigger flares and launch CMEs—and ultimately causing the production of SEPs. A recent review of the topic has been provided by Reames (2015).

Another strong statistical bias toward attributing the gradual events to the ICME driven shocks arises because gradual events are orders of magnitude more intense than impulsive events. As pointed out by Trottet et al. (2015) both flare acceleration and ICME shock acceleration contribute to the production of protons in the energy range of tens of MeV and nearly-relativistic electron populations in large SEP events. Thus impulsive and gradual events cannot easily be distinguished, as the impulsive component can be hidden under the much more intense shock accelerated component of a gradual event. For practical space weather purposes, the strong events are, of course, critical because they are the most harmful to spacecraft and humans in space.

Further challenge with SEPs is their east-west variation. Most SEP events have been measured at one single point in space and the global picture has been composed putting together different events with different characteristics and using numerical modelling. The problem of how SEPs populate a broad range of solar longitudes dates back to the times before the association with ICMEs was suspected. Reinhard and Wibberenz (1974) studied a number of events using widely spaced spacecraft in the late 1960s and early 1970s. They attributed the longitudinal transport of SEPs to a three-step process: first, a rapid spreading of the accelerated population in the corona, followed by drifts in the highly structured high-altitude coronal arcades, and finally by a slower diffusion phase that extended the spatially broadening SEP populations further in the inner heliosphere.

Today the wide longitudinal extent of gradual events is, at least partially, understandable, as ICME-driven shocks are longitudinally wide and expand during the propagation. de Lucas et al. (2011) analyzed all ICME-driven shocks that had been observed during 1974–1986 by multiple spacecraft, including Helios. They found that there was a 50% chance to observe a shock at two locations separated by 90°, after which the probability quickly dropped. Only in four cases the extent was in the range 120°–160°. However, the energetic particle enhancements can be longitudinally even wider than the shock, which could at least partially be attributed to cross-field diffusion in the interplanetary space.

The STEREO mission would have been ideal for further studies of the east-west extent of the shocks and SEPs. Unfortunately, the number of good events during the optimal separation of the two spacecraft was limited due to the long solar minimum 2008–2010. An interesting SEP event took place on November 11, 2011, and was observed by near-Earth spacecraft, both STEREO spacecraft and MESSENGER. Gómez-Herrero et al. (2015) found an almost circumsolar particle enhancement although the data indicated that all particles originated from a single active region on the Sun. Their conclusion was that the data were not consistent with a major role of interplanetary diffusion, as had been suggested by Dresing et al. (2012) to have been the case in an earlier event on January 17, 2010. Instead, in the November 2011 event the population appeared to have quickly spread already close to the Sun.

That the longitudinal extent of SEP events can have so different physical causes is a great challenge for modellers. While the extension of diffusion models to three spatial dimensions already is a formidable task, it may not be enough, as coronal drifts and other transport processes form a complex cocktail that may well vary from event to event.

2.5 Space Weather in the Magnetosphere

Most technological systems sensitive to space weather operate inside the magnetosphere, including those in the atmosphere or on the surface of the Earth. The global scale magnetospheric storms and substorms have been in the focus of STP research since the early days of in situ solar wind and magnetosphere observations (Akasofu and Chapman 1961; Akasofu 1964). It has become clear that the magnetosphere responds to the variations in the solar wind forcing in widely varying temporal and spatial scales. The internal magnetospheric processes and different magnetospheric domains are coupled to each other through complicated direct and feed-back mechanisms that often are non-linear. While there are good motivations to try to describe the magnetospheric activity in terms of different activity parameters, e.g., Dst, Kp, or AE (for definitions, see Mayaud 1980), they only give very coarse description of the magnetospheric dynamics. Space weather forecasts based on indices calculated from solar wind observations remain statistical and are not detailed enough if temporally or spatially localized forecasts are required. During the past 15 years the multi-satellite missions Cluster, THEMIS, and most recently MMS and Van Allen Probes (VAP) have opened new possibilities to deal with spatio-temporal structures and phenomena in the near-Earth solar wind, magnetosheath and magnetosphere. These observations together with improving physics-based numerical simulation tools carry a promise to qualitative improvements in our understanding of magnetospheric physics, but we are not yet there.

The most important drivers of intense magnetospheric activity are ICMEs with the associated shocks and sheaths, fast solar wind, and fast/slow stream interaction regions (SIR). As magnetic reconnection, introduced in the magnetospheric context by Dungey (1961), is a key element in the energy, momentum, and plasma transfer from the solar wind to the magnetosphere, the southward IMF (B_S) is a major contributor to the storm strength. However, it is not sufficient to consider the upstream solar magnetic field only. The magnetic field can have significant evolution in the magnetosheath and finally it is the shocked solar wind that interacts with the magnetopause. The sheath fluctuations are quite different behind the quasi-parallel and quasi-perpendicular sectors of the bow shock (for comprehensive reviews, see Bale et al. 2005; Burgess et al. 2005), and the kinetic effects must be taken into account, at least of ions. For numerical simulations, this calls for hybrid approach where electrons are treated as a fluid and ions are described by quasi-particles, as in the widely applied particle-in-cell simulations (see, e.g., Karimabadi et al. 2014), or in terms of distribution functions, as in the newly-developed Vlasov-hybrid method (von Alfthan et al. 2014).

The peak intensity of intense storms (defined here as Dst < -100 nT) is observed equally often during ICME flux rope and sheath passages and only about 15% of intense storms take place during fast streams or SIRs (Huttunen and Koskinen 2004). The ICME flux ropes and sheaths tend to enhance the geospace response in different ways. Smoothly rotating flux ropes are more efficient drivers of large-scale convection and ring current, whereas the more fluctuating magnetic field in the sheath regions leads to stronger high-latitude effects (e.g., Huttunen et al. 2002; Pulkkinen et al. 2007). Also here the mixture of causes and effects, as well as involved feed-back mechanisms between different processes, make correlation studies non-straightforward.

The main, but not the only, contributor to the low-latitude magnetic deviations on ground, from which the Dst index is calculated, is the ring current. During intense storms the ring current dominates together with enhanced magnetopause current, the latter of which can be eliminated with rather good reliability (O'Brien and McPherron 2000), if the "pure" ring current effect is sought for. During moderate storms the contributions from the tail current become important, which are more difficult to handle (for a comprehensive reviews of the geospace current systems, see Ganushkina et al. 2015, and Ganushkina et al., this issue).

The positive correlation between fast solar wind and radiation belt electron fluxes has been known for a long time. High solar wind speed very quickly leads to enhancements of the lower-energy electron population, whereas the appearance of relativistic electrons can have a substantial delay. The wave–particle interactions needed to lift the energies to the relativistic range take their time (e.g., Horne et al. 2005).

Recently the Van Allen Probes (VAP) of NASA have shed much new light to the radiation belt dynamics (for a thorough review of VAP contributions, see Baker et al., this issue). A lucky coincidence was that a new ultrarelativistic electron belt was formed on September 2, 2012, soon after the VAP launch (Baker et al. 2013). It is becoming increasingly evident that "if you have seen one storm, you have seen one storm" (Geoff Reeves, private communication). In fact, energization by interplanetary shock-induced electric field impulses (e.g., Kanekal et al. 2014) as well as wave–particle interactions through both chorus (e.g., Thorne et al. 2013) and ULF (e.g., Mann et al. 2016) waves have been successfully applied to VAP observations of different storm events. As the characteristics of the various wave modes critically depend on the plasma background where they propagate, the coupling to the low-energy plasmas in the plasmasphere and ionosphere is of decisive importance. During strong magnetospheric activity the plasmapause is pushed closer to the Earth. On the other hand, the plasma outflow from the ionosphere is enhanced, which contributes to the seed particles for the ring current and radiation belt acceleration processes. This is still a poorly understood complex of feed-back processes, which may lead to quite different storm characteristics under otherwise similar the solar wind driving conditions. The VAP results strongly emphasize the importance to reach the heart of the electron belt instead of just skimming the low-altitude end of the radiation belt L-shells. This said, the great value of the long-term statistics from polar orbiting low-altitude satellites, such as SAMPEX, deserves to be acknowledged (e.g., Baker and Kanekal 2008).

It will take time to digest all new VAP information. From a space weather viewpoint an important challenge is to be able to associate the local magnetospheric processes with the solar wind drivers. There is a general tendency, when focusing on the details of magnetospheric processes, to oversimplify the parametrization of the solar and interplanetary drivers. Similarly, when assessing the effects of the interplanetary drivers, the complexity of the magnetospheric response is all too often oversimplified. For example, in studies of ICMEs and their geoefficiency it is tempting to lump the magnetospheric response into one single index, often Dst, or at most into a few of the indices. On the other hand, when trying to understand observations inside the magnetosphere, correlations are often sought only with some particular solar wind parameter, e.g., velocity, dynamic pressure, or magnetic field, or some combination of these, e.g., the epsilon parameter of Akasofu (1981), without taking into account the intrinsic structure and temporal evolution of the driver. An attempt to consider the details of both the driver and the response was presented by Kilpua et al. (2015). They investigated the cause–effect relationship by carefully considering which part of the shock and ICME was driving either the growth or decay of the electron belt during different storms. This was a promising step towards solving the long-standing problem, stated by Reeves et al. (2003), why some storms lead to electron enhancements, some others to depletions.

Finally, the role of substorms in this context is worthy of comment. The term "substorm" is sometimes misleading. Although one or more substorms almost always take place during magnetospheric storms, they are not building blocks of the storms but represent an independent mode of magnetospheric activity. Most substorms occur outside of storm periods and often without any clear trigger in the solar wind parameters. The substorm process intensifies the magnetosphere–ionosphere interaction, strengthens the ionosphere's role as the magnetospheric plasma source and produce seed population for high-energy electrons through wave–particle interactions. The actual role of substorm process in the electron content of the radiation belts is yet to be understood (for further discussion, see Baker et al., this issue).

2.6 Space Weather Effects in the Upper Atmosphere and Ionosphere

The main space weather effects in the upper atmosphere and ionosphere are mediated through the electromagnetic magnetosphere–ionosphere coupling and precipitation of energetic charged particles to the upper atmosphere. While the statistical patterns of ionospheric current systems and magnetic field-aligned currents (FAC) have been established long time ago (Iijima and Potemra 1976), in the space weather context the dynamic changes and often highly localised perturbations are more critical. Again, packaging the ionospheric response in a small number of activity indices is far from sufficient.

In the high-latitude ionosphere the most visible space weather phenomena are the auroral manifestations of magnetospheric storms and substorms. During this process in particular the Region 1 FAC system, connecting the auroral ionosphere to the outer magnetospheric boundary, enhances as does the westward ionospheric electrojet, which is indexed with the electrojet index AL. While AL as well as the eastward electrojet index AU are routinely calculated as the lower (AL) and upper (AU) envelope of the horizontal magnetic field component recordings at a limited number of magnetometer stations below the statistical auroral oval (see Mayaud 1980), more detailed information can be obtained from individual magnetometers when they happen to be located under the most interesting ionospheric activity. Here the longitudinal chains, e.g., IMAGE in the Fennoscandian sector (http://space.fmi.fi/image), and CARISMA in Canada (http://www.carisma.ca) have turned out to be very valuable tools.

An interesting fact about the AL index is that it has the strongest coupling to the solar wind during relatively weak solar activity (see McPherron et al., this issue). In case of very strong activity the magnetosphere does not seem to be capable to take all of the energy the solar wind is offering, which also shows in the saturation of the cross polar cap potential (for a review of a number of proposed explanations for the saturation, see Borovsky et al. 2009). A fresh view on the saturation issue was recently given by Pulkkinen et al. (2016). They analyzed magnetosheath observations of the five-spacecraft THEMIS mission jointly with auroral electrojets calculated from IMAGE magnetometer data and argued that at least a part of the saturation takes place already in the magnetosheath before the solar wind directly interacts with the magnetopause.

While auroral displays caused by electron and proton precipitation during magnetospheric storms and substorms can be considered as the most pleasant face of space weather, the beautiful phenomenon has also its dark side. The enhanced and temporally variable particle precipitation causes perturbations in the ion content of the ionosphere disturbing radio wave communication and satellite navigation systems. Intense high-energy particle fluxes also increase the radiation dose to aircraft crews and passengers, forcing lowering of the cruise altitude to get more atmospheric shielding.

Finally, energetic particles penetrating into the middle atmosphere have effects on the local ion chemistry. The atmospheric effects of strong SEP events and energetic electron precipitation (EEP) are roughly comparable. However, sufficiently strong SEP events are rare, as only a small fraction of all CME-driven shocks are able to accelerate protons to high enough energies to have significant effects in the middle atmosphere. On the other hand, sufficiently energized EEP from the outer radiation belt is a common storm-time phenomenon, in particular during substorm activity, and cause observable changes in atmospheric molecular composition (Andersson et al. 2014). For further discussion of atmospheric space weather matters, see Sojka (this issue).

2.7 Space Weather Effects on Earth

In addition to the communications and positioning systems that can be perturbed by space weather-related ionospheric irregularities and direct noise from strong solar radio bursts, the list of ground-based space weather sensitive technological systems is long, including long conductive transmission lines, gas pipe lines, railway signal systems, prospecting and deep drilling, etc.

A fundamental phenomenon behind the most important ground-based effects is the electromagnetic induction leading to disturbance currents in conductor systems, called geomagnetically induced currents (GIC). The basic physics behind GIC is elementary electrodynamics. Fast changes in ionospheric currents cause fast variations in the magnetic field on ground. These induce an electric field that drives electric currents in the conductive earth. The currents seek their way through the best conductors in the vicinity.

From observed magnetic field variations and accurate conductivity models of both ground and the conductor system in question post-event analysis of GIC events is straightforward. Also nowcasting close enough to a good magnetometer is not a problem, as the calculations do not require large computing power (e.g., Pirjola and Viljanen 1998). While the role of ICMEs as ultimate drivers of large GICs is well-established (Huttunen et al. 2008), the path from the upstream solar wind down to the Earth is long and complicated and a real challenge lies in the forecasting. Rapid enough temporal changes in the ionospheric currents are related to substorms and often highly localized. They cannot be reliably predicted using current forecasting tools.

2.8 Space Climate—Long-term Variations

The Sun is a changing star and the changes determine the long-term space climate under which space weather variations take place. During the 400 years since the time of Galileo the solar cycle as seen in the appearance of the sunspots has shown great variability both in the cycle length and in the peak and bottom sunspot numbers (for a recent review, see Hathaway 2015). While the statistical and physics-based sunspot number models have become increasingly sophisticated, forecasting of the cycle is yet a big challenge. The peak of cycle 23 was not as strong as some of the most advanced models suggested, and the following long (2008–2010) and low valley before cycle 24 was not predicted at all. Also the physical properties of the minimum phase solar wind were intriguing. Based on Ulysses observations McComas et al. (2008) found that during the minimum before cycle 24 the fast solar wind was significantly less dense, cooler and carried less mass and momentum flux than during the previous minimum. Furthermore, the open magnetic flux was reduced and, as discussed by Schwadron et al. (this issue), this tendency has continued through cycle 24.

Cycle 24 is now approaching its end. It is becoming, in terms of sunspot number, the weakest cycle since cycle 14 (for regularly updated sunspot records see, e.g., http://sidc.oma.be/silso/). Is the state of the solar dynamo undergoing some kind of long-term change resembling, e.g., the Maunder minimum (e.g. Eddy 1976), is this a signature of the Gleissberg (1939) cycle, or is it just a question of "statistical fluctuation", remains to be seen. As discussed by Schwadron et al. (this issue), the decreasing solar activity is enhancing the galactic cosmic ray fluxes to the inner heliosphere leading to enhanced levels of radiation doses with consequent hazard risks both in space and at airline altitudes. If this trend continues, it will shorten the time humans can be safely exposed to galactic cosmic rays. This is, evidently, a serious concern for those who are planning for manned missions to Mars.

The terrestrial space climate environment is also affected by the slowly changing geomagnetic field. Currently the dipole moment is decaying 5–6% per century and both the north and south magnetic poles have been moving toward the northwest. As discussed by Mandea (in this issue and references therein) the migration of the north magnetic pole has been accelerating until recently, but is now slowing down. The geomagnetic field is weakest over a large area, called South Atlantic Anomaly (SAA), containing the east cost of South America and the southern part of Atlantic Ocean. Whereas the nominal dipole field at the Earth's surface is about 30 μT at the dipole equator and 60 μT at the poles, the minimum field of SAA is only 22 μT close to Asuncion in Paraguay. The SAA evolves naturally with the decaying and moving dipole moment. In the past the field minimum has been moving in the direction from the southern Africa towards South America. As the magnetic field is weakening above the SAA, more energetic particles can penetrate to lower altitudes with increasing risks to low-altitude satellites and aircraft. Thus the continuous updating of geomagnetic and ionospheric field models is also critical for space weather (for further discussion, see Mandea, this issue).

2.9 Extreme Space Weather

The most severe space weather events are often, but not always, associated with very complex solar activity. A historically important sequence of events, illustrating wide variety of space weather phenomena both on the Sun and in the near-Earth space, took place during the second half of May 1967 (for a recent review, see Knipp et al. 2016, and references therein). A particularly active region moved over the visible solar disc during 17–31 May. It produced several tens of significant flares and, once the activity intensified on 21 May, three major solar radio bursts (SRB) occurred on 21, 23 and 28 May. The burst on the 21st was associated with a large white-light flare and strong X-ray emissions. On 23 May the magnetic complexity enhanced and another white-light flare with solar proton emission took place. The X-ray flare activity and the SRB continued for several hours. The flare was followed with a gradual SEP event peaking at 1 AU on 25 May, suggesting the passage of an ICME-driven shock. The subsequent magnetospheric storm on 25–26 May was one of the strongest in the history of magnetic observations, the Dst index reaching −387 nT. This event took place before the existence of CMEs was discovered and there is no direct evidence that the magnetospheric storm would have been driven by an ICME. The geomagnetic indicators and the observation that the dayside magnetopause was inside geostationary orbit, however, point to an ICME. For further analysis of this event and its wide reaching consequences in geospace we refer to Knipp et al. (2016).

While magnetic and magnetospheric storms are driven by shocks, ICMEs and fast solar wind, during extreme solar activity also the electromagnetic connection from the Sun, in form of X-ray emissions and SRBs, can have serious consequences to communication and satellite navigation systems long before the plasma connection is established. The X-ray-enhanced ionization causes perturbations to signal propagation in the ionosphere, whereas strong radio bursts introduce noise directly at the receivers if the energy flux in the used frequency band is large enough.

The strongest SRB event thus far recorded at GPS frequencies took place in December 2006 during a sequence of X-class flares. The estimated flux of the burst on 6 December was about 10^6 SFU (solar flux unit) at 1.4 GHz, which was an order of magnitude more than in other large events. The noise caused long-duration (about 30 min) reduction of the carrier-to-noise ratio up to 25 dB (Cerruti et al. 2008; Carrano et al. 2009). Another burst exceeding 10^5 SFU took place on 13 December. Loss of lock with GPS satellites was experienced for around 2 hours following the event and for some receivers a complete loss of navigation solution experienced for 6–10 minutes. A curious fact is that these events took place close to solar minimum. The quiet Sun may sometimes be more dangerous than the active.

Evidently, strong space weather events have wide societal consequences already today, and an extreme event of the Carrington class or stronger may cause far reaching hazards. However, forecasting such events is difficult, if not impossible. Riley (2012) estimated the likelihood of extreme events statistically from past events assuming power law distribution of the peak Dst values. He found that an event with Dst < -850 nT has a 12% likelihood to occur within the next decade. For an event exceeding Dst < -1700 nT the likelihood dropped to 1.5%. Assuming log-normal or exponential distributions the likelihoods are a bit lower. Clearly, events of this size are rare but not unlikely.

As discussed by Riley (2012) the Dst exceeded –600 nT only once during the 20th century. That happened on 13 March 1989 when the peak Dst was about −640 nT. This was the storm causing the famous collapse of Hydro-Québec power distribution system. If the 23 July 2012 CME would have hit the Earth, the storm would most likely have been much stronger. The transit time to 1 AU was about the same as in the Carrington event and, as discussed in Sect. 2.3 and the peak IMF as observed by STEREO was larger 100 nT. According to existing models, Dst exceeding -1000 nT would have been expected, e.g., the Dst model by Temerin and Li (2006) predicts -1200 nT. Of course, the Dst models have not been calibrated to so large events.

According to standard magnetopause models based on pressure balance, the July 2012 event would have pushed the dayside magnetopause almost down to the ionosphere. Also this prediction needs to be considered with some care, as we do not really know the strength of induced currents to counteract the push. As noted in Sect. 2.6, the polar cap potential has a tendency to saturate during extreme driving, which is an indication of counteraction by induction. However, it is quite likely that the event could have wiped a significant fraction of plasma and energetic particles away from the magnetosphere.

How serious can the consequences of the worst imaginable space weather event really be? The honest answer is that we do not know. Claims have been voiced that the outcome of a super event could be almost an armageddon of modern civilization. While such claims may give publicity to space weather efforts, scientists should stay calm and concentrate on serious and unbiased investigations of the extreme events and their effects on critical systems. One should not cry wolf too often (Aesop, the 6th century BCE). For further discussion of extreme events, see Riley et al. (this issue).

3 Space Weather Beyond Solar-Terrestrial Physics

Even if we leave the design and maintenance issues of technological space weather sensitive systems outside of our discussion, there are some interesting and challenging topics that are not directly solar-terrestrial physics but where improved STP understanding is needed.

3.1 Space Weather Elsewhere

The details of interaction between solar eruptions, solar wind and ICMEs with other planets depend on distinct plasma and magnetic field conditions at each individual planet. While the basic physics of these interactions is known, very little is known of the details. If we really had to make reliable space weather forecasts at other planets, as may some day be the case with Mars, we would be in much deeper trouble than at the Earth.

Mercury, the planet closest to the Sun, has the most direct influence by the solar wind. The planet's magnetic field is so weak that it does not require a particularly strong event to open the dayside magnetosphere for a direct access for the solar wind down to the surface. As the planet has no atmosphere, and thus no ionosphere, there is no shielding against solar energetic particles, even less to Galactic cosmic rays. The MESSENGER mission has brought a wealth of new observational information of the dynamics of the Hermean magnetosphere. Due to the small scales of the system everything is happening much faster than at the Earth. In response to three ICMEs events discussed by Slavin et al. (2014), Mercury's already dynamic magnetosphere was found to be significantly deformed, with the subsolar distance to the magnetopause reduced from 1.45 $R_☿$ to as little as 1.03–1.12 $R_☿$. Thick, low-β plasma depletion layers were observed, caused by the impact of the ICMEs in the inner magnetosheath adjacent to the subsolar magnetopause, leading to the magnetic pressure being nearly equal on both sides of the magnetopause. This vastly increased the magnetic reconnection rate, which already during normal conditions is much larger at Mercury than at the Earth (DiBraccio et al. 2013). The reconnection process led to a major erosion of the field at the magnetopause, with expected dynamic effects throughout the magnetosphere.

Venus is an opposite case. Venus does not have a magnetic field but its atmosphere is dense and hot. The direct interaction between the solar wind with the ionosphere leads to continuous outflow of plasma to space but not much is known how space weather events modify the escape. Remarkable examples are the so-called Hot Flow Anomalies in the Venusian foreshock region, which are also known to take place in the foreshocks of the Earth, Mercury and Mars (Collinson et al. 2012, and references therein). Hot Flow Anomalies are driven by solar wind discontinuities and in case of Venus they are bigger than the planet and lead to strongly enhanced loss of plasma from the ionosphere of the planet.

Mars is probably the most interesting planet for space weather considerations. Given its complex and fragmented environmental make-up (e.g., Dubinin et al. 2008) the impact of solar wind drivers is correspondingly complex. The planet has a tenuous atmosphere, an ionosphere and some localized magnetic structures. On Mars, the lower and upper atmospheres are strongly coupled and loss of gas from the upper atmosphere to space is an important process that shapes the planetary atmosphere and its dynamics. It has been shown (Jakosky et al. 2015) that, in addition to the expected magnetospheric effects (compression of the bow shock and magnetosheath, formation of aurora), the escape rate of ions is also significantly enhanced during the passage of an ICME.

There are and will be technological systems on the surface of Mars, and recently a very ambitious private plan to send humans to Mars was announced by Elon Musk at the 67th International Astronautical Conference. Before this can become reality, serious space weather hazards during the transit to Mars need to be assessed and taken into account. The very weak shield against both solar and galactic cosmic rays provided by the Martian magnetosphere and atmosphere means that, once on the surface, human explorers will need continuous and reliable space weather services to survive.

Disruptions of the distant plasma tails of comets by ICMEs have been repeatedly observed remotely (e.g. Kuchar et al. 2008). A fortuitous observation by ESA's Rosetta mission to comet 67P/Churyumov–Gerasimenko (Edberg et al. 2016) gave a chance to study a widespread disruption in the near-environment of an active comet caused by an impacting ICME.

The giant gas planets have very strong magnetospheres and thick atmospheres. Consequently their plasma dynamics is much more internally controlled than at the Earth. As discussed in Sect. 2.3, fast ICMEs slow down toward the ambient solar wind speed while propagating through the solar system. This together with the shocks being dissipative weakens the ICME effects at outer planets. Space weather processes are still important as illustrated, e.g., by the Jovian auroras. An interesting group of celestial bodies, again with greatly variable characteristics, are the big moons of Jupiter and Saturn, which move both inside and outside of the planets' magnetospheres. Thus far we have very little observational facts about space weather phenomena there.

The Sun is only one of the more than 100 billion stars in the Milky Way. Many of them are magnetically active, not too different from the Sun, and the biggest sunspots can be identified in the observed light-curves. There are no direct observations of stellar winds yet but there is no reason to doubt their existence. Stellar flares are easier to observe and many magnetically active stars are known to have continuous flare activity similar to our Sun. Present day observations are limited to X- and M-class flares. There are rare observations of 2–3 orders of magnitude larger flares than the largest observed in the Sun. If we again extend the distribution beyond observations, it is possible that such flares could happen in the Sun once in a few thousand years.

3.2 Anthropogenic Space Weather

Although space weather is ultimately driven by solar activity, humans can and have caused space weather events and modified the environment where space weather phenomena take place. The high-altitude nuclear weapon tests 1958–1962 introduced significant new trapped particle populations that remained observable for several years. In particular the 1.4-megaton Starfish Prime in July 1962 at an altitude of 400 km resulted in an artificial radiation belt that has been claimed to have destroyed three satellites and caused problems to at least three more soon after the detonation. In 1962 this was a large fraction of all satellites in orbit.

Another man-made effect is thought to be related to the question: "Why is there a slot region between outer and inner belt electrons?" Radio waves from ground-based VLF transmitters have been proposed to scatter the electrons into the atmospheric loss cone and clean up the region. The recent VAP data supports this interpretation indicating that the VLF waves really form an almost impenetrable barrier against the outer belt electrons at $L = 2.8$ (Foster et al. 2016). For further discussion of anthropogenic space weather issues, see Gombosi et al. (this issue).

3.3 Space Weather in the Public Eye

Regardless of the risks that motivate most space weather activities, we must not forget the beautiful face of space weather. Time after time it is equally fascinating to be out in the silent arctic winter night watching a major auroral event and see the entire sky full of moving colourful displays. Although we know that the auroras are at 100-km altitude, the impression that they sweep the nearby hilltops is unavoidable. The feeling is simply unforgettable.

Because space weather is visible and beautiful, it has great educational potential for physics students and for wider public as well. It is remarkably easy to make a well-received public presentation of space weather. One just needs to collect a few breathtaking pictures and movies of solar eruptions at one end and of auroras at the other, and prepare to tell a few facts that have been discussed above. This works particularly well in those parts of the world where people have seen real auroras with their own eyes. And the number of these people is increasing also elsewhere due to the fairly intense auroral tourism, which is a growing but thus far underexploited form of business.

4 Concluding Remarks

Since the advent of in situ observations in the solar wind and the magnetosphere impressive progress in understanding the physics behind space weather phenomena has taken place. We know what the most relevant solar phenomena driving space weather are, and what their consequences in the near-Earth space can be, but not always are. In fact, our detailed knowledge and modelling capabilities are far from sufficient for reliable forecasting of hazardous space weather conditions in space or on ground. There is both room and need for further research, including new observations, advanced data analysis, and numerical modelling.

Observations of space weather in space, whether remotely of solar processes or in situ of charged particles and electromagnetic fields, are technologically and economically demanding. A particular concern is that the observations are often difficult to repeat, which is generally considered as one of the backbones of the scientific method. New observations and re-analysis of old data continuously modify our understanding of the scientific foundations of space weather. The recent VAP observations provide great illustrations of the need for fresh data with improved instrumentation.

As it is not always possible to directly probe the physical processes, at least not continuously, various proxies are needed, such as empirical estimates of energy input from the solar wind to the magnetosphere, based on known energy output mechanisms and theoretical ideas of the input processes. To obtain the proxies statistical analyses are called for. Too often the "statistics" in this field is based on way too few samples, in particular when dealing with rare space weather events, which in turn are the most important from the practical point of view. An issue are the complicated interdependencies of observable phenomena, which may or may not have causal connections, e.g., SPEs vs. CMEs, SPEs vs. solar flares, or CMEs vs. solar flares. More sophisticated statistical tools than simple linear correlation analyses are definitely called for.

In the field of numerical modelling and forecasting the space weather modellers are nowhere near their colleagues in the atmospheric weather business. There is much room for improved numerical modelling and simulations throughout the road from the Sun to the ionosphere. Some of the challenges are really serious, as physical processes often have their roots in the spatial and temporal scales of electron motion while the modellers are still struggling to include ion kinetic effects in the fluid models. Also the vast range of relevant physical parameters is a serious issue, in particular from the solar surface to the corona where much of space weather originates.

The challenges to improve practical space weather services cannot be met using more efficient computational methods and tools alone. Deep understanding of the underlying physics and innovative ideas to improve the understanding remain basic requirements of all progress in space weather activities. The decisive role of temporally and spatially sufficiently wide observational input for both theoretical and practical aspects of space weather must always be kept in mind.

Acknowledgements We wish to thank the International Space Science Institute for inviting us to assemble the program of this workshop. The contributions of all attendees and the excellent spirit during the sessions are gratefully acknowledged. The efficient and utmost helpful local ISSI staff deserves our greatest thanks. As always.

References

S.-I. Akasofu, The development of the auroral substorm. Planet. Space Sci. **12**, 273–282 (1964)

S.-I. Akasofu, Energy coupling between the solar wind and the mangetosphere. Space Sci. Rev. **28**, 121–190 (1981)

S.-I. Akasofu, S. Chapman, The ring current, geomagnetic disturbance, and the Van Allen radiation belts. J. Geophys. Res. **66**, 1321–1350 (1961)

H. Alfvén, On the theory of comet tails. Tellus **9**, 92–96 (1957)

M.E. Andersson, P.T. Verronen, C.J. Rodger, M.A. Clilverd, A. Seppälä, Missing driver in the Sun–Earth connection from energetic electron precipitation impacts mesospheric ozone. Nat. Commun. **5** (2014). doi:10.1038/ncomms6197

D.N. Baker, S.G. Kanekal, Solar cycle changes, geomagnetic variations, and energetic particle properties in the inner magnetosphere. J. Atmos. Sol.-Terr. Phys. **70**, 195–206 (2008)

D.N. Baker, S.G. Kanekal, V.C. Hoxie, M.G. Henderson, X. Li, H.E. Spence, S.R. Elkington, R.H.W. Friedel, J. Goldstein, M.K. Hudson, G.D. Reeves, R.M. Thorne, C.A. Kletzing, S.G. Claudepierre, A long-lived relativistic electron storage ring embedded in Earth's outer Van Allen belt. Science **340** (2013). doi:10.1126/science.1233518

S.D. Bale, M.A. Balikhin, T.S. Horbury, V.V. Krasnoselskikh, H. Kucharek, E. Möbius, S.N. Walker, A. Balogh, D. Burgess, B. Lembège, E.A. Lucek, M. Scholer, S.J. Schwartz, M.F. Thomsen, Quasi-perpendicular shock structure and processes. Space Sci. Rev. **118**, 161–203 (2005). doi:10.1007/s11214-005-3827-0

L. Biermann, Kometenschweife und solare Korpuscular-strahlung. Z. Astrophys. **29**, 274–286 (1951)

L. Biermann, Solar corpuscular radiation and the interplanetary gas. Observatory **77**, 109–110 (1957)

J.E. Borovsky, B. Lavraud, M.M. Kuznetsova, Polar cap potential saturation, dayside reconnection and changes to the magnetosphere. J. Geophys. Res. **114** (2009). doi:10.1029/2005JA014058

D.A. Bryant, T.L. Cline, U.D. Desai, F.B. McDonald, Explorer 12 observations of solar cosmic rays and Energetic Storm Particles after the solar flare of September 28, 1961. J. Geophys. Res. **67**, 4983–5000 (1962)

D. Burgess, E.A. Lucek, M. Scholer, S.D. Bale, M.A. Balikhin, A. Balogh, T.S. Horbury, V. Krasnoselskikh, H. Kucharek, B. Lembège, E. Möbius, S.J. Schwartz, M. Thomsen, S.N. Walker, Quasi-parallel shock structure and processes. Space Sci. Rev. **118**, 205–222 (2005). doi:10.1007/s11214-005-3832-3

H.V. Cane, R.E. McGuire, T.T. von Rosenvinge, Two classes of solar energetic particle events associated with impulsive and long-duration soft X-ray flares. Astrophys. J. **301**, 448–459 (1986)

C.S. Carrano, C.T. Bridgwood, K.M. Groves, Impacts of the December 2006 solar radio bursts on the performance of GPS. Radio Sci. **44** (2009). doi:10.1029/2008RS004071

A.P. Cerruti, P.M. Kinter Jr., D.E. Gary, A.J. Mannucci, R.F. Meyer, P. Doherty, A.J. Coster, Effect of intense December 2006 solar radio bursts on GPS receivers. Space Weather **6** (2008). doi:10.1029/2007SW000375

S. Chapman, An outline of a theory of magnetic storms. Proc. R. Soc. Lond. Ser. A, Math. Phys. Sci. **95**, 61–83 (1918)

S. Chapman, V.C.A. Ferraro, A new theory of magnetic storms. Terr. Magn. Atmos. Electr. **36**, 77–97 (1931a)

S. Chapman, V.C.A. Ferraro, A new theory of magnetic storms. Terr. Magn. Atmos. Electr. **36**, 171–186 (1931b)

E.W. Cliver, L. Svalgaard, The 1859 solar-terrestrial disturbance and the current limits of extreme space weather activity. Sol. Phys. **224** (2004). doi:10.1007/s11207-015-0683-2

G.A. Collinson, D.G. Sibeck, A. Masters, N. Shane, J.A. Slavin, A.J. Coates, T.L. Zhang, M. Sarantos, S. Boardsen, T.E. Moore, S. Barabash, Hot flow anomalies at venus. J. Geophys. Res. **117** (2012). doi:10.1029/2011JA017277

P. Corona-Romero, J.A. Gonzalez-Esparza, R. Aguilar-Rodriguez, V. De-la-Luz, J.C. Mejia-Ambriz, Kinematics of ICMEs/shocks: blast wave reconstruction using type-II emissions. Sol. Phys. **290**, 407–422 (2015)

A. de Lucas, R. Schwenn, A. dal Lago, E. Marsch, A.L. Clúa de Gonzalez, Interplanetary shock wave extent in the inner heliosphere as observed by multiple spacecraft. J. Atmos. Sol.-Terr. Phys. **73**, 1281–1292 (2011). doi:10.1016/j.jastp.2010.12.011

G.A. DiBraccio, J.A. Slavin, S.A. Boardsen, B.J. Anderson, H. Korth, T.H. Zurbuchen, J.M. Raines, D.N. Baker, R.L. McNutt, S.C. Solomon, MESSENGER observations of magnetopause structure and dynamics at Mercury. J. Geophys. Res. **118** (2013). doi:10.1002/jgra.50123

N. Dresing, R. Gómez-Herrero, A. Klassen, B. Heber, Y. Kartavykh, W. Dröge, The large longitudinal spread of solar energetic particles during the 17 January 2010 solar event. Sol. Phys. **281** (2012). doi:10.1007/s11207-012-0049-y

E. Dubinin, R. Modolo, M. Fraenz, J. Woch, G. Chanteur, F. Duru, F. Akalin, D. Gurnett, R. Lundin, S. Barabash, J.D. Winningham, R. Frahm, Plasma environment of Mars as observed by simultaneous MEX-ASPERA-3 and MEX-MARSIS observations. J. Geophys. Res. **113** (2008). doi:10.1029/2008JA013355

J.W. Dungey, Interplanetary magnetic field and auroral zones. Phys. Rev. Lett. **6**, 47–48 (1961)

N.J.T. Edberg, M. Alho, M. André, D.J. Andrews, E. Behar, J.L. Burch, C.M. Carr, E. Cupido, I.A.D. Engelhardt, A.I. Eriksson, K.-H. Glassmeier, C. Goetz, R. Goldstein, P. Henri, F.L. Johansson, C. Koenders, K. Mandt, C. Möstl, H. Nilsson, E. Odelstad, I. Richter, S. Wedlund, G. Stenberg Wieser, K. Szego, E. Vigren, M. Volwerk, CME impact on comet 67P/Chuyumov–Gerasimenko. Mon. Not. R. Astron. Soc. **462**, 45–56 (2016). doi:10.1093/mnras/stw2112

J.A. Eddy, The Maunder Minimum. Science **192** (1976). doi:10.1126/science.192.4245.1189

S.E. Forbush, On world-wide changes in cosmic-ray intensity. Phys. Rev. **54** (1938). doi:10.1103/PhysRev.54.975

J.C. Foster, P.J. Erickson, D.N. Baker, A.N. Jaynes, E.V. Mishin, J.F. Fennel, X. Li, M.G. Henderson, S.G. Kanekal, Observations of the impenetrable barrier, the plasmapause, and the VLF bubble during the 17 March 2015 storm. J. Geophys. Res. **121**, 5537–5548 (2016). doi:10.1002/2016JA022509

N.Y. Ganushkina, M.W. Liehmohn, S. Dubyagin, I.A. Daglis, I. Dandouras, D.L. De Zeeuw, Y. Ebihara, R. Ilie, R. Katus, M. Kubyshkina, S.E. Milan, S. Ohtani, N. Ostgaard, J.P. Reistad, P. Tenfjord, F. Toffo-

letto, S. Zaharia, O. Amariutei, Defining and resolving current systems in geospace. Ann. Geophys. **33**, 1369–1402 (2015). doi:10.5194/angeo-33-1369-2015
J. Geiss, G. Gloeckler, R. von Steiger, H. Balsiger, L.A. Fisk, A.B. Galvin, F.M. Ipavich, S. Livi, J.F. McKenzie, K.W. Ogilvie, B. Wilken, The southern high-speed stream: results from the SWICS instrument on Ulysses. Science **268**, 1033–1035 (1995). doi:10.1126/science.7754380
W. Gleissberg, A long-periodic fluctuation of the sun-spot numbers. Observatory **62**, 158–159 (1939)
R. Gómez-Herrero, N. Dresing, A. Klassen, B. Heber, D. Lario, N. Agueda, O.E. Malandraki, J.J. Blanco, J. Rodríguez-Pacheco, S. Banjac, Circumsolar energetic particle distribution on 2011 November 3. Astrophys. J. **799**(55) (2015). doi:10.1088/0004-637X/799/1/55
J.T. Gosling, The solar flare myth. J. Geophys. Res. **98**, 18937–18949 (1993)
G.E.. Hale, On the probable existence of a magnetic field in sun-spots. Astrophys. J. **28**, 315–342 (1908)
D.H. Hathaway, The Solar Cycle. Living Rev. Sol. Phys. **12** (2015). doi:10.1007/lrsp-2015-4
R.B. Horne, R.M. Thorne, Y.Y. Shprits, N.P. Meredith, S.A. Glauert, A.J. Smith, S.G. Kanekal, D.N. Baker, M.J. Engebretson, J.L. Posch, M. Spasojevic, U.S. Inan, J.S. Pickett, P.M.E. Decreau, Wave acceleration of electrons in the Van Allen radiation belts. Nature **437**, 227–230 (2005)
H.S. Hudson, B. Haisch, K.T. Strong, Comment on the solar flare myth by J.T. Gosling. J. Geophys. Res. **100** (1995). doi:10.1029/94JA02710
K.E.J. Huttunen, H.E.J. Koskinen, Importance of post-shock streams and sheath regions as drivers of intense magnetospheric storms and high-latitude activity. Ann. Geophys. **22**, 1729–1738 (2004)
K.E.J. Huttunen, H.E.J. Koskinen, R. Schwenn, Variability of magnetospheric storms driven by different solar wind perturbations. J. Geophys. Res. **107** (2002). doi:10.1029/2001JA900171
K.E.J. Huttunen, S.P. Kilpua, A. Pulkkinen, A. Viljanen, E. Tanskanen, Solar wind drivers of large geomagnetically induced currents during the solar cycle 23. Space Weather **6** (2008). doi:10.1029/2007SW000374
T. Iijima, T.A. Potemra, Field-aligned currents in the dayside cusp observed by Triad. J. Geophys. Res. **81**, 5971–5979 (1976)
B.M. Jakosky, J.M. Grebowsky, J.G. Luhmann, J. Connerney, F. Epavier, R. Ergun, J. Halekas, D. Larson, P. Mahaffy, J. McFadden, D.F. Mitchell, N. Schneider, R. Zurek, S. Bougher, D. Brain, Y.J. Ma, C. Mazelle, L. Andersson, D. Andrews, D. Baird, D. Baker, B.J. M., M. Benna, M. Chaffin, P. Chamberlin, Y.-Y. Chaufray, J. Clarke, G. Collinson, G. Delory, R. Dewey, G. DiBraccio, C. Dong, Y. Dong, P. Dunn, M. Elrod, S. England, A. Eriksson, J. Espley, S. Evans, X. Fang, M. Fillingim, K. Fortier, F.C. M., J. Fox, H. Gröller, S. Guzewich, T. Hara, Y. Harada, G. Holsclaw, S.K. Jain, R. Jolitz, F. Leblanc, C.O. Lee, Y. Lee, F. Lefevre, R. Lillis, R. Livi, D. Lo, M. Mayyasi, W. McClintock, T. McEnulty, R. Modolo, F. Fontmessin, M. Morooka, A. Nagy, K. Olsen, W. Peterson, A. Rahmati, S. Ruhunusiri, C.T. Russell, S. Sakai, J.-A. Sauvaud, K. Seki, M. Steckiewicz, M. Stevens, A.I.F. Steward, A. Stiepen, S. Stone, V. Tenishev, E. Thiemann, R. Tolson, D. Toublanc, M. Vogt, T. Weber, P. Withers, T. Woodes, R. Yelle, MAVEN observations of the response of Mars to an interplanetary coronal mass ejection. Science **350** (2015). doi:10.1126/science.aad0210
S.W. Kahler, The role of the big flare syndrome in correlations of solar energetic proton fluxes and associated microwave burst parameters. J. Geophys. Res. **87** (1982). doi:10.1029/JA087iA05p03439
S.W. Kahler, N.R. Sheeley Jr., R.A. Howard, M.J. Koomen, D.J. Michels, R.E. McGuire, T.T. von Rosenvinge, D.V. Reames, Associations between coronal mass ejections and solar energetic proton events. J. Geophys. Res. **89**, 9683–9693 (1984). doi:10.1029/JA089iA11p09683
S.G. Kanekal, D.N. Baker, J.F. Fennell, A. Jones, Q. Schiller, I.G. Richardson, X. Li, D.L. Turner, S. Califf, S.G. Claudepierre, L.B. Willson III, A. Jaynes, J.B. Blake, G.D. Reeves, H.E. Spence, C.A. Kletzing, J.R. Wygant, Prompt acceleration of magnetospheric electrons to ultrarelativistic energies by the 17 March 2015 interplanetary shock. J. Geophys. Res. **21**, 7622–7635 (2014). doi:10.1063/1.4882875
H. Karimabadi, V. Roytershteyn, H.X. Vu, Y.A. Omelchenko, J. Scudder, W. Daughton, A. Dimmock, K. Nykyri, M. Wan, D. Sibeck, M. Tatineni, A. Majumdar, B. Loring, B. Geveci, The link between shocks, turbulence, and magnetic reconnection in collisionless plasmas. Phys. Plasmas **42** (2014). doi:10.1002/2015GL064816
R. Kataoka, D. Shiota, E. Kilpua, K. Keika, Pileup accident hypothesis of magnetic storm on 17 March 2015. Geophys. Res. Lett. **42**, 5155–5161 (2015). doi:10.1002/2015GL064816
E.K.J. Kilpua, H. Hietala, D.L. Turner, H.E.J. Koskinen, T.I. Pulkkinen, J.V. Rodriguez, G.D. Reeves, S.G. Claudepierre, H.E. Spence, Unraveling the drivers of the storm time radiation belt response. Geophys. Res. Lett. **42** (2015). doi:10.1002/2015GL063542
J.A. Klimchuk, Key aspects of coronal heating. Philos. Trans. R. Soc. A **373** (2015). doi:10.1098/rsta.2014.0256
D.J. Knipp, A.C. Ramsay, E.D. Beard, A.L. Boright, W.B. Cade, I.M. Hewins, R.H. McFadden, W.F. Denig, L.M. Kilcommons, M.A. Shea, D.F. Smart, The May 1967 great storm and radio disruption

event: Extreme space weather and extraordinary responses. Space Weather **14** (2016). doi:10.1002/2016SW001423

T.A. Kuchar, A. Buffington, C.N. Arge, P.P. Hick, T.A. howard, B.V. Jackson, J.C. Johnston, D.R. Mizuno, S.J. Tappin, D.F. Webb, Observations of a comet tail disruption induced by the passage of a CME. J. Geophys. Res. **113** (2008). doi:10.1029/2007JA012603

L.J. Lanzerotti, D.J. Thomson, C.G. Maclennan, Engineering issues in space weather, in *Modern Radio Science*, ed. by M.A. Stuchly (Wiley-IEEE Press, Hoboken, 1999), pp. 25–51

F.A. Lindeman, Note on the theory of magnetic storms. Philos. Mag. Ser. 6 **38**, 669–684 (1919)

Y.D. Liu, J.G. Luhmann, P. Kajdič, E.K.J. Kilpua, N. Lugaz, N.V. Nitta, C. Möstl, B. Lavraud, S.D. Bale, C.J. Farrugia, A.B. Galvin, Observations of and extreme storm in interplanetary space caused by successive coronal mass ejections. Nat. Commun. **5** (2014). doi:10.1038/ncomms4481

I.R. Mann, L.G. Ozeke, K.R. Murphy, S.G. Claudepierre, D.L. Turner, D.N. Baker, I.J. Rae, A. Kale, D.K. Milling, A.J. Boyd, H.E. SPence, G.D. Reeves, H.J. Singer, S. Dimitrakoudis, I.A. Daglis, F. Honary, Explaining the dynamics of the ultra-relativistic third van allen radiation belt. Nat. Phys. **12** (2016). doi:10.1038/nphys3799

G.M. Mason, N.V. Nitta, M.E. Wiedenbeck, D.E. Innes, Evidence for a common acceleration mechanism for enrichments of ^{3}He and heavy ions in impulsive SEP events. Astrophys. J. **823** (2016). doi:10.3847/0004-637X/823/2/138

P.N. Mayaud, *Derivation, Meaning, and Use of Geomagnetic Indices*. Geophysical Monograph, vol. 22 (American Geophysical Union, Washington, 1980)

D.J. McComas, S.J. Bame, B.L. Barraclough, W.C. Feldman, H.O. Funsten, J.T. Gosling, P. Riley, R. Skoug, A. Balogh, R. Forsyth, B.E. Goldstein, M. Neugebauer, Ulysses return to the slow solar wind. Geophys. Res. Lett. **25**, 1–4 (1998)

D.J. McComas, R.W. Ebert, H.A. Elliot, B.E. Goldstein, J.T. Gosling, N.A. Schwadron, Skoug, Weaker solar wind from the polar coronal holes and the whole Sun. Geophys. Res. Lett. **35** (2008). doi:10.1029/2008GL034896

M. Neugebauer, C.W. Snyder, The mission of Mariner II: preliminary observations—solar plasma experiment. Science **138**, 1095–1097 (1962)

T.P. O'Brien, R.L. McPherron, An empirical phase space analysis of ring current dynamics: Solar wind control of injection and decay. J. Geophys. Res. **105**, 7707–7719 (2000)

E.N. Parker, Dynamics of the interplanetary gas and magnetic fields. Astrophys. J. **128**, 664–676 (1958)

R. Pirjola, A. Viljanen, Complex image method for calculating electric and magnetic fields produced by an auroral electrojet of finite length. Ann. Geophys. **16**, 1434–1444 (1998)

T.I. Pulkkinen, N. Partamies, K.E.J. Huttunen, G.D. Reeves, H.E.J. Koskinen, Differences in geomagnetic storms driven by magnetic clouds and ICME sheath regions. Geophys. Res. Lett. **34** (2007). doi:10.1029/2006GL027775

T.I. Pulkkinen, A.P. Dimmock, A. Lakka, A. Osmane, E. Kilpua, M. Myllys, E.I. Tanskanen, A. Viljanen, Magnetosheath control of solar wind-magnetosphere coupling efficiency. J. Geophys. Res. **121** (2016). doi:10.1002/2016JA023011

U.R. Rao, K.G. McCracken, R.P. Bukata, Cosmic ray propagation processes: 2. the energetic storm-particle event. J. Geophys. Res. **72**, 4325–4341 (1967)

D.V. Reames, What are the sources of solar energetic particles? Element abundances and source plasma temperatures. Space Sci. Rev. **194**, 303–327 (2015). doi:10.1007/s11214-015-0210-7

G.D. Reeves, K.L. McAdams, R.H.W. Friedel, Acceleration and loss of relativistic electrons during geomagnetic storms. Geophys. Res. Lett. **30** (2003). doi:10.1029/2002GL016513

M.J. Reiner, M.L. Kaiser, J.-L. Bougeret, Coronal and interplanetary propagation of CME/shocks from radio, in situ and white-light observations. Astrophys. J. **663** (2007). doi:10.1086/518683

R. Reinhard, G. Wibberenz, Propagation of flare protons in the solar atmosphere. Sol. Phys. **36**, 473–494 (1974)

P. Riley, On the probability of occurrence of extreme space weather events. Space Weather **10** (2012). doi:10.1029/2011SW000734

H. Schwabe, Sonnen-Beobachtungen im Jahre 1843. Astron. Nachr. **21** (1844). doi:10.1002/asna.18440211505

J.A. Slavin, G.A. DiBraccio, D.J. Gershman, S.M. Imber, G.K. Poh, J.M. Raines, T.H. Zurbuchen, X. Jia, D.N. Baker, K.-H. Glassmeier, S.A. Livi, S.A. Boardsen, T.A. Cassidy, M. Sarantos, T. Sundeberg, A. Masters, C.L. Johnson, R.M. Winslow, B.J. Anderson, H. Korth, R.L. McNutt Jr., S.C. Solomon, MESSENGER observations of Mercury's dayside magnetosphere under extreme solar wind conditions. J. Geophys. Res. **119** (2014). doi:10.1002/2014JA020319

M. Temerin, X. Li, Dst model for 1995–2002. J. Geophys. Res. **111** (2006). doi:10.1029/2005JA011257

R.M. Thorne, W. Li, B. Ni, Q. Ma, J. Bornik, L. Chen, D.N. Baker, H.E. Spence, G.D. Reeves, M.G. Henderson, C.A. Kletzing, W.S. Kurth, G.B. Hospodarsky, J.B. Blake, J.F. Fennell, S.G. Claudepierre,

S.G. Kanekal, Rapid local acceleration of relativistic radiation-belt electrons by magnetospheric chorus. Nature **506** (2013). doi:10.1038/nature12889
R. Tousey, The solar corona. Space Res. **XIII**, 713–730 (1973)
G. Trottet, S. Samwel, K.-L. Klein, T. Dudok de Wit, R. Miteva, Statistical evidence for contributions of flares and coronal mass ejections to major solar energetic particle events. Sol. Phys. **290**, 819–839 (2015). doi:10.1007/s11207-014-0628-1
S. von Alfthan, D. Pokhotelov, Y. Kempf, S. Hoilijoki, I. Honkonen, A. Sandroos, M. Palmroth, Vlasiator: first global hybrid-Vlasov simulations of Earth's foreshock and magnetosheath. J. Atmos. Sol.-Terr. Phys. **120**, 24–35 (2014). doi:10.1016/j.jastp.2014.08.012
V. Yurchyshyn, Y. S., V. Abramenko, H. Wang, N. Gopalswamy, Statistical distributions of speeds of coronal mass ejections. Astrophys. J. **619** (2005). doi:10.1086/426129

Space Sci Rev (2017) 212:1253–1270
DOI 10.1007/s11214-017-0408-y

Space Weather: Historical and Contemporary Perspectives

Louis J. Lanzerotti[1]

Received: 3 June 2017 / Accepted: 4 August 2017 / Published online: 4 September 2017

Abstract A somewhat personalized overview is presented of the effects of solar-terrestrial processes on electrical technologies, beginning with the electrical telegraphs in the mid-eighteenth century to the current era of extensive ground- and space-based commercial and governmental technical systems (including national security) upon which modern society depends. As human technologies increased in diversity over the last century and a half, space weather effects have continued to be of significant importance in the successful design and operations of many of these technologies.

Keywords History · Space weather · Electrical technologies

> "*. . . during the appearance of the aurora borealis of November 17, 1848, the soft iron armatures employed in the electric telegraph between Florence and Pisa remain attached to their electromagnetics, as if the latter were powerfully magnetized, without, however, the apparatus being in action, and without the currents in the battery being set in action.*" (Prescott 1875).

1 Introduction

Within a few decades in the early to mid 19th century, the electrical telegraph had revolutionized communications within countries, and across countries and oceans (Standage 1998). The telegraph had eliminated the line of sight semaphore signaling that had begun to be emplaced across Europe in the late 18th century (Napoleon had planned a line to Boulogne in preparation for his invasion of England), and in the eastern U.S. For optimum

The Scientific Foundation of Space Weather
Edited by Rudolf von Steiger, Daniel Baker, André Balogh, Tamás Gombosi, Hannu Koskinen and Astrid Veronig

✉ L.J. Lanzerotti
ljl@njit.edu

[1] Center for Solar Terrestrial Research, New Jersey Institute of Technology, Newark, NJ 07102, USA

signaling, semaphores were placed in so far as feasible on elevated locations ("telegraph hills"). Such signaling required significant manpower for relaying, and could often be obscured by weather. The electrical telegraph was a communications revolution, much faster than semaphores and not affected by visible weather. But the telegraph could be disturbed by electrical discharges from lightning.

The development of an electric telegraph of some type was of considerable importance to railway companies of the 19th century as railroads expanded across various nations. For example, as written in 1838 with respect to the development of telegraphs, "Probably no subject will contribute more to guard railway travellers [sic] against the accidents of collisions than the perfection of some rapid mode of communications from station to station, by giving timely notice of whatever may be on the road" (Electro-Magnetic Telegraphs 1838). That the electrical telegraph still required considerable development at that time were these hopeful words "And as it is thought that electro-galvanic telegraphs, if they can ever be perfected, bid fair to be among the most useful....". Thus the beginnings of the first technology—for communications—that would be subject to space weather.

Some sixty or more claimants to the concept of the telegraph existed up to about 1838 (Prescott 1875, page 425). But the February 1854 U.S. Supreme Court decision (Silverman 2003) settled the issue once and for all that the telegraph embodiment by Samuel F.B. Morse was the patent that counted: "Writing, printing or recording at a distance ... was never invented, perfected, or put into practical operation, till it was done by Morse." (Standage 1998, page 183).

The electrical telegraph operated under the simple principal that a long wire, with its attached telegraph key, was attached to a battery whose other pole was grounded by a conducting plate buried in the Earth. The wire at the other end was attached to a receiver that in turn was affixed to a wire connected to a conducting plate also buried in the ground. Thus, the Earth provided the return electrical path. However, Morse's original line in 1844 from Washington to Baltimore (along the Baltimore & Ohio Railroad Company tracks), which carried the first words (in Morse code) "What hath God wrought", consisted of two wires, that is, an "entire metallic circuit" (Prescott 1875, page 54). This line therefore did not need the Earth for a return circuit. The biblical phrase that was transmitted was suggested by the mother of Annie Ellsworth, the daughter of Henry Ellsworth, the Commissioner of Patents and a Yale classmate of Morse (Silverman 2003).

Studies of the Earth as a conductor were carried out by a significant number of "philosophers" in the early 19th century, many in the context of studies of various types of electrical telegraphy. Such individuals included Gauss and Steinheil in Germany, Bain and Wheatstone in England, and Matteucci in Italy (Prescott 1875). The conclusion to use the Earth as a return conductor in this first electrical technology—a technology of long conductors—is the underlying reason why what is now called space weather became of significant importance to any technical system that needed long conductors for its implementation.

The following broad overview of space weather and aspects of its historical contexts are influenced by the author's personal experiences and research in the field, as well as by discussions and collaborations with national and international colleagues over many decades. This overview is not intended to be a comprehensive historical accounting, but rather the providing of a number of important, by the author's judgment, technical developments and occurrences as space weather became an ever-more prevalent element in the design and operations of many technologies. A *Resource Letter* on space weather by Baker and Lanzerotti (2016) contains some historical and more contemporary references.

The underlying theme of this overview is

as human technologies increased in number, and especially in diversity, over the last 170 or so years, space weather effects have continued to be of significance in the successful design and operations of many of these technologies—particularly those used for communications and navigation, for both civilian and national security objectives.

2 Technologies Based upon Long Conductors

Long conductors with groundings at both ends are subject to what are now commonly called geomagnetically-induced currents (GICs) or telluric currents or earth currents. Space weather processes originating in the Sun produce these currents. From the earliest electrical technology employing long conductors until about 1940, GICs were the nemesis of, first, telegraph lines and then telephone circuits. Following the ever-more interconnectedness of power systems on all continents beginning in the mid-20th century, and especially the geomagnetic event of Easter Sunday in March 1940, the effects of GICs on power grids have become of very significant interest and concern. Knipp (2015) organized and moderated a collection of contemporary GIC-related publications, largely from American Geophysical Union journals, on GICs, and their especial present-day relevance to electrical power systems (Gaunt 2016).

Unknown currents were observed on the early electrical telegraphs, causing anomalous operations of the receivers. The first documentation of these spontaneous currents was in a publication by William Henry Barlow, an English civil engineer who at the time was the superintendent engineer of the Midland Counties Railway. Barlow (1849) reported that his "attention was strongly drawn to the subject" by his observation that during an aurora on the evening of May 19, 1847, "strong alternating deflections occurred on all the [telegraph] instruments". He further wrote that "... in every case which has come under my observation, the telegraph needles have been deflected whenever aurora has been visible".

Barlow determined to make "a systematic set of observations". He executed an experimental observation program employing spare, unused telegraph wires along the rail lines. First, he determined experimentally that only when both ends of a wire were connected to the ground were these spontaneous currents observed: not when only one end was grounded, or when the wire was floating. Then he took five-minute data on the position of the needle of a galvanometer over a two-week period. A plot of the tabular mean hourly data (from five minute observation points) in Barlow's publication from the Derby galvanometer for the Derby to Birmingham railway line ($\sim$70 km) is shown in Fig. 1.

These data taken in May 1847 show large excursions in the galvanometer deflections during these two weeks (weekend days of data apparently not taken). Large daily excursions are seen during each week, and especially so during the first week. Barlow remarked about the occurrences of these diurnal variations. His measurements on these telegraph lines can be considered to represent the first observations of what is now well known as the ionosphere Sq current system (e.g., Matsushita 1969).

These diurnal variations remained puzzles to telegraph operators and engineers for years. For example, Richard Spelman Culley, an English engineer, listed the times of maximum excursion of diurnal galvanometer readings at stations located from Madras (now Chennai) to Munich to England to Ireland and to Toronto (Culley 1874). He stressed that the times of greatest and least diurnal excursion of the galvanometer needles "differ according to locality", but did not draw any conclusions as to the possible origin of the excursions.

Barlow includes in his paper as a Postscript a communication from Culley dated May 8, 1848. Culley writes that he has "twice this winter foretold an aurora: the connection between

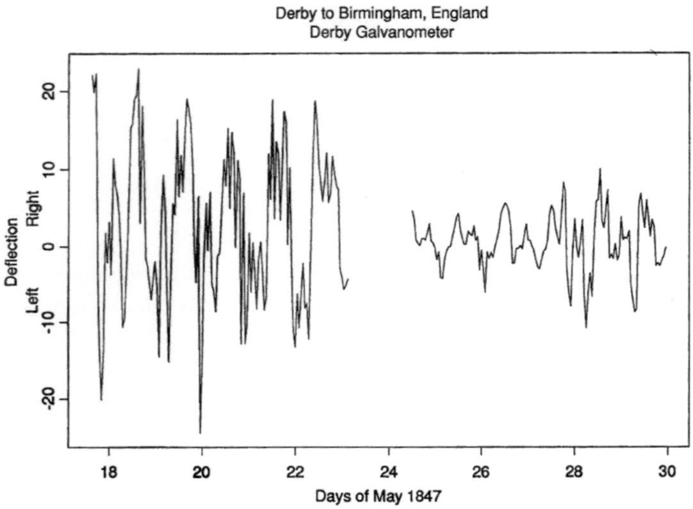

Fig. 1 Hourly mean declination readings of the Derby galvanometer on a telegraph line from Derby to Birmingham of the Midland Counties Railway Company. Adopted from tabular data in Barlow (1849)

this phenomenon and the deflections [of a galvanometer needle] is indisputable." In the section devoted to Earth Currents in his *Handbook*, Culley (1874) writes "The telegraph always foretells an aurora". These remarks by Culley could be considered the initial steps in the "forecasting" of solar-terrestrial processes and of space weather effects on technical systems. Prior to Culley's proposal of a technology (the telegraph) foretelling an aurora, a correlation of auroral sightings and deflections of a compass needle showing magnetic activity was studied in the early 18th century by the Swedish scientists Anders Celcius and his student Olof Hiorter (e.g., Odenwlad 2007).

Much has been written and compiled about the exceptionally large geomagnetic storm interval that followed the observation of the first white light solar flare by Richard Christopher Carrington in 1859 (Carrington 1863), more than a decade after the work of Barlow. Prescott (1875), Superintendent of Electric Telegraph Lines in Boston, reports extensively on the telegraph impacts of the geomagnetic activity during the interval from August 28th through about September 2nd, 1859. A 120 page volume on the event as it affected telegraph systems was written soon afterwards by Clement (1860). Shea and Smart (2006) compiled a set of eight articles on the Carrington event that were published in the *American Journal of Science* in the two-year period following the event.

In more recent times, space weather has become of considerably more interest and importance in the context of solar-terrestrial effects on technological systems. This 1859 geomagnetic storm ("the Carrington event") has served as the reference point for defining great storms, and for identifying impacts from such a great storm on contemporary technologies, including long conductors (e.g., Cliver 2006; National Academies 2008; Tsurutani et al. 2003; Siscoe et al. 2006; Clauer and Siscoe 2006; Baker et al. 2013; Chigomezyo et al. 2013).

Prescott (1860, 1875) records a number of first hand experiences of the telegraph operators during the Carrington event. Other of his observations, and of a number of other operators in the U.S. and Europe, are contained in the eight articles compiled by Shea and Smart (2006). They all make for interesting reading. A flavor of the experiences of operators can be obtained from an exchange that Prescott (1860, 1875) records between operators in

Fig. 2 Old trans-Atlantic telegraph cables entering from Trinity Bay into the historic Heart's Content, Newfoundland, cable station (*red building* in background)

Boston and Portland, Maine. The two operators discuss the huge anomalous current situation they are experiencing. They decide to try to operate without their batteries. The Boston operator is quoted as writing to the Portland operator "We are working with the aid of the aurora alone. How do you receive my writing?" The Boston operator responds "Very well indeed, much better than when the batteries were on; the current is steadier and more reliable." Prescott (1875) concludes by noting that the Boston-Portland line "was worked in this manner more than two hours, when, the aurora having subsided, the batteries were resumed."

The successes of telegraph on land led quickly to extend communications across bodies of water (e.g., Newfoundland to New York), and the Atlantic Ocean. The saga of the American Atlantic cable entrepreneur Cyrus Field, who made his fortune in the paper industry, has been recorded in many volumes (e.g., Standage 1998; Gordon 2003; Rowe 2009). The initial cable laid in 1858, from Valentia, Ireland, to Bull Arm (Trinity Bay), Newfoundland, failed after about one month. Finally, success was achieved in 1866 with the connection from Valentia to Heart's Content (Trinity Bay), Newfoundland (Fig. 2). So successful was it, the Heart's Content cable station was in operation for telegraphy for essentially a century, until 1965.

Thus, the first Atlantic cable missed recording the 1859 Carrington event. Nevertheless, as Rowe (2009) relates for the transatlantic connection, "the beginning of cable service was far from reliable [...] earth's magnetic currents, the aurora borealis [...] sent the [galvanometer] into wild and rapid gyrations." Gordon (2003) relates that as the cable was being laid there was a "tiny whispering effect [of] the earth's magnetic field [...] on the current in the cable."

It is of interest in the context of the history of space weather that the individual who was centrally involved in engineering the solution to the transmission of signals in undersea

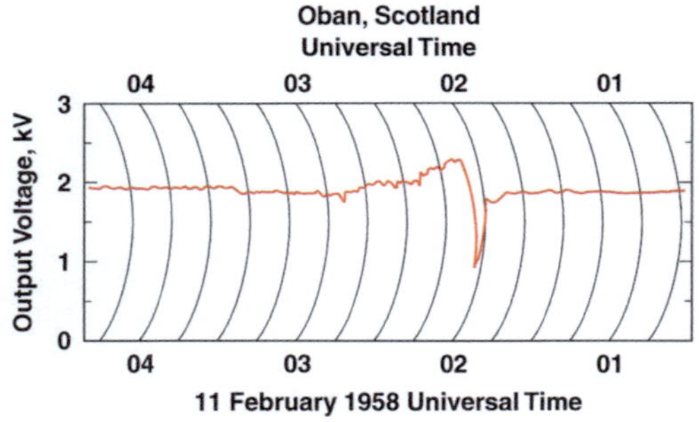

Fig. 3 Voltage at the AT&T Oban, Scotland, telecommunications station on one of the two Oban to Clarenville, Newfoundland, TAT-1 cables during the February 11, 1958, geomagnetic storm. Adapted from Axe (1968)

cables was also the individual who was most influential in arguing against the effect of the sun on geomagnetism and the aurora, and thus on the production of large earth currents that disturbed telegraph systems. Sir William Thomson (Lord Kelvin, Kelvin being the name of the small river near the campus of the University of Glasgow where Thomson spent his career) addressed and provided solutions for the data rate problem for a cable that is laid in the conducting medium of seawater (Lindley 2004). Lord Kelvin's 1892 address to the Royal Society is often cited as the summary basis for his influence in shaping the intellectual opposition to solar influences on the aurora, and thus earth currents (Kelvin 1892).

The huge geomagnetic storm event of May 14–15, 1921, was of particular interest in that Washington, DC, "was cut off telegraphically from the rest of the United States" (Odenwald 2015). *The New York Times* (May 15, 1921) carried a front page story where "The American Telegraph and Telephone Company … reported that the disturbance [on its lines] was unprecedented in the memory of the men thirty years in the service. At times, every wire was 'down' and operators could not get them back into service although they kept trying to communicate with city after city." Anderson (1928) wrote that during the geomagnetic storm "…grounded telegraph circuits were rendered inoperative due to excessive earth potentials which burned out line protectors, broke down condensers and insulation, and even started fires in terminal offices."

As the complexities of technical systems increase, as new technologies are invented and employed, human-built systems can be susceptible to the effects of earth's space environment. This is exemplified by the advance made in communications when cables (long conductors) were laid under the Atlantic to carry not telegraph, but telephone, traffic. The first transatlantic telephone circuits were established in September 1956, nearly a century after the first telegraph cable. This telephone cable from Oban, Scotland, to Clarenville, Newfoundland, consisted of two cables, one carrying voice from Oban to Clarenville and the other carrying voice in the reverse direction. The cable was retired after two decades, in 1978. A year and a half after the start of telephone transmissions, the magnetic storm of 10 February 1958 produced about an hour disruption of telephone traffic on the cable (Axe 1968; Meloni et al. 1983).

Because of the widespread disturbances, including on power lines, that the 1958 magnetic storm event caused, an article in the *New Yorker* magazine was devoted to it (Brooks 1959). Brooks wrote that "At almost the exact moment when the magnetograph traces leaped and the aurora flared up, huge currents in the earth, induced by the heavenly turbulence, manifested themselves not only in power lines in Canada but in cables under the north Atlantic". Voltage excursions on the telephone power supply at Oban exceeded 1 kV to produce the communications disruption (Fig. 3; Axe 1968).

Swedenborg and Wyke (1959) documented effects in the communications systems in Sweden during the February 1958 storm. Effects included voltage swings, spurious alarms,

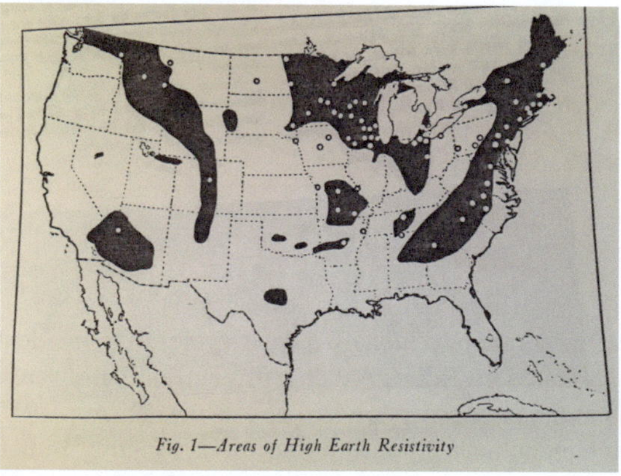

Fig. 4 Areas of high Earth resistivity (*black shaded areas*) with locations shown (*open circles*) of AT&T Long Lines system telecommunications problems during the Easter Sunday, March 24, 1940, geomagnetic storm. Adapted from Germaine (1940)

and even fire in some cables. They wrote "The effects of the magnetic storm of 11th February 1958 stressed the need of an immediate inquiry into the possibilities of organizing an efficient warning service and the means of providing for preventive measures." They also wrote that the "proposed warning service should be organized so as to avoid as far as possible the emission of unnecessary warnings." These features remain today pillars for space weather forecast and modeling services.

Brooks further noted with respect to the February 1958 event "circuit breakers began tripping out in Ontario transformer stations, plunging the Toronto area into a temporary darkness broken only by the strange light of the aurora overhead." This 1958 storm was not the first large and widespread disturbance of long conductor electric power systems. That can be attributed, in so far as I have determined, to the very large Easter Sunday 24 March 1940 geomagnetic storm (e.g., Nicholson 1940; McNish 1940; Davidson 1940) that disrupted telegraph, telephone, and electrical power across a considerable portion of the U.S. The Edison Electric Institute convened a meeting in Chicago in May 1940, just after the March event at which representatives of telephone and power companies made presentations on the event.

Davidson (1940) provided a lengthy list and discussion of impacts on power companies in the U.S. and Canada from the March 1940 storm. Disturbances were recorded by 10 companies in New England, New York, eastern Pennsylvania, and southern and eastern parts of Minnesota, and in Ontario and Quebec. There were 7 cases of voltage dips, 5 cases of tripping transformer banks, and 4 cases of large surges or swings in reactive power.

Germaine (1940), in his discussion of the effects of the March 1940 storm, notes that the highest voltages across telephone cables "as a rule will be experienced in those territories where the earth resistivity is highest. . . ." He presented a map of the continental U.S. showing high resistivity areas (shaded black in Fig. 4) and locations where anomalies were seen in the AT&T Long Lines system (open dots in Fig. 4). A sample of large voltage swings (as much as 500 V overall) for several hours on the line from Minneapolis to Fargo is shown in Fig. 5 (Germaine 1940). The communications impacts of the 1940 event were sufficient to warrant a one-page article in Life Magazine (1940). The article reproduced Germaine's resistivity map as well as contained three pictures of the originating sunspot region moving across the solar disk.

Beginning in the decade preceding the March 1940 event, a cooperative project of the United States Coast and Geodetic Survey, the Carnegie Institution of Washington, and the AT&T Company carried out studies of earth currents at the Tucson magnetic observatory in Arizona (Rooney 1949). This program, lasting from 1931 until 1943, resulted in nearly

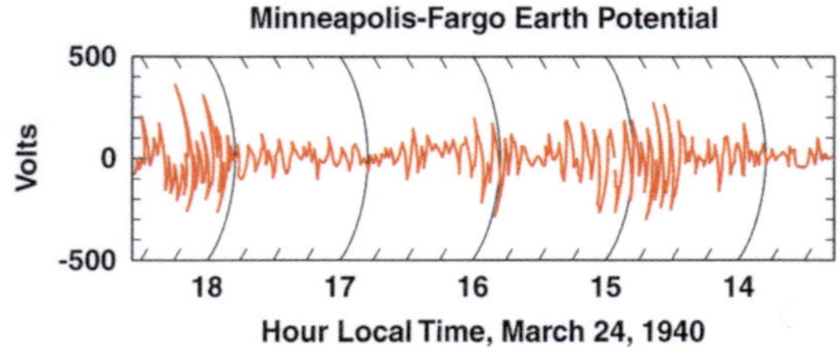

Fig. 5 Minneapolis, Minnesota, to Fargo, North Dakota, Earth potential recorded on AT&T telecommunications line during the March 24, 1940, geomagnetic storm. Adapted from Germaine (1940)

12 years (approximately a solar cycle) of coincident magnetic and telluric current studies. As stated by Rooney, such earth current data were rare in the literature. He further stated that the only exception that he knew were from 1910–1938 at the Ebro, Spain, observatory, which published hourly values in its Monthly Bulletins. The interests of Bell Laboratories, and thus AT&T, in this project undoubtedly were as a result of the GIC problems that continued to occur in long telephone lines. Employed in the Tucson-based study were unused, approximately orthogonal, lines of the Mountain States Telephone and Telegraph Company. One line ran 90 km due east of Tucson and one ran about 60 km a little east of north. It is not clear how the results of these studies may have influenced the operations of long conductor AT&T telephone lines.

Following the 1958 event, at the beginning of the space age, concern for long conductors began to become more centered on electric power grids. For example, the very large solar event of August 1972 (occurring between the last two Apollo flights to the moon) caused power disruptions in regions as widespread as Newfoundland, Wisconsin, South Dakota, and Manitoba, and destroyed a transformer in British Columbia (Odenwald 2015). An AT&T long lines telephone cable from Plano, IL, to Cascade, IA, was rendered out of service (Anderson et al. 1974). Geomagnetic depth sounding studies of the Plano-Cascade region were carried out (Meloni et al. 1985). Design changes were made in telephone cable power systems following this 1972 event. The changes in the powering system would appear to have prevented the outage of the first transatlantic fiber cable (Medford et al. 1989) in the March 1989 geomagnetic storm that caused the electrical power blackout of the province of Quebec (Czech et al. 1989).

A considerable amount of research was conducted in the high resistivity area of northern Minnesota (Fig. 4), stimulated by space weather effects on power systems (e.g., Albertson et al. 1970, 1974; Kappenman 1996). The large geomagnetic storm of March 1989 which blacked out Quebec, and the "Halloween storm" of 2003 produced a large increase in research and development related to geomagnetic storm effects on long conductors, both in the U.S. and in Europe (e.g., Czech et al. 1989; Bolduc 2002; Kappenman 2005; National Academies 2008; Wik et al. 2009; Royal Academy of Engineering 2013).

Pipelines are another class of long conductor that can be affected by anomalous space weather-induced telluric currents. These earth currents can produce enhanced erosion of pipelines if there are breaks in insulation, and can cause disturbances and outage of corrosion-protection electronics (e.g., Campbell 1978, 1980; Viljanen et al. 2006).

While there has been considerable research on individual events, and modeling to address such events, statistics on the occurrence and occurrence frequency of GICs in specific locations are few. Such statistics would be useful in achieving deeper understanding of local ground conductivities in specific areas, and the overhead ionosphere current systems that can develop in geomagnetic disturbances. Studies of this nature include Lanzerotti et al. (1993) and Pulkkinen et al. (2012). The lengthy data set of Rooney (1949) has never been analyzed in a statistical manner.

3 Wireless Technologies

The first wireless telegraph signals using electromagnetic waves were sent by Guglielmo Marconi in 1895. He carried out many experiments from his parent's home in Villa Griffone outside Bologna (Weightman 2003). Marconi was one of many in Europe, the UK, and the U.S. who were intrigued by the possibility to send Morse Code signals by the electromagnetic waves that had been discovered by Heinrich Hertz based upon the theoretical work of electromagnetic forces proposed by James Clerk Maxwell.

Marconi was an astute businessperson as well as talented in exploiting technology. He recognized that his ship-to-shore wireless businesses in England could, and should, be extended to intercontinental distances. He achieved a demonstration of this with the reception of the Morse code letter "S" (dot dot dot) in blizzard conditions on Telegraph Hill, St. John's, Newfoundland, on the 12th and 13th of December 1901 (Weightman 2003). There had been considerable skepticism surrounding Marconi's belief that the wireless signals transmitted from Poldhu on Cornwall would be received across the curvature of Earth. Of course, the reason why this achievement was possible (and that the Poldhu signals did not just propagate off into space) was the existence of the as-yet-to-be discovered ionosphere. And the ionosphere would be the geophysical entity that would be influenced by space weather phenomena. Breit and Tuve (1925) and Appleton and Barnett (1925) physically identified the reflecting layer more than two decades later. Marconi (1909) and Appleton (1947) were awarded Nobel prizes for their work. It is relevant to note that in the final paragraph of his Nobel Prize Lecture, Sir Edward Appleton addressed the "practical side of applications". He discussed the "forecast of what I may call the 'ionosphere weather' some time ahead" for the "most suitable wavelengths at any time of the day, over any distance of transmission, at any part of the world." His phrase "ionosphere weather" is a forerunner to today's "space weather".

Wireless telegraphy was the second human-built system that was, unbeknownst to Marconi and subsequent developers, susceptible to space weather effects in Earth's space environment. The same electrical currents in Earth's ionosphere that produced the "spontaneous" telluric currents that disrupted cable telegraph could also disrupt the propagation and reception of long-distance wireless signals. Wireless engineers noted this at the time, and Marconi wrote (1928) "... times of bad fading [of radio signals] practically always coincide with the appearance of large sun-spots and intense aurora-borealis usually accompanied by magnetic storms...." He importantly noted that these are "... the same periods when cables and land lines experience difficulties or are thrown out of action."

The telephone communications industry early saw the importance of wireless for voice transmission. And also experienced the effects of space weather on transmissions, as illustrated in Fig. 6 (reproduced from Fagen 1975, which contains information on early wireless research in the former Bell Telephone System). The intensities in the signal strength (upper curve) were obtained by averaging the signal strength values from about ten European stations that were broadcasting in the ~15–23 kHz band (very long wavelengths), after reducing them to the signal from Nauen, Germany, as the common base. The relationship of the signal strength to the monthly-averaged sunspot numbers per year for two solar cycles in the early 20th century is clear. The physical reason(s) for such an association was obscure at this time, of course, but such a relationship could provide communications engineers some gross expectations as to signal quality and strength on a year-to-year basis. Such an example of industrial interest in solar effects on wireless telephony can be associated with very early "forecasting" of space weather phenomena affecting technologies.

As described above, Marconi noted that disturbances on wireless transmissions occurred at the same time that wired communications were affected. *The New York Times* (May 15,

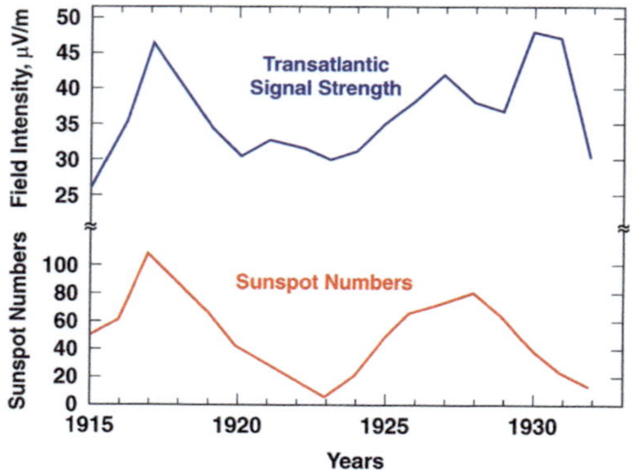

Fig. 6 *Upper trace*: yearly transatlantic wireless signal strength (∼15–23 kHz band) for approximately two solar cycles in the very early days of wireless telegraphy. *Lower trace*: yearly sunspot numbers. Adopted from Fagen (1975)

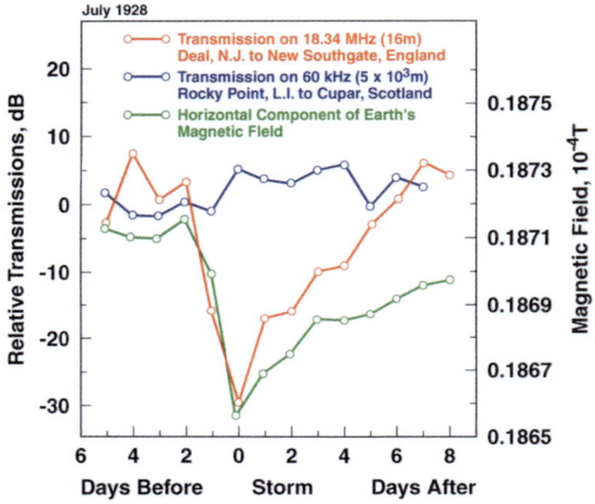

Fig. 7 Transatlantic wireless transmissions on two different frequencies (*blue* and *red*) during a geomagnetic storm in July 1928. Shown *in green* is the horizontal component of Earth's geomagnetic field (location unspecified). Adapted from Bowen (1930)

1921) reported, "Curiously, the disturbance [of May 15, 1921] that tied up the land wires seemed to strengthen the signals of the wireless apparatus. Operators in the special wireless station of *The New York Times* reported that their signals from Berlin and the Lafayette station at Bordeaux were much stronger [during] the period of the greatest disturbance on the land wires." Anderson (1928) provides transmission data from three locations in the eastern United States to Chelmsford, England, during this large event when both wireless and wired communication services were affected.

In fact, in those early days where the ionosphere was so important in facilitating transmissions (even though the ionosphere was not "discovered" or understood), geomagnetic storms would affect different frequencies in quite different ways. As *The New York Times* (January 23, 1938) subhead on a front-page article about the solar event of January 22, 1938 reported "Transoceanic Services Transfer Phone and Other Traffic to Long Wave Lengths as Sun Spot Disturbance Strikes". This is illustrated in Fig. 7 for a geomagnetic storm event in July 1928 (Bowen 1930). Here the transmissions at long wavelengths across the Atlantic were relatively undisturbed while those at the shorter wavelength were basically inoperable.

It is important to recognize that space weather effects did not abate and also did not hinder the commercialization of wired and wireless communications technologies. This is the general trend one finds in technology advances that can be affected by solar-induced processes. Bowen (1937) reviews the developments in transoceanic radiotelephone in the decade since the beginning of commercial service in January 1927. The technological advances were huge. At the time (1937) that he writes, United States subscribers could reach 68 other countries, and the radiotelephone circuits that he illustrated in his Fig. 1 permitted the interconnection of "about ninety-three percent of the world's telephones". He does not once address solar effects in his review, even though he was an early engineer addressing the topic (Bowen 1930). As a side note, Ralph Bowen became research director at Bell Telephone Laboratories, and led the 1948 press conference that announced the invention of the transistor by John Bardeen, Walter Brattain, and William Shockley (the solid state device that revolutionized communications technologies, as well as spawned countless inventions and industries in many fields).

The use of electromagnetic waves in new technologies has exploded since the 1930s. Radar was first employed for offense and defense in WWII. British radars were temporarily blinded (with the blinding initially attributed to enemy jamming) by the large solar radio burst in February 1942 (Hewish 2002). The large solar event of May 1967, including the radio burst content (Castelli et al. 1968) that interfered with radar was recently re-visited by Knipp et al. (2016), who discussed the cold war implications. Solar radio noise and bursts continues to be of design and operational concern for the potential of interference with radar and navigation (GPS) technologies (e.g., Cerruti et al. 2006, 2008).

The ionosphere was found by the COMSAT engineer Taur (1973, 1976) to disturb even the GHz frequencies used for satellite-to-ground signals. The scintillation of these signals has proven to be a significant impediment to communications and navigation, especially in the equatorial and auroral regions (e.g., Basu and Basu 1981). The scintillations and disturbances of the signals by the ionosphere have critical implications for national security interests (e.g., Kelly et al. 2014).

Temporal changes in Earth's natural magnetic field due to geomagnetic disturbances and storms are used for geomagnetic depth sounding of Earth's crust and upper mantle (magnetotelluric technique). Geomagnetic field variations from minutes to many hours are measured and analyzed to investigate a large range of depths into the Earth (Chave and Jones 2012). Exploration for minerals, other resources of economic interest, and even archeological artifacts are often conducted by surveys using magnetic sensing instruments on land and from aircraft. Such surveys can be severely compromised if the geomagnetic conditions are disturbed (e.g., Reeves 1993). Thus, depending upon the objective, natural geomagnetic field variations can be either a source or a disturbance for exploration.

4 Air and Space Technologies

4.1 Spacecraft Radiation Effects

When shortly after WWII Sir Arthur Clarke proposed his crewed geosynchronous communications satellite, he did not envision that Earth's space environment would be anything but benign. Neither did John Pierce at AT&T Bell Laboratories when he proposed the low altitude active communications satellite Telstar-1 (launched in July 1962). The surprising discovery by Van Allen (1959) with an instrument on the Explorer 1 satellite of the intense fluxes of radiation trapped in Earth's magnetic field meant that any technologies that humans placed into space would have to contend with this radiation. Research on Earth's (and the interplanetary medium) radiation environment continues at a high and sophisticated level to this day.

Extensive references are provided in Hastings and Garrett (2004), Baker and Lanzerotti (2016), and Pisacane (2016) on the effects of the space environment on space systems. The

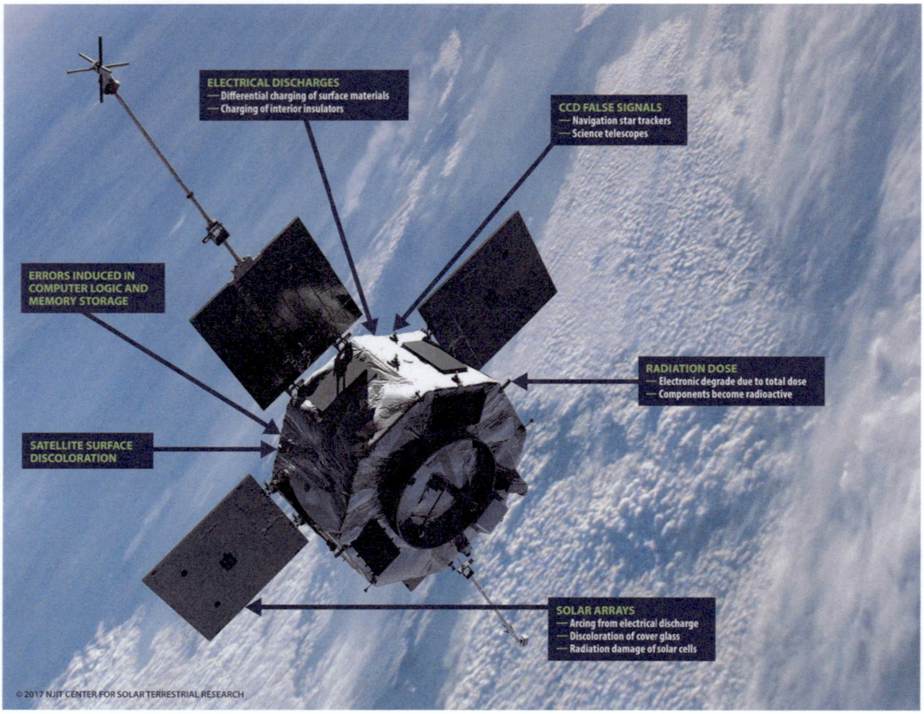

Fig. 8 Space radiation effects on space systems

major dominant radiation effects are illustrated in Fig. 8. Radiation impinging on solar arrays, on spacecraft surface materials, and on insulating materials deep within a spacecraft can produce high levels of charge buildup on the arrays and in the insulators (e.g., Vampola 1987; Lanzerotti et al. 1998; Garrett and Whittlesey 2012). When the charges become too large they will break down and will produce electromagnetic signals that can interfere with the satellite electronics (as an atmospheric lightning discharge produces interference in radio receivers).

Radiation will also discolor spacecraft surface materials (such as thermal insulations) and the cover glasses that are over the solar arrays (e.g., Bouquet and Phillips 1983). Radiation can also cause components on satellites to become radioactive. Errors will be induced in computer logic systems and solid-state memories (e.g., Lohmeyer et al. 2015; Clements et al. 2016). Radiation will induce false signals in scientific optical instruments and in sun and star trackers that are used for spacecraft attitude control (Birnbaum 1996).

As a space vehicle continues to reside within the radiation environment, the impinging radiation will slowly degrade the solar arrays, decreasing the power available to operations of the payload. Solar array degradation as measured on the Van Allen Probes B spacecraft for 1152 mission days post launch (August 30, 2012) is shown in Fig. 9. The degradation of two separate solar array patches on spacecraft B is illustrated. These data and the design of the solar arrays are presented in detail in Butler (2016). The sizes of satellite solar arrays are designed to take into account in so far as feasible such degradation over the time of the designed mission.

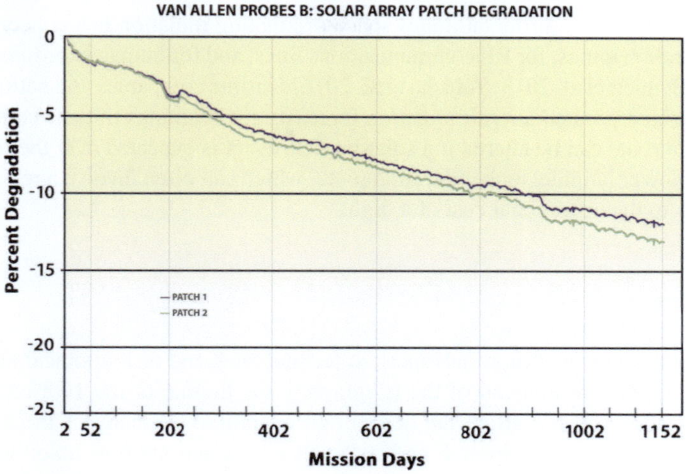

Fig. 9 Solar array patch degradation for two array patches on Van Allen Probes B normalized to Day 2 (August 31, 2012) of the mission. The data plotted are the short circuit current of each solar patch showing the decrease over the first 1100 days of the mission

4.2 Satellite Drag

Variations in the properties (density, composition, temperature, winds) of Earth's atmosphere above a few hundred km under differing solar and geomagnetic conditions are often referred to as the second major understanding of the space age, after the discovery of the radiation belts (Jacchia 1963). Studies of orbits of satellites at low Earth altitudes, and of satellites that have low altitude perigees, yield data that provide some of this atmospheric information (e.g., King-Hele 1992; Knipp et al. 2004; Moe and Moe 2011).

These data are also needed, and are used, in atmospheric models for the prediction of the orbits of satellites (and space debris) around Earth. Predictive orbital knowledge is essential for monitoring space assets of potentially hostile entities. Predictive knowledge of satellite orbits also is essential for spacecraft operators to adjust orbital conditions using on-board fuel. Adjusting satellite orbits can prolong lifetimes, and contribute to de-orbiting at the end of useful life. Predictions of the locations of space debris enable satellite operators (including those operating the International Space Station) to maneuver their satellites to avoid mission-ending collisions.

4.3 Space Radiation at Aircraft Altitudes

Aircraft flying at altitudes typical of jet airplane travel encounter galactic cosmic ray radiation because of reduced magnetic field and atmospheric shielding at flight altitudes (Knipp 2017). The amount of enhanced radiation depends upon the latitude of the flight paths; those at lower latitudes encounter less cosmic ray intensity due to the larger screening of Earth's magnetic field. Following the end of the cold war, aircraft from North America could fly the shorter, great circle, routes over the North Polar Region to reach Asian destinations, and vice versa. Such flight paths encounter more galactic cosmic rays. Further, and importantly, energetic solar particles from solar events can penetrate down polar magnetic field lines to jet flight altitudes. These solar particles can present substantially enhanced radiation conditions for such flights (e.g., Kataoka et al. 2015).

Therefore, at typical airline altitudes space-originating radiation is a concern for its effects on aircraft avionics, for RF communications links, and for human exposure (e.g., Jones et al. 2005; Bennett et al. 2013; Tobiska et al. 2016). Airline companies and national security entities have incorporated aircraft radiation forecasts and warnings in their control centers. A great circle route can be altered if a solar particle event is expected over the polar region. The altered, lower latitude, longer duration route, which can often involve a re-fueling, adds considerably to the operational cost of a flight.

5 Summary

As Fig. 10 illustrates, enormous advances in technologies and technological systems have occurred since the development of the telegraph in the middle of the 19th century. Solar-terrestrial occurrences can affect the design and operations of many of these systems as is illustrated schematically. Ground- and space-based systems are now ubiquitous not only for communications, but for navigation, weather, and critical national security purposes. Contemporary understanding of Earth's space weather environment, from the sun to the top of the atmosphere, has little comparison to Carrington's solar research.

Research on Earth's space environment has of necessity become more sophisticated in order to provide the "space weather" data required to design and operate the ground-based and satellite systems that use contemporary cutting edge technologies (Baker and Lanzerotti 2016). Because of its critical importance, frontier space environment research continues to be of high relevance and importance, and is supported by national governments and industry.

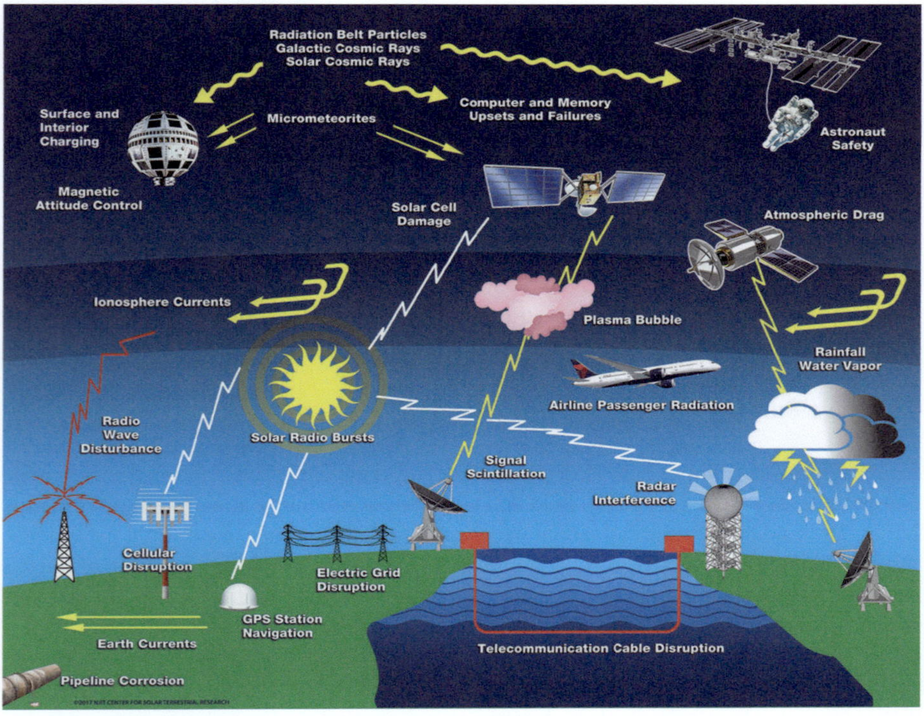

Fig. 10 Solar-terrestrial effects on contemporary technologies

As technologies—whether on the ground or in space—that can be impacted by space processes continue to advance in sophistication, the understanding (and even predictability) of Earth's dynamic space weather environment will need to similarly advance in sophistication by forefront research in order to ensure operational success and mission survival.

References

V.D. Albertson, S.C. Tripathy, R.E. Clayton, Electric power systems and geomagnetic disturbances, in *Proc. Midwest Power Symposium* (1970)

V.D. Albertson, J.M. Thorson Jr., S.A. Miske Jr., The effects of geomagnetic storms on electrical power systems. IEEE Trans. Power Appar. Syst. **PAS-93**, 1031 (1974)

C.W. Anderson III., L.J. Lanzerotti, C.G. Maclennan, Outage of the L4 system and the geomagnetic disturbances of 4 August 1972. Bell Syst. Tech. J. **53**, 1817–1837 (1974). doi:10.1002/j.1538-7305.1974.tb02817.x

C.N. Anderson, Correlation of long wave transatlantic radio transmission with other factors affected by solar activity. Proc. IRE **16**, 267 (1928)

E.V. Appleton, M.A.F. Barnett, Local reflection of wireless waves from the upper atmosphere. Nature **115**, 333 (1925)

G.A. Axe, The effects of the Earth's magnetism on submarine cables. Post Off. Electr. Eng. J. **61**, 37 (1968)

D.N. Baker, L.J. Lanzerotti, Resource letter SW1: Space weather. Am. J. Phys. **84**, 166–180 (2016). doi:10.1119/1.4938403

D.N. Baker, X. Li, A. Pulkkinen, C.M. Ngwira, M.L. Mays, A.B. Galvin, K.D.C. Simunac, A major solar eruptive event in July 2012: Defining extreme space weather scenarios. Space Weather **11**, 1 (2013). doi:10.1002/swe.20097

W.H. Barlow, On spontaneous electrical currents observed in the wires of the electric telegraph. Philos. Trans. R. Soc. Lond. **139**, 61–72 (1849)

S. Basu, S. Basu, Equatorial scintillations: a review. J. Atmos. Terr. Phys. **43**, 473–489 (1981)

L.G.I. Bennett, B.J. Lewis, B.H. Bennett, M.J. McCall, M. Bean, L. Doré, L. Getley, A survey of the cosmic radiation exposure of Air Canada pilots during maximum galactic radiation conditions in 2009. Radiat. Meas. **49**, 103–108 (2013). doi:10.1016/j.radmeas.2012.12.004

M.M. Birnbaum, Spacecraft attitude control using star field trackers. Acta Astronaut. **39**, 763 (1996)

L. Bolduc, GIC observations and studies in the hydro-Quebec power system. J. Atmos. Sol.-Terr. Phys. **64**, 1793 (2002)

F.L. Bouquet, A. Phillips, Simulated space radiation effects on dielectrics and coatings. IEEE Trans. Nucl. Sci. **30**, 4090 (1983)

R. Bowen, Transoceanic telephone service—Short-wave transmission. Bell Syst. Tech. J. **9**, 258 (1930)

R. Bowen, Transoceanic radio telephone development. Bell Syst. Tech. J. **16**, 560–567 (1937)

M.A. Breit, M.A. Tuve, A radio method of estimating the height of the conducting layer. Nature **116**, 357 (1925)

J. Brooks, A reporter at large; The subtle storm, New Yorker (February 19, 1959)

M. Butler, The Van Allen Probes Power System Mission Performance, in *AIAA Propulsion and Energy Forum, 14th International Energy Conversion Engineering Conference*, Salt Lake City, UT (2016)

W.H. Campbell, Induction of auroral zone electric currents within the Alaska pipeline. Pure Appl. Geophys. **16**, 1143 (1978)

W.H. Campbell, Observation of electrical currents in the Alaska oil pipeline resulting from auroral electrojet current sources. Geophys. J. R. Astron. Soc. **61**, 437 (1980)

R.C. Carrington, *Observations of the Spots on the Sun from November 9, 1853, to March 24, 1861, Made at Redhill* (Williams and Norgate, London, 1863)

J.P. Castelli, J. Aarons, G.A. Michael, The great solar radio burst of May 23, 1967. Astrophys. J. **153**, 267–273 (1968)

A.P. Cerruti, P.M. Kintner, D.E. Gary, L.J. Lanzerotti, E.R. dePaula, H.B. Vo, Observed solar radio burst effects on GPS/wide area augmentation system carrier-to-noise ratio. Space Weather **4**, S10006 (2006). doi:10.1029/2006SW000254

A.P. Cerruti, P.M. Kinter, D.E. Gary, A.J. Mannucci, R.F. Meyer, P. Doherty, A.J. Coster, Effect of intense 2006 solar radio bursts on GPS receivers. Space Weather **6**, S10D07 (2008). doi:10.1029/2007SW000375

A.D. Chave, A.G. Jones (eds.), *The Magnetotelluric Method* (Cambridge University Press, Cambridge, 2012)

M.N. Chigomezyo, A. Pulkkinen, M.L. Mays, M.M. Kuznetsova, A.B. Galvin, K. Simunac, D.N. Baker, X. Lin, Y. Zheng, A. Glocer, Simulation of the 23 July 2012 extreme space weather event: What if this extremely rare CME was Earth directed? Space Weather **11**, 671 (2013). doi:10.1002/2013SW000990

C.R. Clauer, G. Siscoe, The great historical geomagnetic storm of 1859: A modern look. Adv. Space Res. **38**, 117 (2006)

K.J. Clement, *Das Große Nordlicht in der Nacht zun 29 Aug. 1859 und die Telegraphenverwirrung in Nordamerica und Europa* (Perthes-Besser und Mauke, Hamburg, 1860)

E.B. Clements, A.K. Carlton, C.J. Joyce, N.A. Schwadron, H.E. Spence, X. Sun, K. Cahoy, Interplanetary space weather effects on lunar reconnaissance orbiter avalanche photodiode performance. Space Weather **14**, 343–350 (2016). doi:10.1002/2016SW001381

E.W. Cliver, The 1859 space weather event: Then and now. Adv. Space Res. **38**, 119 (2006). doi:10.1016/j.asr.2005.07.077

R.S. Culley, *Handbook of Practical Telegraphy* (Longmans, Green, Reader, and Dyer, London, 1874)

P. Czech, C.H. Huynh, A. Dutil, The hydro-Quebec system blackout of March 13, 1989: System response to geomagnetic disturbance, in *Proc. Geomagnetically Induced Currents Conf.*, EPRI, Palo Alto, CA (1989)

W.F. Davidson, The magnetic storm of March 24, 1940—Effects in the power system, Edison Electric Institute Bulletin, p. 365, July 1940

Electro-Magnetic Telegraphs, The Railway Magazine **33**, 365 (1838)

M.D. Fagen, *A History of Science and Engineering in the Bell System* (Bell Telephone Laboratories, Inc., Murray Hill, 1975)

H.B. Garrett, A.C. Whittlesey, *Guide to Mitigating Spacecraft Charging Effects* (Wiley, Hoboken, 2012)

C.T. Gaunt, Why space weather is relevant to electrical power systems. Space Weather **14**, 2–9 (2016). doi:10.1002/2015SW001306

L.W. Germaine, The magnetic storm of March 24, 1940—Effects in the communication system, Edison Electric Institute Bulletin, p. 367, July 1940

J.S. Gordon, *A Thread Across the Ocean: The Heroic Story of the Transatlantic Cable* (Harper, New York, 2003)

D. Hastings, H. Garrett, *Spacecraft-Environment Interactions* (Cambridge University Press, Cambridge, 2004)

A. Hewish, James Stanley Hey, M.B.E. 3 May 1909–27 February 2000. Biogr. Mem. Fellows R. Soc. **48**, 167–178 (2002)

L.G. Jacchia, Variations in the Earth's upper atmosphere as revealed by satellite drag. Rev. Mod. Phys. **35**, 973 (1963)

J.B.L. Jones, R.D. Bentley, R. Hunter, R.H.A. Iles, G.C. Taylor, D.J. Thomas, Space weather and commercial airlines. Adv. Space Res. **36**, 2258 (2005)

J.G. Kappenman, Geomagnetic storms and their impact on power systems. IEEE Power Eng. Rev. **16**, 5 (1996)

J.G. Kappenman, An overview of the impulsive geomagnetic field disturbances and power grid impacts associated with the violent Sun-Earth connection events of 29–31 October 2003 and a comparative evaluation with other contemporary storms. Space Weather **3**, S08C01 (2005)

R. Kataoka, Y. Nakagawa, T. Sato, Radiation dose of aircrews during a solar proton event without ground-level enhancement. Ann. Geophys. **33**, 75–78 (2015). doi:10.5194/angeo-33-75-2015

M.A. Kelly, J.M. Comberiate, E.S. Miller, L.J. Paxton, Progress toward forecasting of space weather effects on UHF SATCOM after Operation Anaconda. Space Weather **12**, 601–611 (2014). doi:10.1002/2014SW001081

W.T. Lord Kelvin, Presidential address to the Royal Society. Nature **47** (1892)

D. King-Hele, *A Tapestry of Orbits* (Cambridge University Press, Cambridge, 1992)

D.J. Knipp, Forward to space weather collection on geomagnetically induced currents: Commentary and research. Space Weather **13**, 742 (2015). doi:10.1002/2015SW001318

D.J. Knipp, Essential science for understanding risks from radiation for air line passengers and crews. Space Weather **15**, 549–552 (2017). doi:10.1002/2017SW001639

D.J. Knipp, W.K. Tobiska, B.A. Emery, Direct and indirect thermospheric heating sources for solar cycles 21–23. Sol. Phys. **224**, 495 (2004). doi:10.1007/s11207-005-6393-4

D.J. Knipp, A.C. Ramsey, E.D. Beard, A.L. Borright, W.B. Cade, I.M. Hewins, R.H. McFadden, W.F. Denig, L.M. Kilcommons, M.A. Shea, D.F. Smart, The May 1967 great storm and radio disruption event: Extreme space weather and extraordinary responses. Space Weather **14**, 614–633 (2016). doi:10.1002/2016SW001423

L.J. Lanzerotti, C.H. Sayres, L.V. Medford, J.S. Kraus, C.G. Maclennan, D.J. Thomson, Statistical examination of induced voltages across long oceanic telecommunications cables, in *Proceedings of the Workshop Solar-Terrestrial Predictions-IV*, Ottawa, Canada, May 18–22, 1992, vol. 1 (NOAA Environ. Res. Lab., Boulder, 1993), p. 224

L.J. Lanzerotti, K. LaFleur, C.G. Maclennan, D.W. Maurer, Geosynchronous spacecraft charging in January 1997. Geophys. Res. Lett. **25**, 2967 (1998)

Life Magazine, Spots on the face of the sun mess up earth's communications, **8**, p. 38, April 8, 1940

D. Lindley, *Degrees Kelvin, a Tale of Genius, Invention, and Tragedy* (Joseph Henry Press, Washington, 2004)

W. Lohmeyer, A. Carlton, F. Wong, M. Bodeau, A. Kennedy, K. Cahoy, Response of geostationary communications satellite solid-state power amplifiers to high-energy electron fluence. Space Weather **13**, 298–315 (2015)

G. Marconi, Radio communication. Proc. IRE **16**, 40 (1928)

S. Matsushita, Dynamo currents, winds, and electric fields. Radio Sci. **4**, 771–780 (1969). doi:10.1029/RS004i009p00771

A.G. McNish, The magnetic storm of March 24, 1940. J. Geophys. Res. **45**, 359 (1940)

L.V. Medford, L.J. Lanzerotti, J.S. Kraus, C.G. Maclennan, Transatlantic Earth potential variations during the March 1989 magnetic storm. Geophys. Res. Lett. **16**, 1145 (1989)

A. Meloni, L.J. Lanzerotti, G.P. Gregori, Induction of currents in long submarine cables by natural phenomena. Rev. Geophys. **21**, 795 (1983)

A. Meloni, L.J. Lanzerotti, L.V. Medford, Geomagnetic anomaly detected at hydromagnetic wave frequencies. J. Geophys. Res. **90**, 3569 (1985). doi:10.1029/JB090iB05p03569

K. Moe, M.M. Moe, Operational models and drag-derived density trends in the thermosphere. Space Weather **9**, S00E10 (2011)

National Academies, *Severe Space Weather Events—Understanding Societal and Economic Impacts: A Workshop Report* (2008)

S.B. Nicholson, The great magnetic storm of March 24, 1940. Publ. Astron. Soc. Pac. **52**, 169 (1940)

S. Odenwald, Solar Storms, Washington, 2015, www.astronomycafe.net

S. Odenwlad, Newspaper reporting of space weather: End of a golden age. Space Weather **5**, S11005 (2007). doi:10.1029/2007SW000344

V.L. Pisacane, *The Space Environment and Its Effects on Space Systems* (AIAA, Reston, 2016)

G.B. Prescott, Observations made at Boston, Mass, and its vicinity. Am. J. Sci. **29**, 92 (1860)

G.B. Prescott, *Theory and Practice of the Electrical Telegraph* (Osgood and Co., Boston, 1875)

A. Pulkkinen, E. Bernaubeu, J. Eichner, C. Beggan, A.W.P. Thomson, Generation of 100-year geomagnetically induced current scenarios. Space Weather **10**, S04003 (2012)

C.V. Reeves, Limitations imposed by geomagnetic variations on high quality aeromagnetic surveys. Explor. Geophys. **24**, 115 (1993)

W.J. Rooney, *Earth Current Results at Tucson Magnetic Observatory (1932–1942)*, vol. 9 (Carnegie Institution of Washington, Washington, 1949), 309 pp., Issue 175

T. Rowe, *Connecting the Continents* (Creative Publishers, St. John's, Newfoundland, 2009)

Royal Academy of Engineering, *Extreme Space Weather: Impacts on Engineered Systems and Infrastructure* (2013)

M.A. Shea, D.F. Smart, Compendium of the eight articles on the "Carrington Event" attributed to or written by Elias Loomis in the American Journal of Science 1859–1861. Adv. Space Res. **38**, 313 (2006). doi:10.1016/j.asr.2006.07.005

K. Silverman, *Lightning Man: The Accursed Life of Samuel F.B. Morse* (Knopf, New York, 2003)

G. Siscoe, N.U. Crooker, C.R. Clauer, Dst of the Carrington storm of 1859. Adv. Space Res. **38**, 173 (2006). doi:10.1016/j.asr.2005.02.102

T. Standage, *The Victorian Internet* (Walker and Co., New York, 1998)

G. Swedenborg, K. Wyke, Disturbances caused by Earth currents to telecommunication plants. Tele **1**, 41 (1959)

R.R. Taur, Ionospheric scintillation at 4 and 6 GZ. COMSAT Tech. Rev. **3**, 145–163 (1973)

R.R. Taur, Simultaneous 1.5- and 4-GHz ionospheric scintillation measurement. Radio Sci. **11**, 1029–1036 (1976)

W.K. Tobiska et al., Global real-time dose measurements using the automated radiation measurements for aerospace safety (ARMAS) system. Space Weather **14**, 1053–1080 (2016). doi:10.1002/2016SW001419

B.T. Tsurutani, W.D. Gonzalez, G.S. Lakhina, S. Alex, The extreme magnetic storm of 1–2 September 1859. J. Geophys. Res. **108**, 1268 (2003). doi:10.1029/2002JA009504

A.L. Vampola, Thick dielectric charging of high-altitude spacecraft. J. Electrost. **20**, 21–30 (1987)

J.A. Van Allen, The geomagnetically trapped corpuscular radiation. J. Geophys. Res. **64**, 1683–1689 (1959)

A. Viljanen, A. Pulkkinen, R. Pirjola, K. Pajunpaa, P. Posio, A. Koistinen, Recordings of geomagnetically induced currents and nowcasting service of the Finnish natural gas pipeline system. Space Weather **4** (2006)

G. Weightman, *Signor Marconi's Magic Box* (DaCapo Press, Perseus Books, Boston, 2003)

M. Wik, R. Pirjola, H. Lundstedt, A. Viljanen, P. Wintoft, A. Pulkkinen, Space weather effects in July 1982 and October 2003 and the effects of geomagnetically induced currents on Swedish technical systems. Ann. Geophys. **27**, 1775 (2009)

Space Sci Rev (2017) 212:1345–1384
DOI 10.1007/s11214-017-0416-y

Origins of the Ambient Solar Wind: Implications for Space Weather

Steven R. Cranmer[1] · **Sarah E. Gibson**[2] · **Pete Riley**[3]

Received: 2 June 2017 / Accepted: 30 August 2017 / Published online: 2 October 2017

Abstract The Sun's outer atmosphere is heated to temperatures of millions of degrees, and solar plasma flows out into interplanetary space at supersonic speeds. This paper reviews our current understanding of these interrelated problems: coronal heating and the acceleration of the ambient solar wind. We also discuss where the community stands in its ability to forecast how variations in the solar wind (i.e., fast and slow wind streams) impact the Earth. Although the last few decades have seen significant progress in observations and modeling, we still do not have a complete understanding of the relevant physical processes, nor do we have a quantitatively precise census of which coronal structures contribute to specific types of solar wind. Fast streams are known to be connected to the central regions of large coronal holes. Slow streams, however, appear to come from a wide range of sources, including streamers, pseudostreamers, coronal loops, active regions, and coronal hole boundaries. Complicating our understanding even more is the fact that processes such as turbulence, stream-stream interactions, and Coulomb collisions can make it difficult to unambiguously map a parcel measured at 1 AU back down to its coronal source. We also review recent progress—in theoretical modeling, observational data analysis, and forecasting techniques that sit at the interface between data and theory—that gives us hope that the above problems are indeed solvable.

Keywords Coronal holes · Coronal streamers · Heliosphere · Solar corona · Solar wind

The Scientific Foundation of Space Weather
Edited by Rudolf von Steiger, Daniel Baker, André Balogh, Tamás Gombosi, Hannu Koskinen and Astrid Veronig

✉ S.R. Cranmer
steven.cranmer@colorado.edu

[1] Department of Astrophysical and Planetary Sciences, Laboratory for Atmospheric and Space Physics, University of Colorado, Boulder, CO 80309, USA

[2] High Altitude Observatory, National Center for Atmospheric Research, 3080 Center Green Drive, Boulder, CO 80027, USA

[3] Predictive Science Inc., 9990 Mesa Rim Road, Suite 170, San Diego, CA 92121, USA

1 Introduction

This paper surveys the current state of understanding about how the solar wind is accelerated along magnetic field lines rooted in the Sun's hot corona. It is based on talks and discussions that took place at a June 2016 workshop devoted to *The Scientific Foundations of Space Weather* at the International Space Science Institute (ISSI) in Bern, Switzerland. A primary goal of this interdisciplinary workshop was to review the causal chain of events that link the Sun and the terrestrial environment, and thus to assess where we stand in our basic physical understanding of this complex system. This paper focuses on the origins of the "ambient" solar wind, by which we mean to exclude eruptive events like coronal mass ejections (CMEs), but to include a wide range of stochastic processes that produce global-scale structure in the heliosphere. This global-scale structure (consisting mainly of fast and slow streams that interact with one another as they expand out from the Sun) evolves on timescales from minutes to years, so it is clear that the term "ambient" is not equivalent to "time-steady."

The ambient solar wind is known to be a driver of geoeffective space weather activity. There are three main ways in which this driving occurs:

1. CMEs, the most dramatic source of space weather, accelerate through a background flow consisting of fast and slow wind streams. CME flux ropes can be accelerated or decelerated by drag-like interactions with the surrounding solar wind (Gopalswamy et al. 2000; Vršnak et al. 2010; Temmer et al. 2011). Large-scale spatial structures in the wind can also distort CMEs, deflect their trajectories, and affect their overall strengths (Riley et al. 1997; Odstrčil and Pizzo 1999; Wang et al. 2004; Isavnin et al. 2014; Zhou and Feng 2017). Thus, being able to predict the properties of the ambient solar wind appears to be a necessary component of predicting CME geoeffectiveness. In this issue, related reviews of the space weather impacts of CMEs and other transient forcing events include Manchester et al. (2017), Green et al. (2017), Eastwood et al. (2017), McPherron et al. (2017), Lester et al. (2017), and Sojka et al. (2017).
2. Sustained high-speed wind streams that intersect the Earth's magnetosphere have been shown to drive geomagnetic activity (see, e.g., Tsurutani et al. 2006; Baker et al. 2017; Ganushkina and Jaynes 2017). The primary impact of a fast stream appears to be the acceleration of additional energetic electrons in the radiation belts (e.g., Iles et al. 2002; Reeves et al. 2003; Jaynes et al. 2015; Kilpua et al. 2015). High-speed streams usually also contain stronger Alfvén waves than the slow wind, and these have been shown (McGregor et al. 2014) to enhance magnetospheric ultralow frequency (ULF) fluctuations associated with storms and radiation belt dynamics.
3. The apparently bimodal structure of the solar wind—i.e., its tendency to produce fast and slow streams—leads to the production of compressions, rarefactions, and shocks when the streams interact with one another. The passage of such corotating interaction regions (CIRs) past the Earth's magnetosphere is known to contribute to geomagnetic storm activity (in this issue, see Kilpua et al. 2017; Klein and Dalla 2017). Although only a small fraction of the most intense storms appear to come from CIRs alone (Gosling et al. 1991; Huttunen et al. 2002; Zhang et al. 2007), they are responsible for the majority of moderate-strength storms, especially at solar minimum (Verbanac et al. 2011; Echer et al. 2013). CIR events, in combination with high-speed wind streams, also provide extra heat to the Earth's ionosphere/thermosphere layers (Sojka et al. 2009), which can enhance spacecraft drag and alter its infrared energy budget.

Despite the apparently modest space-weather impacts from fast streams and CIRs (compared to CMEs) they have the potential for increased significance because they can persist

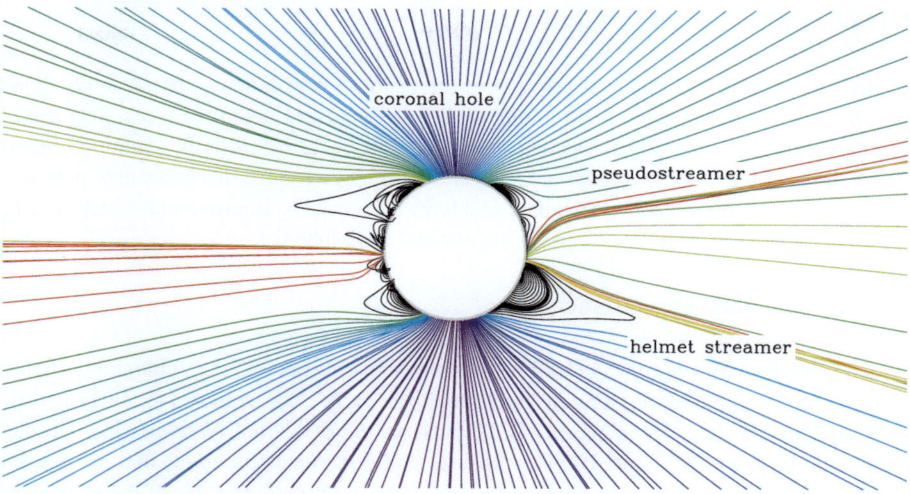

Fig. 1 Closed (*black*) and open (*multi-color*) magnetic field lines traced from a time-steady solution of the polytropic MHD conservation equations, computed by the Magnetohydrodynamics Around a Sphere (MAS) code (Linker et al. 1999). Photospheric boundary conditions were from Carrington Rotation 2058 (June–July 2007). *Colors* of open field lines correspond to the Wang and Sheeley (1990) expansion factor: $f \leq 4$ (*violet*), $f \sim 6$ (*blue*), $f \sim 10$ (*green*), $f \sim 15$ (*gold*), $f \geq 40$ (*red*). Labeled structures are discussed in more detail in Sect. 2.2

over long times and are likely to repeat over multiple solar rotations (see, e.g., Sibeck and Richardson 1997; Borovsky and Denton 2006).

An ongoing topic of debate is whether the solar wind is truly bimodal (i.e., cleanly separable into two distinct source regions). In the half-dozen years around each minimum in the Sun's 11-year activity cycle, there are large unipolar coronal holes at the north and south poles, with mostly closed fields at low latitudes. Figure 1 shows extrapolated field lines from a rotation-averaged magnetohydrodynamic (MHD) model constructed for a representative solar-minimum time period. We have high confidence that the fast solar wind is rooted in the central regions of coronal holes. The slow solar wind appears to be associated with "everywhere else" on the Sun that connects out to the distant heliosphere. Some slow-wind source regions may start as closed magnetic loops and undergo jet-like magnetic reconnection. Other regions may be topologically similar to fast-wind source regions, but with lower levels of momentum and energy deposition. This paper will discuss several unanswered questions about the solar wind's bimodality, magnetic topology, and radial evolution.

The remainder of this paper is organized as follows. Section 2 provides a review of solar wind observations, both remote and *in situ*, as well as a discussion of how coronal structures appear to be connected to their counterparts in the heliosphere. In Sect. 3 we summarize the current state of theoretical solar wind modeling. Section 4 gazes into the crystal ball to speculate about what future improvements are needed, and Sect. 5 concludes with some broader context about the impact of this work on other fields. Because the solar wind has been studied by hundreds of researchers for more than a half-century, this paper cannot be truly comprehensive in its review of the literature. Interested readers are urged to fill in the gaps by surveying other reviews, such as those by Dessler (1967), Holzer and Axford (1970), Hundhausen (1972), Leer et al. (1982), Barnes (1992), Parker (1997), Cranmer (2002, 2009), Marsch (2006), Velli (2010), Abbo et al. (2016), and Chen (2016).

2 Observations of Solar Wind Origins

In order to identify the physical processes responsible for producing the solar wind, we must have accurate empirical measurements of the plasma and field properties. Section 2.1 summarizes *in situ* interplanetary measurements, and Sect. 2.2 describes remote-sensing observations of the coronal origin regions near the Sun. Section 2.3 discusses how periodicities and other correlations between data sets have been used to improve our understanding of "what connects to what" between the corona and heliosphere.

2.1 Interplanetary Measurements

Evidence for the existence of an outflow of "corpuscular radiation" (i.e., charged particles) from the Sun accumulated gradually throughout the early 20th century (see historical reviews by Dessler 1967; Hundhausen 1972). Early *in situ* detections of solar wind particles were made between 1959 and 1961 by Russian and American spacecraft that left Earth's magnetosphere. The continuous, supersonic, and possibly bimodal nature of the solar wind was confirmed by *Mariner 2* on its journey to Venus (Neugebauer and Snyder 1962). Those early data indicated a range of outflow speeds (roughly from 250 to 800 km s^{-1}) that seem to act as an organizing quantity. In other words, many of the other plasma and field quantities measured at 1 AU appear to be correlated with whether one is in a fast or slow stream.

Table 1 summarizes some representative properties of the fast and slow wind regimes as revealed over the past half-century of exploration (see also Schwenn 2006). There is still substantial debate about whether the solar wind plasma can be classified into more than two distinct types—based on, e.g., source regions, acceleration mechanisms, or local plasma physics—and whether or not the wind speed is in fact a reliable indicator of which type is being detected (see, e.g., Wang et al. 2009; Zurbuchen et al. 2012; Stakhiv et al. 2015; Neugebauer et al. 2016). Difficulties arise because much of the solar wind at 1 AU has undergone some kind of processing or mixing (see Sect. 2.3), such that the global magnetic topology and coronal connections are not easy to determine.

Despite the above difficulties, there are many regularities in the *in situ* data. The raw probability distribution of wind speeds u in the ecliptic is usually single-peaked around

Table 1 Properties of slow and fast solar wind streams

Quantity	Slow wind	Fast wind
Radial flow speed	250–450 km s^{-1}	450–800 km s^{-1}
Proton density (1 AU)	5–20 cm^{-3}	2–4 cm^{-3}
Proton temperature (1 AU)	0.03–0.1 MK	0.1–0.3 MK
Electron temperature (1 AU)	0.1–0.15 MK	∼0.1 MK
Freezing-in temperature (corona)	1.4–1.7 MK	1.0–1.3 MK
Helium abundance	0.5–4%	3–5%
Heavy ion abundances	low-FIP enhanced	∼photospheric
Ion/proton temperature ratio	$< m_{\text{ion}}/m_p$	$> m_{\text{ion}}/m_p$
Coulomb collisional age (1 AU)	0.1–10	0.001–0.1
Coronal WSA expansion factor	15–100	3–10
Coronal sources (Sects. 2.2–2.3)	streamers, quiet loops, active regions, coronal hole boundaries, separatrices	coronal hole cores

400 $\mathrm{km\,s^{-1}}$, with a relatively sharp cutoff below about 250 $\mathrm{km\,s^{-1}}$ and a skewed tail toward higher speeds (e.g., Gosling et al. 1971; McGregor et al. 2011b). An interesting exception was in 2008 during the "peculiar solar minimum" when the presence of long-lived, low-latitude coronal holes led to a truly bimodal distribution of solar wind speed at 1 AU (de Toma 2011). Proton and electron densities n are negatively correlated with speed, but the mass flux (i.e., the product nu) has a slight residual trend toward higher values in the slow wind. Le Chat et al. (2012) found that the kinetic energy flux (proportional to nu^3) is very nearly constant as a function of wind speed, latitude, and solar cycle. The radial magnetic flux also tends to be reasonably constant throughout the low- and high-latitude heliosphere (Smith and Balogh 1995), but its overall value does change as a function of global solar activity (Svalgaard and Cliver 2007). Any theoretical model of the solar wind must reproduce these trends and quasi-invariants.

Heliospheric measurements in the ecliptic plane tend to show a preponderance of slow solar wind, with high-speed streams being occasional interlopers. This led to early widespread identification of the slow wind as the "ambient" background state (e.g., Hundhausen 1972). However, there were hints—going back to at least Bame et al. (1977)—that the fast wind was a much better candidate for being the most time-steady and quiescent type of solar wind. The *Ulysses* probe confirmed this picture when it left the ecliptic plane and showed that the fast wind is ubiquitous over large polar coronal holes, which (1) persist over more than half of each solar cycle, and (2) expand out to fill the majority of the heliospheric volume (Goldstein et al. 1996; Marsden 2001; McComas et al. 2008).

In the 1990s, *Ulysses* and *ACE* also began to show that ion composition measurements (i.e., both elemental abundances and ionization states) can be used to reliably distinguish slow and fast wind streams from one another. These composition signatures are established close to the Sun and are subsequently "frozen in" along most of the extent of each wind stream. On the other hand, the wind speed itself continues to evolve dynamically between the Sun and 1 AU as streams interact with one another. Thus, ion composition is suspected to be more reliable as a wind-stream identification tag than the flow speed (see, e.g., Neugebauer et al. 2016; Fu et al. 2017). The ratio of O^{+7} to O^{+6} charge-state number densities tends to be the most widely reported composition signature, mainly because the large oxygen abundance allows for good measurement statistics. However, Landi et al. (2012a) suggested that the relative fractions of carbon ions C^{+4}, C^{+5}, and C^{+6} may be more precise probes of the plasma conditions in the low corona ($r \approx 1.2\,R_\odot$) where the freezing-in occurs.

Figure 2 shows data from the *Ulysses* SWICS (Solar Wind Ion Composition Spectrometer) instrument final archive to illustrate how the traditional O^{+7}/O^{+6} charge-state ratio varies as a function of solar wind speed (Gloeckler et al. 1992; von Steiger et al. 2000). The polar plots show (a) wind speed in $\mathrm{km\,s^{-1}}$, and (b) a scaled ratio with magnitude $3.4 + \log_{10}(O^{+7}/O^{+6})$, as a function of latitude during an orbit near solar minimum. In panels (c)–(d), the equivalent O^{+7}/O^{+6} freezing-in temperature (i.e., the electron temperature corresponding to a given charge-state ratio in coronal equilibrium) was computed from ionization balance curves provided in version 7.1 of CHIANTI (Landi et al. 2012b). Note that panel (c) sometimes indicates abrupt changes in the ionization state at intermediate wind speeds, but panel (d) shows that, statistically speaking, the trend is rather gradual.

Additional clues about the physical origins of fast and slow wind streams come from the kinetic properties of the plasma. It has been known since the first decade of interplanetary exploration (e.g., Sturrock and Hartle 1966) that solar wind parcels are not just expanding adiabatically, but are continuing to undergo changes in their energy budgets at 1 AU and beyond. The relatively slow radial decline in particle temperature $T(r)$ indicates some combination of sustained thermal energy input (a continuation of coronal heating) and strong

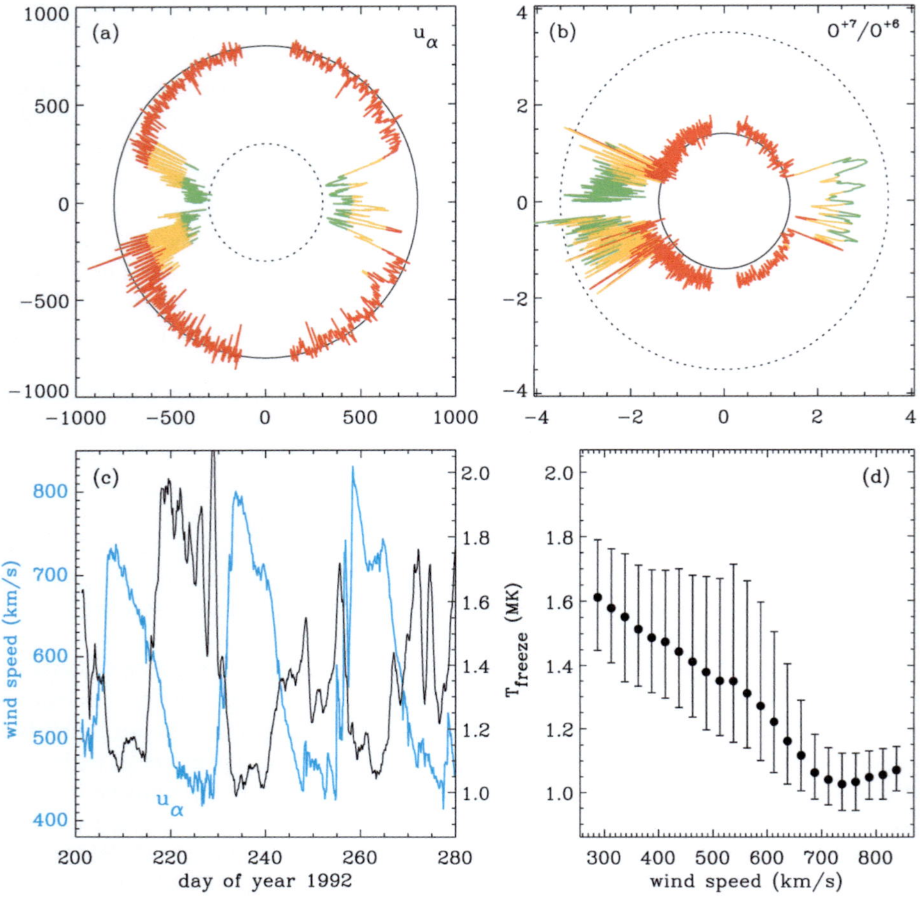

Fig. 2 Polar plots of (**a**) alpha particle wind speeds, and (**b**) ratios of O^{+7} to O^{+6} ion number densities from *Ulysses*/SWICS during its first high-latitude orbit in 1992–1997. Parcels are color-coded by wind speed (*green*: $u < 450$ km s^{-1}, *red*: $u > 650$ km s^{-1}, *yellow*: intermediate) with the same labels applied to data points in panel (**b**). The speed/ion-ratio anticorrelation is also shown in (**c**) as a function of time for several solar rotations in 1992 (i.e., the same period analyzed by Geiss et al. 1995). Panel (**d**) shows the same anticorrelation, collected into 25 km s^{-1} bins over the entire *Ulysses* mission, with each bin's median (*filled circles*) and $\pm 1\sigma$ error bars. In (**c**) and (**d**) the ion ratio was converted to freezing-in temperature (see text)

heat conduction due to the presence of skewness in the velocity distributions. The latter is certainly true for electrons (e.g., Bale et al. 2013), and it has been recently argued to be an important contributor to proton thermodynamics as well (Scudder 2015).

Coulomb collisions in the solar wind appear to be infrequent enough to allow the protons and electrons to evolve away from a common thermal state. Figure 3 illustrates this by showing the dominant trends of proton temperature T_p and electron temperature T_e versus wind speed at 1 AU. The protons appear to be strongly correlated with wind speed (see also Elliott et al. 2012) while the electrons are much less sensitive to local conditions. In the slow wind, it seems possible that stronger electron conduction keeps the coronal T_e high for a larger range of distance, while weaker proton conduction (and a lack of equilibrating collisions) allows the protons to cool off more rapidly (see, e.g., Freeman 1988). In the fast wind, the data show $T_p > T_e$, which suggests sustained heating for the protons. There have

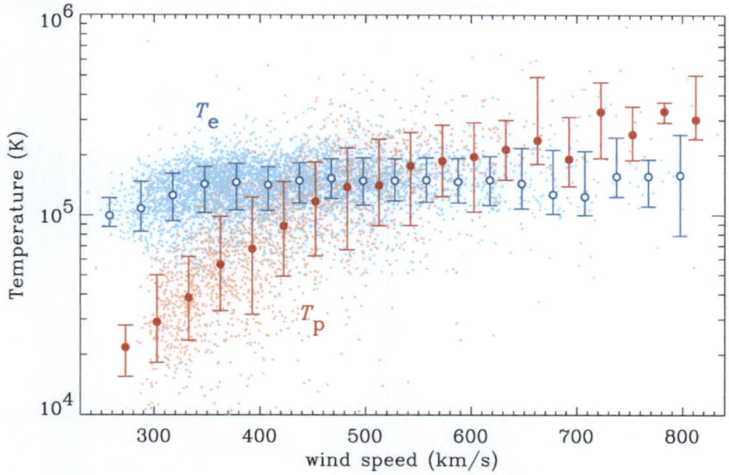

Fig. 3 Hourly averaged proton (*red*) and electron (*blue*) temperatures measured at 1 AU by *ISEE-3* (Newbury et al. 1998) between January 1980 and October 1982. *Small points* indicate individual measurements, and *large symbols with error bars* show median and $\pm 1\sigma$ values within 30 $\mathrm{km\,s^{-1}}$ bins of solar wind speed

been several empirical estimates of heat input rates that indicate the protons receive more "extended coronal heating" than do the electrons (Stawarz et al. 2009; Cranmer et al. 2009; Štverák et al. 2015).

Protons in the inner heliosphere also tend to exhibit thermal anisotropies, with unequal temperatures measured perpendicular and parallel to the background magnetic field (Marsch 2006). In the fast solar wind, the proton magnetic moment $\mu \propto T_{\perp}/B$ has been seen to increase with increasing heliocentric distance (Marsch et al. 1983). This suggests the existence of kinetic wave-particle interactions that transfer thermal energy to only *some* of the proton degrees of freedom. When the proton data at 1 AU are plotted in a two-dimensional plane of the anisotropy ratio ($\mathscr{R} = T_{\perp}/T_{\parallel}$) versus the parallel plasma beta parameter ($\beta_{\parallel}$, parallel gas pressure divided by magnetic pressure), the resulting distribution of data points (see Hellinger et al. 2006; Maruca et al. 2012) provides additional constraints on the nature of wave-particle interactions that energize the protons. Some kinds of simple linear theory—i.e., the damping of a cascading spectrum of ion cyclotron waves (Cranmer 2014b)—predict reasonably correct shapes for the populated region in ($\mathscr{R}, \beta_{\parallel}$) parameter space. However, more physically realistic numerical simulations (e.g., Servidio et al. 2015; Hellinger et al. 2017) may be needed to reproduce all of the relevant details of this region.

Ions heavier than hydrogen are also useful probes of kinetic physics in the collisionless solar wind. Both alpha particles and other minor ion species are heated and accelerated preferentially in comparison to the protons. At 1 AU, these differences appear to be organized by the Coulomb collisional "age" of the solar wind parcel; i.e., parcels that experience the fewest number of collisions between the Sun and 1 AU show the strongest departures from thermal equilibrium (e.g., Kasper et al. 2008). Preferential ion heating appears to be necessary condition for preferential ion acceleration. Geiss et al. (1970) investigated models without extra heating and found that ions tend to flow out more slowly than the protons; in fact, in those models Coulomb friction may help bring ions up to the proton outflow speed, but no faster. Ryan and Axford (1975) and others realized that heating the ions more strongly than the protons—at least proportionally to their masses (to provide comparable

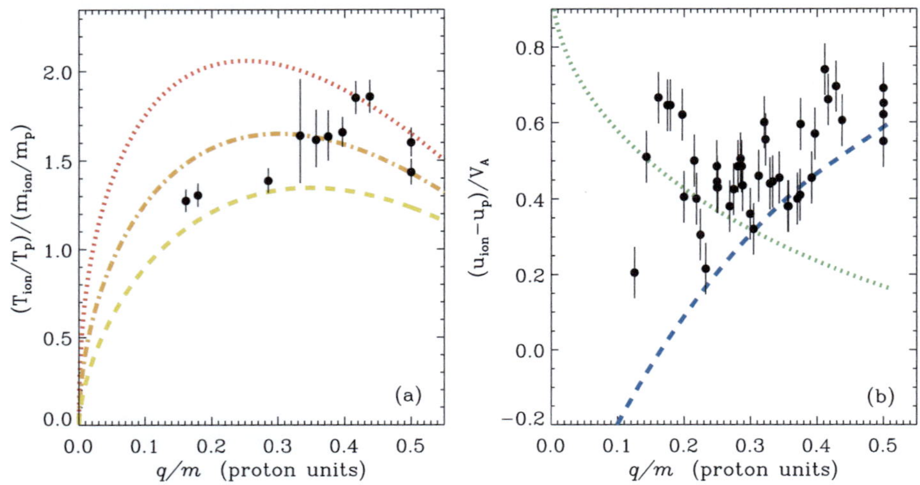

Fig. 4 Heavy-ion preferential heating (**a**) and acceleration (**b**) with respect to solar wind protons at 1 AU. *Both panels* show relative ion-proton quantities versus the charge/mass ratio (q/m) in units of proton charge/mass. (**a**) *Points* show the ratio of ion to proton squared thermal speeds ($v_{\rm th}^2 \propto T/m$) from *ACE*/SWICS (Tracy et al. 2016). (**b**) *Points* show the difference between ion and proton bulk flow speeds in units of the local Alfvén speed, as measured by SWICS and SWEPAM on *ACE* (Berger et al. 2011). For discussion of the model curves, see text

pressure gradients) or even more than that (to accelerate them even faster)—was a natural explanation for the data.

Figure 4 shows recent measurements of preferential ion heating (Tracy et al. 2016) and preferential ion acceleration (Berger et al. 2011) measured at 1 AU for collisionally young plasma that tends to be dominant in the fast wind. The particles measured by *ACE* include multiple ionization stages of He, C, N, O, Ne, Mg, Si, S, Ca, and Fe. Each ion temperature is shown as a squared thermal speed (i.e., T_i/m_i) in units of a similar quantity corresponding to the protons. The ion bulk flow speeds are shown as differences ($u_i - u_p > 0$) in units of the local Alfvén speed $V_{\rm A}$.

There is still no consensus about the identity of the physical processes responsible for the observed ion properties. The model curves shown in Fig. 4 are meant to illustrate the challenges inherent in explaining the data with a single kinetic theory. Curves in Fig. 4a are predictions from ion cyclotron resonance excited by MHD turbulence (see equation 26 of Cranmer 2002). These curves correspond to a turbulent power-law spectrum $P \propto k_{\parallel}^{-\eta}$ with $\eta = 1.57$ (red dotted curve), $\eta = 1.47$ (orange dot-dashed curve), and $\eta = 1.37$ (yellow dashed curve), where $k_{\parallel}$ is the wavenumber of cyclotron resonant fluctuations in the direction parallel to the background magnetic field. The model curves in Fig. 4b are upper and lower limits on the differential ion flow speeds compatible with ion cyclotron resonance. The cyclotron waves were assumed to obey a cold-plasma dispersion relation (e.g., Hollweg and Isenberg 2002) with alpha particles flowing $0.55 V_{\rm A}$ faster than protons, as measured by Berger et al. (2011). The blue dashed curve shows minimum resonant ion speeds for $k_{\parallel} < 0$, and the green dotted curve shows maximum resonant ion speeds for $k_{\parallel} > 0$ (see also McKenzie and Marsch 1982). It is important to note that the relevance of these ion-cyclotron curves to the data has not yet been demonstrated conclusively. However, it may be noteworthy that the rightmost “wedge” region of the plot (below the blue curve and above the green curve, for $q/m > 0.3$) is firmly excluded by both curves and is also more or less empty of data points.

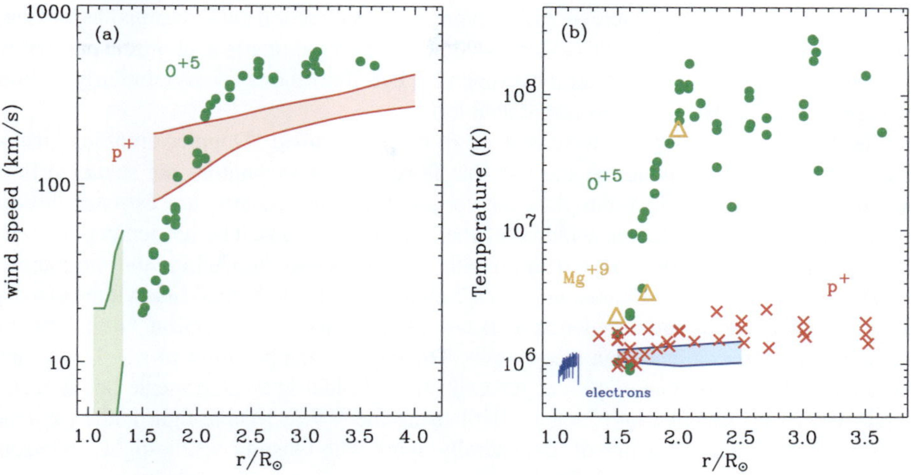

Fig. 5 Off-limb measurements of (**a**) outflow speeds and (**b**) temperatures above polar coronal holes. *Red*: proton flow speeds (Cranmer et al. 1999) and temperatures assembled from various sources (e.g., Antonucci et al. 2000; Cranmer 2009). *Green*: O^{+5} data points from Cranmer et al. (2008), and near-Sun bounded region in outflow speed from Teriaca et al. (2003). *Gold*: Mg^{+9} ion temperatures (Kohl et al. 1999). *Blue*: electron temperatures at $r < 1.3\,R_\odot$ (Landi 2008) and $r > 1.5\,R_\odot$ (Cranmer 2017, in prep). See text for details, and original sources for error bars

2.2 Coronal Measurements

A wide variety of remote observation techniques—direct imaging, spectroscopy, radio sounding, and coronagraphic occultation—have been used to put useful constraints on solar wind origins (Bird and Edenhofer 1990; Kohl et al. 2006; Habbal et al. 2013; Judge et al. 2013; Slemzin et al. 2014). These techniques have been implemented on a number of different platforms—spacecraft, rockets, ground-based observatories, and movable "eclipse-chasing" instruments—each with its own unique advantages and challenges. The combined analysis of data from these different platforms (also including *in situ* particle and field detection) has been a crucial ingredient in the advances made so far in our knowledge about the complex Sun-heliosphere system.

The solar disk contains small-scale features (e.g., bright points, faculae, ephemeral regions) and medium-scale structures (e.g., active regions, filaments) that are associated mainly with closed magnetic loops. Because these features do not appear to be connected continuously to the open heliosphere, much of the work in studying solar wind origins has focused on *large-scale* features such as coronal holes and streamers. The remainder of this subsection describes these features. However, Sect. 3.2 discusses a range of proposed coronal heating processes that includes the dynamical evolution of (temporarily) closed magnetic regions.

Coronal holes are low-density patches of nearly unipolar magnetic flux on the surface that appear to expand out superradially into the heliosphere. The central regions of large coronal holes are known sources of fast solar wind (Wilcox 1968; Krieger et al. 1973; Noci 1973). Because they are associated with tenuous, collisionless plasmas and are long-lived time-steady structures, coronal holes have been ideal hunting grounds for similar kinetic effects as seen in fast wind streams at 1 AU. Figure 5 summarizes the evidence found by the Ultraviolet Coronagraph Spectrometer (UVCS) instrument on the *Solar and Heliospheric*

Observatory (*SOHO*) for preferential ion heating and acceleration above coronal holes (see, e.g., Kohl et al. 2006). Most of this evidence comes from the comparison of proton properties (measured by proxy using the neutral hydrogen H I Lyα line) and O^{+5} ions (similarly probed by the O VI 103.2–103.7 nm resonance doublet).

The flow speeds in Fig. 5a were derived from the so-called "Doppler dimming" technique, which takes advantage of the fact that fewer solar-disk photons are scattered into our line of sight when the atoms (i.e., the coronal scattering centers) are Doppler shifted away from the narrow spectral window of the available photons. The ion temperatures in Fig. 5b were derived from spectral line widths and associated modeling, and are mainly probes of $T_\perp$. These temperatures were corrected to remove nonthermal line widths associated with MHD waves and turbulence. This is a model-dependent correction, but it is based on additional observational data (see below). The electron temperatures at $r > 1.5\,R_\odot$ are preliminary results from an empirical generalization of older hydrostatic scale-height techniques (see, e.g., Lemaire and Stegen 2016) using the UVCS visible-light and Lyα data as constraints. These estimates of T_e generally agree with existing visible-light Thomson scattering results (Reginald et al. 2011).

The remote-sensing data provide evidence for $T_{\text{ion}} \gg T_p > T_e$ in the source regions of fast solar wind, which is reminiscent of the heliospheric data at distances greater than 0.3 AU. The initial reports of heavy ions with temperatures of order 10^8 K (i.e., even hotter than in the solar core), together with $T_\perp > T_\parallel$ and ion flow speeds roughly double those of the protons at $r \approx 3\,R_\odot$, were surprising. There was some skepticism about the uniqueness of these interpretations of the data (e.g., Raouafi and Solanki 2004). However, more rigorous data analysis (Cranmer et al. 2008) has generally upheld those initial results, albeit with some tempering (i.e., the O^{+5} anisotropy ratio $T_\perp/T_\parallel$ was found to be more like 3–10, instead of the earlier claim of $\sim$100).

In addition to coronal holes, the large-scale corona contains a variety of other magnetic features that appear to be connected to the slow solar wind (see Sect. 2.3). When observing above the solar limb, the most striking of these are the bright streamers illustrated in Fig. 1. The magnetic field in streamers appears to be closed at low heights, with surrounding open field lines converging above a cusp-like point at the top. The helmet-like appearance of many streamers has been compared to the 19th century Prussian *pickelhaube*, and it is clear that the solar wind acts to open up the magnetic field above these structures. Streamers are generally assumed to be sources of low-speed solar wind, but the precise topological connections (Sect. 2.3) and mass-release mechanisms (Sect. 3.2) are still being debated.

In recent years, distinctions have been made between: (1) helmet streamers that expand up from a bipolar loop, and thus have a large current sheet between the two opposite-polarity legs, and (2) "pseudostreamers" that are connected to an even number of bipoles, and thus have legs with the same polarity (e.g., Wang et al. 2007; Riley and Luhmann 2012; Rachmeler et al. 2014). There are also differences in the plasma properties between large quiescent equatorial streamers and the brighter, more compact streamers associated with active regions (Liewer et al. 2001; Ko et al. 2002). The relatively high densities seen in all coronal streamers (Gibson et al. 1999; Strachan et al. 2002) appear to indicate rapid Coulomb collisions that generally lead to temperature equilibration ($T_p \approx T_e$). However, the largest streamers do start to exhibit collisionless kinetic effects, such as high O^{+5} temperatures similar to what is seen in coronal holes, above their cusps (Frazin et al. 2003).

Both coronal holes and streamers are intrinsically time-variable. In addition to changes in connectivity that occur as the Sun's magnetic field evolves over multiple solar rotations, the corona is also observed to be full of large-amplitude oscillations (e.g., waves, shocks, and transient eddies). A comprehensive review of oscillation measurement techniques is beyond the scope of this paper (see, e.g., Cranmer 2002, 2004; Nakariakov 2006), but there are several aspects that are relevant to solar wind origins:

1. Sensitive measurements of the off-limb coronal intensity allow *low-frequency density fluctuations* to be tracked in space and time. The tips of most helmet streamers appear to be unstable to the production of blob-like plasmoids that flow out with the slow solar wind (Sheeley et al. 1997, 2009; Wang et al. 2000; Plotnikov et al. 2016). Similar features continue to be detected as 1–2 hour density modulations at larger distances (Viall and Vourlidas 2015), and they appear to be most intense in the heliospheric current sheet (HCS). Above coronal holes, there are appear to be weak, but ubiquitous field-aligned compressive waves with periods of order 10–20 minutes (Ofman et al. 1999; Threlfall et al. 2013; Liu et al. 2015).
2. Radio telescopes probe plasma properties near the Sun by measuring how signals are distorted by changes in the refractive index as they pass through the corona (e.g., Bastian 2001). Interplanetary scintillation (IPS) measurements are sensitive to *high-frequency density fluctuations* (i.e., millisecond timescales), and additional information can be extracted about the coronal magnetic field and the solar wind speed. Global IPS maps of solar wind acceleration show the presence of fast and slow streams (Kojima and Kakinuma 1990; Grall et al. 1996; Imamura et al. 2014), but some information is lost by the integration over long lines of sight. Efimov et al. (2010) detected spatial anisotropy in radio-detected turbulent eddies at heliocentric distances smaller than $\sim 25\, R_\odot$, and isotropy above $\sim 30\, R_\odot$. This is a similar qualitative transition as the one seen in the shapes of larger visible-light structures resolved by heliospheric imagers. However, for the latter, DeForest et al. (2016) found that the transition to isotropy does not occur until at least 60–80 $R_\odot$.
3. A combination of motion-tracking and spectroscopic Doppler-shift techniques allows *transverse Alfvénic fluctuations* to be detected in the solar wind. It is suspected (see Sect. 3.2) that Alfvén waves and turbulence are major players in heating the extended corona and solar wind. Figure 6 shows a summary of inferred velocity amplitudes over polar coronal holes. The associated model curves show predictions for undamped and damped Alfvénic turbulence from Cranmer and van Ballegooijen (2005). Measured amplitudes derived from nonthermal line widths are shown from SUMER/*SOHO* (Banerjee et al. 1998, orange crosses), near-limb EIS/*Hinode* data (Landi and Cranmer 2009, red diamonds), and UVCS/*SOHO* (Esser et al. 1999, green region). Taken together, those data appeared to agree well with the predictions for Alfvén waves that dissipate and heat the corona. More recently, however, the EIS instrument has been used to probe larger heights above the poles; magenta points show data from Hahn and Savin (2013) (see also Hahn et al. 2012; Bemporad and Abbo 2012; Gupta 2017). There is now clearly some "tension" with the model curves and with the inferred UVCS result from Esser et al. (1999).

Figure 6 makes it clear that our knowledge of the global evolution of waves and turbulence in the solar wind is still lacking. The recent EIS data call into question our understanding of where Alfvén waves are damped and how their energy is converted to heat. There is also some inherent uncertainty in interpreting the properties of off-limb emission lines, especially when observing diffuse areas such as coronal holes. If the line of sight contains N independently fluctuating flux tubes, with $N \gg 1$, then many of the desired diagnostics (e.g., Doppler shifts or plane-of-sky swaying motions) are reduced in amplitude by roughly $1/\sqrt{N}$. Monte Carlo forward models (De Pontieu et al. 2007; McIntosh et al. 2011) have proven to be helpful in estimating the magnitude of this effect, but definitive "inversions"

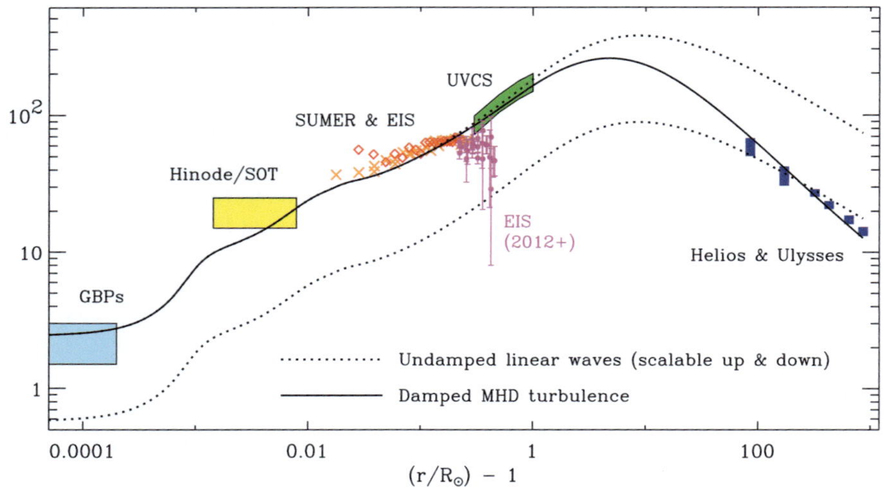

Fig. 6 Height dependence of transverse velocity amplitudes of MHD fluctuations in coronal holes and the fast solar wind. Model curves and the photospheric G-band Bright Point (GBP) data are from Cranmer and van Ballegooijen (2005). Other data, *from left to right*, are from Type II spicule motions observed by *Hinode*/SOT (De Pontieu et al. 2007), nonthermal line broadening from SUMER, EIS, and UVCS (see text), and direct *in situ* measurement from *Helios* and *Ulysses* (Bavassano et al. 2000)

are not yet possible. In Sect. 4, we discuss future efforts to improve upon the existing measurements.

2.3 Periodicities Linking the Sun and Heliosphere

There is not yet a fully-understood one-to-one mapping between observed features in the corona and *in situ* detected structures in the heliosphere. Multi-point measurements made over multiple solar rotations—sometimes extending to multiple solar cycles—have helped us find correlations between large, long-lived structures on the Sun and in the solar wind. Whether or not these correlations are related to physics-based causations is a separate issue, but good correlations provide good starting points for space weather prediction.

For example, the half-century long OMNI database of plasma and field measurements at 1 AU has been shown to be useful for long-baseline studies of all kinds (e.g., O'Brien and McPherron 2000; King and Papitashvili 2005; Lee et al. 2009). Analogous databases for regions near the Sun have run the gamut from careful hand-drawings based on daily images (Harvey and Recely 2002; McIntosh 2003) to automated "big data" feature-extraction systems (Martens et al. 2012; Bobra et al. 2014). Taking inspiration from worldwide events like the 1957 International Geophysical Year, multiple communities came together in coordinated projects—e.g., three "Whole Sun Months" in 1996, 1998, and 1999 (Galvin and Kohl 1999; Riley et al. 1999; Breen et al. 2000) and a "Whole Heliosphere Interval" in 2008 (Gibson et al. 2009, 2011; Riley et al. 2011; Thompson et al. 2011)—to improve our understanding of Sun-heliosphere connectivity.

Figure 7 illustrates the synergistic power of combining multiple databases. Stacking up a solar cycle's worth of OMNI wind speeds versus Carrington longitude reveals the presence of high-speed streams that recur over multiple rotations and fade in and out over time (see also Lee et al. 2009). The occurrences of these streams line up quite well with the presence of

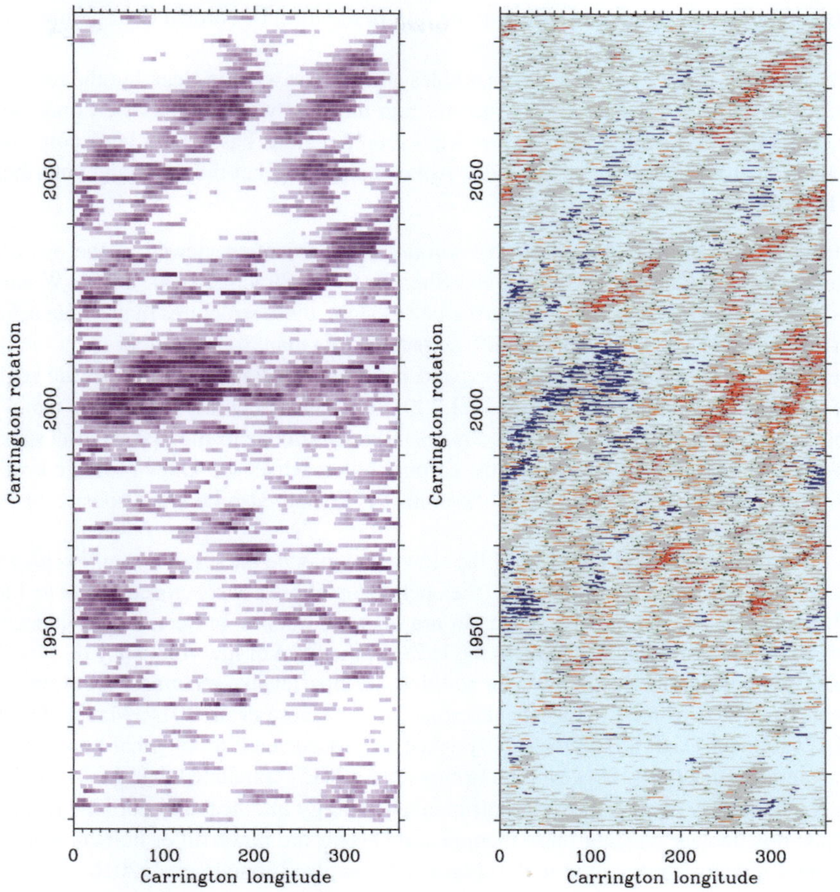

Fig. 7 Carrington rotation stack plots showing (**a**) in-ecliptic OMNI wind speeds and (**b**) surface features from the McIntosh archive, both for the duration of solar cycle 23 (June 1996 to July 2009). In panel (**a**), *white* denotes $u \leq 450\ \mathrm{km\,s^{-1}}$ and increasingly *darker shades of purple* eventually saturate at the *darkest color* for $u \geq 750\ \mathrm{km\,s^{-1}}$. Longitudes have been offset by 50.55°, or 3.83 days, to account for propagation from the Sun to 1 AU at a mean speed of 450 $\mathrm{km\,s^{-1}}$. Panel (**b**) shows equatorial (±20° from equator) features, with *blue* [*red*] showing coronal holes of positive [negative] polarity, *cyan* [*gray*] showing quiet regions with predominantly positive [negative] polarity, *orange* indicating sunspots, and *green* indicating filaments

large equatorial coronal holes as recorded in the McIntosh synoptic image archive (Gibson et al. 2017b). The long-lived coronal holes (blue/red) seen in panel (b) are rotating at a rate somewhat faster than the 27.275 day Carrington rotation, and thus they have a positive slope in this plot. This correlates well with the slopes seen in the fast wind streams indicated in panel (a).

The correlations shown in Fig. 7 do not stop at the solar wind, but indeed extend to the Earth's space environment and upper atmosphere. Clear connections can be found between high-speed solar wind streams and modulations of the aurora and geomagnetic indices, radiation belts, ionosphere, and thermosphere (Gibson et al. 2009; Solomon et al. 2010; Lei et al. 2011). Long-time series analyses over years and decades show periodicities in all of these quantities that may be associated with periodicities in the fast solar wind, and consequently the distribution of open magnetic flux at the Sun in the form of coronal holes (Emery et al. 2011; Love et al. 2012).

The connection between large coronal holes and the fast wind is clear, but the remaining connections between other coronal structures and the slow wind are less well understood. An exact census or mass budget of slow-wind source regions has not yet been constructed (see also Poletto 2013; Kilpua et al. 2016; Abbo et al. 2016), but the following contributors may be significant:

1. Steady flows from the *boundaries of coronal holes* are often viewed as the open-field "legs" of helmet streamers (Wang and Sheeley 1990; Strachan et al. 2002). When the axis of the streamer belt is oblique to the line of sight, these structures may be identifiable in coronagraph images merely as diffuse patches of Quiet Sun. In either case, the open field lines in these regions tend to expand more superradially than the central regions of the large coronal holes. Stakhiv et al. (2015) coined the phrase "boundary wind" for this component, which tends to be compositionally similar to the fast wind despite its lower asymptotic speed. The smallest coronal holes, which are known to be correlated with slow wind speeds at 1 AU (Nolte et al. 1976), may also be close cousins of these boundary-layer type flows.
2. A more time-variable component of the slow wind may be the result of *multi-scale magnetic reconnection* in the corona; i.e., the opening up of previously closed magnetic loops. Theoretical arguments for this scenario are discussed below in Sect. 3.2. Evidence for large-scale intermittent mass loss in the HCS (in the form of low-frequency density fluctuations) was summarized above. In addition, smaller jet-like reconnection events have been suggested to feed mass into the solar wind (Moore et al. 2011; Madjarska et al. 2012; Raouafi et al. 2016), especially when they occur near topological boundaries of magnetic connectivity. However, Paraschiv et al. (2015) concluded that the hot jets seen in X-ray images convert most of their magnetic energy into heat and not kinetic energy. Thus, it is unclear whether these reconnection events are powerful or numerous enough to make a major contribution to the solar wind (see also Lionello et al. 2016).
3. Images and spectra of *active regions* show rapid flows with speeds of at least 100 $\mathrm{km\,s^{-1}}$ along their fanned-out edges (e.g., Harra et al. 2008; Brooks and Warren 2011; Morgan 2013; Zangrilli and Poletto 2016). The slow solar wind associated with these structures may come from small, short-lived coronal holes adjacent to the active regions themselves (Wang et al. 2009). Active-region slow wind tends to be associated with larger expansion factors, stronger magnetic fields, higher mass fluxes, higher O^{+7}/O^{+6} ratios, and larger abundance enhancements of low first ionization potential (FIP) elements than the slow wind associated with streamers.
4. Although there is still some debate, it is becoming increasingly clear that *pseudostreamers* are sources of slow solar wind (Riley and Luhmann 2012; Crooker et al. 2014; Owens et al. 2014). Open field lines near pseudostreamers are topologically complex and "squashed," and the asymptotic speed of their solar wind may depend on small details of their geometric expansion (Wang et al. 2012; Panasenco and Velli 2013; Gibson et al. 2017a). Nevertheless, the narrow HCS generally appears to be surrounded by a web-like band of pseudostreamer separatrix surfaces (Antiochos et al. 2011), and the 20° to 30° width of this band in latitude corresponds closely to the zone of slow solar wind seen by *Ulysses* (see Fig. 2). Wind streams associated with pseudostreamers tend to have charge states and kinetic properties intermediate between those typical of fast and slow wind (Wang et al. 2012; Abbo et al. 2015) and extreme values of the proton mass flux (Zhao et al. 2013).

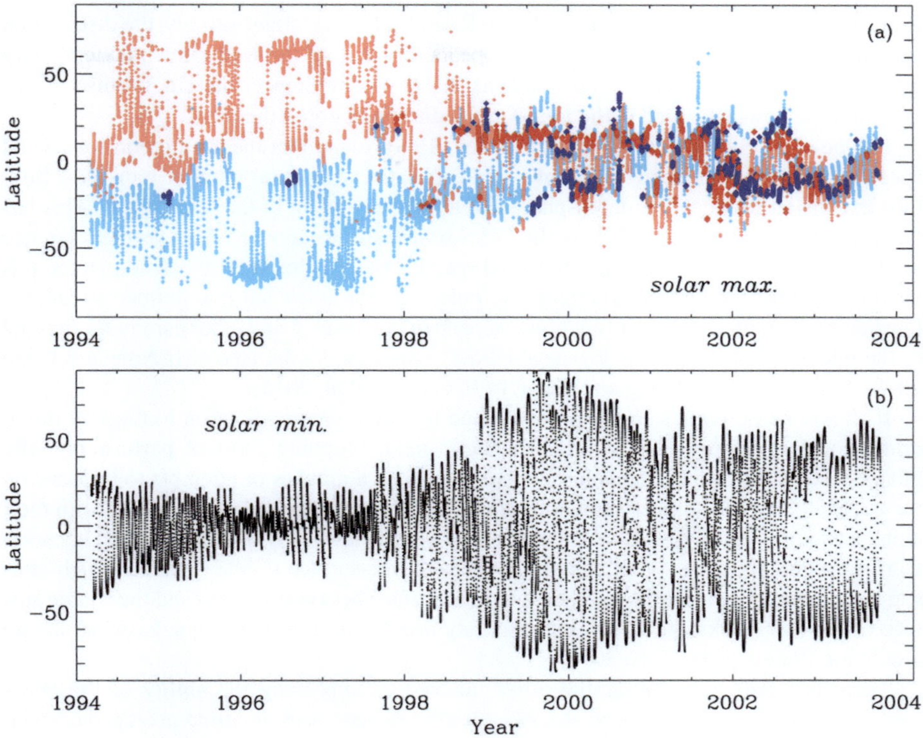

Fig. 8 (**a**) Latitudes of photospheric footpoints of open field lines that connect to the ecliptic plane. The PFSS technique was used to extrapolate synoptic magnetogram data from the Wilcox Solar Observatory (Hoeksema and Scherrer 1986), and a series of 133 sequential Carrington rotations was stacked together in time. *Red* [*blue*] points show footpoints with positive [negative] polarities. *Large points with darker colors* indicate strong photospheric fields ($|B_r| > 5$ G), and *small points with lighter colors* indicate weak fields below this threshold. (**b**) Latitude of the HCS neutral line at the source surface, mapped from the same set of PFSS models as in panel (**a**)

Although the coronal magnetic field is not yet measurable in a routine way, there are several semi-empirical extrapolation models that have been successful in estimating how the photospheric field maps out into the heliosphere. The community's workhorse is the potential-field source-surface (PFSS) technique, which assumes the corona is current-free between the photosphere and a spherical surface in the mid-corona, typically at $r = 2.5\ R_\odot$ (Schatten et al. 1969; Altschuler and Newkirk 1969). Above the so-called source surface, the magnetic field is assumed to be stretched out by the solar wind into a radially-pointing "split monopole" configuration.

The PFSS technique is computationally efficient to implement, and it reproduces a number of large-scale features of the corona as seen with coronagraphs and during eclipses (Riley et al. 2006). Figure 8 shows how PFSS models can also be useful tools for mapping the origins of solar wind streams (see also Luhmann et al. 2002; Liewer et al. 2004; Fazakerley et al. 2016). At solar minimum, it is clear that high-latitude coronal holes have significant "reach" down into the ecliptic plane. However, the persistently low latitude of the HCS also means that slow wind from equatorial streamers must also contribute to the measurement record at 1 AU. At solar maximum, the Sun's dominant dipole field is in the process of being destroyed and reconstituted with opposite polarity, so the tilted HCS tends to spend time at nearly all latitudes (see also Riley et al. 2001). Interestingly, the distribution of photospheric footpoints of open field appears to trace out the well-known butterfly diagram of active regions (see also Gibson et al. 2017b, and references therein, for discussions of similar patterns observed in the long-term evolution of coronal holes).

Levine et al. (1977) and Wang and Sheeley (1990) found that the asymptotic solar wind speed along a field line tends to be inversely correlated with the amount of transverse flux-tube expansion between the photosphere and a reference point in the mid-corona. This has been subsequently formalized using the PFSS source surface at $r = 2.5\,R_\odot$ as the reference point. The anticorrelation between the wind speed u and the flux-tube expansion factor f is most evident for the largest structures like polar coronal holes and low-latitude streamers. Further refinement in the exact functional dependence of u on f and other parameters has led to the widely-used Wang-Sheeley-Arge (WSA) empirical model (see also Arge and Pizzo 2000; Arge et al. 2003; Wang and Sheeley 2006; Riley et al. 2015).

It should be noted that the PFSS technique is only an approximation to the true three-dimensional structure of the coronal magnetic field. Stopping short of performing fully global MHD simulations (see below), there have been a number of attempts to improve on the accuracy of PFSS-like extrapolation methods. Many of these methods, along with their alphabet soup of acronyms, have been reviewed comprehensively by Sun (2012). One noteworthy technique is the so-called current-sheet source-surface (CSSS) model, which adds some complexity by inserting another spherical surface between the Sun and the source surface (Zhao and Hoeksema 1995), but also may provide improvement to solar wind stream prediction (Poduval and Zhao 2014).

There has also been substantial effort devoted to improving the ability of the WSA method to predict the wind speed at 1 AU. Despite its successes, the time-averaged correlation coefficient between the predicted and measured wind speed tends to never exceed $\sim$50% (e.g., McGregor et al. 2011b; Gressl et al. 2014; Riley et al. 2015). Statistical comparisons between multiple models and the observed solar wind (Jian et al. 2015, 2016) show generally similar results for other quantities such as density, temperature, and the magnetic field. Improvements to the original Wang and Sheeley (1990) anticorrelation have been found by including a second parameter, such as the angular distance θ between each field-line footpoint and the nearest coronal hole edge (Arge et al. 2003; Owens et al. 2008; Shen et al. 2012; Riley and Luhmann 2012) or the magnetic field magnitude at the source surface (Suzuki 2006; Fujiki et al. 2015; Wang 2016). In addition to magnetic-field parameters, it is possible that data from EUV images of the chromosphere and low corona can be used to improve these kinds of empirical predictions (Leamon and McIntosh 2007; Luo et al. 2008; Rotter et al. 2015).

Lastly, it is important to note that there is a difference between the largest-scale stream structure of the solar wind, which clearly survives the journey to 1 AU, and smaller-scale structure, which may or may not have a one-to-one correspondence with features on the Sun. There have been many reports of *in situ* "microstreams" that may be the imprints or relics of coronal structures (Thieme et al. 1990; Reisenfeld et al. 1999; Borovsky 2008, 2016). However, there are several stochastic processes that appear to vigorously blend or scramble the plasma and magnetic flux tubes to such a degree that deterministic mappings may not be possible. Section 3.3 discusses these processes in more detail.

3 Physical Processes that Produce the Solar Wind

Empirically based prediction techniques have been successful, but it can be argued that moving beyond correlations into the realm of fundamental physics is required to make substantial new gains in predictive accuracy. Once the key physical processes are identified and characterized, it will be much more straightforward to benchmark, assess, and refine the simulations used for predicting heliospheric conditions at 1 AU. This section reviews recent work along these lines. Section 3.1 begins by outlining the ideas about which most researchers agree, and Sect. 3.2 describes the areas of active debate. Section 3.3 discusses one notable difficulty in choosing between the various model proposals: the fact that wind streams tend to lose their unique connections back to the corona due to a range of dynamical effects.

3.1 Uncontroversial Fundamentals

The Sun's corona is hot. Although Grotrian, Edlén, and others began to understand the high ionization state of coronal emission lines in the 1930s, it was left to Alfvén (1941) to assemble additional lines of evidence and make the definitive case that the corona is comprised of plasma with $T \approx 10^6$ K (see also Peter and Dwivedi 2014). The high gas pressure gradient in such an extended atmosphere led Parker (1958) to determine that the most likely steady state would be a supersonic outflow. This is still the dominant idea in solar wind theory, but there may be other supplementary sources of radial acceleration in addition to the gas pressure gradient (see, e.g., Jacques 1977; Hollweg and Isenberg 2002).

The mechanism by which the coronal plasma is heated is not yet known, but its ultimate energy source is universally understood to be the convection zone. Photospheric granulation enables some kind of upward Poynting flux that delivers kinetic and magnetic energy to the higher layers of the atmosphere. After an undetermined time over which much of this energy is "stored" in the magnetic field, it is converted irreversibly to heat. Some of that thermal energy conducts back down to the chromosphere, and some is extracted by the Parker (1958) mechanism to do work against the Sun's gravitational potential. In steady-state, the power input at the base (i.e., energy flux multiplied by available surface area) should equal the solar wind's kinetic power far above the solar surface,

$$\left[4\pi r^2 f\left(F_{\rm heat} - F_{\rm cond} + \frac{1}{2}\rho u_r^3 - \rho u_r \frac{GM_\odot}{r}\right)\right]_{\rm base} \approx \left[4\pi r^2\left(\frac{1}{2}\rho u_r^3\right)\right]_{r\gg R_\odot} \quad (1)$$

where f is the surface filling factor of magnetic field lines that eventually reach the solar wind, $F_{\rm heat}$ is the energy flux deposited by the still-unidentified source of coronal heating, and $F_{\rm cond}$ is the energy flux conducted back down to the chromosphere (see also Hammer 1982; Hansteen and Leer 1995; Schwadron and McComas 2003; Cranmer and Saar 2011). The equation above neglects enthalpy fluxes and radiative losses, both of which are usually negligible above the transition region. From the standpoint of the supersonic solar wind, the high coronal temperature is only a kind of temporary holding area; i.e., a stopover between the original source of the energy and its eventual destiny as outflowing kinetic energy.

The energy balance shown in Eq. (1) sets the mass loss rate $\dot{M}$ of the wind, but the relative magnitudes of the terms on the left-hand side are still not known. This is an analogous situation to the long-studied problem of heating in static coronal loops (e.g., Rosner et al. 1978), in which the "base pressure" is determined by time-steady energy conservation. Both the wind's $\dot{M}$ and a loop's pressure are measures of how much plasma is drawn up from the relatively vast chromospheric reservoir. A key point is that the mass loss rate is *not* determined by the Parker (1958) solution of the momentum equation. The accelerating flow through the Parker critical point (i.e., the radius at which the wind speed exceeds the sound speed) merely takes whatever mass is supplied at the coronal base and draws it out. Wang (1998) estimated the sphere-averaged value of $\dot{M}$ varies between about $2 \times 10^{-14}\ M_{\odot}\ \mathrm{yr}^{-1}$ (at solar minimum) to $3 \times 10^{-14}\ M_{\odot}\ \mathrm{yr}^{-1}$ (at solar maximum).

In the past, theorists have disagreed about whether the solar wind is more properly described using fluid or kinetic equations (Chamberlain 1960; Jockers 1970; Lemaire and Scherer 1971). The consensus now is that both pictures agree on the basic properties of the outflow (Lemaire and Pierrard 2001; Parker 2010). To some extent, this ought to be the case, because the conservation equations based on fluid moments are derived directly from Liouville's theorem and the associated kinetic transport equations. However, the fluid picture does make closure assumptions about the shapes of the velocity distribution functions, and there remain disagreements about, e.g., the validity of classical heat conduction (Landi and Pantellini 2003) and the available linear wave modes (Verscharen et al. 2017).

It has also been proposed that there are suprathermal particles (i.e., power-law tails that augment the normally Maxwellian velocity distributions) in the solar atmosphere, and that these particles escape preferentially to produce high coronal temperatures (Levine 1974; Scudder 1992). This "velocity filtration" idea has been implemented in kinetic exobase-type models that successfully predict some aspects of the particle measurements at 1 AU (e.g., Meyer-Vernet 1999; Zouganelis et al. 2004; Pierrard and Pieters 2014). Despite this idea being somewhat outside the mainstream of research, we list it here in the subsection about uncontroversial physics. It may or may not be important on the Sun, but it is similar to any other coronal heating theory in that it requires converting some other form of energy (i.e., kinetic or magnetic) into thermal energy. The difference is that this conversion would have to occur down in the chromosphere, where a combination of Coulomb collisions and radiative losses would keep the majority of particles cool.

A final uncontroversial statement to make about the solar wind is that it its fluctuations (e.g., waves, turbulence, shocks, and end-products of magnetic reconnection) are likely to both affect and be affected by the time-averaged properties of the flow. There do not seem to be any theories of coronal heating and solar wind acceleration that do *not* ultimately involve the summed impact from multiple transient or oscillating events. The extent to which terms like "waves" and "turbulence" are useful descriptors of the physics is still being debated, but the variability is ubiquitous and important.

3.2 Controversial Alternatives

The exact chain of events by which the corona is heated and the solar wind is accelerated is not yet known. It has proven exceedingly difficult to distinguish between competing theoretical models because the basic energy conversion processes appear to be acting on spatial and time scales unresolved by existing observations. It is also probably the case that different mechanisms are dominant in different source-regions of the solar wind (e.g., active regions versus coronal holes), and that in some regions multiple mechanisms may be contributing at comparable levels.

With the above caveats in mind, we have sorted the proposed physical models into three broad categories:

1. If solar wind field lines are open to interplanetary space—and if they remain open on timescales comparable to the time it takes plasma to accelerate into the corona—then the main sources of energy must be injected at the footpoints. Thus, in *wave/turbulence-driven (WTD) models*, the convection-driven jostling of the flux-tube is assumed to gen-

erate wave-like fluctuations that propagate up into the extended corona (Coleman 1968; Hollweg 1986; Velli et al. 1991; Matthaeus et al. 1999; Suzuki and Inutsuka 2006; Cranmer et al. 2007; Ofman 2010; van Ballegooijen et al. 2011; Chandran et al. 2011; Matsumoto and Suzuki 2012; Perez and Chandran 2013; Lionello et al. 2014; Tenerani and Velli 2017; van Ballegooijen and Asgari-Targhi 2017). The coronal heating comes from wave dissipation, whose physical origin is still a subject of debate. Fast and slow wind streams come from the fact that flux tubes with different expansion factors have different radial distributions of the heating rate and different locations of the Parker critical point (see, e.g., Leer and Holzer 1980; Cranmer 2005).
2. Near the Sun, all open magnetic flux tubes are observed to exist in the vicinity of closed loops. The complex distribution of mixed-polarity loop footpoints—which is evolving continuously via emergence, cancellation, diffusion, splitting, and merging—has been called the Sun's "magnetic carpet" (Title and Schrijver 1998). It is natural to propose a class of *reconnection/loop-opening (RLO) models*, in which the mass and energy in some coronal loops is fed into the open regions that connect to the solar wind. Some have suggested that RLO-type energy interchange primarily occurs at the scale of the supergranular network (Axford and McKenzie 1992; Fisk et al. 1999; Fisk 2003; Schwadron et al. 2006; Yang et al. 2013; Karpen et al. 2017), and others favor larger-scale reconnection events near global null points and streamer cusps (Suess et al. 1996; Einaudi et al. 1999; Wang et al. 2000). The idea of a so-called S-web, or separatrix-web (Antiochos et al. 2011; Edmondson 2012; Higginson et al. 2017) involves a continuous range of scales between the two, with the complex topological rearrangement helping to energize the slow wind.
3. The upper chromosphere is filled with a range of narrow features known variously as spicules, jets, fibrils, surges, and mottles. Some have suggested that much of the corona's mass and energy may be injected directly from these structures (Pneuman 1986; Loucif 1994; De Pontieu et al. 2009; Moore et al. 2011; Tian et al. 2014; Raouafi et al. 2016). Strictly speaking, this idea could be considered a subset of either the WTD or RLO models, depending on whether the spicules and jets are driven by waves (Sterling and Hollweg 1984; Kudoh and Shibata 1999; Cranmer and Woolsey 2015) or by reconnection (e.g., Uchida 1969; Pariat et al. 2016). Still, the direct chromospheric source of the mass appears to distinguish this idea—here called *chromospheric mass supply (CMS)*—from the other two, in which the processes giving rise to the corona and solar wind are generally located up in the corona itself.

The above list of processes does not include some that have been applied mostly to closed loops that do not connect directly to the solar wind. For example, the classical idea of direct-current (DC) heating—in which the corona evolves in a succession of twisted and quasi-static states via small-scale current dissipation events (Parker 1972; Heyvaerts and Priest 1984; van Ballegooijen 1986)—does not sit solidly within the WTD, RLO, or CMS categories. These DC mechanisms are often associated with *nanoflares*: tiny episodic bursts of energy that may dominate the coronal heating (Parker 1988; Parnell and Jupp 2000; Joulin et al. 2016). It is fair to say that all three of the above model categories can produce nanoflare-like intermittent heating. In many cases, however, the predicted small-scale bursts are expected to be highly dynamic and not quasi-static; this is consistent with observations of "braiding" by high-resolution imagers (e.g., Cirtain et al. 2013; van Ballegooijen et al. 2014). No matter the source of these bursts, they may generate nonthermal electrons that propagate out into the heliosphere (e.g., Che and Goldstein 2014).

Observations have been used to attempt to support or rule out the above processes, but no consensus has been reached. For example, Roberts (2010) claimed there to be insufficient WTD energy present to heat the corona. Cranmer and van Ballegooijen (2010) similarly claimed the RLO model cannot energize the plasma in open-field regions (see also Karachik and Pevtsov 2011; Lionello et al. 2016). Klimchuk and Bradshaw (2014) concluded that CMS-type processes cannot produce sufficient heat to power the corona. However, none of these ideas has been ruled out conclusively because (1) the relevant energy-release events cannot yet be measured directly, and (2) most models still employ free parameters that can be adjusted to improve agreement with the existing data.

As mentioned above, it may also be possible for aspects of more than one model to be present. Stakhiv et al. (2016) suggested that both closed loops and open flux tubes produce solar wind—with closed loops contributing more to the slow wind and open flux tubes contributing more to the fast wind—and that once the plasma is released, both types share a single WTD acceleration mechanism. There are other ways that the WTD, RLO, or CMS models can exhibit intermingled characteristics. If there is a turbulent cascade that produces WTD-type heating, the ultimate energy dissipation may be best describable by small-scale reconnection events that reconfigure the magnetic topology (e.g., Matthaeus and Velli 2011). On the other hand, the proposed RLO reconnection events in the magnetic carpet may generate MHD waves (Lynch et al. 2014; Karpen et al. 2017) that go on to dissipate and heat the plasma. Lastly, even if the evolving S-web does not produce sufficient reconnection to heat the global corona, it may be that the mere presence of the sharp transverse gradients and separatrices can act as an additional source of shear-driven waves (see, e.g., Lee and Roberts 1986; Kaghashvili 1999, 2007; Evans et al. 2012).

How can we as a community progress toward the goal of identifying and characterizing the physical processes at work in the solar corona? It is worthwhile listing some promising paths forward (see Sect. 4), but there have also been unsuccessful attempts to rule out models. For example, the prevalence of enhanced low-FIP elements and high freezing-in temperatures (e.g., high O^{+7}/O^{+6} ratios) in the slow wind has been used as evidence for RLO-type processes. Closed coronal loops exhibit similar composition patterns to the slow wind, so it is natural to connect them together. However, time-steady WTD models have been shown to predict variations with wind speed in abundances and charge states that follow the measured patterns in interplanetary space (Cranmer et al. 2007; Jin et al. 2012; Cranmer 2014a; Oran et al. 2015). Thus, for now, it appears that solar wind composition is not a useful discriminator between the main theoretical categories.

3.3 Does Radial Evolution Mask Bimodality?

It is still not clear whether the dominant acceleration processes in the corona produce *bimodal* (i.e., cleanly separated) fast and slow wind streams, or if they produce a continuous distribution of states by varying one of more parameters. In some versions of the WTD model, a slow variation of the superradial flux-tube expansion factor can produce a bimodal jump in the wind speed (Cranmer 2005). This occurs because there are often multiple possible radii for the Parker (1958) critical point, and the global time-steady solution can undergo a rapid transition from one of those radii to another, depending on small changes in the expansion factor (see also Kopp and Holzer 1976). If the overall radial distribution of coronal heating remains more or less unchanged, a wind with a lower critical point tends to have a higher asymptotic speed, and a wind with a higher critical point tends to have a lower speed (Leer and Holzer 1980; Pneuman 1980).

Whether or not the corona produces a bimodal or broad/continuous distribution of wind speeds, it remains difficult to use *in situ* interplanetary data to make definitive conclusions

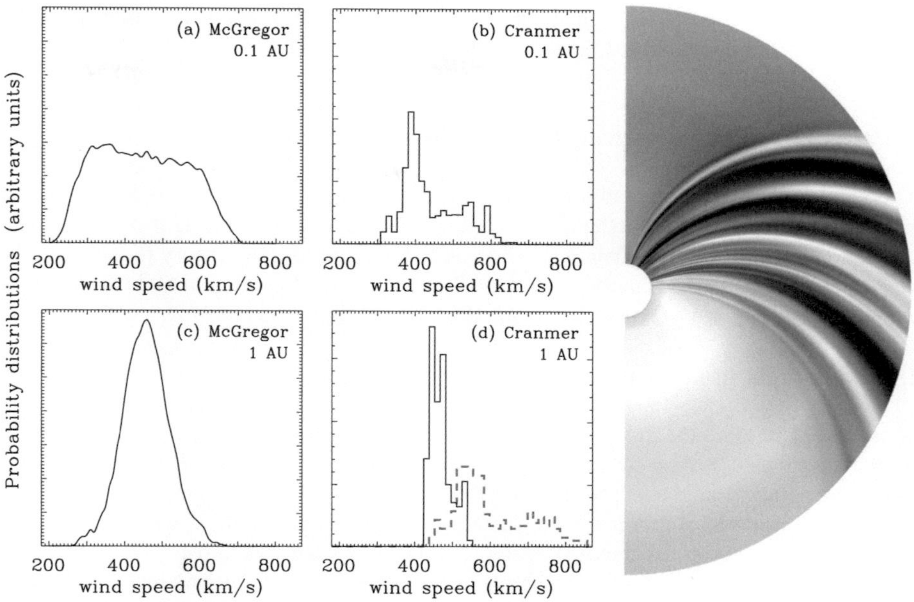

Fig. 9 Comparison of solar wind-speed probability distributions in near-ecliptic regions of semi-empirical three-dimensional models: (**a**), (**c**) WSA-ENLIL models (McGregor et al. 2011a) assembled over 3 years during the 1995–1997 solar minimum, sampled at 0.1 AU and 1 AU; (**b**), (**d**) coordinated set of ZEPHYR models (Cranmer et al. 2013) for a quiet region observed with SOLIS in 2003, sampled at 0.1 AU and 1 AU. The *red dashed curve* is the distribution of wind speeds at 1 AU for separate ZEPHYR-model flux tubes computed *without* stream interactions in the ecliptic plane. *Right-hand panel* shows the simulated CIR pattern for the 2003 quiet region. *Grayscale levels* correspond to density variations, with the large-scale radial dropoff removed

about this issue. In the ecliptic plane, the solar wind becomes mixed—in ways that usually increase its randomness and stochasticity—in at least three distinct ways. (1) MHD turbulence gives rise to an effective random walk of field lines and a potential loss of identity for initial plasma parcels that become shredded in space and time (Matthaeus et al. 1998; Greco et al. 2012). (2) The Sun's rotation produces stream-stream interactions that create CIR spiral structures and ever-greater longitudinal blending with increasing distance (Burlaga and Szabo 1999; Riley 2007). (3) Coulomb collisions, whose randomizing effects accumulate with radial distance, tend to erase the field-aligned kinetic effects discussed in Sect. 2 (e.g., Kasper et al. 2008). The main impact of these processes is to produce ambiguity when mapping wind streams back from interplanetary space to the Sun. This ambiguity becomes more apparent for smaller scales in longitude and latitude, and for more distant mappings from the outer heliosphere (e.g., Elliott et al. 2012).

Figure 9 illustrates the second of the three mixing processes listed above. Model-based reconstructions of radial and longitudinal evolution of the solar wind are shown from McGregor et al. (2011a) and Cranmer et al. (2013). High-contrast flux-tube structure near the Sun appears to be eroded rapidly by CIR stream interactions. The most extreme solar wind parcels (e.g., the highest and lowest speeds) tend to disappear and leave behind a single-peaked distribution dominated by moderate speeds of order 400–450 $\mathrm{km\,s^{-1}}$. The red dashed curve in Fig. 9d shows what the distribution of speeds would have been if stream-stream interactions in the ecliptic were ignored (i.e., the slight bimodality at 0.1 AU would have been preserved at 1 AU). Thus, the presence of stream-stream interactions tends make any intrinsic coronal bimodality extremely difficult to detect at 1 AU.

4 Paths Forward to Improved Space Weather Prediction

Although past observational and theoretical work has improved our understanding of the basic processes at work in generating the solar wind, much more needs to be done. The subsections below describe some of what is on the horizon for pure theory (Sect. 4.1), observations and *in situ* measurements (Sect. 4.2), and empirical prediction tools that attempt to make use of the best of what models and data have to offer (Sect. 4.3).

4.1 Theoretical Improvements

In order to determine the quantitative contributions of WTD, RLO, and CMS processes to accelerating the solar wind, each of these models must be developed to the point of eliminating their free parameters. The difficulty in doing this lies in the large range of relevant spatial and time scales important to the physics—e.g., the Sun can exhibit wavelike variability with periods from milliseconds (Bastian et al. 1998) to years (McIntosh et al. 2017). This reality demands adaptive, multi-scale modeling techniques that go beyond many traditional plasma-in-a-box type simulations. These models must also strive to contain as many of the proposed physical processes as possible. The true impact of any one process on the system may not be made clear until it is allowed to interact with the others in a realistic way.

Although there are still many physics questions that can be answered by models with limited spatial extent, the community's ultimate goal is to develop and improve *global three-dimensional simulations* of the entire corona and heliosphere. In the last decade, several of these models have made the transition from polytropic energy equations and prescribed heating functions to more physics-based (usually WTD) ways of computing the coronal heating. Additional recent developments include solving multi-fluid energy equations instead of single-fluid MHD (Usmanov et al. 2012, 2016; van der Holst et al. 2014; Oran et al. 2015) and assimilating time-dependent photospheric data instead of using static synoptic maps (Feng et al. 2015; Linker et al. 2016; Yalim et al. 2017). Aspects of the simulations that have been shown to be especially important for space weather prediction include resolving the sharp HCS (i.e., avoiding unphysical diffusion associated with too coarse a grid; Stevens et al. 2012) and using Monte Carlo ensembles instead of single models (Riley et al. 2013; Owens et al. 2017).

Because of the need to compare model predictions with actual observational data (see below), it is important to include *forward modeling* in the theorist's toolbox. It has long been a goal to "invert" the data; i.e., to extract information from images and spectra that allow us to solve for the three-dimensional distributions of plasma parameters such as temperature and density. However, the corona is optically thin, highly structured, and time-variable. Thus, attempts at data inversion are often nonunique and fraught with uncertainty (see, e.g., Judge and McIntosh 1999; De Moortel and Bradshaw 2008). It is a much safer procedure to take the theoretical model output, simulate what observers would see, and make direct comparisons at the level of the data. Several sets of sophisticated software tools are being developed to enable this kind of forward modeling (e.g., Nita et al. 2015; Gibson 2015; Gibson et al. 2016; Van Doorsselaere et al. 2016). Methodologies are also being developed to iteratively optimize models to match observations, thus achieving the ultimate goal of inversion (e.g., Dalmasse et al. 2016). A benefit of these approaches is that they provide information about which observables are the most influential in validating or falsifying a given model.

4.2 Observational Improvements

Despite the difficulties described above regarding forward and inverse modeling, we need to remain on the lookout for new ways in which observations can put tight constraints on theory. A clear source of "low-hanging fruit" is to pursue better measurements of the ingredients of proposed physical mechanisms. For example, testing the WTD model requires knowing the amplitudes and damping rates of fluctuations that propagate along open field lines (i.e., filling in the gaps in Fig. 6). Testing the CMS model requires knowing how much mass and Poynting flux comes up through the photosphere, chromosphere, and transition region—and does not come back down again. Testing the RLO model requires measuring how much plasma and magnetic energy gets processed through magnetic reconnection events that convert closed to open field lines.

Global models of the solar wind depend on photospheric magnetic field maps as lower boundary conditions. Traditional synoptic maps, built up from longitudinal magnetograms, are problematic because (1) they do not contain information about magnetic currents in the solar atmosphere, (2) they ignore variability on timescales shorter than a solar rotation, and (3) the north and south poles are poorly resolved (Sun et al. 2011; Petrie 2017). Vector magnetograms are improving the situation (e.g., Liu et al. 2017), but it is still the case that magnetograms from different observatories tend to produce markedly different predictions for the ecliptic plane at 1 AU (Jian et al. 2011; Riley et al. 2014). The ideal solution would be to have telescopes on multiple spacecraft, positioned throughout the heliosphere, so that all 4π steradians of the Sun can be monitored continuously (Roelof et al. 2004; Liewer et al. 2008; Strong et al. 2012). A continuous view of the solar poles is particularly compelling, both for space-weather monitoring and for establishing the nature of the solar dynamo. There are also mission concepts that would provide new insights while stopping short of full 4π coverage, such as an early-warning system at the Earth-Sun L5 point (e.g., Lavraud et al. 2016; Pevtsov et al. 2016), or a two-spacecraft system that would improve upon *STEREO's* initial exploration of stereoscopic imaging (Strugarek et al. 2015). Sustained multi-vantage observations of the photospheric magnetic field, the inner boundary of the heliosphere, has transformative potential.

Having the ability to make direct and routine measurements of the coronal magnetic field would complement the photospheric data (Lin et al. 2004; Judge et al. 2013). Most proposed coronal heating mechanisms are magnetic in nature, and they often depend on the properties of twisted, non-potential structures that are difficult to extrapolate up from photospheric boundary conditions. Indirect methods such as coronal seismology (e.g., De Moortel and Nakariakov 2012) have been helpful, but direct measurements of the field magnitude and direction could put more stringent constraints on models. Observations of linear polarization in infrared coronal emission lines by the *Coronal Multichannel Spectropolarimeter* (Tomczyk et al. 2008) have demonstrated the power of such observations for establishing the topologies of magnetic structures such as flux ropes and pseudostreamers (Dove et al. 2011; Bąk-Stęślicka et al. 2013; Rachmeler et al. 2013, 2014; Gibson et al. 2017a). The *Daniel K. Inouye Solar Telescope* (DKIST; Tritschler et al. 2016) will provide a more direct measure of the line-of-sight coronal field strength via high-precision measurements of circular polarization, and explore new regimes of coronal magnetometry. Future programs, such as the *Coronal Solar Magnetism Observatory* (COSMO; Tomczyk et al. 2016), would enable full-Sun, synoptic measurements of the coronal magnetic field. Ultimately, a vantage away from the Earth-Sun line would be beneficial for monitoring Earth-directed space weather; the proposed balloon-borne *Waves and Magnetism in the Solar Atmosphere* (WAMIS; Ko et al. 2016) and the *massively-multiplexed Coronal Spectropolarimetric Magnetometer* (mxCSM; Lin 2016) represent development in this direction.

Additional information about solar wind acceleration is likely to be learned by detecting scattered light from plasma parcels in the extended corona ($r \approx 2$–$30\,R_\odot$). This range of heights is where coronal "structures" evolve into solar wind streams. It is also where the plasma becomes collisionless; i.e., where departures from thermal equilibrium (Fig. 5) start to become useful probes of the physics. It would be beneficial for next-generation ultraviolet coronagraph spectrometers (e.g., Kohl et al. 2008; Strachan et al. 2016) to be developed, in order to follow up on the successes of UVCS and extend the remote-sensing field of view to larger heights and more ions.

To monitor solar wind acceleration at slightly larger distances, there are opportunities for improving both the analysis techniques and instrumentation for space-based visible-light heliospheric imagers (Rouillard et al. 2011; DeForest et al. 2011, 2016; DeForest and Howard 2015) and ground-based radio arrays (Manoharan et al. 2017). Large-scale collaborations such as HELCATS (Heliospheric Cataloguing, Analysis, and Techniques Service; Plotnikov et al. 2016; Rouillard et al. 2017) are building a broad range of analysis tools and observational databases. New remote-sensing heliospheric observations have the potential to improve our knowledge of how parcels accelerate in both isolated wind streams and CIRs (as well as CMEs), and also to reveal how coronal waves evolve into turbulent eddies.

Of course, future improvements in solar wind measurements must also include *in situ* particle and field detection. We anticipate that the *Parker Solar Probe* (Fox et al. 2016) and *Solar Orbiter* (Müller et al. 2013), both scheduled for launch in 2018, will revolutionize our conception of the inner heliosphere. Specifically, these missions will explore regions in which all three mixing processes discussed in Sect. 3.3 have not yet had time to smear out the unique field-line mappings back to the corona. As these missions are starting to explore the inner heliosphere, India's *Aditya-L1* mission is expected to launch around 2020. Its combination of remote-sensing and *in situ* instruments (Ghosh et al. 2016; Venkata et al. 2017) should extend the multi-diagnostic capabilities pioneered by *SOHO* and *ACE* at Earth-Sun L1. Lastly, the *Turbulence Heating Observer* (THOR; Vaivads et al. 2016) spacecraft is expected to launch in 2026, and it will be dedicated to a complete characterization of micro-scale turbulent dissipation in the solar wind and Earth's magnetosphere.

4.3 Improvements in Empirical Prediction Tools

As discussed in Sect. 2.3, empirically based techniques to predict solar wind conditions at 1 AU (e.g., PFSS and WSA) continue to be tested and upgraded. Comparisons are being made between potential-based field extrapolations and global MHD models (Jian et al. 2015, 2016), but both techniques still have problems reproducing the full range of observed solar-wind variability. Ongoing improvements in model validation are being made by collaborative efforts such as the Community Coordinated Modeling Center (CCMC; Hesse et al. 2010), the Space Weather Modeling Framework (SWMF; Tóth et al. 2005), the European Heliospheric Forecasting Information Asset (EUHFORIA; Pomoell et al. 2017), and the NOAA Space Weather Prediction Center (SWPC; Berger et al. 2015).

Eventually, forecasts based on solutions to conservation equations ought to replace empirical recipes such as PFSS, but at present that appears to be too computationally expensive. A promising middle ground may be to use extrapolations based on *magnetofrictional evolution*, which stops short of full MHD but still manages to include the time-dependent development of non-potential coronal currents (Yeates 2014; Edwards et al. 2015; Fisher et al. 2015). For the actual prediction of solar wind properties along open field lines, there are new semi-empirical tools (see, e.g., Woolsey and Cranmer 2014; Pinto and Rouillard 2017) that use the output of physics-based models (rather than WSA-type correlations) but are also designed for computational speed. For solar-cycle-length predictions, it may be important to couple corona/wind models to flux-transport models of the photospheric fields (e.g., Merkin et al. 2016; Weinzierl et al. 2016) or even to more self-consistent simulations of the interior dynamo.

Space weather forecasting also requires efficient *data assimilation* (or *data incorporation*, as discussed by Schrjiver et al. 2015) for it to be accurate. Unfortunately, there is a stark contrast with terrestrial weather forecasting, in that (1) the space-based data are so sparse that the "upwind conditions" are not really known with sufficient precision, and (2) the corona/heliosphere physics that produces these upwind conditions is still not understood. There are also differences in how data products are used. Photospheric magnetograms are straightforward drivers of the models because they can be used as lower boundary conditions. However, it is more difficult to use data measured in the corona and inner heliosphere, because those regions are inside the computational grids of most MHD simulations. Just straightforwardly inserting newly measured plasma properties into an MHD model would overconstrain the equations and produce discontinuities in the conserved quantities. One promising path forward is to run multiple simulations in parallel—with randomly varied initial/boundary conditions or different choices for unknown coronal heating processes—and then to choose the model (or linear combination of models) that best matches the data (see Doxas et al. 2007). The optimization method of Dalmasse et al. (2016) discussed above also suggests possible solutions to this problem. This type of *ensemble modeling* is now being applied to a range of solar wind and ICME prediction methods (Lee et al. 2013; Riley et al. 2013; Cash et al. 2015; Owens et al. 2017).

5 Conclusions

The objective of this paper has been to review our present-day understanding of the origins of ambient solar wind streams. We also examined a wide range of possible ways to improve space-weather forecasts related to the ambient wind. There are still some fundamental questions about solar wind acceleration for which we do not yet have answers: What are the physical processes that dominate coronal heating? Where does most of the slow wind come from on the Sun? Do fast and slow wind streams originate from two separate mechanisms? However, it is nevertheless the case that forecasts are improving. This is due in part to the successes of empirical correlation techniques, which help point the way to identifying the "best" physical processes to include in simulations. As these processes are identified and characterized, we also expect improvements in models of CMEs, which have proposed heating mechanisms (see, e.g., Liu et al. 2006; Murphy et al. 2011) very similar to those discussed in Sect. 3 for the ambient wind.

Much of the discussion in this paper has been focused on the idea of applying insights from fundamental research to the practical goals of space weather forecasting. This phase of work has been called "research-to-operations" (R2O). However, there are also benefits that flow in the opposite direction; i.e., "operations-to-research" (O2R; Steenburgh et al. 2014). Increased knowledge about our well-studied Sun and heliosphere feeds back into astrophysical studies of Sun-like stars (e.g., Haisch and Schmitt 1996; Schmelz 2003; Brun et al. 2015) and even more distant objects such as supermassive black holes (Quataert and Gruzinov 1999; Li et al. 2017) and galaxy clusters (Parrish et al. 2012). For decades, the Sun has been considered a testbed for studying fundamental physical processes that cannot be reproduced in terrestrial laboratories. Because the solar wind unifies studies of waves, turbulence, and magnetic reconnection, it also feeds into interdisciplinary studies of universal processes in heliophysics (see Davila et al. 2009; Raulin et al. 2010).

Acknowledgements The authors would like to thank Ruedi von Steiger, André Balogh, Dan Baker, Tamás Gombosi, Hannu Koskinen, and Astrid Veronig for convening the fantastic 2016 ISSI workshop on the scientific foundations of space weather. SRC's work was supported by NASA grants NNX15AW33G and NNX16AG87G, NSF grants 1540094 (SHINE) and 1613207 (AAG), and start-up funds from the Department of Astrophysical and Planetary Sciences at the University of Colorado Boulder. PR's work was supported through a grant from NASA's Living With a Star (LWS) Program. The National Center for Atmospheric Research (NCAR) is supported by the National Science Foundation. This research made extensive use of NASA's Astrophysics Data System (ADS).

References

L. Abbo, R. Lionello, P. Riley, Y.-M. Wang, Coronal pseudo-streamer and bipolar streamer observed by SOHO/UVCS in March 2008. Sol. Phys. **290**, 2043–2054 (2015)

L. Abbo, L. Ofman, S.K. Antiochos, V.H. Hansteen, L. Harra, Y.-K. Ko, G. Lapenta, B. Li, P. Riley, L. Strachan, R. von Steiger, Y.-M. Wang, Slow solar wind: Observations and modeling. Space Sci. Rev. **201**, 55–108 (2016)

H. Alfvén, On the solar corona. Arkiv Math. Astron. Fysik (Band 27A) **25**, 1–23 (1941)

M.D. Altschuler, G. Newkirk, Magnetic fields and the structure of the solar corona: I: Methods of calculating coronal fields. Sol. Phys. **9**, 131–149 (1969)

S.K. Antiochos, Z. Mikić, V.S. Titov, R. Lionello, J.A. Linker, A model for the sources of the slow wind. Astrophys. J. **731**, 112 (2011)

E. Antonucci, M.A. Dodero, S. Giordano, Fast solar wind velocity in a polar coronal hole during solar minimum. Sol. Phys. **197**, 115–134 (2000)

C.N. Arge, V.J. Pizzo, Improvement in the prediction of solar wind conditions using near-real time solar magnetic field updates. J. Geophys. Res. **105**, 10465–10480 (2000)

C.N. Arge, D. Odstrčil, V.J. Pizzo, L.R. Mayer, Improved method for specifying solar wind speed near the Sun, in *Solar Wind Ten*, ed. by M. Velli, R. Bruno. AIP Conf. Proc., vol. 679 (AIP, New York, 2003), pp. 190–193

W.I. Axford, J.F. McKenzie, The origin of high speed solar wind streams, in *Solar Wind Seven*, ed. by E. Marsch, R. Schwenn (Pergamon, New York, 1992), pp. 1–5

U. Bąk-Stęślicka, S.E. Gibson, Y. Fan, C. Bethge, B. Forland, L.A. Rachmeler, The magnetic structure of solar prominence cavities: New observational signature revealed by coronal magnetometry. Astrophys. J. Lett. **770**, L28 (2013)

D.N. Baker, P.J. Erickson, J.F. Fennell, J.C. Foster, A.N. Jaynes, P.T. Verronen, Space weather effects in the Earth's radiation belts. Space Sci. Rev. (2017), this issue

S.D. Bale, M. Pulupa, C. Salem, C.H.K. Chen, E. Quataert, Electron heat conduction in the solar wind: Transition from Spitzer-Härm to the collisionless limit. Astrophys. J. Lett. **769**, L22 (2013)

S.J. Bame, J.R. Asbridge, W.C. Feldman, J.T. Gosling, Evidence for a structure-free state at high solar wind speeds. J. Geophys. Res. **82**, 1487–1492 (1977)

D. Banerjee, L. Teriaca, J.G. Doyle, K. Wilhelm, Broadening of Si VIII lines observed in the solar polar coronal holes. Astron. Astrophys. **339**, 208–214 (1998)

A. Barnes, Acceleration of the solar wind. Rev. Geophys. **30**, 43–55 (1992)

T.S. Bastian, Radio wave propagation in the corona and the interplanetary medium. Astrophys. Space Sci. **277**, 107–116 (2001)

T.S. Bastian, A.O. Benz, D.E. Gary, Radio emission from solar flares. Annu. Rev. Astron. Astrophys. **36**, 131–188 (1998)

B. Bavassano, E. Pietropaolo, R. Bruno, On the evolution of outward and inward Alfvénic fluctuations in the polar wind. J. Geophys. Res. **105**, 15959–15964 (2000)

A. Bemporad, L. Abbo, Spectroscopic signature of Alfvén waves damping in a polar coronal hole up to 0.4 solar radii. Astrophys. J. **751**, 110 (2012)

L. Berger, R.F. Wimmer-Schweingruber, G. Gloeckler, Systematic measurements of ion-proton differential streaming in the solar wind. Phys. Rev. Lett. **106**, 151103 (2011)

T. Berger, R. Viereck, H. Singer, T. Onsager, D. Biesecker, R. Rutledge, S. Hill, R. Akmaev, G. Milward, T. Fuller-Rowell, Characteristics of operational space weather forecasting: Observations and models. Presented at 1st Joint AAS/AGU Triennial Earth-Sun Summit, 112.04 (2015)

M.K. Bird, P. Edenhofer, Remote sensing observations of the solar corona, in *Physics of the Inner Heliosphere I*, ed. by R. Schwenn, E. Marsch (Springer, Berlin, 1990), pp. 13–97

M.G. Bobra, X. Sun, J.T. Hoeksema, M. Turmon, Y. Liu, K. Hayashi, G. Barnes, K.D. Leka, The Helioseismic and Magnetic Imager (HMI) vector magnetic field pipeline: SHARPs: Space-weather HMI active region patches. Sol. Phys. **289**, 3549–3578 (2014)

J.E. Borovsky, Flux tube texture of the solar wind: Strands of the magnetic carpet at 1 AU. J. Geophys. Res. **113**, A08110 (2008)

J.E. Borovsky, The plasma structure of coronal hole solar wind: Origins and evolution. J. Geophys. Res. **121**, 5055–5087 (2016)

J.E. Borovsky, M.H. Denton, Differences between CME-driven storms and CIR-driven storms. J. Geophys. Res. **111**, A07S08 (2006)

A.R. Breen, B.J. Thompson, M. Kojima, D.A. Biesecker, A. Canals, R.A. Fallows, J.A. Linker, A.J. Lazarus, A. Lecinski, Z. Mikić, P.J. Moran, P.J.S. Williams, Measurements of the solar wind over a wide range of heliocentric distances: A comparison of results from the first three Whole Sun Months. J. Atmos. Sol.-Terr. Phys. **62**, 1527–1543 (2000)

D.H. Brooks, H.P. Warren, Establishing a connection between active region outflows and the solar wind: Abundance measurements with EIS/Hinode. Astrophys. J. Lett. **727**, L13 (2011)

A.S. Brun, R.A. García, G. Houdek, D. Nandy, M. Pinsonneault, The solar-stellar connection. Space Sci. Rev. **196**, 303–356 (2015)

L.F. Burlaga, A. Szabo, Fast and slow flows in the solar wind near the ecliptic at 1 AU. Space Sci. Rev. **87**, 137–140 (1999)

M.D. Cash, D.A. Biesecker, V. Pizzo, C.A. Koning, G. Millward, C.N. Arge, C.J. Henney, D. Odstrčil, Space Weather **13**, 611–625 (2015)

J.W. Chamberlain, Interplanetary gas. II. Expansion of a model solar corona. Astrophys. J. **131**, 47–56 (1960)

B.D.G. Chandran, T.J. Dennis, E. Quataert, S.D. Bale, Incorporating kinetic physics into a two-fluid solar-wind model with temperature anisotropy and low-frequency Alfvén-wave turbulence. Astrophys. J. **743**, 197 (2011)

H. Che, M.L. Goldstein, The origin of non-Maxwellian solar wind electron velocity distribution function: Connection to nanoflares in the solar corona. Astrophys. J. Lett. **795**, L38 (2014)

C.H.K. Chen, Recent progress in astrophysical plasma turbulence from solar wind observations. J. Plasma Phys. **82**, 535820602 (2016)

J.W. Cirtain, L. Golub, A.R. Winebarger, B. De Pontieu, K. Kobayashi, R.L. Moore, R.W. Walsh, K.E. Korreck, M. Weber, P. McCauley, A. Title, S. Kuzin, C.E. DeForest, Energy release in the solar corona from spatially resolved magnetic braids. Nature **493**, 501–503 (2013)

P.J. Coleman, Turbulence, viscosity, and dissipation in the solar-wind plasma. Astrophys. J. **153**, 371–388 (1968)

S.R. Cranmer, Coronal holes and the high-speed solar wind. Space Sci. Rev. **101**, 229–294 (2002)

S.R. Cranmer, Observational aspects of wave acceleration in open magnetic regions, in *SOHO-13: Waves, Oscillations, and Small-Scale Events in the Solar Atmosphere*, ed. by H. Lacoste (ESA, Noordwijk, 2004), pp. 353–362

S.R. Cranmer, Why is the fast solar wind fast and the slow solar wind slow? A survey of geometrical models, in *Connecting Sun and Heliosphere*, ed. by B. Fleck, T.H. Zurbuchen, H. Lacoste. ESA, vol. SP-592 (ESA, Noordwijk, 2005), pp. 159–164

S.R. Cranmer, Coronal holes. Living Rev. Sol. Phys. **6**, 3 (2009)

S.R. Cranmer, Suprathermal electrons in the solar corona: Can nonlocal transport explain heliospheric charge states? Astrophys. J. Lett. **791**, L31 (2014a)

S.R. Cranmer, Ensemble simulations of proton heating in the solar wind via turbulence and ion cyclotron resonance. Astrophys. J. Suppl. Ser. **213**, 16 (2014b)

S.R. Cranmer, S.H. Saar, Testing a predictive theoretical model for the mass loss rates of cool stars. Astrophys. J. **741**, 54 (2011)

S.R. Cranmer, A.A. van Ballegooijen, On the generation, propagation, and reflection of Alfvén waves from the solar photosphere to the distant heliosphere. Astrophys. J. Suppl. Ser. **156**, 265–293 (2005)

S.R. Cranmer, A.A. van Ballegooijen, Can the solar wind be driven by magnetic reconnection in the Sun's magnetic carpet? Astrophys. J. **720**, 824–847 (2010)

S.R. Cranmer, L.N. Woolsey, Driving solar spicules and jets with magnetohydrodynamic turbulence: Testing a persistent idea. Astrophys. J. **812**, 71 (2015)

S.R. Cranmer, J.L. Kohl, G. Noci, E. Antonucci, G. Tondello, M.C.E. Huber, L. Strachan, A.V. Panasyuk, L.D. Gardner, M. Romoli, S. Fineschi, D. Dobrzycka, J.C. Raymond, P. Nicolosi, O.H.W. Siegmund, D. Spadaro, C. Benna, A. Ciaravella, S. Giordano, S.R. Habbal, M. Karovska, X. Li, R. Martin, J.G. Michels, A. Modigliani, G. Naletto, R.H. O'Neal, C. Pernechele, G. Poletto, P.L. Smith, R.M. Suleiman, An empirical model of a polar coronal hole at solar minimum. Astrophys. J. **511**, 481–501 (1999)

S.R. Cranmer, A.A. van Ballegooijen, R.J. Edgar, Self-consistent coronal heating and solar wind acceleration from anisotropic magnetohydrodynamic turbulence. Astrophys. J. Suppl. Ser. **171**, 520–551 (2007)

S.R. Cranmer, A.V. Panasyuk, J.L. Kohl, Improved constraints on the preferential heating and acceleration of oxygen ions in the extended solar corona. Astrophys. J. **678**, 1480–1497 (2008)
S.R. Cranmer, W.H. Matthaeus, B.A. Breech, J.C. Kasper, Empirical constraints on proton and electron heating in the solar wind. Astrophys. J. **702**, 1604–1614 (2009)
S.R. Cranmer, A.A. van Ballegooijen, L.N. Woolsey, Connecting the Sun's high-resolution magnetic carpet to the turbulent heliosphere. Astrophys. J. **767**, 125 (2013)
N.U. Crooker, R.L. McPherron, M.J. Owens, Comparison of interplanetary signatures of streamers and pseudostreamers. J. Geophys. Res. **119**, 4157–4163 (2014)
K. Dalmasse, D. Nychka, S.E. Gibson, N. Flyer, Y. Fan, ROAM: A radial-basis-function optimization approximation method for diagnosing the three-dimensional coronal magnetic field. Front. Astron. Space Sci. **3**, 24 (2016)
J.M. Davila, N. Gopalswamy, B.J. Thompson, Universal processes in heliophysics, in *Universal Heliophysical Processes*, ed. by N. Gopalswamy, D.F. Webb. IAU Symp., vol. 257 (Cambridge University Press, Cambridge, 2009), pp. 11–16
I. De Moortel, S.J. Bradshaw, Forward modelling of coronal intensity perturbations. Sol. Phys. **252**, 101–119 (2008)
I. De Moortel, V.M. Nakariakov, Magnetohydrodynamic waves and coronal seismology: An overview of recent results. Philos. Trans. R. Soc. A **370**, 3193–3216 (2012)
B. De Pontieu, S.W. McIntosh, M. Carlsson, V.H. Hansteen, T.D. Tarbell, C.J. Schrijver, A.M. Title, R.A. Shine, S. Tsuneta, Y. Katsukawa, K. Ichimoto, Y. Suematsu, T. Shimizu, S. Nagata, Chromospheric Alfvén waves strong enough to power the solar wind. Science **318**, 1574–1577 (2007)
B. De Pontieu, S.W. McIntosh, V.H. Hansteen, C.J. Schrijver, Observing the roots of solar coronal heating in the chromosphere. Astrophys. J. Lett. **701**, L1–L6 (2009)
G. de Toma, Evolution of coronal holes and implications for high-speed solar wind during the minimum between cycles 23 and 24. Sol. Phys. **274**, 195–217 (2011)
C.E. DeForest, T.A. Howard, Feasibility of heliospheric imaging from near Earth. Astrophys. J. **804**, 126 (2015)
C.E. DeForest, T.A. Howard, S.J. Tappin, Observations of detailed structure in the solar wind at 1 AU with STEREO/HI-2. Astrophys. J. **738**, 103 (2011)
C.E. DeForest, W.H. Matthaeus, N.M. Viall, S.R. Cranmer, Fading coronal structure and the onset of turbulence in the young solar wind. Astrophys. J. **828**, 66 (2016)
A.J. Dessler, Solar wind and interplanetary magnetic field. Rev. Geophys. Space Phys. **5**, 1–41 (1967)
J.B. Dove, S.E. Gibson, L.A. Rachmeler, S. Tomczyk, P. Judge, A ring of polarized light: Evidence for twisted coronal magnetism in cavities. Astrophys. J. Lett. **731**, L1 (2011)
I. Doxas, W. Horton, J. Lyon, M. Wiltberger, R.S. Weigel, Branch prediction and speculative execution: A magnetospheric data assimilation scheme for space weather forecasting. Space Weather **5**, S11001 (2007)
J. Eastwood, R. Nakamura, M. Hesse, Space weather driven dynamics of the magnetosphere. Space Sci. Rev. (2017), this issue
E. Echer, B.T. Tsurutani, W.D. Gonzalez, Interplanetary origins of moderate (-100 nT $<$ Dst ≤ -50 nT) geomagnetic storms during solar cycle 23 (1996–2008). J. Geophys. Res. **118**, 385–392 (2013)
J.K. Edmondson, On the role of interchange reconnection in the generation of the slow solar wind. Space Sci. Rev. **172**, 209–225 (2012)
S.J. Edwards, A.R. Yeates, F.-X. Bocquet, D.H. Mackay, Influence of non-potential coronal magnetic topology on solar-wind models. Sol. Phys. **290**, 2791–2808 (2015)
A.I. Efimov, T. Imamura, K.-I. Oyama, K. Noguchi, L.N. Samoznaev, A.S. Nabatov, M.K. Bird, I.V. Chashei, Properties of solar wind turbulence from radio occultation experiments with the NOZOMI spacecraft. Astron. Rep. **54**, 1032–1041 (2010)
G. Einaudi, P. Boncinelli, R.B. Dahlburg, J.T. Karpen, Formation of the slow solar wind in a coronal streamer. J. Geophys. Res. **104**, 521–534 (1999)
H.A. Elliott, C.J. Henney, D.J. McComas, C.W. Smith, B.J. Vasquez, Temporal and radial variation of the solar wind temperature–speed relationship. J. Geophys. Res. **117**, A09102 (2012)
B.A. Emery, I.G. Richardson, D.S. Evans, F.J. Rich, G.R. Wilson, Solar rotational periodicities and the semiannual variation in the solar wind, radiation belt, and aurora. Sol. Phys. **274**, 399–425 (2011)
R. Esser, S. Fineschi, D. Dobrzycka, S.R. Habbal, R.J. Edgar, J.C. Raymond, J.L. Kohl, M. Guhathakurta, Plasma properties in coronal holes derived from measurements of minor ion spectral lines and polarized white light intensity. Astrophys. J. Lett. **510**, L63–L67 (1999)
R.M. Evans, M. Opher, R. Oran, B. van der Holst, I.V. Sokolov, R. Frazin, T.I. Gombosi, A. Vásquez, Coronal heating by surface Alfvén wave damping: Implementation in a global magnetohydrodynamics model of the solar wind. Astrophys. J. **756**, 155 (2012)

A.N. Fazakerley, L.K. Harra, L. van Driel-Gesztelyi, An investigation of the sources of Earth-directed solar wind during Carrington Rotation 2053. Astrophys. J. **823**, 145 (2016)
X. Feng, X. Ma, C. Xiang, Data-driven modeling of the solar wind from 1 R_s to 1 AU. J. Geophys. Res. **120**, 10159–10174 (2015)
G.H. Fisher, W.P. Abbett, D.J. Bercik, M.D. Kazachenko, B.J. Lynch, B.T. Welsch, J.T. Hoeksema, K. Hayashi, Y. Liu, A.A. Norton, A. Sainz Dalda, X. Sun, M.L. DeRosa, M.C.M. Cheung, The coronal global evolutionary model: Using HMI vector magnetogram and Doppler data to model the buildup of free magnetic energy in the solar corona. Space Weather **13**, 369–373 (2015)
L.A. Fisk, Acceleration of the solar wind as a result of the reconnection of open magnetic flux with coronal loops. J. Geophys. Res. **108**, 1157 (2003)
L.A. Fisk, N.A. Schwadron, T.H. Zurbuchen, Acceleration of the fast solar wind by the emergence of new magnetic flux. J. Geophys. Res. **104**, 19765–19772 (1999)
N.J. Fox, M. Velli, S.D. Bale, R. Decker, A. Driesman, R.A. Howard, J.C. Kasper, J. Kinnison, M. Kusterer, D. Lario, M.K. Lockwood, D.J. McComas, N.E. Raouafi, A. Szabo, The solar probe plus mission: Humanity's first visit to our star. Space Sci. Rev. **204**, 7–48 (2016)
R.A. Frazin, S.R. Cranmer, J.L. Kohl, Empirically determined anisotropic velocity distributions and outflows of O^{5+} ions in a coronal streamer at solar minimum. Astrophys. J. **597**, 1145–1157 (2003)
J.W. Freeman, Estimates of solar wind heating inside 0.3 AU. Geophys. Res. Lett. **15**, 88–91 (1988)
H. Fu, M.S. Madjarska, L.D. Xia, B. Li, Z.H. Huang, Z. Wangguan, Charge states and FIP bias of the solar wind from coronal holes, active regions, and quiet Sun. Astrophys. J. **836**, 169 (2017)
K. Fujiki, M. Tokumaru, T. Iju, K. Hakamada, M. Kojima, Relationship between solar-wind speed and coronal magnetic-field properties. Sol. Phys. **290**, 2491–2505 (2015)
A.B. Galvin, J.L. Kohl, Whole Sun Month at solar minimum: An introduction. J. Geophys. Res. **104**, 9673–9678 (1999)
N. Ganushkina, A. Jaynes, Space weather effects produced by ring current particles. Space Sci. Rev. (2017), this issue
J. Geiss, P. Hirt, H. Leutwyler, On acceleration and motion of ions in corona and solar wind. Sol. Phys. **12**, 458–483 (1970)
J. Geiss, G. Gloeckler, R. von Steider, Origin of the solar wind from composition data. Space Sci. Rev. **72**, 49–60 (1995)
A. Ghosh, S. Chatterjee, A.R. Khan, D. Tripathi, A.N. Ramaprakash, D. Banerjee, P. Chordia, A.M. Gandorder, N. Krivova, D. Nandy, C. Rajarshi, S.K. Solanki, S. Sriram, The solar ultraviolet imaging telescope onboard Aditya-L1. Proc. SPIE **9905**, 990503 (2016)
S.E. Gibson, Data–model comparison using FORWARD and CoMP, in *Polarimetry: From the Sun to Stars and Stellar Environments*, ed. by K. Nagendra, S. Bagnulo, R. Centeno, M. Martínez. IAU Symp., vol. 305 (Cambridge University Press, Cambridge, 2015), pp. 245–250
S.E. Gibson, A. Fludra, F. Bagenal, D. Biesecker, G. del Zanna, B. Bromage, Solar minimum streamer densities and temperatures using Whole Sun Month coordinated data sets. J. Geophys. Res. **104**, 9691–9700 (1999)
S.E. Gibson, J.U. Kozyra, G. de Toma, B.A. Emery, T. Onsager, B.J. Thompson, If the Sun is so quiet, why is the Earth ringing? A comparison of two solar minimum intervals. J. Geophys. Res. **114**, A09105 (2009)
S.E. Gibson, G. de Toma, B. Emery, P. Riley, L. Zhao, Y. Elsworth, R.J. Leamon, J. Lei, S. McIntosh, R.A. Mewaldt, B.J. Thompson, D. Webb, The Whole Heliosphere Interval in the context of a long and structured solar minimum: An overview from Sun to Earth. Sol. Phys. **274**, 5–27 (2011)
S.E. Gibson, T. Kucera, S. White, J. Dove, Y. Fan, B. Forland, L. Rachmeler, C. Downs, K. Reeves, FORWARD: A toolset for multiwavelength coronal magnetometry. Front. Astron. Space Sci. **3**, 8 (2016)
S.E. Gibson, K. Dalmasse, L.A. Rachmeler, M.L. DeRosa, S. Tomczyk, G. de Toma, J. Burkepile, M. Galloy, Magnetic nulls and super-radial expansion in the solar corona. Astrophys. J. **840**, L13 (2017a)
S.E. Gibson, D. Webb, I.M. Hewins, R.H. McFadden, B.A. Emery, W. Denig, P.S. McIntosh, Beyond sunspots: Studies using the McIntosh archive of global solar magnetic field patterns, in *Living Around Active Stars*, ed. by D. Nandi, A. Valio, P. Petit. IAU Symp., vol. 328 (2017b), in press
G. Gloeckler, J. Geiss, H. Balsiger, P. Bedini, J.C. Cain, J. Fischer, L.A. Fisk, A.B. Galvin, F. Gliem, D.C. Hamilton, J.V. Hollweg, F.M. Ipavich, R. Joos, S. Livi, R.A. Lundgren, U. Mall, J.F. McKenzie, K.W. Ogilvie, F. Ottens, W. Rieck, E.O. Tums, R. von Steiger, W. Weiss, B. Wilken, The solar wind ion composition spectrometer. Astron. Astrophys. Suppl. Ser. **92**, 267–289 (1992)
B.E. Goldstein, M. Neugebauer, J.L. Phillips, S. Bame, J.T. Gosling, D. McComas, Y.-M. Wang, N.R. Sheeley, S.T. Suess, Ulysses plasma parameters: Latitudinal, radial, and temporal variations. Astron. Astrophys. **316**, 296–303 (1996)
N. Gopalswamy, A. Lara, R.P. Lepping, M.L. Kaiser, D. Berdichevsky, O.C.St. Cyr, Interplanetary acceleration of coronal mass ejections. Geophys. Res. Lett. **27**, 145–148 (2000)

J.T. Gosling, R.T. Hansen, S.J. Bame, Solar wind speed distributions: 1962–1970. J. Geophys. Res. **76**, 1811–1815 (1971)
J.T. Gosling, D.J. McComas, J.L. Phillips, S.J. Bame, Geomagnetic activity associated with Earth passage of interplanetary shock disturbances and coronal mass ejections. J. Geophys. Res. **96**, 7831–7839 (1991)
R.R. Grall, W.A. Coles, M.T. Klinglesmith, A.R. Breen, P.J.S. Williams, J. Markkanen, R. Esser, Nature **379**, 429–432 (1996)
A. Greco, W.H. Matthaeus, R. D'Amicis, S. Servidio, P. Dmitruk, Evidence for nonlinear development of magnetohydrodynamic scale intermittency in the inner heliosphere. Astrophys. J. **749**, 105 (2012)
L. Green, B. Vršnak, T. Török, S. Krucker, C. Manchester, A. Veronig, The origin and predictability of solar eruptions. Space Sci. Rev. (2017), this issue
C. Gressl, A.M. Veronig, M. Temmer, D. Odstrčil, J.A. Linker, Z. Mikić, P. Riley, Comparative study of MHD modeling of the background solar wind. Sol. Phys. **289**, 1783–1801 (2014)
G.R. Gupta, Spectroscopic evidence of Alfvén wave damping in the off-limb solar corona. Astrophys. J. **836**, 4 (2017)
S.R. Habbal, H. Morgan, M. Druckmüller, A. Ding, J.F. Cooper, A. Daw, E.C. Sittler, Probing the fundamental physics of the solar corona with lunar solar occultation observations. Sol. Phys. **285**, 9–24 (2013)
M. Hahn, D.W. Savin, Observational quantification of the energy dissipated by Alfvén waves in a polar coronal hole: Evidence that waves drive the fast solar wind. Astrophys. J. **776**, 78 (2013)
M. Hahn, E. Landi, D.W. Savin, Evidence of wave damping at low heights in a polar coronal hole. Astrophys. J. **753**, 36 (2012)
B. Haisch, J.H.M.M. Schmitt, Advances in solar-stellar astrophysics. Publ. Astron. Soc. Pac. **108**, 113–129 (1996)
R. Hammer, Energy balance of stellar coronae. I. Methods and examples. Astrophys. J. **259**, 767–778 (1982)
V.H. Hansteen, E. Leer, Coronal heating, densities, and temperatures and solar wind acceleration. J. Geophys. Res. **100**, 21577–21594 (1995)
L.K. Harra, T. Sakao, C.H. Mandrini, H. Hara, S. Imada, P.R. Young, L. van Driel-Gesztelyi, D. Baker, Outflows at the edges of active regions: Contribution to solar wind formation? Astrophys. J. Lett. **676**, L147 (2008)
K.L. Harvey, F. Recely, Polar coronal holes during cycles 22 and 23. Sol. Phys. **211**, 31–52 (2002)
P. Hellinger, P. Trávníček, J.C. Kasper, A.J. Lazarus, Solar wind proton temperature anisotropy: Linear theory and WIND/SWE observations. Geophys. Res. Lett. **33**, L09101 (2006)
P. Hellinger, S. Landi, L. Matteini, A. Verdini, L. Franci, Mirror instability in the turbulent solar wind. Astrophys. J. **838**, 158 (2017)
M. Hesse, M. Kuznetsova, M. Maddox, A. Pulkkinen, J.S. Shim, D. Berrios, L. Rastaetter, P. MacNeice, S. Taktakishvili, A. Chulaki, L. Moiseev, S. Bakshi, K. Patel, Space weather models and their validation and verification at the CCMC. Presented at COSPAR Scientific Assembly, PSW1-003-10 (2010)
J. Heyvaerts, E.R. Priest, Coronal heating by reconnection in DC current systems: A theory based on Taylor's hypothesis. Astron. Astrophys. **137**, 63–78 (1984)
A.K. Higginson, S.K. Antiochos, C.R. DeVore, P.F. Wyper, T.H. Zurbuchen, Dynamics of coronal hole boundaries. Astrophys. J. **837**, 113 (2017)
J.T. Hoeksema, P.H. Scherrer, An atlas of photospheric magnetic field observations and computed coronal magnetic fields: 1976–1985. Sol. Phys. **105**, 205–211 (1986)
J.V. Hollweg, Transition region, corona, and solar wind in coronal holes. J. Geophys. Res. **91**, 4111–4125 (1986)
J.V. Hollweg, P.A. Isenberg, Generation of the fast solar wind: A review with emphasis on the resonant cyclotron interaction. J. Geophys. Res. **107**, 1147 (2002)
T.E. Holzer, W.I. Axford, The theory of stellar winds and related outflows. Annu. Rev. Astron. Astrophys. **8**, 31–60 (1970)
A.J. Hundhausen, *Coronal Expansion and Solar Wind* (Springer, Berlin, 1972)
E.K.J. Huttunen, H.E.J. Koskinen, R. Schwenn, Variability of magnetospheric storms driven by different solar wind perturbations. J. Geophys. Res. **107**, 1121 (2002)
R.H.A. Iles, A.N. Fazakerley, A.D. Johnstone, N.P. Meredith, P. Bühler, The relativistic electron response in the outer radiation belt during magnetic storms. Ann. Geophys. **20**, 957–965 (2002)
T. Imamura, M. Tokumaru, H. Isobe, D. Shiota, H. Ando, M. Miyamoto, T. Toda, B. Häusler, M. Pätzold, A. Nabatov, A. Asai, Outflow structure of the quiet Sun corona probed by spacecraft radio scintillations in strong scattering. Astrophys. J. **788**, 117 (2014)
A. Isavnin, A. Vourlidas, E.K.J. Kilpua, Three-dimensional evolution of flux-rope CMEs and its relation to the local orientation of the heliospheric current sheet. Sol. Phys. **289**, 2141–2156 (2014)
S.A. Jacques, Momentum and energy transport by waves in the solar atmosphere and solar wind. Astrophys. J. **215**, 942–951 (1977)

A.N. Jaynes, D.N. Baker, H.J. Singer, J.V. Rodriguez, T.M. Loto'aniu, A.F. Ali, S.R. Elkington, X. Li, S.G. Kanekal, S.G. Claudepierre, J.F. Fennell, W. Li, R.M. Thorne, C.A. Kletzing, H.E. Spence, G.D. Reeves, Source and seed populations for relativistic electrons: Their roles in radiation belt changes. J. Geophys. Res. **120**, 7240–7254 (2015)
L.K. Jian, C.T. Russell, J.G. Luhmann, P.J. MacNeice, D. Odstrčil, P. Riley, J.A. Linker, R.M. Skoug, J.T. Steinberg, Comparison of observations at ACE and Ulysses with Enlil model results: Stream interaction regions during Carrington Rotations 2016–2018. Sol. Phys. **273**, 179–203 (2011)
L.K. Jian, P.J. MacNeice, A. Taktakishvili, D. Odstrčil, B. Jackson, H.-S. Yu, P. Riley, I.V. Sokolov, R.M. Evans, Validation for solar wind prediction at Earth: Comparison of coronal and heliospheric models installed at the CCMC. Space Weather **13**, 316–338 (2015)
L.K. Jian, P.J. MacNeice, M.L. Mays, A. Taktakishvili, D. Odstrčil, B. Jackson, H.-S. Yu, P. Riley, I.V. Sokolov, Validation for global solar wind prediction using Ulysses comparison: Multiple coronal and heliospheric models installed at the Community Coordinated Modeling Center. Space Weather **14**, 592–611 (2016)
M. Jin, W.B. Manchester, B. van der Holst, J.R. Gruesbeck, R.A. Frazin, E. Landi, A.M. Vásquez, P.L. Lamy, A. Llebaria, A. Fedorov, G. Tóth, T.I. Gombosi, A global two-temperature corona and inner heliosphere model: A comprehensive validation study. Astrophys. J. **745**, 6 (2012)
K. Jockers, Solar wind models based on exospheric theory. Astron. Astrophys. **6**, 219–239 (1970)
V. Joulin, E. Buchlin, J. Solomon, C. Guennou, Energetic characterisation and statistics of solar coronal brightenings. Astron. Astrophys. **591**, A148 (2016)
P.G. Judge, S.W. McIntosh, Non-uniqueness of atmospheric modeling. Sol. Phys. **190**, 331–350 (1999)
P.G. Judge, S. Habbal, E. Landi, From forbidden coronal lines to meaningful coronal magnetic fields. Sol. Phys. **288**, 467–480 (2013)
E.K. Kaghashvili, On the acceleration of the solar wind: Role of the inhomogeneous flow. Astrophys. J. **512**, 969–974 (1999)
E.K. Kaghashvili, Alfvén wave driven compressional fluctuations in shear flows. Phys. Plasmas **14**, 044502 (2007)
N.V. Karachik, A.A. Pevtsov, Solar wind and coronal bright points inside coronal holes. Astrophys. J. **735**, 47 (2011)
J.T. Karpen, C.R. DeVore, S.K. Antiochos, E. Pariat, Reconnection-driven coronal-hole jets with gravity and solar wind. Astrophys. J. **834**, 62 (2017)
J.C. Kasper, A.J. Lazarus, S.P. Gary, Hot solar-wind helium: Direct evidence for local heating by Alfvén-cyclotron dissipation. Phys. Rev. Lett. **101**, 261103 (2008)
E.K.J. Kilpua, H. Hietala, D.L. Turner, H.E.J. Koskinen, T.I. Pulkkinen, J.V. Rodriguez, G.D. Reeves, S.G. Claudepierre, H.E. Spence, Unraveling the drivers of the storm time radiation belt response. Geophys. Res. Lett. **42**, 3076–3084 (2015)
E.K.J. Kilpua, M.S. Madjarska, N. Karna, T. Wiegelmann, C. Farrugia, W. Yu, K. Andreeova, Sources of the slow solar wind during the solar cycle 23/24 minimum. Sol. Phys. **291**, 2441–2456 (2016)
E.K.J. Kilpua, A. Balogh, R. von Steiger, Y.D. Liu, Geoeffective properties of solar transients and stream interaction regions. Space Sci. Rev. (2017), this issue. doi:10.1007/s11214-017-0411-3
J.H. King, N.E. Papitashvili, Solar wind spatial scales in and comparisons of hourly wind and ACE plasma and magnetic field data. J. Geophys. Res. **110**, A02104 (2005)
L. Klein, S. Dalla, Acceleration and propagation of solar energetic particles. Space Sci. Rev. (2017), this issue. doi:10.1007/s11214-017-0382-4
J.A. Klimchuk, S.J. Bradshaw, Are chromospheric nanoflares a primary source of coronal plasma? Astrophys. J. **791**, 60 (2014)
Y.-K. Ko, J.C. Raymond, J. Li, A. Ciaravella, J. Michels, S. Fineschi, R. Wu, Solar and heliospheric observatory ultraviolet coronagraph spectrometer and Yohkohl soft X-ray telescope observations of the high-temperature corona above an active region complex. Astrophys. J. **578**, 979–995 (2002)
Y.-K. Ko, J.D. Moses, J.M. Laming, L. Strachan, S.T. Beltran, S. Tomczyk, S.E. Gibson, F. Auchère, R. Casini, S. Fineschi, M. Knoelker, C. Korendyke, S.W. McIntosh, M. Romoli, J. Rybak, D.G. Socker, A. Vourlidas, Q. Wu, Waves and magnetism in the solar atmosphere (WAMIS). Front. Astron. Space Sci. **3**, 1 (2016)
J.L. Kohl, R. Esser, S.R. Cranmer, S. Fineschi, L.D. Gardner, A.V. Panasyuk, L. Strachan, R.M. Suleiman, R.A. Frazin, G. Noci, EUV spectral line profiles in polar coronal holes from 1.3 to 3.0 solar radii. Astrophys. J. Lett. **510**, L59–L62 (1999)
J.L. Kohl, G. Noci, S.R. Cranmer, J.C. Raymond, Ultraviolet spectroscopy of the extended solar corona. Astron. Astrophys. Rev. **13**, 31–157 (2006)
J.L. Kohl, R. Jain, S.R. Cranmer, L.D. Gardner, A.K. Pradhan, J.C. Raymond, L. Strachan, Next generation UV coronagraph instrumentation for solar cycle 24. J. Astrophys. Astron. **29**, 321–327 (2008)

M. Kojima, T. Kakinuma, Solar cycle dependence of global distribution of solar wind speed. Space Sci. Rev. **53**, 173–222 (1990)

R.A. Kopp, T.E. Holzer, Dynamics of coronal hole regions, I, Steady polytropic flows with multiple critical points. Sol. Phys. **49**, 43–56 (1976)

A.S. Krieger, A.F. Timothy, E.C. Roelof, A coronal hole and its identification as the source of a high velocity solar wind stream. Sol. Phys. **29**, 505–525 (1973)

T. Kudoh, K. Shibata, Alfvén wave model of spicules and coronal heating. Astrophys. J. **514**, 493–505 (1999)

E. Landi, The off-disk thermal structure of a polar coronal hole. Astrophys. J. **685**, 1270–1276 (2008)

E. Landi, S.R. Cranmer, Ion temperatures in the low solar corona: Polar coronal holes at solar minimum. Astrophys. J. **691**, 794–805 (2009)

S. Landi, F. Pantellini, Kinetic simulations of the solar wind from the subsonic to the supersonic regime. Astron. Astrophys. **400**, 769–778 (2003)

E. Landi, R.L. Alexander, J.R. Gruesbeck, J.A. Gilbert, S.T. Lepri, W.B. Mancheser, T.H. Zurbuchen, Carbon ionization states as a diagnostic of the solar wind. Astrophys. J. **744**, 100 (2012a)

E. Landi, G. Del Zanna, P.R. Young, K.P. Dere, H.E. Mason, CHIANTI: An atomic database for emission lines. XII. Version 7 of the database. Astrophys. J. **744**, 99 (2012b)

B. Lavraud, Y. Liu, K. Segura, J. He, G. Qin, M. Temmer, J.-C. Vial, M. Xiong, J.A. Davies, A.P. Rouillard, R. Pinto, F. Auchère, R.A. Harrison, C. Eyles, W. Gan, P. Lamy, L. Xia, J.P. Eastwood, L. Kong, J. Wang, R.F. Wimmer-Schweingruber, S. Zhang, Q. Zong, J. Soucek, J. An, L. Prech, A. Zhang, P. Rochus, V. Bothmer, M. Janvier, M. Maksimovic, C.P. Escoubet, E.K.J. Kilpua, J. Tappin, R. Vainio, S. Poedts, M.W. Dunlop, N. Savani, N. Gopalswamy, S.D. Bale, G. Li, T. Howard, C. DeForest, D. Webb, N. Lugaz, S.A. Fuselier, K. Dalmasse, J. Tallineau, D. Vranken, J.G. Fernández, A small mission concept to the Sun-Earth Lagrangian L5 point for innovative solar, heliospheric, and space weather science. J. Atmos. Sol.-Terr. Phys. **146**, 171–185 (2016)

G. Le Chat, K. Issautier, N. Meyer-Vernet, The solar wind energy flux. Sol. Phys. **279**, 197–205 (2012)

R.J. Leamon, S.W. McIntosh, Empirical solar wind forecasting from the chromosphere. Astrophys. J. **659**, 738–742 (2007)

M.A. Lee, B. Roberts, On the behavior of hydromagnetic surface waves. Astrophys. J. **301**, 430–439 (1986)

C.O. Lee, J.G. Luhmann, X.P. Zhao, Y. Liu, P. Riley, C.N. Arge, C.T. Russell, I. de Pater, Effects of the weak polar fields of solar cycle 23: Investigation using OMNI for the STEREO mission period. Sol. Phys. **256**, 345–363 (2009)

C.O. Lee, C.N. Arge, D. Odstrčil, G. Millward, V. Pizzo, J.M. Quinn, C.J. Henney, Ensemble modeling of CME propagation. Sol. Phys. **285**, 349–368 (2013)

E. Leer, T.E. Holzer, Energy addition in the solar wind. J. Geophys. Res. **85**, 4681–4688 (1980)

E. Leer, T.E. Holzer, T. Flå, Acceleration of the solar wind. Space Sci. Rev. **33**, 161–200 (1982)

J. Lei, J.P. Thayer, W. Wang, R.L. McPherron, Impact of CIR storms on thermosphere density variability during the solar minimum of 2008. Sol. Phys. **274**, 427–437 (2011)

J.F. Lemaire, V. Pierrard, Kinetic models of solar and polar winds. Astrophys. Space Sci. **277**, 169–180 (2001)

J.F. Lemaire, M. Scherer, Kinetic models of the solar wind. J. Geophys. Res. **76**, 7479–7490 (1971)

J.F. Lemaire, K. Stegen, Improved determination of the location of the temperature maximum in the corona. Sol. Phys. **291**, 3659–3683 (2016)

M. Lester, J. Huba, J. Foster, Space weather effects: Ionosphere. Space Sci. Rev. (2017), this issue

R.H. Levine, A new theory of coronal heating. Astrophys. J. **190**, 457–466 (1974)

R.H. Levine, M.D. Altschuler, J.W. Harvey, Solar sources of the interplanetary magnetic field and solar wind. J. Geophys. Res. **82**, 1061–1065 (1977)

Y.-P. Li, Q. Yuan, Q.D. Wang, P.F. Chen, J. Neilsen, T. Fang, S. Zhang, J. Dexter, Statistical and theoretical studies of flares from Sagittarius A*, in *Multi-Messenger Astrophysics of the Galactic Centre*, ed. by R. Crocker, S. Longmore, G. Bicknell. IAU Symp., vol. 322 (Cambridge University Press, Cambridge, 2017), pp. 31–38

P.C. Liewer, J.R. Hall, M. De Jong, D.G. Socker, R.A. Howard, P.C. Crane, P. Reiser, N. Rich, A. Vourlidas, Determination of three-dimensional structure of corona streamers and relationship to the solar magnetic field. J. Geophys. Res. **106**, 15903–15916 (2001)

P.C. Liewer, M. Neugebauer, T. Zurbuchen, Characteristics of active-region sources of solar wind near solar maximum. Sol. Phys. **223**, 209–229 (2004)

P.C. Liewer, D. Alexander, J. Ayon, L. Floyd, G. Garbe, B. Goldstein, D. Hassler, A. Kosovichev, R. Mewaldt, N. Murphy, M. Neugebauer, A. Sandman, D. Socker, S. Suess, R. Ulrich, M. Velli, A. Vourlidas, T.H. Zurbuchen, Solar polar imager: Observing solar activity from a new perspective, in *NASA Space Science Vision Missions*, ed. by M.S. Allen. Progress in Astronautics and Aeronautics, vol. 224 (AIAA, Reston, 2008), pp. 1–37

H. Lin, mxCSM: A 100-slit, 6-wavelength wide-field coronal spectropolarimeter for the study of the dynamics and the magnetic fields of the solar corona. Front. Astron. Space Sci. **3**, 9 (2016)
H. Lin, J.R. Kuhn, R. Coulter, Coronal magnetic field measurements. Astrophys. J. Lett. **613**, L177–L180 (2004)
J.A. Linker, Z. Mikić, D.A. Biesecker, R.J. Forsyth, S.E. Gibson, A.J. Lazarus, A. Lecinski, P. Riley, A. Szabo, B.J. Thompson, Magnetohydrodynamic modeling of the solar corona during Whole Sun Month. J. Geophys. Res. **104**, 9809–9830 (1999)
J.A. Linker, R.M. Caplan, C. Downs, R. Lionello, P. Riley, Z. Mikić, C.J. Henney, C.N. Arge, T. Kim, N. Pogorelov, An empirically driven time-dependent model of the solar wind. J. Phys. Conf. Ser. **719**, 012012 (2016)
R. Lionello, M. Velli, C. Downs, J.A. Linker, Z. Mikić, A. Verdini, Validating a time-dependent turbulence-driven model of the solar wind. Astrophys. J. **784**, 120 (2014)
R. Lionello, T. Török, V.S. Titov, J.E. Leake, Z. Mikić, J.A. Linker, M.G. Linton, The contribution of coronal jets to the solar wind. Astrophys. J. Lett. **831**, L2 (2016)
Y. Liu, J.D. Richardson, J.W. Belcher, J.C. Kasper, H.A. Elliott, Thermodynamic structure of collision-dominated expanding plasma: Heating of interplanetary coronal mass ejections. J. Geophys. Res. **111**, A01102 (2006)
J. Liu, S.W. McIntosh, I. De Moortel, Y. Wang, On the parallel and perpendicular propagating motions visible in polar plumes: An incubator for (fast) solar wind acceleration. Astrophys. J. **806**, 273 (2015)
Y. Liu, J.T. Hoeksema, X. Sun, K. Hayashi, Vector magnetic field synoptic charts from the Helioseismic and Magnetic Imager (HMI). Sol. Phys. **292**, 29 (2017)
M.L. Loucif, Giant macrospicules as possible sources of the fast solar wind. Astron. Astrophys. **281**, 95–107 (1994)
J.J. Love, E.J. Rigler, S.E. Gibson, Geomagnetic detection of the sectorial solar magnetic field and the historical peculiarity of minimum 23–24. Geophys. Res. Lett. **39**, L04102 (2012)
J.G. Luhmann, Y. Li, C.N. Arge, P.R. Gazis, R. Ulrich, Solar cycle changes in coronal holes and space weather cycles. J. Geophys. Res. **107**, 1154 (2002)
B. Luo, Q. Zhong, S. Liu, J. Gong, A new forecasting index for solar wind velocity based on EIT 284 Å observations. Sol. Phys. **250**, 159–170 (2008)
B.J. Lynch, J.K. Edmondson, Y. Li, Interchange reconnection Alfvén wave generation. Sol. Phys. **289**, 3043–3058 (2014)
M.S. Madjarska, Z. Huang, J.G. Doyle, S. Subramanian, Coronal hole boundaries evolution at small scales, III, EIS and SUMER views. Astron. Astrophys. **545**, A67 (2012)
C. Manchester, Y. Liu, P. Riley, E. Kilpua, T. Török, N. Lugaz, B. Vršnak, Physical processes of CME propagation. Space Sci. Rev. (2017), this issue
P.K. Manoharan, C.R. Subrahmanya, J.N. Chengalur, Space weather and solar wind studies with OWFA. J. Astrophys. Astron. **38**, 16 (2017)
E. Marsch, Kinetic physics of the solar corona and solar wind. Living Rev. Sol. Phys. **3**, 1 (2006)
E. Marsch, K.-H. Mühlhäuser, H. Rosenbauer, R. Schwenn, On the equation of state of solar wind ions derived from Helios measurements. J. Geophys. Res. **88**, 2982–2992 (1983)
R.G. Marsden, The heliosphere after Ulysses. Astrophys. Space Sci. **277**, 337–347 (2001)
P.C.H. Martens, G.D.R. Attrill, A.R. Davey, A. Engell, S. Farid, P.C. Grigis, J. Kasper, K. Korreck, S.H. Saar, A. Savcheva, Y. Su, P. Testa, M. Wills-Davey, P.N. Bernasconi, N.-E. Raouafi, V.A. Delouille, J.F. Hochedez, J.W. Cirtain, C.E. DeForest, R.A. Angryk, I. De Moortel, T. Wiegelmann, M.K. Georgoulis, R.T.J. McAteer, R.P. Timmons, Computer vision for the Solar Dynamics Observatory (SDO). Sol. Phys. **275**, 79–113 (2012)
B.A. Maruca, J.C. Kasper, S.P. Gary, Instability-driven limits on helium temperature anisotropy in the solar wind: Observations and linear Vlasov analysis. Astrophys. J. **748**, 137 (2012)
T. Matsumoto, T.K. Suzuki, Connecting the Sun and the solar wind: The first 2.5-dimensional self-consistent MHD simulation under the Alfvén wave scenario. Astrophys. J. **749**, 8 (2012)
W.H. Matthaeus, M. Velli, Who needs turbulence: A review of turbulence effects in the heliosphere and on the fundamental process of reconnection. Space Sci. Rev. **160**, 145–168 (2011)
W.H. Matthaeus, C.W. Smith, S. Oughton, Dynamical age of solar wind turbulence in the outer heliosphere. J. Geophys. Res. **103**, 6495–6502 (1998)
W.H. Matthaeus, G.P. Zank, S. Oughton, D.J. Mullan, P. Dmitruk, Coronal heating by magnetohydrodynamic turbulence driven by reflected low-frequency waves. Astrophys. J. **523**, L93–L96 (1999)
D.J. McComas, R.W. Ebert, H.A. Elliott, B.E. Goldstein, J.T. Gosling, N.A. Schwadron, R.M. Skoug, Weaker solar wind from the polar coronal holes and the whole Sun. Geophys. Res. Lett. **35**, L18103 (2008)
S.L. McGregor, W.J. Hughes, C.N. Arge, D. Odstrčil, N.A. Schwadron, The radial evolution of solar wind speeds. J. Geophys. Res. **116**, A03106 (2011a)

S.L. McGregor, W.J. Hughes, C.N. Arge, M.J. Owens, D. Odstrčil, The distribution of solar wind speeds during solar minimum: Calibration for numerical solar wind modeling constraints on the source of the slow solar wind. J. Geophys. Res. **116**, A03101 (2011b)
S.L. McGregor, M.K. Hudson, W.J. Hughes, Modeling magnetospheric response to synthetic Alfvénic fluctuations in the solar wind: ULF wave fields in the magnetosphere. J. Geophys. Res. **119**, 8801–8812 (2014)
P.S. McIntosh, Patterns and dynamics of solar magnetic fields and He I coronal holes in cycle 23, in *Solar Variability as an Input to the Earth's Environment*, ed. by A. Wilson. ESA, vol. SP-535 (ESA, Noordwijk, 2003), pp. 807–818
S.W. McIntosh, B. de Pontieu, M. Carlsson, V. Hansteen, P. Boerner, M. Goossens, Alfvénic waves with sufficient energy to power the quiet solar corona and fast solar wind. Nature **475**, 477–480 (2011)
S.W. McIntosh, W.J. Cramer, M. Pichardo Marcano, R.J. Leamon, The detection of Rossby-like waves on the Sun. Nature Astron. **1**, 0086 (2017)
J.F. McKenzie, E. Marsch, Resonant wave acceleration of minor ions in the solar wind. Astrophys. Space Sci. **81**, 295–314 (1982)
B. McPherron, B. Anderson, A. Vijanen, Dynamic magnetospheric forcing, ionospheric current systems. Space Sci. Rev. (2017), this issue
V.G. Merkin, J.G. Lyon, D. Lario, C.N. Arge, C.J. Henney, Time-dependent magnetohydrodynamic simulations of the inner heliosphere. J. Geophys. Res. **121**, 2866–2890 (2016)
N. Meyer-Vernet, How does the solar wind blow? A simple kinetic model. Eur. J. Phys. **20**, 167–176 (1999)
R.L. Moore, A.C. Sterling, J.W. Cirtain, D.A. Falconer, Solar X-ray jets, type-II spicules, granule-size emerging bipoles, and the genesis of the heliosphere. Astrophys. J. Lett. **731**, L18 (2011)
H. Morgan, An observation of solar active region expansion into the heliosphere. Mon. Not. R. Astron. Soc. **433**, L74–L78 (2013)
D. Müller, R.G. Marsden, O.C.St. Cyr, H.R. Gilbert, Solar orbiter: Exploring the Sun-heliosphere connection. Sol. Phys. **285**, 25–70 (2013)
N.A. Murphy, J.C. Raymond, K.E. Korreck, Plasma heating during a coronal mass ejection observed by the solar and heliospheric observatory. Astrophys. J. **735**, 17 (2011)
V.M. Nakariakov, Coronal waves and oscillations, in *Solar Activity and Its Magnetic Origin*, ed. by V. Bothmer, A. Hady. IAU Symp., vol. 233 (Cambridge University Press, Cambridge, 2006), pp. 464–471
M. Neugebauer, C.W. Snyder, Solar plasma experiment. Science **138**, 1095–1097 (1962)
M. Neugebauer, D. Reisenfeld, I.G. Richardson, Comparison of algorithms for determination of solar wind regimes. J. Geophys. Res. **121**, 8215–8227 (2016)
J.A. Newbury, C.T. Russell, J.L. Phillips, S.P. Gary, Electron temperature in the ambient solar wind: Typical properties and a lower bound at 1 AU. J. Geophys. Res. **103**, 9553–9566 (1998)
G.M. Nita, G.D. Fleishman, A.A. Kuznetsov, E.P. Kontar, D.E. Gary, Three-dimensional radio and X-ray modeling and data analysis software: Revealing flare complexity. Astrophys. J. **799**, 236 (2015)
G. Noci, Energy budget in coronal holes. Sol. Phys. **28**, 403–407 (1973)
J.T. Nolte, A.S. Krieger, A.F. Timothy, R.E. Gold, E.C. Roelof, G. Vaiana, A.J. Lazarus, J.D. Sullivan, P.S. McIntosh, Coronal holes as sources of solar wind. Sol. Phys. **46**, 303–322 (1976)
T.P. O'Brien, R.L. McPherron, An empirical phase space analysis of ring current dynamics: Solar wind control of injection and decay. J. Geophys. Res. **105**, 7707–7720 (2000)
D. Odstrčil, V.J. Pizzo, Three-dimensional propagation of coronal mass ejections in a structured solar wind flow, 2: CME launched adjacent to the streamer belt. J. Geophys. Res. **104**, 493–504 (1999)
L. Ofman, Wave modeling of the solar wind. Living Rev. Sol. Phys. **7**, 4 (2010)
L. Ofman, V.M. Nakariakov, C.E. DeForest, Slow magnetosonic waves in coronal plumes. Astrophys. J. **514**, 441–447 (1999)
R. Oran, E. Landi, B. van der Holst, S.T. Lepri, A.M. Vásquez, F.A. Nuevo, R. Frazin, W. Manchester, I. Sokolov, T.I. Gombosi, A steady-state picture of solar wind acceleration and charge state composition derived from a global wave-driven MHD model. Astrophys. J. **806**, 55 (2015)
M.J. Owens, H.E. Spence, S.L. McGregor, W.J. Hughes, J.M. Quinn, C.N. Arge, P. Riley, J. Linker, D. Odstrčil, Metrics for solar wind prediction models: Comparison of empirical, hybrid, and physics-based schemes with 8 years of L1 observations. Space Weather **6**, S08001 (2008)
M.J. Owens, N.U. Crooker, M. Lockwood, Solar cycle evolution of dipolar and pseudostreamer belts and their relation to the slow solar wind. J. Geophys. Res. **119**, 36–46 (2014)
M.J. Owens, P. Riley, T.S. Horbury, Probabilistic solar wind and geomagnetic forecasting using an analogue ensemble or 'similar day' approach. Sol. Phys. **292**, 69 (2017)
O. Panasenco, M. Velli, Coronal pseudostreamers: Source of fast or slow solar wind, in Solar Wind 13, in *Proceedings of the Thirteenth International Solar Wind Conference*, ed. by G. Zank, J. Borovsky, R. Bruno, J. Cirtain, S. Cranmer, H. Elliott, J. Giacalone, W. Gonzalez, G. Li, E. Marsch, E. Moebius, N. Pogorelov, J. Spann, O. Verkhoglyadova. AIP Conf. Proc., vol. 1539 (AIP, New York, 2013), pp. 50–53

A.R. Paraschiv, A. Bemporad, A.C. Sterling, Physical properties of solar polar jets: A statistical study with Hinode XRT data. Astron. Astrophys. **579**, A96 (2015)
E. Pariat, K. Dalmasse, C.R. DeVore, S.K. Antiochos, J.T. Karpen, A model for straight and helical solar jets, II, Parametric study of the plasma beta. Astron. Astrophys. **596**, A36 (2016)
E.N. Parker, Dynamics of the interplanetary gas and magnetic fields. Astrophys. J. **128**, 664–676 (1958)
E.N. Parker, Topological dissipation and the small-scale fields in turbulent gases. Astrophys. J. **174**, 499–510 (1972)
E.N. Parker, Nanoflares and the solar X-ray corona. Astrophys. J. **330**, 474–479 (1988)
E.N. Parker, Reflections on macrophysics and the Sun. Sol. Phys. **176**, 219–247 (1997)
E.N. Parker, Kinetic and hydrodynamic representations of coronal expansion and the solar wind, in *Twelfth International Solar Wind Conference*, ed. by M. Maksimovic, K. Issautier, N. Meyer-Vernet, M. Moncuquet. AIP Conf. Proc., vol. 1216 (AIP, New York, 2010), pp. 3–7
C.E. Parnell, P.E. Jupp, Statistical analysis of the energy distribution of nanoflares in the Quiet Sun. Astrophys. J. **529**, 554–569 (2000)
I.J. Parrish, M. McCourt, E. Quataert, P. Sharma, The effects of anisotropic viscosity on turbulence and heat transport in the intracluster medium. Mon. Not. R. Astron. Soc. **422**, 704–718 (2012)
J.C. Perez, B.D.G. Chandran, Direct numerical simulations of reflection-driven, reduced magnetohydrodynamic turbulence from the Sun to the Alfvén critical point. Astrophys. J. **776**, 124 (2013)
H. Peter, B.N. Dwivedi, Discovery of the Sun's million-degree hot corona. Front. Astron. Space Sci. **1**, 2 (2014)
G. Petrie, High-resolution vector magnetograms of the Sun's poles from Hinode: Flux distributions and global coronal modeling. Sol. Phys. **292**, 13 (2017)
A.A. Pevtsov, L. Bertello, P. MacNeice, G. Petrie, What if we had a magnetograph at Lagrangian L5. Space Weather **14**, 1026–1031 (2016)
V. Pierrard, M. Pieters, Coronal heating and solar wind acceleration for electrons, protons, and minor ions obtained from kinetic models based on kappa distributions. J. Geophys. Res. **119**, 9441–9455 (2014)
R.F. Pinto, A.P. Rouillard, A multiple flux-tube solar wind model. Astrophys. J. **838**, 89 (2017)
I. Plotnikov, A.P. Rouillard, J.A. Davies, V. Bothmer, J.P. Eastwood, P. Gallagher, R.A. Harrison, E. Kilpua, C. Möstl, C.H. Perry, L. Rodriguez, B. Lavraud, V. Génot, R.F. Pinto, E. Sanchez-Diaz, Long-term tracking of corotating density structures using heliospheric imaging. Sol. Phys. **291**, 1853–1875 (2016)
G.W. Pneuman, The physical structure of coronal holes: Influence of magnetic fields and coronal heating. Astron. Astrophys. **81**, 161–166 (1980)
G.W. Pneuman, Driving mechanisms for the solar wind. Space Sci. Rev. **43**, 105–138 (1986)
B. Poduval, X.P. Zhao, Validating solar wind prediction using the current sheet source surface model. Astrophys. J. Lett. **782**, L22 (2014)
G. Poletto, Sources of solar wind over the solar activity cycle. J. Adv. Res. **4**, 215–220 (2013)
J. Pomoell, E. Kilpua, C. Verbeke, E. Lumme, S. Poedts, E. Palmerio, A. Isavnin, Modeling the Sun-to-Earth evolution of the magnetic structure of coronal mass ejections with EUHFORIA, in *19th EGU General Assembly* (2017), p. 11747
E. Quataert, A. Gruzinov, Turbulence and particle heating in advection-dominated accretion flows. Astrophys. J. **520**, 248–255 (1999)
L.A. Rachmeler, S.E. Gibson, J.B. Dove, C.R. DeVore, Y. Fan, Polarimetric properties of flux ropes and sheared arcades in coronal prominence cavities. Sol. Phys. **288**, 617–636 (2013)
L.A. Rachmeler, S.J. Platten, C. Bethge, D.B. Seaton, A.R. Yeates, Observations of a hybrid double-streamer/pseudostreamer in the solar corona. Astrophys. J. Lett. **787**, L3 (2014)
N.-E. Raouafi, S.K. Solanki, Effect of the electron density stratification on off-limb O VI profiles: How large is the velocity distribution anisotropy in the solar corona. Astron. Astrophys. **427**, 725–733 (2004)
N.-E. Raouafi, S. Patsourakos, E. Pariat, P.R. Young, A.C. Sterling, A. Savcheva, M. Shimojo, F. Moreno-Insertis, C.R. DeVore, V. Archontis, T. Török, H. Mason, W. Curdt, K. Meyer, K. Dalmasse, Y. Matsui, Solar coronal jets: Observations, theory, and modeling. Space Sci. Rev. **201**, 1–53 (2016)
J.-P. Raulin, J.M. Davila, T. Bogdan, K. Yumoto, J. Leibacher, The future of IHY campaigns: Transition to the international space weather initiative. Highlights Astron. **15**, 501–503 (2010)
G.D. Reeves, K.L. McAdams, R.H.W. Friedel, T.P. O'Brien, Acceleration and loss of relativistic electrons during geomagnetic storms. Geophys. Res. Lett. **30**, 1529 (2003)
N.L. Reginald, J.M. Davila, O.C.St. Cyr, D.M. Rabin, M. Guhathakurta, D.M. Hassler, H. Gashut, Electron temperatures and flow speeds of the low solar corona: MACS results from the total solar eclipse of 29 March 2006 in Libya. Sol. Phys. **270**, 235–251 (2011)
D.B. Reisenfeld, D.J. McComas, J.T. Steinberg, Evidence of a solar origins for pressure balance structures in the high-latitude solar wind. Geophys. Res. Lett. **26**, 1805–1808 (1999)
P. Riley, Modeling corotating interaction regions: From the Sun to 1 AU. J. Atmos. Sol.-Terr. Phys. **69**, 32–42 (2007)

P. Riley, J.G. Luhmann, Interplanetary signatures of unipolar streamers and the origin of the slow solar wind. Sol. Phys. **277**, 355–373 (2012)
P. Riley, J.T. Gosling, V.J. Pizzo, A two-dimensional simulation of the radial and latitudinal evolution of a solar wind disturbance driven by a fast, high-pressure coronal mass ejection. J. Geophys. Res. **102**, 14677–14686 (1997)
P. Riley, J.T. Gosling, D.J. McComas, V.J. Pizzo, J.G. Luhmann, D. Biesecker, R.J. Forsyth, J.T. Hoeksema, A. Lecinski, B.J. Thompson, Relationship between Ulysses plasma observations and solar observations during the Whole Sun Month. J. Geophys. Res. **104**, 9871–9880 (1999)
P. Riley, J.A. Linker, Z. Mikić, An empirically-driven global MHD model of the solar corona and inner heliosphere. J. Geophys. Res. **106**, 15889–15901 (2001)
P. Riley, J.A. Linker, Z. Mikić, R. Lionello, S.A. Ledvina, J.G. Luhmann, A comparison between global solar magnetohydrodynamic and potential field source surface model results. Astrophys. J. **653**, 1510–1516 (2006)
P. Riley, R. Lionello, J.A. Linker, Z. Mikić, J. Luhmann, J. Wijaya, Global MHD modeling of the solar corona and inner heliosphere for the Whole Heliosphere Interval. Sol. Phys. **274**, 361–377 (2011)
P. Riley, J.A. Linker, Z. Mikić, On the application of ensemble modeling techniques to improve ambient solar wind models. J. Geophys. Res. **118**, 600–607 (2013)
P. Riley, M. Ben-Nun, J.A. Linker, Z. Mikić, L. Svalgaard, J. Harvey, L. Bertello, T. Hoeksema, Y. Liu, R. Ulrich, A multi-observatory inter-comparison of line-of-sight synoptic magnetograms. Sol. Phys. **289**, 769–792 (2014)
P. Riley, J.A. Linker, C.N. Arge, On the role played by magnetic expansion factor in the prediction of solar wind speed. Space Weather **13**, 154–169 (2015)
D.A. Roberts, Demonstrations that the solar wind is not accelerated by waves or turbulence. Astrophys. J. **711**, 1044–1050 (2010)
E.C. Roelof, G.B. Andrews, P.C. Liewer, D. Moses, Telemachus: A mission for a polar view of solar activity. Adv. Space Res. **34**, 467–471 (2004)
R. Rosner, W.H. Tucker, G.S. Vaiana, Dynamics of the quiescent solar corona. Astrophys. J. **220**, 643–665 (1978)
T. Rotter, A.M. Veronig, M. Temmer, B. Vršnak, Real-time solar wind prediction based on SDO/AIA coronal hole data. Sol. Phys. **290**, 1355–1370 (2015)
A.P. Rouillard, N.R. Sheeley, T.J. Cooper, J.A. Davies, B. Lavraud, E.K.J. Kilpua, R.M. Skoug, J.T. Steinberg, A. Szabo, A. Opitz, J.-A. Sauvaud, The solar origin of small interplanetary transients. Astrophys. J. **734**, 7 (2011)
A.P. Rouillard, B. Lavraud, V. Genot, M. Bouchemit, N. Dufourg, I. Plotnikov, R.F. Pinto, E. Sanchez-Diaz, M. Lavarra, M. Penou, C. Jacquey, N. Andre, S. Caussarieu, J.-P. Toniutti, D. Popescu, E. Buchlin, S. Caminade, P. Alingery, J.A. Davies, D. Odstrčil, L. Mays, A propagation tool to connect remote-sensing observations with in-situ measurements of heliospheric structures. Planet. Space Sci. (2017), submitted. arXiv:1702.00399
J.M. Ryan, W.I. Axford, The behaviour of minor species in the solar wind. Z. Geophys. **41**, 221–232 (1975)
K.H. Schatten, J.M. Wilcox, N.F. Ness, A model of interplanetary and coronal magnetic fields. Sol. Phys. **6**, 442–455 (1969)
J.T. Schmelz, Why stellar astronomers should be interested in the Sun. Adv. Space Res. **32**, 895–904 (2003)
C.J. Schrjiver, K. Kauristie, A.D. Aylward, C.M. Denardini, S.E. Gibson, A. Glover, N. Gopalswamy, M. Grande, M. Hapgood, D. Heynderickx, N. Jakowski, V.V. Kalegaev, G. Lapenta, J.A. Linker, S. Liu, C.H. Mandrini, I.R. Mann, T. Nagatsuma, D. Nandy, T. Obara, T.P. O'Brien, T. Onsager, H.J. Opgenoorth, M. Terkildsen, C.E. Valladares, N. Vilmer, Understanding space weather to shield society: A global road map for 2015–2025 commissioned by COSPAR and ILWS. Adv. Space Res. **55**, 2745–2807 (2015)
N.A. Schwadron, D.J. McComas, Solar wind scaling law. Astrophys. J. **599**, 1395–1403 (2003)
N.A. Schwadron, D.J. McComas, C.E. DeForest, Relationship between solar wind and coronal heating: Scaling laws from solar X-rays. Astrophys. J. **642**, 1173–1176 (2006)
R. Schwenn, Space weather: The solar perspective. Living Rev. Sol. Phys. **3**, 2 (2006)
J.D. Scudder, Why all stars should possess circumstellar temperature inversions. Astrophys. J. **398**, 319–349 (1992)
J.D. Scudder, Radial variation of the solar wind proton temperature: Heat flow or addition. Astrophys. J. **809**, 126 (2015)
S. Servidio, F. Valentini, D. Perrone, A. Greco, F. Califano, W.H. Matthaeus, P. Veltri, A kinetic model of plasma turbulence. J. Plasma Phys. **81**, 325810107 (2015)
N.R. Sheeley, Y.-M. Wang, S.H. Hawley, G.E. Brueckner, K.P. Dere, R.A. Howard, M.J. Koomen, C.M. Korendyke, D.J. Michels, S.E. Paswaters, D.G. Socker, O.C.St. Cyr, D. Wang, P.L. Lamy, A. Llebaria,

R. Schwenn, G.M. Simnett, S. Plunkett, D.A. Biesecker, Measurements of flow speeds in the corona between 2 and 30 $R_{\odot}$. Astrophys. J. **484**, 472–478 (1997)
N.R. Sheeley, D.D.H. Lee, K.P. Casto, Y.-M. Wang, N.B. Rich, The structure of streamer blobs. Astrophys. J. **694**, 1471–1480 (2009)
F. Shen, X. Feng, C. Xiang, Improvement to the global distribution of coronal plasma and magnetic field on the source surface using expansion factor f_s and angular distance θ_b. J. Atmos. Sol.-Terr. Phys. **77**, 125–131 (2012)
D.G. Sibeck, J.D. Richardson, Toward forecasting space weather in the heliosphere. J. Geophys. Res. **102**, 14721–14734 (1997)
V.A. Slemzin, F.F. Goryaev, S.V. Kuzin, Spectroscopic diagnostics of the solar coronal plasma. Plasma Phys. Rep. **40**, 855–892 (2014)
E.J. Smith, A. Balogh, Ulysses observations of the radial magnetic field. Geophys. Res. Lett. **22**, 3317–3320 (1995)
J.J. Sojka, R.L. McPherron, A.P. van Eyken, M.J. Nicholls, C.J. Heinselman, J.D. Kelly, Observations of ionospheric heating during the passage of solar coronal hole fast streams. Geophys. Res. Lett. **36**, L19105 (2009)
J.J. Sojka et al., Space weather effects in the atmosphere. Space Sci. Rev. (2017), this issue
S.C. Solomon, T.N. Woods, L.V. Didkovsky, J.T. Emmert, L. Qian, Anomalously low solar extreme-ultraviolet irradiance and thermospheric density during solar minimum. Geophys. Res. Lett. **37**, L16103 (2010)
M. Stakhiv, E. Landi, S.T. Lepri, R. Oran, T.H. Zurbuchen, On the origin of mid-latitude fast wind: Challenging the two-state solar wind paradigm. Astrophys. J. **801**, 100 (2015)
M. Stakhiv, S.T. Lepri, E. Landi, P. Tracy, T.H. Zurbuchen, On solar wind origin and acceleration: Measurements from ACE. Astrophys. J. **829**, 117 (2016)
J.E. Stawarz, C.W. Smith, B.J. Vasquez, M.A. Forman, B.T. MacBride, The turbulence cascade and proton heating in the solar wind at 1 AU. Astrophys. J. **697**, 1119–1127 (2009)
R.A. Steenburgh, D.A. Biesecker, G.H. Millward, From predicting solar activity to forecasting space weather: Practical examples of research-to-operations and operations-to-research. Sol. Phys. **289**, 675–690 (2014)
A.C. Sterling, J.V. Hollweg, Alfvénic resonances on solar spicules. Astrophys. J. **285**, 843–850 (1984)
M.L. Stevens, J.A. Linker, P. Riley, W.J. Hughes, Underestimates of the magnetic flux in coupled MHD model solar wind solutions. J. Atmos. Sol.-Terr. Phys. **83**, 22–31 (2012)
L. Strachan, R. Suleiman, A.V. Panasyuk, D.A. Biesecker, J.L. Kohl, Empirical densities, kinetic temperatures, and outflow velocities in the equatorial streamer belt at solar minimum. Astrophys. J. **571**, 1008–1014 (2002)
L. Strachan, J.M. Laming, Y.-K. Ko, C.M. Korendyke, S.T. Beltran, D.G. Socker, C. Brown, E. Provornikova, The Ultraviolet Spectro-Coronagraph pathfinder mission for the detection of coronal suprathermal seed particles. SPD meeting 47, abstract 301.04, Am. Astron. Soc. (2016)
K. Strong, J. Saba, D. Pesnell, J. Luhmann, F. Hill, T. Duvall, 4PI: A global understanding of the solar cycle, in White paper 262 submitted to *Solar and Space Physics: A Science for a Technological Society* (National Academies Press, Washington, 2012)
A. Strugarek, N. Janitzek, A. Lee, P. Löschl, B. Seifert, S. Hoilijoki, E. Kraaikamp, A. Isha Mrigakshi, T. Philippe, S. Spina, M. Bröse, S. Massahi, L. O'Halloran, V. Pereira Blanco, C. Stausland, P. Escoubet, G. Kargl, A space weather mission concept: Observatories of the solar corona and active regions (OSCAR). J. Space Weather Space Clim. **5**, A4 (2015)
P.A. Sturrock, R.E. Hartle, Two-fluid model of the solar wind. Phys. Rev. Lett. **16**, 628–631 (1966)
Š. Štverák, P. Trávníček, P. Hellinger, Electron energetics in the expanding solar wind via Helios observations. J. Geophys. Res. **120**, 8177–8193 (2015)
S.T. Suess, A.-H. Wang, S.T. Wu, Volumetric heating in coronal streamers. J. Geophys. Res. **101**, 19957–19966 (1996)
X. Sun, The magnetic solar photosphere and corona: Observation, modeling, and prediction. Ph.D. dissertation, Stanford University (2012)
X. Sun, Y. Liu, J.T. Hoeksema, K. Hayashi, X. Zhao, A new method for polar field interpolation. Sol. Phys. **270**, 9–22 (2011)
T.K. Suzuki, Forecasting solar wind speeds. Astrophys. J. Lett. **640**, L75–L78 (2006)
T.K. Suzuki, S.-I. Inutsuka, Solar winds driven by nonlinear low-frequency Alfvén waves from the photosphere: Parametric study for fast/slow winds and disappearance of solar winds. J. Geophys. Res. **111**, A06101 (2006)
L. Svalgaard, E.W. Cliver, A floor in the solar wind magnetic field. Astrophys. J. Lett. **661**, L203–L206 (2007)

M. Temmer, T. Rollett, C. Möstl, A.M. Veronig, B. Vršnak, D. Odstrčil, Influence of the ambient solar wind flow on the propagation behavior of interplanetary coronal mass ejections. Astrophys. J. **743**, 101 (2011)

A. Tenerani, M. Velli, Evolving waves and turbulence in the outer corona and inner heliosphere: The accelerating expanding box. Astrophys. J. **843**, 26 (2017)

L. Teriaca, G. Poletto, M. Romoli, D.A. Biesecker, The nascent solar wind: Origin and acceleration. Astrophys. J. **588**, 566–577 (2003)

K.M. Thieme, E. Marsch, R. Schwenn, Spatial structures in high-speed streams as signatures of fine structures in coronal holes. Ann. Geophys. **8**, 713–723 (1990)

B.J. Thompson, S.E. Gibson, P.C. Schroeder, D.F. Webb, C.N. Arge, M.M. Bisi, G. de Toma, B.A. Emery, A.B. Galvin, D.A. Haber, B.V. Jackson, E.A. Jensen, R.J. Leamon, J. Lei, P.K. Manoharan, M.L. Mays, P.S. McIntosh, G.J.D. Petrie, S.P. Plunkett, L. Qian, P. Riley, S.T. Suess, M. Tokumaru, B.T. Welsch, T.N. Woods, A snapshot of the Sun near solar minimum: The Whole Heliosphere Interval. Sol. Phys. **274**, 29–56 (2011)

J. Threlfall, I. De Moortel, S.W. McIntosh, C. Bethge, First comparison of wave observations from CoMP and AIA/SDO. Astron. Astrophys. **556**, A124 (2013)

H. Tian, E.E. DeLuca, S.R. Cranmer, B. De Pontieu, H. Peter, J. Martínez-Sykora, L. Golub, S. McKillop, K.K. Reeves, M.P. Miralles, P. McCauley, S. Saar, P. Testa, M. Weber, N. Murphy, J. Lemen, A. Title, P. Boerner, N. Hurlburt, T.D. Tarbell, J.P. Wuelser, L. Kleint, C. Kankelborg, S. Jaeggli, M. Carlsson, V. Hansteen, S.W. McIntosh, Prevalence of small-scale jets from the networks of the solar transition region and chromosphere. Science **346**, 1255711 (2014)

A.M. Title, C.J. Schrijver, The Sun's magnetic carpet, in *Tenth Cambridge Workshop on Cool Stars, Stellar Systems, and the Sun*, ed. by R. Donahue, J. Bookbinder. ASP Conf. Ser., vol. 154 (ASP, San Francisco, 1998), pp. 345–358

S. Tomczyk, G.L. Card, T. Darnell, D.F. Elmore, R. Lull, P.G. Nelson, K.V. Streander, J. Burkepile, R. Casini, P.G. Judge, An instrument to measure coronal emission line polarization. Sol. Phys. **247**, 411–428 (2008)

S. Tomczyk, E. Landi, J.T. Burkepile, R. Casini, E.E. DeLuca, Y. Fan, S.E. Gibson, H. Lin, S.W. McIntosh, S.C. Solomon, G. de Toma, A.G. de Wijn, J. Zhang, Scientific objectives and capabilities of the coronal solar magnetism observatory. J. Geophys. Res. **121**, 7470–7487 (2016)

G. Tóth, I.V. Sokolov, T.I. Gombosi, D.R. Chesney, C.R. Clauer, D.L. De Zeeuw, K.C. Hansen, K.J. Kane, W.B. Manchester, R.C. Oehmke, K.G. Powell, A.J. Ridley, I.I. Roussev, Q.F. Stout, O. Volberg, R.A. Wolf, S. Sazykin, A. Chan, B. Yu, J. Kóta, Space weather modeling framework: A new tool for the space science community. J. Geophys. Res. **110**, A12226 (2005)

P.J. Tracy, J.C. Kasper, J.M. Raines, P. Shearer, J.A. Gilbert, T.H. Zurbuchen, Constraining solar wind heating processes by kinetic properties of heavy ions. Phys. Rev. Lett. **116**, 255101 (2016)

A. Tritschler, T.R. Rimmele, S. Berukoff, R. Casini, J.R. Kuhn, H. Lin, M.P. Rast, J.P. McMullin, W. Schmidt, F. Wöger, D.K. Inouye (the DKIST Team), Solar telescope: High-resolution observing of the dynamic Sun. Astron. Nachr. **337**, 1064–1069 (2016)

B.T. Tsurutani, W.D. Gonzalez, A.L.C. Gonzalez, F.L. Guarnieri, N. Gopalswamy, M. Grande, Y. Kamide, Y. Kasahara, G. Lu, I. Mann, R. McPherron, F. Soraas, V. Vasyliunas, Corotating solar wind streams and recurrent geomagnetic activity: A review. J. Geophys. Res. **111**, A07S01 (2006)

Y. Uchida, A mechanism for the acceleration of solar chromospheric spicules. Publ. Astron. Soc. Jpn. **21**, 128–140 (1969)

A.V. Usmanov, M.L. Goldstein, W.H. Matthaeus, Three-dimensional magnetohydrodynamic modeling of the solar wind including pickup protons and turbulence transport. Astrophys. J. **754**, 40 (2012)

A.V. Usmanov, M.L. Goldstein, W.H. Matthaeus, A four-fluid MHD model of the solar wind/interstellar medium interaction with turbulence transport and pickup protons as separate fluid. Astrophys. J. **820**, 17 (2016)

A. Vaivads, A. Retinò, J. Soucek, Y.V. Khotyaintsev, F. Valentini, C.P. Escoubet, O. Alexandrova, M. André, S.D. Bale, M. Balikhin, D. Burgess, E. Camporeale, D. Caprioli, C.H.K. Chen, E. Clacey, C.M. Cully, J. De Keyser, J.P. Eastwood, A.N. Fazakerley, S. Eriksson, M.L. Goldstein, D.B. Graham, S. Haaland, M. Hoshino, H. Ji, H. Karimabadi, H. Kucharek, B. Lavraud, F. Marcucci, W.H. Matthaeus, T.E. Moore, R. Nakamura, Y. Narita, Z. Nemecek, C. Norgren, H. Opgenoorth, M. Palmroth, D. Perrone, J.-L. Pinçon, P. Rathsman, H. Rothkaehl, F. Sahraoui, S. Servidio, L. Sorriso-Valvo, R. Vainio, Z. Vörös, R.F. Wimmer-Schweingruber, Turbulence heating observer: Satellite mission proposal. J. Plasma Phys. **82**, 905820501 (2016)

A.A. van Ballegooijen, Cascade of magnetic energy as a mechanism of coronal heating. Astrophys. J. **311**, 1001–1014 (1986)

A.A. van Ballegooijen, M. Asgari-Targhi, Direct and inverse cascades in the acceleration region of the fast solar wind. Astrophys. J. **835**, 10 (2017)

A.A. van Ballegooijen, M. Asgari-Targhi, S.R. Cranmer, E. DeLuca, Heating of the solar chromosphere and corona by Alfvén wave turbulence. Astrophys. J. **736**, 3 (2011)

A.A. van Ballegooijen, M. Asgari-Targhi, M.A. Berger, On the relationship between photospheric footpoint motions and coronal heating in solar active regions. Astrophys. J. **787**, 87 (2014)

B. van der Holst, I.V. Sokolov, X. Meng, M. Jin, W.B. Manchester, G. Tóth, T.I. Gombosi, Alfvén Wave Solar Model (AWSoM): Coronal heating. Astrophys. J. **782**, 81 (2014)

T. Van Doorsselaere, P. Antolin, D. Yuan, V. Reznikova, N. Magyar, Forward modelling of optically thin coronal plasma with the FoMo model. Front. Astron. Space Sci. **3**, 4 (2016)

M. Velli, Solar wind acceleration: Mechanisms and scaling laws, in *Twelfth International Solar Wind Conference*, ed. by M. Maksimovic, K. Issautier, N. Meyer-Vernet, M. Moncuquet. AIP Conf. Proc., vol. 1216 (AIP, New York, 2010), pp. 14–19

M. Velli, R. Grappin, A. Mangeney, Waves from the Sun. Geophys. Astrophys. Fluid Dyn. **62**, 101–121 (1991)

S.N. Venkata, B.R. Prasad, R.K. Nalla, J. Singh, Scatter studies for visible emission line coronagraph on board Aditya-L1 mission. J. Astron. Telesc. Instrum. Syst. **3**, 014002 (2017)

G. Verbanac, B. Vršnak, S. Živković, T. Hojsak, A.M. Veronig, M. Temmer, Solar wind high-speed streams and related geomagnetic activity in the declining phase of solar cycle 23. Astron. Astrophys. **533**, A49 (2011)

D. Verscharen, C.H.K. Chen, R.T. Wicks, On kinetic slow modes, fluid slow modes, and pressure-balanced structures in the solar wind. Astrophys. J. **840**, 106 (2017)

N.M. Viall, A. Vourlidas, Periodic density structures and the origin of the slow solar wind. Astrophys. J. **807**, 176 (2015)

R. von Steiger, N.A. Schwadron, L.A. Fisk, J. Geiss, G. Gloeckler, S. Hefti, B. Wilken, R.F. Wimmer-Schweingruber, T.H. Zurbuchen, Composition of quasi-stationary solar wind flows from Ulysses/solar wind ion composition spectrometer. J. Geophys. Res. **105**, 27217–27238 (2000)

B. Vršnak, T. Žic, T.V. Falkenberg, C. Möstl, S. Vennerstrom, D. Vrbanec, The role of aerodynamic drag in propagation of interplanetary coronal mass ejections. Astron. Astrophys. **512**, A43 (2010)

Y.-M. Wang, Cyclic magnetic variations of the Sun, in *Tenth Cambridge Workshop on Cool Stars, Stellar Systems, and the Sun*, ed. by R. Donahue, J. Bookbinder. ASP Conf. Ser., vol. 154 (ASP, San Francisco, 1998), pp. 131–151

Y.-M. Wang, Role of the coronal Alfvén speed in modulating the solar-wind helium abundance. Astrophys. J. Lett. **833**, L21 (2016)

Y.-M. Wang, N.R. Sheeley, Solar wind speed and coronal flux-tube expansion. Astrophys. J. **355**, 726–732 (1990)

Y.-M. Wang, N.R. Sheeley, Sources of the solar wind at Ulysses during 1990–2006. Astrophys. J. **653**, 708–718 (2006)

Y.-M. Wang, N.R. Sheeley, D.G. Socker, R.A. Howard, N.B. Rich, The dynamical nature of coronal streamers. J. Geophys. Res. **105**, 25133–25142 (2000)

Y. Wang, C. Shen, S. Wang, P. Ye, Deflection of coronal mass ejection in the interplanetary medium. Sol. Phys. **222**, 329–343 (2004)

Y.-M. Wang, N.R. Sheeley, N.B. Rich, Coronal pseudostreamers. Astrophys. J. **658**, 1340–1348 (2007)

Y.-M. Wang, Y.-K. Ko, R. Grappin, Slow solar wind from open regions with strong low-coronal heating. Astrophys. J. **691**, 760–769 (2009)

Y.-M. Wang, R. Grappin, E. Robbrecht, N.R. Sheeley, On the nature of the solar wind from coronal pseudostreamers. Astrophys. J. **749**, 182 (2012)

M. Weinzierl, A.R. Yeates, D.H. Mackay, C.J. Henney, C.N. Arge, A new technique for the photospheric driving of non-potential solar coronal magnetic field simulations. Astrophys. J. **823**, 55 (2016)

J.M. Wilcox, The interplanetary magnetic field: Solar origin and terrestrial effects. Space Sci. Rev. **8**, 258–328 (1968)

L.N. Woolsey, S.R. Cranmer, Turbulence-driven coronal heating and improvements to empirical forecasting of the solar wind. Astrophys. J. **787**, 160 (2014)

M.S. Yalim, N. Pogorelov, Y. Liu, A data-driven MHD model of the global solar corona within Multi-Scale Fluid-Kinetic Simulation Suite (MS-FLUKSS). J. Phys. Conf. Ser. **837**, 012015 (2017)

L. Yang, J. He, H. Peter, C.-Y. Tu, W. Chen, L. Zhang, E. Marsch, L. Wang, X. Feng, L. Yan, Injection of plasma into the nascent solar wind via reconnection driven by supergranular advection. Astrophys. J. **770**, 6 (2013)

A.R. Yeates, Coronal magnetic field evolution from 1996 to 2012: Continuous non-potential simulations. Sol. Phys. **289**, 631–648 (2014)

L. Zangrilli, G. Poletto, Evolution of active region outflows throughout an active region lifetime. Astron. Astrophys. **594**, A40 (2016)

J. Zhang, I.G. Richardson, D.F. Webb, N. Gopalswamy, E. Huttunen, J.C. Kasper, N.V. Nitta, W. Poomvises, B.J. Thompson, C.-C. Wu, S. Yashiro, A.N. Zhukov, Solar and interplanetary sources of major geomagnetic storms (Dst ≤ -100 nT) during 1996–2005. J. Geophys. Res. **112**, A10102 (2007)

X. Zhao, J.T. Hoeksema, Prediction of the interplanetary magnetic field strength. J. Geophys. Res. **100**, 19–33 (1995)

L. Zhao, S.E. Gibson, L.A. Fisk, Association of solar wind proton flux extremes with pseusostreamers. J. Geophys. Res. **118**, 2834–2841 (2013)

Y. Zhou, X. Feng, Numerical study of the propagation characteristics of coronal mass ejections in a structured ambient solar wind. J. Geophys. Res. **122**, 1451–1462 (2017)

I. Zouganelis, M. Maksimovic, N. Meyer-Vernet, H. Lamy, K. Issautier, A transonic collisionless model of the solar wind. Astrophys. J. **606**, 542–554 (2004)

T.H. Zurbuchen, R. von Steiger, J. Gruesbeck, E. Landi, S.T. Lepri, L. Zhao, V. Hansteen, Sources of solar wind at solar minimum: Constraints from composition data. Space Sci. Rev. **172**, 41–55 (2012)

Space Sci Rev (2018) 214:56
https://doi.org/10.1007/s11214-018-0487-4

Evolution of the Sunspot Number and Solar Wind *B* Time Series

Edward W. Cliver[1] · **Konstantin Herbst**[2]

Received: 28 April 2017 / Accepted: 7 February 2018 / Published online: 8 March 2018

Abstract The past two decades have witnessed significant changes in our knowledge of long-term solar and solar wind activity. The sunspot number time series (1700-present) developed by Rudolf Wolf during the second half of the 19th century was revised and extended by the group sunspot number series (1610–1995) of Hoyt and Schatten during the 1990s. The group sunspot number is significantly lower than the Wolf series before ∼1885. An effort from 2011–2015 to understand and remove differences between these two series via a series of workshops had the unintended consequence of prompting several alternative constructions of the sunspot number. Thus it has been necessary to expand and extend the sunspot number reconciliation process. On the solar wind side, after a decade of controversy, an ISSI International Team used geomagnetic and sunspot data to obtain a high-confidence time series of the solar wind magnetic field strength (B) from 1750-present that can be compared with two independent long-term (> ∼600 year) series of annual B-values based on cosmogenic nuclides. In this paper, we trace the twists and turns leading to our current understanding of long-term solar and solar wind activity.

Keywords Sunspot number · solar wind magnetic field strength

1 Introduction

Schwabe's discovery of the "decennial" sunspot cycle based on his 1826–1843 observations (Schwabe 1844) led Wolf to compile sunspot observations for earlier years and thus extend the 11-yr solar cycle back to 1700. More recently, the impulse to extend solar and

The Scientific Foundation of Space Weather
Edited by Rudolf von Steiger, Daniel Baker, André Balogh, Tamás Gombosi, Hannu Koskinen and Astrid Veronig

✉ E.W. Cliver
ecliver@nso.edu

[1] National Solar Observatory, Boulder, CO, USA

[2] Christian-Albrechts-Universität zu Kiel, Kiel, Germany

solar-terrestrial time series back in time, or to re-examine existing long-term series, has been manifest in attempts to determine solar wind B prior to the space age and to recalibrate the sunspot number. Both efforts involve considerable uncertainty. For the sunspot number, the primary challenge is to normalize differences in telescopes, visual acuity, observing conditions, and reporting practices between modern sunspot counters and those from earlier epochs. The challenge for solar wind reconstruction is to deduce values of solar wind parameters from indirect observations, viz., sunspot number, geomagnetic variability, and concentrations of cosmogenic nuclides in tree rings (^{14}C) and polar ice cores (^{10}Be). Research on long-term reconstructions of both the sunspot number and solar wind B was spurred by the Lockwood et al. (1999) report of a doubling of the Sun's coronal magnetic field during the 20th century. This paper led to a prolonged contentious debate on long-term solar wind B that was recently largely resolved (Owens et al. 2016a, 2016b) while, at the same time, an equally animated discussion sprang up surrounding the sunspot number time series.

In this paper, we trace the evolution of research on the sunspot number (Sect. 2) and long-term solar wind B (Sect. 3), including recent developments for the Maunder Minimum (Sect. 4). Section 5 contains a summary and a list of lessons learned.

2 Sunspot Number

2.1 Schwabe and Wolf

Sunspots were first observed telescopically in 1610 and 1611 by Galileo and others but it is fair to say that little progress was made in their understanding before Schwabe's (1844) report of the sunspot cycle. Within the following 20 years, Schwabe's discovery triggered several key advances: Earth's magnetic variability was found to track the 11-yr solar cycle (Sabine 1852; Wolf 1852a, 1852b; Gautier 1852); Carrington discovered the latitude variation of sunspots over the cycle (1858) and used sunspots to deduce the Sun's differential rotation (1859a, 1863); and the first solar flare and first confirmed solar-terrestrial event were recorded (Carrington 1859b; Hodgson 1859; Stewart 1861; see Bartels 1937). Prior to Schwabe's discovery, the prevailing view of sunspots among astronomers was that voiced by Delambre (1814; see Clerke 1902), the Director of the Paris Observatory, "It is true that the sunspots are more puzzling than really useful."

With von Humboldt's help (Hufbauer 1991), Schwabe changed this perception and set Swiss astronomer Rudolf Wolf on the scientific direction that he would take until his death in 1893. Wolf's definition of the sunspot number (1851, 1856),

$$\mathrm{WSN} = k(10 \times Ng + Ns) \tag{1}$$

where WSN = Wolf sunspot number for a given day,[1] k = normalization factor for a secondary observer relative to the primary Zürich observer, Ng = Number of sunspot groups, Ns = Number of sunspots. Wolf intended his relative sunspot number to be an indicator of the total spotted area of the Sun. In the 1856 article, he wrote, "I was relying on the fact that the level of spottedness must be given primarily by the number of groups, but also by the number of spots as a reflection of the size of a group"... [I] "selected the number 10 that

[1]The Wolf number is also referred to as the Zürich number because of the long period during which it was made in Switzerland, and, more recently, as the international sunspot number following the transfer of the curatorship to the Royal Observatory of Belgium (ROB) in 1981 (Clette et al. 2007; Stenflo 2016).

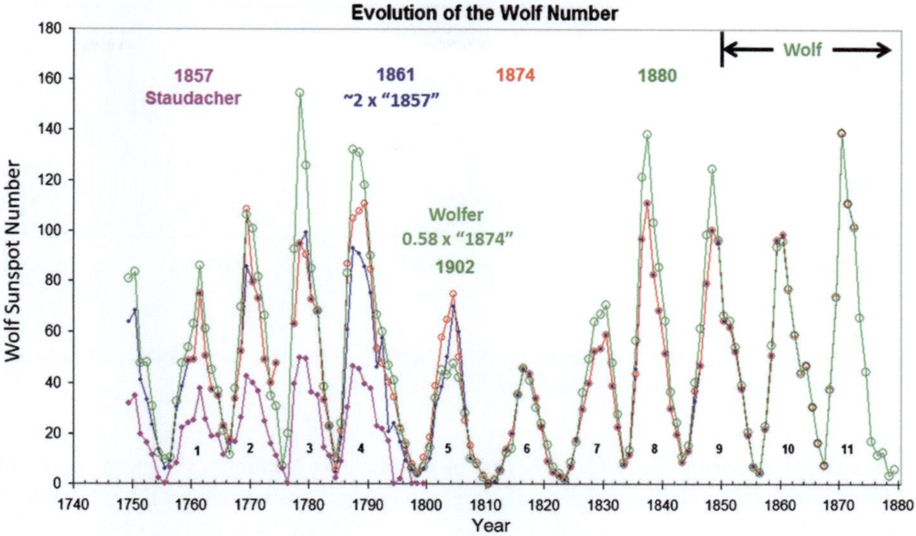

Fig. 1 Evolution of Wolf's time series, with revisions made (primarily to the pre-1848 record) by Wolf in 1861, 1874, and 1880, and by Wolfer for cycle 5 in 1902 (adapted from Svalgaard 2010)

seemed to fit a large number of cases and also was more convenient than other values near it [for ease of calculation]"... "I had preferred that my relative number would have been proportional to the actual area of the spots, but that would have necessitated too much work with data that I most often didn't have". In practice, it turned out that on average a group has approximately 10 spots, e.g., see Fig. 60 in Clette et al. (2014), so the WSN gives approximately equal weight to Ng and Ns. Over time, Wolf realized that the historical records of sunspot counts he acquired, predominantly from Europe, needed to be scaled up to be commensurate with his own series of observations initiated in late 1848. Thus in 1861, 1874, and 1880, he made (almost exclusively) upward adjustments to the pre-1848 series with only minor adjustments (in 1880) for years after 1848 (Fig. 1, from Svalgaard 2010). It appears that at least some of these adjustments were made with reference to the daily variation of geomagnetic activity (Loomis 1873; Hoyt and Schatten 1998a, 1998b; Cliver et al. 2015). The last adjustment to the Wolf series, a downward revision to cycle 5 (1798–1809) was made by Wolf's successor early in the 20th century (Wolfer 1902).

2.2 The Maunder Minimum

The next key development regarding the sunspot number time series was Eddy's (1976) rediscovery of the prolonged sunspot drought during the second half of the 17th century, which he termed the Maunder Minimum (MM). Wolf had been stymied in his effort to extend the sunspot number time series back in time by the relative absence of reported sunspots before 1700. Subsequently Spörer (1887, 1889) and Maunder (1890, 1894, 1922) called attention to this episode of apparently anomalous behavior, but few astronomers paid attention until Eddy's landmark paper. Eddy bolstered the case for an interval of unusually low sunspot activity from 1645–1715 with cosmogenic nuclide data (^{14}C) and chose the alliterative name "Maunder Minimum" (vs. Spörer Minimum; see the transcribed interview of Jack Eddy by Spencer Weart at https://history.aip.org/climate/eddy_int.htm) to describe it. In addition, he extended Wolf's annual time series of sunspot numbers back to 1610,

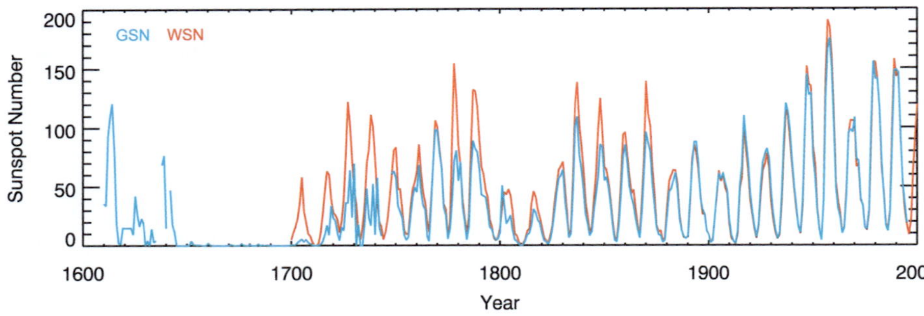

Fig. 2 Comparison of the Wolf sunspot number (WSN) with the Hoyt and Schatten (1998a, 1998b) group sunspot number (GSN), 1610–1995 (adapted from Clette et al. 2015)

with many years during the 1645–1700 core of the Maunder Minimum (Vaquero and Trigo 2015) having sunspot numbers of zero. The highest average sunspot number Eddy calculated for any year during this ~50 year period was 11 (in 1684) vs. annual values ranging from ~150–190 at the maxima of cycles 18, 19, 21, and 22, spanning the ~1945–1995 interval. Eddy's paper and supporting work, notably that of Ribes and Nesme-Ribes (1993), Beer et al. (1998), and Berggren et al. (2009) established the Maunder Minimum, the persistence of the 11-yr cycle during the MM, and long-term variability in the envelope of 11-yr solar maxima as basic tenets of solar physics.

2.3 The Group Sunspot Number

In his landmark paper, Eddy (1976) commented that, "Past counts of sunspot numbers are readily available from the year 1700 [...], and workers in solar and terrestrial studies often use the record as though it were of uniform quality. In fact, it is not. Thus it is advisable from time to time, to review the origin and pedigree of past sunspot numbers, and to recognize the uncertainty in much of the early record."

In 1994, Hoyt et al. took up Eddy's challenge in *Geophysical Research Letters*, titling their paper: "The one hundredth year of Rudolf Wolf's death: Do we have the correct reconstruction of solar activity?" Their answer, given in Hoyt and Schatten (1998a, 1998b), was in the negative. Their revised sunspot number time series, based solely on the daily number of sunspot groups (Ng in Eq. (1)), was significantly lower than the Wolf series before 1882 (see blue line in Fig. 2). The new Hoyt and Schatten "group" sunspot number (GSN) was defined as

$$\mathrm{GSN} = 12.08k' \times Ng \tag{2}$$

where the factor of 12.08 scaled group counts to the Wolf number and k' was the normalization factor used to scale secondary observers to the primary Royal Greenwich Observatory (RGO; Willis et al. 2013a, 2013b, 2016a, 2016b; Erwin et al. 2013) reference series (1874–1976). In deriving the GSN, Hoyt and Schatten (1998a, 1998b) greatly expanded the data base that Wolf and his successors at Zürich had compiled, achieving, e.g., an increase of 80% in the number of individual daily observations of sunspot groups for years before 1874. Moreover, they digitized the entire data base, making it accessible to all through the World Data Centers (https://www.ngdc.noaa.gov/stp/space-weather/solar-data/solar-indices/sunspot-numbers/group/). The new GSN time series was used to support the notion that the recent period of intense solar activity from ~1945–1995 was the strongest in several thousand years (Solanki et al. 2004; Usoskin et al. 2007; Usoskin 2017) and represented a Modern Grand Maximum, a counterpoint to the Maunder Minimum and an earlier Grand Minimum (Spörer, 1390–1550: Usoskin 2017) identified in the ^{14}C record by Eddy (1976).

2.4 The Sunspot Number Workshops

For more than a decade following the publication of the Hoyt and Schatten (1998a, 1998b) GSN, the solar community lived with two disparate solar activity time series, both widely used, with no consensus as to which was more correct. The choice of which sunspot number to use became a free parameter in solar and solar-terrestrial physics. Some argued that because the WSN and GSN had different prescriptions, it should not be surprising that their time series differed. This prompted the rejoinder: Why should series which agree reasonably well after ~1885 (see Fig. 2) part company for earlier times? Clearly further scrutiny was in order. In 2011, Frédéric Clette, who had shortly before become the Director of the World Data Center responsible for the production of the sunspot number, initiated an effort to address this unacceptable situation. A series of sunspot number workshops from 2011–2014 (Cliver et al. 2013a, 2015; Clette et al. 2014), co-organized by Clette, Ed Cliver of the US Air Force Research Laboratory, and Leif Svalgaard of Stanford University, examined the two discordant time series in Fig. 2 in an attempt to understand the cause of the differences between them and achieve a reconciliation, if possible.

In the course of the workshops, discontinuities were discovered in both the WSN and GSN time series. The Wolf series exhibited four inhomogeneities: (1) ca. 1850, at the join of the Schwabe and Wolf series (Leussu et al. 2013); (2) ~1865, change of observers/telescopes at Zürich (Clette and Lefèvre 2016); (3) after 1946, reflecting the change in directorship at Zürich from Brunner to Waldmeier near that time (Lockwood et al. 2014c; Clette et al. 2014; Clette and Lefèvre 2016; Lockwood et al. 2016a, 2016d, 2016e; Friedli 2016; Svalgaard et al. 2017); and (4) after 1980, resulting from the transfer of production of the sunspot number from Zürich to Brussels in that year (Clette et al. 2014, 2016a). The third of these discontinuities, uncovered by Svalgaard (2010, 2012), was caused by Waldmeier's decision to break from previous procedure by weighting individual sunspot numbers according to their size. Thus the standard definition of the WSN given in Eq. (1) and all solar physics textbooks or monographs (e.g., Zirin 1988; Foukal 2004) was not followed after 1946.[2] Equally surprising, the post-1980 inhomogeneity was due to a complex drift in the group counts of the Locarno station, the reference observer for the international sunspot number, that was not apparent in the group count time series of other long-term observers during this period. The GSN was found to be too low before ~1885 because of an inhomogeneity in the early years of the group count time series from RGO that Hoyt and Schatten (1998a, 1998b) used as a reference observer (Clette et al. 2014; Cliver and Ling 2016; Cliver 2017; cf., Willis et al. 2016a, 2016b; Lockwood et al. 2016e) and the flawed and somewhat opaque procedure Hoyt and Schatten used to normalize observers who did not overlap with RGO (Cliver and Ling 2016; Cliver 2017).

The end products of the sunspot number workshops, viz., a revised WSN series and an independently-derived GSN time series, are shown in Fig. 3(a). The modified WSN series,

[2]This practice may have been implemented on a trial basis before Waldmeier (Friedli 2016; Svalgaard et al. 2017) but the effect is not apparent until after 1946. The weighting was stopped with the publication of the new WSN (Clette et al. 2015). Thus the formula in Eq. (1) is correct again.

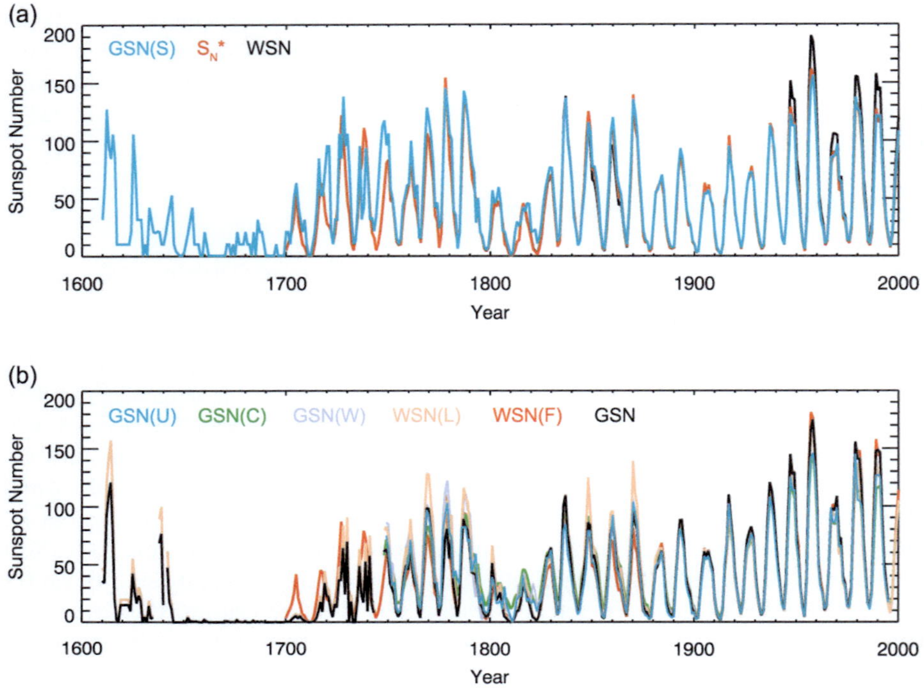

Fig. 3 (**a**) Comparison of the two new time series produced during the sunspot number workshops—S_N* (= 0.6 × S_N), where S_N is the corrected WSN series of Clette and Lefèvre (2016), and GSN(S), the newly constructed GSN series of Svalgaard and Schatten (2016)—plotted along with the original WSN series that they resemble. (**b**) Sunspot number time series produced independently of the sunspot number workshops—GSN(U) Usoskin et al. (2016), GSN(C) Chatzistergos et al. (2017), GSN(W) Willamo et al. (2017), WSN(L) Lockwood et al. (2014c, 2014d; revised in Lockwood et al. 2016d, 2016e), WSN(F) Friedli (2016)—plotted with the Hoyt and Schatten (1998a, 1998b) GSN time series they resemble. All series in (**a**) and (**b**) are scaled to WSN over the 1916–1946 interval

designated S_N (*version 2.0*; Clette and Lefèvre 2016 (red line)), employed corrections for the inhomogeneities 1–4 noted above. The independently-derived GSN-type series was based on a "backbone" approach that used multiple reference observers in order to minimize the uncertainty that results from the daisy-chaining needed to link distant secondary observers to a single primary observer (Svalgaard and Schatten 2016 (light-blue line)). The two new series in Fig. 3(a) track each other reasonably well and adhere more closely to the original WSN (black line) than to the Hoyt and Schatten GSN. The reconciliation sought by the sunspot number workshops was achieved but it proved to be short-lived.

2.5 Unintended Consequences and the Path Forward

The new S_N series was published on the website of the World Data Center for the Solar Index and Long-term Solar Observations (WDC-SILSO; http://sidc.oma.be/silso/) at ROB (Clette et al. 2015) and a Topical Issue in *Solar Physics* (Clette et al. 2016b) called for papers to both document and comment on the new time series. In a classic case of unintended consequences, the results exceeded expectations of the sunspot number workshop organizers. Criticism of the new GSN of Svalgaard and Schatten (e.g., Lockwood et al. 2016a, 2016b, 2016c, 2016d) was accompanied by the introduction of alternative GSN

(Usoskin et al. 2016) and Wolf-type (i.e., based on the formula in Eq. (1)) time series (Friedli 2016) which more closely resembled the Hoyt and Schatten GSN than the original WSN. Earlier, Lockwood et al. (2014c, 2014d; revised in Lockwood et al. 2016d, 2016e) had introduced a new Wolf-type series that was somewhat intermediate between the two new time series that resulted from the sunspot number workshops (Clette and Lefèvre 2016; Svalgaard and Schatten 2016) and those of Usoskin et al. (2016) and Friedli (2016). Detailed comparisons/critiques of these various new sunspot number series are given in Lockwood et al. (2016e) and Cliver (2016). Recently, Chatzistergos et al. (2017) introduced another GSN time series that is similar to that of Usoskin et al. (2016) which itself has been updated by Willamo et al. (2017). Figure 3(b) shows the time series of Usoskin et al. (2016; light blue trace), Chatzistergos et al. (2017; green), Willamo et al. (2017; light purple), Lockwood et al. (2016d, 2016e; light orange), and Friedli (2016; red), along with that of the original Hoyt and Schatten (1998a, 1998b; black) GSN.

Comparison of Fig. 3(a) and Fig. 3(b) indicate that the circumstance of the existence of two discordant time series (WSN and GSN) for long-term solar activity that motivated the sunspot number workshops has been exacerbated. Now, instead of two disparate series, the solar community has, to first order, two *classes* of series, with both WSN- and GSN-type series in each class. Moreover, the various new time series differ widely in their choices of standard or reference observers (e.g., RGO vs. Wolfer) and the techniques used to scale secondary observers to the standard observer (e.g., linear regression vs. a non-linear non-parametric probability distribution function). A key schism among the new proposed series involved the central problem of comparing of secondary observers with non-overlapping primary or reference observers. In the traditional daisy-chaining approach of Hoyt and Schatten (1998a, 1998b), secondary observer C who does not overlap with primary observer A is scaled to A through observer B who does; then D is linked to C, and so forth. In the backbone method (Svalgaard 2013; Clette et al. 2014; Svalgaard and Schatten 2016; Chatzistergos et al. 2017), the error propagation/accumulation inherent in daisy-chaining is reduced by designating several primary or backbone observers, scaling all overlapping secondary observers to them, and then linking the backbones together. In effect, the number of links between non-overlapping observers is reduced in such a scheme, providing fewer opportunities for error to creep in. Alternatively, the innovative active day fraction method proposed by Usoskin et al. (2016; and refined by Willamo et al. 2017) eliminates daisy-chaining altogether by scaling all secondary observers (regardless of overlap or non-overlap) directly to RGO by a comparison of the fraction of days per month the secondary observer reported (non-zero) spot groups (a measure of observer quality) with the corresponding fraction for RGO. The Usoskin et al. (2016), Willamo et al. (2017), and Chatzistergos et al. (2017) time series employ advanced non-linear techniques to scale secondary observers to primary observers vs. the combination of linear and non-linear regression used by Svalgaard and Schatten (2016) and criticized in, e.g., Lockwood et al. (2016c; cf., Svalgaard and Schatten 2017).

The situation facing the solar community in 2016 was thus scientifically complicated and, on a human level, becoming increasingly contentious. The danger was that the proliferation of new disparate series, if left unaddressed in a systematic fashion, would render the sunspot number meaningless as a measure of solar activity. How to proceed?

Matters came to a head at the Space Climate 6 Symposium in Levi, Finland in April 2016. Following a lively scientific session devoted to the sunspot number, developers of the new sunspot series held an informal meeting and agreed to work together to examine the causes of their differences and to reconcile them, in so far as possible—in other words, to continue, on a broader scale, the effort that was begun under the sunspot number workshops. The end

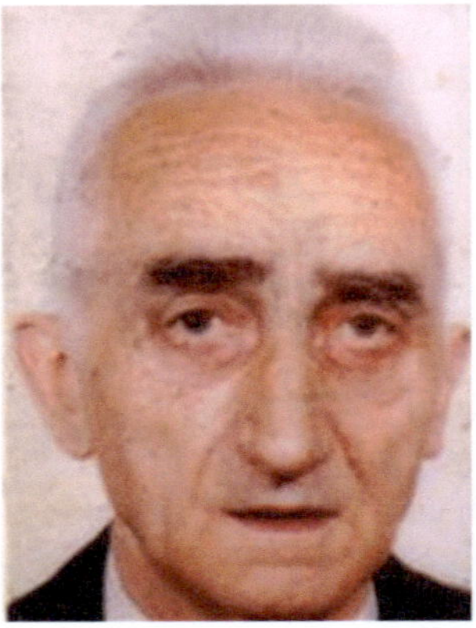

Fig. 4 Pierre-Nöel Mayaud (1923–2006), creator of the geomagnetic *aa* time series. Mayaud was a Jesuit who wrote the "bible" (Mayaud 1980) on geomagnetic indices. After he left geophysics, he became an authority on the interaction between the Catholic church and Galileo

goal is the creation of community vetted and accepted versions of both the GSN and WSN time series (with stated uncertainties, e.g., Dudok de Wit et al. 2016) and the establishment of a procedure for publication of further revisions as warranted. This work is underway, first involving small teams that are examining each series separately, to be followed by larger group meetings that will identify/implement best practices for series construction, with a targeted release date of 2019 for the new versions of the two series.

3 Solar Wind *B* Time Series

Fortunately, solar scientists have recent experience in dealing with a situation similar to that which has arisen for the sunspot number. The long-term time series in question previously was that for the solar wind magnetic field strength (B). We will spend some time recounting the evolution of that time series. It contains the various elements—conflicting findings, error, insight, miscommunications, confusion, emerging consensus and resolution—characteristic of scientific debate. The positive outcome of the work on solar wind B underwrites the present effort to reconcile the various sunspot number time series. For recent independent accounts of the evolution of thinking on long-term solar wind activity, see Usoskin (2017) and Lockwood et al. (2017).

3.1 Doubling of the Sun's Open Magnetic Flux During the 20th Century

Because geomagnetic activity is driven by the solar wind, solar wind parameters—first available directly from space observations in the early 1960s (Neugebauer and Snyder 1962; Snyder et al. 1963)—can be extracted from geomagnetic observations for earlier years. In 1972, P.-N. Mayaud, pictured in Fig. 4, published a new long-term (1868–1967) geomagnetic index, termed *aa* (Mayaud 1972), based on sequences of geomagnetic stations in the UK and Australia, that significantly extended the objective, easily accessible, record of geomagnetic activity, more than doubling the length of *ap* series (available from 1932) at the time. Following Russell (1975), Feynman and Crooker (1978) used the empirically-derived dependence of *aa* ($\sim V^2 B_S$; where B_S is the magnitude of the southward-pointing component of B and V is the solar wind speed) on solar wind parameters to constrain the range of B_S to 0.65–1.75 nT and V to 240–400 km s^{-1} for the solar minimum year of 1901, at the depth of a minimum of the ~100-yr Gleissberg cycle. Because *aa* depends on both B and V, a second relationship involving either B and/or V is needed for further specificity.

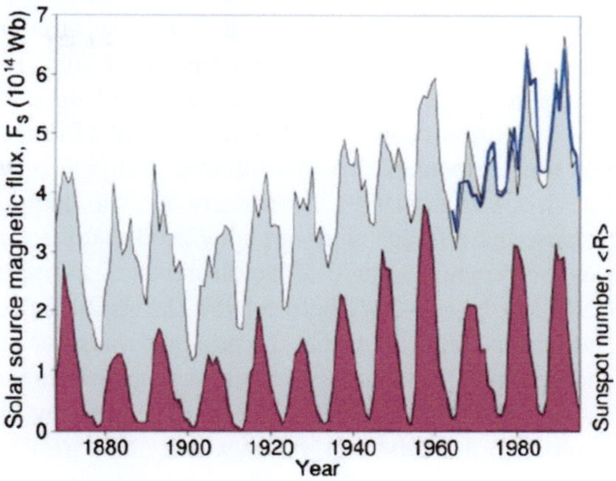

Fig. 5 Top: Modeled open solar flux (gray-shaded plot) from 1868–1996 with observed values (dark blue line). Bottom: sunspot number. (From Lockwood et al. 1999.)

Approximately two decades after Feynman and Crooker's pioneering paper, Lockwood et al. (1999) used Sargent's (1986) recurrence index, which is a measure of the 27-day repeatability of geomagnetic activity, as a proxy for solar wind speed and thus were able to deduce from the *aa* index an annual time series for the open solar open flux (*OSF*), based on the radial component Br of the solar wind magnetic field, from 1868 to the present.[3] Lockwood et al. (1999) found a doubling (increase by 131% from 1901 to 1992) of the coronal magnetic field over the 20th century (see Fig. 5). In the same paper, these authors reported a 41% increase in *OSF* between the solar minima in 1964 and 1996. In a key development, Solanki et al. (2000) introduced a model for the long-term time evolution of the open solar flux with which they were able to reproduce the doubling of the *OSF* during the 20th century reported by Lockwood et al. (1999). However, Arge et al. (2002), using a potential field source surface model and solar magnetogram data from three different observatories found no evidence for an increase in the open flux between the minima of 1976 and 1996. They wrote, "Thus the Lockwood et al. claim that open solar flux has increased by 41% from 1964 to 1995 would require that the entire increase must have occurred over the twelve-year period between 1964 to 1976, which does not seem credible." At the same time, Wang and Sheeley (2002) obtained a record of the *OSF* similar to that of Arge et al. (2002) but noted that the fact that the open flux was higher on average during strong cycles 21 and 22 than in weaker cycles 20 and 23 tended to support the finding of Lockwood et al. (1999) of a secular increase in the open flux over the last century, given the overall increase in sunspot activity since 1900.

Although both Svalgaard (1977) and Cliver and Ling (2002) noted evidence in the geomagnetic record that was consistent with a doubling of the open solar flux during the 20th century, they disputed the Lockwood et al. (1999) result in a paper submitted to *Nature* in 2003 entitled "No Doubling of the Sun's Coronal Magnetic Field During the Last 100 years". The "no doubling" paper was based largely on a newly-derived inter-hourly variability (*IHV*) long-term geomagnetic index (Svalgaard et al. 2004; first presented at CoSpaR 2002). For a single station, *IHV* is defined to be the sum of the [unsigned] differences between each hourly-averaged value and the next of the horizontal component of the geomagnetic field for the 6 h around local midnight, divided by 6. A key attribute

[3]Lockwood et al. (1999) also employed the "Ulysses result" (Balogh et al. 1995; Smith and Balogh 1995; Lockwood et al. 2004; Lockwood and Owens 2009) that *Br* (normalized to 1 AU) is independent of latitude when averaged over a rotation cycle or more, thus permitting the determination of *OSF* from a point measurement of *Br*.

of the *IHV* index is its objectivity and ease of construction from early geomagnetic data which often is readily available only in the form of hourly values (Svalgaard et al. 2004; Svalgaard and Cliver 2005). Comparison of the *IHV* index with *aa* indicated that the *aa* index was significantly too low (by ~5–10 nT) for the early years of the 20th century and was thus responsible for the inferred doubling of the coronal magnetic field reported by Lockwood et al. (1999). Fortunately, the "no doubling" paper was rejected for publication because the referees were not convinced that the *aa* index rather than the new *IHV* index was the principal source of the discrepancy. Subsequently, it was found that *aa* was artificially low by ~2–3 nT before 1957 (Jarvis 2005; Lockwood et al. 2006, 2009a; Svalgaard and Cliver 2007b), but that the principal cause of the difference between *IHV* and *aa* that underpinned the no doubling claim was an error in the interpretation of the raw geomagnetic data used to construct *IHV*. This error, detected by Svalgaard, was corrected in a note added in proof to the 2004 *IHV* paper which read: "*IHV* calculated for Cheltenham before 1915 should be reduced by 30% because it was based on a single reading per hour rather than on the average for an hour. This increases the variability and thus *IHV*. Values after 1915 are not affected." This change in recording from "spot" to hourly-averaged values at individual stations, occurring for British stations in 1912 and all American stations in 1915, is a common feature in long-term geomagnetic archives (Svalgaard and Cliver 2005, Sect. 2.1; 2007b, Appendix A2; Love et al. 2010).

The early geomagnetic data contained other station-unique artefacts, e.g., the records in the World Data Centers for the long-running Eskdalemuir (ESK) station were affected by two-point averaging of observed hourly values from 1911–1931 (Martini and Mursula 2006; Svalgaard and Cliver 2007b, Appendix A4.1; Macmillan and Clarke 2011). The net effect of this "post-processing" was to inflate 20th century increases in *IHV* and *aa* indices obtained by Mursula et al. (2004) and Clilverd et al. (2005), respectively, that were based in part on ESK observations.

3.2 Interdiurnal Variability of Geomagnetic Activity

In his 1977 paper on geomagnetic activity and solar wind parameters, Svalgaard wrote, "It would be very desirable to infer the interplanetary magnetic field strength independently from the solar wind speed from geomagnetic data." Twenty-six years later, in 2003, he achieved this goal with the realization that the geomagnetic u-index is not sensitive to the solar wind speed, but only to the IMF magnitude. The daily u-index (at a given station) developed by Bartels (1932), who termed it the interdiurnal variability, was the difference, regardless of sign, between successive daily means of the magnetic horizontal intensity. Svalgaard et al. (2003) retained the interdiurnal variability name but used the descriptor *IDV* to distinguish it from the u-index because *IDV* is based on only the single hour (taken to start 1 hour after the UT hour closest to local midnight) for consecutive days.[4] As shown in Fig. 6, *IDV* "has the useful property that its yearly averages are highly correlated with the solar wind magnetic field strength (B) and are independent of solar wind speed (V)" (Svalgaard et al. 2003). Thus the *IDV* index yields solar wind B directly and, in conjunction with *IHV*, solar wind V from hourly data for the second half of the 19th century. Svalgaard and Cliver (2005) showed that *IDV* is highly-correlated, on average, with the negative component of the *Dst* index.

Svalgaard and Cliver (2005) used *IDV* to construct a time series for B from 1872–2004. Based on this time series and a proportional relationship between Br and B ($Br = 0.53B$)

[4]For *IDV*, as well as *IHV*, restriction to night time hours removes contamination by the EUV-driven regular ionospheric variation.

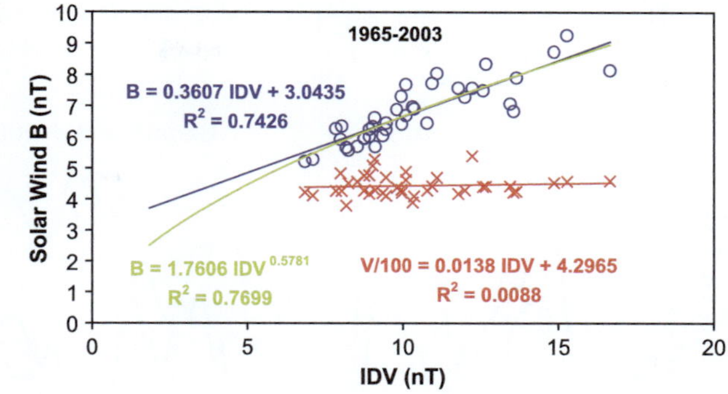

Fig. 6 Scatter plot showing the correlation between solar wind magnetic field strength at 1 AU (B) and the *IDV* geomagnetic index from 1965–2003 and the negligible dependence of *IDV* on solar wind speed (V) during this period (adapted from Svalgaard and Cliver 2005)

in Lockwood et al. (1999), they reported that, "solar cycle average B increased by ~25% from the 1900s to the 1950s and has been lower since", consistent with a Gleissberg-type (~100-yr) cycle (Gleissberg 1939; Feynman and Ruzmaikin 2011; Hathaway 2015; Le Mouël et al. 2017) during the 20th century. This paper prompted a comment by Lockwood et al. (2006; response in Svalgaard and Cliver 2006) who argued that the estimate of ~25% was too low and determined an increase of 38% based on the ratio of 11-yr running means in 1956 (near sunspot cycle maximum) and 1903 (near minimum) for the B values given in Table 3 of Svalgaard and Cliver (2005). Moreover, Lockwood et al. argued that for a more valid averaging interval for B of one-day vs. the one hour averages used in Lockwood et al. (1999), the proportional relationship between B and Br should be replaced by a linear relationship with an offset ($Br = 0.55B–0.91$), implying a 54% increase in smoothed Br from 1903–1956 based on the Svalgaard and Cliver B series. Lockwood et al. (2006) questioned the method used by Svalgaard and Cliver (2005) to treat gaps in the observed solar wind data and their B vs. *IDV* regression analysis. Using an alternative technique for data gaps and a least median of squares regression method for a correlation between *IDV* and Br, Lockwood et al. (2006) obtained an 85% increase in 11-yr averaged Br from 1903 to 1956.

3.3 Stumbling in the Dark

The exchange between Lockwood et al. (2006) and Svalgaard and Cliver (2006) was accompanied by a related debate on the long-term reconstruction of solar activity based on cosmogenic radionuclides. In that field, Solanki et al. (2004), from an analysis of the GSN and the record of ^{14}C concentration in tree rings, calculated that the interval of solar activity from ~1940–2000 was the strongest in the last ~8000 years. Following Usoskin et al. (2007), this ~60-yr interval came to be called the Modern Grand Maximum (MGM). In contrast, Muscheler et al. (2005) used a ^{14}C-based reconstruction of the solar modulation parameter (empirically related to B; e.g., Fig. 12 in Cliver et al. 2013b) and deduced high levels of this parameter during the second half of the 18th century that were comparable to those during the MGM. They noted that Bard et al. (2000) had earlier reported similar results based on ^{10}Be measurements from the South Pole. In response, Solanki et al. (2005) argued that the Muscheler et al. (2005) result was in conflict with other independent proxy time series for long-term solar activity such as that based on the ^{44}Ti cosmogenic nuclide (Bonino et al. 1995).

In short, long-term behavior of solar wind activity was a controversial topic in the middle of the last decade, with no resolution in sight. This scientific turbulence and confusion continued during 2007. From comparisons of their *IDV*-based B with the yearly-averaged

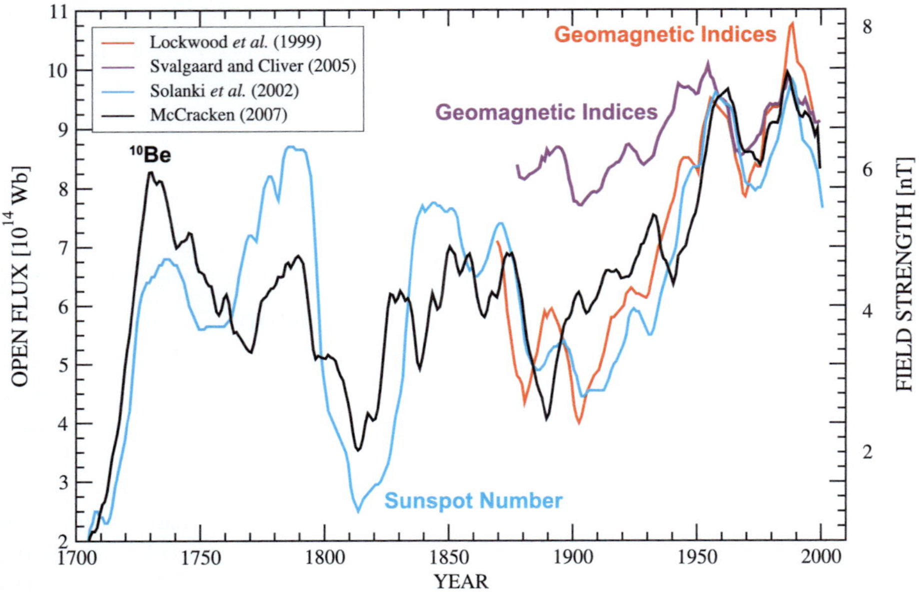

Fig. 7 Reconstructions of solar open flux (*OSF*) and solar wind B: *OSF* from Lockwood et al. (1999) based on geomagnetic data; B from Svalgaard and Cliver (2005) based on *IDV; OSF* from Solanki et al. (2002) based on the sunspot number; and B from McCracken (2007) based on ^{10}Be. All plots are 11-yr running averages. (Adapted from McCracken 2007.)

sunspot number, Svalgaard and Cliver (2007a) concluded that there was a "floor" in annual averages of solar wind B of 4.6 nT below which B would not drop. In the same month (April) that the paper was published, monthly-averaged B dropped below 4.6 nT and remained there for 30 of the next 32 months. Annual averages of B were 4.5 nT, 4.2 nT, and 3.9 nT for 2007, 2008, and 2009, respectively. In May 2007, Rouillard used a newly devised, mid-latitude range index (similar to *aa*, *ap*, and *IHV*), designated the m-index (for median), in conjunction with a corrected (for the pre-1957 deficit discussed in Sect. 3.1) *aa* index aa_C, to obtain time series of both B and V. Using a Bayesian least squares regression technique, they found an increase of 87% in the open solar flux between 11-year running means in 1903 and 1956. Their Fig. 6 indicated an anomalous spike in V near the 1901 minimum of cycle 14 which would lead to an under-estimate of the open flux at this time (and a corresponding over-estimate in the increase of this parameter between cycles 14 and 19). Generally, however, their series for B resembled that of Svalgaard and Cliver (2005). In September, however, McCracken (2007) published a long-term time series for solar wind B based on ^{10}Be that re-emphasized the differences between the time series of Lockwood et al. (1999) and Svalgaard and Cliver (2005). Figure 7, taken from McCracken's paper, contains plots of 11-yr running averages of time series for: solar open flux from Lockwood et al. (1999) based on geomagnetic data, solar wind B from Svalgaard and Cliver (2005) based on *IDV*, solar open flux from Solanki et al. (2002) based on the sunspot number, and solar wind B from McCracken (2007) based on ^{10}Be. McCracken wrote that, "The agreement between curves 1 [Lockwood et al. 1999], 3 [Solanki et al. 2002], and 4 [McCracken 2007] in Fig. 7 provides confidence in the overall validity of these three independent methods." He added, "It is important, however, that the disagreement between curve 2 [Svalgaard and Cliver 2005] and the other three be resolved."

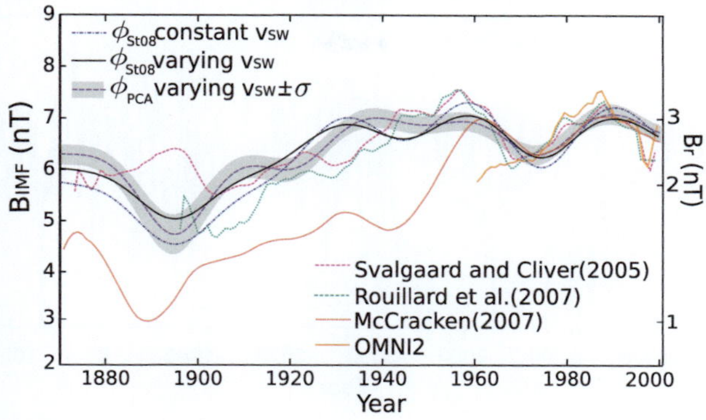

Fig. 8 Comparison of three reconstructions of solar wind B (left-hand y-axis) from Steinhilber et al. (2010; key in upper left corner) with other reconstructions and observation (lower right corner). Calculated Br is given on the right-hand y-axis. The Steinhilber et al. plots are 40-yr running means (Φ_{St08}) and 25-yr running means (Φ_{PCA}). The comparison plots (Svalgaard and Cliver 2005; Rouillard et al. 2007; McCracken 2007; OMNI2) are 11-yr running means. (From Steinhilber et al. 2010.)

3.4 Approaching Consensus

Lockwood et al. (2009a) used the aa_{C} index and the m-index to derive a time series for B that, other than for a few years in cycle 14 (from 1905, when the reconstruction began, to ~1910) was quite similar to the Svalgaard and Cliver (2005) series. In this reconstruction, they employed a correction for Br that makes allowance for the kinematic effect of longitudinal structure in solar wind flow speed (Lockwood et al. 2009b). Subsequently, a ^{10}Be-based reconstruction of B from Steinhilber et al. (2010) reproduced in Fig. 8 represented a marked departure from the McCracken (2007) time series for this parameter and agreed reasonably well with that of Svalgaard and Cliver (2005). Things were beginning to converge, at least for the post-1870 time period. Regarding the longer-term level of solar activity, Steinhilber et al. (2010) published the long-term reconstruction of B shown in Fig. 9 which indicated that the Svalgaard and Cliver (2005) floor at 4.6 nT, already violated by direct observations from 2007–2009, was also breached on numerous occasions in the past, including the Maunder (1645–1715) and Spörer (1390–1550) Grand Minima.

Svalgaard and Cliver (2010) updated their *IDV*-based B series and called attention to the increasing convergence of the various geomagnetic-based time series of B (see Fig. 10(a)). Lockwood and Owens (2011) concurred, reminding, however, that Lockwood et al. (1999) computed the variation of the open flux rather than B. Figure 10(b), taken from their paper, shows that a proper reconstruction of B from the *OSF*, taking into account the role of V (Eq. 5 in Lockwood et al. 1999), agreed more closely with subsequent reconstructions than had been indicated in Fig. 10(a). As further evidence of convergence, Cliver and Ling (2011) used the B values from Svalgaard and Cliver (2010) and *IHV*-based V values from Svalgaard and Cliver (2007b) to deduce an increase of 84% in the open solar flux between the sunspot minimum years of 1901 and 1954, comparable to the values obtained by Lockwood et al. (2006) and Rouillard et al. (2007) from 11-yr running means for 1903 and 1956.

3.5 ISSI International Team

Bolstered by the results shown in Fig. 8 and Fig. 10, Svalgaard, Lockwood, and Beer proposed to ISSI in 2011 that an International Team be convened to systematically examine

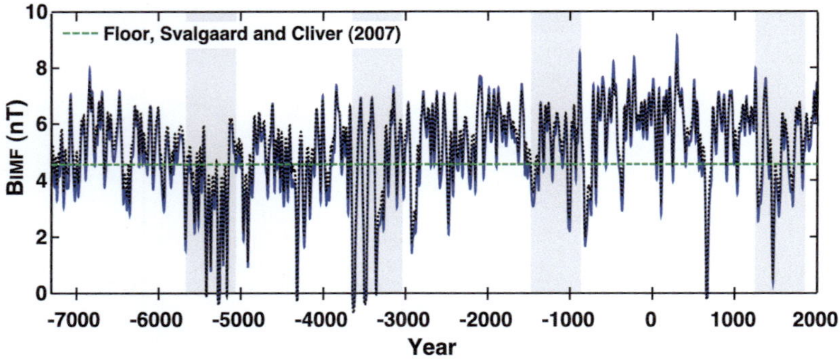

Fig. 9 Reconstruction of 40-yr running means of solar wind B for the past 9300 years. The solid-blue (dotted-black) curve assumes a constant (varying) solar wind speed. The dashed-green line shows the floor of Svalgaard and Cliver (2007a). Note the multiple excursions to $B = \sim 0$ nT, including during the Spörer minimum, ca. 1500 (adapted from Steinhilber et al. 2010)

geomagnetic, solar, and cosmogenic radionuclide data to derive a consensus long-term series for solar wind B. Quoting from the proposal abstract,

> *After a decade of vigorous research, reasonable agreement has been achieved between IMF strength (and open flux) estimates based on geomagnetic data and the inversion of the paleo-cosmic radiation data for the last* ∼100 years. [...] *The* [*primary*] *purpose of the workshop is* ...] to [...] *extend/substantiate the geomagnetic-based reconstruction of solar wind parameters from* ∼*1840–2010 and to resolve the remaining discrepancies among the geomagnetic-, cosmic-ray-, and sunspot-based reconstructions* [...]

The principal results of the resulting ISSI team meetings are summarized here.

3.5.1 Long-Term B Series Based on the IDV Index and on the Sunspot Number

Owens et al. (2016a, 2016b) obtained three separate time series for B based on the three types of data considered (geomagnetic, sunspot, and cosmogenic radionuclides), and their correlations with directly observed solar wind B, B[OBS]. Of these three time series, the one based on geomagnetic data (1845–2013) is considered to be the "gold standard". It is a composite of *IDV*-based B series from Lockwood et al. (2013a, 2013b, 2014a, 2014b; see Fig. 11(a)) and (Svalgaard 2014; Fig. 11(b)). The Svalgaard et al. series considers data from all stations at corrected geomagnetic latitudes $\lesssim |50°|$ for only the hour taken to start 1 hour after the UT hour closest to local midnight. The Lockwood et al. approach considers only a single chain of magnetometers but all 24 hourly-observations, similar to Bartels' u-index. As a measure of the robustness of *IDV*, the results produced by both methods are very similar as can be seen in (Fig. 11(c)). Following Lockwood et al. (2013b, 2014b), the B[GEO] composite and uncertainty bands in Fig. 11(c) are based on a Monte Carlo technique that considered ∼17,000 realizations of correlations of the two *IDV*-based series with actual observations, weighted approximately 3:2 toward the Svalgaard method. A similar approach was used to obtain a sunspot number based time series for B from 1749 to present (Fig. 12). In that case, time series for B, based on two separate methods, were constructed from B[OBS] for the space age and the sunspot number time series of Lockwood et al. (2014c, 2014d; Fig. 12(a)), Clette and Lefèvre (2016; Fig. 12(b)), Svalgaard and Schatten (2016; Fig. 12(c)), and

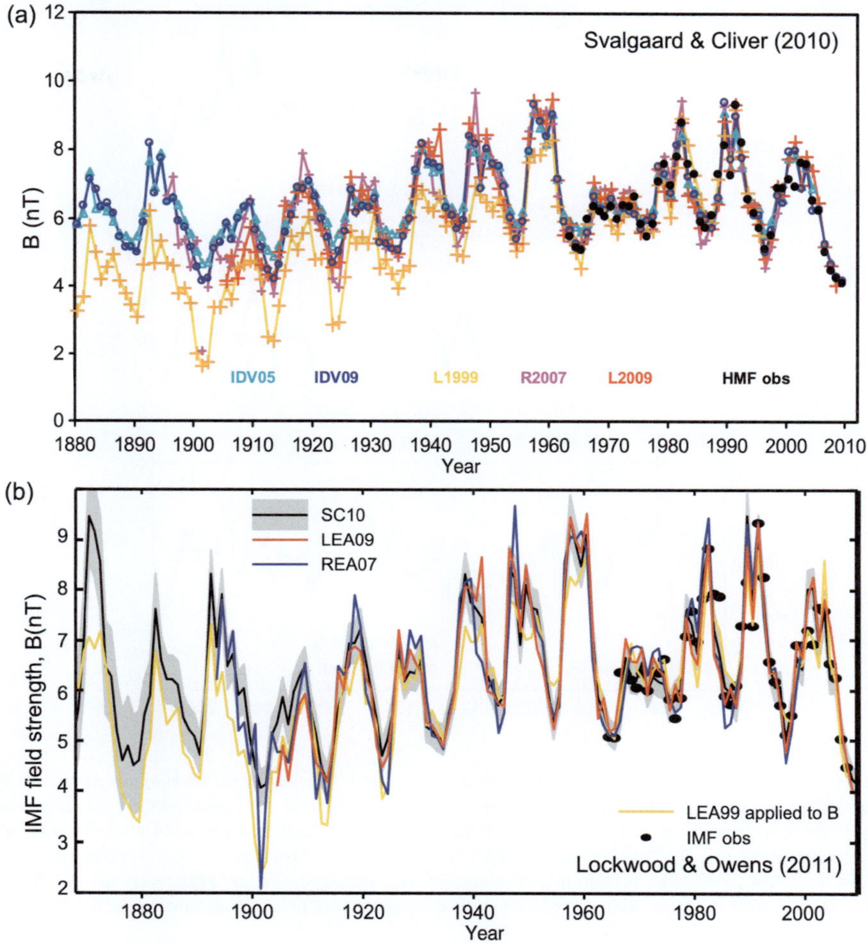

Fig. 10 (**a**) Figure from Svalgaard and Cliver (2010) showing emerging consensus in reconstructions of solar wind B. The plots of annual averages of B are from Lockwood et al. (1999; L1999), Svalgaard and Cliver (2005; IDV05), Rouillard et al. (2007; R2007), Lockwood et al. (2009a; L2009), and Svalgaard and Cliver (2010; IDV09). Filled black circles are annual means of directly observed B. (**b**) A figure similar to (**a**) from Lockwood and Owens (2011). Plots are from Rouillard et al. (2007; REA07), Lockwood et al. (2009a; LEA09), and Svalgaard and Cliver (2010; SC10). The yellow curve is the correct reconstruction of solar wind B based on Lockwood et al. (2009) (cf., yellow curve in panel (**a**)), and solid black circles are annual means of observed B

Usoskin et al. (2016; Fig. 12(d)). In the first method, designated B[SSN$^{1/2}$], B is assumed to be linearly correlated to the square root of sunspot number (Wang and Sheeley 2003; Wang et al. 2005; Svalgaard and Cliver 2005). The second method (B[OL2012]) used the model of Owens and Lockwood (2012) which assumes that open solar flux generation varies with the coronal mass ejection (CME) rate (with the sunspot number used as a proxy for the CME rate before the advent of coronagraph observations) and employs an *OSF* loss term that varies with the observed tilt angle of the heliospheric current sheet over the solar cycle and is assumed to be invariant from cycle to cycle. The resulting eight B series were then compared with *IDV*-based B (1845–2013) and a composite [B(SSN)], shown in Fig. 12(e),

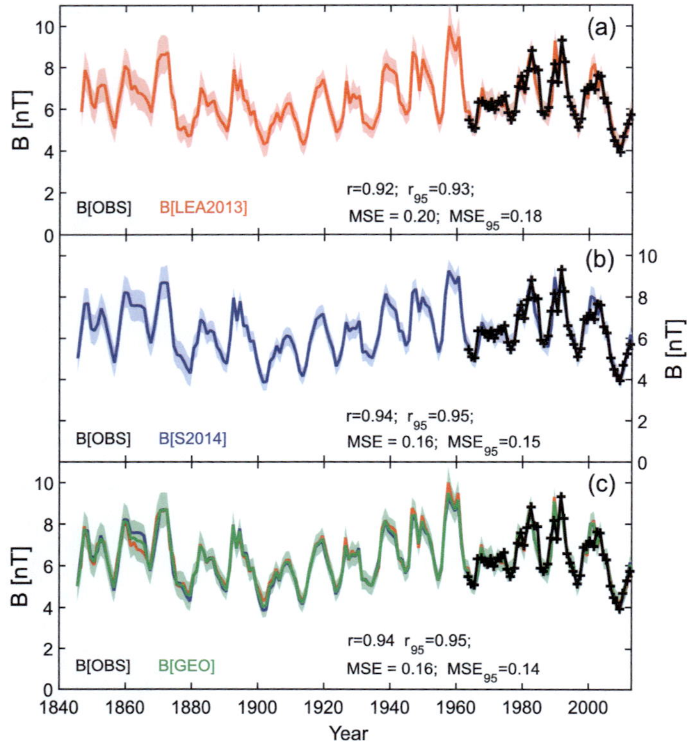

Fig. 11 Annual means of near-Earth solar wind B, 1845–2013. Black symbols show observed B, B[OBS] from OMNI (https://omniweb.gsfc.nasa.gov/) for the 1963–2013 interval. (**a**) Geomagnetic-based reconstruction of B from (Lockwood et al. (2013b; B[LEA2013], red). (**b**) Geomagnetic based reconstruction of B from Svalgaard (2014; B[S2014], blue). (**c**) B[GEO] (green), a composite based on a weighted average of B[LEA2013] and B[S2014]. Correlation coefficients (r) and mean square errors (MSE) relative to B[OBS] are given. Shaded areas represent 90% confidence intervals. Subscripts "95" indicate values over the 1995–2013 interval. (From Owens et al. 2016a.)

was formed with slightly higher weights given to the series of Svalgaard and Schatten (2016) and Clette and Lefèvre (2016).

3.5.2 Long-Term Series of Solar Wind B Based on Cosmogenic Radionuclides

In Owens et al. (2016b) two independent cosmogenic based series for B (1750–$\sim$1990) were compared, using different methodologies, the first from McCracken and Beer (2015) and the second from Usoskin and colleagues (Usoskin et al. 2002; Alanko-Huotari et al. 2006; Kovaltsov and Usoskin 2010; Usoskin et al. 2015). Both series employ ^{10}Be concentrations from the Dye 3 and North GRIP Greenland ice cores. A key aspect of the McCracken and Beer (2015) time series is the correction (upward adjustment of B) necessitated by several large solar cosmic ray enhancements, either observed (1942–1956) or inferred (1816–1938), as well as by high-altitude nuclear explosions in 1962. It includes clusters of upper limit values of 2.5 nT for years near the depths of the Spörer, Maunder, and Dalton (ca. 1800)

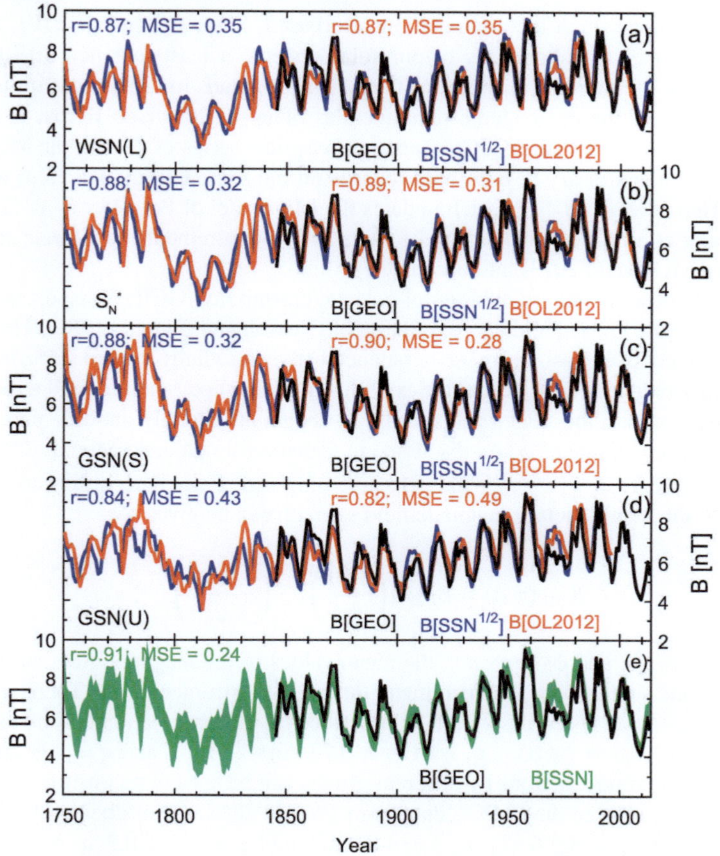

Fig. 12 Sunspot number based reconstructions of *B* over the period 1750–2013 compared with the 1845–2013 B[GEO] composite from Fig. 11(c; black lines). Blue lines show the results of the B[SSN 1/2] reconstruction method, while red lines show reconstructions based on B[OL2012]. (**a–d**) Estimates of *B* based on: (**a**) WSN(L), (**b**) S_N*, (**c**) GSN(S), and (**d**) GSN(U). (**e**) B[SSN], weighted composite of the eight individual *B* estimates. Correlation coefficients (r) and mean square errors (MSE) relative to B[GEO; 1945–2013] are given (adapted from Owens et al. 2016a)

minima.[5] Figure 3 in Owens et al. (2016b) shows that the Usoskin series lays consistently below the McCracken and Beer (2015) series from 1750 to ∼1900.

The McCracken and Beer (2015) annual time series for *B* extends from 1391–1983 and can be compared over this interval with the long-term (0–2000) series of the modulation potential (ϕ) obtained by Muscheler et al. (2016) based on annual measurements of ^{14}C concentrations in tree rings to 1950 and neutron monitor measurements thereafter. The modulation potential has units of MV and is a measure of heliospheric resistance to incoming cosmic rays (see, e.g., Herbst et al. 2010, 2017; Beer et al. 2012). The time series of magnetic field strength by McCracken and Beer (2015) and of the solar modulation potential by Muscheler et al. (2016) are based on numerical models to describe the ^{10}Be and ^{14}C measurements, respectively. The results strongly depend on the Local Interstellar

[5] Spörer (1412.0, 1420.0, 1421.0, 1436.0, 1442.1, 1448.4, 1450.3, 1451.1, 1560.1, 1461.9, 1531.6); Maunder (1681.0, 1692.7, 1693.6, 1694.6, 1696.0, 1697.0, 1697.8, 1698.7, 1704.6, 1705.9); Dalton (1809, 1810).

Spectrum (LIS) on which the computations are based. Since the end of 2013, after Voyager 1 crossed the outer boundary of our solar system, at least four new LIS have been proposed in the literature. The choice of the LIS model leads to differences in the production rate values of the cosmogenic radionuclides of up to 20% (see Herbst et al. 2017). Thus, in order to compare the McCracken and Beer time series of B with the Muscheler et al. ϕ-series, it is necessary to base both records on the same LIS model. Here we use the LIS from Herbst et al. (2017), which updates the LIS model of Potgieter et al. (2014) used by Muscheler et al. (2016), by taking the Voyager 1 measurements (see Stone et al. 2013; Webber and McDonald 2013) into account.

In order to describe the modulation of galactic cosmic ray (GCR) transport in the heliosphere, we use the Force-Field approximation of Gleeson and Axford (1968). The transport can be described as a quasi-steady-state balance between radially inward diffusion through interplanetary magnetic field irregularities with a diffusion coefficient κ, and a radially outward convection with the solar wind speed $\mathrm{v_{sw}}$, depending on only one free parameter, the solar modulation parameter ϕ. As discussed in Caballero-Lopez and Moraal (2004), ϕ can be given by $\phi(\mathrm{r},\ \mathrm{t}) = \int_r^{r_b} (\frac{v_{sw}(r',t)}{3\kappa(r',t)})dr'$. By assuming $\kappa \propto {B_{IMF}}^{-\alpha}$ according to Steinhilber et al. (2010) the interplanetary magnetic field strength can be given by:

$$B_{IMF}(\phi,\mathrm{t}) = \mathrm{B_{IMF,0}}\left(\frac{\phi(\mathrm{t})}{\phi_0}\right)^{\frac{1}{\alpha}} \cdot \left(\frac{v_{sw}(t)}{\mathrm{v_{sw,0}}}\right)^{-\frac{1}{\alpha}}, \tag{3}$$

where $\mathrm{B_{IMF,0}}$, $\mathrm{v_{sw,0}}$, and ϕ_0 represent the mean values of the magnetic field strength, solar wind speed, and solar modulation parameter during the instrumental era. Thereby, the time-dependent solar wind speed $v_{sw}(t)$ can be described by a linear relation between solar wind speed and $\phi : v_{sw}(t) = \mathrm{m}\phi(\mathrm{t}) + \mathrm{q}$, with $\mathrm{m} = 0.23\ \mathrm{km\,s^{-1}\,MV^{-1}}$ and $\mathrm{q} = 303$ km/s. Using this equation, ϕ-reconstructions by different groups can be used to reconstruct the temporal variation of B_{IMF}. Based on OMNI2 data from 1963 to 2008, Steinhilber et al. (2010) used $\phi_0 = 615$ MV, $\mathrm{B_{IMF,0}} = 6.6$ nT, $\mathrm{v_{sw,0}} = 446$ km/s and $\alpha = 1.8 \pm 0.3$ in order to convert the ϕ-reconstructions of Steinhilber et al. (2008) into B. Note, however, that all parameters strongly depend on the time interval over which the measurements were taken, and that in particular ϕ_0 and α also strongly depend on the LIS model on which the ϕ-reconstructions are based. Thus, only B_{IMF} records which include the most recent measurements and which are based on same LIS model should be compared with each other.

We update the investigations by Steinhilber et al. (2010) by (a) including OMNI2 data from 2008 to 2016 and (b) converting ϕ-reconstructions to the LIS model of Herbst et al. (2017), in the following denoted as ϕ_{HE17}, in order to achieve comparability among the existing B_{IMF} records.

The following steps were performed:

1. Based on the measurements from 1963 to 2016 the mean values of the interplanetary magnetic field and the solar wind speed change to $\mathrm{B_{IMF,0}} = 6.4$ nT and $\mathrm{v_{sw,0}} = 438$ km/s, respectively. The mean value of the updated solar modulation parameter is given by $\phi_0 = \phi_{\mathrm{HE17}} = 678$ MV. The exponent α is the only free parameter in the model by Steinhilber et al. (2010). By fitting Eq. (3) to the OMNI2 observations via least square regression we found α to be in the order of 1.25 ± 0.3. The connection between ϕ and B_{IMF} (measurements: filled triangles; computations using Eq. (3): red line) is shown in Fig. 13(a).
2. To validate our results, we compare a B time series derived from a ϕ-reconstruction based on the Oulu Neutron Monitor measurements (see e.g., Usoskin et al. 2011) with a corresponding series based on the ϕ-reconstruction of Muscheler et al. (2016). In order

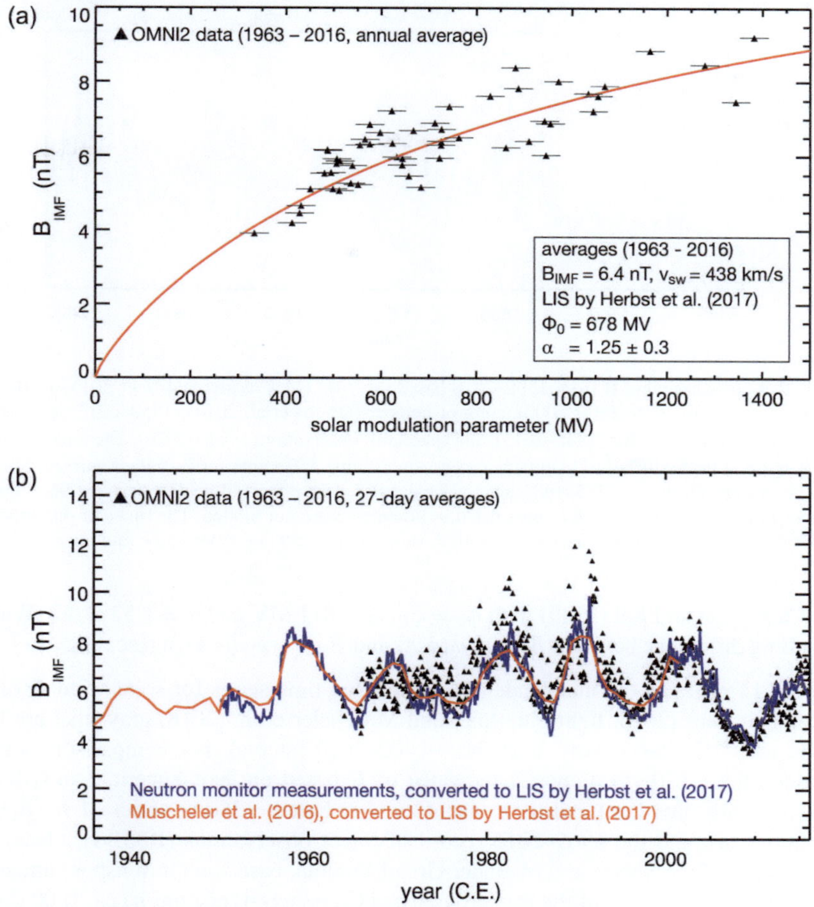

Fig. 13 (**a**) Scatter plot of B_{IMF} (IMF = Interplanetary Magnetic Field) as a function of the solar modulation parameter ϕ for 1963–2016, with red regression line (see Eq. (3)). (**b**) Comparison of the temporal variation of B_{IMF} based on monthly NM data (Usoskin et al. 2011; blue line) and annual ^{14}C data (Muscheler et al. 2016; red line). Black triangles represent 27-day averages of the OMNI2 B_{IMF} data for 1963–2017

to do so, we first converted the two ϕ-records, which depend on different LIS models, to ϕ_{HE17}. The ϕ-record of Usoskin et al. (2011) is based on the LIS model of Usoskin et al. (2005, ϕ_{US05}), while the record of Muscheler et al. (2016) is based on the LIS model of Potgieter et al. (2014, ϕ_{PG14}). To make the conversions, we used the linear regression functions $\phi_{HE17} = 1.025 \cdot \phi_{US05} + 24.18$ MV and $\phi_{HE17} = 1.025 \cdot ((\phi_{PG14} - 28.0\ \text{MV})/1.029) + 24.18$ MV (see Herbst et al. 2017). Based on these new ϕ-values and Eq. (3), a reconstruction of B_{IMF} for both records can be performed. The good agreement between the measured B_{IMF} time series (filled triangles; 27-day averaged OMNI2 data) and those calculated from the Oulu NM data (Usoskin et al. 2011; blue line) and ^{14}C measurements (Muscheler et al. 2016; red line) is displayed in Fig. 13(b).

3. The ^{10}Be based B record of McCracken and Beer (2015) uses the LIS model by Castagnoli and Lal (1980). To scale this record to reconstructions based on the LIS by Herbst et al. (2017), we performed step 1) and 2) for a B_{IMF} reference proxy based on the LIS

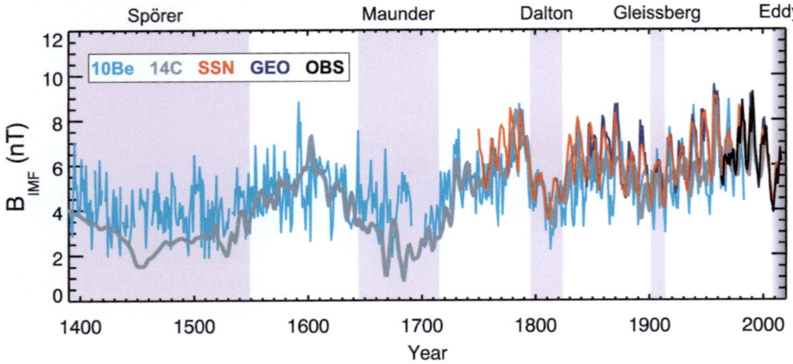

Fig. 14 Plots of solar wind B from 1390–2016 based on ^{10}Be (McCracken and Beer 2015; 1391–1983), ^{14}C (Muscheler et al. 2016; 1390–2002), sunspot number (Owens et al. 2016a; 1750–2013); geomagnetic *IDV* index (Owens et al. 2016a; 1845–2013), and spacecraft observations (1965–2016). The trace for the ^{14}C based *B* series is made thicker to facilitate comparison. For the ^{10}Be time series, values corrected for solar cosmic ray events during the 1775–1983 interval (and also for the effects of nuclear tests in 1962) are used, and upper limit values of 2.5 nT for years listed in Footnote 5 are not plotted. The times of the Spörer and Maunder Grand Minima are indicated along with those of subsequent Gleissberg-type minima

by Castagnoli and Lal (1980) with $\phi_0 = \phi_{CL80} = 611$ MV and $\alpha = 1.52 \pm 0.3$. With the resulting difference between $B_{IMF,CL80}(\phi, \mathrm{t})$ and $B_{IMF,\ HE17}(\phi, \mathrm{t})$, a rescaling is possible.

Figure 14 contains the independently-constructed time series for solar wind *B* of McCracken and Beer (2015; light-blue line) and Muscheler et al. (2016; gray) that are based on ^{10}Be and ^{14}C, respectively, over the ~1400–2000 interval. For comparison, we have also added the ~170-yr composite sequence of *B* based on geomagnetic data (B[GEO], Fig. 11(c); dark blue line) and the sunspot number based ~260-yr series of *B* (B[SSN], Fig. 12(e); red); and the 1965–2016 record of direct observations (B[OBS]; black). The nominal dates of the Spörer and Maunder Grand Minima, based on the sunspot number, are indicated along with those of the less-pronounced Gleissberg-type minima ca. 1800 (Dalton minimum, 1798–1823), 1900 (Gleissberg minimum, 1900–1913), and 2010 (*Eddy* minimum; 2008–?). We have thickened the trace of the ^{14}C based series to increase its visibility. Compared to the other plotted series, its peak values are significantly lower than those of the other series from ~1850–1950. Also noteworthy are the higher values of *B* for the McCracken and Beer (2015) series during the two Grand Minima. The large year-to-year variability of the ^{10}Be based series reflects in part the ~20% uncertainty for any given year (Owens et al. 2016b). As McCracken and Beer (2015) point out, increasing the number of long-term ice cores from two to five, with two of the new cores coming from the Antarctic, will result in a considerable reduction of noise in the ^{10}Be-based record for *B*. The more gradual year-to-year variation of ^{14}C-based *B* reflects the extended and complex path from creation to sequestration of ^{14}C in tree rings vs. the more rapid deposition of ^{10}Be in ice cores (Beer et al. 2012).

3.5.3 Composite B Series (1391–1983)

In Fig. 15, we have concatenated the following time series for solar wind *B*: direct observations (1965–2016), geomagnetic-based (1845–1964; Owens et al. 2016a); sunspot-based (1749–1844; Owens et al. 2016a); cosmogenic-nuclide based (1391–1748; McCracken and Beer 2015; Muscheler et al. 2016). This figure represents the state of the art in our knowledge of the last ~600 years of solar wind *B*. Uncertainties increase for each link of the

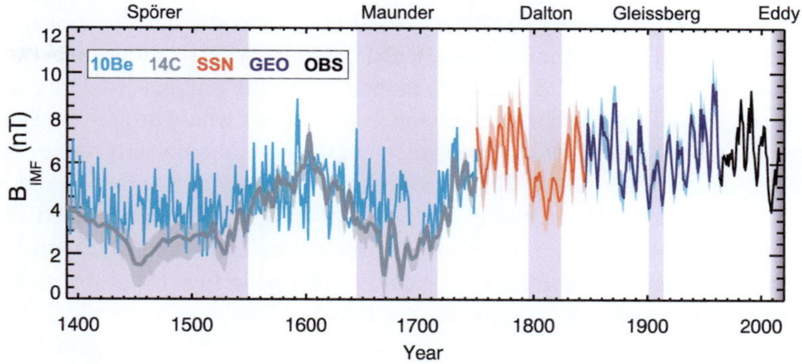

Fig. 15 Composite plot of solar wind B, with uncertainties, from 1390–2016 based on cosmogenic ^{10}Be (McCracken and Beer 2015; 1391–1748), ^{14}C (Muscheler et al. 2016; 1390–1748), sunspot number (Owens et al. 2016a; 1749–1844), geomagnetic data (Owens et al. 2016a; 1845–1964), and from observations in space near-Earth (1965–2016). Upper limit values of 2.5 nT in the McCracken and Beer (2015) for ~20 years (total) during the Spörer and Maunder Minima are not plotted (see Footnote 5). The times of the Spörer, Maunder, Dalton, Gleissberg, and (proposed) Eddy minima are indicated

chain as one goes back in time—from directly observed B, through time series based on geomagnetic data, sunspot number, and cosmogenic nuclides, in turn.

4 Recent Developments on the Maunder Minimum and the Floor in Solar Wind *B*

The low inferred values of sunspot activity and solar wind *B* during the Maunder Minimum mark it as a particularly important epoch, accounting for much of the observed/inferred range of long-term solar activity since ~1400 (Fig. 15). Recently, Zolotova and Ponyavin (2015, 2016) have argued that the Maunder Minimum may not be as grand as currently thought. They suggested that 11-yr peak annual averages of daily group counts for this period might range from ~3–8 in contrast to the values of <0.2 obtained by Hoyt and Schatten (1998a, 1998b) for the core years from 1645–1700. In other words, the Maunder Minimum might have looked more like a typical Gleissberg minimum, then a "Grand" minimum. The thrust of their argument was that, in general, the scientists who reported spots were not interested in tabulating their numbers but focused instead on round spots as possible evidence of intra-Mercurial planets (which partially motivated Schwabe's observations), or large spots to provide insight on the nature of spots, or reported only haphazardly, e.g., as an occasional by-product of meridian transit observations. They also raised the issue of the inhibiting effect of contemporary theological support for Aristotelian views of an immaculate Sun. These arguments were rebutted by Usoskin et al. (2015) with a variety of data including "[...] the fraction of sunspot active days, the latitudinal extent of sunspot positions, auroral sightings at high latitudes, cosmogenic radionuclide data [including ^{44}Ti from meteorites for which interpretations of measurements do not need to take atmospheric transport and deposition processes into account as is the case for ^{10}Be and ^{14}C] as well as solar eclipse observations [...]" in addition to considerations of "[...]peculiar features of the Sun (very strong hemispheric asymmetry of sunspot location, unusual differential rotation and the lack of the K-corona) that imply a special mode of solar activity during the Maunder minimum."

The low end of the range of inferred *B* values during the Maunder Minimum is of interest in relation to the putative floor in solar wind *B* of Svalgaard and Cliver (2007a). Cliver and

Ling (2011) used "(i) a relationship between solar polar-field strength and yearly averages of B for the last four 11-year minima (B_{MIN}), and (ii) a precursor relationship between peak sunspot number for cycles 14–23 and B_{MIN} at their preceding minima" to derive a reduced floor value of ~2.8 nT to which the solar wind in the ecliptic would drop if the Sun's 11-yr cycle ceased. This value is in reasonable agreement with minimum yearly values in Fig. 15 from McCracken and Beer (2014, 2015) for both the Maunder and Spörer Grand Minima, but it lies above those of Muscheler et al. (2016) as well as recently modeled values for B during the Maunder Minimum. Even for the McCracken and Beer (2015) series, it must be noted that there are concentrations of years (not plotted in Fig. 15) at the depths of the Maunder and Spörer minima with upper limit values of 2.5 nT.

In an output from the ISSI team meetings, Riley et al. (2015) used a global thermodynamic MHD model to deduce that during the deepest part of the Maunder Minimum, the corona was "devoid of any large-scale structure, driven by a photospheric field composed of only ephemeral regions, and likely substantially reduced in strength." Model-based estimates of solar wind Br for the two most extreme scenarios considered, viz., no solar polar field at sunspot minimum (absence of the 11-yr cycle) combined with a photosphere consisting of only small-scale ephemeral flux regions having "parasitic" magnetic polarities with average strength of either ± 10 G or ± 3.3 G, are 0.29 nT and 0.08 nT, respectively. For an assumed solar wind speed of ~300 km s^{-1} (Cliver and Ling 2011; Cliver and von Steiger 2017; Owens et al. 2017), and using the formula in Fig. 5 of Cliver and Ling (2011) or the relationship in Fig. 11 of Lockwood and Owens (2011), these Br values translate into B values between ~0.5 to 2.0 nT. Modeled 11-yr minimum B values during the Maunder Minimum of ~1.5 nT and ~1.0 nT by Lockwood and Owens (2014; based on the *OSF* continuity equation of Owens and Lockwood 2012) and Owens et al. (2017; based on Owens and Lockwood 2012, and the three-dimensional magnetohydrodynamic model of the solar corona of Linker et al. 1999), respectively, fall within this range of values. These time series for B show the continuation of the cycle during the Maunder Minimum deduced by Beer et al. (1998) and Berggren et al. (2009), however, so a true floor level presumably would be lower since the cycle requires the presence of solar polar fields at 11-yr minima that will add to solar wind B at these times (Svalgaard et al. 2005). In fact, in another recent determination of B for the Maunder Minimum that begins with a continuity equation for the open solar flux, Rahmanifard et al. (2017), determined that any floor in solar wind B during this period could not be $\gtrsim 1.5$ nT and could be as low zero nT.

Observationally, we note the absence in Fig. 14 of the downward 40-yr spike to ~0 nT in B ca. 1500 seen in Fig. 9 for the Spörer Minimum, along with similar downward spikes in the modulation potential to values less than zero for both the Maunder and Spörer minima in Fig. S13 of Steinhilber et al. (2012).

5 Conclusions

5.1 Summary

We reviewed the evolution of thinking on time series for the sunspot number and solar wind B. The first of these has a research history going back over 400 years, the second of only ~50 years. Both fields have undergone intense development over the past two decades.

Key developments in sunspot research during this time include the development of the group sunspot number (GSN) by Hoyt and Schatten (1998a, 1998b), the sunspot number workshops of 2011–2015 (Cliver et al. 2013a, 2015; Clette et al. 2014), and independent work on sunspot number time series (e.g., Lockwood et al. 2014c, 2014d, 2016d, 2016e; Usoskin et al. 2016; Willamo et al. 2017; Chatzistergos et al. 2017) that they triggered. Recent progress includes: (1) uncovering of inhomogeneities in both WSN (e.g., Svalgaard 2010, 2012; Leussu et al. 2013; Clette et al. 2014; Clette and Lefèvre 2016) and GSN (Clette et al. 2014; Cliver and Ling 2016; Cliver 2017), (2) a new group count data base (Vaquero et al. 2016), and (3) the introduction of novel reconstruction methods (e.g., Usoskin et al. 2016). This story is ongoing, with work in progress on reconciling, in so far as possible given the difference in the base formulae for the two series (Eqs. (1) and (2)), the recent proliferation of sunspot number time series. This effort will lead to new versions of both the Wolf and group sunspot number.

The key advances on construction of a long-term solar wind B time series include the seminal paper by Lockwood et al. (1999) that re-energized the field of space climate, the development of the long-term interdiurnal variability index of geomagnetic index by Svalgaard and colleagues (Svalgaard et al. 2003; Svalgaard and Cliver 2005), the ISSI International Team (2012–2013) on long-term solar and solar wind reconstruction (Owens et al. 2016a, 2016b), and the development of long-term high-time resolution series by McCracken and Beer (2015; from 1391–1983 for solar wind B) and Muscheler et al. (2016; from 0–2000 for modulation potential Φ), based on cosmogenic nuclides. For the solar wind B time series, we gave a detailed account of the path from Lockwood et al. (1999) to the present, as a case study of what one scientist memorably called "the truly gritty, uncertain and fun nature of scientific research" (Cliver and van Driel-Gesztelyi 2010).

Despite initial resistance, the work of both Lockwood and Svalgaard found relatively rapid acceptance. The Lockwood et al. (1999) report that the underlying open solar flux (the minimum-to-minimum variation) could change by a factor of two over 100 years was shown to be conservative as a comparison of directly-observed annual Br values at cycle minima in 1986 (2.42 nT) and 2009 (1.25 nT) (Lockwood et al. 2009a; Cliver and Ling 2011), revealed a "reverse direction rise" of $\sim$95% (or $\sim$52% fall) over two solar cycles. Whether such changes in the *OSF* as that from 1986–2009 or the approximate doubling of the OSF during the first half of the 20th century represent random secular variations or are manifestations of longer cyclic (e.g., Gleissberg-type) behavior remains to be determined. Svalgaard's development of the IDV index from Bartels' u index, and his discovery of its relationship to near-Earth *IMF* (Svalgaard et al. 2003), became the basis for the reference geomagnetic-based long-term time series of solar wind B (Owens et al. 2016a).

The Maunder Minimum remains a lively field of research on both solar and solar wind activity. In regard to the Sun, the key question involves the level/duration of low sunspot activity: Was the Maunder a Grand Minimum (Usoskin et al. 2015), or more in the run of a Gleissberg-type centennial variation (Zolotova and Ponyavin 2015, 2016), or something in between? For the solar wind, a prominent question involves the existence of a floor, with proposed levels dropping from the original $\sim$4.6 nT value of Svalgaard and Cliver (2007a) to the possibility of no floor recently modeled by Rahmanifard et al. (2017).

5.2 Lessons Learned

The efforts to obtain high-confidence long-term time series for solar wind B (Sect. 2) and the sunspot number (Sect. 1) suggest the following guidelines for the ongoing work on the sunspot number:

(1) *Nullius in verba* (take nobody's word for it), the motto of the Royal Society, applies. This is a scientific truism but it was violated in the case of the Hoyt and Schatten (1998a, 1998b) group sunspot number, which was not independently vetted and found wanting (Cliver and Ling 2016; Cliver 2017) until nearly two decades after its introduction. The

lesson of the need for scrutiny is now being applied to all six of the newly modified or conceived sunspot number time series.

(2) For widely-used time series, the entire community must be involved. This was not the case for the sunspot number workshops and it only became so after the publication of the new time series by SILSO. If there is a general lesson here, it is the need to truly work, i.e., to publish, together. That way, all have ownership of the process, and the result.

(3) Reaching consensus takes time, approximately 15 years were required to reach the present state for solar wind B, and six so far for the sunspot number, with an estimate of two-to-three more needed for completion.

(4) There is always room for improvement. Error bars can be reduced as evidence accumulates. Versioning is a healthy and humbling concept. Continuous monitoring is required to preserve homogeneity. Formulate a prescribed mechanism for implementing and sanctioning updates. All of these precepts and practices apply for the nascent 3.0 versions of the S_N and GSN time series. The value of a single accepted time series, be it solar wind B or sunspot number, is akin to that of an established paradigm for a field in that it provides a framework against which new ideas and data can be measured.

(5) While ancillary or proxy data such as geomagnetic based solar B can be used to corroborate and guide the reconciliation of the various new sunspot number time series, it should be used in their actual derivation solely as a last resort, i.e., if there is no other way through a gap in the sunspot record. Before 1610, the proxy cosmogenic radionuclide data is the only proven way to reliably extend the solar activity record. A key end goal of the current sunspot recalibration effort is to provide guidelines and constraints for a reliable millennial-scale record of solar activity based on the paleo-cosmic-ray record.

Acknowledgements E.W.C. thanks Daniel Baker, André Balogh, Tamás Gombosi, Hannu Koskinen, Rudolf von Steiger, and Astrid Veronig for organizing the ISSI Workshop on The Scientific Foundation of Space Weather and the invitation to speak on long-term solar variability. He is grateful to ISSI for supporting both this Workshop and, previously, the Svalgaard, Lockwood, and Beer International Team on Long-term reconstruction of Solar and Solar Wind Parameters. He thanks Leif Svalgaard for the translation of the Wolf (1856) article. K.H. thanks Jürg Beer, Ken McCracken, Bernd Heber, and Raimund Muscheler for their support and many insightful discussions over the past years.

References

K. Alanko-Huotari, K. Mursula, I. Usoskin, G. Kovaltsov, Global heliospheric parameters and cosmic-ray modulation: an empirical relation for the last decades. Sol. Phys. **238**, 391 (2006). https://doi.org/10.1007/s11207-006-0233-z

C.N. Arge, E. Hildner, V.J. Pizzo, J.W. Harvey, Two solar cycles of nonincreasing magnetic flux. J. Geophys. Res. **107**, 1319 (2002). https://doi.org/10.1029/2001JA000503

A. Balogh, E.J. Smith, B.T. Tsurutani, D.J. Southwood, R.J. Forsyth, T.S. Horbury, The heliospheric magnetic field over the South polar region of the Sun. Science **286**, 1007 (1995). https://doi.org/10.1126/science.268.5213.1007

E. Bard, G. Raisbeck, F. Yiou, J. Jouzel, Solar irradiance during the last 1200 years based on cosmogenic nuclides. Tellus B **52**, 985 (2000). https://doi.org/10.1034/j.1600-0889.2000.d01-7.x

J. Bartels, Terrestrial-magnetic activity and its relations to solar phenomena. Terr. Magn. Atmos. Electr. **37**, 1 (1932). https://doi.org/10.1029/TE037i001p00001

J. Bartels, Solar eruptions and their ionospheric effects—a classical observation and its new interpretation. Terr. Magn. Atmos. Electr. **42**, 235 (1937). https://doi.org/10.1029/TE042i003p00235

J. Beer, S. Tobias, N. Weiss, An active Sun throughout the Maunder Minimum. Sol. Phys. **181**, 237 (1998). https://doi.org/10.1023/A:1005026001784

J. Beer, K.G. McCracken, R. von Steiger, *Cosmogenic Radionuclides: Theory and Applications in the Terrestrial and Space Environments* (Springer, Berlin, 2012). p. 54 ff., p. 227 ff., p. 267 ff.

G. Bonino, G. Cini Castagnoli, N. Bhandari, C. Taricco, Behavior of the heliosphere over prolonged solar quiet periods by ^{44}Ti measurements in meteorites. Science **270**, 1648 (1995). https://doi.org/10.1126/science.270.5242.1648

A.-M. Berggren, J. Beer, G. Possnert, A. Aldahan, P. Kubik, M. Christl, S.J. Johnsen et al., A 600-year annual ^{10}Be record from the NGRIP ice core, Greenland. Geophys. Res. Lett. **36**, L11801 (2009). https://doi.org/10.1029/2009GL038004

R.A. Caballero-Lopez, H. Moraal, Limitations of the force field equation to describe cosmic ray modulation. J. Geophys. Res. **109**, A01101 (2004). https://doi.org/10.1029/2003JA010098

R.C. Carrington, On the distribution of the solar spots in latitudes since the beginning of the year 1854, with a map. Mon. Not. R. Astron. Soc. **19**, 1 (1858). https://doi.org/10.1093/mnras/19.1.1

R.C. Carrington, On certain phenomena in the motions of solar spots. Mon. Not. R. Astron. Soc. **19**, 81 (1859a). https://doi.org/10.1093/mnras/19.3.81

R.C. Carrington, Description of a singular appearance seen in the Sun on September 1, 1859. Mon. Not. R. Astron. Soc. **20**, 13 (1859b). https://doi.org/10.1093/mnras/20.1.13

R.C. Carrington, *Observations of the Spots on the Sun from November 9, 1853 to March 24, 1861* (Williams and Norgate, London, 1863)

G. Castagnoli, D. Lal, Solar modulation effects in terrestrial production of carbon-14. Radiocarbon **22**, 133 (1980)

T. Chatzistergos, I.G. Usoskin, G.A. Kovaltsov, N.A. Krivova, S.K. Solanki, New reconstruction of the sunspot group number since 1739 using the direct calibration and 'backbone' methods. Astron. Astrophys. **602**, A69 (2017). https://doi.org/10.1051/0004-6361/201630045

A.M. Clerke, *A Popular History of Astronomy During the Nineteenth Century*, 4th edn. (Adam and Charles Black, London, 1902), p. 125

F. Clette, L. Lefèvre, The new sunspot number: assembling all corrections. Sol. Phys. **291**, 2629 (2016). https://doi.org/10.1007/s11207-016-1014-y

F. Clette, D. Berghmans, P. Vanlommel, R.A.M. Van der Linden, A. Koeckelenbergh, L. Wauters, From the Wolf number to the International Sunspot Index: 25 years of SIDC. Adv. Space Res. **40**, 919 (2007). https://doi.org/10.1016/j.asr.2006.12.045

F. Clette, L. Svalgaard, J.M. Vaquero, E.W. Cliver, Revisiting the sunspot number: a 400-year perspective on the solar cycle. Space Sci. Rev. **186**, 35 (2014). https://doi.org/10.1007/s11214-014-0074-2

F. Clette, E.W. Cliver, L. Lefèvre, L. Svalgaard, J.M. Vaquero, Revision of the sunspot number(s). Space Weather **13**, 529 (2015). https://doi.org/10.1002/2015SW001264

F. Clette, E.W. Cliver, L. Lefèvre, L. Svalgaard, J.M. Vaquero, J.W. Leibacher, Preface to topical issue: recalibration of the sunspot number. Sol. Phys. **291**, 2479 (2016a). https://doi.org/10.1007/s11207-016-1017-8

F. Clette, L. Lefèvre, M. Cagnotti, S. Cortesi, A. Bulling, The revised Brussels–Locarno sunspot number (1981–2015). Sol. Phys. **291**, 2733 (2016b). https://doi.org/10.1007/s11207-016-0875-4

E.W. Cliver, Comparison of new and old sunspot number time series. Sol. Phys. **291**, 2891 (2016). https://doi.org/10.1007/s11207-016-0929-7

E.W. Cliver, Sunspot number recalibration: the $\sim$1840–1920 anomaly in the observer normalization factors of the group sunspot number. J. Space Weather Space Clim. **7**, A12 (2017). https://doi.org/10.1051/swsc/2017010

E.W. Cliver, A.G. Ling, Secular change in geomagnetic indices and the solar open magnetic flux during the first half of the twentieth century. J. Geophys. Res. **107**, 1303 (2002). https://doi.org/10.1029/2001JA000505

E.W. Cliver, A.G. Ling, The floor in the solar wind magnetic field revisited. Sol. Phys. **274**, 285 (2011). https://doi.org/10.1007/s11207-010-9657-6

E.W. Cliver, A.G. Ling, The discontinuity circa 1885 in the group sunspot number. Sol. Phys. **291**, 2763 (2016). https://doi.org/10.1007/s11207-015-0841-6

E. Cliver, L. van Driel-Gesztelyi, Solar physics memoir series reinstituted. Sol. Phys. **267**, 233 (2010). https://doi.org/10.1007/s11207-010-9667-4

E.W. Cliver, R. von Steiger, Minimal magnetic states of the Sun and the solar wind: implications for the origin of the slow solar wind. Space Sci. Rev. **210**, 227 (2017). https://doi.org/10.1007/s11214-015-0224-1

E.W. Cliver, F. Clette, L. Svalgaard, Recalibrating the sunspot number (SSN): the SSN workshops. Cent. Eur. Astrophys. Bull. **37**, 401 (2013a)

E.W. Cliver, I.G. Richardson, A.G. Ling, Solar drivers of 11-yr and long-term cosmic ray modulation. Space Sci. Rev. **176**, 3 (2013b). https://doi.org/10.1007/s11214-011-9746-3

E.W. Cliver, F. Clette, L. Svalgaard, J.M. Vaquero, Recalibrating the sunspot number (SN): the 3rd and 4th SN workshops. Cent. Eur. Astrophys. Bull. **39**, 1 (2015)

M.A. Clilverd, E. Clarke, T. Ulich, J. Linthe, H. Rishbeth, Reconstructing the long-term aa index. J. Geophys. Res. **110**, A07205 (2005). https://doi.org/10.1029/2004JA010762

J.B.J. Delambre, *Astronomie Théorique et Pratique*, vol. 3 (1814), p. 20
T. Dudok de Wit, L. Lefèvre, F. Clette, Uncertainties in the sunspot numbers: estimation and implications. Sol. Phys. **291**, 2709 (2016). https://doi.org/10.1007/s11207-016-0970-6
E.H. Erwin, H.E. Coffey, W.F. Denig, D.M. Willis, R. Henwood, M.N. Wild, The Greenwich photo-heliographic results (1874–1976): initial corrections to the printed publications. Sol. Phys. **288**, 157 (2013). https://doi.org/10.1007/s11207-013-0310-z
J.A. Eddy, The Maunder Minimum. Science **192**, 1189 (1976). https://doi.org/10.1126/science.192.4245.1189
J. Feynman, N.U. Crooker, The solar wind at the turn of the century. Nature **275**, 626 (1978). https://doi.org/10.1038/275626a0
J. Feynman, A. Ruzmaikin, The Sun's strange behavior: Maunder Minimum or Gleissberg cycle? Sol. Phys. **272**, 351 (2011). https://doi.org/10.1007/s11207-011-9828-0
P.V. Foukal, *Solar Astrophysics* (Wiley-VCH, Weinheim, 2004), p. 365
T.K. Friedli, The construction of the Wolf series from 1749 to 1980. Solar Phys. (submitted, 2016)
A. Gautier, Relation entre les taches du Soleil et les phénomènes magnétiques. Arch. Sci. **21**, 194 (1852)
L.J. Gleeson, W.I. Axford, Solar modulation of galactic cosmic rays. Astrophys. J. **154**, 1011 (1968). https://doi.org/10.1086/149822
W. Gleissberg, A long-periodic fluctuation of the sun-spot numbers. Observatory **62**, 158 (1939)
D.H. Hathaway, The solar cycle. Living Rev. Sol. Phys. **12**, 4 (2015). https://doi.org/10.1007/lrsp-2015-4
K. Herbst, A. Kopp, B. Heber, F. Steinhilber, H. Fichtner, K. Scherer, D. Matthiä, On the importance of the local interstellar spectrum for the solar modulation parameter. J. Geophys. Res. **115**, D00I20 (2010). https://doi.org/10.1029/2009JD012557
K. Herbst, R. Muscheler, B. Heber, The new local interstellar spectra and their influence on the production rates of the cosmogenic radionuclides ^{10}Be and ^{14}C. J. Geophys. Res. **122**, 23 (2017). https://doi.org/10.1002/2016JA023207
R. Hodgson, On a curious appearance seen in the Sun. Mon. Not. R. Astron. Soc. **20**, 15 (1859). https://doi.org/10.1093/mnras/20.1.15
D.V. Hoyt, K.H. Schatten, Group sunspot numbers: a new solar activity reconstruction. Sol. Phys. **179**, 189 (1998a). https://doi.org/10.1023/A:1005007527816
D.V. Hoyt, K.H. Schatten, Group sunspot numbers: a new solar activity reconstruction. Sol. Phys. **181**, 491 (1998b). https://doi.org/10.1023/A:1005056326158
D. Hoyt, K.H. Schatten, E. Nesme-Ribes, The one hundredth year of Rudolf Wolf's death: do we have the correct reconstruction of solar activity? Geophys. Res. Lett. **21**, 2067 (1994). https://doi.org/10.1029/94GL01698
K. Hufbauer, *Exploring the Sun: Solar Science Since Galileo* (Johns Hopkins University Press, Baltimore, 1991), p. 46
M.J. Jarvis, Observed tidal variation in the lower thermosphere through the 20th century and the possible implication of ozone depletion. J. Geophys. Res. **110**, A04303 (2005). https://doi.org/10.1029/2004JA010921
G.A. Kovaltsov, I.G. Usoskin, A new 3D numerical model of cosmogenic nuclide ^{10}Be production in the atmosphere. Earth Planet. Sci. Lett. **291**, 182 (2010). https://doi.org/10.1016/j.epsl.2010.01.011
J.-L. Le Mouël, F. Lopes, V. Courtillot, Identification of Gleissberg cycles and a rising trend in a 315-year-long series of sunspot numbers. Sol. Phys. **292**, #43 (2017)
R. Leussu, I.G. Usoskin, R. Arlt Rainer, K. Mursula, Inconsistency of the Wolf sunspot number series around 1848. Astron. Astrophys. **559**, A28 (2013). https://doi.org/10.1051/0004-6361/201322373
J.A. Linker, Z. Mikić, D.A. Biesecker, R.J. Forsyth, S.E. Gibson, A.J. Lazarus, A. Lecinski et al., Magnetohydrodynamic modeling of the solar corona during Whole Sun Month. J. Geophys. Res. **104**, 9809 (1999). https://doi.org/10.1029/1998JA900159
M. Lockwood, M. Owens, The accuracy of using the Ulysses result of the spatial invariance of the radial heliospheric field to compute the open solar flux. Astrophys. J. **701**, 964 (2009). https://doi.org/10.1088/0004-637X/701/2/964
M. Lockwood, M.J. Owens, Centennial changes in the heliospheric magnetic field and open solar flux: the consensus view from geomagnetic data and cosmogenic isotopes and its implications. J. Geophys. Res. **116**, A04109 (2011). https://doi.org/10.1029/2010JA016220
M. Lockwood, M.J. Owens, Implications of the recent low solar minimum for the solar wind during the Maunder Minimum. Astrophys. J. Lett. **781**, L7 (2014). https://doi.org/10.1088/2041-8205/781/1/L7
M. Lockwood, R. Stamper, M.N. Wild, A doubling of the Sun's coronal magnetic field during the past 100 years. Nature **399**, 437 (1999). https://doi.org/10.1038/20867
M. Lockwood, R.B. Forsyth, A. Balogh, D.J. Mcomas, Open solar flux estimates from near-Earth measurements of the interplanetary magnetic field: comparison of the first two perihelion passes of the Ulysses spacecraft. Ann. Geophys. **22**, 1395 (2004). https://doi.org/10.5194/angeo-22-1395-2004

M. Lockwood, A.P. Rouillard, I. Finch, R. Stamper, Comment on "the IDV index: its derivation and use in inferring long-term variations of the interplanetary magnetic field strength" by Leif Svalgaard and Edward W. Cliver. J. Geophys. Res. **111**, A09109 (2006). https://doi.org/10.1029/2006JA011640

M. Lockwood, A.P. Rouillard, I.D. Finch, The rise and fall of open solar flux during the current grand solar maximum. Astrophys. J. **700**, 937 (2009a). https://doi.org/10.1088/0004-637X/700/2/937

M. Lockwood, M. Owens, A.P. Rouillard, Excess open solar magnetic flux from satellite data: 2. A survey of kinematic effects. J. Geophys. Res. **114**, A11104 (2009b). https://doi.org/10.1029/2009JA014450

M. Lockwood, L. Barnard, H. Nevanlinna, M.J. Owens, R.G. Harrison, A.P. Rouillard, C.J. Davis, Reconstruction of geomagnetic activity and near-Earth interplanetary conditions over the past 167 yr—Part 1: A new geomagnetic data composite. Ann. Geophys. **31**, 1957 (2013a). https://doi.org/10.5194/angeo-31-1957-2013

M. Lockwood, L. Barnard, H. Nevanlinna, M.J. Owens, R.G. Harrison, A.P. Rouillard, C.J. Davis, Reconstruction of geomagnetic activity and near-Earth interplanetary conditions over the past 167 yr—Part 2: A new reconstruction of the interplanetary magnetic field. Ann. Geophys. **31**, 1979 (2013b). https://doi.org/10.5194/angeo-31-1979-2013

M. Lockwood, H. Nevanlinna, M. Vokhmyanin, D. Ponyavin, S. Sokolov, L. Barnard, M. Owens, R. Harrison, A. Rouillard, C. Scott, Reconstruction of geomagnetic activity and near-Earth interplanetary conditions over the past 167 yr—Part 3: Improved representation of solar cycle 11. Ann. Geophys. **32**, 367 (2014a). https://doi.org/10.5194/angeo-32-367-2014

M. Lockwood, H. Nevanlinna, L. Barnard, M.J. Owens, R.G. Harrison, A.P. Rouillard, C.J. Scott, Reconstruction of geomagnetic activity and near-Earth interplanetary conditions over the past 167 yr—Part 4: Near-Earth solar wind speed, IMF, and open solar flux. Ann. Geophys. **32**, 383 (2014b). https://doi.org/10.5194/angeo-32-383-2014

M. Lockwood, M.J. Owens, L. Barnard, Centennial variations in sunspot number, open solar flux, and streamer belt width: 1. Correction of the sunspot number record since 1874. J. Geophys. Res. **119**, 5172 (2014c). https://doi.org/10.1002/2014JA019970

M. Lockwood, M.J. Owens, L. Barnard, Centennial variations in sunspot number, open solar flux, and streamer belt width: 2. Comparison with the geomagnetic data. J. Geophys. Res. **119**, 5183 (2014d). https://doi.org/10.1002/2014JA019972

M. Lockwood, C.J. Scott, M.J. Owens, L. Barnard, D.M. Willis, Tests of sunspot number sequences: 1. Using ionosonde data. Sol. Phys. **291**, 2785 (2016a). https://doi.org/10.1007/s11207-016-0855-8

M. Lockwood, M.J. Owens, L. Barnard, C.J. Scott, I.G. Usoskin, H. Nevanlinna, Tests of sunspot number sequences: 2. Using geomagnetic and auroral data. Sol. Phys. **291**, 2811 (2016b). https://doi.org/10.1007/s11207-016-0913-2

M. Lockwood, M.J. Owens, L. Barnard, I.G. Usoskin, Tests of sunspot number sequences: 3. Effects of regression procedures on the calibration of historic sunspot data. Sol. Phys. **291**, 2829 (2016c). https://doi.org/10.1007/s11207-015-0829-2

M. Lockwood, M.J. Owens, L. Barnard, Tests of sunspot number sequences: 4. Discontinuities around 1946 in various sunspot number and sunspot-group-number reconstructions. Sol. Phys. **291**, 2843 (2016d). https://doi.org/10.1007/s11207-016-0967-1

M. Lockwood, M.J. Owens, L. Barnard, I.G. Usoskin, An assessment of sunspot number data composites over 1845–2014. Astrophys. J. **824**, 54 (2016e). https://doi.org/10.3847/0004-637X/824/1/54

M. Lockwood, M. Owens, L.A. Barnard, C.J. Scott, C.E. Watt, Space Climate and Space Weather over the past 400 years: 1. The power input to the magnetosphere. J. Space Weather Space Clim. **7**, A25 (2017). https://doi.org/10.1051/swsc/2017019

E. Loomis, Comparison of the mean daily range of the magnetic declination and the number of auroras observed each year, with the extent of black spots on the Sun. Am. J. Sci. Arts **105**, 245 (1873)

J.J. Love, V.C. Tsai, J.L. Gannon, Averaging and sampling for magnetic-observatory hourly data. Ann. Geophys. **28**, 2079 (2010). https://doi.org/10.5194/angeo-28-2079-2010

S. Macmillan, E. Clarke, Resolving issues concerning Eskdalemuir geomagnetic hourly values. Ann. Geophys. **29**, 283 (2011). https://doi.org/10.5194/angeo-29-283-2011

D. Martini, K. Mursula, Correcting the geomagnetic IHV index of the Eskdalemuir observatory. Ann. Geophys. **24**, 3411 (2006). https://doi.org/10.5194/angeo-24-3411-2006

E.W. Maunder, Professor Spoerer's researches on sun-spots. Mon. Not. R. Astron. Soc. **50**, 251 (1890)

E.W. Maunder, A prolonged sunspot minimum. Knowledge **17**, 173 (1894)

E.W. Maunder, The prolonged sunspot minimum. J. Br. Astron. Assoc. **32**, 140 (1922)

P.-N. Mayaud, The aa indices: a 100-year series characterizing the magnetic activity. J. Geophys. Res. **77**, 6870 (1972). https://doi.org/10.1029/JA077i034p06870

P.-N. Mayaud, *Derivation, Meaning, and Use of Geomagnetic Indices* (American Geophysical Union, Washington, 1980)

K.G. McCracken, Heliomagnetic field near Earth, 1428–2005. J. Geophys. Res. **112**, A09106 (2007). https://doi.org/10.1029/2006JA012119

K.G. McCracken, J. Beer, Comparison of the extended solar minimum of 2006–2009 with the Spoerer, Maunder, and Dalton Grand Minima in solar activity in the past. J. Geophys. Res. **119**, 2379 (2014). https://doi.org/10.1002/2013JA019504

K.G. McCracken, J. Beer, The annual cosmic-radiation intensities 1391–2014; the annual heliospheric magnetic field strengths 1391–1983, and identification of solar cosmic-ray events in the cosmogenic record 1800–1983. Sol. Phys. **290**, 3051 (2015). https://doi.org/10.1007/s11207-015-0777-x

K. Mursula, D. Martini, A. Karinen, Did open solar magnetic field increase during the last 100 years? A reanalysis of geomagnetic activity. Sol. Phys. **224**(1–2), 85 (2004). https://doi.org/10.1007/s11207-005-4981-y

R. Muscheler, F. Joos, S.A. Müller, I. Snowball, How unusual is today's solar activity? Nature **436**, E3 (2005). https://doi.org/10.1038/nature04045

R. Muscheler, F. Adolphi, K. Herbst, A. Nilsson, The revised sunspot record in comparison to cosmogenic radionuclide-based solar activity reconstructions. Sol. Phys. **291**, 3025 (2016). https://doi.org/10.1007/s11207-016-0969-z

M. Neugebauer, C.W. Snyder, Solar plasma experiment. Science **138**, 1095 (1962). https://doi.org/10.1126/science.138.3545.1095-a

M.J. Owens, M. Lockwood, Cyclic loss of open solar flux since 1868: the link to heliospheric current sheet tilt and implications for the Maunder Minimum. J. Geophys. Res. **117**, A04102 (2012). https://doi.org/10.1029/2011JA017193

M.J. Owens, E. Cliver, K.G. McCracken, J. Beer, L. Barnard, M. Lockwood, A. Rouillard et al., Near-Earth heliospheric magnetic field intensity since 1750: 1. Sunspot and geomagnetic reconstructions. J. Geophys. Res. **121**, 6048 (2016a). https://doi.org/10.1002/2016JA022529

M.J. Owens, E. Cliver, K.G. McCracken, J. Beer, L. Barnard, M. Lockwood, A. Rouillard et al., Near-Earth heliospheric magnetic field intensity since 1750: 2. Cosmogenic radionuclide reconstructions. J. Geophys. Res. **121**, 6064 (2016b). https://doi.org/10.1002/2016JA022550

M.J. Owens, M. Lockwood, P. Riley, Global solar wind variations over the last four centuries. Nature Sci. Rep. **7**, 41548 (2017)

M.S. Potgieter, E.E. Vos, M. Boezio, N. De Simone, V. Di Felice, V. Formato, Modulation of galactic protons in the heliosphere during the unusual solar minimum of 2006 to 2009. Sol. Phys. **289**, 391–406 (2014). https://doi.org/10.1007/s11207-013-0324-6

F. Rahmanifard, N.A. Schwadron, C.W. Smith, K.G. McCracken, K.A. Duderstadt, N. Lugaz, M.L. Goelzer, Inferring the heliospheric magnetic field back through Maunder Minimum. Astrophys. J. **837**, 165 (2017). https://doi.org/10.3847/1538-4357/aa6191

J.C. Ribes, E. Nesme-Ribes, The solar sunspot cycle in the Maunder Minimum AD1645 to AD1715. Astron. Astrophys. **276**, 549 (1993)

P. Riley, R. Lionello, J.A. Linker, E. Cliver, A. Balogh, J. Beer, P. Charbonneau et al., Inferring the structure of the solar corona and inner heliosphere during the Maunder Minimum using global thermodynamic magnetohydrodynamic simulations. Astrophys. J. **802**, 105 (2015). https://doi.org/10.1088/0004-637X/802/2/105

A.P. Rouillard, M. Lockwood, I. Finch, Centennial changes in the solar wind speed and in the open solar flux. J. Geophys. Res. **112**, A05103 (2007). https://doi.org/10.1029/2006JA012130

C.T. Russell, On the possibility of deducing interplanetary and solar parameters from geomagnetic records. Sol. Phys. **42**, 259 (1975). https://doi.org/10.1007/BF00153301

E. Sabine, On periodical laws discoverable in the mean effects of the larger magnetic disturbances. Philos. Trans. R. Soc. Lond. **142**, 103 (1852)

H.H. Sargent III, The 27-day recurrence index, in *Solar wind–magnetosphere coupling*, ed. by Y. Kamide, J.A. Slavin (Dordrecht, Boston, 1986), p. 143. https://doi.org/10.1007/978-90-277-2303-1_11

S.H. Schwabe, Sonnenbeobachtungen im Jahre 1843. Von Herrn Hofrath Schwabe in Dessau. Astron. Nachr. **21**, 233 (1844)

E.J. Smith, A. Balogh, Ulysses observations of the radial magnetic field. Geophys. Res. Lett. **22**, 3317 (1995). https://doi.org/10.1029/95GL02826

C.W. Snyder, M. Neugebauer, U.R. Rao, The solar wind velocity and its correlation with cosmic-ray variations and with solar and geomagnetic activity. J. Geophys. Res. **68**, 6361 (1963). https://doi.org/10.1029/JZ068i024p06361

S.K. Solanki, M. Schüssler, M. Fligge, Evolution of the Sun's large-scale magnetic field since the Maunder Minimum. Nature **408**, 445 (2000). https://doi.org/10.1038/35044027

S.K. Solanki, M. Schüssler, M. Fligge, Secular variation of the Sun's magnetic flux. Astron. Astrophys. **383**, 706 (2002). https://doi.org/10.1051/0004-6361:20011790

S.K. Solanki, I.G. Usoskin, B. Kromer, M. Schüssler, J. Beer, Unusual activity of the Sun during recent decades compared to the previous 11,000 years. Nature **431**, 1084 (2004). https://doi.org/10.1038/nature02995

S.K. Solanki, I.G. Usoskin, B. Kromer, M. Schüssler, J. Beer, Climate: how unusual is today's solar activity? (reply). Nature **436**, 4 (2005). https://doi.org/10.1038/nature04046

F.W.G. Spörer, On the periodicity of sunspots since the year 1618, especially with respect to the heliographic latitude of the same, and reference to a significant disturbance of this periodicity during a long period. Vierteljahrsschr. Astron. Ges. (Leipzig) **22**, 323 (1887)

F.W.G. Spörer, Sur les différences que présentent l'hémisphère sud du Soliel. Bull. Astron. **6**, 60 (1889)

F. Steinhilber, J.A. Abreu, J. Beer, Solar modulation during the Holocene. Astrophys. Space Sci. Trans. **4**, 1 (2008). https://doi.org/10.5194/astra-4-1-2008

B. Stewart, On the great magnetic disturbance which extended from August 28 to September 7, 1859, as recorded by photography at the Kew observatory. Philos. Trans. R. Soc. Lond. **151**, 423 (1861)

F. Steinhilber, J.A. Abreu, J. Beer, K.G. McCracken, Interplanetary magnetic field during the past 9300 years inferred from cosmogenic radionuclides. J. Geophys. Res. **115**, A01104 (2010). https://doi.org/10.1029/2009JA014193

F. Steinhilber, J.A. Abreu, J. Beer, I. Brunner, M. Christl, H. Fischer, U. Heikkila et al., 9,400 years of cosmic radiation and solar activity from ice cores and tree rings. Proc. Natl. Acad. Sci. **109**, 5967 (2012). https://doi.org/10.1073/pnas.1118965109

J.O. Stenflo, Transition of the sunspot number from Zurich to Brussels in 1980: a personal perspective. Sol. Phys. **291**, 2487 (2016). https://doi.org/10.1007/s11207-015-0837-2

E.C. Stone, A.C. Cummings, F.B. McDonald, B.C. Heikkila, N. Lal, W.R. Webber, Voyager 1 observes low-energy galactic cosmic rays in a region depleted of heliospheric ions. Science **341**, 150 (2013). https://doi.org/10.1126/science.1236408

L. Svalgaard, Geomagnetic activity: dependence on solar wind parameters, in *Coronal Holes and High Speed Wind Streams*, ed. by J.B. Zirker (Colorado Associated University Press, Boulder, 1977), p. 371

L. Svalgaard, Updating the historical sunspot record, in *Understanding a Peculiar Solar Minimum*, ed. by S.R. Cranmer, J.T. Hoeksema, J.L. Kohl. ASP Conference Series, vol. 428 (Astron. Society Pacific, San Francisco, 2010), p. 297

L. Svalgaard, How well do we know the sunspot number? Comparative magnetic minima, characterizing quiet times in the Sun and stars, in *Proc. IAU Symp.*, vol. 286, ed. by C.H. Mandrini, D.F. Webb (Cambridge University Press, Cambridge, 2012), p. 27

L. Svalgaard, Solar activity—past, present, future. J. Space Weather Space Clim. **3**, A24 (2013). https://doi.org/10.1051/swsc/2013046

L. Svalgaard, Correction of errors in scale values for magnetic elements for Helsinki. Ann. Geophys. **32**, 633 (2014). https://doi.org/10.5194/angeo-32-633-2014

L. Svalgaard, E.W. Cliver, The IDV index: its derivation and use in inferring long-term variations of the interplanetary magnetic field strength. J. Geophys. Res. **110**, A12103 (2005). https://doi.org/10.1029/2005JA011203

L. Svalgaard, E.W. Cliver, Reply to the comment by M. Lockwood et al. on "The IDV index: its derivation and use in inferring long-term variations of the interplanetary magnetic field". J. Geophys. Res. **111**, A09110 (2006). https://doi.org/10.1029/2006JA011678

L. Svalgaard, E.W. Cliver, A floor in the solar wind magnetic field. Astrophys. J. Lett. **661**, L203 (2007a). https://doi.org/10.1086/518786

L. Svalgaard, E.W. Cliver, Interhourly variability index of geomagnetic activity and its use in deriving the long-term variation of solar wind speed. J. Geophys. Res. **112**, A10111 (2007b). https://doi.org/10.1029/2007JA012437

L. Svalgaard, E.W. Cliver, Heliospheric magnetic field 1835–2009. J. Geophys. Res. **115**, A09111 (2010). https://doi.org/10.1029/2009JA015069

L. Svalgaard, K.H. Schatten, Reconstruction of the sunspot group number: the backbone method. Sol. Phys. **291**, 2653 (2016). https://doi.org/10.1007/s11207-015-0815-8

L. Svalgaard, K.H. Schatten, On the sunspot group number reconstruction: the Backbone method revisited (2017). arXiv:1704.07061

L. Svalgaard, E.W. Cliver, P. Le Sager, Determination of interplanetary magnetic field strength, solar wind speed and EUV irradiance, 1890–2003, in *Solar Variability as an Input to the Earth's Environment, ESA SP-535*, ed. by A. Wilson (ESA Publications Division, Noordwijk, 2003), p. 15

L. Svalgaard, E.W. Cliver, P. Le Sager, IHV: a new long-term geomagnetic index. Adv. Space Res. **34**, 436 (2004). https://doi.org/10.1016/j.asr.2003.01.029

L. Svalgaard, E.W. Cliver, Y. Kamide, Sunspot cycle 24: smallest cycle in 100 years? Geophys. Res. Lett. **32**, L01104 (2005). https://doi.org/10.1029/2004GL021664

L. Svalgaard, M. Cagnotti, S. Cortesi, The effect of sunspot weighting. Sol. Phys. **292**, #34 (2017). https://doi.org/10.1007/s11207-016-1024-9

I.G. Usoskin, A history of solar activity over millennia. Living Rev. Sol. Phys. **14**, 3 (2017). https://doi.org/10.1007/s41116-017-0006-9

I.G. Usoskin, K. Mursula, S.K. Solanki, M. Schüssler, G.A. Kovaltsov, A physical reconstruction of cosmic ray intensity since 1610. J. Geophys. Res. **107**, 1374 (2002). https://doi.org/10.1029/2002JA009343

I.G. Usoskin, K. Alanko-Huotari, G.A. Kovaltsov, K. Mursula, Heliospheric modulation of cosmic rays: monthly reconstruction for 1951–2004. J. Geophys. Res. **110**(A12), A12108 (2005). https://doi.org/10.1029/2005JA011250

I.G. Usoskin, S.K. Solanki, G.A. Kovaltsov, Grand minima and maxima of solar activity: new observational constraints. Astron. Astrophys. **471**, 301 (2007). https://doi.org/10.1051/0004-6361:20077704

I.G. Usoskin, G.A. Bazilevskaya, G.A. Kovaltsov, Solar modulation parameter for cosmic rays since 1936 reconstructed from ground-based neutron monitors and ionization chambers. J. Geophys. Res. **116**(A2), A02104 (2011). https://doi.org/10.1029/2010JA016105

I.G. Usoskin, R. Arlt, E. Asvestari, E. Hawkins, M. Käpylä, G.A. Kovaltsov, N. Krivova et al., The Maunder minimum (1645–1715) was indeed a grand minimum: a reassessment of multiple datasets. Astron. Astrophys. **581**, A95 (2015). https://doi.org/10.1051/0004-6361/201526652

I.G. Usoskin, G.A. Kovaltsov, M. Lockwood, K. Mursula, M. Owens, S.K. Solanki, A new calibrated sunspot group series since 1749: statistics of active day fractions. Sol. Phys. **291**, 2685 (2016). https://doi.org/10.1007/s11207-015-0838-1

J.M. Vaquero, R.M. Trigo, Redefining the limit dates for the Maunder Minimum. New Astron. **34**, 120 (2015). https://doi.org/10.1016/j.newast.2014.06.002

J.M. Vaquero, L. Svalgaard, V.M.S. Carrasco, F. Clette, L. Lefèvre, M.C. Gallego, R. Arlt et al., A revised collection of sunspot group numbers. Sol. Phys. **291**, 3061 (2016). https://doi.org/10.1007/s11207-016-0982-2

Y.-M. Wang, N.R. Sheeley Jr., On the fluctuating component of the Sun's large-scale magnetic field. Astrophys. J. **590**, 1111 (2003). https://doi.org/10.1086/375026

Y.-M. Wang, J.L. Lean, N.R. Sheeley Jr., Modeling the Sun's magnetic field and irradiance since 1713. Astrophys. J. **625**, 522 (2005). https://doi.org/10.1086/429689

W.R. Webber, F.B. McDonald, Recent Voyager 1 data indicate that on 25 August 2012 at a distance of 121.7 AU from the Sun, sudden and unprecedented intensity changes were observed in anomalous and galactic cosmic rays. Geophys. Res. Lett. **40**, 1665–1668 (2013). https://doi.org/10.1002/grl.50383

T. Willamo, I.G. Usoskin, G.A. Kovaltsov, Updated sunspot group number reconstruction for 1749–1996 using the active day fraction method. Astron. Astrophys. **601**, A109 (2017). https://doi.org/10.1051/0004-6361/201629839

D.M. Willis, H.E. Coffey, R. Henwood, E.H. Erwin, D.V. Hoyt, M.N. Wild, W.F. Denig, The Greenwich photo-heliographic results (1874–1976): summary of the observations, applications, datasets, definitions and errors. Sol. Phys. **288**, 117 (2013a). https://doi.org/10.1007/s11207-013-0311-y

D.M. Willis, R. Henwood, M.N. Wild, H.E. Coffey, W.F. Denig, E.H. Erwin, D.V. Hoyt, The Greenwich photo-heliographic results (1874–1976): procedures for checking and correcting the sunspot digital datasets. Sol. Phys. **288**, 141 (2013b). https://doi.org/10.1007/s11207-013-0312-x

D.M. Willis, M.N. Wild, J.S. Warburton, Re-examination of the daily number of sunspot groups for the royal observatory, Greenwich (1874–1885). Sol. Phys. **291**, 2519 (2016a). https://doi.org/10.1007/s11207-016-0856-7

D.M. Willis, M.N. Wild, G.M. Appleby, L.T. Macdonald, The Greenwich photo-heliographic results (1874–1885): observing telescopes, photographic processes and solar images. Sol. Phys. **291**, 2553 (2016b). https://doi.org/10.1007/s11207-016-0894-1

R. Wolf, Sonnenflecken Beobachtungen in der zweiten Hälfte des Jahres 1850. Mitt. Nat.forsch. Ges. Bern **207**, 89 (1851)

R. Wolf, Sonnenflecken – Beobachtungen in der ersten Hälfte des Jahres 1852; Entdeckung des Zusammen-hanges zwischen den Declinationsvariationen der Magnetnadel und den Sonnenflecken. Mitt. Nat.forsch. Ges. Bern **224**, 179 (1852a)

R. Wolf, Liaison entre les taches du Soleil et les variations en declinaison de l'aiguille aimantée. Compt. Rend. **35**, 364 (1852b)

R. Wolf, Beobachtungen der Sonnenflecken in den Jahren 1849–1855. Astron. Mitteil. Eidgn. Sternw. Zür. **1**, 3 (1856)

A. Wolfer, Die Wolf'schen Tafeln der Sonnenflecken – Relativzahlen von 1749 bis zur Gegenwart, und der Maximums- und Minimumsepochen von 1610–1894, neu herausgegeben mit Berichtigungen und Ergänzungen. Astron. Mitt. Zür. **47**(93), 80 (1902)

H. Zirin, *Astrophysics of the Sun* (Cambridge University Press, New York, 1988), p. 303

N.V. Zolotova, D.I. Ponyavin, The Maunder minimum is not as grand as it seemed to be. Astrophys. J. **800**, 42 (2015). https://doi.org/10.1088/0004-637X/800/1/42

N.V. Zolotova, D.I. Ponyavin, How deep was the Maunder minimum? Sol. Phys. **291**, 2869 (2016). https://doi.org/10.1007/s11207-016-0908-z

Space Sci Rev (2018) 214:46
https://doi.org/10.1007/s11214-017-0462-5

The Origin, Early Evolution and Predictability of Solar Eruptions

Lucie M. Green[1] · Tibor Török[2] · Bojan Vršnak[3] · Ward Manchester IV[4] · Astrid Veronig[5]

Received: 15 August 2017 / Accepted: 15 December 2017 / Published online: 7 February 2018

Abstract Coronal mass ejections (CMEs) were discovered in the early 1970s when space-borne coronagraphs revealed that eruptions of plasma are ejected from the Sun. Today, it is known that the Sun produces eruptive flares, filament eruptions, coronal mass ejections and failed eruptions; all thought to be due to a release of energy stored in the coronal magnetic field during its drastic reconfiguration. This review discusses the observations and physical mechanisms behind this eruptive activity, with a view to making an assessment of the current capability of forecasting these events for space weather risk and impact mitigation. Whilst a wealth of observations exist, and detailed models have been developed, there still exists a need to draw these approaches together. In particular more realistic models are encouraged in order to asses the full range of complexity of the solar atmosphere and the criteria for

The Scientific Foundation of Space Weather
Edited by Rudolf von Steiger, Daniel Baker, André Balogh, Tamás Gombosi, Hannu Koskinen and Astrid Veronig

✉ L.M. Green
lucie.green@ucl.ac.uk

T. Török
tibor@predsci.com

B. Vršnak
bvrsnak@gmail.com

W. Manchester IV
chipm@umich.edu

A. Veronig
astrid.veronig@uni-graz.at

1 Mullard Space Science Laboratory, UCL, Holmbury St. Mary, RH5 6NT, UK

2 Predictive Science Inc., 9990 Mesa Rim Rd, Ste 170, San Diego, CA 92121, USA

3 Hvar Observatory, Faculty of Geodesy, University of Zagreb, Kaciceva 26, 10000 Zagreb, Croatia

4 Department of Climate and Space Sciences and Engineering, University of Michigan, Ann Arbor, MI 48109, USA

5 Institute of Physics/IGAM, University of Graz, Universitaetsplatz 5, 8010 Graz, Austria

which an eruption is formed. From the observational side, a more detailed understanding of the role of photospheric flows and reconnection is needed in order to identify the evolutionary path that ultimately means a magnetic structure will erupt.

Keywords Sun · CME · Space weather

1 Introduction

Our Sun is a dynamic star, exhibiting a range of large-scale eruptive activity that, over the last roughly 150 years of study, has led to various activity categories being developed. These include eruptive flares, filament eruptions, coronal mass ejections (CMEs) and failed eruptions. The modern understanding is that this list represents different manifestations of a disruption and reconfiguration of the coronal magnetic field, which ultimately leads to an ejection of magnetised plasma that may, or may not, propagate into the heliosphere. In this review we use the general term CME to encapsulate all eruptive activity types, and focus on the physical processes that are involved.

Coronal mass ejections are the sudden release of 10^{15}–10^{16} g of chromospheric and coronal plasma into the heliosphere. Their speed ranges from between a few tens to a few thousands $\mathrm{km\,s^{-1}}$, with an average value of $\sim$490 $\mathrm{km\,s^{-1}}$ (Webb and Howard 2012). Fast CMEs (that have speeds greater than the background solar wind through which they are propagating) can drive shocks ahead of them. CMEs are identified through the use of coronagraph data in which they are seen via the detection of Thomson scattered photons in outward moving plasma structures that have a higher electron density than the background corona (Fig. 1). Their plane-of-sky configuration varies from event to event and includes amorphous blobs, a "3-part" structure consisting of a bright front, dark cavity and bright central core (Illing and Hundhausen 1985), and helical shapes. The plasma traces out the structure of the magnetic field and the ejected plasma is carried away from the Sun as it is "frozen-in" to the magnetic field. Indeed studying the magnetic field is central to understanding CMEs since the energy to power CMEs, which can be up to the order of 10^{32} ergs, can only come from the conversion of free magnetic energy into other forms (Forbes 2000). The free magnetic energy is the excess energy stored in the coronal field above the energy of the current-free potential field configuration. The free magnetic energy is that available to be released and is stored in the form of electric currents in the pre-eruptive non-potential magnetic structures (for a review see, e.g., Low 1996; Priest and Forbes 2002; Vršnak 2006; Aulanier 2014; Schmieder et al. 2015).

When CMEs are formed at the solar limb many can be seen to originate from coronal streamers (Hundhausen 1993). However, for the majority of CMEs, the eruption includes a low-lying sheared core field, which may or may not contain filament plasma, and that cannot be observed with a coronagraph. Therefore instruments that image the solar disk must be used to study the origin of CMEs in addition to coronagraph data. It is these low altitude observations that showed CMEs are closely related to solar flares (e.g. Vršnak 2008; Schmieder et al. 2015). During a solar flare, free magnetic energy is converted into energetic particles, heat and electromagnetic radiation and the close temporal and spatial CME-flare relationship shows that they are the product of the same disruption and reconfiguration of the coronal magnetic field (e.g. Temmer et al. 2010, and references therein). A relationship between CME occurrence and flare size has been noted, with observations revealing that the likelihood of a flare having an associated CME increases with flare magnitude and that all flares above X5 in the GOES classification occur in concert with a CME (Andrews 2003;

Fig. 1 A coronal mass ejection imaged by the SOHO/LASCO C2 coronagraph on 4th January 2002. The CME is revealed due to Thomson scattered photospheric light and the ejection can be seen in the *upper left* of the image to have a bright core in addition to the surmounting structure. The *black disk* is a result of the occulting disk, which has a radius of $1.7 R_{sun}$

Yashiro et al. 2005). However, it should be pointed out that CMEs and flares do not occur on a one-to-one basis.

CMEs are common events and their occurrence rate varies with the phase of the solar cycle. On average one CME is formed per day at cycle minimum with around five per day at cycle maximum (Webb and Howard 2012) when active regions are more abundant on the solar disk. Active regions can be the source of CMEs during their emergence phase, when current carrying magnetic field is being brought into the corona (Wang et al. 1994; Leka et al. 1996), and through the decay phase when fragmentation and flux dispersal occur, redistributing the magnetic field into the quiet sun. However, CME characteristics can be very different at different evolutionary stages of an active region. From compact, fast CMEs that occur during the emergence phase to quiescent filament eruptions that occur when active regions are at the end of their lives, and finally streamer blowouts associated with the largest-scale coronal structures/or quiescent filaments. The eruptive structure that forms the CME can come from within an active region and also between active regions. Two examples of how the CME *rate* varies with an active region's evolutionary phase were presented in Démoulin et al. (2002) and Green et al. (2002). These studies followed NOAA 7978 and NOAA AR 8100, respectively, for several months and showed that these active regions were CME productive throughout their lifetime (whereas flare activity disappeared as the decay phase progressed). The findings indicate that CMEs do not seem to be so dependent on the magnetic field conditions that are necessary for flares to occur. Flares, especially major flares, occurred when each active region was young and flux emergence had recently taken place. CMEs on the other hand occurred at all evolutionary stages, including during the decay phase when there was no new major flux emergence and the energy density of the magnetic field had reduced. These studies revealed that CMEs are not as dependent on strong field and magnetic complexity as are flares. Still, active regions that have large and complex magnetic field configurations can be the origin of many CMEs during their emergence phase. For example, NOAA active region 9236 produced 17 CMEs in 5 days (Gopalswamy et al. 2005) and NOAA active region 8100 produced 15 CMEs in 7 days (Green et al. 2002).

However, these large and dramatic regions should not distract from the fact that CMEs are formed in magnetic structures that span a range of spatial scales. The size-scale of the pre-eruption structure can be as large as 100s thousands of km long (polar crown filaments) but the structure can be as small as only 10s of thousands of km across (Mandrini et al. 2005).

Some CMEs detected in coronagraph data show no signatures in the lower atmosphere, earning them the name "stealth CMEs". They originate at altitudes of $>0.1 R_\odot$, where $R_\odot$ is the solar radius, (Robbrecht et al. 2009) in magnetic structures where the field strength is weak and the plasma density low, meaning that observational signatures are less likely to be seen (Howard and Harrison 2013). A study of the Sun at solar minimum suggests that stealth CMEs may make up 30% of all CMEs that occur (Ma et al. 2010). Kilpua et al. (2014) found in their study after solar maximum in 2014 that ten out of the 16 studied events could be classed as stealth CMEs.

Taking into account both the remote sensing and in situ data it has been realised that a CME in the inner corona can be composed of some or all of the following:

- a shock that moves ahead of the CME
- a frontal structure with high plasma density ($n_e \approx 10^{14}$ m^{-3}), coronal temperature ($\sim$2 MK) and high magnetic field strength (10^{-4} T)
- a cavity with low plasma density ($n_e \approx 10^{13}$ m^{-3}), coronal temperature ($\sim$2 MK) and field strength of few 10^{-4} T
- a prominence core with high plasma density ($n_e \approx 10^{17}$ m^{-3}), low temperature ($\sim$80 000 K) and high field strength (few 10^{-3} T)
- a post-eruption arcade with a plasma temperature of $\sim$10 MK and a few 10^{-4} T field strength

Studies of CMEs have found a new relevance in the last three decades through the realisation that they can drive severe space weather at Earth (Gosling 1993; Green and Baker 2015). Parameters that are important for determining the level to which a CME will be geo-effective upon its impact with the magnetosphere include the strength of the southward component of the CME's magnetic field (B_z), the speed and plasma density of the CME (V_{CME}, ρ) and the CME's dynamic pressure ($\rho\, V_{\mathrm{CME}}^2$). This means that fast CMEs with a strong and sustained southward magnetic field component and/or a high plasma density are likely to be the ones that have the most significant effect. Historical case studies of such events include the Carrington event of 1859 when a CME reached the Earth in just 17.5 hours (Cliver and Svalgaard 2004) giving it an average velocity along the Sun–Earth line of $\approx$2300 km s^{-1}, and the March 1989 event which was exceptionally bright in coronagraph data, implying it had a high plasma density although it also had a much lower average Sun–Earth line speed of around 770 km s^{-1} (Feynman and Hundhausen 1994). These CMEs produced geomagnetic storms with a disturbance storm time (*Dst*) index estimated to be -850 nT (Siscoe et al. 2006) and -548 nT (Cliver and Svalgaard 2004) respectively. It is worth noting that the *Dst* value during the 1859 event has been a topic of controversy, with Tsurutani et al. (2003) proposing that -1760 nT may have been reached.

Especially important in the creation of severe geomagnetic storms appear to be CMEs that follow one another in quick succession (Liu et al. 2014; Dumbović et al. 2015; Temmer and Nitta 2015; Vennerstrom et al. 2016). The Carrington event is one example of such a case. In these events, the preceding CME may "pre-condition" the space between the Sun and the Earth so that the following CME is embedded in a faster solar wind stream and so experiences less drag; allowing it to maintain a high velocity out to 1 AU (Temmer et al. 2017).

The impact of a CME extends beyond its direct interaction with the magnetosphere as shocks driven by fast CMEs (roughly above around 800 km s^{-1}, Kahler 2001) can accelerate particles, producing an additional space weather effect when these solar energetic particles (SEPs) arrive at the Earth (Reames 2013). Meaning that when CME predictions are being developed, speed and spatial extent of the CME are important quantities to predict too. As are the upstream wind conditions ahead of the shock that can affect the shock geometry.

The very significant role that CMEs play in the creation of space weather has naturally led to a desire to predict their time of arrival at Earth. Once a CME has been launched, it can take from less than a day to more than three days before it arrives. Since the fastest CMEs can be the most geo-effective, the ability to predict their occurrence before they happen would significantly improve forecasting lead times. In this review paper we draw together the wide-ranging observational (Sect. 2.1) and theoretical (Sect. 2.2) work on CMEs to assess the current status of understanding how eruptive structures are formed and how energy is injected into the magnetic field, over timescales of weeks to hours before the eruption. We focus on processes that occur during the energy build-up phase that bring the magnetic field to a point where a "driving" mechanism occurs to drive the rapid expansion of the structure. From this, we look at current and future directions of predicting CMEs based on both observations and modelling (Sect. 3).

2 Origin and Evolution of Solar Eruptions

2.1 Observations

Coronal mass ejections are observed using a wide range of imaging and spectroscopic data, spanning wavelengths from radio to X-ray and altitudes from the photosphere to the outer corona. These observations reveal that in many cases, the onset of a CME will be marked by the eruption of a filament/prominence in the lower solar atmosphere. Indeed, over 70% of CMEs begin this way (Munro et al. 1979). The filament may be seen as a bright core in coronagraph data in CMEs that have a 3-part structure. Lower coronal observations of CME source regions can also show the rise of plasma structures that are emitting at EUV and/or soft X-ray wavelengths. Such rising structures have been seen in *Yohkoh*/SXT data (e.g. Moore et al. 2001; Green and Kliem 2014), *SDO*/AIA channels (e.g. Liu et al. 2010; Zharkov et al. 2011) and *STEREO*/EUVI data (Bein et al. 2011; Patsourakos et al. 2013). In the wake of a CME, the reconfiguration results in a post-eruption arcade, which is also known as a flare arcade. When the CME source region is located at the solar limb, hot plasma structures referred to as plasmoids or flux ropes have been observed (Shibata et al. 1995; Cheng et al. 2011; Reeves and Golub 2011). Once the eruption is well underway, the footpoints of the expanding and erupting magnetic volume can sometimes be identified through EUV or soft X-ray regions which have dimmed due to the reduction in plasma density (Rust and Hildner 1976; Sterling and Hudson 1997; Zarro et al. 1999). Figure 2 provides examples of the variety of lower corona signatures that indicate the occurrence of a CME.

In this section we review observational aspects of the coupled evolution of the photosphere and the corona in the long-term (Sect. 2.1.1) and short-term (Sect. 2.1.2) leading up to the eruption and during the rapid acceleration phase of the eruption itself (Sect. 2.1.3).

2.1.1 Long-Term Evolution Prior to Eruption

The energy required to power a CME appears to be built-up in the coronal field over days/weeks prior to the eruption, in an energy storage and release process. This injection of

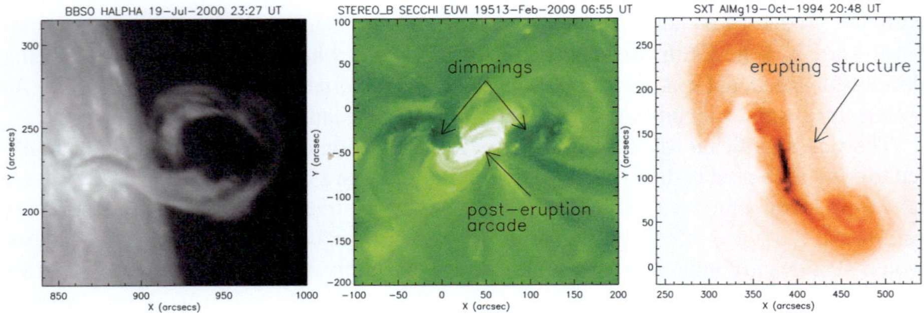

Fig. 2 Lower coronal signatures of the occurrence of a coronal mass ejection (CME). There are a variety of observational manifestations of CMEs including filament eruptions (*left panel*), post-eruption arcades (also called flare arcades, *middle panel*), dimming regions (*middle panel*) and rising EUV/soft X-ray structures (*right panel*)

energy is driven by both the emergence of flux (which is non-potential, Leka et al. 1996) and photospheric motions due to convection and differential rotation that can stress the coronal field. In some cases sunspot rotation is observed (e.g., Brown et al. 2003) and/or rotational motions of same-polarity magnetic fragments, one about the other (James et al. 2017).

Development of non-linear force-free field (NLFFF) modelling techniques have provided methods to reconstruct the coronal magnetic field using photospheric vector magnetograms. In turn the evolution of the energy in the coronal field can be analysed from these extrapolated configurations. The analysis of the magnetically complex NOAA active region 11158 by Sun et al. (2012) followed the energy stored in the coronal field over a period of five days, during which time the region produced multiple flares and CMEs. The study found a rapid increase in the free magnetic energy of the coronal field lasting around four hours as the flux emergence began. This was followed by a more gradual injection of free energy into the coronal configuration during the following ~two days. The energy was mostly accumulated at low heights (less than 6 Mm) along a polarity inversion line where a low-lying filament formed, and where later eruptions originated. The extrapolated coronal field configuration shows that around the time of an X-class flare approximately 3.4×10^{31} ergs of energy is released. The NLFFF modelling studies by Gibb et al. (2014), Mackay et al. (2011) followed the evolution of two short-lived small bipolar active regions during their emergence and decay phases using a magnetofrictional method. The simulated coronal field in these studies captured the energy build-up prior to the CMEs that the regions produce. Free energy is found to be built-up during the flux emergence phase and also during the decay phase when the regions start to undergo flux cancellation at their internal polarity inversion lines (flux cancellation is described and discussed at the end of this section).

As well as energetics, the quantity of magnetic helicity has been considered in the evolution of active regions that produce CMEs. Magnetic helicity is the quantity that encapsulates how twisted, sheared and interlinked the magnetic field is. Magnetic helicity is approximately conserved even during non-ideal magnetohydrodynamic (MHD) processes (Berger 1984). For example, magnetic reconnection serves to transport magnetic helicity to larger spatial scales, opposite to energy, which cascades to smaller scales where it can dissipate. Magnetic helicity is injected into the corona via flux emergence, photospheric motions and torsional and shear Alfvén waves and this, along with its conservation property, has led to the suggestion that CMEs may act as a valve to prevent magnetic helicity from endlessly accumulating in the solar atmosphere (Heyvaerts and Priest 1984; Low 1996). Indeed, active regions during their entire lifetime can be the source of many CMEs, which eject a significant amount of magnetic helicity (Démoulin et al. 2002; Green et al. 2002). It has even been proposed that CMEs are the unavoidable product of magnetic helicity accumulation in the force-free magnetic field of the corona (Zhang et al. 2006). The proposal of an intimate link between this topological quantity and a CME spurred a wave of research in this area. New methods to quantify magnetic helicity injection into the solar atmosphere via flux emergence and photospheric flows provided the tools to be able to make these investigations (Démoulin and Berger 2003).

Many studies of the evolution of magnetic helicity in active regions in the time periods around the occurrence of flares and CMEs have been carried out. Nindos and Andrews (2004) studied the coronal magnetic helicity evolution of active regions that were the source of 133 M-class and X-class flares. They found that the magnetic helicity contained in the coronal field of an active region was larger in regions that produced flares that had associated CMEs than for those regions that had CME-less flares. Other studies have looked at magnetic helicity flux, even proposing that a threshold helicity injection rate must be exceeded for a CME to occur (LaBonte et al. 2007). However, in contrast to this, other work has found that major changes in the magnetic helicity content of an active region can occur both before and after a CME (Smyrli et al. 2010). Helicity injection prior to a CME in the Smyrli et al. (2010) study was seen to be due to flux emergence, whereas the changes after the CME were proposed to be due to the torque imbalance between the helicity-depleted corona and the sub-photospheric portions of the flux as suggested in Longcope and Welsch (2000), Démoulin et al. (2002), Green et al. (2002). Whether or not magnetic helicity is a quantity that can be used to predict CME occurrence will be discussed further in Sect. 3.1.

The role of flux emergence and photospheric shear flows in injecting sufficient helicity into the corona before a CME were studied by Nindos and Zhang (2002) using NOAA active region 9165. The study finds that the horizontal flows did not inject sufficient magnetic helicity into the corona to account for that ejected by CMEs. Instead, flux emergence is invoked as the dominant helicity source. However, studies of smaller and shorter-lived bipolar active regions suggest that during the decay phase of these short-lived regions the photospheric motions caused by convection (super-granular flows) may play an important role in injecting helicity into the corona (Savcheva 2012; Gibb et al. 2014). All the above studies indicate that the magnetic helicity in the corona can be injected via variety of mechanisms, and the mechanism that dominates may depend on the evolutionary stage of the region. Next, we focus on how a ubiquitous process observed in the photosphere known as flux cancellation might prime the corona, ready for an eruption.

Flux cancellation is a common phenomenon that occurs all over the Sun (Livi et al. 1985) and is the name given to an observational event consisting of the approach of two opposite polarity fragments that collide and then disappear from the photospheric magnetic field measurement (Martin et al. 1985). It's importance for this review paper is that flux cancellation is often observed in the time leading up to a CME (Martin et al. 1985), which suggests that it may be important in the creation of an eruptive magnetic field configuration and its destabilisation. The physical interpretation of flux cancellation is that when opposite polarities collide they undergo reconnection in the photosphere or chromosphere (e.g. Yurchyshyn and Wang 2001; Litvinenko and Martin 1999). These opposite polarities represent the footpoints of two sheared loops and the low-altitude reconnection reconfigures the magnetic field so that a long loop forms in the corona and also a very short loop that submerges below the photosphere due to its tension force overcoming buoyancy. Shear and free magnetic energy are concentrated along polarity inversion lines where flux cancellation is ongoing for extended periods of time (Welsch 2006), hence it relevance for CME studies.

In some small bipolar active regions, an eruptive configuration is formed via flux cancellation along the internal polarity inversion line during the decay phase of the region, when the flux is fragmenting. This is well observed in bipolar active regions that form S-shaped "sigmoidal" emission structures in soft X-ray or EUV wavebands over a couple of days (e.g. Green et al. 2011). Both the coronal evolution and the photospheric evolution in these cases match well the scenario proposed by van Ballegooijen and Martens (1989) in which a sheared arcade is transformed into a flux rope. This is also in line with the theoretical findings that sigmoids are created by hot plasma along field lines in the vicinity of a quasi-separatrix layer formed in flux rope topology (Titov and Démoulin 1999).

In summary, the evolution of an active region prior to a CME involves the emergence of magnetic flux into the corona that is then modified by photospheric flows, which eventually disperse the field over an ever wider area. During the lifetime of the active region flux cancellation may be occurring that is able to modify the configuration of the magnetic field and allow free magnetic energy to buildup along polarity inversion lines. CMEs can occur at any point during this evolutionary sequence although the frequency of CMEs is highest at times of new flux emergence (Green et al. 2003).

2.1.2 Short-Term Evolution Prior to Eruption

Collated observations of the evolution of CME source regions in the tens of minutes before an eruption reveal that the magnetic structure evolves through distinct dynamic phases. The majority of CMEs exhibit a slow-rise phase, a phase of rapid acceleration and then a phase where the erupting structure propagates into the heliosphere (Zhang et al. 2001; Vršnak 2001). The slow rise phase occurs over timescales of minutes to hours and can be observationally identified through the upward motion of a filament or the expansion of coronal loops. The phase of rapid acceleration then occurs and signals that the eruption is in progress. During the propagation phase the CME interacts with the ambient solar wind (Vršnak et al. 2008). Here we describe the short-term evolution in the hours or minutes leading up to the phase of rapid acceleration.

It has been realised for many years that the magnetic field configuration associated with a sigmoid is highly likely to erupt (Canfield et al. 1999), however an exact observational definition of a sigmoid has not yet been reached. The term is used to describe continuous S-shaped structures and also collections of loops that appear S-shaped overall. In the cases where a continuous S-shaped sigmoid is observed, an eruption follows around within around 14 hours (Green and Kliem 2014). Given the finding that continuous S-shaped sigmoids that form along polarity inversion lines with sustained flux cancellation are indicative of a low altitude flux rope, the data indicate that these flux ropes can form and remain stable on the Sun for several hours. Sigmoids are best observed near disk centre, so that projection effects do not mask their shape. There is a related set of observations at the solar limb that indicate the formation of flux ropes at a higher altitude, formed by coronal reconnection. These flux ropes appear to form from around 7 hours to 10 minutes before their eruption (Patsourakos et al. 2013; Cheng et al. 2011). A recent study identified such a flux rope near disk centre, allowing the photospheric magnetic field evolution to be observed and revealing that the flux rope formation was driven by flux emergence which exhibited rotational motions of the fragments in the leading polarity, one about the other (James et al. 2017). These photospheric motions and the corresponding evolution of the coronal field led to magnetic reconnection in the corona and the formation of the flux rope.

As the time of the eruption approaches observations reveal that coronal structures begin to rise, albeit slowly, before the main phase of acceleration. Filaments darken and start their slow ascent from around 3 hours before their successful eruption (Martin 1980). More compact active region structures complete their slow rise phase more rapidly, taking as little as just a few minutes (e.g. Zharkov et al. 2011) before the structure undergoes rapid acceleration.

2.1.3 Evolution of the Eruption in the Lower Corona

Over many decades, observations of the onset and evolution of the eruption of a CME in the lower corona have been drawn together to form the standard model for a CME or an eruptive flare. This model is also known as the CSHKP model after the authors of the early papers (Carmichael 1964; Sturrock 1966; Hirayama 1974; Kopp and Pneuman 1976). For a review of some of the many observational papers that have further supported this model see McKenzie (2002). The standard model of an eruption begins with the presence of a core field, either a sheared arcade or magnetic flux rope, embedded in an arcade field. Such sheared structures and flux ropes are well supported by observations from the photospheric vector magnetic field to EUV and soft X-ray emission structures in the corona. At the start of the eruption, the core field starts to rise, stretching the overlying arcade. A current sheet forms in the arcade below the core field, and reconnection sets in. Observations that support the occurrence of reconnection include the reconfiguration of the coronal field, indications of non-thermal particles from the formation of flare ribbons and hard X-ray emission, and reconnection inflow (McKenzie 2002). These observational signatures indicate that reconnection is closely related in time to the rapid acceleration phase of the CME. Indeed, the reconnection helps facilitate the eruption as it builds poloidal flux around the erupting core field and cuts the tethers of the overlying field. Flare loops form below the reconnection region as the downward product of the reconnection. Some CMEs show a clear rotation of the erupting structure as the eruption begins, indicating the presence of a flux rope and a conversion of twist into writhe (deformation of the axis of the flux rope, Green et al. 2007).

Not all CMEs that are initiated make it to a full eruption (Rust 2003). Instead, their rapid ascension is arrested in the lower corona. This sub-set of events are known as failed eruptions or confined eruptions. The failure of the eruption might be due to the reconnection that occurs below the erupting core field not sufficiently cutting the tethers of the overlying arcade field (Moore et al. 2001) or the field strength in the overlying arcade not dropping off sufficiently rapidly with height (Török and Kliem 2005). However, failed eruptions still serve a useful purpose in CME studies as they allow an investigation of the triggers versus the drivers of an eruption. In a failed eruption, presumably the trigger occurs, but the physical processes needed to drive the CME into the heliosphere do not (see Sect. 2.2.2).

2.2 Theory and Simulations

From a theoretical point of view, there is little doubt that large-scale solar eruptions are magnetically driven, since they originate in the corona where the plasma beta (ratio of gas to magnetic pressure) is small. They are powered by the free magnetic energy that is stored in the corona in non-potential magnetic fields, i.e., in volumetric electric currents. These currents are believed to emerge from the convection zone (e.g., Leka et al. 1996) or be produced by photospheric flows that shear and twist the coronal magnetic field (e.g., Klimchuk and Sturrock 1992). Eruptions occur when the slow, continuous stressing of the coronal field by flux emergence and surface flows reaches some threshold above which magnetic equilibrium cannot be maintained and the field violently erupts (Forbes 2000). While this overall picture is well established, the detailed processes that govern the evolution prior to and during an eruption are not yet fully understood. In this subsection, we thus focus on the following three questions: How is the free energy built up in the corona? (Sect. 2.2.1) How are solar eruptions initiated? (Sect. 2.2.2) What are the physical mechanisms that can propel the magnetic field and plasma into interplanetary space? (Sect. 2.2.3) We will also discuss potential connections between individual eruptions (leading to "sympathetic" events) in Sect. 2.2.4.

2.2.1 Modelling the Build-up of Free Magnetic Energy

Non-potential magnetic fields are at the epicenter of solar eruptive behavior. Eruptions may originate from large-scale relatively weak magnetic fields as in the case of streamer blowouts or quiescent prominence eruptions (Hundhausen 1993; Lynch et al. 2010) or they may originate from strong active region magnetic fields. Whatever the circumstances, solar eruptions originate from strongly non-potential fields where the photospheric magnetic field is nearly parallel to the polarity inversion line (PIL), a so-called magnetic shear (e.g. Forbes 2000; Schrijver 2009), which stands in contrast to a potential field that runs perpendicular to the PIL and has no free energy. There is enormous evidence for the existence of highly sheared magnetic fields associated with CMEs and large flares. At the photosphere, magnetic shear in CME-productive active regions is directly measurable with vector magnetographs (e.g. Hagyard et al. 1984; Zirin and Wang 1993; Falconer et al. 2002; Yang et al. 2004; Liu et al. 2005; Sun et al. 2012) and the associated shear flows can be observed along the photospheric PIL (e.g. Strous et al. 1996; Yang et al. 2004; Deng et al. 2006; Sun et al. 2012).

With such observations in mind, MHD simulations of solar eruptions have long invoked shear flows prescribed as boundary conditions to energize magnetic fields to produce large-scale solar eruptions. Many examples exist including simulations by Steinolfson (1991), Mikic and Linker (1994), Wolfson (1995), Antiochos et al. (1999), Amari et al. (2003a), Lynch et al. (2008), van der Holst et al. (2009), which are successful in reproducing many characteristics of CMEs, filament eruptions and flares. CME-like eruptions have also been modelled with prescribed rotational flows that twist up the field (e.g., Amari et al. 1996; Török and Kliem 2003; Aulanier et al. 2005, see also Table 1). A significant limitation of these numerical models is that because of the enormous differences in plasma conditions, they all treat the corona as being disconnected from the solar interior. While this simplification greatly reduces computational costs, it also removes any possibility of elucidating the physical processes that by their very nature transport free magnetic energy and magnetic helicity from the solar interior to the corona.

In contrast, simulations of magnetic flux emerging from the convection zone to the corona set the stage to understand how active region magnetic fields are organized and energized. Global-scale simulations of the solar convection zone have shown how magnetic fields are amplified by the dynamo to produce buoyant loops that capture active region characteristics such as Joy's law and sunspot asymmetries (e.g. Spruit and van Ballegooijen 1982; Abbett et al. 2001; Fan and Fang 2014; Nelson et al. 2014). However, constraints in timescale and numerical resolution prevent these simulations from reaching beyond 95% of the solar radius. To incorporate the upper layers of the solar atmosphere, fully compressible MHD models are applied on small scales in Cartesian domains (e.g. Shibata et al. 1989; Matsumoto et al. 1993) that illustrate magnetic flux buoyantly rising through solar photosphere to the corona. Simulations that followed (Manchester 2001; Fan 2001; Magara and Longcope 2003; Török et al. 2014) found that under these circumstances, shear flows develop spontaneously where horizontal flows reverse direction across emerging loop structures. An analytical model by Manchester and Low (2000) first showed that these shear flows are in response to the Lorentz force, which occurs as magnetic fields rise through the stratified atmosphere and are deformed by the intense pressure gradient of the surrounding plasma. The shearing motions are of the form of large-amplitude shear Alfvén waves where magnetic tension drives the flows that draw the magnetic field along the direction parallel to the polarity inversion line and transport magnetic flux and energy from the convection zone into the corona. Manchester (2003, 2007), Manchester et al. (2004a) found that such shear flows are capable of driving eruptions of magnetic arcades and emerging flux ropes and proposed this mechanism as an energy source and initiation mechanism for CMEs.

Early simulations of flux emergence treated the solar interior as a simple polytropic layer, which roughly captured the temperature structure and density but was free of convection. In recent years, realistic solar models with granular convective and coronal atmospheres have been developed with the inclusion of radiative transfer and heat conduction. Examples include Stein and Nordlund (2006), Abbett (2007), Rempel et al. (2009), Nordlund et al. (2009), Martínez-Sykora et al. (2009), Cheung et al. (2010), Fang et al. (2010), Kitiashvili et al. (2010), which illustrate the complex and dynamic interaction between magnetic fields and granular convection. Work in this vein began with quiet sun magneto-convection (e.g. Stein and Nordlund 2006; Abbett 2007; Rempel 2014), which shows the development of a near-surface dynamo and formation of kiloGauss flux tubes in convective downdrafts. In related work, Martínez-Sykora et al. (2009), Kitiashvili et al. (2010) found strong vortex flows in downdrafts at the vertices of convective granules where the twisting magnetic field produces strong outflows representative of spicules. The inclusion of strong large-scale vertical magnetic fields in the presence of convective flows leads to the formation of sunspots as shown by Rempel et al. (2009), Cheung et al. (2010), Rempel and Cheung (2014).

Following this work, came realistic simulations of flux emergence and active region formation in a fully convective layer (e.g. Fang et al. 2010, 2012b; Cheung et al. 2010; Rempel and Cheung 2014). In the convection zone, high-plasma beta magnetic flux is severely distorted as distinct parts of the flux system rise and fall with the large-scale convective flows. As flux passes through the photosphere, it is shredded by the granular convection only to later reorganize as like polarities coalesce into kiloGauss pores and sunspots (Fang et al. 2012b; Cheung et al. 2010; Rempel and Cheung 2014). Even with this complex structure the Lorentz force continues to drive both shear and rotational flows that transport magnetic flux, helicity and free magnetic energy from the convection zone just as found in earlier simulations. In the case of Fang et al. (2012a), it was also found that while rotational flows provide free energy in the area of sunspots, shear flows dominate the energy build-up at the PIL. Fang et al. (2012a) further found that converging motions, flux cancellation and tether-cutting reconnection can combine to build-up the magnetic shear and free energy in the corona necessary for eruptive events. In this case, nearly 40% of the magnetic energy is free energy, which is near the amount required to produce eruptions (Amari et al. 2003a; Aulanier et al. 2010; Moore et al. 2012).

Recent observations are proving that the ubiquitous shear and rotational flows so strongly associated with CMEs produced during an active region's emergence phase are driven by the Lorentz force. Analysis of high-resolution vector magnetograms (Georgoulis et al. 2012) indicate that photospheric electric currents generate the Lorentz force that drives shear flows along active region PILs during this phase. Similar analysis by Su et al. (2008) has revealed that the Lorentz force is the driver of rotational flows in sunspots. In these cases, the flows are shown to produce a build-up of magnetic energy with magnetic stress passing from the convection zone into the corona to form non-potential fields. These observational results both verify prior numerical simulations and confirm a fundamental mechanism for energizing coronal magnetic fields before eruption.

Simulation results of the development of shear flows, sheared magnetic fields, Poynting flux and an erupting flux rope are shown in Fig. 3. Panel (a) shows the results of a

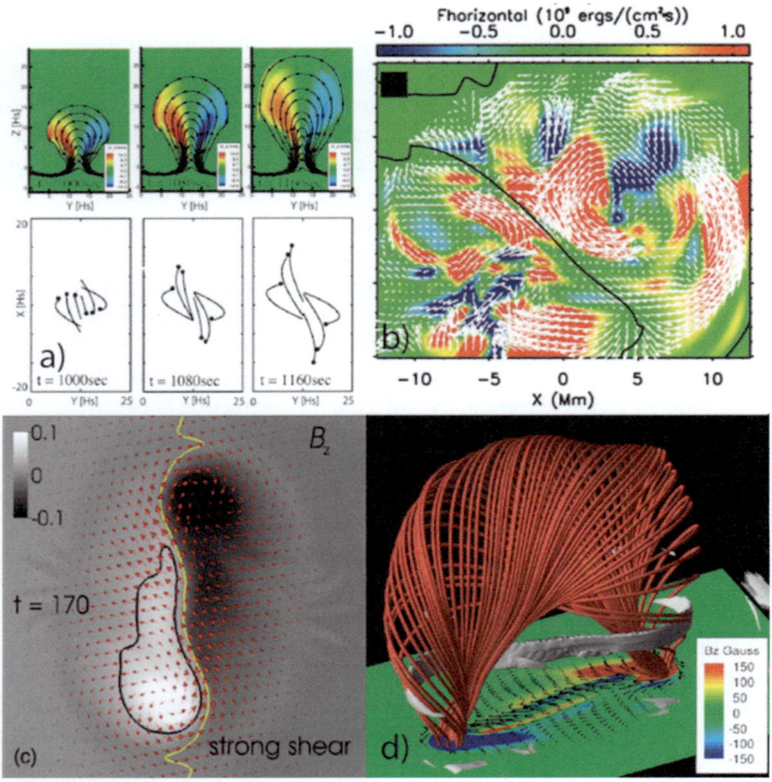

Fig. 3 Simulations of shear and rotational flows, energy transport, and magnetic flux-rope eruption. Panel (**a**) Manchester (2001) shows a 2.5D simulation of flux emerging, with the *top row* showing the shear velocity shown over the cross section of the emerging loop, and the *bottom row* showing the legs of the loops where *dots* are at the foot points. Panel (**b**) Fang et al. (2012a) shows the Poynting flux associated with horizontal flows passing though the corona. *White arrows* show the velocity vectors. Panel (**c**) Török et al. (2014) shows late-phase signatures of an emerged magnetic flux rope at the photosphere. *Red arrows* show the horizontal magnetic field direction while the *yellow line* shows the PIL. Note the magnetic shear with the alignment of the magnetic field with the PIL. Panel (**d**) Manchester et al. (2004a) shows the eruption of the flux rope resulting from shear flows. At the photosphere, the vertical magnetic field strength is shown in *color* and the horizontal direction is shown with *vectors*. Here, the current sheet is shown with a *gray iso-surface*, forming a sigmoid structure that runs nearly parallel to the highly sheared PIL

2.5 D simulation of flux emergence that shows the development of sheared magnetic fields (Manchester 2001). Panel (b) shows the Poynting flux associated with horizontal flows passing into the corona (Fang et al. 2012a) while Panel (c) shows the development of a highly sheared magnetic field resulting from flux emergence (Török et al. 2014). Panel (d) shows the magnetic field of an erupting magnetic flux rope, where the eruption is triggered by shear flows driven by the Lorentz force.

2.2.2 Initiation Mechanisms of Solar Eruptions

As described in the previous sections, solar eruptions are powered by the free magnetic energy that is stored in current-carrying magnetic fields in the corona. Prior to an eruption,

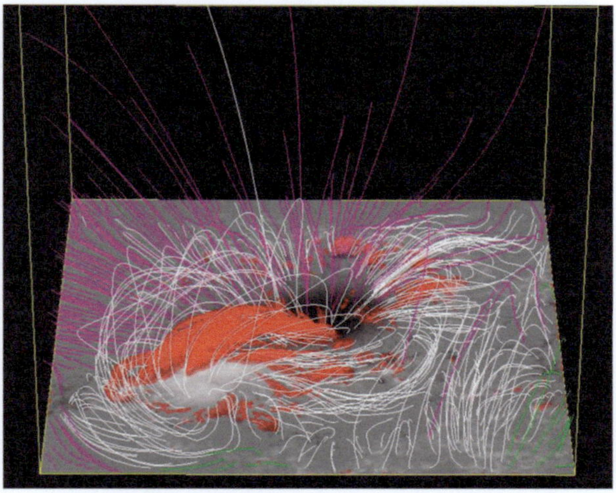

Fig. 4 Non-linear force-free field extrapolation of NOAA AR 10930 before it produced an X3.4 flare and CME (from Schrijver et al. 2008a). *Red iso-surfaces* show electric currents. *Coloured (white)* field lines are open (closed). Closed field lines arch over the currents, stabilizing the sheared or twisted core magnetic field

the magnetic configuration essentially consists of two parts. One part is the current-carrying (sheared and/or twisted) core field, which is always located above a polarity inversion line and which may or may not harbor a filament. This field naturally seeks to expand and must therefore be stabilized by the other part of the configuration, namely by the largely current-free (potential) ambient field that surrounds the core field (often referred to as the "strapping field"), as illustrated in Fig. 4. The idea underlying many eruption-models is that an eruption can be initiated by any physical mechanism that is capable of bringing this configuration to a state at which stable equilibria cease to exist and the system violently erupts, thereby releasing some of the stored free energy. The initiation mechanism could be one that either increases the current in the core field or decreases the stabilizing magnetic tension of the ambient field, or both (see, e.g., Aulanier 2014).

However, even though eruptions have been observed extensively for many years, the specific build-up and trigger mechanisms, and the mechanisms that drive CMEs remain rather elusive. This is, to a large extent, due to the persistent lack of routine magnetic field measurements in the corona. Analytical models and numerical simulations provide a useful tool for our attempts to close this gap. These approaches typically employ idealized initial configurations and a reduced set of the MHD equations, both of which are strong simplifications of reality. However, such a simplified approach has two significant advantages. On the one hand, it often allows one to isolate in the model the specific mechanism of interest. On the other hand, it keeps the computational costs moderate and thus provides the possibility of performing extensive parametric studies. Such studies are useful because they allow, for example, an assessment of physical quantities that are difficult or impossible to measure in the corona, such as magnetic twist (e.g., Kliem et al. 2012). In the past decades, idealized analytical models and numerical simulations have been widely used to test and study suggested eruption mechanisms, and have thereby greatly increased our understanding of how solar eruptions work. We note that significantly more complex MHD simulations of eruptions exist as well. These simulations include observed data and more of the physics, and they are aiming to reproduce specific eruptions with as much realism as presently possible (see Manchester et al. in this issue).

Table 1 shows a compilation of eruption models of the so-called "storage and release" type, together with references and a compilation of review papers on the subject.[1] Storage-and-release models assume that the free energy required to power an eruption is slowly (i.e., quasi-statically) accumulated in the corona over hours or days, and then rapidly released during the eruption. "Directly driven" models, such as thermal blasts during flares or flux injection from the solar interior, appear outdated and are not considered here (for discussions of their validity see, e.g., Forbes 2000, Klimchuk 2001, Linker et al. 2003; but see also Chen 2017). Furthermore, the table does not include eruption scenarios that have, to the best of our knowledge, not yet been tested using numerical simulations. Such scenarios are, for instance, the triggering of eruptions by the merging of magnetic flux ropes (e.g., Pevtsov et al. 1996) or by the unloading of prominence mass (e.g., Low 1996; Seaton et al. 2011), as well as the conjecture that CMEs occur in order to avoid an over-accumulation of magnetic helicity in the corona (e.g., Low 1994).

The mechanisms are grouped according to their predominant role in the eruption process, i.e., whether they act as a "trigger" or as a "driver".[2] By driver we mean here any mechanism that can account for the rapid acceleration and huge expansion of plasma and magnetic field observed during eruptions, i.e., a mechanism that can produce a CME. As a trigger, on the other hand, we consider mechanisms that are capable of initiating an eruption, but can by themselves not produce a CME. The role of a trigger is rather to bring the system to a state at which a driver can take over and complete the eruption.

Such a strict distinction between trigger and driver is somewhat artificial, but we use it here for the following reasons. First, rise profiles of eruptions sometimes clearly show two distinct phases that cannot be fitted with one functional form (see Sect. 2.1.2 and e.g., Sterling et al. 2007a,b; Schrijver et al. 2008b), indicating the occurrence of two distinct physical mechanisms. Second, it has been demonstrated in MHD simulations that certain mechanisms (such as tether-cutting and magnetic breakout) do not seem sufficient for producing a full eruption (e.g., Aulanier et al. 2010; Karpen et al. 2012). Finally, some mechanisms do not appear to be capable of producing a CME, just by their nature. The helical kink instability, for example, is expected to lead to a (relatively moderate) deformation of the shape of an unstable flux rope, but not to its huge expansion.

We note that some of the processes listed as a trigger mechanism in Table 1 (flux emergence, sunspot rotation, flux cancellation) typically act on time-scales of days, much longer than the typical duration of the slow rise phase preceding eruptions. Their preliminary role is to accumulate free energy in the corona (see Sect. 2.2.1). However, as demonstrated by the numerical simulations referenced in the table, they can also, on shorter time scales, induce changes to the system that push it towards an eruption (an example would be flux emergence into an existing pre-eruptive configuration). In that sense it appears justified to consider these processes not just as build-up, but also as trigger mechanisms.

At present, it seems that there may exist only two driving mechanisms in solar eruptions, namely the torus instability and the "flare reconnection" (for a somewhat different view, see Aulanier 2014). The latter is the reconnection that takes place across the vertical current sheet below a CME and produces the flare signatures. The torus instability is an ideal MHD instability, closely related to the processes known as "flux-rope catastrophe" and "loss of equilibrium" (see Démoulin and Aulanier 2010 and Kliem et al. 2014a for a detailed discussion). In this instability, the driver responsible for the acceleration of the ejecta is the hoop

[1] This table is not intended to be complete, more mechanisms and articles may be found in the literature. See Aulanier (2014) for the differences between "flux decrease" and "flux dispersion" models.

[2] Any such grouping is somewhat subjective. Aulanier (2014), for example, considers magnetic breakout rather than flare-reconnection as the second driving mechanism next to the torus instability.

Table 1 Compilation of suggested physical mechanisms of the "storage-and-release" type for the triggering and driving of solar eruptions. "FE" stands for flux emergence. Note that (i) we do not distinguish here between 2D and 3D simulations; (ii) we do not include simulations that model CMEs by starting with an "out-of-equilibrium" flux-rope configuration (e.g., Roussev et al. 2003; Manchester et al. 2004b; Tóth et al. 2007; Cohen et al. 2009; Lugaz et al. 2011; Pagano et al. 2013), even though they are compatible with the storage-and-release paradigm; and that (iii) each article is referenced only once, even if the model or scenario it describes involves more than one mechanism. The *bottom row* contains related review articles

Mechanism	References
TRIGGER:	
Sunspot rotation	Amari et al. (1996), Tokman and Bellan (2002), Török and Kliem (2003), Aulanier et al. (2005), Rachmeler et al. (2009)
Twisting overlying field	Török et al. (2013)
Shearing of arcade	Mikic et al. (1988), Biskamp and Welter (1989), Mikic and Linker (1994), Choe and Lee (1996a,b), Amari et al. (1996a), Jacobs et al. (2006, 2009), Roussev et al. (2007), Shiota et al. (2008), Downs et al. (2011)
Reversed shear	Kusano et al. (2004)
Self-induced shear flows	Manchester (2003, 2007), Manchester et al. (2004a)
Magnetic breakout	Antiochos et al. (1999), MacNeice et al. (2004), Lynch et al. (2004, 2008), van der Holst et al. (2007, 2009), Masson et al. (2013)
Tether cutting	Moore and Roumeliotis (1992), Moore et al. (2001), Aulanier et al. (2010)
Converging flows / Flux cancellation	Inhester et al. (1992), Amari et al. (2003a), Roussev et al. (2004), Zuccarello et al. (2012), Mikić et al. (2013)
Flux decrease / dispersion	Lin et al. (1998), Linker et al. (2001, 2003), Amari et al. (2000, 2003b), Titov et al. (2008), Reeves et al. (2010)
FE close/below flux rope	Chen and Shibata (2000), Lin et al. (2001), Shiota et al. (2005), Dubey et al. (2006)
FE into potential arcade	Zuccarello et al. (2008), Jacobs and Poedts (2012), Roussev et al. (2012)
FE into sheared arcade	Notoya et al. (2007), Kusano et al. (2012)
Helical kink instability	Sakurai (1976), Hood and Priest (1979), Gerrard et al. (2001), Fan and Gibson (2003), Török et al. (2004), Török and Kliem (2005)
Flux transfer/feeding	Zhang et al. (2014), Kliem et al. (2014b)
Tilt instability	Keppens et al. (2014)
Double-arc instability	Ishiguro and Kusano (2017)
DRIVER:	
Torus instability / Flux-rope catastrophe / Loss of equilibrium	van Tend and Kuperus (1978), Priest and Forbes (1990), Forbes and Isenberg (1991), Lin et al. (1998), Kliem and Török (2006), Török and Kliem (2007), Fan and Gibson (2007), Olmedo and Zhang (2010), Démoulin and Aulanier (2010), Kliem et al. (2014a)
Flare-reconnection	Lin and Forbes (2000), Vršnak (2008), Temmer et al. (2010), Karpen et al. (2012)
Review articles:	Forbes (2000), Chen (2001), Klimchuk (2001), Low (2001), Priest and Forbes (2002), Lin et al. (2003), Linker et al. (2003), Zhang and Low (2005), Forbes et al. (2006), Moore and Sterling (2006), Mikić and Lee (2006), Roussev (2008), Vršnak (2008), Amari and Aly (2009), Linton and Moldwin (2009), Schrijver (2009), Aulanier et al. (2010), Chen (2011), Jacobs and Poedts (2011), Kleimann (2012), Schmieder et al. (2013, 2015), Aulanier (2014), Inoue (2016), Chen (2017)

force or Lorentz-self force, which results from the curvature of a current-carrying flux rope (e.g., Titov and Démoulin 1999). The flare reconnection, on the other hand, can drive the eruption in several ways. For instance, by transferring restraining overlying flux into flux of the erupting sheared core or flux rope, or by producing highly bent field lines in a rope that accelerates it by means of a slingshot effect (e.g., Linton et al. 2001; Mackay and van Ballegooijen 2006; Archontis and Török 2008). These two processes (instability and reconnection) often occur simultaneously and are closely coupled (e.g., Bein et al. 2012; Vršnak 2016, see also Sect. 2.2.3). Their respective contributions to the acceleration of the ejecta are presently not well known; they likely depend on various parameters and may therefore differ from case to case. Disentangling the contributions is not trivial and poses a challenge for modellers. As we will see in the next section, it appears that at least the most impulsive eruptions are driven predominantly by reconnection.

As apparent from Table 1, many more trigger mechanism than driving mechanisms seem to exist. Rather than describing all the trigger mechanisms that have been suggested, we only give a few general remarks here, noting that more detailed explanations can be found in the references compiled in Table 1. First of all, trigger mechanisms should not be thought of as acting in isolation or being mutually exclusive. It is likely that several of them work together, simultaneously or successively, in starting an eruption (e.g., Williams et al. 2005). Unfortunately it is often difficult, if not impossible, to pin down which particular mechanism or combination of mechanisms triggered a specific eruption. Most mechanisms cannot be observed directly, so indirect evidence such as EUV brightening or the slow rise of a filament must be used. Also, the same observational manifestations may be produced by different mechanisms. This lack of detailed knowledge greatly hampers our capabilities to forecast the onset of eruptions (see Sect. 3), as different trigger mechanisms may have very different onset conditions. Moreover, even if one could predict that a certain trigger mechanism will take place, it would still be hard to predict whether an eruption will produce a CME or remains confined (e.g., Moore et al. 2001; Török and Kliem 2005; Guo et al. 2010) as we know little about the conditions under which trigger mechanisms can succeed in switching on driving mechanisms. In fact, it seems that, at least in some events, several confined eruptions are required before a full eruption can occur (e.g., Panesar et al. 2015; Chintzoglou et al. 2015; Liu et al. 2016).

Overall, observations and simulations have helped us to develop a relatively clear qualitative picture of how a typical solar eruption works. Slow magnetic reconnection acting below (potentially also above) a sheared or twisted core field results in a slow rise of this field. During this slow rise, the core field successively detaches from the photosphere, reconnection occurs at increasingly larger heights in the corona, and the field successively transforms from a structure that was initially probably more like an arcade into a flux rope. Once the rope has been lifted to a sufficiently large height, the torus instability sets in and starts to rapidly accelerate the rope upward. More or less simultaneously, the flare reconnection sets in and supports the rapid acceleration. What is still needed, though, is a better understanding of the onset conditions, efficiency, and interplay of the underlying mechanisms. In the next sub-section, we will use some general considerations and an analytical approach to dig deeper into the physics behind driving mechanisms and the acceleration of CMEs.

2.2.3 Eruption Driving Mechanisms and Physical Background

The free magnetic energy contained in the coronal field can be expressed in the form $W = LI^2$, where I is the total electric current flowing through the system and L is the self-inductance that includes information on the entire magnetic-field geometry (e.g., Garren and Chen 1994; Chen 1989; Chen and Krall 2003; Jackson et al. 2004; Žic et al. 2007, and references therein). Depending on the desired complexity of the model, an otherwise intricate coronal magnetic configuration is represented in the analytical approach as a simple line current, a flux rope, a flux rope embedded in a magnetic arcade, etc. In the process of an eruption, the free magnetic energy is explosively released and transformed into kinetic and gravitational potential energy. In addition, if the eruption is accompanied by a flare, part of the energy is consumed for various thermal and non-thermal processes (heating, convective flows, non-thermal particle populations, etc.). Finally, as the speed of the eruption increases, progressively more and more energy is spent for the work against the "aerodynamic" drag. Let's now go on to consider the maximum velocity and acceleration that a CME can attain.

Bearing in mind energy conservation, one can relate the kinetic energy density and the magnetic field energy density as $\rho v^2/2 < B^2/2\mu$, where ρ, v, B, and μ are the plasma mass-density, flow speed, magnetic field strength, and permeability, respectively. This relation can be rewritten as $v < B/\sqrt{\rho\mu}$, i.e., $v < v_{\mathrm{A}}$, where v_{A} is the Alfvén speed *within* the erupting structure. Consequently, it is expected that eruptions from active regions should attain higher speeds than those occurring in quiet-sun regions, which is fully consistent with observations (e.g., Vršnak et al. 2005; Bein et al. 2011, 2012, and references therein).

The CME acceleration is driven by the Lorentz force, or formulated in a more simplified way, by the gradient of the magnetic pressure $\nabla(B^2/2\mu)$. Taking an order of magnitude estimate for the magnetic pressure as $\nabla(B^2/2\mu) \approx B^2/2d\mu \approx f_L$, where d is the size of the structure and f_L is the Lorentz force *per* unit volume, the acceleration can be expressed as $a = f_L/\rho = B^2/2d\rho\mu$, i.e., $a \approx v_{\mathrm{A}}^2/2d$. This implies that the acceleration of a CME is limited by the intrinsic Alfvén speed, and that the acceleration should be larger in compact eruptions than in extended ones. Indeed, in eruptions of active-region structures, where the Alfvén speed is high and the source region is rather compact, the acceleration is usually much larger than in eruptions from quiet-sun regions (e.g. eruptions of quiescent prominences, stealth CMEs; see, e.g., Vršnak et al. 2005; Bein et al. 2012; Howard and Harrison 2013). Note that the relation $a \approx v_{\mathrm{A}}^2/2d$ implies that in the case of very compact sources, $d < 100$ Mm, accelerations can achieve values on the order of 10 km s^{-2}, such as is observed in the most impulsive events (Vršnak et al. 2007; Bein et al. 2011).

Using these estimates for the maximum velocity and acceleration, one can also roughly estimate the time and distance over which an eruption will be accelerated. The acceleration time can be expressed as $t_a \approx v/a$, implying $t_a \approx d/v_{\mathrm{A}} = t_A$, where t_A represents the Alfvén travel time over the eruptive structure. The acceleration, a, speed, v, and the acceleration distance, d_a, can be related as $a \approx \sqrt{2ad_a}$, implying $d_a \approx v^2/2a = d$, i.e., the acceleration distance should be comparable with the size of the eruptive structure. Indeed, both types of relationships $t_a \propto d$ and $d_a \propto d$, as well as $t_a \propto 1/a$, are found in observations (Vršnak et al. 2007; Bein et al. 2011). Furthermore, it is well known that eruptions originating from active regions (smaller size, higher Alfvén speed) are more impulsive (stronger acceleration over a shorter acceleration time) than those launched from quiet-sun regions (Zhang and Dere 2006; Vršnak et al. 2007; Bein et al. 2011), consistent with $t_a \approx d/v_{\mathrm{A}}$ and $a \approx v_{\mathrm{A}}^2/2d$. Characteristic values for active-region and quiescent-prominence eruptions, based on the presented order-of-magnitude relations, are displayed in Table 2.

Let us now consider some general aspects related to the evolution of an eruption. The self-inductance L is primarily dependent on the size d of the current-carrying structure, $L \propto d$ (Jackson et al. 2004). Thus, as an erupting structure expands, the self-inductance increases, and it can be approximately taken that d is proportional to the heliospheric distance r, i.e., $d \propto r$. Under this assumption $L \propto r$. On the other hand, the self-inductance of the magnetic structure is by definition related to the electric current as $\Phi = LI$, where Φ is the magnetic flux associated with the current I (see, e.g., Batygin and Toptygin 1962; Žic et al. 2007).

Table 2 Typical values for active region (AR) and quiescent prominence (QP) eruptions. The size of the CME source structure is given by d, the Alfvén speed within the erupting structure by v_{A}, velocity by v, acceleration by a, acceleration duration by t_a and distance over which the structure is accelerated by d_a

	d [Mm]	v_{A} [km s^{-1}]	v [km s^{-1}]	a [m s^{-2}]	t_a [s]	d_a [Mm]
AR	100	1000	1000	10000	100	100
QP	1000	400	400	160	2500	1000

In the absence of dissipative processes that could lead to magnetic reconnection, the flux Φ stays preserved, implying that the electric current has to decrease as $I \propto 1/r$. This means that the free magnetic energy of the system, $W = LI^2$, decreases, being transformed into the kinetic energy and the work done against drag and gravity.

In line with the magnetic origin of an eruption, the CME acceleration is driven by the Lorentz force, or formulated in a simplified way, by the Ampère force $\boldsymbol{F} = \boldsymbol{I} \times \boldsymbol{B}$. Consequently, in the absence of reconnection, the driving force of a CME rapidly decreases with distance from the Sun, because both the electric current I and the magnetic field B protruding through the magnetic structure decrease (for the dependence of the magnetic field on height see, e.g., Dulk and McLean 1978; Gary 2001; Vršnak et al. 2002, 2004a; Liu 2008). Indeed, observations indicate that the driving force decreases approximately as $r^{-\alpha}$, where the exponent ranges between $\alpha \approx 1$ and 2 (Vršnak 2001, 2006; Vršnak et al. 2004c). In other words, the inductive effects and the decrease of the magnetic field with height, act to prevent a very strong acceleration happening over a large time and distance range.

The general physical principles that govern the eruption process described here should be incorporated in any quantitative model, no matter what the specific magnetic configuration is. Most often, the eruptive structure is represented by a semi-toroidal magnetic flux rope that is fixed at both ends in the photosphere. It is worth noting that the flux-rope concept has been in use since before the discovery of CMEs, i.e., in early models of solar flares and eruptive prominences (e.g., Gold and Hoyle 1960; Ellison et al. 1961; Gold 1962; Byrne et al. 1964; Hyder 1967; Anzer and Tandberg-Hanssen 1970). Gradually, flux-rope models developed from very simple ones (e.g., Kuperus and Raadu 1974; Mouschovias and Poland 1978; Sakurai 1976; Anzer 1978; van Tend and Kuperus 1978; Anzer and Pneuman 1982; Pneuman 1980; Vršnak 1984; Steele and Priest 1989) to more complex forms, where various relevant effects have been sequentially taken into account (e.g., Chen 1989; Vršnak 1990; Titov and Démoulin 1999; Török and Kliem 2003). In addition, the development of numerical techniques has provided more demanding studies, beyond those provided by the analytical approach.

Now we focus on the implications of the basic physical principles that have just been discussed as applied to the flux rope concept of an eruption. To do so, let us consider the forces acting on a semi-toroidal magnetic flux rope, sketched in Fig. 5a. In such a configuration, several effects have to be considered. First, the curvature of the flux rope axis causes the Lorentz self-force (also known as "hoop force"). The contribution of the self-force to the equation of motion was firstly estimated by applying the "virtual work" principle (e.g., Shafranov 1966; Garren and Chen 1994). Hereafter, we will follow the procedure proposed by Vršnak (1984) and Vršnak (1990), where the self-force is estimated considering directly the Lorentz force components related to the magnetic pressure gradient and tension (sketched in the upper right and left corners of Fig. 5a).

Next, one has to take into account a diamagnetic effect, which comes from the influence of eddy electric currents induced at the solar "surface". In the simplest way, this effect can

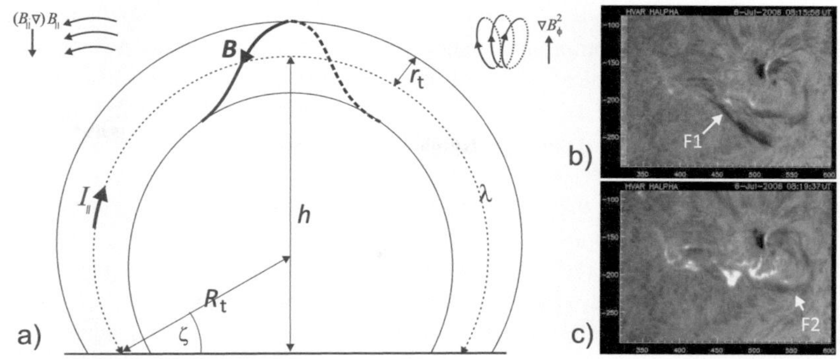

Fig. 5 **a**) Cartoon of a semi-toroidal flux rope with definition of symbols used in the text. In the *upper-left* and *upper-rigth* corner the effects of magnetic tension and pressure gradient are sketched, respectively. **b**) & **c**) Eruption of July 6, 2016: Filament F1 in a slow-rise phase (*b*), indicating a presence of a metastable flux rope, gradually evolving towards the loss-of-equilibrium state; The eruption of the filament F1 caused a flare below it (*c*) and destabilized the nearby (also metastable) filament F2 that erupted too, causing a westwards extension of the flare (Hvar Observatory filtergrams; x- and y-coordinates are shown in arcsec)

be roughly estimated by considering the effect of a line-current located at the height h above the conducting surface, as done by Kuperus and Raadu (1974). Furthermore, the flux rope concept includes an overlying coronal field (B_c), and the influence of this field must be taken into account. The effect depends on the orientation of the background field—in the case of the Kuperus–Raadu configuration (Kuperus and Raadu 1974), the force is directed downwards, whereas in the case of the Kippenhahn–Schlüter configuration (Kippenhahn and Schlüter 1957), the force is directed upwards.

Finally, one has to consider the effect of gravity. Taking into account buoyancy, the acceleration reads $a_g = -g(\rho - \rho_w)/\rho$, where g is the local gravitational acceleration and ρ and ρ_w represent the density of the CME and the ambient plasma, respectively. Note that in the case $\rho \approx \rho_w$ the gravitation can be neglected, and if $\rho < \rho_w$ then a_g becomes positive (i.e., acts upwards) (Low et al. 1982).

Let us now consider the stability of a flux-rope and the conditions under which it will erupt. Putting all the previously mentioned effects together leads to an expression for the net acceleration (for details of the derivation see Vršnak 2008, 2016):

$$a = \frac{C_L}{\Lambda}\left[\frac{1}{2\tilde{R}_t} - \frac{1}{\tilde{R}_t X^2} + \frac{1}{H}\right] - \frac{C_c}{\Lambda^2 \tilde{R}_t} - C_g\, g, \tag{1}$$

where $\tilde{R}_t$, Λ, and H represent the major radius of the semi-toroidal flux rope, the length of its axis, and the height of the axis summit, respectively, all normalized with respect to the footpoint half-separation (following the notation presented in Fig. 5a, one can write $\tilde{R}_t \equiv R_t/d$, $\Lambda \equiv \lambda/d$, $H \equiv h/d$, and $\tilde{r}_t \equiv r_t/d$). The parameter X represents the ratio of the poloidal and longitudinal magnetic field components, $X = B_\phi/B_\parallel$, at the surface of the flux rope (Fig. 5a). The first three terms on the right-hand side of (1), written within the square brackets, represent the effects of the magnetic pressure gradient, magnetic tension, and diamagnetic effect, respectively. The fourth term is due to the background field, whereas the last term considers gravity.

In (1), we took into account the previously inferred relation $I \propto 1/\lambda$, i.e., we express the current as $I = I_0\lambda_0/\lambda$, where I_0 is the value of the axial current when the flux-rope axis is at the height $h = d$ (i.e., $R_t = h = d$, $\lambda = R_t\pi$, $\tilde{R}_t = H = 1$, $\Lambda = \pi$) and we considered that the mass of the flux rope is $M = const$. The constant $C_L = \mu I_0^2 \pi / 4M$ determines the overall "strength" of the eruption, and it shows that the CME acceleration is proportional to I_0^2. Taking into account typical values of $M = 10^{12}$–10^{13} kg (e.g., Vourlidas et al. 2010; Webb and Howard 2012) and $I_0 = 10^{10}$–10^{11} A (e.g., Hofmann et al. 1992; Filippov et al. 2015, and references therein), it can be estimated that the value of C_L should typically range between 100 and 1000 m s^{-2}.

Furthermore, we assumed that B_c (background coronal magnetic field surrounding the rope) is uniform over the rope. So the flux conservation $B_c r_t \lambda = const.$, implies $B_c = B_{c0} r_{0t} \lambda_0 / r_t \lambda$. We note that in the absence of reconnection the eruptive structure should expand in a self-similar way, i.e., $r_t \propto R_t$ (Vršnak 2008), which means $B_c = B_{c0} R_{0t} \lambda_0 / R_t \lambda = B_{c0} \pi / \tilde{R}_t \Lambda$. Furthermore, we introduced a constant $C_c = I_0 B_{c0} \pi^3 d / M$, where we have taken into account $I\lambda = I_0 \lambda_0$, i.e., $I = I_0 \pi / \Lambda$. Taking typical values of $d = 100$ Mm and $B_{c0} = 0.1$–1 G (i.e., 10^{-5}–10^{-4} T), together with the previously used values for I_0 and M, one finds C_c could be as large as 5×10^4 m s^{-2}. In the case where the effect of a decrease of the background magnetic field with height is considered, B_{c0} should be replaced by, e.g., $B_c = B_{c0}(R_{0t}/R_t)^n = B_{c0} \tilde{R}_t^{-n}$, as is often used in treatments of the torus instability (e.g., Kliem and Török 2006; Démoulin and Aulanier 2010; Olmedo and Zhang 2010, and references therein). Under this approximation, the last term in (1) should be replaced by $C_c / \Lambda^2 \tilde{R}_t^{1+n}$.

To solve (1), one has to specify the relationship between the geometrical parameters $\tilde{R}_t$, H, and Λ. Using Fig. 5a, where we have included the angle ζ that represents the inclination of a line connecting a flux-rope footpoint and the center of curvature of the rope axis, one finds the parametric relationships $\tilde{R}_t = 1/\cos\zeta$, $H = (1 + \sin\zeta)/\cos\zeta$, and $\Lambda = (\pi + 2\zeta)/\cos\zeta$. Taking into account that the poloidal-to-axial field ratio can be expressed as $X \equiv B_\phi / B_\parallel = 2\pi \tilde{r}_t / \Lambda$, (1) constitutes an explicit expression for the acceleration as a function of the angle ζ, and from that, one can establish a functional dependence $a(H)$.

Various solutions to (1) are given in Fig. 6. Several $a(H)$ curves are shown, following an increasing C_L (Fig. 6a and b) and X (Fig. 6c and d) sequence. The increase of C_L and/or X can be caused by various evolutionary processes related to the photospheric and subphotospheric motions. For example, twisting motions at the flux-rope footpoints would lead to a transport of poloidal flux into the rope (thus increasing X) and increase the electric current (thus increasing C_L). A similar effect may also be expected in the case of reconnection below the rope, which tends to increase the poloidal flux in the rope (see e.g., Vršnak 2016). The value of C_L could also be increased by emerging flux, since the increase of the total magnetic flux encircled by the current circuit associated with the rope would lead to an enhancement of the current. Finally, C_L might be increased by a mass leakage through the flux-rope legs, since $C_L \propto 1/M$.

The intercepts of the $a(H)$ curves with the x-axis are equilibrium points ($a = 0$). If the slope of the $a(H)$ curve, $\partial a / \partial H \equiv \omega^2$, at the equilibrium point is characterized by $\omega^2 < 0$ the equilibrium is stable (points denoted in Fig. 6b as H_1, H_2, H_3, and H_4). If the rope is moved from this position, it will oscillate around it, with a period of $P = 2\pi/\omega$. Note that from the observational point of view, oscillations of coronal structures are especially well visualized in the case of prominence oscillations initiated by disturbances coming from distant eruptions, a phenomenon being reported as "winking filaments" a half of century ago by Ramsey and Smith (1966). Inspecting Figs. 6b and 6d, one finds that as the value of C_L or X increases, the equilibrium position shifts to larger heights (see the insets in Figs. 6b and 6d) and the oscillation period increases (the slope $\partial a / \partial H$ becomes less steep).

The x-axis intercepts of $a(H)$ curves characterized by $\partial a / \partial H > 0$ represent unstable equilibria. If the rope is displaced from a stable equilibrium to an unstable one (e.g., by

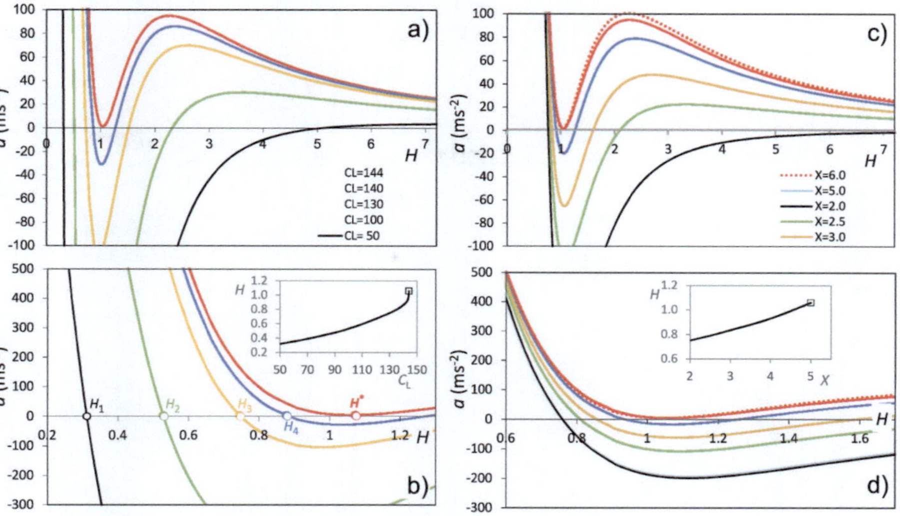

Fig. 6 Various solutions of (1), representing stable, metastable, and unstable ropes. **a)** $a(H)$ for various values of C_{L} (written in the *legend*), in combination with $X = 5$, $r_{\mathrm{t}}/R_{\mathrm{t}} = 0.1$, $C_{\mathrm{c}} = 1.3 \times 10^4\ \mathrm{m\,s^{-2}}$. **b)** Enlarged part of *a*) around equilibrium points; in the *inset* the equilibrium height H_e is shown as a function of C_{L}. **c)** $a(H)$ for various values of X (written in the *legend*), in combination with $C_{\mathrm{L}} = 144\ \mathrm{m\,s^{-2}}$, $r_{\mathrm{t}}/R_{\mathrm{t}} = 0.1$, $C_{\mathrm{c}} = 1.3 \times 10^4\ \mathrm{m\,s^{-2}}$; the *red-dotted curve* represents the case of $X = 6.0$ with $C_{\mathrm{L}} = 143\ \mathrm{m\,s^{-2}}$. **d)** Enlarged part of *c*) around equilibrium points; in the *inset* the equilibrium height H_e is shown as a function of X

a disturbance from a distant eruption), it will erupt. Thus, ropes that are characterized by $a(H)$ curves that have both types of x-axis intercepts are in a metastable state. Note that with increasing C_{L} or X, the distance between stable and unstable equilibria decreases, i.e., it becomes easier to push the rope from the stable to the unstable point and trigger the eruption. Finally, it should be noted that the $a(H)$ curve achieves the metastable form only for a sufficiently large value of X. This is illustrated in Fig. 6d, where the $a(H)$ curve for $X = 2$ (black curve) is shown, which is characterized by only one (stable) equilibrium point. The curve calculated for $X = 2.5$ (green curve) already contains the unstable equilibrium at $H \approx 2$. Thus, a value of X at which the magnetic structure transforms from stable to metastable configuration lies between 2 and 2.5; a more detailed analysis shows that for the considered combination of parameters ($C_{\mathrm{L}} = 144\ \mathrm{m\,s^{-2}}$, $r_{\mathrm{t}}/R_{\mathrm{t}} = 0.1$, $C_{\mathrm{c}} = 1.3 \times 10^4\ \mathrm{m\,s^{-2}}$) this transition occurs at $X = 2.2$.

The increase of C_{L} or X leads to another essential physical situation. As the $a(H)$ curves gradually shift "upwards", at a given critical value of C_{L}^* or X^*, the stable and unstable equilibrium points merge (see the red curve in Fig. 6), after which there is no equilibrium possible. The local minimum in the $a(H)$ curve just touches the x-axis (in Fig. 6b this critical height is denoted as H^*), i.e., the whole $a(H)$ curve is located in the $a \geq 0$ region. Consequently, the rope erupts, dynamically trying to find a new equilibrium. In the case when the gravity is neglected ($C_{\mathrm{g}} = 0$) a new equilibrium does not exist, and the rope erupts. Note that the critical value C_{L}^* is lower for a higher value of X. This is illustrated in Fig. 6c, where the loss-of-equilibrium curve for $X = 5$ (full-red line; $C_{\mathrm{L}}^* = 144\ \mathrm{m\,s^{-2}}$) is compared with the analogous curve for $X = 6$ (dotted-red line; $C_{\mathrm{L}}^* = 143\ \mathrm{m\,s^{-2}}$). An example of a metastable flux rope containing filament plasma, gradually evolving towards the loss-of-equilibrium stage and eruption, is presented in Fig. 5b. Filament F1 was slowly rising until

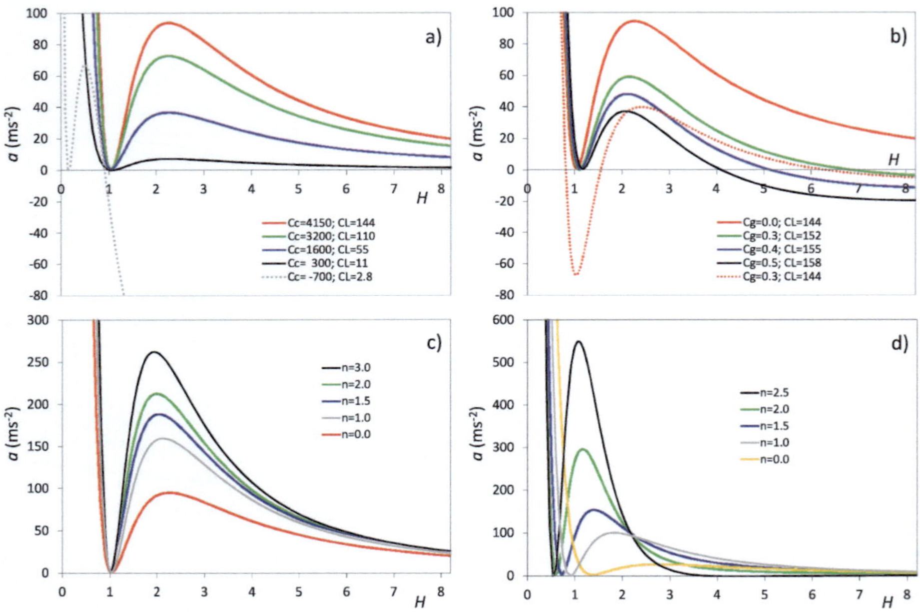

Fig. 7 Flux ropes in the loss-of-equilibrium state: **a**) $a(H)$ solutions for different levels of the background-field effect (values of C_c expressed in $\mathrm{m\,s^{-2}}$ are written in the *legend*, together with the corresponding critical values of C_L^*); *gray-dotted line* represents Kippenhahn–Schlüter configuration with $C_g = 1$ ($IB_c < 0$, i.e., $C_c < 0$). **b**) The gravity effect (values of C_g are written in the *legend*, together with the corresponding critical values of C_L^*) calculated using $d = 70$ Mm, $X = 5$, and $C_c = 13000\ \mathrm{m\,s^{-2}}$. **c**) The effect of decreasing background field (values of the decay index n are given in the *legend*), calculated using $X = 5$, $C_L^* = 144\ \mathrm{m\,s^{-2}}$, $C_c = 13000\ \mathrm{m\,s^{-2}}$, and , $C_g = 0$. **d**) The torus-instability effect (values of n are given in the *legend*), calculated using the same set of parameters as in *c*)

reaching a critical height, after which it erupted (Fig. 5c). Besides causing a flare below it, the eruption destabilized and triggered the eruption of the nearby metastable flux rope that was embedding the filament F2 (for details see Miklenic et al. 2009).

For a flux rope characterized by a given value of the poloidal-to-axial field ratio X, a critical value C_L^* at which the eruption starts depends on the relative contribution of the background magnetic field and the gravity. In other words, it depends on the values of C_c and C_g, which is illustrated in Figs. 7a and 7b, respectively.

In Fig. 7a we present four loss-of-equilibrium $a(H)$ curves calculated for different values of $C_c > 0$ and the corresponding values of C_L^*, applying fixed values $X = 5$ and $C_g = 0$. The graph illustrates that the peak acceleration a_m is larger for a larger value of C_c, i.e., larger C_L^*, and that $a_m \to 0$ for $C_c \to 0$. The value of a_m increases if larger values of X or lower values of C_g are applied. Note that in all cases shown in Fig. 7a the eruption starts at the same height, very close to $H = 1$. This is consistent with conclusions drawn by (e.g., Chen 1989; Vršnak 1990) and confirmed by observations (e.g., Vršnak et al. 1991; Chen and Krall 2003). This property is a consequence of the hoop-force behaviour, achieving a maximum at $H = 1$, since at this height the radius of curvature attains minimum, $\tilde{R}_t = 1$.

For comparison, besides the four $a(H)$ curves based on the Kuperus–Raadu configuration ($C_c > 0$) shown in Fig. 7a, one example of the Kippenhahn–Schlüter case is presented (gray-dotted curve; $C_c = -700\ \mathrm{m\,s^{-2}}$). It reveals a completely different behaviour than the Kuperus–Raadu configuration. First, the equilibrium point is shifted to much lower heights than in the Kuperus–Raadu case. Furthermore, the metastable form of the $a(H)$ curve is possible only for $C_g > 0$, and in that case there is always an upper-equilibrium position present (the x-axis intercept of the $a(H)$ curve located at $H \approx 1$ in the presented case). Thus, the Kippenhahn–Schlüter configuration can result only in failed eruptions, and will not be considered hereafter.

To illustrate the effect of gravity, Fig. 7b shows four loss-of-equilibrium $a(H)$ curves for different values of C_g and the corresponding values of C_L^*, applying fixed values $d =$ 70 Mm, $X = 5$, and $C_c = 1.3\ \mathrm{km\,s^{-2}}$. The three curves calculated with $C_g \neq 0$ show an upper equilibrium point, unlike the curve calculated with $C_g = 0$. Thus, in $C_g \neq 0$ cases, the eruption should stop at an upper equilibrium. As a matter of fact, after loosing equilibrium, the rope should accelerate upwards, reach maximum speed at the upper equilibrium, after which it should be decelerated until being stopped. Then, the restoring force at the upper equilibrium would start accelerating it downwards—in the absence of drag the rope should oscillate around the upper equilibrium position. Taking into account the drag effect, the oscillations would be damped until the rope settles at the upper equilibrium (for observations see, e.g., Vršnak et al. 1990; Mrozek 2008).

In Fig. 7b we also show the case $C_g = 0.3\ \mathrm{m\,s^{-2}}$ with $C_L^* = 144\ \mathrm{m\,s^{-2}}$ (red-dotted curve), to compare it with the curve calculated for the same $C_L^* = 144\ \mathrm{m\,s^{-2}}$ and $C_g = 0$ (red-full curve). The graph shows that the gravity effect shifts the $a(H)$ curve downwards, and since the effect is weaker at larger heights due to decreasing g, this leads to the formation of the upper equilibrium position. In this respect, let us also note that the gravity effect weakens with increasing d, i.e., it is smaller for larger ropes. For example, for d two times larger than the applied value of $d = 70$ Mm, the upper equilibrium disappears in the $C_g = 0.3\ \mathrm{m\,s^{-2}}$ case.

The effect of the decreasing background magnetic field is illustrated in Fig. 7c, where we show $a(H)$ curves calculated using again $C_L^* = 144\ \mathrm{m\,s^{-2}}$, $C_c = 1.3\ \mathrm{km\,s^{-2}}$, $X = 5$, and $C_g = 0$, but now applying several different values of the exponent n in the dependence $B_c = B_{0c}\tilde{R}_t^{-n}$. The red curve, representing the case $n = 0$, is the same as the red curves in Figs. 7a and 7b. The graph shows that the peak acceleration increases with increasing n, i.e., it is larger for eruptions occurring in coronal regions where the decrease of the magnetic field is steeper. The rate of decrease is dependent on the flux distribution at the photosphere. Furthermore, the acceleration peak systematically shifts to lower heights.

Although the effects presented here are similar to those governing the torus instability, they do not describe the same physical mechanism. If one would like to consider the torus-instability effect, the pressure-gradient term in the hoop force, i.e., the first term in (1), should be replaced by a new term, estimated using a new difference between the magnetic field strength at the inner and outer boundary of the torus. The difference can be estimated from the derivative of $B_c = B_{0c}\tilde{R}_t^{-n}$, and from that one gets the pressure-gradient term of the form $2nC_L/\Lambda\tilde{R}_t^{n+1}$. Note that this approach is different from that usually applied (e.g., Kliem and Török 2006), where the equation of motion is derived employing the virtual-work principle. Consequently, the equation of motion is of a somewhat different form than that used in torus instability studies. For example, the "standard" torus instability equation of motion includes explicitly the self-inductance, whereas it is herein included implicitly. In addition, (1) includes the diamagnetic effect, Ampére force, and gravity, which are not usually considered in torus instability studies.

The torus instability effect is illustrated in Fig. 7d, where $a(H)$ curves for several values of n are presented. Comparing this graph with the graphs presented in Fig. 7a–c, one finds a considerably different behaviour of the flux rope. First, the loss-of-equilibrium height H^* is increasing with decreasing n, and at the same time, the peak acceleration a_m decreases.

Second, in the case of $n > 1$ the $a(H)$ profile is more sharply peaked than profiles shown in Fig. 7a–c, and the peak accelerations are higher. Thus, eruptions driven by the torus instability should be more impulsive than those driven by the hoop force that represents a form of the kink instability. Finally, note that an eruption is not possible if the value of n is too low, due to the trend $a_{\mathrm{m}} \to 0$ and $H^* \to \infty$ for decreasing n. For example, for the set of parameters used for Fig. 7d, the critical value of n is ≈ 0.8.

The analysis presented here shows that under the conditions considered it is difficult to get accelerations higher than 1000 m s^{-2}. For example, even taking C_{L} as high as 1000 m s^{-2} in the hoop-force option (at the same time requiring extremely large $C_{\mathrm{c}} = 100$ km s^{-2}), one gets a peak acceleration up to 600 m s^{-2}. This limitation is related to the conservation of magnetic flux associated with the electric current in the eruptive structure, in conjunction with the inductivity effects. As previously demonstrated, these effects cause a decrease of the current as the size of an eruption increases, which in turn leads to a rapid decrease of the driving Lorentz force. Thus, in order to get a really high acceleration, the flux-conservation constraint has to be broken. This can be attributed either to an abundant flux emergence, as proposed by Chen (1989), or by eruption-related reconnection resulting in the flare energy release, as suggested by Vršnak (2008). Since in the former case the emergence rate should be much higher than ever observed, the latter option seems to be more appropriate (Kunkel and Chen 2010). It is also important to note that since the open field configuration holds more energy than the closed state, it has been postulated that magnetic reconnection is necessary for a successful eruption to occur (Aly 1984; Sturrock 1991).

The role of reconnection in the dynamics of a flux rope eruption was analysed by Vršnak (2008) and Vršnak (2016), providing an insight into the physics of the CME/flare relationship. Reconnection results in two effects that can significantly enhance and prolong the CME acceleration. First, reconnection supplies the rope with additional poloidal flux (see Fig. 4 in Vršnak 2008), which enhances the hoop force, providing higher peak accelerations. Furthermore, the poloidal-flux supply reduces the inductive decrease of the axial current in the later phases of the eruption, i.e., the acceleration phase is prolonged. The second important effect of reconnection is the weakening of the magnetic tension of the arcade field overlying the flux rope by detaching this field from the photosphere.

In Vršnak (2008) and Vršnak (2016) the reconnection effects were analysed assuming that reconnection starts simultaneously with the force imbalance, attains maximum rate when the rope is at a certain height, and ends at twice this height. Various combinations of the height range and peak reconnection rate were considered to find out how these parameters affect the eruption kinematics (see Fig. 2 in Vršnak 2016). It was demonstrated that the acceleration time profile is tightly synchronized with the reconnection-rate time profile, whereas the velocity time profile is synchronized with the cumulative reconnected flux. This explains the observed synchronization of the CME acceleration phase and the impulsive phase of the associated flare (Kahler et al. 1988; Neupert et al. 2001; Zhang et al. 2001, 2004; Shanmugaraju et al. 2003; Vršnak et al. 2004b; Zhang and Dere 2006; Maričić et al. 2007; Temmer et al. 2008, 2010). It was also shown that the peak acceleration is higher for higher peak reconnection rate. Consequently the maximum speed of the eruption is higher for larger total reconnected flux, consistent with observations (Qiu and Yurchyshyn 2005). It was also demonstrated that the highest observed accelerations, reaching ≈ 10 km s^{-2}, could be achieved if the total reconnected flux is a few times larger than the initial poloidal flux of the rope, corresponding to $\approx 10^{22}$ Mx. Such amounts are frequently observed in flares (e.g., Qiu et al. 2004, 2007; Qiu and Yurchyshyn 2005; Miklenic et al. 2007, 2009).

The sequence of events presented here can be brought together in a relatively simple eruption scenario. If the coronal structure contains a flux rope and this rope is characterized by weak electric current and/or low degree of twist (i.e., low values of C_L and X, respectively), it is stable. If the rope is displaced from the equilibrium position, it will oscillate around the equilibrium position. If C_L or X (or both) start to increase due to photospheric twisting, shearing, or flux emergence, the structure gradually evolves through a series of quasi-equilibrium states. The equilibrium height in the $a(H)$ curve increases, which is reflected in a slow rise of the structure. At a given amount of twist, the structure becomes metastable, and if pushed strongly enough to reach the point of unstable equilibrium, it will erupt.

If the structure is sufficiently sheared or twisted, the gradual rise of the structure caused by increasing C_L and X ends up with a loss of equilibrium, and consequently, an eruption. The threshold value of X (the ratio of poloidal to axial field) is lower for structures characterized by larger C_L. It is important to note that observations of the flux swept out by flare ribbons, compared to in situ flux measurements of the associated magnetic cloud, indicate that the pre-eruption structure is weakly twisted, consistent with EUV and soft X-ray observations. The poloidal flux is added via the flare reconnection (Qiu et al. 2007; Hu et al. 2014). Under certain circumstances, when the $a(H)$ curve shows the presence of an upper equilibrium, the eruption might be halted, showing damped oscillations around the upper equilibrium position. Otherwise, the eruption continues into the heliosphere. The acceleration is stronger, and speed higher, for larger values of C_L and X. Since C_L is related to the current ($C_L \propto I^2$) and X with the self-inductance ($L \propto X$), this is equivalent to saying that the acceleration and speed are higher for structures containing larger amount of free energy $W = LI^2$. Furthermore, since C_L can be also expressed as $C_L \propto B^2/\rho$, we find that acceleration and speed are higher for structures characterized by a higher Alfvén speed. This is consistent with the order-of-magnitude relationships anticipated at the beginning of this section.

To conclude this sub-section, the analysis presented demonstrates that events characterized by $a_{max} > 1\ \mathrm{km\,s^{-2}}$ can occur only if the eruption is accompanied by strong magnetic reconnection below the erupting rope. Indeed, the most powerful/impulsive eruptions are associated with powerful/impulsive flares, supporting such a conclusion. Note that the reconnection process can go on for a long time, on the order of a day, as evidenced by growing post-eruption loop systems (e.g., Veronig et al. 2006; Vršnak et al. 2006, and references therein) and signatures of current sheets in the wake of the CME (e.g., Ko et al. 2003; Webb et al. 2003; Lin et al. 2005; Ciaravella and Raymond 2008; Vršnak et al. 2009). Finally, the acceleration can be additionally enhanced by a transport of twist into the expanded summit of the rope (Jockers 1978; Browning and Priest 1983) and by writhing of the flux-rope axis that decreases the radius of curvature (e.g., Török et al. 2010).

2.2.4 Sympathetic Eruptions

The previous discussions in this section focussed on single, isolated eruptions. However, solar eruptions can occur relatively close to each other in both space and time, and may thus interact low in the corona or later on in interplanetary space. Such interactions are discussed in Manchester et al. in this Volume; see also the recent review by Lugaz et al. (2016). Here we focus on a special type of interaction, namely those in which one eruption appears to initiate another one. Such events were first mentioned more than 80 years ago (Richardson 1936), and are now commonly referred to as *sympathetic* eruptions.

Sympathetic eruptions occur within a relatively short period of time (tens of minutes to several hours) at different locations on the Sun. The individual eruptions may originate in one, typically large and complex source region (e.g., Liu et al. 2009) or in different source regions that are often, but not always, adjacent.[3] Research on sympathetic eruptions revolves around two main questions. Is the close timing of individual eruptions purely coincidental, or does some causal link exist between them? And if it does, what is the physical nature of such a link?

Prior to the launch of SDO, research on the timing of flare/CME events focused mainly on the flares (Fritzova-Svestkova et al. 1976; Pearce and Harrison 1990; Bumba and Klvana 1993; Bagalá et al. 2000; Biesecker and Thompson 2000; Wang et al. 2001; Moon et al. 2002; Wheatland and Craig 2006). Many of these studies considered waiting-times (i.e., the time interval between two successive events) and the location of the event, both of which are easier to establish for flares than for CMEs (Moon et al. 2003). These predominantly statistical studies have provided strong evidence that a causal relationship between sympathetic eruptions does indeed exist. This was supported by detailed case studies (e.g., Bumba and Klvana 1993; Wang et al. 2001). Sympathetic CMEs were studied less frequently though (e.g., Simnett and Hudson 1997; Moon et al. 2003; Cheng et al. 2005; Jiang et al. 2008), and the evidence for their existence appeared to be weaker than for flares (Moon et al. 2003). However, since flares and CMEs are often merely different observational manifestations of the same eruption process, this discrimination is somewhat artificial. The physical connections between sympathetic eruptions were attributed to, e.g., large-scale convective motions that led to simultaneous flux emergence in the flaring regions (Bumba and Klvana 1993) or waves triggered by a previous eruption (e.g., Ramsey and Smith 1966, Khan and Hudson 2000; see also Mullan 1976 for sympathetic stellar flares). Later it was suggested that these connections are of a magnetic nature, such as surges or jets produced by an adjacent eruption (Wang et al. 2001; Jiang et al. 2008) or changes in the background field due to expansion or reconnection (e.g., DeVore and Antiochos 2005; Zuccarello et al. 2009; Jiang et al. 2011; Yang et al. 2012).

The launch of SDO spurred a new series of investigations. This was triggered by the complex sympathetic eruptions that took place on 1–2 August 2010 and that were observed in unprecedented detail by AIA. Schrijver and Title (2011) analyzed the coronal magnetic field during the time period of the eruptions (using potential field extrapolations), and found evidence for connections between all source regions involved via structural features such as separators and quasi-separatrix layers (QSLs; Priest and Forbes 1992; Démoulin et al. 1996). They concluded that the structural properties of the large-scale coronal field play an important role in the genesis of sympathetic eruptions (see also Schrijver et al. 2013). Other cases of sympathetic eruptions observed by AIA can be found in Shen et al. (2012, 2013), Joshi et al. (2016), and Wang et al. (2016).

The 1–2 August 2010 events also inspired new modelling efforts (for earlier attempts see, e.g., Ding et al. 2006; Peng and Hu 2007). Figure 8 shows an MHD simulation by Török et al. (2011), which was designed to qualitatively reproduce a sub-set of these events (see also Lynch and Edmondson 2013). The simulation suggests two purely magnetic mechanisms for the links between successive eruptions. Figure 8 panel (a) shows the initial configuration, consisting of two flux ropes (F2, F3) located in the lobes of a pseudostreamer and one rope (F1) next to it. Pseudostreamers or pseudostreamer-like configurations appear to be prone to producing sympathetic eruptions; they often harbor twin-filaments that erupt successively (Panasenco and Velli 2010), which was the case also in the 1–2 August 2010 events. In the simulation, the eruption of F1 triggers the formation of a thin current

[3] Successive eruptions that start from essentially the same location and display very similar morphological properties are typically referred to as *homologous* eruptions, which we do not consider here (for details on such events see, e.g., Duchlev et al. 2016; Lugaz et al. 2016, and references therein).

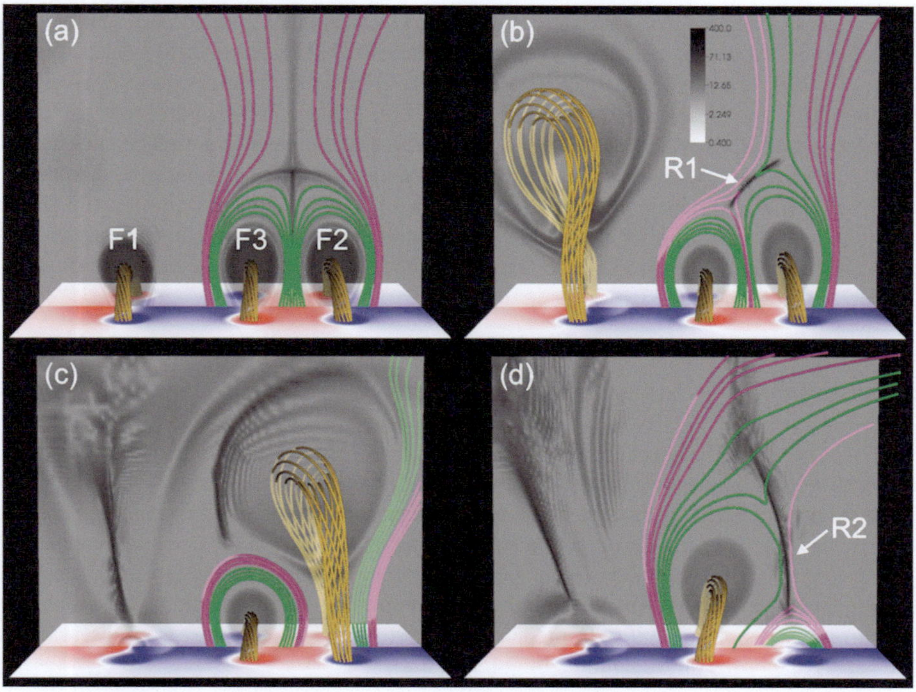

Fig. 8 MHD simulation of sympathetic flux-rope eruptions by Török et al. (2011). The eruption of F1 triggers the subsequent eruptions of F2 and F3 due to the reconnection events R1 and R2, respectively

layer around the separator line above the pseudostreamer lobes, across this current sheet breakout-like reconnection (R1) transfers flux from the right lobe to the left one (panel (b)). This reduction of flux, and thus of stabilizing magnetic tension, triggers the torus instability in F2, which subsequently erupts (panel (c)). Finally, the flare-reconnection (R2) below F2 redistributes flux from the left lobe to the growing flare arcade (panel (d)). This new flux reduction now destabilizes F3, completing the sequence of eruptions. Note that these two similar mechanisms can work independently, and their applicability is not restricted to pseudostreamers. They can work also in a quadrupolar configuration, as hypothesized by DeVore and Antiochos (2005). What is required for their occurrence is merely a flux system that contains adjacent lobes overlaid by a separator line and a sufficiently strong perturbation to set things in motion. An important and yet largely unexplored question is under which conditions such reconnections will be actually strong enough to trigger a subsequent eruption (see Török et al. 2011). Very recently, Jin et al. (2016) undertook the first steps in answering this question, by investigating and quantifying the effects CMEs can have on adjacent and remote magnetic fields.

The simulation shown in Fig. 8 constituted a significant step forward in understanding the physical links between sympathetic eruptions. However, it was designed as a "proof of concept" model, using a simplified magnetic configuration. Employing instead a potential field source surface (PFSS) model based on observed magnetograms, Titov et al. (2012) performed a detailed topological analysis of the coronal magnetic field during the time of the 1–2 August 2010 events, and found support for the occurrence of the mechanisms suggested in Török et al. (2011). In order to further substantiate this scenario, MHD simulations with more realistic magnetic fields need to be employed to test whether sympathetic eruptions can occur also under less idealized and symmetric conditions (Mikic et al. 2011). Such simulations will allow us also to investigate the long-range effects of eruptions (e.g., Schrijver and Higgins 2015), as recently demonstrated by Jin et al. (2016). Furthermore, whether other mechanisms, such as waves, can serve as a possible link between eruptions can be tested too. At present, the observational and numerical studies that have been performed since the launch of SDO strongly support the conjecture that the mechanisms at work in sympathetic eruptions are magnetic in nature. These mechanisms, by means of reconnection, act predominantly by reducing the magnetic tension in the source region of an eruption, allowing the core field of the region to erupt.

3 Predictability of Solar Eruptions

Significant progress has been made in understanding the physics that underlies CME occurrence, as well as the observational characteristics of eruptions. So, where do we stand in our ability to predict CMEs for use in space weather forecasting? There are two basic goals and approaches: (1) to predict the *occurrence* and timing of an eruption (before it has happened) and (2) to predict the erupting magnetic field configuration that leaves the Sun, and its arrival time at Earth which is related to the *impact* of an eruption. In this section, we summarise the current status of our ability to predict the occurrence of a CME based on observations (Sect. 3.1) and based on MHD modelling (Sect. 3.2).

3.1 Predictions Based on Observations

The evolution of the coronal and chromospheric plasma structures in CME source regions can sometimes provide an indication of an impending eruption, as summarised in Green and Baker (2015). However, the timescales over which observations reveal that a CME might be imminent can vary. For example, filaments darken, broaden and slowly rise in the minutes prior to their eruption (for a review see Martin 1980) whereas streamers brighten over around a day before they lift off as a CME (Illing and Hundhausen 1986). Ultimately, a variety of observations may need to be utilised to both identify that a magnetic structure is evolving toward an eruptive configuration, and give a warning of when the eruption is imminent. Here we summarise the large body of observational work that is relevant to CME prediction.

The observational approach to predicting CMEs broadly involves two aspects; identifying the existence of a potentially eruptive magnetic field configuration and identifying the path to its destabilisation. Perhaps the longest-known eruptive magnetic field configuration is that which supports a filament. Feynman and Martin (1995) investigated the role that flux emergence may play in the destabilisation of filaments and found that $19/30$ of the eruptions studied were observed to have new flux emergence in the vicinity of the filament. The flux emerged in a time period of a few days before the CME. Filaments that did not have nearby flux emergence did not erupt. The study showed that the orientation of the newly emerging flux was important, with bipoles that were orientated in such a way as to be favourable for reconnection being key for CME occurrence.

Sigmoids are another coronal configuration that is associated with eruptive activity. Sigmoids are S-shaped coronal emission structures seen in soft X-ray and EUV observations (see Sect. 2.1.1), which form in a sub-set of CME source regions. These configurations are one of the most successful indicators of an eruption (Canfield et al. 1999) and have become synonymous with CME occurrence. Given the observation that sigmoidal active regions are highly likely to produce a CME, the question can be asked "how much time elapses between the sigmoid formation and the CME?". Green and Kliem (2014) found that around 5 to 14 hours elapsed between formation and eruption in the small sample of sigmoid regions studied. This study used the time of the first observation of a continuous S-shaped emission structure (rather than an overall S shape formed in a collection of loops) to indicate that a sigmoid had formed. Canfield et al. (2007) found 107 active regions with overall sigmoidal shape in a study of all the active regions imaged in partial-frame SXT data. This means that around 6% of active regions observed in the partial frame *Yohkoh*/SXT data during 1991 to 2001 were sigmoidal. Although Sterling (2000) has a slightly more optimistic finding, stating that "These observations indicate that the pre-eruption sigmoid patterns are more prominent in SXRs than in EUV, and that sigmoid precursors are present in over 50% of CMEs." So, whilst sigmoids are a useful proxy, not all active regions evolve to produce a sigmoidal configuration. In addition, sigmoids are more likely to be seen in active regions when they are near disk centre and viewed from above, when projection effects do not mask the S shape. As discussed in Sects. 2.1.1 and 2.1.2, a sigmoidal emission structure can indicate the presence of a flux rope magnetic field configuration.

The observation of a continuous sigmoid therefore gives a only few hours warning that the region will produce a CME. Sigmoids can form in decaying bipolar active regions though, that appear to show a systematic evolution toward this shape during their lifetime and this can be exploited to give a longer lead time for CME prediction. For sigmoids that form in decaying bipolar active regions, the process of formation takes a couple of days. During this time the coronal structures evolve due to flux cancellation in the photosphere. Flux cancellation can take the coronal field through three observational stages; 1) increase of shear along the polarity inversion line, 2) the formation of two sets of J-shaped loops and 3) the formation of a sigmoid (Green and Kliem 2009, 2014). Accurately identifying stage 2 would increase the CME forecast time by up to a day. Sigmoids are relatively easy to identify when present in small and bipolar regions and viewed from above. They could be harder to identify in complex multipolar active regions with several polarity inversion lines and strong EUV and soft X-ray emission from adjacent structures. However, it is still possible to observe sigmoids in these environments (e.g. Zharkov et al. 2011) where they also seem to be associated with flux cancellation.

The interpretation that some sigmoids indicate a flux rope topology has led to studies of flux cancellation that probe the amount of flux that builds into the rope. This can potentially be used to investigate the stability of the structure. The flux cancellation associated with the formation of a sigmoid was studied in detail for the small bipolar NOAA active region 10977 (Green et al. 2011). The amount of flux cancelled was measured from the time that the active region had fully emerged to the time of the first eruption from the region. The results showed that at the time of the eruption the ratio of flux in the rope to that in the overlying field of the active region had an upper value of 1:1.5. Another study, in an active region that wasn't sigmoidal found that at the time of a filament eruption the ratio of the flux contained in the rope to the flux of the overlying arcade field was 1:0.9 (Yardley et al. 2016). However, there were previous eruptions from this region and the study wasn't able to track the flux cancellation during the entire time during which the filament was forming, and so may represent a lower limit to the flux that was actually contained in the rope. These observational results can be compared to a series of studies that use the flux rope insertion method (van Ballegooijen 2004) to investigate the stability of flux ropes embedded in an overlying arcade. Stable flux ropes have been produced that have a ratio of flux in the rope to flux in the overlying arcade ranging from 1.15 to 1.9 (2008; Savcheva and van Ballegooijen 2009; Su et al. 2009; Savcheva 2012). The Savcheva (2012) study compared the modelled results to the observational deductions of flux rope flux, which found that around 60% to 100% of the flux involved in the flux cancellation built into the rope.

Flux cancellation at the photosphere may be an indicator of the formation of an eruptive configuration but cancellation is a ubiquitous process that takes place all over the photosphere. The challenge is to recognise when and where this process is important for CME occurrence, which can only be done when studied in concert with the evolution of the coronal field. Another aspect is that a small number of sigmoids appear more transiently in active regions. These sigmoids appear to be formed via reconnection in the corona (rather than via flux cancellation and reconnection in the photosphere or chromosphere). Sigmoids formed in this way have been observed around 2 hours prior to their eruption (Cheng et al. 2014; James et al. 2017) although they tend to flash into view and fade, only to reappear again around eruption onset making them harder to use for CME forecasting.

An alternative and complementary approach is to study the occurrence and predictability of solar flares in order to make a CME forecast. This approach is relevant due to the relationship between flares and CMEs (see Sect. 2.1.3) and has the added advantage that the datasets used to study flares include observations of the photosphere, which are available over a much longer time period than X-ray and EUV imaging, and coronagraph observations are. Sammis et al. (2000) analysed eight years of active region observations in order to establish whether a relation between an active region's magnetic classification and the occurrence of large flares (for a description of the active region classes see Hale et al. 1919). They find that active regions classified as $\beta\gamma\delta$ produce significantly more large flares than other regions of comparable size. Each active region of magnetic class $\beta\gamma\delta$ and a size greater than 1000 μh revealed a probability of nearly 40% of producing flares of GOES class X1 or greater. Flares of this size have around a 90% chance of being associated to a CME (Yashiro et al. 2005). More recently, surface magnetic parameters were combined with the amount of stored (free) magnetic energy reconstructed from local non-linear force-free field modelling, and tested against the flare productivity of active regions (Jing et al. 2010). Though this study revealed that the free magnetic energy clearly differentiates flaring active regions from non-flaring active regions, the temporal variation of an region's free magnetic energy did not exhibit a clear pre-flare pattern.

Studies that apply the physical knowledge of an active region's morphology and magnetic properties, to parameterize the solar data in order to quantify flare/CME productivity, mostly use the photospheric line-of-sight magnetic field. However, vector field measurements can also be used. For the parameterizations based on the magnetic properties of active regions, one can further distinguish between those that relate directly to the observed photospheric field (such as magnetic flux, length and strength of main polarity inversion line, etc.) and those that are derived after reconstruction of the coronal field distribution from the vector photospheric data (such as total magnetic energy, free magnetic energy, etc.), see Barnes et al. (2016). In general, it seems that quantities that are integrated over active region scales are stronger determinants for upcoming flare/CME activity than are localized variations on small scales (e.g. review by Schrijver 2009). In addition, the flaring history of an active region plays a decisive role (Falconer et al. 2012). There exist statistical methods that predict flaring probabilities using solely the flaring history of the active region under study. This type of statistical model makes use of the information that the past history of flare occurrence in an active region is an important indicator of its future flare productivity and that the solar flare size distribution follows a power-law (Wheatland 2004).

One of the key features used to describe the magnetic characteristics of an active region with respect to its flare/CME productivity is the polarity-inversion line. Schrijver (2007) conducted a statistical study of the magnetic properties of active regions associated with almost 300 M- and X-class flares, in comparison with 2,500 randomly selected active regions.

The main finding of the study was that active regions that are the source of large flares all have a pronounced polarity inversion line with a strong gradient across it. Calculating the unsigned flux within 15 Mm of such strong field polarity inversion lines is shown to be a good indicator for the occurrence of a large flare.

A series of papers led by Falconer used the polarity inversion line along with a photospheric proxy for the free magnetic energy stored in the coronal magnetic field to investigate whether there exists a relation to the region's CME productivity. The work used vector magnetic field data to derive various parameters that describe the global non-potentiality of the active region's magnetic field. These parameters are the length of the strongly sheared field across the polarity inversion line, the net electric current and the degree of twist. They showed that all three parameters are significantly correlated with the active region's CME productivity (Falconer et al. 2002). Note that in their study, they focussed on active regions with one well-defined, dominant polarity-inversion line. In a follow up study, Falconer et al. (2006) extended this work to a larger sample of active regions, and also extended the set of parameters derived from the vector field data. Three of the six parameters calculated characterize the total non-potentiality of the active region (length of the strong-shear segment of polarity inversion line, length of the main polarity inversion line with a strong field-gradient across it, and the total net electric current). Two of them describe the overall twist in the active region (the net-current α and best constant α derived from a linear force-free field) using the twist parameter α, which is determined from the net current divided by the magnetic flux. One parameter describes the active region size (the total magnetic flux). They find that the magnetic twist and the size are not correlated, i.e. they are separate quantities describing the active region's characteristics, but both show a strong relation to the active region's CME productivity, suggesting that a combination of total flux and total twist would provide the best metric for forecasting. In addition, they find that the total free magnetic energy in an active region is more relevant to its CME productivity than is the overall twist (or its large-scale helicity) alone.

Georgoulis and Rust (2007) carried out a study where only a single metric was used to provide a probability forecast of flaring occurrence in an active region. The authors defined a quantity called the "effective connected magnetic field", B_{eff}, that is calculated by connectivity matrices of field-line lengths and fluxes between positive and negative flux elements. They derived B_{eff} for 298 active regions (93 X- and M-flaring, 205 non-flaring) from MDI/SOHO data over a 10 yr period during solar cycle 23, finding that B_{eff} provides a robust measure to distinguish flaring from non-flaring regions.

Leka and Barnes (2003a,b) conducted a thorough study into whether an active region will be flare-productive or flare-quiet. They derived various parameters, together with their statistical moments from vector magnetic field data. These parameters include the magnetic flux distribution, horizontal spatial gradients of the field, vertical current, twist parameter α, current helicity, shear angles, and photospheric excess magnetic energy. Leka and Barnes (2007) examined a set of 496 active regions observed in more than 1200 magnetograms. Statistical tests based on linear discriminant analysis are applied to the numerous photospheric magnetic parameters listed above, and compared to epochs of non-flaring. The most accurate predictor for the occurrence of large flares (defined as larger than M1), is found to be the total excess photospheric magnetic energy (in excess of the potential field). Other top performers identified are parameters that measure global properties of the field, i.e. ones that are integrated over the full active region. The analysis of Leka and Barnes (2007) shows that a number of parameters are important to quantify an active region as flare-productive, but many of these quantities are strongly correlated. For instance, large active regions as measured by the total magnetic flux also tend to have large vertical currents, significant excess energy, and significant current helicity. Thus, the choice of which few variables to use is not unique. The discriminant analysis in Leka and Barnes (2007) gives a success rate of 80% for the occurrence of flares of GOES class C1 or larger. This is compared to a success rate of 70% when all regions are uniformly classified as "flare quiet". For flares of class M1 and larger, the success rate obtained is 93%, which is only slightly larger than the rate of 91% that is obtained for the uniform flare-quiet labelling. This is an intrinsic problem related to the low occurrence rate of large flares, as low event occurrences typically lead to large false alarm rates. Based on these outcomes, the authors conclude that the state of the photospheric magnetic field at any given time does not have a large influence on whether an active region will be flare productive or not. This is in line with the conclusion in Schrijver (2009), that so far none of these various measures (either single or combined) has been identified as being particularly well correlated with the flare/CME activity of an active region.

In order to directly compare the relative success of existing flare prediction algorithms, an interagency "All-Clear workshop" was held in Colorado in 2009. The outcomes of the workshop are summarised in Barnes et al. (2016). The workshop brought together people working on a number of existing algorithms. Previously, it had been difficult to compare the outcome of different flare forecasting algorithms, as they use different analysis techniques and data sets. The analysis carried out for the All-Clear workshop was based on a common data set, using line-of-sight magnetic field and continuum intensity maps from MDI/SOHO, and applied standard verification metrics to evaluate the performance of the different methods. In total, 11 different algorithms were applied including the ones that are described above, i.e. the methods of Georgoulis and Rust (2007), Falconer et al. (2002, 2008), Leka and Barnes (2003a,b, 2007), Schrijver (2007) and Wheatland (2004). In addition, a number of further algorithms were included, which are now described. The method of Colak and Qahwaji (2008, 2009) is based on a feature-recognition technique to automatically derive the McIntosh classification from white-light images and subsequently uses machine learning to make a probability forecast based on the sunspot class. The method by Bloomfield et al. (2012) uses historical flaring rates related to the McIntosh class of the source active region to make forecasts using Poisson probabilities. Yuan et al. (2010, 2011) apply machine learning techniques to provide flare forecasts from characteristic magnetic parameters, such as the total unsigned magnetic flux, length of the strong-gradient polarity inversion line, and total magnetic energy dissipation. The method by McAteer et al. (2010) is based on the fractal dimension of the magnetic flux concentrations in an active region to determine its flare productivity. The algorithm by Higgins et al. (2011) and Qahwaji et al. (2008) uses automatic characterization of various magnetic parameters of the active region to apply flare forecasts based on cascade correlation network techniques. The main outcome of the comprehensive workshop study detailed in Barnes et al. (2016) is that no method was clearly outperforming the others. This may be due to the fact that there are strong correlations between the different parameters that are used in the different prediction schemes. For the prediction of flares of class M and larger, the set of methods studied tends to give a slightly positive skill score, but none of the methods achieved large skill score values. All skill scores were $\lesssim 0.2$. For context, a perfect forecast would have a skill score of 1 and forecasts that have positive skill score values indicate that the method is more successful than the reference method being used.

An interesting case study against which to examine and test ideas around predicting flares/CMEs based on observations is NOAA active region 12192. This region appeared on the solar disk in October 2014 and it hosted the largest sunspot seen by that point in solar cycle 24. From its size and complexity ($\beta\gamma\delta$) it was expected to be the source of a significant amount of free magnetic energy and high-intensity flares, therefore catching the attention of space weather forecasters. Indeed the region produced six GOES X-class flares during its disk passage but no successful eruptions occurred (Thalmann et al. 2015; Sun et al. 2015). Although the magnetic field of NOAA region 12192 was complex there was a separation of the main positive and negative magnetic polarities with time. No convergence or collision of polarities was seen during its first disk passage. Small bipoles between the main spots (called "serpentine field") were observed indicating that the flux emergence process was still taking place. Overall, the active region configuration was that of one main bipole, with at least 2 new bipolar emergences on the periphery of the sunspot region or near the centre, later in time. The region was studied by Sun et al. (2015) who compared NOAA active region 12192 to two other flare-productive regions (11429 and 11158). They constructed NLFFF extrapolations and calculated the free energy in the coronal field, the decay index of the coronal field and the squashing factor. They find that the region's core field had a low level of non-potentiality and that the overlying field was relatively strong. The lack of colliding polarities (and therefore opportunities for magnetic reconnection) may be the reason that a strongly non-potential field didn't build up along any polarity inversion line in the active region and the lack of flux cancellation would have resulted in no tether cutting (which removes overlying restraining field).

The suggestion that CMEs may act as a valve for magnetic helicity that has accumulated in the corona appears to lead intuitively to the idea that CMEs might occur once a threshold in this quantity has been reached. Can this be used to forecast the occurrence of a CME? Although this idea is by no means accepted or proven, some work has indicated that magnetic helicity is a necessary but not sufficient condition for an eruption (Amari et al. 2003b) whilst others show that magnetic helicity is not necessary for CMEs (Zuccarello et al. 2009). It is worth noting that CMEs originate from a wide range of coronal magnetic structures, from bright points with a flux (ϕ) of around 10^{20} Mx, to large active regions, where $\phi = 10^{22}$ Mx, giving a potentially large range in magnetic helicity values (since helicity scales with ϕ^2). Therefore, there will not be a one-size-fits-all value for magnetic helicity in order to create an eruptive configuration. Instead, the helicity of an eruptive configuration will vary from case-to-case. A study by Pariat et al. (2017) takes a different approach into the potential utility of magnetic helicity to forecast and eruption. Pariat et al. (2017) study the evolution of stable and eruptive magnetic configurations using the flux emergence simulations of Leake et al. (2013, 2014). The relative magnetic helicity of the computed field is decomposed into that of the current carrying component and that of the potential component. They find that the eruptive configurations are not necessarily those with the highest magnetic helicity value but that an eruptivity indicator exists in the ratio of these two components of the relative magnetic helicity. Opening, for the first time, an avenue for investigating how magnetic helicity may be used for CME forecasting.

3.2 Predictions Based on MHD Models

MHD models provide a potentially powerful future tool in the quest for a predictive capability of CMEs. However, at present, the MHD numerical models described in Sect. 2.2 cannot be used for predicting the time of an eruption. This is in part because most of these models do not provide quantitative onset criteria that could then be applied to observations or magnetic-field extrapolations of potential CME source regions. Even those that do (i.e. the kink and torus instability models) can at present only provide rough estimates of the onset criteria. This is because the exact thresholds of these instabilities depends on the detailed structure of the magnetic-field configuration, which is not sufficiently well known. Moreover, exact thresholds have so far only been derived for rather idealized model configurations rather than those exhibited by the real Sun. A complementary approach is to create a simulation that is driven by line-of-sight magnetograms. Using real data provides an opportunity to create a realistic simulation so that the coronal field can be monitored for its eruptive potential. This is the approach used by Gibb et al. (2014) in a study of NOAA active region 10977 that employs a magnetofrictional method driven by line-of-sight magnetograms. The study found that a *realistic* magnetic field line reconstruction was not possible around the time that the active region erupted suggesting that the approach captured well the destabilisation of the coronal field. Since the top boundary of the simulation box is closed, the flux rope that had formed and started to erupt wasn't able to escape the computational volume. That is, the computed field was trying to relax by ejecting the flux rope but the boundary conditions did not allow this to happen. Consequently, this led to the formation of highly twisted structures in the computational volume.

Apart from the eruption onset, one would also like to forecast the *arrival time* and *impact* at the Earth of the CME, once the eruption is on its way. CMEs typical take from 0.7 to 4 days to travel from the Sun to the Earth. This is an amount of time that allows an MHD model, based on solar data, to be run. In contrast, predicting the effects of energetic-particle events created by CME-induced shocks is much more challenging, as these particles can reach the Earth in less than an hour. We therefore consider here only to what extent MHD models can be employed for predicting the time and impact of space-weather disturbances caused by CMEs. The most important quantities for the latter are the speed, duration, plasma density, and magnetic field strength and orientation of the CME. Sophisticated models should also include the effects of the shock and the sheath that precede the CME flux-rope in fast events.

As discussed in Jin et al. (2017), present CME forecasting models can be divided into different categories (see also Siscoe and Schwenn 2006; Messerotti et al. 2009). The first categories contain empirical forecasting models (e.g., Gopalswamy et al. 2001; Schwenn et al. 2005), kinematic models such as the cone model (e.g., Zhao et al. 2002) and analytical models such as the drag-model (Vršnak et al. 2013). These models have been widely used to predict the arrival time, orientation, and speed of CMEs (and of associated shocks), but they cannot provide predictions of the plasma density and the magnetic field (though see, for example, Möstl et al. 2014 for attempts to predict the maximum magnetic field strength in an CME from its predicted speed). To this end, the use of models of the third category, namely of MHD simulations, appears inevitable.

However, MHD simulations that employ observed data and continuously model an eruption from its onset in the low corona out to 1 AU (see Manchester et al. in this issue) are limited to using data from the Earth's point-of-view. Therefore, the Earth-directed component of the CME velocity is not accurately determined. Moreover, these simulations cannot yet be performed in a constantly evolving and accurate solar wind. For this, in situ solar wind measurements are needed that provide real data to adjust the models to. This limits the capability of these MHD models, but efforts are being made to develop such models for future operational use (e.g., Jin et al. 2017). Presently, ENLIL (Odstrcil 2003) is the only MHD model that is used for operational forecasting (at NOAA and the UK's Met Office Space Weather Operations Centre for example). ENLIL produces estimates of the solar wind speed, density, and temperature and magnetic field, and provides 1–4 day advance warning of disturbances that will produce geomagnetic storms, such as CMEs. To be applicable in practice, simplifications had to be made to the model. For example, ENLIL simulates the solar wind plasma and magnetic field only beyond 30 $R_\odot$, i.e., the corona is neglected. To feed the model, the solar wind plasma flow and radial magnetic field at 30 $R_\odot$ are provided by the semi-empirical WSA model (Wang and Sheeley 1990; Arge and Pizzo 2000). If an eruption is to be simulated, the cone model is used to set up the location, direction, and speed of the CME (e.g., Odstrcil et al. 2005). Also, while the solar wind magnetic field is included in the model, CMEs are initiated without an internal magnetic field. Therefore, the model cannot yet be used to predict the southward component of the magnetic field during CMEs. However, observations of the CME source region can be used to give first order approximation of the magnetic configuration of the erupting structure. For example, through the detection of a flux rope and determination of the orientation of its axis and its chiral properties (e.g. Palmerio et al. 2017).

3.3 Bringing Observational and Theoretical Predictions Closer Together

There is much to be learnt from the complementary approaches of observational CME studies and theoretical work, including numerical simulations. In terms of the most promising approaches to predicting CME occurrence ahead of time, combining current observational and theoretical knowledge, there appear to be three main avenues:

(i) Flux imbalance between a flux rope and overlying arcade
(ii) Torus instability due to the rate of decay of the arcade overlying a flux rope
(iii) Helicity proxy derived from the current carrying and potential field components

For (i), observations of flux cancellation can be used to investigate the flux built into the rope versus that in the overlying field. However, more work is needed to understand how the shear and spatial extent of the cancellation in the active region influence what proportion of cancelled flux is built into the rope, and what the flux imbalance criteria is. For (ii), the gradient of the overlying field can be investigated using a potential field extrapolation, but simulations so far describe rather idealised configurations and the exact stability criteria varies on a case-to-case basis. For (iii), the theoretical knowledge is still developing, but early indications suggest that there might be a critical value for the ratio of helicity in the current carrying field component to the helicity in potential component. Still, none of the above are able to give the desired several days notice of the occurrence of a CME. In addition, approaches (i) and (ii) are relevant for the flux rope concept of CMEs rather than a sheared arcade.

4 Summary

This review aims to provide a summary of the wide range of observational and theoretical work on coronal mass ejections (CMEs). These eruptions are known to be a driver of space weather, including the most severe space weather events. In response to this, an increasing number of developed and developing nations provide space weather forecasts to help mitigate the risk of economic and societal disruption resulting from the technology affected by space weather. The severe impact that results from the arrival of a CME of high velocity, southward directed magnetic field and enhanced plasma density means that there is a significant interest in forecasting these eruptions ahead of time.

Currently the physical processes at play after CME onset can be brought together in the standard CHSKP model. This captures and explains CME observations well, once the eruption is underway. However, the energy required to power a CME (and any associated flare) cannot be supplied to the corona on the timescale of the dynamic event and instead it must be built up and stored in the coronal magnetic field in the hours or days beforehand. We need to extend the standard model further back in time to understand the physical processes that create an eruptive magnetic field configuration and the associated currents, as well as the processes that trigger and drive the eruption. CME occurrence requires more than just estimating the free magnetic energy stored in the coronal field—the free energy is a necessary but not sufficient condition. The exact magnetic field configuration and its stability need to be better characterised.

Many flare/CME forecasts currently focus on using a snapshot of the configuration of the photosheric magnetic field to provide a probabilistic forecast. Improvements to this technique are limited by the physical origins of a CME being related to the stability of the coronal field, when a trigger switches a driver on, which is unlikely to be captured in the photospheric data. CME triggers themselves can be monitored over many days using both photospheric and coronal data. But this must be supplemented with data to reveal the likelihood of a CME occurring, using either a NLFFF extrapolation or an eruptivity signature(s) determined using atmospheric observations. With no regular direct measurements of the coronal magnetic field, observations are limited to inferences obtained from plasma structures.

To move on from probabilistic forecasting, to the capability of being able to predict CME onset time, we first need a clear understanding of the criteria that encapsulates the eruptive status of the magnetic field. At what point does a CME driver switch on? Parametric modelling studies that can identify the loss of stability criteria are required and this could be determined from more realistic magnetic field configurations and their evolution. Developing the capability to conduct data driven simulations, relevant for active regions from their birth to their dispersal and decay into the background, field will greatly help this aim.

Acknowledgements LMG acknowledges support through a Royal Society University Research Fellowship and through the Leverhulme Trust Research Project Grant 2014-051. T.T. acknowledges support by NASA's LWS and H-SR programs and by NSF grants AGS-1249270, AGS-1348577, and AGS-1135432 (Sun-2-Ice). BV acknowledges the support by the Croatian Science Foundation under the project 6212 "Solar and Stellar Variability". AV acknowledges support by the Austrian Science Fund (FWF): P27292-N20. We acknowledge the International Space Science Institute (ISSI) in Bern Switzerland for their generous support and for travel and accommodation.

References

W.P. Abbett, The magnetic connection between the convection zone and corona in the quiet sun. Astrophys. J. **665**, 1469–1488 (2007). https://doi.org/10.1086/519788

W.P. Abbett, G.H. Fisher, Y. Fan, The effects of rotation on the evolution of rising omega loops in a stratified model convection zone. Astrophys. J. **546**, 1194–1203 (2001). https://doi.org/10.1086/318320. arXiv:astro-ph/0008501

J.J. Aly, On some properties of force-free magnetic fields in infinite regions of space. Astrophys. J. **283**, 349–362 (1984). https://doi.org/10.1086/162313

T. Amari, J.J. Aly, Magnetic flux ropes: fundamental structures for eruptive phenomena, in *Universal Heliophysical Processes, IAU Symposium*, vol. 257, ed. by N. Gopalswamy, D.F. Webb (2009), pp. 211–222. https://doi.org/10.1017/S1743921309029329

T. Amari, J.F. Luciani, J.J. Aly, M. Tagger, Plasmoid formation in a single sheared arcade and application to coronal mass ejections. Astron. Astrophys. **306**, 913 (1996a)

T. Amari, J.F. Luciani, J.J. Aly, M. Tagger, Very fast opening of a three-dimensional twisted magnetic flux tube. Astrophys. J. Lett. **466**, L39–L42 (1996). https://doi.org/10.1086/310158

T. Amari, J.F. Luciani, Z. Mikic, J. Linker, A twisted flux rope model for coronal mass ejections and two-ribbon flares. Astrophys. J. Lett. **529**, L49–L52 (2000). https://doi.org/10.1086/312444

T. Amari, J.F. Luciani, J.J. Aly, Z. Mikic, J. Linker, Coronal mass ejection: initiation, magnetic helicity, and flux ropes. I. Boundary motion-driven evolution. Astrophys. J. **585**, 1073–1086 (2003a). https://doi.org/10.1086/345501

T. Amari, J.F. Luciani, J.J. Aly, Z. Mikic, J. Linker, Coronal mass ejection: initiation, magnetic helicity, and flux ropes. II. Turbulent diffusion-driven evolution. Astrophys. J. **595**, 1231–1250 (2003b). https://doi.org/10.1086/377444

M.D. Andrews, A search for CMEs associated with big flares. Sol. Phys. **218**, 261–279 (2003). https://doi.org/10.1023/B:SOLA.0000013039.69550.bf

S.K. Antiochos, C.R. DeVore, J.A. Klimchuk, A model for solar coronal mass ejections. Astrophys. J. **510**, 485–493 (1999). https://doi.org/10.1086/306563. arXiv:astro-ph/9807220

U. Anzer, Can coronal loop transients be driven magnetically. Sol. Phys. **57**, 111–118 (1978). https://doi.org/10.1007/BF00152048

U. Anzer, G.W. Pneuman, Magnetic reconnection and coronal transients. Sol. Phys. **79**, 129–147 (1982)

U. Anzer, E. Tandberg-Hanssen, A model for quiescent prominences with helical structure. Sol. Phys. **11**, 61–67 (1970). https://doi.org/10.1007/BF00156551

V. Archontis, T. Török, Eruption of magnetic flux ropes during flux emergence. Astron. Astrophys. **492**, L35–L38 (2008). https://doi.org/10.1051/0004-6361:200811131. arXiv:0811.1134

C.N. Arge, V.J. Pizzo, Improvement in the prediction of solar wind conditions using near-real time solar magnetic field updates. J. Geophys. Res. **105**, 10,465–10,480 (2000). https://doi.org/10.1029/1999JA000262

G. Aulanier, The physical mechanisms that initiate and drive solar eruptions, in *IAU Symposium*, vol. 300, ed. by B. Schmieder, J.M. Malherbe, S.T. Wu (2014), pp. 184–196. https://doi.org/10.1017/S1743921313010958. arXiv:1309.7329

G. Aulanier, P. Démoulin, R. Grappin, Equilibrium and observational properties of line-tied twisted flux tubes. Astron. Astrophys. **430**, 1067–1087 (2005). https://doi.org/10.1051/0004-6361:20041519

G. Aulanier, T. Török, P. Démoulin, E.E. DeLuca, Formation of torus-unstable flux ropes and electric currents in erupting sigmoids. Astrophys. J. **708**, 314–333 (2010). https://doi.org/10.1088/0004-637X/708/1/314

L.G. Bagalá, C.H. Mandrini, M.G. Rovira, P. Démoulin, Magnetic reconnection: a common origin for flares and AR interconnecting arcs. Astron. Astrophys. **363**, 779–788 (2000)

G. Barnes, K.D. Leka, C.J. Schrijver, T. Colak, R. Qahwaji, O.W. Ashamari, Y. Yuan, J. Zhang, R.T.J. McAteer, D.S. Bloomfield, P.A. Higgins, P.T. Gallagher, D.A. Falconer, M.K. Georgoulis, M.S. Wheatland, C. Balch, T. Dunn, E.L. Wagner, A comparison of flare forecasting methods. I. Results from the "All-Clear" workshop. Astrophys. J. **829**, 89 (2016). https://doi.org/10.3847/0004-637X/829/2/89. arXiv:1608.06319

V.V. Batygin, I.N. Toptygin, *Problems in Electrodynamics* (Academic Press, New York, 1962)

B.M. Bein, S. Berkebile-Stoiser, A.M. Veronig, M. Temmer, N. Muhr, I. Kienreich, D. Utz, B. Vršnak, Impulsive acceleration of coronal mass ejections. I. Statistics and coronal mass ejection source region characteristics. Astrophys. J. **738**, 191 (2011). https://doi.org/10.1088/0004-637X/738/2/191. arXiv:1108.0561

B.M. Bein, S. Berkebile-Stoiser, A.M. Veronig, M. Temmer, B. Vršnak, Impulsive acceleration of coronal mass ejections. II. Relation to soft X-ray flares and filament eruptions. Astrophys. J. **755**, 44 (2012). https://doi.org/10.1088/0004-637X/755/1/44. arXiv:1206.2144

M.A. Berger, Rigorous new limits on magnetic helicity dissipation in the solar corona. Geophys. Astrophys. Fluid Dyn. **30**, 79–104 (1984). https://doi.org/10.1080/03091928408210078

D.A. Biesecker, B.J. Thompson, Sympathetic flaring with BATSE, GOES, and EIT data. J. Atmos. Sol.-Terr. Phys. **62**, 1449–1455 (2000). https://doi.org/10.1016/S1364-6826(00)00085-7

D. Biskamp, H. Welter, Magnetic arcade evolution and instability. Sol. Phys. **120**, 49–77 (1989). https://doi.org/10.1007/BF00148534

D.S. Bloomfield, P.A. Higgins, R.T.J. McAteer, P.T. Gallagher, Toward reliable benchmarking of solar flare forecasting methods. Astrophys. J. Lett. **747**, L41 (2012). https://doi.org/10.1088/2041-8205/747/2/L41. arXiv:1202.5995

M.G. Bobra, A.A. van Ballegooijen, E.E. DeLuca, Modeling nonpotential magnetic fields in solar active regions. Astrophys. J. **672**, 1209–1220 (2008). https://doi.org/10.1086/523927

D.S. Brown, R.W. Nightingale, D. Alexander, C.J. Schrijver, T.R. Metcalf, R.A. Shine, A.M. Title, C.J. Wolfson, Observations of rotating sunspots from TRACE. Sol. Phys. **216**, 79–108 (2003). https://doi.org/10.1023/A:1026138413791

P.K. Browning, E.R. Priest, The structure of twisted magnetic flux tubes. Astrophys. J. **266**, 848–865 (1983). https://doi.org/10.1086/160833

V. Bumba, M. Klvana, Questions concerning the existence of sympathetic flares. Astrophys. Space Sci. **199**, 45–52 (1993). https://doi.org/10.1007/BF00612976

F.N. Byrne, M.A. Ellison, J.H. Reid, A survey of solar flare phenomena. Space Sci. Rev. **3**, 319–341 (1964). https://doi.org/10.1007/BF00230516
R.C. Canfield, H.S. Hudson, D.E. McKenzie, Sigmoidal morphology and eruptive solar activity. Geophys. Res. Lett. **26**, 627–630 (1999). https://doi.org/10.1029/1999GL900105
R.C. Canfield, M.D. Kazachenko, L.W. Acton, D.H. Mackay, J. Son, T.L. Freeman, Yohkoh SXT full-resolution observations of sigmoids: structure, formation, and eruption. Astrophys. J. Lett. **671**, L81–L84 (2007). https://doi.org/10.1086/524729
H. Carmichael, A process for flares. NASA Spec. Publ. **50**, 451 (1964)
J. Chen, Effects of toroidal forces in current loops embedded in a background plasma. Astrophys. J. **338**, 453–470 (1989). https://doi.org/10.1086/167211
J. Chen, Physics of coronal mass ejections: a new paradigm of solar eruptions. Space Sci. Rev. **95**, 165–190 (2001)
J. Chen, Physics of erupting solar flux ropes: coronal mass ejections (CMEs)—recent advances in theory and observation. Phys. Plasmas **24**(9), 090501 (2017). https://doi.org/10.1063/1.4993929
J. Chen, J. Krall, Acceleration of coronal mass ejections. J. Geophys. Res. **108**, 1410 (2003). https://doi.org/10.1029/2003JA009849
P.F. Chen, Coronal mass ejections: models and their observational basis. Living Rev. Sol. Phys. **8**, 1 (2011). https://doi.org/10.12942/lrsp-2011-1
P.F. Chen, K. Shibata, An emerging flux trigger mechanism for coronal mass ejections. Astrophys. J. **545**, 524–531 (2000). https://doi.org/10.1086/317803
J. Cheng, C. Fang, P. Chen, M. Ding, Two sympathetic homologous CMEs on 2002 May 22. Chin. J. Astron. Astrophys. **5**, 265–272 (2005). https://doi.org/10.1088/1009-9271/5/3/006
X. Cheng, J. Zhang, Y. Liu, M.D. Ding, Observing flux rope formation during the impulsive phase of a solar eruption. Astrophys. J. Lett. **732**, L25 (2011). https://doi.org/10.1088/2041-8205/732/2/L25. arXiv:1103.5084
X. Cheng, M.D. Ding, J. Zhang, X.D. Sun, Y. Guo, Y.M. Wang, B. Kliem, Y.Y. Deng, Formation of a double-decker magnetic flux rope in the sigmoidal solar active region 11520. Astrophys. J. **789**, 93 (2014). https://doi.org/10.1088/0004-637X/789/2/93. arXiv:1405.4923
M.C.M. Cheung, M. Rempel, A.M. Title, M. Schüssler, Simulation of the formation of a solar active region. Astrophys. J. **720**, 233–244 (2010). https://doi.org/10.1088/0004-637X/720/1/233. arXiv:1006.4117
G. Chintzoglou, S. Patsourakos, A. Vourlidas, Formation of magnetic flux ropes during a confined flaring well before the onset of a pair of major coronal mass ejections. Astrophys. J. **809**, 34 (2015). https://doi.org/10.1088/0004-637X/809/1/34. arXiv:1507.01165
G.S. Choe, L.C. Lee, Evolution of solar magnetic arcades. I. Ideal MHD evolution under footpoint shearing. Astrophys. J. **472**, 360 (1996a). https://doi.org/10.1086/178069
G.S. Choe, L.C. Lee, Evolution of solar magnetic arcades. II. Effect of resistivity and solar eruptive processes. Astrophys. J. **472**, 372 (1996b). https://doi.org/10.1086/178070
A. Ciaravella, J.C. Raymond, The current sheet associated with the 2003 November 4 coronal mass ejection: density, temperature, thickness, and line width. Astrophys. J. **686**, 1372–1382 (2008). https://doi.org/10.1086/590655
E.W. Cliver, L. Svalgaard, The 1859 solar-terrestrial disturbance and the current limits of extreme space weather activity. Sol. Phys. **224**, 407–422 (2004). https://doi.org/10.1007/s11207-005-4980-z
O. Cohen, G.D.R. Attrill, W.B.Manchester IV, M.J. Wills-Davey, Numerical simulation of an EUV coronal wave based on the 2009 February 13 CME event observed by STEREO. Astrophys. J. **705**, 587 (2009). https://doi.org/10.1088/0004-637X/705/1/587. arXiv:0909.3095
T. Colak, R. Qahwaji, Automated McIntosh-based classification of sunspot groups using MDI images. Sol. Phys. **248**, 277–296 (2008). https://doi.org/10.1007/s11207-007-9094-3
T. Colak, R. Qahwaji, Automated solar activity prediction: a hybrid computer platform using machine learning and solar imaging for automated prediction of solar flares. Space Weather **7**, S06001 (2009). https://doi.org/10.1029/2008SW000401
P. Démoulin, G. Aulanier, Criteria for flux rope eruption: non-equilibrium versus torus instability. Astrophys. J. **718**, 1388–1399 (2010). https://doi.org/10.1088/0004-637X/718/2/1388. arXiv:1006.1785
P. Démoulin, M.A. Berger, Magnetic energy and helicity fluxes at the photospheric level. Sol. Phys. **215**, 203–215 (2003). https://doi.org/10.1023/A:1025679813955
P. Démoulin, J.C. Henoux, E.R. Priest, C.H. Mandrini, Quasi-separatrix layers in solar flares. I. Method. Astron. Astrophys. **308**, 643–655 (1996)
P. Démoulin, C.H. Mandrini, L. van Driel-Gesztelyi, B.J. Thompson, S. Plunkett, Z. Kovári, G. Aulanier, A. Young, What is the source of the magnetic helicity shed by CMEs? The long-term helicity budget of AR 7978. Astron. Astrophys. **382**, 650–665 (2002). https://doi.org/10.1051/0004-6361:20011634

N. Deng, Y. Xu, G. Yang, W. Cao, C. Liu, T.R. Rimmele, H. Wang, C. Denker, Multiwavelength study of flow fields in flaring super active region NOAA 10486. Astrophys. J. **644**, 1278–1291 (2006). https://doi.org/10.1086/503600

C.R. DeVore, S.K. Antiochos, Magnetic free energies of breakout coronal mass ejections. Astrophys. J. **628**, 1031–1045 (2005). https://doi.org/10.1086/431141

J.Y. Ding, Y.Q. Hu, J.X. Wang, Catastrophic behavior of multiple coronal flux rope system. Sol. Phys. **235**, 223–234 (2006). https://doi.org/10.1007/s11207-006-0092-7

C. Downs, I.I. Roussev, B. van der Holst, N. Lugaz, I.V. Sokolov, T.I. Gombosi, Studying extreme ultraviolet wave transients with a digital laboratory: direct comparison of extreme ultraviolet wave observations to global magnetohydrodynamic simulations. Astrophys. J. **728**, 2–17 (2011). https://doi.org/10.1088/0004-637X/728/1/2

G. Dubey, B. van der Holst, S. Poedts, Initiation of CMEs by magnetic flux emergence. J. Astrophys. Astron. **27**, 159–166 (2006). https://doi.org/10.1007/BF02702518

P. Duchlev, K. Koleva, M.S. Madjarska, M. Dechev, Homologous prominence non-radial eruptions: a case study. New Astron. **48**, 66–73 (2016). https://doi.org/10.1016/j.newast.2016.05.001. arXiv:1605.02299

G.A. Dulk, D.J. McLean, Coronal magnetic fields. Sol. Phys. **57**, 279–295 (1978). https://doi.org/10.1007/BF00160102

M. Dumbović, A. Devos, B. Vršnak, D. Sudar, L. Rodriguez, D. Ruždjak, K. Leer, S. Vennerstrøm, A. Veronig, Geoeffectiveness of coronal mass ejections in the SOHO era. Sol. Phys. **290**, 579–612 (2015). https://doi.org/10.1007/s11207-014-0613-8. arXiv:1410.3303

M.A. Ellison, S.M.P. McKenna, J.H. Reid, Flares associated with the 1960 November event and the flare nimbus phenomenon. Mon. Not. R. Astron. Soc. **122**, 491 (1961). https://doi.org/10.1093/mnras/122.6.491

D.A. Falconer, R.L. Moore, G.A. Gary, Correlation of the coronal mass ejection productivity of solar active regions with measures of their global nonpotentiality from vector magnetograms: baseline results. Astrophys. J. **569**, 1016–1025 (2002). https://doi.org/10.1086/339161

D.A. Falconer, R.L. Moore, G.A. Gary, Magnetic causes of solar coronal mass ejections: dominance of the free magnetic energy over the magnetic twist alone. Astrophys. J. **644**, 1258–1272 (2006). https://doi.org/10.1086/503699

D.A. Falconer, R.L. Moore, G.A. Gary, Magnetogram measures of total nonpotentiality for prediction of solar coronal mass ejections from active regions of any degree of magnetic complexity. Astrophys. J. **689**, 1433–1442 (2008). https://doi.org/10.1086/591045

D.A. Falconer, R.L. Moore, A.F. Barghouty, I. Khazanov, Prior flaring as a complement to free magnetic energy for forecasting solar eruptions. Astrophys. J. **757**, 32 (2012). https://doi.org/10.1088/0004-637X/757/1/32

Y. Fan, The emergence of a twisted Ω-tube into the solar atmosphere. Astrophys. J. Lett. **554**, L111–L114 (2001). https://doi.org/10.1086/320935

Y. Fan, F. Fang, A simulation of convective dynamo in the solar convective envelope: maintenance of the solar-like differential rotation and emerging flux. Astrophys. J. **789**, 35 (2014). https://doi.org/10.1088/0004-637X/789/1/35. arXiv:1405.3926

Y. Fan, S.E. Gibson, The emergence of a twisted magnetic flux tube into a preexisting coronal arcade. Astrophys. J. Lett. **589**, L105–L108 (2003). https://doi.org/10.1086/375834

Y. Fan, S.E. Gibson, Onset of coronal mass ejections due to loss of confinement of coronal flux ropes. Astrophys. J. **668**, 1232–1245 (2007). https://doi.org/10.1086/521335

F. Fang, W. Manchester, W.P. Abbett, B. van der Holst, Simulation of flux emergence from the convection zone to the corona. Astrophys. J. **714**, 1649–1657 (2010). https://doi.org/10.1088/0004-637X/714/2/1649. arXiv:1003.6118

F. Fang, W. Manchester IV, W.P. Abbett, B. van der Holst, Buildup of magnetic shear and free energy during flux emergence and cancellation. Astrophys. J. **754**, 15 (2012a). https://doi.org/10.1088/0004-637X/754/1/15. arXiv:1205.3764

F. Fang, W. Manchester IV, W.P. Abbett, B. van der Holst, Dynamic coupling of convective flows and magnetic field during flux emergence. Astrophys. J. **745**, 37 (2012b). https://doi.org/10.1088/0004-637X/745/1/37. arXiv:1111.1679

J. Feynman, A.J. Hundhausen, Coronal mass ejections and major solar flares: the great active center of March 1989. J. Geophys. Res. **99**, 8451–8464 (1994). https://doi.org/10.1029/94JA00202

J. Feynman, S.F. Martin, The initiation of coronal mass ejections by newly emerging magnetic flux. J. Geophys. Res. **100**, 3355–3367 (1995). https://doi.org/10.1029/94JA02591

B. Filippov, O. Martsenyuk, A.K. Srivastava, W. Uddin, Solar magnetic flux ropes. J. Astrophys. Astron. **36**, 157–184 (2015). https://doi.org/10.1007/s12036-015-9321-5. arXiv:1501.02562

T.G. Forbes, A review on the genesis of coronal mass ejections. J. Geophys. Res. **105**, 23,153–23,166 (2000). https://doi.org/10.1029/2000JA000005

T.G. Forbes, P.A. Isenberg, A catastrophe mechanism for coronal mass ejections. Astrophys. J. **373**, 294–307 (1991). https://doi.org/10.1086/170051

T.G. Forbes, J.A. Linker, J. Chen, C. Cid, J. Kóta, M.A. Lee, G. Mann, Z. Mikić, M.S. Potgieter, J.M. Schmidt, G.L. Siscoe, R. Vainio, S.K. Antiochos, P. Riley, CME theory and models. Space Sci. Rev. **123**, 251–302 (2006). https://doi.org/10.1007/s11214-006-9019-8

L. Fritzova-Svestkova, R.C. Chase, Z. Svestka, On the occurrence of sympathetic flares. Sol. Phys. **48**, 275–286 (1976). https://doi.org/10.1007/BF00151996

D.A. Garren, J. Chen, Lorentz self-forces on curved current loops. Phys. Plasmas **1**, 3425–3436 (1994). https://doi.org/10.1063/1.870491

G.A. Gary, Plasma beta above a solar active region: rethinking the paradigm. Sol. Phys. **203**, 71–86 (2001)

M.K. Georgoulis, D.M. Rust, Quantitative forecasting of major solar flares. Astrophys. J. Lett. **661**, L109–L112 (2007). https://doi.org/10.1086/518718

M.K. Georgoulis, V.S. Titov, Z. Mikić, Non-neutralized electric current patterns in solar active regions: origin of the shear-generating Lorentz force. Astrophys. J. **761**, 61 (2012). https://doi.org/10.1088/0004-637X/761/1/61. arXiv:1210.2919

C.L. Gerrard, T.D. Arber, A.W. Hood, R.A.M. Van der Linden, Numerical simulations of kink instability in line-tied coronal loops. Astron. Astrophys. **373**, 1089–1098 (2001). https://doi.org/10.1051/0004-6361:20010678

G.P.S. Gibb, D.H. Mackay, L.M. Green, K.A. Meyer, Simulating the formation of a sigmoidal flux rope in AR10977 from SOHO/MDI magnetograms. Astrophys. J. **782**, 71 (2014). https://doi.org/10.1088/0004-637X/782/2/71

T. Gold, Magnetic storms. Space Sci. Rev. **1**, 100–114 (1962). https://doi.org/10.1007/BF00174637

T. Gold, F. Hoyle, On the origin of solar flares. Mon. Not. R. Astron. Soc. **120**, 89 (1960)

N. Gopalswamy, A. Lara, S. Yashiro, M.L. Kaiser, R.A. Howard, Predicting the 1-AU arrival times of coronal mass ejections. J. Geophys. Res. **106**, 29,207–29,218 (2001). https://doi.org/10.1029/2001JA000177

N. Gopalswamy, S. Yashiro, Y. Liu, G. Michalek, A. Vourlidas, M.L. Kaiser, R.A. Howard, Coronal mass ejections and other extreme characteristics of the 2003 October–November solar eruptions. J. Geophys. Res. Space Phys. **110**, A09S15 (2005). https://doi.org/10.1029/2004JA010958

J.T. Gosling, The solar flare myth. J. Geophys. Res. **98**, 18,937–18,950 (1993). https://doi.org/10.1029/93JA01896

L. Green, D. Baker, Coronal mass ejections: a driver of severe space weather. Weather **70**(1), 31–35 (2015). https://doi.org/10.1002/wea.2437

L.M. Green, B. Kliem, Flux rope formation preceding coronal mass ejection onset. Astrophys. J. Lett. **700**, L83–L87 (2009). https://doi.org/10.1088/0004-637X/700/2/L83. arXiv:0906.4794

L.M. Green, B. Kliem, Observations of flux rope formation prior to coronal mass ejections, in *Nature of Prominences and Their Role in Space Weather, IAU Symposium*, vol. 300, ed. by B. Schmieder, J.M. Malherbe, S.T. Wu (2014), pp. 209–214. https://doi.org/10.1017/S1743921313010983. arXiv:1312.4388

L.M. Green, M.C. López fuentes, C.H. Mandrini, P. Démoulin, L. Van Driel-Gesztelyi, J.L. Culhane, The magnetic helicity budget of a CME-prolific active region. Sol. Phys. **208**, 43–68 (2002). https://doi.org/10.1023/A:1019658520033

L.M. Green, P. Démoulin, C.H. Mandrini, L. Van Driel-Gesztelyi, How are emerging flux, flares and CMEs related to magnetic polarity imbalance in MDI data? Sol. Phys. **215**, 307–325 (2003). https://doi.org/10.1023/A:1025678917086 arXiv:astro-ph/0304092

L.M. Green, B. Kliem, T. Török, L. van Driel-Gesztelyi, G.D.R. Attrill, Transient coronal sigmoids and rotating erupting flux ropes. Sol. Phys. **246**, 365–391 (2007). https://doi.org/10.1007/s11207-007-9061-z

L.M. Green, B. Kliem, A.J. Wallace, Photospheric flux cancellation and associated flux rope formation and eruption. Astron. Astrophys. **526**, A2 (2011). https://doi.org/10.1051/0004-6361/201015146. arXiv:1011.1227

Y. Guo, M.D. Ding, B. Schmieder, H. Li, T. Török, T. Wiegelmann, Driving mechanism and onset condition of a confined eruption. Astrophys. J. Lett. **725**, L38–L42 (2010). https://doi.org/10.1088/2041-8205/725/1/L38

M.J. Hagyard, D. Teuber, E.A. West, J.B. Smith, A quantitative study relating observed shear in photospheric magnetic fields to repeated flaring. Sol. Phys. **91**, 115–126 (1984). https://doi.org/10.1007/BF00213618

G. Hale, F. Ellerman, S. Nicholson, A. Joy, The magnetic polarity of sun-spots. Astrophys. J. **49**, 153–178 (1919). https://doi.org/10.1086/142452

J. Heyvaerts, E.R. Priest, Coronal heating by reconnection in DC current systems—a theory based on Taylor's hypothesis. Astron. Astrophys. **137**, 63–78 (1984)

P.A. Higgins, P.T. Gallagher, R.T.J. McAteer, D.S. Bloomfield, Solar magnetic feature detection and tracking for space weather monitoring. Adv. Space Res. **47**, 2105–2117 (2011). https://doi.org/10.1016/j.asr.2010.06.024. arXiv:1006.5898

T. Hirayama, Theoretical model of flares and prominences. I: evaporating flare model. Sol. Phys. **34**, 323–338 (1974). https://doi.org/10.1007/BF00153671

A. Hofmann, V. Ruzdjak, B. Vrsnak, A study on electric currents in a solar active region—a dynamo process at a place of repeated flaring. Hvar Obs. Bull. **16**, 29 (1992)

A.W. Hood, E.R. Priest, Kink instability of solar coronal loops as the cause of solar flares. Sol. Phys. **64**, 303–321 (1979). https://doi.org/10.1007/BF00151441

T.A. Howard, R.A. Harrison, Stealth coronal mass ejections: a perspective. Sol. Phys. **285**, 269–280 (2013). https://doi.org/10.1007/s11207-012-0217-0

Q. Hu, J. Qiu, B. Dasgupta, A. Khare, G.M. Webb, Structures of interplanetary magnetic flux ropes and comparison with their solar sources. Astrophys. J. **793**, 53 (2014). https://doi.org/10.1088/0004-637X/793/1/53. arXiv:1408.1470

A.J. Hundhausen, Sizes and locations of coronal mass ejections—SMM observations from 1980 and 1984–1989. J. Geophys. Res. **98**, 13,177 (1993). https://doi.org/10.1029/93JA00157

C.L. Hyder, A phenomenological model for disparitions brusques followed by flarelike chromospheric brightenings, I: the model, its consequences, and observations in quiet solar regions. Sol. Phys. **2**, 49–74 (1967). https://doi.org/10.1007/BF00155892

R.M.E. Illing, A.J. Hundhausen, Observation of a coronal transient from 1.2 to 6 solar radii. J. Geophys. Res. **90**, 275–282 (1985). https://doi.org/10.1029/JA090iA01p00275

R.M.E. Illing, A.J. Hundhausen, Disruption of a coronal streamer by an eruptive prominence and coronal mass ejection. J. Geophys. Res. **91**, 10,951–10,960 (1986). https://doi.org/10.1029/JA091iA10p10951

B. Inhester, J. Birn, M. Hesse, The evolution of line-tied coronal arcades including a converging footpoint motion. Sol. Phys. **138**, 257–281 (1992)

S. Inoue, Magnetohydrodynamics modeling of coronal magnetic field and solar eruptions based on the photospheric magnetic field. Prog. Earth Planet. Sci. **3**, 19 (2016). https://doi.org/10.1186/s40645-016-0084-7

N. Ishiguro, K. Kusano, Double arc instability in the solar corona. Astrophys. J. **843**, 101 (2017). https://doi.org/10.3847/1538-4357/aa799b. arXiv:1706.06112

B.V. Jackson, A. Buffington, P.P. Hick, R.C. Altrock, S. Figueroa, P.E. Holladay, J.C. Johnston, S.W. Kahler, J.B. Mozer, S. Price, R.R. Radick, R. Sagalyn, D. Sinclair, G.M. Simnett, C.J. Eyles, M.P. Cooke, S.J. Tappin, T. Kuchar, D. Mizuno, D.F. Webb, P.A. Anderson, S.L. Keil, R.E. Gold, N.R. Waltham, The Solar Mass-Ejection Imager (SMEI) mission. Sol. Phys. **225**, 177–207 (2004). https://doi.org/10.1007/s11207-004-2766-3

C. Jacobs, S. Poedts, Models for coronal mass ejections. J. Atmos. Sol.-Terr. Phys. **73**, 1148–1155 (2011). https://doi.org/10.1016/j.jastp.2010.12.002

C. Jacobs, S. Poedts, A numerical study of the response of the coronal magnetic field to flux emergence. Sol. Phys. **280**, 389–405 (2012). https://doi.org/10.1007/s11207-012-9941-8

C. Jacobs, S. Poedts, B. van der Holst, The effect of the solar wind on CME triggering by magnetic foot point shearing. Astron. Astrophys. **450**, 793–803 (2006). https://doi.org/10.1051/0004-6361:20054670

C. Jacobs, I.I. Roussev, N. Lugaz, S. Poedts, The internal structure of coronal mass ejections: are all regular magnetic clouds flux ropes? Astrophys. J. Lett. **695**, L171–L175 (2009). https://doi.org/10.1088/0004-637X/695/2/L171

A.W. James, L.M. Green, E. Palmerio, G. Valori, H.A.S. Reid, D. Baker, D.H. Brooks, L. van Driel-Gesztelyi, E.K.J. Kilpua, On-disc observations of flux rope formation prior to its eruption. Sol. Phys. **292**, 71 (2017). https://doi.org/10.1007/s11207-017-1093-4. arXiv:1703.10837

Y. Jiang, Y. Shen, B. Yi, J. Yang, J. Wang, Magnetic interaction: a transequatorial jet and interconnecting loops. Astrophys. J. **677**, 699–703 (2008). https://doi.org/10.1086/529417

Y. Jiang, J. Yang, J. Hong, Y. Bi, R. Zheng, Sympathetic filament eruptions connected by coronal dimmings. Astrophys. J. **738**, 179 (2011). https://doi.org/10.1088/0004-637X/738/2/179

M. Jin, C.J. Schrijver, M.C.M. Cheung, M.L. DeRosa, N.V. Nitta, A.M. Title, A numerical study of long-range magnetic impacts during coronal mass ejections. Astrophys. J. **820**, 16 (2016). https://doi.org/10.3847/0004-637X/820/1/16. arXiv:1603.04900

M. Jin, W.B. Manchester, B. van der Holst, I. Sokolov, G. Tóth, R.E. Mullinix, A. Taktakishvili, A. Chulaki, T.I. Gombosi, Data-constrained coronal mass ejections in a global magnetohydrodynamics model. Astrophys. J. **834**, 173 (2017). https://doi.org/10.3847/1538-4357/834/2/173. arXiv:1605.05360

J. Jing, C. Tan, Y. Yuan, B. Wang, T. Wiegelmann, Y. Xu, H. Wang, Free magnetic energy and flare productivity of active regions. Astrophys. J. **713**, 440–449 (2010). https://doi.org/10.1088/0004-637X/713/1/440

K. Jockers, Transport of twist in force-free magnetic flux tubes. Astrophys. J. **220**, 1133–1136 (1978). https://doi.org/10.1086/155998

N.C. Joshi, B. Schmieder, T. Magara, Y. Guo, G. Aulanier, Chain reconnections observed in sympathetic eruptions. Astrophys. J. **820**, 126 (2016). https://doi.org/10.3847/0004-637X/820/2/126. arXiv:1602.07792

S.W. Kahler, The correlation between solar energetic particle peak intensities and speeds of coronal mass ejections: effects of ambient particle intensities and energy spectra. J. Geophys. Res. **106**, 20,947–20,956 (2001). https://doi.org/10.1029/2000JA002231

S.W. Kahler, R.L. Moore, S.R. Kane, H. Zirin, Filament eruptions and the impulsive phase of solar flares. Astrophys. J. **328**, 824–829 (1988). https://doi.org/10.1086/166340

J.T. Karpen, S.K. Antiochos, C.R. DeVore, The mechanisms for the onset and explosive eruption of coronal mass ejections and eruptive flares. Astrophys. J. **760**, 81 (2012). https://doi.org/10.1088/0004-637X/760/1/81

R. Keppens, O. Porth, C. Xia, Interacting tilt and kink instabilities in repelling current channels. Astrophys. J. **795**, 77 (2014). https://doi.org/10.1088/0004-637X/795/1/77. arXiv:1409.4543

J.I. Khan, H.S. Hudson, Homologous sudden disappearances of transequatorial interconnecting loops in the solar corona. Geophys. Res. Lett. **27**, 1083 (2000). https://doi.org/10.1029/1999GL010730

E.K.J. Kilpua, M. Mierla, A.N. Zhukov, L. Rodriguez, A. Vourlidas, B. Wood, Solar sources of interplanetary coronal mass ejections during the solar cycle 23/24 minimum. Sol. Phys. **289**, 3773–3797 (2014). https://doi.org/10.1007/s11207-014-0552-4

R. Kippenhahn, A. Schlüter, Eine Theorie der solaren Filamente. Mit 7 Textabbildungen. Z. Astrophys. **43**, 36 (1957)

I.N. Kitiashvili, A.G. Kosovichev, A.A. Wray, N.N. Mansour, Mechanism of spontaneous formation of stable magnetic structures on the Sun. Astrophys. J. **719**, 307–312 (2010). https://doi.org/10.1088/0004-637X/719/1/307. arXiv:1004.2288

J. Kleimann, 4 π models of CMEs and ICMEs (Invited review). Sol. Phys. **281**, 353–367 (2012). https://doi.org/10.1007/s11207-012-9994-8. arXiv:1203.6187

B. Kliem, T. Török, Torus instability. Phys. Rev. Lett. **255**, 002 (2006). https://doi.org/10.1103/PhysRevLett.96.255002. arXiv:physics/0605217

B. Kliem, T. Török, W.T. Thompson, A parametric study of erupting flux rope rotation. Modeling the "Cartwheel CME" on 9 April 2008. Sol. Phys. **281**, 137–166 (2012). https://doi.org/10.1007/s11207-012-9990-z. arXiv:1112.3389

B. Kliem, J. Lin, T.G. Forbes, E.R. Priest, T. Török, Catastrophe versus instability for the eruption of a toroidal solar magnetic flux rope. Astrophys. J. **789**, 46 (2014a). https://doi.org/10.1088/0004-637X/789/1/46. arXiv:1404.5922

B. Kliem, T. Török, V.S. Titov, R. Lionello, J.A. Linker, R. Liu, C. Liu, H. Wang, Slow rise and partial eruption of a double-decker filament. II. A double flux rope model. Astrophys. J. **792**, 107 (2014b). https://doi.org/10.1088/0004-637X/792/2/107. arXiv:1407.2272

J.A. Klimchuk, *Theory of Coronal Mass Ejections*. Geophysical Monograph Series, vol. 125 (American Geophysical Union, Washington, 2001). https://doi.org/10.1029/GM125p0143

J.A. Klimchuk, P.A. Sturrock, Three-dimensional force-free magnetic fields and flare energy buildup. Astrophys. J. **385**, 344–353 (1992). https://doi.org/10.1086/170943

Y.K. Ko, J.C. Raymond, J. Lin, G. Lawrence, J. Li, A. Fludra, Dynamical and physical properties of a post-coronal mass ejection current sheet. Astrophys. J. **594**, 1068–1084 (2003). https://doi.org/10.1086/376982

R.A. Kopp, G.W. Pneuman, Magnetic reconnection in the corona and the loop prominence phenomenon. Sol. Phys. **50**, 85–98 (1976). https://doi.org/10.1007/BF00206193

V. Kunkel, J. Chen, Evolution of a coronal mass ejection and its magnetic field in interplanetary space. Astrophys. J. Lett. **715**, L80–L83 (2010). https://doi.org/10.1088/2041-8205/715/2/L80

M. Kuperus, M.A. Raadu, The support of prominences formed in neutral sheets. Astron. Astrophys. **31**, 189 (1974)

K. Kusano, T. Maeshiro, T. Yokoyama, T. Sakurai, The trigger mechanism of solar flares in a coronal arcade with reversed magnetic shear. Astrophys. J. **610**, 537–549 (2004). https://doi.org/10.1086/421547

K. Kusano, Y. Bamba, T.T. Yamamoto, Y. Iida, S. Toriumi, A. Asai, Magnetic field structures triggering solar flares and coronal mass ejections. Astrophys. J. **760**, 31 (2012). https://doi.org/10.1088/0004-637X/760/1/31. arXiv:1210.0598

B.J. LaBonte, M.K. Georgoulis, D.M. Rust, Survey of magnetic helicity injection in regions producing X-class flares. Astrophys. J. **671**, 955–963 (2007). https://doi.org/10.1086/522682

J.E. Leake, M.G. Linton, T. Török, Simulations of emerging magnetic flux. I. The formation of stable coronal flux ropes. Astrophys. J. **778**, 99 (2013). https://doi.org/10.1088/0004-637X/778/2/99. arXiv:1308.6204

J.E. Leake, M.G. Linton, S.K. Antiochos, Simulations of emerging magnetic flux. II. The formation of unstable coronal flux ropes and the initiation of coronal mass ejections. Astrophys. J. **787**, 46 (2014). https://doi.org/10.1088/0004-637X/787/1/46. arXiv:1402.2645

K.D. Leka, G. Barnes, Photospheric magnetic field properties of flaring versus flare-quiet active regions. I. Data, general approach, and sample results. Astrophys. J. **595**, 1277–1295 (2003a). https://doi.org/10.1086/377511

K.D. Leka, G. Barnes, Photospheric magnetic field properties of flaring versus flare-quiet active regions. II. Discriminant analysis. Astrophys. J. **595**, 1296–1306 (2003b). https://doi.org/10.1086/377512
K.D. Leka, G. Barnes, Photospheric magnetic field properties of flaring versus flare-quiet active regions. IV. A statistically significant sample. Astrophys. J. **656**, 1173–1186 (2007). https://doi.org/10.1086/510282
K.D. Leka, R.C. Canfield, A.N. McClymont, L. van Driel-Gesztelyi, Evidence for current-carrying emerging flux. Astrophys. J. **462**, 547 (1996). https://doi.org/10.1086/177171
J. Lin, T.G. Forbes, Effects of reconnection on the coronal mass ejection process. J. Geophys. Res. **105**, 2375–2392 (2000). https://doi.org/10.1029/1999JA900477
J. Lin, T.G. Forbes, P.A. Isenberg, P. Démoulin, The effect of curvature on flux-rope models of coronal mass ejections. Astrophys. J. **504**, 1006–1019 (1998). https://doi.org/10.1086/306108
J. Lin, T.G. Forbes, P.A. Isenberg, Prominence eruptions and coronal mass ejections triggered by newly emerging flux. J. Geophys. Res. **106**, 25,053–25,074 (2001). https://doi.org/10.1029/2001JA000046
J. Lin, W. Soon, S.L. Baliunas, Theories of solar eruptions: a review. New Astron. Rev. **47**, 53–84 (2003)
J. Lin, Y.K. Ko, L. Sui, J.C. Raymond, G.A. Stenborg, Y. Jiang, S. Zhao, S. Mancuso, Direct observations of the magnetic reconnection site of an eruption on 2003 November 18. Astrophys. J. **622**, 1251–1264 (2005). https://doi.org/10.1086/428110
J.A. Linker, R. Lionello, Z. Mikić, T. Amari, Magnetohydrodynamic modeling of prominence formation within a helmet streamer. J. Geophys. Res. **106**, 25,165–25,176 (2001). https://doi.org/10.1029/2000JA004020
J.A. Linker, Z. Mikić, R. Lionello, P. Riley, T. Amari, D. Odstrcil, Flux cancellation and coronal mass ejections. Phys. Plasmas **10**, 1971–1978 (2003)
M.G. Linton, M.B. Moldwin, A comparison of the formation and evolution of magnetic flux ropes in solar coronal mass ejections and magnetotail plasmoids. J. Geophys. Res. Space Phys. **114**, A00B09 (2009). https://doi.org/10.1029/2008JA013660
M.G. Linton, R.B. Dahlburg, S.K. Antiochos, Reconnection of twisted flux tubes as a function of contact angle. Astrophys. J. **553**, 905–921 (2001). https://doi.org/10.1086/320974
Y.E. Litvinenko, S.F. Martin, Magnetic reconnection as the cause of a photospheric canceling feature and mass flows in a filament. Sol. Phys. **190**, 45–58 (1999). https://doi.org/10.1023/A:1005284116353
C. Liu, N. Deng, Y. Liu, D. Falconer, P.R. Goode, C. Denker, H. Wang, Rapid change of δ spot structure associated with seven major flares. Astrophys. J. **622**, 722–736 (2005). https://doi.org/10.1086/427868
C. Liu, J. Lee, M. Karlický, D. Prasad Choudhary, N. Deng, H. Wang, Successive solar flares and coronal mass ejections on 2005 September 13 from NOAA AR 10808. Astrophys. J. **703**, 757–768 (2009). https://doi.org/10.1088/0004-637X/703/1/757. arXiv:0908.0487
R. Liu, C. Liu, S. Wang, N. Deng, H. Wang, Sigmoid-to-flux-rope transition leading to a loop-like coronal mass ejection. Astrophys. J. Lett. **725**, L84–L90 (2010). https://doi.org/10.1088/2041-8205/725/1/L84. arXiv:1011.1181
R. Liu, B. Kliem, V.S. Titov, J. Chen, Y. Wang, H. Wang, C. Liu, Y. Xu, T. Wiegelmann, Structure, stability, and evolution of magnetic flux ropes from the perspective of magnetic twist. Astrophys. J. **818**, 148 (2016). https://doi.org/10.3847/0004-637X/818/2/148. arXiv:1512.02338
Y. Liu, Magnetic field overlying solar eruption regions and kink and torus instabilities. Astrophys. J. Lett. **679**, L151 (2008). https://doi.org/10.1086/589282
Y.D. Liu, J.G. Luhmann, P. Kajdič, E.K.J. Kilpua, N. Lugaz, N.V. Nitta, C. Möstl, B. Lavraud, S.D. Bale, C.J. Farrugia, A.B. Galvin, Observations of an extreme storm in interplanetary space caused by successive coronal mass ejections. Nat. Commun. **5**, 3481 (2014). https://doi.org/10.1038/ncomms4481. arXiv:1405.6088
S.H.B. Livi, J. Wang, S.F. Martin, The cancellation of magnetic flux. I—on the quiet sun. Aust. J. Phys. **38**, 855–873 (1985). https://doi.org/10.1071/PH850855
D.W. Longcope, B.T. Welsch, A model for the emergence of a twisted magnetic flux tube. Astrophys. J. **545**, 1089–1100 (2000). https://doi.org/10.1086/317846
B.C. Low, Magnetohydrodynamic processes in the solar corona: flares, coronal mass ejections, and magnetic helicity. Phys. Plasmas **1**, 1684–1690 (1994). https://doi.org/10.1063/1.870671
B.C. Low, Solar activity and the corona. Sol. Phys. **167**, 217–265 (1996). https://doi.org/10.1007/BF00146338
B.C. Low, Coronal mass ejections, magnetic flux ropes, and solar magnetism. J. Geophys. Res. **106**, 25,141–25,164 (2001). https://doi.org/10.1029/2000JA004015
B.C. Low, R.H. Munro, R.R. Fisher, The initiation of a coronal transient. Astrophys. J. **254**, 335–342 (1982). https://doi.org/10.1086/159737
N. Lugaz, C. Downs, K. Shibata, I.I. Roussev, A. Asai, T.I. Gombosi, Numerical investigation of a coronal mass ejection from an anemone active region: reconnection and deflection of the 2005 August 22 eruption. Astrophys. J. **738**, 127 (2011). https://doi.org/10.1088/0004-637X/738/2/127. arXiv:1106.5284

N. Lugaz, M. Temmer, Y. Wang, C.J. Farrugia, The Interaction of Successive Coronal Mass Ejections: A Review (2016). ArXiv e-prints. arXiv:1612.02398

B.J. Lynch, J.K. Edmondson, Sympathetic magnetic breakout coronal mass ejections from pseudostreamers. Astrophys. J. **764**, 87 (2013). https://doi.org/10.1088/0004-637X/764/1/87. arXiv:1212.6677

B.J. Lynch, S.K. Antiochos, P.J. MacNeice, T.H. Zurbuchen, L.A. Fisk, Observable properties of the breakout model for coronal mass ejections. Astrophys. J. **617**, 589–599 (2004). https://doi.org/10.1086/424564

B.J. Lynch, S.K. Antiochos, C.R. DeVore, J.G. Luhmann, T.H. Zurbuchen, Topological evolution of a fast magnetic breakout CME in three dimensions. Astrophys. J. **683**, 1192–1206 (2008). https://doi.org/10.1086/589738

B.J. Lynch, Y. Li, A.F.R. Thernisien, E. Robbrecht, G.H. Fisher, J.G. Luhmann, A. Vourlidas, Sun to 1 AU propagation and evolution of a slow streamer-blowout coronal mass ejection. J. Geophys. Res. Space Phys. **115**, A07106 (2010). https://doi.org/10.1029/2009JA015099

S. Ma, G.D.R. Attrill, L. Golub, J. Lin, Statistical study of coronal mass ejections with and without distinct low coronal signatures. Astrophys. J. **722**, 289–301 (2010). https://doi.org/10.1088/0004-637X/722/1/289

D.H. Mackay, A.A. van Ballegooijen, Models of the large-scale corona. I. Formation, evolution, and liftoff of magnetic flux ropes. Astrophys. J. **641**, 577–589 (2006). https://doi.org/10.1086/500425

D.H. Mackay, L.M. Green, A. van Ballegooijen, Modeling the dispersal of an active region: quantifying energy input into the corona. Astrophys. J. **729**, 97 (2011). https://doi.org/10.1088/0004-637X/729/2/97. arXiv:1102.5296

P. MacNeice, S.K. Antiochos, A. Phillips, D.S. Spicer, C.R. DeVore, K. Olson, A numerical study of the breakout model for coronal mass ejection initiation. Astrophys. J. **614**, 1028–1041 (2004). https://doi.org/10.1086/423887

T. Magara, D.W. Longcope, Injection of magnetic energy and magnetic helicity into the solar atmosphere by an emerging magnetic flux tube. Astrophys. J. **586**, 630–649 (2003). https://doi.org/10.1086/367611

W. Manchester, Buoyant disruption of magnetic arcades with self-induced shearing. J. Geophys. Res. Space Phys. **108**, 1162 (2003). https://doi.org/10.1029/2002JA009252

W. Manchester, B.C. Low, Magnetostatic atmospheres possessing identical invariants of ideal magnetohydrodynamics. Phys. Plasmas **7**, 1263–1279 (2000). https://doi.org/10.1063/1.873937

W. Manchester IV, The role of nonlinear Alfvén waves in shear formation during solar magnetic flux emergence. Astrophys. J. **547**, 503–519 (2001). https://doi.org/10.1086/318342

W. Manchester IV, Solar atmospheric dynamic coupling due to shear motions driven by the Lorentz force. Astrophys. J. **666**, 532–540 (2007). https://doi.org/10.1086/520493

W. Manchester IV, T. Gombosi, D. DeZeeuw, Y. Fan, Eruption of a buoyantly emerging magnetic flux rope. Astrophys. J. **610**, 588–596 (2004a). https://doi.org/10.1086/421516

W.B. Manchester, T.I. Gombosi, I. Roussev, A. Ridley, D.L. de Zeeuw, I.V. Sokolov, K.G. Powell, G. Tóth, Modeling a space weather event from the Sun to the Earth: CME generation and interplanetary propagation. J. Geophys. Res. Space Phys. **109**, A02107 (2004b). https://doi.org/10.1029/2003JA010150

C.H. Mandrini, S. Pohjolainen, S. Dasso, L.M. Green, P. Démoulin, L. van Driel-Gesztelyi, C. Copperwheat, C. Foley, Interplanetary flux rope ejected from an X-ray bright point. The smallest magnetic cloud source-region ever observed. Astron. Astrophys. **434**, 725–740 (2005). https://doi.org/10.1051/0004-6361:20041079

D. Maričić, B. Vršnak, A.L. Stanger, A.M. Veronig, M. Temmer, D. Roša, Acceleration phase of coronal mass ejections: II. Synchronization of the energy release in the associated flare. Sol. Phys. **241**, 99–112 (2007). https://doi.org/10.1007/s11207-007-0291-x

S.F. Martin, Preflare conditions, changes and events. Sol. Phys. **68**, 217–236 (1980). https://doi.org/10.1007/BF00156861

S.F. Martin, S.H.B. Livi, J. Wang, The cancellation of magnetic flux. II—in a decaying active region. Aust. J. Phys. **38**, 929–959 (1985). https://doi.org/10.1071/PH850929

J. Martínez-Sykora, V. Hansteen, B. DePontieu, M. Carlsson, Spicule-like structures observed in three-dimensional realistic magnetohydrodynamic simulations. Astrophys. J. **701**, 1569–1581 (2009). https://doi.org/10.1088/0004-637X/701/2/1569. arXiv:0906.4446

S. Masson, S.K. Antiochos, C.R. DeVore, A model for the escape of solar-flare-accelerated particles. Astrophys. J. **771**, 82 (2013). https://doi.org/10.1088/0004-637X/771/2/82. arXiv:1301.0654

R. Matsumoto, T. Tajima, K. Shibata, M. Kaisig, Three-dimensional magnetohydrodynamics of the emerging magnetic flux in the solar atmosphere. Astrophys. J. **414**, 357–371 (1993). https://doi.org/10.1086/173082

R.T.J. McAteer, P.T. Gallagher, P.A. Conlon, Turbulence, complexity, and solar flares. Adv. Space Res. **45**, 1067–1074 (2010). https://doi.org/10.1016/j.asr.2009.08.026. arXiv:0909.5636

D.E. McKenzie, Signatures of reconnection in eruptive flares [Invited], in *Multi-Wavelength Observations of Coronal Structure and Dynamics*, ed. by P.C.H. Martens, D. Cauffman (2002), p. 155

M. Messerotti, F. Zuccarello, S.L. Guglielmino, V. Bothmer, J. Lilensten, G. Noci, M. Storini, H. Lundstedt, Solar weather event modelling and prediction. Space Sci. Rev. **147**, 121–185 (2009). https://doi.org/10.1007/s11214-009-9574-x

Z. Mikić, M.A. Lee, An introduction to theory and models of CMEs, shocks, and solar energetic particles. Space Sci. Rev. **123**, 57–80 (2006). https://doi.org/10.1007/s11214-006-9012-2

Z. Mikic, J.A. Linker, Disruption of coronal magnetic field arcades. Astrophys. J. **430**, 898–912 (1994). https://doi.org/10.1086/174460

Z. Mikic, D.C. Barnes, D.D. Schnack, Dynamical evolution of a solar coronal magnetic field arcade. Astrophys. J. **328**, 830–847 (1988). https://doi.org/10.1086/166341

Z. Mikic, T. Török, V.S. Titov, J.A. Linker, R. Lionello, P. Riley, *MHD Modeling of the Sympathetic Eruptions Observed on August 1, 2010. AGU Fall Meeting Abstracts* (2011)

Z. Mikić, T. Török, V. Titov, J.A. Linker, R. Lionello, C. Downs, P. Riley, The challenge in making models of fast CMEs, in *American Institute of Physics Conference Series*, ed. by G.P. Zank, J. Borovsky, R. Bruno, J. Cirtain, S. Cranmer, H. Elliott, J. Giacalone, W. Gonzalez, G. Li, E. Marsch, E. Moebius, N. Pogorelov, J. Spann, O. Verkhoglyadova. American Institute of Physics Conference Series, vol. 1539 (2013), pp. 42–45. https://doi.org/10.1063/1.4810985

C.H. Miklenic, A.M. Veronig, B. Vršnak, A. Hanslmeier, Reconnection and energy release rates in a two-ribbon flare. Astron. Astrophys. **461**, 697–706 (2007). https://doi.org/10.1051/0004-6361:20065751

C.H. Miklenic, A.M. Veronig, B. Vršnak, Temporal comparison of nonthermal flare emission and magnetic-flux change rates. Astron. Astrophys. **499**, 893–904 (2009). https://doi.org/10.1051/0004-6361/200810947. arXiv:0910.1701

Y. Moon, G.S. Choe, Y.D. Park, H. Wang, P.T. Gallagher, J. Chae, H.S. Yun, P.R. Goode, Statistical evidence for sympathetic flares. Astrophys. J. **574**, 434–439 (2002). https://doi.org/10.1086/340945

Y. Moon, G.S. Choe, H. Wang, Y.D. Park, Sympathetic coronal mass ejections. Astrophys. J. **588**, 1176–1182 (2003). https://doi.org/10.1086/374270

R.L. Moore, G. Roumeliotis, Triggering of eruptive flares—destabilization of the preflare magnetic field configuration, in *IAU Colloq. 133: Eruptive Solar Flares*, ed. by Z. Svestka, B.V. Jackson, M.E. Machado. Lecture Notes in Physics, vol. 399 (Springer, Berlin, 1992), p. 69. https://doi.org/10.1007/3-540-55246-4_79

R.L. Moore, A.C. Sterling, *Initiation of Coronal Mass Ejections*. Geophysical Monograph Series, vol. 165 (American Geophysical Union, Washington, 2006), p. 43. https://doi.org/10.1029/165GM07

R.L. Moore, A.C. Sterling, H.S. Hudson, J.R. Lemen, Onset of the magnetic explosion in solar flares and coronal mass ejections. Astrophys. J. **552**, 833–848 (2001). https://doi.org/10.1086/320559

R.L. Moore, D.A. Falconer, A.C. Sterling, The limit of magnetic-shear energy in solar active regions. Astrophys. J. **750**, 24 (2012). https://doi.org/10.1088/0004-637X/750/1/24

C. Möstl, K. Amla, J.R. Hall, P.C. Liewer, E.M. De Jong, R.C. Colaninno, A.M. Veronig, T. Rollett, M. Temmer, V. Peinhart, J.A. Davies, N. Lugaz, Y.D. Liu, C.J. Farrugia, J.G. Luhmann, B. Vršnak, R.A. Harrison, A.B. Galvin, Connecting speeds, directions and arrival times of 22 coronal mass ejections from the Sun to 1 AU. Astrophys. J. **787**, 119 (2014). https://doi.org/10.1088/0004-637X/787/2/119. arXiv:1404.3579

T.C. Mouschovias, A.I. Poland, Expansion and broadening of coronal loop transients—a theoretical explanation. Astrophys. J. **220**, 675–682 (1978). https://doi.org/10.1086/155951

T. Mrozek, TRACE and RHESSI observations of a failed eruption of a magnetic flux rope and oscillating coronal loops. Cent. Eur. Astrophys. Bull. **32**, 85–92 (2008)

D.J. Mullan, Sympathetic stellar flares and electron precipitation as probes of coronal structure in flare stars. Astrophys. J. **204**, 530–538 (1976). https://doi.org/10.1086/154199

R.H. Munro, J.T. Gosling, E. Hildner, R.M. MacQueen, A.I. Poland, C.L. Ross, The association of coronal mass ejection transients with other forms of solar activity. Sol. Phys. **61**, 201–215 (1979). https://doi.org/10.1007/BF00155456

N.J. Nelson, B.P. Brown, A. Sacha Brun, M.S. Miesch, J. Toomre, Buoyant magnetic loops generated by global convective dynamo action. Sol. Phys. **289**, 441–458 (2014). https://doi.org/10.1007/s11207-012-0221-4. arXiv:1212.5612

W.M. Neupert, B.J. Thompson, J.B. Gurman, S.P. Plunkett, Eruption and acceleration of flare-associated coronal mass ejection loops in the low corona. J. Geophys. Res. **106**, 25,215–25,226 (2001). https://doi.org/10.1029/2000JA004012

A. Nindos, M.D. Andrews, The association of big flares and coronal mass ejections: what is the role of magnetic helicity? Astrophys. J. Lett. **616**, L175–L178 (2004). https://doi.org/10.1086/426861

A. Nindos, H. Zhang, Photospheric motions and coronal mass ejection productivity. Astrophys. J. Lett. **573**, L133–L136 (2002). https://doi.org/10.1086/341937

Å. Nordlund, R.F. Stein, M. Asplund, Solar surface convection. Living Rev. Sol. Phys. **6**, 2 (2009). https://doi.org/10.12942/lrsp-2009-2

S. Notoya, T. Yokoyama, K. Kusano, T. Sakurai, T. Miyagoshi, H. Isobe, T. Yamamoto, Three-dimensional filament eruption driven by an emerging flux, in *New Solar Physics with Solar-B Mission*, ed. by K. Shibata, S. Nagata, T. Sakurai. Astronomical Society of the Pacific Conference Series, vol. 369 (2007), p. 381

D. Odstrcil, Modeling 3-D solar wind structure. Adv. Space Res. **32**, 497–506 (2003). https://doi.org/10.1016/S0273-1177(03)00332-6

D. Odstrcil, V.J. Pizzo, C.N. Arge, Propagation of the 12 May 1997 interplanetary coronal mass ejection in evolving solar wind structures. J. Geophys. Res. Space Phys. **110**, A02106 (2005). https://doi.org/10.1029/2004JA010745

O. Olmedo, J. Zhang, Partial torus instability. Astrophys. J. **718**, 433–440 (2010). https://doi.org/10.1088/0004-637X/718/1/433

P. Pagano, D.H. Mackay, S. Poedts, Magnetohydrodynamic simulations of the ejection of a magnetic flux rope. Astron. Astrophys. **554**, A77 (2013). https://doi.org/10.1051/0004-6361/201220947

E. Palmerio, E.K.J. Kilpua, A.W. James, L.M. Green, J. Pomoell, A. Isavnin, G. Valori, Determining the intrinsic CME flux rope type using remote-sensing solar disk observations. Sol. Phys. **292**, 39 (2017). https://doi.org/10.1007/s11207-017-1063-x. arXiv:1701.08595

O. Panasenco, M.M. Velli, Magnetic Structure of Twin Filaments Inside Pseudostreamers, AGU, Fall Meeting, Abstracts (2010), p. A1663

N.K. Panesar, A.C. Sterling, D.E. Innes, R.L. Moore, Destabilization of a solar prominence/filament field system by a series of eight homologous eruptive flares leading to a CME. Astrophys. J. **811**, 5 (2015). https://doi.org/10.1088/0004-637X/811/1/5. arXiv:1508.01952

E. Pariat, J.E. Leake, G. Valori, M.G. Linton, F.P. Zuccarello, K. Dalmasse, Relative magnetic helicity as a diagnostic of solar eruptivity. Astron. Astrophys. **601**, A125 (2017). https://doi.org/10.1051/0004-6361/201630043. arXiv:1703.10562

S. Patsourakos, A. Vourlidas, G. Stenborg, Direct evidence for a fast coronal mass ejection driven by the prior formation and subsequent destabilization of a magnetic flux rope. Astrophys. J. **764**, 125 (2013). https://doi.org/10.1088/0004-637X/764/2/125. arXiv:1211.7211

G. Pearce, R.A. Harrison, Sympathetic flaring. Astron. Astrophys. **228**, 513–516 (1990)

Z. Peng, Y. Hu, Interaction between adjacent sheared magnetic arcades in the solar corona. Astrophys. J. **668**, 513–519 (2007). https://doi.org/10.1086/520806

A.A. Pevtsov, R.C. Canfield, H. Zirin, Reconnection and helicity in a solar flare. Astrophys. J. **473**, 533 (1996). https://doi.org/10.1086/178164

G.W. Pneuman, The physical structure of coronal holes—influence of magnetic fields and coronal heating. Astron. Astrophys. **81**, 161–166 (1980)

E.R. Priest, T.G. Forbes, Magnetic field evolution during prominence eruptions and two-ribbon flares. Sol. Phys. **126**, 319–350 (1990). https://doi.org/10.1007/BF00153054

E.R. Priest, T.G. Forbes, Magnetic flipping—reconnection in three dimensions without null points. J. Geophys. Res. **97**, 1521–1531 (1992). https://doi.org/10.1029/91JA02435

E.R. Priest, T.G. Forbes, The magnetic nature of solar flares. Astron. Astrophys. Rev. **10**, 313–377 (2002). https://doi.org/10.1007/s001590100013

R. Qahwaji, T. Colak, M. Al-Omari, S. Ipson, Automated prediction of CMEs using machine learning of CME—flare associations. Sol. Phys. **248**, 471–483 (2008). https://doi.org/10.1007/s11207-007-9108-1

J. Qiu, V.B. Yurchyshyn, Magnetic reconnection flux and coronal mass ejection velocity. Astrophys. J. Lett. **634**, L121–L124 (2005). https://doi.org/10.1086/498716

J. Qiu, H. Wang, C.Z. Cheng, D.E. Gary, Magnetic reconnection and mass acceleration in flare-coronal mass ejection events. Astrophys. J. **604**, 900–905 (2004). https://doi.org/10.1086/382122

J. Qiu, Q. Hu, T.A. Howard, V.B. Yurchyshyn, On the magnetic flux budget in low-corona magnetic reconnection and interplanetary coronal mass ejections. Astrophys. J. **659**, 758–772 (2007). https://doi.org/10.1086/512060

L.A. Rachmeler, C.E. DeForest, C.C. Kankelborg, Reconnectionless CME eruption: putting the Aly–Sturrock conjecture to rest. Astrophys. J. **693**, 1431–1436 (2009). https://doi.org/10.1088/0004-637X/693/2/1431. arXiv:0812.3199

H.E. Ramsey, S.F. Smith, Flare-initiated filamei it oscillations. Astron. J. **71**, 197 (1966)

D.V. Reames, The two sources of solar energetic particles. Space Sci. Rev. **175**, 53–92 (2013). https://doi.org/10.1007/s11214-013-9958-9. arXiv:1306.3608

K.K. Reeves, L. Golub, Atmospheric imaging assembly observations of hot flare plasma. Astrophys. J. Lett. **727**, L52 (2011). https://doi.org/10.1088/2041-8205/727/2/L52

K.K. Reeves, J.A. Linker, Z. Mikić, T.G. Forbes, Current sheet energetics, flare emissions, and energy partition in a simulated solar eruption. Astrophys. J. **721**, 1547–1558 (2010). https://doi.org/10.1088/0004-637X/721/2/1547

M. Rempel, Numerical simulations of quiet sun magnetism: on the contribution from a small-scale dynamo. Astrophys. J. **789**, 132 (2014). https://doi.org/10.1088/0004-637X/789/2/132. arXiv:1405.6814

M. Rempel, M.C.M. Cheung, Numerical simulations of active region scale flux emergence: from spot formation to decay. Astrophys. J. **785**, 90 (2014). https://doi.org/10.1088/0004-637X/785/2/90. arXiv:1402.4703

M. Rempel, M. Schüssler, M. Knölker, Radiative magnetohydrodynamic simulation of sunspot structure. Astrophys. J. **691**, 640–649 (2009). https://doi.org/10.1088/0004-637X/691/1/640. arXiv:0808.3294

R.S. Richardson, Annu. Rep. Dir. Mt. Wilson Obs. **114**, 871 (1936)

E. Robbrecht, S. Patsourakos, A. Vourlidas, No trace left behind: STEREO observation of a coronal mass ejection without low coronal signatures. Astrophys. J. **701**, 283–291 (2009). https://doi.org/10.1088/0004-637X/701/1/283. arXiv:0905.2583

I. Roussev, Eruptive events in the solar atmosphere: new insights from theory and 3-D numerical modelling. Contemp. Phys. **49**, 237–254 (2008). https://doi.org/10.1080/00107510802366658

I.I. Roussev, T.G. Forbes, T.I. Gombosi, I.V. Sokolov, D.L. DeZeeuw, J. Birn, A three-dimensional flux rope model for coronal mass ejections based on a loss of equilibrium. Astrophys. J. Lett. **588**, L45–L48 (2003). https://doi.org/10.1086/375442

I.I. Roussev, I.V. Sokolov, T.G. Forbes, T.I. Gombosi, M.A. Lee, J.I. Sakai, A numerical model of a coronal mass ejection: shock development with implications for the acceleration of GeV protons. Astrophys. J. Lett. **605**, L73–L76 (2004). https://doi.org/10.1086/392504

I.I. Roussev, N. Lugaz, I.V. Sokolov, New physical insight on the changes in magnetic topology during coronal mass ejections: case studies for the 2002 April 21 and August 24 events. Astrophys. J. Lett. **668**, L87–L90 (2007). https://doi.org/10.1086/522588

I.I. Roussev, K. Galsgaard, C. Downs, N. Lugaz, I.V. Sokolov, E. Moise, J. Lin, Explaining fast ejections of plasma and exotic X-ray emission from the solar corona. Nat. Phys. **8**, 845–849 (2012). https://doi.org/10.1038/nphys2427

D.M. Rust, The helicial flux rope structure of solar filaments. Adv. Space Res. **32**, 1895–1903 (2003). https://doi.org/10.1016/S0273-1177(03)90623-5

D.M. Rust, E. Hildner, Expansion of an X-ray coronal arch into the outer corona. Sol. Phys. **48**, 381–387 (1976). https://doi.org/10.1007/BF00152003

T. Sakurai, Magnetohydrodynamic interpretation of the motion of prominences. Publ. Astron. Soc. Jpn. **28**, 177–198 (1976)

I. Sammis, F. Tang, H. Zirin, The dependence of large flare occurrence on the magnetic structure of sunspots. Astrophys. J. **540**, 583–587 (2000). https://doi.org/10.1086/309303

A. Savcheva, A. van Ballegooijen, Nonlinear force-free modeling of a long-lasting coronal sigmoid. Astrophys. J. **703**, 1766–1777 (2009). https://doi.org/10.1088/0004-637X/703/2/1766

A.S. Savcheva, Photospheric flux cancellation and the build-up of sigmoidal flux ropes on the Sun. Astrophys. J. **759**, 105 (2012). https://doi.org/10.1088/0004-637X/759/2/105

B. Schmieder, P. Démoulin, G. Aulanier, Solar filament eruptions and their physical role in triggering coronal mass ejections. Adv. Space Res. **51**, 1967–1980 (2013). https://doi.org/10.1016/j.asr.2012.12.026. arXiv:1212.4014

B. Schmieder, G. Aulanier, B. Vršnak, Flare-CME models: an observational perspective (invited review). Sol. Phys. (2015). https://doi.org/10.1007/s11207-015-0712-1

C.J. Schrijver, A characteristic magnetic field pattern associated with all major solar flares and its use in flare forecasting. Astrophys. J. Lett. **655**, L117–L120 (2007). https://doi.org/10.1086/511857

C.J. Schrijver, Driving major solar flares and eruptions: a review. Adv. Space Res. **43**, 739–755 (2009). https://doi.org/10.1016/j.asr.2008.11.004. arXiv:0811.0787

C.J. Schrijver, P.A. Higgins, A statistical study of distant consequences of large solar energetic events. Sol. Phys. **290**, 2943–2950 (2015). https://doi.org/10.1007/s11207-015-0785-x. arXiv:1509.05680

C.J. Schrijver, A.M. Title, Long-range magnetic couplings between solar flares and coronal mass ejections observed by SDO and STEREO. J. Geophys. Res. Space Phys. **A04**, 108 (2011). https://doi.org/10.1029/2010JA016224

C.J. Schrijver, M.L. DeRosa, T. Metcalf, G. Barnes, B. Lites, T. Tarbell, J. McTiernan, G. Valori, T. Wiegelmann, M.S. Wheatland, T. Amari, G. Aulanier, P. Démoulin, M. Fuhrmann, K. Kusano, S. Régnier, J.K. Thalmann, Nonlinear force-free field modeling of a solar active region around the time of a major flare and coronal mass ejection. Astrophys. J. **675**, 1637–1644 (2008a). https://doi.org/10.1086/527413. arXiv:0712.0023

C.J. Schrijver, C. Elmore, B. Kliem, T. Török, A.M. Title, Observations and modeling of the early acceleration phase of erupting filaments involved in coronal mass ejections. Astrophys. J. **674**, 586–595 (2008b). https://doi.org/10.1086/524294. arXiv:0710.1609

C.J. Schrijver, A.M. Title, A.R. Yeates, M.L. DeRosa, Pathways of large-scale magnetic couplings between solar coronal events. Astrophys. J. **773**, 93 (2013). https://doi.org/10.1088/0004-637X/773/2/93. arXiv:1305.0801

R. Schwenn, A. dal Lago, E. Huttunen, W.D. Gonzalez, The association of coronal mass ejections with their effects near the Earth. Ann. Geophys. **23**, 1033–1059 (2005). https://doi.org/10.5194/angeo-23-1033-2005

D.B. Seaton, M. Mierla, D. Berghmans, A.N. Zhukov, L. Dolla, SWAP-SECCHI observations of a mass-loading type solar eruption. Astrophys. J. Lett. **727**, L10 (2011). https://doi.org/10.1088/2041-8205/727/1/L10. arXiv:1011.1144

V.D. Shafranov, Plasma equilibrium in a magnetic field. Rev. Plasma Phys. **2**, 103 (1966)

A. Shanmugaraju, Y.J. Moon, M. Dryer, S. Umapathy, On the kinematic evolution of flare-associated CMEs. Sol. Phys. **215**, 185–201 (2003)

C. Shen, G. Li, X. Kong, J. Hu, X.D. Sun, L. Ding, Y. Chen, Y. Wang, L. Xia, Compound twin coronal mass ejections in the 2012 May 17 GLE event. Astrophys. J. **763**, 114 (2013). https://doi.org/10.1088/0004-637X/763/2/114

Y. Shen, Y. Liu, J. Su, Sympathetic partial and full filament eruptions observed in one solar breakout event. Astrophys. J. **750**, 12 (2012). https://doi.org/10.1088/0004-637X/750/1/12. arXiv:1202.4867

K. Shibata, T. Tajima, R.S. Steinolfson, R. Matsumoto, Two-dimensional magnetohydrodynamic model of emerging magnetic flux in the solar atmosphere. Astrophys. J. **345**, 584–596 (1989). https://doi.org/10.1086/167932

K. Shibata, S. Masuda, M. Shimojo, H. Hara, T. Yokoyama, S. Tsuneta, T. Kosugi, Y. Ogawara, Hot-plasma ejections associated with compact-loop solar flares. Astrophys. J. Lett. **451**, L83 (1995). https://doi.org/10.1086/309688

D. Shiota, H. Isobe, P.F. Chen, T.T. Yamamoto, T. Sakajiri, K. Shibata, Self-consistent magnetohydrodynamic modeling of a coronal mass ejection, coronal dimming, and a giant cusp-shaped arcade formation. Astrophys. J. **634**, 663–678 (2005). https://doi.org/10.1086/496943. arXiv:astro-ph/0508478

D. Shiota, K. Kusano, T. Miyoshi, N. Nishikawa, K. Shibata, A quantitative MHD study of the relation among arcade shearing, flux rope formation, and eruption due to the tearing instability. J. Geophys. Res. Space Phys. **113**, A03S05 (2008). https://doi.org/10.1029/2007JA012516

G.M. Simnett, H.S. Hudson, The evolution of a rapidly-expanding active region loop into a trans-equatorial coronal mass ejection, in *Correlated Phenomena at the Sun, in the Heliosphere and in Geospace*, ed. by A. Wilson. ESA Special Publication, vol. 415 (1997), p. 437

G. Siscoe, R. Schwenn, CME disturbance forecasting. Space Sci. Rev. **123**, 453–470 (2006). https://doi.org/10.1007/s11214-006-9024-y

G. Siscoe, N.U. Crooker, C.R. Clauer, Dst of the Carrington storm of 1859. Adv. Space Res. **38**, 173–179 (2006). https://doi.org/10.1016/j.asr.2005.02.102

A. Smyrli, F. Zuccarello, P. Romano, F.P. Zuccarello, S.L. Guglielmino, D. Spadaro, A.W. Hood, D. Mackay, Trend of photospheric magnetic helicity flux in active regions generating halo coronal mass ejections. Astron. Astrophys. **521**, A56 (2010). https://doi.org/10.1051/0004-6361/200913275

H.C. Spruit, A.A. van Ballegooijen, Stability of toroidal flux tubes in stars. Astron. Astrophys. **106**, 58–66 (1982)

C.D.C. Steele, E.R. Priest, The eruption of a prominence and coronal mass ejection which drive reconnection. Sol. Phys. **119**, 157–195 (1989). https://doi.org/10.1007/BF00146219

R.F. Stein, Å. Nordlund, Solar small-scale magnetoconvection. Astrophys. J. **642**, 1246–1255 (2006). https://doi.org/10.1086/501445

R.S. Steinolfson, Coronal evolution due to shear motion. Astrophys. J. **382**, 677–687 (1991). https://doi.org/10.1086/170755

A.C. Sterling, H.S. Hudson, Yohkoh SXT observations of X-ray "dimming" associated with a halo coronal mass ejection. Astrophys. J. Lett. **491**, L55–L58 (1997). https://doi.org/10.1086/311043

A.C. Sterling, L.K. Harra, R.L. Moore, New evidence for the role of emerging flux in a solar filament's slow rise preceding its CME-producing fast eruption. Astrophys. J. **669**, 1359–1371 (2007a). https://doi.org/10.1086/520829

A.C. Sterling, R.L. Moore, T.E. Berger, M. Bobra, J.M. Davis, P. Jibben, R. Kano, L.L. Lundquist, D. Myers, N. Narukage, T. Sakao, K. Shibasaki, R.A. Shine, T.D. Tarbell, M. Weber, Hinode observations of the onset stage of a solar filament eruption. Publ. Astron. Soc. Jpn. **59**, 823 (2007b). https://doi.org/10.1093/pasj/59.sp3.S823

L.H. Strous, G. Scharmer, T.D. Tarbell, A.M. Title, C. Zwaan, Phenomena in an emerging active region. I. Horizontal dynamics. Astron. Astrophys. **306**, 947 (1996)

P.A. Sturrock, Model of the high-energy phase of solar flares. Nature **211**, 695–697 (1966). https://doi.org/10.1038/211695a0

P.A. Sturrock, Maximum energy of semi-infinite magnetic field configurations. Astrophys. J. **380**, 655–659 (1991). https://doi.org/10.1086/170620

J. Su, Y. Liu, J. Liu, X. Mao, H. Zhang, H. Li, X. Wang, W. Xie, Lorentz force: a possible driving force for sunspot rotation. Sol. Phys. **252**, 55–71 (2008). https://doi.org/10.1007/s11207-008-9236-2

Y. Su, A. van Ballegooijen, B.W. Lites, E.E. Deluca, L. Golub, P.C. Grigis, G. Huang, H. Ji, Observations and nonlinear force-free field modeling of active region 10953. Astrophys. J. **691**, 105–114 (2009). https://doi.org/10.1088/0004-637X/691/1/105

X. Sun, J.T. Hoeksema, Y. Liu, T. Wiegelmann, K. Hayashi, Q. Chen, J. Thalmann, Evolution of magnetic field and energy in a major eruptive active region based on SDO/HMI observation. Astrophys. J. **748**, 77 (2012). https://doi.org/10.1088/0004-637X/748/2/77. arXiv:1201.3404

X. Sun, M.G. Bobra, J.T. Hoeksema, Y. Liu, Y. Li, C. Shen, S. Couvidat, A.A. Norton, G.H. Fisher, Why is the great solar active region 12192 flare-rich but CME-poor? Astrophys. J. Lett. **804**, L28 (2015). https://doi.org/10.1088/2041-8205/804/2/L28. arXiv:1502.06950

M. Temmer, N.V. Nitta, Interplanetary propagation behavior of the fast coronal mass ejection on 23 July 2012. Sol. Phys. **290**, 919–932 (2015). https://doi.org/10.1007/s11207-014-0642-3. arXiv:1411.6559

M. Temmer, A.M. Veronig, B. Vršnak, J. Rybák et al., Acceleration in fast halo CMEs and synchronized flare HXR bursts. Astrophys. J. **673**, L95–L98 (2008). https://doi.org/10.1086/527414

M. Temmer, A.M. Veronig, E.P. Kontar, S. Krucker, B. Vršnak, Combined STEREO/RHESSI study of coronal mass ejection acceleration and particle acceleration in solar flares. Astrophys. J. **712**, 1410–1420 (2010). https://doi.org/10.1088/0004-637X/712/2/1410. arXiv:1002.3080

M. Temmer, M.A. Reiss, L. Nikolic, S.J. Hofmeister, A.M. Veronig, Preconditioning of interplanetary space due to transient CME disturbances. Astrophys. J. **835**, 141 (2017). https://doi.org/10.3847/1538-4357/835/2/141. arXiv:1612.06080

J.K. Thalmann, Y. Su, M. Temmer, A.M. Veronig, The confined X-class flares of solar active region 2192. Astrophys. J. Lett. **801**, L23 (2015). https://doi.org/10.1088/2041-8205/801/2/L23. arXiv:1502.05157

V.S. Titov, P. Démoulin, Basic topology of twisted magnetic configurations in solar flares. Astron. Astrophys. **351**, 707–720 (1999)

V.S. Titov, Z. Mikic, J.A. Linker, R. Lionello, 1997 May 12 coronal mass ejection event. I. A simplified model of the preeruptive magnetic structure. Astrophys. J. **675**, 1614–1628 (2008). https://doi.org/10.1086/527280. arXiv:0711.3801

V.S. Titov, Z. Mikic, T. Török, J.A. Linker, O. Panasenco, 2010 August 1–2 sympathetic eruptions. I. Magnetic topology of the source-surface background field. Astrophys. J. **759**, 70 (2012). https://doi.org/10.1088/0004-637X/759/1/70. arXiv:1209.5797

M. Tokman, P.M. Bellan, Three-dimensional model of the structure and evolution of coronal mass ejections. Astrophys. J. **567**, 1202–1210 (2002). https://doi.org/10.1086/338699

T. Török, B. Kliem, The evolution of twisting coronal magnetic flux tubes. Astron. Astrophys. **406**, 1043–1059 (2003). https://doi.org/10.1051/0004-6361:20030692

T. Török, B. Kliem, Confined and ejective eruptions of kink-unstable flux ropes. Astrophys. J. Lett. **630**, L97–L100 (2005). https://doi.org/10.1086/462412. arXiv:astro-ph/0507662

T. Török, B. Kliem, Numerical simulations of fast and slow coronal mass ejections. Astron. Nachr. **328**, 743 (2007). https://doi.org/10.1002/asna.200710795. arXiv:0705.2100

T. Török, B. Kliem, V.S. Titov, Ideal kink instability of a magnetic loop equilibrium. Astron. Astrophys. **413**, L27–L30 (2004). https://doi.org/10.1051/0004-6361:20031691. arXiv:astro-ph/0311198

T. Török, M.A. Berger, B. Kliem, The writhe of helical structures in the solar corona. Astron. Astrophys. **516**, A49 (2010). https://doi.org/10.1051/0004-6361/200913578. arXiv:1004.3918

T. Török, O. Panasenco, V.S. Titov, Z. Mikić, K.K. Reeves, M. Velli, J.A. Linker, G. De Toma, A model for magnetically coupled sympathetic eruptions. Astrophys. J. Lett. **739**, L63 (2011). https://doi.org/10.1088/2041-8205/739/2/L63. arXiv:1108.2069

T. Török, M. Temmer, G. Valori, A.M. Veronig, L. van Driel-Gesztelyi, B. Vršnak, Initiation of coronal mass ejections by sunspot rotation. Sol. Phys. **286**, 453–477 (2013). https://doi.org/10.1007/s11207-013-0269-9. arXiv:1401.2922

T. Török, J.E. Leake, V.S. Titov, V. Archontis, Z. Mikić, M.G. Linton, K. Dalmasse, G. Aulanier, B. Kliem, Distribution of electric currents in solar active regions. Astrophys. J. Lett. **782**, L10 (2014). https://doi.org/10.1088/2041-8205/782/1/L10. arXiv:1401.2931

G. Tóth, D.L. de Zeeuw, T.I. Gombosi, W.B. Manchester, A.J. Ridley, I.V. Sokolov, I.I. Roussev, Sun-to-thermosphere simulation of the 28–30 October 2003 storm with the space weather modeling framework. Space Weather **5**, S06003 (2007). https://doi.org/10.1029/2006SW000272

B.T. Tsurutani, W.D. Gonzalez, G.S. Lakhina, S. Alex, The extreme magnetic storm of 1–2 September 1859. J. Geophys. Res. Space Phys. **108**, 1268 (2003). https://doi.org/10.1029/2002JA009504

A.A. van Ballegooijen, Observations and modeling of a filament on the Sun. Astrophys. J. **612**, 519–529 (2004). https://doi.org/10.1086/422512

A.A. van Ballegooijen, P.C.H. Martens, Formation and eruption of solar prominences. Astrophys. J. **343**, 971–984 (1989). https://doi.org/10.1086/167766

B. van der Holst, C. Jacobs, S. Poedts, Simulation of a breakout coronal mass ejection in the solar wind. Astrophys. J. Lett. **671**, L77–L80 (2007). https://doi.org/10.1086/524732

B. van der Holst, W. Manchester IV, I.V. Sokolov, G. Tóth, T.I. Gombosi, D. DeZeeuw, O. Cohen, Breakout coronal mass ejection or streamer blowout: the bugle effect. Astrophys. J. **693**, 1178 (2009). https://doi.org/10.1088/0004-637X/693/2/1178

W. van Tend, M. Kuperus, The development of coronal electric current systems in active regions and their relation to filaments and flares. Sol. Phys. **59**, 115–127 (1978)

S. Vennerstrom, L. Lefevre, M. Dumbović, N. Crosby, O. Malandraki, I. Patsou, F. Clette, A. Veronig, B. Vršnak, K. Leer, T. Moretto, Extreme geomagnetic storms—1868–2010. Sol. Phys. **291**, 1447–1481 (2016). https://doi.org/10.1007/s11207-016-0897-y

A.M. Veronig, M. Karlický, B. Vršnak, M. Temmer, J. Magdalenić, B.R. Dennis, W. Otruba, W. Pötzi, X-ray sources and magnetic reconnection in the X3.9 flare of 2003 November 3. Astron. Astrophys. **446**, 675–690 (2006). https://doi.org/10.1051/0004-6361:20053112

A. Vourlidas, R.A. Howard, E. Esfandiari, S. Patsourakos, S. Yashiro, G. Michalek, Comprehensive analysis of coronal mass ejection mass and energy properties over a full solar cycle. Astrophys. J. **722**, 1522–1538 (2010). https://doi.org/10.1088/0004-637X/722/2/1522. arXiv:1008.3737

B. Vršnak, Dynamics of solar coronal eruptions. J. Geophys. Res. **106**, 25,249–25,260 (2001). https://doi.org/10.1029/2000JA004007

B. Vršnak, The oscillating loop prominence of July 17, 1981. Sol. Phys. **94**, 289–297 (1984)

B. Vršnak, Eruptive instability of cylindrical prominences. Sol. Phys. **129**, 295–312 (1990)

B. Vršnak, Dynamics of solar coronal eruptions. J. Geophys. Res. **106**, 25,249–25,260 (2001). https://doi.org/10.1029/2000JA004007

B. Vršnak, Forces governing coronal mass ejections. Adv. Space Res. **38**, 431–440 (2006). https://doi.org/10.1016/j.asr.2005.03.090

B. Vršnak, Processes and mechanisms governing the initiation and propagation of CMEs. Ann. Geophys. **26**, 3089–3101 (2008)

B. Vršnak, Solar eruptions: the CME-flare relationship. Astron. Nachr. **337**, 21 (2016). https://doi.org/10.1002/asna.201612424

B. Vršnak, V. Ruždjak, R. Brajša, F. Zloch, Oscillatory motions in an active prominence. Sol. Phys. **127**, 119–128 (1990)

B. Vršnak, V. Ruždjak, B. Rompolt, Stability of prominences exposing helical-like patterns. Sol. Phys. **136**, 151–167 (1991)

B. Vršnak, J. Magdalenić, H. Aurass, G. Mann, Band-splitting of coronal and interplanetary type II bursts. II. Coronal magnetic field and Alfvén velocity. Astron. Astrophys. **396**, 673–682 (2002). https://doi.org/10.1051/0004-6361:20021413

B. Vršnak, J. Magdalenić, P. Zlobec, Band-splitting of coronal and interplanetary type II bursts. III. Physical conditions in the upper corona and interplanetary space. Astron. Astrophys. **413**, 753–763 (2004a). https://doi.org/10.1051/0004-6361:20034060

B. Vršnak, D. Maričić, A.L. Stanger, A. Veronig, Coronal mass ejection of 15 May 2001: II. Coupling of the CME acceleration and the flare energy release. Sol. Phys. **225**, 355–378 (2004b). https://doi.org/10.1007/s11207-004-4995-x

B. Vršnak, D. Ruždjak, D. Sudar, N. Gopalswamy, Kinematics of coronal mass ejections between 2 and 30 solar radii. What can be learned about forces governing the eruption? Astron. Astrophys. **423**, 717–728 (2004c). https://doi.org/10.1051/0004-6361:20047169

B. Vršnak, D. Sudar, D. Ruždjak, The CME-flare relationship: are there really two types of CMEs? Astron. Astrophys. **435**, 1149–1157 (2005). https://doi.org/10.1051/0004-6361:20042166

B. Vršnak, M. Temmer, A. Veronig, M. Karlický, J. Lin, Shrinking and cooling of flare loops in a two-ribbon flare. Sol. Phys. **234**, 273–299 (2006). https://doi.org/10.1007/s11207-006-0093-6

B. Vršnak, D. Maričić, A.L. Stanger, A.M. Veronig, M. Temmer, D. Roša, Acceleration phase of coronal mass ejections: I. Temporal and spatial scales. Sol. Phys. **241**, 85–98 (2007). https://doi.org/10.1007/s11207-006-0290-3

B. Vršnak, D. Vrbanec, J. Čalogović, Dynamics of coronal mass ejections. The mass-scaling of the aerodynamic drag. Astron. Astrophys. **490**, 811–815 (2008). https://doi.org/10.1051/0004-6361:200810215

B. Vršnak, G. Poletto, E. Vujić, A. Vourlidas, Y.K. Ko, J.C. Raymond, A. Ciaravella, T. Žic, D.F. Webb, A. Bemporad, F. Landini, G. Schettino, C. Jacobs, S.T. Suess, Morphology and density structure of post-CME current sheets. Astron. Astrophys. **499**, 905–916 (2009). https://doi.org/10.1051/0004-6361/200810844. arXiv:0902.3705

B. Vršnak, T. Žic, D. Vrbanec, M. Temmer, T. Rollett, C. Möstl, A. Veronig, J. Čalogović, M. Dumbović, S. Lulić, Y.J. Moon, A. Shanmugaraju, Propagation of interplanetary coronal mass ejections: the drag-based model. Sol. Phys. **285**, 295–315 (2013). https://doi.org/10.1007/s11207-012-0035-4

H. Wang, J. Chae, V. Yurchyshyn, G. Yang, M. Steinegger, P. Goode, Inter-active region connection of sympathetic flaring on 2000 February 17. Astrophys. J. **559**, 1171–1179 (2001). https://doi.org/10.1086/322377

R. Wang, Y.D. Liu, I. Zimovets, H. Hu, X. Dai, Z. Yang, Sympathetic solar filament eruptions. Astrophys. J. Lett. **827**, L12 (2016). https://doi.org/10.3847/2041-8205/827/1/L12. arXiv:1608.01067

T. Wang, A. Xu, H. Zhang, Evolution of vector magnetic fields and vertical currents and their relationship with solar flares in AR 5747. Sol. Phys. **155**, 99–112 (1994). https://doi.org/10.1007/BF00670733

Y.M. Wang, N.R. Sheeley Jr., Solar wind speed and coronal flux-tube expansion. Astrophys. J. **355**, 726–732 (1990). https://doi.org/10.1086/168805

D.F. Webb, T.A. Howard, Coronal mass ejections: observations. Living Rev. Sol. Phys. **9**, 3 (2012). https://doi.org/10.12942/lrsp-2012-3

D.F. Webb, J. Burkepile, T.G. Forbes, P. Riley, Observational evidence of new current sheets trailing coronal mass ejections. J. Geophys. Res. Space Phys. **108**, 1440 (2003). https://doi.org/10.1029/2003JA009923

B.T. Welsch, Magnetic flux cancellation and coronal magnetic energy. Astrophys. J. **638**, 1101–1109 (2006). https://doi.org/10.1086/498638

M.S. Wheatland, A Bayesian approach to solar flare prediction. Astrophys. J. **609**, 1134–1139 (2004). https://doi.org/10.1086/421261. arXiv:astro-ph/0403613

M.S. Wheatland, I.J.D. Craig, Including flare sympathy in a model for solar flare statistics. Sol. Phys. **238**, 73–86 (2006). https://doi.org/10.1007/s11207-006-0206-2

D.R. Williams, T. Török, P. Démoulin, L. van Driel-Gesztelyi, B. Kliem, Eruption of a kink-unstable filament in NOAA active region 10696. Astrophys. J. Lett. **628**, L163–L166 (2005). https://doi.org/10.1086/432910. arXiv:astro-ph/0507661

R. Wolfson, Shear-induced opening of the coronal magnetic field. Astrophys. J. **443**, 810–817 (1995). https://doi.org/10.1086/175571

G. Yang, Y. Xu, W. Cao, H. Wang, C. Denker, T.R. Rimmele, Photospheric shear flows along the magnetic neutral line of active region 10486 prior to an X10 flare. Astrophys. J. Lett. **617**, L151–L154 (2004). https://doi.org/10.1086/427210

J. Yang, Y. Jiang, R. Zheng, Y. Bi, J. Hong, B. Yang, Sympathetic filament eruptions from a bipolar helmet streamer in the Sun. Astrophys. J. **745**, 9 (2012). https://doi.org/10.1088/0004-637X/745/1/9

S.L. Yardley, L.M. Green, D.R. Williams, L. van Driel-Gesztelyi, G. Valori, S. Dacie, Flux cancellation and the evolution of the eruptive filament of 2011 June 7. Astrophys. J. **827**, 151 (2016). https://doi.org/10.3847/0004-637X/827/2/151. arXiv:1606.08264

S. Yashiro, N. Gopalswamy, S. Akiyama, G. Michalek, R.A. Howard, Visibility of coronal mass ejections as a function of flare location and intensity. J. Geophys. Res. Space Phys. **110**, A12S05 (2005). https://doi.org/10.1029/2005JA011151

Y. Yuan, F.Y. Shih, J. Jing, H.M. Wang, Automated flare forecasting using a statistical learning technique. Res. Astron. Astrophys. **10**, 785–796 (2010). https://doi.org/10.1088/1674-4527/10/8/008

Y. Yuan, F.Y. Shih, J. Jing, H. Wang, Solar flare forecasting using sunspot-groups classification and photospheric magnetic parameters, in *Physics of Sun and Star Spots, IAU Symposium*, vol. 273, ed. by D. Prasad Choudhary, K.G. Strassmeier (2011), pp. 446–450. https://doi.org/10.1017/S1743921311015742

V.B. Yurchyshyn, H. Wang, Magnetic flux cancellation observed in the sunspot moat. Sol. Phys. **202**, 309–318 (2001). https://doi.org/10.1023/A:1012200827156

D.M. Zarro, A.C. Sterling, B.J. Thompson, H.S. Hudson, N. Nitta, SOHO EIT observations of extreme-ultraviolet "dimming" associated with a halo coronal mass ejection. Astrophys. J. Lett. **520**, L139–L142 (1999). https://doi.org/10.1086/312150

J. Zhang, K.P. Dere, A statistical study of main and residual accelerations of coronal mass ejections. Astrophys. J. **649**, 1100–1109 (2006). https://doi.org/10.1086/506903

J. Zhang, K.P. Dere, R.A. Howard, M.R. Kundu, S.M. White, On the temporal relationship between coronal mass ejections and flares. Astrophys. J. **559**, 452–462 (2001). https://doi.org/10.1086/322405

J. Zhang, K.P. Dere, R.A. Howard, A. Vourlidas, A study of the kinematic evolution of coronal mass ejections. Astrophys. J. **604**, 420–432 (2004). https://doi.org/10.1086/381725

M. Zhang, B.C. Low, The hydromagnetic nature of solar coronal mass ejections. Annu. Rev. Astron. Astrophys. **43**, 103–137 (2005). https://doi.org/10.1146/annurev.astro.43.072103.150602

M. Zhang, N. Flyer, B.C. Low, Magnetic field confinement in the corona: the role of magnetic helicity accumulation. Astrophys. J. **644**, 575–586 (2006). https://doi.org/10.1086/503353. arXiv:astro-ph/0603011

Q. Zhang, R. Liu, Y. Wang, C. Shen, K. Liu, J. Liu, S. Wang, A Prominence Eruption Driven by Flux Feeding from Chromospheric Fibrils (2014). ArXiv e-prints. arXiv:1405.6833

X.P. Zhao, S.P. Plunkett, W. Liu, Determination of geometrical and kinematical properties of halo coronal mass ejections using the cone model. J. Geophys. Res. Space Phys. **107**, 1223 (2002). https://doi.org/10.1029/2001JA009143

S. Zharkov, L.M. Green, S.A. Matthews, V.V. Zharkova, 2011 February 15: sunquakes produced by flux rope eruption. Astrophys. J. Lett. **741**, L35 (2011). https://doi.org/10.1088/2041-8205/741/2/L35. arXiv:1110.2005

T. Žic, B. Vršnak, M. Skender, The magnetic flux and self-inductivity of a thick toroidal current. J. Plasma Phys. **73**, 741–756 (2007). https://doi.org/10.1017/S0022377806006209

H. Zirin, H. Wang, Narrow lanes of transverse magnetic field in sunspots. Nature **363**, 426–428 (1993). https://doi.org/10.1038/363426a0

F. Zuccarello, P. Romano, F. Farnik, M. Karlicky, L. Contarino, V. Battiato, S.L. Guglielmino, M. Comparato, I. Ugarte-Urra, The X17.2 flare occurred in NOAA 10486: an example of filament destabilization caused by a domino effect. Astron. Astrophys. **493**, 629–637 (2009). https://doi.org/10.1051/0004-6361:200809887

F.P. Zuccarello, A. Soenen, S. Poedts, F. Zuccarello, C. Jacobs, Initiation of coronal mass ejections by magnetic flux emergence in the framework of the breakout model. Astrophys. J. Lett. **689**, L157–L160 (2008). https://doi.org/10.1086/595821

F.P. Zuccarello, C. Jacobs, A. Soenen, S. Poedts, B. van der Holst, F. Zuccarello, Modelling the initiation of coronal mass ejections: magnetic flux emergence versus shearing motions. Astron. Astrophys. **507**, 441–452 (2009). https://doi.org/10.1051/0004-6361/200912541

F.P. Zuccarello, Z. Meliani, S. Poedts, Numerical modeling of the initiation of coronal mass ejections in active region NOAA 9415. Astrophys. J. **758**, 117 (2012). https://doi.org/10.1088/0004-637X/758/2/117

Space Sci Rev (2017) 212:1159–1219
DOI 10.1007/s11214-017-0394-0

The Physical Processes of CME/ICME Evolution

Ward Manchester IV[1] · Emilia K.J. Kilpua[2] · Ying D. Liu[3,4] · Noé Lugaz[5] · Pete Riley[6] · Tibor Török[6] · Bojan Vršnak[7]

Received: 1 February 2017 / Accepted: 26 June 2017 / Published online: 3 August 2017

Abstract As observed in Thomson-scattered white light, coronal mass ejections (CMEs) are manifest as large-scale expulsions of plasma magnetically driven from the corona in the most energetic eruptions from the Sun. It remains a tantalizing mystery as to how these erupting magnetic fields evolve to form the complex structures we observe in the solar wind at Earth. Here, we strive to provide a fresh perspective on the post-eruption and interplanetary evolution of CMEs, focusing on the physical processes that define the many complex interactions of the ejected plasma with its surroundings as it departs the corona and

The Scientific Foundation of Space Weather
Edited by Rudolf von Steiger, Daniel Baker, André Balogh, Tamás Gombosi, Hannu Koskinen and Astrid Veronig

✉ W. Manchester IV
chipm@umich.edu

E.K.J. Kilpua
Emilia.Kilpua@helsinki.fi

Y.D. Liu
liuxying@spaceweather.ac.cn

N. Lugaz
noe.lugaz@unh.edu

P. Riley
pete@predsci.com

T. Török
tibor@predsci.com

B. Vršnak
bvrsnak@gmail.com

1 Department of Climate and Space Sciences and Engineering, University of Michigan, Ann Arbor, MI 48109, USA

2 Department of Physics, University of Helsinki, P.O. Box 64, Helsinki, Finland

3 State Key Laboratory of Space Weather, National Space Science Center, Chinese Academy of Sciences, Beijing 100190, China

4 University of Chinese Academy of Sciences, Beijing 100049, China

propagates through the heliosphere. We summarize the ways CMEs and their interplanetary CMEs (ICMEs) are rotated, reconfigured, deformed, deflected, decelerated and disguised during their journey through the solar wind. This study then leads to consideration of how structures originating in coronal eruptions can be connected to their far removed interplanetary counterparts. Given that ICMEs are the drivers of most geomagnetic storms (and the sole driver of extreme storms), this work provides a guide to the processes that must be considered in making space weather forecasts from remote observations of the corona.

Keywords Sun · Coronal mass ejections · Solar wind · Space weather

1 Introduction

Coronal Mass Ejections were first observed from space with the coronagraph onboard NASA's Seventh Orbiting Solar Observatory (OSO-7) on 14 December 1971, Tousey (1973) as bright transients expelled through the coronagraph field of view over a period of minutes to hours. Subsequent space-borne coronagraphs such as the Coronagraph/Polarimeter (or C/P) on the Solar Maximum Mission (SMM) (MacQueen et al. 1980) and the Large Angle Spectrometric Coronagraph (LASCO) onboard the Solar and Heliosphere Observatory (SOHO) (Brueckner et al. 1995) have observed thousands of CMEs from which their characteristics are documented over more than two solar cycles (e.g., Hundhausen 1993; St. Cyr et al. 2000). An analysis of these data has led to a very detailed understanding of the structure and evolution of CMEs in the corona, which has been summarized in several reviews (e.g., Hundhausen et al. 1984; Kahler 1987; Hundhausen 1987; Kahler 1992; Gosling 1993a). Related reviews also discuss the pre-event conditions leading to CMEs (Gopalswamy et al. 2006), while the theoretic underpinnings of CME initiation are treated in e.g., Forbes (2000), Forbes et al. (2006). The structure of interplanetary CMEs (ICMEs) is well described in Gopalswamy (2006) and by Kilpua et al. in this issue. For the purpose of this paper, we define CMEs as transients that occur within the field-of-view of classical coronagraphs, which may extend to maximum of 30 $R_\odot$ (solar radii) in the case of the LASCO C3 coronagraph. Transients beyond this range, and certainly beyond orbit of Mercury, we classify as ICMEs.

A fresh look at the literature on CMEs and ICMEs is timely given the enormous advances that have occurred in the past decade. We have just passed the ten year anniversary of the launch of the twin Solar Terrestrial Relations Observatory (STEREO) spacecraft (Kaiser et al. 2008), which provide continuous multi-viewpoint white-light observations of CMEs from Sun to Earth (Howard et al. 2008). STEREO also provides *in situ* measurements including both the plasma and the magnetic field made with the In-situ Measurements of Particles and CME Transients (IMPACT) instrument (Luhmann et al. 2008; Galvin et al. 2008). With these capabilities, STEREO was ideally designed to connect solar eruptions with their solar wind disturbances. Of similar importance is the introduction of massively-parallel supercomputers in the late 1990's, which have allowed the first realistic three-dimensional magnetohydrodynamic (MHD) simulations of specific CME events (e.g., Odstrcil et al. 2005; Lugaz et al. 2007; Tóth et al. 2007; Manchester et al. 2008; Taktakishvili et al. 2009). A review by Webb and Howard (2012) summarizes observational studies of CMEs utilizing data

[5] Space Science Center and Department of Physics, University of New Hampshire, Durham, NH 03824, USA

[6] Predictive Science Inc., 9990 Mesa Rim Rd, Ste 170, San Diego, CA 92121, USA

[7] Hvar Observatory, Faculty of Geodesy, University of Zagreb, Kaciceva 26, 10000 Zagreb, Croatia

from STEREO among many other spacecraft while a companion paper by Chen (2011) summarizes recent advances in numerical modeling of CME initiation. We complement these works by reviewing the results of both observational studies and numerical simulations to garner a more complete understanding of the physical processes governing the evolution of CMEs in the corona and ICMEs in the solar wind. We do so by following the sequence of events that affect the structure and velocity of CMEs as they depart the closed fields of the corona to pass through the depths of interplanetary space.

As we take inventory of relevant physical processes, we first consider the structure and appearance of CMEs and ICMEs from the low corona to 1 AU in Sect. 2. We then cover those processes that are most pronounced in close proximity to the Sun, where CMEs make their way through the highly structured magnetic field of the low corona. In this regard, we consider deflections and rotations (in Sects. 3 and 4, respectively). In Sect. 5, we examine the kinematic evolution followed by a discussion of the impact of ICMEs on the surrounding heliosphere in Sect. 6. We then cover the processes that affect the magnetic structure of ICMEs in Sect. 7, most significantly reconnection, which causes magnetic ejecta to erode. In Sect. 8, we examine the charge state composition of CMEs with a special emphasis on filament material. We consider the effects of CME-CME interaction and the formation of complex ejecta in Sect. 9, and in Sect. 10, we describe a Sun-to-Earth simulation of the Bastille Day CME event. Finally, in Sect. 11, we summarize the salient points of CME and ICME evolution. As we elucidate these processes, we endeavor to first introduce them as they were discovered, and then describe supporting work in a historical narrative. Understanding the evolution of CMEs and ICMEs in the corona and solar wind is fundamental to explaining and potentially predicting the complex structures observed throughout the heliosphere. This fundamental goal of heliospheric physics has great practical value in space weather forecasting capability, which is addressed by Kilpua et al. in this issue. Similarly the topic of solar energetic particles (SEPs) will be covered in two complementary texts, one by Dalla and Klein and another by Schwadron et al.

2 Structure of CMEs to ICMEs

Manifest in coronagraph images, CMEs are seen in Thomson-scattered white-light, where the brightness reflects the electron density near the plane of the sky (Billings 1966). While CMEs may show many morphologies, the simplest configuration is a three-part structure: a bright leading loop enclosing a dark low-density cavity, which contains a high-density core (e.g., Hundhausen 1993; Howard et al. 1997). This three-part structure can usually be traced directly to a progenitor of the same form that may exist at small-scale confined to an active region or exist at global-scale contained within a helmet streamer. The entire system, with a mass in the range of 10^{15}–10^{16} g (Colaninno and Vourlidas 2009) may lift out of the corona gradually over a period of hours with speeds less than 100 $\mathrm{km\,s^{-1}}$, or they may be impulsively ejected with speeds approaching 3000 $\mathrm{km\,s^{-1}}$ (Hundhausen et al. 1994), with kinetic energy approaching 10^{33} ergs. The journey continues beyond the view of classical coronagraphs where the corresponding ejecta in the solar wind are identified as ICMEs that may or may not bare a clear connection to a coronal counterpart (e.g., Gopalswamy et al. 1998).

Disturbances in the solar wind have long been associated with various eruptive phenomena, such as flares, eruptive prominences, type II radio bursts, and later with coronal mass ejections. Since the 1970's efforts have been made to predict the arrival time and impact speed of the eruption-related solar wind disturbances (e.g., De Young and Hundhausen

1971, 1973; Steinolfson and Dryer 1978; Wu et al. 1979). However, at that time, CMEs were not recognized as a major source of interplanetary disturbances and research was primarily focused on the propagation of flare-driven shocks. With time, the CME and more precisely, the ICME ejecta and sheath came to be seen as the source of the most significant solar wind disturbances and consequently the source of non-recurrent geomagnetic storms (e.g., Gosling 1993b). This transformation in understanding was reflected in the evolution of modeling efforts from flare-driven shocks (also called blast waves) to CMEs (Wu et al. 1981; Wei 1982; Dryer et al. 1984; Dryer and Smart 1984; Smart and Shea 1985; Smith and Dryer 1990; Wei and Dryer 1991; Farrugia et al. 1993; Osherovich et al. 1993; Gosling and Riley 1996; Vandas et al. 1996, e.g., and references therein; see also reviews by Pizzo 1985 and Dryer et al. 1988).

The connection between CME and ICME was made abundantly more clear through STEREO observations (e.g., Davis et al. 2009; Möstl et al. 2009; Wood et al. 2009; Liu et al. 2010a). An example of a particularly well observed CME/ICME pair is found in the 12–18 December 2008 event, which serves to highlight a range of physical processes governing the evolution of these phenomena (e.g., Davis et al. 2009; Byrne et al. 2010; Liu et al. 2010a,b; Lugaz et al. 2010; DeForest et al. 2011, 2013; Howard and DeForest 2012). In this case, STEREO data is utilized to make a convincing connection between the solar eruption and the disturbance observed at Earth. The CME is induced by a prominence eruption in the northern hemisphere (see left panel of Fig. 1), which started between 03–04 UT on 12 December 2008. The prominence material (visible in EUVI at 304 Å) is well aligned with the CME core. The CME slowly rotates and expands toward the ecliptic plane, and seems fully developed in COR2. The basic structure of the CME remains organized out to at least the field of view of HI1. In HI2 of STEREO A, we see a dark cavity bracketed by structures with enhanced densities. The time-elongation maps shown in Fig. 1 are produced by stacking running difference intensities of COR2, HI1 and HI2 within a slit along the ecliptic plane. Two features corresponding to the CME leading and trailing edges can be identified up to 50° elongation for both STEREO A and B. Intermittent ones between the two tracks, probably associated with the CME core, are also seen but later disappear presumably due to the expansion of the ICME. Figure 2 shows the plasma and magnetic signatures of the corresponding ICME, which passed Wind on 17 December 2008. The shaded region identifies the magnetic ejecta of the CME.

2.1 ICMEs and Magnetic Clouds

The observed structure and evolution of CMEs/ICMEs, as illustrated in Figs. 1, 2 and 3, can be described by physical models, which can reproduce observed properties and explain and predict their evolution. For CMEs with a common three-part structure, the magnetic field is commonly taken to be of the form of a twisted flux rope contained within an interior plasma cavity (e.g., Gibson et al. 2010). The core of the structure is typically considered to be filament material that was supported by the magnetic field above the system's photospheric polarity inversion line (PIL) prior to the eruption. While long-standing, it is worth noting recent work by Howard et al. (2017) questions the filament-core connection for some events. Flux ropes have often been invoked as a theoretical construction corresponding to the pre-event plasma cavity, which contains the free energy necessary to drive CMEs (e.g., Low 2001; Török and Kliem 2003; Kliem et al. 2004; Fuller et al. 2008). The pre-event magnetic field supporting the filament can also be well described by highly sheared magnetic arcades crossing the PIL (e.g., Mikić et al. 1988; Steinolfson 1991; Antiochos et al. 1999; Amari et al. 2003; Manchester 2003; Lynch et al. 2008; van der Holst et al. 2009).

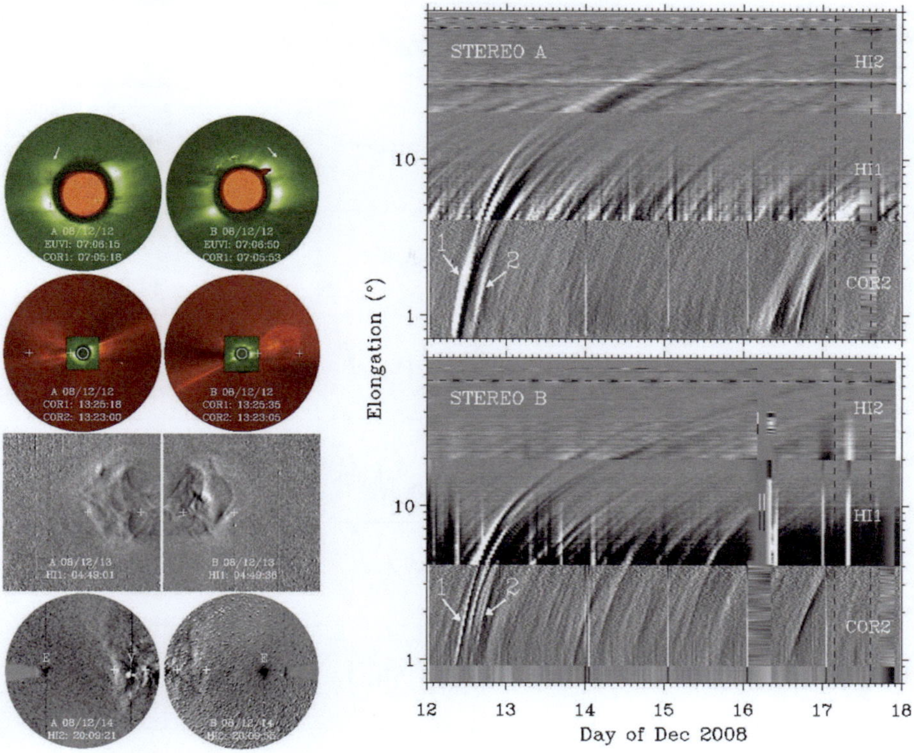

Fig. 1 STEREO observations of the 12 December 2008 CME/ICME event. The *left column* shows the CME/ICME evolution observed by STEREO A (*left*) and STEREO B (*right*) near simultaneously. From top to bottom, the *panels* display the composite images of EUVI at 304 Å and COR1 showing the nascent CME (indicated by the *arrow*), combined COR1 and COR2 images of the fully developed CME, and running difference images from HI1 and HI2 when the ICME is far away from the Sun. The *crosses* mark the locations of the CME leading and trailing edges obtained from the time-elongation map. The positions of the Earth and Venus are labeled as E and V. *Right*: Time-elongation maps constructed from running difference images of COR2, HI1 and HI2 along the ecliptic plane for STEREO A (*upper*) and B (*lower*). The *arrows* indicate two tracks associated with the CME. The *vertical dashed lines* show the MC interval observed at Wind, and the *horizontal dashed line* marks the elongation angle of the Earth. Adapted from Liu et al. (2010a)

Upon eruption, these arcades invariably neck off and reconnect to form erupting magnetic flux ropes attached to the Sun at both ends. Regardless of the simulated CME initiation process, the magnetic structure expelled from the corona is almost universally considered to be a twisted structure that can be characterized as a flux rope. It is here that we begin our evaluation of the physical processes that guide the evolution of CMEs from the low corona to interplanetary space.

Flux rope models have been shown to self-consistently reproduce many observed properties of CMEs, including the three-part density structure (e.g., Gibson and Low 1998; Wu et al. 2001; Manchester et al. 2004b; Wood and Howard 2009). In a similar vein, the graduated cylindrical shell (GCS) model (Thernisien et al. 2009; Vourlidas et al. 2011; Colaninno and Vourlidas 2015) provides a geometric representation of the CME cavity that is consistent with an idealized flux rope. Parameters for the model are determined by fitting two or three nearly simultaneous multi-viewpoint observations derived from STEREO Sun Earth Connection Coronal and Heliospheric Investigation (SECCHI), SOHO/LASCO C2 and C3

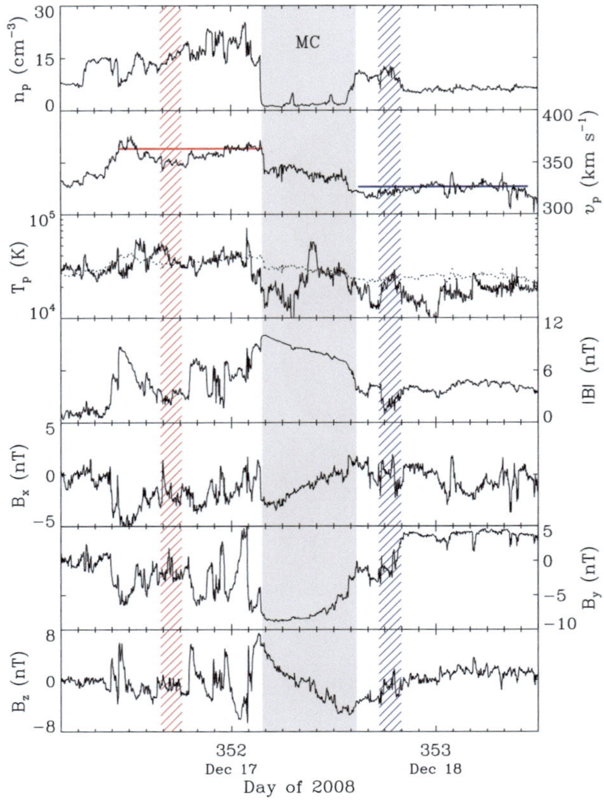

Fig. 2 The MC observed at Wind corresponding to the 12 December 2008 CME (after Liu et al. 2010a). From top to bottom, the *panels* show the proton density, bulk speed, proton temperature, and magnetic field strength and components, respectively. The *shaded region* indicates the MC interval, and the *hatched area* shows the predicted arrival times (with uncertainties) of the ICME leading and trailing edges. The *horizontal lines* mark the corresponding predicted velocities at 1 AU. The *dotted line* denotes the expected proton temperature from the observed speed

and Solar Dynamics Observatory (SDO) Atmospheric Imaging Assembly (AIA). The GCS model is capable of describing a wide range of quantities including the bulk velocity mass distribution and three-dimensional trajectory of the CME (e.g., Shi et al. 2015). Figure 3 shows application of the GCS model to the 16 June 2010 CME event, which is a relatively slow CME occurring with the eruption of a quiescent filament. Here, coronagraph images from STEREO/SECCHI and LASCO coronagraphs are shown along with GCS model represented as green circular lines describing the location of a three-dimensional flux rope in the shape of a crescent as seen in the COR2-A field of view of Fig. 3. The model in this case shows evidence for a variety of physical processes we will discuss in following sections, including super-radial expansion and rotation within the first 5 $R_\odot$. In several examples (e.g., Liu et al. 2010b; Vourlidas et al. 2011; Nieves-Chinchilla et al. 2012; Isavnin et al. 2014; Shi et al. 2015; Schmidt et al. 2016), the GCS model has shown how the early 3-D evolution of CMEs can be well described as magnetic flux ropes that are prone to both deflection and rotation.

Flux ropes ejected from the solar corona during CMEs may travel through interplanetary space largely intact, and careful examination suggests that they can be connected to the magnetic structures observed at 1 AU (e.g., Yurchyshyn et al. 2007; Démoulin 2008; Möstl et al. 2008; Davis et al. 2009; Liu et al. 2010a,b, 2011; Howard and DeForest 2012; Manchester et al. 2014a; Hu et al. 2016). The magnetic fields associated with ICMEs often retain a coherent structure resembling a flux rope. Referred to as magnetic clouds (MCs) (Burlaga 1981, 1988; Lepping et al. 1990; Burlaga et al. 1995), these ICMEs are characterized by high magnetic field strength with a smooth rotation (e.g., south to north or east to

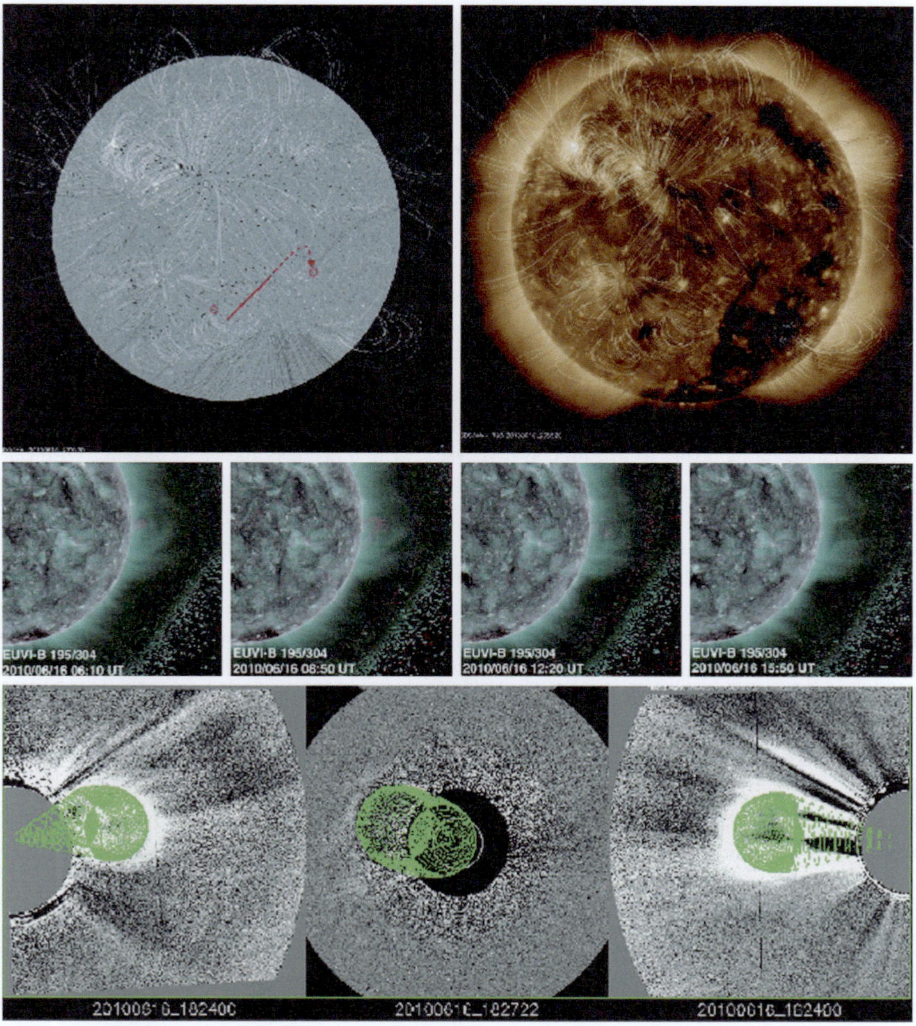

Fig. 3 Solar observations of the source region and CME of 16 June 2010. *Top left*, the HMI synoptic magnetic field is shown with PFSS coronal extrapolation. The *red line* indicates the source of the CME *red circles* giving the foot points of the erupting flux rope. *Top right*, AIA 193 Å image of Sun. *Middle Row*: STEREO EUVI-B images in 195 & 304 Å show a quiescent filament lifting off as the CME erupts. *Bottom Row*: GCS model fitting to the CME event of 16 June 2010. The left, center and right panels are simultaneous data from STEREO COR2-B, SOHO LASCO C2, and STEREO COR2-A, respectively. The images have been over plotted (*green*) with the GCS model represented by a grid of points on the surface of the model flux rope

west) of the field direction, low ion temperature, low plasma beta (typically less than 0.1). The rotation of the field is suggestive of a flux rope geometry (e.g., Lepping et al. 1990; Hu and Sonnerup 2002; Liu et al. 2008a), while the occasional presence of counter-streaming electrons suggest the magnetic field remains attached to the Sun at both ends (e.g., Gosling et al. 2001). The charge state composition of MCs shows elevated ionization states, which are suggestive of flare heated material being ejected with the CME (e.g., Neugebauer and Goldstein 1997; Lepri and Zurbuchen 2004; Zurbuchen and Richardson 2006). Magnetic clouds are also distinguished by their large-scale, and their passage past the Earth that may last 7 to 48 hours, with an average of approximately 21 hours (Lepping et al. 2006). The time and speed indicates that near Earth, the average radial width of MCs is about 0.22 AU and about 1.33 AU at 10 AU (Liu et al. 2006a). MCs are particularly likely to be detected in the near-Earth solar wind when a CME originates within 30° of disk center (Gopalswamy et al. 2001a), indicative of the longitudinal size of these truly global-scale heliospheric disturbances. It should be noted that apart from size, the signatures of ICMEs do not usually occur simultaneously and few ICMEs have all of them.

The relative proportion of ICMEs that appear as MC events has historically shown great variation through the solar cycle. At solar minimum, nearly all ICMEs at Earth can be identified as MCs (Cane and Richardson 2003; Richardson and Cane 2004b), while at solar maximum only $\approx 15\%$ can be. Averaged over the solar cycle, MCs comprised $\approx 30\%$ of ICMEs (Gosling 1990). The cycle dependence reflects several different aspects of CMEs including their place of origin, orientation and mutual interaction. At solar minimum, a greater majority of CMEs originate from streamer blowouts and quiescent filament eruptions. These eruptions are more prone to produce slow CMEs, which are less likely to interact with one another owing to lower eruption rates. Also at solar minimum, CMEs erupt more often at low latitude, providing a greater opportunity for near-central impacts for observing spacecraft, which are more likely to register MC signatures. In contrast, at solar maximum, high latitude eruptions result in off-center ICME *in situ* measurements which are less likely to register the field line rotations of a flux rope. Also at solar maximum, more CMEs originate from active regions where high eruption rates are prone to produce complex interacting ICMEs, as will be discussed in Sect. 9. However, more recent analysis suggest that nearly all ICMEs have flux rope structure (e.g., Owens et al. 2005; Gopalswamy et al. 2013; Mäkelä et al. 2013), and that even plasma-dense ejecta may be fit with flux ropes (Marubashi et al. 2015). ICME-related signatures can also continue well beyond the MC boundaries (e.g., Richardson and Cane 2010b; Kilpua et al. 2013). Manchester and Zurbuchen (2006) found that open field lines deflected around the ejected flux rope can have plasma and magnetic characteristics of a MC at latitudes extending beyond the ejected flux rope.

3 CME and ICME Deflection

3.1 Characteristics and Causes of CME Deflection

CME deflection is the departure from a radial trajectory that commonly occurs with significant in-course changes in direction (e.g., Gosling et al. 1987; Vandas et al. 1996; Wang et al. 2004; Gui et al. 2011; Lugaz et al. 2011; Shen et al. 2011; Kay et al. 2013, 2016; Rollett et al. 2014; Möstl et al. 2015). Figure 4 shows a clear example of deflection for the 2 November 2008 event where the CME's change in latitude is obvious when comparing the STEREO-B COR1 to COR2 images. A survey by Isavnin et al. (2014) found a maximum total CME deflection (from Sun to Earth) in latitude was 49° and almost 30° in longitude. These deflections can be attributed to two primary causes: First, magnetic forces produced by the background corona (e.g., MacQueen et al. 1986; Kilpua et al. 2009; Shen et al. 2011), including the active region of origin (Möstl et al. 2015). Second, the background solar wind flow pattern can inhibit the latitudinal expansion of the CME in the corona (e.g., Cremades et al. 2006) and the wind also interacts with ICMEs farther out in the heliosphere (e.g., Wang et al. 2004). Magnetic forces control the deflection low in the corona, while the importance of kinematic interactions increases at larger heliospheric distances. In the case of kinematic interactions, the deflecting forces are also ultimately magnetic in nature (see e.g., discussion

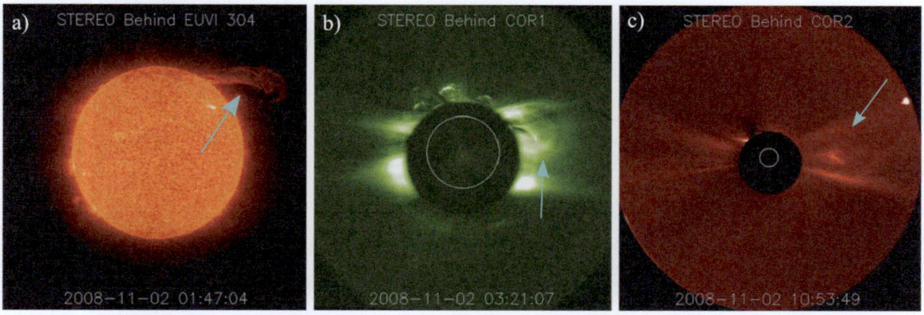

Fig. 4 (**a**): A high-latitude prominence eruption on 2 November 2008 seen by STEREO-B EUVI at 304 Å wavelength and the corresponding CME in (**b**): STEREO-B COR1 (**c**): STEREO-B COR2. This CME deflected quickly to the ecliptic and was observed at STEREO-A as a well-defined magnetic cloud a few days later. The event is studied in detail in Kilpua et al. (2009). Panels (**a**)–(**c**) adapted from Kilpua et al. (2009)

in Isavnin et al. 2014). This deflection occurs because of ICME interactions with the ambient solar wind that pileup plasma and drape magnetic field at the edges of the CME ejecta, as will be discussed in detail in Sects. 5 and 6.

Shen et al. (2011) argue that CMEs tend to deflect toward the region of the lower magnetic energy density by the combined effect of the magnetic pressure and tension forces. The authors presented a theoretical method to account for this effect, which was used statistically by Gui et al. (2011) to confirm the deflection toward the magnetic energy minimum. Kay et al. (2015) demonstrated this deflection property using the Forecasting CMEs Altered Trajectory (ForeCAT) model (see also Kay et al. 2013). This tool propagates CMEs using a drag-based empirical model that takes into account magnetic forces as well as CME expansion. The results of Kay et al. (2015) show a wide range of deflection and also illustrate circumstances when forces are insufficiently strong to fully deflect CMEs toward the energy minimum.

Consider CME deflection from the corona/heliosphere system in the minimum energy state, which has open flux extending from coronal holes separated by a streamer belt with a heliospheric current sheet (HCS) extension. The magnetic field in coronal holes is typically stronger than that found in the surrounding closed flux systems, which provides a magnetic gradient that readily deflects CMEs. The importance of coronal holes for deflection is featured in many studies (e.g., Cremades et al. 2006). Gopalswamy et al. (2009a) and Mohamed et al. (2012) estimated the magnitude and direction of the resultant magnetic force exerted by all coronal holes present on the solar disk during the time of the CME eruption (defined as the CHIP parameter in Mohamed et al. 2012) and compared this with CME trajectories and *in situ* observations. It was found that CMEs tend to move away from the coronal holes and the CMEs that erupt close to disk center but had large CHIP parameters produced generally complex ICMEs or driverless shocks in the near-Earth solar wind. In addition, several recent studies have demonstrated that CMEs are also deflected by strong magnetic fields in the CME source active region (e.g., Kay et al. 2015; Möstl et al. 2015; Wang et al. 2015). Figure 5 shows examples of deflection in both circumstances: CMEs originating from a low-latitude active region and from a coronal hole boundary depicted in Fig. 5a, b, respectively. These results, calculated with ForeCAT, indicate that deflections in latitude and longitude may reach 30° to 40°, respectively, with the magnitude being inversely related to CME speed and mass. The work of Lugaz et al. (2011) on CME deflection presents a slightly different configuration, that of an anemone active region, where the Lorentz force drives the deflection.

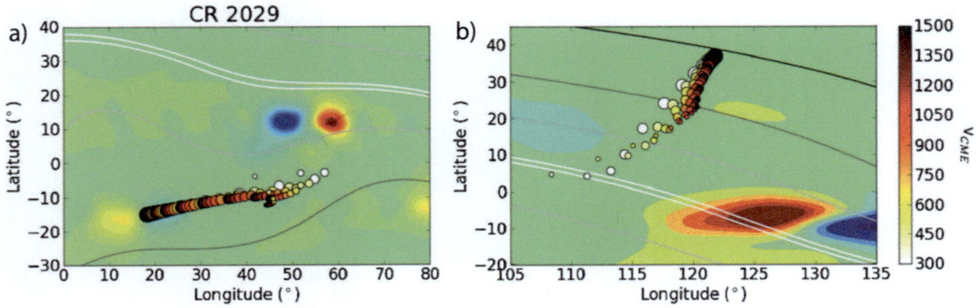

Fig. 5 FOREcat model results (adapted from Kay et al. 2015) shown for CMEs originating from two locations: (**a**) low-latitude active region and (**b**) coronal hole boundary. Here, the *dots* represent the deflected location of the CMEs where the size of the dots are proportional to mass ranging from 10^{14} to 10^{15} g, and the speed of the CMEs ranging from 300 to 1500 $\mathrm{km\,s^{-1}}$ is indicated by *color*. It is shown that deflection in latitude and longitude may reach 30° to 40°, respectively, with deflection being inversely related to CME speed and mass

The CME deflection in latitude is constrained by the location of the streamer belt/HCS and this deflection occurs predominantly close to the Sun in the neighborhood of the streamers. However, longitudinal deflections are largely controlled by kinematic interactions (e.g., Gosling et al. 1987; Wang et al. 2004, 2014b) that occur at larger distances in the corona and heliosphere. The main source of longitudinal deflection is caused by interaction of ICMEs with the Parker spiral structured solar wind that occurs when there is sufficient speed difference with the ICME (Wang et al. 2004). The resulting magnetic forces and the direction of the deflection are different for slow and fast ICME populations: Slow ICMEs are deflected westward when they are pushed from behind by the faster wind, while fast ICMEs deflect eastward as they are decelerated by the slower wind head. We note however that ambient Parker spiral field by itself may not be able to deflect a ICME by more than a few degrees as even for a slow ICME its kinetic energy density is about two orders of magnitude higher than the magnetic energy density of the Parker field. Finally, interactions between multiple CMEs/ICMEs, in particular when they collide, can also cause longitudinal deflections (e.g., Lugaz et al. 2012; Shen et al. 2012; Liu et al. 2012, 2014a).

3.2 Deflection Dependence on the Background Corona

Based on the above-described studies, the rate and amount of CME/ICME deflection is controlled by the strength and distribution of the background magnetic field, and the mass, size, and speed of the CME/ICME relative to the solar wind. Hence, both the global configuration of the Sun's magnetic field and intrinsic CME properties have crucial importance on the degree and direction of deflection, which we now further quantify. For example, Xie et al. (2009) showed that during solar minimum slow CMEs deflected toward the ecliptic and the streamer belt by strong polar magnetic fields, while fast CMEs are deflected less, sometimes also away from the streamer belt away from the streamer belt, confirming earlier results by MacQueen et al. (1986). Similarly Wang et al. (2011) found 62% of CMEs deflected towards equator with an average deflection angle of 22°, and only 5% of CMEs deflected towards poles, with an average deflection angle of 16°. Also a case study by Kilpua et al. (2009) found that slower and wider CMEs deflected toward the equator while the faster and narrower CME propagated radially from its source active region. It was suggested that slow and wider CMEs cannot penetrate through the overlying coronal fields, but are channeled toward the streamer belt. These findings are consistent with the ForeCAT model (Kay et al. 2015; Kay and Opher 2015) showing that slow, wide and low-mass CMEs deflect most as shown in Fig. 5.

The rate of CME deflection is clearly fastest close to the Sun where magnetic forces dominate. The analysis of 14 CMEs by Isavnin et al. (2014) showed that about 60% of the total evolution from the Sun to Earth orbit takes place in the corona, i.e., within about first 20–30 $R_\odot$ from the Sun, in particular, the rate being highest in the low corona ($< 5\ R_\odot$). Kay et al. (2015) came to similar conclusions using the ForeCAT model; the majority of CME deflection occurs within 10 $R_\odot$ from the Sun. Kay et al. (2015) also pointed out that deflections are confined closer to the Sun in the case of stronger background magnetic fields. For strong fields studied in their paper, the deflections occurred primarily below 2 $R_\odot$ from the solar surface. This result is consistent with Gui et al. (2011) who found a positive correlation with the rate of the deflection and the strength of the magnetic energy density gradient. The intrinsic magnetic polarity of the CME flux rope relative to the background coronal magnetic field determines whether and where the magnetic reconnection can occur and consequently deflects the CME (e.g., see simulation work by Chané et al. 2005; Zuccarello et al. 2012a; Zhou and Feng 2013). If the CMEs intrinsic field is parallel to the ambient field, the CME deflects equator-ward while when the fields are antiparallel, the CME is likely to deflect poleward (Zhou and Feng 2013).

3.3 Characteristics and Examples of CME/ICME Deflection

There are several examples in the literature where deflections of CMEs have changed the expected ICME impact at Earth and the expected geoeffectivity. For example, Zhou et al. (2006) showed that intrinsically high-latitude CMEs can drive strong storms; nearly 30% of the Earth-encountered ICMEs associated with high-latitude polar crown filament disappearances investigated in their study caused at least moderate space weather effects. CMEs have a strong tendency to deflect from high latitudes toward the equator in particular near solar minimum (e.g., Plunkett et al. 2001; Cremades et al. 2006; Kilpua et al. 2009; Byrne et al. 2010; Isavnin et al. 2014). This behavior is expected, as at this time, the global magnetic field of the Sun is relatively close to a dipole field and two large polar coronal holes dominate the field structure. These coronal holes can effectively guide CMEs toward the ecliptic, consistent also with the suggestion by Shen et al. (2011) as the minimum energy region streamer belt/HCS is relatively flat and confined close to the equator during slow solar activity period. Near solar maximum the configuration of the Sun's global magnetic field and the distribution of magnetic energy density is more complex than near solar minimum and CMEs deflect less and more randomly. Poleward deflection occurs mainly during the times of solar maximum when large low-latitude polar coronal holes are present.

As discussed in the beginning of this section, longitudinal deflection may cause CMEs/ICMEs to deviate from the Sun-Earth line or toward it. According to study by Gopalswamy et al. (2009b), almost ten percent of large geospace storms are caused by CMEs that originate close to the limb of the Sun. In such cases, the ICME sheath is typically the primary driver of the storm (see also Huttunen et al. 2002), but cases have been reported where clear ejecta signatures upstream of the Earth and geomagnetic activity have been associated with limb CMEs (e.g., Schwenn et al. 2005; Cid et al. 2012; Wang et al. 2014b; Liu et al. 2016). For example, the CME studied by Wang et al. (2014b) was initially heading toward the STEREO-B, but it deflected in the heliosphere and arrived at the Earth instead, which was located about 35° away from the STEREO-B at that time. An opposite case is featured e.g., in Möstl et al. (2015), Wang et al. (2015) where the authors studied a CME

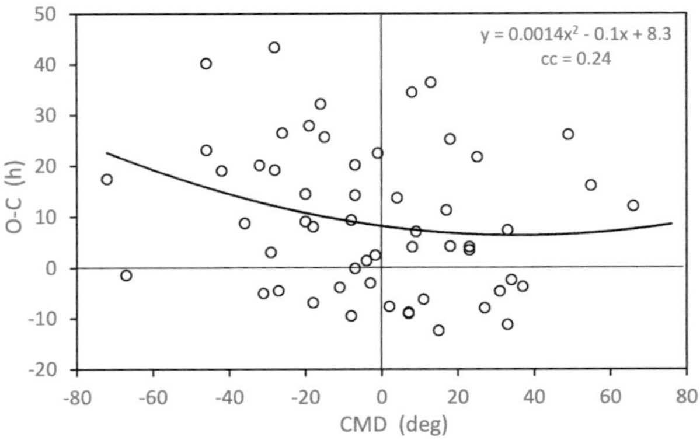

Fig. 6 Difference between the observed and calculated transit times ($O-C$) of 55 fast CMEs ($v > 500\ \text{km}\,\text{s}^{-1}$) presented as a function of central meridian distance (*CMD*). The calculated transit times are based on the Drag-Based Model (DBM) using the solar wind speed of $w = 450\ \text{km}\,\text{s}^{-1}$ and the drag parameter $\Gamma = 0.2$ (for details see Vršnak et al. 2013)

that originated from an active region near the solar disk center and significant geomagnetic response was expected. However, multi-spacecraft observations and modeling demonstrated that this CME deflected almost 40° in longitude and caused only minimal space weather effects. A similar event illustrating the magnitude of CME/ICME longitudinal deflection as reported by Mays et al. (2015). Gopalswamy et al. (2009a) also demonstrated that some halo CMEs that erupt near disk center appear to be associated with driverless shocks near Earth orbit (i.e., an interplanetary shock that is not followed by the discernible driver or a slow-fast stream interaction region) due to significant deflection in longitude away from the Sun-Earth line. In these cases the ICME shock and sheath is detected *in situ* at Earth, but the driving ejecta is missed completely.

Finally, we illustrate the eastward-deflection effect on fast CMEs by presenting in Fig. 6 the dependence of the difference between the observed and calculated transit times ($O-C$) on the central meridian distance (*CMD*) of their source region. The sample contains 55 fast CMEs, whose transit times are calculated using the Drag-Based Model (DBM) by using the solar wind speed of $w = 450\ \text{km}\,\text{s}^{-1}$ and the drag parameter $\Gamma = 0.2$ (for details see Vršnak et al. 2013). Note that $O-C > 0$ means that the ICME arrived later than expected for the chosen DBM parameters (in the presented case the presumed solar wind speed is too high, causing underestimation of transit times and leading to the average $O-C$ of 9.6 h). The scatter-plot includes the quadratic least squares fit, characterized by the correlation coefficient of $cc = 0.24$ and the F-test statistical significance of $P > 99\%$. The fit shows that $O-C$ values are on average larger for CMEs launched closer to the solar limb, meaning that such CMEs are slower than those launched close to the disc center, perhaps indicating that the flank speed is lower than at the nose of the CME. Furthermore, the effect is larger for the eastern-hemisphere events than for western ones, demonstrating the eastward deflection of fast events.

4 CME Rotation

Erupting prominences (or filaments) frequently exhibit a rotation about their rise direction as they ascend in the corona, which leads to a deviation from their original orientation on the solar surface. Strongly rotating filaments display a characteristic "inverse γ" shape, which develops when the legs of the filament cross along the line of sight of the observer (Fig. 7a, b). The direction of the rotation is determined by the sign of helicity of the source region, though there seem to be exceptions (Muglach et al. 2009). When viewed from above,

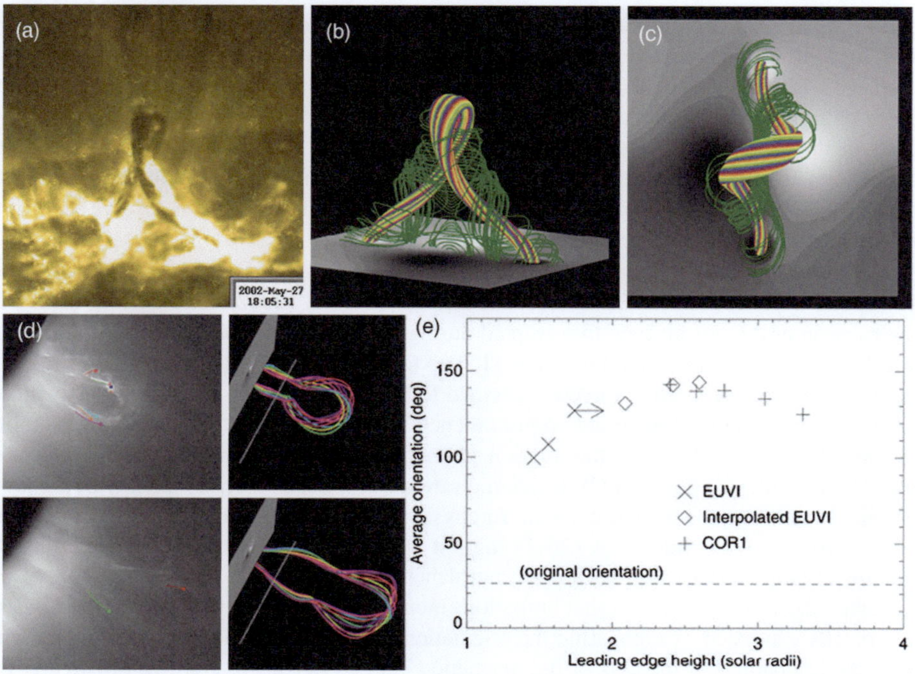

Fig. 7 (**a**): Confined filament eruption on 27 May 2002 observed by *TRACE* in 195 Å. The filament exhibits an "inverse γ" shape, suggestive of a rotation about its rise direction. (**b**): MHD simulation of the eruption by Török and Kliem (2005). *Colored field lines* outline the core of a kink-unstable flux rope with positive helicity (right-handed twist), surrounded by *green ambient field lines*. (**c**): Top view on the simulation. The flux-rope core has rotated clockwise by almost 90° with respect to its initial orientation. (**d**): Fully erupting prominence on 9 April 2008 observed by STEREO *Ahead* at 10:55 and 11:25 UT. The *colored strands* were used for a 3D reconstruction. The *right panels* show field lines from a simulation of the event Kliem et al. (2012). (**e**): Prominence rotation *vs.* heliocentric height of the prominence's leading edge. Panels (**a**)–(**c**) adapted from Green et al. (2007). Panels (**d**)–(**e**) adapted from Thompson et al. (2012)

clockwise (anti-clockwise) rotation is typically observed for filaments erupting from source regions with positive (negative) helicity (Green et al. 2007; Fig. 7c). This suggests the conversion of twist into writhe in a kink-unstable magnetic flux rope as a possible mechanism of the rotation (Kliem et al. 2012). Figure 7 shows a confined (or failed) eruption, but significant rotations occur also in cases where the filament fully erupts as part of a CME (Fig. 7d). While filament material outlines only a small fraction of a CME, it is believed to be located close to the axis of the CME flux rope. Thus, the observed rotation of a filament implies that the whole CME flux rope rotates as well, even though outer flux surfaces of the rope may rotate at a smaller rate.

Since the magnetic orientation of an ICME upon arrival at Earth is one of the main parameters that determine its geo-effectiveness (as discussed by Kilpua et al. in this issue), it is important to understand the physical mechanisms that cause the rotation, and to quantify the total rotation that CMEs/ICMEs undergo during their travel to Earth, particularly in light of the desire to develop methods to reliably predict the sign and magnitude of B_z at 1 AU. However, obtaining reliable measurements of the total rotation directly from the observations is difficult for several reasons. First, complete observational coverage of the CME and ICME propagation from Sun to Earth is not always available. Second, rotation is hard to recognize in coronagraph images, especially if the CME is observed above the solar limb and no filament is present, though measurements could be obtained for halo CMEs using flux-rope fitting and three-dimensional (3D) reconstruction techniques (e.g., Yurchyshyn et al. 2007; Liu et al. 2010b; Vourlidas et al. 2011). Third, most of the rotation may take place before the CME enters a coronagraphs' field of view, which requires EUV observations of the associated filament eruption, with the filament material exhibiting a coherent shape for a sufficiently long time. Even if the latter condition is met, reliable measurements of the rotation can only be obtained if the filament either erupts toward the observer or if 3D reconstructions of its shape can be made.

A rare example of direct rotation measurements is shown in Fig. 7d and e. Using observations from both STEREO spacecraft, Thompson et al. (2012) were able to measure the rotation of the prominence that erupted during the "Cartwheel" CME on 9 April 2008, which showed a strong rotation of about 115° up to a height of about 1.5 $R_\odot$ above the surface. Afterward, the rotation direction seemed to reverse slowly. In a similar investigation of an erupting quiescent polar-crown prominence, Thompson (2011) obtained a rotation of at least 90°. Also in this case, the rotation saturated before the prominence became invisible. These results suggest that CME rotation saturates (or even reverses) already in the low corona. In contrast, the studies by, e.g., Yurchyshyn et al. (2007), Liu et al. (2010b), Lynch et al. (2010), and Vourlidas et al. (2011) suggest that CME/ICMEs still exhibit a significant rotation at larger coronal heights and in interplanetary space.

Given the various observational limitations mentioned above, the total rotation of CMEs is typically estimated by comparing the orientation of the pre-eruptive structure on the Sun with the orientation of the axis of the magnetic cloud at 1 AU. Pre-eruptive orientation is inferred from the polarity inversion line of the source region or from filament, sigmoid or coronal loop observations, while the orientation at 1 AU is obtained by fitting 3D flux-rope models to the 1D *in situ* data, given that the circumstances allow such fitting. Such comparisons of orientation, albeit hampered by some uncertainty, suggest that in many, if not most, cases the total rotation of the CME/ICME remains relatively small. However, as reviewed by Démoulin (2008), cases with total rotations larger than 30° are not uncommon, and very large values of about 120° or more have been reported (e.g., Rust et al. 2005; Dasso et al. 2007; Liu et al. 2008b, 2016; Isavnin et al. 2014; Vemareddy et al. 2016).

Numerical simulations and theoretical considerations have been employed to understand the physical causes of CME rotation and to explain the wide range of rotations observed. Török and Kliem (2003) showed that erupting flux ropes undergo a significant rotation (of more than 90°) when driven by vortex flows at their foot points (used to model rotating sunspots), and suggested that the strong rotation occurred due to the development of the ideal MHD kink instability. Starting from analytical flux-rope models, Fan and Gibson (2004) and Török and Kliem (2005) reported similarly large rotations (up to 120°) as a result of the same instability. Using a similar flux-rope model, Fan (2016) obtained an even larger rotation of almost 180°, i.e., a full reversal of the magnetic field vector at the front of the flux rope, in a recent simulation of the 13 December 2006 event.

Isenberg and Forbes (2007) suggested that the presence of an external shear field surrounding a pre-eruptive flux rope (i.e., of an ambient field component pointing along the axis of the rope) provides a different mechanism for the rotation of a flux rope, once the rope leaves its equilibrium state and rises in the corona. The Lorentz forces that arise from the interaction of the flux-rope current with the external shear field cause a rotation that acts in the same direction as the kink instability. Kliem et al. (2012) confirmed the suggestion by Isenberg and Forbes in a series of MHD simulations, in which they varied the twist within a flux rope, the strength of the external shear field, and the decrease of the ambient potential field with height (by changing the distance between the photospheric polarities that generate that field). Changing these parameters, they obtained flux-rope rotations in a wide range of about (35–170)°. Their study revealed a number of interesting results: (1) For small distances between the polarities, the dominant contribution to the rotation comes from the external shear, even for strongly kink-unstable flux ropes; (2) a moderate rotation in erupting flux ropes is always present due to the conversion of flux-rope twist into writhe (Török et al. 2010), even in the absence of a shear field and of the kink instability; (3) the amount of rotation depends on the slope of the ambient field: the slower the field drops with height, the more the rope rotates. This effect is, however, relatively weak as long as the distance between the polarities is smaller or comparable to the distance of the flux-rope foot points (as it is typically the case on the Sun), but becomes significant if the former distance dominates.

In contrast to these simulations, in which the eruption starts from a fully-developed flux rope, Lynch et al. (2009) modeled CME rotation starting from a sheared arcade, using the breakout model (Antiochos et al. 1999; Lynch et al. 2008). Initially, the arcade rose slowly, without a clear sign of rotation. When it reached a height of about 2 (heliocentric) $R_{\odot}$, flare reconnection set in and a twisted flux rope was formed. From that point on, the rope rotated at a constant rate of about 30° $R_{\odot}^{-1}$, and reached a rotation angle of about 50° at a height of 3.5 $R_{\odot}$. The twist in the rope was low, clearly below the threshold of the kink instability. The authors explained the rotation by the effect of the tension force associated with the sigmoidal shape of the erupting field lines, which acted to straighten out those field lines once the eruption was underway. As the field lines became straight, the rotation still continued due to the angular momentum imparted in the early phase of the eruption. Lynch et al. (2009) did not report how the rotation evolved beyond 3.5 $R_{\odot}$, so it is not clear at which point a saturation of the rotation may have occurred. Their simulation results agree qualitatively very well with the case observed by Vourlidas et al. (2011), where apparently little or no rotation of the CME occurred below 2 $R_{\odot}$, and an almost constant rotation (at a rate twice as large as in the simulation) was seen above that height.

Other mechanisms that have been suggested based on observational or numerical studies are the straightening of an initially strong S-shape of a flux rope during its eruption (Török et al. 2010; Kliem et al. 2012); reconnection of an erupting flux rope with the ambient magnetic field (Jacobs et al. 2009; Shiota et al. 2010; Cohen et al. 2010; Thompson 2011; Lugaz et al. 2011); and the alignment of the CME flux rope with the heliospheric current sheet (e.g., Yurchyshyn 2008).

Overall, these studies reveal a substantial number of mechanisms that can cause CMEs to rotate about their direction of propagation. Moreover, as discussed in detail in Kliem et al. (2012), several mechanisms can contribute simultaneously or successively to the rotation in a complicated, parameter-dependent manner, making the quantitative prediction of CME rotation a very challenging task. Specifically, predictions of the sign of B_z at 1 AU merely based on the pre-eruptive orientation of the ejecta have to be taken with care, given the significant number of events that exhibit a large total rotation. Lynch et al. (2009) suggested that the amount of total rotation my be predicted by the degree of "sigmoidality" of the pre-eruptive field, but this approach needs to be tested using observations, and it will likely underestimate the total rotation if other mechanisms such as the kink instability are involved. If most of the rotation typically occurs low in the corona, as suggested by some observations and simulations, then measurements of the rotation of erupting filaments may be utilized to improve (shorter-term) B_z predictions, but the practicability of such an approach needs to be tested as well. Attempts to systematically predict the rotation (and deflection) of CMEs in the corona, based on the properties of the background magnetic field and using analytical flux-rope models, have been recently developed and tested with two observed events by Kay et al. (2015, 2016).

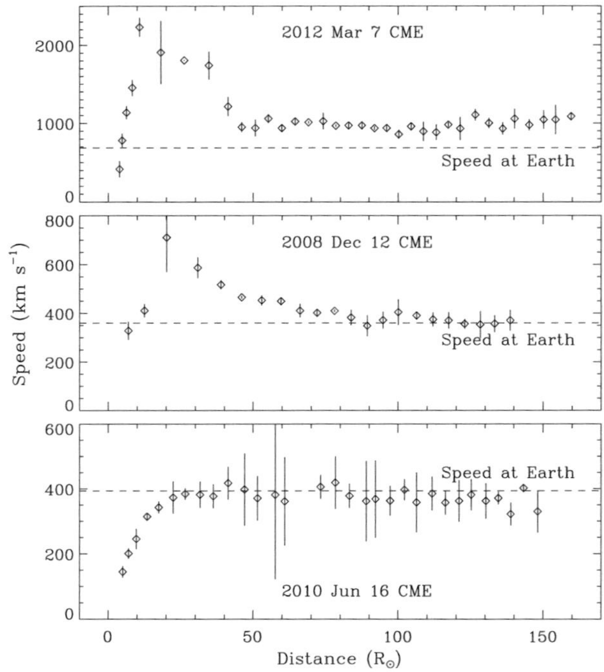

Fig. 8 Comparison of Sun-to-Earth propagation profiles between a typical fast ICME (*upper*), a typical intermediate-speed one (middle) and a typical slow one (*lower*). The *horizontal dashed line* indicates the observed speed at the Earth. Adapted from Liu et al. (2016)

5 CME/ICME Kinematics

5.1 A Brief Overview of CME Kinematics

Here, we examine the details of when, where, and how CMEs/ICMEs accelerate/decelerate in interplanetary space and quantify the kinematics behavior. To that end, observations by the STEREO spacecraft have provided unprecedented opportunities to observed kinematic behavior of ICMEs with multiple views that enable accurate measurements of ICMEs over a large distance. Liu et al. (2010a) have developed a triangulation technique to determine CME/ICME Sun-to-Earth kinematics based on the wide-angle imaging observations from STEREO. The technique initially assumes a relatively compact CME structure simultaneously seen by the two spacecraft. It has no free parameters and can give CME/ICME kinematics (both propagation direction and radial velocity) as a function of distance from the Sun continuously out to 1 AU. This capability is key to probing CME propagation and interaction with the inner heliosphere. Later, Lugaz et al. (2010) and Liu et al. (2010b) realize that the same idea can be applied by assuming CME geometry as a spherical front attached to the Sun. In this case, what is seen by a spacecraft is the segment tangent to the line of sight. The triangulation concept has proven to be a useful tool for determining CME Sun-to-Earth kinematics and connecting imaging observations with *in situ* signatures (e.g., Liu et al. 2010a,b, 2011, 2012, 2013; Möstl et al. 2010; Lugaz et al. 2010; Harrison et al. 2012; Temmer et al. 2012; Davies et al. 2013; Mishra and Srivastava 2013). Details of when, where, and how CMEs accelerate/decelerate in interplanetary space can be quantified using the kinematics derived with the triangulation method.

Figure 8 shows the comparison of Sun-to-Earth propagation profiles characteristic of slow, fast and intermediate-speed CMEs. The 7 March 2012 CME, with a peak speed of more than 2000 $\mathrm{km\,s^{-1}}$ shows a typical three-phase speed profile developed by Liu et al.

(2013): an impulsive acceleration for a CME up to 10–15 $R_\odot$, followed by a rapid deceleration for the ICME out to about 50 $R_\odot$, and thereafter a nearly constant speed (or gradual deceleration). The predicted speed at the Earth overestimates the observed speed by about 250 km s^{-1}, which may be partly due to the large longitudinal separation between the two STEREO spacecraft (227°, i.e., both observing the CME from behind the Sun). Note that the observed speed at the Earth is the average solar wind speed in the sheath between the shock and ejecta, which is usually a little smaller than the shock speed at 1 AU. Hence, a better agreement may be achieved if the shock speed (rather than the sheath speed) is used for the comparison. The 12 December 2008 CME with a peak speed of about 700 km s^{-1} has a similar profile but different cessation distances for the acceleration and deceleration. Compared with the 7 March 2012 CME, its speed first increased with a lower rate up to about 20 $R_\odot$, then decreased out to 80–90 $R_\odot$, and thereafter became roughly constant, similar to the evolution described by Wood and Howard (2009) for a CME of similar speed. The cessation distance of ICME rapid deceleration (about 50 and 85 $R_\odot$ for the 7 March 2012 and 12 December 2008 CMEs, respectively) is much shorter than the average cessation distance of 0.76 AU (163 $R_\odot$) inferred indirectly by Gopalswamy et al. (2001b). The Sun-to-Earth propagation profile of the 16 June 2010 CME/ICME exhibited only two phases: an acceleration with an even slower rate up to 25–30 $R_\odot$ followed by a nearly invariant speed at about 390 km s^{-1}. The predicted speeds at the Earth for the latter two cases are well confirmed by the *in situ* measurements at 1 AU; the longitudinal separation between the two STEREO spacecraft is 86.3° for the 12 December 2008 CME and 143.6° for the 16 June 2010 event.

In examining the question of CME/ICME kinematics and momentum transfer between the ejecta and the solar wind, there are many aspects that must be considered: First, the ambient solar wind is often highly variable in time and space (e.g., Temmer et al. 2011; Kilpua et al. 2012; Rollett et al. 2012; Liu et al. 2015, 2016). Second, that the Lorentz force driving the CME can sometimes significantly contribute to the CME dynamics up to large distances from the Sun (e.g., Vršnak et al. 2004; Temmer et al. 2011). Third, CME-CME interactions including the affects of CME preconditioning the solar wind, (e.g., Lugaz et al. 2005b; Temmer et al. 2012; Maričić et al. 2014; Rollett et al. 2014; Liu et al. 2012, 2014a,b), which will be taken up in Sect. 9. All of these phenomena affect the CME/ICME propagation and significantly contribute to the uncertainties in predicting the arrival time and impact speed of ICMEs at Earth.

Generally, the CME/ICME propagation can be divided into several phases. The CME initiation phase is most often characterized by swelling and slow rising motion of the pre-eruptive structure, which is usually interpreted as an evolution through a series of quasi-equilibrium states (see, e.g., Vršnak 2008, and references therein). When the slowly evolving structure reaches an unstable state (so called "loss of equilibrium" mechanism of Forbes and Isenberg 1991), the structure starts to accelerate, driven by some form of ideal MHD instabilities such as the kink or torus modes (e.g., and references therein Török and Kliem 2003, 2005; Kliem et al. 2004; Kliem and Török 2006; Démoulin and Aulanier 2010; Olmedo and Zhang 2010; Török et al. 2010). Then follows the take-off stage, which is most often characterized by accelerations on the order of 100 m s^{-2}, and lasts for $\sim$ 1 h, so the majority of CMEs achieve velocities in the range 100–1000 km s^{-1} (e.g., Vršnak et al. 2007; Bein et al. 2011; Yashiro et al. 2004). However, sometimes the take-off phase is characterized by an extremely impulsive acceleration, achieving peak values on the order of 10 km s^{-2}, but lasting only for several minutes (e.g., Vršnak et al. 2007; Bein et al. 2011). In such events the gradual pre-eruptive evolution is frequently not observed. In fact, in some events a confined-flare type of process leads to fast reconfiguration of the pre-eruptive structure into an unstable configuration and erupts immediately after being formed (Aurass et al. 1999).

On the other side of the "spectrum" of event types are very gradual events, characterized by weak accelerations that result in velocities on the order of 100 km s^{-1}.

The take-off phase imparts the initial velocity, which can be categorized as slow (speeds below 400 km s^{-1}) intermediate (speeds between 400 and 1000 km s^{-1}) and fast (speeds above 1000 km s^{-1}), and it is this speed that largely determines the kinematical evolution in the ensuing "main acceleration" phase. Most generally, in the upper corona and heliosphere, CMEs that are slower than the ambient solar wind continuously accelerate, whereas those that are faster than the wind decelerate (Lindsay et al. 1999; Gopalswamy et al. 2001b; Jones et al. 2007) such that CMEs tend to approach the speed of the ambient solar wind. This acceleration/deceleration tendency is already observable in the upper corona (Moon et al. 2002; Vršnak et al. 2004) and was found by comparing the distribution of coronagraphic CME velocities with the ICME speeds measured *in situ* at 1 AU (Gopalswamy et al. 2000, 2001b). In the heliospheric range, the velocity profile was directly confirmed by various types of measurements, e.g., by white-light heliospheric imagers such as those made by Coriolis instrument (e.g., Reiner et al. 2005b,a; Tappin 2006; Webb et al. 2006; Howard et al. 2013) onboard the Solar Mass Ejection Imager (SMEI) (Eyles et al. 2003; Jackson et al. 2004) and STEREO-HI (e.g., Liu et al. 2013, 2016), by tracking interplanetary radio type II bursts (e.g., Reiner et al. 2005a,b; Liu et al. 2008b), by analyzing radio interplanetary scintillation (e.g., Manoharan et al. 2001; Manoharan 2006), or by comparing coronagraphic and 1 AU CME/ICME speeds (e.g., Manoharan and Mujiber Rahman 2011).

5.2 Empirical Models of CME/ICME Kinematics

In parallel with progress of numerical propagation models, a number of empirical methods and analytical physics-based models were developed. Although, quite clearly, the future of space weather forecasting lies in numerical modeling, the performance of empirical and analytical methods is currently comparable to, or even slightly better than that of numerical methods. On the other hand, these methods are easy to handle and could be promptly adjusted to the new data incoming in the course of the CME/ICME propagation, i.e., the predictions could be easily refreshed instantaneously with receiving the newest input data. Furthermore, the empirical and physics-based analytical models educe our comprehension of the CME/ICME dynamics, and thus, could help in advancing the numerical methods. In this section, we review a physical background of this approach to the space weather forecasting, which is generally based on empirical studies of the CME/ICME kinematics and/or theoretical considerations of their dynamics.

Results of various statistical studies of the CME/ICME Sun-Earth transit times, reflecting the described general characteristics of the CME/ICME kinematics, can be used to establish empirical methods for predicting the ICME arrival. The simplest one, based on a statistical study of a sample of geoeffective ICMEs, was proposed by Brueckner et al. (1998), saying: "In many cases, the travel time between the explosion on the Sun and the maximum geomagnetic activity is about 80 hours." To explain this, so-called "Brückner's 80-hour rule", corresponding to the Sun-Earth average speed of 500 km s^{-1}, one has to bear in mind that the Sun-Earth travel time for a typical slow solar wind of the velocity $\simeq 400$ km s^{-1} is about 100 h. On the other hand, most of geoeffective ICMEs are relatively fast, and for the speed of 1000 km s^{-1}, the transit time is $\simeq 40$ h. Thus, it can be concluded that the 80-hour rule is a consequence of deceleration of fast CMEs in the ambient solar wind.

A somewhat more detailed empirical method, based on the correlation of the Sun-Earth transit times, tt, and the coronagraphic speeds of CMEs, V_{CME} is illustrated in Fig. 9a. A similar plot is presented in Vršnak and Žic (2007), and for additional examples see also

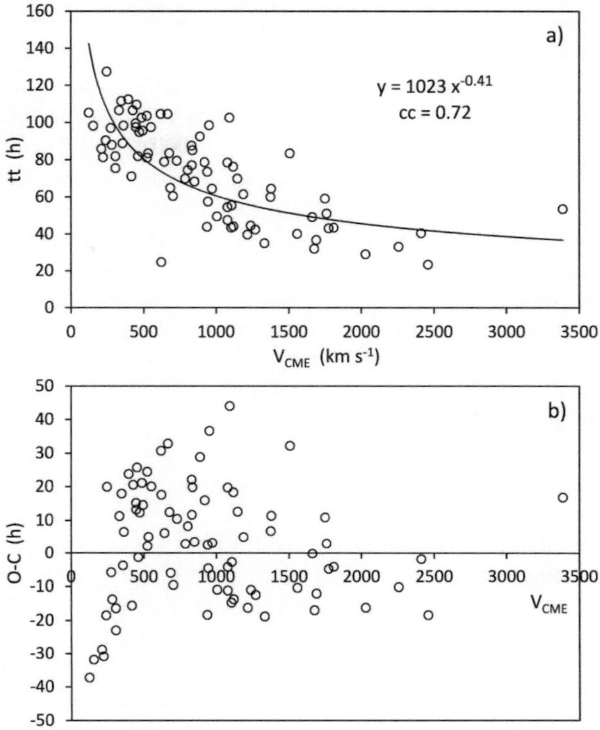

Fig. 9 (**a**) Sun-Earth transit times of 78 CMEs presented as a function of their speed in the coronagraphic field of view, together with the power-law fit. (**b**) Differences $O - C$ between observed transit times and the values based on the power-law fit

Manoharan et al. (2004), Schwenn et al. (2005), Manoharan and Mujiber Rahman (2011). In Fig. 9a, the arrival time for a CME of a given coronagraphic speed can be predicted by employing the presented power-law fit. In Fig. 9b, the residuals $O - C$ are presented to show the span of the prediction errors as a function of V_{CME}. Inspecting the graph one finds that the errors can be as large as $\simeq 40$ h and tend to be larger for slower CMEs (however the relative errors are similar for the whole range of V_{CME}). The standard deviation of the $O - C$ distribution is 18 h. The errors can be somewhat reduced by taking into account the source-region location, the CME mass, the associated-flare importance, etc., but not significantly. Note that the same procedure can be applied by using, e.g., the coronal shock velocities inferred from the radio type II burst dynamic spectra.

The scatter-plot in Fig. 9a shows that the power-law fit is characterized by $tt \propto V^{-0.41}$, which is distinctively different from $tt \propto V^{-1}$ that would be expected in the case of $V_{\mathrm{CME}} =$ const. This means that slow/fast CMEs have shorter/longer transit times than they would have if propagating at constant speed. Again, this can be attributed to the adjustment of the CME/ICME propagation to the ambient solar wind.

The statistical relationships between the arrival time and various CME parameters measured from coronagraphic observations can be used to employ more sophisticated prediction methods. For example, Sudar et al. (2016) applied the "neural network" approach to determine the most probable transit time using the coronagraphic CME speed and the central meridian distance of the source-region as the input parameters. The analysis, involving a sample of 153 CME-ICME pairs, showed that the average tt error is $\simeq 12$ h. The $tt(V_{\mathrm{CME}})$ dependence showed a typical drag-like trend, with acceleration turning to deceleration at $V_{\mathrm{CME}} \simeq 500\ \mathrm{km\,s^{-1}}$, consistent with the "Brückner's 80-hour rule" (note that similar holds for the scatter-plot shown in Fig. 9a). Furthermore, the results clearly demonstrate larger *tt* for larger *CMD*, as well as the eastward/westward deflection of fast/slow CMEs. Note that an analogous technique can be applied to forecast the geoeffectiveness of CMEs (e.g., Valach et al. 2009; Uwamahoro et al. 2012; Dumbović et al. 2015, 2016).

The acceleration/deceleration tendency of slow/fast CMEs was quantified by Lindsay et al. (1999) and Gopalswamy et al. (2000, 2001b) by defining a linear relationship between the CME acceleration expressed in $\mathrm{m\,s^{-2}}$ and the CME speed expressed in $\mathrm{km\,s^{-1}}$ as $a = 2.193 - 0.0054V_{\mathrm{CME}}$, which can be also expressed as $a = -0.0054(V_{\mathrm{CME}} - 406)$. The range over which the deceleration occurs was estimated to $r < 0.76$ AU. Later on, Manoharan et al. (2004), Manoharan (2006), and Gopalswamy (2009) extended the $a(V_{\mathrm{CME}})$ relationship to 2nd-degree polynomial forms. The described kinematical forecasting technique, employing various forms of the $a(V_{\mathrm{CME}})$ relationship, was applied also by, e.g., Manoharan et al. (2004), Gopalswamy et al. (2005), Reiner et al. (2005b), Manoharan and Mujiber Rahman (2011), Salas-Matamoros and Klein (2015), where also forecasting of the CME-driven shocks was included.

5.3 Physics-Based Kinematic Models

The statistical tendency showing that fast CMEs decelerate, whereas slow CMEs accelerate, as demonstrated by Lindsay et al. (1999), and confirmed by Gopalswamy et al. (2001b), Moon et al. (2002), Vršnak et al. (2004), Yashiro et al. (2004), indicates that CMEs tend to adjust their velocity to the ambient solar wind, which could be interpreted as a consequence of "aerodynamic" drag (see, e.g., Cargill et al. 1996; Vršnak 2001; Vršnak et al. 2004, 2008; Cargill 2004; Owens and Cargill 2004; Manoharan 2006). In particular, Vršnak et al. (2004) have shown that the $a(V_{\mathrm{CME}})$ relationship for the CMEs observed in the LASCO field of view has a quadratic form, providing a strong evidence for the aerodynamic drag effect. Furthermore, they showed that the Lorentz force is decaying rapidly with the height, implying that the drag becomes a dominant force in the heliospheric dynamics of CMEs.

Following observational facts and theoretical considerations, the so called Drag-Based Model (DBM; Vršnak and Gopalswamy 2002; Vršnak et al. 2010, 2013) was developed to describe analytically the dynamics/kinematics of the heliospheric propagation of CMEs. DBM is based on the assumption that the aerodynamic drag is a dominant force that governs the CME propagation in the interplanetary space. Thus, CME/ICME dynamics are described by the equation of motion of the form $a \equiv \ddot{r} \equiv \dot{v} = -\gamma(v - w)|v - w|$, where $a(t)$, $v(t)$, and $r(t)$ are the instantaneous CME acceleration, speed, and position, w is the ambient solar wind speed (generally depending on r, but often used to be constant). The drag parameter γ defines the drag "effectiveness", usually being expressed as $\gamma = c_{\mathrm{d}} A \rho_w / M$, where c_{d} is the dimensionless drag coefficient, A is the CME cross-sectional area, ρ_w is the ambient solar wind density, and M is the total CME mass (for details see, e.g., Cargill 2004; Vršnak et al. 2013).

The described equation of motion, as well as its modifications (e.g., Tappin 2006; Borgazzi et al. 2009; Lara and Borgazzi 2009; Shanmugaraju and Vršnak 2014; Shi et al. 2015), was employed to study various aspects of the CME/ICME heliospheric propagation. For example, Vršnak and Žic (2007) and Shanmugaraju and Vršnak (2014) studied the role of the solar wind speed on the CME/ICME transit times, showing that the wind speed is an essential parameter for the CME/ICME dynamics, which results in "Brückner's 80-hour rule". Temmer et al. (2011) analyzed the effects of variable/structured solar wind and showed that it significantly affects the CME/ICME kinematics, but also, they demonstrated that in some cases the driving Lorentz force might contribute to CME/ICME dynamics beyond a distance of 100 $R_\odot$. Vršnak et al. (2008) studied the role of the CME mass, and found that in the range of distances covered by LASCO, the drag affects more strongly the propagation of light CMEs, and inferred that the Lorentz force is usually stronger in more massive events. Temmer et al. (2012) and Rollett et al. (2014) employed DBM to investigate how the CME-CME interaction affects the kinematics, whereas Wang et al. (2016b) used it to infer the degree of deflection of a CME/ICME event. Falkenberg et al. (2010) and Vršnak et al. (2010, 2014) compared the DBM and ENLIL ICME arrival-time predictions, and found that results of the two models to be similar accurate.

After being proposed in its initial form by Vršnak and Gopalswamy (2002), the formulation of the basic form of DBM, suited for space-weather forecasting, was completed by Vršnak et al. (2013). In this basic option of DBM, it is taken, approximately, that the CME/ICME cross section increases as $A \propto r^2$, the solar wind density decreases as $\rho_w \propto r^{-2}$ (consistent with the $w = \text{const.}$ approximation), and the mass and drag coefficient are constant. Under this approximation one finds $\gamma = \text{const.}$, which simplifies the equation of motion, providing explicit solutions $r(t)$ and $v(t)$. Obviously, the main advantage of so-defined DBM option is its practical use in real-time forecasting of the ICME arrival time and impact speed. However, serious intrinsic drawbacks are related to approximations applied to the DBM: (1) c_d does not vary with distance (e.g., Cargill 2004), (2) ICME virtual-mass effect is not included (e.g., Bein et al. 2013; Feng et al. 2015; Cargill 2004; Tappin 2006), (3) solar wind properties are taken to be constant in time (e.g., Temmer et al. 2011), (4) and (5) the geometry of the ICME leading edge and location of the source region are not considered, and finally, (6) and (7) the roles of ICME over-expansion (e.g., Riley and Crooker 2004) and the ICME-driven shock (Liu et al. 2013, 2016) are omitted. Some of these drawbacks were removed by Žic et al. (2015), who included CME geometry and source-region location factors, and also took into account more general forms for $\rho_w(r)$ and $w(r)$. Finally, it should be noted that there are also various forms of empirical forecasting methods that rely on relationships between CME-driven shocks and CMEs, CME-associated flares, or type II radio bursts. Analytical models also exit as well as hybrid approaches that combine components of empirical and analytical models (e.g., Feng and Zhao 2006; Zhao and Dryer 2014; Zhao and Feng 2015; Zhao et al. 2016, and references therein).

6 CME/ICME-Driven Heliospheric Disturbances

6.1 Mass Increase in CMEs/ICMEs

The momentum transfer between ICMEs and the solar wind can be described as "aerodynamic" drag, however this process is strongly affected by the expanding nature of CMEs and ICMEs as shown by Siscoe and Odstrcil (2008). ICMEs tend to expand only slightly faster than the spherically diverging flow of the surrounding solar wind. Consequently, the ejecta has a nearly constant (Sun-centered) angular size while the transverse size increases linearly with distance from the Sun. Siscoe and Odstrcil (2008) found the expansion speed of the ICME flux rope can be greater than the non-radial deflection speed of the solar wind, so that impacted solar wind is unable to move completely around the ICME, and accumulates in the sheath. The deflection flow pattern around an ICME is shown in Fig. 10a, which depicts the non-radial velocity and the trajectory of solar wind parcels ahead of a moderately fast simulated ICME. The impacted solar wind is unable to move around the flux rope, and as a result, the plasma is swept up in the ICME sheath, increasing the total mass of the disturbance.

This snow-plow effect has been difficult to discern in near-sun coronagraph observations because the swept up mass increase is difficult to distinguish from the sustained outflow of dense plasma with the CME (Bein et al. 2013). However, far from the Sun, the

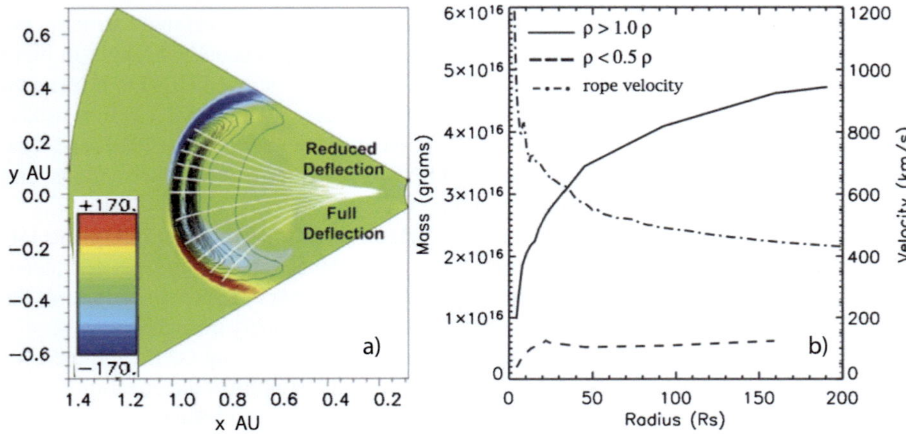

Fig. 10 Solar wind swept up by ICMEs. (**a**): simulated flow pattern around an ICME from Siscoe and Odstrcil (2008). *Color contours* show the non-radial velocity while *white lines* show the trajectory of solar wind parcels ahead of the ICME. The azimuthal deflection speed is insufficient to allow the impacted solar wind to move around the flux rope, and as a result the plasma is swept up in the ICME sheath. (**b**): plots of the mass and velocity of a simulated ICME as a function of distance from the Sun taken from Lugaz et al. (2005a). The *solid line* shows the mass of plasma above ambient density, the *dashed line* shows the mass of low density material while the *dot-dash line* shows speed of the ICME. Mass and speed follow equal and opposite trends indicating momentum conservation as the solar wind is compressed at the shock and then piles up in the sheath region

snow-plow affect has recently been measured in STEREO HI observations (DeForest et al. 2013), which confirms prior predictions made with numerical simulations (Manchester et al. 2004b), which established a large mass increase occurring with CME deceleration close to the Sun. Here, the speed of the CME was found to follow an exponential-like decay that asymptotically approaches the speed of the background solar wind, thus verifying the empirical relationship established by Sheeley et al. (1999) (see also Fig. 6). A study by Lugaz et al. (2005a) explicitly calculated this ICME mass from both 3-D structure and from synthetic coronagraph images, the results of which are shown in Fig. 10a. Here, ICME mass and velocity are plotted as functions of distance from the Sun. The CME/ICME mass increased by more than a factor of four from the low corona to 1 AU with the majority (factor of three) of the increase occurring within 30 $R_{\odot}$. This plasma is compressed and accelerated at the CME-driven shock where a significant amount accumulates in the sheath region between the flux rope and the shock as shown in Fig. 11. As a result of momentum conservation, ICME velocity follows an equal and opposite decrease in magnitude.

Manchester et al. (2008) found similar mass increases in their simulation of the 28 October 2003 CME event. In this case, the mass is derived from analysis of synthetic LASCO C2 and C3 coronagraph images, which showed the mass more than doubled to over 4×10^{16} g. Analysis of ICME data from SMEI have found similar mass increases, and in the case of the 28 October 2003 event, the mass reached the extraordinary level of 13.6×10^{16} g when the ICME reached 1 AU as reported by Jackson et al. (2006). Much of the mass increase is related to the extreme nature of this event, which reached Earth in approximately 20 hours.

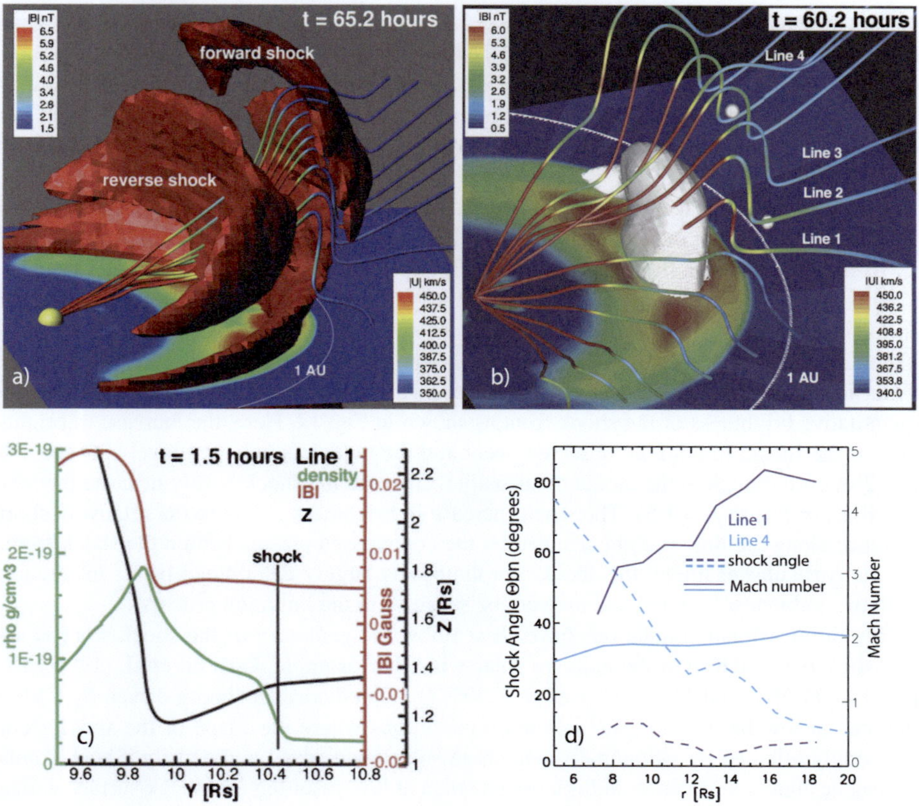

Fig. 11 Simulated CME-driven heliospheric disturbances. (**a**): forward and reverse shocks shown with red iso-surfaces, open magnetic field lines colored to denote field strength, and the equatorial plane colored to show flow speed. (**b**): an iso-surface at 25 nT shows the center of the flux rope above the equatorial plane. (**c**): the shock structure is shown along a selected magnetic field line ((**b**): Line 1). Here, density (*green*) and magnetic field strength (*red*) are plotted as functions of radius along the field line shown in (*black*) passing through the shock. (**d**): Mach number and shock geometry (θ_{BN}) plotted as functions of distance for low and high latitude field lines labeled, respectively (Lines 1 and 4 shown in (**b**))

6.2 CME-Driven Shocks

High initial speed in conjunction with CME/ICME expansion drive fast-mode waves into the surrounding medium, which may steepen into shocks. These disturbances can propagate far beyond the expanding magnetic driver and are typically observed as forward shocks proceeding ejecta (e.g., Reiner et al. 1998; Wang et al. 2001; Manchester et al. 2008; Wood et al. 2012; Temmer and Nitta 2015; Hu et al. 2016; Liu et al. 2017). CME-driven shocks were originally identified in coronagraph images as faint arcs observed at the outer edges of fast CMEs. An early example is provided by Sime and Hundhausen (1987), who observed a bright loop at the front of a fast CME identified as a shock from its high speed (1070 $\mathrm{km\,s^{-1}}$), absence of deflections preceding the loop, as well as the absence of stationary legs, and continuously expanding arc. Perhaps the most compelling observational evidence for shocks appearing in LASCO images is presented in Raymond et al. (2000), Mancuso et al. (2002), and Vourlidas et al. (2003). In the case of Raymond et al. (2000) and Mancuso et al. (2002), shocks were observed simultaneously in the low corona ($r < 3\ R_{\odot}$) by LASCO, the Ultraviolet Coronagraph Spectrometer (UVCS), and associated type II radio bursts. UVCS gave clear spectroscopic evidence for the presence of shock fronts, while type II radio bursts indicated the presence of shock-accelerated electrons. A connection between type II radio bursts and CME-driven shocks has been established by many studies (e.g., Reiner et al. 1998; Liu et al. 2009, 2013; Reames 2013; Gopalswamy et al. 2013; Nitta et al. 2014; Gopalswamy et al. 2016)

MHD simulations also provide predictions of the appearance of CME-driven shocks in a variety of observing modes. Pagano et al. (2008) modeled the spectral line signatures of CME-driven shocks as they appear in UVCS, and Lugaz et al. (2007), Manchester et al. (2008) simulated the appearance of coronal shocks in synthetic coronagraph images derived from simulations of specific CME events. In the case of Manchester et al. (2008) the simulated shock is directly compared to LASCO observations of the 28 October 2003 event (Vourlidas et al. 2003), which shows agreement with both the general morphology and the quantitative brightness of the shock front as shown in Fig. 12. Here, the increase in brightness at the shock, in both the observed event and the model, is only at a level between 1% and 2%, even though in the model the density increase at the shock is very near the theoretical limit of 5 (for $\gamma = 1.5$). The small increase in brightness is due to the relatively short distance along the line-of-sight spanned by the compressed plasma behind the shock front. In this particular example, the shock morphology is largely determined by its interaction with the ambient solar wind and may not be sensitive to the initiation process.

Multidimensional simulations have shown how the geometry of the shock front is affected by the structure of the ambient solar wind. For example, Odstrčil et al. (1996), Riley et al. (1997), and Manchester et al. (2005) simulated coronal shocks driven by CMEs initiated at low latitude in solar minimum conditions where the shape of the shock front is strongly affected by latitudinal gradients in solar wind speed and density. Shock fronts propagate significantly faster at high latitude than at low, resulting in a shock surface with a saddle-like structure that is outward-concave in latitude while outward-convex in longitude, as shown in Fig. 11a. Examination of a shock in this configuration by Manchester et al. (2005) revealed that in the concave region, the fast-mode shock deflection away from the shock normal necessarily drives converging flows that lead to a significant density enhancement behind the shock as seen in the line plots in Fig. 11c In complement, line plots in Fig. 11d show the shock Mach number, compression ratio and shock angle (between B and the shock normal, θ_{BN}) as functions of distance from Sun center.

6.3 Flux Rope Over-Expansion and Forward-Reverse Shock Pairs

Flux ropes erupting from the Sun have high internal magnetic pressures that drive CME expansion, a process that may persist well beyond 1 AU. Departures from force-free equilibrium are common in ICMEs, providing an overpressure that can drive expansion (Liu et al. 2006a). In fact, ICME expansion is often observed *in situ* as declining speed profiles (e.g., Farrugia et al. 1993; Liu et al. 2006a), which are consistent with over-expansion or may be a residual effect of initiation. At high helio-latitudes, measurements by the Ulysses spacecraft led to the identification of a new class of disturbances produced by CMEs coasting through the high-speed wind (Gosling et al. 1994). Bounded by a pair of forward and reverse shocks, these events were also defined by low internal densities, temperatures, and hence, pressures. It was inferred, and confirmed by numerical calculations (e.g., Riley et al. 1997, 2003, 2004; Gosling et al. 1998; Reisenfeld et al. 2003) that the initial high internal pressure drove magnetosonic waves into the surrounding plasma in all directions, which then steepened into shock waves. The shocks are divided into two distinct modes: forward

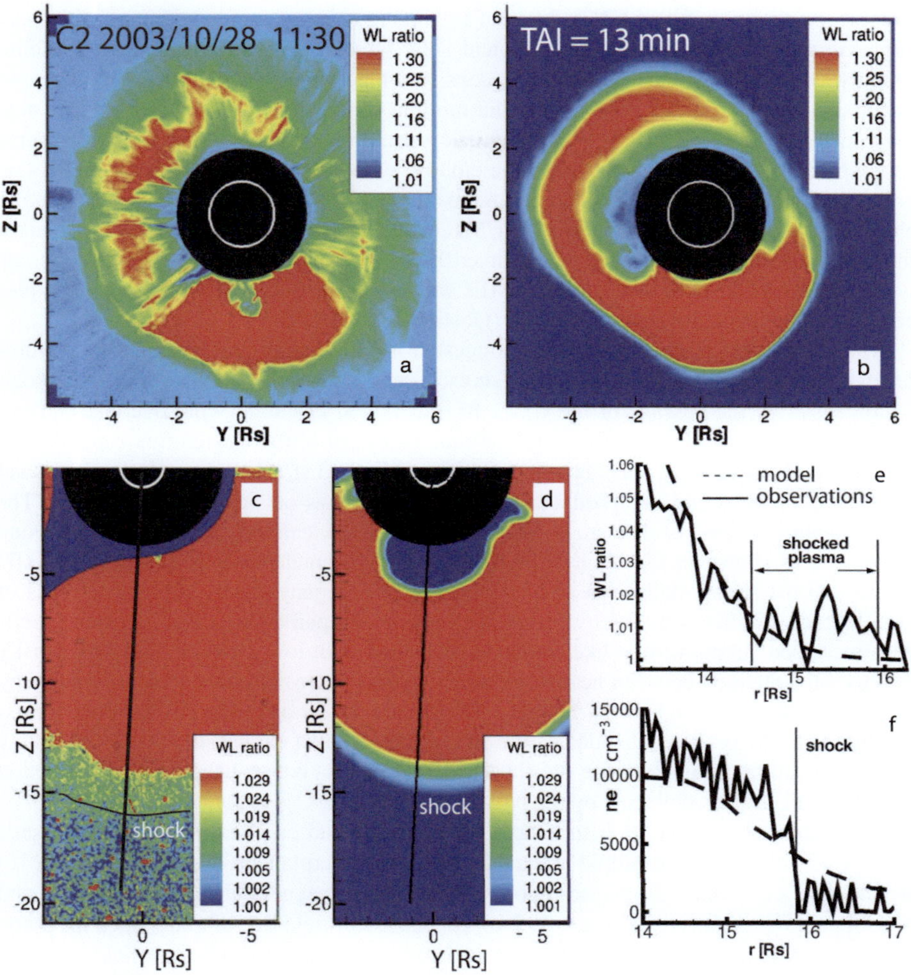

Fig. 12 Simulation of the 28 October 2003 CME event adapted from Manchester et al. (2008). (**a**) and (**b**): quantitative comparison between the observed (LASCO C2) and simulated coronagraph images, respectively. The *color images* show brightness relative to (divided by that of) the pre-event corona. (**c**) and (**d**): close-ups of the shock front in the observed (LASCO C3) and simulated coronagraph images, respectively, where color contour levels are adjusted to highlight the faint shock front. (**e**) and (**f**): line plots of the total brightness and electron density, respectively, where observed values are shown with a *solid line* and model results are shown with a *dashed line*. The model shows quantitative agreement with the observed event in both magnitude and morphology of the brightness

shocks that travel ahead of the ejecta, and in contrast, reverse shocks that propagate back toward the Sun in the rest frame of the ejecta, while actually being carried away from the Sun by the solar wind. In the case of a reverse shock, fast low-density plasma passes through the shock to produce slow, dense, hot shocked plasma on the far side (away from the Sun) of shock. Similar processes occur at low latitudes, but because the ejecta is usually immersed within slow and denser material, the interaction is more asymmetric (Gosling 2000) and occurs without the companion reverse shock.

CMEs associated with forward-reverse shock pairs have many distinguishing properties. First, all of these CMEs are found at high heliographic latitude embedded in high-speed streams. Furthermore, these CMEs have an expanding velocity profile with a central speed approximately the same as the fast ambient solar wind. These CMEs commonly exhibit roughly symmetric increases in pressure, density, temperature, and magnetic field strength down stream of the shock pairs, while the minima of these plasma quantities occur near the centers of the disturbances. The magnetic structures associated with forward-reverse shock pairs have the features of magnetic clouds with shock-enhanced field strengths that are typically a factor of 2 or 3 greater than the ambient solar wind field. Counter streaming electrons suggest closed field lines in the CME, while the ion charge state of CME plasma is more typical of the ambient solar wind rather than a more highly ionized ejecta. While this is the common and ideal structure of CMEs with forward-reverse shocks, some examples are much more irregular and asymmetric (Gosling et al. 1998).

Manchester and Zurbuchen (2006) suggested an alternative explanation for forward-reverse shock pairs in which the reverse shock forms as a result of deflections of the solar wind caused by the passage of the CME. In this model, fast solar wind overtakes slower plasma forming a reverse shock at high-latitude poleward of the ejected flux rope. Apparent signatures of ICMEs, such as enhanced magnetic field strength, low plasma beta, and field direction rotations, are produced by the plasma flows outside of the flux rope. The salient feature of this mechanism is that the reverse shock can only form laterally beyond the ejecta, as shown in Fig. 11a. Further analysis by Manchester and Zurbuchen (2007) shows that the stream-deflection mechanism reproduces many observed characteristics of high-latitude ICMEs bounded by forward-reverse shock pairs, including: magnetic fields with increased strength, rope like rotations, entrained with low pressure, and low density plasma, all contained between nearly symmetric pressure/density enhancements created by the forward and reverse shocks. In this case, stream deflections naturally produce an expanding velocity profile imbedded in the fast wind, and also explain the fast-wind charge state composition found between the shocks, all as a direct consequence of the interaction of solar wind streams deflected by the passage of an ICME.

Apart from shocks, large disturbances in the solar wind can be driven by the passage of the CME ejecta. One example is the draping of the interplanetary magnetic field (IMF) around the ejecta, which forms strong out-of-the-ecliptic magnetic fields (e.g., Gosling and McComas 1987; Odstrčil and Pizzo 1999a). In the case of field line draping, open magnetic field lines are bent around the ICME ejecta and compressed to have enhanced field strengths that may be two or three times as great as found in the ambient solar wind, as was found in Pioneer observations examined by McComas et al. (1988). A CME simulation showing such draping is shown in Fig. 11. Here, open field lines are colored to show field strength, which is particularly enhanced on the Sun-ward side of the flux rope. Another important disturbance is found trailing fast ICMEs, the large rarefactions that form in the trailing wake of ICMEs (e.g., Gosling and McComas 1987; Riley and Gosling 2007; Liu et al. 2014a). Here, the ICME outruns the material following it, forming an expansion wave (or rarefaction region) of low density (Riley et al. 1999). Sunward of fast CMEs, the magnetic field is also distorted away from the nominal Parker Spiral configuration into a near radial configuration (e.g., Riley and Gosling 2007; Liu et al. 2014a).

7 Magnetic Evolution

7.1 MC/ICME Distortion

Although suspected, the substantial distortion that ICMEs undergo was not fully realized until the Ulysses mission, when combined with Geotail, provided the first multi-spacecraft *in situ* measurements of the same CME at significantly different helio-latitudes and distances from the Sun (Hammond et al. 1995). Liu et al. (2006b) similarly analyzed Ulysses *in situ* observations to find that MCs are highly flattened. At solar minimum, MCs have curved cross sections that are outward concave, while at solar maximum the MCs are outward convex. The results are indicative that the MCs are deformed by their interaction with the ambient solar wind. White-light observations of ICMEs at large elongation made with SMEI found a relatively small number of ICMEs to be in the shape of outward-concave arcs, consistent with deformation solar wind velocity shear (Kahler and Webb 2007). Liu et al. (2008c) found that at solar minimum eight out of eleven MCs were outward concave as determined by the *in situ* measurement of the elevation angle of the MC normal with respect to the solar equatorial plane. Finally, as reported by Savani et al. (2010), coronagraphs onboard the STEREO-B observed the distortion of the leading edge of the CME/ICME event of 14 November 2007, which began close to the Sun and progressively increased as the CME moved to large elongation.

Subsequent numerical simulations provided ever-increasingly more accurate explanations for how these deformations evolve as the flux rope transits the space from the Sun to the Earth and beyond (e.g., Riley et al. 1997, 2003; Odstrčil and Pizzo 1999b; Manchester et al. 2004a). These simulations show the deformation occurring from the exchange of momentum between the ejecta and the surrounding medium. Early on, radial expansion of the ejecta naturally provides a flow that evolves the erupting flux rope in a nearly self-similar fashion (e.g., Gibson and Low 1998). This simple evolution breaks downs and the ICME flattens as the solar wind inhibits the expansion of the MC whose radial extent becomes more nearly constant, while simultaneously, the spherically diverging flow causes the transverse extent to increase linearly with distance from the Sun. The aspect ratio of the ejecta, (transverse versus radial size) begins to increases with time, which simulations show to be particularly noticeable beyond 40 $R_{\odot}$.

Simulations also illustrate the deformation of CMEs by bimodal ambient solar wind typical of solar minimum (e.g., Gosling 1999; Odstrčil and Pizzo 1999b; Riley and Crooker 2004; Manchester et al. 2004a). At the largest scales, ICMEs are deformed into a highly flattened, concave-outward structure by the latitudinal velocity gradient (speed increasing with latitude) that serves to carry the high-latitude portion of the flux rope to greater distances. An example of a 3-D numerical simulation of a CME in a bimodal wind is shown in Fig. 13. Here, the ejected flux rope is shown from two perspectives at 4.5 and 60 hours after initiation in the top and bottom rows, respectively. Close to the Sun, the flux rope reflects its original shape greatly expanded. As the flux rope approaches 1 AU, it appears flattened and outward-convex with field lines connected to the IMF. At solar maximum, the solar wind velocity is relatively slow and unstructured so MC are deformed by aerodynamic drag to a convex shape, which is particularly true for fast CMEs (e.g., Manchester et al. 2008).

7.2 Flux Rope Erosion

Magnetic reconnection is ubiquitous in the solar wind, from large-scale structures such as the heliospheric current sheet (e.g., Gosling et al. 2005, 2006) to the smallest scale associated with magnetic turbulence. So it should be no surprise that there are clear signatures that reconnection occurs with magnetic ejecta in interplanetary space (Farrugia et al. 2001). Reconnection between the IMF and the ICME field is most prone to occur where more-nearly-opposed fields are brought into contact, which for a highly twisted horizontal flux rope would tend to be either at the front or rear of the magnetic flux rope. Dasso et al. (2006) point out that such reconnection will reduce the magnetic flux of a rope, resulting in

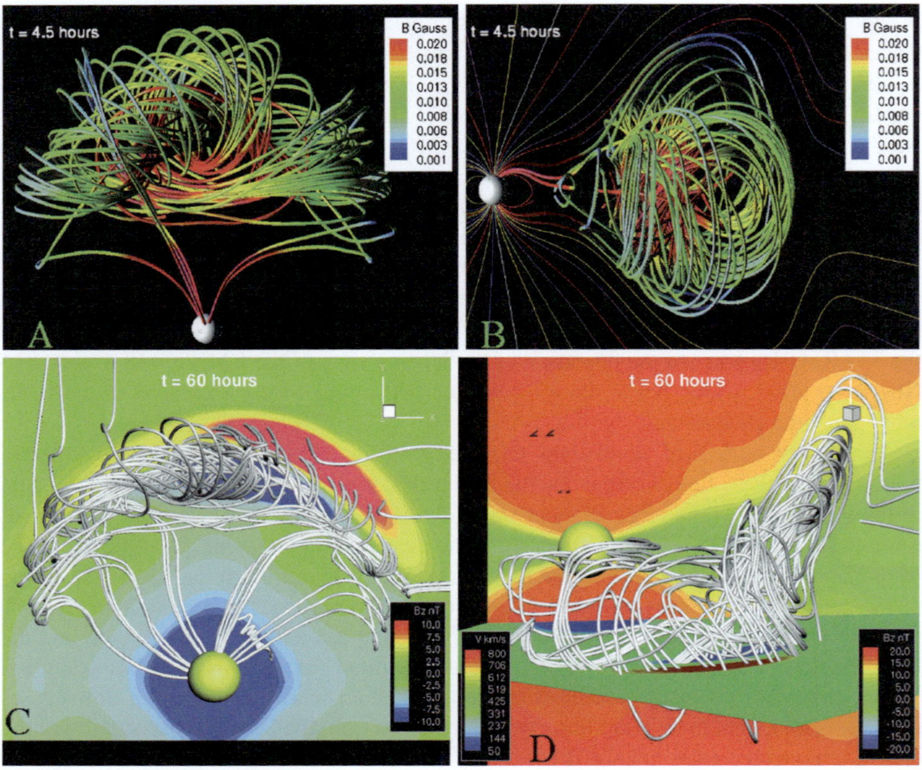

Fig. 13 Flux rope structure and evolution in a simulated CME. The magnetic field of a CME simulated with a spheromak flux rope is shown at 4.5 hours (*top row*) and 60 (*bottom row*) hours after initiation. (**a**) and (**c**): simulation results seen viewed from above ($+z$ axis). (**b**) and (**d**): side and oblique views of the simulated CME. Close to the Sun, the flux rope evolves in a nearly self-similar fashion. Far from the Sun, the flux rope is deformed to an outward-convex crescent, while the outer field lines have reconnected with the IMF

an imbalance of poloidal flux, a process called flux rope erosion. A schematic of this process is shown in Fig. 14 adapted from Ruffenach et al. (2012). Here, Fig. 14a shows the flux distribution of a cylindrical flux rope prior to reconnection with the IMF, where as Fig. 14b shows the outer layers of the flux rope peeled away by reconnection, leaving an imbalance in poloidal flux.

The analyses of MCs in many studies (e.g., Chané et al. 2005; Dasso et al. 2006, 2007; Ruffenach et al. 2012, 2015; Lavraud et al. 2013, 2014) have revealed the predicted features of flux rope erosion. Dasso et al. (2006) examined the 21 January 2005 ICME and found a large imbalance in B_z that they attribute to magnetic reconnection. Examination of the 9 November 2004 9 ICME by Dasso et al. (2007) revealed a long lasting tail (more than a day) of flux that is interpreted to be the result of magnetic reconnection at the front (and not at the rear) of the MC. A similar tail was found with the 18–20 October 1995 ICME, where the remaining magnetic field is interpreted as a partially reconnected flux rope by Dasso et al. (2006). Ruffenach et al. (2012) and Lavraud et al. (2014) closely examined several ICME events and found evidence (including reconnection jets) that magnetic reconnection occurred on the front side of the MC flux ropes, leading to their erosion. Ruffenach et al. (2012) examined the 19–21 November 2007 event with multi-spacecraft including STEREO A, B, ACE, Wind and THEMIS to establish that the axis of the rope was near the

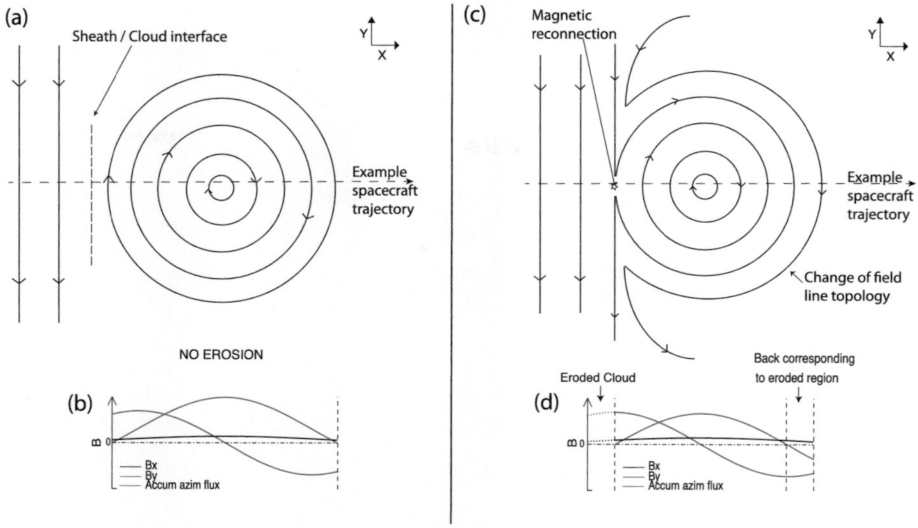

Fig. 14 Schematic of MC erosion adapted from Ruffenach et al. (2012). The examples on the left/right show a flux rope with no erosion/erosion and balanced/unbalanced poloidal flux B_y component

equatorial plane, proving that flux imbalance is not the result of high flux rope inclination (Lepping and Wu 2007). Ruffenach et al. (2015) analyzed the magnetic flux content of 50 MCs and found erosion at the front or rear in similar proportions with an average of 40% of the total azimuthal magnetic flux removed by reconnection. 30% of the magnetic clouds showed signs of reconnection.

Numerical simulations have also been used to study magnetic erosion in CMEs (e.g., Schmidt and Cargill 2003; Lugaz et al. 2013; Manchester et al. 2014a,b). The earliest examples are two-dimensional numerical experiments conducted by Schmidt and Cargill (2003), where the cylindrical flux rope was made to move out through open flux and through the current sheet. It was found that reconnection occurs on the lead/trailing edges of the flux rope in the case of fast/slow CMEs, and that the reconnection rate increases with the relative speed of the rope. An example from this simulation is shown in Fig. 15a, where the flux rope is initially traveling at 1.5 times the solar wind speed through the current sheet. Reconnection is manifest by the formation of small islands on opposite sides of the flux rope.

The magnetic reconnection rate is greatest close to the Sun where field strengths and Alfvén speeds are highest, with the rate decreasing with distance from the Sun. This evolutionary trend was shown in Manchester et al. (2014b), where the poloidal flux of the ejected rope is integrated and plotted as a function of time. Here, 14% of the total flux is lost during the first hour after initiation followed by additional 6% loss as the ICME propagates to 1 AU. A more extreme example of CME-IMF reconnection is found in the simulation of the 15 May 2005 ICME (Manchester et al. 2014a) where the poloidal field of the ejected rope is in the opposite direction of the surrounding streamer belt. Figure 15b provides a global view of the reconnection between the MC and the IMF occurring primarily at the north and south extremities of the flux rope. As the ICME approaches 1 AU, the MC is largely reconfigured by reconnection, yet the portion passing Earth retains the B_z orientations of the original rope and reproduces *in situ* observations of total field strength and B_z. As the result of reconnection, a long duration of B_z-south trails behind the MC. Lugaz et al. (2013) examines simulations of reconnection between multiple CMEs, which will be discussed in Sect. 9.

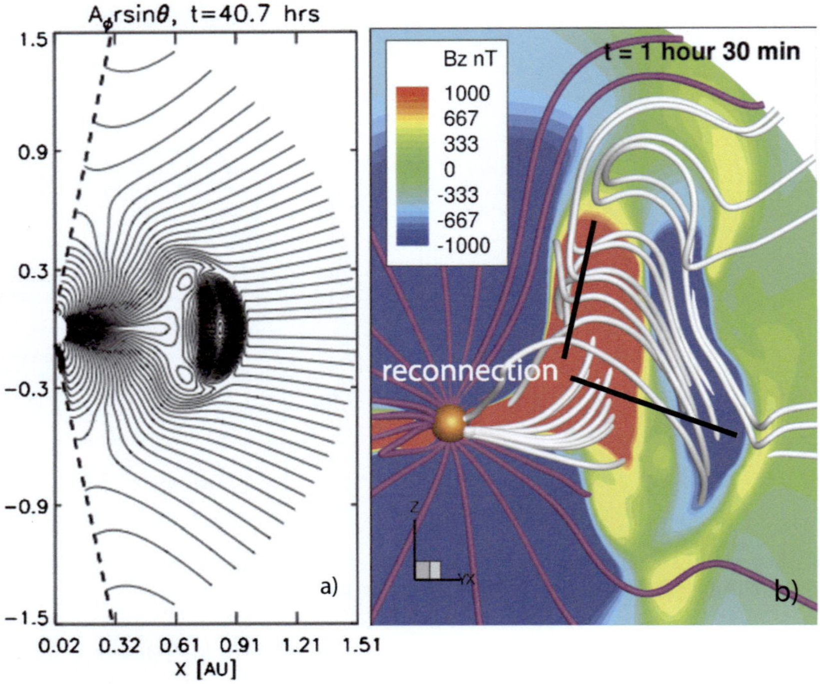

Fig. 15 Simulations of MC erosion. (**a**): results of the 2-D simulation of Schmidt and Cargill (2003). Here, magnetic reconnection with the IMF occurs near the top and bottom front of the flux rope resulting in the formation of a pair of magnetic islands. (**b**): the results of the 3-D simulation of Manchester et al. (2014a). In this case, reconnection occurs at both the front and back sides of the flux rope

A magnetic flux rope can also be distorted and the flux redistributed to produce localized flux imbalances and give the appearance of flux rope erosion. Manchester et al. (2014b) provides a CME simulation exhibiting an extreme case of flux redistribution that mimics flux rope erosion. In this case, as a fast ICME propagates through the solar wind, it undergoes strong deceleration, as previously discussed in Sect. 5. As a result, there is a tendency for denser filament material to move forward through the decelerating flux rope. Manchester et al. (2014b) have found that such movement is accompanied by azimuthal flows that can transport poloidal flux away from the nose of the ICME producing a large imbalance in B_z, giving the appearance of flux rope erosion, while, in fact, the majority of the flux remains intact. In this example, reconnection did occur with the IMF, but only removed less than twenty percent of the toroidal flux of the rope at 1 AU. Evidence of this behavior may be found in the 21 January 2005 ICME. *In situ* observations of this event show very dense cold plasma with a charge state composition consistent with filament material following immediately behind the ICME sheath, and most significantly the MC is almost entirely unbalanced in poloidal flux.

8 Plasma Evolution

Plasma that escapes the corona is heated to temperatures near 1.5 MK, which is sufficiently high to partially ionize all heavy elements. The freeze-in location of each ion charge state

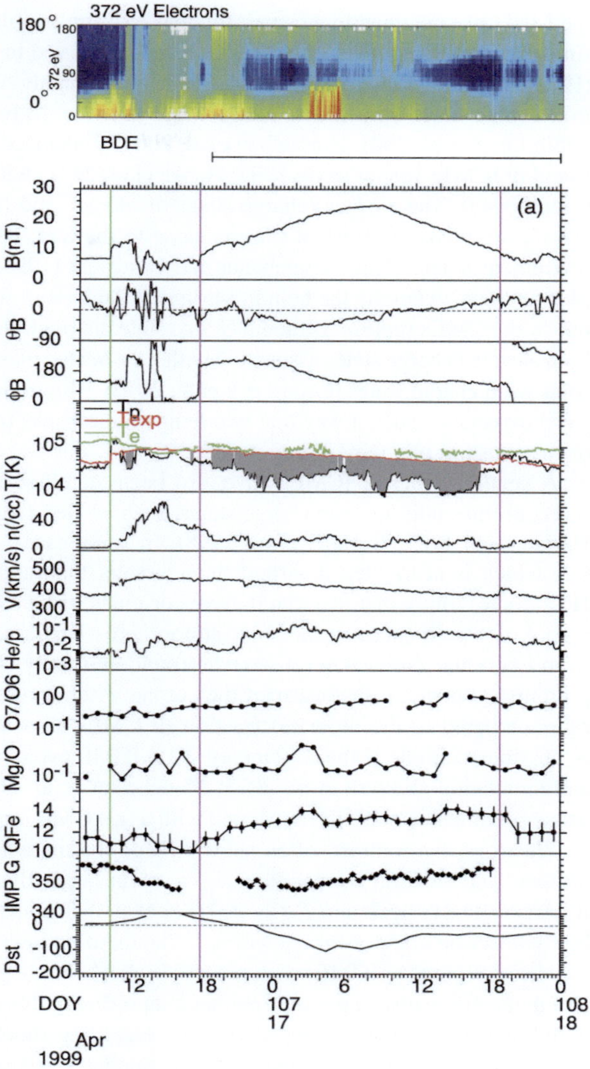

Fig. 16 Solar wind plasma and magnetic field parameters across an ICME (after Zurbuchen and Richardson 2006). The *vertical green line* marks the arrival time of the ICME-driven shock, and *two vertical pink lines* indicate the ICME interval. This event can be classified as a magnetic cloud, showing an enhanced, smoothly-rotating magnetic field. Other ICME signatures are also present: (bidirectional electrons) BDEs (with the interval indicated by the *horizontal bar*), depressed proton temperatures, declining solar wind speed profiles, He/proton abundance enhancements, enhanced oxygen and iron charge states and Mg/O ratio, cosmic ray depressions ($\gtrsim 60$ MeV particle count rate), and geomagnetic storms (indicated by the D_{st} index)

(typically within 4 $R_\odot$) depends both on ionization and recombination rates, which are functions of plasma density and temperature. The *in situ* charge state composition, therefore, complements remote observations and provides a unique and very important measure of the heating and expansion of the solar wind near the Sun (e.g., Fenimore 1980; Galvin 1997; Henke et al. 1998; Gloeckler et al. 1998; Zurbuchen et al. 2004; Lepri and Zurbuchen 2004; Reinard 2005; Zurbuchen and Richardson 2006). Used in combination, ion freeze-in states can be used to map out the heating and expansion rates of CMEs and as well as the ambient solar wind (e.g., Buergi and Geiss 1986; Esser and Edgar 2001; Lynch et al. 2011; Jin et al. 2012)

Numerous studies have shown the value of ion charge states as identifiers of ICMEs, finding that MCs are more likely to exhibit a higher composition of elevated charge states than non-MC ICMEs (e.g., Henke et al. 1998). Figure 16 (adapted from Zurbuchen and Richardson 2006) provides an example of ICME charge state signatures shown in conjunction with the MHD plasma quantities, supra-thermal electron populations and magnetospheric DST. Enhanced helium abundance (He^{2+}) has been observed in ICMEs since the early space era (Hirshberg et al. 1972; Borrini et al. 1982), and roughly 50%–70% ICMEs have enhanced iron charge states up to Fe^{16+} and average (Q)Fe > 12 (e.g., Bame et al. 1979; Fenimore 1980; Lepri et al. 2001; Zurbuchen et al. 2004). Enhanced charge states of oxygen are also found in ICMEs (e.g., Galvin 1997; Henke et al. 2001) with $O^{7+}/O^{6+} > 1$ (Richardson and Cane 2004a). The elevated charge states of oxygen and iron (Lepri and Zurbuchen 2004) signify an intense heating of plasma close to the Sun, while the cause of the helium enhancement is less clear. Neugebauer and Goldstein (1997) ascribe it to a "sludge removal phenomenon" whereby the helium settling at the base of the solar wind flux tube is cleared out by the CME eruption. Bimodal charge state distributions are also found in ICMEs where both elevated charge states associated with flare heated plasma are found along with ambient solar wind charge states (Gruesbeck et al. 2012). This condition suggest that either the material expands in such a way that two temperature peaks form (e.g., Gruesbeck et al. 2011) or two plasmas mix (e.g., Lepri et al. 2012).

A small number of ICMEs have also been identified with intervals of highly elevated levels of unusually low ion charge states, such as He^{+} (e.g., Gosling et al. 1980; Illing and Hundhausen 1985; Burlaga et al. 1998). In some cases, the He^{+}/He^{++} ratio may reach 0.1, which is nearly ten thousand times greater than ambient levels. The singly charged He particles imply that these cold events originate from a low temperature region, possibly a cool, dense filament in the solar atmosphere. This assertion has found further support from Lepri and Zurbuchen (2010) who found unusually cold material that they attribute to prominence material that escaped the corona without further heating. Such events remain rare compared to the observed fraction of CMEs associated with prominence eruptions. A possible example is the 21 January 2005 ICME event, where both charge state analysis and simulation (Kozyra et al. 2013; Manchester et al. 2014b) strongly suggest that cold dense prominence material associated with a fast CME impacted the Earth.

Electrons, by contrast, often show strange thermodynamic behavior across the ICME plasma. The electron temperature, T_e, is often larger than the proton temperature within ICMEs, with a typical ratio $T_e/T_p > 3$ (Liu et al. 2005), so the thermal pressure is dominated by the electron component. However, T_e tends to be lower than the proton temperature in the sheath regions of ICMEs with preceding shocks (Skoug et al. 2000), analogous to planetary magnetosheaths. A possible explanation is that protons resonate at their gyro-frequency with magnetic fluctuations induced at the preceding shock and thus gain energy from the fluctuations, while electrons cannot do so. Another explanation is that the shocks are supersonic relative to the protons, but subsonic relative to the electrons, and as a consequence, the protons are preferentially shock heated (Kosovichev and Stepanova 1991; Manchester et al. 2012; Jin et al. 2013). Another feature of electron thermodynamics is that the polytropic index $\gamma_e < 1$ inside ICMEs (Gosling 1999), the physical cause of which remains a mystery.

The radial evolution of T_α/T_p and $v_{\alpha p}$ from 0.3 to 5.4 AU is shown in Fig. 17. The differential streaming in the solar wind (lower panel) shows a fairly steady decrease with distance, probably due to the accumulative effect of Coulomb collisions. In contrast to the solar wind, the ICME $v_{\alpha p}$ decreases to less than 10 $\mathrm{km\,s^{-1}}$ by 0.4 AU, which indicates the importance of Coulomb collisions inside ICME plasma. The temperature ratio T_α/T_p of ICMEs is larger than that of the solar wind inside 1 AU but roughly the same magnitude beyond 1 AU. The little variation of the ICME temperature ratio with distance implies the existence of a local mechanism responsible for heating the alphas. One may consider that the free energy associated with the differential streaming is converted into the thermal energy of the plasma. It is not likely that the free energy preferentially goes to alphas through

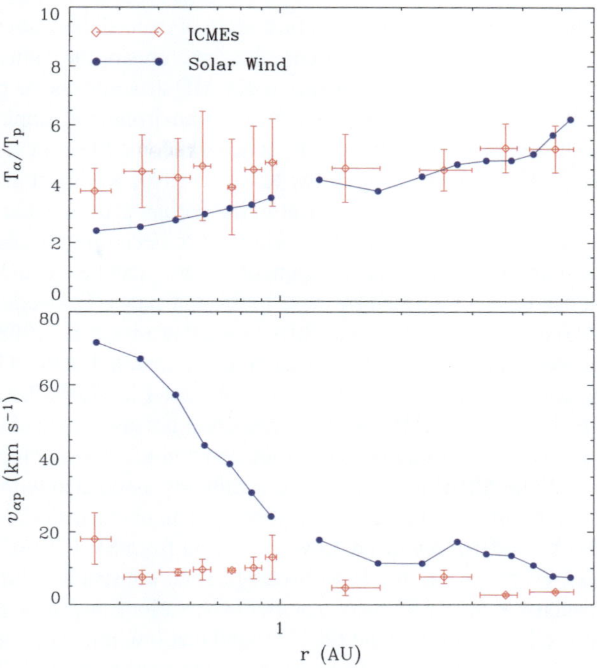

Fig. 17 Radial variations of the temperature ratio and differential speed between alphas and protons for ICMEs and solar wind (after Liu et al. 2006a). For ICMEs, the *horizontal bars* indicate the bounds of the bins while the *error bars* show the standard deviation of the parameters within the bins. The solar wind levels are represented by the average values over 0.1 AU bins within 1 AU and 0.5 AU bins beyond. All the data are within ±20° in latitude

Coulomb collisions. In addition, the differential speed within ICMEs is very small outside about 0.4 AU, but the temperature ratio is conserved at much larger distances. These results, in particular the collisional nature of the ICME plasma which has been ignored in previous studies, have significant consequences for modeling the realistic thermodynamics of ICMEs. The readers are directed to Liu et al. (2006a) for more detailed discussions.

Recent advancements in techniques for analyzing the solar wind composition measured have provided the most comprehensive ionic composition data to date, which set solid constraints on the heating and acceleration of the solar wind and ICMEs. These observations are even more powerful by combining them with remote observations of coronal plasma, which reveals details about CME expansion and solar wind source regions (e.g., Gruesbeck et al. 2011; Jin et al. 2012). These data are further complemented by recent simulations (Lynch et al. 2011) of ICME and solar wind (Jin et al. 2012) charge state composition.

Filaments (or more specially the filament channel) are at the heart of the CME energy source, where highly sheared magnetic fields run nearly parallel to the PIL in an energized state (Manchester 2001, 2007, 2008; Fang et al. 2010, 2012). Here, filaments form as cold dense concentrations of plasma suspended in the corona at chromospheric temperatures with densities 100 times greater than the surrounding corona. Filament material is typically taken to be the source of the dense core of the canonical three-part structure that is often wholly expelled in CMEs (Hundhausen 1993; Gibson and Low 1998). Gilbert et al. (2000) found 94% of erupting filaments were in association with CMEs and of those, 84% were found to have a portion of the filament escape the corona and contribute to the ejected CME plasma. However, connecting the components of the CME three-part density structure to *in situ* counterparts is often not possible. While a pileup of plasma is commonly found ahead of the flux rope, signatures of a cavity are not always clear and filament material is most often not detectable. ICME expansion tends to produce densities (and kinetic temperatures) lower than that of the surrounding solar wind (Liu et al. 2006a). However, there are often density fluctuations within the cloud that attain or exceed the ambient value (Burlaga 1981; Burlaga et al. 1982), which may obscure the signatures of the cavity.

A key to distinguishing different CME structures is the plasma charge state composition, which is determined by the ionization environment. Rapid decreases in mass density and collision rates cause the charge state to become fixed, or frozen in, beyond a few solar radii (e.g., Hundhausen et al. 1968; Ko et al. 1999; Kilpua et al. 2013). Given its relatively cool temperatures, filament material in the corona is dominated by heavy ion charge states such as C^{2-3+}, O^{2-4+} and Fe^{4-6+}, which have been directly observed with SOHO/UVCS (e.g., Lee and Raymond 2012). Filament material can be ejected into interplanetary space where it can be distinguished by these ionization states (Schwenn et al. 1980; Burlaga et al. 1998; Skoug et al. 1999; Yao et al. 2010). Gopalswamy et al. (1998) provides a clear example of the identification of low charge state filament material in the 07–11 February 1997 CME/ICME event. Here, filament material was observed in SOHO/LASCO C2 coronagraph images at the base of the CME, while corresponding dense filament material was found *in situ* in the Sun-ward side of a magnetic cloud containing relatively high abundances of Fe^{5+} and Fe^{6+}.

While filament eruptions are commonly associated with CMEs, it is surprising that convincing *in situ* charge state evidence of filament material is rarely found. Lepri and Zurbuchen (2010) reviewed 10 years of data from SWICS/ACE and found only 4% of ICMEs (11 events of the total) with charge state consistent with filament material. The relative paucity of *in situ* filament material observations may have been resolved by recent observations. Lee and Raymond (2012) found that low-ionization filament material (as observed by UVCS) covers only a few percent of the total area of CMEs, which is consistent with the low rate of *in situ* detection. This explanation is further corroborated by remarkable observations by Wood et al. (2016), which found that unlike most plasma in the CME, the filament material does not expand to fill space but fragments into high-density parcels of plasma that ballistically propagate through the heliosphere. Furthermore, in the case of fast ICMEs that strongly decelerate, the dense filament material maintains a high velocity and moves radially outward through the body of the CME, just as first predicted by Manchester et al. (2014b).

9 CME-CME Interaction

ICME properties may be strongly influenced as they propagate between the Sun and the Earth, in particular when they interact with other ICMEs or with corotating interaction regions (CIRs) and fast wind streams (Prise et al. 2015; Winslow et al. 2016; Liu et al. 2015, 2016). Two of the main sources of successive CMEs are homologous and sympathetic eruptions. Solar observations have revealed that recurrent CMEs occur from the same active region, often associated with homologous flares (Schmieder et al. 1984; Moon et al. 2003; Liu et al. 2014a,b; Wang et al. 2014a). Sympathetic CMEs, when the eruption of one CME triggers another, is another source of successive CMEs in relatively close angular and temporal separation (see Fig. 18a), revealed by means of observations and simulations to occur regularly (Schrijver and Title 2011; Török et al. 2011; Wang et al. 2016a).

During their Sun-to-Earth propagation, the interaction of successive ICMEs may take a variety of forms: (i) multiple CME-driven shock waves may interact without the associated ejecta interacting, (ii) one shock wave may interact with a preceding magnetic ejecta, or (iii) the successive magnetic ejecta may interact and/or reconnect. Early detections of multiple flares and complex ejecta in the interplanetary space revealed in the 1970s and 1980s that CME-CME interaction is a frequent occurrence. For example, the series of seven flares in 72 hours in early August 1972 have been extensively analyzed (Dryer et al. 1976; Intriligator

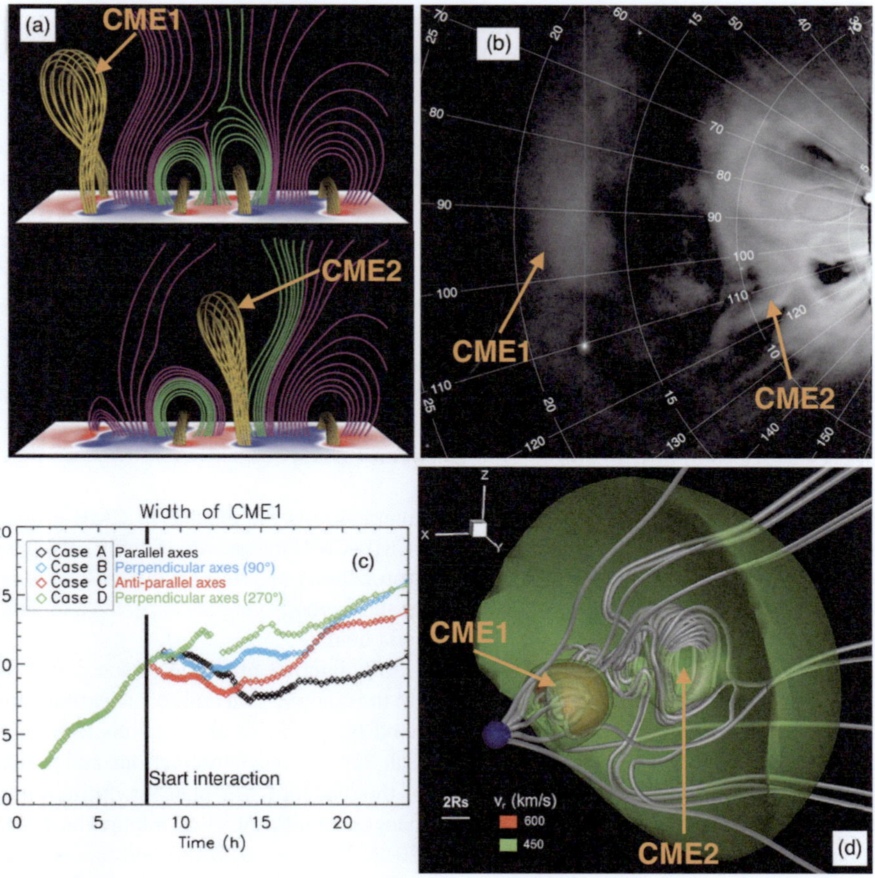

Fig. 18 MHD simulations and observations of CME-CME interaction. (**a**): Simulations of sympathetic eruptions from Török et al. (2011). (**b**): STEREO-A/HI1observations of two CMEs propagating in the inner heliosphere. The two CMEs have similar directions and interact and deflect each other on their way to Earth Lugaz et al. (2012). (**c**): Radial size of simulated CMEs before and during their interaction depending on their initial relative orientation, as adapted from Lugaz et al. (2013). (**d**): 3-D view of a MHD simulation of two interacting CMEs from Shen et al. (2013)

1976; Ivanov 1982). Another example is the study by Burlaga et al. (1987), in which the authors describe a variety of compound streams resulting from the interaction of a transient with another transient or with a solar wind stream. The authors concluded that "the interaction between two fast flows, is in general a nonlinear process, and hence, a compound stream is more than a linear superposition of its parts". An early numerical investigation is that of Vandas et al. (1997), who simulated the interaction of a shock wave with a magnetic cloud in 2.5-D MHD. The authors noted that the shock propagation results in a radial compression of the magnetic cloud, a change of its aspect ratio, and acceleration as well as heating of the cloud.

Direct observations of CME-CME interaction became possible in the mid-1990s with the larger field-of-view of LASCO/C3 (up to 32 $R_\odot \sim 0.15$ AU), which yielded the first reported white-light observations of CME-CME interaction (Gopalswamy et al. 2001c). Interest in this topic grew following this initial report (e.g., see Reiner et al. 2003), and from debates about the association of interacting CMEs with large solar energetic particle (SEP) events (Gopalswamy et al. 2002; Richardson et al. 2003). In solar cycle 24 (2008–present), observations with high spatial and temporal resolution by SDO have returned the study of sympathetic eruptions to central stage. The development of heliospheric imaging with the Solar Mass Ejection Imager (SMEI, Eyles et al. 2003; Jackson et al. 2004) and the Heliospheric Imagers (HIs, Eyles et al. 2009) onboard STEREO have led to a large increase in the number of published cases of ICME-ICME interactions being remotely observed in the past 10 years (see Fig. 18b). From the first remote detection in 2001 to routine remote observations in the late 2000s, numerical simulations have been used to fill the gap between the upper corona and the near-Earth space (Wu et al. 2002; Odstrcil et al. 2003; Schmidt and Cargill 2004; Lugaz et al. 2005b; Xiong et al. 2006). In the past decade, the combination of remote observations, *in situ* measurements and numerical simulations has resulted in a deeper understanding of the physical processes occurring during CME-CME interaction.

9.1 Compression and Reconnection Between CMEs

One of the essential aspects of CME-CME interaction is the change in CME properties, such as their speed, size, expansion rate, etc. CME-CME interaction directly affects space weather forecasting, by modifying the hit/miss probability and the expected arrival time of a CME (changes in the CME speed and direction) and by modifying the expected geomagnetic responses (changes in the CME internal magnetic field).

Here, we first discuss the changes in the CME internal properties, such as radial extent and magnetic field strength, associated with the compression and reconnection between CMEs. The radial size of a CME can be derived from remote observations (Savani et al. 2009; Nieves-Chinchilla et al. 2012; Lugaz et al. 2012). These observations and numerical simulations confirmed that the radial extent of the leading CME plateaus during the main phase (i.e. when the speed of both CMEs changes significantly) of interaction (Schmidt and Cargill 2004; Lugaz et al. 2005b, 2012, 2013; Xiong et al. 2006) and this is typically associated with a "pancaking" of the leading CME (Vandas et al. 1997). In Lugaz et al. (2005b), the authors discussed how the shock propagation through the leading CME is the main way in which the expansion slows. *In situ* measurements at 1 AU of shocks propagating inside CMEs (Lugaz et al. 2015) revealed the compression of the leading CME by the overtaking shock and CME (Wang et al. 2003b; Farrugia and Berdichevsky 2004; Liu et al. 2012, 2014c). It is, as yet, unclear whether this compression changes for cases when the overtaking CME drives a shock or not. What clearly changes is the resulting expansion of the leading CME after the end of the main interaction phase (i.e. after the shock exited the ejecta). In Xiong et al. (2006), the authors found that the leading CME over-expands to return to its expected size; as such the compression is only a temporary state. This was confirmed by a statistical study of the radial size of magnetic ejecta at different heliocentric distances (Gulisano et al. 2010). In one case study, similar results were found: remote observations indicated compression but, a day after the interaction ended, when the CME impacted Earth, the *in situ* measurements indicated a typical CME size (Lugaz et al. 2012). In numerical simulations with two magnetic ejecta, the rate of over-expansion is found to depend on the rate of reconnection between the two ejecta (see Fig. 18c); as such, it depends on the relative orientation of the two magnetic ejecta (Schmidt and Cargill 2004; Lugaz et al. 2013), but also probably on their density. Associated with ICME compression is the enhanced magnetic field inside the ICMEs as revealed by *in situ* measurements. Liu et al. (2014a) find that an extremely enhanced field strength, larger than 100 nT, can be produced by CME-CME interactions.

CME-CME "cannibalism", i.e. the total coalescence of two CMEs, was reported based on LASCO images by Gopalswamy et al. (2001c). However, because coronagraphs and white-light imagers only give direct information about a CME density structure, it is unclear whether this coalescence is associated with a total reconnection of one CME's magnetic flux or not. Known magnetic reconnection rates and CME magnetic flux makes it unlikely that total reconnection is a common occurrence until beyond Earth's orbit. The potential full coalescence of two ejecta into one was discussed in a few studies (Odstrcil et al. 2003; Schmidt and Cargill 2004; Chatterjee and Fan 2013) but has not been investigated in detail with realistic reconnection rates. *In situ* measurements of transients resulting from the interaction of two CMEs sometimes contain a period of weaker and more turbulent magnetic field, higher in density and temperature, resulting in a higher plasma β (Wang et al. 2003a). These periods have been explained by ongoing or past reconnection between the two magnetic ejecta; however multi-spacecraft measurements or a successful Walén test (Paschmann and Sonnerup 2008) would be required to assert that reconnection is occurring. No such case has been reported yet.

In Lugaz et al. (2013), the authors considered the interaction of two CMEs with axial fields perpendicular to one another by means of MHD simulations, and found that the resulting ejecta appeared as one of a single, extended but seemingly isolated, CME with a complex magnetic field rotation. Such events can sometimes be measured at 1 AU (Dasso et al. 2009; Lugaz and Farrugia 2014) and may be related to these instances of CME-CME cannibalism.

9.2 Momentum Transfer Between CMEs

Many studies have investigated the momentum transfer between CMEs during their interaction, both through remote observations and numerical simulations (Schmidt and Cargill 2004; Lugaz et al. 2005b; Liu et al. 2012; Shen et al. 2012; Mishra et al. 2015a,b; Colaninno and Vourlidas 2015). CMEs are large-scale magnetized plasma structures propagating in solar wind, and therefore, the momentum transfer between CMEs is complex for a variety of factors: (i) depending on the speed of the CMEs, the interaction between two CMEs may involve zero, one or two CME-driven shocks, some of which may dissipate during the interaction, (ii) the interaction takes at least one Alfvén crossing time of a CME, i.e. typically 8–24 hours, (iii) the CME speeds can change significantly, even at large distances from the Sun, due to their interaction with the solar wind (Manchester et al. 2004a; Temmer et al. 2011; Liu et al. 2013), (iv) CME-CME interaction is inherently a three-dimensional process, and the changes in kinematics may differ greatly depending on the CME part that is considered (Shen et al. 2012; Temmer et al. 2014). Using numerical simulations (see Fig. 18d), it is somewhat possible to control for some of these effects, for example, by performing simulations with or without interaction but with identical CME properties, and by knowing the velocity field in the entire 3-D domain (Shen et al. 2013, 2016).

As CME-CME interaction involves a faster CME overtaking a slower CME, the end result is to homogenize the speed, as was noted from *in situ* measurements in Burlaga et al. (2002), Farrugia and Berdichevsky (2004) and through simulations by Schmidt and Cargill (2004) and Lugaz et al. (2005b), among others. It has been found to occur independently of the relative speeds of the two CMEs. Based on this finding, one of the main issues related to momentum transfer is to understand what determines the final speed of the transient that was formed by the CME-CME interaction. In an early work, Wang et al. (2005) found that, in the absence of CME-driven shock waves, the final speed is determined by that of the slower ejecta, whereas Schmidt and Cargill (2004) and Lugaz et al. (2005b) found that when the

CMEs drive shocks, the final speed is primarily determined by that of the faster ejecta, as the shock's propagation through the first magnetic ejecta accelerates it to a speed similar to that of the second ejecta. Work combining remote observations and numerical simulations lead us to conclude that the final speed of the transient depends on the relative masses of the CMEs, as well as their approach speed (Poedts et al. 2003; Shen et al. 2016).

In addition to changes in velocity, CME-CME interaction may result in the deflection of one CME by another (Xiong et al. 2009; Lugaz et al. 2012; Shen et al. 2012). Combining these works, it appears that the deflection can reach up to 15° when the two CMEs are initially about 15–20° apart. Such angular separations are quite frequent between successive CMEs, as it corresponds to a delay of about one day for two CME originating from the same active region (due to solar rotation). This change in direction must be taken into account when deriving the changes in velocity, as done in Shen et al. (2012) and Mishra et al. (2016).

9.3 Changes in the Shock Properties

In addition to changes in the CME properties, any fast forward shock propagating inside the overtaken magnetic ejecta encounters highly-varying and unusual upstream conditions, affecting the shock properties. Most of what is known about the changes in shock properties was learned from numerical simulations; however, there have been many reported detections of shocks propagating inside a magnetic cloud or magnetic ejecta at 1 AU (Wang et al. 2003c; Collier et al. 2007; Richardson and Cane 2010a; Liu et al. 2012, 2014c; Lugaz et al. 2015, 2016).

Vandas et al. (1997) noted that a shock propagates faster inside a magnetic cloud due to the enhanced fast magnetosonic speed inside, which may result in shock-shock merging close to the nose of the magnetic cloud but two distinct shocks in the flanks. Odstrcil et al. (2003) noted that associated with this acceleration, the density jump becomes smaller. Lugaz et al. (2005b) performed an in-depth analysis of the changes in the shock properties, dividing the interaction into four main phases: (i) before any physical interaction, when the shock propagates faster than an identical isolated shock due to the smaller density in the solar wind, (ii) during the shock propagation inside the magnetic cloud, when the shock speed in a rest frame increases and its compression ratio decreases, confirming the findings of Odstrcil et al. (2003), (iii) during the shock propagation inside the dense sheath when the shock decelerates, as pointed out by Vandas et al. (1997), and (iv) the shock-shock merging when, as predicted by MHD theory, a stronger shock forms followed by a contact discontinuity. If the shock is weak or slow enough, it may dissipate as it propagates into the region of higher magnetosonic speed inside the magnetic cloud (Xiong et al. 2006; Lugaz et al. 2007). High spatial resolution is necessary to resolve weak shocks in MHD simulations, and low resolution may affect the prediction of shock dissipation. The merging or dissipation of shocks was noted by Farrugia and Berdichevsky (2004), when Helios measured four shocks at 0.67 AU and ISEE-3 measured only two shocks later on at 1 AU.

For Solar Cycle 23, Lugaz et al. (2015) identified in total 49 shocks that were propagating within magnetic ejecta. Most such shocks occur toward the back of the ejecta, and shocks tend to be slower as they get closer to the CME front. This can be interpreted as an indication that a number of shocks dissipate inside a CME before exiting it. The two main reasons are that CMEs tend to be expanding and have a decreasing speed profile and that the peak Alfvén speed typically occurs close to the center of the magnetic ejecta. The latter reason means that shocks become weaker as they approach the center of the ejecta. The former reason implies that shocks propagate into higher and higher upstream speeds as they move from the back to the front of the CME. Lugaz et al. (2015) reported cases when the speed at the front of the first CME exceeds the speed of the overtaking shock, i.e. because of the CME expansion, the shock cannot overtake the front of the CME.

10 Modeling CMEs from Sun to Earth: The Bastille Day Event

MHD simulations of CMEs can be roughly divided into two classes. On the one hand, there are models that consider idealized configurations and are mainly used to test suggested ideas and to investigate specific aspects of CMEs, such as their initiation mechanisms. This kind of simulation uses a simplified set of the MHD equations and numerical domains that typically cover only the lower corona (see Green et al. in this Volume). On the other hand, there are simulations that are designed predominantly to reproduce observed eruptions. Two examples are shown in Figs. 12 and 15, respectively. These semi-realistic simulations vary widely in complexity, depending on the physics included, the phases of the CME/ICME evolution modeled, and the extent of observational constraints and real data included. Many models use observed magnetograms as boundary condition for the magnetic field at the lower boundary, and recently observed surface flows have been included as well (Jiang et al. 2016). Some simulations restrict the calculation to the evolution in the corona (e.g., Roussev et al. 2007; Cohen et al. 2009; Zuccarello et al. 2012b; Amari et al. 2014; Fan 2016), while others model the propagation of the ejecta to 1 AU but do not include the coronal evolution. Instead, the latter models start the simulation in the inner heliosphere (typically at around 20–30 $R_\odot$) and use for the initial ICME some idealized model whose parameters are chosen guided by coronagraph observations (e.g., Shiota and Kataoka 2016). Some of these models, such as ENLIL (Odstrcil 2003), which is the only MHD model that is currently used for operational space weather predictions (see Green et al. in this Volume), simplify even further by neglecting the magnetic field of the ICME and initiating its propagation by specifying a cone of constant velocity at the inner boundary of the model.

The currently most advanced simulations model CMEs all the way from the Sun to Earth (e.g., Manchester et al. 2008). The most recent ones employ a sophisticated treatment of the energy transfer in the corona, often referred to as "thermodynamic MHD" (Lionello et al. 2009; Downs et al. 2010; van der Holst et al. 2010). Such a treatment is required for a realistic description of the plasma properties in the corona, in particular if one wants to reproduce simultaneously the emission in active regions, the quiet sun, and coronal holes (Lionello et al. 2009). Some models also allow for a separate treatment of the electron and ion temperatures (van der Holst et al. 2014). CMEs are then launched in such a thermodynamic MHD environment, often using an analytical magnetic flux-rope model such as the one by Titov and Démoulin (1999), and propagated into the corona (Lugaz et al. 2011; Downs et al. 2012; Jin et al. 2013) or, by coupling different MHD codes (Tóth et al. 2007), further out to 1 AU and beyond (Lionello et al. 2013; Manchester et al. 2014a; Jin et al. 2017b). A long-term aim is to use these sophisticated simulations for space-weather predictions by modeling observed CMEs, especially the strongest, most geo-effective events, in real-time, i.e., while the eruption is still on its way to Earth (see, e.g., Jin et al. 2017a). In what follows, we briefly describe the properties and limitations of these modeling efforts, using as an example a recent simulation of the well-known "Bastille Day" eruption. Other examples can be found in the articles mentioned above.

The 14 July 2000 ("Bastille Day") event, one of the largest eruptions of solar cycle 23, occurred in NOAA active region (AR) 9077. It was associated with a fast halo CME, an X5.7 flare, and an intense radiation storm that resulted in one of the 16 GLE events of cycle 23. The CME traveled in the corona at a speed of $\approx 1700\ \mathrm{km\,s^{-1}}$ (Andrews 2001). The

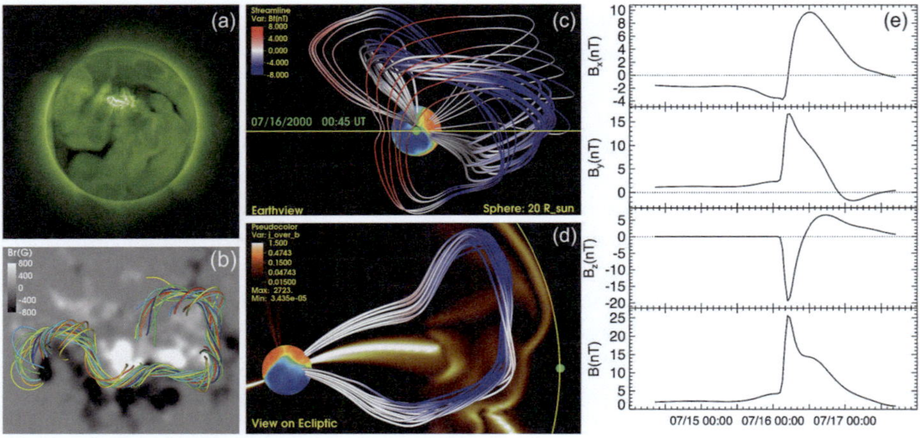

Fig. 19 Coronal and heliospheric MHD simulation of the "Bastille Day" event. (**a**): Synthetic SOHO/EIT 195 Å emission image before the eruption. (**b**): Stable flux rope in the eruption's source region. The SOHO/MDI magnetogram is shown in *gray-scale*. (**c**): Field lines of the ICME flux-rope as viewed from Earth. *Thick* (*thin*) *lines* outline the flux-rope core (front). Field lines are colored by B_θ, corresponding to $-B_z$. The *sphere shows* B_r at the inner boundary of the heliospheric run; the *green ball* indicates Earth. (**d**): Same as (**c**), viewed from above the ecliptic plane. Thin field lines are omitted for clarity. Electric currents are visualized by j/B (in code units) in the ecliptic plane. (**e**): Synthetic measurements of the magnetic field components at 1 AU, taken about 20° north of Earth. Panels (**c**)–(**e**) are adapted from Linker et al. (2016)

shock wave driven by the interplanetary CME/ICME and a large magnetic cloud reached Earth in about 28 and 33 hours, respectively, triggering a large geomagnetic storm (Lepping et al. 2001). Strong eruptions such as the Bastille Day event are perfectly suited to test the capabilities of MHD simulations in reproducing the properties of "extreme" events.

A simulation of this event was performed using the 3D-MHD code MAS developed at Predictive Science Inc. (e.g., Mikić et al. 1999; Lionello et al. 1999, 2013). Here, we focus on the main properties of the simulation; a more detailed description can be found in Linker et al. (2016) and in a forthcoming publication (T. Török et al., in preparation).

To provide the background environment for the CME, first an MHD solution of the global corona was calculated, using a spherical grid that covers 1 to 20 $R_\odot$, and includes the sonic and Alfvénic critical points. The surface radial magnetic field, B_r, was specified by combining a line-of-sight (LOS) SOHO/MDI synoptic map (10/7/2000-6/8/2000) with a LOS MDI magnetogram measured about 30 minutes prior to the flare onset. After calculating a global potential magnetic field based on this boundary condition, a thermodynamic MHD model of the corona is produced, using the equations and coronal-heating specification described by Lionello et al. (2009). The system is relaxed for about 64 hours, until a steady state, including the solar wind, is obtained. The inclusion of thermal conduction, radiative cooling, and empirical coronal heating provides realistic density and temperature distributions, which allows one to produce synthetic EUV or soft X-rays satellite images that can be directly compared to the observations. Figure 19a shows, as an example, a synthetic SOHO/EIT image of the corona obtained from the simulation.

The AR was then energized by inserting several instances of the modified Titov-Démoulin (TDm) flux-rope model (Titov et al. 2014) along the PIL. The original TD model of a force-free coronal flux rope (Titov and Démoulin 1999) has been widely used as an initial condition for CME simulations. Typically, a TD rope out of magnetic equilibrium is used to initiate a CME (e.g., Manchester et al. 2008; Lugaz et al. 2011). This approach is convenient as it reduces the complexity of the computation and reproduces many observed CME properties sufficiently well, but neglects that in reality CMEs always arise from magnetic fields in an initial state of force balance. Thus, starting simulations from a stable magnetic equilibrium is important for understanding the energy storage and release of CMEs. The TDm model facilitates the construction of such configurations by taking into account the background coronal magnetic field, and was therefore, used in the simulation described here. To account for the highly curved and elongated PIL, along which the strength of the background field strongly varies, a chain of seven overlapping TDm ropes was inserted (Fig. 19b). The axial field orientation was chosen to produce a left-handed flux rope, as suggested by the observations (e.g., Yurchyshyn et al. 2001). The model was further modified to preserve the observed B_r distribution when inserting the ropes, and the parameters of each TDm rope were chosen such that the merged flux rope is stable with respect to the kink and torus instabilities (e.g. Török et al. 2004; Kliem and Török 2006). The resulting configuration is close to being force-balanced but is not in thermal equilibrium. Flows and plasma condensations appeared along the flux-rope field lines that possibly arose from thermal non-equilibrium (e.g., Mikić et al. 2013a). The system was relaxed for about one hour, during which time the numerical dissipation (enhanced by the thermal flows) leads to some loss of magnetic energy.

After this additional relaxation, the eruption of the flux rope was triggered by imposing sub-Alfvénic flows at the inner boundary that converge toward the PIL (e.g., Linker et al. 2003; Mikić et al. 2013b). These flows were imposed for about 50 minutes, at which point the configuration destabilized. The flows were not directly derived from the observations, but adjusted such that the eruption starts at the western end of the AR and proceeds toward the East, in qualitative agreement with the observations. The full eruption released 1.3×10^{33} ergs in about 4 minutes; about 31% (4×10^{32} ergs) was converted into kinetic energy. The simulated CME reached a speed of about 1500 $\mathrm{km\,s^{-1}}$, close to the observed speed. The striking morphology of the flare arcade and the halo CME are qualitatively well reproduced in the simulation (see Linker et al. 2016). The simulation also self-consistently (i.e., without further boundary-driving) reproduces a second, "sympathetic" eruption, that was observed four hours after the main eruption in the western part of the AR (T. Török et al., in preparation).

Finally, in order to model the propagation of the ICME to Earth, the coronal simulation was coupled to the heliospheric version of MAS (Lionello et al. 2013), which uses a simpler set of equations and, is therefore, computationally more efficient. The coupling is done by setting the outer boundary of the coronal simulation as the inner boundary of the heliospheric simulation, using time-dependent boundary conditions as described in Lionello et al. (2013). The heliospheric calculation uses a $486 \times 264 \times 583$ nonuniform spherical mesh extending from 19 to 230 $R_\odot$.

The observed ICME decelerated during its propagation, and arrived at Earth at about 19 UT on 15/7/2000 with a speed of ≈ 1100 $\mathrm{km\,s^{-1}}$ and a maximum field strength of ≈ 50 nT (e.g., Smith et al. 2001). The magnetic-field measurements at 1 AU suggest a left-handed flux rope passed the spacecraft below its axis (Yurchyshyn et al. 2001). Figures 19c, d show the simulated ICME, with field lines colored by the magnitude of B_θ; positive B_θ (red) is in the direction of negative B_z at Earth. It can be seen that a negative B_z arises from the azimuthal fields wrapping the core of the flux rope, and that the flux rope connects back to the inner boundary of the simulation domain. Figure 19e shows that the simulated ICME arrives at 1 AU seven hours later than the real event, with a speed of ~ 900 $\mathrm{km\,s^{-1}}$ and a maximum field strength of ~ 27 nT. The simulated ICME is qualitatively consistent with the flux rope inferred from the observations: a full rotation of B_z from negative to positive, and $B_x, B_y > 0$. However, it passes Earth about 15–20° northward of the real event. At Earth's location, the correct (negative) sign of B_z is reproduced, but the rotation of B_z from negative to positive is not seen, and the field strengths are weaker.

This example illustrates that state-of-the-art CME simulations are capable of qualitatively reproducing many important features of extreme events, such as the rapid, strong energy release, the observed emission and white-light morphology, and the magnetic structure of the ICME. The quantitative differences in the case shown here are relatively small, given that the simulation still uses various simplifications, such as a steady-state corona and heliosphere, and does not employ "trial-and-error" runs to match the observations (such as the CME speed) quantitatively. Yet, the discrepancies are still significant in light of the potential future application of such simulations for space-weather predictions. Further improvements, such as data-driven, time-dependent models of the background corona, and more case studies are required to assess to which accuracy state-of-the-art CME simulations can match the observations at 1 AU.

11 Conclusions

3-D numerical simulations and wide-angle heliospheric imaging of CMEs and ICMEs now routinely performed by STEREO, have resulted in a new era of study of CME/ICME evolution and their interaction in the heliosphere. Upon eruption, CMEs strongly interact with the surrounding environment from the active region magnetic field, the streamer belt, the solar wind and even other CMEs. A multitude of physical processes govern the transfer of energy, mass, momentum and magnetic flux to and from CMEs and ICMEs as they travel from low corona to interplanetary space, most of which can be well described in the context of extended MHD. CME deflection has an early impact on CMEs, where the trajectory departs from the expected radial direction and is most prevalent in the low corona. Here, magnetic forces can drive deflections in longitude and latitude that can reach 30–40° for low-mass and slow-speed CMEs. Meanwhile, patterns of CME impact on Earth suggest a consistent eastward deflection related to spiral structure of the ambient solar wind. Apart from being deflected, CMEs can also rotate about their rise direction as they propagate in the corona and interplanetary space. Like deflection, rotations are largest in the low corona, but are less predictable given that a number of mechanisms, including reconnection of the ejecta with complex background magnetic fields, can contribute to CME rotation in a complicated manner. At present, it is unclear what determines the total amount of rotation in CMEs, and further efforts are required in order to gain a better understanding of this important problem.

Once free of the low corona where strong acceleration occurs, ICME evolution is foremost dominated by kinematic processes as ICMEs ballistically propagate through the solar wind. As a first approximation, this process is described by aerodynamic drag where slow CMEs are accelerated and fast CMEs decelerated by the interaction with the solar wind. These processes are sufficiently well represented by empirical and numerical models that they can typically predict CME arrival times at Earth to within eight hours or less. Consider next the reaction that ICME propagation has on the heliosphere. CMEs traveling faster than the solar wind drive forward propagating shocks that can extend far beyond the ejecta, and whose geometry can be largely modified by the structure of the ambient solar wind. Simulations show that radial expansion of the ejecta can drive high-latitude reverse shocks, which may also be caused by CME-deflected high-speed streams. The ejecta also expands laterally with the diverging solar wind at a rate that is faster than the flow deflected around the ejecta. As a result, plasma accumulates in a sheath surrounding the flux rope and adds substantially to the overall mass of the disturbance. The interplanetary magnetic field can similarly be compressed as it drapes around the ejecta, leading to significant enhancements in the overall flux and field strength. The IMF can also reconnect with the ejecta, eroding the magnetic rope by removing poloidal flux. Reconnection is common with 30% of flux ropes showing signs of erosion with an average of 40% of the flux removed at 1 AU. In addition to MHD quantities, charge state composition is measured both *in situ* and spectroscopically, which gives measures of the temperature and thermodynamic behavior of the low corona where the charge state freezes in. These data are particularly important in identifying cold filament material and flare heated plasma.

Many recent investigations have focused on the relation between CME-CME and ICME-ICME interaction and the resulting structures at 1 AU, as well as their exchange of momentum. CME-CME interaction is one of the ways in which large changes in the properties of CMEs may occur during their Sun-to-Earth propagation, in addition to interaction with the solar wind. It often results in a compression and acceleration of the leading CME, although this might only be a temporary effect; it can also result in a partial or total coalescence of the two interacting CMEs (CME-CME "cannibalism"). There remains significant debate about momentum exchange between CMEs and the resulting final speed. In many instances, after their interaction, the two CMEs propagate with a uniform speed profile, similar to that of an isolated CME, but how this speed relates to that of the two CMEs before the interaction is still an area of active research. Other potential effects are a deflection of the two CMEs, and the "disappearance" of shock waves, either through shock-shock merging or as the trailing shock dissipates as it propagates inside the leading magnetic ejecta. About 15% of shocks at 1 AU are found to propagate inside a preceding magnetic ejecta, a clear example of CME-CME interaction in the process of happening.

Numerical simulations have evolved to a state where a semi-realistic modeling of observed eruptions has become possible. State-of-the-art simulations employ measured magnetograms, a realistic treatment of the energy transfer in the corona, and propagate ICMEs to 1 AU or beyond. Using the example of the well-known Bastille Day event, we described the capability of such simulations to reproduce many important observational features of eruptions, which makes them extremely valuable for improving our physical understanding of these complex events. Further improvements in computational power and in the accuracy of the simulations are required, though, before they can be used for forecasting the impacts of CMEs on space weather close to Earth.

Together, these results indicate general pictures and universal features, that regulate the Sun-to-Earth propagation of different types of CMEs. Although exceptions from the norms can occur, depending on the coronal and interplanetary conditions where a CME/ICME propagates, the results are representative of many events. Any theory/model of CME/ICME Sun-to-Earth propagation and space weather forecasting should be guided and constrained by these results.

Acknowledgements W. M. was supported by NASA grant NNX16AL12G and NSF grant AGS1322543. E. K. acknowledges the University of Helsinki 3-year project and Academy of Finland project 1267087. Y. L. was supported by the Recruitment Program of Global Experts of China, NSFC under grant 41374173 and the Specialized Research Fund for State Key Laboratories of China. N. L. was supported by NASA grants NNX13AH94G and NNX15AB87G and NSF grant AGS1435785. T. T. acknowledges support by NASA's LWS and HSR programs and by NSF grants AGS1249270, AGS1348577, and AGS1135432 (Sun-2-Ice). B. V. acknowledges support by Croatian Science Foundation under the project 6212 Solar and Stellar Variability. We acknowledge the International Space Science Institute (ISSI) in Bern Switzerland for their generous support and for travel and accomodations.

References

T. Amari, J.F. Luciani, J.J. Aly, Z. Mikic, J. Linker, Coronal mass ejection: initiation, magnetic helicity, and flux ropes. I. Boundary motion-driven evolution. Astrophys. J. **585**, 1073–1086 (2003). doi:10.1086/345501

T. Amari, A. Canou, J.J. Aly, Characterizing and predicting the magnetic environment leading to solar eruptions. Nature **514**, 465 (2014)

M.D. Andrews, Lasco and EIT observations of the Bastille day 2000 solar storm. Sol. Phys. **204**, 179–196 (2001). doi:10.1023/A:1014215923912

S.K. Antiochos, C.R. DeVore, J.A. Klimchuk, A model for solar coronal mass ejections. Astrophys. J. **510**, 485–493 (1999)

H. Aurass, B. Vršnak, A. Hofmann, V. Ruždjak, Flares in sigmoidal coronal structures a case study. Sol. Phys. **190**, 267–293 (1999). doi:10.1023/A:1005261709955

S.J. Bame, J.R. Asbridge, W.C. Feldman, E.E. Fenimore, J.T. Gosling, Solar wind heavy ions from flare-heated coronal plasma. Sol. Phys. **62**, 179–201 (1979). doi:10.1007/BF00150143

B.M. Bein, S. Berkebile-Stoiser, A.M. Veronig, M. Temmer, N. Muhr, I. Kienreich, D. Utz, B. Vršnak, Impulsive acceleration of coronal mass ejections. I. Statistics and coronal mass ejection source region characteristics. Astrophys. J. **738**, 191 (2011). arXiv:1108.0561. doi:10.1088/0004-637X/738/2/191

B.M. Bein, M. Temmer, A. Vourlidas, A.M. Veronig, D. Utz, The height evolution of the "true" coronal mass ejection mass derived from STEREO COR1 and COR2 observations. Astrophys. J. **768**, 31 (2013). arXiv:1303.3372. doi:10.1088/0004-637X/768/1/31

D.W. Billings, *A Guide to the Solar Corona* (Academic, San Diego, 1966)

A. Borgazzi, A. Lara, E. Echer, M.V. Alves, Dynamics of coronal mass ejections in the interplanetary medium. Astron. Astrophys. **498**, 885–889 (2009). doi:10.1051/0004-6361/200811171

G. Borrini, J.T. Gosling, S.J. Bame, W.C. Feldman, Helium abundance enhancements in the solar wind. J. Geophys. Res. **87**, 7370–7378 (1982). doi:10.1029/JA087iA09p07370

G.E. Brueckner, R.A. Howard, M.J. Koomen, C.M. Korendyke, D.J. Michels, J.D. Moses, D.G. Socker, K.P. Dere, P.L. Lamy, A. Llebaria, M.V. Bout, R. Schwenn, G.M. Simnett, D.K. Bedford, C.J. Eyles, The Large Angle Spectroscopic Coronagraph (LASCO). Sol. Phys. **162**, 357–402 (1995). doi:10.1007/BF00733434

G.E. Brueckner, J.P. Delaboudiniere, R.A. Howard, S.E. Paswaters, O.C. St. Cyr, R. Schwenn, P. Lamy, G.M. Simnett, B. Thompson, D. Wang, Geomagnetic storms caused by coronal mass ejections (CMEs): March 1996 through June 1997. Geophys. Res. Lett. **25**, 3019–3022 (1998). doi:10.1029/98GL00704

A. Buergi, J. Geiss, Helium and minor ions in the corona and solar wind—dynamics and charge states. Sol. Phys. **103**, 347–383 (1986). doi:10.1007/BF00147835

L.F. Burlaga, Magnetic loop behind an interplanetary shock: Voyager, Helios and MP-8 observations. J. Geophys. Res. **86**, 6673 (1981)

L.F. Burlaga, Magnetic clouds and force-free fields with constant alpha. J. Geophys. Res. **93**, 7217–7224 (1988)

L.E. Burlaga, L.W. Klein, N.R. Sheeley Jr., D.J. Michels, R.A. Howard, M.J. Koomen, R. Schwenn, H. Rosenbauer, A magnetic cloud and a coronal mass ejection. Geophys. Res. Lett. **9**, 1317 (1982)

L.F. Burlaga, K.W. Behannon, L.W. Klein, Compound streams, magnetic clouds, and major geomagnetic storms. J. Geophys. Res. **92**, 5725–5734 (1987). doi:10.1029/JA092iA06p05725

L.F. Burlaga, N.F. Ness, F.B. McDonald, Magnetic fields and cosmic rays in the distant heliosphere at solar maximum: Voyager 2 observations near 32 AU during 1990. J. Geophys. Res. **100**(A8), 14,763–14,771 (1995)

L. Burlaga, R. Fitzenreiter, R. Lepping, K. Ogilvie, A. Szabo, A. Lazarus, J. Steinberg, G. Gloeckler, R. Howard, D. Michels, C. Farrugia, R.P. Lin, D.E. Larson, A magnetic cloud containing prominence material: January 1997. J. Geophys. Res. **103**, 277–285 (1998)

L.F. Burlaga, S.P. Plunkett, O.C. St. Cyr, Successive CMEs and complex ejecta. J. Geophys. Res. **107**, 1 (2002)

J.P. Byrne, S.A. Maloney, R.T.J. McAteer, J.M. Refojo, P.T. Gallagher, Propagation of an Earth-directed coronal mass ejection in three dimensions. Nat. Commun. **1**, 74 (2010). arXiv:1010.0643. doi:10.1038/ncomms1077

H.V. Cane, I.G. Richardson, Interplanetary coronal mass ejections in the near-Earth solar wind during 1996–2002. J. Geophys. Res. **108**, 1156 (2003). doi:10.1029/2002JA009817

P.J. Cargill, On the aerodynamic drag force acting on interplanetary coronal mass ejections. Sol. Phys. **221**, 135–149 (2004). doi:10.1023/B:SOLA.0000033366.10725.a2

P.J. Cargill, J. Chen, D.S. Spicer, S.T. Zalesak, Magnetohydrodynamic simulations of the motion of magnetic flux tubes through a magnetized plasma. J. Geophys. Res. **101**, 4855–4870 (1996). doi:10.1029/95JA03769

E. Chané, C. Jacobs, B. van der Holst, S. Poedts, D. Kimpe, On the effect of the initial magnetic polarity and of the background wind on the evolution of CME shocks. Astron. Astrophys. **432**, 331–339 (2005). doi:10.1051/0004-6361:20042005

P. Chatterjee, Y. Fan, Simulation of homologous and cannibalistic coronal mass ejections produced by the emergence of a twisted flux rope into the solar corona. Astrophys. J. Lett. **778**, L8 (2013)

P.F. Chen, Coronal mass ejections: models and their observational basis. Living Rev. Sol. Phys. **8**, 1 (2011). doi:10.12942/lrsp-2011-1

C. Cid, H. Cremades, A. Aran, C. Mandrini, B. Sanahuja, B. Schmieder, M. Menvielle, L. Rodriguez, E. Saiz, Y. Cerrato, S. Dasso, C. Jacobs, C. Lathuillere, A. Zhukov, Can a halo CME from the limb be geoeffective? J. Geophys. Res. **117**(A16), A11102 (2012). doi:10.1029/2012JA017536

O. Cohen, G.D.R. Attrill, W.B. Manchester IV, M.J. Wills-Davey, Numerical simulation of an EUV coronal wave based on the 2009 February 13 CME event observed by STEREO. Astrophys. J. **705**, 587 (2009). arXiv:0909.3095. doi:10.1088/0004-637X/705/1/587

O. Cohen, G.D.R. Attrill, N.A. Schwadron, N.U. Crooker, M.J. Owens, C. Downs, T.I. Gombosi, Numerical simulation of the 12 May 1997 CME event: the role of magnetic reconnection. J. Geophys. Res. **115**, A10104 (2010). arXiv:1006.4337. doi:10.1029/2010JA015464

R.C. Colaninno, A. Vourlidas, First determination of the true mass of coronal mass ejections: a novel approach to using the two STEREO viewpoints. Astrophys. J. **698**, 852–858 (2009). arXiv:0903.4344. doi:10.1088/0004-637X/698/1/852

R.C. Colaninno, A. Vourlidas, Using multiple-viewpoint observations to determine the interaction of three coronal mass ejections observed on 2012 March 5. Astrophys. J. **815**, 70 (2015). doi:10.1088/0004-637X/815/1/70

M.R. Collier, R.P. Lepping, D.B. Berdichevsky, A statistical study of interplanetary shocks and pressure pulses internal to magnetic clouds. J. Geophys. Res. **112**, A06102 (2007). doi:10.1029/2006JA011714

H. Cremades, V. Bothmer, D. Tripathi, Properties of structured coronal mass ejections in Solar Cycle 23. Adv. Space Res. **38**, 461–465 (2006). doi:10.1016/j.asr.2005.01.095

S. Dasso, C.H. Mandrini, P. Démoulin, M.L. Luoni, A new model-independent method to compute magnetic helicity in magnetic clouds. Astron. Astrophys. **455**, 349–359 (2006). doi:10.1051/0004-6361:20064806

S. Dasso, M.S. Nakwacki, P. Démoulin, C.H. Mandrini, Progressive transformation of a flux rope to an ICME. Comparative analysis using the direct and fitted expansion methods. Sol. Phys. **244**, 115–137 (2007). arXiv:0706.2889. doi:10.1007/s11207-007-9034-2

S. Dasso, C.H. Mandrini, B. Schmieder, H. Cremades, C. Cid, Y. Cerrato, E. Saiz, P. Démoulin, A.N. Zhukov, L. Rodriguez, A. Aran, M. Menvielle, S. Poedts, Linking two consecutive nonmerging magnetic clouds with their solar sources. J. Geophys. Res. **114**, A02109 (2009). doi:10.1029/2008JA013102

J.A. Davies, C.H. Perry, R.M.G.M. Trines, R.A. Harrison, N. Lugaz, C. Möstl, Y.D. Liu, K. Steed, Establishing a stereoscopic technique for determining the kinematic properties of solar wind transients based on a generalized self-similarly expanding circular geometry. Astrophys. J. **777**, 167 (2013). doi:10.1088/0004-637X/777/2/167

C.J. Davis, J.A. Davies, M. Lockwood, A.P. Rouillard, C.J. Eyles, R.A. Harrison, Stereoscopic imaging of an Earth-impacting solar coronal mass ejection: a major milestone for the STEREO mission. Geophys. Res. Lett. **36**, L08102 (2009). doi:10.1029/2009GL038021

D.S. De Young, A.J. Hundhausen, Two-dimensional simulation of flare-associated disturbances in the solar wind. J. Geophys. Res. **76**, 2245 (1971). doi:10.1029/JA076i010p02245

D.S. De Young, A.J. Hundhausen, Simulation of driven flare-associated disturbances in the solar wind. J. Geophys. Res. **78**, 3633 (1973). doi:10.1029/JA078i019p03633

C.E. DeForest, T.A. Howard, S.J. Tappin, Observations of detailed structure in the solar wind at 1 AU with STEREO/HI-2. Astrophys. J. **738**, 103 (2011). arXiv:1104.1615. doi:10.1088/0004-637X/738/1/103

C.E. DeForest, T.A. Howard, D.J. McComas, Tracking coronal features from the low corona to Earth: a quantitative analysis of the 2008 December 12 coronal mass ejection. Astrophys. J. **769**, 43 (2013). doi:10.1088/0004-637X/769/1/43

P. Démoulin, A review of the quantitative links between CMEs and magnetic clouds. Ann. Geophys. **26**, 3113–3125 (2008). doi:10.5194/angeo-26-3113-2008

P. Démoulin, G. Aulanier, Criteria for flux rope eruption: non-equilibrium versus torus instability. Astrophys. J. **718**, 1388–1399 (2010). arXiv:1006.1785. doi:10.1088/0004-637X/718/2/1388

C. Downs, I.I. Roussev, B. van der Holst, N. Lugaz, I.V. Sokolov, T.I. Gombosi, Toward a realistic thermodynamic magnetohydrodynamic model of the global solar corona. Astrophys. J. **712**, 1219–1231 (2010). doi:10.1088/0004-637X/712/2/1219

C. Downs, I.I. Roussev, B. van der Holst, N. Lugaz, I.V. Sokolov, Understanding SDO/AIA observations of the 2010 June 13 EUV wave event: direct insight from a global thermodynamic MHD simulation. Astrophys. J. **750**, 134 (2012). doi:10.1088/0004-637X/750/2/134

M. Dryer, D.F. Smart, Dynamical models of coronal transients and interplanetary disturbances. Adv. Space Res. **4**, 291–301 (1984). doi:10.1016/0273-1177(84)90573-8

M. Dryer, R.S. Steinolfson, J.D. Mihalov, J.H. Wolfe, J. Chao, Z.K. Smith, Interplanetary disturbances caused by the August 1972 solar flares as observed by Pioneer 9. J. Geophys. Res. **81**, 4651–4663 (1976). doi:10.1029/JA081i025p04651

M. Dryer, Z.K. Smith, S.T. Wu, J.F. Wang, G. Gislason, S.M. Han, D.F. Smart, M.A. Shea, Magnetohydrodynamic modelling of interplanetary disturbances between the Sun and Earth. Astrophys. Space Sci. **105**, 187–208 (1984). doi:10.1007/BF00651218

M. Dryer, Z.K. Smith, S.T. Wu, The role of magnetohydrodynamics in heliospheric space plasma physics research. Astrophys. Space Sci. **144**, 407–425 (1988). doi:10.1007/BF00793195

M. Dumbović, A. Devos, B. Vršnak, D. Sudar, L. Rodriguez, D. Ruždjak, K. Leer, S. Vennerstrøm, A. Veronig, Geoeffectiveness of coronal mass ejections in the SOHO era. Sol. Phys. **290**, 579–612 (2015). arXiv:1410.3303. doi:10.1007/s11207-014-0613-8

M. Dumbović, B. Vršnak, J. Čalogović, Forbush decrease prediction based on remote solar observations. Sol. Phys. **291**, 285–302 (2016). arXiv:1510.03282. doi:10.1007/s11207-015-0819-4

R. Esser, R.J. Edgar, Differential flow speeds of ions of the same element: effects on solar wind ionization fractions. Astrophys. J. **563**, 1055–1062 (2001). doi:10.1086/323987

C.J. Eyles, G.M. Simnett, M.P. Cooke, B.V. Jackson, A. Buffington, P.P. Hick, N.R. Waltham, J.M. King, P.A. Anderson, P.E. Holladay, The Solar Mass Ejection Imager (SMEI). Sol. Phys. **217**, 319–347 (2003). doi:10.1023/B:SOLA.0000006903.75671.49

C.J. Eyles, R.A. Harrison, C.J. Davis, N.R. Waltham, B.M. Shaughnessy, H.C.A. Mapson-Menard, D. Bewsher, S.R. Crothers, J.A. Davies, G.M. Simnett, R.A. Howard, J.D. Moses, J.S. Newmark, D.G. Socker, J.P. Halain, J.M. Defise, E. Mazy, P. Rochus, The heliospheric imagers onboard the STEREO mission. Sol. Phys. **254**, 387–445 (2009). doi:10.1007/s11207-008-9299-0

T.V. Falkenberg, B. Vršnak, A. Taktakishvili, D. Odstrcil, P. MacNeice, M. Hesse, Investigations of the sensitivity of a coronal mass ejection model (ENLIL) to solar input parameters. Space Weather **8**, S06,004 (2010). doi:10.1029/2009SW000555

Y. Fan, Modeling the initiation of the 2006 December 13 coronal mass ejection in AR 10930: the structure and dynamics of the erupting flux rope. Astrophys. J. **824**, 93 (2016). arXiv:1604.05687. doi:10.3847/0004-637X/824/2/93

Y. Fan, S.E. Gibson, Numerical simulations of three-dimensional coronal magnetic fields resulting from the emergence of twisted magnetic flux tubes. Astrophys. J. **609**, 1123–1133 (2004). doi:10.1086/421238

F. Fang, W.B. Manchester IV, W.P. Abbett, B. van der Holst, Simulation of flux emergence from the convection zone to the corona. Astrophys. J. **714**, 1649–1657 (2010). doi:10.1088/0004-637X/714/2/1649

F. Fang, W.B. Manchester IV, W.P. Abbett, B. van der Holst, Buildup of magnetic shear and free energy during flux emergence and cancellation. Astrophys. J. **754**, 15 (2012). doi:10.1088/0004-637X/754/1/15

C. Farrugia, D. Berdichevsky, Evolutionary signatures in complex ejecta and their driven shocks. Ann. Geophys. **22**, 3679–3698 (2004)

C.J. Farrugia, L.F. Burlaga, V.A. Osherovich, I.G. Richardson, M.P. Freeman, R.P. Lepping, A.J. Lazarus, A study of an expanding interplanetary magnetic cloud and its interaction with the Earth's magnetosphere—the interplanetary aspect. J. Geophys. Res. **98**, 7621–7632 (1993)

C.J. Farrugia, S. Mühlbachler, H.K. Biernat, R.B. Torbert, Dayside erosion during intervals of tenuous solar wind. J. Geophys. Res. **106**, 25,517–25,528 (2001). doi:10.1029/2001JA000087

X. Feng, X. Zhao, A new prediction method for the arrival time of interplanetary shocks. Sol. Phys. **238**, 167–186 (2006). doi:10.1007/s11207-006-0185-3

L. Feng, Y. Wang, F. Shen, C. Shen, B. Inhester, L. Lu, W. Gan, Why does the apparent mass of a coronal mass ejection increase? Astrophys. J. **812**, 70 (2015). arXiv:1509.02246. doi:10.1088/0004-637X/812/1/70

E.E. Fenimore, Solar wind flows associated with hot heavy ions. Astrophys. J. **235**, 245–257 (1980). doi:10.1086/157628

T.G. Forbes, A review on the genesis of coronal mass ejections. J. Geophys. Res. **105**(A10), 23,153–23,165 (2000)

T.G. Forbes, P.A. Isenberg, A catastrophe mechanism for coronal mass ejection. Astrophys. J. **373**, 294–307 (1991)

T.G. Forbes, J.A. Linker, J. Chen, C. Cid, J. Kóta, M.A. Lee, G. Mann, Z. Mikić, M.S. Potgieter, J.M. Schmidt, G.L. Siscoe, R. Vainio, S.K. Antiochos, P. Riley, CME theory and models. Space Sci. Rev. **123**, 251–302 (2006). doi:10.1007/s11214-006-9019-8

J. Fuller, S.E. de Gibson, G. Toma, Y. Fan, Observing the unobservable? Modeling coronal cavity densities. Astrophys. J. **678**, 515–530 (2008). doi:10.1086/533527

A.B. Galvin, Minor ion composition in CME-related solar wind, in *Washington DC American Geophysical Union Geophysical Monograph Series*, vol. 99 (1997), pp. 253–260. doi:10.1029/GM099p0253

A.B. Galvin, L.M. Kistler, M.A. Popecki, C.J. Farrugia, K.D.C. Simunac, L. Ellis, E. Möbius, M.A. Lee, M. Boehm, J. Carroll, A. Crawshaw, M. Conti, P. Demaine, S. Ellis, J.A. Gaidos, J. Googins, M. Granoff, A. Gustafson, D. Heirtzler, B. King, U. Knauss, J. Levasseur, S. Longworth, K. Singer, S. Turco, P. Vachon, M. Vosbury, M. Widholm, L.M. Blush, R. Karrer, P. Bochsler, H. Daoudi, A. Etter, J. Fischer, J. Jost, A. Opitz, M. Sigrist, P. Wurz, B. Klecker, M. Ertl, E. Seidenschwang, R.F. Wimmer-Schweingruber, M. Koeten, B. Thompson, D. Steinfeld, The Plasma and Suprathermal Ion Composition (PLASTIC) investigation on the STEREO observatories. Space Sci. Rev. **136**, 437–486 (2008). doi:10.1007/s11214-007-9296-x

S. Gibson, B.C. Low, A time-dependent three-dimensional magnetohydrodynamic model of the coronal mass ejection. Astrophys. J. **493**, 460–473 (1998). doi:10.1086/305107

S.E. Gibson, T.A. Kucera, D. Rastawicki, J. Dove, G. de Toma, J. Hao, S. Hill, H.S. Hudson, C. Marqué, P.S. McIntosh, L. Rachmeler, K.K. Reeves, B. Schmieder, D.J. Schmit, D.B. Seaton, A.C. Sterling, D. Tripathi, D.R. Williams, M. Zhang, Three-dimensional morphology of a coronal prominence cavity. Astrophys. J. **724**, 1133–1146 (2010). doi:10.1088/0004-637X/724/2/1133

H.R. Gilbert, T.E. Holzer, J.T. Burkepile, A.J. Hundhausen, Active and eruptive prominences and their relationship to coronal mass ejections. Astrophys. J. **537**, 503–515 (2000). doi:10.1086/309030

G. Gloeckler, P. Bedini, L.A. Fisk, J.C. Cain, E.O. Tums, P. Bochsler, J. Fischer, R.F. Wimmer-Schweingruber, J. Geiss, R. Kallenbach, Investigation of the composition of solar and interstellar matter using solar wind and pickup ion measurements with SWICS and SWIMS on the ACE spacecraft. Space Sci. Rev. **86**, 497–539 (1998). doi:10.1023/A:1005036131689

N. Gopalswamy, Consequences of coronal mass ejections in the heliosphere. Sun Geosph. **1**(2), 5–12 (2006)

N. Gopalswamy, Coronal mass ejections and space weather, in *Climate and Weather of the Sun-Earth System (CAWSES): Selected Papers from the 2007 Kyoto Symposium*, ed. by T. Tsuda, R. Fujii, K. Shibata, M.A. Geller (2009), pp. 77–120

N. Gopalswamy, Y. Hanaoka, T. Kosugi, R.P. Lepping, J.T. Steinberg, S. Plunkett, R.A. Howard, B.J. Thompson, J. Gurman, G. Ho, N. Nitta, H.S. Hudson, On the relationship between coronal mass ejections and magnetic clouds. Geophys. Res. Lett. **25**, 2485–2488 (1998). doi:10.1029/98GL50757

N. Gopalswamy, A. Lara, R.P. Lepping, M.L. Kaiser, D. Berdichevsky, O.C. St. Cyr, Interplanetary acceleration of coronal mass ejections. Geophys. Res. Lett. **27**, 145–148 (2000). doi:10.1029/1999GL003639

N. Gopalswamy, A. Lara, M.L. Kaiser, J.L. Bougeret, Near-Sun and near-Earth manifestations of solar eruptions. J. Geophys. Res. **106**, 25,261–25,278 (2001a)

N. Gopalswamy, A. Lara, S. Yashiro, M.L. Kaiser, R.A. Howard, Predicting the 1-AU arrival times of coronal mass ejections. J. Geophys. Res. **106**, 29,207–29,218 (2001b). doi:10.1029/2001JA000177

N. Gopalswamy, S. Yashiro, M.L. Kaiser, R.A. Howard, J.L. Bougeret, Radio signatures of coronal mass ejection interaction: coronal mass ejection cannibalism? Astrophys. J. Lett. **548**, L91–L94 (2001c)

N. Gopalswamy, S. Yashiro, G. Michaek, M.L. Kaiser, R.A. Howard, D.V. Reames, R. Leske, T. von Rosenvinge, Interacting coronal mass ejections and solar energetic particles. Astrophys. J. **572**, L103–L108 (2002)

N. Gopalswamy, A. Lara, P.K. Manoharan, R.A. Howard, An empirical model to predict the 1-AU arrival of interplanetary shocks. Adv. Space Res. **36**, 2289–2294 (2005). doi:10.1016/j.asr.2004.07.014

N. Gopalswamy, Z. Mikić, D. Maia, D. Alexander, H. Cremades, P. Kaufmann, D. Tripathi, Y.M. Wang, The Pre-CME Sun. Space Sci. Rev. **123**, 303–339 (2006). doi:10.1007/s11214-006-9020-2

N. Gopalswamy, P. Mäkelä, H. Xie, S. Akiyama, S. Yashiro, CME interactions with coronal holes and their interplanetary consequences. J. Geophys. Res. **114**, A00A22 (2009a). doi:10.1029/2008JA013686

N. Gopalswamy, S. Yashiro, H. Xie, S. Akiyama, P. Mäkelä, Large geomagnetic storms associated with limb coronal mass ejections. Adv. Geosci. **21** (2009b)

N. Gopalswamy, H. Xie, P. Mäkelä, S. Yashiro, S. Akiyama, W. Uddin, A.K. Srivastava, N.C. Joshi, R. Chandra, P.K. Manoharan, K. Mahalakshmi, V.C. Dwivedi, R. Jain, A.K. Awasthi, N.V. Nitta, M.J. Aschwanden, D.P. Choudhary, Height of shock formation in the solar corona inferred from observations of type II radio bursts and coronal mass ejections. Adv. Space Res. **51**, 1981–1989 (2013). arXiv:1301.0893. doi:10.1016/j.asr.2013.01.006

N. Gopalswamy, S. Yashiro, N. Thakur, P. Mäkelä, H. Xie, S. Akiyama, The 2012 July 23 backside eruption: an extreme energetic particle event? Astrophys. J. **833**, 216 (2016). arXiv:1610.05790. doi:10.3847/1538-4357/833/2/216

J.T. Gosling, Coronal mass ejections and magnetic flux ropes in interplanetary space, in *Physics of Magnetic Flux Ropes*, ed. by C.T. Russell, E.R. Priest, L.C. Lee. AGU Geophys. Monograph Ser., vol. 58 (AGU, Washington, D.C., 1990), pp. 343–364
J.T. Gosling, Coronal mass ejections—the link between solar and geomagnetic activity. Phys. Fluids, B Plasma Phys. **5**, 2638–2645 (1993a). doi:10.1063/1.860701
J.T. Gosling, The solar flare myth. J. Geophys. Res. **98**, 18,937–18,950 (1993b). doi:10.1029/93JA01896
J.T. Gosling, On the determination of electron polytrope indices within coronal mass ejections in the solar wind. J. Geophys. Res. **104**, 19,851–19,858 (1999). doi:10.1029/1999JA900254
J.T. Gosling, Coronal mass ejections, in *Proc. 26th International Cosmic Ray Conference*, ed. by B.L. Dingus, D. Kieda, M. Salamon. AIP Conference Proceedings, vol. 516 (2000), pp. 59–79
J.T. Gosling, D.J. McComas, Field line draping about fast coronal mass ejecta: a source of strong out-of-the-ecliptic interplanetary magnetic clouds. Geophys. Res. Lett. **14**, 355 (1987)
J.T. Gosling, P. Riley, The acceleration of slow coronal mass ejections in the high-speed solar wind. Geophys. Res. Lett. **23**, 2867–2870 (1996). doi:10.1029/96GL02843
J.T. Gosling, J.R. Asbridge, S.J. Bame, W.C. Feldman, R.D. Zwickl, Observations of large fluxes of He^+ in the solar wind following an interplanetary shock. J. Geophys. Res. **85**, 3431–3434 (1980). doi:10.1029/JA085iA07p03431
J.T. Gosling, M.F. Thomsen, S.J. Bame, R.D. Zwickl, The eastward deflection of fast coronal mass ejecta in interplanetary space. J. Geophys. Res. **92**, 12,399–12,406 (1987). doi:10.1029/JA092iA11p12399
J.T. Gosling, S.J. Bame, D.J. McComas, J.L. Phillips, E.E. Scime, V.J. Pizzo, B.E. Goldstein, A. Balogh, A forward-reverse shock pair in the solar wind driven by over-expansion of a coronal mass ejections: Ulysses observations. Geophys. Res. Lett. **21**(3), 237–240 (1994)
J.T. Gosling, P. Riley, D.J. McComas, V.J. Pizzo, Overexpanding coronal mass ejections at high heliographic latitudes: observations and simulations. J. Geophys. Res. **103**, 1941 (1998)
J.T. Gosling, R.M. Skoug, W.C. Feldman, Solar wind electron halo depletions at 90 degree pitch angle. Geophys. Res. Lett. **28**, 4155–4158 (2001). doi:10.1029/2001GL013758
J.T. Gosling, R.M. Skoug, D.J. McComas, C.W. Smith, Direct evidence for magnetic reconnection in the solar wind near 1 AU. J. Geophys. Res. **A01**, 107 (2005). doi:10.1029/2004JA010809
J.T. Gosling, D.J. McComas, R.M. Skoug, C.W. Smith, Magnetic reconnection at the heliospheric current sheet and the formation of closed magnetic field lines in the solar wind. Geophys. Res. Lett. **33**, L17102 (2006). doi:10.1029/2006GL027188
L.M. Green, B. Kliem, T. Török, L. van Driel-Gesztelyi, G.D.R. Attrill, Transient coronal sigmoids and rotating erupting flux ropes. Sol. Phys. **246**, 365–391 (2007). doi:10.1007/s11207-007-9061-z
J.R. Gruesbeck, S.T. Lepri, T.H. Zurbuchen, S.K. Antiochos, Constraints on coronal mass ejection evolution from in situ observations of ionic charge states. Astrophys. J. **730**, 103 (2011). doi:10.1088/0004-637X/730/2/103
J.R. Gruesbeck, S.T. Lepri, T.H. Zurbuchen, Two-plasma model for low charge state interplanetary coronal mass ejection observations. Astrophys. J. **760**, 141 (2012). doi:10.1088/0004-637X/760/2/141
B. Gui, C. Shen, Y. Wang, P. Ye, J. Liu, S. Wang, X. Zhao, Quantitative analysis of CME deflections in the corona. Sol. Phys. **271**, 111–139 (2011). arXiv:1105.3382. doi:10.1007/s11207-011-9791-9
A.M. Gulisano, P. Démoulin, S. Dasso, M.E. Ruiz, E. Marsch, Global and local expansion of magnetic clouds in the inner heliosphere. Astron. Astrophys. **509**, A39 (2010). doi:10.1051/0004-6361/200912375
C.M. Hammond, G.K. Crawford, J.T. Gosling, H. Kojima, J.L. Phillips, H. Matsumoto, A. Balogh, L.A. Frank, S. Kokubun, T. Yamamoto, Latitudinal structure of a coronal mass ejection inferred from ULYSSES and GEOTAIL observations. Geophys. Res. Lett. **22**, 1169–1172 (1995). doi:10.1029/95GL01016
R.A. Harrison, J.A. Davies, C. Möstl, Y. Liu, M. Temmer, M.M. Bisi, J.P. de Eastwood, C.A. Koning, N. Nitta, T. Rollett, C.J. Farrugia, R.J. Forsyth, B.V. Jackson, E.A. Jensen, E.K.J. Kilpua, D. Odstrcil, D.F. Webb, An analysis of the origin and propagation of the multiple coronal mass ejections of 2010 August 1. Astrophys. J. **750**, 45 (2012). doi:10.1088/0004-637X/750/1/45
T. Henke, J. Woch, U. Mall, S. Livi, B. Wilken, R. Schwenn, G. Gloeckler, R. von Steiger, R.J. Forsyth, A. Balogh, Differences in the O^{7+}/O^{6+} ratio of magnetic cloud and non-cloud coronal mass ejections. Geophys. Res. Lett. **25**, 3465–3468 (1998). doi:10.1029/98GL02632
T. Henke, J. Woch, R. Schwenn, U. Mall, G. Gloeckler, R. von Steiger, R.J. Forsyth, A. Balogh, Ionization state and magnetic topology of coronal mass ejections. J. Geophys. Res. **106**, 10,597–10,614 (2001). doi:10.1029/2000JA900176
J. Hirshberg, S.J. Bame, D.E. Robbins, Solar flares and solar wind helium enrichments: July 1965 July 1967. Sol. Phys. **23**, 467–486 (1972). doi:10.1007/BF00148109
T.A. Howard, C.E. DeForest, Inner heliospheric flux rope evolution via imaging of coronal mass ejections. Astrophys. J. **746**, 64 (2012). doi:10.1088/0004-637X/746/1/64

R.A. Howard, G.E. Brueckner, O.C. St. Cyr, D.A. Biesecker, K.P. Dere, M.J. Koomen, C.M. Korendyke, P.L. Lamy, A. Llebaria, M.V. Bout, D.J. Michels, J.D. Moses, S.E. Paswaters, S.P. Plunkett, R. Schwenn, G.M. Simnett, D.G. Socker, S.J. Tappin, D. Wang, Observations of CMEs from SOHO/LASCO, in *Washington DC American Geophysical Union Geophysical Monograph Series*, vol. 99 (1997), pp. 17–26. doi:10.1029/GM099p0017

R.A. Howard, J.D. Moses, A. Vourlidas, J.S. Newmark, D.G. Socker, S.P. Plunkett, C.M. Korendyke, J.W. Cook, A. Hurley, J.M. Davila, W.T. Thompson, O.C. St. Cyr, E. Mentzell, K. Mehalick, J.R. Lemen, J.P. Wuelser, D.W. Duncan, T.D. Tarbell, C.J. Wolfson, A. Moore, R.A. Harrison, N.R. Waltham, J. Lang, C.J. Davis, C.J. Eyles, H. Mapson-Menard, G.M. Simnett, J.P. Halain, J.M. Defise, E. Mazy, P. Rochus, R. Mercier, M.F. Ravet, F. Delmotte, F. Auchere, J.P. Delaboudiniere, V. Bothmer, W. Deutsch, D. Wang, N. Rich, S. Cooper, V. Stephens, G. Maahs, R. Baugh, D. McMullin, T. Carter, Sun Earth Connection Coronal and Heliospheric Investigation (SECCHI). Space Sci. Rev. **136**, 67–115 (2008). doi:10.1007/s11214-008-9341-4

T.A. Howard, M.M. Bisi, A. Buffington, J.M. Clover, M.P. Cooke, C.J. Eyles, P.P. Hick, P.E. Holladay, B.V. Jackson, J.C. Johnston, S.W. Kahler, T.A. Kuchar, D.R. Mizuno, A.J. Penny, S.D. Price, R.R. Radick, G.M. Simnett, S.J. Tappin, N.R. Waltham, D.F. Webb, The Solar Mass Ejection Imager and its heliospheric imaging legacy. Space Sci. Rev. **180**, 1–38 (2013). doi:10.1007/s11214-013-9992-7

T.A. Howard, C.E. DeForest, U.G. Schneck, C.R. Alden, Challenging some contemporary views of coronal mass ejections. II. The case for absent filaments. Astrophys. J. **834**, 86 (2017). doi:10.3847/1538-4357/834/1/86

Q. Hu, B.U.Ö. Sonnerup, Reconstruction of magnetic clouds in the solar wind: orientations and configurations. J. Geophys. Res. **107**, 1142 (2002). doi:10.1029/2001JA000293

H. Hu, Y.D. Liu, R. Wang, C. Möstl, Z. Yang, Sun-to-Earth characteristics of the 2012 July 12 coronal mass ejection and associated geo-effectiveness. Astrophys. J. **829**, 97 (2016). arXiv:1607.06287. doi:10.3847/0004-637X/829/2/97

A.J. Hundhausen, The origin and propagation of coronal mass ejections, in *Solar Wind Six*, ed. by V.J. Pizzo, T.E. Holzer, D.G. Sime (NCAR, Boulder, 1987), pp. 181–214

A.J. Hundhausen, Sizes and locations of coronal mass ejections—SMM observations from 1980 and 1984–1989. J. Geophys. Res. **98**, 13,177 (1993)

A.J. Hundhausen, H. Gilbert, S. Bame, Ionization state of the interplanetary plasma. J. Geophys. Res. **73**, 5485 (1968)

A.J. Hundhausen, C.B. Sawyer, L. House, R.M.E. Illing, W.J. Wagner, Coronal mass ejections observed during the solar maximum mission—latitude distribution and rate of occurrence. J. Geophys. Res. **89**, 2639–2646 (1984). doi:10.1029/JA089iA05p02639

A.J. Hundhausen, J.T. Burkpile, O.C. St. Cyr, Speeds of coronal mass ejections: SMM observations from 1980 and 1984–1989. J. Geophys. Res. **99**(A4), 6543–6552 (1994)

K.E.J. Huttunen, H.E.J. Koskinen, T.I. Pulkkinen, A. Pulkkinen, M. Palmroth, E.G.D. Reeves, H.J. Singer, April 2000 magnetic storm: solar wind driver and magnetospheric response. J. Geophys. Res. **107**, 1440 (2002). doi:10.1029/2001JA009154

R.M.E. Illing, A.J. Hundhausen, Observation of a coronal transient from 1.2 to 6 solar radii. J. Geophys. Res. **90**, 275–282 (1985). doi:10.1029/JA090iA01p00275

D.S. Intriligator, The August 1972 solar-terrestrial events—solar wind plasma observations. Space Sci. Rev. **19**, 629–660 (1976). doi:10.1007/BF00210644

A. Isavnin, A. Vourlidas, E.K.J. Kilpua, Three-dimensional evolution of flux-rope CMEs and its relation to the local orientation of the heliospheric current sheet. Sol. Phys. **289**, 2141–2156 (2014). arXiv:1312.0458. doi:10.1007/s11207-013-0468-4

P.A. Isenberg, T.G. Forbes, A three-dimensional line-tied magnetic field model for solar eruptions. Astrophys. J. **670**, 1453–1466 (2007). doi:10.1086/522025

K.G. Ivanov, A study of some interplanetary shock wave tendencies. Space Sci. Rev. **32**, 49–63 (1982). doi:10.1007/BF00225176

B.V. Jackson, A. Buffington, P.P. Hick, R.C. Altrock, S. Figueroa, P.E. Holladay, J.C. Johnston, S.W. Kahler, J.B. Mozer, S. Price, R.R. Radick, R. Sagalyn, D. Sinclair, G.M. Simnett, C.J. Eyles, M.P. Cooke, S.J. Tappin, T. Kuchar, D. Mizuno, D.F. Webb, P.A. Anderson, S.L. Keil, R.E. Gold, N.R. Waltham, The Solar Mass-Ejection Imager (SMEI) mission. Sol. Phys. **225**, 177–207 (2004). doi:10.1007/s11207-004-2766-3

B. Jackson, A. Buffington, P. Hick, D. Webb, Preliminary three-dimensional analysis of the heliospheric response to the 28 October 2003 CME using SMEI white-light observations. J. Geophys. Res. **111**, A04S91 (2006)

C. Jacobs, I.I. Roussev, N. Lugaz, S. Poedts, The internal structure of coronal mass ejections: are all regular magnetic clouds flux ropes? Astrophys. J. Lett. **695**, L171–L175 (2009). doi:10.1088/0004-637X/695/2/L171

C. Jiang, S.T. Wu, X. Feng, Q. Hu, Data-driven magnetohydrodynamic modelling of a flux-emerging active region leading to solar eruption. Nat. Commun. **7**, 11522 (2016). doi:10.1038/ncomms11522

M. Jin, W.B. Manchester IV, B. van der Holst, J.R. Gruesbeck, R.A. Frazin, E. Landi, A.M. Vasquez, P.L. Lamy, A. Llebaria, A. Fedorov, G. Toth, T. Gombosi, A global two-temperature corona and inner heliosphere model: a comprehensive validation study. Astrophys. J. **7745**, 6 (2012). doi:10.1088/0004-637X/745/1/6

M. Jin, W.B. Manchester IV, B. van der Holst, R. Oran, I. Sokolov, G. Toth, T. Gombosi, Y. Liu, X. Sun, Numerical simulations of coronal mass ejection on 2011 march 7: one-temperature and two-temperature model comparison. Astrophys. J. **773**, 50 (2013). doi:10.1088/0004-637X/773/1/50

M. Jin, W.B. Manchester IV, B. van der Holst, I. Sokolov, G. Tóth, R.E. Mullinix, A. Taktakishvili, A. Chulaki, T.I. Gombosi, Data-constrained coronal mass ejections in a global magnetohydrodynamics model. Astrophys. J. **834**, 173 (2017a). arXiv:1605.05360. doi:10.3847/1538-4357/834/2/173

M. Jin, W.B. Manchester IV, B. van der Holst, I. Sokolov, G. Tóth, A. Vourlidas, C.A. de Koning, Chromosphere to 1 AU simulation of the 2011 March 7th event: a comprehensive study of coronal mass ejection propagation. Astrophys. J. **834**, 172 (2017b). arXiv:1611.08897. doi:10.3847/1538-4357/834/2/172

R.A. Jones, A.R. Breen, R.A. Fallows, A. Canals, M.M. Bisi, G. Lawrence, Interaction between coronal mass ejections and the solar wind. J. Geophys. Res. **112**, A08107 (2007). doi:10.1029/2006JA011875

S. Kahler, Coronal mass ejections. Rev. Geophys. **25**, 663–675 (1987). doi:10.1029/RG025i003p00663

S.W. Kahler, Solar flares and coronal mass ejections. Annu. Rev. Astron. Astrophys. **30**, 113 (1992)

S.W. Kahler, D.F. Webb, V arc interplanetary coronal mass ejections observed with the Solar Mass Ejection Imager. J. Geophys. Res. **112**, A09103 (2007). doi:10.1029/2007JA012358

M.L. Kaiser, T.A. Kucera, J.M. Davila, O.C. St. Cyr, M. Guhathakurta, E. Christian, The STEREO mission: an introduction. Space Sci. Rev. **136**, 5–16 (2008). doi:10.1007/s11214-007-9277-0

C. Kay, M. Opher, The heliocentric distance where the deflections and rotations of solar coronal mass ejections occur. Astrophys. J. Lett. **811**, L36 (2015). arXiv:1509.04948. doi:10.1088/2041-8205/811/2/L36

C. Kay, M. Opher, R.M. Evans, Forecasting a coronal mass ejection's altered trajectory: ForeCAT. Astrophys. J. **775**, 5 (2013). arXiv:1307.7603. doi:10.1088/0004-637X/775/1/5

C. Kay, M. Opher, R.M. Evans, Global trends of CME deflections based on CME and solar parameters. Astrophys. J. **805**, 168 (2015). arXiv:1410.4496. doi:10.1088/0004-637X/805/2/168

C. Kay, M. Opher, R.C. Colaninno, A. Vourlidas, Using ForeCAT deflections and rotations to constrain the early evolution of CMEs. Astrophys. J. **827**, 70 (2016). arXiv:1606.03460. doi:10.3847/0004-637X/827/1/70

E.K.J. Kilpua, J. Pomoell, A. Vourlidas, R. Vainio, J. Luhmann, Y. Li, P. Schroeder, A.B. Galvin, K. Simunac, STEREO observations of interplanetary coronal mass ejections and prominence deflection during solar minimum period. Ann. Geophys. **27**, 4491–4503 (2009). doi:10.5194/angeo-27-4491-2009

E.K.J. Kilpua, M. Mierla, L. Rodriguez, A.N. Zhukov, N. Srivastava, M.J. West, Estimating travel times of coronal mass ejections to 1 AU using multi-spacecraft coronagraph data. Sol. Phys. **279**, 477–496 (2012). doi:10.1007/s11207-012-0005-x

E.K.J. Kilpua, A. Isavnin, A. Vourlidas, H.E.J. Koskinen, L. Rodriguez, On the relationship between interplanetary coronal mass ejections and magnetic clouds. Ann. Geophys. **31**, 1251–1265 (2013). doi:10.5194/angeo-31-1251-2013

B. Kliem, T. Török, Torus instability. Phys. Rev. Lett. **255**, 002 (2006). arXiv:physics/0605217. doi:10.1103/PhysRevLett.96.255002

B. Kliem, V.S. Titov, T. Török, Formation of current sheets and sigmoidal structure by the kink instability of a magnetic loop. Astron. Astrophys. **413**, L23–L26 (2004). arXiv:astro-ph/0311199. doi:10.1051/0004-6361:20031690

B. Kliem, T. Török, W.T. Thompson, A parametric study of erupting flux rope rotation. Modeling the "Cartwheel CME" on 9 April 2008. Sol. Phys. **281**, 137–166 (2012). arXiv:1112.3389. doi:10.1007/s11207-012-9990-z

Y.K. Ko, G. Gloeckler, C.M.S. Cohen, A.B. Galvin, Solar wind ionic charge states during the Ulysses pole-to-pole pass. J. Geophys. Res. **104**(17), 005 (1999). 020. doi:10.1029/1999JA900112

A.G. Kosovichev, T.V. Stepanova, Numerical simulation of shocks in the heliosphere. Sov. Astron. **35**, 646 (1991)

J.U. Kozyra, W.B. Manchester IV, C.P. Escoubet, S.T. Lepri, M.W. Liemohn, W.D. Gonzalez, M.W. Thomsen, B.T. Tsurutani, Earth's collision with a solar filament on 21 January 2005: overview. J. Geophys. Res. **118**, 5967–5978 (2013). doi:10.1002/jgra.50567

A. Lara, A.I. Borgazzi, Dynamics of interplanetary CMEs and associated type II bursts, in *Universal Heliophysical Processes*, IAU Symposium, vol. 257, ed. by N. Gopalswamy, D.F. Webb (2009), pp. 287–290. doi:10.1017/S1743921309029421

B. Lavraud, E. Larroque, E. Budnik, V. Génot, J.E. Borovsky, M.W. Dunlop, C. Foullon, H. Hasegawa, C. Jacquey, K. Nykyri, A. Ruffenach, M.G.G.T. Taylor, I. Dandouras, H. Rème, Asymmetry of magnetosheath flows and magnetopause shape during low Alfvén Mach number solar wind. J. Geophys. Res. **118**, 1089–1100 (2013). doi:10.1002/jgra.50145

B. Lavraud, A. Ruffenach, A.P. Rouillard, P. Kajdic, W.B. Manchester IV, N. Lugaz, Geo-effectiveness and radial dependence of magnetic cloud erosion by magnetic reconnection. J. Geophys. Res. **119**, 26–35 (2014). doi:10.1002/2013JA019154

J.Y. Lee, J.C. Raymond, Low ionization state plasma in coronal mass ejections. Astrophys. J. **758**, 116 (2012). doi:10.1088/0004-637X/758/2/116

R.P. Lepping, C.C. Wu, On the variation of interplanetary magnetic cloud type through Solar Cycle 23: wind events. J. Geophys. Res. **112**(A11), A10103 (2007). doi:10.1029/2006JA012140

R.P. Lepping, L.F. Burlaga, J.A. Jones, Magnetic field structure of interplanetary magnetic clouds at 1 AU. J. Geophys. Res. **95**, 11,957–11,965 (1990)

R.P. Lepping, D.B. Berdichevsky, L.F. Burlaga, A.J. Lazarus, J. Kasper, M.D. Desch, C.C. Wu, D.V. Reames, H.J. Singer, C.W. Smith, K.L. Ackerson, The Bastille day magnetic clouds and upstream shocks: near-Earth interplanetary observations. Sol. Phys. **204**, 285–303 (2001). doi:10.1023/A:1014264327855

R.P. Lepping, D.B. Berdichevsky, C.C. Wu, A. Szabo, T. Narock, F. Mariani, A.J. Lazarus, A.J. Quivers, A summary of WIND magnetic clouds for years 1995–2003: model-fitted parameters, associated errors and classifications. Ann. Geophys. **24**, 215–245 (2006). doi:10.5194/angeo-24-215-2006

S.T. Lepri, T.H. Zurbuchen, Iron charge state distributions as an indicator of hot ICMEs: possible sources and temporal and spatial variations during solar maximum. J. Geophys. Res. **109**, A01112 (2004). doi:10.1029/2003JA009954

S.T. Lepri, T.H. Zurbuchen, Direct observational evidence of filament material within interplanetary coronal mass ejections. Astrophys. J. Lett. **723**, L22–L27 (2010). doi:10.1088/2041-8205/723/1/L22

S.T. Lepri, T.H. Zurbuchen, L.A. Fisk, I.G. Richardson, H.V. Cane, G. Gloeckler, Iron charge distribution as an identifier of interplanetary coronal mass ejections. J. Geophys. Res. **106**, 29,231–29,238 (2001) doi:10.1029/2001JA000014

S.T. Lepri, J.M. Laming, C.E. von Rakowski, R. Steiger, Spatially dependent heating and ionization in an ICME observed by both ACE and Ulysses. Astrophys. J. **760**, 105 (2012). doi:10.1088/0004-637X/760/2/105

G.M. Lindsay, J.G. Luhmann, C.T. Russell, J.T. Gosling, Relationships between coronal mass ejection speeds from coronagraph images and interplanetary characteristics of associated interplanetary coronal mass ejections. J. Geophys. Res. **104**, 12,515–12,524 (1999) doi:10.1029/1999JA900051

J.A. Linker, Z. Mikić, R. Lionello, P. Riley, T. Amari, D. Odstrcil, Flux cancellation and coronal mass ejections. Phys. Plasmas **10**, 1971–1978 (2003)

J. Linker, T. Torok, C. Downs, R. Lionello, V. Titov, R.M. Caplan, Z. Mikić, P. Riley, MHD simulation of the Bastille day event. AIP Conf. Ser. **1720**, 020002 (2016). doi:10.1063/1.4943803

R. Lionello, Z. Mikić, J.A. Linker, Stability of algorithms for waves with large flows. J. Comput. Phys. **152**, 346–358 (1999). doi:10.1006/jcph.1999.6250

R. Lionello, J.A. Linker, Z. Mikic, Multispectral emission of the sun during the first Whole Sun Month: magnetohydrodynamic simulations. Astrophys. J. **690**(1), 902–912 (2009). doi:10.1088/0004-637X/690/1/902

R. Lionello, C. Downs, J.A. Linker, T. Török, P. Riley, Z. Mikić, Magnetohydrodynamic simulations of interplanetary coronal mass ejections. Astrophys. J. **777**, 76 (2013). doi:10.1088/0004-637X/777/1/76

Y. Liu, J.D. Richardson, J.W. Belcher, A statistical study of the properties of interplanetary coronal mass ejections from 0.3 to 5.4 AU. Planet. Space Sci. **53**, 3–17 (2005). doi:10.1016/j.pss.2004.09.023

Y. Liu, J.D. Richardson, J.W. Belcher, J.C. Kasper, H.A. Elliott, Thermodynamic structure of collision-dominated expanding plasma: heating of interplanetary coronal mass ejections. J. Geophys. Res. **111**, A01102 (2006a). doi:10.1029/2005JA011329

Y. Liu, J.D. Richardson, J.W. Belcher, C. Wang, Q. Hu, J.C. Kasper, Constraints on the global structure of magnetic clouds: transverse size and curvature. J. Geophys. Res. **111**(A10), A12S03 (2006b). arXiv:physics/0606003. doi:10.1029/2006JA011890

Y. Liu, J.G. Luhmann, K.E.J. Huttunen, R.P. Lin, S.D. Bale, C.T. Russell, A.B. Galvin, Reconstruction of the 2007 May 22 magnetic cloud: how much can we trust the flux-rope geometry of CMEs? Astrophys. J. Lett. **677**, L133 (2008a). doi:10.1086/587839

Y. Liu, J.G. Luhmann, R. Müller-Mellin, P.C. Schroeder, L. Wang, R.P. Lin, S.D. Bale, Y. Li, M.H. Acuña, J.A. Sauvaud, A comprehensive view of the 2006 December 13 CME: from the Sun to interplanetary space. Astrophys. J. **689**, 563–571 (2008b). arXiv:0802.2423. doi:10.1086/592031

Y. Liu, W.B. Manchester IV, J.D. Richardson, J.G. Luhmann, R.P. Lin, S.D. Bale, Deflection flows ahead of ICMEs as an indicator of curvature and geoeffectiveness. J. Geophys. Res. **113**, A00B03 (2008c). doi:10.1029/2007JA012996

Y. Liu, J.G. Luhmann, S.D. Bale, R.P. Lin, Relationship between a coronal mass ejection-driven shock and a coronal metric type II burst. Astrophys. J. Lett. **691**, L151 (2009)

Y. Liu, J.A. Davies, J.G. Luhmann, A. Vourlidas, S.D. Bale, R.P. Lin, Geometric triangulation of imaging observations to track coronal mass ejections continuously out to 1 AU. Astrophys. J. Lett. **710**, L82–L87 (2010a). arXiv:1001.1352. doi:10.1088/2041-8205/710/1/L82

Y. Liu, A. Thernisien, J.G. Luhmann, A. Vourlidas, J.A. Davies, R.P. Lin, S.D. Bale, Reconstructing coronal mass ejections with coordinated imaging and in situ observations: global structure, kinematics, and implications for space weather forecasting. Astrophys. J. **722**, 1762–1777 (2010b). arXiv:1009.1414. doi:10.1088/0004-637X/722/2/1762

Y. Liu, J.G. Luhmann, S.D. Bale, R.P. Lin, Solar source and heliospheric consequences of the 2010 April 3 coronal mass ejection: a comprehensive view. Astrophys. J. **734**(2), 84 (2011). http://stacks.iop.org/0004-637X/734/i=2/a=84

Y.D. Liu, J.G. Luhmann, C. Möstl, J.C. Martinez-Oliveros, S.D. Bale, R.P. Lin, R.A. Harrison, M. Temmer, D.F. Webb, D. Odstrcil, Interactions between coronal mass ejections viewed in coordinated imaging and in situ observations. Astrophys. J. Lett. **746**, L15 (2012). doi:10.1088/2041-8205/746/2/L15

Y.D. Liu, J.G. Luhmann, N. Lugaz, C. Möstl, J.A. Davies, S.D. Bale, R.P. Lin, On Sun-to-Earth propagation of coronal mass ejections. Astrophys. J. **769**, 45 (2013). doi:10.1088/0004-637X/769/1/45

Y.D. Liu, J.G. Luhmann, P. Kajdič, E.K.J. Kilpua, N. Lugaz, N.V. Nitta, C. Möstl, B. Lavraud, S.D. Bale, C.J. Farrugia, A.B. Galvin, Observations of an extreme storm in interplanetary space caused by successive coronal mass ejections. Nat. Commun. **5**, 3481 (2014a). arXiv:1405.6088. doi:10.1038/ncomms4481

Y.D. Liu, J.D. Richardson, C. Wang, J.G. Luhmann, Propagation of the 2012 March coronal mass ejections from the Sun to heliopause. Astrophys. J. Lett. **788**, L28 (2014b). arXiv:1405.6086. doi:10.1088/2041-8205/788/2/L28

Y.D. Liu, Z. Yang, R. Wang, J.G. Luhmann, J.D. Richardson, N. Lugaz, Sun-to-Earth characteristics of two coronal mass ejections interacting near 1 AU: formation of a complex ejecta and generation of a two-step geomagnetic storm. Astrophys. J. Lett. **793**, L41 (2014c). doi:10.1088/2041-8205/793/2/L41

Y.D. Liu, H. Hu, R. Wang, Z. Yang, B. Zhu, Y.A. Liu, J.G. Luhmann, J.D. Richardson, Plasma and magnetic field characteristics of solar coronal mass ejections in relation to geomagnetic storm intensity and variability. Astrophys. J. **809**, L34 (2015). arXiv:1508.01267. doi:10.1088/2041-8205/809/2/L34

Y.D. Liu, H. Hu, C. Wang, J.G. Luhmann, J.D. Richardson, Z. Yang, R. Wang, On Sun-to-Earth propagation of coronal mass ejections: II. Slow events and comparison with others. Astrophys. J. Suppl. Ser. **222**(2), 23 (2016). http://stacks.iop.org/0067-0049/222/i=2/a=23

Y.D. Liu, H. Hu, B. Zhu, J.G. Luhmann, A. Vourlidas, Structure, propagation and expansion of a CME-driven shock in the heliosphere: a revisit of the 2012 July 23 extreme storm. Astrophys. J. Lett. **834**, 158 (2017). doi:10.3847/1538-4357/834/2/158

B.C. Low, Coronal mass ejections, magnetic flux ropes, and solar magnetism. J. Geophys. Res. **106**, 25,141–25,163 (2001)

N. Lugaz, C.J. Farrugia, A new class of complex ejecta resulting from the interaction of two CMEs and its expected geo-effectiveness. J. Geophys. Res. (2014). doi:10.1002/2013GL058789

N. Lugaz, W.B. Manchester IV, T.I. Gombosi, The evolution of CME density structures. Astrophys. J. **627**, 1019–1030 (2005a)

N. Lugaz, W.B. Manchester IV, T.I. Gombosi, Numerial simulation of the interaction of two coronal mass ejections from Sun to Earth. Astrophys. J. **634**, 651–662 (2005b)

N. Lugaz, W.B. Manchester IV, I.I. Roussev, G. Tóth, T.I. Gombosi, Numerical investigation of the homologous coronal mass ejection events from active region 9236. Astrophys. J. **659**, 788–800 (2007). doi:10.1086/512005

N. Lugaz, J.N. Hernandez-Charpak, I.I. Roussev, C.J. Davis, A. Vourlidas, J.A. Davies, Determining the azimuthal properties of coronal mass ejections from multi-spacecraft remote-sensing observations with STEREO SECCHI. Astrophys. J. **715**(1), 493 (2010). http://stacks.iop.org/0004-637X/715/i=1/a=493

N. Lugaz, C. Downs, K. Shibata, I.I. Roussev, A. Asai, T.I. Gombosi, Numerical investigation of a coronal mass ejection from an Anemone active region: reconnection and deflection of the 2005 August 22 eruption. Astrophys. J. **738**, 127 (2011). arXiv:1106.5284. doi:10.1088/0004-637X/738/2/127

N. Lugaz, C.J. Farrugia, J.A. Davies, C. Möstl, C.J. Davis, I.I. Roussev, M. Temmer, The deflection of the two interacting coronal mass ejections of 2010 May 23–24 as revealed by combined in situ measurements and heliospheric imaging. Astrophys. J. **759**, 68 (2012). doi:10.1088/0004-637X/759/1/68

N. Lugaz, C.J. Farrugia, W.B. Manchester IV, N. Schwadron, The interaction of two coronal mass ejections: influence of relative orientation. Astrophys. J. **778**, 20 (2013)

N. Lugaz, C.J. Farrugia, C.W. Smith, K. Paulson, Shocks inside CMEs: a survey of properties from 1997 to 2006. J. Geophys. Res. **120**, 2409–2427 (2015). doi:10.1002/2014JA020848

N. Lugaz, C.J. Farrugia, R.M. Winslow, N. Al-Haddad, E.K.J. Kilpua, P. Riley, Factors affecting the geoeffectiveness of shocks and sheaths at 1 AU. J. Geophys. Res. **121**(A10), 10 (2016). arXiv:1610.07885. doi:10.1002/2016JA023100

J.G. Luhmann, D.W. Curtis, P. Schroeder, J. McCauley, R.P. Lin, D.E. Larson, S.D. Bale, J.A. Sauvaud, C. Aoustin, R.A. Mewaldt, A.C. Cummings, E.C. Stone, A.J. Davis, W.R. Cook, B. Kecman, M.E. von Wiedenbeck, T. Rosenvinge, M.H. Acuna, L.S. Reichenthal, S. Shuman, K.A. Wortman, D.V. Reames, R. Mueller-Mellin, H. Kunow, G.M. Mason, P. Walpole, A. Korth, T.R. Sanderson, C.T. Russell, J.T. Gosling, STEREO IMPACT investigation goals, measurements, and data products overview. Space Sci. Rev. **136**, 117–184 (2008). doi:10.1007/s11214-007-9170-x

B.J. Lynch, S.K. Antiochos, C.R. DeVore, J.G. Luhmann, T.H. Zurbuchen, Topological evolution of a fast magnetic breakout CME in three dimensions. Astrophys. J. **683**, 1192–1206 (2008). doi:10.1086/589738

B.J. Lynch, S.K. Antiochos, Y. Li, J.G. Luhmann, C.R. DeVore, Rotation of coronal mass ejections during eruption. Astrophys. J. **697**, 1918–1927 (2009). doi:10.1088/0004-637X/697/2/1918

B.J. Lynch, Y. Li, A.F.R. Thernisien, E. Robbrecht, G.H. Fisher, J.G. Luhmann, A. Vourlidas, Sun to 1 AU propagation and evolution of a slow streamer-blowout coronal mass ejection. J. Geophys. Res. **115**, A07106 (2010). doi:10.1029/2009JA015099

B.J. Lynch, A.A. Reinard, T. Mulligan, K.K. Reeves, C.E. Rakowski, J.C. Allred, Y. Li, J.M. Laming, P.J. MacNeice, J.A. Linker, Ionic composition structure of coronal mass ejections in axisymmetric magnetohydrodynamic models. Astrophys. J. **740**, 112 (2011). doi:10.1088/0004-637X/740/2/112

R.M. MacQueen, A. Csoeke-Poeckh, E. Hildner, L. House, R. Reynolds, A. Stanger, H. Tepoel, W. Wagner, The high altitude observatory coronagraph/polarimeter on the solar maximum mission. Sol. Phys. **65**, 91–107 (1980). doi:10.1007/BF00151386

R.M. MacQueen, A.J. Hundhausen, C.W. Conover, The propagation of coronal mass ejection transients. J. Geophys. Res. **91**, 31–38 (1986). doi:10.1029/JA091iA01p00031

P. Mäkelä, N. Gopalswamy, H. Xie, A.A. Mohamed, S. Akiyama, S. Yashiro, Coronal hole influence on the observed structure of interplanetary CMEs. Sol. Phys. **284**, 59–75 (2013). arXiv:1301.2176. doi:10.1007/s11207-012-0211-6

W.B. Manchester IV, The role of nonlinear Alfven waves in shear formation during solar magnetic flux emergence. Astrophys. J. **547**, 503–519 (2001). doi:10.1086/318342

W.B. Manchester IV, Buoyant disruption of magnetic arcades with self-induced shearing. J. Geophys. Res. **108**, 1162 (2003). doi:10.1029/2002JA009252

W.B. Manchester IV, Solar atmospheric dynamic coupling due to shear motions driven by the Lorentz force. Astrophys. J. **666**, 532–540 (2007). doi:10.1086/520493

W.B. Manchester IV, Shear flows driven by the Lorentz force: an energy source for coronal mass ejections and flares, in *Subsurface and Atmospheric Influences on Solar Activity*, ed. by R. Howe, R.W. Komm, K.S. Balasubramaniam, G.J.D. Petrie. Astronomical Society of the Pacific Conference Series, vol. 383 (2008), pp. 91–98

W.B. Manchester IV, T.H. Zurbuchen, Are high-latitude forward-reverse shock pairs driven by overexpansion? J. Geophys. Res. **111**, A05,101 (2006). doi:10.1029/2005JA011461

W.B. Manchester IV, T.H. Zurbuchen, Reply to comment by P. Riley and J.T. Gosling on "Are high-latitude forward-reverse shock pairs driven by overexpansion?". J. Geophys. Res. **112**, A07,103 (2007). doi:10.1029/2007JA012272

W. Manchester IV, T. Gombosi, A. Ridley, I. Roussev, D.D. Zeeuw, I. Sokolov, K. Powell, G. Tóth, Modeling a space weather event from the Sun to the Earth: CME generation and interplanetary propagation. J. Geophys. Res. **109**, A02107 (2004a). doi:10.1029/2003JA010150

W.B. Manchester IV, T. Gombosi, I. Roussev, D. De Zeeuw, I. Sokolov, K. Powell, G. Tóth, M. Opher, Three-dimensional MHD simulation of a flux rope driven CME. J. Geophys. Res. **109**, A01,102 (2004b). doi:10.1029/2002JA009672

W.B. Manchester IV, T.I. Gombosi, D.L. De Zeeuw, I.V. Sokolov, I.I. Roussev, K.G. Powell, J. Kóta, G. Tóth, T.H. Zurbuchen, Coronal mass ejection shock and sheath structures relevant to particle acceleration. Astrophys. J. **622**, 1225–1239 (2005). doi:10.1086/427768

W.B. Manchester IV, A. Vourlidas, G. Tóth, N. Lugaz, I.I. Roussev, I.V. Sokolov, T.I. Gombosi, D.L.D. Zeeuw, M. Opher, Three-dimensional MHD simulation of the 2003 October 28 coronal mass ejection: comparison with LASCO coronagraph observations. Astrophys. J. **684**, 1448–1460 (2008). doi:10.1086/590231

W.B. Manchester IV, B. van der Holst, G. Tóth, T.I. Gombosi, The coupled evolution of electrons and ions in coronal mass ejection-driven shocks. Astrophys. J. **756**, 1 (2012). doi:10.1088/0004-637X/756/1/81

W.B. Manchester IV, B. van der Holst, B. Lavraud, Flux rope evolution in interplanetary coronal mass ejections: the 13 May 2005 event. Plasma Phys. Control. Fusion **56**, 1–11 (2014a). doi:10.1088/0741-3335/56/6/064006

W.B. Manchester IV, J.U. Kozyra, S.T. Lepri, B. Lavraud, Simulation of magnetic cloud erosion during propagation. J. Geophys. Res. **119**, 1–16 (2014b). doi:10.1002/2014JA019882

S. Mancuso, J.C. Raymond, J. Kohl, Y.K. Ko, M. Uzzo, R. Wu, UVCS/SOHO observations of a CME-driven shock: consequences on ion heating mechanisms behind a coronal shock. Astron. Astrophys. **383**, 267–274 (2002). doi:10.1051/0004-6361:20011721

P.K. Manoharan, Evolution of coronal mass ejections in the inner heliosphere: a study using white-light and scintillation images. Sol. Phys. **235**, 345–368 (2006). doi:10.1007/s11207-006-0100-y

P.K. Manoharan, A. Mujiber Rahman, Coronal mass ejections: propagation time and associated internal energy. J. Atmos. Sol.-Terr. Phys. **73**, 671–677 (2011). arXiv:1011.3687. doi:10.1016/j.jastp.2011.01.017

P.K. Manoharan, M. Tokumaru, M. Pick, P. Subramanian, F.M. Ipavich, K. Schenk, M.L. Kaiser, R.P. Lepping, A. Vourlidas, Coronal mass ejection of 2000 July 14 flare event: imaging from near-Sun to Earth environment. Astrophys. J. **559**, 1180–1189 (2001). doi:10.1086/322332

P.K. Manoharan, N. Gopalswamy, S. Yashiro, A. Lara, G. Michalek, R.A. Howard, Influence of coronal mass ejection interaction on propagation of interplanetary shocks. J. Geophys. Res. **109**, A06109 (2004). doi:10.1029/2003JA010300

D. Maričić, B. Vršnak, M. Dumbović, T. Žic, D. Roša, D. Hržina, S. Lulić, I. Romštajn, I. Bušić, K. Salamon, M. Temmer, T. Rollett, A. Veronig, N. Bostanjyan, A. Chilingarian, B. Mailyan, K. Arakelyan, A. Hovhannisyan, N. Mujić, Kinematics of interacting ICMEs and related Forbush decrease: case study. Sol. Phys. **289**, 351–368 (2014). doi:10.1007/s11207-013-0314-8

K. Marubashi, S. Akiyama, S. Yashiro, N. Gopalswamy, K.S. Cho, Y.D. Park, Geometrical relationship between interplanetary flux ropes and their solar sources. Sol. Phys. **290**, 1371–1397 (2015). doi:10.1007/s11207-015-0681-4

M.L. Mays, B.J. Thompson, L.K. Jian, R.C. Colaninno, D. Odstrcil, C. Möstl, M. Temmer, N.P. Savani, G. Collinson, A. Taktakishvili, P.J. MacNeice, Y. Zheng, Propagation of the 7 January 2014 CME and resulting geomagnetic non-event. Astrophys. J. **812**, 145 (2015). arXiv:1509.06477. doi:10.1088/0004-637X/812/2/145

D.J. McComas, J.T. Gosling, D. Winterhalter, E.J. Smith, Interplanetary magnetic field draping about fast coronal mass ejecta in the outer heliosphere. J. Geophys. Res. **93**, 2519–2526 (1988). doi:10.1029/JA093iA04p02519

Z. Mikić, D. Barnes, D.D. Schnack, Dynamical evolution of a solar coronal magnetic field arcade. Astrophys. J. **328**, 830–847 (1988)

Z. Mikić, J. Linker, D. Schnack, R. Lionello, A. Tarditi, Magnetohydrodynamic modeling of the global solar corona. Phys. Plasmas **6**, 2217–2224 (1999)

Z. Mikić, R. Lionello, Y. Mok, J.A. Linker, A.R. Winebarger, The importance of geometric effects in coronal loop models. Astrophys. J. **773**, 94 (2013a). doi:10.1088/0004-637X/773/2/94

Z. Mikić, T. Török, V. Titov, J.A. Linker, R. Lionello, C. Downs, P. Riley, The challenge in making models of fast CMEs, in *American Institute of Physics Conference Series*, ed. by G.P. Zank, J. Borovsky, R. Bruno, J. Cirtain, S. Cranmer, H. Elliott, J. Giacalone, W. Gonzalez, G. Li, E. Marsch, E. Moebius, N. Pogorelov, J. Spann, O. Verkhoglyadova. American Institute of Physics Conference Series, vol. 1539 (2013b), pp. 42–45. doi:10.1063/1.4810985

W. Mishra, N. Srivastava, Estimating the arrival time of Earth-directed coronal mass ejections at in situ spacecraft using COR and HI observations from STEREO. Astrophys. J. **772**(1), 70 (2013). http://stacks.iop.org/0004-637X/772/i=1/a=70

W. Mishra, N. Srivastava, D. Chakrabarty, Evolution and consequences of interacting CMEs of 9–10 November 2012 using STEREO/SECCHI and in situ observations. Sol. Phys. **290**, 527–552 (2015a). doi:10.1007/s11207-014-0625-4

W. Mishra, N. Srivastava, T. Singh, Kinematics of interacting CMEs of 25 and 28 September 2012. J. Geophys. Res. **120**, 10 (2015b). doi:10.1002/2015JA021415

W. Mishra, Y. Wang, N. Srivastava, On understanding the nature of collisions of coronal mass ejections observed by STEREO. Astrophys. J. **831**, 99 (2016). arXiv:1607.07692. doi:10.3847/0004-637X/831/1/99

A.A. Mohamed, N. Gopalswamy, S. Yashiro, S. Akiyama, P. Mäkelä, H. Xie, H. Jung, The relation between coronal holes and coronal mass ejections during the rise, maximum, and declining phases of Solar Cycle 23. J. Geophys. Res. **117**, A01103 (2012). doi:10.1029/2011JA016589

Y.J. Moon, G.S. Choe, H. Wang, Y.D. Park, N. Gopalswamy, G. Yang, S. Yashiro, A statistical study of two classes of coronal mass ejections. Astrophys. J. **581**, 694–702 (2002). doi:10.1086/344088

Y.J. Moon, G.S. Choe, H. Wang, Y.D. Park, Sympathetic coronal mass ejections. Astrophys. J. **588**, 1176–1182 (2003). doi:10.1086/374270

C. Möstl, C. Miklenic, C.J. Farrugia, M. Temmer, A. Veronig, A.B. Galvin, B. Vršnak, H.K. Biernat, Two-spacecraft reconstruction of a magnetic cloud and comparison to its solar source. Ann. Geophys. **26**, 3139–3152 (2008). doi:10.5194/angeo-26-3139-2008

C. Möstl, C.J. Farrugia, M. Temmer, C. Miklenic, A.M. Veronig, A.B. Galvin, M. Leitner, H.K. Biernat, Linking remote imagery of a coronal mass ejection to its in situ signatures at 1 AU. Astrophys. J. **705**, L180–L185 (2009). arXiv:0910.1188. doi:10.1088/0004-637X/705/2/L180

C. Möstl, M. Temmer, T. Rollett, C.J. Farrugia, Y. Liu, A.M. Veronig, M. Leitner, A.B. Galvin, H.K. Biernat, STEREO and Wind observations of a fast ICME flank triggering a prolonged geomagnetic storm on 5–7 April 2010. Geophys. Res. Lett. **37**, L24103 (2010). arXiv:1010.4150. doi:10.1029/2010GL045175

C. Möstl, T. Rollett, R.A. Frahm, Y.D. Liu, D.M. Long, R.C. Colaninno, M.A. Reiss, M. Temmer, C.J. Farrugia, A. Posner, M. Dumbović, M. Janvier, P. Démoulin, P. Boakes, A. Devos, E. Kraaikamp, M.L. Mays, B. Vršnak, Strong coronal channelling and interplanetary evolution of a solar storm up to Earth and Mars. Nat. Commun. **6**, 7135 (2015). arXiv:1506.02842. doi:10.1038/ncomms8135

K. Muglach, Y.M. Wang, B. Kliem, Evidence for mixed helicity in erupting filaments. Astrophys. J. **703**, 976–981 (2009). arXiv:0907.4446. doi:10.1088/0004-637X/703/1/976

M. Neugebauer, R. Goldstein, Particle and field signatures of coronal mass ejections in the solar wind, in *Washington DC American Geophysical Union Geophysical Monograph Series*, vol. 99 (1997), 245–251. doi:10.1029/GM099p0245

T. Nieves-Chinchilla, R. Colaninno, A. Vourlidas, A. Szabo, R.P. Lepping, S.A. Boardsen, B.J. Anderson, H. Korth, Remote and in situ observations of an unusual Earth-directed coronal mass ejection from multiple viewpoints. J. Geophys. Res. **117**, A06106 (2012). doi:10.1029/2011JA017243

N.V. Nitta, W. Liu, N. Gopalswamy, S. Yashiro, The relation between large-scale coronal propagating fronts and type II radio bursts. Sol. Phys. **289**, 4589–4606 (2014). arXiv:1409.4754. doi:10.1007/s11207-014-0602-y

D. Odstrcil, Modeling 3-D solar wind structures. Adv. Space Res. **32**(4), 497–506 (2003)

D. Odstrčil, V.J. Pizzo, Distortion of the interplanetary magnetic field by three-dimensional propagation of coronal mass ejections in a structured solar wind. J. Geophys. Res. **104**, 28,225–28,239 (1999a)

D. Odstrčil, V.J. Pizzo, Three-dimensional propagation of CMEs in a structured solar wind flow, 1, CME launched within the streamer belt. J. Geophys. Res. **104**, 483–492 (1999b)

D. Odstrčil, M. Dryer, Z. Smith, Propagation of an interplanetary shock along the heliospheric plasma sheet. J. Geophys. Res. **101**, 19,973–19,984 (1996)

D. Odstrcil, M. Vandas, V.J. Pizzo, P. MacNeice, Numerical simulation of interacting magnetic flux ropes, in *Solar Wind Ten*, AIP Conf. Proc., vol. 679 (2003), pp. 699–702

D. Odstrcil, V.J. Pizzo, C.N. Arge, Propagation of the 12 May 1997 interplanetary coronal mass ejection in evolving solar wind structures. J. Geophys. Res. **A02**, 106 (2005). doi:10.1029/2004JA010745

O. Olmedo, J. Zhang, Partial torus instability. Astrophys. J. **718**, 433–440 (2010). doi:10.1088/0004-637X/718/1/433

V.A. Osherovich, C.J. Farrugia, L.F. Burlaga, Dynamics of aging magnetic clouds. Adv. Space Res. **13**, 57–62 (1993). doi:10.1016/0273-1177(93)90391-N

M. Owens, P. Cargill, Predictions of the arrival time of coronal mass ejections at 1 AU: an analysis of the causes of errors. Ann. Geophys. **22**, 661–671 (2004). doi:10.5194/angeo-22-661-2004

M.J. Owens, P.J. Cargill, C. Pagel, G.L. Siscoe, N.U. Crooker, Characteristic magnetic field and speed properties of interplanetary coronal mass ejections and their sheath regions. J. Geophys. Res. **110**, A01105 (2005). doi:10.1029/2004JA010814

P. Pagano, J.C. Raymond, F. Reale, S. Orlando, Modeling magnetohydrodynamics and non-equilibrium SoHO/UVCS line emission of CME shocks. Astron. Astrophys. **481**, 835–844 (2008). arXiv:0801.2705. doi:10.1051/0004-6361:20079088

G. Paschmann, B.U.O. Sonnerup, Proper frame determination and Walen test, in *ISSI Scientific Reports Series*, vol. 8 (2008), pp. 65–74

V.J. Pizzo, Interplanetary shocks on the large scale—a retrospective on the last decade's theoretical efforts, in *Washington DC American Geophysical Union Geophysical Monograph Series*, vol. 35 (1985), pp. 51–68

S.P. Plunkett, B.J. Thompson, O.C. St. Cyr, R.A. Howard, Solar source regions of coronal mass ejections and their geomagnetic effects. J. Atmos. Sol.-Terr. Phys. **63**, 389–402 (2001). doi:10.1016/S1364-6826(00)00166-8

S. Poedts, B. van der Holst, I. Chattopadhyay, D. Banerjee, T. van Lier, R. Keppens, Simulation of shock waves in the interplanetary medium, in *Solar Variability as an Input to the Earth's Environment*, ed. by A. Wilson. ESA Special Publication, vol. 535 (2003), pp. 603–612

A.J. Prise, L.K. Harra, S.A. Matthews, C.S. Arridge, N. Achilleos, Analysis of a coronal mass ejection and corotating interaction region as they travel from the Sun passing Venus, Earth, Mars, and Saturn. J. Geophys. Res. **120**, 1566–1588 (2015). doi:10.1002/2014JA020256

J.C. Raymond, B.J. Thompson, O.C. St. Cyr, N. Gopalswamy, S. Kahler, M. Kaiser, A. Lara, A. Ciaravella, M. Romoli, R. O'Neal, SOHO and radio observations of a CME shock wave. Geophys. Res. Lett. **27**, 1439–1442 (2000). doi:10.1029/1999GL003669

D.V. Reames, The two sources of solar energetic particles. Space Sci. Rev. **175**, 53–92 (2013). arXiv:1306.3608. doi:10.1007/s11214-013-9958-9

A. Reinard, Comparison of interplanetary CME charge state composition with CME-associated flare magnitude. Astrophys. J. **620**, 501–505 (2005). doi:10.1086/426109

M.J. Reiner, M.L. Kaiser, J. Fainberg, R.G. Stone, A new method for studying remote type II radio emissions from coronal mass ejection-driven shocks. J. Geophys. Res. **103**, 29,651–29,664 (1998). doi:10.1029/98JA02614

M.J. Reiner, A. Vourlidas, O.C. St. Cyr, J.T. Burkepile, R.A. Howard, M.L. Kaiser, N.P. Prestage, J.L. Bougeret, Constraints on coronal mass ejection dynamics from simultaneous radio and white-light observations. Astrophys. J. **590**, 533–546 (2003). doi:10.1086/374917

M.J. Reiner, B.V. Jackson, D.F. Webb, D.R. Mizuno, M.L. Kaiser, J.L. Bougeret, Coronal mass ejection kinematics deduced from white light (Solar Mass Ejection Imager) and radio (Wind/WAVES) observations. J. Geophys. Res. **110**, A09S14 (2005a). doi:10.1029/2004JA010943

M.J. Reiner, M.L. Kaiser, J.L. Bougeret, CME kinematics in interplanetary space, in *Solar Wind 11/SOHO 16, Connecting Sun and Heliosphere*, ed. by B. Fleck, T.H. Zurbuchen, H. Lacoste. ESA Special Publication, vol. 592 (2005b), p. 307

D.B. Reisenfeld, J.T. Gosling, R.J. Forsyth, P. Riley, O.C. St. Cyr, Properties of high-latitude CME-driven disturbances during Ulysses second northern polar passage. Geophys. Res. Lett. **30**, 8031 (2003). doi:10.1029/2003GL017155

I.G. Richardson, H.V. Cane, Identification of interplanetary coronal mass ejections at 1 AU using multiple solar wind plasma composition anomalies. J. Geophys. Res. **109**, A09104 (2004a). doi:10.1029/2004JA010598

I.G. Richardson, H.V. Cane, The fraction of interplanetary coronal mass ejections that are magnetic clouds: evidence for a solar cycle variation. Geophys. Res. Lett. **31**, L18804 (2004b). doi:10.1029/2004GL020958

I.G. Richardson, H.V. Cane, Interplanetary circumstances of quasi-perpendicular interplanetary shocks in 1996–2005. J. Geophys. Res. **115**, A07103 (2010a). doi:10.1029/2009JA015039

I.G. Richardson, H.V. Cane, Near-Earth interplanetary coronal mass ejections during Solar Cycle 23 (1996–2009): catalog and summary of properties. Sol. Phys. **264**, 189–237 (2010b). doi:10.1007/s11207-010-9568-6

I.G. Richardson, G.R. Lawrence, D.K. Haggerty, T.A. Kucera, A. Szabo, Are CME "interactions" really important for accelerating major solar energetic particle events? Geophys. Res. Lett. **30**, 8014 (2003). doi:10.1029/2002GL016424

P. Riley, N.U. Crooker, Kinematic treatment of coronal mass ejection evolution in the solar wind. Astrophys. J. **600**, 1035–1042 (2004). doi:10.1086/379974

P. Riley, J.T. Gosling, On the origin of near-radial magnetic fields in the heliosphere: numerical simulations. J. Geophys. Res. **112**, A06115 (2007). doi:10.1029/2006JA012210

P. Riley, J.T. Gosling, V.J. Pizzo, A two-dimensional simulation of the radial and latitudinal evolution of a solar wind disturbance driven by a fast, high-pressure coronal mass ejection. J. Geophys. Res. **102**, 14,577 (1997)

P. Riley, J.T. Gosling, D.J. McComas, V.J. Pizzo, J.G. Luhmann, D. Biesecker, R.J. Forsyth, J.T. Hoeksema, A. Lecinski, B.J. Thompson, Relationship between Ulysses plasma observations and solar observations during the Whole Sun Month campaign. J. Geophys. Res. **104**, 9871–9880 (1999). doi:10.1029/1998JA900078

P. Riley, J.A. Linker, Z. Mikić, D. Odstrcil, T.H. Zurbuchen, D. Lario, R.P. Lepping, Using an MHD simulation to interpret the global context of a coronal mass ejection observed by two spacecraft. J. Geophys. Res. **108**, 1272 (2003). doi:10.1029/2002JA009760

P. Riley, J.A. Linker, Z. Mikic, D. Odstrcil, Magnetohydrodynamic modeling of interplanetary CMEs. IEEE Trans. Plasma Sci. **32**, 1415–1424 (2004)

T. Rollett, C. Möstl, M. Temmer, A.M. Veronig, C.J. Farrugia, H.K. Biernat, Constraining the kinematics of coronal mass ejections in the inner heliosphere with in-situ signatures. Sol. Phys. **276**, 293–314 (2012). arXiv:1110.0300. doi:10.1007/s11207-011-9897-0

T. Rollett, C. Möstl, M. Temmer, R.A. Frahm, J.A. Davies, A.M. Veronig, B. Vršnak, U.V. Amerstorfer, C.J. Farrugia, T. Žic, T.L. Zhang, Combined multipoint remote and in situ observations of the asymmetric evolution of a fast solar coronal mass ejection. Astrophys. J. Lett. **790**, L6 (2014). arXiv:1407.4687. doi:10.1088/2041-8205/790/1/L6

I.I. Roussev, N. Lugaz, I.V. Sokolov, New physical insight on the changes in magnetic topology during coronal mass ejections: case studies for the 2002 April 21 and August 24 events. Astrophys. J. Lett. **668**, L87–L90 (2007). doi:10.1086/522588

A. Ruffenach, B. Lavraud, M.J. Owens, J.A. Sauvaud, N.P. Savani, A.P. Rouillard, P. Démoulin, C. Foullon, A. Opitz, A. Fedorov, C.J. Jacquey, V. Génot, P. Louarn, J.G. Luhmann, C.T. Russell, C.J. Farrugia, A.B. Galvin, Multispacecraft observation of magnetic cloud erosion by magnetic reconnection during propagation. J. Geophys. Res. **117**, A09101 (2012). doi:10.1029/2012JA017624

A. Ruffenach, B. Lavraud, C.J. Farrugia, P. Démoulin, S. Dasso, M.J. Owens, J.A. Sauvaud, A.P. Rouillard, A. Lynnyk, C. Foullon, N.P. Savani, J.G. Luhmann, A.B. Galvin, Statistical study of magnetic cloud erosion by magnetic reconnection. J. Geophys. Res. **120**, 43–60 (2015). doi:10.1002/2014JA020628

D.M. Rust, B.J. Anderson, M.D. Andrews, M.H. Acuña, C.T. Russell, P.W. Schuck, T. Mulligan, Comparison of interplanetary disturbances at the NEAR spacecraft with coronal mass ejections at the Sun. Astrophys. J. **621**, 524–536 (2005). doi:10.1086/427401

C. Salas-Matamoros, K.L. Klein, On the statistical relationship between CME speed and soft X-ray flux and fluence of the associated flare. Sol. Phys. **290**, 1337–1353 (2015). arXiv:1503.08613. doi:10.1007/s11207-015-0677-0

N.P. Savani, A.P. Rouillard, J.A. Davies, M.J. Owens, R.J. Forsyth, C.J. Davis, R.A. Harrison, The radial width of a coronal mass ejection between 0.1 and 0.4 AU estimated from the Heliospheric Imager on STEREO. Ann. Geophys. **27**, 4349–4358 (2009)

N.P. Savani, M.J. Owens, A.P. Rouillard, R.J. Forsyth, J.A. Davies, Observational evidence of a coronal mass ejection distortion directly attributable to a structured solar wind. Astrophys. J. Lett. **714**, L128–L132 (2010). doi:10.1088/2041-8205/714/1/L128

J.M. Schmidt, P.J. Cargill, Magnetic reconnection between magnetic cloud and the solar wind. J. Geophys. Res. **108**(A1), 1023 (2003). doi:10.1029/2002JA009325

J. Schmidt, P. Cargill, A numerical study of two interacting coronal mass ejections. Ann. Geophys. **22**, 2245–2254 (2004)

J.M. Schmidt, I.H. Cairns, H. Xie, O.C. St. Cyr, N. Gopalswamy, CME flux rope and shock identifications and locations: comparison of white light data, Graduated Cylindrical Shell model, and MHD simulations. J. Geophys. Res. **121**, 1886–1906 (2016). doi:10.1002/2015JA021805

B. Schmieder, G. Simon, M.J. Martres, P. Mein, N. Mein, E. Tandberg-Hanssen, Recurrent mass ejections observed in H-alpha and CIV. Adv. Space Res. **4**, 27–30 (1984). doi:10.1016/0273-1177(84)90154-6

C.J. Schrijver, A.M. Title, Long-range magnetic couplings between solar flares and coronal mass ejections observed by SDO and STEREO. J. Geophys. Res. **116**, A04108 (2011). doi:10.1029/2010JA016224

R. Schwenn, H. Rosenbauer, K.H. Muehlhaeuser, Singly-ionized helium in the driver gas of an interplanetary shock wave. Geophys. Res. Lett. **7**, 201–204 (1980). doi:10.1029/GL007i003p00201

R. Schwenn, A. dal Lago, E. Huttunen, W.D. Gonzalez, The association of coronal mass ejections with their effects near the Earth. Ann. Geophys. **23**, 1033–1059 (2005)

A. Shanmugaraju, B. Vršnak, Transit time of coronal mass ejections under different ambient solar wind conditions. Sol. Phys. **289**, 339–349 (2014). doi:10.1007/s11207-013-0322-8

N.R. Sheeley, J.H. Walters, Y.M. Wang, R.A. Howard, Continuous tracking of coronal outflows: two kinds of coronal mass ejections. J. Geophys. Res. **104**, 24,739–24,767 (1999)

C. Shen, Y. Wang, B. Gui, P. Ye, S. Wang, Kinematic evolution of a slow CME in corona viewed by STEREO-B on 8 October 2007. Sol. Phys. **269**, 389–400 (2011). arXiv:1101.3205. doi:10.1007/s11207-011-9715-8

C. Shen, Y. Wang, S. Wang, Y. Liu, R. Liu, A. Vourlidas, B. Miao, P. Ye, J. Liu, Z. Zhou, Super-elastic collision of large-scale magnetized plasmoids in the heliosphere. Nat. Phys. **8**, 923–928 (2012). doi:10.1038/nphys2440

F. Shen, C. Shen, Y. Wang, X. Feng, C. Xiang, Could the collision of CMEs in the heliosphere be super-elastic? Validation through three-dimensional simulations. Geophys. Res. Lett. **40**, 1457–1461 (2013). doi:10.1002/grl.50336

F. Shen, Y. Wang, C. Shen, X. Feng, Turn on the super-elastic collision nature of coronal mass ejections through low approaching speed. Sci. Rep. **6**, 19576 (2016). doi:10.1038/srep19576

T. Shi, Y. Wang, L. Wan, X. Cheng, M. Ding, J. Zhang, Predicting the arrival time of coronal mass ejections with the graduated cylindrical shell and drag force model. Astrophys. J. **806**, 271 (2015). arXiv:1505.00884. doi:10.1088/0004-637X/806/2/271

D. Shiota, R. Kataoka, Magnetohydrodynamic simulation of interplanetary propagation of multiple coronal mass ejections with internal magnetic flux rope (SUSANOO-CME). Space Weather **14**, 56–75 (2016). doi:10.1002/2015SW001308

D. Shiota, K. Kusano, T. Miyoshi, K. Shibata, Magnetohydrodynamic modeling for a formation process of coronal mass ejections: interaction between an ejecting flux rope and an ambient field. Astrophys. J. **718**, 1305–1314 (2010). arXiv:1006.0307. doi:10.1088/0004-637X/718/2/1305

D.G. Sime, A.J. Hundhausen, The coronal mass ejection of July 6, 1980—a candidate for interpretation as a coronal shock wave. J. Geophys. Res. **92**, 1049–1055 (1987)

G. Siscoe, D. Odstrcil, Ways in which ICME sheaths differ from magnetosheaths. J. Geophys. Res. **113**, A00B07 (2008). doi:10.1029/2008JA013142

R.M. Skoug, S.J. Bame, W.C. Feldman, J.T. Gosling, D.J. McComas, J.T. Steinberg, R.L. Tokar, P. Riley, L.F. Burlaga, N.F. Ness, C.W. Smith, A prolonged He^+ enhancement within a coronal mass ejection in the solar wind. Geophys. Res. Lett. **26**, 161–164 (1999). doi:10.1029/1998GL900207

R.M. Skoug, W.C. Feldman, J.T. Gosling, D.J. McComas, C.W. Smith, Solar wind electron characteristics inside and outside coronal mass ejections. J. Geophys. Res. **105**(23), 069 (2000). 084. doi:10.1029/2000JA000017

D.F. Smart, M.A. Shea, A simplified model for timing the arrival of solar flare-initiated shocks. J. Geophys. Res. **90**, 183–190 (1985). doi:10.1029/JA090iA01p00183

Z. Smith, M. Dryer, MHD study of temporal and spatial evolution of simulated interplanetary shocks in the ecliptic plane within 1 AU. Sol. Phys. **129**, 387–405 (1990). doi:10.1007/BF00159049

C.W. Smith, N.F. Ness, L.F. Burlaga, R.M. Skoug, D.J. McComas, T.H. Zurbuchen, G. Gloeckler, D.K. Haggerty, R.E. Gold, M.I. Desai, G.M. Mason, J.E. Mazur, J.R. Dwyer, M.A. Popecki, E. Möbius, C.M.S. Cohen, R.A. Leske, ACE observations of the Bastille day 2000 interplanetary disturbances. Sol. Phys. **204**, 227–252 (2001). doi:10.1023/A:1014265108171

O.C. St. Cyr, S.P. Plunkett, D.J. Michels, S.E. Paswaters, M.J. Koomen, G.M. Simnett, B.J. Thompson, J.B. Gurman, R. Schwenn, D.F. Webb, E. Hildner, P.L. Lamy, Properties of coronal mass ejections: SOHO LASCO observations from January 1996 to June 1998. J. Geophys. Res. **105**, 18,169–18,186 (2000). doi:10.1029/1999JA000381

R.S. Steinolfson, Coronal evolution due to shear motion. Astrophys. J. **382**, 677–687 (1991). doi:10.1086/170755

R.S. Steinolfson, M. Dryer, Numerical simulation of MHD shock waves in the solar wind. J. Geophys. Res. **83**, 1576–1582 (1978). doi:10.1029/JA083iA04p01576

D. Sudar, B. Vršnak, M. Dumbović, Predicting coronal mass ejections transit times to Earth with neural network. Mon. Not. R. Astron. Soc. **456**, 1542–1548 (2016). arXiv:1511.07620. doi:10.1093/mnras/stv2782

A. Taktakishvili, M. Kuznetsova, P. MacNeice, M. Hesse, L. Rastätter, A. Pulkkinen, A. Chulaki, D. Odstrcil, Validation of the coronal mass ejection predictions at the Earth orbit estimated by ENLIL heliosphere cone model. Space Weather **7**, S03004 (2009). doi:10.1029/2008SW000448

S.J. Tappin, The deceleration of an interplanetary transient from the Sun to 5 AU. Sol. Phys. **233**, 233–248 (2006). doi:10.1007/s11207-006-2065-2

M. Temmer, N.V. Nitta, Interplanetary propagation behavior of the fast coronal mass ejection on 23 July 2012. Sol. Phys. **290**, 919–932 (2015). arXiv:1411.6559. doi:10.1007/s11207-014-0642-3

M. Temmer, T. Rollett, C. Möstl, A.M. Veronig, B. Vršnak, D. Odstrčil, Influence of the ambient solar wind flow on the propagation behavior of interplanetary coronal mass ejections. Astrophys. J. **743**, 101–112 (2011). doi:10.1088/0004-637X/743/2/101

M. Temmer, B. Vršnak, T. Rollett, B. Bein, C.A. de Koning, Y. Liu, E. Bosman, J.A. Davies, C. Möstl, T. Žic, A.M. Veronig, V. Bothmer, R. Harrison, N. Nitta, M. Bisi, O. Flor, J. Eastwood, D. Odstrcil, R. Forsyth, Characteristics of kinematics of a coronal mass ejection during the 2010 August 1 CME-CME interaction event. Astrophys. J. **749**, 57 (2012). arXiv:1202.0629. doi:10.1088/0004-637X/749/1/57

M. Temmer, A.M. Veronig, V. Peinhart, B. Vršnak, Asymmetry in the CME-CME interaction process for the events from 2011 February 14–15. Astrophys. J. **785**, 85 (2014). doi:10.1088/0004-637X/785/2/85

A. Thernisien, A. Vourlidas, R.A. Howard, Forward modeling of coronal mass ejections using STEREO/SECCHI data. Sol. Phys. **256**, 111–130 (2009). doi:10.1007/s11207-009-9346-5

W.T. Thompson, Strong rotation of an erupting quiescent polar crown prominence. J. Atmos. Sol.-Terr. Phys. **73**, 1138–1147 (2011). doi:10.1016/j.jastp.2010.07.005

W.T. Thompson, B. Kliem, T. Török, 3D reconstruction of a rotating erupting prominence. Sol. Phys. **276**, 241–259 (2012). arXiv:1112.3388. doi:10.1007/s11207-011-9868-5

V.S. Titov, P. Démoulin, Basic topology of twisted magnetic configurations in solar flares. Astron. Astrophys. **351**, 707–720 (1999)

V.S. Titov, T. Török, Z. Mikic, J.A. Linker, A method for embedding circular force-free flux ropes in potential magnetic fields. Astrophys. J. **790**, 163 (2014). doi:10.1088/0004-637X/790/2/163

T. Török, B. Kliem, The evolution of twisting coronal magnetic flux tubes. Astron. Astrophys. **406**, 1043–1059 (2003). doi:10.1051/0004-6361:20030692

T. Török, B. Kliem, Confined and ejective eruptions of kink-unstable flux ropes. Astrophys. J. Lett. **630**, L97–L100 (2005). arXiv:astro-ph/0507662. doi:10.1086/462412

T. Török, B. Kliem, V.S. Titov, Ideal kink instability of a magnetic loop equilibrium. Astron. Astrophys. **413**, L27–L30 (2004). arXiv:astro-ph/0311198. doi:10.1051/0004-6361:20031691

T. Török, M.A. Berger, B. Kliem, The writhe of helical structures in the solar corona. Astron. Astrophys. **516**, A49 (2010). arXiv:1004.3918. doi:10.1051/0004-6361/200913578

T. Török, O. Panasenco, V.S. Titov, Z. Mikić, K.K. Reeves, M. Velli, J.A. Linker, G. De Toma, A model for magnetically coupled sympathetic eruptions. Astrophys. J. Lett. **739**, L63 (2011). doi:10.1088/2041-8205/739/2/L63

G. Tóth, D.L.D. Zeeuw, T.I. Gombosi, W.B. Manchester IV, A.J. Ridley, I.V. Sokolov, I.I. Roussev, Sun to thermosphere simulation of the October 28–30, 2003 storm with the Space Weather Modeling Framework. Space Weather **5**, S06,003 (2007). doi:10.1029/2006SW000272

R. Tousey, The solar corona, in *Space Research Conference*, Space Research Conference, vol. 2, ed. by M.J. Rycroft, S.K. Runcorn (1973), pp. 713–730

J. Uwamahoro, L.A. McKinnell, J.B. Habarulema, Estimating the geoeffectiveness of halo CMEs from associated solar and IP parameters using neural networks. Ann. Geophys. **30**, 963–972 (2012). doi:10.5194/angeo-30-963-2012

F. Valach, M. Revallo, J. Bochníček, P. Hejda, Solar energetic particle flux enhancement as a predictor of geomagnetic activity in a neural network-based model. Space Weather **7**, S04004 (2009). doi:10.1029/2008SW000421

B. van der Holst, W.B. Manchester IV, I. Sokolov, G. Tóth, T. Gombosi, D.D. Zeeuw, O. Cohen, Breakout coronal mass ejection or streamer blowout: the bugle effect. Astrophys. J. **693**, 1178–1187 (2009)

B. van der Holst, W.B. Manchester IV, R. Frazin, A. Vásquez, G. Tóth, T. Gombosi, A data-driven, two-temperature solar wind model with Alfvén waves. Astrophys. J. **725**, 1373–1383 (2010). doi:10.1088/0004-637X/725/1/1373

B. van der Holst, I. Sokolov, X. Meng, M. Jin, W.B. Manchester IV, G. Tóth, T.I. Gombosi, Alfvén wave solar model (AWSOM): coronal heating. Astrophys. J. **782**, 81 (2014). doi:10.1088/0004-637X/782/2/81

M. Vandas, S. Fischer, M. Dryer, Z. Smith, T. Detman, Simulation of magnetic cloud propagation in the inner heliosphere in two dimensions 2. A loop parallel to the ecliptic plane and the role of helicity. J. Geophys. Res. **101**, 2505–2510 (1996). doi:10.1029/95JA02446

M. Vandas, S. Fischer, P. Pelant, M. Dryer, Z. Smith, T. Detman, Propagation of a spheromac: 1. Some comparisons of cylindrical and spherical magnetic clouds. J. Geophys. Res. **102**, 24,183 (1997)

P. Vemareddy, C. Möstl, T. Amerstorfer, W. Mishra, C. Farrugia, M. Leitner, Comparison of magnetic properties in a magnetic cloud and its solar source on 2013 April 11–14. Astrophys. J. **828**, 12 (2016). arXiv:1607.03811. doi:10.3847/0004-637X/828/1/12

A. Vourlidas, S.T. Wu, A. Wang, P. Subramanian, R.A. Howard, Direct detection of a CME-associated shock in LASCO white light images. Astrophys. J. **598**, 1392 (2003)

A. Vourlidas, R. Colaninno, T. Nieves-Chinchilla, G. Stenborg, The first observation of a rapidly rotating coronal mass ejection in the middle corona. Astrophys. J. **733**, L23 (2011). doi:10.1088/2041-8205/733/2/L23

B. Vršnak, Deceleration of coronal mass ejections. Sol. Phys. **202**, 173–189 (2001)

B. Vršnak, Processes and mechanisms governing the initiation and propagation of CMEs. Ann. Geophys. **26**, 3089–3101 (2008)

B. Vršnak, N. Gopalswamy, Influence of the aerodynamic drag on the motion of interplanetary ejecta. J. Geophys. Res. **107**, SSH2 (2002). doi:10.1029/2001JA000120

B. Vršnak, T. Žic, Transit times of interplanetary coronal mass ejections and the solar wind speed. Astron. Astrophys. **472**, 937–943 (2007). doi:10.1051/0004-6361:20077499

B. Vršnak, D. Ruždjak, D. Sudar, N. Gopalswamy, Kinematics of coronal mass ejections between 2 and 30 solar radii. What can be learned about forces governing the eruption? Astron. Astrophys. **423**, 717–728 (2004). doi:10.1051/0004-6361:20047169

B. Vršnak, D. Maričić, A.L. Stanger, A.M. Veronig, M. Temmer, D. Roša, Acceleration phase of coronal mass ejections: I. Temporal and spatial scales. Sol. Phys. **241**, 85–98 (2007). doi:10.1007/s11207-006-0290-3

B. Vršnak, D. Vrbanec, J. Čalogović, Dynamics of coronal mass ejections. The mass-scaling of the aerodynamic drag. Astron. Astrophys. **490**, 811–815 (2008). doi:10.1051/0004-6361:200810215

B. Vršnak, T. Žic, T.V. Falkenberg, C. Möstl, S. Vennerstrom, D. Vrbanec, The role of aerodynamic drag in propagation of interplanetary coronal mass ejections. Astron. Astrophys. **512**, A43 (2010). doi:10.1051/0004-6361/200913482

B. Vršnak, T. Žic, D. Vrbanec, M. Temmer, T. Rollett, C. Möstl, A. Veronig, J. Čalogović, M. Dumbović, S. Lulić, Y.J. Moon, A. Shanmugaraju, Propagation of interplanetary coronal mass ejections: the drag-based model. Sol. Phys. **285**, 295–315 (2013). doi:10.1007/s11207-012-0035-4

B. Vršnak, M. Temmer, T. Žic, A. Taktakishvili, M. Dumbović, C. Möstl, A.M. Veronig, M.L. Mays, D. Odstrčil, Heliospheric propagation of coronal mass ejections: comparison of numerical WSA-ENLIL+cone model and analytical drag-based model. Astrophys. J. Suppl. Ser. **213**, 21 (2014). doi:10.1088/0067-0049/213/2/21

C. Wang, J.D. Richardson, L. Burlaga, Propagation of the Bastille day 2000 CME shock in the outer heliosphere. Sol. Phys. **204**, 413–423 (2001). doi:10.1023/A:1014293527951

Y.M. Wang, P.Z. Ye, S. Wang, Multiple magnetic clouds: several examples during March-April 2001. J. Geophys. Res. **108**, 6 (2003a)

Y.M. Wang, P.Z. Ye, S. Wang, M. Xiong, Theoretical analysis on the geoeffectiveness of a shock overtaking a preceding magnetic cloud. Sol. Phys. **216**, 295–310 (2003b). doi:10.1023/A:1026150630940

Y.M. Wang, P.Z. Ye, S. Wang, X.H. Xue, An interplanetary cause of large geomagnetic storms: fast forward shock overtaking preceding magnetic cloud. Geophys. Res. Lett. **30**, 33 (2003c)

Y. Wang, C. Shen, S. Wang, P. Ye, Deflection of coronal mass ejection in the interplanetary medium. Sol. Phys. **222**, 329–343 (2004). doi:10.1023/B:SOLA.0000043576.21942.aa

Y.M. Wang, H. Zheng, S. Wang, P. Ye, MHD simulation of the formation and propagation of multiple magnetic clouds in the heliosphere. Astron. Astrophys. **434**, 309–316 (2005)

Y. Wang, C. Chen, B. Gui, C. Shen, P. Ye, S. Wang, Statistical study of coronal mass ejection source locations: understanding CMEs viewed in coronagraphs. J. Geophys. Res. **116**, A04104 (2011). arXiv:1101.0641. doi:10.1029/2010JA016101

R. Wang, Y.D. Liu, Z. Yang, H. Hu, Magnetic field restructuring associated with two successive solar eruptions. Astrophys. J. **791**, 84 (2014a). arXiv:1407.4004. doi:10.1088/0004-637X/791/2/84

Y. Wang, B. Wang, C. Shen, F. Shen, N. Lugaz, Deflected propagation of a coronal mass ejection from the corona to interplanetary space. J. Geophys. Res. **119**, 5117–5132 (2014b). arXiv:1406.4684. doi:10.1002/2013JA019537

R. Wang, Y.D. Liu, X. Dai, Z. Yang, C. Huang, H. Hu, The role of active region coronal magnetic field in determining coronal mass ejection propagation direction. Astrophys. J. **814**, 80 (2015). arXiv:1510.06177. doi:10.1088/0004-637X/814/1/80

R. Wang, Y.D. Liu, I. Zimovets, H. Hu, X. Dai, Z. Yang, Sympathetic solar filament eruptions. Astrophys. J. Lett. **827**, L12 (2016a). arXiv:1608.01067. doi:10.3847/2041-8205/827/1/L12

Y. Wang, Q. Zhang, J. Liu, C. Shen, F. Shen, Z. Yang, T. Zic, B. Vrsnak, D.F. Webb, R. Liu, S. Wang, J. Zhang, Q. Hu, B. Zhuang, On the propagation of a geoeffective coronal mass ejection during 15–17 March 2015. J. Geophys. Res. **121**, 7423–7434 (2016b). arXiv:1607.07750. doi:10.1002/2016JA022924

D.F. Webb, T.A. Howard, Coronal mass ejections: observations. Living Rev. Sol. Phys. **9**, 3 (2012). doi:10.12942/lrsp-2012-3

D.F. Webb, D.R. Mizuno, A. Buffington, M.P. Cooke, C.J. Eyles, C.D. Fry, L.C. Gentile, P.P. Hick, P.E. Holladay, T.A. Howard, J.G. Hewitt, B.V. Jackson, J.C. Johnston, T.A. Kuchar, J.B. Mozer, S. Price, R.R. Radick, G.M. Simnett, S.J. Tappin, Solar Mass Ejection Imager (SMEI) observations of Coronal Mass Ejections (CMEs) in the heliosphere. J. Geophys. Res. **111**(A10), A12101 (2006). doi:10.1029/2006JA011655

F.S. Wei, The blast wave propagating in a moving medium with variable density. Chin. J. Space Sci. **2**, 63–72 (1982)

F. Wei, M. Dryer, Propagation of solar flare-associated interplanetary shock waves in the heliospheric meridional plane. Sol. Phys. **132**, 373–394 (1991). doi:10.1007/BF00152294

R.M. Winslow, N. Lugaz, N.A. Schwadron, C.J. Farrugia, W. Yu, J.M. Raines, M.L. Mays, A.B. Galvin, T.H. Zurbuchen, Longitudinal conjunction between MESSENGER and STEREO A: development of ICME complexity through stream interactions. J. Geophys. Res. **121**, 6092–6106 (2016). doi:10.1002/2015JA022307

B.E. Wood, R.A. Howard, An empirical reconstruction of the 2008 April 26 coronal mass ejection. Astrophys. J. **702**, 901–910 (2009). doi:10.1088/0004-637X/702/2/901

B.E. Wood, R.A. Howard, S.P. Plunkett, D.G. Socker, Comprehensive observations of a solar minimum coronal mass ejection with the solar terrestrial relations observatory. Astrophys. J. **694**, 707–717 (2009). arXiv:0811.3226. doi:10.1088/0004-637X/694/2/707

B.E. Wood, C.C. Wu, A.P. Rouillard, R.A. Howard, D.G. Socker, A coronal hole's effects on coronal mass ejection shock morphology in the inner heliosphere. Astrophys. J. **755**, 43 (2012). arXiv:1206.3584. doi:10.1088/0004-637X/755/1/43

B.E. Wood, R.A. Howard, M.G. Linton, Imaging prominence eruptions out to 1 AU. Astrophys. J. **816**, 67 (2016). arXiv:1512.06748. doi:10.3847/0004-637X/816/2/67

S.T. Wu, S.M. Han, M. Dryer, Two-dimensional, time-dependent MHD description of interplanetary disturbances simulation of high speed solar wind interactions. Planet. Space Sci. **27**, 255–264 (1979). doi:10.1016/0032-0633(79)90068-0

S.T. Wu, R.S. Steinolfson, M. Dryer, E. Tandberg-Hanssen, Magnetohydrodynamic models of coronal transients in the meridional plane. IV—effect of the solar wind. Astrophys. J. **243**, 641–643 (1981). doi:10.1086/158629

S.T. Wu, M.D. Andrews, S.P. Plunkett, Numerical Magnetohydrodynamic (MHD) modeling of Coronal Mass Ejections (CMEs). Space Sci. Rev. **95**, 191–213 (2001)

S.T. Wu, A.H. Wang, N. Gopalswamy, MHD modelling of CME and CME interactions in a bi-modal solar wind: a preliminary analysis of the 20 January 2001 two CMEs interaction event, in *ESA SP-505: SOLMAG 2002*. Proceedings of the Magnetic Coupling of the Solar Atmosphere Euroconference (2002), pp. 227–230

H. Xie, O.C. St. Cyr, N. Gopalswamy, S. Yashiro, J. Krall, M. Kramar, J. Davila, On the origin, 3D structure and dynamic evolution of CMEs near solar minimum. Sol. Phys. **259**, 143–161 (2009). doi:10.1007/s11207-009-9422-x

M. Xiong, H. Zheng, Y. Wang, S. Wang, Magnetohydrodynamic simulation of the interaction between interplanetary strong shock and magnetic cloud and its consequent geoeffectiveness. J. Geophys. Res. **111**, A08105 (2006). doi:10.1029/2005JA011593

M. Xiong, H. Zheng, S. Wang, Magnetohydrodynamic simulation of the interaction between two interplanetary magnetic clouds and its consequent geoeffectiveness: 2. Oblique collision. J. Geophys. Res. **A11**, 101 (2009). doi:10.1029/2009JA014079

S. Yao, E. Marsch, C.Y. Tu, R. Schwenn, Identification of prominence ejecta by the proton distribution function and magnetic fine structure in interplanetary coronal mass ejections in the inner heliosphere. J. Geophys. Res. **115**, A05103 (2010). doi:10.1029/2009JA014914

S. Yashiro, N. Gopalswamy, G. Michalek, O.C. St. Cyr, S.P. Plunkett, N.B. Rich, R.A. Howard, A catalog of white light coronal mass ejections observed by the SOHO spacecraft. J. Geophys. Res. **A07**, 105 (2004). doi:10.1029/2003JA010282

V. Yurchyshyn, Relationship between EIT posteruption arcades, coronal mass ejections, the coronal neutral line, and magnetic clouds. Astrophys. J. Lett. **675**, L49–L52 (2008). arXiv:0710.1292. doi:10.1086/533413

V.B. Yurchyshyn, H. Wang, P.R. Goode, Y. Deng, Orientation of the magnetic fields in interplanetary flux ropes and solar filaments. Astrophys. J. **563**, 381–388 (2001). doi:10.1086/323778

V. Yurchyshyn, Q. Hu, R.P. Lepping, B.J. Lynch, J. Krall, Orientations of LASCO Halo CMEs and their connection to the flux rope structure of interplanetary CMEs. Adv. Space Res. **40**, 1821–1826 (2007). arXiv:astro-ph/0703627. doi:10.1016/j.asr.2007.01.059

X. Zhao, M. Dryer, Current status of CME/shock arrival time prediction. Space Weather **12**, 448–469 (2014). doi:10.1002/2014SW001060

X.H. Zhao, X.S. Feng, Influence of a CME's initial parameters on the arrival of the associated interplanetary shock at Earth and the shock propagational model version 3. Astrophys. J. **809**, 44 (2015). doi:10.1088/0004-637X/809/1/44

X. Zhao, Y.D. Liu, B. Inhester, X. Feng, T. Wiegelmann, L. Lu, Comparison of CME/shock propagation models with heliospheric imaging and in situ observations. Astrophys. J. **830**, 48 (2016). arXiv:1607.05533. doi:10.3847/0004-637X/830/1/48

Y.F. Zhou, X.S. Feng, MHD numerical study of the latitudinal deflection of coronal mass ejection. J. Geophys. Res. **118**, 6007–6018 (2013). doi:10.1002/2013JA018976

G. Zhou, Y. Wang, J. Wang, Coronal mass ejections associated with polar crown filaments. Adv. Space Res. **38**, 466–469 (2006). doi:10.1016/j.asr.2005.03.110

T. Žic, B. Vršnak, M. Temmer, Heliospheric propagation of coronal mass ejections: drag-based model fitting. Astrophys. J. Suppl. Ser. **218**, 32 (2015). arXiv:1506.08582. doi:10.1088/0067-0049/218/2/32

F.P. Zuccarello, A. Bemporad, C. Jacobs, M. Mierla, S. Poedts, F. Zuccarello, The role of streamers in the deflection of coronal mass ejections: comparison between STEREO three-dimensional reconstructions and numerical simulations. Astrophys. J. **744**, 66 (2012a). doi:10.1088/0004-637X/744/1/66

F.P. Zuccarello, Z. Meliani, S. Poedts, Numerical modeling of the initiation of coronal mass ejections in active region NOAA 9415. Astrophys. J. **758**, 117 (2012b). doi:10.1088/0004-637X/758/2/117

T.H. Zurbuchen, I.G. Richardson, In-situ solar wind and magnetic field signatures of interplanetary coronal mass ejections. Space Sci. Rev. **123**, 31–43 (2006). doi:10.1007/s11214-006-9010-4

T.H. Zurbuchen, G. Gloeckler, F. Ipavich, J. Raines, C.W. Smith, L.A. Fisk, On the fast coronal mass ejections in October/November 2003: ACE-SWICS results. Geophys. Res. Lett. **L11**, 805(2004). doi:10.1029/2004GL019461

Space Sci Rev (2017) 212:1107–1136
DOI 10.1007/s11214-017-0382-4

Acceleration and Propagation of Solar Energetic Particles

Karl-Ludwig Klein[1] · **Silvia Dalla**[2]

Received: 22 December 2016 / Accepted: 17 May 2017 / Published online: 27 July 2017

Abstract Solar Energetic Particles (SEPs) are an important component of Space Weather, including radiation hazard to humans and electronic equipment, and the ionisation of the Earth's atmosphere. We review the key observations of SEPs, our current understanding of their acceleration and transport, and discuss how this knowledge is incorporated within Space Weather forecasting tools. Mechanisms for acceleration during solar flares and at shocks driven by Coronal Mass Ejections (CMEs) are discussed, as well as the timing relationships between signatures of solar eruptive events and the detection of SEPs in interplanetary space. Evidence on how the parameters of SEP events are related to those of the parent solar activity is reviewed and transport effects influencing SEP propagation to near-Earth locations are examined. Finally, the approaches to forecasting Space Weather SEP effects are discussed. We conclude that both flare and CME shock acceleration contribute to Space Weather relevant SEP populations and need to be considered within forecasting tools.

Keywords Sun: particle emission · Solar-terrestrial relations · Space weather

1 Introduction

Solar energetic particle (SEP) events are distinct enhancements in space of particle fluxes of electrons, protons and heavy ions at energies well above the average thermal energy in

The Scientific Foundation of Space Weather
Edited by Rudolf von Steiger, Daniel Baker, André Balogh, Tamás Gombosi, Hannu Koskinen and Astrid Veronig

Further affiliations for K.-L. Klein: PSL Research University, Univ. P & M Curie, Univ. Paris-Diderot.

✉ K.-L. Klein
ludwig.klein@obspm.fr

S. Dalla
sdalla@uclan.ac.uk

1 LESIA—Observatoire de Paris, CNRS, Meudon, France

2 Jeremiah Horrocks Institute, University of Central Lancashire, Preston PR1 2HE, UK

the corona, which is a few hundreds of eV. The energy of solar energetic protons can reach GeV in some events. The events last from a few hours to several days. Space weather effects of SEPs are due to the interaction of the particles with electronics, the Earth's atmosphere and living beings. Strong SEP events are a potential space weather hazard because they may affect space borne electronics and generate substantial particle radiation. Energetic ions are themselves primary radiation and they generate secondaries through nuclear reactions with the Earth's atmosphere and any exposed materials. This leads to enhanced ionisation and sometimes to modifications of the local chemistry of the high polar atmosphere of the Earth. SEP events are a major obstacle to human spaceflight outside the Earth's magnetosphere.

In this paper we discuss the acceleration and propagation of SEPs from a space weather perspective. Scientific understanding of these topics underpins many forecasting tools that aim to predict the occurrence of SEP events and their impact.

Two key questions regarding SEPs within a Space Weather framework are: Which solar events are responsible for the production of SEPs most dangerous for Space Weather? What type of events can give rise to extreme SEP enhancements? In this chapter we review the observational evidence and modelling efforts related to SEPs, with the aim of addressing the above questions.

Since SEP events usually have well-defined onset time and intensity enhancements up to several orders of magnitude, they can be related to individual events of solar activity. Historically flares and filament eruptions were the only known major transient features in the solar atmosphere. SEP events were considered as a consequence of flares until the 1970s. Since the discovery of CMEs in the 1970s a bimodal framework for the interpretation of SEP events has emerged, based on their classification into impulsive and gradual (Reames 1999). Impulsive SEP events are short ($\leq$1 day), low intensity and numerous (estimated as about 1000/year in periods of high activity). Gradual events are long (several days at energies of a few MeV/nuc), rather rare (a few tens per year), and orders of magnitude more intense in protons than impulsive SEP events. Within this scenario, the latter type of events, characterised by the largest proton fluences and therefore of most relevance to Space Weather, are ascribed to acceleration by CME-driven shocks as they propagate through the heliosphere. There is some debate as to the role played by "flare acceleration" in these events. The question for space weather forecasting is whether it suffices to consider the parameters of CMEs in order to provide an accurate SEP forecast, or whether flare information must also be taken into account. One can immediately see that the science related to SEP acceleration should inform Space Weather, for example through statistical studies of links between solar events and SEP occurrence and parameters.

The propagation of energetic particles through interplanetary space is also key to determining whether or not SEPs will be observed, for example, near Earth and potentially produce Space Weather effects. The classic description of SEP transport, embedded within the two-class paradigm, is spatially one-dimensional, in the sense that particles are assumed to remain tied to the magnetic field line on which they were initially injected. The detection of particle flux at one location therefore requires direct magnetic connection to the acceleration region. At the same time, measurements from a number of spacecraft have shown that in many events particles have easy access to wide regions in longitude and latitude. Is this observation the result of wide CME driven accelerating fronts in the heliosphere, as postulated within the two-class paradigm, or could other physical processes be at play? Again scientific understanding is required to assess and forecast the Space Weather effects of SEPs.

We begin in Sect. 2 by reviewing key SEP observations, on which our current understanding is based and which need to be explained by any overarching SEP theory. In Sect. 3 we briefly touch upon mechanisms for particle acceleration that are thought to operate at the

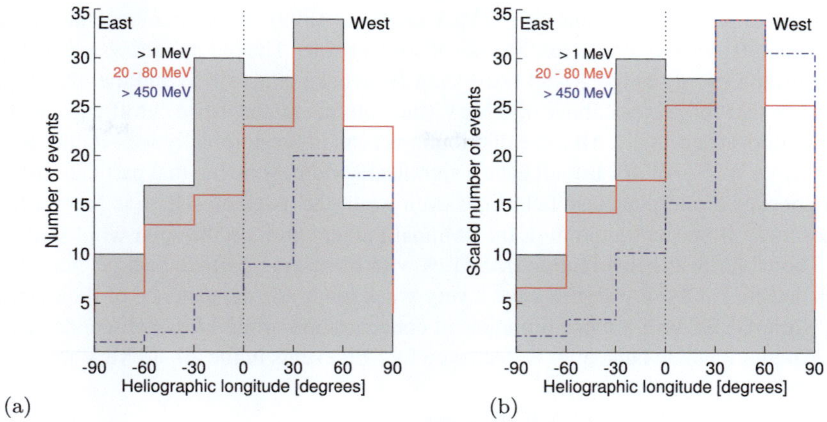

Fig. 1 Heliolongitude distribution of the flares associated with SEP events in different energy ranges: gradual events at energies >1 MeV (*black line* and *grey background shading*; numbers rebinned into 30°-wide intervals, after Fig. 2.3 of Reames 1999); 20–80 MeV (*red solid line*) from Van Hollebeke et al. (1975); >450 MeV (*blue dashed–dotted line*) from the list of ground level enhancements (GLEs) at www.nmdb.eu. Panel (**a**) gives the actual event numbers, (**b**) the numbers scaled to the maximum of the distribution of >1 MeV events

Sun, and in Sect. 4 we discuss the links and timings between remote sensing signatures of acceleration and SEP detection in interplanetary space. Section 5 summarises studies of statistical relationships between solar activity and SEP occurrence and parameters. In Sect. 6 the particle transport through the corona and interplanetary space is discussed. Approaches to forecasting SEPs within a Space Weather perspective are reviewed in Sect. 7 and finally Sect. 8 presents our conclusions and discusses future perspectives.

2 Key Observations

A large amount of observations have been gathered on SEPs over the past decades. It is beyond the scope of this article to review all available experimental information, but in this Section we will focus on the main features of SEP observations. The reader is referred to reviews by Reames (1999), Klecker et al. (2006) and Desai and Giacalone (2016) for additional information.

2.1 Location of Parent Active Region

SEP enhancements associated with solar flares and Coronal Mass Ejections (CMEs) can last from just a few hours to several days. Whether or not an SEP event will be detected, for example near Earth, depends on the location at the Sun of the parent event: eruptions at Western locations on the solar disk have a much higher likelihood of resulting in an SEP event at Earth, due to the curvature of the Parker spiral interplanetary magnetic field. However, in many events particles can be detected at locations widely separated, either in longitude (*e.g.*, Richardson et al. 2014) or in latitude (*e.g.*, Dalla et al. 2003), from the parent region. While this was thought initially to apply only to gradual events, it has been shown that impulsive events can also be detected over wide longitudinal ranges (Wiedenbeck et al. 2013). In some cases particles can spread over 360° in longitude (Gómez-Herrero et al. 2015).

Observed longitude distributions of the flares associated with major SEP events are displayed in Fig. 1, for three separate SEP energy ranges. The black histogram and grey shading show gradual SEP proton events at $\sim$1 MeV (Reames 1999) (where numbers have been re-binned into 30°-wide intervals for the sake of comparison). The red line shows the histogram for the proton energy range 20–80 MeV (Van Hollebeke et al. 1975) and the blue dashed–dotted line that for protons above 450 MeV (the nominal atmospheric cutoff energy of neutron monitors at sea level). All three distributions peak in the nominally well-connected longitude range W30°–W60°. But all have a significant width, showing that particles can reach Earth-connected interplanetary field lines even when the associated flare is several tens of degrees away from the footpoint of the nominal Parker spiral on the solar wind source surface. The width of the distribution decreases with increasing particle energy: gradual SEP events detected at MeV energies have a very weak longitude dependence of the associated flare, while GLEs have a more pronounced concentration around the well-connected field lines. An intermediate behaviour is displayed by SEP events in the 20–80 MeV range.

2.2 Particle Intensities and Anisotropies

Profiles of particle intensities versus time vary depending on the location of the parent event, with SEPs from Western events displaying a fast rise and generally shorter duration, and those from Eastern events having a much more gradual rise phase (Cane et al. 1988). Peak intensities show a strong dependence on the longitudinal separation from the source active region and tend to be largest at the spacecraft best connected to the region (*e.g.*, Richardson et al. 2014). In some events intensities show a large peak at energies up to $\sim$100 MeV at the time when a CME-driven interplanetary shock passes the Earth. The phenomenon is termed Energetic Storm Particle (ESP) event. ESP events are not addressed here in detail. Recent reviews of observations and theory are given in Lee et al. (2012) and Desai and Giacalone (2016).

Anisotropies, describing the degree to which particles are beamed when arriving at the detecting spacecraft, are an important property of SEP events that can be obtained from sectored particle instruments. Anisotropies tend to be larger at the beginning of an event, and usually they quickly fall to zero, indicating isotropy of the particle distribution. In a recent study of electron anisotropies detected by STEREO, it was shown that in many events anisotropies are large at the spacecraft best connected to the flare region while they are small at less well connected spacecraft, although other types of multi-spacecraft anisotropy signatures are also possible (Dresing et al. 2014).

The energy range over which SEPs are observed at 1 AU varies considerably between different events and is of course dependent on instrumentation and background fluxes. In the largest events, the presence of protons of energies up to tens of GeVs can be detected through neutron monitors at the Earth's surface, in so-called Ground Level Enhancements (GLEs). Measurements of solar γ-rays by the FERMI/LAT instrument have shown that the acceleration of protons of energies above 300 MeV is a common occurrence in the solar corona (see Sect. 4.3) and can even be observed in M-class flares

2.3 Elemental Abundances and Charge States

Measured properties of heavy ion SEP populations provide important information on their acceleration and propagation. Both depend on the mass-to-charge ratio m/q. Heavy ion SEPs detected at 1 AU are typically partially ionised.

Heavy ion properties provide some key criteria for separating events into gradual and impulsive within the two-class scenario (Reames 1999). In the initial version of this scenario impulsive events are characterised by values of the event-averaged Fe/O around 1 (about 10

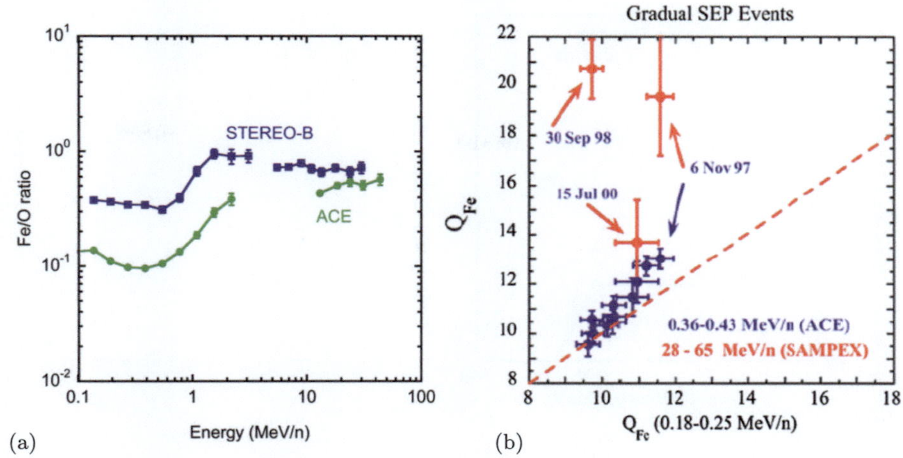

Fig. 2 (**a**) Energy dependence of Fe/O ratio for the event of 2013 April 11 (Cohen et al. 2014). ©AAS. Reproduced with permission. (**b**) Compilation of energy dependence of Fe charge states for a number of SEP events (Klecker et al. 2007). ©Springer. Reproduced with permission

times larger than typical coronal values), while gradual events are less rich in iron, with values around 0.1. Impulsive events show large enhancements in the event-averaged $^3He/^4He$ ratio (about 1000 times larger than typical coronal values), while the ratio is typically orders of magnitudes lower in gradual events. Charge states of Fe in impulsive events are typically higher ($Q_{\mathrm{Fe}} \simeq 20$) than in gradual events ($Q_{\mathrm{Fe}} \simeq 12$) (Reames 1999).

The above criteria were established in the 1990s using SEP data for heavy ions in the $\sim$1 MeV/nuc energy range. It was realised later that charge states are in fact strongly dependent on the energy of the ions. In the common impulsive SEP events, charge states at energies below 1 MeV/nuc were found to increase with increasing energy (Klecker et al. 2007; DiFabio et al. 2008). In several large ('gradual') events the charge states were shown to increase with energy above 1 MeV/nuc such as to approach values at several tens of MeV/nuc that were considered as typical of 'impulsive' SEP events in the 1990s (Cohen et al. 1999; Mazur et al. 1999; Möbius et al. 1999; Mewaldt et al. 2006; Klecker et al. 2007; Rouillard et al. 2012). The left-hand panel in Fig. 2 shows the Fe/O ratio during an event as a function of particle energy, observed by two spacecraft at different heliolongitudes. At both spacecraft the Fe/O ratio increases with increasing energy above $\sim$0.7 MeV/nuc. In addition, the ratio is different at different vantage points. The right-hand panel shows event-integrated Fe charge states in two energy ranges (blue and red crosses, respectively) as a function of charge states in the energy range 0.18–0.25 MeV/nuc for several gradual SEP events. Each cross designates a different event. Events with energy-independent charge states are on the inclined dashed line. The blue crosses are mostly close to this line, showing that charge states in the ranges (0.36–0.43) MeV/nuc are the same as those in the (0.18–0.25) MeV/nuc range. However, at energies above 20 MeV the average charge of the Fe ions is significantly higher, with charge states typical of the classical value for impulsive events on 1998 Sep. 30 and 1997 Nov. 06. Elemental abundances also display a time variation in individual events. For example the Fe/O ratio often decays over time from values considered typical of impulsive events, to much lower values (Tylka et al. 2013; Mason et al. 2006). Other ionic ratios also display a time dependence and this is dependent on their m/q values (Zelina et al. 2017).

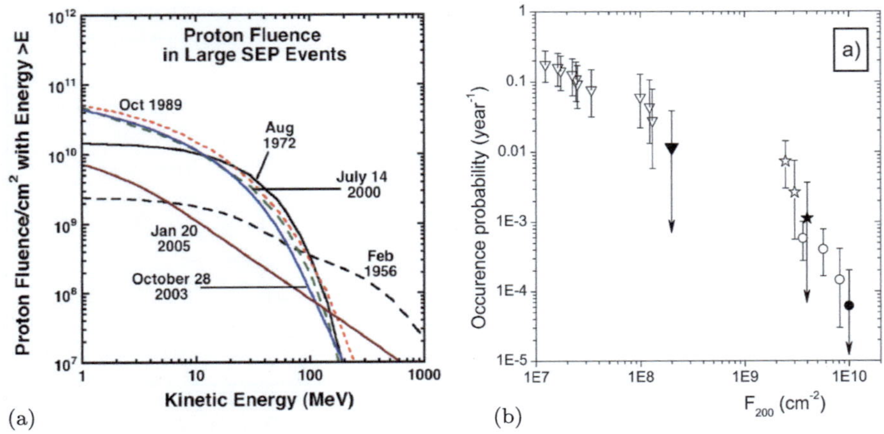

Fig. 3 (**a**) Fluence spectra of strong SEP events observed from spacecraft (reproduced from Mewaldt et al. 2007, with the permission of AIP Publishing). (**b**) Cumulative occurrence probability of yearly fluences at energies above 200 MeV inferred from neutron monitor observations of relativistic SEP events and from radionuclides in ice cores (Kovaltsov et al. 2014). ©Springer. Reproduced with permission

2.4 Extreme SEP Events

Studies of space weather effects from SEPs can look back to about 50 years of space observations at energies up to a few hundreds of MeV, and to 70 years of ground-based measurements of relativistic SEP events, up to some GeV or even a few tens of GeV. Fluence spectra of strong events are shown in Fig. 3(a). The spectra can be divided in two categories: (1) events with high fluence at several tens of MeV, with steep spectra towards higher energies (typical examples: August 1972, October 1989, July 2000), and (2) relativistic particle events, called ground-level enhancements (GLEs) that have spectra extending into the GeV range, but are much weaker at energies of tens of MeV than the first category. The distinction is not bimodal, since the events of Aug. 1972, Oct. 1989 and Jul. 2000 also were GLEs. But high particle intensities at tens of MeV and several GeV do not appear to be correlated, and extreme event scenarios must be elaborated separately for the two categories. This is important since the two relevant energy ranges define different potential space weather hazards: impact on satellites in exposed orbits and on ionisation of the high polar atmosphere at tens of MeV, and tropospheric effects through the formation of atmospheric cascades at several GeV. The reason for the difference is that the extreme fluences at tens of MeV are not dominated by acceleration at the Sun, but near Earth, when it encounters a CME-driven interplanetary shock. The high particle fluxes come from the proximity of the accelerator, rather than its intrinsic strength. GeV particles are accelerated closer to the Sun.

The probability of occurrence of SEP events at energies of tens of MeV has been studied, *e.g.*, by Xapsos et al. (1999). From an analysis of SEP events in cycles 20 to 22, considering only events occurring in a time interval from 2.5 years before to 4.5 years after the sunspot maximum, these authors inferred a power-law with a rollover at high fluences and a smooth approach of zero at some maximum fluence. The extrapolated maximum fluence at energies above 30 MeV is found to be near $1.3 \cdot 10^{10}\ \mathrm{cm}^{-2}$, about 1.6 times the highest fluence observed. There are probably some differences in the calibration with Fig. 3(a), which shows slightly lower fluences than Fig. 1 of Xapsos et al. (1999). But this does not affect the above ratio. Worst case scenarios for prescribed mission durations are also presented by these authors. More recent analyses of extreme events relevant to space missions are, *e.g.*, Jiggens et al. (2014).

The data base for relativistic solar particle events goes beyond the space age. Direct measurements date back to the 1940s, showing events that are much stronger than during the space age (McCracken 2007). If transient enhancements in the traces of cosmogenic radionuclides in ice cores and tree rings, such as ^{10}Be and ^{14}C, can be related to past solar events, the historic view covers the last 10,000 years. This has been exploited by Kovaltsov et al. (2014) to establish the cumulative occurrence probability of yearly ion fluence at energies above 200 MeV (Fig. 3(b)). The result is that once every 100 years a yearly fluence can be expected of about 5 times or more the strongest yearly fluence observed by a network of neutron monitors so far, that is the fluence in 1956, with the strongest GLE ever seen—GLE1956Feb23 (GLE 5). This estimate does not depend crucially on the interpretation of signatures in ice cores or tree rings. Those suggest that the distribution function derived from observed GLEs rolls over at a fluence near $2 \cdot 10^9$ cm^{-2}, which is about 25 times the 1956 fluence. There is still some debate, however, if the high-fluence branch in the plot is really related to solar events. Furthermore, the fluences are yearly values, not individual events. No signature of the Carrington event in 1859 is identified in these terrestrial archives.

3 Acceleration Processes

The association of SEP events with flares and CMEs leads to the distinction of two groups of acceleration mechanisms. One is related to magnetic reconnection and turbulence in active regions, exemplified by solar flares and particularly the hard X-ray, gamma-ray and microwave signatures during the impulsive flare phase. The other is particle acceleration at large-scale shock waves, exemplified by interplanetary shocks. Energetic storm particle events accelerated when such a shock passes near Earth are a prime example.

The neat distinction between the two types of processes may be difficult in practice. Magnetic reconnection, for instance, is not restricted to the impulsive flare phase, but may also occur in the downstream region of CMEs, where current sheets are formed, and at the interface between the outward propagating CME and the ambient corona. Shock acceleration may also occur in reconnection regions, at the interface between the reconnection jet and its coronal environment (see Warmuth et al. 2009, and references therein).

3.1 Particle Acceleration in Solar Flares

3.1.1 Observational Evidence on the Location of the Acceleration Region

Hard X-ray imaging with Yohkoh and RHESSI frequently shows configurations suggesting magnetic loops with a particle acceleration region near or above the loop top. The RHESSI image of the 2005 Jan 20 flare (Fig. 4(a)) is a prime example: the red contours outline a thermal X-ray source, which traces the upper part of a coronal loop. The blue contours are the sources of hard X-ray emission from the chromospheric footpoints. They project onto flare ribbons seen in UV (the two elongated grey bands), which outline the part of the chromosphere heated by energy deposition during the flare. This source morphology is generally interpreted as a signature of energy release near or above the loop top, which heats the plasma in the coronal loop and accelerates electrons that escape from the primary acceleration site as beams. Because of their high energy they interact very little with the dilute coronal plasma and precipitate into the dense chromospheric footpoints, where they

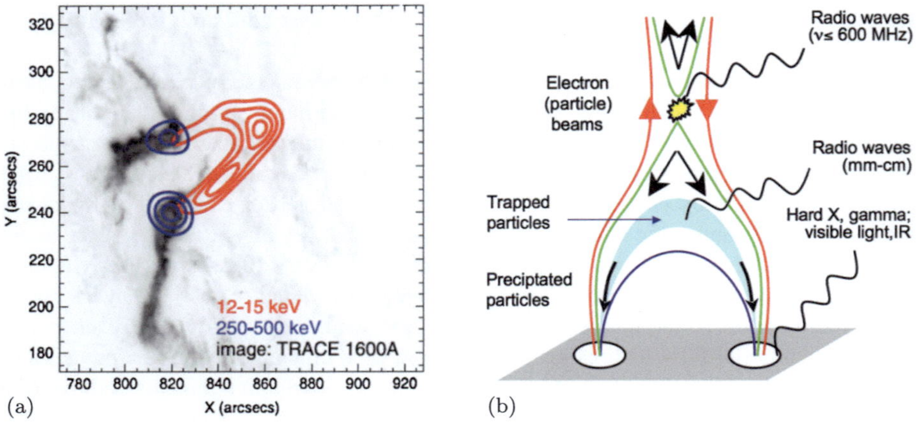

Fig. 4 (**a**) Contours of hard X-ray emission in two spectral bands (RHESSI) superposed on a TRACE image of chromospheric flare ribbons in UV (Krucker et al. 2008b). ©AAS. Reproduced with permission. (**b**) Cartoon scenario of particle populations and related electromagnetic emissions during a flare

lose their energy instantaneously through collisions with the ambient medium, while simultaneously transforming a small amount of energy into hard X-ray emission. The RHESSI image is a snapshot: during the impulsive phase of the event the X-ray sources occur in an irregular temporal succession at neighbouring places, but the hard X-ray footpoints are always located on the UV ribbons. A standard cartoon scenario is shown in Fig. 4(b). The acceleration is ascribed to energy release above the loop top, probably related to magnetic reconnection. The upward field lines may be part of a plasmoid that is ejected upward, or they may be open to the high corona. The energy release may equally well be related to magnetic reconnection with another closed magnetic structure, as in the classical scenario in Heyvaerts et al. (1977). The observation (Sui et al. 2004) of pairs of hard X-ray sources at flare loop tops with temperature gradients pointing to a region between them corroborates the idea that the basic flare energy release occurs near or above loop tops in active regions.

From the energy-dependent timing of hard X-ray intensity peaks and from the observation that decimetric radio emission shows bidirectional electron beams, with downward-directed beams at high frequency (high ambient electron density) and upward directed beams at low frequency, Aschwanden and coworkers concluded that the typical acceleration region is placed at an altitude of about 1.5 times the half-length of the magnetic loop above the photosphere, with an ambient density of about $(1\text{–}10) \cdot 10^9\ \mathrm{cm}^{-3}$ (see Sects. 3.3 and 3.6 of Aschwanden 2002, and references therein). Recent analyses show that similar bidirectional streams can also show up at higher frequencies, and hence higher ambient electron densities: Tan et al. (2016) derive electron densities in the acceleration region in the range $(1\text{–}10) \cdot 10^{10}\ \mathrm{cm}^{-3}$.

An alternative estimate of the thermal electron density in the acceleration region can be obtained from the Fe charge states measured in SEPs, Q_{Fe}. High values of $Q_{\mathrm{Fe}} \simeq 20$ (to be compared with $Q_{\mathrm{Fe}} \simeq 12$ in the quiet corona) are observed in impulsive SEP events and at high energies in some gradual SEP events: if these charge states are to be explained by collisional stripping in the acceleration region, the product of the election density n_e and the residence time in the acceleration region τ must be $(1\text{–}10) \cdot 10^{10}\ \mathrm{cm}^{-3}\ \mathrm{s}$ (Kocharov et al. 2000; Kartavykh et al. 2007). This is consistent with the high electron densities inferred by Tan et al. (2016), given that the time scales of hard X-ray emission suggest acceleration times $\tau < 1$ s for the electrons. The acceleration time of interacting protons at energies of tens of MeV is unlikely to be significantly higher, since similar time scales of hard X-ray emission and nuclear gamma-ray emission have been found in some flares where the gamma rays were observed with high time resolution (Kane et al. 1986; Trottet et al. 1993).

3.1.2 Current Sheets, Magnetic Reconnection and Particle Acceleration

Magnetic reconnection is widely believed to be the fundamental process explaining both the restructuring of coronal magnetic field configurations and the conversion of the magnetically stored flare energy. Particle acceleration can be due to the direct electric field in the diffusion region, turbulence created during the process, or the reconnection shock that may form as the outflow jets from the diffusion region impinge onto underlying magnetic field structures. These processes are not expected to be neatly separated in a realistic scenario. Current sheets related to solar flares have spatial scales that are largely below the spatial resolution of any imager. They are expected to be highly fragmented and highly dynamic, with the formation and coalescence of magnetic islands, and the retraction of reconnected magnetic field lines (Kliem et al. 2000). The interaction of particles wandering around within a sea of magnetic islands can lead to more efficient particle acceleration than a single reconnection region as considered in many earlier studies (*e.g.*, Heerikhuisen et al. 2002). The fragmentation of current sheets is also physically more realistic than reconnection in a single large-scale structure. A detailed discussion is given in Cargill et al. (2012). Multiple current sheets can also be formed throughout a magnetic loop that is twisted by motions of the photospheric plasma at one of its footpoints (Gordovskyy et al. 2013). Such a scenario with distributed sites throughout the volume of the loop may resolve problems with understanding the stable transport of intense electron beams over macroscopic distances from the coronal acceleration sites to the chromospheric footpoints where the hard X-rays are emitted. Recent reviews of particle acceleration in solar flares can be found in Holman et al. (2011), Fletcher et al. (2011) and Vilmer et al. (2011).

Scenarios of CME eruption include current sheets either as the fundamental driver of the eruption or as a consequence. In coronographic observations bright ray-shaped features that form in the aftermath of the ejected structure are often interpreted as current sheets. The visible features are dense, and might better be called plasma sheets. But judging from their geometry they must contain current sheets. UV coronographic spectroscopy shows they start hot, at temperatures of several tens of MK, and cool down over time scales of several hours. This is accompanied by the appearance of new loop structures over an extended range of temperatures and wavelengths, and by signatures of sunward-retracting structures, which is reminiscent of the depolarization observed in magnetic substorms. Reviews of recent work are given in Raymond et al. (2012) and Lin et al. (2015). These post-eruptive current sheets are likely regions of long-lasting electron acceleration seen in metric radio bursts (Aurass et al. 2009; Benz et al. 2011; Démoulin et al. 2012a, 2012b), and were suggested to also be a site of long-lasting acceleration of high-energy ions that escape to the interplanetary space (Carmichael 1964; Akimov et al. 1996; Klein et al. 2014).

3.1.3 Stochastic Acceleration

Stochastic acceleration generalises Fermi's idea that charged particles gain energy by encounters with randomly moving magnetic obstacles, because energy-increasing head-on collisions occur more often than energy-reducing trailing collisions. Instead of macroscopic magnetic obstacles, any fluctuating electric field in different types of plasma waves and in turbulence can provide the acceleration. A review of stochastic acceleration especially in solar flares is given in Petrosian (2012).

Turbulence is expected to arise in many situations where plasma flows occur. In the solar atmosphere such plasma flows occur in magnetic reconnection, but also at the interface between jets or CMEs and the ambient corona. In a flare scenario as shown by the cartoon in Fig. 4(b) magnetic energy released during magnetic reconnection or other processes is expected to create high levels of turbulence on MHD scales, which will cascade to smaller scales. Charged particles may be trapped in this turbulence around the loop top, and be accelerated. This could explain why loop top sources are commonly seen in hard X-ray images (Krucker et al. 2008). They are especially observed whenever the bright footpoints, which dominate the hard X-ray morphology of flares on the disk, are occulted by the solar disk.

The cyclotron frequency of particles with charge $q = Qe$ and mass $m = Am_p$, where m_p is the mass of the proton, is $\Omega_c = \frac{Q}{A}\Omega_{cp}$. $Q/A = 1$ (protons), 2/3 (^{3}He), 1/2 (^{4}He and heavier elements, if they are fully ionised). In the corona, H and He are expected to be fully ionised, but the highest charge states of Fe observed in SEP events are $Q \simeq 20$. This corresponds to $Q/A \leq 0.4$. As MHD waves cascade towards smaller spatial scales, i.e. increasingly high frequencies, they are hence expected to first resonate with ion species having low cyclotron frequencies, which may then be expected to be more efficiently accelerated than species with higher Q/A ratio. This expected behaviour fits well the observed ordering of enhanced abundances by the Q/A-ratio in small impulsive SEP events displayed in Fig. 4.4 of Reames (1999). For this reason stochastic acceleration is considered as the archetypal "flare acceleration" process in the SEP literature.

However, the abundance enhancements due to the cascade of MHD waves cannot account for the highly efficient acceleration of ^{3}He, which has a high cyclotron frequency. Interpretations of ^{3}He enrichments include the resonance with waves produced by electron beams (Miller et al. 1997; Aschwanden 2002) or the presence of multiple resonances of ^{3}He with waves when the ^{4}He population is accounted for in the dispersion relations of the waves (Petrosian 2012).

3.2 Shock Acceleration

Shocks are ubiquitous in the corona, including CMEs and jets. Shock acceleration mechanisms are reviewed in Jones and Ellison (1991), Krauss-Varban (2010), Lee et al. (2012) and Desai and Giacalone (2016). Below we briefly discuss shock drift acceleration, which is mainly expected at quasi-perpendicular shocks (i.e. in regions of the shock where the normal on the shock front makes a large angle with the upstream magnetic field), and diffusive shock acceleration.

3.2.1 Shock-Drift Acceleration

In the shock-drift acceleration process particles gain energy from the convective electric field $\mathbf{E} = -\mathbf{V} \times \mathbf{B}$, where $\mathbf{V}$ is the inflow speed of the plasma, $\mathbf{B}$ the upstream magnetic field in the rest frame of an oblique shock. In the upstream region, electrons and ions $\mathbf{E} \times \mathbf{B}$-drift towards the shock front. Because the magnetic field is compressed by the oblique shock, they undergo a gradient drift along the shock front. This drift is directed along the electric field for positively charged particles, and opposite to the electric field for negatively charged particles. Hence electrons, protons and ions gain energy. Depending on the energy and pitch angle before the first encounter with the shock, the energy gained from the drift along the convection electric field may be such that the particle is again injected into the upstream medium and may escape. Besides a gradient drift, particles also undergo a curvature drift while their guiding centre travels along the magnetic field, which is curved in the shock transition. In a planar fast shock, the angle between a field line and the shock normal is larger downstream than upstream, and the curvature drift is opposite to the gradient drift. The curvature drift hence leads to energy losses. But the drift speed decreases as the shock becomes more oblique, so that in quasi-perpendicular shocks the gradient drift, and hence the energy gain, dominates. The acceleration process is widely advocated for electrons in type II radio bursts (Holman and Pesses 1983; Mann and Klassen 2005). But it also applies to SEP-acceleration, especially when particles encounter the shock repeatedly, for instance when they are trapped in the upstream region of a shock that propagates into increasing ambient magnetic field strengths or in a curved ambient magnetic field (Sandroos and Vainio 2006).

3.2.2 Diffusive Shock Acceleration

When ions are reflected at shock waves and stream into the upstream region, they have a beam-like distribution and are therefore likely to generate waves, parallel-propagating Alfvén waves as well as obliquely propagating fast magnetosonic waves. When these waves grow to sufficient amplitudes, they can scatter subsequent ions back to the shock. Since the shock propagates faster than these waves, ions find themselves confined between approaching scattering centers downstream and upstream, and gain energy by bouncing back and forth through the shock font, until they eventually escape. The process requires that particles be able to escape into the upstream medium after the initial reflection to interact with the waves. This means that they must stream away from the shock front at a minimum speed $V/\cos\theta_{\mathrm{Bn}}$, where V is the speed of the shock, and θ_{Bn} the angle between the shock normal and the upstream magnetic field vector. The diffusive shock acceleration process is therefore expected to work best at quasi-parallel shocks. Numerical simulations indicate that it can accelerate protons to relativistic energies (Afanasiev et al. 2015). When the shock becomes more oblique, an increasing fraction of the reflected particles will be unable to escape, and will be convected into the downstream region and thermalised. This is the reason why the injection energy into the diffusive acceleration process is expected to increase with increasing shock angle.

On the other hand, quasi-perpendicular shocks accelerate particles faster and to higher energies. There is hence a theoretical distinction between (1) fast acceleration to high energies at quasi-perpendicular shocks, out of a suprathermal seed population, and (2) slower acceleration to lower energies at quasi-parallel shocks, which can accelerate ions out of the thermal background. In simple scenarios predominantly perpendicular shocks are expected to exist on the flanks of a CME, while shocks near the nose are expected to be quasi-parallel when the surrounding magnetic field is more or less radial, i.e. almost everywhere in the high corona and the interplanetary medium (see Fig. 4 of Rouillard et al. 2016). However, realistic shocks are expected to have rippled surfaces so that local geometries may be quasi-parallel or quasi-perpendicular irrespective of the location on the shock front.

3.2.3 Seed Particle Population

The idea that a CME-driven shock would accelerate particles out of the ambient coronal and interplanetary plasma was a prominent justification for the bimodal impulsive-gradual SEP scenario. It seems a natural explanation why the elemental abundances and Fe charge states in gradual SEP events were about coronal at energies up to about 1 MeV/nuc (*e.g.*, Reames 1999), which was the limit of the energy range accessible to in situ measurements of abundances and charge states at that time. This line of reasoning was challenged subsequently by the observation that both charge states and abundances of gradual SEP events depend on energy, and come closer to those observed in impulsive SEP events at energies above 10 MeV (see Sect. 2). One way of addressing this problem within a model of shock acceleration is to invoke seed populations that already contain the overabundant elements and the high charge states. In a study of impulsive SEP events from several active regions, Wang et al. (2006) noticed numerous releases of ^{3}He-rich particle populations related to coronal jets, and observed enhanced levels of ^{3}He throughout the two to three days during which the spacecraft was magnetically connected with the parent active regions. This finding corroborates the idea (Mason et al. 1999; Mewaldt et al. 2012) that coronal acceleration episodes during flares or minor activity populate the high corona with enhanced levels of "flare suprathermals" out of which CME-shocks accelerate particles on occasion to the high energies observed in gradual SEP events. The observed abundances and charge states would then rather reflect the seed population than the acceleration process.

The argument that quasi-perpendicular shocks need a seed population with higher energy than quasi-parallel shocks was used to explain energy-dependent differences in the composition between large SEP events (Tylka and Lee 2006). In these authors' view quasi-perpendicular shocks accelerate suprathermal seed populations injected by previous flares, while quasi-parallel shocks would rather accelerate particles out of the thermal background. This clear distinction depends much on the pre-existing turbulence in the acceleration region. Giacalone (2005) argues that a sufficiently high level of turbulence allows particle acceleration out of the thermal background both at quasi-parallel and quasi-perpendicular shocks. The injection problem as a function of shock angle is discussed, *e.g.*, in Caprioli et al. (2015) and Desai and Giacalone (2016).

4 Timing Relationships Between Interacting Particles and SEPs

4.1 A Solar Energetic Particle Event Observed by HELIOS Near 0.4 AU

The most direct indication of a physical relationship between SEPs in space and non thermal particles interacting with the solar atmosphere is the identification of common time structures. But SEP time profiles observed near 1 AU are usually smeared out by transport in the turbulent interplanetary magnetic field (*e.g.*, Dröge 2000—see Sect. 6). *Helios* observed a few SEP events while within 0.4 AU from the Sun. One of the rare events with a well-defined time structure is shown in Fig. 5(a) (from Kallenrode and Wibberenz 1991). It was observed when the spacecraft was 0.38 AU from the Sun. The abscissa gives the time when the particles were released at the Sun, computed by subtracting $\sim$500 s from the photon arrival time at Earth and the travel time of the particles along the nominal Parker spiral to *Helios*. The two curves in the bottom panel show that deka-MeV protons started to be released about 10 min after near relativistic electrons. The apparent new rise in the electron profile at that time may be due to cross-talking protons. The important feature is that the releases of the two particle populations are accompanied by distinct episodes of particle acceleration in the corona, as shown by the double-peaked profile of the radio emission at 8.8 GHz, which is gyro-synchrotron emission of near-relativistic electrons (100 keV to a few MeV) in flaring loops in the low corona. The hard X-ray time profile shows a similar

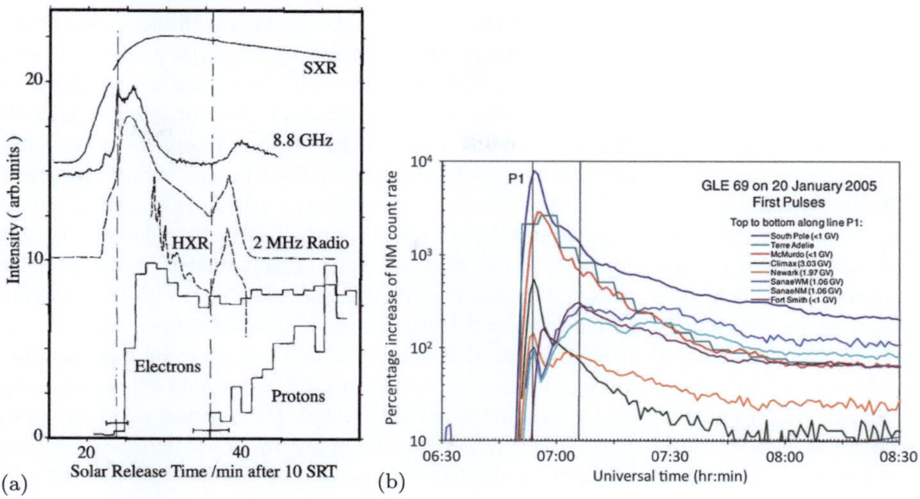

Fig. 5 Time structure of SEP events: (**a**) *Helios* SEP measurements (electrons 300–800 keV, protons 27–37 MeV) and associated radio and X-ray emission (HXR: hard X-rays, SXR: soft X-rays) in the corona on 1980 June 08. The abscissa ("solar release time") gives the time when the particles (photons, electrons, protons) were released at the Sun (adapted from Kallenrode and Wibberenz 1991; ©AAS, reproduced with permission). (**b**) Neutron monitor profiles measured during he GLE on 2005 January 20 (Moraal and McCracken 2012). ©Springer. Reproduced with permission

structure, but the SMM spacecraft was in the Earth's shadow when the burst began. The radio profile at 2 MHz (ISEE 3) has also two bursts (type III bursts). They are produced by electron beams streaming along open magnetic field lines to the interplanetary space. The simultaneous occurrence of microwave bursts from electrons accelerated in a flare and type III bursts hence demonstrates that electrons accelerated during the two microwave bursts can escape from the flaring active region. The succession of an initial electron-rich SEP phase and a second proton-rich phase is in this case related to two distinct episodes of electron acceleration in the corona. The first, electron-rich phase is connected to a burst during the rise of the soft X-ray burst, shown in the top curve of Fig. 5, hence to the impulsive flare phase. The proton-rich acceleration episode occurs after the soft X-ray peak, in the so-called gradual flare phase. The pronounced difference in the e/p-ratio suggests that the two episodes of particle acceleration did not occur in the same region or involved different mechanisms, or both. The association of each of the two SEP releases with a microwave and a hard X-ray burst argues for closely related particle acceleration processes occurring in the flaring active region and its surroundings.

4.2 Time Structure of SEPs and Transport Modelling

When observed near 1 AU, particle events with little pitch angle scattering in the turbulent interplanetary magnetic field are rare. But occasionally electrons or relativistic protons may display long mean free paths (see Fig. 4 in Dröge 2000). Agueda et al. (2008) developed a model of interplanetary electron transport where the time profiles of the intensity and anisotropy of electrons at energies of a few tens to a few hundreds of keV observed in space are reproduced by a series of elementary releases in the corona, at a solar radius above the photosphere. The 1D interplanetary transport model considers focussing along the magnetic field by the overall decrease of the magnetic field strength with increasing heliocentric distance, and pitch angle scattering by the interplanetary magnetic field turbulence. These authors showed that short, impulsive SEP events are well represented by short releases of near-relativistic electrons at the Sun. The inferred releases are accompanied by type III bursts at decametre-to-kilometre wavelengths, which reveal electron beams travelling from the high corona to 1 AU. In electron events of long duration the situation is more complex, with electron release near the Sun that can last several hours (Agueda et al. 2014). These long-duration releases are not accompanied by type III bursts, but by long-duration broadband bursts at dm–m wavelengths (type IV), which are presumably related to reconnection in current sheets behind the rising CME, and by type II bursts related to shock waves in the corona (metre wavelengths) and interplanetary space (kilometre wavelengths).

Evidence of successive distinguishable SEP releases within a given event has been discussed in recent systematic analyses of ground-level events (GLEs), which are produced by relativistic protons with typical energies in the few GeV range. These primary particles generate atmospheric nuclear cascades that can be detected in ground-based particle detectors, neutron monitors, neutron telescopes or muon telescopes. GLEs often have a double-peaked structure, with an initial fast rise and an anisotropic particle population—called the 'prompt component'—followed by a more gradual and less anisotropic 'delayed component' (see Chap. 7.3 of Miroshnichenko 2001). McCracken et al. (2012) demonstrate that the sequence of an anisotropic impulsive peak, and a less anisotropic gradual peak occurring 7–15 min later is a common occurrence when the parent active region is magnetically connected to the Earth, while the absence of the impulsive peak is typical of poorly connected activity near to or east of the central solar meridian or well beyond the western limb.

A prominent case illustrating this double-peaked structure is the GLE of 2005 Jan. 20. It displayed a well-defined, rapidly rising time profile at the beginning and a distinct second peak a few minutes later. The difference is clearly seen in the neutron monitor time profiles plotted in Fig. 5(b). In a detailed timing comparison, Klein et al. (2014) identified microwave and type III counterparts to the two proton peaks similar to the case of 1980 June 08 in Fig. 5(a). The second GLE peak was in addition found to occur during a metre wave type IV burst, which is generally ascribed to electron acceleration in the post-CME current sheet. Using the time profile of the microwave burst as the injection profile of relativistic protons at the Sun and modelling the interplanetary transport of the protons in the focussed transport model developed by Agueda et al. (2008), Klein et al. (2015) showed that a consistent timing could be found with the neutron monitor profiles. This gives support to the idea that different acceleration processes in the low corona successively release the relativistic protons that are observed at Earth. An alternative possibility, advocated by Moraal and McCracken (2012) for the GLE and by Kallenrode and Wibberenz (1991) for the *Helios* observations, is the successive acceleration of the first SEP population in the impulsive flare phase and of the second SEP population at the CME shock. In this scenario it is less clear why there is a timing relationship between SEP acceleration at the CME shock, which at that time is at a heliocentric distance of a few solar radii (*e.g.*, Klein et al. 2014), and radiative signatures of renewed electron acceleration in the corona behind the CME. More complex shock acceleration scenarios in the corona were devised that could account for such a situation (Pomoell et al. 2008). But the direct contribution from acceleration processes during magnetic restructuring of the low corona that are expected in the impulsive flare phase, when the CME lifts off, and in the post-CME corona, is a plausible alternative in the light of these observations. The escape of particles accelerated in a current sheet behind the CME is not fully understood. Reconnection of the closed CME structures, supposedly containing the accelerated particles, with neighbouring open field lines could be essential here, as modelled by Masson et al. (2013).

4.3 Relativistic Nucleons in the Solar Atmosphere

The consistent timing between post-impulsive radio or hard X-ray emissions produced by energetic electrons and SEPs suggests that protons are accelerated in the post-CME corona, sometimes up to relativistic energies. Events where radiative diagnostics of energetic ions were reported (Kanbach et al. 1993; Kocharov et al. 1994; Akimov et al. 1996; Hudson and Ryan 1995) have long been considered as exceptional, because the detectors of nuclear gamma-ray emission were not sensitive enough or were contaminated by protons after the impulsive flare phase. FERMI/LAT changed this situation in discovering that high-energy gamma-ray emission, attributed to the decay of pions produced by protons with energy above 300 MeV, is by no means an extreme phenomenon. Pion-decay gamma-rays were detected in a number of GOES M class flares. On occasions this emission is observed during several hours (Ackermann et al. 2014). The persistence of the gamma-ray source in the vicinity of an active region shown by some imaging observations (Ajello et al. 2014) during long-duration emission is consistent with acceleration in the low corona. Stochastic acceleration was advocated by Hudson and Ryan (1995). However, pion-decay gamma-rays were also observed from flares behind the limb (Pesce-Rollins et al. 2015). This shows that relativistic protons can radiate far from the parent active region. This observation was ascribed to shock acceleration high in the corona. One then needs to understand how the shock maintains a magnetic connection with a localised region in the low solar atmosphere while travelling over many solar radii out into the heliosphere, and how the downward streaming protons overcome magnetic mirroring. The origin of the high energy gamma-ray events, especially those lasting several hours, and their relationship with SEPs is not yet understood.

5 Association Between SEP Events and Solar Activity

Another attempt to identify the solar origin of SEP events uses statistical relationships. SEP events were only ascribed to flares at a time when CMEs had not been discovered as an independent phenomenon. Since the 1980s CMEs took an increasing importance in the discussion about the origin of SEP events. The ability of CME-driven shocks to accelerate particles all along their way from the Sun to the Earth and beyond is a natural explanation of enhanced SEP intensities that persist over more than a day at energies up to a few tens of MeV (see Kahler 1992; Reames 1999). In an attempt to identify candidate solar events that might produce space-weather relevant SEP events, we discuss in the following whether there are preferential associations or statistical correlations between SEP events and either CMEs or flares.

5.1 Confined and Eruptive Solar Activity

Fast CMEs tend to be accompanied by intense flares and vice versa. However, about 10% of the X class solar flares of cycle 23 were not accompanied by CMEs (Wang and Zhang 2007), they were so-called confined flares. It turned out that these X-ray bursts were not accompanied by SEP events either, even those which were located in the western solar hemisphere, and which therefore were *a priori* magnetically connected with the Earth (Klein et al. 2010, 2011). Besides CMEs, the confined flares also lacked radio counterparts at long wavelengths (metre waves and longer), which indicates that flare-accelerated electrons did not get access to extended magnetic structures in the corona and open field lines towards the Heliosphere. Since the particles accelerated in and around the flare remained confined, and since no CME or CME shock was associated with these flares, no SEP event could be expected. This is confirmed by the impressive series of X-class flares without CMEs that was observed from one active region during its disk passage in October 2014 (Thalmann et al. 2015). It produced six X-class flares (X1.0–X3.1), among them four while in the western solar hemisphere (October 22–27), with no CME. The GOES particle intensity measurements did not exceed background during this time (http://www.solarmonitor.org). As shown in Thalmann et al. (2015), the flares occurred in an environment with strong surrounding and overlying magnetic fields, like in the events analyzed in Wang and Zhang (2007). Electron acceleration accompanied the flares, as shown by the hard X-ray emission. But none of the X-class flares was accompanied by conspicuous type III bursts at decameter and longer waves (Wind/WAVES observations), and none of the three bursts that occurred during observing hours of the Nançay radio observatory was accompanied by a radio burst between 1 GHz and 30 MHz (see http://radio-monitoring.obspm.fr).

5.2 SEP Events Without Flares?

So while confined flares as defined above do not produce SEP events, one can ask the opposite question: do CMEs that have no other particle acceleration signature in the corona, such as hard X-rays or radio emission at cm-to-dm wavelengths, produce SEP events? Kahler et al. (1986) studied a filament eruption and CME without an impulsive flare or radio emission from the low corona, but which was associated with a type II burst at hectometric wavelengths, i.e. a shock wave at heliocentric distances beyond a few solar radii. The CME was accompanied by an SEP event with a steep spectrum, seen up to about 80 MeV, and also by near-relativistic electrons. This is widely considered as a prime example that a CME shock is a sufficient condition to produce SEPs, although it was also pointed out (Cane et al. 2002) that besides the type II burst a type III burst, signalling electron acceleration in the lower corona, was also observed in association with a weak SXR burst. Whatever the interpretation of the filament-associated SEP events, there are at best very few SEP events associated with a CME and no alternative signature of particle acceleration in the corona (see also Marqué et al. 2006).

5.3 Statistical Relationships Between SEP Peak Intensities and Parameters of Solar Activity

Besides considering the common occurrence of SEP events, flares and CMEs, one can search for a preferential correlation between parameters describing the importance of the SEPs on the one hand and of the eruptive solar activity on the other hand. Many years of work led always to the same significant, but noisy type of correlation between the SEP peak intensity, at energies of a few MeV to a few tens of MeV, and the peak flux of the SXR burst on the one hand, the CME speed on the other (*e.g.*, Kahler 2001; Gopalswamy et al. 2003; Cane et al. 2010; Miteva et al. 2013; Richardson et al. 2014). Comparable correlations were found with CME speed and with SXR peak flux. A correlation between the flux of protons above 10 MeV in space and the gamma-ray fluence in the 4–7 MeV range, where nuclear lines are superposed on the bremsstrahlung continuum, was found by Chertok (1990).

Kahler (2001) pointed out that the correlation between SEP intensity and CME speed could be blurred by the contribution of other parameters. He found that higher peak SEP intensities were observed when the pre-event background was higher. He argued that this was evidence that high SEP intensities result when there is a suprathermal seed population for shock acceleration in the high corona and the interplanetary space. CME speed by itself is also not as useful a parameter to characterise the strength of a shock as is the Mach number, which cannot be measured directly in the corona. In a systematic study of fast CMEs, Gopalswamy et al. (2008) showed that those which lacked type II emission at decametric and longer waves had no SEP event. The authors concluded that this pointed to a varying Alfvén speed in the corona, and that even CMEs with speeds as high as 1000 $\mathrm{km\,s^{-1}}$ did not necessarily drive a shock. The existence of a type II burst, however, proved the presence of the shock, and strengthened the idea that the shock revealed by the radio emission also accelerated the SEPs. Clearly, the conventional correlation analysis oversimplifies the problem.

Another reason to explain the failure of classical correlation studies in establishing preferential associations of SEP importance with flare importance or CME speed is the fact that the solar activity parameters are themselves correlated (Kahler 1982; Trottet et al. 2015). More sophisticated statistical methods then suggest that SEP peak intensities are indeed correlated significantly both with CME speed and with flare parameters, especially SXR fluence (Trottet et al. 2015). This argues for a mixed scenario, where processes related with both flares and CMEs contribute to SEP acceleration.

This conclusion is confirmed by a detailed study of the dependence on SEP energy. Most statistical studies were limited to a single energy range. When considering SEP peak intensities in different energy ranges between 1 MeV and 100 MeV, Dierckxsens et al. (2015) discovered a systematic trend: at energies well below 20 MeV, the correlation coefficient with the SEP peak intensity is higher for CME speed than for SXR peak flux, but at energies above 20 MeV the inverse is true. This is new evidence for a mixed acceleration scenario, where different acceleration processes dominate at different SEP energies. This is consistent with the result discussed above that SEP events associated with the eruption of quiescent filaments have steep energy spectra (Kahler et al. 1986; Gopalswamy et al. 2015).

5.4 Comparisons of the Numbers of Particles Interacting in the Solar Atmosphere and Particles in SEP Events at 1 AU

Quantitative comparisons between the spectra and numbers of interacting particles in the solar atmosphere and SEPs are important further clues, but depend strongly on the models employed. Collisional bremsstrahlung at hard X-rays and nuclear gamma-ray emission are in principle understood, and can be modelled. But simple models for the energy spectrum, the angular distribution, and the nuclear abundances of the non-thermal and target particle populations have to be assumed. In SEP events, single-point measurements often have to be used to estimate SEP intensity over an extended range in heliocentric longitudes and latitudes, and the total number is inferred using an assumption of the spatial extent. While multi-spacecraft observations bring supplementary constraints (Lario et al. 2013; Dresing et al. 2014), one still has to suppose a simple smooth variation of the particle intensity with heliolongitude. This may be unrealistic, if particles are released into discrete channels of the interplanetary space, implying an irregular variation of the intensity with spacecraft position, as shown for protons by flux dropouts in impulsive SEP events (Mazur et al. 2000), and for electrons of low energy emitting type III radio bursts (Buttighoffer 1998) as well as for near-relativistic electrons (Klassen et al. 2016). So any quantitative comparison between interacting protons and SEPs at 1 AU cannot be done to better accuracy than an order of magnitude.

Such comparisons were conducted for interacting protons and protons in SEPs at energies of a few tens of MeV. In a study of several events of the SMM era, Ramaty et al. (1993) found an event-to-event variation of the ratio of protons at energies above 30 MeV inferred from gamma-ray modelling and from in situ measurements in a broad range between about 0.01 and 100 (see also Emslie et al. 2005). The number of protons above 30 MeV in the large SEP event on 1990 May 24 was found to be smaller than the number of protons interacting in the low solar atmosphere (Debrunner et al. 1997). For electrons in the energy range between a few tens of keV and a few MeV the situation is different: the number of electrons detected in space is always found to be orders of magnitude smaller than the number required for hard X-ray emission by collisional bremsstrahlung, although a correlation seems to exist between the energy spectra of the two populations (Ramaty et al. 1993; Dröge 1996; Krucker et al. 2007).

6 Particle Transport

Once acceleration has taken place, propagation from the source region distributes particles within the heliosphere, and in some cases they can reach near-Earth space, where they may produce Space Weather effects. The magnetic fields of the corona and interplanetary space are key to transport. The coronal fields have a complex geometry including closed and open magnetic field configurations. The average interplanetary magnetic field can be described as a Parker spiral, superimposed on which are a variety of transient structures as well as turbulence. The latter produces scattering of energetic particles and meandering of the magnetic field lines.

6.1 Coronal Magnetic Field

The longitude distribution of parent flares of SEP events peaks at Western heliolongitudes (Fig. 1), as one would expect considering that the footpoint of the Parker spiral for an Earth observer is located in this region. However, the distributions are broad, and multi-spacecraft observations discussed in Sect. 2 show that in individual cases the longitudinal injection cones may exceed 180°.

The width of the heliolongitude distribution tends to be larger than the range over which the heliolongitude of the Earth-connected Parker spiral on the solar wind source surface varies due to varying solar wind speed. One important contribution to broad particle injection cones is the geometry of the coronal magnetic field above active regions. While active regions are largely composed of closed magnetic fields, there are also open field lines. A well-known illustration of this fact are type III radio bursts, which are produced by electron beams travelling outward throughout the corona and the interplanetary space, and are a very frequent form of radio emission related with solar flares (*e.g.*, Saint-Hilaire et al. 2013). Other evidence comes from the extrapolation of photospheric magnetic field measurements to the source surface using models such as the Potential–Field–Source–Surface (PFSS) model of Schrijver and DeRosa (2003). The open field lines form very narrow bundles at altitudes below, say, 0.5 $R_{\odot}$ above the photosphere, due to the pressure of neighbouring closed magnetic field structures. But the closed magnetic flux diminishes with increasing height, and the open field lines are bound to spread, connecting a tiny part of an active region with a longitude range as large as 90° at the source surface (*e.g.*, Liewer et al. 2004). This spreading was conjectured in the interpretation of early SEP observations (Fan et al. 1968; Reinhard and Wibberenz 1974). Using radio observations of electron beams in the corona, Klein et al. (2008) showed that it explains the observation of simple electron events at Earth related to parent activity as far as 50° away from the nominal Parker spiral longitude. It is, however, not sufficient to explain very broad injection cones observed in some simple impulsive SEP events observed by STEREO (Wiedenbeck et al. 2013). But the spread of open magnetic field lines cannot be ignored in the discussion of particle transport in the corona.

6.2 Classical 1D Models of Interplanetary Transport

The classical description of particle transport in the interplanetary medium relies on a 1D focussed transport equation, describing the effects of focussing along the magnetic field and scattering due to magnetic field fluctuations (Roelof 1969). The only spatial variable retained in the model is the distance measured along a Parker spiral magnetic field line, so that particles are assumed to remain bound to the field line on which they were injected, for their entire propagation. A revised formalism that includes a description of adiabatic deceleration was later developed (Ruffolo 1995) and this forms the basis of several current SEP models.

Within a 1D formalism, the value of the scattering mean free path λ is a key parameter in determining the overall shape of SEP intensity profiles at 1 AU. Many studies used the focussed transport equation to fit measured intensity profiles (and in some cases anisotropies) and derive a value of λ (*e.g.*, Kallenrode 1993). These types of studies generally found regimes of strong scattering for protons, with typical parallel mean free paths of $\sim$0.1 AU, with weak dependence on rigidity (Bieber et al. 1994). For electrons, the radial mean free path is typically between 0.1 and 0.5 AU, showing a tendency to decrease with increasing particle rigidity (Agueda et al. 2014).

The 1D assumption is embedded within the two-class paradigm, in which the wide extent in longitude of many large events is explained as efficient acceleration over a wide portion of a CME-driven interplanetary shock (Reames 1999). The role of scattering in shaping intensity profiles of SEPs accelerated by interplanetary shocks is less clear than for the case of a compact injection region at the Sun. For the former case, in fact, acceleration can continue as the shock has left the corona, resulting in a time extended injection. Models that aim to account for an extended injection from a shock as well as propagation, need to consider that a given observer will be magnetically connected to different parts of the shock over time (Lario et al. 1998; Bain et al. 2016).

6.3 3D Models of Interplanetary Transport

Interaction with turbulence in interplanetary space may produce particle diffusion across the magnetic field, in addition to the effects on propagation along the field, already included in classic 1D models.

For this reason, extensions of the focussed transport equation to include scattering across the magnetic field have been developed (Zhang et al. 2009; Dröge et al. 2010). Here particles may slowly diffuse away from the magnetic field line on which they were injected, requiring a 3D description. Parallel and perpendicular mean free paths are an input to the models.

The need for 3D modelling was also emphasized by studies based on the test particle approach, where the full equations of motion are integrated numerically for an energetic particle population. When this methodology was applied to SEPs for the case of a Parker spiral configuration, it was found that guiding centre drifts associated with the gradient and curvature of the magnetic field produce transport in both longitude and latitude, in the absence of perpendicular scattering (Marsh et al. 2013; Dalla et al. 2013). One important consequence of drift is the associated deceleration (Dalla et al. 2015).

Drifts are particularly strong for SEP heavy ions due to their large m/q, which results from being partially ionised. When heavy ions are injected from a compact area at the Sun, they are found to propagate efficiently to locations that are not magnetically well connected to the injection region. This drift-dominated propagation produces a decrease over time of the Fe/O ratio (Dalla et al. 2017) similar to that observed in SEP events (Mason et al. 2006), for not magnetically well connected observers. The reason for the observed decrease in Fe/O is that Fe drifts more than O due to its larger mass-to-charge ratio, so that events tend to be Fe rich early on.

In addition, a 3D model including drift naturally produces an energy dependence of the 1 AU charge states for an observer without direct connection to the injection region (Dalla et al. 2017), qualitatively similar to SEP observations (Mazur et al. 1999). In fact, due to the dependence of drift velocity on the product of m/q by energy per nucleon, for particles at low energy a low charge state is the only way to ensure significant drift across the field. At higher particle energies, higher charge states are also able to drift efficiently, resulting in a larger measured event charge state. The timescales over which drift effects become important depend on the specific particle species, energy and charge-to-mass ratio. For heavy ions they are important over timescales of the order of an hour.

In addition to scattering of energetic particles, turbulence in the interplanetary magnetic field produces meandering of the magnetic field lines, so that the magnetic connection between a given location in space and the source of an SEP event may differ considerably from what is predicted by a simple Parker spiral model. Laitinen et al. (2016) incorporated field line meandering into a model of SEP propagation in the heliosphere and showed that it can explain the fast onset of SEP events over wide longitudinal ranges, as can be observed experimentally. They also showed that the early SEP propagation across the magnetic field cannot be described by a diffusive approach.

7 Approaches to SEP Forecasting

Forecasting SEP events has attracted a large amount of effort in recent years. Producing forecasts with long lead times (of the order of days) relies on the analysis of solar active regions and their magnetic configuration. Short lead time forecasts, on the other hand, are issued when a solar eruptive event, typically including a solar flare and/or a CME, is detected. Here we focus on the latter scenario, where the aim is to predict whether or not the observed eruption will produce SEPs at Earth, and in the case of positive forecast to estimate the impact the SEP event is likely to have. In the following some recent models are briefly described. An earlier overview is given in Vainio et al. (2009).

7.1 Empirical Models

Empirical models are based on known statistical relationships between various types and magnitudes of solar events and SEPs, as reviewed in Sect. 5. The PROTONS system, the SEP prediction tool in use at the NOAA Space Weather Prediction Center (SWPC), makes use of an algorithm that takes into account the GOES X-ray peak flux of the flare, as well as information on the flare location (Balch 2008). Another empirical tool is the Proton Prediction System (PPS), also based on flare parameters (Kahler et al. 2007). The SEP forecasting tool developed within the *COronal Mass Ejections and Solar Energetic Particles: forecasting the space weather impact* (COMESEP) alert system makes use of empirical statistical relationships to forecast the occurrence of an event (Dierckxsens et al. 2015), while it uses a physics based model to forecast event parameters (Marsh et al. 2015). The SEP prediction model of the University of Malaga (UMASEP model) tracks simultaneously the time profiles of soft X-rays and protons observed by the GOES spacecraft (Núñez 2011). It predicts the occurrence, peak time and peak intensity of an SEP event. This model has been incorporated within the SEPsFLAREs forecasting system (García-Rigo et al. 2016). Methods that forecast proton arrival based on the detection of relativistic electrons or relativistic protons, which travel to 1 AU faster than ions, have also been proposed (Posner 2007; Souvatzoglou et al. 2014).

Besides the above-mentioned models a number of empirical relationships have been discovered that may be exploited in future operational models.

- The examination of hard X-ray bursts observed by the *Solar Maximum Mission* (SMM) showed that those associated with SEP events exhibit a distinct spectral hardening (*i.e.* flattening) throughout the event (Kiplinger 1995). This is a distinctive feature, since in typical hard X-ray bursts the photon spectrum starts soft, hardens as the intensity rises, and softens again in the decay phase (Grigis and Benz 2004). The SMM-finding was confirmed by RHESSI observations (Grayson et al. 2009).
- Chertok et al. (2009) noticed that SEP events at energies above 100 MeV were accompanied by microwave bursts with spectra that had peaks at or above about 15 GHz, whereas on average microwave peak frequencies are around 10 GHz. A possible interpretation is that SEP events are associated with particularly strong electron acceleration, leading to high densities of radio-emitting electrons in the flaring active region.
- When examining the soft X-ray bursts associated with SEP events, it was noticed that apparently cooler events, which means events that were relatively faint in the 0.05–0.4 nm channel of the GOES monitoring instrument as compared to the 0.1–0.8 nm channel, are more strongly associated with SEPs than the others (see Garcia 2004, and references therein).

Both the continuous spectral hardening of hard X-ray bursts and the low temperature of the SXR bursts associated with SEP events may translate the association of SEP events with extended coronal structures that exist during CMEs. But it is not clear why continuous hardening of the hard X-ray spectrum indicates escaping SEPs, especially since the effect is usually ascribed to energy-dependent energy losses or energy-dependent precipitation of a trapped electron population.

7.2 Physics Based Models

Physics based models aim to provide an SEP forecast by modelling the relevant acceleration and transport processes. This is complicated by the fact that only limited information about the acceleration is known within a real time forecasting context, so that a number of assumptions are usually necessary to produce predicted SEP intensity profiles.

The *Solar Particle Engineering Code* (SOLPENCO) forecasting framework solves a 1D focussed transport equation coupled with MHD modelling of the shock accelerating the SEPs (Aran et al. 2006). Here the fact that over time an observer is connected to different portions of the shock, having varying acceleration efficiencies is taken into account.

An alternative approach uses the output of ENLIL simulations to obtain information on the location and properties of CME driven interplanetary shocks (Bain et al. 2016). Particles injected from these shocks are propagated in 1D along the magnetic field within a scatter-free assumption (Luhmann et al. 2010).

The *Energetic Particle Radiation Environment Module* (EPREM) within the *Earth–Moon–Mars Radiation Environment Module* (EMMREM) framework couples a 1D focussed transport equation with a convection-diffusion equation to describe transport in 3D (Schwadron et al. 2010).

The SPARX model, based on the test particle approach, solves for SEP trajectories in 3D to forecast time profiles of particle intensities at 1 AU (Marsh et al. 2015).

In recent years, new numerical models have also attempted to couple realistic simulations of CME propagation in the corona with particle simulations to describe the associated particle acceleration (Manchester et al. 2005; Kozarev et al. 2013).

8 Discussion

The manifestations of energetic particles of solar origin in the Earth's plasma environment are a complex interplay between acceleration at the Sun and in the interplanetary space, and of transport in the turbulent interplanetary magnetic field. As of today it is impossible to predict the occurrence and properties of SEP events from properties of an active region or of the coronal magnetic field before a solar eruptive event occurs. The earliest observable signature related to an upcoming particle event is electromagnetic emission. The earliest electromagnetic emissions are related to the impulsive flare phase. Forecasting schemes of SEP events that use electromagnetic emissions, mostly soft X-rays because of their continuous availability from the GOES monitoring observations, therefore rely on the hypothesis of at least a statistical, if not physical, connection between the energy release in a flare and the importance of the associated SEP event.

Within the two-class scenario for SEP events, where impulsive and gradual events are distinguished, events with the largest space-weather impact are of the gradual type, i.e. they are associated with coronal and interplanetary shocks driven by fast CMEs. It is widely considered that it is these CME shocks that accelerate the SEPs. This conclusion appears to be in conflict with the use of flare-related emission in SEP forecasting schemes. An important question is therefore whether acceleration in flares, that in most cases happens in conjunction with sufficiently fast CMEs, is also a source of space-weather relevant SEPs. Furthermore, the liftoff of CMEs is also followed by the formation of current sheets and by signatures of magnetic reconnection that persist well beyond the impulsive flare phase. So there is a number of possible processes that may contribute to the acceleration of particles released into the interplanetary space. The term "flare acceleration" should not be restricted to magnetic reconnection processes and turbulence in the impulsive flare phase, but considered in a more general sense.

The energy range of SEPs spans many orders of magnitude, from hundreds of keV to possibly tens of GeV. Different space weather aspects refer to different particle energy ranges: tens of MeV for the impact on space missions outside the Earth's magnetosphere and in polar orbits, on human spaceflight outside the magnetosphere, and on radio communications through the polar ionosphere of the Earth; >1 GeV for the production of atmospheric particle cascades and their impact on aviation. Many observables of SEP events, including the longitude distribution of the parent activity, the elemental abundances and the Fe charge states, depend on the particle energy. As a result, it is not possible to define a sharp separation between 'impulsive' and 'gradual' SEP events that holds across the entire energy spectrum. The relative contribution of acceleration at CME shocks on the one hand and in small-scale plasma processes involving magnetic reconnection on the other is hence likely to depend on the energy of the particles being considered. At which energies the distinction is made is presently an open question and needs more research. The work of Dierckxsens et al. (2015) suggests a stronger dependence of SEP intensities on CME speed than on soft X-ray flux at energies below 10 MeV, and the opposite at energies above 20 MeV. The transition in the range 10–20 MeV is a statistical average, not a sharp physical limit.

The complexity of particle acceleration models has considerably increased in recent years, in order to cope with the observations. The role of fragmentation of the acceleration region in reconnecting current sheets has been emphasised by physical considerations and numerical modelling. As far as shock acceleration is concerned, the importance of the seed population has been highlighted by observations that show sometimes enhanced levels of ^{3}He and other elements in the interplanetary medium. Shock acceleration out of this seed population would then create energetic particle populations with elemental abundances that contain the peculiar enhancements of the seed population. This points to the importance of particularly active regions in the production of major solar events, where numerous flares and CMEs occur before the eruptive event at the origin of the SEPs, and which apparently release the suprathermal seed particle populations over times that are much longer than the electromagnetic emissions of the flares themselves.

But in addition to the question of particle acceleration, understanding the transport of SEPs is the other key to successful forecasting. While classical models of SEP propagation have made use of the 1D approximation, which assumes particles remain tied to the magnetic field line on which they are injected, the importance of using a 3D description has recently been emphasised. Within the standard two-class scenario, properties such as elemental abundances and charge states of heavy ions are ascribed purely to the acceleration process. Describing particle motion in 3D and accounting for the effects of drifts opens up the possibility that such properties may be partly the result of propagation, for the case of an observer that is not well connected to the source. In addition to guiding centre drifts in the spatially varying interplanetary magnetic field, 3D propagation may also be caused by processes such as scattering by turbulence and field line meandering. An accurate description of the complex magnetic fields of the solar corona is also of great importance to understanding SEP escape and propagation.

A number of institutions and projects are attempting to make SEP forecasts and improve their reliability. At present, a number of factors are limiting the efficacy of SEP forecasts. On the observational side, although real time flare observations are well established, little information is available on CME shocks, including their extent and Mach numbers. As a result it is difficult to obtain observational information on SEP injection in a real time forecast scenario. On the physics modelling side, a balance needs to be struck between correctly including the relevant physics of acceleration and transport and computational efficiency, so that useful information can be obtained over timescales relevant for forecasts.

A major limitation on our understanding of SEP acceleration and propagation is the exclusive availability of in situ measurements near 1 AU. Effects of particle acceleration and particle transport are then mixed and difficult to disentangle. We expect that the upcoming Solar Orbiter and Solar Probe Plus missions will provide invaluable new measurements of SEPs. By going closer to the Sun, these missions will allow detection of SEP properties closer to the acceleration region, therefore minimising any modifications introduced by transport effects. In addition they will be able to detect shocks and the associated populations of particles and plasma waves much closer to the Sun, where their particle acceleration efficiencies are expected to be much higher. The new insight gained on particle acceleration and transport by these missions will enable a major step foreward in our forecasting abilities.

Acknowledgements The authors are grateful to ISSI for organizing the workshop on the scientific foundations of space weather, as well as a number of earlier team meetings that helped us shape our ideas on solar energetic particles. S.D. acknowledges support from the UK Science and Technology Facilities Council (STFC) (grant ST/M00760X/1) and the Leverhulme Trust (grant RPG-2015-094). Research at Paris Observatory received funding from the French space agency *CNES* and the European Union's Horizon 2020 research and innovation programme under grant agreement No 637324 (HESPERIA project).

References

M. Ackermann, M. Ajello, A. Albert, A. Allafort, L. Baldini et al., High-energy gamma-ray emission from solar flares: summary of Fermi Large Area Telescope detections and analysis of two M-class flares. Astrophys. J. **787**, 15 (2014). doi:10.1088/0004-637X/787/1/15

A. Afanasiev, M. Battarbee, R. Vainio, Self-consistent Monte Carlo simulations of proton acceleration in coronal shocks: effect of anisotropic pitch-angle scattering of particles. Astron. Astrophys. **584**, A81 (2015). doi:10.1051/0004-6361/201526750

N. Agueda, K.L. Klein, N. Vilmer, R. Rodríguez-Gasén, O.E. Malandraki et al., Release timescales of solar energetic particles in the low corona. Astron. Astrophys. **570**, A5 (2014). doi:10.1051/0004-6361/201423549

N. Agueda, R. Vainio, D. Lario, B. Sanahuja, Injection and interplanetary transport of near-relativistic electrons: modeling the impulsive event on 2000 May 1. Astrophys. J. **675**, 1601–1613 (2008). doi:10.1086/527527

M. Ajello, A. Albert, A. Allafort, L. Baldini, G. Barbiellini et al., Impulsive and long duration high-energy gamma-ray emission from the very bright 2012 March 7 solar flares. Astrophys. J. **789**, 20 (2014). doi:10.1088/0004-637X/789/1/20

V.V. Akimov, P. Ambrož, A.V. Belov, A. Berlicki, I.M. Chertok et al., Evidence for prolonged acceleration based on a detailed analysis of the long-duration solar gamma-ray flare of June 15, 1991. Sol. Phys. **166**, 107–134 (1996)

A. Aran, B. Sanahuja, D. Lario, SOLPENCO: a solar particle engineering code. Adv. Space Res. **37**, 1240–1246 (2006). doi:10.1016/j.asr.2005.09.019

M.J. Aschwanden, Particle acceleration and kinematics in solar flares. Space Sci. Rev. **101**, 1–227 (2002)

H. Aurass, F. Landini, G. Poletto, Coronal current sheet signatures during the 17 May 2002 CME-flare. Astron. Astrophys. **506**, 901–911 (2009). doi:10.1051/0004-6361/200912229

H.M. Bain, M.L. Mays, J.G. Luhmann, Y. Li, L.K. Jian, D. Odstrcil, Shock connectivity in the August 2010 and July 2012 solar energetic particle events inferred from observations and ENLIL modeling. Astrophys. J. **825**, 1 (2016). doi:10.3847/0004-637X/825/1/1

C.C. Balch, Updated verification of the Space Weather Prediction Center's solar energetic particle prediction model. Space Weather **6**, S01001 (2008). doi:10.1029/2007SW000337

A.O. Benz, M. Battaglia, N. Vilmer, Location of decimetric pulsations in solar flares. Sol. Phys. **273**, 363–375 (2011). doi:10.1007/s11207-011-9760-3

J.W. Bieber, W.H. Matthaeus, C.W. Smith, W. Wanner, M.B. Kallenrode, G. Wibberenz, Proton and electron mean free paths: the Palmer consensus revisited. Astrophys. J. **420**, 294–306 (1994). doi:10.1086/173559

A. Buttighoffer, Solar electron beams associated with radio type III bursts: propagation channels observed by Ulysses between 1 and 4 AU. Astron. Astrophys. **335**, 295–302 (1998)

H.V. Cane, W.C. Erickson, N.P. Prestage, Solar flares, type III radio bursts, coronal mass ejections and energetic particles. J. Geophys. Res. Space Phys. **107**, 1315 (2002). doi:10.1029/2001JA000320

H.V. Cane, D.V. Reames, T.T. von Rosenvinge, The role of interplanetary shocks in the longitude distribution of solar energetic particles. J. Geophys. Res. Space Phys. **93**, 9555–9567 (1988). doi:10.1029/JA093iA09p09555

H.V. Cane, I.G. Richardson, T.T. von Rosenvinge, A study of solar energetic particle events of 1997–2006: their composition and associations. J. Geophys. Res. Space Phys. **115**, A08101 (2010). doi:10.1029/2009JA014848

D. Caprioli, A.R. Pop, A. Spitkovsky, Simulations and theory of ion injection at non-relativistic collisionless shocks. Astrophys. J. Lett. **798**, L28 (2015). doi:10.1088/2041-8205/798/2/L28

P.J. Cargill, L. Vlahos, G. Baumann, J.F. Drake, Å. Nordlund, Current fragmentation and particle acceleration in solar flares. Space Sci. Rev. **173**, 223–245 (2012). doi:10.1007/s11214-012-9888-y

H. Carmichael, A process for flares. NASA Spec. Publ. **50**, 451 (1964)

I.M. Chertok, On the correlation between the solar gamma-ray line emission, radio bursts and proton fluxes in the interplanetary space. Astron. Nachr. **311**, 379–381 (1990)

I.M. Chertok, V.V. Grechnev, N.S. Meshalkina, On the correlation between spectra of solar microwave bursts and proton fluxes near the Earth. Astron. Rep. **53**, 1059–1069 (2009). doi:10.1134/S1063772909110110
C.M.S. Cohen, G.M. Mason, R.A. Mewaldt, M.E. Wiedenbeck, The longitudinal dependence of heavy-ion composition in the 2013 April 11 solar energetic particle event. Astrophys. J. **793**, 35 (2014). doi:10.1088/0004-637X/793/1/35
C.M.S. Cohen, R.A. Mewaldt, R.A. Leske, A.C. Cummings, E.C. Stone et al., New observations of heavy-ion-rich solar particle events from ACE. Geophys. Res. Lett. **26**, 2697–2700 (1999). doi:10.1029/1999GL900560
S. Dalla, A. Balogh, S. Krucker, A. Posner, R. Müller-Mellin et al., Properties of high heliolatitude solar energetic particle events and constraints on models of acceleration and propagation. Geophys. Res. Lett. **30**, 8035 (2003). doi:10.1029/2003GL017139
S. Dalla, M.S. Marsh, M. Battarbee, Solar energetic particle drifts and the energy dependence of 1 AU charge states. Astrophys. J. **834**, 167 (2017). doi:10.3847/1538-4357/834/2/167
S. Dalla, M.S. Marsh, J. Kelly, T. Laitinen, Solar energetic particle drifts in the Parker spiral. J. Geophys. Res. Space Phys. **118**, 5979–5985 (2013). doi:10.1002/jgra.50589
S. Dalla, M.S. Marsh, T. Laitinen, Drift-induced deceleration of solar energetic particles. Astrophys. J. **808**, 62 (2015). doi:10.1088/0004-637X/808/1/62
S. Dalla, M.S. Marsh, P. Zelina, T. Laitinen, Time dependence of Fe/O ratio within a 3D solar energetic particle propagation model including drift. Astron. Astrophys. **598**, A73 (2017). doi:10.1051/0004-6361/201628618
H. Debrunner, J.A. Lockwood, C. Barat, R. Buetikofer, J.P. Dezalay et al., Energetic neutrons, protons, and gamma rays during the 1990 May 24 solar cosmic-ray event. Astrophys. J. **479**, 997–1011 (1997)
P. Démoulin, A. Vourlidas, M. Pick, A. Bouteille, Erratum: Initiation and development of the white-light and radio coronal mass ejection on 2001 April 15. Astrophys. J. **754**, 156 (2012a). doi:10.1088/0004-637X/754/2/156
P. Démoulin, A. Vourlidas, M. Pick, A. Bouteille, Initiation and development of the white-light and radio coronal mass ejection on 2001 April 15. Astrophys. J. **750**, 147 (2012b). doi:10.1088/0004-637X/750/2/147
M. Desai, J. Giacalone, Large gradual solar energetic particle events. Living Rev. Sol. Phys. **13**, 3 (2016). doi:10.1007/s41116-016-0002-5
M. Dierckxsens, K. Tziotziou, S. Dalla, I. Patsou, M.S. Marsh et al., Relationship between solar energetic particles and properties of flares and CMEs: statistical analysis of solar cycle 23 events. Sol. Phys. **290**, 841–874 (2015). doi:10.1007/s11207-014-0641-4
R. DiFabio, Z. Guo, E. Möbius, B. Klecker, H. Kucharek, G.M. Mason, M. Popecki, Energy-dependent charge states and their connection with ion abundances in impulsive solar energetic particle events. Astrophys. J. **687**, 623–634 (2008). doi:10.1086/591833
W. Dröge, Particle scattering by magnetic fields. Space Sci. Rev. **93**, 121–151 (2000)
N. Dresing, R. Gómez-Herrero, B. Heber, A. Klassen, O. Malandraki et al., Statistical survey of widely spread out solar electron events observed with STEREO and ACE with special attention to anisotropies. Astron. Astrophys. **567**, A27 (2014). doi:10.1051/0004-6361/201423789
W. Dröge, Energetic solar electron spectra and gamma-ray observations, in *High Energy Solar Physics*, ed. by R. Ramaty, N. Mandzhavidze, X.M. Hua. American Institute of Physics Conference Series, vol. 374 (1996), pp. 78–85
W. Dröge, Y.Y. Kartavykh, B. Klecker, G.A. Kovaltsov, Anisotropic three-dimensional focused transport of solar energetic particles in the inner heliosphere. Astrophys. J. **709**, 912–919 (2010). doi:10.1088/0004-637X/709/2/912
A.G. Emslie, B.R. Dennis, G.D. Holman, H.S. Hudson, Refinements to flare energy estimates: a followup to "Energy partition in two solar flare/CME events" by A. G. Emslie et al. J. Geophys. Res. Space Phys. **110**, A11,103 (2005). doi:10.1029/2005JA011305
C.Y. Fan, M. Pick, R. Pyle, J.A. Simpson, D.R. Smith, Protons associated with centers of solar activity and their propagation in interplanetary magnetic field regions corotating with the Sun. J. Geophys. Res. Space Phys. **73**, 1555–1582 (1968)
L. Fletcher, B.R. Dennis, H.S. Hudson, S. Krucker, K. Phillips et al., An observational overview of solar flares. Space Sci. Rev. **159**, 19–106 (2011). doi:10.1007/s11214-010-9701-8
H.A. Garcia, Forecasting methods for occurrence and magnitude of proton storms with solar soft X-rays. Space Weather **S02**, 002 (2004). doi:10.1029/2003SW000001
A. García-Rigo, M. Núñez, R. Qahwaji, O. Ashamari, P. Jiggens et al., Prediction and warning system of SEP events and solar flares for risk estimation in space launch operations. J. Space Weather Space Clim. **6**(27), A28 (2016). doi:10.1051/swsc/2016021
J. Giacalone, The efficient acceleration of thermal protons by perpendicular shocks. Astrophys. J. Lett. **628**, L37–L40 (2005). doi:10.1086/432510

R. Gómez-Herrero, N. Dresing, A. Klassen, B. Heber, D. Lario et al., Circumsolar energetic particle distribution on 2011 November 3. Astrophys. J. **799**, 55 (2015). doi:10.1088/0004-637X/799/1/55
N. Gopalswamy, P. Mäkelä, S. Akiyama, S. Yashiro, H. Xie et al., Large solar energetic particle events associated with filament eruptions outside of active regions. Astrophys. J. **806**, 8 (2015). doi:10.1088/0004-637X/806/1/8
N. Gopalswamy, S. Yashiro, A. Lara, M.L. Kaiser, B.J. Thompson et al., Large solar energetic particle events of cycle 23: a global view. Geophys. Res. Lett. **30**, 8015 (2003). doi:10.1029/2002GL016435
N. Gopalswamy, S. Yashiro, H. Xie, S. Akiyama, E. Aguilar-Rodriguez et al., Radio-quiet fast and wide coronal mass ejections. Astrophys. J. **674**, 560–569 (2008). doi:10.1086/524765
M. Gordovskyy, P.K. Browning, E.P. Kontar, N.H. Bian, Effect of collisions and magnetic convergence on electron acceleration and transport in reconnecting twisted solar flare loops. Sol. Phys. **284**, 489–498 (2013). doi:10.1007/s11207-012-0124-4
J.A. Grayson, S. Krucker, R.P. Lin, A statistical study of spectral hardening in solar flares and related solar energetic particle events. Astrophys. J. **707**, 1588–1594 (2009). doi:10.1088/0004-637X/707/2/1588
P.C. Grigis, A.O. Benz, The spectral evolution of impulsive solar X-ray flares. Astron. Astrophys. **426**, 1093–1101 (2004). doi:10.1051/0004-6361:20041367
J. Heerikhuisen, Y.E. Litvinenko, I.J.D. Craig, Proton acceleration in analytic reconnecting current sheets. Astrophys. J. **566**, 512–520 (2002). doi:10.1086/337957
J. Heyvaerts, E.R. Priest, D.M. Rust, An emerging flux model for the solar flare phenomenon. Astrophys. J. **216**, 123–137 (1977). doi:10.1086/155453
G.D. Holman, M.J. Aschwanden, H. Aurass, M. Battaglia, P.C. Grigis et al., Implications of X-ray observations for electron acceleration and propagation in solar flares. Space Sci. Rev. **159**, 107–166 (2011). doi:10.1007/s11214-010-9680-9
G.D. Holman, M.E. Pesses, Solar type II radio emission and the shock drift acceleration of electrons. Astrophys. J. **267**, 837–843 (1983)
H. Hudson, J. Ryan, High-energy particles in solar flares. Annu. Rev. Astron. Astrophys. **33**, 239–282 (1995). doi:10.1146/annurev.aa.33.090195.001323
P. Jiggens, M.A. Chavy-Macdonald, G. Santin, A. Menicucci, H. Evans, A. Hilgers, The magnitude and effects of extreme solar particle events. J. Space Weather Space Clim. **4**(27), A20 (2014). doi:10.1051/swsc/2014017
F.C. Jones, D.C. Ellison, The plasma physics of shock acceleration. Space Sci. Rev. **58**, 259–346 (1991). doi:10.1007/BF01206003
S. Kahler, Solar flares and coronal mass ejections. Annu. Rev. Astron. Astrophys. **30**, 113–141 (1992)
S.W. Kahler, The role of the big flare syndrome in correlations of solar energetic proton fluxes and associated microwave burst parameters. J. Geophys. Res. Space Phys. **87**, 3439–3448 (1982). doi:10.1029/JA087iA05p03439
S.W. Kahler, The correlation between solar energetic particle peak intensities and speeds of coronal mass ejections: effects of ambient particle intensities and energy spectra. J. Geophys. Res. Space Phys. **106**, 20,947–20,956 (2001). doi:10.1029/2000JA002231
S.W. Kahler, E.W. Cliver, H.V. Cane, R.E. McGuire, R.G. Stone, N.R. Sheeley, Solar filament eruptions and energetic particle events. Astrophys. J. **302**, 504–510 (1986)
S.W. Kahler, E.W. Cliver, A.G. Ling, Validating the proton prediction system (PPS). J. Atmos. Sol.-Terr. Phys. **69**, 43–49 (2007). doi:10.1016/j.jastp.2006.06.009
M.B. Kallenrode, Particle propagation in the inner heliosphere. J. Geophys. Res. Space Phys. **98**, 19 (1993). doi:10.1029/93JA02079
M.B. Kallenrode, G. Wibberenz, Particle injection following solar flares on 1980 May 28 and June 8—evidence for different injection time histories in impulsive and gradual events? Astrophys. J. **376**, 787–796 (1991)
G. Kanbach, D.L. Bertsch, C.E. Fichtel, R.C. Hartman, S.D. Hunter et al., Detection of a long-duration solar gamma-ray flare on June 11, 1991 with EGRET on COMPTON-GRO. Astron. Astrophys. Suppl. Ser. **97**, 349–353 (1993)
S.R. Kane, E.L. Chupp, D.J. Forrest, G.H. Share, E. Rieger, Rapid acceleration of energetic particles in the 1982 February 8 solar flare. Astrophys. J. Lett. **300**, L95–L98 (1986). doi:10.1086/184610
Y.Y. Kartavykh, W. Dröge, B. Klecker, G.M. Mason, E. Möbius et al., Evidence of a two-temperature source region in the ^{3}He-rich solar energetic particle event of 2000 May 1. Astrophys. J. **671**, 947–954 (2007). doi:10.1086/522687
A.L. Kiplinger, Comparative studies of hard X-ray spectral evolution in solar flares with high-energy proton events observed at Earth. Astrophys. J. **453**, 973–986 (1995)
A. Klassen, N. Dresing, R. Gómez-Herrero, B. Heber, R. Müller-Mellin, Unexpected spatial intensity distributions and onset timing of solar electron events observed by closely spaced STEREO spacecraft. Astron. Astrophys. **593**, A31 (2016). doi:10.1051/0004-6361/201628734

B. Klecker, H. Kunow, H.V. Cane, S. Dalla, B. Heber et al., Energetic particle observations. Space Sci. Rev. **123**, 217–250 (2006). doi:10.1007/s11214-006-9018-9
B. Klecker, E. Möbius, M.A. Popecki, Ionic charge states of solar energetic particles: a clue to the source. Space Sci. Rev. **130**, 273–282 (2007). doi:10.1007/s11214-007-9207-1
K.L. Klein, N. Agueda, R. Bütikofer, On the origin of relativistic solar particle events: interplanetary transport modelling and radio emission, in *34th International Cosmic Ray Conference*, ed. by R. Caballero, J.C. D'Olivo, G. Medina-Tanco, L. Nellen, F.A. Sánchez, J.F. Valdés-Galicia. Proceedings of Science (2015), pp. 121–124
K.L. Klein, S. Krucker, G. Lointier, A. Kerdraon, Open magnetic flux tubes in the corona and the transport of solar energetic particles. Astron. Astrophys. **486**, 589–596 (2008). doi:10.1051/0004-6361:20079228
K.L. Klein, S. Masson, C. Bouratzis, V. Grechnev, A. Hillaris, P. Preka-Papadema, The relativistic solar particle event of 2005 January 20: origin of delayed particle acceleration. Astron. Astrophys. **572**, A4 (2014). doi:10.1051/0004-6361/201423783
K.L. Klein, G. Trottet, A. Klassen, Energetic particle acceleration and propagation in strong CME-less flares. Sol. Phys. **263**, 185–208 (2010). doi:10.1007/s11207-010-9540-5
K.L. Klein, G. Trottet, S. Samwel, O. Malandraki, Particle acceleration and propagation in strong flares without major solar energetic particle events. Sol. Phys. **269**, 309–333 (2011). doi:10.1007/s11207-011-9710-0
B. Kliem, M. Karlický, A.O. Benz, Solar flare radio pulsations as a signature of dynamic magnetic reconnection. Astron. Astrophys. **360**, 715–728 (2000)
L. Kocharov, G.A. Kovaltsov, J. Torsti, V.M. Ostryakov, Evaluation of solar energetic Fe charge states: effect of proton-impact ionization. Astron. Astrophys. **357**, 716–724 (2000)
L.G. Kocharov, G.A. Kovaltsov, G.E. Kocharov, E.I. Chuikin, I.G. Usoskin et al., Electromagnetic and corpuscular emission from the solar flare of 1991 June 15: continuous acceleration of relativistic particles. Sol. Phys. **150**, 267–283 (1994)
G.A. Kovaltsov, I.G. Usoskin, E.W. Cliver, W.F. Dietrich, A.J. Tylka, Fluence ordering of solar energetic proton events using cosmogenic radionuclide data. Sol. Phys. **289**, 4691–4700 (2014). doi:10.1007/s11207-014-0606-7
K.A. Kozarev, R.M. Evans, N.A. Schwadron, M.A. Dayeh, M. Opher et al., Global numerical modeling of energetic proton acceleration in a coronal mass ejection traveling through the solar corona. Astrophys. J. **778**, 43 (2013). doi:10.1088/0004-637X/778/1/43
D. Krauss-Varban, Particle acceleration in shocks, in *Heliophysics: Space Storms and Radiation: Causes and Effects*, ed. by C.J. Schrijver, G.L. Siscoe (Cambridge University Press, Cambridge, 2010), pp. 209–231
S. Krucker, M. Battaglia, P.J. Cargill, L. Fletcher, H.S. Hudson et al., Hard X-ray emission from the solar corona. Astron. Astrophys. Rev. **16**, 155–208 (2008). doi:10.1007/s00159-008-0014-9
S. Krucker, G.J. Hurford, A.L. MacKinnon, A.Y. Shih, R.P. Lin, Coronal γ-ray bremsstrahlung from solar flare-accelerated electrons. Astrophys. J. Lett. **678**, L63–L66 (2008b). doi:10.1086/588381
S. Krucker, E.P. Kontar, S. Christe, R.P. Lin, Solar flare electron spectra at the sun and near the Earth. Astrophys. J. Lett. **663**, L109–L112 (2007). doi:10.1086/519373
T. Laitinen, A. Kopp, F. Effenberger, S. Dalla, M.S. Marsh, Solar energetic particle access to distant longitudes through turbulent field-line meandering. Astron. Astrophys. **591**, A18 (2016). doi:10.1051/0004-6361/201527801
D. Lario, A. Aran, R. Gómez-Herrero, N. Dresing, B. Heber et al., Longitudinal and radial dependence of solar energetic particle peak intensities: STEREO, ACE, SOHO, GOES, and MESSENGER observations. Astrophys. J. **767**, 41 (2013). doi:10.1088/0004-637X/767/1/41
D. Lario, B. Sanahuja, A.M. Heras, Energetic particle events: efficiency of interplanetary shocks as 50 keV $<$ E $<$ 100 MeV proton accelerators. Astrophys. J. **509**, 415–434 (1998). doi:10.1086/306461
M.A. Lee, R.A. Mewaldt, J. Giacalone, Shock acceleration of ions in the Heliosphere. Space Sci. Rev. **173**, 247–281 (2012). doi:10.1007/s11214-012-9932-y
P.C. Liewer, M. Neugebauer, T. Zurbuchen, Characteristics of active-region sources of solar wind near solar maximum. Sol. Phys. **223**, 209–229 (2004). doi:10.1007/s11207-004-1105-z
J. Lin, N.A. Murphy, C. Shen, J.C. Raymond, K.K. Reeves et al., Review on current sheets in CME development: theories and observations. Space Sci. Rev. **194**, 237–302 (2015). doi:10.1007/s11214-015-0209-0
J.G. Luhmann, S.A. Ledvina, D. Odstrcil, M.J. Owens, X.P. Zhao et al., Cone model-based SEP event calculations for applications to multipoint observations. Adv. Space Res. **46**, 1–21 (2010). doi:10.1016/j.asr.2010.03.011
W.B. Manchester IV., T.I. Gombosi, D.L. De Zeeuw, I.V. Sokolov, I.I. Roussev et al., Coronal mass ejection shock and sheath structures relevant to particle acceleration. Astrophys. J. **622**, 1225–1239 (2005). doi:10.1086/427768
G. Mann, A. Klassen, Electron beams generated by shock waves in the solar corona. Astron. Astrophys. **441**, 319–326 (2005). doi:10.1051/0004-6361:20034396

C. Marqué, A. Posner, K.L. Klein, Solar energetic particles and radio-silent fast coronal mass ejections. Astrophys. J. **642**, 1222–1235 (2006)

M.S. Marsh, S. Dalla, M. Dierckxsens, T. Laitinen, N.B. Crosby SPARX: a modeling system for Solar Energetic Particle Radiation Space Weather forecasting. Space Weather **13**, 386–394 (2015). doi:10.1002/2014SW001120

M.S. Marsh, S. Dalla, J. Kelly, T. Laitinen, Drift-induced perpendicular transport of solar energetic particles. Astrophys. J. **774**, 4 (2013). doi:10.1088/0004-637X/774/1/4

G.M. Mason, M.I. Desai, C.M.S. Cohen, R.A. Mewaldt, E.C. Stone, J.R. Dwyer, The role of interplanetary scattering in western hemisphere large solar energetic particle events. Astrophys. J. Lett. **647**, L65–L68 (2006). doi:10.1086/507469

G.M. Mason, J.E. Mazur, J.R. Dwyer, ^{3}He enhancements in large solar energetic particle events. Astrophys. J. Lett. **525**, L133–L136 (1999)

S. Masson, S.K. Antiochos, C.R. DeVore, A model for the escape of solar-flare-accelerated particles. Astrophys. J. **771**, 82 (2013). doi:10.1088/0004-637X/771/2/82

J.E. Mazur, G.M. Mason, J.R. Dwyer, J. Giacalone, J.R. Jokipii, E.C. Stone, Interplanetary magnetic field line mixing deduced from impulsive solar flare particles. Astrophys. J. **532**, L79–L82 (2000)

J.E. Mazur, G.M. Mason, M.D. Looper, R.A. Leske, R.A. Mewaldt, Charge states of solar energetic particles using the geomagnetic cutoff technique: SAMPEX measurements in the 6 November 1997 solar particle event. Geophys. Res. Lett. **26**, 173–176 (1999). doi:10.1029/1998GL900075

K.G. McCracken, High frequency of occurrence of large solar energetic particle events prior to 1958 and a possible repetition in the near future. Space Weather **5**, 07004 (2007). doi:10.1029/2006SW000295

K.G. McCracken, H. Moraal, M.A. Shea, The high-energy impulsive ground-level enhancement. Astrophys. J. **761**, 101 (2012). doi:10.1088/0004-637X/761/2/101

R.A. Mewaldt, C.M.S. Cohen, D.K. Haggerty, G.M. Mason, M.L. Looper et al., Radiation risks from large solar energetic particle events, in *Turbulence and Nonlinear Processes in Astrophysical Plasmas*, ed. by D. Shaikh, G.P. Zank. American Institute of Physics Conference Series, vol. 932 (2007), pp. 277–282. doi:10.1063/1.2778975

R.A. Mewaldt, C.M.S. Cohen, G.M. Mason, *The Source Material for Large Solar Energetic Particle Events*. American Geophysical Union Geophysical Monograph Series, vol. 165 (2006). Washington D.C.

R.A. Mewaldt, G.M. Mason, C.M.S. Cohen, R. Gómez-Herrero, D.K. Haggerty et al., Evolution of suprathermal seed particle and solar energetic particle abundances, in *Physics of the Heliosphere: A 10 Year Retrospective*, ed. by J. Heerikhuisen, G. Li, N. Pogorelov, G. Zank. American Institute of Physics Conference Series, vol. 1436 (2012), pp. 206–211. doi:10.1063/1.4723609

J.A. Miller, P.J. Cargill, A.G. Emslie, G.D. Holman, B.R. Dennis et al., Critical issues for understanding particle acceleration in impulsive solar flares. J. Geophys. Res. Space Phys. **102**, 14,631–14,660 (1997). doi:10.1029/97JA00976

L.I. Miroshnichenko, *Solar Cosmic Rays* (Kluwer Acad. Publ., Dordrecht/Boston/London, 2001)

R. Miteva, K.L. Klein, O. Malandraki, G. Dorrian, Solar energetic particle events in the 23rd solar cycle: interplanetary magnetic field configuration and statistical relationship with flares and CMEs. Sol. Phys. **282**, 579–613 (2013). doi:10.1007/s11207-012-0195-2

E. Möbius, M. Popecki, B. Klecker, L.M. Kistler, A. Bogdanov et al., Energy dependence of the ionic charge state distribution during the November 1997 solar energetic particle event. Geophys. Res. Lett. **26**, 145–148 (1999). doi:10.1029/1998GL900131

H. Moraal, K.G. McCracken, The time structure of ground level enhancements in solar cycle 23. Space Sci. Rev. **171**, 85–95 (2012). doi:10.1007/s11214-011-9742-7

M. Núñez, Predicting solar energetic proton events ($E > 10$ MeV). Space Weather **9**, 07003 (2011). doi:10.1029/2010SW000640

M. Pesce-Rollins, N. Omodei, V. Petrosian, W. Liu, F. Rubio da Costa et al., First detection of >100 MeV gamma-rays associated with a behind-the-limb solar flare. Astrophys. J. Lett. **805**, L15 (2015). doi:10.1088/2041-8205/805/2/L15

V. Petrosian, Stochastic acceleration by turbulence. Space Sci. Rev. **173**, 535–556 (2012). doi:10.1007/s11214-012-9900-6

J. Pomoell, R. Vainio, R. Kissmann, MHD modeling of coronal large-amplitude waves related to CME lift-off. Sol. Phys. **253**, 249–261 (2008). doi:10.1007/s11207-008-9186-8

A. Posner, Up to 1-hour forecasting of radiation hazards from solar energetic ion events with relativistic electrons. Space Weather **5**, 05001 (2007). doi:10.1029/2006SW000268

R. Ramaty, N. Mandzhavidze, B. Kozlovsky, J.G. Skibo, Acceleration in solar flares: interacting particles versus interplanetary particles. Adv. Space Res. **13**(9), 275–284 (1993). doi:10.1016/0273-1177(93)90490-3

J.C. Raymond, S. Krucker, R.P. Lin, V. Petrosian, Observational aspects of particle acceleration in large solar flares. Space Sci. Rev. **173**, 197–221 (2012). doi:10.1007/s11214-012-9897-x

D.V. Reames, Particle acceleration at the Sun and in the heliosphere. Space Sci. Rev. **90**, 413–491 (1999). doi:10.1023/A:1005105831781
R. Reinhard, G. Wibberenz, Propagation of flare protons in the solar atmosphere. Sol. Phys. **36**, 473–494 (1974)
I.G. Richardson, T.T.von Rosenvinge, H.V. Cane, E.R. Christian, C.M.S. Cohen et al., >25 MeV proton events observed by the High Energy Telescopes on the STEREO A and B spacecraft and/or at Earth during the first seven years of the STEREO mission. Sol. Phys. **289**, 3059–3107 (2014). doi:10.1007/s11207-014-0524-8
E.C. Roelof, Propagation of solar cosmic rays in the interplanetary magnetic field, in *Lectures in High-Energy Astrophysics*, ed. by H. Ögelmann, J. Wayland. NASA, vol. SP-199 (1969), pp. 111–136
A.P. Rouillard, I. Plotnikov, R.F. Pinto, M. Tirole, M. Lavarra et al., Deriving the properties of coronal pressure fronts in 3D: application to the 2012 May 17 ground level enhancement. Astrophys. J. **833**, 45 (2016). doi:10.3847/1538-4357/833/1/45
A.P. Rouillard, N.R. Sheeley, A. Tylka, A. Vourlidas, C.K. Ng et al., The longitudinal properties of a solar energetic particle event investigated using modern solar imaging. Astrophys. J. **752**, 44 (2012). doi:10.1088/0004-637X/752/1/44
D. Ruffolo, Effect of adiabatic deceleration on the focused transport of solar cosmic rays. Astrophys. J. **442**, 861–874 (1995). doi:10.1086/175489
P. Saint-Hilaire, N. Vilmer, A. Kerdraon, A decade of solar type III radio bursts observed by the Nançay Radioheliograph 1998–2008. Astrophys. J. **762**, 60 (2013). doi:10.1088/0004-637X/762/1/60
A. Sandroos, R. Vainio, Particle acceleration at shocks propagating in inhomogeneous magnetic fields. Astron. Astrophys. **455**, 685–695 (2006). doi:10.1051/0004-6361:20054754
C.J. Schrijver, M.L. DeRosa, Photospheric and heliospheric magnetic fields. Sol. Phys. **212**, 165–200 (2003). doi:10.1023/A:1022908504100
N.A. Schwadron, L. Townsend, K. Kozarev, M.A. Dayeh, F. Cucinotta et al., Earth–Moon–Mars radiation environment module framework. Space Weather **8**, S00E02 (2010). doi:10.1029/2009SW000523
G. Souvatzoglou, A. Papaioannou, H. Mavromichalaki, J. Dimitroulakos, C. Sarlanis, Optimizing the real-time ground level enhancement alert system based on neutron monitor measurements: introducing GLE Alert Plus. Space Weather **12**, 633–649 (2014). doi:10.1002/2014SW001102
L. Sui, G.D. Holman, B.R. Dennis, Evidence for magnetic reconnection in three homologous solar flares observed by RHESSI. Astrophys. J. **612**, 546–556 (2004). doi:10.1086/422515
B. Tan, H. Mészárosová, M. Karlický, G. Huang, C. Tan, Microwave type III pair bursts in solar flares. Astrophys. J. **819**, 42 (2016). doi:10.3847/0004-637X/819/1/42
J.K. Thalmann, Y. Su, M. Temmer, A.M. Veronig, The confined X-class flares of solar active region 2192. Astrophys. J. Lett. **801**, L23 (2015). doi:10.1088/2041-8205/801/2/L23
G. Trottet, S. Samwel, K.L. Klein, T. Dudok de Wit, R. Miteva, Statistical evidence for contributions of flares and coronal mass ejections to major solar energetic particle events. Sol. Phys. **290**, 819–839 (2015). doi:10.1007/s11207-014-0628-1
G. Trottet, N. Vilmer, C. Barat, J.P. Dezalay, R. Talon et al., Temporal and spectral characteristics of the June 11, 1991 gamma-ray flare. Astron. Astrophys. Suppl. Ser. **97**, 337–339 (1993)
A.J. Tylka, M.A. Lee, A model for spectral and compositional variability at high energies in large, gradual solar particle events. Astrophys. J. **646**, 1319–1334 (2006). doi:10.1086/505106
A.J. Tylka, O.E. Malandraki, G. Dorrian, Y.K. Ko, R.G. Marsden et al., Initial Fe/O enhancements in large, gradual, solar energetic particle events: observations from Wind and Ulysses. Sol. Phys. **285**, 251–267 (2013). doi:10.1007/s11207-012-0064-z
R. Vainio, L. Desorgher, D. Heynderickx, M. Storini, E. Flückiger et al., Dynamics of the Earth's particle radiation environment. Space Sci. Rev. **147**, 187–231 (2009). doi:10.1007/s11214-009-9496-7
M.A.I. Van Hollebeke, L.S. Ma Sung, F.B. McDonald, The variation of solar proton energy spectra and size distribution with heliolongitude. Sol. Phys. **41**, 189–223 (1975). doi:10.1007/BF00152967
N. Vilmer, A.L. MacKinnon, G.J. Hurford, Properties of energetic ions in the solar atmosphere from γ-ray and neutron observations. Space Sci. Rev. **159**, 167–224 (2011). doi:10.1007/s11214-010-9728-x
Y. Wang, J. Zhang, A comparative study between eruptive X-class flares associated with coronal mass ejections and confined X-class flares. Astrophys. J. **665**, 1428–1438 (2007). doi:10.1086/519765
Y.M. Wang, M. Pick, G.M. Mason, Coronal holes, jets, and the origin of ^{3}He-rich particle events. Astrophys. J. **639**, 495–509 (2006). doi:10.1086/499355
A. Warmuth, G. Mann, H. Aurass, Modelling shock drift acceleration of electrons at the reconnection outflow termination shock in solar flares. Observational constraints and parametric study. Astron. Astrophys. **494**, 677–691 (2009). doi:10.1051/0004-6361:200810101
M.E. Wiedenbeck, G.M. Mason, C.M.S. Cohen, N.V. Nitta, R. Gómez-Herrero, D.K. Haggerty, Observations of solar energetic particles from ^{3}He-rich events over a wide range of heliographic longitude. Astrophys. J. **762**, 54 (2013). doi:10.1088/0004-637X/762/1/54

M.A. Xapsos, G.P. Summers, J.L. Barth, E.G. Stassinopoulos, E.A. Burke, Probability model for worst case solar proton event fluences. IEEE Trans. Nucl. Sci. **46**, 1481–1485 (1999). doi:10.1109/23.819111

P. Zelina, S. Dalla, C.M.S. Cohen, R.A. Mewaldt, Time evolution of elemental ratios in solar energetic particle events. Astrophys. J. **835**, 71 (2017). doi:10.3847/1538-4357/aa5274

M. Zhang, G. Qin, H. Rassoul, Propagation of solar energetic particles in three-dimensional interplanetary magnetic fields. Astrophys. J. **692**, 109–132 (2009). doi:10.1088/0004-637X/692/1/109

Space Sci Rev (2017) 212:1069–1106
DOI 10.1007/s11214-017-0381-5

Particle Radiation Sources, Propagation and Interactions in Deep Space, at Earth, the Moon, Mars, and Beyond: Examples of Radiation Interactions and Effects

Nathan A. Schwadron[1] · John F. Cooper[2] · Mihir Desai[3] · Cooper Downs[4] · Matt Gorby[1] · Andrew P. Jordan[1] · Colin J. Joyce[1] · Kamen Kozarev[5] · Jon A. Linker[4] · Zoran Mikíc[4] · Pete Riley[4] · Harlan E. Spence[1] · Tibor Török[4] · Lawrence W. Townsend[6] · Jody K. Wilson[1] · Cary Zeitlin[7]

Received: 16 October 2016 / Accepted: 16 May 2017 / Published online: 3 July 2017

Abstract Particle radiation has significant effects for astronauts, satellites and planetary bodies throughout the Solar System. Acute space radiation hazards pose risks to human and robotic exploration. This radiation also naturally weathers the exposed surface regolith of the Moon, the two moons of Mars, and other airless bodies, and contributes to chemical evolution of planetary atmospheres at Earth, Mars, Venus, Titan, and Pluto. We provide a select review of recent areas of research covering the origin of SEPs from coronal mass ejections low in the corona, propagation of events through the solar system during the anomalously weak solar cycle 24 and important examples of radiation interactions for Earth, other planets and airless bodies such as the Moon.

Keywords Cosmic rays · Energetic particles · Radiation

1 Introduction—Radiation Hazards and Interactions

Particle radiation has far-reaching effects throughout our solar system. There are wide-ranging sources of particle radiation including but not limited to solar flares, shocks or

The Scientific Foundation of Space Weather
Edited by Rudolf von Steiger, Daniel Baker, André Balogh, Tamás Gombosi, Hannu Koskinen and Astrid Veronig

✉ N.A. Schwadron
nschwadron@unh.edu

1 8 College Road, Durham, NH 03824, USA

2 Heliospheric Physics Laboratory, Code 672, NASA Goddard Space Flight Center, Greenbelt, MD 20771, USA

3 Southwest Research Institute, 6220 Culebra Rd., San Antonio, TX 78238, USA

4 Predictive Science Inc., 9990 Mesa Rim Rd, Suite 170, San Diego, CA 92121, USA

5 Harvard-Smithsonian Center for Astrophysics, 60 Garden Street, Cambridge, MA 02138, USA

6 Nuclear Engineering Department, U. Tenn. Knoxville, Knoxville, TN 37996, USA

7 Leidos, Houston, TX, USA

compressions driven by coronal mass ejections, the termination shock surrounding our solar system and sources beyond our heliosphere. Galactic Cosmic Rays (GCRs) accelerated at supernova shocks continually penetrate inside the solar system, representing a high-energy particle population that affects all bodies in the solar system. The largest solar-generated events (Solar Energetic Particle, SEP events) are associated with complex sunspot groups (active regions on the Sun) with strong, stressed magnetic fields. Highly energetic protons are accelerated very low in the corona by compression regions or shocks driven by coronal mass ejections (CMEs). These energetic protons travel near the speed of light and can arrive at Earth and Mars minutes after the eruptive event. Whether these particles reach Earth, the Moon, Mars (or any other point) depends on their transport in interplanetary space and their magnetic connection to the particle source. The size, duration, and radial gradients of SEP events determines how they interact with the planets throughout the solar system and other bodies.

The inner planets, the asteroid belt, and Jupiter remain within the range of significant influence by SEP events but this connection begins to fade at the orbit of Saturn and beyond, where ions accelerated at the termination shock, the Anomalous Cosmic Rays (ACRs), and in interstellar space beyond the heliosphere, the galactic cosmic ray (GCR) ions and electrons, become the dominant particle radiation species. Finally, there are competing radiation effects from solar and stellar ultraviolet light emissions which may be particularly important for atmospheric chemistry, e.g. the ozone layer at Earth, early atmospheric loss at Mars, and organic aerosols in the atmospheres of Titan and Pluto. In this work we are mainly concerned with particle radiation sources, interplanetary propagation, and interaction effects.

Our understanding of SEP activity has been biased toward the large events by having evolved during the Modern Maximum of solar activity from the 1930s onward during the last century until the more recent decline in the current solar cycle 24. It became clear that the largest events were not directly from solar flares at the Sun but from acceleration by interplanetary shocks and compression regions or shocks low in the corona. We discuss here contemporaneous observations during the largest events in solar cycle 24 (Russell et al. 2013; Joyce et al. 2015) that show the longitudinal distributions of ionizing radiation events broadly distributed from sources near the Sun, yet highly isolated during the passage of the CME shocks or compression regions that accelerate particles throughout interplanetary space.

The current evolution of the Sun between solar cycles 23 and 24 and during cycle 24 remains highly anomalous compared to previous periods of the space age. The Sun has been abnormally quiet over a relatively long solar minimum when GCRs achieved the highest flux levels observed in the space age (Mewaldt et al. 2010; Schwadron et al. 2014a), and the power, pressure, flux and magnetic flux of solar wind were at the lowest levels (McComas et al. 2008; Schwadron et al. 2008; Connick et al. 2011). Even observations of the global heliosphere show remarkably rapid changes (McComas et al. 2010) caused by dropping solar wind pressure. This new period of solar evolution presents a unique opportunity to study the doses of GCRs when their fluxes remain at high levels. The period is also important for studying SEP events during low solar activity. A fundamental question remains as to how large SEP events are created, how they are related to coronal mass ejections and active region eruptions, and what factors control the differential fluxes incident at Earth and other observers during SEP events. The drop in solar activity represents a unique opportunity to test our understanding of SEP acceleration.

In past and the present work we have used observations from the Cosmic Ray Telescope for the Effects of Radiation (CRaTER, Spence et al. 2010) on the Lunar Reconnaissance Orbiter (LRO) to examine the implications of these highly unusual solar conditions for human space exploration and for planetary bodies throughout the inner solar system. While the conditions of the space environment are not a show-stopper for long-duration human missions (e.g., to the Moon, an asteroid, or Mars), galactic cosmic ray radiation remains a significant and worsening factor that limits mission durations. If the heliospheric magnetic field continues to weaken over time, as is likely, then allowable mission durations will continue to decrease. As galactic cosmic ray fluxes rise, the doses also increase down to aircraft altitudes. The changing solar conditions therefore affect Earth and other planets with atmospheres that absorb primary incident radiation and create cascades of secondary particles.

Solar energetic particle events and galactic cosmic rays cause significant, yet under-appreciated effects of space weathering: the general process of planetary surface and atmospheric modification by space environmental components including micrometeoroid bombardment, solar ultraviolet irradiation, solar wind and magnetospheric plasma, and more energetic charged and neutral particles including GCRs. Plasma-surface interaction processes (Johnson 1990) produce a thin layer of outer material covering and sometimes obscuring the materials of greatest interest for understanding origins and evolution of the affected body (e.g., Moon, planet, asteroid, comet). Within a series of weathered layers at increasing depth, products of GCR interactions extending to meter and greater depths may be covered by other more highly processed products of lower energy particle interactions. Outermost layers are eroded by plasma and energetic ion sputtering but residual materials are processed into volatile-poor refractory materials, thus for example accounting for the low surface albedo of cometary nuclei (Johnson et al. 1987) and on Kuiper Belt and Oort Cloud Objects (Moore et al. 2003; Cooper et al. 2003; Hudson and Moore 1999; Hudson et al. 2008). Radiation causes both chemical change of lunar regolith (Schwadron et al. 2012) and leads to dielectric breakdown, resulting in significant changes in regolith porosity and measured plane albedo (e.g., Jordan et al. 2013, 2015). Finally, radiation also penetrates into atmospheres causing a cascade of chemical reactions (e.g., Bazilevskaya et al. 2008; Duderstadt et al. 2016) and production of isotopes (Beer et al. 2012). The effects of radiation potentially extend to Earth's climate and biosphere (Schwadron et al. 2011).

No introduction to radiation effects in the solar system would be complete without reference to one of the most extreme radiation environments, the orbital zone of the ocean moon Europa in the Jovian magnetosphere. Radiation effects are literally imprinted into the surface chemistry of Europa, as revealed by near-infrared spectral mapping of sulfate hydrate deposits (McCord et al. 1998a,b, 1999, 2001; Carlson et al. 1999, 2005; Dalton et al. 2005; Hibbitts et al. 2000) concentrated on the trailing hemisphere where energetic electrons have maximum surface fluxes (Paranicas et al. 2001, 2009). Irradiation produces oxidants and simple organics that could be critical for life (e.g., Chyba 2000; Cooper et al. 2001) but also obscure and even destroy chemical signatures of prebiotic and biological evolution on potentially habitable worlds such as Europa and Enceladus. Radiation-induced chemistry in surface ices has also been proposed (Cooper et al. 2009) as an energy source via magnetospheric electron irradiation for gas-driven cryovolcanism on Enceladus. A similar process on far more intensely irradiated Europa might account for the apparent confirmation (Sparks et al. 2016) of active plumes on that ocean moon. The common chemical product driving such activity on Europa and Enceladus could be oxygen, perhaps produced in the initial and relatively stable form of hydrogen peroxide (Cooper et al. 2001) at cryogenic temperatures.

Space weathering may also be a factor in the surface albedos and compositional colors of Kuiper Belt Objects (KBOs) and their own satellites as suggested by Cooper et al. (2003). Thin outer layers of highly processed and refractory materials could cover thicker layers of moderately irradiated ices, e.g. the latter having red colors as compared to grey to black colors for the outermost radiation crusts. Plasma and meteoritic impact erosion could

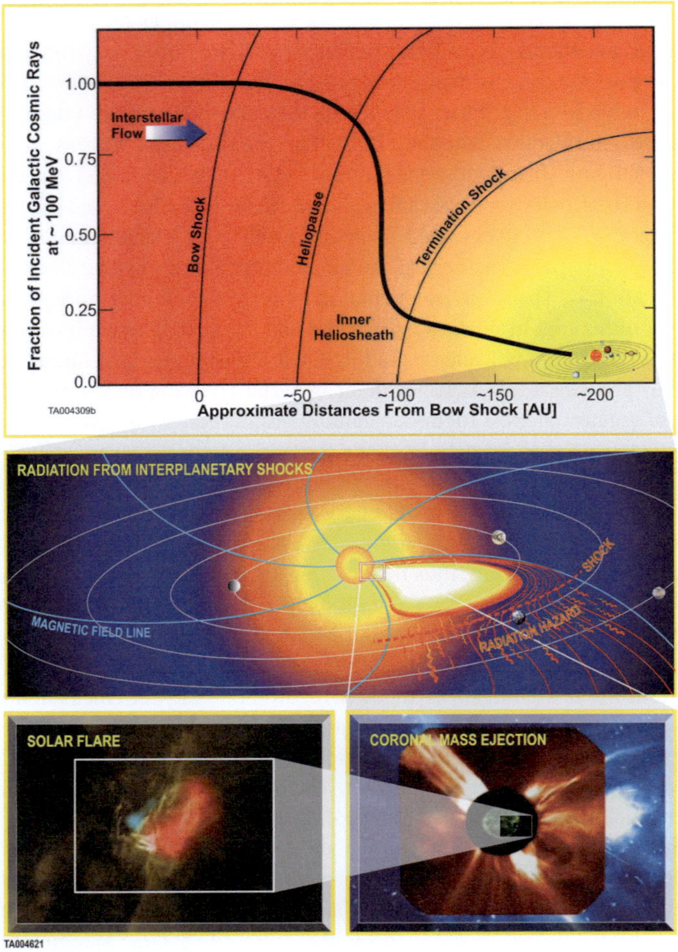

Fig. 1 Radiation hazards posed by space radiation: (*top*) galactic cosmic rays from outside our solar system and from (*middle*) shocks and (*bottom left*) flares often driven by (*bottom right*) coronal mass ejections (reproduced from Schwadron et al. 2010)

thin the outer crusts and expose the underlying redder ices. The KBO Eris has an extremely bright surface, like Enceladus, and could similarly be resurfaced by gas-driven cryovolcanic ejecta. There could, for example, be a gas-generating interaction between oxidized species produced by penetrating GCR interactions with more pristine ices and reduced species composing surface refractory materials.

Particle radiation poses hazards for electronic systems and human explorers (Cucinotta 1999; Cucinotta et al. 2010; Schwadron et al. 2010, 2014a, Fig. 1). For example, it is unclear if long missions outside of low-Earth orbit can be accomplished with acceptable risk (Cucinotta et al. 2010; Schwadron et al. 2014a). The hazards and effects caused by particle radiation in space depicted in Fig. 1 pose a serious challenge to human and robotic exploration missions to the Moon, Mars and beyond. The hazards and effects include the following:

– *Galactic cosmic rays (GCRs)*, which are always present in the near Earth space environment and throughout the solar system, originate from beyond our heliosphere and produce chronic but not acute exposures. GCRs are very difficult to shield against beyond the Earth's protective atmosphere and magnetosphere. At ground level on Earth we live under the shielding equivalent of a swimming pool ten meters in depth, while the geomagnetic field reduces ground level radiation (e.g., secondary neutrons) by another factor of four at the magnetic equator. A ten-meter thick layer of water covering an area of ten square meters, the size of a small habitat for one person, would weigh 100 metric tons, near maximum capacity for currently envisaged launch vehicles. It is more likely, therefore, that astronauts will make use of natural shielding materials at their destinations than that artificial materials will be launched from Earth. For the same habitable area an artificial magnetic field of 60 T, far too strong for habitation by humans, would be required to provide magnetic shielding equivalent to that at the Earth's magnetic equator. Otherwise, astronauts operating in space under typical shielding of a few-10 g/cm^2 of aluminum by spacecraft structures could reach their career limit of radiation exposure from GCRs in one or several years (Cucinotta et al. 2001; Schwadron et al. 2010). Current research in this area is focused on understanding the constraints imposed by GCRs and how they vary with mission transit time, shielding type and thickness, and on developing better techniques to shield against GCRs. The intensities of GCRs vary with the solar cycle with the largest intensities occurring near solar minimum when interplanetary field strengths (e.g., Le Roux and Potgieter 1995) are weakest and there are the fewest number of interplanetary disturbances from transient disturbances such as coronal mass ejections (e.g., Owens and Crooker 2006; Schwadron et al. 2008). GCRs are modulated by the outflowing solar wind and its embedded magnetic field; the modulation is therefore weakest when the interplanetary field strength is low (e.g., Potgieter et al. 2001) and the associated intensities of GCRs are commensurately high.

- *Solar energetic particle (SEP) events* (which we define to include ions; also solar particle events, SPEs) are also dangerous to astronauts outside of Earth's protective layers (the atmosphere and magnetosphere). Current research in this area focuses on developing the ability to predict when and where SEP events will occur and finding ways to adequately shield against SEP-associated particle radiation. *Ground Level Enhancements (GLEs)* represent the largest subset of SEP events, which have very large fluxes of high energy particles ($\sim$ GeV) leading to the production of showers of secondary particles in the Earth's atmosphere with sufficient intensity to be detected by ground-level neutron monitors, above the background of galactic cosmic rays. The association of GLE events with both solar flares and CMEs is well known. Energetic storm particle (ESP) events are associated with shocks driven by coronal mass ejections (CMEs), which efficiently accelerate energetic particles throughout the shock's propagation through the inner heliosphere and likely beyond.
- Unique radiation environments at each planet and their satellites. We have thoroughly characterized the locations of the radiation belts at Earth, which allows us to reduce the hazard they pose by rapidly transiting them. Human and robotic exploration of other planets and satellites requires that we adequately characterize planetary radiation environments and develop appropriate mitigation strategies and adequate shielding. Shielding is often considered the solution to space radiation hazards. Very high energy radiation (e.g., >100 MeV), however, produces secondary penetrating particles such as neutrons and nuclear fragments in shielding material. Some types of shielding material may actually increase the radiation hazard (Wilson et al. 1999). Recent work has shown that relatively lightweight material with high hydrogen content (e.g., the tissue-equivalent plastic between detectors in the Cosmic Ray Telescope for the Effects of Radiation, CRaTER, on the Lunar Reconnaissance Orbiter, LRO) allows significant dose reductions (Zeitlin et al.

2013) and has significant advantages for shielding galactic cosmic rays, and likely SEPs as well. The radiation hazard is not sufficiently well characterized to determine if long missions outside of low-Earth orbit can be accomplished with acceptable risk (Cucinotta et al. 2001, 2010; Schwadron et al. 2014a).

This paper reviews selected recent work on the effects and hazards posed by SEPs and GCR radiation throughout the heliosphere. A comprehensive review is beyond the scope of the paper. We do not cover topics such as the influence of GCRs on aerosols and clouds, ozone destruction of SEPs, and related topics. Section 2 focuses on the origin (Sect. 2.1), propagation (Sect. 2.2) and evolution (Sect. 2.3) of SEP and GCR particle radiation. Section 3 highlights examples of the effects radiation interactions at the Moon (Sect. 3.1), exemplifying radiation interactions at an airless body, at other planetary bodies (Sect. 3.2) and at Earth (Sect. 3.3), exemplifying radiation interactions in the presence of an atmosphere. We conclude in Sect. 4. Appendix A provides an acronym list and Appendix B discusses components of the Earth Moon Mars Radiation Environment Modules along with the Predictions of Radiation from Release, EMMREM, and Data Incorporating the CRaTER, COSTEP and other SEP measurements (PREDICCS) online system.

2 Origin, Propagation and Evolution of SEP and GCR Particle Radiation in the Heliosphere

In this section we discuss the origin and evolution of particle radiation in the heliosphere. As discussed in the introduction the primary sources of particle radiation are solar energetic particle events and galactic cosmic radiation. Section 2.1 discusses the origin of large solar energetic particle events from the low corona and their propagation through the heliosphere. Section 2.2 discusses the propagation of SEPs and GCRs through the solar system. Section 2.3 discusses the evolution of both SEP and GCR particle radiation. We have considerably broadened our understanding of the evolution of particle radiation as our observations have extended through the deep and prolonged solar minimum between cycles 23 and 24 and the remarkably weak activity in cycle 24. The observations become increasingly important as our Sun and heliosphere may be headed toward a prolonged period of low solar activity such a Dalton minimum, or a Gleissberg minimum.

2.1 Origin of Solar Energetic Particle Events from the Low Corona

Particle acceleration is associated with flare acceleration and the shocks driven by coronal mass ejections (Aschwanden 2006). Generally, the sites of particle acceleration and the evolution of SEP events depend strongly on the sources and evolution of solar disturbances. Rapid reconfiguration of coronal magnetic fields through magnetic reconnection is associated with solar flares causing impulsive SEP events. CMEs drive compressions and shocks that accelerate energetic particles as the CME driver propagates out through the corona and out into interplanetary space. Moderate speed CMEs accelerate gradually with radius, leading to the formation of gradual SEP events from energetic particles that are accelerated from shocks that form typically beyond 10 R_s. The largest and fastest CMEs form powerful compressions and shocks low in the corona (<5 R_s) that more rapidly accelerate high energy particles often up to the GeV energies required for GLEs. It is difficult to disentangle energetic particles from flares and from CME drivers. Magnetic reconfiguration and reconnection are inherently associated with CME initiation (Antiochos et al. 1999; Forbes et al.

2006; Chen 1989). Further, the flares sources, which are typically rich in ^{3}He, serve as the pre-accelerated seed populations of SEPs accelerated at shocks and compressions revealed by heavy ion and ^{3}He enhancements observed at high energies in SEP events (Mason et al. 1984, 1999, 2002; Desai et al. 2003, 2006).

Recent work focusing on *large* SEP events (frequently associated with GLEs) demonstrates the overriding importance of CMEs, which drive shocks and compressions *low in the corona* that, in turn, accelerate energetic particles. Gopalswamy et al. (2005) studied CME coronal heights at times of energetic particle release. They found that particle acceleration typically begins ∼1.4–8.7 R_s. These results are highly consistent with timing studies based on *in situ* observations (Reames 2009a,b). Remote observations confirm that CME-driven shock waves or compressions occur low in the corona below 2 R_s (Kozarev et al. 2011; Ma et al. 2011; Veronig et al. 2010; Liu et al. 2010). The region between 1–10 R_s remains difficult to study due to the complexity of the magnetic structure, the uncertainties related to CME initiation and the difficulty in describing the underlying particle acceleration (e.g., Kozarev et al. 2013; Schwadron et al. 2015).

The passage of shocks beyond 1 AU observers has allowed significant insight into particle acceleration in interplanetary space. This area of research has facilitated detailed study of ESP events. Particle-in-cell simulations provide significant insights into the underlying physical mechanisms associated with diffusive shock acceleration (Giacalone 2005). Global shock acceleration and particle propagation models typically focus on interplanetary space (Verkhoglyadova et al. 2008; Zhang et al. 2009), and the development of gradual events between ∼20 R_s and 1 AU where shocks have sufficient time to become strong.

The environment closer to the Sun, below 10 R_s has remained a difficult problem due to the complexity of competing physical processes that control this plasma environment and the lack of direct observations within this region of space. Simulations and theories of particle acceleration in the low corona have shown that it is an environment conducive to accelerating high energy particles very rapidly. Solar energetic protons can be rapidly accelerated up to very high energies of 10 GeV for strong shocks (Roussev et al. 2004; Manchester et al. 2005; Kocharov et al. 2005; Tsurutani et al. 2003). Schwadron et al. (2014b) and Schwadron et al. (2015) show results of sophisticated MHD models where the powerful compressions and shocks on the flanks and expanding CMEs close to the Sun (<5 R_s) can accelerate particles to more than 300 MeV/nuc in 10's of seconds. This likely relates to the formation of prompt GLE events arising from the compressions and shocks driven by accelerating CMEs low in the corona.

Schwadron et al. (2014b, 2015) applied the Energetic Particle Radiation Environment Model (EPREM, Schwadron et al. 2010) for describing particle acceleration arising from a modeled CME low in the corona. The model combines an energetic particle solver that describes the evolution of the particle distribution based on the focused transport equation and simulated magnetohydrodynamic (MHD) fields. EPREM uses a methodology for energetic particle modeling based on a concept first introduced by Kóta et al. (2005a) and Kóta et al. (2005b) in which the focused transport equation is solved in a Lagrangian system co-moving with the plasma. The same approach involving EPREM and coupling to an MHD model was used by Kozarev et al. (2013) to describe particle acceleration from the low corona during CME acceleration. The acceleration of the CME was modeled with the Michigan Block-Adaptive-Tree-Solarwind-Roe-Upwind-Scheme (BATS-R-US) MHD code (Tóth et al. 2005).

In Schwadron et al. (2014b), energetic particle acceleration was simulated from a very rapidly accelerating CME, which was initiated through eruption of a modified Titov (Titov

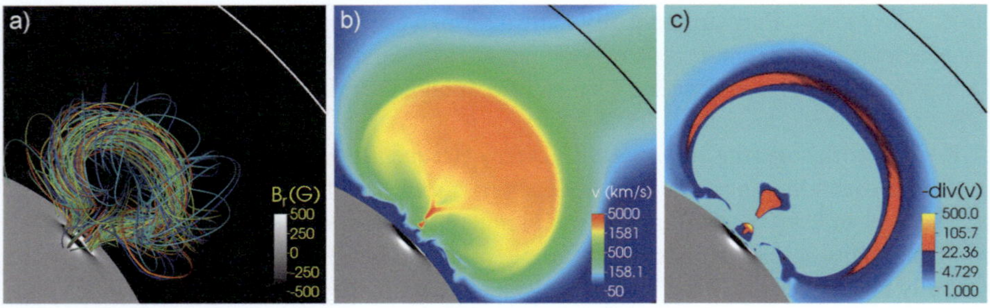

Fig. 2 Eruption of a rapidly accelerating CME (reproduced from Schwadron et al. 2015): the eruption begins with a strong flux rope low in the corona (panel **a**); as the flux rope loses equilibrium, the CME eruption occurs, leading to a large outward speed of the plasma within the CME (panel **b**). The rapid acceleration of the CME causes a large divergence of the plasma velocity (panel **c**), particularly on the flanks of the CME. The CME-driven shock forms at a height of ∼1.4 solar radii. The compression (negative divergence in velocity) is stronger at the flanks of the CME than at its front (Ontiveros and Vourlidas 2009; Hudson 2011). Strong compression is also found in the wake of the CME (panel **c**) due to reconnection outflows (e.g., Shibata and Magara 2011)

and Démoulin 1999) flux rope (Titov et al. 2014). Strong (flare) reconnection jets occur below the erupting rope, which helps the CME to rapidly accelerate to a velocity of $\sim 3000\ \mathrm{km\,s^{-1}}$ low in the corona ($r < 2\ \mathrm{R}_s$; Fig. 2b), after which it slows down and finally travels with an almost constant speed of $\sim 1000\ \mathrm{km\,s^{-1}}$ at heights $r > 3\ \mathrm{R}_s$. This simulation is coupled to a newly developed heliospheric MHD code and the eruption is propagated to 1 AU (Lionello et al. 2013). The CME arrives at 1 AU with a speed of $\sim 700\ \mathrm{km\,s^{-1}}$, moving faster than the surrounding slow solar wind. The most pronounced acceleration of the CME occurred very close to the Sun, within 2 R_s, causing extremely strong compression on the flanks of the CME, as shown in Fig. 2. The simulations of Schwadron et al. (2014b) suggest the reconnection jets during CME initiation help provide the energy for the outward expansion that drives the compressions on the flanks of the CME. In other words, it is likely that reconnection plays a fundamental role in both creating the compressions that accelerate energetic particles while also providing the flare-related seed populations that feed into this acceleration process.

Results from the Schwadron et al. (2015) study shown in Fig. 3 demonstrate that the size of the shock or compression limits the break energy. This implies that larger shocks or compressions should lead to increased fluxes at higher energies. Conversely, small shocks or compressions lead to decreased break energies. This result may explain why large CMEs tend to produce higher energy particle events.

The theoretical investigations of the origins of SEP events in the low corona highlight several important components with observational implications:

1. Magnetic reconnection during the initial acceleration of CMEs appears to play a significant role in energizing the plasma within the CME, and causes rapid CME expansion. The expansion, in turn, drives strong compressions and shocks low in the corona that rapidly accelerate high energy particles within minutes of the formation of the compressions or shocks. This leads to the association of high SEP fluxes during the initial acceleration of the CME. It also causes broad longitudinal distributions of SEPs (see Fig. 4) from the CME as it moves low within the corona.
2. We expect very different signatures of SEPs from different spatial regions connected to the CME (Reames 1999). Retrograde observers should have stronger magnetic connections to the CME, which would cause prompt increases in SEP fluxes during CME

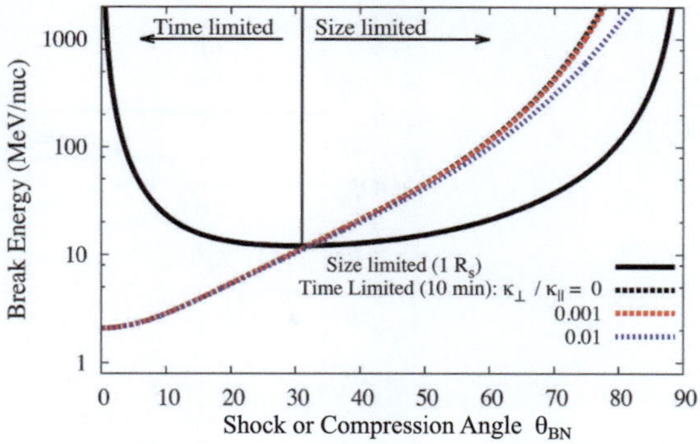

Fig. 3 Break energy of accelerated particles as a function of the magnetic field shock-normal angle (reproduced from Schwadron et al. 2015). Size-limited shocks and compressions limit break energies for large shock angles (>30°), whereas time-limited acceleration limits break energies for lower shock angles (<30°). Note generally that larger shock angles lead to higher break energies

acceleration in the low corona. In contrast, prograde observers should have the weakest magnetic connection to the CME and show more gradual acceleration and lower SEP fluxes.

3. The largest fluxes of SEPs should be observed from observers that are directly in the path of the CME. In this case, we expect not only the fluxes of promptly accelerated particles from the CME shocks and compressions within the corona, but also the build up of fluxes from the shocks driven by the CME as it moves out through the interplanetary medium.

A critical resource in understanding the evolution of CMEs close to the Sun is coronagraph measurements such as those provided by COR1, the innermost coronagraph of the Sun Earth Connection Coronal and Heliospheric Investigation (SECCHI) instrument suite aboard the twin Solar Terrestrial Relations Observatory (STEREO) spacecraft. The paired COR1 telescopes observe the K-corona from 1.4 to 4 R_s in a waveband 22.5 nm wide centered on the H-α line at 656 nm (Thompson and Reginald 2008).

Another essential resource in understanding the evolution of CMEs in the low corona is the Mauna Loa Solar Observatory (MLSO) operated by the National Center for Atmospheric Research (NCAR). MLSO contains a suite of instruments for studying both short-term (minutes-hours) and long-term (years) changes on the Sun from the photosphere, through the chromosphere and into the corona. The MK4 white light K-coronameter (Cho et al. 2007; Lee et al. 2008) records images of the low corona (1.12 to 2.9 R_s). The MK4 instituted improvements over the earlier MK3 K-coronameter, which operated from 1980 to 1999, and provides white light views into the very low corona (down to 1.05 R_s) where CMEs form. These observations enhance existing and new observatories on the ground (e.g. SOLIS, ATST, COSMO, FASR) and in space (e.g. SDO, Hinode, STEREO, SOHO, RHESSI, and ACE) by providing information on the early evolution of CMEs in the region where the largest SEP events, GLEs, are spawned. The European Space Agency's Proba-3 pair of coronagraph-occulter satellites will be designed to image the lower corona down to 1.05 R_s (Zhukov 2016; Llorente et al. 2013). Alternative approaches using planetary objects as occulters for in-space coronagraph imaging of the lower corona below 1.4 R_s have been suggested by Habbal et al. (2013) for the Moon and by Kim et al. (2017) for Mars.

Remote observations provide important information about the coronal plasma response to solar eruptions, which are accompanied by large-scale wave-like disturbances visible in vari-

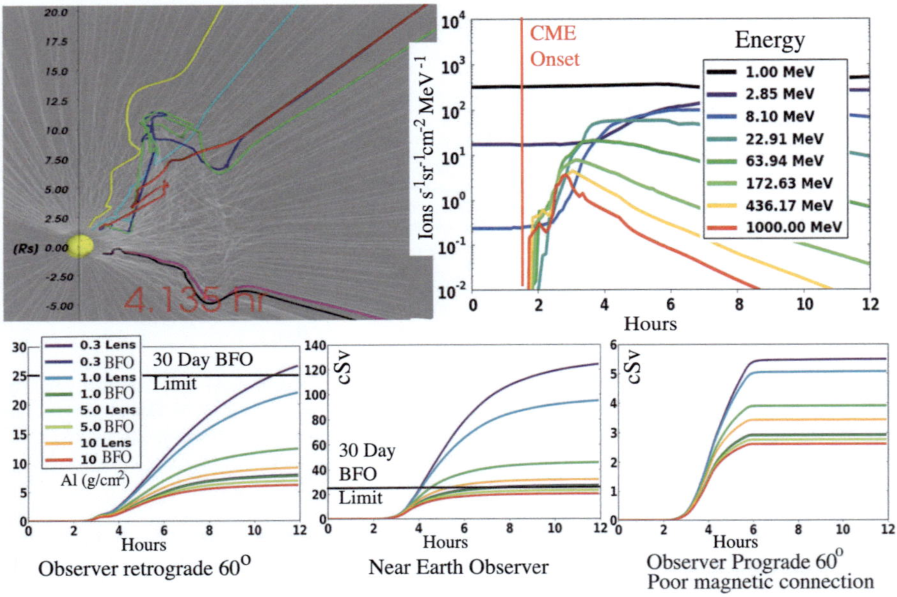

Fig. 4 Energetic particles are accelerated over a broad latitudinal and longitudinal spread from the CME released following destabilization (in Fig. 2). The figure is reproduced from Schwadron et al. (2014b). The colored magnetic field *lines* (*left top*) show strong distortions by the plasma expansion, driven by reconnection jets. Using coupled MAS-EPREM simulations, we link coronal conditions, CMEs and associated shocks and transients to solar energetic particles (*top right*), solar wind conditions, and time-dependent radiation exposure (*bottom panels*). Shown in the *right top panel* are the results for particle differential energy fluxes at 1 AU from the event shown in Fig. 2. The *bottom panels* shows the resulting integrated dose equivalents for Lens and Blood Forming Organs (BFO) behind different levels of shielding. We find 10's of cSv even for well-shielded (10 g/cm^2 Al) BFO dose equivalents, indicating a radiation hazard that approaches the 30-Day Limit (25 cSv) in roughly 2 hours after CME initiation. The CME-onset begins at $t = 1.43$ hours, and then strong compression and energetic particle acceleration begins at 1.45 hours. Integrated doses at prograde (*bottom-right*) and retrograde observers (*bottom-left*) are shown in addition to the near-Earth observer. The retrograde observer is well connected magnetically to the CME driver, which results in higher and more prolonged SEP fluxes at this observer (Schwadron et al. 2014b)

ous spectral ranges. Moreton waves (Moreton 1960) observed in H-α were initially proposed (Uchida 1968) to be a chromospheric trail associated with a coronal fast-mode MHD wave. Observations of the low corona in extreme ultraviolet (EUV) with EIT (Delaboudinière et al. 1995) on SOHO revealed large-scale wave-like disturbances either quasi-stationary or propagating over large distances up to the whole disk along the solar surface or expanding above the limb. These "EIT waves" or "EUV waves" are registered by EUV telescopes such as EIT, TRACE, STEREO/EUVI, SDO/AIA (Wills-Davey and Attrill 2009; Gallagher and Long 2011).

EUV waves are likely associated with MHD fast-mode waves in the lower corona (e.g., Thompson et al. 1999; Warmuth et al. 2001) and plasma compression in bases of coronal loops in their successive stretching by an expanding CME (e.g., Delannée 2000). Schmidt and Ofman (2010) presented the first 3D MHD modeling of an EIT wave from an eruption-driven shock wave. There is an expected association of particle acceleration on the flanks of CMEs from large-scale compressions and EIT waves driven by shocks in the low corona.

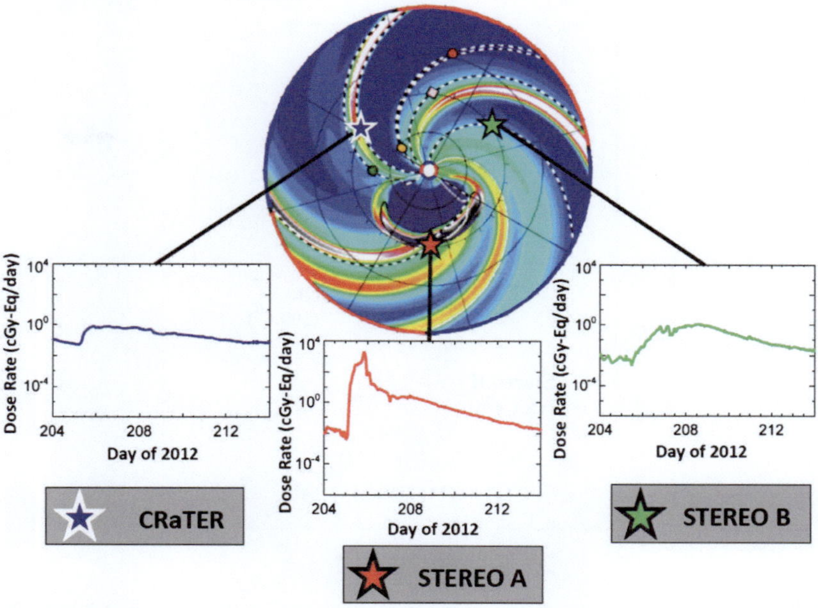

Fig. 5 Demonstration of the characterizations observed by Reames (1999) in the July 23 2012 event (reproduced from Joyce et al. 2015). Shown is the ENLIL MHD model output at the time the shock arrives at 1 AU along with the dose rates observed at CRaTER (*left*, *blue*), derived at STEREO A (*middle*, *red*) and STEREO B (*right*, *green*), representing the observers located at the eastern flank (retrograde), center and western flank (prograde) of the shock front respectively

2.2 Propagation of SEPs and GCRs Through the Solar System

In this subsection, we consider the propagation of SEPs from the Sun through the heliosphere. We also discuss GCR propagation through the heliosphere. This section collects recent work summarizing a very broad topic, with an emphasis on the propagation profiles, which provides utility for assessing radiation interactions at planetary bodies throughout the solar system.

The first topic considered is the radiation environment at 1 AU. With the advent of STEREO in addition to observations at LRO, ACE, Wind and SOHO, we have had unique opportunities to observe how radiation processes evolve at 1 AU over a broad longitudinal distribution. The July 23 SEP event of 2012 (Russell et al. 2013; Joyce et al. 2015) provided a unique opportunity to simultaneously observe the evolution of a large SEP event from an observer (STEREO A) directly in the path of a large CME, in addition to one observer (LRO/CRaTER) on the eastern flank (retrograde) and another observer on the western flank (retrograde). Figure 5 shows time profiles for this event from STEREO A, STEREO B and from LRO/CRaTER. Figure 6 shows proton flux spectra from the two Stereo spacecraft and from other spacecraft orbiting upstream of Earth at the L1 Lagrange point, the strongest fluxes clearly being at STEREO A and the weakest at STEREO B. Note that 10-MeV intensities are 10^3 and 10^5 times lower at Earth and STEREO B, respectively. Other events, e.g. October 1989, had higher proton fluxes above 10 MeV near Earth but the July 23rd event was notable for its extremes at the three different vantage points. While the multi-point average of these fluxes, represented approximately by those near Earth in Fig. 6, was not particular high for applications to long-term space weathering effects at planetary surfaces

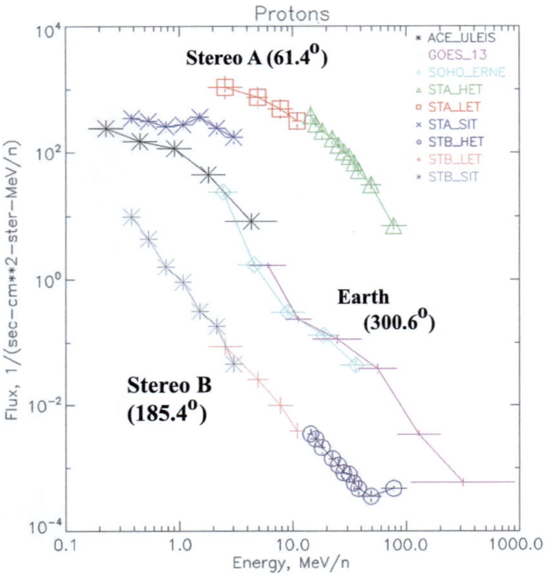

Fig. 6 Proton flux spectra from the "Carrington" coronal Mass ejection (CME) event of July 23, 2012, as plotted by the Multi-Source Spectral Plot service of the Virtual Energetic Particle Observatory (VEPO) at http://vepo.gsfc.nasa.gov/. Sources: Advanced Composition Explorer (ACE) Ultra Low Energy Ion Spectrometer (ULEIS); Geostationary Satellite (GOES) 13 Space Environment Monitor (SEM); Solar Heliospheric Observatory (SOHO) Energetic and Relativistic Nuclei and Electron (ERNE) sensor; Solar Terrestrial Relations Observatory A and B (STA, STB) High Energy Telescope (HET), Low Energy Telescope (LET), and Solar Isotope Telescope (SIT). Note that STA SIT data apparently saturated at extremely high intensity of this event. Locations of STEREO-A, Earth, and STEREO-B are given in Solar Ecliptic Longitude (°)

of the inner solar system, the peak fluxes in the direction of STEREO A could have been a major radiation concern for human astronauts voyaging into deep space at that longitude enroute to Mars or another destination.

Accumulated doses for the July 2012 SEP event as computed by the Earth Moon Radiation Environment Module (EMMREM) at STEREO A are shown in Fig. 7. Doses are computed for skin/eye as well as Blood Forming Organ (BFO) for various levels of shielding. Also indicated on the plot are the 30-day dose limits (NASA 2007) for skin, eye and BFO, as well as the dose accumulated for the CRaTER GCR rate.

The July 23 2012 event generated by far the largest dose rates for the observer (STEREO A) directly in the path of the CME. Figure 7 shows that the largest increase in dose occurred slightly after day 205 (2012) when the shock passed the spacecraft. In other words, the passage of the CME through the low corona did not result in the highest fluxes of energetic particles. It was the CME shock itself and the sheath behind it that contained the largest fluxes of energetic particles and resulted in the most significant radiation hazard. The delayed increase in energetic particle fluxes observed at STEREO A is also consistent with the relatively low doses observed at CRaTER and STEREO B. Both these observers on the East and West flanks of the CME failed to connect to the strong shock formed as the CME pushed out through the interplanetary medium.

When assessing effects for various bodies in the solar system it is important to address how both SEP and GCR radiation evolves with radial distance and also over the solar cycle. The results discussed here are currently being compiled as a part of paper that focuses on the radiation environment with implications for planetary bodies throughout the solar

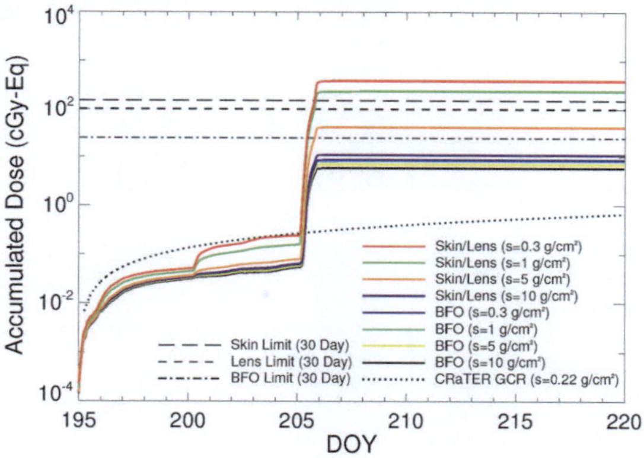

Fig. 7 Dose rate for proxies of skin/lens dose and blood-forming organ (BFO) dose for shielding thicknesses, $s = 0.3, 1, 5, 10\ \mathrm{g/cm^2}$ and BFO for four different levels of shielding which correspond to spacesuit, heavy spacesuit, spacecraft and heavy protective shielding (reproduced from Joyce et al. 2015). Average background GCR dose rate measured by CRaTER during this time is also shown. NASA 30 day dose limits (NASA 2007) for skin and eye dose are exceeded for both 0.3 and 1 $\mathrm{g/cm^2}$ Al shielding; the heaviest shielding reduces the total accumulated dose by more than an order of magnitude. The BFO limit is not exceeded for any level of shielding, though heavier shielding is less effective at reducing the total dose

system (Spence et al. 2017). SEPs originate from the Sun and accordingly SEP fluxes drop off with radial distance (e.g., Dayeh et al. 2010). In contrast, GCRs come from outside the heliosphere and are modulated by the solar wind implying that GCR fluxes increase with radial distance through the heliosphere. A data and model-based projection of the radial evolution of SEPs and GCRs is summarized in Fig. 8. The empirical model of Webber and Lockwood (2004) is used to quantify how the intensity of GCR protons varies with radial distance through the solar system and as a function of solar cycle. Webber and Lockwood (2004) used GCR measurements from the IMP, Voyager, and Pioneer spacecraft to construct their model, which included data collected over several solar cycles over a broad range of heliocentric distances near the ecliptic plane. They found that the observed GCR intensity profiles were well approximated mathematically with a simple radial dependence characterized by a gradient of several percent per AU (only marginally better fits were accomplished with a mathematical form that exhibited a weak dependence of that gradient with radial distance). We use their empirical fits (possessing a radial gradient of 3.2%/AU), and the typical intensities measured at Earth during solar minimum and maximum, to estimate the commensurate intensity of the 120 to 230 MeV GCR protons at Mercury, the Moon, in the Mars system, at Ceres, and in the Pluto system; this energy range is representative of those with important effects at planetary surfaces.

We find a factor of $\sim$8 decrease in GCR fluxes during solar maximum compared to solar minimum, a factor of $\sim$3.5 between Mercury and Pluto and a factor of $\sim$30 between environment extremes (Mercury at solar max versus Pluto at solar minimum). This represents significant variability in the radiation environment, which causes large differences between the effects of radiation at the planets and over time.

Figure 8 also shows the fluence associated with the March 2012 SEP event. The projection is made using 1 AU (GOES) observations and the EPREM model to determine how fluxes vary radially out through the solar system (Spence et al. 2017). The projection assumes a 1% ratio of perpendicular to parallel diffusion, and a scaling of the parallel mean free with rigidity to the power 2/3, $\lambda_{\parallel} = \lambda_0 (R/R_0)^{2/3}$ where R is rigidity, $R_0 = 1$ GV and

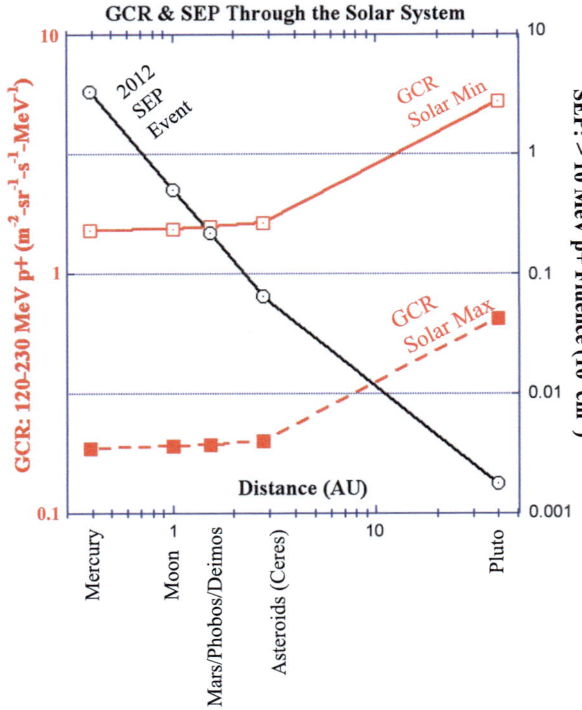

Fig. 8 Variability in GCR fluxes and SEP fluxes throughout the solar system. Shown here are fluxes of GCR protons from ~120–230 MeV (*red*) based on the empirical model of Webber and Lockwood (2004). Also shown is the flux of the March 2012 SEP event, projected throughout the solar system. The results demonstrate the large variability in the radiation environment throughout the solar system, with inner planets experiencing higher levels of SEP radiation and outer planets experiencing higher levels of GCRs

$\lambda_0 = 0.4$ AU. As in the case of GCRs, we find a large variation in SEP fluences throughout the solar system; in this case the fluence drops by more than 1800 times through the solar system. The large variability in SEP events over the solar cycle, and the large changes in fluence throughout the solar system creates radically different radiation environments at the different planets and for asteroids, comets and small bodies throughout the solar system. The radial dependence of GCRs and SEPs becomes important in this context for understanding how radiation effects bodies throughout the solar system.

2.3 Evolution of SEP and GCR Radiation

A holistic view SEP and GCR radiation interactions at planets and moons depends in part on understanding the temporal evolution of radiation. In Fig. 8, we indicate temporal variability of GCRs over the solar cycle. There is a well known source of variability as the interplanetary magnetic field becomes stronger near solar maximum due predominantly to the increased frequency (~3 CMEs/day) of coronal mass ejections, which drag out new magnetic flux into the heliosphere (Owens and Crooker 2006; Schwadron et al. 2008). Conversely, near solar minimum the reduced CME frequency (~1 CMEs/3 days) leads to a weaker magnetic field. The magnetic field intensity directly causes variation in cosmic ray diffusion. Stronger magnetic fields reduce the diffusion coefficient and thereby reduce the flux of GCRs, whereas weaker magnetic fields increase the diffusion coefficient and the GCR flux. Conversely, solar activity increases the frequency of solar flares and coronal mass ejections, which accelerate SEPs. Therefore, GCR fluxes and SEP events vary out of phase through the solar cycles with GCR fluxes largest near solar minimum, whereas SEP events are most frequent at solar maximum.

The evolution of GCRs and SEPs over the course of many solar cycles is also thought to vary largely with solar activity. Periods of extreme lows in solar activity such as the Maunder minimum should have infrequent SEPs and large fluxes of GCRs. Recent observations provide a rare glimpse into the extreme lows of solar activity conditions that are drivers of longer term solar and heliospheric variability. The deep and prolonged solar minimum between solar cycles 23 and 24 and the activity in solar cycle 24 have differed significantly from previous cycles during the last 100 years (Schwadron et al. 2011, 2014; McComas et al. 2013). The fast solar wind was somewhat slower, while the solar wind in general was less dense, cooler, and had significantly lower momentum and mass fluxes (McComas et al. 2008). In addition, the solar wind had significantly weaker heliospheric magnetic fields (Smith and Balogh 2008) compared to earlier cycles within the space age. As the activity level of solar cycle 24 rose, the mass flux of solar wind remained extremely low (McComas et al. 2013) and the magnetic flux of the solar wind remained at much lower levels than observed at previous solar maxima in the space age (Smith et al. 2013). As a result, solar cycle 24 is the weakest solar maximum of the space age, which continues the anomalous trends observed in the deep cycle 23–24 minimum. Conditions during the cycle 23–24 solar minimum were similar to conditions at the beginning of the Dalton Minimum (Goelzer et al. 2013). These recent changes suggest that the next solar minimum may continue to show a decline in sunspot numbers, and cause, as a result, further reductions in magnetic flux and solar wind particle flux.

Figure 9 from Schwadron et al. (2014a) shows the evolution of the dose rate from GCR throughout the space age and out beyond 2020. The reduction of solar activity and weakening magnetic fields throughout the interplanetary medium caused increasing fluxes and doses of galactic cosmic rays in successive solar minima starting in 1988 and continuing in successive solar minima through 2009. A major question is whether the weak solar activity observed over the last decade will continue into the future. The solar maximum of cycle 24 has continued to be extremely weak, leading McComas et al. (2013) to coin the phrase, the "mini-solar maximum" of cycle 24. Goelzer et al. (2013) examined conditions of previous solar cycles, concluding that the observed activity of cycle 23 and 24 appeared most like the Dalton minimum (1790–1830) or the Gleissberg minimum (1890–1920). The projections of Goelzer et al. (2013) based on the Dalton and Gleissberg minima were used to develop a new model for the galactic cosmic rays (Schwadron et al. 2014a). The projections suggest that galactic cosmic ray fluxes will continue to increase in the next cycle. These projections demonstrate the vast importance of changes in solar activity for regulating the radiation environment of the solar system.

A second major consequence of reduced solar activity is that SEP events have reduced fluxes and frequency. Figure 10 shows an example highlighting the reduced fluxes of SEP events in cycle 24. The blood forming organ (BFO) doses in SEP events are associated with the fluxes of high energy particles that can penetrate deeply into the body. In this case, we have modeled doses behind relatively thick Al shielding, $10\ \mathrm{g\,cm^{-2}}$. The modeling was done here with the EMMREM and data from PREDICCS (Predictions of radiation from REleASE, EMMREM, and Data Incorporating CRaTER, COSTEP, and other SEP measurements Joyce et al. 2013). The probability of a large SEP event for up to 1 year missions fails to exceed the 30 day-BFO limit. The results show that while solar energetic particle events in cycle 24 present some hazard, accumulated doses for astronauts behind $10\ \mathrm{g/cm^2}$ shielding are well below current dose limits (NASA 2007).

Galactic cosmic radiation presents a more significant challenge: the time to 3% Risk of Exposure Induced Death (REID) in interplanetary space was less than 400 days for a 30 year old male and less than 300 days for a 30 year old female in the last cycle 23–24 minimum

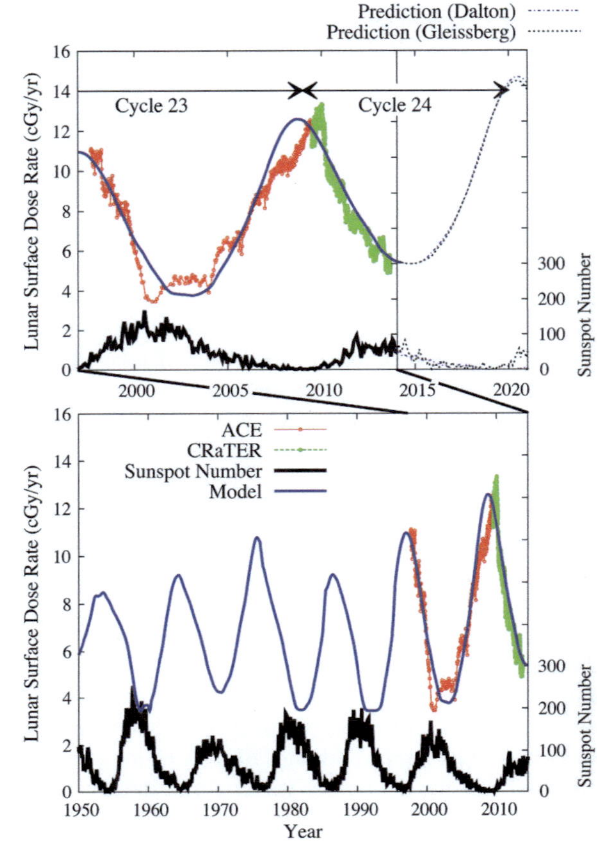

Fig. 9 Evolving and increasing radiation levels in space (reproduced from Schwadron et al. 2014a). ACE dose rates (*red points*) are derived from fits to CRIS spectra (O'Neill 2006), CRaTER measurements (*green points*) are from the zenith facing D1/D2 detectors, commonly used as proxies for lens dose rates behind 0.3 g/cm^2 Al shielding (Schwadron et al. 2012). Sunspot number predictions (*upper panel, black dashed lines*) show are based on a Gleissberg-like and a Dalton-like minimum. Dose predictions (*solid blue line* and the *upper black* and *blue dashed curves*) are from a sunspot-based model of the heliospheric magnetic field and the correlated variation in modulation of GCRs (Schwadron et al. 2014a). The dose rates are projected to the lunar surface. The *bottom panel* is the same as *top panel* but for a longer time span

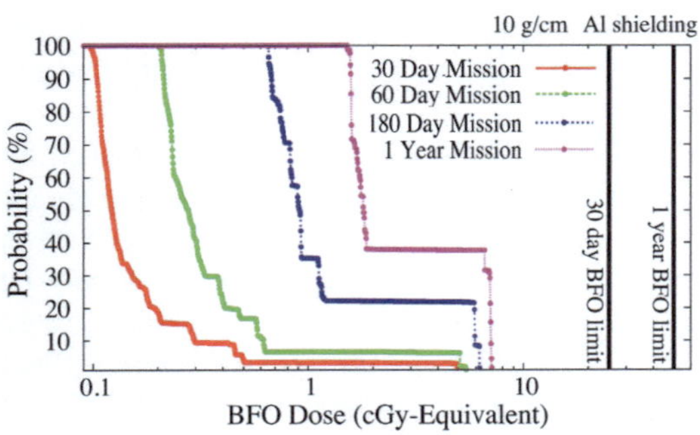

Fig. 10 Probability of an SEP event (%) versus the event's integrated BFO dose for 30 day to 1 year missions (reproduced from Schwadron et al. 2014a). The data used covers events for the period from July 2011 through April 2014. We use the PREDICCS database (Joyce et al. 2013, http://prediccs.sr.unh.edu) to build up statistics for the probability of SEP events of varying integrated dose behind spacecraft shielding (10 g/cm^2). The database currently provides doses for the period from July 2011 through April 2014. The PREDICCS doses are derived from proton spectra and use dose in 10 g/cm^2 water as a proxy for the Blood Forming Organ (BFO) dose

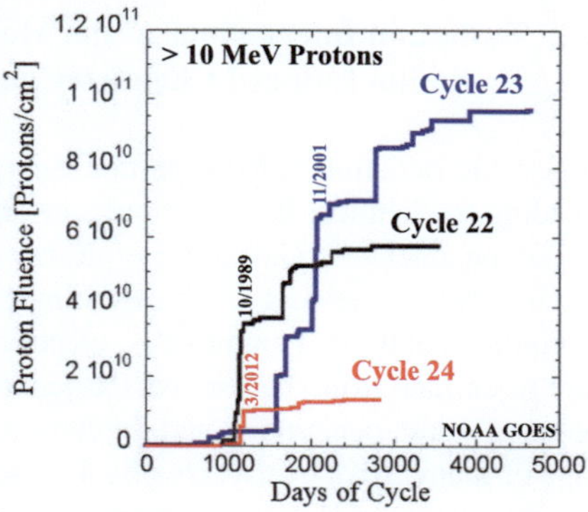

Fig. 11 SEP proton fluences in three successive solar cycles highlighting the effects of dropping solar activity for lowering the fluxes of SEP events (adapted from Mewaldt et al. 2015)

(Schwadron et al. 2014a). The time to 3% REID is estimated to be ~20% lower in the coming cycle 24–25 minimum (Fig. 9). If the heliospheric magnetic field continues to weaken over time, as is likely, then allowable mission durations will decrease correspondingly.

We compare the proton fluences from SEP events (>10 MeV) from the last three solar cycles in Fig. 11 (Mewaldt et al. 2015). The results reinforce the remarkably low activity in cycle 24, which has created weak SEP events with low fluences. Weakening solar activity has had wide-spread manifestations in declining interplanetary field strengths, dropping solar wind flux, elevated galactic cosmic ray fluxes and dropping SEP fluences.

3 Examples of Radiation Interactions at the Moon, at Other Airless Bodies Throughout the Solar System and at Earth

In this section, we explore examples of the effects of radiation interactions at an airless body, the Moon, and at Earth, which has a thick atmosphere. In the case of airless bodies, radiation interacts directly with the surface of the body—the regolith. Radiation interactions cause chemical changes of the regolith surface and lead to space weathering via a number of chemical breakdown processes, which are now being explored. The lunar regolith (Sect. 3.1) provides the nearest planetary body in which to study these space weathering processes actively.

The processes active at the Moon are also active at other airless bodies throughout the solar system (Sect. 3.2). As we expand and generalize our understanding of radiation interactions at other planetary bodies, we address important issues that touch on the habitability of bodies throughout the solar system.

We further discuss radiation interactions at Earth, which highlights the associated physical processes in the presence of a thick atmosphere (Sect. 3.3). Specifically, we find that atmospheric interactions lead to a cascade of secondary particles generated through collisions initiated by the parent population of the source radiation. The radiation dose drops as we move more deeply through the atmosphere. We also observe significant effects due to the interaction of charged particle radiation with the Earth's magnetic field, which causes attenuation as a function of magnetic latitude (geomagnetic cutoff). This example also demonstrates the significance of radiation interactions in view of the large population of air travelers that move through high altitudes and are exposed to increased levels of radiation compared with bodies on the Earth's surface.

3.1 Radiation Interactions at the Moon: A Laboratory for Understanding Radiation Induced Chemistry and Space Weathering at an Airless Body

Energetic radiation at MeV particle energies and above penetrates deeply within surfaces, leading to chemical modification, and space weathering of the physical media with which radiation interacts. Galactic cosmic ray particles strike regolith with very high energies (∼100 MeV to several GeV); resulting collisions create a spectrum of secondary particles (protons, neutrons, gamma-rays, electrons, positrons and other elementary particles) that are re-emitted from the surface. Particle radiation at energies below 100 MeV from SEPs is generally less penetrating and, in contrast to GCR radiation, infrequent, but the time average of many SEPs over years of solar cycles can have significant space weathering effects. Accumulation of Helium-3 isotopes is a well-known effect of solar wind interactions at keV energies with the lunar surface from SEP events enriched in this isotope and generally also in ions heavier than helium up through iron. The lunar surface also reflects small fractions of solar wind protons (Saito et al. 2008) and neutral hydrogen (McComas et al. 2009; Saul et al. 2013) at sub-keV energies.

An important application of radiation interactions at the Moon is the detection of near-surface deposits of water ice via absorption of low-energy neutrons produced by cosmic ray interactions. Elphic et al. (2007) estimate from neutron measurements in the polar regions that 150 million tons of water ice could reside there in permanently shadowed regions. Widespread global surface production of water group molecules (H_2O, OH) has been identified from the ensemble of remote observations by Cassini (Clark 2009), Chandrayaan-1 (Pieters et al. 2009), and Deep Impact (Sunshine et al. 2009). One potential source of these molecules is combination of implanted hydrogen with hydroxyl groups (Starukhina and Shkuratov 2000) in oxygen-containing minerals and impact or volcanic glasses. That is, solar wind protons could drive production of water group volatiles that eventually migrate to the cold trapping regions of the lunar poles and become buried in the meters-deep impact regolith there. Additionally, the lunar polar water could arise from meteoritic impact vaporization and subsequent polar migration of exogenous water.

Figure 12 shows the fraction of dose due to more energetic radiation that is re-emitted from the Moon during quiet periods when GCRs represent the dominant incident flux at the Moon, as reported by Spence et al. (2013). In this case, we analyzed 160 days of CRaTER observations, spanning the period when LRO first entered its mapping orbit on 16 September 2009 through 6 March 2010, just before the first Forbush decrease of the LRO mission. Near the Moon, we found that GCR accounts for ∼91.4% of the total absorbed dose, with GCR protons accounting for ∼42.8%, GCR alpha particles for ∼18.5%, and GCR heavy ions for ∼30.1%. The remaining ∼8.6% of the dose at Lunar Reconnaissance Orbiter altitudes (∼50 km) arises from secondary lunar species, primarily "albedo" protons (3.1%) and electrons (2.2%). Other lunar nuclear evaporation species contributing to the dose rate are positrons (1.5%), gammas (1.1%), and neutrons (0.7%).

Relative contributions of these same species to the total dose equivalent rate in skin, a quantity of more direct biological relevance, favor those with comparatively high quality factors. Consequently, the primary GCR heavy ion components dominate the estimated effective skin dose. Finally, we note that when considering the lunar radiation environment, although the Moon blocks approximately half of the sky, thus essentially halving the absorbed dose rate near the Moon relative to deep space, the secondary radiation created by the presence of the Moon adds back a small, but measurable, absorbed dose (∼8%) that can and should be now accounted for quantitatively in radiation risk assessments at the Moon and other similar exploration targets.

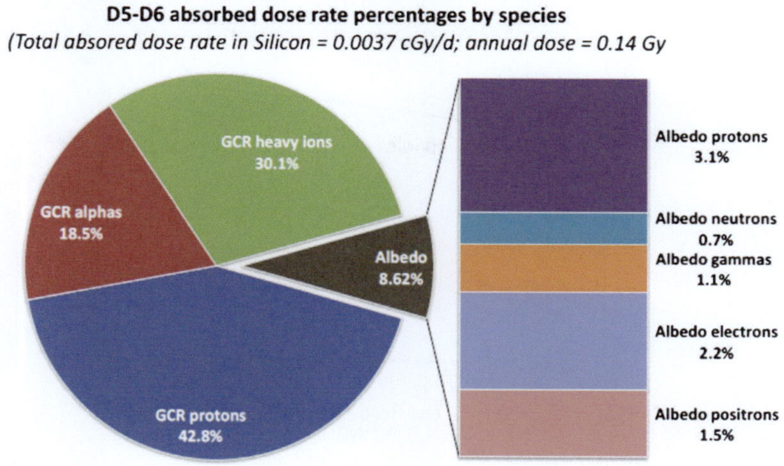

Fig. 12 Relative percentage contribution of primary GCR and albedo species to the total absorbed dose rate (and dose) at LRO altitudes derived from the Looper et al. (2013) Geant4-derived and validated model (reproduced from Spence et al. 2013). Note that the radiation re-emitted from the Moon accounts from ~8.6% of the radiation on the lunar surface

Incident GCR radiation at the Moon is also a significant source of chemical change. Schwadron et al. (2012) found the maximum CRaTER-observed GCR dose rate (~11.7 cGy/yr where Gy indicates energy deposition per unit mass, 1 Gy = 1 J/kg), equivalent to an energy deposition ~88 eV/molecule in water over four billion years. At this dosage level up to one O_2 or H_2O_2 product molecule (Johnson et al. 2004) could be produced per ten target H_2O molecules. This large dose from GCR radiation is significant for the older terrains of the Moon and modifies the molecular composition and physical structure (e.g., density, color, crystallinity) of water ice within the regolith, leading to loss of molecular hydrogen, and production of more complex molecules linking carbon and other elements in the irradiated ice. Space weathering by GCRs and lower energy particles down to plasma energies is important for chemical evolution of the regolith on the Moon and other airless bodies throughout the solar system.

The rates of chemical change induced by radiation (Fig. 13) at the Moon may have significant implications for recent observations. For example, the results of Jordan et al. (2013) have direct bearing on observations of the composition of the regolith made following the impact of the Lunar Crater Observation and Sensing Satellite (LCROSS) on October 2009. The satellite's collision within the PSR within Cabeus crater ejected a plume that was directly observed by LAMP. LAMP observed FUV emissions from the fluorescence of sunlight within the plume by molecular hydrogen (H_2) and other constituents. Jordan et al. (2013) found that GCRs and SEPs convert at least 1–7% of the original water molecules into H_2 within the PSR regolith. This chemical conversion of water molecules likely accounted for ~10%–100% of the H_2 measured by LAMP.

Energetic charged particles can cause deep dielectric charging, the creation of subsurface electric fields in materials that are electrically insulating. If the particles have sufficient fluxes and the material has a sufficiently low electrical conductivity, then the deep dielectric charging can create electric fields strong enough to cause dielectric breakdown, vaporizing tiny (diameters of microns), electrically conductive channels through the material (Budenstein 1980). The conditions for dielectric breakdown have been well-characterized, because breakdown is the leading cause of anomalies on spacecraft orbiting near Earth's radiation

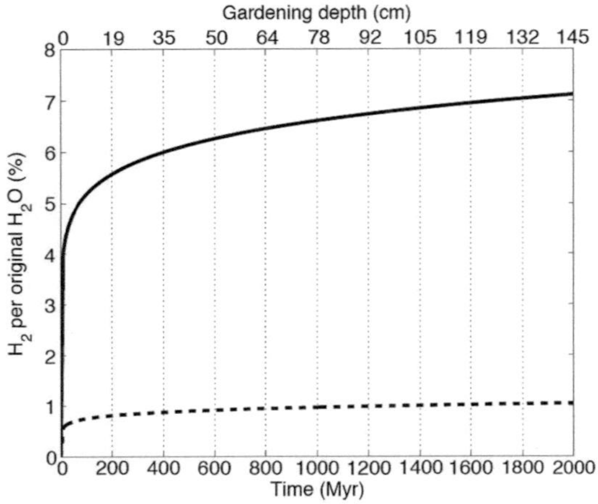

Fig. 13 An example of the effects of incident GCR radiation for chemical modification of the regolith (reproduced from Jordan et al. 2013). Shown is the number (as a percentage) of H_2 molecules created by GCRs and SEPs with respect to the original number of water molecules as a function of gardening time (*lower axis*) and depth (*upper axis*). The gardening time refers to the timescale over which lunar regolith is modified by bombardment of dust and small objects. The *dashed line* and the *solid line* refer to different limiting scenarios based on rates of molecular change induced by radiation

belts (Koons et al. 1999). It also occurred dozens of times in Voyager 1's electronics as the spacecraft passed through Jupiter's radiation belts (Leung et al. 1986).

Dielectric breakdown is an explosive process that both melts and vaporizes material, like meteoroid impacts. Consequently, it may also be a space weathering process on solar system bodies that lack atmospheres to protect their surface (Campins and Krider 1989). The necessary conditions are low electrical conductivity (conductivity and temperature are inversely related for most materials) and exposure to high fluxes of energetic charged particles, like radiation belt particles or SEPs.

On the Moon, PSRs are very cold and thus have very low conductivities. Furthermore, large SEP events can have fluences greater than the threshold needed to cause dielectric breakdown. Consequently, it has been predicted that breakdown can occur in PSRs (Jordan et al. 2014). Based on SEP data compiled over the last ~50 years, sufficiently large SEP events have occurred approximately each year—two of the most recent being in 2012 (Jordan et al. 2015).

Lunar regolith, however, is not exposed to SEPs indefinitely. Meteoroid impacts thoroughly mix, or garden, the top layer of regolith, burying previously exposed regolith with fresh regolith (the top layer, though it varies significantly, typically extends down to about a meter). On average, then, all regolith in the gardened zone has been exposed to SEPs for about 1 Myr, though not all at the same time (Jordan et al. 2013). If the yearly occurrence rate of large SEP events were constant well into the past, as indicated by Apollo samples, then gardened regolith in PSRs has experienced breakdown-causing events about a million times, as shown in Fig. 14 (Jordan et al. 2015). At this rate, dielectric breakdown caused by SEP events is predicted to cause space weathering (melting and vaporizing in this case) comparable to meteoroid impacts (Jordan et al. 2017). And because the nightside of the Moon also reaches low temperatures (below 100 K), low-latitude regolith may also have experienced significant dielectric breakdown weathering, though less than PSRs (Jordan et al. 2017).

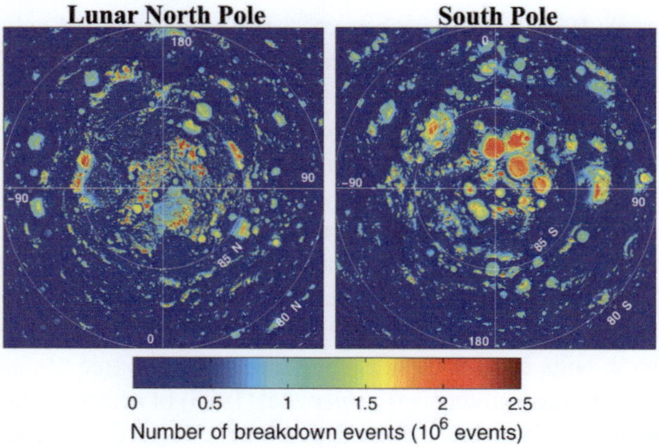

Fig. 14 Permanently shadowed craters on the Moon experience a large number of SEP events that may cause dielectric breakdown (adapted from Fig. 8 in Jordan et al. 2015). The resulting breakdown weathering may melt and vaporize as much regolith as do hypervelocity meteoroid impacts, suggesting that dielectric breakdown could be a significant source of space weathering

Consequently, there may be measurable effects of breakdown weathering. For example, because breakdown is explosive and tends to occur along mineralogical boundaries, it may enhance comminution of the regolith (Jordan et al. 2015). There may thus be a latitudinal trend in regolith grain sizes, with polar regions having smaller grains. Also, if breakdown does occur at lower latitudes, its byproducts may exist in Apollo or Luna samples returned to Earth. Work is ongoing to determine how melting and vaporization by breakdown differs from melting and vaporing by hypervelocity impacts (Shusterman et al. 2016); this will enable future work to determine if breakdown-weathered soil was brought to Earth by those mission.

3.2 Radiation Induced Chemistry and Space Weathering at Other Planetary Bodies of the Inner and Outer Solar System

Most current research regarding the space weathering induced by SEPs has focused on the Moon due to its accessibility for direct measurement. However, as demonstrated by Fig. 9, SEP event fluences remain large throughout the inner heliosphere. Mercury, the Moon, Phobos and Deimos and asteroids (e.g., Ceres) are all objects that will experience the effects of chemical change and dielectric breakdown. Due to its proximity near the Sun, Mercury may be the most strongly affected and cold icy regions at the poles may have been heavily processed by energetic particles from the solar wind and the Mercury magnetosphere (Delitsky et al. 2017).

Other extremes of space weathering by energetic particles can be found at the surfaces of the Galilean moons within the Jovian magnetosphere. These moons are irradiated at far higher intensities than even during the largest SEP events at Mercury. The "ocean moon" Europa has a surface and thin atmosphere enriched in molecular oxygen that likely arises from sputtering and chemical processing of surface ice by energetic electrons, protons, and heavier ions (Cooper et al. 2001), so much that this oxygen could be a resource for life in that Moon's subsurface ocean (Chyba 2000; Cooper et al. 2001; Hand et al. 2006). The surface of Europa is visibly imprinted by space weathering in at least two other respects: (1) sulfur plasma ions from volcanically active Io implant onto the trailing hemisphere surface and form hydrated compounds, and (2) these compounds are irradiated by Jovian magnetospheric electrons with highest surface energy fluxes in that hemisphere (Paranicas et al. 2009). The surface of Ganymede is partially protected against electron irradiation by the Moon's dipole magnetic field (Cooper et al. 2001) but still shows signs of space weathering from accumulation of molecular oxygen at the trailing hemisphere surface (Spencer et al. 1995) and from changes in surface albedo and color at the polar cap boundary between closed and open magnetic field lines (Fatemi et al. 2016). The least active Galilean moon, Callisto, shows concentration of trapped CO_2 at the trailing hemisphere surface (Hibbitts et al. 2000) that is likely related to irradiation processing. Although the inner Galilean moon, Io, receives the most irradiation at its position deepest within the Jovian magnetosphere, it is also the most active moon of the four and is constantly resurfaced by volcanic ejecta so space weathering products are not visible. In fact, the ionic emissions from the Io atmospheric are primary agents of space weathering for the other moons.

In the Saturn system galactic cosmic ray interactions with the main rings (Cooper 1983) produce secondary neutrons that decay into trapped protons and electrons within the radiation belts beyond these rings. This neutron source was constrained in part by direct Pioneer 11 measurements (Cooper et al. 1985) of neutrons, gamma rays, protons, and electrons near the main rings. The trapped energetic particles then interact with the Saturn moons, the narrow G ring, and the diffuse E ring. There is likely a source of molecular oxygen in the rings from the GCR irradiation but the primary source in the ring region appears to be photo-sputtering from irradiation by solar ultraviolet photons (Johnson et al. 2006). A strong seasonal (equinox to solstice) variation has been observed in the oxygen ions of the inner magnetosphere (Elrod et al. 2012, 2014), inwards of the orbit of the cryovolcanically active moon Enceladus, that is consistent with seasonal variation in main ring illumination by the Sun and not with a more constant production rate by cosmic ray interactions (Cooper et al. 2017). Enceladus is the dominant source of water group species for the magnetosphere via ionization of the gas plume molecules emanating from the south pole (Waite et al. 2006, 2009) but not evidently of molecular oxygen.

The thick atmosphere of Titan is chemically reduced with less than two percent methane relative to molecular nitrogen in the stratosphere and five percent near the surface (Raulin and Owen 2002). Oxygen abundance in the form primarily of carbon monoxide is several orders of magnitude lower than for methane. Sittler et al. (2009) have noted that a primary influx of oxygen into the Titan upper atmosphere could come from magnetospheric oxygen ions. Hydrocarbon hazes in the middle to lower atmosphere originate from a combination of solar ultraviolet, magnetospheric particle, and cosmic ray irradiation, increasing in importance from the first to the last with increasing depth towards the surface.

The magnetospheric plasma environments of Uranus and Neptune, explored by the Voyager 2 spacecraft in 1986 and 1989, are distinguished from those of Jupiter and Saturn by a general paucity of heavy ions. No volcanic activity coupling its neutral gas emissions to the magnetospheres have been found in those systems. Trapped energetic electrons and protons are present in both systems and presumably produce some radiolysis products on interaction with the moon surfaces. Due to large tilt angles between the rotational and magnetic equators, respectively 60° and 47° at Uranus and Neptune, the moons of these ice giants pass through large changes in magnetic coordinates and surface irradiation fluxes (e.g., Cooper 1990; Cooper and Stone 1991). Even so, the moons of Uranus show signs of space weathering interactions with some related compositional asymmetries on the trailing and leading hemispheres (Grundy et al. 2006). Darkness of material in the thin rings and arcs of these planets are thought to arise from space weathering effects of trapped and solar wind irradiation as these features sweep over wide ranges in magnetic coordinates.

If Planet 9 truly existed at about ten Earth masses on a 200–1200 AU elongated orbit, as suggested by the dynamical model of Batygin and Brown (2016), it would likely be another ice giant, possibly with its own magnetosphere, ring, and moon systems. Far beyond the influence of the heliospheric solar wind, the Planet 9 system would mainly interact with the local interstellar magnetic field, plasma, and energetic particle environment now being explored by the Voyager 1 and 2 spacecraft (Stone et al. 2013; Cummings et al. 2016). The available measurements, also including neutral interstellar gas flow measurements from the Interstellar Boundary Explorer (IBEX), indicate a relatively cold (8000 K) interstellar plasma environment (e.g., McComas et al. 2015), far colder than the 10^6 K temperature of the solar wind within the heliosphere, but relatively high intensities of galactic cosmic ray protons, ions, and electrons. Ultraviolet photon, plasma electron, and cosmic ray particle irradiation could ionize neutral gas from the atmosphere of Planet 9 to form a trapped population of hot pickup ions within a corotating planetary magnetic field and magnetosphere. These ions and associated trapped electrons could then produce space weathering effects on any rings and moons of that system.

Beyond the giant planets we find the small to large bodies of the Kuiper Belt and the Oort Cloud. Among the 10^3 known objects we now count Pluto as a dwarf planet in the same class as the large asteroid Ceres. The Kuiper Belt Objects (KBOs) mainly orbit at 30–50 AU, well within the supersonic solar wind and energetic particle environment of the heliosphere, but some are on highly elliptical orbits passing out of the heliosphere into the local interstellar medium (Cooper et al. 2006). Since the extreme objects on the highly elliptical orbits would spend most of their time near aphelion, space weathering of these objects, as also for the putative Planet 9, would largely reflect the properties of the interstellar medium.

The plasma and radiation environment of the majority of the KBOs near 40 AU was earlier modeled by Cooper et al. (2003), this model continues to be updated with past measurements from the Pioneer 10 and 11, and Voyager 1 and 2 spacecraft passing through this region. Upcoming public availability of plasma and energetic particle data from the New Horizons spacecraft will further improve the environmental model for KBO space weathering. Thus far, the 2003 model suggests that space weathering occurs in several layers on the diverse KBO surfaces: (1) high intensities of low-energy less-penetrating particles produce an outer radiation crust of low albedo like on the nucleus of Halley's's Comet (Johnson et al. 1987), (2) deeply-penetrating high-energy particles produce an underlying mantle to meters in depth of moderately irradiated material that can appear red in color, (3) deeper materials retain the original color and composition of primordial formation. One subsequent update to the 2003 model is that Enceladus-like cryovolcanic activity (Porco et al. 2006) could eject material from layers (2) and (3) to partially or wholly cover layer (1) and provide the appearance of a brighter and younger surface. Another update is that the heliospheric boundary regions may been compressed at times when the Sun passed through dense molecular clouds, such that the high intensities of the heliosheath region (Cooper et al. 2006) would have swept over the KBOs for 10^5–10^6 years. This suggests the possibility that the history of the Sun's multi-billion year journey through the Galaxy has been recorded in outer layers of cold classical (circular orbit, low inclination, least subject to collisional disruption) KBOs like tree-rings have recorded climate change on Earth.

3.3 Radiation Interactions with Earth's Atmosphere: Effects at Airline Altitudes

Earth's atmosphere represents an important example atmospheric interaction causing the attenuation of incident radiation and the creation of a large population of secondary radiation products. Recent findings indicating the significant rise GCR fluxes due to dropping solar activity highlight the importance of accurate radiation modeling for the planning of future human exploration missions. Atmospheric radiation modeling may also gain increased significance given the growing interest in supersonic passenger transport at altitudes much higher than traditional commercial aircraft. At higher altitudes the reduced amount of atmospheric shielding increases radiation exposure. Atmospheric shielding of radiation is also an important consideration for potential future missions to Mars.

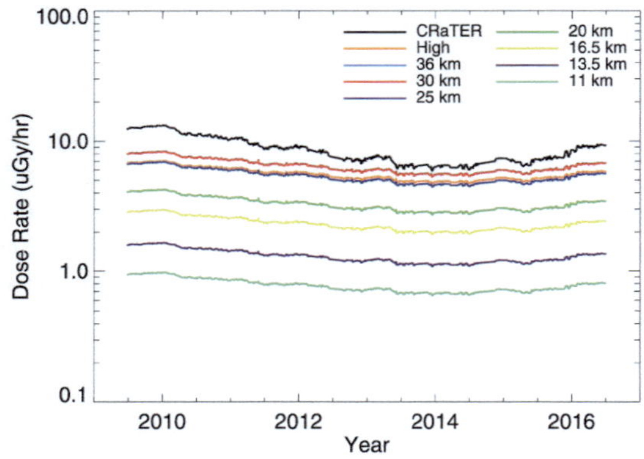

Fig. 15 GCR dose rates in the Earth's atmosphere over the course of the LRO mission to date (reproduced from Joyce et al. 2016). The model dose rates shown are computed using the Badhwar–O'Neil model together with the computed modulation potential from Joyce et al. (2016) and updated EMMREM-generated dose lookup tables. Also shown for comparison is the CRaTER GCR dose rate (*black curve*) measured over this period. We also plot the dose rate for zero atmospheric shielding (labeled "High") to isolate the effect of the magnetosphere on the radiation level. Observe that the magnetic field alone reduces the dose rate by about 30% on average relative to the CRaTER dose rate and that the atmospheric shielding at 36 and 30 km (5 and 12 g/cm^2, respectively), serves to increase the dose by about 30%. This suggests the presence of a Pfotzer maximum in the radiation due to particles somewhere above 36 km, with the dose rate does not falling below the zero atmosphere levels until below an altitude of 25 km

Recent work by Joyce et al. (2016) provides an analysis of the radiation impact of GCRs in Earth's atmosphere using measurements made by CRaTER together with the Badhwar–O'Neil GCR radiation model (O'Neill 2006) and data lookup tables computed by the high-charge-and energy transport model (HZETRN) (Nealy et al. 2007) and the high-energy transport code (HETC-HEDS) (Townsend et al. 2005). These components of EMMREM allow predictions of the total ionizing dose resulting from protons, heavy ions and neutrons in addition to gamma rays at airline altitudes and throughout the atmosphere.

Joyce et al. (2016) validated the EMMREM's atmospheric radiation model by comparing predictions to data from the Automated Radiation Measurements for Aviation Safety (ARMAS) project. The ARMAS data (http://sol.spacenvironment.net/~ARMAS/Level_2_3_Data.html) are based on measurements taken aboard commercial and research aircraft flights from 2013 to 2016.

Figure 15 shows GCR dose rates as a function of atmospheric height. At 20 km, we observe a dose rate of ~3.5 μSv/hr, which implies that over ~30 hours of time in flight at this altitude, crew and passengers would acquire a radiation dose ~0.1 mSv, the equivalent of a chest X-ray. At typical airline altitudes we observe doses ~0.7 μSv/hr. A total of 1430 flight hours is required to reach the maximum recommended 1 mSv effective dose in 1 year (ICRP 2007). We find a radiation-imposed ceiling of ~240 flights in 1 year with an average flight time of 5.9 hours.

In the study of Joyce et al. (2016), the geomagnetic cutoff model of Nymmik et al. (2009) was incorporated in order to take into account the shielding provided by the Earth's magnetic field, thus enabling the model to produce dose rates for any latitudinal or longitudinal position in the Earth's atmosphere. Using the computed rigidity, Joyce et al. (2016) compute radiation levels using only particles with magnetic rigidities greater than the geomagnetic cutoff. Using this modified flux spectrum together with dose tables generated by the HETC-HEDS code, dose rates were determined as a function of geographic position at various altitudes in the Earth's atmosphere, extending down to commercial airline altitudes (11 km).

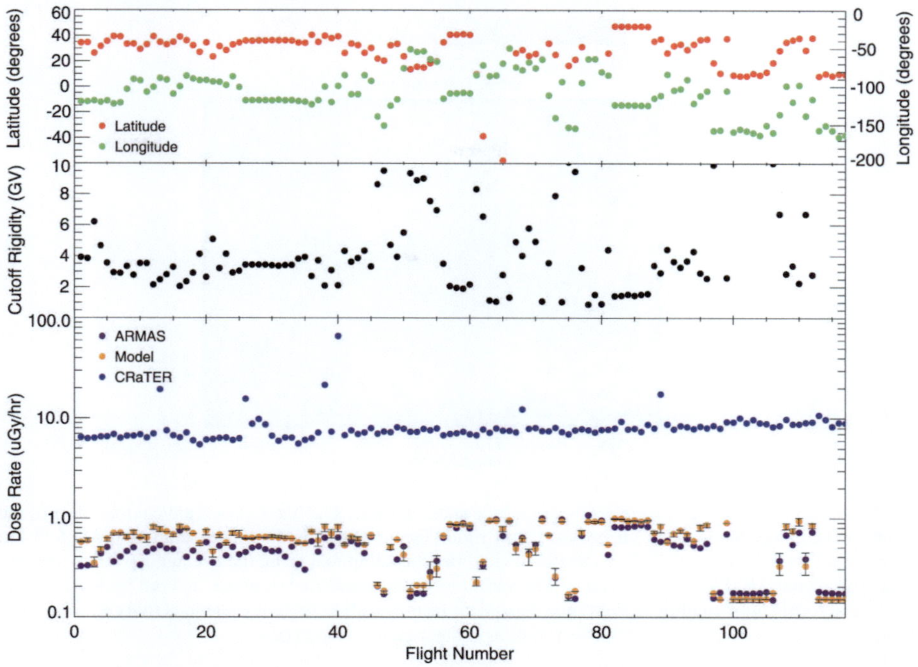

Fig. 16 Dose rates measured aboard commercial aircraft during 117 flights spanning from mid-2013 to early 2016, as provided by the ARMAS project (reproduced from Joyce et al. 2016). Modeled dose rates are also shown and are found to be 30% greater than the measurements on average. The average latitude and longitude as well as the cutoff rigidity used by the model are plotted to explain how variations seen in the airline dose rates. The dose rate measured by CRaTER during each flight is also shown for comparison

The particles accounted for in the computation include protons, heavy ions from He to Fe, and secondary neutrons, pions and muons, with all particle energy losses being deposited locally at the site of the collision. Π_0 particles were not transported, with all their energy being deposited locally at the cite of their production.

Figure 16 shows the comparison between the model and the level 3 data from the Automated Radiation Measurements for Aviation Safety (ARMAS) project (version v5.35) for dose rates measured on 117 airline flights spanning from 18 June 2013 to 18 February 2016. In order to be comparable to the model's lowest altitude output (11 km), data measured below 8 km were not used for this comparison. The 117 flights used have been numbered in chronological order. The ARMAS dose rate data is presented in terms of personal dose equivalent at a depth of 10 mm into tissue, generally considered a sufficiently precise assessment of the effective dose (ICRP 2007). Quality factors are estimated based on the cutoff rigidity. Figure 17 shows that most of the significant variations in the airline dose rates are due to differences in geographic position thats determines the cutoff rigidity.

While Earth's atmosphere and magnetosphere play a significant role in attenuating radiation, we still observe large levels of GCR radiation throughout the atmosphere. The effects of secondary radiation products including gamma rays and neutrons play an important role deep in the atmosphere where collisions with primary GCR nuclei create a cascade of lower elementary particles and gamma rays. Nonetheless, the effective shielding provided by Earth's atmosphere is very significant in attenuating incident radiation, which leaves the Earth's surface relatively unaffected by the space weathering effects induced by radiation.

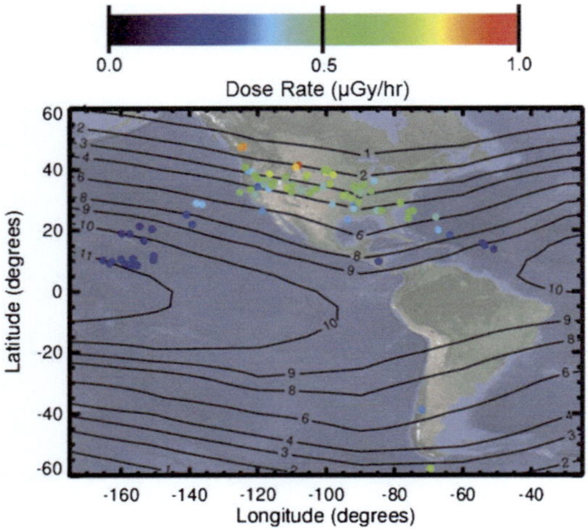

Fig. 17 ARMAS dose rates as a function of latitude and longitude. The *black curves* are contour plots of the geomagnetic cutoff rigidity (GV) using the cutoff rigidities computed by the model at an altitude of 11 km (reproduced from Joyce et al. 2016). The cutoff rigidities are computed using the average Kp index over the time span of the ARMAS measurements. The variations in the measured dose are ordered by location and the typical local cutoff rigidity, though time dependent factors such as or geomagnetic activity and the GCR modulation are also responsible for some variation in the measured dose rates

4 Conclusions

We have discussed recent work on the effects of radiation throughout the solar system. The Sun is currently exhibiting activity at much lower levels than has existed in almost 100 years prior. This weakening activity has lead to higher GCR fluxes than previously observed during the space age.

Large solar radiation events produced at the Sun are associated with the eruption of coronal mass ejections, which drive strong expansions low in the corona, the sites of rapid particle acceleration. The events from low in the corona naturally show broad longitudinal distributions. However, particle acceleration continues as CME shocks develop and are driven throughout interplanetary space. The largest SEP events are typically associated with the passage of CME shocks past the observer.

Radiation effects the surfaces of airless bodies such as the Moon. The space weathering induced by particle radiation can affect the chemical composition of the regolith, and large particle events may induce dielectric breakdown, affecting the porosity and plane albedo of the regolith. In some cases, the effects of dielectric breakdown rival meteoroid impacts in terms of space weathering of the regolith.

Radiation induced chemical changes and space weathering affect planetary bodies throughout the solar system. There are critical links between radiation, habitability, and life in the solar system and beyond. Important questions remain: did the late bombardment period of four billion years ago bring processed prebiotic materials to Earth and other planets and moons from spawning grounds of the outer solar system? Do radiation-produced oxidants contribute to habitability and even extant life at planetary bodies such as Europa, Enceladus, and Titan? The lessons learned by studying the Moon, and other planets are important as we also search for signs of habitability and life in extrasolar planetary systems.

Galactic cosmic rays and solar energetic particles penetrate through the atmospheres of Earth and other planets, producing a cascade of secondary radiation. Near Earth, airline crews and passengers receive doses from particle radiation. The doses from galactic cosmic rays increases during solar minima, particularly as solar activity weakens. In general, the widespread biological effects of radiation represent a concern that limits mission durations for human exploration, and the net exposure for airline crews and passengers, particularly as higher altitude airline travel becomes more heavily utilized.

Thus, we review the widespread sources and effects of particle radiation within the solar system. Particle radiation represents an energy source that affects biological systems, atmospheres and the surfaces of airless bodies. As we observe major changes in the Sun and their manifestations in the radiation environment, we gain a deeper perspective of the importance of the radiation environment for biological systems and space weathering. The research in this area must proceed as its effects may hold essential keys for understanding the continued habitability of our planet, the planets throughout the solar system, and, by extension, planets around other stars.

Acknowledgements We thank all those who made CRaTER possible. CRaTER is primarily funded by the LRO program (Contract No. NNG11PA03C). This work was also funded EMMREM (grant number NNX07AC14G), C-SWEPA (NASA grant number NNX13AI75G), Sun-2-Ice (NSF grant number AGS1135432) projects, and DoSEN (NASA grant NNX13AC89G), DREAM (NASA grant NNX10AB17A) and DREAM2 (NASA grant NNX14AG13A). CRaTER and VEPO data are respectively available at http://prediccs.sr.unh.edu/craterweb/ and http://vepo.gsfc.nasa.gov/. We thank the International Space Science Institute, which made possible this paper as a part of the ISSI workshop, the Scientific Foundation of Space Weather, and International Team 353, Radiation Interactions at Planetary Bodies.

Appendix A: Acronyms

Table 1 Acronyms and Models used throughout the paper

Acronym/Model	Definition
ACE	Anomalous Composition Explorer
ARMAS	Automated Radiation Measurements for Aviation Safety
ATST	Advanced Technology Solar Telescope
AU	Astronomical Unit
BATS-R-US	Block-Adaptive-Tree-Solarwind-Roe-Upwind-Scheme
BFO	Blood Forming Organ
CME	Coronal Mass Ejection
COSMO	COronal Solar Magnetism Observatory
COSTEP	Comprehensive Suprathermal and Energetic Particle Analyzer
CRaTER	Cosmic Ray Telescope for the Effects of Radiation
CRIS	Cosmic Ray Isotope Spectrometer
EIT	(SOHO) Extreme ultraviolet Imaging Telescope
EMMREM	Earth Moon Mars Radiation Environment Modules (Schwadron et al. 2010)
EPREM	Energetic Particle Radiation Environment Model
ESP	Energetic Storm Particles
EUV	Extreme Ultraviolet
FASR	Frequency Agile Solar Radiotelescope
GOES	Geostationary Satellite
GCR	Galactic Cosmic Ray
GLE	Ground Level Event
HETC-HEDS	High Energy Transport Code for Human Exploration and Development in Space

Table 1 (*Continued*)

Acronym/Model	Definition
HZETRN	high-charge-and energy transport model
LAMP	LRO Lyman Alpha Mapping Project
LCROSS	Lunar Crater Observation and Sensing Satellite
LET	Low Energy Telescope
LET	Linear Energy Transfer
LRO	Lunar Reconnaissance Orbiter
MAS	Magnetohydrodynamics-Around-a-Sphere
MLSO	Mauna Loa Solar Observatory
NCAR	National Center for Atmospheric Research
PREDICCS	Predictions of radiation from REleASE, EMMREM, and Data Incorporating CRaTER, COSTEP, and other SEP measurement
PSR	Permanently Shaded Regions
REID	Risk of Exposure Induced Death
RHESSI	Reuven Ramaty High Energy Solar Spectroscopic Imager
SDO	Solar Dynamics Observatory
SECCHI	Sun Earth Connection Coronal and Heliospheric Investigation
SEP	Solar Energetic Particle
SIT	Solar Isotope Telescope
SOLIS	Synoptic Optical Long-Term Investigations of the Sun
SOHO	Solar and Heliospheric Observatory
STP	Standard Temperature and Pressure
STEREO	Solar Terrestrial Relations Observatory
ULEIS	Ultra Low Energy Ion Spectrometer
UV	Ultraviolet

Appendix B: Components of EMMREM and PREDICCS

Throughout parts of this paper, we have used results from the Earth–Moon–Mars Radiation Environment Module (EMMREM, Fig. 18) and PREDICCS (Predictions of Radiation from Release, EMMREM, and Data Incorporating the CRaTER, COSTEP and other SEP measurements, http://prediccs.sr.unh.edu). EMMREM includes a suite of modules is designed to predict the radiation environment at Earth, the Moon, Mars and throughout the interplanetary medium in the inner heliosphere (Schwadron et al. 2010). PREDICCS is an online system that provides characterization of the radiation environment of the inner heliosphere in near real-time. PREDICCS utilizes data from satellites (ACE, Wind, GOES) in conjunction with EMMREM to produce dose rate and particle flux data at the Earth, Moon and Mars (Joyce et al. 2013). Validation has been performed through detailed comparisons between dose rates produced by PREDICCS with those measured by CRaTER.

The EMMREM modules consists of four primary pieces:

- The Energetic Particle Radiation Environment Module (EPREM) solves for the propagation and acceleration of energetic particles in the evolving magnetic fields of the inner heliosphere with input based on observations from satellites.
- The Baryon Transport Module (BRYNTRN) is a deterministic, coupled proton-neutron space radiation transport model that transports incident protons and their secondary prod-

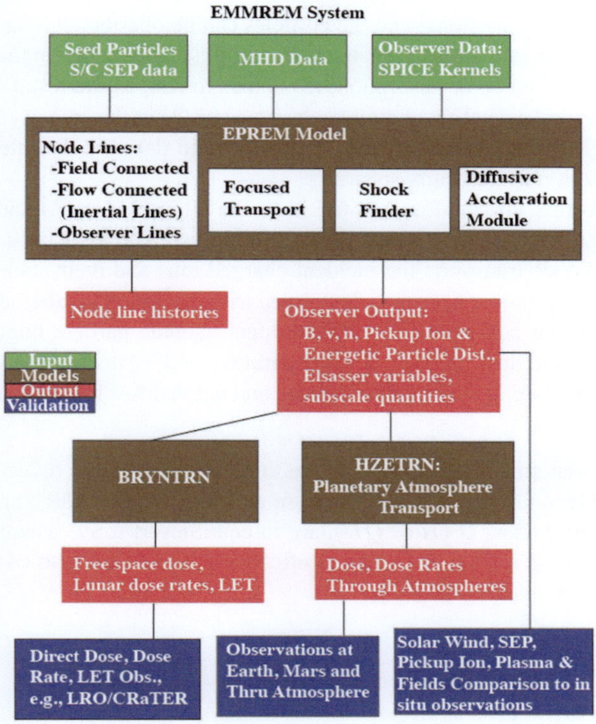

Fig. 18 The Earth–Moon–Mars Radiation Environment Model (EMMREM) provides a link from the Space Science to Space Exploration programs by characterizing time-dependent radiation exposure from observed and simulated particle radiation events. EMMREM provides a series of primary and secondary transport modules for the prediction and validation of particle radiation environment. Key components of EMMREM include: the Energetic Particle Radiation Environment Module (EPREM) solves for the propagation and acceleration of energetic particles in the evolving magnetic fields of the inner heliosphere with input based on observations from satellites; the Baryon Transport Module (BRYNTRN), which is a deterministic, coupled proton-neutron space radiation transport model that transports incident protons and their secondary products (protons, neutrons, deuterons, tritons, helions, and alphas) through shields of arbitrary composition and thickness (Wilson et al. 1991); the High-charge (Z) and Energy (HZE) Transport (HZETRN) model (Nealy et al. 2007) and the high-energy transport code (HETC-HEDS) (Townsend et al. 2005) to estimate associated dose rate from galactic cosmic rays (Wilson and Badavi 1986; Wilson et al. 1991, 2003; Shinn et al. 1991; Cucinotta 1993; Nealy et al. 2006)

ucts (protons, neutrons, deuterons, tritons, helions, and alphas) through shields of arbitrary composition and thickness (Wilson et al. 1991). BRYNTRN is used primarily in the modeling of solar proton events.

– The High Energy Transport Code for Human Exploration and Development in Space (HETC-HEDS) is a Monte Carlo based algorithm that was specially designed to address space radiation problems, and, in particular, with the secondary particle distributions that are produced when the high-energy GCRs or SPEs interact with shielding and/or body organs (e.g., bone marrow or the central nervous system). HETC-HEDS has been extensively verified and validated, particularly against available laboratory beam data. HETC-HEDS models the nuclear elastic and inelastic interactions, decay and atomic interactions (Townsend et al. 2005). It accomplishes this using a Monte Carlo approach for computing the trajectories of particle cascades that result from the nuclear collisions. The particles included in HETC-HEDS are: protons, neutrons, π^+, π^-, μ^+, μ^-, and heavy ions. The code is fully three-dimensional and the particles can be distributed in angle, energy, and space. Each particle in the cascade is tracked until it: (1) escapes from the system; (2) undergoes nuclear collision or absorption; (3) comes to rest, as the result of energy losses from ionization and excitation of atomic electrons; or (4) in the case of pions and muons, decays. HETC-HEDS has been used in EMMREM to describe the interaction of GCR distributions with planetary atmospheres.

- GCR and SEP flux distributions are fed in to the High-charge (Z) and Energy (HZE) Transport (HZETRN) model (Nealy et al. 2007) to estimate associated GCR dose rate. The HZETRN code transports the incident charged ions and their nuclear reaction secondary particles (protons, neutrons, deuterons, tritons, ^{3}He, ^{4}He, and heavier ions) generated from nuclear collisions. The code outputs, include particle fluences, dose, dose equivalent, effective dose, and linear energy transfer (LET) distributions. The HZETRN model results are used to derive effective dose and organ doses for the assumed aluminum shield configurations.

In model calculations, organ doses (D) are in units of centiGray (cGy) where 1 cGy $=$ 1 rad and 100 cGy $=$ Gy $=$ 1 J/kg. Organ dose equivalents (H), which are the product of dose with a quality factor, Q $(H = QD)$, are in centiSievert (cSv) where 1 cSv $=$ 1 rem and 100 cSv $=$ 1 Sv $=$ 1 J/kg. The units of effective dose (E) are also cSv. For HZETRN effective dose is calculated from

$$E = \sum_T w_T H_T \qquad (1)$$

where H_T is the organ dose equivalent for the organ specified by T (e.g., skin, eye, lens, etc.). The tissue (organ) weighting factors w_T are the proportionate detriment of the organ when the whole body is irradiated and are tabulated in Table 5.1 of National Council on Radiation Protection and Measurements (NCRP) [1993].

References

S.K. Antiochos, C.R. DeVore, J.A. Klimchuk, A model for solar coronal mass ejection. Astrophys. J. **510**, 485 (1999)

M. Aschwanden, The localization of particle acceleration sites in solar flares and CMEs. Space Sci. Rev. **124**, 361–372 (2006). doi:10.1007/s11214-006-9095-9

K. Batygin, M.E. Brown, Evidence for a distant giant planet in the solar system. Astron. J. **151**, 22 (2016). doi:10.3847/0004-6256/151/2/22

G.A. Bazilevskaya, I.G. Usoskin, E.O. Flückiger, R.G. Harrison, L. Desorgher, R. Bütikofer, M.B. Krainev, V.S. Makhmutov, Y.I. Stozhkov, A.K. Svirzhevskaya, N.S. Svirzhevsky, G.A. Kovaltsov, Cosmic ray induced ion production in the atmosphere. Space Sci. Rev. **137**, 149–173 (2008). doi:10.1007/s11214-008-9339-y

J. Beer, K. McCracken, R. von Steiger, *Cosmogenic Radionuclides* (2012). doi:10.1007/978-3-642-14651-0

P.P. Budenstein IEEE Trans. Electr. Insul. **3**, 225 (1980)

H. Campins, E.P. Krider, Surface discharges on natural dielectrics in the solar system. Science **245**, 622–624 (1989). doi:10.1126/science.245.4918.622

R.W. Carlson, R.E. Johnson, M.S. Anderson, Sulfuric acid on Europa and the radiolytic sulfur cycle. Science **286**, 97–99 (1999). doi:10.1126/science.286.5437.97

R.W. Carlson, M.S. Anderson, R. Mehlman, R.E. Johnson, Distribution of hydrate on Europa: further evidence for sulfuric acid hydrate. Icarus **177**, 461–471 (2005). doi:10.1016/j.icarus.2005.03.026

J. Chen, Effects of toroidal forces in current loops embedded in a background plasma. Astrophys. J. **338**, 453–470 (1989). doi:10.1086/167211

K.-S. Cho, J. Lee, Y.-J. Moon, M. Dryer, S.-C. Bong, Y.-H. Kim, Y.D. Park, A study of CME and type II shock kinematics based on coronal density measurement. Astron. Astrophys. **461**, 1121–1125 (2007). doi:10.1051/0004-6361:20064920

C.F. Chyba, Energy for microbial life on Europa. Nature **403**, 381–382 (2000). doi:10.1038/35000281
R.N. Clark, Detection of adsorbed water and hydroxyl on the Moon. Science **326**, 562 (2009). doi:10.1126/science.1178105
D.E. Connick, C.W. Smith, N.A. Schwadron, Interplanetary magnetic flux depletion during protracted solar minima. Astrophys. J. **727**, 8 (2011). doi:10.1088/0004-637X/727/1/8. http://adsabs.harvard.edu/abs/2011ApJ...727....8C
J.F. Cooper, Nuclear cascades in Saturn's rings—cosmic ray albedo neutron decay and origins of trapped protons in the inner magnetosphere. J. Geophys. Res. **88**, 3945–3954 (1983). doi:10.1029/JA088iA05p03945
J.F. Cooper, Satellite sweeping of electrons at Neptune and Uranus. Geophys. Res. Lett. **17**, 1665–1668 (1990). doi:10.1029/GL017i010p01665
J.F. Cooper, E.C. Stone, Electron signatures of satellite sweeping in the magnetosphere of Uranus. J. Geophys. Res. **96**, 7803–7821 (1991). doi:10.1029/90JA02629
J.F. Cooper, J.H. Eraker, J.A. Simpson, The secondary radiation under Saturn's A–B–C rings produced by cosmic ray interactions. J. Geophys. Res. **90**, 3415–3427 (1985). doi:10.1029/JA090iA04p03415
J.F. Cooper, R.E. Johnson, B.H. Mauk, H.B. Garrett, N. Gehrels, Energetic ion and electron irradiation of the icy Galilean satellites. Icarus **149**, 133–159 (2001). doi:10.1006/icar.2000.6498
J.F. Cooper, E.R. Christian, J.D. Richardson, C. Wang, Proton irradiation of Centaur, Kuiper belt, and Oort cloud objects at plasma to cosmic ray energy. Earth Moon Planets **92**, 261–277 (2003). doi:10.1023/B:MOON.0000031944.41883.80
J.F. Cooper, M.E. Hill, J.D. Richardson, S.J. Sturner, Proton irradiation environment of solar system objects in the heliospheric boundary regions, in *Physics of the Inner Heliosheath*, ed. by J. Heerikhuisen, V. Florinski, G.P. Zank, N.V. Pogorelov American Institute of Physics Conference Series, vol. 858, 2006, pp. 372–380. doi:10.1063/1.2359353
J.F. Cooper, P.D. Cooper, E.C. Sittler, S.J. Sturner, A.M. Rymer, Old Faithful model for radiolytic gas-driven cryovolcanism at Enceladus. Planet. Space Sci. **57**, 1607–1620 (2009). doi:10.1016/j.pss.2009.08.002
J.F. Cooper, P. Kollman, E.C. Sittler Jr., R.E. Johnson, E. Roussos, Plasma, neutral atmosphere, and energetic radiation environments of planetary rings, in *Planetary Ring Systems*, (Cambridge University Press, Cambridge, 2017), in press
F. Cucinotta, Issues in risk assessment from solar particle events. Radiat. Meas. **30**, 261 (1999)
F.A. Cucinotta, Calculations of cosmic-ray helium transport in shielding materials. NASA Tech. Paper 3354 **3354** (1993)
F.A. Cucinotta, W. Schimmerling, J.W. Wilson, L.E. Peterson, P. Saganti, G.D. Badhwar, J.F. Dicello, Space radiation cancer risks and uncertainties for mars missions. Radiat. Res. **156**, 682 (2001)
F.A. Cucinotta, S. Hu, N.A. Schwadron, K. Kozarev, L.W. Townsend, M.-H.Y. Kim, Space radiation risk limits and Earth–Moon–Mars environmental models. Space Weather **8**, S00E09 (2010). doi:10.1029/2010SW000572
A.C. Cummings, E.C. Stone, B.C. Heikkila, N. Lal, W.R. Webber, G. Jóhannesson, I.V. Moskalenko, E. Orlando, T.A. Porter, Galactic cosmic rays in the local interstellar medium: Voyager 1 observations and model results. Astrophys. J. **831**, 18 (2016). doi:10.3847/0004-637X/831/1/18
J.B. Dalton, O. Prieto-Ballesteros, J.S. Kargel, C.S. Jamieson, J. Jolivet, R. Quinn, Spectral comparison of heavily hydrated salts with disrupted terrains on Europa. Icarus **177**, 472–490 (2005). doi:10.1016/j.icarus.2005.02.023
M.A. Dayeh, M.I. Desai, K. Kozarev, N.A. Schwadron, L.W. Townsend, M. PourArsalan, C. Zeitlin, R.B. Hatcher, Modeling proton intensity gradients and radiation dose equivalents in the inner heliosphere using EMMREM: May 2003 solar events. Space Weather **8**, S00E07 (2010)
J.-P. Delaboudinière, G.E. Artzner, J. Brunaud, A.H. Gabriel, J.F. Hochedez, F. Millier, X.Y. Song, B. Au, K.P. Dere, R.A. Howard, R. Kreplin, D.J. Michels, J.D. Moses, J.M. Defise, C. Jamar, P. Rochus, J.P. Chauvineau, J.P. Marioge, R.C. Catura, J.R. Lemen, L. Shing, R.A. Stern, J.B. Gurman, W.M. Neupert, A. Maucherat, F. Clette, P. Cugnon, E.L. van Dessel, EIT: Extreme-ultraviolet Imaging Telescope for the SOHO mission. Sol. Phys. **162**, 291–312 (1995). doi:10.1007/BF00733432
C. Delannée, Another view of the EIT wave phenomenon. Astrophys. J. **545**, 512–523 (2000). doi:10.1086/317777
M.L. Delitsky, D.A. Paige, M.A. Siegler, E.R. Harju, D. Schriver, R.E. Johnson, P. Travnicek, Ices on Mercury: chemistry of volatiles in permanently cold areas of Mercury's north polar region. Icarus **281**, 19–31 (2017). doi:10.1016/j.icarus.2016.08.006
M.I. Desai, G.M. Mason, J.R. Dwyer, J.E. Mazur, R.E. Gold, S.M. Krimigis, C.W. Smith, R.M. Skoug, Evidence for a suprathermal seed population of heavy ions accelerated by interplanetary shocks near 1 AU. Astrophys. J. **588**, 1149–1162 (2003). doi:10.1086/374310
M.I. Desai, G.M. Mason, J.E. Mazur, J.R. Dwyer, Solar cycle variations in the composition of the suprathermal heavy-ion population near 1 AU. Astrophys. J. Lett. **645**, 81–84 (2006). doi:10.1086/505935

K.A. Duderstadt, J.E. Dibb, N.A. Schwadron, H.E. Spence, S.C. Solomon, V.A. Yudin, C.H. Jackman, C.E. Randall, Nitrate ion spikes in ice cores not suitable as proxies for solar proton events. J. Geophys. Res., Atmos. **121**, 2994–3016 (2016). doi:10.1002/2015JD023805

R.C. Elphic, V.R. Eke, L.F.A. Teodoro, D.J. Lawrence, D.B.J. Bussey, Models of the distribution and abundance of hydrogen at the lunar south pole. Geophys. Res. Lett. **34**, 13204 (2007). doi:10.1029/2007GL029954

M.K. Elrod, W.-L. Tseng, R.J. Wilson, R.E. Johnson, Seasonal variations in Saturn's plasma between the main rings and Enceladus. J. Geophys. Res. Space Phys. **117**, 03207 (2012). doi:10.1029/2011JA017332

M.K. Elrod, W.-L. Tseng, A.K. Woodson, R.E. Johnson, Seasonal and radial trends in Saturn's thermal plasma between the main rings and Enceladus. Icarus **242**, 130–137 (2014). doi:10.1016/j.icarus.2014.07.020

S. Fatemi, A.R. Poppe, K.K. Khurana, M. Holmström, G.T. Delory, On the formation of Ganymede's surface brightness asymmetries: kinetic simulations of Ganymede's magnetosphere. Geophys. Res. Lett. **43**(10), 4745–4754 (2016). doi:10.1002/2016GL068363

T.G. Forbes, J.A. Linker, J. Chen, C. Cid, J. Kóta, M.A. Lee, G. Mann, Z. Mikić, M.S. Potgieter, J.M. Schmidt, G.L. Siscoe, R. Vainio, S.K. Antiochos, P. Riley, CME theory and models. Space Sci. Rev. **123**, 251–302 (2006). doi:10.1007/s11214-006-9019-8

P.T. Gallagher, D.M. Long, Large-scale bright fronts in the solar corona: a review of "EIT waves". Space Sci. Rev. **158**, 365–396 (2011). doi:10.1007/s11214-010-9710-7

J. Giacalone, The efficient acceleration of thermal protons by perpendicular shocks. Astrophys. J. Lett. **628**, 37–40 (2005). doi:10.1086/432510

M.L. Goelzer, C.W. Smith, N.A. Schwadron, K.G. McCracken, An analysis of heliospheric magnetic field flux based on sunspot number from 1749 to today and prediction for the coming solar minimum. J. Geophys. Res. Space Phys. **118**, 7525–7531 (2013). doi:10.1002/2013JA019404

N. Gopalswamy, E. Aguilar-Rodriguez, S. Yashiro, S. Nunes, M.L. Kaiser, R.A. Howard, Type II radio bursts and energetic solar eruptions. J. Geophys. Res. Space Phys. **110**(A9), 12 (2005). doi:10.1029/2005JA011158

W.M. Grundy, L.A. Young, J.R. Spencer, R.E. Johnson, E.F. Young, M.W. Buie, Distributions of H_2O and CO_2 ices on Ariel, Umbriel, Titania, and Oberon from IRTF/SpeX observations. Icarus **184**, 543–555 (2006). doi:10.1016/j.icarus.2006.04.016

S.R. Habbal, H. Morgan, M. Druckmüller, A. Ding, J.F. Cooper, A. Daw, E.C. Sittler, Probing the fundamental physics of the solar corona with lunar solar occultation observations. Sol. Phys. **285**, 9–24 (2013). doi:10.1007/s11207-012-0115-5

K.P. Hand, C.F. Chyba, R.W. Carlson, J.F. Cooper, Clathrate hydrates of oxidants in the ice shell of Europa. Astrobiology **6**, 463–482 (2006). doi:10.1089/ast.2006.6.463

C.A. Hibbitts, T.B. McCord, G.B. Hansen, Distributions of CO_2 and SO_2 on the surface of Callisto. J. Geophys. Res. **105**, 22541–22558 (2000). doi:10.1029/1999JE001101

H.S. Hudson, Global properties of solar flares. Space Sci. Rev. **158**, 5–41 (2011). doi:10.1007/s11214-010-9721-4

R.L. Hudson, M.H. Moore, Laboratory studies of the formation of methanol and other organic molecules by water+carbon monoxide radiolysis: relevance to comets, icy satellites, and interstellar ices. Icarus **140**, 451–461 (1999). doi:10.1006/icar.1999.6144

R.L. Hudson, M.E. Palumbo, G. Strazzulla, M.H. Moore, J.F. Cooper, S.J. Sturner, Laboratory studies of the chemistry of transneptunian object surface materials, in *The Solar System Beyond Neptune*, ed. by M.A. Barucci, H. Boehnhardt, D.P. Cruikshank, A. Morbidelli, R. Dotson (2008), pp. 507–523. http://adsabs.harvard.edu/abs/2008ssbn.book..507H

ICRP, *International Commission on Radiological Protection ICRP Publication 103: The 2007 Recommendations of the International Commission on Radiological Protection*. Ann. ICRP, vol. 37 (Sage, Thousand Oaks, 2007)

R.E. Johnson, *Energetic Charged-Particle Interactions with Atmospheres and Surfaces*. Physics and Chemistry in Space Planetology, vol. 19 (1990)

R.E. Johnson, J.F. Cooper, L.J. Lanzerotti, G. Strazzulla, Radiation formation of a non-volatile comet crust. Astron. Astrophys. **187**, 889 (1987)

R.E. Johnson, R.W. Carlson, J.F. Cooper, C. Paranicas, M.H. Moore, M.C. Wong, Radiation effects on the surfaces of the Galilean satellites, in *Jupiter. The Planet, Satellites and Magnetosphere*, ed. by F. Bagenal, T.E. Dowling, W.B. McKinnon (2004), pp. 485–512. http://adsabs.harvard.edu/abs/2004jpsm.book..485J

R.E. Johnson, J.G. Luhmann, R.L. Tokar, M. Bouhram, J.J. Berthelier, E.C. Sittler, J.F. Cooper, T.W. Hill, H.T. Smith, M. Michael, M. Liu, F.J. Crary, D.T. Young, Production, ionization and redistribution of O_2 in Saturn's ring atmosphere. Icarus **180**, 393–402 (2006). doi:10.1016/j.icarus.2005.08.021

A.P. Jordan, T.J. Stubbs, C.J. Joyce, N.A. Schwadron, H.E. Spence, J.K. Wilson, The formation of molecular hydrogen from water ice in the lunar regolith by energetic charged particles. J. Geophys. Res., Planets **118**, 1257–1264 (2013). doi:10.1002/jgre.20095

A.P. Jordan, T.J. Stubbs, J.K. Wilson, N.A. Schwadron, H.E. Spence, C.J. Joyce, Deep dielectric charging of regolith within the Moon's permanently shadowed regions. J. Geophys. Res., Planets **119**, 1806–1821 (2014). doi:10.1002/2014JE004648

A.P. Jordan, T.J. Stubbs, J.K. Wilson, N.A. Schwadron, H.E. Spence, Dielectric breakdown weathering of the Moon's polar regolith. J. Geophys. Res., Planets **120**, 210–225 (2015). doi:10.1002/2014JE004710

A.P. Jordan, T.J. Stubbs, J.K. Wilson, N.A. Schwadron, H.E. Spence, The rate of dielectric breakdown weathering of lunar regolith in permanently shadowed regions. Icarus **283**, 352–358 (2017). doi:10.1016/j.icarus.2016.08.027. Lunar Reconnaissance Orbiter—Part II. http://www.sciencedirect.com/science/article/pii/S0019103516305358

C.J. Joyce, N.A. Schwadron, J.K. Wilson, H.E. Spence, J.C. Kasper, M. Golightly, J.B. Blake, J. Mazur, L.W. Townsend, A.W. Case, E. Semones, S. Smith, C.J. Zeitlin, Validation of PREDICCS using LRO/CRaTER observations during three major solar events in 2012. Space Weather **11**, 350–360 (2013). doi:10.1002/swe.20059

C.J. Joyce, N.A. Schwadron, L.W. Townsend, R.A. Mewaldt, C.M.S. Cohen, T.T. Rosenvinge, A.W. Case, H.E. Spence, J.K. Wilson, M. Gorby, M. Quinn, C.J. Zeitlin, Analysis of the potential radiation hazard of the 23 July 2012 SEP event observed by STEREO a using the EMMREM model and LRO/CRaTER. Space Weather **13**, 560–567 (2015). doi:10.1002/2015SW001208

C.J. Joyce, N.A. Schwadron, L.W. Townsend, W.C. deWet, J.K. Wilson, H.E. Spence, W.K. Tobiska, K. Shelton-Mur, A. Yarborough, J. Harvey, A. Herbst, A. Koske-Phillips, F. Molina, S. Omondi, C. Reid, D. Reid, J. Shultz, B. Stephenson, M. McDevitt, T. Phillips, Atmospheric radiation modeling of galactic cosmic rays using LRO/CRaTER and the EMMREM model with comparisons to balloon and airline based measurements. Space Weather **14**, 659–667 (2016). doi:10.1002/2016SW001425

I.S. Kim, L.P. Nasonova, D.V. Lisin, V.V. Popov, N.L. Krusanova, Imaging the structure of the low k-corona. J. Geophys. Res. Space Phys. **122**(1), 77–88 (2017). 2016JA022623. doi:10.1002/2016JA022623

L. Kocharov, M. Lytova, R. Vainio, T. Laitinen, J. Torsti, Modeling the shock aftermath source of energetic particles in the solar corona. Astrophys. J. **620**, 1052–1068 (2005). doi:10.1086/427162

H.C. Koons, J.E. Mazur, R.S. Selesnick, J.B. Blake, J.F. Fennell, The Impact of the Space Environment on Space Systems. NASA STI/Recon Technical Report N (1999)

J. Kóta, W.B. Manchester, T.I. Gombosi, SEP acceleration at realistic CMEs: two sites of acceleration? Int. Cosm. Ray Conf., **1**, 125 (2005a)

J. Kóta, W.B. Manchester, J.R. Jokipii, D.L. de Zeeuw, T.I. Gombosi, Simulation of SEP acceleration and transport at CME-driven shocks, in *The Physics of Collisionless Shocks: 4th Annual IGPP International Astrophysics Conference*, ed. by G. Li, G.P. Zank, C.T. Russell American Institute of Physics Conference Series, vol. 781, 2005b, pp. 201–206. doi:10.1063/1.2032697

K.A. Kozarev, K.E. Korreck, V.V. Lobzin, M.A. Weber, N.A. Schwadron, Off-limb solar coronal wavefronts from SDO/AIA extreme-ultraviolet observations—implications for particle production. Astrophys. J. Lett. **733**, 25 (2011). doi:10.1088/2041-8205/733/2/L25

K.A. Kozarev, R.M. Evans, N.A. Schwadron, M.A. Dayeh, M. Opher, K.E. Korreck, B. van der Holst, Global numerical modeling of energetic proton acceleration in a coronal mass ejection traveling through the solar corona. Astrophys. J. **778**, 43 (2013). doi:10.1088/0004-637X/778/1/43

J.A. Le Roux, M.S. Potgieter, The simulation of complete 11 and 12 year modulation cycles for cosmic rays in the heliosphere using a drift model with global merged interaction regions. Astrophys. J. **442**, 847–851 (1995). doi:10.1086/175487

K.-S. Lee, Y.-J. Moon, K.-S. Kim, J.-Y. Lee, K.-S. Cho, G.S. Choe, Comparison of SOHO UVCS and MLSO MK4 coronameter densities. Astron. Astrophys. **486**, 1009–1013 (2008). doi:10.1051/0004-6361:20078976

P. Leung, A.C. Whittlesey, H.B. Garrett, P.A. Robinson Jr., Environment-induced electrostatic discharges as the cause of Voyager 1 power-on resets. J. Spacecr. Rockets **23**, 323 (1986). doi:10.2514/3.25805

R. Lionello, C. Downs, J.A. Linker, T. Török, P. Riley, Z. Mikić, Magnetohydrodynamic simulations of interplanetary coronal mass ejections. Astrophys. J. **777**, 76 (2013). doi:10.1088/0004-637X/777/1/76

W. Liu, N.V. Nitta, C.J. Schrijver, A.M. Title, T.D. Tarbell, First SDO AIA observations of a global coronal EUV "Wave": multiple components and "ripples". Astrophys. J. Lett. **723**, 53–59 (2010). doi:10.1088/2041-8205/723/1/L53

J.S. Llorente, A. Agenjo, C. Carrascosa, C. de Negueruela, A. Mestreau-Garreau, A. Cropp, A. Santovincenzo, Proba-3: precise formation flying demonstration mission. Acta Astronaut. **82**(1), 38–46 (2013). doi:10.1016/j.actaastro.2012.05.029. 6th International Workshop on Satellite Constellation and Formation Flying. http://www.sciencedirect.com/science/article/pii/S0094576512002202

M.D. Looper, J.E. Mazur, J.B. Blake, H.E. Spence, N.A. Schwadron, M.J. Golightly, A.W. Case, J.C. Kasper, L.W. Townsend, The radiation environment near the lunar surface: CRaTER observations and Geant4 simulations. Space Weather **11**, 142–152 (2013). doi:10.1002/swe.20034

S. Ma, J.C. Raymond, L. Golub, J. Lin, H. Chen, P. Grigis, P. Testa, D. Long, Observations and interpretation of a low coronal shock wave observed in the EUV by the SDO/AIA. Astrophys. J. **738**, 160 (2011). doi:10.1088/0004-637X/738/2/160

W.B. Manchester IV, T.I. Gombosi, D.L. De Zeeuw, I.V. Sokolov, I.I. Roussev, K.G. Powell, J. Kóta, G. Tóth, T.H. Zurbuchen, Coronal mass ejection shock and sheath structures relevant to particle acceleration. Astrophys. J. **622**, 1225–1239 (2005). doi:10.1086/427768

G.M. Mason, G. Gloeckler, D. Hovestadt, Temporal variations of nucleonic abundances in solar flare energetic particle events. II. Evidence for large-scale shock acceleration. Astrophys. J. **280**, 902–916 (1984). doi:10.1086/162066

G.M. Mason, C.M.S. Cohen, A.C. Cummings, J.R. Dwyer, R.E. Gold, S.M. Krimigis, R.A. Leske, J.E. Mazur, R.A. Mewaldt, E. Möbius, M. Popecki, E.C. Stone, T.T. von Rosenvinge, M.E. Wiedenbeck, Particle acceleration and sources in the November 1997 solar energetic particle events. Geophys. Res. Lett. **26**, 141–144 (1999)

G.M. Mason, M.E. Wiedenbeck, J.A. Miller, J.E. Mazur, E.R. Christian, C.M.S. Cohen, A.C. Cummings, J.R. Dwyer, R.E. Gold, S.M. Krimigis, R.A. Leske, R.A. Mewaldt, P.L. Slocum, E.C. Stone, T.T. von Rosenvinge, Spectral properties of He and heavy ions in ^{3}He-rich solar flares. Astrophys. J. **574**, 1039–1058 (2002). doi:10.1086/341112

D.J. McComas, R.W. Ebert, H.A. Elliott, B.E. Goldstein, J.T. Gosling, N.A. Schwadron, R.M. Skoug, Weaker solar wind from the polar coronal holes and the whole Sun. Geophys. Res. Lett. **35**, 18103 (2008). doi:10.1029/2008GL034896

D.J. McComas, F. Allegrini, P. Bochsler, P. Frisch, H.O. Funsten, M. Gruntman, P.H. Janzen, H. Kucharek, E. Möbius, D.B. Reisenfeld, N.A. Schwadron, Lunar backscatter and neutralization of the solar wind: first observations of neutral atoms from the Moon. Geophys. Res. Lett. **36**, 12104 (2009). doi:10.1029/2009GL038794

D.J. McComas, M. Bzowski, P. Frisch, G.B. Crew, M.A. Dayeh, R. DeMajistre, H.O. Funsten, S.A. Fuselier, M. Gruntman, P. Janzen, M.A. Kubiak, G. Livadiotis, E. Möbius, D.B. Reisenfeld, N.A. Schwadron, Evolving outer heliosphere: large-scale stability and time variations observed by the interstellar boundary explorer. J. Geophys. Res. Space Phys. **115**(A14), 9113 (2010). doi:10.1029/2010JA015569

D.J. McComas, N. Angold, H.A. Elliott, G. Livadiotis, N.A. Schwadron, R.M. Skoug, C.W. Smith, Weakest solar wind of the space age and the current "mini" solar maximum. Astrophys. J. **779**, 2 (2013). doi:10.1088/0004-637X/779/1/2. http://adsabs.harvard.edu/abs/2013ApJ...779....2M

D.J. McComas, M. Bzowski, P. Frisch, S.A. Fuselier, M.A. Kubiak, H. Kucharek, T. Leonard, E. Möbius, N.A. Schwadron, J.M. Sokół, P. Swaczyna, M. Witte, Warmer local interstellar medium: a possible resolution of the Ulysses-IBEX enigma. Astrophys. J. **801**, 28 (2015). doi:10.1088/0004-637X/801/1/28

T.B. McCord, G.B. Hansen, R.N. Clark, P.D. Martin, C.A. Hibbitts, F.P. Fanale, J.C. Granahan, M. Segura, D.L. Matson, T.V. Johnson, R.W. Carlson, W.D. Smythe, G.E. Danielson, Non-water-ice constituents in the surface material of the icy Galilean satellites from the Galileo near-infrared mapping spectrometer investigation. J. Geophys. Res. **103**, 8603–8626 (1998a). doi:10.1029/98JE00788

T.B. McCord, G.B. Hansen, F.P. Fanale, R.W. Carlson, D.L. Matson, T.V. Johnson, W.D. Smythe, J.K. Crowley, P.D. Martin, A. Ocampo, C.A. Hibbitts, J.C. Granahan, Salts on Europa's surface detected by Galileo's near infrared mapping spectrometer. Science **280**, 1242 (1998b). doi:10.1126/science.280.5367.1242

T.B. McCord, G.B. Hansen, D.L. Matson, T.V. Johnson, J.K. Crowley, F.P. Fanale, R.W. Carlson, W.D. Smythe, P.D. Martin, C.A. Hibbitts, J.C. Granahan, A. Ocampo, Hydrated salt minerals on Europa's surface from the Galileo near-infrared mapping spectrometer (NIMS) investigation. J. Geophys. Res., Planets **104**(E5), 11827–11851 (1999). doi:10.1029/1999JE900005

T.B. McCord, T.M. Orlando, G. Teeter, G.B. Hansen, M.T. Sieger, N.G. Petrik, L. Van Keulen, Thermal and radiation stability of the hydrated salt minerals epsomite, mirabilite, and natron under Europa environmental conditions. J. Geophys. Res., Planets **106**(E2), 3311–3319 (2001). doi:10.1029/2000JE001282

R.A. Mewaldt, A.J. Davis, K.A. Lave, R.A. Leske, E.C. Stone, M.E. Wiedenbeck, W.R. Binns, E.R. Christian, A.C. Cummings, G.A. de Nolfo, M.H. Israel, A.W. Labrador, T.T. von Rosenvinge, Record-setting cosmic-ray intensities in 2009 and 2010. Astrophys. J. Lett. **723**, 1–6 (2010). doi:10.1088/2041-8205/723/1/L1. http://adsabs.harvard.edu/abs/2010ApJ...723L...1M

R. Mewaldt, C. Cohen, G. Mason, T. von Rosenvinge, G. Li, C.W. Smith, A. Vourlidas, An investigation of the causes of solar-cycle variations in SEP fluences and composition, in *34th International Cosmic Ray Conference*, ICRC2015, vol. 34 (2015), p. 30

M.H. Moore, R.L. Hudson, R.F. Ferrante, Radiation products in processed ices relevant to Edgeworth–Kuiper-belt objects. Earth Moon Planets **92**, 291–306 (2003). doi:10.1023/B:MOON.0000031946.53696.f6

G.E. Moreton, Hα observations of flare-initiated disturbances with velocities ~1000 km/sec. Astron. J. **65**, 494 (1960). doi:10.1086/108346

NASA, *NASA Space Flight Human System Standard Volume 1: Crew Health, NASA-STD-3001*, vol. NASA-STD-3001 (NASA Headquarters, Washington, 2007)

NCRP, *Limitation of Exposure to Ionizing Radiation*. NCRP Report 116 (National Council on Radiation Protection and Measurements, Bethesda, 1993)

J.E. Nealy, F.A. Cucinotta, J.W. Wilson, F.F. Badavi, N. Zapp, E. Semones, S.A. Walker, G. de Angelis, S.R. Blattnig, Pre-engineering spaceflight validation of environmental models and the 2005 HZETRN simulation code, in *36th COSPAR Scientific Assembly*. COSPAR Meeting, vol. 36, 2006, p. 1399

J.E. Nealy, F.A. Cucinotta, J.W. Wilson, F.F. Badavi, T.P. Dachev, B.T. Tomov, S.A. Walker, G. De Angelis, S.R. Blattnig, W. Atwell, Pre-engineering spaceflight validation of environmental models and the 2005 HZETRN simulation code. Adv. Space Res. **40**, 1593–1610 (2007). doi:10.1016/j.asr.2006.12.029

R.A. Nymmik, M.I. Panasyuk, V.V. Petrukhin, B.Y. Yushkov, A method of calculation of vertical cutoff rigidity in the geomagnetic field. Cosm. Res. **47**, 191–197 (2009). doi:10.1134/S0010952509030010

P.M. O'Neill, Badhwar O'Neill galactic cosmic ray model update based on advanced composition explorer (ACE) energy spectra from 1997 to present. Adv. Space Res. **37**, 1727–1733 (2006). doi:10.1016/j.asr.2005.02.001

V. Ontiveros, A. Vourlidas, Quantitative measurements of coronal mass ejection-driven shocks from LASCO observations. Astrophys. J. **693**, 267–275 (2009). doi:10.1088/0004-637X/693/1/267

M.J. Owens, N.U. Crooker, Coronal mass ejections and magnetic flux buildup in the heliosphere. J. Geophys. Res. Space Phys. **111**, 10104 (2006). doi:10.1029/2006JA011641. http://adsabs.harvard.edu/abs/2006JGRA..11110104O

C. Paranicas, R.W. Carlson, R.E. Johnson, Electron bombardment of Europa. Geophys. Res. Lett. **28**, 673–676 (2001). doi:10.1029/2000GL012320

C. Paranicas, J.F. Cooper, H.B. Garrett, R.E. Johnson, S.J. Sturner, Europa's radiation environment and its effects on the surface, in *Europa*, ed. by R.T. Pappalardo, W.B. McKinnon, K.K. Khurana, with the assistance of René, Dotson with 85 collaborating authors. The University of Arizona Space Science Series (University of Arizona Press, Tucson, 2009), p. 529. ISBN: 9780816528448

C.M. Pieters, J.N. Goswami, R.N. Clark, M. Annadurai, J. Boardman, B. Buratti, J.-P. Combe, M.D. Dyar, R. Green, J.W. Head, C. Hibbitts, M. Hicks, P. Isaacson, R. Klima, G. Kramer, S. Kumar, E. Livo, S. Lundeen, E. Malaret, T. McCord, J. Mustard, J. Nettles, N. Petro, C. Runyon, M. Staid, J. Sunshine, L.A. Taylor, S. Tompkins, P. Varanasi, Character and spatial distribution of OH/H_2O on the surface of the Moon seen by M^3 on Chandrayaan-1. Science **326**, 568 (2009). doi:10.1126/science.1178658

C.C. Porco, P. Helfenstein, P.C. Thomas, A.P. Ingersoll, J. Wisdom, R. West, G. Neukum, T. Denk, R. Wagner, T. Roatsch, S. Kieffer, E. Turtle, A. McEwen, T.V. Johnson, J. Rathbun, J. Veverka, D. Wilson, J. Perry, J. Spitale, A. Brahic, J.A. Burns, A.D. Del Genio, L. Dones, C.D. Murray, S. Squyres, Cassini observes the active South pole of Enceladus. Science **311**, 1393–1401 (2006). doi:10.1126/science.1123013

M.S. Potgieter, R.A. Burger, S.E.S. Ferreira, Modulation of cosmic rays in the heliosphere from solar minimum to maximum: a theoretical perspective. Space Sci. Rev. **97**, 295–307 (2001). doi:10.1023/A:1011837303094

F. Raulin, T. Owen, Organic chemistry and exobiology on Titan. Space Sci. Rev. **104**, 379–395 (2002)

D.V. Reames, Particle acceleration at the Sun and in the heliosphere. Space Sci. Rev. **90**, 413–491 (1999). doi:10.1023/A:1005105831781

D.V. Reames, Solar energetic-particle release times in historic ground-level events. Astrophys. J. **706**, 844–850 (2009a). doi:10.1088/0004-637X/706/1/844

D.V. Reames, Solar release times of energetic particles in ground-level events. Astrophys. J. **693**, 812–821 (2009b). doi:10.1088/0004-637X/693/1/812

I.I. Roussev, I.V. Sokolov, T.G. Forbes, T.I. Gombosi, M.A. Lee, J.I. Sakai, A numerical model of a coronal mass ejection: shock development with implications for the acceleration of GeV protons. Astrophys. J. Lett. **605**, 73–76 (2004). doi:10.1086/392504

C.T. Russell, R.A. Mewaldt, J.G. Luhmann, G.M. Mason, T.T. von Rosenvinge, C.M.S. Cohen, R.A. Leske, R. Gomez-Herrero, A. Klassen, A.B. Galvin, K.D.C. Simunac, The very unusual interplanetary coronal mass ejection of 2012 July 23: a blast wave mediated by solar energetic particles. Astrophys. J. **770**, 38 (2013). doi:10.1088/0004-637X/770/1/38

Y. Saito, S. Yokota, T. Tanaka, K. Asamura, M.N. Nishino, M. Fujimoto, H. Tsunakawa, H. Shibuya, M. Matsushima, H. Shimizu, F. Takahashi, T. Mukai, T. Terasawa, Solar wind proton reflection at the lunar surface: low energy ion measurement by MAP-PACE onboard SELENE (KAGUYA). Geophys. Res. Lett. **35**, 24205 (2008). doi:10.1029/2008GL036077

L. Saul, P. Wurz, A. Vorburger, D.F. Rodríguez M., S.A. Fuselier, D.J. McComas, E. Möbius, S. Barabash, H. Funsten, P. Janzen, Solar wind reflection from the lunar surface: the view from far and near. Planet. Space Sci. **84**, 1–4 (2013). doi:10.1016/j.pss.2013.02.004

J.M. Schmidt, L. Ofman, Global simulation of an extreme ultraviolet imaging telescope wave. Astrophys. J. **713**, 1008–1015 (2010). doi:10.1088/0004-637X/713/2/1008

N.A. Schwadron, M. Owens, N.U. Crooker, The heliospheric magnetic field over the hale cycle. Astrophys. Space Sci. Trans. **4**, 19–26 (2008). doi:10.5194/astra-4-19-2008. http://adsabs.harvard.edu/abs/2008ASTRA...4...19S

N.A. Schwadron, H.E. Spence, R. Came, Does the space environment affect the ecosphere? Eos **92**, 297–298 (2011). doi:10.1029/2011EO360001

N.A. Schwadron, L. Townsend, K. Kozarev, M.A. Dayeh, F. Cucinotta, M. Desai, M. Golightly, D. Hassler, R. Hatcher, M.-Y. Kim, A. Posner, M. PourArsalan, H.E. Spence, R.K. Squier, Earth–Moon–Mars radiation environment module framework. Space Weather **8**, S00E02 (2010). doi:10.1029/2009SW000523

N.A. Schwadron, C.W. Smith, H.E. Spence, J.C. Kasper, K. Korreck, M.L. Stevens, B.A. Maruca, K.K. Kiefer, S.T. Lepri, D. McComas, Coronal electron temperature from the solar wind scaling law throughout the space age. Astrophys. J. **739**, 9 (2011). doi:10.1088/0004-637X/739/1/9. http://adsabs.harvard.edu/abs/2011ApJ...739....9S

N.A. Schwadron, T. Baker, B. Blake, A.W. Case, J.F. Cooper, M. Golightly, A. Jordan, C. Joyce, J. Kasper, K. Kozarev, J. Mislinski, J. Mazur, A. Posner, O. Rother, S. Smith, H.E. Spence, L.W. Townsend, J. Wilson, C. Zeitlin, Lunar radiation environment and space weathering from the cosmic ray telescope for the effects of radiation (CRaTER). J. Geophys. Res., Planets **117** (2012). doi:10.1029/2011JE003978. http://adsabs.harvard.edu/abs/2012JGRE..117.0H13S

N.A. Schwadron, M.L. Goelzer, C.W. Smith, J.C. Kasper, K. Korreck, R.J. Leamon, S.T. Lepri, B.A. Maruca, D. McComas, M.L. Steven, Coronal electron temperature in the protracted solar minimum, the cycle 24 mini maximum, and over centuries. J. Geophys. Res. Space Phys. (2014). doi:10.1002/2013JA019397. http://onlinelibrary.wiley.com/doi/10.1002/2013JA019397/abstract

N.A. Schwadron, J.B. Blake, A.W. Case, C.J. Joyce, J. Kasper, J. Mazur, N. Petro, M. Quinn, J.A. Porter, C.W. Smith, S. Smith, H.E. Spence, L.W. Townsend, R. Turner, J.K. Wilson, C. Zeitlin, Does the worsening galactic cosmic radiation environment observed by CRaTER preclude future manned deep space exploration? Space Weather **12**, 622–632 (2014a). doi:10.1002/2014SW001084

N.A. Schwadron, M. Gorby, T. Török, C. Downs, J. Linker, R. Lionello, Z. Mikić, P. Riley, J. Giacalone, B. Chandran, K. Germaschewski, P.A. Isenberg, M.A. Lee, N. Lugaz, S. Smith, H.E. Spence, M. Desai, J. Kasper, K. Kozarev, K. Korreck, M. Stevens, J. Cooper, P. MacNeice, Synthesis of 3-D coronal-solar wind energetic particle acceleration modules. Space Weather **12**, 323–328 (2014b). doi:10.1002/2014SW001086

N.A. Schwadron, M.A. Lee, M. Gorby, N. Lugaz, H.E. Spence, M. Desai, T. Török, C. Downs, J. Linker, R. Lionello, Z. Mikić, P. Riley, J. Giacalone, J.R. Jokipii, J. Kota, K. Kozarev, Particle acceleration at low coronal compression regions and shocks. Astrophys. J. **810**, 97 (2015). doi:10.1088/0004-637X/810/2/97

K. Shibata, T. Magara, Solar flares: magnetohydrodynamic processes. Living Rev. Sol. Phys. **8**, 6 (2011)

J.L. Shinn, J.W. Wilson, M. Weyland, F.A. Cucinotta, Improvements in computational accuracy of BRYNTRN (A Baryon transport code). NASA Tech. Paper 3093 **3093** (1991)

M.L. Shusterman, N.R. Izenberg, C.A. Hibbitts, A.P. Jordan, T.J. Stubbs, J.K. Wilson, Weathering effects of dielectric breakdown in the lunar polar regions, in *47th Lunar and Planetary Science Conference* (2016)

E.C. Sittler, A. Ali, J.F. Cooper, R.E. Hartle, R.E. Johnson, A.J. Coates, D.T. Young, Heavy ion formation in Titan's ionosphere: magnetospheric introduction of free oxygen and a source of Titan's aerosols? Planet. Space Sci. **57**, 1547–1557 (2009). doi:10.1016/j.pss.2009.07.017

C.W. Smith, N.A. Schwadron, C.E. DeForest, Decline and recovery of the interplanetary magnetic field during the protracted solar minimum. Astrophys. J. **775**, 59 (2013). doi:10.1088/0004-637X/775/1/59

E.J. Smith, A. Balogh, Decrease in heliospheric magnetic flux in this solar minimum: recent Ulysses magnetic field observations. Geophys. Res. Lett. **35**, 22103 (2008). doi:10.1029/2008GL035345. http://adsabs.harvard.edu/abs/2008GeoRL..3522103S

W.B. Sparks, K.P. Hand, M.A. McGrath, E. Bergeron, M. Cracraft, S.E. Deustua, Probing for evidence of plumes on Europa with HST/STIS. Astrophys. J. **829**, 121 (2016). doi:10.3847/0004-637X/829/2/121

H.E. Spence, A.W. Case, M.J. Golightly, T. Heine, B.A. Larsen, J.B. Blake, P. Caranza, W.R. Crain, J. George, M. Lalic, A. Lin, M.D. Looper, J.E. Mazur, D. Salvaggio, J.C. Kasper, T.J. Stubbs, M. Doucette, P. Ford, R. Foster, R. Goeke, D. Gordon, B. Klatt, J. O'Connor, M. Smith, T. Onsager, C. Zeitlin, L.W. Townsend, Y. Charara, CRaTER: the cosmic ray telescope for the effects of radiation experiment on the lunar reconnaissance orbiter mission. Space Sci. Rev. **150**, 243–284 (2010). doi:10.1007/s11214-009-9584-8. http://adsabs.harvard.edu/abs/2010SSRv..150..243S

H.E. Spence, M.J. Golightly, C.J. Joyce, M.D. Looper, N.A. Schwadron, S.S. Smith, L.W. Townsend, J. Wilson, C. Zeitlin, Relative contributions of galactic cosmic rays and lunar proton "albedo" to dose and dose rates near the Moon. Space Weather **11**, 643–650 (2013). doi:10.1002/2013SW000995

H.E. Spence, N.A. Schwadron, C.J. Joyce, S.S. Smith, J. Wilson, A. Jordan, Space weathering throughout the solar system (2017 in work)

J.R. Spencer, W.M. Calvin, M.J. Person, Charge-coupled device spectra of the Galilean satellites: molecular oxygen on Ganymede. J. Geophys. Res., Planets **100**(E9), 19049–19056 (1995). doi:10.1029/95JE01503

L.V. Starukhina, Y.G. Shkuratov, NOTE: the lunar poles: water ice or chemically trapped hydrogen? Icarus **147**, 585–587 (2000). doi:10.1006/icar.2000.6476

E.C. Stone, A.C. Cummings, F.B. McDonald, B.C. Heikkila, N. Lal, W.R. Webber, Voyager 1 observes low-energy galactic cosmic rays in a region depleted of heliospheric ions. Science **341**, 150–153 (2013). doi:10.1126/science.1236408

J.M. Sunshine, T.L. Farnham, L.M. Feaga, O. Groussin, F. Merlin, R.E. Milliken, M.F. A'Hearn, Temporal and spatial variability of lunar hydration as observed by the deep impact spacecraft. Science **326**, 565 (2009). doi:10.1126/science.1179788

B.J. Thompson, J.B. Gurman, W.M. Neupert, J.S. Newmark, J.-P. Delaboudinière, O.C.S. Cyr, S. Stezelberger, K.P. Dere, R.A. Howard, D.J. Michels, SOHO/EIT observations of the 1997 April 7 coronal transient: possible evidence of coronal Moreton waves. Astrophys. J. Lett. **517**, 151–154 (1999). doi:10.1086/312030

W.T. Thompson, N.L. Reginald, The radiometric and pointing calibration of SECCHI Cor1 on STEREO. Sol. Phys. **250**(2), 443–454 (2008). doi:10.1007/s11207-008-9228-2

V.S. Titov, P. Démoulin, Basic topology of twisted magnetic configurations in solar flares. Astron. Astrophys. **351**, 707–720 (1999)

V.S. Titov, T. Török, Z. Mikic, J.A. Linker, A method for embedding circular force-free flux ropes in potential magnetic fields. Astrophys. J. **790**, 163 (2014). doi:10.1088/0004-637X/790/2/163

G. Tóth, I.V. Sokolov, T.I. Gombosi, D.R. Chesney, C.R. Clauer, D.L. de Zeeuw, K.C. Hansen, K.J. Kane, W.B. Manchester, R.C. Oehmke, K.G. Powell, A.J. Ridley, I.I. Roussev, Q.F. Stout, O. Volberg, R.A. Wolf, S. Sazykin, A. Chan, B. Yu, J. Kóta, Space weather modeling framework: a new tool for the space science community. J. Geophys. Res. Space Phys. **110**(A9), 12226 (2005). doi:10.1029/2005JA011126

L.W. Townsend, T.M. Miller, T.A. Gabriel, Hetc radiation transport code development for cosmic ray shielding applications in space. Radiat. Prot. Dosim. **116**(1–4), 135–139 (2005). doi:10.1093/rpd/nci091

B. Tsurutani, S.T. Wu, T.X. Zhang, M. Dryer, Coronal Mass Ejection (CME)-induced shock formation, propagation and some temporally and spatially developing shock parameters relevant to particle energization. Astron. Astrophys. **412**, 293–304 (2003). doi:10.1051/0004-6361:20031413

Y. Uchida, Propagation of hydromagnetic disturbances in the solar corona and Moreton's wave phenomenon. Sol. Phys. **4**, 30–44 (1968). doi:10.1007/BF00146996

O.P. Verkhoglyadova, G. Li, G.P. Zank, Q. Hu, Modeling a mixed SEP event with the PATH model: December 13, 2006, in *American Institute of Physics Conference Series*, ed. by G. Li, Q. Hu, O. Verkhoglyadova, G.P. Zank, R.P. Lin, J. Luhmann American Institute of Physics Conference Series, vol. 1039, 2008, pp. 214–219. doi:10.1063/1.2982448

A.M. Veronig, N. Muhr, I.W. Kienreich, M. Temmer, B. Vršnak, First observations of a dome-shaped large-scale coronal extreme-ultraviolet wave. Astrophys. J. Lett. **716**, 57–62 (2010). doi:10.1088/2041-8205/716/1/L57

J.H. Waite Jr., W.S. Lewis, B.A. Magee, J.I. Lunine, W.B. McKinnon, C.R. Glein, O. Mousis, D.T. Young, T. Brockwell, J. Westlake, M.-J. Nguyen, B.D. Teolis, H.B. Niemann, R.L. McNutt, M. Perry, W.-H. Ip, Liquid water on Enceladus from observations of ammonia and ^{40}Ar in the plume. Nature **460**, 487–490 (2009). doi:10.1038/nature08153

J.H. Waite, M.R. Combi, W.-H. Ip, T.E. Cravens, R.L. McNutt, W. Kasprzak, R. Yelle, J. Luhmann, H. Niemann, D. Gell, B. Magee, G. Fletcher, J. Lunine, W.-L. Tseng, Cassini ion and neutral mass spectrometer: Enceladus plume composition and structure. Science **311**, 1419–1422 (2006). doi:10.1126/science.1121290

A. Warmuth, B. Vršnak, H. Aurass, A. Hanslmeier, Evolution of two EIT/Hα Moreton waves. Astrophys. J. Lett. **560**, 105–109 (2001). doi:10.1086/324055

W.R. Webber, J.A. Lockwood, Heliocentric radial intensity profiles of galactic cosmic rays measured by the IMP, Voyager, and Pioneer spacecraft in solar 11-year modulation cycles of opposite magnetic polarity. J. Geophys. Res. Space Phys. **109**(A18), 11101 (2004). doi:10.1029/2004JA010642

M.J. Wills-Davey, G.D.R. Attrill, EIT waves: a changing understanding over a solar cycle. Space Sci. Rev. **149**, 325–353 (2009). doi:10.1007/s11214-009-9612-8

J.W. Wilson, F.F. Badavi, Methods of galactic heavy ion transport. Radiat. Res., **108**, 231–237 (1986). ISSN 0033-7587

J.W. Wilson, L.W. Townsend, W.S. Schimmerling, G.S. Khandelwal, F.S. Khan, J.E. Nealy, F.A. Cucinotta, L.C. Simonsen, J.L. Shinn, J.W. Norbury, Transport methods and interactions for space radiations. NASA STI/Recon Technical Report N **92**, 15956 (1991)

J.W. Wilson, F.A. Cucinotta, J. Miller, J.L. Shinn, S.A. Thibeault, R.C. Singleterry, L.C. Simonsen, M.H. Kim, Materials for shielding astronauts from the hazards of space radiations. Mater. Res. Soc. Symp. Proc. **551**, 3 (1999)

J.W. Wilson, J.E. Nealy, G. de Angelis, M.S. Clowdsley, F.F. Badavi, Deep space environment and shielding, in *Space Technology and Applications International Forum—STAIF 2003*, ed. by M.S. El-Genk American Institute of Physics Conference Series, vol. 654 (2003), pp. 993–1010. doi:10.1063/1.1541395

C. Zeitlin, A.W. Case, H.E. Spence, N.A. Schwadron, M. Golightly, J.K. Wilson, J.C. Kasper, J.B. Blake, M.D. Looper, J.E. Mazur, L.W. Townsend, Y. Iwata, Measurements of galactic cosmic ray shielding with the CRaTER instrument. Space Weather **11**, 284–296 (2013). doi:10.1002/swe.20043

M. Zhang, G. Qin, H. Rassoul, Propagation of solar energetic particles in three-dimensional interplanetary magnetic fields. Astrophys. J. **692**, 109–132 (2009). doi:10.1088/0004-637X/692/1/109

A. Zhukov, The PROBA-3 mission, in *41st COSPAR Scientific Assembly*. COSPAR Meeting, vol. 41 (2016)

Space Sci Rev (2017) 212:1271–1314
DOI 10.1007/s11214-017-0411-3

Geoeffective Properties of Solar Transients and Stream Interaction Regions

E.K.J. Kilpua[1] · A. Balogh[2] · R. von Steiger[3] · Y.D. Liu[4]

Received: 28 March 2017 / Accepted: 14 August 2017 / Published online: 6 September 2017

Abstract Interplanetary Coronal Mass Ejections (ICMEs), their possible shocks and sheaths, and co-rotating interaction regions (CIRs) are the primary large-scale heliospheric structures driving geospace disturbances at the Earth. CIRs are followed by a faster stream where Alfvénic fluctuations may drive prolonged high-latitude activity. In this paper we highlight that these structures have all different origins, solar wind conditions and as a consequence, different geomagnetic responses. We discuss general solar wind properties of sheaths, ICMEs (in particular those showing the flux rope signatures), CIRs and fast streams and how they affect their solar wind coupling efficiency and the resulting magnetospheric activity. We show that there are two different solar wind driving modes: (1) Sheath-like with turbulent magnetic fields, and large Alfvén Mach (M_A) numbers and dynamic pressure, and (2) flux rope-like with smoothly varying magnetic field direction, and lower M_A numbers and dynamic pressure. We also summarize the key properties of interplanetary shocks for space weather and how they depend on solar cycle and the driver.

Keywords Interplanetary coronal mass ejections · Interplanetary shocks · Space weather · Solar wind · Magnetic storms · Magnetosphere

The Scientific Foundation of Space Weather
Edited by Rudolf von Steiger, Daniel Baker, André Balogh, Tamás Gombosi, Hannu Koskinen and Astrid Veronig

✉ E.K.J. Kilpua
emilia.kilpua@helsinki.fi

Y.D. Liu
liuxying@spaceweather.ac.cn

[1] Department of Physics, University of Helsinki, Helsinki, Finland

[2] Imperial College London, London, UK

[3] International Space Science Institute, Bern, Switzerland

[4] State Key Laboratory of Space Weather, National Space Science Center, Chinese Academy of Sciences, Beijing 100190, China

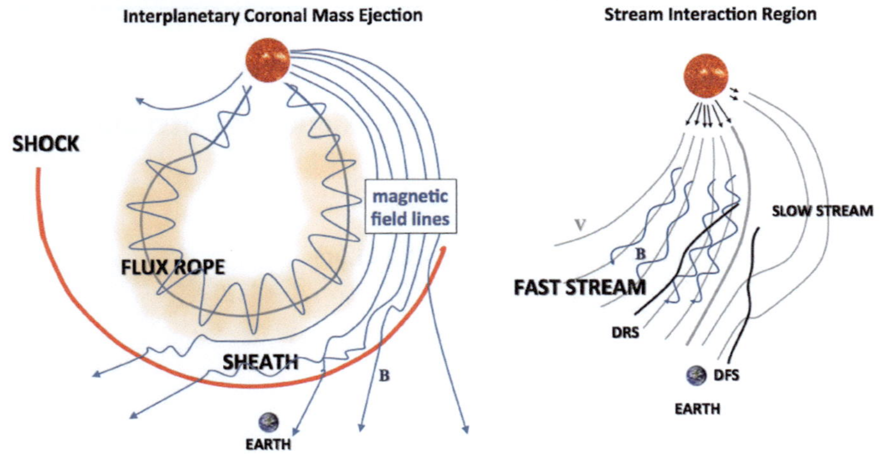

Fig. 1 Schematic of (*Left*) an Interplanetary Coronal Mass Ejection (ICME) and the associated shock (*red arc* in the figure) and sheath. The ICME shown here features the flux rope or magnetic cloud structure, see text for details. Note that not all ICMEs are fast enough to have shocks and clear sheaths. (*Right*) Co-rotating Interaction Region (CIR). Near Earth orbit CIRs are typically bounded by a developing fast forward–fast reverse shock pair (DFS and DRS)

1 Introduction

The large-scale propagating structures in interplanetary space that originate from solar wind streams and transient solar eruptions cause planetary-scale disturbances in the geomagnetic field (for a recent review, see Badruddin and Falak 2016). These magnetospheric/magnetic storms can last from hours to weeks and they are the periods when most severe space weather effects occur. During magnetic storms electric currents that flow in the magnetosphere and ionosphere, and the field-aligned currents coupling these regions are strongly intensified and relativistic electron fluxes in the Van Allen radiation belts may experience dramatic changes. Overall, geomagnetic storms are very complex phenomena, even if their typical morphology is well understood (e.g., Gonzalez et al. 1994; Pulkkinen 2007). Given their important role in space weather scenarios, they have been, and remain, extensively studied.

The most important large-scale solar wind structures that drive magnetic storms (e.g., Gosling et al. 1991; Crooker 2000; Huttunen et al. 2002; Zhang et al. 2007; Echer et al. 2008b; Richardson and Cane 2012) are shown schematically in Fig. 1. On left is an interplanetary counterpart of a coronal mass ejection (CME). CMEs are gigantic magnetized plasma clouds that are expelled on a daily basis from the Sun (e.g., Webb and Howard 2012). When detected in-situ CMEs are called interplanetary CME (ICMEs) based on a set of characteristic solar wind magnetic field, plasma and compositional signatures, see e.g., Zurbuchen and Richardson (2006) and Wimmer-Schweingruber et al. (2006). When CMEs are launched, their speed often exceeds (at least initially) the speed of the ambient solar wind ahead. Farther in the heliosphere, if the speed difference between the ICME and the preceding plasma is larger than the local magnetosonic speed a fast forward shock propagates supersonically into the upstream solar wind ahead of the ICME. Fast forward shocks are the most frequently observed shock waves in space plasmas, and also of greatest interest to space weather studies. Their impact on Earth may change dramatically the state of the magnetosphere (e.g., Samsonov et al. 2007). The ICME sheaths, i.e., the region extending from the shock to the ICME leading edge, are relatively little studied when compared to the ICME itself, but as we will demonstrate later, they are highly important drivers of space weather disturbances (see e.g. recent review, Kilpua et al. 2017, and references therein). Sheaths pile-up gradually over several days the CME travels and expands through interplanetary space (e.g., Siscoe and Odstrcil 2008). Other key large-scale structures, Co-rotating Interaction Regions (CIRs) shown on right in Fig. 1, are compression regions that form when high-speed streams from the Sun follow and catch the slower streams ahead (e.g., Belcher and Davis 1971; Balogh et al. 1999; Gosling and Pizzo 1999). The above-discussed structures often arrive in rapid sequences or are merged together. Such interacting events can lead to particularly intense and long-lasting geospace disturbances.

Whether a given large-scale structure can drive a storm depends on its solar wind conditions. The most critical prerequisite is that the interplanetary magnetic field (IMF) is southward for a sufficiently long-time to facilitate magnetic reconnection between the interplanetary magnetic field (IMF) and the geomagnetic field (Dungey 1961; Burton et al. 1975). The rate of this reconnection is often approximated with "solar wind driving electric field", i.e., $E_Y \approx V B_S$, where V is the solar wind speed (typically very close to the Sun–Earth direction), and B_S is the southward component of the IMF (e.g., Vasyliunas 1975). In reality, solar wind—magnetospheric coupling efficiency is more complex than being a simple linear function of the driving electric field (e.g., Gonzalez et al. 1990; Pulkkinen 2007; Borovsky and Birn 2014). One of the key focuses of this review is to discuss in detail the importance of other solar wind parameters as well as the role of the Earth's magnetosheath in controlling the solar wind—magnetosphere coupling. The main focus will be on how coupling efficiency varies depending on the large-scale solar wind structure driving the storm.

The magnetosphere exhibits two basic modes or time-scales (e.g., Akasofu 1981); a directly driven mode that is related to large-scale magnetospheric convection driven by dayside reconnection and the loading-unloading mode that is related to energy storage and release process in the magnetotail, i.e., to the substorm cycle. During magnetic storms both modes are usually in action. Occasionally, periods of steady magnetospheric convection may arise when reconnection at the dayside and in the tail are balanced (e.g., Sergeev et al. 1996; Walach and Milan 2015), or alternatively, extreme substorms may occur even without sustained large-scale convection (e.g., Huttunen and Koskinen 2004; Tsurutani et al. 2015). Relevant questions related to large-scale solar wind structures and the sequences at which they arrive while interacting are also the pre-conditioning of the magnetosphere by previous activity and how long memory the system has.

Figure 2 demonstrates that the large-scale structures have clear differences in their solar wind properties. The sheath and CIR are associated with large-amplitude magnetic field fluctuations and as they are both compressed regions, they have considerably higher magnetic field and dynamic pressure than the preceding quiet solar wind. In turn, the ICME has distinctly smoother magnetic field when compared to the preceding sheath (or to CIR). The root-mean-square of the magnetic field (Fig. 2b) is clearly depressed and the magnetic field vectors rotate smoothly (Fig. 2c). These are signatures of a magnetic cloud, first introduced by Burlaga et al. (1981) as solar wind structures with (1) enhanced magnetic fields, (2) long-lasting rotation of the magnetic field direction, and (3) depressed proton temperatures. Magnetic clouds can be approximated as force-free ($\mathbf{J} \times \mathbf{B} = 0$) magnetic flux ropes (e.g., Goldstein 1983; Lepping et al. 1990). However, a significant number of ICMEs do not show clear magnetic cloud signatures due to deformation during their interplanetary journey (see Manchester et al. 2017, this issue) and/or the flux rope being sampled far from its centre (e.g., Cane et al. 1997; Liu et al. 2005; Jian et al. 2006b; Richardson and Cane 2010; Kilpua et al. 2011). The ICME in Fig. 2 drove a clear shock that is associated with an abrupt jump in the magnetic field and plasma parameters. The speed profile stays almost flat during the sheath and then declines monotonically from the leading edge of the magnetic cloud to its trailing edge. This indicates that the magnetic cloud was still expanding while it moved past

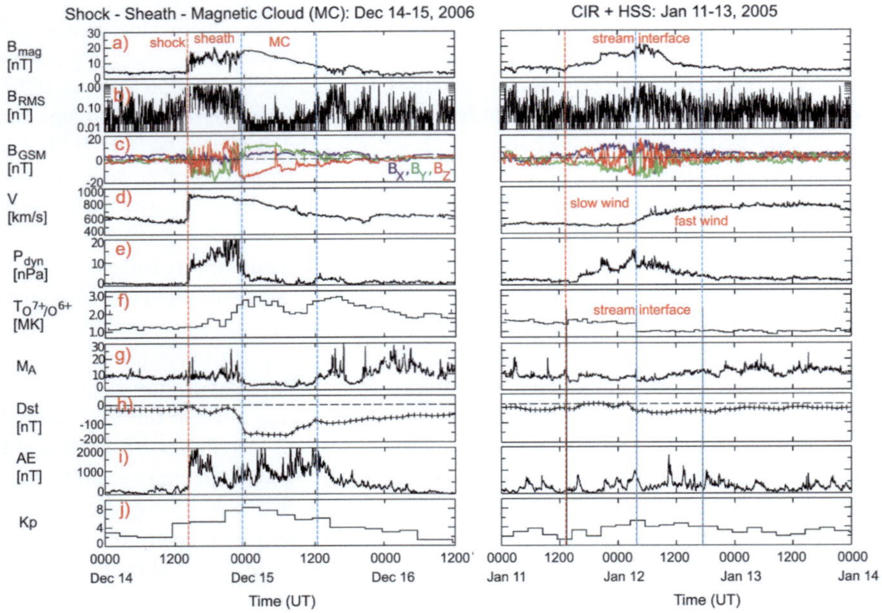

Fig. 2 Solar wind measurements and geomagnetic response during (*Left*) ICME on December 14–15, 2006 (see also a comprehensive study by Liu et al. 2008a) and (*Right*) CIR on January 11–13, 2005. The panels show from top to bottom: (**a**) magnetic field magnitude, (**b**) root-mean-square (RMS) of the magnetic field magnitude, (**c**) magnetic field components in the GSM coordinates, (**d**) solar wind speed, (**e**) Alfvén Mach number (the ratio of the solar wind speed to the Alfvén speed), (**f**) O^{7+}/O^{6+} ratio (CIR ratio is multiplied by a factor of ten), and geomagnetic indices, (**g**) Dst, (**h**) AE, and (**i**) Kp. The data (1-minute for solar wind and AE, 1-hour for Dst and Kp) is from the OMNI database. The *red dashed line* shows the ICME shock, and the magnetic cloud is bounded by a pair of *solid black lines*. In the CIR plot the *vertical lines* present in chronological order the start of the CIR, stream interface, and the end of the CIR. The times are taken from the ACE CIR catalog compiled by L. Jian

the Earth. This is a common feature of ICMEs near the orbit of the Earth (e.g., Klein and Burlaga 1982).

Within the CIR the speed increases gradually and the stream interface (e.g., Gosling et al. 1978) separates the denser and slower solar wind from the faster and more tenuous solar wind. However, due to the dynamic evolution around the stream-stream interfaces, the kinetic parameters of the solar wind (speed, density and temperature) cannot indicate equally clearly the actual stream interface between slow and fast wind (Wimmer-Schweingruber et al. 1997). The solar wind streams of different speeds and origin can be clearly separated by the oxygen ion charge state ratio (O^{7+}/O^{6+}; e.g., Geiss et al. (1995)). This transition is clearly illustrated in Fig. 2f. The charge states "freeze-in" at the distance of a few solar radii from the Sun when the propagation time of the solar wind is short compared to the recombination-ionisation time-scale (e.g., Zhao et al. 2009). At the interface between the streams, the freezing in temperature drops abruptly from 1.5 MK, corresponding to the slow wind, to ~1 MK, corresponding to the fast wind. Note that the freezing-in temperature is high throughout the ICME in Fig. 2, showing the high coronal temperatures of the plasma, in excess of 2 MK within the magnetic cloud. Increased O^{7+}/O^{6+} ratios are known to be a good indicator of ICME-related plasma in the solar wind (e.g., Henke et al. 2001) The continued enhanced freezing-in temperature, depressed field fluctuations and low temperature (data not shown) reveal also a second much weaker ICME following shortly after the magnetic cloud.

Three bottom panels of Fig. 2 show geomagnetic activity indices Dst, AE and Kp. These indices are derived from different sets of magnetometer stations and they reflect variations in different magnetospheric and ionospheric current systems. The 1-hour Dst primarily monitors changes in the equatorial ring current, 1-minute AE is the auroral electrojet index and 3-hour Kp gives a more global activity level. Note that for Dst the more negative the index is, the large the storm. Dst values below -50 nT are generally considered as moderate storms and the intense storm threshold is -100 nT (e.g., Gonzalez et al. 1994). The NOAA Space Weather scales[1] define $\mathrm{Kp} = 5$ as the small storm limit, and the extreme storms have the maximum possible Kp value 9. For further information of geomagnetic indices see e.g., Mayaud (1980). As the contributing current systems react differently to different solar forcing conditions and magnetospheric modes described above, there is no one-to-one correspondence between the indices (e.g., Huttunen et al. 2002; Huttunen and Koskinen 2004). In these events the sheath and the magnetic cloud both drove significant geomagnetic activity, while CIR and HSS were associated to a weaker level activity. In addition, the small ICME following the main magnetic cloud did not cause any significant response to geomagnetic indices.

The arrival of the ICME shock caused a positive excursions in Dst (panel 2g). Storm Sudden Commencements (SSCs) are precursors in the geomagnetic field of many large geomagnetic storms and indicate the first impact of a traveling disturbance on the magnetosphere (Joselyn and Tsurutani 1990). SSCs are generally associated with shocks, but they can also arise from sudden enhancements in the solar wind dynamic pressure (e.g., Wang et al. 2009; Xie et al. 2015). The lack of a shock or a clear pressure pulse preceding a geoeffective ICME or a CIR leads to a storm without an SSC, and indeed, about half of magnetic storms occur without clear SSCs (Park et al. 2015). SSCs are also associated with geomagnetically induced currents (GICs), in particular those that can occur at mid- and low-latitudes (e.g., Zhang et al. 2015). Note that Sudden Impulse (SI) refers to a case when the shock or a pressure pulse impacts on the magnetosphere, but no storm follows.

Above, we have briefly presented the main large-scale interplanetary structures that drive geomagnetic activity and highlighted the complexity of the magnetospheric response to varying solar wind driving conditions. In this review we focus in describing the space weather relevant properties of these drivers and their characteristic geomagnetic responses. We start by historical outlook on finding solar wind drivers for "recurrent" and "sporadic" geomagnetic storms in Sect. 2. In Sect. 3 we first discuss in a statistical manner solar wind properties in different large-scale structures, followed by a review on their ability to drive magnetic storms. Section 4 is focused on solar wind—magnetosphere coupling efficiency. In Sect. 5 we discuss how interactions between different large-scale drivers affect their space weather relevant parameters/properties and modify the resulting geospace response and Sect. 6 is devoted to solar cycle variations. Finally in Sect. 7 we conclude and also discuss the status and future challenges of long-lead time and near-real time space weather predictions in the context of these drivers.

2 Historical Outlook

Geomagnetic storms have been related to the Sun and solar activity for a long time. This connection was realized well before the direct physical links could be identified and the causal phenomenological chains observed, from the Sun through interplanetary space to the

[1]http://www.swpc.noaa.gov/noaa-scales-explanation.

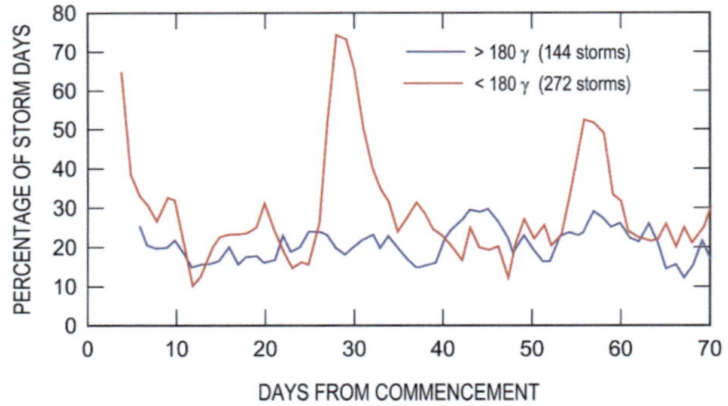

Fig. 3 The recurrence characteristics of (*red line*) weak and (*blue line*) strong geomagnetic storms for the period 1874–1927. For storm definition, see the text. Based on data in Greaves and Newton (1929)

magnetosphere and a plethora of terrestrial effects. The Carrington flare in 1859 and the subsequent major geomagnetic disturbance were among the first well-documented solar-terrestrial events (e.g., Carrington 1859; Cliver and Svalgaard 2004). The further works started to establish the connection between the sunspots numbers and perturbations in the geomagnetic field (e.g., Ellis 1898; Maunder 1904). Many theories of the nature of phenomena driving perturbations in the geomagnetic field emerged during the early twentieth century, but could not be explored in more detail before the space age.

Already the early studies identified two primary types of geomagnetic storms; moderate storms that have approximately 27-day recurrence and stronger and sporadic storms. In a study of 426 geomagnetic storms for the period 1874–1927, Greaves and Newton (1929) investigated the recurrence characteristics of five magnitude categories determined by the changes in declination and in the horizontal and vertical components of the geomagnetic field using Greenwich magnetic records. Figure 3 shows the recurrence characteristics for weaker and stronger storms based on this study. Weak storms (red line) are defined here those that have the horizontal or vertical geomagnetic field intensity less than 180 nT and large storms (blue line) those that have horizontal or vertical intensity exceeding 180 nT. Weaker storms have a clear recurrence signal at about 27 days, the solar synodic period, while for the two strongest categories, no recurrence was noted. For more recent study on this with similar results see, e.g., Borovsky and Denton (2006).

We now know that intense and sporadic magnetic storms are caused mostly by CMEs (e.g., Webb et al. 2000; Schwenn et al. 2005), although prior to their identification in 1970s with space-borne coronagraphs (Tousey 1973; Gosling et al. 1975), solar flares were thought to be the responsible drivers. The landmark paper by Gosling (1993) highlighted the role of CMEs in producing major transient disturbances in interplanetary space. Distinguishing the true driver was not straightforward as flares and CMEs often occur almost simultaneously and they are understood as manifestations of similar magnetic destabilization and energy release process. The early theories were indeed based on the idea that flares release "plasma clouds" or "streams of corpuscles" that propagate to the Earth and cause perturbations in the terrestrial magnetic field (e.g., Lindeman 1911; Chapman and Ferraro 1929). Plasma clouds were also suggested to explain decreases in cosmic ray intensities (e.g., Forbush 1937; Morrison 1956) and it was soon realized that such clouds should form either "magnetic tongues" of extended loops with both field lines rooted to the Sun (Cocconi et al. 1958) or "magnetic bubbles" that are completely detached from the Sun (Piddington 1958; Gold 1962). In-situ observations confirmed the existence of ICMEs soon after beginning of space era, including the likely existence of "magnetic bottles" (for the first identifications see e.g., Hirshberg et al. 1970; Gosling et al. 1973). Rapidly accumulating joint chronograph and in-situ

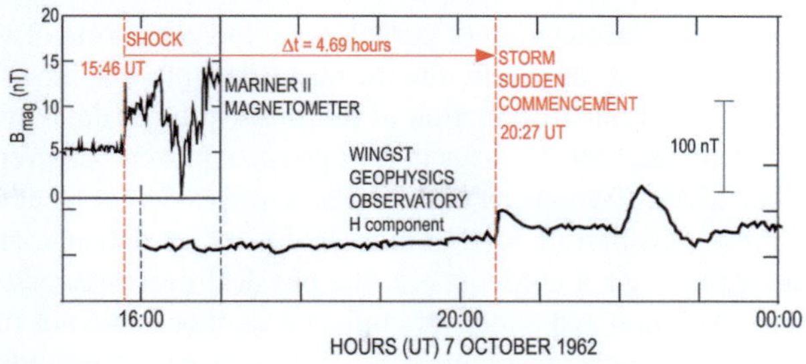

Fig. 4 The first observation of a propagating interplanetary shock wave (*inset*) by the magnetometer on Mariner II, from Sonett et al. (1964). The observation of the shock wave was followed 4.69 hours later by a Storm Sudden Commencement at Earth. The main plot shows the horizontal component of the geomagnetic field (based on Zelwer et al. 1967)

observations, e.g., by the Solwind coronagraph onboard the P78-1 spacecraft and the Helios spacecraft established a more detailed linkage between CMEs and ICMEs (e.g, Burlaga et al. 1982; Schwenn 1983; Sheeley et al. 1985).

Finding the cause for recurrent storms was not straightforward either. An explanation was put forward by Bartels (1932) who assumed that distinct regions on the Sun called M (or magnetic) regions caused the 27-day recurrence. However, no distinguishing features could be identified on the Sun, but it was noted that M-regions appeared to avoid active regions with sunspots (Allen 1944). Although the discovery of the M-regions was often claimed, and the search was slowly narrowing to the actual cause (Toman 1958; Saemundsson 1962), it was only the discovery of coronal holes with Skylab X-ray observations and associations of fast solar wind streams with them (Krieger et al. (1973) and Nolte et al. (1976), see also Cranmer (2002)) that resolved this puzzle. Coronal holes are long-lasting features on the Sun that can maintain their shape for several solar rotations. As a consequence, CIRs and related geomagnetic storms repeat on a 27-days cycles.

Another historically important phenomena related to connecting geomagnetic storms and solar wind drivers were SSC signatures in the geomagnetic field, discussed in Introduction (for definitions and history of the concepts, see e.g., Joselyn and Tsurutani 1990; Curto et al. 2007). The linkage between interplanetary shocks and SSCs was initially suggested by Gold (1955). This was before the discovery of the solar wind, the existence of interplanetary shocks and even before the main characteristics of the magnetosphere were understood. His suggestion was confirmed when the first interplanetary shock wave (Sonett et al. 1964) and the associated SSC (Zelwer et al. 1967) were identified (see Fig. 4). A strong statistical association was found between SSCs and interplanetary shocks by Chao and Lepping (1974) who studied 93 SSCs at solar maximum in 1968–1971 and found that 81 of these could be associated with solar events, although, as this was before the concept of CME was introduced, the association was with flaring activity. Of the 93 SSCs, interplanetary data were available for 48; of these 41 could be shown to be shock events. Later, over the next solar maximum between 1978 and 1981, Smith et al. (1986) found that the association between SSCs and interplanetary shocks could be unambiguously identified in about 85% of the cases.

At first, the whole concept of the interplanetary shock was puzzling. In space plasmas, contrary to hydrodynamics or even to laboratory MHD, the interaction is collisionless in the sense that the mean free paths of the particles that form the plasma are many orders of magnitude larger than the scale on which the interaction takes place. The necessary dissipation process cannot simply be located at the shock surface; it is now well known that the dissipation process involves the acceleration of particles and the generation of waves that carry information about the shock far upstream into the unshocked plasma, modifying its properties that in turn affect the eventual formation of the shock wave Balogh and Riley (1997). The formation of collisionless shock waves in astrophysical plasmas is a very complex process (e.g., Treumann 2009). Phenomena at these shock waves do not meet the assumptions that would allow the application of MHD or any hydro- or gasdynamic analogy for their comprehensive description. As a consequence, the physical processes associated with the formation, structure, evolution and effects of collisionless shocks are not fully understood, even though many detailed phenomenological details have been successfully described. Basic physical considerations that have led, historically, to the definition and evolution of shock waves have been reviewed by Salas (2007).

3 Characteristic Properties and Geoeffectivity of Large-Scale Drivers

In introduction we emphasized that large-scale interplanetary structures have distinct solar wind conditions. In this section we will first study these differences statistically and then proceed on discussing the ability of different structures in driving geospace storms.

3.1 Solar Wind Conditions in Geoeffective Large-Scale Drivers

In Fig. 5 we plot the probability distributions of several space weather relevant solar wind parameters and geomagnetic indices for geoeffective (Dst at least −50 nT reached) ICMEs, sheaths, CIRs, and HSSs using the event list in Kilpua et al. (2015c). The HSS was defined as a 12-hour period following the CIR. We summarize the key features below (see also studies by Liu et al. 2006; Guo et al. 2010 and Myllys et al. 2016 that compare distributions of several solar wind parameters in sheaths and ICMEs). We note that some of the possible extreme values are excluded from the plot and as many strong events (e.g., Halloween storms in 2003) lack standard solar wind measurements.

The magnetic field distribution (panel 5a) peaks at the largest value for ICMEs, but differences are not large when compared to sheaths and CIRs, whose distributions also peak close to 10 nT. However, ICMEs and sheaths have clearly the longest strong-field tails. In turn, the HSS magnetic field distribution is clearly focused towards weaker magnetic fields; it peaks around 5 nT, which is close to the nominal solar wind value. Similar characteristics are also visible in the next panel that shows distributions for the north-south IMF component. Strongest southward fields occur predominantly in ICMEs and sheaths. The distribution of the root-mean-square of the magnetic field in Fig. 5c shows that sheaths are clearly the most turbulent of the investigated large-scale structures, followed by CIRs and then by HSSs, while ICMEs have the smoothest fields. The obvious tendency for lower fluctuation levels in ICMEs than in sheaths was shown statistically also in Kilpua et al. (2013a) where the authors conducted a superposed epoch analysis of the magnetic field fluctuation power in the Ultra Low Frequency (ULF) range.

As expected, the HSSs have on average the highest speeds, but their distribution has large variations, reflecting that not all CIRs are followed by particularly fast wind. What matters is that there is a sufficient speed gradient to form a compressive CIR. In addition, CIRs are three-dimensional structures and in some cases the spacecraft crosses only the outskirts of the CIR and may miss most of the fast stream. The speed distributions for ICMEs and sheaths peak around 400 km/s, while CIRs have the widest distribution extending to higher speeds. This is expected as the speed in CIRs generally increases gradually from the slow

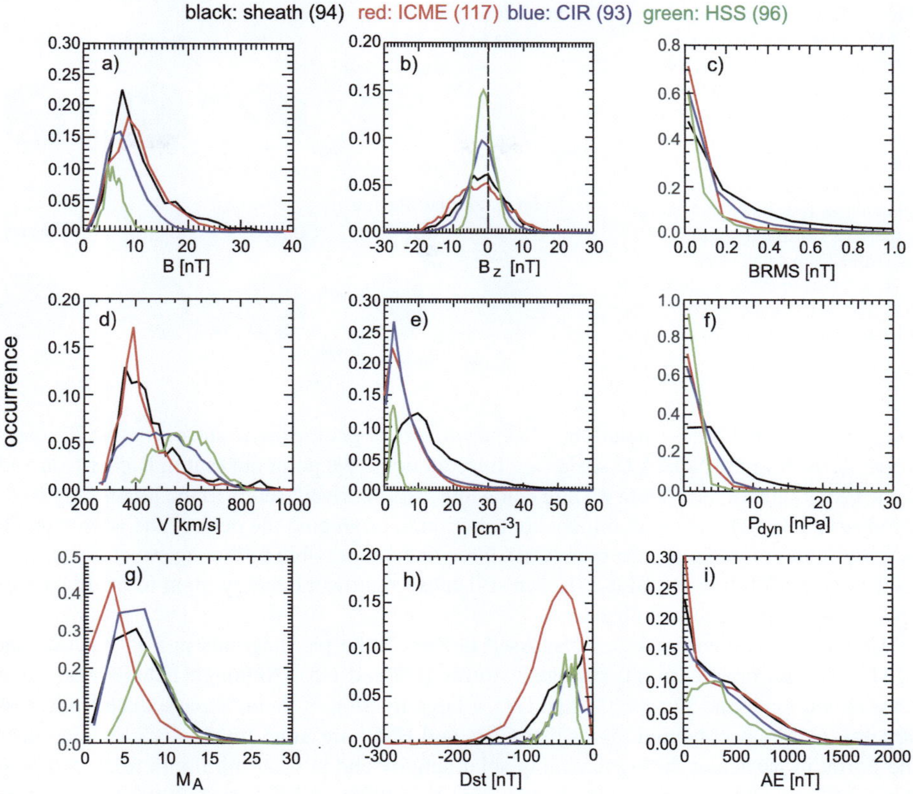

Fig. 5 Probability distributions of selected solar wind parameters and geomagnetic indices within sheaths (*black*), ICMEs (*red*), CIRs (*blue*) and HSS (*green*). To calculate the HSS distributions we used a twelve-hour interval after the CIR end time. The panels give: (**a**) magnetic field magnitude, (**b**) IMF north-south component, (**c**) root-mean-square of the magnetic field, (**d**) solar wind speed, (**e**) solar wind density, (**f**) solar dynamic pressure, (**g**) solar wind Alfvén Mach number, (**h**) Dst index, (**i**) AE index. The data are 5-minute OMNI data (Dst 1-hour). The numbers in parenthesis show the number of events in each structure category

solar wind towards the fast wind (see also Fig. 2). It should be noted that the most extreme speeds (>800 km/s) are, however, related almost solely to ICMEs and sheaths.

The next panel, density and dynamic pressure distributions, highlights that sheaths are particularly compressed regions. It is also clear from the figure that HSSs have the lowest densities and dynamic pressure of all investigated structures. The Alfvén Mach number (M_A) distributions are very similar for sheaths and CIRs. ICMEs have clear tendency towards lower M_A and it is seen that practically all of the lowest M_A conditions in the solar wind ($M_A < 4$) are related to ICMEs. The HSSs lack completely low M_A values, but their high-M_A tail matches to that of CIRs and sheaths.

3.2 Large-Scale Structures as Drivers of Magnetic Storms and High-Latitude Activity

A comprehensive survey of the different sizes of magnetic storms, as well as their solar wind origins was carried out by Richardson and Cane (2012) over four solar cycles (1963–2011). Figure 6 shows the importance of ICMEs, slow interstream solar wind, and fast streams

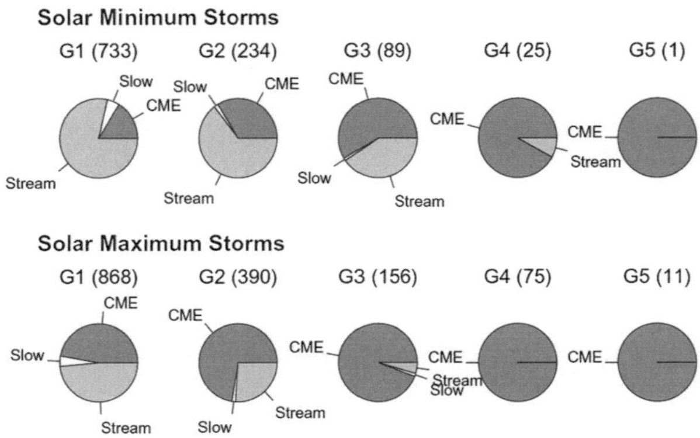

Fig. 6 The importance of ICMEs, slow interstream solar wind, and fast solar wind from coronal holes (including the CIRs) as drivers of geomagnetic storms over more than four full solar cycles (1964–2011). The geomagnetic storm definition shown here follows NOAA storm scale. The storm magnitude increases from small storms (G1, Kp = 5) to extreme storms (G5, Kp = 9). Figure is from Richardson and Cane (2012)

wind from coronal holes (including CIRs) as drivers of geomagnetic storms in five Kp categories following the NOAA G scale (see Introduction). We point out that in Richardson and Cane (2012) and in the papers we cite below the "storm driver" refers to the large-scale solar wind structure that had the dominant contribution in producing the peak of the geomagnetic activity (maximum Kp or minimum Dst). Note that the possible following structure(s), e.g., a CIR and HSS following an ICME, can still cause significant energy input to the magnetosphere also in the recovery phase.

Consistent with early studies discussed in Sect. 2 the pie diagrams in Fig. 6 show that ICMEs caused practically all strongest storms (G4 and G5). Although ICMEs also drive weaker and moderate storms (G1 and G2), a large fraction of them, are associated with fast streams. One reason behind this is that CIRs and HSSs are more frequently identified near the Earth, in particular during declining solar activity and at solar minimum (e.g., compare the ICME and CIR statistics published in Jian et al. 2006a and Jian et al. 2006b). The statistical analysis by Kilpua et al. (2015b) over 13 solar cycles supports the above conclusions; the occurrence of extreme storms correlates with the evolution of the toroidal magnetic field of the Sun that is connected to the formation of sunspots and active regions, where the majority of CMEs arise. In turn, the occurrence of weaker storms correlates with the evolution of the poloidal component of the Sun's magnetic field. This again emphasizes the linkage between the CIRs/HSSs and weaker storms, as poloidal field determines the distribution of coronal holes.

The Dst distributions in Fig. 5 also demonstrate that ICMEs and sheaths produced the majority of the most depressed Dst values; for Dst < -150 nT there are practically no contributions from HSSs and CIRs. These results are in agreement with the above-described Richardson and Cane (2012) study and with several other statistical studies. For example, Echer et al. (2008b) investigated the drivers of 90 intense storms (Dst < -100 nT) that occurred during Solar Cycle 23. The authors found that ICMEs were involved in the clear majority (87%) of these storms, while CIRs caused only about 13% of storms. Zhang et al. (2007) arrived to the same 13% contribution for CIRs in their investigation of the causes of 88 intense storms during 1996–2005, i.e., the period covering most of Solar Cycle 23. Nikolaeva et al. (2011) studied the effectiveness of different large-scale solar wind structures in driving 382 storms with Dst < -50 nT that occurred over multiple solar cycles (1976–2010). If we consider only those 202 storms in their study that could be clearly associated either with CIR or with an ICME, it was found that CIRs caused almost one-third (30%) of the storms. Direct comparison between different studies is not straightforward, but Nikolaeva et al. (2011) result is in agreement with Richardson and Cane (2012) and shows that the contribution from CIRs increases with decreasing storm threshold also when Dst is used

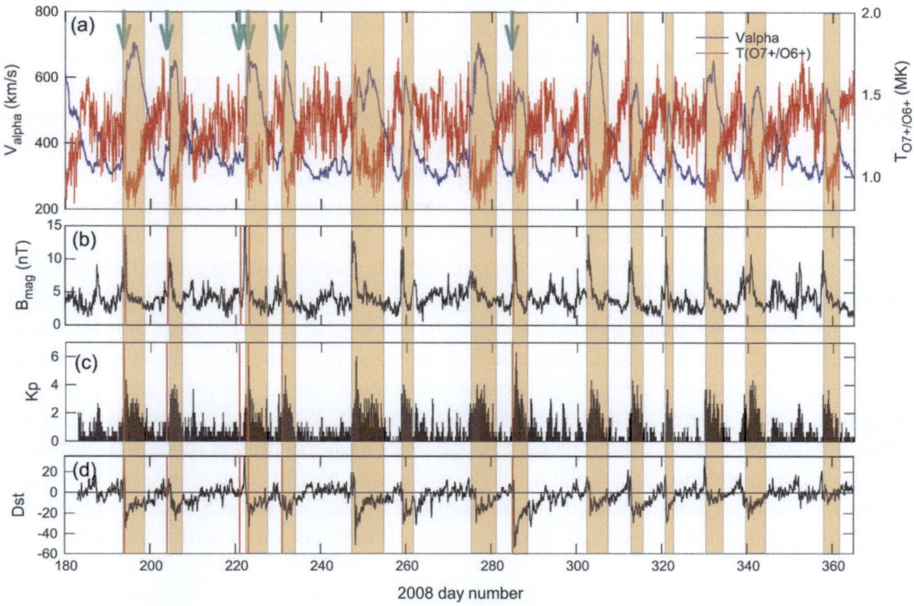

Fig. 7 A series of 14 Stream Interaction regions (SIRs) observed at 1 AU by ACE in the second half of 2008, at a time of very low solar activity, together with the signatures of their geomagnetic effects as measured by the Kp and Dst indices. Panel (**a**) shows the speed of the He^{2+} ions, equivalent to the solar wind speed (*blue line*) and the freezing-in coronal temperature (*red line*) calculated from the ratio of the Fe ions Fe^{7+}/Fe^{6+} measured by the SWICS instrument on ACE. Panel (**b**) is the magnitude of the magnetic field measured by the MAG instrument on ACE. Panels (**c**) and (**d**) show, respectively, the geomagnetic indices Kp and Dst. The *green arrows* indicate the passage of interplanetary shock waves. The *shaded areas* indicate the SIRs, identified (on this scale) by the anti-correlated (high) solar wind speed and (low) freezing-in coronal temperatures. Note that in all 14 cases, there is a corresponding geomagnetic signature, indicating low level stormy conditions

as the storm proxy. In turn, Echer et al. (2008a) found that all (11 in total) super-intense ($Dst < -250$ nT) storms during Solar Cycle 23 were related to ICMEs and sheaths, i.e., none of them were driven by a CIR. Although ICMEs and their sheaths clearly dominate as causes of intense geomagnetic storms, we note that CIRs and HSS have an important role in enhancing the geoeffectivity of some ICMEs (Sect. 5) and in prolonging the recovery phase of the storm (see below).

The geomagnetic effects of CIRs are further illustrated in Fig. 7, showing a series of 14 CIRs in the second half of 2008, observed at 1 AU by the ACE spacecraft. The CIRs were identified by the presence of HSSs, preceded by a compression in the magnetic field. The sudden drop in the freezing-in temperature at the leading edges of the HSSs indicates the stream interfaces. Note that only five of the CIRs were preceded by a shock wave, the others had not steepened enough to form a shock at 1 AU. All shown 14 CIRs were accompanied by a recognizable geomagnetic signature, but usually the minimum of the Dst did not exceed the moderate storm limit ($Dst < -50$ nT). For most of the cases the Kp index reached 3+ to 4+, i.e., the NOAA scale small storm limit ($Kp = 5$) was not exceeded. The likelihood of CIRs/HSSs in driving moderate storms increases in the declining phase, as we will discuss in Sect. 6.1.

As sheaths and ICMEs have different origin and solar wind conditions, it is also important to investigate their separated contributions as storm drivers. Now we consider only the subset of storms that can be traced to an eruption of a CME at the Sun. In Echer et al. (2008b) study 47% of such intense storms were caused by ICMEs, 31% were pure sheath storms, and 22% had contribution both from the sheath and ICME. Zhang et al. (2007) found that 69% of storms were ICMEs-driven, and 31% were sheath-driven. Nikolaeva et al. (2011) arrived to very similar percentages for CME induced moderate storms. In their study 68% of Dst < −50 nT storms were associated to ICMEs, and 32% to sheaths. Note that Nikolaeva et al. (2011) and Zhang et al. (2007) did not report the percentage of the events with significant contribution from both the sheath and ICME.

Some studies however highlight the role of sheaths as storm drivers. One of the earliest statistical studies that demonstrated the relevance of sheath fields was conducted by Tsurutani et al. (1988). The authors showed that for 10 storms that occurred near Solar Cycle 21 maximum the sheaths had equal importance to ICMEs. A statistical analysis by Huttunen et al. (2002) found that in 1996–1999 two-third of intense storms when defined by Kp were either pure sheath storms or the sheath had a big influence. When Dst was used as the proxy of the geomagnetic activity the importance of ICMEs as storm drivers increased. When the analysis was extended to cover Solar Cycle 23 maximum by Huttunen and Koskinen (2004) sheaths drove the largest fraction also of intense Dst storms. As we will discuss in Sect. 6.1, the relative importance of different structures as storm drivers varies with solar activity cycle.

Differences between large-scale structures as drivers of auroral activity and substorms is less systematically studied. Auroral region activity can be studied using geomagnetic indices (e.g., AE, AL and AU, see Mayaud 1980) that are derived from high-latitude magnetometer stations. Substorms manifest themselves with various signatures from which one of the clearest one are is the rapid injections of energetic charged particles to geosynchronous orbit when the geomagnetic tail reconfigures (e.g., Koskinen 2011 and references therein). These injections are followed by slow decay and when they occur at about 2–4 periodicity they are called as sawtooth oscillations. These oscillations occur both during fluctuating and smooth IMF and roughly with the same periodicity (Huang et al. 2004). However, Brambles et al. (2013) found the evidence using a global multi-fluid simulation that sawtooth events during CIRs are externally triggered, while ICME related sawtooths are internally triggered. Farrugia et al. (1993) studied the substorm activity during ∼18 hours of southward IMF associated with the MC passage. The authors identified over 20 substorms during this interval that repeated with a rapid 50 min periodicity, even when solar wind driving diminished due to decreasing magnetic field magnitude and solar wind speed in the end part of the MC. This occurrence frequency agrees with the model developed by Klimas et al. (1992), but is substantially shorter than 2–3 hours reported in several other studies (e.g., Borovsky et al. 1993; Baker et al. 1991; Huang et al. 2004; Borovsky and Yakymenko 2017). As pointed out by Huang et al. (2004), at some periods resonant state of the magnetosphere may exist where solar wind oscillations are comparable with the substorm cycle. Too fast oscillations prevent enough energy to accumulated and if oscillations are too long the energy will be released through internally triggered processes.

Statistical studies show that substorms induced by ICMEs are on average stronger and they occur at lower geomagnetic latitudes than substorms related to CIRs and HSSs (e.g., Holappa et al. 2014; Myllys et al. 2015). Huttunen and Koskinen (2004) showed that sheaths and CIRs can cause considerable high-latitude activity and strong substorms even though the ring current is not significantly enhanced. Similarly, in Nikolaeva et al. (2011) study sheaths were associated with the largest average AE values for a given solar wind electric field than ICMEs and CIRs.

If one considers cumulative effects rather than more instantaneous strong high-latitude activity, the role of HSSs is emphasized. Tsurutani et al. (2006) remarked that recovery phases of CIR-storms can extend up to 27 days. This is attributed to Alfvénic fluctuations in the following HSSs (e.g., Tsurutani and Gonzalez 1987; Balogh et al. 1995; Tsurutani et al. 1995; Lucek and Balogh 1998) that ultimately originate from the Sun and propagate along open field lines into the heliosphere (e.g., Tomczyk and McIntosh 2009). Combined with high solar wind speed these Alfvénic fluctuations cause intermittent but prolonged dayside reconnection that drives efficiently high-latitude geomagnetic activity, called also as High-Intensity, Long-Duration, Continuous AE Activity (HILDCAA) (Tsurutani and Gonzalez 1987). Tanskanen et al. (2005) further showed that during the years when HSSs were frequent substorms deposited approximately twice as much magnetic energy to the auroral ionosphere than during those years when only few HSSs occurred.

3.2.1 ICMEs and Magnetic Clouds

Not all ICMEs are equally geoeffective. Some of them drive extreme storms, while the others pass the Earth practically unnoticed. Near the Earth orbit ICME speeds vary from typical slow solar wind speeds up to few thousands kilometers per second and their magnetic fields range from only a few nT to up about 100 nT. What we observe in-situ is determined by intrinsic CME properties, the evolution and interaction of the CME with the ambient solar wind (see Manchester et al. 2017, this issue, and the references therein), and the path the observing spacecraft (and the Earth) makes through the ICME.

Table 1 shows the Dst response of ICMEs based on their magnetic complexity. The events shown are taken from the Richardson and Cane ICME list[2] and they are divided to MCs, non-MCs and MC-like events. MC-like events show rotation of the field, but they lack other specific MC signatures. Table 1 clearly shows that MCs are the most geoeffective subset of ICMEs. They are associated with clearly stronger average Dst disturbance than non-MCs and MC-like events and a higher fraction of them caused moderate and intense storms. This result is expected as the lack of clear MC signature in ICMEs is generally attributed to cases where the spacecraft crosses the flux rope farther from the centre where the field is typically weaker (see Introduction). Non-MC events likely represent interacting ICMEs, so called complex ejecta (see Sect. 5), where individual MC characteristics are lost. Note that in Table 1 the sheath and MC contributions have not been separated, i.e., part of the storms may have been driven by the sheath alone or the sheath had a significant contribution. However, whatever the geoeffective structure is, the results show that ICMEs that contain MCs are clearly the most likely to drive significant geomagnetic activity.

The results shown in Table 1 are also consistent with statistical studies we discussed in Sect. 3.2. In Zhang et al. (2007) study from those 41 intense storms in their study that were driven by ICMEs, 30 (73%) were MCs. Nikolaeva et al. (2011) also showed that non-MC ICMEs and CIRs were associated with clearly lower average electric fields and geomagnetic responses than the other investigated structures, emphasising again that MCs are associated with stronger geomagnetic activity than non-MC ICMEs and CIRs.

The strong potential of MCs to drive magnetic storms stems from their enhanced magnetic fields and smooth field rotation creating particular favorable conditions for long-lasting and strong magnetospheric convection. Several studies have also indicated a clear relationship between the speed of an MC and its magnetic field (e.g., Gonzalez et al. 1998; Owens et al. 2005; Richardson and Cane 2010), i.e., stronger MCs tend to be faster, which enhances their geoeffectivity. According to Gonzalez and Tsurutani (1987) intense magnetic storms (Dst < -100 nT) require that the solar wind driving electric field E_Y is equal or larger than

[2] http://www.srl.caltech.edu/ACE/ASC/DATA/level3/icmetable2.htm.

Table 1 The geoeffectivity of magnetic clouds (MC), cloud-like events (MC-like) and non-magnetic cloud (non-MC) ICMEs. MC-like events have at least weak rotation in the field, but they lack other MC characteristics. The events and Dst values are taken from the Richardson and Cane online ICME catalog for the period from 1995 to 2016. The first column (N) gives the number of events in each driver category and the second column (< Dst >) shows the averages of the resulting storm Dst minima. The last two columns show the percentages of events (number of such events in parenthesis) in each driver category that caused Dst < -100 nT and Dst < -200 nT storms

	N	⟨Dst⟩	Dst < -100 nT	Dst < -200 nT
MC	170	−86.7	34% (57)	7% (12)
MC-like	161	−51.8	11% (17)	4% (6)
non-MC	147	−41.6	11% (16)	1% (2)

5 mV/m at least for 3 hours. As MCs have on average strong magnetic fields, also slow MCs can create large enough electric fields to drive intense magnetic storms (Tsurutani et al. 2004).

The geomagnetic response of a given MC depends also strongly on its flux rope type (e.g., Huttunen et al. 2005), i.e., how the IMF B_Z rotates within the cloud (e.g., Bothmer and Schwenn 1998; Mulligan et al. 1998). In bipolar MCs the field rotates either from north to south (SN) or south to north (SN), see Fig. 2 for an example of an SN-type MC. Such clouds are also called low-inclination as their axes are more or less aligned with respect to the ecliptic plane. In turn in unipolar or high-inclination MCs (N- and S-type) the field rotates in a way that B_Z maintains its sign.

It is clear that MCs with fully northward fields (i.e., N-type MCs) cannot drive magnetic storms, while MCs with fully southward fields (S-type MCs) provide the most long-lasting southward IMF and they typically drive intense storms. Figure 8 shows examples of both S-type and N-type MCs and their geomagnetic responses. The unipolar S-type MC on 19–20 November, 2003, shown on left in Fig. 8, drove the largest Dst storm of Solar Cycle 23 (−422 nT). In turn, although the sheath of the N-type MC shown on right caused some substorm activity (enhanced AE and Kp values), the MC itself did not cause any significant geomagnetic activity. In particular the AE index shows that the activity in the auroral region was minimal during the whole passage of the MC.

N-type MCs may also lead to false space weather alerts. Even if a CME would be fast and associated with a strong flare, no storm follows if its magnetic field in the MC is northward. One interesting example of such a case was the CME on August 1972 that was associated to a major flare and that had one of the fastest Sun-to-Earth transit times ever reported (Tsurutani et al. 1992), but no significant geomagnetic activity followed due to northward fields.

Bipolar/low-inclination MCs have also distinct differences in their geospace responses (e.g., Kilpua et al. 2012). For example, the expansion of the CME can attenuate significantly the geoeffectivity of an NS type cloud as the magnetic field and speed are weaker in the trailing part of the cloud where the field is directed southward. In turn, compression of the tail part by a fast stream can enhance significantly geomagnetic response of an NS-type MC (discussed in more detail in Sect. 5).

3.2.2 Shocks and Sheath Regions

As discussed in Sect. 3.2 ICME sheaths are important drivers of magnetic storms and their importance increases when high-latitude activity is considered. It is also interesting to note

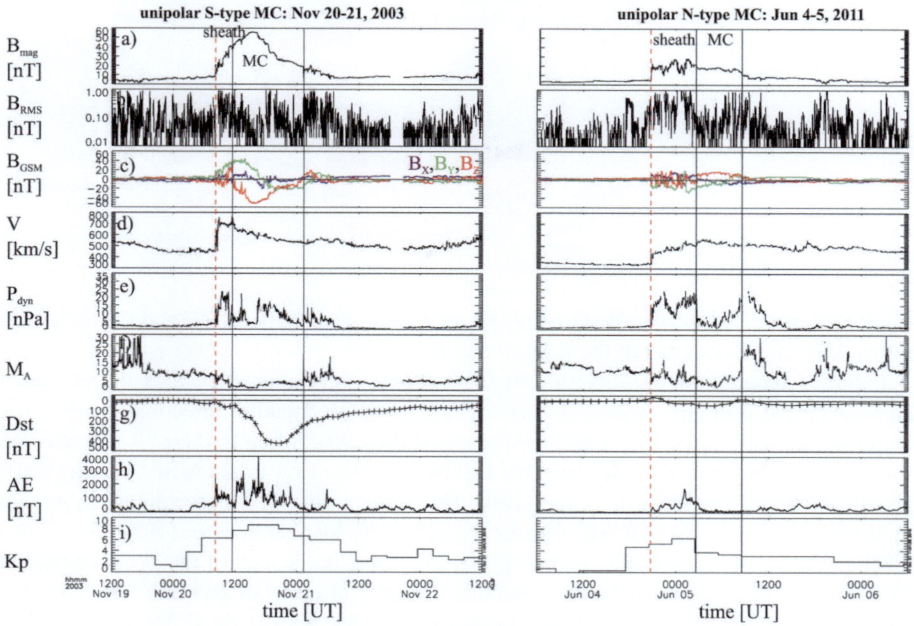

Fig. 8 Unipolar magnetic clouds (MCs) observed in the near-Earth solar wind. (*Left*) S-type MC observed on November 20–21, 2003. (*Right*) N-type MC observed on June 4–5, 2011. The panels are the same as in Fig. 2. The *dashed line* indicates the fast forward shock and the pair of *solid lines* bound the MC

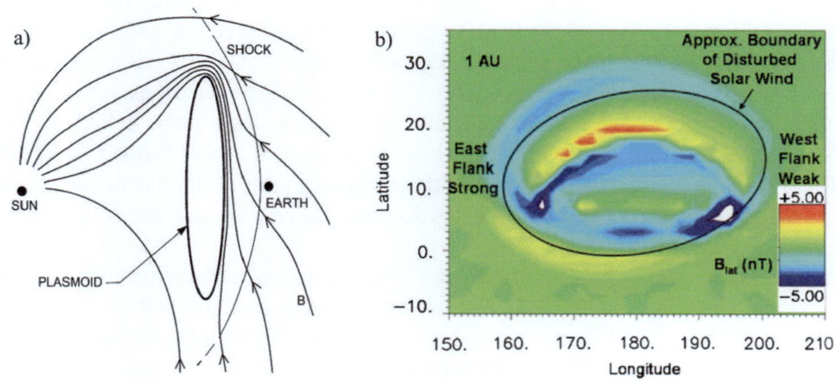

Fig. 9 Direction of the out-of-ecliptic fields in CME sheaths (**a**) for radial IMF from Gosling and McComas (1987). Note that this is a meridional cut, (**b**) Parker spiral-type solar wind from the global MHD simulation by (Siscoe et al. 2007). The strength of the out-of-ecliptic IMF component is shown for a view towards the Sun. The relevant part to focus here are the outer boundary and outskirts of the ellipse that corresponds to the ejecta

that Nikolaeva et al. (2011) showed that on average, sheath regions preceding MCs (only eight included for the whole data set) were associated clearly to the largest driving electric fields and strongest storms and substorms from all investigated driver categories.

There are two basic mechanisms that create strong southward magnetic fields in sheaths (e.g., Gonzalez et al. 1994; Kataoka et al. 2005; Liu et al. 2008b): (1) processes related to the

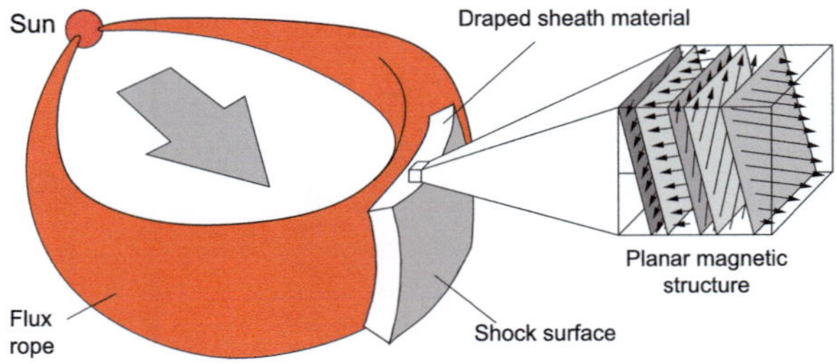

Fig. 10 Formation of planar magnetic structures in an ICME sheath. Figure is from Jones et al. (2002)

shock transition, i.e., the compression of the pre-existing southward IMF and the deflection of the IMF, and (2) draping of the IMF around the CME.

Figure 9 shows how the direction of the out-of-ecliptic magnetic fields in the sheath depends on the upstream field direction and the CME propagation direction. Panel (a) is from Gosling and McComas (1987) for the radial IMF. According to their "Bz forecasting rule" geoeffective sheaths occur when the IMF points away (towards) the Sun and the CME passes southward (northward) of the Earth. The applicability of this technique to predict the dominant sign of B_Z in ICME sheaths was tested by McComas et al. (1989) for 17 fast ICMEs. The authors found that the rule predicted correctly the sign of the B_Z perturbation for 13 events (76%). It was however noted that from the initial set of 26 fast ICMEs nine did not exhibit clear change in the average B_Z upstream of the shock to the ICME leading edge, hence decreasing the amount of events for which this technique can be applied. The Gosling and McComas (1987) scenario was extended by Siscoe et al. (2007) to take into account the Parker spiral type configuration. The results from their global MHD simulation are shown in Fig. 9b. The plot shows that the strongest out-of-ecliptic magnetic fields are found at the eastern flank of the CMEs (the area outside the ellipse that corresponds the ICME). This implies that the eastern flank of the ICME would be more geoeffective, and as suggested by the authors, could explain the known greater geoeffectiveness of CMEs originating from the western hemisphere (e.g., Wang et al. 2002). The magnetic field variations related to shock processes and draping effects occur typically parallel to a single plane called as "planar magnetic structures" (e.g., Jones and Balogh 2000; Jones et al. 2002; Palmerio et al. 2016) (see Fig. 10). Palmerio et al. (2016) showed that the planar parts of the sheath are associated with considerably larger out-of-ecliptic fields than non-planar parts, and are, hence, expected to be the most geoeffective.

The shock angle (i.e., the angle between the shock normal and the upstream magnetic field) has a big influence on how geoeffective the following structure will be. For example, Jurac et al. (2002) studied 107 fast forward shocks observed by Wind in the near-Earth solar wind. The authors showed that regardless of the shock driver 40% of quasi-perpendicular fast forward shocks were associated with an intense storm within 48 hours from the shock passage compared to only 10–15% for quasi-parallel shocks. Quasi-perpendicular and -parallel shocks have different internal structure, because the shock angle (θ_{Bn}) controls the behavior of the particles incident to the shock. Quasi-parallel shocks feature an extended foreshock and wave activity, and as a consequence, the shock transition is more gradual than at quasi-perpendicular shocks (e.g., Bale et al. 2005; Burgess et al. 2005; Lucek et al. 2008; Kilpua et al. 2015a). Examples of quasi-perpendicular and parallel shocks are shown in Fig. 11.

In addition, the shock impact angle (i.e., the angle between the shock normal and the Sun–Earth line) affects the geomagnetic response. Frontal shock collisions compress the whole magnetosphere symmetrically, while inclined collisions lead to an asymmetric collision (e.g., Samsonov et al. 2015). According to global MHD simulations performed by

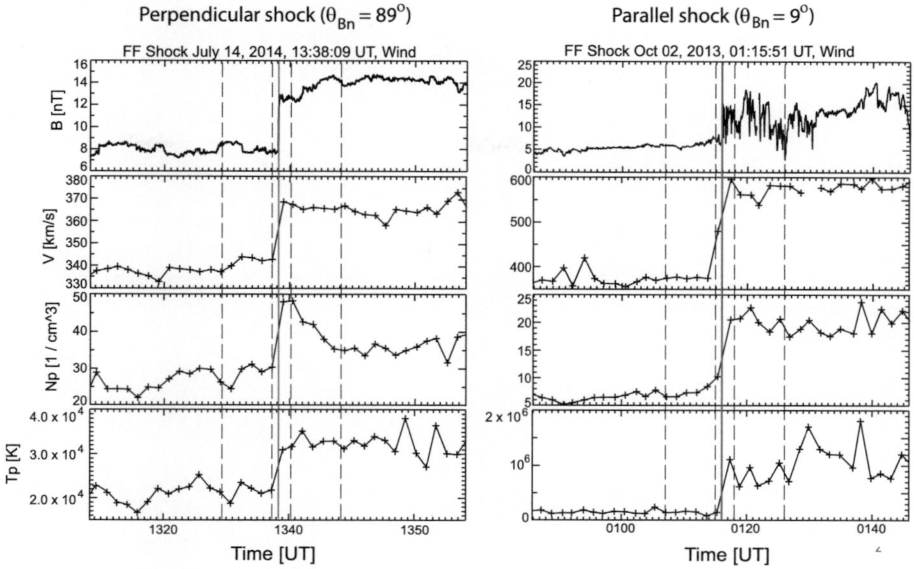

Fig. 11 Examples of (*Left*) quasi-perpendicular and (*Right*) quasi-parallel shocks detected by Wind. The shocks are from the Heliospheric Shock Database (ipshocks.fi)

Oliveira and Raeder (2014) inclined shocks are associated to overall weaker compression of the magnetosphere and weaker response of the field-aligned currents coupling the magnetosphere and ionosphere than head-on shock impacts. The authors also reported that despite otherwise similar parameters the head-on case in their study triggered a substorm, while the inclined shock did not. The stronger auroral response for frontal shocks was also evident in the extensive experimental study by Oliveira and Raeder (2015) where the authors used the geomagnetic index derived from more than 300 SuperMAG magnetometer stations. The authors suggest that symmetric compression of the magnetosphere associated with frontal shocks creates favorable conditions for the release of magnetic energy stored in the magnetotail. We note that CIRs are generally associated with larger impact angle, i.e., being more inclined than ICME-driven shocks (e.g., Jurac et al. 2002; Kilpua et al. 2015a).

Figure 12 in turn features an exceptional shock wave observed by the STEREO-A spacecraft on 23 July 2012 (Liu et al. 2014a). Riley et al. (2016) determined that the shock was quasi-parallel, with an extremely high speed of 3300 km/s, reaching STEREO-A (0.96 AU) in 18.6 hours, and unusually high sonic Mach number ($M_S = 28$), and that the shock normal was pointing very close to the radial direction. They also concluded that what was observed by STEREO-A was the result of the interaction of two exceptional CMEs, propagating into interplanetary space that had been pre-conditioned by a previous CME, so as to influence significantly the formation, evolution and propagation of the shock wave and the CME ejecta. The observed event was thus caused by compound and exceptional conditions, both in the corona and in interplanetary space. Given the extraordinary parameter regime of the event, the STEREO-A plasma instruments could not make all the measurements that would have been necessary for a full description of the observations. Russell et al. (2013) noted that this shock actually rather resembled a subsonically driven compressional disturbance due to intense flux of very high energy particles related to the strong CME. The authors also argued that this could have likely weakened the shock-related space weather consequences. As was concluded by Riley et al. (2016), it remains unclear what role high

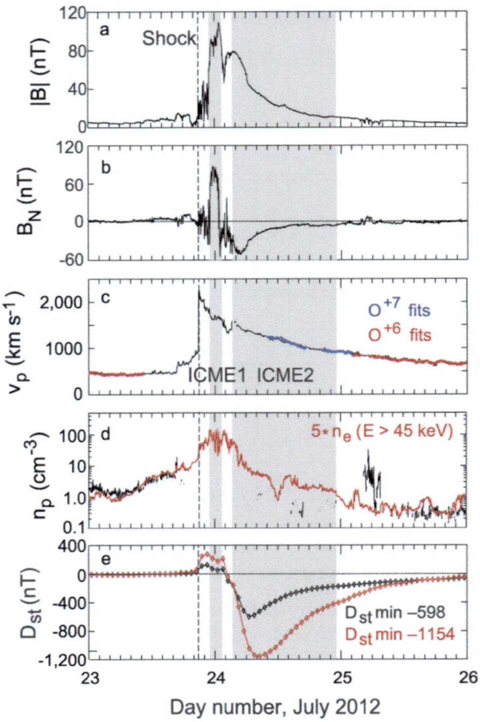

Fig. 12 The exceptional shock wave observed by the STEREO-A spacecraft on 23 July 2012. The panels show (**a**) the magnetic field magnitude, (**b**) the magnetic field normal component, solar wind (**c**) speed and (**d**) density, and (**e**) simulated Dst index. Because of the difficulties of processing the data from the plasma instrument due to the exceptional speeds, densities and temperatures this plot is a composite reconstruction of the best available data (see Liu et al. 2014a)

energy particles played in the observations. This exceptional shock wave illustrates the potentially extreme events that can be caused by CMEs, with a major impact on a large region of the inner heliosphere and, if Earth-directed, causing extreme disturbances in the near-Earth space as well as on the ground.

3.2.3 CIRs and High-Speed Streams

Similarly to ICMEs not all CIRs that pass the Earth are geoeffective. Alves et al. (2006) showed that approximately one-third of CIRs during 1964–2003 were followed by at least a moderate (Dst < -50 nT) magnetic storm. The leading shock enhances the ability of a CIR to drive a storm; according to Zhang et al. (2008) a clear majority (85%) of shock-driving CIRs produced a storm.

The evolution of a CIR structure at the Earth orbit at three spacecraft is shown in Fig. 13: ACE sunward of the Earth at L1, STEREO-A and B, close to 1 AU, but to the west and the east of the Sun–Earth line by about 23°, respectively (see the orbits on the left hand side in Fig. 13). The front of this corotating structure was observed first at STEREO-B, then at the Earth (as observed by ACE), followed by STEREO-A. At STEREO-B there is no evidence of a shock forming (although the solar wind velocity increases sharply) and there is an increase in the magnetic field magnitude, indicating the compression. At ACE, the magnetic field has already steepened considerably, ready to form the forward shock. At STEREO-A, both forward and reverse shocks were identified. As the evolution of this CIR structure varies in longitude, at least at 1 AU the geoeffectivity of a CIR may vary in longitude in the interplanetary space. Panel (d) in Fig. 13 shows the oxygen freezing-in temperature at ACE clearly identifying the stream interface between the slow and fast wind

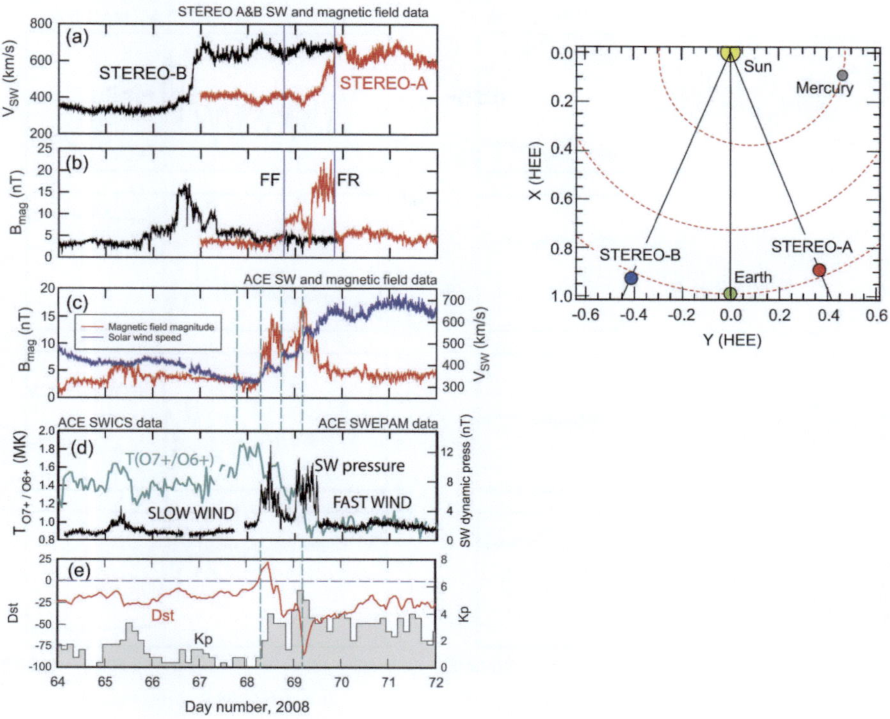

Fig. 13 (*Left*) The evolution of a CIR near 1 AU as seen by STEREO-B, ACE and STEREO-A. Te panels show: (**a**)–(**b**) the solar wind speed and magnetic field magnitude at the two STEREO spacecraft, respectively, (**c**) the magnetic field and solar wind speed profiles observed at 1 AU by ACE, (**d**) oxygen freezing-in temperature measured at ACE, (*black line*). In addition, panel (**e**) shows the geomagnetic response to the CIR, in terms of the disturbance indices Dst and Kp. (*Right*) The positions of the three spacecraft that observed the CIR are shown on 8 March 2008

streams, not so clearly seen in the kinetic parameters of the solar wind. The oxygen freezing-in temperature profile shows that in this case the slow wind ahead of the interaction region was a compound stream, corroborated by the solar wind dynamic pressure. Even though the compression within CIR at ACE was (apparently) not enough to form either a forward or a reverse shock, the solar wind pressure increase was sufficient to trigger a moderate geomagnetic storm, with $Dst = -85$ nT and a maximum $Kp = 6$. A search of geomagnetic data records for the time did not show the occurrence of a SSC, confirming that the CIR leading edge had not formed a shock wave.

Figure 14 shows the close-up the magnetic field measurements for the CIR shocks in Fig. 13 as detected by STEREO-A. The panels show the 32 Hz and 8 Hz magnetic field data in the left and right panels, respectively, for the forward and reverse shocks, over equal, 60 seconds, intervals. The left panels in Fig. 14 illustrate the very prominent wave field in the upstream region, known to be whistler waves generated at the shock. In the right panels the data show clearly how the waves decay when moving away from the shock in the upstream region. The downstream waves in the left hand panel are likely to be magnetosonic waves (Russell et al. 2009). An overview of the generally low Mach number shock waves and the waves associated with both the upstream and downstream regions, as observed by the STEREO mission, was given by (Blanco-Cano et al. 2016).

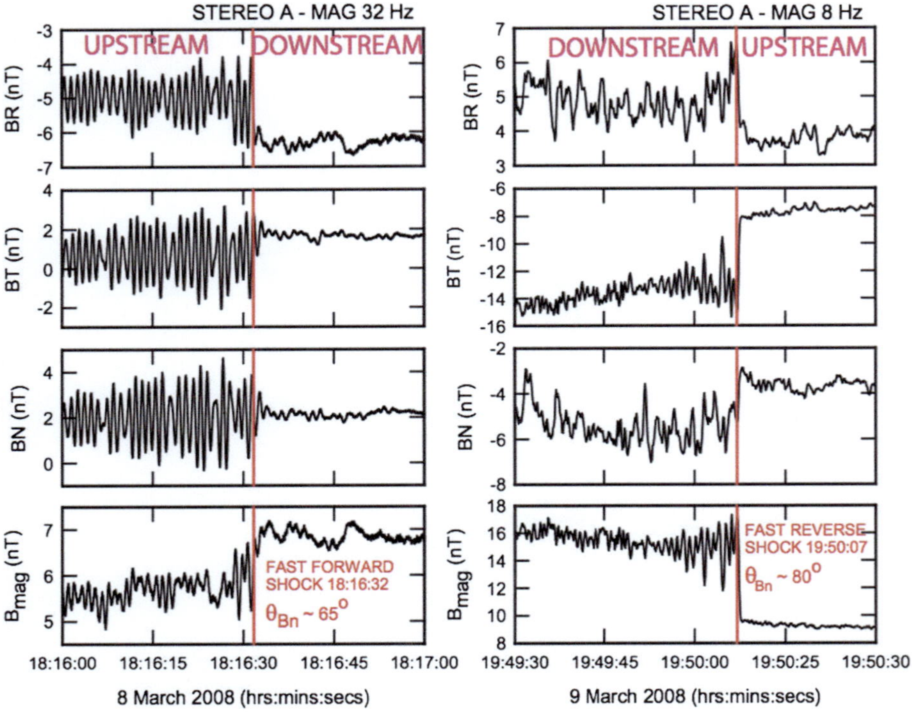

Fig. 14 A close-up of the shock shown in Fig. 13 at STEREO-A. The panels show (**a**)–(**c**) the magnetic field components in the RTN system, (**d**) the magnetic field magnitude. In the *left panels* the frequency of the measurements in 32 Hz and in the *right panels* 8 Hz

The most geoeffective part of a CIR is the region after the stream interface. The superposed epoch analysis using multiple epoch times by Kilpua et al. (2015c) showed that this part has strongest southward fields, speed and magnetic field fluctuation levels. For CIRs that were followed by slower streams the activity declined soon after the interface, while for CIRs that were followed by fast streams the activity continued at relatively high-levels until the end of the CIR and then declined slowly during the following HSS. Considering the effects of the HSS, the Alfvénicity is a critical parameter for causing prolonged geomagnetic activity (see discussion in Sect. 3.2). Snekvik et al. (2013) showed that HSSs with strong Alfvénicity are the most geoeffective as such streams tend to have clearly higher speeds and magnetic fields and slightly higher dynamic pressure than mostly non-Alfvénic streams.

It is also clear that CIRs that embed ICMEs are generally more geoeffective than "pure" CIRs. For example, Zhang et al. (2008) found that from 345 CIRs identified during Solar Cycle 23 157 were "pure" CIRs and they caused mainly weak and moderate storms, while CIRs with ICMEs could drive significantly stronger responses (see Sect. 5). Many CIRs also embed small ICME-like transients, i.e., solar wind periods that feature many ICME signatures, such as smooth/coherent field rotation and enhanced fields, but that are significantly shorter in duration (a few hours on average) than typical large-scale ICMEs (e.g., Moldwin et al. 2000; Kilpua et al. 2009; Yu et al. 2014). Entrainment of such transients, when they have southward fields, may enhance the geoeffectiveness of a CIR. The origin of these transients is not clear, but at least part of them are associated to streamer blobs (e.g., Sheeley and Rouillard 2010; Rouillard et al. 2011) that are released in quasi-continuous manner, likely via magnetic reconnection, from the Sun (e.g., Sheeley et al. 1997). More studies are needed to clarify how these blobs modify the global CIR structures and their geoffectiveness.

CIRs can be associated either with helmet streamers or pseudo-streamers (e.g., Crooker et al. 2012). Helmet streamers are loop structures that separate two coronal holes of the same polarity, while pseudo-streamers separate coronal holes of the same polarity (Wang et al. 2007). Borovsky and Denton (2013) showed that although storms associated with helmet streamer CIRs and pseudo streamer CIRs have similar overall magnitudes, they feature some obvious differences in their characteristics (see their Table 1). The most significant differences are that helmet streamer CIRs have higher densities and dynamic pressure than pseudo-streamer CIRs that also tend to peak earlier concomitant with the magnetic field, and clearly longer duration HSS following them. In addition, the authors emphasized that the helmet streamer CIRs are often preceded by "a calm before the storm", i.e., a particularly low geomagnetic activity period that pre-conditiones the magnetosphere by building-up the plasmasphere and the plasma sheet. This is because the helmet streamer CIRs tend to have considerably lower solar wind speed, density and dynamic pressure preceding them than the pseudo-streamer CIRs. Interesting questions are also whether the above-discussed streamer blobs occur from pseudo-streamers and whether they have similar frequencies and properties as the blobs emanating from helmet streamers.

4 Solar Wind Magnetosphere Coupling Efficiency During Large-Scale Drivers

In this section we discuss the current knowledge on how the solar wind—magnetosphere coupling efficiency depends on different solar wind parameters and the type of the associated large-scale driver.

4.1 Definition of the Coupling Efficiency

The solar wind—magnetosphere coupling efficiency is defined as the ratio of the input to output energies. One of the simplest energy input estimate, or the so-called "coupling function", is the solar wind driving electric field E_Y. As we mentioned in Introduction, the true coupling efficiency is however more complex. Over the years several coupling functions have been developed, such as the widely used epsilon parameter (e.g., Akasofu 1979, 1981; Koskinen and Tanskanen 2002). For a comprehensive review of early coupling functions see Gonzalez et al. (1990). Examples of more recent coupling functions are the ones formulated by Newell et al. (2007) and Borovsky (2008). The energy consumed in the ring current and in the ionosphere as Joule heating and via auroral precipitation can be estimated in terms of geomagnetic indicates (e.g., Yermolaev et al. 2010; Myllys et al. 2016) or functions based on them (Ahn et al. 1983; Knipp et al. 2004), using global numerical simulations (e.g., Palmroth et al. 2004), or if available, using actual satellite measurements (e.g., Turner et al. 2006).

The studies comparing coupling efficiency for different storm drivers (e.g., Turner et al. 2009; Yermolaev et al. 2010; Myllys et al. 2016) show that ICMEs, despite being the key drivers of intense geomagnetic storms, have on average the weakest coupling efficiency. This can be traced to their characteristic plasma and magnetic field conditions we highlighted in Sect. 3.1.

4.2 Influence of Various Solar Wind Parameters

The solar wind Alfvén Mach number (M_A; i.e., the ratio of the solar wind speed to the Alfvén speed), is particularly important parameter for controlling the coupling efficiency. Firstly, M_A controls how efficiently the plasma and magnetic field are compressed at the bow shock (Lopez et al. 2004). The larger the Alfvén Mach number is, the more compression at the bow shock occurs. The tendency for low M_A in ICMEs when compared to sheaths and CIRs should therefore lead to weaker bow shock and weaker compression of the plasma and magnetic field (e.g., Turc et al. 2015; Lugaz et al. 2016). The Alfvén Mach number has also an important role in controlling the saturation of the polar cap potential. The saturation means that although the solar wind driving electric field E_Y would intensify, the magnetosphere cannot dissipate energy more effectively. Saturation generally occurs during intense solar wind driving (e.g., Reiff et al. 1981; Hairston et al. 2005; Raeder and Lu 2005), but when M_A is low the saturation can occur even during the moderate driving conditions (e.g., Lavraud and Borovsky 2008; Lopez et al. 2010; Myllys et al. 2016). These results imply that during ICMEs saturation may occur already during relatively low solar wind driving when compared to sheaths an CIRs.

Another important effect attributed to low Alfvén Mach number solar wind conditions is the resulting low plasma beta in the magnetosheath (e.g., Lavraud and Borovsky 2008). Based on this Lopez et al. (2010) presented a force magnitude balance hypothesis for the M_A control of the coupling efficiency (see also Koskinen 2011). Variations in the plasma beta change the relative magnitude of the pressure force and the magnetic force in the magnetosheath that affects how flows are diverged. When the magnetic force dominates the plasma deviates increasingly and the fraction of the solar wind electric field that can impinge the magnetopause and the associated reconnection will decrease. This effect again decreases the coupling efficiency of low M_A ICMEs when compared to high M_A structures. However, Lavraud and Borovsky (2008) mentioned that increased magnetic tension forces in the magnetosheath accelerate plasma effectively around the magnetopause. The authors also pointed out that high velocity shear at the magnetopause creates favorable conditions for non-linear Kelvin-Helmholtz vortices (Lavraud and Borovsky 2008) that may increase the role of viscous interactions in transferring the energy (e.g., Hasegawa et al. 2004). As the flow speeds increase for low M_A conditions this effect could in turn enhance coupling efficiency for ICME-driven storms.

Rapid IMF fluctuations in sheaths, CIRS and HSSs can also enhance viscous interactions at the magnetopause as demonstrated e.g., by Borovsky and Funsten (2003) and Osmane et al. (2015) and lead to larger geoeffectivity. Turbulent conditions in the upstream solar wind could increase turbulence in the low-latitude boundary layer where the majority of the energy transfer occurs. This is also the region where field-aligned currents arise to the ionosphere and the vorticity increases the strength of these currents (e.g., Koskinen 2011). Large variations in IMF have also been suggested to trigger externally substorms (e.g., McPherron et al. 1986; Lyons et al. 1997), although more recent studies suggest that this association is rather a coincidence than causality (e.g., Morley and Freeman 2007; Newell and Liou 2011).

The role of solar wind dynamic pressure has also been highlighted in several studies. Increased solar wind dynamic pressure evidently compresses the whole magnetosphere (e.g., Shue et al. 1998), increasing magnetopause currents (e.g., Burton et al. 1975). Global MHD simulations by Palmroth et al. (2004) found evidence that increasing solar wind dynamic pressure increases the field aligned currents coupling the magnetosphere to ionosphere, and consequently the ionospheric Joule heating. Myllys et al. (2017) showed that solar wind dynamic pressure seems to inhibit polar cap potential from saturation during intense solar wind driving electric fields. It also interesting to note that both simulations and observations suggest that coupling efficiency is higher when the solar wind speed makes the driving electric field or dynamic pressure large (Palmroth et al. 2010; Pulkkinen et al. 2015; Myllys et al. 2016). The above findings again give support that sheaths have stronger coupling efficiency than ICMEs.

4.3 Bow Shock Configuration and Structures

The bow shock configuration has also an important role in controlling the coupling efficiency. The way the plasma and magnetic field change at the bow depends strongly on whether the shock is quasi-perpendicular or quasi-parallel. This is defined by the upstream solar wind magnetic field. The transition through the bow shock and the magnetosheath has been studied in detail for MCs by Turc et al. (2015) who showed that their magnetic structure alters significantly behind quasi-parallel regime of the bow shock. This alteration can even reverse the sign of B_Z. However, the magnetic fields in MCs are oriented in a way that they tend to be associated with quasi-perpendicular sub-solar bow shock configurations (e.g., Turc et al. 2016). It is yet unclear how the bow shock responds to turbulent drivers (sheaths, CIRs and HSSs) as due to rapid IMF fluctuations the bow shock configuration is expected to vary over relatively short time scales. As discussed in the previous section the shock angle controls the physics of the shock, and hence, the magnetosheath has very different properties behind quasi-perpendicular and quasi-parallel sections of the bow shock (e.g., Walsh et al. 2012; Dimmock et al. 2015).

We also note that the vicinity of the bow shock and magnetosheath have complex internal structure. In particular the foreshock associated with the quasi-parallel bow shock and the magnetosheath behind are populated by various localized structures, such as cavitons, shocklets, Short Large Amplitude Magnetic Structures (SLAMS), hot flow anomalies, spontaneous hot flow anomalies and jets (e.g., Hietala et al. 2009; Blanco-Cano et al. 2011; Omidi et al. 2013; Zhang et al. 2013). It is expected that these structures affect how upstream solar wind plasma and magnetic field change before reaching the magnetopause. However, is not yet clear to what extent they can regulate coupling efficiency and how solar wind conditions and the type of the large-scale solar wind driver control their formation and evolution.

4.4 Plasma Sheet Density and Ring Current Composition

Additional factor that we have not discussed so far, but that may significantly affect the storm magnitude are the density in the Earth's inner plasma sheet and the composition of the ring current. The importance of both of these may vary depending on the large-scale solar wind structure driving the storm. Firstly, the plasma sheet density is known to correlate with the solar wind density, and in particular the sustained northward IMF periods are associated with a "super-dense" plasma sheet (Terasawa et al. 1997; Borovsky et al. 1997). The later convection of this high density plasma inwards toward the Earth may build up ring current and intensify the storm (e.g., Farrugia et al. 2006; Lavraud et al. 2006). The importance of plasma sheet density for contributing to the ring current has been also demonstrated in several simulation studies (e.g., Liemohn et al. 2001; Kozyra et al. 2002; Jordanova et al. 2009). We note that a sheath with primarily northward fields and high densities followed by an MC with southward fields can provide particularly favorable conditions for first creating a super-dense plasma sheet and subsequent sustained convection building a strong ring current. Another possible structure is helmet streamer related CIRs that are generally preceded by high densities (Sect. 3.2.3).

While the quiet time ring current is carried mainly by protons, the contribution from oxygen ions increases dramatically during magnetic storms and with increasing storm magnitude (Daglis et al. 1999). As shown in Kamide et al. (1998b) O^+ comprises only about 10% of the ring current energy density during quiet times, but during extreme storms the contribution can increase up to about 70%. The H^+ ions in the ring current originate both from the solar wind and from the ionosphere, while O^+ is primarily of ionospheric origin. As ICMEs generally drive stronger storms than CIRs, it would be expected that ICME related storms would have larger O^+ content than CIR storms. This was indeed demonstrated statistically by Denton et al. (2006) using LANL/MPA data. In addition, we suggest that high solar wind densities in sheaths could result in particularly oxygen-rich ring current as it was shown by Daglis and Kozyra (2002) that dynamic pressure pulses are related to the enhanced O^+ outflow from the ionosphere.

4.5 Energy Partitioning and the Large-Scale Driver Type

Different structures lead also to different storm energy partitioning: CIR- and HSS-driven storms dissipate more energy into the ionosphere through Joule heating and particle precipitation, while ICMEs dissipate more energy into the ring current (e.g., Turner et al. 2009; Guo et al. 2011; Hajra et al. 2014). These studies did not separate contributions from ICMEs and their sheaths. It is, however, expected that turbulent sheaths dissipate more energy to the ionosphere, similar to CIRs and HSSs. This was also suggested by Huttunen et al. (2002) and Huttunen and Koskinen (2004) who demonstrated that ICMEs cause relatively stronger response to the ring current index Dst, while sheaths produce a stronger response to high-latitude auroral indices and to Kp.

As suggested by Kamide et al. (1998a) the ring current intensifies primarily due to prolonged convection and associated plasma transport during the southward IMF, not due to successive substorm injections. In addition, Tsurutani et al. (2003) reported the lack of substorm expansion phases for steady southward IMF intervals that are typical conditions during MCs. As a consequence, MCs are likely drivers of steady magnetospheric convection events, discussed in Introduction. Simulation studies that do not take into account substorm injections typically underestimate the modeled Dst, and this underestimation is clearly larger (by 20–30%) for CIR-driven storms when compared to ICME-driven storms (e.g., Liemohn et al. 2010; Jordanova et al. 2009; Cramer et al. 2013). This implies that substorm injections may play on average larger role in building up the ring current for CIR than ICME related storms. Another possible causes are for example the significant contribution from other current systems than the ring current to Dst (e.g., tail current and auroral currents). Hajra et al. (2014) also compared energy partitioning for CIRs and HSSs that drove HILDCAAs (see Sect. 3.2). The authors found that in HSSs more energy went into Joule heating than during CIRs-storms (67% and 49%, respectively), while both had comparable share to the ring current.

5 Interacting Large-Scale Structures

In previous sections we have examined geo-effective properties of ICMEs, shocks/sheaths and CIRs separately. Individually these structures usually lead to a one-step geomagnetic storm sequence with or without a sudden commencement followed by a main phase and then a recovery phase. A combination of these structures, however, is often seen in the solar wind that can give rise to a multi-step, enhanced geomagnetic storm (e.g., Burlaga et al. 1987;

Gonzalez et al. 1999; Echer and Gonzalez 2004; Zhang et al. 2007). Zhang et al. (2007) divided their 77 ICME-related intense storms (Dst < -100 nT) to single and multiple types, depending on whether the storm was caused by a single ICME associated with a single CME at the Sun, or by multiple ICMEs and CMEs. While they found that the majority of intense ICME-driven storms were single type (69%), almost one-third of them are still caused by multiple ICMEs. We also note that Zhang et al. (2007) classification did separate ICMEs interacting with CIRs and HSSs.

The "perfect storm" scenario proposed by Liu et al. (2014a, 2015) refers to a combination of circumstances that can result in a particular aggravation of the situation. There are many combinations of circumstances that can occur to make an event more geo-effective, including pileup of events, preconditioning of the upstream solar wind for CME propagation, shock enhancement of pre-existing southward magnetic fields, and following high-speed streams causing compression. The "perfect storm" scenario can be frequent enough for us to worry about because complex structures with these combinations are common in the solar wind.

Burlaga et al. (2001, 2002) defined complex ejecta as fast, transient flows resulting from interactions between successive CMEs. Such interactions are a frequent phenomenon in interplanetary space especially near solar maximum (e.g., Liu et al. 2013). The readers are directed to Manchester et al. 2017, this issue (and references therein) for discussions of the physics about CME-CME interactions. Complex ejecta, in general, do not have well-ordered magnetic fields, but may have a sustained period of enhanced southward magnetic fields produced by CME-CME interactions (e.g., Farrugia and Berdichevsky 2004; Wang et al. 2003; Liu et al. 2014a; Lugaz and Farrugia 2014). As sustained southward magnetic field is a key factor in geomagnetic storm generation, complex ejecta can be very geo-effective given their prolonged duration. They have been considered as a trigger of two-step geomagnetic storms (e.g., Farrugia and Berdichevsky 2004; Farrugia et al. 2006; Liu et al. 2014b), in addition to the sheath-ejecta mechanism (Tsurutani et al. 1988). Complex ejecta are also often associated with a high solar wind densities and consequently high Alfvén Mach numbers, due to interactions between CMEs that are likely to enhance solar wind—magnetosphere coupling, see previous section.

An extreme example of complex ejecta formed by CME-CME interactions is the 2012 July 23 event (Liu et al. 2014a), see Fig. 12. The peak solar wind speed downstream of the shock measured in situ at STEREO A was about 2250 $\mathrm{km\,s^{-1}}$, and the magnetic field inside the ejecta was as high as 109 nT (e.g., Russell et al. 2013; Baker et al. 2013; Liu et al. 2014a). Both the peak solar wind speed and the maximum magnetic field strength are among the largest ever measured near 1 AU. Liu et al. (2014a) suggest that the extremely enhanced ejecta magnetic field was caused by the in-transit interaction between two closely launched eruptions. They also indicate that the unusually high speed at STEREO A was due to preconditioning of the upstream solar wind by another earlier CME, which produced a low-density environment with radial magnetic fields for the propagation of the 2012 July 23 complex CME. Further analyses of the shock properties and propagation behavior give results consistent with this pre-conditioning idea (Temmer and Nitta 2015; Cash et al. 2015; Riley et al. 2016). If the event had hit the Earth, it would have generated a geomagnetic storm comparable to or even larger than the Carrington event (Baker et al. 2013; Liu et al. 2014a).

A shock overtaking a preceding ejecta can enhance the speed and the pre-existing southward magnetic field inside the ejecta, an idea for increased geo-effectiveness dating back several decades (Burlaga 1991; Lepping et al. 1997; Vandas et al. 1997). This could be a special case of complex ejecta (e.g., Liu et al. 2012, 2014b; Möstl et al. 2012; Harrison et al.

2012; Webb et al. 2013), but the shock may or may not be associated with a following CME. Shock compression of pre-existing southward magnetic fields can lead to a two-step geomagnetic storm (e.g., Liu et al. 2014b). A statistical analysis indicates that 39% (19 out of 49 shocks) of shocks propagating inside ICMEs are associated with an intense geomagnetic storm (minimum $D_{st} < -100$ nT; Lugaz et al. 2015).

The geo-effectiveness of a CME can also be enhanced when the CME is compressed from behind by a CIR or a fast solar wind stream (e.g., Fenrich and Luhmann 1998; Webb et al. 2000; Kilpua et al. 2012; Liu et al. 2015; Kataoka et al. 2015). Similar to CME-CME interactions, this kind of compression can help in maintaining a strong ejecta magnetic field as well as a relatively high speed. As discussed in Sect. 3.2 whether this compression can enhance the storm magnitude depends on the ICME "flux rope type". Tail-part compression for NS-type MCs enhances southward magnetic fields, density and speed, and consequently solar wind dynamic pressure and M_A. All these can intensify the coupling efficiency and storm magnitude, see examples e.g., from Fenrich and Luhmann (1998) and Kilpua et al. (2012). In addition, the northward fields in the leading part of the cloud (and in the sheath) can contribute to the above-discussed preconditioning of the plasma sheet density and intensify the subsequent storm. Sheaths may also pre-condition the magnetosphere favorably for SN-type clouds. Storms driven by SN-type MCs with geoeffective sheaths often feature two-step Dst development (e.g., Kamide et al. 1998c), while for NS-type MCs two separate magnetic storms are expected as the magnetosphere has time to return to its quiet-time state during the northward fields in the leading part of the MC. Interactions between a CME and the following SIR or another CME can also deflect the CME in interplanetary space and affect its ability to drive magnetospheric storm.

However, it is not obvious how much preceding intensification of the ring current would contribute to the storm intensification. Kozyra et al. (2002) performed a kinetic ring current drift-loss simulation of a strong magnetic storm on 4–6 June 1991 whose main phase was created by three consecutive southward B_Z excursions. The authors found that the previous intensifications of the ring current were quickly swept by the enhanced magnetospheric convection associated with the next B_Z excursion. This is because during sustained southward IMF the ions in the inner magnetosphere that constitute the ring current are mainly on open drift paths and they are lost at the dayside magnetopause (e.g., Liemohn et al. 2001). Only the weakening or northward turning of the IMF will trap these particles on closed drift paths. However, Kozyra et al. (2002) study emphasised the importance of the plasma sheet density variations in subsequent intensifications in altering the ring current development. This can be an important factor in understanding and forecasting the magnetospheric response for interacting structures.

In turn, magnetic flux erosion by magnetic reconnection at the MC boundaries may reduce significantly its expected geomagnetic response. Whether the reconnection can occur depends on the magnetic shear, i.e., the relative orientation of the magnetic field between the flux rope and the ambient solar wind. Figure 15 shows a parametric study by Lavraud et al. (2014) of the modeled Dst response for SN-type MCs having different levels of erosion. It is clear that the Dst response quickly weakens with the increasing erosion. The authors also noted that the erosion may be absent even when the magnetic shear would be favorable if the current sheet at the MC boundary is too thick. However, it seems that erosion is a relatively common process: The detailed statistical analysis of 50 MCs by Ruffenach et al. (2015) showed that nearly 30% of them showed reconnection signatures at their boundaries and the average erosion was significant, 40% of the total azimuthal flux.

The combinations can be more complex involving multiple solar wind structures. For example, the 2015 March 17 intense geomagnetic storm may have resulted from the interaction between two successive CMEs (Liu et al. 2015) or with the interaction of a CME

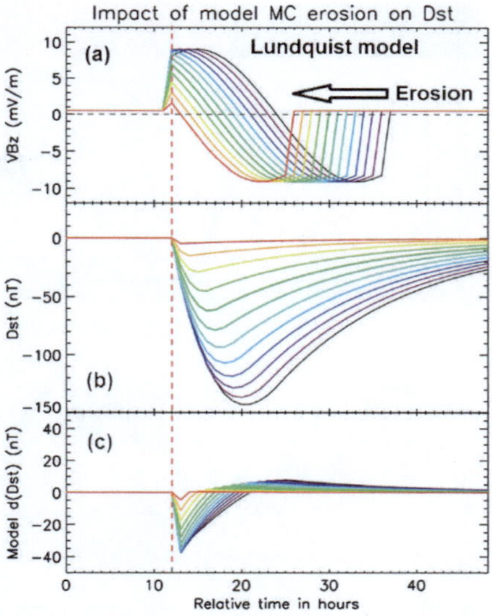

Fig. 15 The panels show (**a**) solar wind driving electric field from the Lundquist flux rope model with maximum magnetic field of 23 nT and speed 400 km/s, (**b**) Dst response as modeled using the O'Brien and McPherron (2000) model, (**c**) the rate of change in Dst. The colors depict level of erosion. *Black* is the non-eroded case, while the *red* is the most eroded case. Figure is from Lavraud et al. (2014)

with the heliospheric plasma sheet (Kataoka et al. 2015) (or having the both aspects), plus the compression by a high-speed stream from behind. Wang et al. (2016) also point out that the CME associated with the main magnetic cloud deflected significantly toward the Earth, which contributed to the unexpected geoeffectiveness of this event. Another intense geomagnetic storm, which occurred on 2015 June 22, was caused by a combination of several preceding shocks with a following CME (Liu et al. 2015). These preceding shocks amplified the southward magnetic field, although fluctuating, as well as the solar wind density, which created favorable conditions for a major geomagnetic storm. A sheath-ejecta-ejecta and a sheath-sheath-ejecta combinations are proposed for the multi-step development of geomagnetic storms based on the studies of the 2015 March 17 and June 22 events (Liu et al. 2015). All these findings reinforce the idea and importance of the "perfect storm" scenario for severe space weather.

6 Solar Cycle and Seasonal Variations

To understand long-term variations in space weather it is important to understand how the frequencies and magnitudes of storms associated with different drivers vary with solar cycle. CIR- and ICME-driven storms exhibit clearly different solar cycle trends that are connected to their different origins, discussed in Sect. 2. Properties of interplanetary shocks and sheaths, and consequently their geomagnetic effects, vary also strongly with solar cycle.

6.1 Solar Cycle Variations

The number of recurrent, generally small to medium CIR/HSS-driven magnetic storms peaks in the declining phase of solar activity cycle (e.g., Tsurutani et al. 1995; Gonzalez et al. 1999; Borovsky and Denton 2006; Zhang et al. 2008; Richardson and Cane 2012). As CIRs arise when slow and fast streams interact, their solar cycle variations are connected to the variations in the occurrence and latitudinal distribution of coronal holes and the helmet streamer structure. As discussed earlier the coronal hole characteristics are controlled by

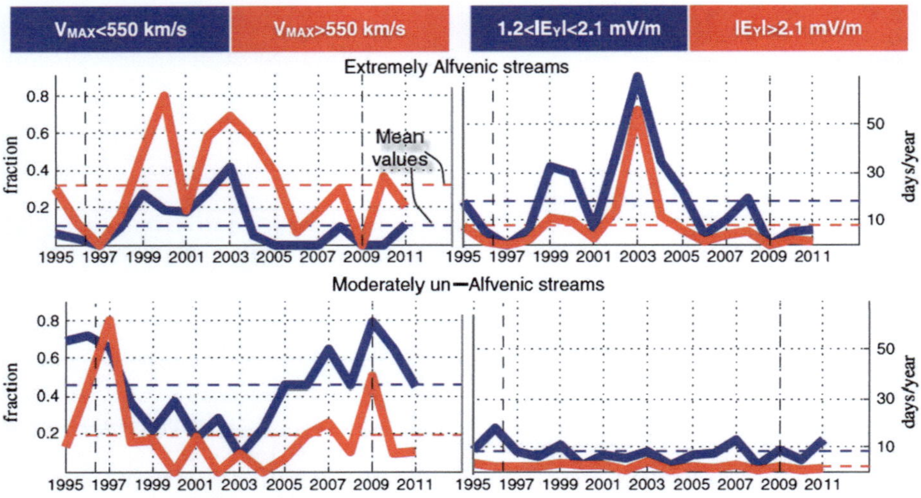

Fig. 16 Yearly distribution of solar wind speeds and electric field for (*Top*) extremely Alfvénic streams and (*Bottom*) moderately non-Alfvénic streams. Figure is from Snekvik et al. (2013)

the poloidal (meridional) component of the Sun's global magnetic field that maximises in the late declining phase and at solar minimum. CIRs do not show clear solar cycle trend in their properties, but they have a tendency towards lower speed gradients, magnetic fields, dynamic pressure and durations at solar minimum than at other times, in particular in the late declining phase (see Fig. 5 in Jian et al. 2006b). Hence, although CIRs are frequent also at solar minimum they drive fewer magnetic storms as they are weaker at this time.

As discussed in Sect. 3.2 the geoeffectivity of HSSs is largely controlled by Alfvénicity. Snekvik et al. (2013) showed that the Alfvénicity of HSSs varies substantially with solar cycle. Figure 16 shows that near solar maximum and early declining phase most fast streams are extremely Alfvénic, while in the late declining phase and at solar minimum fast streams are largely non-Alfvénic. At solar minimum also the number of slow-HSSs increases (i.e. streams that have considerable speed gradient to form CIRs, but whose peak speed is < 550 km/s). The figure also shows that Alfvénic streams produce most significant driving electric fields. As discussed in Snekvik et al. (2013) these variations are presumably related to the variations in the sources of the fast wind, i.e., whether they originate from mid/low-latitude coronal holes or from equatorward extensions of polar coronal holes.

In turn, ICME-related storms follow more closely the sunspot number (SSN). This overall correlation is expected as the occurrence of the fast and strong CMEs from solar active regions follows the evolution of the toroidal component of the Sun's global magnetic field (e.g., Gopalswamy et al. 2004; Webb and Howard 2012). CMEs originate also outside active regions, from helmet streamers and in association with quiet Sun filament disappearances (e.g., Subramanian and Dere 2001; Webb and Howard 2012; Liu et al. 2016) and hence, CMEs are observed in all phases of the solar activity cycle. Solar cycle trends in the ICME occurrence rate and properties naturally reflects those for CMEs (e.g., Webb and Howard 2012), and as a consequence, ICMEs are generally faster and stronger, and more frequent near solar maximum than near solar minimum (e.g., Cane and Richardson 2003; Liu et al. 2005; Jian et al. 2006b). However, the variations in the near ecliptic ICME rate diverge occasionally from the variations in the CME rate. For example, surprisingly many ICMEs were observed near the ecliptic during solar minimum (Kilpua et al. 2011). This is because at this time the helmet streamer belt was confined close to the equator and large polar coronal holes effectively deflected high-latitude CMEs towards lower latitudes (see Manchester et al. 2017, this issue). At solar maximum times the deflection is more random and CME sources are distributed more evenly at all latitudes (e.g., von Steiger et al. 2005; Webb and Howard 2012). Gonzalez et al. (2011) separated also the effect from MCs and sheaths when they studied the importance of different large-scale solar wind structures as drivers of intense magnetic storms (Dst < -100 nT) during Solar Cycle 23. They found that in the rising phase MCs and ICME sheaths were the most important storm drivers, near solar maximum sheaths drove most storms, and in the declining phase MCs and CIRs caused the majority of the intense activity.

However, the exact correlation between the ICME-related storms and the SSN is more complex; rather than peaking exactly at the SSN maximum intense storms have a double-peaked occurrence distribution, with the largest number of storms occurring a few years before and after the SSN peak (e.g., Newton and Milsom 1954; Le et al. 2013). In the Le et al. (2013) study the majority of storms in all investigated magnitude categories occurred in the declining solar activity phase. Kilpua et al. (2015b) and Gonzalez et al. (1990) also found that intense storms occur most frequently in the declining phase, but in these studies the most extreme storms occurred at or relatively close to solar maximum. In addition, the authors did not report that drastic difference between the descending and ascending phases as in Le et al. (2013). These partly contradicting results might stem from how solar maximum is defined, and the fact that many solar cycles feature multi-peaked SSN maxima. In particular, the likelihood for interactions between multiple CMEs increases clearly near solar maximum. As discussed in Sect. 5 such interactions combined to "preconditioning" of the heliosphere by previous CMEs is a particularly likely scenario to lead to the most extreme space weather events ("perfect storm scenario" by Liu et al. (2014a)). The bias towards the declining phase for intense (Dst < -100 nT) storms may also result from the CIRs and fast streams enhancing the geomagnetic response of the preceding ICME. We also note that double-peaked solar cycles have been explained to result from the asymmetric evolution between the northern and southern solar hemispheres. Gopalswamy et al. (2015), however, highlighted that there is necessarily no direct link to variations in the CME rates and properties. The authors studied the properties of CMEs and ICME-driven storms during the multi-peaked Solar Cycle 24 and they found that more geomagnetic storms and more energetic CMEs occurred during the first weaker SSN peak (2012) than during the subsequent stronger peak (2014).

The dependence between the yearly mean SSN and the yearly number of Storm Sudden Commencements (SSCs) for a large data set between 1858 and 2014 is shown in Fig. 17. The figure reveals a clear positive correlation, which presumably arises from the fact that a large fraction SSCs are related to fast forward shocks (see Introduction) whose variations follow closely those in SSN. The clear correlation between the yearly number of fast forward shocks detected in the near-Earth solar wind with the SSN is illustrated in the bottom panel of Fig. 18 that is based on Kilpua et al. (2015a). The comparison of the top panels shows that the ICME-associated forward shocks dominate the distribution and are the cause of the solar cycle dependence. In turn, reverse shocks are almost exclusively associated with CIRs. While there is a clear correlation between SSCs and SSN, as discussed in Introduction, SSCs are seen to occur also without interplanetary shocks, evidently driven by solar wind dynamic pressure pulses at the leading edges of ICMEs and CIRs without an accompanying shock wave. As shown by Zuo et al. (2015) the variations in the number of solar wind dynamic pressure pulses are also roughly in phase with solar cycle, consistent with the results in Fig. 17.

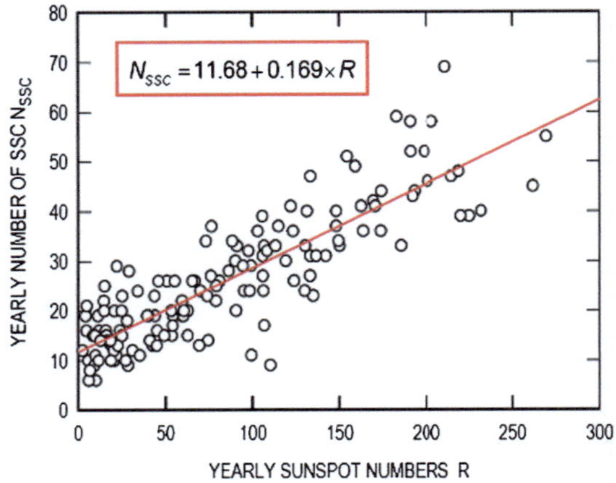

Fig. 17 The correlation between the yearly sunspot number and Storm Sudden Commencements (SSCs) between 1858 and 2014. The red line represents the best linear fit to the data. (Sunspot numbers from: http://www.sidc.be/silso/datafilesand SSCs from ftp://ftp.ngdc.noaa.gov/STP/SOLAR_DATA/SUDDEN_COMMENCEMENTS/SCYEAR

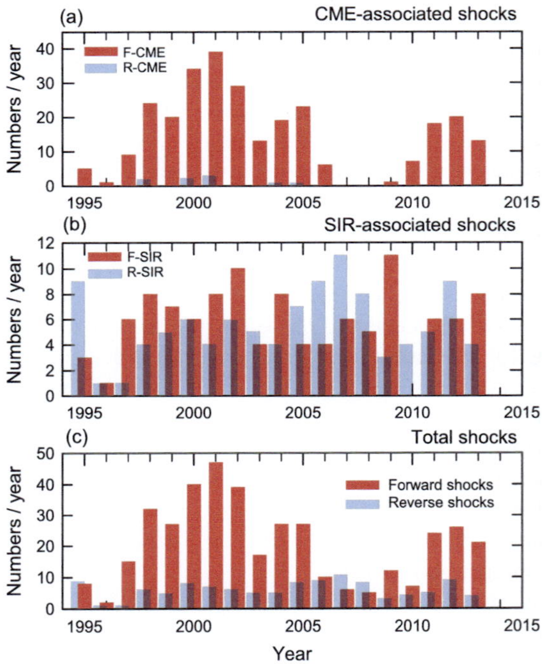

Fig. 18 Annual number of (**a**) CME-associated shocks, (**b**) CIR associated shocks, and (**c**) the total number of shocks. Red histograms show the fast forward shocks and *blue panels* fast reverse shocks. Note the very different scales between panels (**a**) and (**c**) on one hand and panel (**b**) on the other. Figure based on data in Kilpua et al. (2015a)

In Sect. 3.2 we highlighted that the flux rope type of the MC is highly important for its space weather response. The occurrence of bipolar MCs has a clear trend with the 22-year solar magnetic cycle (e.g., Bothmer and Schwenn 1998; Huttunen et al. 2005; Li et al. 2011). The SN-type (NS-type) MCs dominate from the late declining phase of even (odd) numbered solar cycles to the next rising phase. At solar maximum a mixture of SN- and NS-type clouds are observed because the Sun's global magnetic field changes the sign. For unipolar clouds no similar solar cycle trend has been reported. However, Kilpua et al. (2012) noted that during the minimum and early rising phases of Solar Cycle 24 a large fraction of unipolar MCs had their fields northward when compared to previous similar solar cycle phases, which may have contributed to low geomagnetic activity during the early rising phase of cycle 24. On average, the SN and NS type clouds have similar magnitude in their geospace responses, but as discussed in Sect. 3.2 their geoefficiency is controlled in a different way with the kinematic evolution of the CME and the ambient solar wind structure. One additional, interesting solar cycle trend related to MCs, is their fraction from all ICMEs; near solar minimum roughly all ICME are MCs, while at solar maximum the fraction of MCs decreases dramatically to about 10–20% (e.g., Cane and Richardson 2003; Huttunen et al. 2005; Jian et al. 2006a).

6.2 Seasonal Variations

Geomagnetic activity has also seasonal variations that are commonly connected to the changes in the relative orientation between the Sun and the Earth, often called as Russell and McPherron effect (Russell and McPherron 1973). CIR-driven storms have a clear seasonal effect and the majority of CIR-storms occur during "favoured" IMF directions, i.e., close to fall and spring equinoxes (e.g., Alves et al. 2006; Richardson et al. 2006; Zhang et al. 2008). For example, Alves et al. (2006) reported that roughly 50% of CIRs are followed by at least a moderate storms ($Dst < -50$ nT) during equinoxes, compared to before mentioned roughly 30% geoeffectiveness of CIRs when all time periods are considered. According to Richardson et al. (2006) the majority of cases when CIRs cause intense ($Dst < -100$ nT) storms occur during equinoxes. In turn, large ICME-driven storms are usually caused by strongly out-of-ecliptic IMF that has southward component over pro-longed periods, and as a consequence, for ICME driven storms the dipole tilt effect has only a minor contribution (Lockwood et al. 2015). A partly contrary view to seasonal variations has been suggested by Crooker (2000). The authors argue that about two-thirds of the semi-annual variations are attributed to the decreasing coupling efficiency at the solstices, rather than creation of stronger southward IMF at the equinoxes.

6.3 Variations with the Strength of the Solar Cycle

Kilpua et al. (2015b) also examined the relationship of strong magnetic storms with the strength of the solar cycle (as measured by the sunspot number). They found that there is a positive correlation between the strength of solar cycles and magnetic storms. The correlation however decreases with increasing storm magnitude and becomes negligible for the most extreme storms. Hence, there can be some extreme storms even during weak solar cycles (e.g., Carrington storm in 1859, see Sect. 2), but in these cases also the large storms avoid the solar cycle minimum. To investigate long-term driving trends of geomagnetic activity by different large-scale solar wind structures, one viable option is to use the principal component and the independent component analysis of local geomagnetic indices (Holappa et al. 2014)

7 Discussion and Future Outlook

In this article we have reviewed geoeffective properties of large-scale interplanetary structures; Interplanetary coronal mass ejections (ICMEs), shocks/sheaths, Corotating Interaction Regions (CIRs), and High-Speed Streams (HSS) and their combinations. We have demonstrated that these structures are in many respects very different considering their origin, solar wind properties, solar wind—magnetosphere coupling efficiency, geospace responses and solar cycle trends. We discuss below some of the key points highlighted in this review, in particular in the context of understanding and predicting space weather.

We highlight that sheaths and ICMEs, in particular their subset magnetic clouds (MCs), represent very distinct modes of solar wind forcing of the magnetosphere, and therefore, it is of paramount importance to separate these structures. Sheaths are characterized by turbulent magnetic fields and large Alfvén Mach (M_A) numbers and dynamic pressure, while MCs exhibit distinctly smoother driving and generally low dynamic pressure and low M_A conditions.

Based on the existing literature is seems that the "sheath-like" solar wind conditions are associated to more efficient solar wind—magnetosphere coupling. However, due to combination of strong fields and high speeds, and potential to provide sustained southward magnetic fields, MCs are key drivers of intense magnetic storms, in particular when the storm level is determined by Dst. In turn, sheaths deposit most energy in the high-latitude ionosphere. CIRs have more sheath-like conditions, but due to weaker fields and more rapidly fluctuating fields they drive mostly weak to moderate activity, while ICME sheaths cause frequently intense storms. As the large-scale structures often arrive in sequences (e.g., a shock, sheath, MC followed by a CIR and a high-speed stream), the Earth's magnetosphere can encounter rapidly changing forcing conditions and responses.

Two basic steps can be identified when predicting the space weather response of large-scale interplanetary structures: (1) Long-lead time (1–3 days) prediction of their solar wind conditions in the near-Earth solar wind upstream of the bow shock, and (2) Prediction of the details of geomagnetic activity using the near-Earth solar wind observations that are typically available from the Lagrangian point L1 (e.g., Wind, ACE, DSCOVR), i.e., providing lead-times of approximately of 30 min–1 hour.

Some aspects pertaining to Step (1) have improved significantly lately, e.g., numerical heliospheric simulations using information based on the multi-spacecraft CME modeling provide now more accurate CME impact time and details as well as estimates on the overall magnetic field magnitude and plasma parameters. However, the status of long lead-time predictions is currently very limited and in particular the key parameter, the magnetic field north-south component, cannot be measured in the corona, nor predicted in any practical way for space weather purposes. In addition, as discussed in this paper, recent studies also emphasize that to estimate the geospace response it is pivotal to have a detailed knowledge of several other solar wind parameters, including fluctuation levels. Predicting these well in advance would likely require going beyond the MHD-based simulations as kinetic processes can play a significant role in producing variations in particular in turbulent sheath-like drivers and also in regulating the global structure of drivers. Or at least, a very high-resolution grids in MHD simulations are needed.

While the basic processes how the solar wind couples to the magnetosphere are now relatively well understood, recent studies underline the complexity of this coupling. One big issue is the transition through the bow shock and the subsequent propagation through the magnetosheath. As discussed here, the bow shock configuration and the upstream solar wind conditions play a major role in controlling how plasma and magnetic field change. While ICMEs tend to produce quasi-perpendicular bow shock that largely preserves their magnetic field characteristics, it is currently an open issue how the bow shock responds to fluctuating drivers (sheaths, CIRs, and HSSs). Another significant open question is how solar wind conditions control the magnetosheath conditions and conditions in the low-latitude boundary layer through which the majority of the energy transfer occurs and where field-aligned currents arise to the ionosphere. In the future we need more physics-based understanding of details of these coupling processes as a function of upstream solar wind driving conditions and understanding how local and kinetic scale processes and structures regulate global magnetosheath conditions and reconnection details at the magnetopause.

Interactions between the structures introduces a further challenge, and the details of resulting complex structures are largely beyond the current modeling capabilities. As discussed here, there are several processes how interaction between the structures can lead to enhancements of geoeffective properties and solar wind—magnetosphere coupling efficiency and aggravation of the resulting magnetic storm.

Many of the issues discussed above and in particular those involved in interacting structures pertain to "grand challenges in space physics" discussed e.g., by von Steiger (2013), including complexity of space weather, magnetic reconnection, boundaries and collisionless shocks. In particular, this paper emphasizes that although significant progress has been made in many of the above discussed problems, individually, one of the greatest challenges is to unify different aspects into an overall picture.

Acknowledgements Y. D. Liu was supported by the Recruitment Program of Global Experts of China, NSFC under grant 41374173 and the Specialized Research Fund for State Key Laboratories of China. EK acknowledges Academy of Finland project 1267087, UH three-year grant project 490162 and HELCATS project 400931. This project has received funding from the European Research Council (ERC) under the European Union's Horizon 2020 research and innovation programme (SolMAG 724391).

References

B.-H. Ahn, S.-I. Akasofu, Y. Kamide, The Joule heat production rate and the particle energy injection rate as a function of the geomagnetic indices AE and AL. J. Geophys. Res. **88**, 6275–6287 (1983). doi:10.1029/JA088iA08p06275

S.-I. Akasofu, Interplanetary energy flux associated with magnetospheric substorms. Planet. Space Sci. **27**, 425–431 (1979). doi:10.1016/0032-0633(79)90119-3

S.-I. Akasofu, Energy coupling between the solar wind and the magnetosphere. Space Sci. Rev. **28**, 121–190 (1981). doi:10.1007/BF00218810

C.W. Allen, Relation between magnetic storms and solar activity. Mon. Not. R. Astron. Soc. **104**, 13 (1944). doi:10.1093/mnras/104.1.13

M.V. Alves, E. Echer, W.D. Gonzalez, Geoeffectiveness of corotating interaction regions as measured by Dst index. J. Geophys. Res. Space Phys. **111**, A07S05 (2006). doi:10.1029/2005JA011379

A. Badruddin, Z. Falak, Study of the geoeffectiveness of coronal mass ejections, corotating interaction regions and their associated structures observed during Solar Cycle 23. Astrophys. Space Sci. **361**, 253 (2016). doi:10.1007/s10509-016-2839-4

D.N. Baker, A.J. Klimas, D.A. Roberts, Examination of time-variable input effects in a nonlinear analogue magnetosphere model. Geophys. Res. Lett. **18**, 1631–1634 (1991). doi:10.1029/91GL01048

D.N. Baker, X. Li, A. Pulkkinen, C.M. Ngwira, M.L. Mays, A.B. Galvin, K.D.C. Simunac, A major solar eruptive event in July 2012: defining extreme space weather scenarios. Space Weather **11**, 585–591 (2013). doi:10.1002/swe.20097

S.D. Bale, M.A. Balikhin, T.S. Horbury, V.V. Krasnoselskikh, H. Kucharek, E. Möbius, S.N. Walker, A. Balogh, D. Burgess, B. Lembège, E.A. Lucek, M. Scholer, S.J. Schwartz, M.F. Thomsen, Quasi-perpendicular shock structure and processes. Space Sci. Rev. **118**, 161–203 (2005). doi:10.1007/s11214-005-3827-0

A. Balogh, P. Riley, *Overview of Heliospheric Shocks, in Cosmic Winds and the Heliosphere* (University of Arizona Press, Tucson, 1997)

A. Balogh, E.J. Smith, B.T. Tsurutani, D.J. Southwood, R.J. Forsyth, T.S. Horbury, The heliospheric magnetic field over the South polar region of the Sun. Science **268**, 1007–1010 (1995). doi:10.1126/science.268.5213.1007

A. Balogh, V. Bothmer, N.U. Crooker, R.J. Forsyth, G. Gloeckler, A. Hewish, M. Hilchenbach, R. Kallenbach, B. Klecker, J.A. Linker, E. Lucek, G. Mann, E. Marsch, A. Posner, I.G. Richardson, J.M. Schmidt, M. Scholer, Y.-M. Wang, R.F. Wimmer-Schweingruber, M.R. Aellig, P. Bochsler, S. Hefti, Z. Mikić, The solar origin of corotating interaction regions and their formation in the inner heliosphere. Space Sci. Rev. **89**, 141–178 (1999). doi:10.1023/A:1005245306874

J. Bartels, Terrestrial-magnetic activity and its relations to solar phenomena. Terr. Magn. Atmos. Electr. (J. Geophys. Res.) **37**, 1 (1932). doi:10.1029/TE037i001p00001

J.W. Belcher, L. Davis Jr., Large-amplitude Alfvén waves in the interplanetary medium, 2. J. Geophys. Res. **76**, 3534 (1971). doi:10.1029/JA076i016p03534
X. Blanco-Cano, P. Kajdič, N. Omidi, C.T. Russell, Foreshock cavitons for different interplanetary magnetic field geometries: simulations and observations. J. Geophys. Res. Space Phys. **116**, 09101 (2011). doi:10.1029/2010JA016413
X. Blanco-Cano, P. Kajdič, E. Aguilar-Rodríguez, C.T. Russell, L.K. Jian, J.G. Luhmann, Interplanetary shocks and foreshocks observed by STEREO during 2007–2010. J. Geophys. Res. Space Phys. **121**, 992–1008 (2016). doi:10.1002/2015JA021645
J.E. Borovsky, The rudiments of a theory of solar wind/magnetosphere coupling derived from first principles. J. Geophys. Res. Space Phys. **113**, 08228 (2008). doi:10.1029/2007JA012646
J.E. Borovsky, J. Birn, The solar wind electric field does not control the dayside reconnection rate. J. Geophys. Res. Space Phys. **119**, 751–760 (2014). doi:10.1002/2013JA019193
J.E. Borovsky, M.H. Denton, Differences between CME-driven storms and CIR-driven storms. J. Geophys. Res. Space Phys. **111**, A07S08 (2006). doi:10.1029/2005JA011447
J.E. Borovsky, M.H. Denton, The differences between storms driven by helmet streamer CIRs and storms driven by pseudostreamer CIRs. J. Geophys. Res. Space Phys. **118**, 5506–5521 (2013). doi:10.1002/jgra.50524
J.E. Borovsky, H.O. Funsten, Role of solar wind turbulence in the coupling of the solar wind to the Earth's magnetosphere. J. Geophys. Res. Space Phys. **108**, 1246 (2003). doi:10.1029/2002JA009601
J.E. Borovsky, K. Yakymenko, Substorm occurrence rates, substorm recurrence times, and solar wind structure. J. Geophys. Res. Space Phys. **122**, 2973–2998 (2017). doi:10.1002/2016JA023625
J.E. Borovsky, R.J. Nemzek, R.D. Belian, The occurrence rate of magnetospheric-substorm onsets—random and periodic substorms. J. Geophys. Res. **98**, 3807–3813 (1993). doi:10.1029/92JA02556
J.E. Borovsky, M.F. Thomsen, D.J. McComas, The superdense plasma sheet: plasmaspheric origin, solar wind origin, or ionospheric origin? J. Geophys. Res. **102**, 22089–22106 (1997). doi:10.1029/96JA02469
V. Bothmer, R. Schwenn, The structure and origin of magnetic clouds in the solar wind. Ann. Geophys. **16**, 1–24 (1998). doi:10.1007/s00585-997-0001-x
O.J. Brambles, W. Lotko, B. Zhang, J. Ouellette, J. Lyon, M. Wiltberger, The effects of ionospheric outflow on ICME and SIR driven sawtooth events. J. Geophys. Res. Space Phys. **118**, 6026–6041 (2013). doi:10.1002/jgra.50522
D. Burgess, E.A. Lucek, M. Scholer, S.D. Bale, M.A. Balikhin, A. Balogh, T.S. Horbury, V.V. Krasnoselskikh, H. Kucharek, B. Lembège, E. Möbius, S.J. Schwartz, M.F. Thomsen, S.N. Walker, Quasi-parallel shock structure and processes. Space Sci. Rev. **118**, 205–222 (2005). doi:10.1007/s11214-005-3832-3
L.F.E. Burlaga, in *Magnetic Clouds*, ed. by R. Schwenn, E. Marsch (1991), p. 152
L. Burlaga, E. Sittler, F. Mariani, R. Schwenn, Magnetic loop behind an interplanetary shock—Voyager, Helios, and IMP 8 observations. J. Geophys. Res. **86**, 6673–6684 (1981). doi:10.1029/JA086iA08p06673
L.F. Burlaga, L. Klein, N.R. Sheeley Jr., D.J. Michels, R.A. Howard, M.J. Koomen, R. Schwenn, H. Rosenbauer, A magnetic cloud and a coronal mass ejection. Geophys. Res. Lett. **9**, 1317–1320 (1982). doi:10.1029/GL009i012p01317
L.F. Burlaga, K.W. Behannon, L.W. Klein, Compound streams, magnetic clouds, and major geomagnetic storms. J. Geophys. Res. **92**, 5725–5734 (1987). doi:10.1029/JA092iA06p05725
L.F. Burlaga, R.M. Skoug, C.W. Smith, D.F. Webb, T.H. Zurbuchen, A. Reinard, Fast ejecta during the ascending phase of solar cycle 23: ACE observations, 1998–1999. J. Geophys. Res. **106**, 20957–20978 (2001). doi:10.1029/2000JA000214
L.F. Burlaga, S.P. Plunkett, O.C. St. Cyr, Successive CMEs and complex ejecta. J. Geophys. Res. Space Phys. **107**, 1266 (2002). doi:10.1029/2001JA000255
R.K. Burton, R.L. McPherron, C.T. Russell, An empirical relationship between interplanetary conditions and Dst. J. Geophys. Res. **80**, 4204–4214 (1975). doi:10.1029/JA080i031p04204
H.V. Cane, I.G. Richardson, Interplanetary coronal mass ejections in the near-Earth solar wind during 1996-2002. J. Geophys. Res. Space Phys. **108**, 1156 (2003). doi:10.1029/2002JA009817
H.V. Cane, I.G. Richardson, G. Wibberenz, Helios 1 and 2 observations of particle decreases, ejecta, and magnetic clouds. J. Geophys. Res. Space Phys. **102**, 7075–7086 (1997). doi:10.1029/97JA00149
R.C. Carrington, Description of a singular appearance seen in the Sun on September 1, 1859. Mon. Not. R. Astron. Soc. **20**, 13–15 (1859). doi:10.1093/mnras/20.1.13
M.D. Cash, D.A. Biesecker, V. Pizzo, C.A. Koning, G. Millward, C.N. Arge, C.J. Henney, D. Odstrcil, Ensemble modeling of the 23 July 2012 coronal mass ejection. Space Weather **13**, 611–625 (2015). doi:10.1002/2015SW001232
J.K. Chao, R.P. Lepping, A correlative study of ssc's, interplanetary shocks, and solar activity. J. Geophys. Res. **79**, 1799 (1974). doi:10.1029/JA079i013p01799
S. Chapman, V.C.A. Ferraro, The electrical state of solar streams of corpuscles. Mon. Not. R. Astron. Soc. **89**, 470 (1929). doi:10.1093/mnras/89.5.470

E.W. Cliver, L. Svalgaard, The 1859 solar-terrestrial disturbance and the current limits of extreme space weather activity. Sol. Phys. **224**, 407–422 (2004). doi:10.1007/s11207-005-4980-z

G.T. Cocconi, K. Greisen, S. Hayakawa, P. Morrison, The cosmic ray flare effect. Nuovo Cimento **8**, 161–168 (1958)

W.D. Cramer, N.E. Turner, M.-C. Fok, N.Y. Buzulukova, Effects of different geomagnetic storm drivers on the ring current: CRCM results. J. Geophys. Res. Space Phys. **118**, 1062–1073 (2013). doi:10.1002/jgra.50138

S.R. Cranmer, Coronal holes and the high-speed solar wind. Space Sci. Rev. **101**, 229–294 (2002)

N.U. Crooker, Solar and heliospheric geoeffective disturbances. J. Atmos. Sol.-Terr. Phys. **62**, 1071–1085 (2000). doi:10.1016/S1364-6826(00)00098-5

N.U. Crooker, C.-L. Huang, S.M. Lamassa, D.E. Larson, S.W. Kahler, H.E. Spence, Heliospheric plasma sheets. J. Geophys. Res. Space Phys. **109**, 03107 (2004). doi:10.1029/2003JA010170

N.U. Crooker, S.K. Antiochos, X. Zhao, M. Neugebauer, Global network of slow solar wind. J. Geophys. Res. Space Phys. **117**, 04104 (2012). doi:10.1029/2011JA017236

J.J. Curto, T. Araki, L.F. Alberca, Evolution of the concept of sudden storm commencements and their operative identification. Earth Planets Space **59**, i–xii (2007)

I.A. Daglis, J.U. Kozyra, Outstanding issues of ring current dynamics. J. Atmos. Sol.-Terr. Phys. **64**, 253–264 (2002). doi:10.1016/S1364-6826(01)00087-6

I.A. Daglis, R.M. Thorne, W. Baumjohann, S. Orsini, The terrestrial ring current: origin, formation, and decay. Rev. Geophys. **37**, 407–438 (1999). doi:10.1029/1999RG900009

M.H. Denton, J.E. Borovsky, R.M. Skoug, M.F. Thomsen, B. Lavraud, M.G. Henderson, R.L. McPherron, J.C. Zhang, M.W. Liemohn, Geomagnetic storms driven by ICME- and CIR-dominated solar wind. J. Geophys. Res. Space Phys. **111**, A07S07 (2006). doi:10.1029/2005JA011436

A.P. Dimmock, K. Nykyri, H. Karimabadi, A. Osmane, T.I. Pulkkinen, A statistical study into the spatial distribution and dawn-dusk asymmetry of dayside magnetosheath ion temperatures as a function of upstream solar wind conditions. J. Geophys. Res. Space Phys. **120**, 2767–2782 (2015). doi:10.1002/2014JA020734

J.W. Dungey, Interplanetary magnetic field and the auroral zones. Phys. Rev. Lett. **6**, 47–48 (1961). doi:10.1103/PhysRevLett.6.47

E. Echer, W.D. Gonzalez, Geoeffectiveness of interplanetary shocks, magnetic clouds, sector boundary crossings and their combined occurrence. Geophys. Res. Lett. **31**, 09808 (2004). doi:10.1029/2003GL019199

E. Echer, W.D. Gonzalez, B.T. Tsurutani, Interplanetary conditions leading to superintense geomagnetic storms (Dst ≤ -250 nT) during Solar Cycle 23. Geophys. Res. Lett. **35**, L06S03 (2008a). doi:10.1029/2007GL0317

E. Echer, W.D. Gonzalez, B.T. Tsurutani, A.L.C. Gonzalez, Interplanetary conditions causing intense geomagnetic storms (Dst ≤ -100 nT) during Solar Cycle 23 (1996–2006). J. Geophys. Res. **113**, A05221 (2008b). doi:10.1029/2007JA012744

W. Ellis, On the relation between the diurnal range of magnetic declination and horizontal force and the period of solar spot frequency. Proc. R. Soc. Lond. Ser. I **63**, 64–78 (1898)

C. Farrugia, D. Berdichevsky, Evolutionary signatures in complex ejecta and their driven shocks. Ann. Geophys. **22**, 3679–3698 (2004). doi:10.5194/angeo-22-3679-2004

C.J. Farrugia, M.P. Freeman, L.F. Burlaga, R.P. Lepping, K. Takahashi, The Earth's magnetosphere under continued forcing—substorm activity during the passage of an interplanetary magnetic cloud. J. Geophys. Res. **98**, 7657–7671 (1993). doi:10.1029/92JA02351

C.J. Farrugia, V.K. Jordanova, M.F. Thomsen, G. Lu, S.W.H. Cowley, K.W. Ogilvie, A two-ejecta event associated with a two-step geomagnetic storm. J. Geophys. Res. Space Phys. **111**, 11104 (2006). doi:10.1029/2006JA011893

F.R. Fenrich, J.G. Luhmann, Geomagnetic response to magnetic clouds of different polarity. Geophys. Res. Lett. **25**, 2999–3002 (1998). doi:10.1029/98GL51180

S.E. Forbush, On the effects in cosmic-ray intensity observed during the recent magnetic storm. Phys. Rev. **51**, 1108–1109 (1937). doi:10.1103/PhysRev.51.1108.3

J. Geiss, G. Gloeckler, R. von Steiger, Origin of the solar wind from composition data. Space Sci. Rev. **72**, 49–60 (1995). doi:10.1007/BF00768753

T. Gold, Discussion on shock waves and rarefied gas dynamics, in *Gas Dynamics of Cosmic Clouds*. IAU Symposium, vol. 2 (1955), pp. 97–105

T. Gold, Magnetic storms. Space Sci. Rev. **1**, 100–114 (1962). doi:10.1007/BF00174637

H. Goldstein, On the field configuration in magnetic clouds, in *NASA Conference Publication*. NASA Conference Publication, vol. 228 (1983)

W.D. Gonzalez, B.T. Tsurutani, Criteria of interplanetary parameters causing intense magnetic storms (Dst of less than -100 nT). Planet. Space Sci. **35**, 1101–1109 (1987). doi:10.1016/0032-0633(87)90015-8

W.D. Gonzalez, A.L.C. Gonzalez, B.T. Tsurutani, Dual-peak solar cycle distribution of intense geomagnetic storms. Planet. Space Sci. **38**, 181–187 (1990). doi:10.1016/0032-0633(90)90082-2

W.D. Gonzalez, J.A. Joselyn, Y. Kamide, H.W. Kroehl, G. Rostoker, B.T. Tsurutani, V.M. Vasyliunas, What is a geomagnetic storm? J. Geophys. Res. **99**, 5771–5792 (1994). doi:10.1029/93JA02867

W.D. Gonzalez, A.L.C. de Gonzalez, A. Dal Lago, B.T. Tsurutani, J.K. Arballo, G.K. Lakhina, B. Buti, C.M. Ho, S.-T. Wu, Magnetic cloud field intensities and solar wind velocities. Geophys. Res. Lett. **25**, 963–966 (1998)

W.D. Gonzalez, B.T. Tsurutani, A.L. Clúa de Gonzalez, Interplanetary origin of geomagnetic storms. Space Sci. Rev. **88**, 529–562 (1999). doi:10.1023/A:1005160129098

W.D. Gonzalez, E. Echer, B.T. Tsurutani, A.L. Clúa de Gonzalez, A. Dal Lago, Interplanetary origin of intense, superintense and extreme geomagnetic storms. Space Sci. Rev. **158**, 69–89 (2011). doi:10.1007/s11214-010-9715-2

J.A. González-Esparza, A. Balogh, R.J. Forsyth, M. Neugebauer, E.J. Smith, J.L. Phillips, Interplanetary shock waves and large-scale structures: Ulysses' observations in and out of the ecliptic plane. J. Geophys. Res. **101**, 17057–17072 (1996). doi:10.1029/96JA00685

N. Gopalswamy, S. Nunes, S. Yashiro, R.A. Howard, Variability of solar eruptions during cycle 23. Adv. Space Res. **34**, 391–396 (2004). doi:10.1016/j.asr.2003.10.054

N. Gopalswamy, P. Makela, S. Akiyama, S. Yashiro, N. Thakur, CMEs during the two activity peaks in cycle 24 and their space weather consequences. Sun Geosph. **10**, 111–118 (2015)

J.T. Gosling, The solar flare myth. J. Geophys. Res. **98**, 18937–18950 (1993). doi:10.1029/93JA01896

J.T. Gosling, D.J. McComas, Field line draping about fast coronal mass ejecta—a source of strong out-of-the-ecliptic interplanetary magnetic fields. Geophys. Res. Lett. **14**, 355–358 (1987). doi:10.1029/GL014i004p00355

J.T. Gosling, V.J. Pizzo, Formation and evolution of corotating interaction regions and their three dimensional structure. Space Sci. Rev. **89**, 21–52 (1999). doi:10.1023/A:1005291711900

J.T. Gosling, V. Pizzo, S.J. Bame, Anomalously low proton temperatures in the solar wind following interplanetary shock waves—evidence for magnetic bottles? J. Geophys. Res. **78**, 2001 (1973). doi:10.1029/JA078i013p02001

J.T. Gosling, E. Hildner, R.M. MacQueen, R.H. Munro, A.I. Poland, C.L. Ross, Direct observations of a flare related coronal and solar wind disturbance. Sol. Phys. **40**, 439–448 (1975). doi:10.1007/BF00162390

J.T. Gosling, J.R. Asbridge, S.J. Bame, W.C. Feldman, Solar wind stream interfaces. J. Geophys. Res. **83**, 1401–1412 (1978). doi:10.1029/JA083iA04p01401

J.T. Gosling, D.J. McComas, J.L. Phillips, S.J. Bame, Geomagnetic activity associated with Earth passage of interplanetary shock disturbances and coronal mass ejections. J. Geophys. Res. **96**, 7831–7839 (1991). doi:10.1029/91JA00316

W.M.H. Greaves, H.W. Newton, On the recurrence of magnetic storms. Mon. Not. R. Astron. Soc. **89**, 641–646 (1929). doi:10.1093/mnras/89.7.641

J. Guo, X. Feng, J. Zhang, P. Zuo, C. Xiang, Statistical properties and geoefficiency of interplanetary coronal mass ejections and their sheaths during intense geomagnetic storms. J. Geophys. Res. Space Phys. **115**, 09107 (2010). doi:10.1029/2009JA015140

J. Guo, X. Feng, B.A. Emery, J. Zhang, C. Xiang, F. Shen, W. Song, Energy transfer during intense geomagnetic storms driven by interplanetary coronal mass ejections and their sheath regions. J. Geophys. Res. Space Phys. **116**, 05106 (2011). doi:10.1029/2011JA016490

M.R. Hairston, K.A. Drake, R. Skoug, Saturation of the ionospheric polar cap potential during the October-November 2003 superstorms. J. Geophys. Res. Space Phys. **110**, 0926 (2005). doi:10.1029/2004JA010864

R. Hajra, E. Echer, B.T. Tsurutani, W.D. Gonzalez, Solar wind-magnetosphere energy coupling efficiency and partitioning: HILDCAAs and preceding CIR storms during solar cycle 23. J. Geophys. Res. Space Phys. **119**, 2675–2690 (2014). doi:10.1002/2013JA019646

R.A. Harrison, J.A. Davies, C. Möstl, Y. Liu, M. Temmer, M.M. Bisi, J.P. Eastwood, C.A. de Koning, N. Nitta, T. Rollett, C.J. Farrugia, R.J. Forsyth, B.V. Jackson, E.A. Jensen, E.K.J. Kilpua, D. Odstrcil, D.F. Webb, An analysis of the origin and propagation of the multiple coronal mass ejections of 2010 August 1. Astrophys. J. **750**, 45 (2012). doi:10.1088/0004-637X/750/1/45

H. Hasegawa, M. Fujimoto, T.-D. Phan, H. Rème, A. Balogh, M.W. Dunlop, C. Hashimoto, R. TanDokoro, Transport of solar wind into Earth's magnetosphere through rolled-up Kelvin-Helmholtz vortices. Nature **430**, 755–758 (2004). doi:10.1038/nature02799

T. Henke, J. Woch, R. Schwenn, U. Mall, G. Gloeckler, R. von Steiger, R.J. Forsyth, A. Balogh, Ionization state and magnetic topology of coronal mass ejections. J. Geophys. Res. **106**, 10597–10614 (2001). doi:10.1029/2000JA900176

H. Hietala, T.V. Laitinen, K. Andréeová, R. Vainio, A. Vaivads, M. Palmroth, T.I. Pulkkinen, H.E.J. Koskinen, E.A. Lucek, H. Rème, Supermagnetosonic jets behind a collisionless quasiparallel shock. Phys. Rev. Lett. **103**(24), 245001 (2009). doi:10.1103/PhysRevLett.103.245001

J. Hirshberg, A. Alksne, D.S. Colburn, S.J. Bame, A.J. Hundhausen, Observation of a solar flare induced interplanetary shock and helium-enriched driver gas. J. Geophys. Res. **75**, 1 (1970). doi:10.1029/JA075i001p00001

L. Holappa, K. Mursula, T. Asikainen, A new method to estimate annual solar wind parameters and contributions of different solar wind structures to geomagnetic activity. J. Geophys. Res. Space Phys. **119**, 9407–9418 (2014). doi:10.1002/2014JA020599

C.-S. Huang, G. Le, G.D. Reeves, Periodic magnetospheric substorms during fluctuating interplanetary magnetic field B_z. Geophys. Res. Lett. **31**, 14801 (2004). doi:10.1029/2004GL020180

K. Huttunen, H. Koskinen, Importance of post-shock streams and sheath region as drivers of intense magnetospheric storms and high-latitude activity. Ann. Geophys. **22**, 1729–1738 (2004). doi:10.5194/angeo-22-1729-2004

K.E.J. Huttunen, H.E.J. Koskinen, R. Schwenn, Variability of magnetospheric storms driven by different solar wind perturbations. J. Geophys. Res. Space Phys. **107**, 1121 (2002). doi:10.1029/2001JA900171

K.E.J. Huttunen, R. Schwenn, V. Bothmer, H.E.J. Koskinen, Properties and geoeffectiveness of magnetic clouds in the rising, maximum and early declining phases of solar cycle 23. Ann. Geophys. **23**, 625–641 (2005). doi:10.5194/angeo-23-625-2005

L. Jian, C.T. Russell, J.G. Luhmann, R.M. Skoug, Properties of interplanetary coronal mass ejections at one AU during 1995–2004. Sol. Phys. **239**, 393–436 (2006a)

L. Jian, C.T. Russell, J.G. Luhmann, R.M. Skoug, Properties of stream interactions at one AU during 1995–2004. Sol. Phys. **239**, 337–392 (2006b)

G.H. Jones, A. Balogh, Context and heliographic dependence of heliospheric planar magnetic structures. J. Geophys. Res. **105**, 12713–12724 (2000). doi:10.1029/2000JA900003

G.H. Jones, A. Rees, A. Balogh, R.J. Forsyth, The draping of heliospheric magnetic fields upstream of coronal mass ejecta. Geophys. Res. Lett. **29**, 1463 (2002). doi:10.1029/2001GL014110

V.K. Jordanova, H. Matsui, P.A. Puhl-Quinn, M.F. Thomsen, K. Mursula, L. Holappa, Ring current development during high speed streams. J. Atmos. Sol.-Terr. Phys. **71**, 1093–1102 (2009). doi:10.1016/j.jastp.2008.09.043

J.A. Joselyn, B.T. Tsurutani, Geomagnetic sudden impulses and storm sudden commencements—a note on terminology. EOS Trans. **71**, 1808 (1990). doi:10.1029/90EO00350

S. Jurac, J.C. Kasper, J.D. Richardson, A.J. Lazarus, Geomagnetic disturbances and their relationship to interplanetary shock parameters. Geophys. Res. Lett. **29**, 1463 (2002). doi:10.1029/2001GL014034

Y. Kamide, W. Baumjohann, I.A. Daglis, W.D. Gonzalez, M. Grande, J.A. Joselyn, R.L. McPherron, J.L. Phillips, E.G.D. Reeves, G. Rostoker, A.S. Sharma, H.J. Singer, B.T. Tsurutani, V.M. Vasyliunas, Current understanding of magnetic storms: storm-substorm relationships. J. Geophys. Res. **103** 17705–17728 (1998a)

Y. Kamide, J.-H. Shue, X. Li, G. Lu, M.J. Brittnacher, G.K. Parks, G.D. Reeves, Internally and externally triggered substorms: a case study of the January 10, 1997 events, in *Substorms-4*, 1998b

Y. Kamide, N. Yokoyama, W. Gonzalez, B.T. Tsurutani, I.A. Daglis, A. Brekke, S. Masuda, Two-step development of geomagnetic storms. J. Geophys. Res. **103**, 6917–6922 (1998c)

R. Kataoka, S. Watari, N. Shimada, H. Shimazu, K. Marubashi, Downstream structures of interplanetary fast shocks associated with coronal mass ejections. Geophys. Res. Lett. **32**, 12103 (2005). doi:10.1029/2005GL022777

R. Kataoka, D. Shiota, E. Kilpua, K. Keika, Pileup accident hypothesis of magnetic storm on 17 March 2015. Geophys. Res. Lett. **42**, 5155–5161 (2015). doi:10.1002/2015GL064816

E.K.J. Kilpua, J.G. Luhmann, J. Gosling, Y. Li, H. Elliott, C.T. Russell, L. Jian, A.B. Galvin, D. Larson, P. Schroeder, K. Simunac, G. Petrie, Small solar wind transients and their connection to the large-scale coronal structure. Sol. Phys. **256**, 327–344 (2009). doi:10.1007/s11207-009-9366-1

E.K.J. Kilpua, L.K. Jian, Y. Li, J.G. Luhmann, C.T. Russell, Multipoint ICME encounters: pre-STEREO and STEREO observations. J. Atmos. Sol.-Terr. Phys. **73**, 1228–1241 (2011). doi:10.1016/j.jastp.2010.10.012

E.K.J. Kilpua, Y. Li, J.G. Luhmann, L.K. Jian, C.T. Russell, On the relationship between magnetic cloud field polarity and geoeffectiveness. Ann. Geophys. **30**, 1037–1050 (2012). doi:10.5194/angeo-30-1037-2012

E.K.J. Kilpua, H. Hietala, H.E.J. Koskinen, D. Fontaine, L. Turc, Magnetic field and dynamic pressure ULF fluctuations in coronal-mass-ejection-driven sheath regions. Ann. Geophys. **31**, 1559–1567 (2013a)

E.K.J. Kilpua, A. Isavnin, A. Vourlidas, H.E.J. Koskinen, L. Rodriguez, On the relationship between interplanetary coronal mass ejections and magnetic clouds. Ann. Geophys. **31**, 1251–1265 (2013b)

E.K.J. Kilpua, E. Lumme, K. Andreeova, A. Isavnin, H.E.J. Koskinen, Properties and drivers of fast interplanetary shocks near the orbit of the Earth (1995–2013). J. Geophys. Res. **120**, 4112–4125 (2015a)

E.K.J. Kilpua, N. Olspert, A. Grigorievskiy, M.J. Käpylä, E.I. Tanskanen, H. Miyahara, R. Kataoka, J. Pelt, Y.D. Liu, Statistical study of strong and extreme geomagnetic disturbances and solar cycle characteristics. Astrophys. J. **806**, 272 (2015b)
E.K.J. Kilpua, H. Hietala, D.L. Turner, H.E.J. Koskinen, T.I. Pulkkinen, J.V. Rodriguez, G.D. Reeves, S.G. Claudepierre, H.E. Spence, Unraveling the drivers of the storm time radiation belt response. Geophys. Res. Lett. **42**, 3076–3084 (2015c)
E.K.J. Kilpua, H.E.J. Koskinen, T.I. Pulkkinen, Coronal mass ejections and their sheath regions in interplanetary space. Living Rev. Sol. Phys. (2017)
L.W. Klein, L.F. Burlaga, Interplanetary magnetic clouds at 1 AU. J. Geophys. Res. **87**, 613–624 (1982). doi:10.1029/JA087iA02p00613
A.J. Klimas, D.N. Baker, D.A. Roberts, D.H. Fairfield, J. Buechner, A nonlinear dynamical analogue model of geomagnetic activity. J. Geophys. Res. **97**, 12 (1992). doi:10.1029/92JA00794
D.J. Knipp, W.K. Tobiska, B.A. Emery, Direct and indirect thermospheric heating sources for solar cycles 21–23. Sol. Phys. **224**, 495–505 (2004). doi:10.1007/s11207-005-6393-4
H.E.J. Koskinen, *Physics of Space Storms: From the Solar Surface to the Earth* (Springer, Berlin, 2011)
H.E.J. Koskinen, E.I. Tanskanen, Magnetospheric energy budget and the epsilon parameter. J. Geophys. Res. Space Phys. **107**, 1415 (2002). doi:10.1029/2002JA009283
J.U. Kozyra, M.W. Liemohn, C.R. Clauer, A.J. Ridley, M.F. Thomsen, J.E. Borovsky, J.L. Roeder, V.K. Jordanova, W.D. Gonzalez, Multistep Dst development and ring current composition changes during the 4–6 June 1991 magnetic storm. J. Geophys. Res. Space Phys. **107**, 1224 (2002). doi:10.1029/2001JA000023
A.S. Krieger, A.F. Timothy, E.C. Roelof, A coronal hole and its identification as the source of a high velocity solar wind stream. Sol. Phys. **29**, 505–525 (1973). doi:10.1007/BF00150828
B. Lavraud, J.E. Borovsky, Altered solar wind-magnetosphere interaction at low Mach numbers: coronal mass ejections. J. Geophys. Res. Space Phys. **113**, 0008 (2008). doi:10.1029/2008JA013192
B. Lavraud, M.F. Thomsen, J.E. Borovsky, M.H. Denton, T.I. Pulkkinen, Magnetosphere preconditioning under northward IMF: evidence from the study of coronal mass ejection and corotating interaction region geoeffectiveness. J. Geophys. Res. Space Phys. **111**, 09208 (2006). doi:10.1029/2005JA011566
B. Lavraud, A. Ruffenach, A.P. Rouillard, P. Kajdic, W.B. Manchester, N. Lugaz, Geo-effectiveness and radial dependence of magnetic cloud erosion by magnetic reconnection. J. Geophys. Res. Space Phys. **119**, 26–35 (2014). doi:10.1002/2013JA019154
G.-M. Le, Z.-Y. Cai, H.-N. Wang, Z.-Q. Yin, P. Li, Solar cycle distribution of major geomagnetic storms. Res. Astron. Astrophys. **13**, 739–748 (2013). doi:10.1088/1674-4527/13/6/013
R.P. Lepping, L.F. Burlaga, J.A. Jones, Magnetic field structure of interplanetary magnetic clouds at 1 AU. J. Geophys. Res. **95**, 11957–11965 (1990). doi:10.1029/JA095iA08p11957
R.P. Lepping, L.F. Burlaga, A. Szabo, K.W. Ogilvie, W.H. Mish, D. Vassiliadis, A.J. Lazarus, J.T. Steinberg, C.J. Farrugia, L. Janoo, F. Mariani, The wind magnetic cloud and events of October 18-20, 1995: interplanetary properties and as triggers for geomagnetic activity. J. Geophys. Res. **102**, 14049–14064 (1997). doi:10.1029/97JA00272
Y. Li, J.G. Luhmann, B.J. Lynch, E.K.J. Kilpua, Cyclic reversal of magnetic cloud poloidal field. Sol. Phys. **270**, 331–346 (2011)
M.W. Liemohn, J.U. Kozyra, M.F. Thomsen, J.L. Roeder, G. Lu, J.E. Borovsky, T.E. Cayton, Dominant role of the asymmetric ring current in producing the stormtime Dst. J. Geophys. Res. **106**, 10883–10904 (2001). doi:10.1029/2000JA000326
M.W. Liemohn, M. Jazowski, J.U. Kozyra, N. Ganushkina, M.F. Thomsen, J.E. Borovsky, CIR versus CME drivers of the ring current during intense magnetic storms. Proc. R. Soc. Lond. Ser. A **466**, 3305–3328 (2010). doi:10.1098/rspa.2010.0075
F.A. Lindeman, Note on the theory of magnetic storms. Philos. Mag. **38**, 669–684 (1911)
Y. Liu, J.D. Richardson, J.W. Belcher, A statistical study of the properties of interplanetary coronal mass ejections from 0.3 to 5.4 AU. Planet. Space Sci. **53**, 3–17 (2005). doi:10.1016/j.pss.2004.09.023
Y. Liu, J.D. Richardson, J.W. Belcher, J.C. Kasper, R.M. Skoug, Plasma depletion and mirror waves ahead of interplanetary coronal mass ejections. J. Geophys. Res. Space Phys. **111**, 09108 (2006). doi:10.1029/2006JA011723
Y. Liu, J.G. Luhmann, R. Müller-Mellin, P.C. Schroeder, L. Wang, R.P. Lin, S.D. Bale, Y. Li, M.H. Acuña, J.-A. Sauvaud, A comprehensive view of the 2006 December 13 CME: from the Sun to interplanetary space. Astrophys. J. **689**, 563–571 (2008a). doi:10.1086/592031
Y. Liu, W.B. Manchester, J.D. Richardson, J.G. Luhmann, R.P. Lin, S.D. Bale, Deflection flows ahead of ICMEs as an indicator of curvature and geoeffectiveness. J. Geophys. Res. **113**, A00B03 (2008b)
Y.D. Liu, J.G. Luhmann, C. Möstl, J.C. Martinez-Oliveros, S.D. Bale, R.P. Lin, R.A. Harrison, M. Temmer, D.F. Webb, D. Odstrcil, Interactions between coronal mass ejections viewed in coordinated imaging and in situ observations. Astrophys. J. Lett. **746**, 15 (2012). doi:10.1088/2041-8205/746/2/L15

Y.D. Liu, J.G. Luhmann, N. Lugaz, C. Möstl, J.A. Davies, S.D. Bale, R.P. Lin, On Sun-to-Earth propagation of coronal mass ejections. Astrophys. J. **769**, 45 (2013). doi:10.1088/0004-637X/769/1/45

Y.D. Liu, J.G. Luhmann, P. Kajdič, E.K.J. Kilpua, N. Lugaz, N.V. Nitta, C. Möstl, B. Lavraud, S.D. Bale, C.J. Farrugia, A.B. Galvin, Observations of an extreme storm in interplanetary space caused by successive coronal mass ejections. Nat. Commun. **5**, 3481 (2014a)

Y.D. Liu, Z. Yang, R. Wang, J.G. Luhmann, J.D. Richardson, N. Lugaz, Sun-to-Earth characteristics of two coronal mass ejections interacting near 1 AU: formation of a complex ejecta and generation of a two-step geomagnetic storm. Astrophys. J. Lett. **793**, L41 2014b)

Y.D. Liu, H. Hu, R. Wang, Z. Yang, B. Zhu, Y.A. Liu, J.G. Luhmann, J.D. Richardson, Plasma and magnetic field characteristics of solar coronal mass ejections in relation to geomagnetic storm intensity and variability. Astrophys. J. Lett. **809**, 34 (2015). doi:10.1088/2041-8205/809/2/L34

Y.D. Liu, H. Hu, C. Wang, J.G. Luhmann, J.D. Richardson, Z. Yang, R. Wang, On sun-to-Earth propagation of coronal mass ejections: II. Slow events and comparison with others. Astrophys. J. Suppl. Ser. **222**, 23 (2016). doi:10.3847/0067-0049/222/2/23

M. Lockwood, M.J. Owens, L.A. Barnard, S. Bentley, C.J. Scott, C.E. Watt, On the origins and timescales of geoeffective IMF. Space Weather **14** 406–432 (2015). doi:10.1002/2016SW001375

R.E. Lopez, M. Wiltberger, S. Hernandez, J.G. Lyon, Solar wind density control of energy transfer to the magnetosphere. Geophys. Res. Lett. **31**, 08804 (2004). doi:10.1029/2003GL018780

R.E. Lopez, R. Bruntz, E.J. Mitchell, M. Wiltberger, J.G. Lyon, V.G. Merkin, Role of magnetosheath force balance in regulating the dayside reconnection potential. J. Geophys. Res. Space Phys. **115**, 12216 (2010). doi:10.1029/2009JA014597

E.A. Lucek, A. Balogh, The identification and characterization of Alfvénic fluctuations in ULYSSES data at midlatitudes. Astrophys. J. **507**, 984–990 (1998). doi:10.1086/306372

E.A. Lucek, T.S. Horbury, I. Dandouras, H. RèMe, Cluster observations of the Earth's quasi-parallel bow shock. J. Geophys. Res. Space Phys. **113**, 0702 (2008). doi:10.1029/2007JA012756

N. Lugaz, C.J. Farrugia, A new class of complex ejecta resulting from the interaction of two CMEs and its expected geoeffectiveness. Geophys. Res. Lett. **41**, 769–776 (2014). doi:10.1002/2013GL058789

N. Lugaz, C.J. Farrugia, C.W. Smith, K. Paulson, Shocks inside CMEs: a survey of properties from 1997 to 2006. J. Geophys. Res. Space Phys. **120**, 2409–2427 (2015). doi:10.1002/2014JA020848

N. Lugaz, C.J. Farrugia, C.-L. Huang, R.M. Winslow, H.E. Spence, N.A. Schwadron, Earth's magnetosphere and outer radiation belt under sub-Alfvénic solar wind. Nat. Commun. **7**, 13001 (2016)

L.R. Lyons, G.T. Blanchard, J.C. Samson, R.P. Lepping, T. Yamamoto, T. Moretto, Coordinated observations demonstrating external substorm triggering. J. Geophys. Res. **102**, 27039–27052 (1997). doi:10.1029/97JA02639

W. Manchester, E.K.J. Kilpua, Y.D. Liu, N. Lugaz, P. Riley, T. Török, B. Vršnak, The physical processes of CME/ICME evolution. Space Sci. Rev. (2017, this issue). doi:10.1007/s11214-017-0394-0. http://adsabs.harvard.edu/abs/2017SSRv..tmp...90M. Provided by the SAO/NASA Astrophysics Data System

E.W. Maunder, Magnetic disturbance and associated sun-spots. Mon. Not. R. Astron. Soc. **65**, 2 (1904). doi:10.1093/mnras/65.1.2

P.N. Mayaud, *Derivation, Meaning, and Use of Geomagnetic Indices*. Geophysical Monograph, vol. 22 (American Geophysical Union, Washington, 1980)

D.J. McComas, J.T. Gosling, S.J. Bame, E.J. Smith, H.V. Cane, A test of magnetic field draping induced BZ perturbations ahead of fast coronal mass ejecta. J. Geophys. Res. **94**, 1465–1471 (1989). doi:10.1029/JA094iA02p01465

R.L. McPherron, T. Terasawa, A. Nishida, Solar wind triggering of substorm expansion onset. J. Geomagn. Geoelectr. **38**, 1089–1108 (1986)

M.B. Moldwin, S. Ford, R. Lepping, J. Slavin, A. Szabo, Small-scale magnetic flux ropes in the solar wind. Geophys. Res. Lett. **27**, 57–60 (2000). doi:10.1029/1999GL010724

S.K. Morley, M.P. Freeman, On the association between northward turnings of the interplanetary magnetic field and substorm onsets. Geophys. Res. Lett. **34**, 08104 (2007). doi:10.1029/2006GL028891

P. Morrison, Solar origin of cosmic-ray time variations. Phys. Rev. **101**, 1397–1404 (1956). doi:10.1103/PhysRev.101.1397

C. Möstl, C.J. Farrugia, E.K.J. Kilpua, L.K. Jian, Y. Liu, J.P. Eastwood, R.A. Harrison, D.F. Webb, M. Temmer, D. Odstrcil, J.A. Davies, T. Rollett, J.G. Luhmann, N. Nitta, T. Mulligan, E.A. Jensen, R. Forsyth, B. Lavraud, C.A. de Koning, A.M. Veronig, A.B. Galvin, T.L. Zhang, B.J. Anderson, Multi-point shock and flux rope analysis of multiple interplanetary coronal mass ejections around 2010 August 1 in the inner heliosphere. Astrophys. J. **758**, 10 (2012). doi:10.1088/0004-637X/758/1/10

T. Mulligan, C.T. Russell, J.G. Luhmann, Solar cycle evolution of the structure of magnetic clouds in the inner heliosphere. Geophys. Res. Lett. **25**, 2959–2962 (1998). doi:10.1029/98GL01302

M. Myllys, N. Partamies, L. Juusola, Latitude dependence of long-term geomagnetic activity and its solar wind drivers. Ann. Geophys. **33**, 573–581 (2015). doi:10.5194/angeo-33-573-2015
M. Myllys, E.K.J. Kilpua, B. Lavraud, T.I. Pulkkinen, Solar wind-magnetosphere coupling efficiency during ejecta and sheath-driven geomagnetic storms. J. Geophys. Res. Space Phys. **121**, 4378–4396 (2016). doi:10.1002/2016JA022407
M. Myllys, E.K.J. Kipua, B. Lavraud, Interplay of solar wind parameters and physical mechanisms producing the saturation of the cross polar cap potential. Geophys. Res. Lett. **44**, 3019–3027 (2017). doi:10.1002/2017GL072676
P.T. Newell, K. Liou, Solar wind driving and substorm triggering. J. Geophys. Res. Space Phys. **116**, 03229 (2011). doi:10.1029/2010JA016139
P.T. Newell, T. Sotirelis, K. Liou, C.-I. Meng, F.J. Rich, A nearly universal solar wind-magnetosphere coupling function inferred from 10 magnetospheric state variables. J. Geophys. Res. Space Phys. **112**, 01206 (2007). doi:10.1029/2006JA012015
H.W. Newton, A.S. Milsom, The distribution of great and small geomagnetic storms in the sunspot cycle. J. Geophys. Res. **59**, 203–214 (1954). doi:10.1029/JZ059i002p00203
N.S. Nikolaeva, Y.I. Yermolaev, I.G. Lodkina, Dependence of geomagnetic activity during magnetic storms on the solar wind parameters for different types of streams. Geomagn. Aeron. **51**, 49–65 (2011). doi:10.1134/S0016793211010099
J.T. Nolte, A.S. Krieger, A.F. Timothy, R.E. Gold, E.C. Roelof, G. Vaiana, A.J. Lazarus, J.D. Sullivan, P.S. McIntosh, Coronal holes as sources of solar wind. Sol. Phys. **46**, 303–322 (1976). doi:10.1007/BF00149859
T.P. O'Brien, R.L. McPherron, An empirical phase space analysis of ring current dynamics: solar wind control of injection and decay. J. Geophys. Res. **105**, 7707–7720 (2000). doi:10.1029/1998JA000437
D.M. Oliveira, J. Raeder, Impact angle control of interplanetary shock geoeffectiveness. J. Geophys. Res. Space Phys. **119**, 8188–8201 (2014). doi:10.1002/2014JA020275
D.M. Oliveira, J. Raeder, Impact angle control of interplanetary shock geoeffectiveness: a statistical study. J. Geophys. Res. Space Phys. **120**, 4313–4323 (2015). doi:10.1002/2015JA021147
N. Omidi, H. Zhang, D. Sibeck, D. Turner, Spontaneous hot flow anomalies at quasi-parallel shocks: 2. Hybrid simulations. J. Geophys. Res. Space Phys. **118**, 173–180 (2013). doi:10.1029/2012JA018099
A. Osmane, A.P. Dimmock, R. Naderpour, T.I. Pulkkinen, K. Nykyri, The impact of solar wind ULF B_z fluctuations on geomagnetic activity for viscous timescales during strongly northward and southward IMF. J. Geophys. Res. Space Phys. **120**, 9307–9322 (2015). doi:10.1002/2015JA021505
M.J. Owens, P.J. Cargill, C. Pagel, G.L. Siscoe, N.U. Crooker, Characteristic magnetic field and speed properties of interplanetary coronal mass ejections and their sheath regions. J. Geophys. Res. Space Phys. **110**, 01105 (2005). doi:10.1029/2004JA010814
E. Palmerio, E.K.J. Kilpua, N.P. Savani, Planar magnetic structures in coronal mass ejection-driven sheath regions. Ann. Geophys. **34**, 313–322 (2016). doi:10.5194/angeo-34-313-2016
M. Palmroth, T.I. Pulkkinen, P. Janhunen, D.J. McComas, C.W. Smith, H.E.J. Koskinen, Role of solar wind dynamic pressure in driving ionospheric Joule heating. J. Geophys. Res. Space Phys. **109**, 11302 (2004). doi:10.1029/2004JA010529
M. Palmroth, H.E.J. Koskinen, T.I. Pulkkinen, P.K. Toivanen, P. Janhunen, S.E. Milan, M. Lester, Magnetospheric feedback in solar wind energy transfer. J. Geophys. Res. Space Phys. **115**, 0010 (2010). doi:10.1029/2010JA015746
W. Park, J. Lee, Y. Yi, N. Ssessanga, S. Oh, Storm sudden commencements without interplanetary shocks. J. Astron. Space Sci. **32**, 181–187 (2015). doi:10.5140/JASS.2015.32.3.181
J.H. Piddington, Interplanetary magnetic field and its control of cosmic-ray variations. Phys. Rev. **112**, 589–596 (1958). doi:10.1103/PhysRev.112.589
T. Pulkkinen, Space weather: terrestrial perspective. Living Rev. Sol. Phys. **4**, 1 (2007). doi:10.12942/lrsp-2007-1
T.I. Pulkkinen, A.P. Dimmock, A. Osmane, K. Nykyri, Solar wind energy input to the magnetosheath and at the magnetopause. Geophys. Res. Lett. **42**, 4723–4730 (2015). doi:10.1002/2015GL064226
J. Raeder, G. Lu, Polar cap potential saturation during large geomagnetic storms. Adv. Space Res. **36**, 1804–1808 (2005). doi:10.1016/j.asr.2004.05.010
P.H. Reiff, R.W. Spiro, T.W. Hill, Dependence of polar cap potential drop on interplanetary parameters. J. Geophys. Res. **86**, 7639–7648 (1981). doi:10.1029/JA086iA09p07639
I.G. Richardson, H.V. Cane, Near-Earth interplanetary coronal mass ejections during solar cycle 23 (1996–2009): catalog and summary of properties. Sol. Phys. **264**, 189–237 (2010). doi:10.1007/s11207-010-9568-6
I.G. Richardson, H.V. Cane, Solar wind drivers of geomagnetic storms during more than four solar cycles. J. Space Weather Space Clim. **2**(27), 01 (2012). doi:10.1051/swsc/2012001

I.G. Richardson, D.F. Webb, J. Zhang, D.B. Berdichevsky, D.A. Biesecker, J.C. Kasper, R. Kataoka, J.T. Steinberg, B.J. Thompson, C.-C. Wu, A.N. Zhukov, Major geomagnetic storms (Dst ≤ -100 nT) generated by corotating interaction regions. J. Geophys. Res. Space Phys. **111**, 0709 (2006). doi:10.1029/2005JA011476

P. Riley, R.M. Caplan, J. Giacalone, D. Lario, Y. Liu, Properties of the fast forward shock driven by the July 23 2012 extreme coronal mass ejection. Astrophys. J. **819**, 57 (2016). doi:10.3847/0004-637X/819/1/57

A.P. Rouillard, N.R. Sheeley Jr., T.J. Cooper, J.A. Davies, B. Lavraud, E.K.J. Kilpua, R.M. Skoug, J.T. Steinberg, A. Szabo, A. Opitz, J.-A. Sauvaud, The solar origin of small interplanetary transients. Astrophys. J. **734**, 7 (2011). doi:10.1088/0004-637X/734/1/7

A. Ruffenach, B. Lavraud, C.J. Farrugia, P. Démoulin, S. Dasso, M.J. Owens, J.-A. Sauvaud, A.P. Rouillard, A. Lynnyk, C. Foullon, N.P. Savani, J.G. Luhmann, A.B. Galvin, Statistical study of magnetic cloud erosion by magnetic reconnection. J. Geophys. Res. Space Phys. **120**, 43–60 (2015). doi:10.1002/2014JA020628

C.T. Russell, R.L. McPherron, Semiannual variation of geomagnetic activity. J. Geophys. Res. **78**, 92 (1973). doi:10.1029/JA078i001p00092

C.T. Russell, L.K. Jian, X. Blanco-Cano, J.G. Luhmann, STEREO observations of upstream and downstream waves at low Mach number shocks. Geophys. Res. Lett. **36**, 03106 (2009). doi:10.1029/2008GL036991

C.T. Russell, R.A. Mewaldt, J.G. Luhmann, G.M. Mason, T.T. von Rosenvinge, C.M.S. Cohen, R.A. Leske, R. Gomez-Herrero, A. Klassen, A.B. Galvin, K.D.C. Simunac, The very unusual interplanetary coronal mass ejection of 2012 July 23: a blast wave mediated by solar energetic particles. Astrophys. J. **770**, 38 (2013). doi:10.1088/0004-637X/770/1/38

T. Saemundsson, Statistics of geomagnetic storms and solar activity. Mon. Not. R. Astron. Soc. **123**, 299 (1962). doi:10.1093/mnras/123.4.299

M.D. Salas, The curious events leading to the theory of shock waves. Shock Waves **16**, 477–487 (2007). doi:10.1007/s00193-007-0084-z

A.A. Samsonov, D.G. Sibeck, J. Imber, MHD simulation for the interaction of an interplanetary shock with the Earth's magnetosphere. J. Geophys. Res. Space Phys. **112**, 12220 (2007). doi:10.1029/2007JA012627

A.A. Samsonov, V.A. Sergeev, M.M. Kuznetsova, D.G. Sibeck, Asymmetric magnetospheric compressions and expansions in response to impact of inclined interplanetary shock. Geophys. Res. Lett. **42**, 4716–4722 (2015). doi:10.1002/2015GL064294

R. Schwenn, Direct correlations between coronal transients and interplanetary disturbances. Space Sci. Rev. **34**, 85–99 (1983). doi:10.1007/BF00221199

R. Schwenn, A. dal Lago, E. Huttunen, W.D. Gonzalez, The association of coronal mass ejections with their effects near the Earth. Ann. Geophys. **23**, 1033–1059 (2005). doi:10.5194/angeo-23-1033-2005

V.A. Sergeev, R.J. Pellinen, T.I. Pulkkinen, Steady magnetospheric convection: a review of recent results. Space Sci. Rev. **75**, 551–604 (1996). doi:10.1007/BF00833344

N.R. Sheeley Jr., A.P. Rouillard, Tracking streamer blobs into the heliosphere. Astrophys. J. **715**, 300–309 (2010). doi:10.1088/0004-637X/715/1/300

N.R. Sheeley Jr., R.A. Howard, D.J. Michels, M.J. Koomen, R. Schwenn, K.H. Muehlhaeuser, H. Rosenbauer, Coronal mass ejections and interplanetary shocks. J. Geophys. Res. **90**, 163–175 (1985). doi:10.1029/JA090iA01p00163

N.R. Sheeley, Y.-M. Wang, S.H. Hawley, G.E. Brueckner, K.P. Dere, R.A. Howard, M.J. Koomen, C.M. Korendyke, D.J. Michels, S.E. Paswaters, D.G. Socker, O.C. St. Cyr, D. Wang, P.L. Lamy, A. Llebaria, R. Schwenn, G.M. Simnett, S. Plunkett, D.A. Biesecker, Measurements of flow speeds in the corona between 2 and 30 $R_{\odot}$. Astrophys. J. **484**, 472–478 (1997)

J.-H. Shue, P. Song, C.T. Russell, J.T. Steinberg, J.K. Chao, G. Zastenker, O.L. Vaisberg, S. Kokubun, H.J. Singer, T.R. Detman, H. Kawano, Magnetopause location under extreme solar wind conditions. J. Geophys. Res. **103**, 17691–17700 (1998). doi:10.1029/98JA01103

G. Siscoe, D. Odstrcil, Ways in which ICME sheaths differ from magnetosheaths. J. Geophys. Res. Space Phys. **113**, 0007 (2008). doi:10.1029/2008JA013142

G. Siscoe, P.J. MacNeice, D. Odstrcil, East-West asymmetry in coronal mass ejection geoeffectiveness. Space Weather **5**, 04002 (2007). doi:10.1029/2006SW000286

E.J. Smith, J.A. Slavin, R.D. Zwickl, S.J. Bame, Shocks and storm sudden commencements, in *Solar Wind Magnetosphere Coupling*, ed. by Y. Kamide, J.A. Slavin. Astrophysics and Space Science Library, vol. 126 (1986), pp. 345–365

K. Snekvik, E.I. Tanskanen, E.K.J. Kilpua, An automated identification method for Alfvénic streams and their geoeffectiveness. J. Geophys. Res. Space Phys. **118**, 5986–5998 (2013). doi:10.1002/jgra.50588

C.P. Sonett, D.S. Colburn, L. Davis, E.J. Smith, P.J. Coleman, Evidence for a collision-free magnetohydrodynamic shock in interplanetary space. Phys. Rev. Lett. **13**, 153–156 (1964). doi:10.1103/PhysRevLett.13.153

P. Subramanian, K.P. Dere, Source regions of coronal mass ejections. Astrophys. J. **561**, 372–395 (2001). doi:10.1086/323213

E.I. Tanskanen, J.A. Slavin, A.J. Tanskanen, A. Viljanen, T.I. Pulkkinen, H.E.J. Koskinen, A. Pulkkinen, J. Eastwood, Magnetospheric substorms are strongly modulated by interplanetary high-speed streams. Geophys. Res. Lett. **32**, 16104 (2005). doi:10.1029/2005GL023318

M. Temmer, N.V. Nitta, Interplanetary propagation behavior of the fast coronal mass ejection on 23 July 2012. Sol. Phys. **290**, 919–932 (2015). doi:10.1007/s11207-014-0642-3

T. Terasawa, M. Fujimoto, T. Mukai, I. Shinohara, Y. Saito, T. Yamamoto, S. Machida, S. Kokubun, A.J. Lazarus, J.T. Steinberg, R.P. Lepping, Solar wind control of density and temperature in the near-Earth plasma sheet: WIND/GEOTAIL collaboration. Geophys. Res. Lett. **24**, 935–938 (1997). doi:10.1029/96GL04018

K. Toman, Identification of M-regions on the Sun. Nature **181**, 641–642 (1958). doi:10.1038/181641a0

S. Tomczyk, S.W. McIntosh, Time-distance seismology of the solar corona with CoMP. Astrophys. J. **697**, 1384–1391 (2009). doi:10.1088/0004-637X/697/2/1384

R. Tousey, The solar corona, in *Space Research Conference*, ed. by M.J. Rycroft, S.K. Runcorn. Space Research Conference, vol. 2 (1973), pp. 713–730

R.A. Treumann, Fundamentals of collisionless shocks for astrophysical application, 1. Non-relativistic shocks. Astron. Astrophys. Rev. **17**, 409–535 (2009). doi:10.1007/s00159-009-0024-2

B.T. Tsurutani, W.D. Gonzalez, The cause of high-intensity long-duration continuous AE activity (HILDCAAS)—interplanetary Alfven wave trains. Planet. Space Sci. **35**, 405–412 (1987). doi:10.1016/0032-0633(87)90097-3

B.T. Tsurutani, E.J. Smith, W.D. Gonzalez, F. Tang, S.I. Akasofu, Origin of interplanetary southward magnetic fields responsible for major magnetic storms near solar maximum (1978–1979). J. Geophys. Res. **93**, 8519–8531 (1988). doi:10.1029/JA093iA08p08519

B.T. Tsurutani, Y.T. Lee, W.D. Gonzalez, F. Tang, Great magnetic storms. Geophys. Res. Lett. **19**, 73–76 (1992). doi:10.1029/91GL02783

B.T. Tsurutani, W.D. Gonzalez, A.L.C. Gonzalez, F. Tang, J.K. Arballo, M. Okada, Interplanetary origin of geomagnetic activity in the declining phase of the solar cycle. J. Geophys. Res. **100**, 21717–21734 (1995). doi:10.1029/95JA01476

B.T. Tsurutani, X.-Y. Zhou, W.D. Gonzalez, A lack of substorm expansion phases during magnetic storms induced by magnetic clouds, in *Disturbances in Geospace: The Storm-Substorm Relationship*, ed. by A. Surjalal Sharma, Y. Kamide, G.S. Lakhina. Washington DC American Geophysical Union Geophysical Monograph Series, vol. 142 (2003), p. 23. doi:10.1029/142GM03

B.T. Tsurutani, W.D. Gonzalez, X.-Y. Zhou, R.P. Lepping, V. Bothmer, Properties of slow magnetic clouds. J. Atmos. Sol.-Terr. Phys. **66**, 147–151 (2004). doi:10.1016/j.jastp.2003.09.007

B.T. Tsurutani, W.D. Gonzalez, A.L.C. Gonzalez, F.L. Guarnieri, N. Gopalswamy, M. Grande, Y. Kamide, Y. Kasahara, G. Lu, I. Mann, R. McPherron, F. Soraas, V. Vasyliunas, Corotating solar wind streams and recurrent geomagnetic activity: a review. J. Geophys. Res. Space Phys. **111**, 0701 (2006). doi:10.1029/2005JA011273

B.T. Tsurutani, R. Hajra, E. Echer, J.W. Gjerloev, Extremely intense (SML ≤ -2500 nT) substorms: isolated events that are externally triggered? Ann. Geophys. **33**, 519–524 (2015). doi:10.5194/angeo-33-519-2015

L. Turc, D. Fontaine, P. Savoini, R. Modolo, 3D hybrid simulations of the interaction of a magnetic cloud with a bow shock. J. Geophys. Res. Space Phys. **120**, 6133–6151 (2015). doi:10.1002/2015JA021318

L. Turc, C.P. Escoubet, D. Fontaine, E.K.J. Kilpua, S. Enestam, Cone angle control of the interaction of magnetic clouds with the Earth's bow shock. Geophys. Res. Lett. **43**, 4781–4789 (2016). doi:10.1002/2016GL068818

N.E. Turner, E.J. Mitchell, D.J. Knipp, B.A. Emery, Energetics of magnetic storms driven by corotating interaction regions: a study of geoeffectiveness, in *Recurrent Magnetic Storms: Corotating Solar Wind*, ed. by R. McPherron, W. Gonzalez, G. Lu, H.A. José, S. Natchimuthukonar Gopalswamy. Geophysical Monograph Series, vol. 167 (2006, American Geophysical Union, Washington), p. 113. doi:10.1029/167GM11

N.E. Turner, W.D. Cramer, S.K. Earles, B.A. Emery, Geoefficiency and energy partitioning in CIR-driven and CME-driven storms. J. Atmos. Sol.-Terr. Phys. **71**, 1023–1031 (2009). doi:10.1016/j.jastp.2009.02.005

M. Vandas, S. Fischer, M. Dryer, Z. Smith, T. Detman, A. Geranios, MHD simulation of an interaction of a shock wave with a magnetic cloud. J. Geophys. Res. **102**, 22295–22300 (1997). doi:10.1029/97JA01675

V.M. Vasyliunas, Theoretical models of magnetic field line merging. I. Rev. Geophys. Space Phys. **13**, 303–336 (1975). doi:10.1029/RG013i001p00303

R. von Steiger, Space physics—grand challenges for the 21st century. Front. Phys. **1**, 6 (2013). doi:10.3389/fphy.2013.00006

R. von Steiger, T.H. Zurbuchen, A. Kilchenmann, Latitude distribution of interplanetary coronal mass ejections during solar maximum, in *Solar Wind 11/SOHO 16, Connecting Sun and Heliosphere*, ed. by B. Fleck, T.H. Zurbuchen, H. Lacoste. ESA Special Publication, vol. 592 (2005), p. 317

M.-T. Walach, S.E. Milan, Are steady magnetospheric convection events prolonged substorms? J. Geophys. Res. Space Phys. **120**, 1751–1758 (2015). doi:10.1002/2014JA020631

B.M. Walsh, D.G. Sibeck, Y. Wang, D.H. Fairfield, Dawn-dusk asymmetries in the Earth's magnetosheath. J. Geophys. Res. Space Phys. **117**, 12211 (2012). doi:10.1029/2012JA018240

Y.M. Wang, P.Z. Ye, S. Wang, G.P. Zhou, J.X. Wang, A statistical study on the geoeffectiveness of Earth-directed coronal mass ejections from March 1997 to December 2000. J. Geophys. Res. Space Phys. **107**, 1340 (2002). doi:10.1029/2002JA009244

Y.M. Wang, P.Z. Ye, S. Wang, Multiple magnetic clouds: several examples during March–April 2001. J. Geophys. Res. Space Phys. **108**, 1370 (2003). doi:10.1029/2003JA009850

Y.-M. Wang, N.R. Sheeley Jr., N.B. Rich, Coronal pseudostreamers. Astrophys. J. **658**, 1340–1348 (2007). doi:10.1086/511416

C. Wang, J.B. Liu, H. Li, Z.H. Huang, J.D. Richardson, J.R. Kan, Geospace magnetic field responses to interplanetary shocks. J. Geophys. Res. Space Phys. **114**, 05211 (2009). doi:10.1029/2008JA013794

Y. Wang, Q. Zhang, J. Liu, C. Shen, F. Shen, Z. Yang, T. Zic, B. Vrsnak, D.F. Webb, R. Liu, S. Wang, J. Zhang, Q. Hu, B. Zhuang, On the propagation of a geoeffective coronal mass ejection during 15–17 March 2015. J. Geophys. Res. Space Phys. **121**, 7423–7434 (2016). doi:10.1002/2016JA022924

D.F. Webb, T.A. Howard, Coronal mass ejections: observations. Living Rev. Sol. Phys. **9**, 3 (2012). doi:10.1007/lrsp-2012-3. http://www.livingreviews.org/lrsp-2012-3

D.F. Webb, E.W. Cliver, N.U. Crooker, O.C.S. Cry, B.J. Thompson, Relationship of halo coronal mass ejections, magnetic clouds, and magnetic storms. J. Geophys. Res. **105**, 7491–7508 (2000). doi:10.1029/1999JA000275

D.F. Webb, C. Möstl, B.V. Jackson, M.M. Bisi, T.A. Howard, T. Mulligan, E.A. Jensen, L.K. Jian, J.A. Davies, C.A. de Koning, Y. Liu, M. Temmer, J.M. Clover, C.J. Farrugia, R.A. Harrison, N. Nitta, D. Odstrcil, S.J. Tappin, H.-S. Yu, Heliospheric imaging of 3D density structures during the multiple coronal mass ejections of late July to early August 2010. Sol. Phys. **285**, 317–348 (2013). doi:10.1007/s11207-013-0260-5

R.F. Wimmer-Schweingruber, R. von Steiger, R. Paerli, Solar wind stream interfaces in corotating interaction regions: SWICS/Ulysses results. J. Geophys. Res. **102**, 17407–17418 (1997). doi:10.1029/97JA00951

R.F. Wimmer-Schweingruber, N.U. Crooker, A. Balogh, V. Bothmer, R.J. Forsyth, P. Gazis, J.T. Gosling, T. Horbury, A. Kilchenmann, I.G. Richardson, J.D. Richardson, P. Riley, L. Rodriguez, R.V. Steiger, P. Wurz, T.H. Zurbuchen, Understanding interplanetary coronal mass ejection signatures. Space Sci. Rev. **123**, 177–216 (2006). doi:10.1007/s11214-006-9017-x. Report of Working Group B

Y.Q. Xie, P.B. Zuo, X.S. Feng, Y. Zhang, Properties of solar wind dynamic pressure pulses at 1 AU during the deep minimum between solar cycles 23 and 24. Sol. Phys. **290**, 1835–1849 (2015). doi:10.1007/s11207-015-0700-5

Y.I. Yermolaev, N.S. Nikolaeva, I.G. Lodkina, M.Y. Yermolaev, Specific interplanetary conditions for CIR-, sheath-, and ICME-induced geomagnetic storms obtained by double superposed epoch analysis. Ann. Geophys. **28**, 2177–2186 (2010). doi:10.5194/angeo-28-2177-2010

W. Yu, C.J. Farrugia, N. Lugaz, A.B. Galvin, E.K.J. Kilpua, H. Kucharek, C. Möstl, M. Leitner, R.B. Torbert, K.D.C. Simunac, J.G. Luhmann, A. Szabo, L.B. Wilson, K.W. Ogilvie, J.-A. Sauvaud, A statistical analysis of properties of small transients in the solar wind 2007-2009: STEREO and wind observations. J. Geophys. Res. Space Phys. **119**, 689–708 (2014). doi:10.1002/2013JA019115

R. Zelwer, P.C.B. Fernando, S.H. Ward, Interplanetary magnetic field data and corresponding geomagnetic effects for the storm of October 7, 1962. J. Geophys. Res. **72**, 3471–3482 (1967). doi:10.1029/JZ072i013p03471

J. Zhang, I.G. Richardson, D.F. Webb, N. Gopalswamy, E. Huttunen, J.C. Kasper, N.V. Nitta, W. Poomvises, B.J. Thompson, C.-C. Wu, S. Yashiro, A.N. Zhukov, Solar and interplanetary sources of major geomagnetic storms ($Dst \leq -100$ nT) during 1996–2005. J. Geophys. Res. Space Phys. **112**, 10102 (2007). doi:10.1029/2007JA012321

Y. Zhang, W. Sun, X.S. Feng, C.S. Deehr, C.D. Fry, M. Dryer, Statistical analysis of corotating interaction regions and their geoeffectiveness during solar cycle 23. J. Geophys. Res. Space Phys. **113**, 08106 (2008). doi:10.1029/2008JA013095

H. Zhang, D.G. Sibeck, Q.-G. Zong, N. Omidi, D. Turner, L.B.N. Clausen, Spontaneous hot flow anomalies at quasi-parallel shocks: 1. observations. J. Geophys. Res. Space Phys. **118**, 3357–3363 (2013). doi:10.1002/jgra.50376

J.J. Zhang, C. Wang, T.R. Sun, C.M. Liu, K.R. Wang, GIC due to storm sudden commencement in low-latitude high-voltage power network in China: observation and simulation. Space Weather **13**, 643–655 (2015). doi:10.1002/2015SW001263

L. Zhao, T.H. Zurbuchen, L.A. Fisk, Global distribution of the solar wind during solar cycle 23: ACE observations. Geophys. Res. Lett. **36**, 14104 (2009). doi:10.1029/2009GL039181

P. Zuo, X. Feng, Y. Xie, Y. Wang, X. Xu, A statistical survey of dynamic pressure pulses in the solar wind based on WIND observations. Astrophys. J. **808**, 83 (2015). doi:10.1088/0004-637X/808/1/83

T.H. Zurbuchen, I.G. Richardson, In-situ solar wind and magnetic field signatures of interplanetary coronal mass ejections. Space Sci. Rev. **123**, 31–43 (2006). doi:10.1007/s11214-006-9010-4

Space Sci Rev (2017) 212:1221–1252
DOI 10.1007/s11214-017-0399-8

The Scientific Foundations of Forecasting Magnetospheric Space Weather

J.P. Eastwood[1] · R. Nakamura[2] · L. Turc[3] · L. Mejnertsen[1] · M. Hesse[4]

Received: 29 March 2017 / Accepted: 7 July 2017 / Published online: 15 August 2017

Abstract The magnetosphere is the lens through which solar space weather phenomena are focused and directed towards the Earth. In particular, the non-linear interaction of the solar wind with the Earth's magnetic field leads to the formation of highly inhomogenous electrical currents in the ionosphere which can ultimately result in damage to and problems with the operation of power distribution networks. Since electric power is the fundamental cornerstone of modern life, the interruption of power is the primary pathway by which space weather has impact on human activity and technology. Consequently, in the context of space weather, it is the ability to predict geomagnetic activity that is of key importance. This is usually stated in terms of geomagnetic storms, but we argue that in fact it is the substorm phenomenon which contains the crucial physics, and therefore prediction of substorm occurrence, severity and duration, either within the context of a longer-lasting geomagnetic storm, but potentially also as an isolated event, is of critical importance. Here we review the physics of the magnetosphere in the frame of space weather forecasting, focusing on recent results, current understanding, and an assessment of probable future developments.

Keywords Space weather · Magnetosphere · Geomagnetic storm · Substorm · Computer simulation

The Scientific Foundation of Space Weather
Edited by Rudolf von Steiger, Daniel Baker, André Balogh, Tamás Gombosi, Hannu Koskinen and Astrid Veronig

✉ J.P. Eastwood
jonathan.eastwood@imperial.ac.uk

1 The Blackett Laboratory, Imperial College London, London, UK

2 Space Research Institute, Austrian Academy of Sciences, Graz, Austria

3 Department of Physics, University of Helsinki, Helsinki, Finland

4 Birkeland Centre for Space Science, Department of Physics and Technology, University of Bergen, Bergen, Norway

1 Magnetospheric Space Weather—Initial Considerations

It is well known that space weather pertains to the conditions in all regions of space that can harm or disrupt human activity and technology both in space and on the ground (Hapgood 2011, 2012; Schrijver et al. 2015). Space weather is distinct from solar-terrestrial physics, and heliophysics more generally, because of its strong emphasis on societal impact (National Research Council 2008; Cannon et al. 2013). Since space weather is defined by its downstream effects, it is common practice to distinguish between three specific phenomena: radio blackouts, solar radiation storms and geomagnetic storms, which are monitored by various agencies at the national and international level.

In this article we aim to examine different aspects of the scientific foundations behind forecasting magnetospheric space weather. Although in some sense each of the three phenomena mentioned above can be considered 'magnetospheric' space weather (since they all generate impacts that occur inside the magnetosphere) only geomagnetic storms arise as a fundamental consequence of magnetospheric dynamics. Therefore here we will restrict the discussion to geomagnetic activity (of which storms are a manifestation), the underpinning science, and forecasting.

Before discussing the physics that controls magnetospheric activity, it is important to establish why geomagnetic storms are of interest for space weather as this immediately provides further focus and defines the processes that are relevant for magnetospheric space weather forecasting. It is well established that the primary channel by which space weather can generate a major socio-economic impact is through prolonged loss of power over an extended spatial area (Eastwood et al. 2017 and references therein). In the context of geomagnetic storms, the direct risk is from Ground Induced Currents (GICs). Ionospheric current systems, driven by magnetospheric activity, can cause rapidly varying magnetic field signatures at the Earth's surface; in turn these induce geo-electric fields, which then drive GICs that affect power grids in different ways: damage to physical infrastructure; the introduction of voltage instability which may trigger a blackout (without infrastructure damage); and interference with protection systems (Cannon et al. 2013). The most notable historical event with documented impact is the March 1989 geomagnetic storm, which is the largest of the space age but still less severe than the Carrington event, widely considered a reasonable worse-case scenario (Bolduc 2002; Erinmez et al. 2002; Cliver and Dietrich 2013).

Consequently, viewed through the lens of the socio-economic impact, forecasting magnetospheric space weather ultimately revolves around understanding the magnetospheric physics that controls the formation of GICs. Accurate forecasting of this, and magnetospheric space weather more generally depends on several factors. We will examine three in particular:

- The translation of L1 measurements to the nose of the magnetosphere—are the correct boundary conditions being applied to the magnetosphere?
- The physics that controls the behaviour of the magnetosphere—what is the current understanding of key processes based on recent spacecraft observations?
- Appropriateness of modelling approaches—in particular the use of global MHD simulations to capture dynamics and processes that are relevant for magnetospheric space weather forecasting.

In the final part of this review article we consider future developments, in light of present understanding and modelling capabilities. But first, we review fundamental aspects of the solar wind magnetosphere interaction, providing an overview of magnetospheric dynamics and discussing the relevance of substorms and storms for space weather.

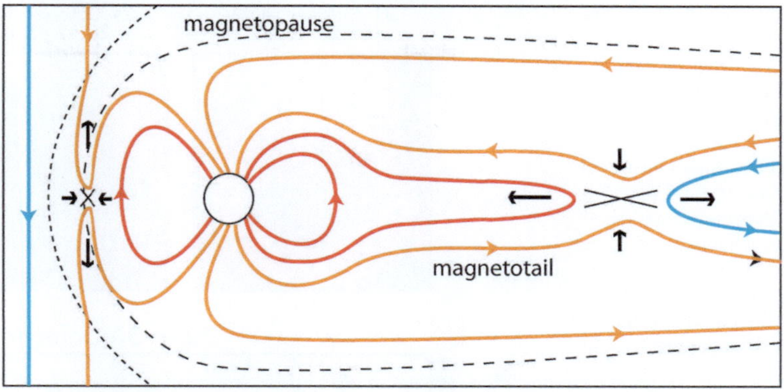

Fig. 1 A schematic of plasma circulation in the Earth's magnetosphere for southward IMF conditions. As originally proposed by Dungey, magnetic reconnection at the dayside magnetopause and in the magnetotail results in the concept of the open magnetosphere

2 The Solar Wind Magnetosphere Interaction

The dynamics of the Earth's magnetosphere are controlled by the solar wind Interplanetary Magnetic Field (IMF) (Dungey 1961). In particular, the stress applied by the solar wind on the Earth's magnetic field results in reconnection at the dayside magnetopause, creating open magnetic flux that is carried by the solar wind toward the night-side. The accumulated stress of the open magnetic field lines on the night-side leads to the formation of an elongated region called the magnetotail (see Fig. 1). Reconnection takes place between the oppositely directed magnetic fields at the center of the cross tail current sheet and the closed magnetic flux of the plasma sheet is transported back again toward the dayside completing the magnetospheric convection cycle. In this way, the global structures and long-time scale ($>$ hours) properties of the Earth's magnetotail are controlled by magnetic flux transport as a consequence of the solar wind-magnetosphere interaction. The reconnection region in the magnetotail responsible for the magnetospheric convection is considered to be nominally located at a distance of about 100 to 200 Re downtail and is called the distant neutral line (e.g., Walker et al. 1999).

Whilst the original Dungey (1961) picture captures the essence of magnetospheric behavior, and is well known, there are two key features fundamental to real magnetospheric dynamics which it does not capture, and which are perhaps not as well known in the wider space weather user community: the Expanding/Contracting Polar Cap paradigm and the Near Earth Neutral Line model.

2.1 The Expanding/Contracting Polar Cap Paradigm

It should be emphasized that the rate of dayside and nightside reconnection are not in balance (Cowley and Lockwood 1992). This means that the amount of open flux in the magnetosphere (orange field lines in Fig. 1) varies with time (e.g., Milan 2015 and references therein). This in particular means that energy can be stored in the magnetic field of the magnetotail, and subsequently released by magnetotail reconnection. The ionospheric polar cap represents the region of magnetic field in the magnetosphere that is 'open', i.e. connected to the Earth at only one end. Dayside reconnection will increase the size of the polar cap, as it will reconnect solar wind plasma and magnetospheric field lines to form open field lines.

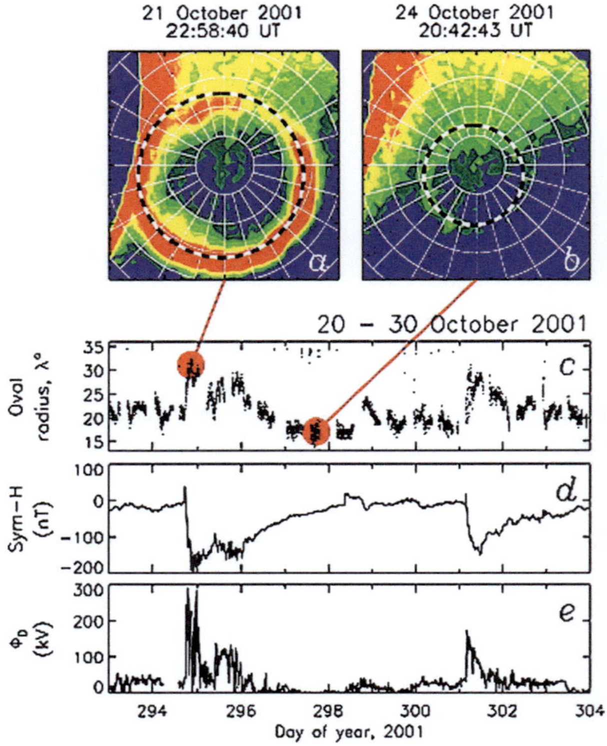

Fig. 2 Example of how the size of the auroral oval changes in response to unbalanced dayside and nightside reconnection. The top two panels show the auroral oval at the start and end of a geomagnetic storm, as measured using the Sym-H index in panel (**d**). Panel (**e**) shows the rate of solar wind magnetosphere coupling. (Figure 2 from Milan 2009; Copyright 2009 by the American Geophysical Union, reproduced with permission.)

Nightside reconnection of open field lines in the northern and southern lobes will reduce the size of the polar cap (see Fig. 2) (Milan 2009). Consequently, if the rate of reconnection at the dayside and in the magnetotail do not match, the polar cap will change in size, and can expand or contract. Whilst the reality is more complex, with other modifying phenomena at work, a very basic prediction is that the polar cap should grow in size before a substorm onset, because energy is being deposited into the lobes by magnetopause reconnection, and it should reduce in size as a substorm occurs, because nightside reconnection is returning closed flux to the inner magnetosphere. This overall picture is known as the expanding/contracting polar cap paradigm. In summary, it may be interpreted as the extension of the open magnetosphere model to time dependent behaviour where the reconnection rates in the magnetotail and on the dayside do not match.

2.2 Formation of the Near-Earth Neutral Line (NENL)

It is now known that internal processes in the magnetotail lead to the formation of thin current sheets also in the closed tail field regions closer to Earth. Consequently reconnection may also take place closer to Earth (closer to 30 Re, down to about ~ 12 Re) in a more explosive manner, resulting in the loss of the plasmasheet plasma tailward and drastic changes in the magnetotail configuration (Hones 1977; Baker et al. 1996). The overall dynamics of the magnetotail are quite well-defined in space and time, with the formation of a near-Earth neutral line leading to plasmoid release downtail (Slavin et al. 1984). This is illustrated in Fig. 3. A key feature of this process is that the inner edge of the magnetotail becomes more dipolar relative to the originally stretched tail current sheet. It is also important to note that

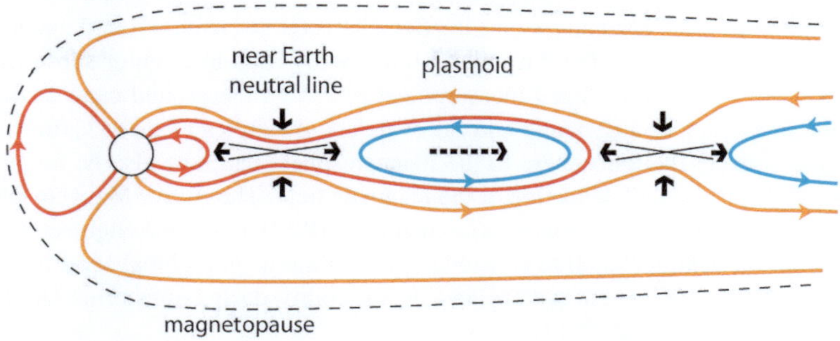

Fig. 3 Formation of the near Earth neutral line. This leads to the development of a plasmoid which is ejected downtail

this picture represents a cut through the magnetotail. Whilst the X-line and plasmoid does extend out of the page, it is limited in size (see, e.g., Eastwood and Kiehas 2015 and references therein). Consequently the dipolar region is approximately confined to the central region of the magnetotail, with consequences for magnetosphere-ionosphere coupling.

2.3 Magnetospheric Substorms and Storms

The elemental behaviour of the magnetosphere is the substorm (McPherron 1970; Rostoker et al. 1980), which has both a magnetospheric and auroral manifestation. The basic nature of substorms can be understood in terms of the overall Dungey model, modified by the expanding/contracting polar cap model, and the near Earth neutral line model. Substorms therefore involve both transient/localized signatures and also large-scale changes in the tail current sheet configuration and energization of the plasma. The THEMIS mission in particular was designed to study the chain of events that lead up to the triggering of a substorm (Angelopoulos et al. 2008) and more recent work on magnetospheric substorms and magnetotail dynamics has been reviewed by Sergeev et al. (2012).

Magnetospheric substorms are relevant for magnetospheric space weather because the reconfiguration of the magnetotail is associated with dynamic auroral emission—the auroral substorm (Akasofu 1964). An auroral substorm has a well-defined behaviour in space and time, which is directly associated with a reconfiguration of currents coupling the magnetosphere to the ionosphere, as well as the release of energy from the magnetotail, where it is stored in the magnetotail lobes.

There are several current systems that connect the magnetosphere and ionosphere (e.g. review by Milan et al. 2017). For example the region 1 field-aligned currents connect the magnetopause to the polar ionosphere, and hence define the open/closed field line boundary that is associated with the auroral oval (see Fig. 2). Region 2 field-aligned currents link the inner magnetosphere and the ring current to the ionosphere at lower latitudes. Field aligned currents are closed in the ionosphere by Pedersen currents in the auroral regions. During substorms, the magnetic field in the magnetotail near the Earth becomes more dipolar. This reduces the cross-tail current in a region transverse to the Sun-Earth line associated with the dipolarization. To ensure current closure, a so-called substorm current wedge is formed, whereby field aligned currents arise that connect the magnetotail with the ionosphere. This occurs in the expansion phase of the substorm. In the ionosphere a corresponding substorm electrojet forms.

The strength of these current systems is often characterised using ground magnetic field measurements. For example, the AE index is commonly used to monitor substorm occurrence and strength. However, fixed measurement points on the ground cannot completely capture the properties of a dynamic auroral oval. The dynamics of these various currents systems that connect the ionosphere to the magnetosphere can now also be monitored by satellites and the associated deflection of the magnetic field. The Active Magnetosphere and Planetary Electrodynamics Response Experiment (AMPERE) used the engineering magnetometers from the Iridium satellite constellation to monitor on a global level the region 1 and region 2 Birkeland current pattern and its variability during substorms (Anderson et al. 2000, 2002; Clausen et al. 2013).

Geomagnetic storms arise when conditions conducive to dayside reconnection persist for long periods, leading to enhanced magnetospheric circulation, injection of energetic particles into the inner magnetosphere, and enhancements of both radiation belt fluxes and the ring current (Gonzalez et al. 1994). Coronal Mass Ejections and Corotating Interaction Regions are the two drivers of geomagnetic storm activity (Gosling et al. 1991; Richardson et al. 2001, 2006). To understand the severity and determine the occurrence rate of geomagnetic storms, different indices are used. Space weather forecast providers often characterise storm severity according to the Kp index ($Kp_{max} = 9$), which is designed to capture the severity of the global perturbation to the quiet magnetic field on the surface of the Earth (time resolution of 3 hours). Whilst the Kp index already rather coarse (not least in its measure of severity and its time resolution), for space weather end-users it is then translated into a five-step scale of severity from G1 to G5. The physical measure of a G5 event would be a Kp index of 9 and it is expected that there would be on average 4 such storm days per solar cycle.

On the other hand, in the scientific literature the Dst index is commonly used to measure the strength of geomagnetic storms, and is used to construct statistics about the severity of geomagnetic storms. Dst, which captures the strength of the ring current via its perturbation to the equatorial magnetic field on the surface of the Earth, provides information about the occurrence of geomagnetic storms on a global level, and therefore its statistics are a necessary part of understanding geomagnetic storm likelihood. The Dst index is usually near 0, but during a storm it becomes negative. Dst $= -30$ to -50 is a weak storm, Dst $= -50$ to -100 is moderate and Dst < -100 is considered intense (Gonzalez et al. 1994; Vasyliūnas 2011).

2.4 Relevance of Substorms and Storms for Magnetospheric Space Weather

As mentioned above in Sect. 2, it is the formation of GICs that are of primary concern for magnetospheric space weather. Geomagnetic storms are usually understood as the primary phenomenon that is considered relevant in this regard; hence the classification of space weather events used by NOAA and UKMO shown in Sect. 1. Much effort has been engaged in understanding the likelihood of a Carrington-class event based on geomagnetic storm indices (e.g., Love 2012; Riley 2012).

However, since geomagnetic indices reduce magnetospheric dynamics to a single number, although they may document occurrence, it is not clear that they adequately capture the likely *impact* (Hapgood 2011). In particular, based on even a crude understanding of the magnetospheric system, one can immediately deduce that GIC production is likely to be highly inhomogenous in both space and time, but organised to some extent by the location of the auroral oval and the substorm current wedge. In fact intense GIC events occur on regional scales, as shown by the analysis of geoelectric fields for 1 in 100 year scale events

(Pulkkinen et al. 2015). Dst (and by extension Kp) must be used with caution (Kamide 2006) as unexceptional storms can nevertheless have notable impact, for example the storm of 4–5 August 1972 which had a Dst of -125 nT (Lanzerotti 1992). Furthermore, since it is an hourly index, Dst naturally smooths over many important details and arguably may hide both problems and physics that must be addressed to improve characterisation of space weather impacts.

It is increasingly recognised in the space weather community that current levels of magnetospheric space weather monitoring and warning, based on the global Kp index, are not fit for purpose, and better measures that more accurately define the threat to the power grid and other similar systems must provide localised, regional information. Therefore there is considerable current effort in developing forecasts that are capable of doing this (Pulkkinen et al. 2013). In particular, better physical understanding of the variation and severity of parameters such as the AE index, dB/dt and the geoelectric field is required to fully understand space weather impact statistics.

This leads to a new perspective whereby in fact intense substorms provide the major space weather risk to power distribution networks, and whilst these may be associated with geomagnetic storms, they may also arise individually. Consequently, forecasting effort is also increasingly focused on improving the predictability of intense substorms at geomagnetic latitudes where power distribution systems are located. Since the precise relationship between geomagnetic storms and substorms is complex and still not well understood, this represents a current challenge to scientific understanding of magnetospheric space weather.

In this article we focus on the following 3 areas where better knowledge of the physical processes at work is key to improving magnetospheric space weather forecasting

- Transition from L1 to the magnetopause and the magnetosheath—i.e. the way in which the solar wind may evolve from L1 to the bow shock, and the way in which it is processed in the magnetosheath. Current knowledge and recent results are discussed in Sect. 3.
- Solar wind magnetosphere coupling—what is the physics that controls how solar wind plasma, momentum and energy all enter the magnetosphere? Where does this occur and on what timescale? This is examined in more detail in Sect. 4.1.
- Magnetotail energy release—formation of bursty bulk flows in the near Earth magnetotail that is associated with particle injection and auroral activity. Considerable progress has been made with spacecraft missions such as THEMIS and Cluster, discussed in Sect. 4.2.

All of these have an impact on the development of magnetospheric modelling and forecasting. This is examined further in Sect. 5.

3 Input to the Magnetosphere: Solar Wind Driving and the Magnetosheath

3.1 Measuring the Inflowing Solar Wind

The best location to measure the inflowing solar wind for operational magnetospheric space weather purposes is the L1 Sun-Earth Lagrange point. Historically, operational data has been provided by ACE (Stone et al. 1998) and is now provided by DSCOVR. A potential challenge associated with the measurement is that spacecraft in fact orbit around the L1 point, and so do not lie exactly on the Sun-Earth line. This may mean that simple models that convect the solar wind from the upstream measurement point to Earth may be incorrect because solar wind phase fronts are unlikely to be perpendicular to the flow (Collier et al.

1998). Algorithms have been developed to account for this (e.g., Weimer and King 2008 and references therein) but different techniques can differ considerably in their prediction of transit time (Case and Wild 2012). A more accurate assessment of the solar wind conditions just upstream of the Earth is also made difficult by the fact that there can be variability in the solar wind perpendicular to the flow direction, that on large scales phase fronts are not planar and moreover that the solar wind evolves between L1 and the Earth (Kessel et al. 1999; Lepping et al. 2003; Tsurutani et al. 2005). Nevertheless, recent analysis shows that for operational modelling of the propagation delays from L1 to the Earth, interestingly there is no significant difference between the assumption of perpendicular and tilted solar wind phase fronts (Cash et al. 2016).

Before the solar wind actually reaches the magnetosphere, it first crosses the bow shock, which slows down the incoming supermagnetosonic flow to submagnetosonic speeds. Downstream of the bow shock, the magnetosheath plasma is denser, hotter, more turbulent, and exhibits stronger magnetic fields. The presence of the magnetosheath influences solar wind-magnetosphere coupling, since it is the magnetosheath plasma that interacts with the magnetosphere at the magnetopause.

3.2 Influence of the Magnetosheath

Firstly, it should be noted that the conditions observed in the solar wind are not necessarily representative of those encountered just upstream of the magnetopause. This may lead to an incorrect prediction of their impact on the magnetosphere. The IMF orientation can be significantly modified inside the magnetosheath (Coleman 2005; Longmore et al. 2006; Turc et al. 2014b). In particular, it has been shown that the magnetic field Bz component can have opposite signs upstream and downstream of the bow shock (Šafránková et al. 2009; Turc et al. 2017). This is exemplified in Fig. 4, which shows the Bz along the magnetopause for an IMF with a northward Bz, obtained from a magnetosheath model (Turc et al. 2014a). As evidenced by the blue areas, Bz turns south in some parts of the magnetosheath. This can in turn influence where reconnection occurs. The modification of the solar wind properties across the bow shock may also explain the saturation of the polar cap potential during strong magnetospheric driving. The polar cap potential is related to the solar wind electric field (and therefore Poynting flux) and is the potential difference across the open field line region in the ionosphere. Observationally, the polar cap potential is proportional to the solar wind electric field up to a certain limit where it then saturates, no longer responding linearly (e.g., Pulkkinen et al. 2016). Pulkkinen et al. (2016) show that in fact the ionospheric response correlates linearly with the Poynting flux normal to the magnetopause, whereas it saturates when compared to the solar wind Poynting flux. This suggests that the saturation is related to the processing of the solar wind through the bow shock and magnetosheath, and supports the fact that the conditions just upstream of the magnetopause are a better proxy of the magnetospheric driving.

It is also worth noting that the magnetosheath properties vary spatially. Many studies have shown evidence of pronounced dawn-dusk asymmetries in the magnetosheath parameters, such as the magnetic field strength and plasma velocity, the ion density and temperature (Dimmock and Nykyri 2013; Dimmock et al. 2015, 2016; Paularena et al. 2001; Longmore et al. 2005; Walsh et al. 2012). These asymmetries are due for the most part to the conditions encountered at the bow shock, and in particular to the angle Θ_{Bn} between the IMF and the normal to the shock's surface, which is defined locally. This implies that the dawn-dusk asymmetries are associated with the Parker-spiral IMF, but that other IMF orientations would result in a different distribution of the magnetosheath parameters. In addition to the

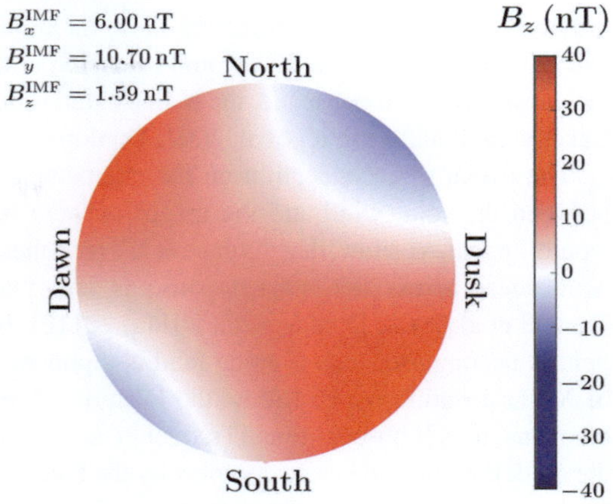

Fig. 4 Magnetic field Bz component along the magnetopause for a weakly northward IMF (Turc et al. 2014a) semi-analytical magnetosheath model based on ideal MHD

bow shock processes, the draping of the field lines along the magnetosphere also contributes to modifying spatially the magnetic field direction inside the magnetosheath. The spatial variation of the magnetosheath properties can affect the onset of magnetic reconnection and the development of the Kelvin-Helmholtz instability (e.g., Borovsky et al. 2008; Nykyri 2013).

Additionally, the bow shock and magnetosheath properties can be significantly modified during low Alfvén Mach number conditions, which are often associated with ICMEs and magnetic clouds. This may alter the solar wind-magnetosphere coupling during extreme space weather events (Lopez et al. 2004, 2011; Lavraud et al. 2007; Lavraud and Borovsky 2008; Farrugia et al. 2013; Turc et al. 2016).

4 Magnetospheric Dynamics: Recent Observations and Current Understanding

4.1 Solar Wind Coupling at the Magnetopause

At the magnetopause, the fundamental challenge for magnetospheric space weather forecasting is to understand when and where reconnection will occur, the physics by which reconnection occurs (so that it may be modelled accurately), and its temporal variability. It is also important to recognise other mechanisms transporting plasma across the magnetopause, specifically the Kelvin Helmholtz instability which may be important when the effects of magnetic reconnection are absent or reduced, for example during northward IMF conditions. We note that there are many recent reviews of both magnetic reconnection in space (e.g., Fuselier and Lewis 2011; Hesse et al. 2011; Eastwood et al. 2013; Paschmann et al. 2013) and the physics of the magnetopause (e.g., Lavraud et al. 2011; Hasegawa 2012; Eastwood et al. 2015b). Here we focus on aspects that are of importance for understanding magnetospheric space weather.

4.1.1 Where and When Will Reconnection Occur?

The concept of the open magnetosphere was cemented by the discovery of reconnection jets at the magnetopause by ISEE (Paschmann et al. 1979). Subsequent satellite measurements provided considerable experimental data, including measurements of oppositely directed exhausts either side of the X-line and extension of the X-line across a substantial fraction of the dayside magnetopause, i.e. the X-line extends out of the plane of Fig. 1 (Phan et al. 2000; Baker et al. 2002; Fuselier et al. 2002; Dunlop et al. 2011a, 2011b).

The extent of reconnection on the magnetopause fundamentally controls the coupling between the solar wind and the magnetosphere (Cowley and Owen 1989). Observations would appear to show that 'component' reconnection (i.e. reconnection where the magnetic shear across the magnetopause is less than 180°) prevails (Retinò et al. 2005; Trenchi et al. 2008; Dunlop et al. 2011a, 2011b). In component reconnection the magnetic field is decomposed into a guide field component and a reconnecting component. Models of X-line location which follow the location of maximum rotation in the field across the boundary match observations (Trattner et al. 2007a, 2007b, 2012). It should be noted that the modelling is made more complex by the fact that the local geometry varies with position on the curved magnetopause surface.

A further complicating factor is the asymmetry of the magnetopause: the different conditions on either side of the current sheet, in particular density, field strength and temperature (e.g., review by Eastwood et al. 2013 and references therein). An important result is the prediction that the X-line and the stagnation point are separated if the reconnection is asymmetric (Cassak and Shay 2007). In combination with the guide field and flow shear this could alter when and where reconnection is likely to occur (Tanaka et al. 2010). The 'beta-shear' condition says that reconnection is suppressed for an increasingly wide range of magnetic shear angles as the change in plasma beta ($\Delta\beta$) increases (Swisdak et al. 2003, 2010), following earlier suggestions that magnetosheath plasma beta was a controlling factor (Paschmann et al. 1986). In particular, reconnection is predicted to be suppressed if $\Delta\beta$ is too large according to the following formula

$$\Delta\beta > 2\left(\frac{L}{\lambda_i}\right)\tan\left(\frac{\theta}{2}\right)$$

Where L can be considered as the width of the magnetopause, λ_i is the ion inertial length and θ is the rotation of the magnetic field across the magnetopause. A small value of θ is equivalent to the statement that the guide field is large. This means that if $\Delta\beta$ is large (i.e. if the magnetosheath plasma beta is high, since the magnetospheric plasma beta is typically low), then reconnection will not occur if the guide field is too large. This relatively simple relationship may therefore be important in producing more refined analytic/empirical models of solar wind magnetosphere coupling, or could be introduced as a heuristic rule into models. This effect has been tested in the solar wind and at the magnetopause, and appears to be a good measure of reconnection occurrence (Phan et al. 2010, 2013). Figure 5a shows 49 magnetopause crossings where magnetic reconnection was observed. The majority of the data points are above the line which is the region where reconnection is not suppressed. In contrast, Fig. 5c shows 24 magnetopause crossings where no reconnection was observed. In these cases, the data points are predominantly below the line. Figures 5b and 5d show similar data but for events restricted in magnetic local time near the subsolar point, where the division in behaviour is even more apparent.

Finally a new area of work is the impact of magnetospheric plasma on reconnection at the magnetopause. In particular, plasmaspheric drainage plumes (caused by enhanced convection electric fields eroding the plasmasphere) can extend to the dayside magnetopause, where they influence reconnection and may reduce the reconnection rate (Borovsky et al.

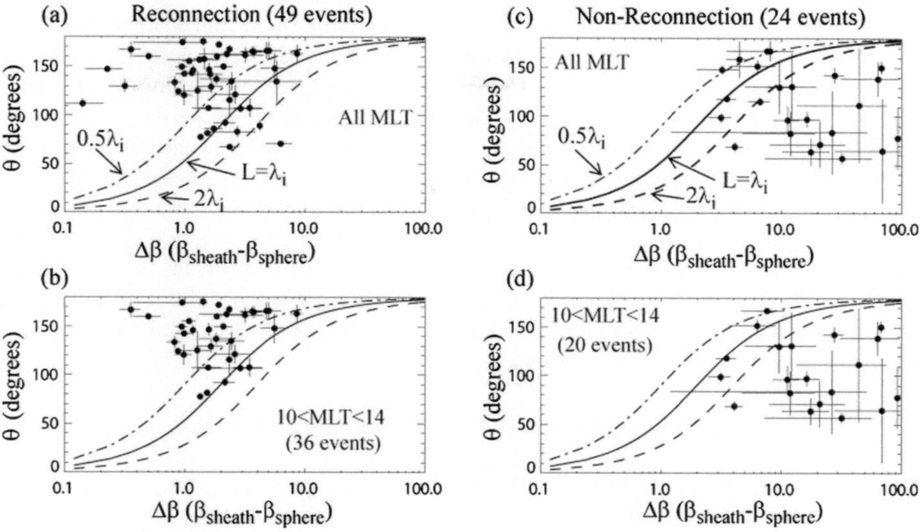

Fig. 5 (Adapted from Phan et al. 2013, Fig. 2; Copyright 2012 by the American Geophysical Union, reproduced with permission.) Results of a statistical survey of reconnection (*left*) and nonreconnection (*right*) events. (**a**, **c**) Scatter plot of magnetic shear versus $\Delta\beta$ across the magnetopause for all magnetic local time (MLT); (**b**, **d**) magnetic shear versus $\Delta\beta$ in the vicinity of the subsolar region ($10 < \mathrm{MLT} < 14$)

2013; Walsh et al. 2014). This suggests that it is important to account for these cold populations in modelling the solar wind magnetosphere coupling at the magnetopause for space weather forecasting purposes.

4.1.2 Physics of Reconnection and the Diffusion Region

Whilst magnetic reconnection has large-scale magnetospheric consequences, reconnection itself occurs within the small-scale diffusion region where the plasma decouples from the magnetic field and the frozen in field theorem no longer applies. In fact, the diffusion region is structured: because of their larger mass, the volume over which the ions are not frozen in (the ion diffusion region), is much larger than the electron diffusion region (EDR) where the electrons decouple and reconnection actually occurs. The length scales for these regions is roughly comparable to the relevant particle's gyroradius (or more precisely the inertial length) (Vasyliunas 1975). At the magnetopause the EDR region is only a few kilometres in size.

The physics of the diffusion region has been the subject of much recent investigation to understand precisely how the plasma and the magnetic field decouples. A key signature of the ion diffusion region which has received considerable attention is the formation of Hall magnetic fields due to the different motion of the ions and electrons (see e.g. recent reviews by Fuselier and Lewis 2011; Paschmann et al. 2013; Goldman et al. 2016). The structure and properties of the ion diffusion region has been resolved by spacecraft missions such as Cluster, Geotail and THEMIS which typically measure the plasma on timescales of a few seconds.

More recent observations from the Magnetospheric Multiscale mission (MMS) have focussed on the electron diffusion region (EDR). MMS is a four spacecraft mission (Burch et al. 2016a) and each spacecraft carries an identical payload of instruments to measure the electric and magnetic fields, the plasma population and composition, and energetic particles. In particular, the MMS fast particle instrument (FPI) can measure the electron and ion all sky distribution function at 30 ms and 150 ms resolution respectively (Pollock et al. 2016), meaning that it is able to probe in detail electron physics and the properties of the EDR. The first major success of the MMS mission has been to record multiple encounters with the EDR, and measure the properties of the electrons there. Prior to the launch of MMS it was suggested that crescent-shaped features should be expected to occur in the plane perpendicular to the magnetic field and MMS has shown that electron orbits in the vicinity of the EDR are coherent, being observed in multiple events (Hesse et al. 2014; Burch et al. 2016b; Burch and Phan 2016 and references therein). This new knowledge about diffusion region physics can help guide the development of new models and theory that accurately captures reconnection physics in the magnetospheric context.

4.1.3 Temporal Variability of Reconnection and Transient Behaviour

Whilst overall magnetic reconnection can be steady at the magnetopause, temporal variations are often observed. In particular, flux transfer events are bipolar signatures in the component of the magnetic field along the magnetopause normal (Russell and Elphic 1978). They are associated with reconnection, which produces the normal magnetic field component and various different models have been proposed to explain their production, including patchy reconnection, multiple X-line reconnection and time-variable reconnection (Russell and Elphic 1978; Lee and Fu 1985; Scholer 1988; Southwood et al. 1988). Of these various production methods, multiple X-line reconnection has received recent attention, being observed in simulations and spacecraft data (Raeder 2006; Omidi and Sibeck 2007; Hasegawa et al. 2010; Oieroset et al. 2014; Pfau-Kempf et al. 2016).

Flux transfer events are therefore important in the context of magnetospheric space weather because they contribute to the transfer of solar wind plasma and energy into the magnetosphere. A key question in this regard is their magnetic topology. If they are connected to the magnetosphere at only one end, then they will indeed transport flux into the magnetotail. However, they may be connected to the magnetosphere at one end, both ends, or neither end. Consequently, flux rope topology is crucial to investigate, and evidence suggests that this may be highly complex (Pu et al. 2013). An alternative approach is to use multi-spacecraft observations of FTE geometry to constrain their shape, size and therefore formation mechanism (Fear et al. 2008, 2012). Together with analysis of FTE transport along the tail magnetopause (Eastwood et al. 2012), this may elucidate how FTEs contribute to flux transport and whether they possibly modulate the rate at which the magnetosphere reaches substorm onset conditions.

4.1.4 Other Plasma Transport Mechanisms: The Kelvin-Helmholtz Instability

It is important to recognise that other mechanisms contribute to the transport of plasma into the magnetosphere, in particular the Kelvin-Helmholtz instability. It arises in association with the large velocity shear on the flank magnetopause and causes magnetopause surface waves to steepen into vortices and break (Hasegawa et al. 2004). It will occur most easily if the field is perpendicular to the flow, and is therefore important for northward IMF conditions, although it has been observed for southward IMF as well (Hwang et al. 2011). In the nonlinear stage microscale instabilities or reconnection can occur, enabling plasma to be transported across the boundary; this is now being resolved by MMS (Eriksson et al. 2016; Stawarz et al. 2016). Ultimately, the KHI is important for space weather because it allows solar wind plasma to enter on the flanks during northward IMF and in fact mass transport across the magnetopause appears to be enhanced for northward IMF (Palmroth et al. 2006). Better quantification of the plasma transport rate from new measurements and simulations will improve knowledge of cold dense plasma sheet formation and magnetospheric 'preconditioning' prior to intervals of southward IMF driving geomagnetic activity.

4.2 Magnetotail Dynamics

Magnetic flux transport in the near Earth tail region is dominated by high-speed plasma flows called Bursty Bulk Flows (BBFs), which are considered to be the outflow of the near-Earth reconnection. Associated electromagnetic field disturbances transport energy closer in towards the Earth's inner magnetosphere, are responsible for energetic particle injection and also for enhanced auroral precipitation. These energetic particle signatures of intense substorms are identified as important space weather phenomena that lead to spacecraft anomalies such as surface charging and deep dielectric charging (Bodeau 2015; Loto'aniu et al. 2015).

Ongoing multi-point spacecraft measurements have significantly contributed to enhancing our understanding in this area from the ability to determine spatial gradients and temporal evolution to the non-adiabatic/adiabatic acceleration processes of the particles related to the transient electric fields. For example, observations by THEMIS and Cluster have enabled advances in our understanding of substorm dynamics and current sheet dynamics relevant to near-Earth magnetotail reconnection (e.g. reviews by Sergeev et al. 2012; Petrukovich et al. 2016). The newly launched MMS spacecraft have started to/will resolve the physical properties of near-Earth magnetotail reconnection and associated disturbances down to electron scales. Whilst there are a number of important physical processes occurring in the magnetotail (reconnection itself, current sheet instabilities, waves, turbulence, etc.) in the following section we highlight the interaction between the reconnection jet and the Earth's dipole region, as this leads to two tail phenomena that are key for magnetospheric space weather: energetic particle injection and auroral precipitation.

4.2.1 Reconnection Jets, Bursty Bulk Flows, Dipolarization Fronts and Dipolarization Flux Bundles

In recent years, a plethora of different terms have arisen to identify different features of magnetotail dynamics. Here we provide a brief summary:

- As mentioned above, **Bursty Bulk Flows** (BBFs) are high speed flows in the plasma sheet. Although their occurrence rate is low, BBFs contribute a significant part of the total magnetic flux transport in the near-Earth magnetotail (Angelopoulos et al. 1999; Schödel et al. 2001). BBFs have 10-min time scales, with embedded velocity peaks of 1-min duration, which are called **flow bursts** (Baumjohann et al. 1989; Angelopoulos et al. 1992). They are localized in the dawn-dusk direction with scales of about 1–5 RE (Sergeev et al. 1996; Angelopoulos et al. 1997; Nakamura et al. 2001, 2004).
- The **Dipolarization Front** (DF) is a thin (ion scale, about 800–2000 km) boundary often observed on the leading edge of a BBF where the Bz component sharply increases (Nakamura et al. 2002).
- The **Dipolarization Flux Bundle** (DFB) (Liu et al. 2013) corresponds to the entire Bz enhancement following the DF.

The most popular model describing flow burst/BBFs and their connection to the ionosphere is the plasma bubble model, where flow burst is a plasma-depleted and dipolarized magnetic flux tube (Pontius and Wolf 1990). Multipoint spacecraft measurements have observed the Earthward propagation of the DF from the midtail to the near-Earth region (Runov et al. 2011). As it propagates Earthward flux piles up and tailward pressure gradient force develops that leads to braking of the flows in the near-Earth dipolar region (Panov et al. 2010). In the flow braking region both the Earthward propagating dipolarization front and subsequent tailward progressing dipolarization can be detected (Nakamura et al. 2009). The plasma bubble model predicts that the bubble penetrates Earthward until the flux tube plasma content matches with that of the ambient dipolar magnetic field, a prediction observationally confirmed by statistical study (Dubyagin et al. 2011). Tailward flow bouncing and oscillatory signatures were also reported in this region indicating that the flow braking is a dynamic process (Panov et al. 2013a, 2013b).

The combined Flow bursts/DF/DFB entity can be regarded as a region of enhanced electric field so that different ions and electrons in the plasma sheet, depending on their pitch angle and energy can enter this region (or, alternatively, enter as the dipolarization front passes the ambient plasma region) for a limited time in a localized region of space as the front propagates Earthward, and are accelerated accordingly. Such a concept of an Earthward propagating electric field pulse has been considered as a source mechanism of injection (Quinn and Southwood 1982; Li et al. 1993). Recent multi-point spacecraft observations have allowed the physical processes around the boundary to be revealed in a more quantitative way. Since the dipolarization front thickness is usually ion-scales (typically less than 5 ion inertial lengths or 3 ion gyroradii) or even smaller, the Hall-electric field and the electron-pressure gradient play a role in forming the front structure and 1-2 electron inertial length structure dominates the dissipation at the front (Fu et al. 2012; Angelopoulos et al. 2013; Balikhin et al. 2014).

Not all, but a large fraction of the fronts were shown to be tangential discontinuities meaning that at least for those boundaries the dipolarization front is simply a thin boundary dividing two different plasma regions (tenuous, hot BBF plasma and denser, colder ambient plasma) without net bulk flow across the boundary (Schmid et al. 2011). Particles that are trapped in the collapsing field line of BBF/DFB can gain energy from Fermi/betatron acceleration as was simulated and observed behind the dipolarization front (Ashour-Abdalla et al. 2011; Fu et al. 2012; Birn et al. 2013). On the other hand, other cases of dipolarization fronts were reported in which ambient ions are passing the front layer, accelerated by the electric field at the dipolarization front and then reflected back ahead of the front (Zhou et al. 2014; Eastwood et al. 2015a). While the occurrence of rapid flux transport rate observations significantly drops between 10 to 15 Re (Schödel et al. 2001), the dipolarization front grows (Bz amplitude increases) closer in to the Earth, and a large dawn-to-dusk electric field in the flow braking region is both predicted from simulations and reported from observational studies (Birn et al. 2011; Liu et al. 2014; Schmid et al. 2016). Energetic particle injections accompanied by DFBs that were found from the midtail reconnection region (30 Re) to inside geosynchronous orbit also suggest the importance of the transient/localized electric field enhancements of BBF and flow braking for the particle acceleration process (Gabrielse et al. 2014; Liu et al. 2016).

Note that the total energy gained from this process is, however, limited by the strength of the dawn-to-dusk (induced) electric field and the width of the dipolarization front. For a typically localized dipolarization front case, the above mechanisms explain acceleration of few keV to hundreds of keV for electrons, and a few tens of keV to hundreds of keV for protons (Birn et al. 2013). These particles are considered as the seed population of the radiation belt particles that are subsequently further accelerated by different mechanisms such as by waves in the inner magnetosphere. Recent observation of a global injection front accompanied by MeV particles detected inside 4 Re during storm time conditions (Dai et al. 2015), indicates that the maximum energy may be enhanced during those active events. Such observations suggest that for an intense substorm case, DF/DFB/BBF related acceleration may also contribute significantly to particle acceleration in the inner magnetosphere/radiation belt.

4.2.2 BBF Size and Magnetotail Flux Transport

From considerations of global flux balance (lobe flux reduction and associated closed flux transport), it is required that BBF should have an effective dawn-dusk size of about 10 RE (Angelopoulos et al. 2013). This is larger than is typically observed. Large-scale size in the azimuthal direction is also expected from the size of the substorm current wedge. Furthermore, a typical time-scale for the substorm expansion phase is about 30 minutes, which is again longer than that of typical DFBs. How then do BBF/DF/DFB/flow bursts, that are shown to be transient in time and localized in size in a number of statistical studies, contribute to the larger scale evolution of the magnetosphere during substorms? One explanation is that large-scale substorm effects are the integrated effects of multiple BBFs that take place in space and time. Multiple flow bursts and associated auroral precipitation (so called auroral streamers, e.g., Kauristie et al. 2003) have been considered as supporting evidence for this. (Temporal and spatial relationships between these BBF-associated field-aligned currents and the substorm current wedge are further discussed by Kepko et al. 2015.) Although only for a limited number of cases, multipoint studies have shown evidence for a dipolarization front extended to 10 RE-scale during an intense substorm in the flow braking region and during storm time intense substorm in the inner magnetosphere (Nakamura et al. 2013; Dai et al. 2015). The longer time-scale of the substorm current wedge or auroral electroject as compared to BBFs/DFBs, on the other hand, could also be due to the accumulated effects of the flow bursts that modify the pressure distribution in the inner magnetosphere and thereby sustain the substorm current wedge (Kepko et al. 2015). In such cases the substorm time-scale would correspond to the time-scale of the pressure redistribution process that is expected to be longer than the BBF time scale.

4.2.3 The Effects of Preconditioning on Magnetotail Dynamics

Magnetotail disturbances are modified by both the preconditioning of the magnetotail and the solar wind driver. In this section we briefly overview the characteristics of the magnetotail for different density/temperature conditions and two different magnetospheric states: steady magnetospheric convection (SMC), and geomagnetic storms. Not only the background magnetic field configuration, such as the location of the thin current sheet, but also the plasma properties, such as plasma density and temperatures, are important factors for determining the magnetotail disturbance as well as the magnetosphere-ionosphere coupling. Sergeev et al. (2014) showed that the solar wind electric field is associated with a larger auroral electrojet when the plasma sheet is hotter and has lower density than for the colder and denser plasma sheet. This difference could be due to stronger field-aligned acceleration of precipitating electrons for higher temperature and lower density as expected from Knight's relationship (Knight 1973). The hot, low density plasma sheet is expected to be produced by the lobe plasma entering and heating by reconnection. On the other hand, the source of the cold and dense plasma sheet is thought to be the solar wind/magnetosheath plasma entering from the flank, as discussed in Sect. 4.1.4. The processes responsible for the entry are diffusion at the low latitude boundary layer, mixing of plasma due to Kelvin-Helmholtz instabilities and associated reconnection, and poleward-of-cusp reconnection during northward IMF Bz (Song and Russell 1992; Terasawa et al. 1997; Hasegawa et al. 2004). It is important to note that although the response time of the reconnection process is less than 1 hour, the mixing process of cold-dense plasma sheet from the flank can take longer ($>$ 3 hours).

Although minor in the number density, the effect of ionospheric heavy ion outflow can further contribute to changes in the time scale (slowing) and changes in the spatial scale of the current sheet due to larger inertial scales (see Kronberg et al. 2014 for more detail regarding the effects of heavy ions on magnetospheric processes). Hence, for understanding the solar wind-magnetosphere-ionosphere coupling, the background plasma conditions of the magnetosphere, produced by the different mechanisms of solar wind-magnetosphere interaction, as well as the ionosphere need to be taken into account. In particular, since the response time scales of these different processes to the solar wind drivers are different, that places significant constraints in predicting the magnetospheric response based on modeling.

The general effect of larger dawn-to-dusk solar wind electric field, created by southward IMF, is to move the near-Earth reconnection region closer to the Earth (Nagai et al. 2005). However, the response of the magnetotail to southward IMF conditions also depends on the nature of the preconditioning and the mode of magnetosphere-ionosphere coupling can differ from the simple substorm cycle discussed before, such as during steady magnetospheric convection (SMC). During SMC the enhanced, stable convection, persists longer than a typical recovery phase, with no substorm expansions (Pytte et al. 1978). The average total pressure in the inner magnetosphere is higher during SMC events than for other types of activity (Kissinger et al. 2012). This higher pressure region extends to larger radial distances, and causes fast Earthward flows to divert toward the dawn or dusk flanks and continue to the dayside leaving the inner magnetosphere relatively quiet.

This pattern of flux transport is in contrast to substorms, during which flows are directed toward the inner magnetosphere. The most extreme case of southward IMF Bz is the storm-time substorm, when significant loading of the magnetotail leads to: the reconnection site moving closer to Earth; enhanced ionospheric outflows causing a significant population of oxygen in the tail current sheet configuration; and MeV injection fronts observed inside 5 Re (Baumjohann et al. 1996; Miyashita et al. 2005; Kistler et al. 2006; Dai et al. 2015; Baker et al. 2016). The mode of the magnetotail disturbance, therefore, can be significantly different depending on the different level of IMF southward and resultant preconditioning of the magnetotail.

5 Modelling and Predicting Magnetospheric Space Weather

As discussed in the previous sections, the processes that cause adverse magnetospheric space weather effects are largely a consequence of solar wind–magnetosphere coupling. Although the processes at work encompass a wide range of spatial and temporal scales, significant success in understanding magnetospheric physics has been received through the use of computer simulations that employ magnetohydrodynamic-based codes. These simulations solve the equations of MHD on a 3-dimensional grid that encompasses the whole magnetosphere. This has the advantage of efficiently describing the large scale structure of the magnetosphere, the transport of plasma, and the coupling between different regions. Large scale communication is accomplished via MHD waves—i.e. magnetosonic and Alfven waves.

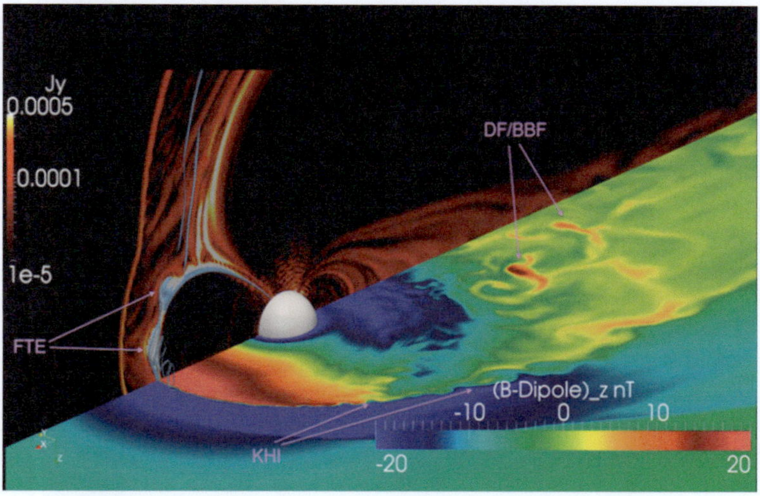

Fig. 6 Lyon-Fedder-Mobarry global magnetosphere model showing flux transfer events (FTEs) at the dayside magnetopause, dipolarization fronts (DFs) and bursty bulk flows (BBFs) in the magnetotail, and Kelvin-Helmholtz instability (KHI) at the flank of the magnetosphere. Color contours show the dawn-to-dusk current (Jy) in the noon-midnight (X–Z) plane and northward magnetic field (Bz) in the equatorial (X–Y) plane (from Sitnov et al. 2016)

They also capture large scale current systems, however, we note that codes typically use the 'v B' approach (as opposed to 'E J') (Vasyliunas 2005), and the current is derived from the curl of the magnetic field. The size of the simulation domain is typically sufficient to include the inflowing solar wind and the magnetotail, and the magnetosphere is modelled as a whole, from the solar wind input at the sunward edge to an inner boundary (usually at 2 RE to 5 RE), and then to the magnetotail. So-called global MHD simulations (GMHD) have been available for a number of years, and have become increasingly popular given the recent increase in affordable computing power. An example of such a simulation is shown in Fig. 6.

5.1 Global Modeling

GMHD codes used to simulate the Earth's magnetosphere include: the Open Geospace General Circulation Model (OpenGGCM), the Block-Adaptive-Tree-Solarwind-Roe-Upwind-Scheme (BATS-R-US), the Grand Unified Magnetosphere–Ionosphere Coupling Simulation (GUMICS) and the Lyon-Fedder-Mobarry (LFM) model (Powell et al. 1999; Lyon et al. 2004; Raeder et al. 2008; Janhunen et al. 2012). It is important to note that given the complexity of the magnetosphere, these simulations in fact form one component of a more complex modelling architecture. In all cases, the core GMHD code must be coupled to a module capable of capturing the ionospheric (and thermospheric) dynamics via field aligned currents. For example the Coupled-Magnetosphere Ionosphere Thermosphere (CTIM) model, the Thermosphere-Ionosphere-Electrodynamic Global Circulation Model (TIEGCM) and the Magnetosphere Ionosphere Coupler Solver (MIX) are used in conjunction with LFM (Roble and Ridley 1994; Wang et al. 2004; Wiltberger et al. 2004; Merkin and Lyon 2010). OpenGGCM is also used with CTIM. BATS-R-US is used with the Rice Convection Model and the Ridley Ionosphere Model as part of the Space Weather

Modeling Framework (SWMF), which brings together simulation models ultimately capable of capturing the whole chain of space weather physical processes between the Sun and the Earth (Wolf et al. 1982; Toffoletto et al. 2003; Ridley et al. 2004; Toth et al. 2005, 2012). Despite the growing complexity of these codes, it is of course acknowledged that they do not contain all the relevant physics. In addition, capturing the onset of magnetotail reconnection leading to a substorm may well depend on numerical resistivity, which arises from small inconsistencies in the calculation due to rounding errors. Furthermore, instabilities will not be correctly described at the most fundamental level. Although code-coupling attempts to adequately describe the inner magnetosphere, radiation belts and ionospheric outflow, this is not without its own difficulty or complexity, and this remains a current challenge (Eastwood et al. 2015b). Nevertheless, GMHD simulations are widely used since they can run in real time with current computational resources, and they provide a route to a first-principle physics-based model of substorms. It is also possible to derive magnetospheric diagnostics relevant for space weather, such as: the size of the auroral oval and polar cap boundary; the location and shape of the magnetopause; and upper atmospheric heating (by calculating Joule heating) (Gordeev et al. 2015). Furthermore, they can in principle be used to predict the ground dB/dt which is crucial for space weather forecasting in the magnetosphere (discussed in more detail below) (Pulkkinen et al. 2013).

The utility of these simulations is therefore multi-faceted. Firstly, they enable exploration of the basic physics that controls magnetospheric dynamics; here we mention a few illustrative examples: the analysis of Palmroth et al. (2006) using the GUMICS code elucidated the fact that mass transport across the magnetopause may be enhanced for northward IMF; investigating the magnetospheric drivers and energy input under different IMF conditions (Raeder et al. 2001; Palmroth et al. 2003; Pulkkinen et al. 2007; Sergeev et al. 2008; Yu and Ridley 2009); exploring the temperature asymmetry in the magnetosheath under different solar wind conditions (Dimmock et al. 2015); modelling the effects of shocks on the magnetosphere (Andréeová et al. 2008; Oliveira and Raeder 2014); and performing large scale statistical studies (Juusola et al. 2014; Facskó et al. 2016).

Secondly, they can be used to provide large scale context and interpretation of sparse in situ spacecraft measurements (e.g., Raeder et al. 2008). This is a research approach that has proved particularly fruitful, to the extent that the Community Coordinated Modeling Center (CCMC) at NASA's Goddard Space Flight Center now provides simulation runs on demand to the general space physics community (MacNeice et al. 2012). The growth of the CCMC in particular has stimulated the use of these codes as a general tool that can be employed by scientists outside the core development team, rather like how spacecraft data is now easily used by scientists who are not part of the instrument team. Specific examples of this approach are numerous and we refer the reader to the CCMC website.[1]

5.2 Benchmarking and Forecast Accuracy

GMHD models are being transitioned into the operational domain for the purposes of space weather modelling and forecasting (Pulkkinen et al. 2010). In doing this, it is necessary to have confidence that the models and simulations are sufficiently accurate and this can be achieved through the use of appropriate metrics (e.g., Lopez et al. 2007). A recent example of this is the benchmarking performed by Gordeev et al. (2015). In this paper, several key global variables were identified which as well as characterising the global magnetosphere, are potentially relevant for space weather forecasting. The first set of metrics concern the

[1] https://ccmc.gsfc.nasa.gov/.

Table 1 Summary of skill scores for four different GMHD models, adapted from Table 3 of Gordeev et al. (2015; Copyright 2015 by the American Geophysical Union, reproduced with permission). The slope and CC (correlation coefficient) refer to the gradient and uncertainty of the best fit line when a regression is performed against the relevant empirical model. The PE (prediction efficiency) is a measure of the forecast skill. In all cases, values equal to unity indicate accurate performance. Grey boxes highlight the best performing model for each metric, according to the correlation coefficient

Parameter	BATS-R-US			GUMICS			Open GGCM			LFM		
	Slope	CC	PE	Slope	CC	PE	Slope	CC	PE	Slope	CC	PE
Subsolar m'pause	1.01	0.95	+0.88	0.96	0.85	+0.31	0.90	0.77	−1.47	0.92	0.91	−0.43
Terminator m'pause	1.03	0.64	−0.02	0.97	0.93	+0.62	1.11	0.47	−3.20	1.00	0.44	−0.95
Tail Mpause	1.03	0.77	+0.22	0.93	0.80	−0.55	1.13	0.73	−3.19	0.99	0.73	−0.04
Tail B field	0.83	0.95	+0.49	0.75	0.94	+0.01	0.93	0.97	+0.86	0.78	0.86	+0.02
Field aligned currents	0.37	0.86	−0.44	0.05	0.48	−2.15	2.01	0.68	−5.15	0.98	0.93	+0.87

size of the magnetosphere, as defined by the distance to the subsolar magnetopause, and the radius of the magnetosphere at the terminator and 15 RE downtail. The second set of metrics is focussed on the magnetotail, its pressure balance and its magnetic flux content. These are defined by the plasma sheet thermal pressure and the lobe magnetic field. The remaining metrics are related to magnetospheric dynamics and substorms, and include the polar cap potential drop (to capture the rate of global convection), the total field aligned current, and more recently in Gordeev et al. (2017), the extent to which substorm loading occurs. Each of these can be validated against an empirical model, such as the Shue et al. (1998) magnetopause model in the case of the distance to the sub-solar magnetopause.

Gordeev et al. (2015) then defined a series of standard solar wind inputs, with densities, temperatures and velocities consistent with the statistical properties of the solar wind. In this study, 19 different synthetic solar wind profiles were identified, of 4 hour duration. The initial two hours contained northward IMF, which may then vary according to the specific run. These input conditions were applied to four models running at the CCMC—OpenGGCM, GUMICS, BATS-R-US and LFM. The output was analysed at 5 minute resolution, and various metrics such as the prediction efficiency were computed (Pulkkinen et al. 2011).

This analysis reveals that different models have different strengths and weaknesses. For example, the BATS-R-US model is found to best agree with empirical models of the subsolar magnetopause location, as shown in Fig. 7. On the other hand, the high-latitude tail magnetopause is best captured by GUMICS. The strength of the lobe field is most accurately predicted by the OpenGGCM simulation, and the LFM model is best performing against the field-aligned current metric. Table 1 below reproduces a summary of the results.

Gordeev et al. (2015, 2017) note that in fact all four models perform reasonably well and capture at least in a qualitative sense the correct behaviour that is expected to be observed for a given change in solar wind conditions. However, this analysis highlights the fact that although simulations are, thanks to CCMC, now available 'on demand', their use and interpretation requires care and expert insight into the nature of the code itself. Furthermore, it raises the potentially important point that the choice of model to be used for space weather forecasting may depend on the purpose for which the forecast is required. For example, the real-time modelling of the magnetopause location is important to predict whether it moves within geostationary orbit. Spacecraft charging will be altered in the higher density plasma of the magnetosheath and furthermore the magnetic field orientation will change, which could affect older spacecraft that use the magnetic field as part of their Attitude and Orientation Control System (AOCS).

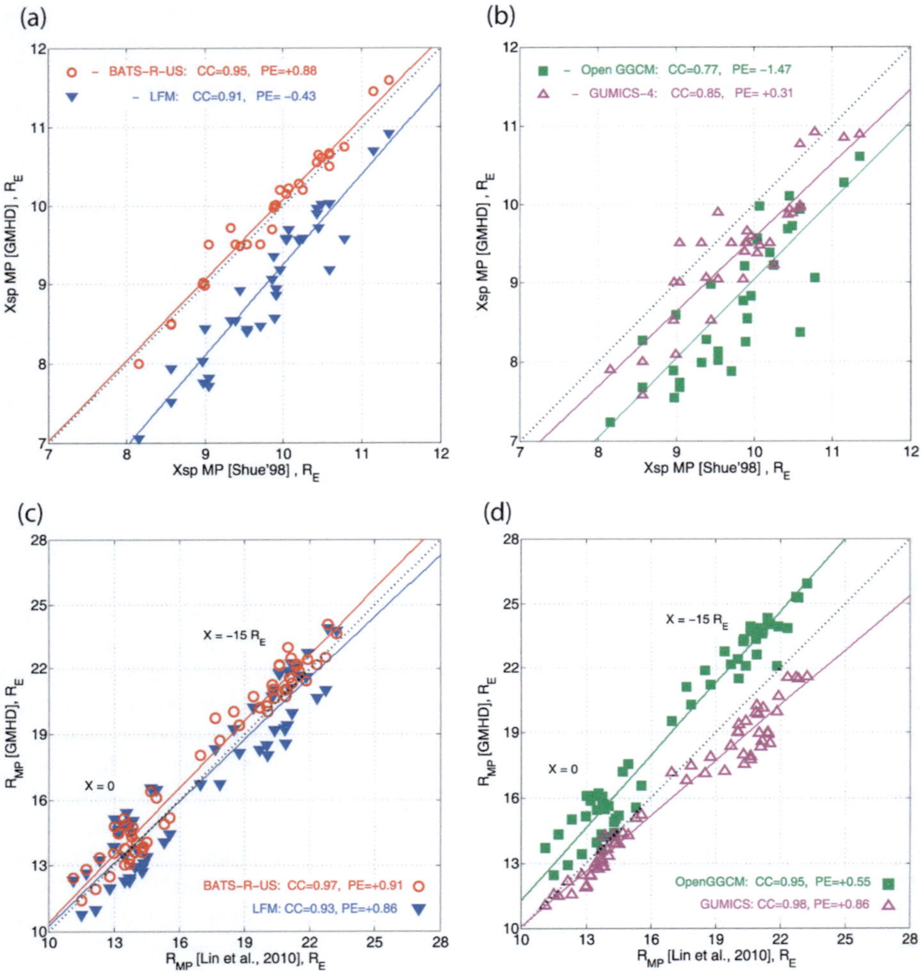

Fig. 7 **(a–b)** The performance of different GMHD models in identifying the location of the subsolar magnetopause, as benchmarked against the Shue et al. (1998) empirical magnetopause model. It is found that in this test, the BATS-R-US code is best performing. **(c–d)** Comparable plots for the combined location of the high-latitude terminator and tail magnetopause. Separate analysis of these two regions is shown in Table 1. (Figure adapted from Gordeev et al. 2015; Copyright 2015 by the American Geophysical Union, reproduced with permission.)

As explained above, the most important parameter for space weather modelling is dB/dt, the rate of change of the magnetic field on the ground. The ability of different magnetospheric models to predict this parameter has been specifically analysed by Pulkkinen et al. (2013). They identified 6 baseline geomagnetic storm events including the 2003 Halloween storms. For these events, the driving solar wind data from L1 was available, as were measurements of dB/dt on the ground at 12 geomagnetic observatories. Three different global MHD models were evaluated: LFM-MIX, OpenGGCM, and the SWMF and were also compared with two empirical models. In each case, the L1 data were used to drive the simulations, and the ground magnetic field perturbation was predicted at 6 of the magnetometer stations. The simulations were all performed at CCMC. The simulation/modelling outputs were then

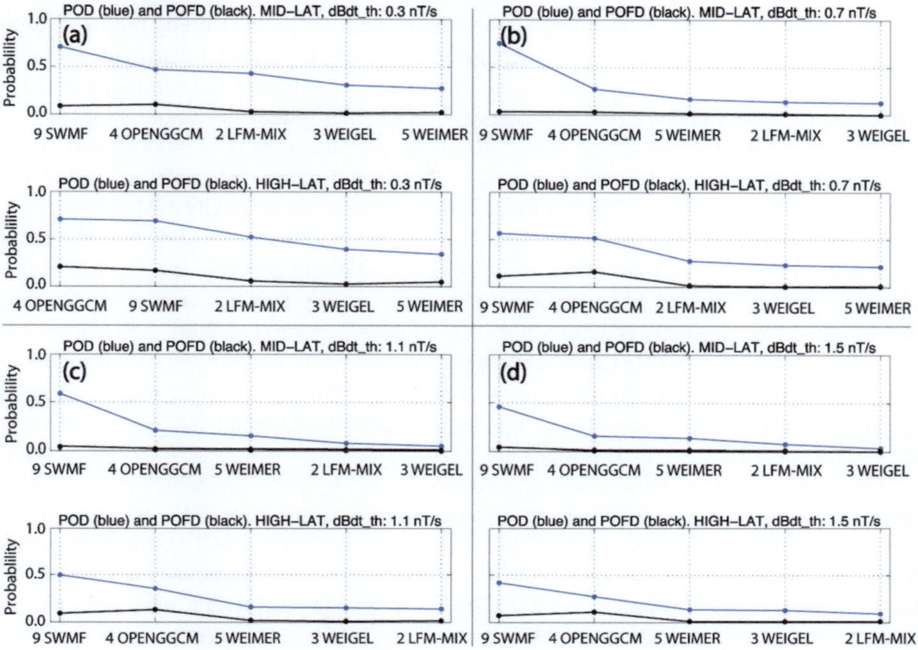

Fig. 8 The probability of detection and probability of false detection of dB/dt is shown for different dB/dt thresholds, at high- and mid-latitudes (adapted from Fig. 5 of Pulkkinen et al. 2013; Copyright 2013 by the American Geophysical Union, reproduced with permission)

compared to the observed data using a series of different metrics. These were the probability of detection (POD), the probability of false detection (POFD) and the Heidke Skill Score (HSS; $1 \geq \text{HSS} > -\infty$). If $\text{HSS} = 0$ then the prediction is as good as random, and if $\text{HSS} = 1$ then the prediction is perfect (Lopez et al. 2007). A score of 0.5 therefore indicates a 50% chance of event detection. Example results are shown in Fig. 8 which shows the POD and POFD of dB/dt by different models and simulations. The data is split according to the location of the measurement on the ground (high- and mid-latitude) and the strength of the dB/dt signal (0.3, 0.7, 1.1 and 1.5 nT/s). It can be seen that in general the SWMF is on average the best performing according to this metric. However, it was found that the POD and HSS was less than 0.5 for a dB/dt threshold of 1.5 nT/s. Nevertheless, it was concluded by Pulkkinen et al. (2013) that "users satisfied with more rough characterisation of dB/dt activation over the storm periods may be able to use the models for generating actionable information."

6 Future Developments

6.1 Future Developments: Spacecraft Measurement

Spacecraft measurements for magnetospheric space weather forecasting can be divided into two types. On the one hand, data is required to advance understanding of the underlying physics. In this regard, at the time of writing MMS is primed to make considerable progress. The combination of multiple operating missions (Cluster, Geotail, MMS, THEMIS, etc.)

allows the magnetosphere to be studied more comprehensively. Outstanding problems relevant for space weather prediction include: how to accurately translate L1 data to the subsolar magnetopause; how to accurately model the location, onset, duration and variability of magnetopause reconnection; how to adequately capture magnetospheric preconditioning; how to include multiscale processes, both temporal and spatial, in magnetotail models; and to understand where, when and how magnetotail reconnection starts and evolves, to establish whether localized plasma jets are due to localized reconnection or locally evolved structure, and whether multiple activations are spatial or temporal in nature. Nevertheless, the measurements are sparse, and this represents a major obstacle to progress in magnetospheric physics as a whole. So-called constellation missions represent an obvious next step forward and an intriguing possibility is to use CubeSat platforms (Spence et al. 2001, 2004; Schwartz et al. 2009; Eastwood 2015). This would be complemented by missions that exploit remote sense. One potentially transformative idea is to remote sensing the outer magnetospheric boundaries through soft X-ray charge exchange (Branduardi-Raymont et al. 2012), which is the goal of the ESA/CAS SMILE mission.

The other activity pertains to operational monitoring, by platforms which provide measurements to be used in operational space weather forecasting. This is likely to make use of dedicated platforms—for example DSCOVR at L1, and GOES in GEO. Improving lead times for magnetospheric space weather monitoring requires measurements to be made in the solar wind further upstream than L1. This could be achieved with a solar sail mission or inner heliosphere measurements (Eastwood et al. 2015c; Kubicka et al. 2016). Within the magnetosphere, improving the density of measurements can also be achieved by using hosted payloads. An example of this within the ESA space situational awareness programme is SOSMAG which will be hosted on GEO-KOMPSAT-2A (Auster et al. 2016).

6.2 Future Developments: Modelling and Forecasting

At the time of writing, NOAA's Space Weather Prediction Center is using the SWMF model operationally.[2] In the future, the availability of multiple magnetospheric models at the operational level is desirable to provide diversity of forecast methodology. Linked to this is the use of ensemble forecasting (Knipp 2016); this requires the ability to run predictions faster than real time. Finally, another option for improving forecasts would be to perform data assimilation, but this is challenging because of sparse measurements in the magnetosphere.

Looking to the future, modelling of strong storms requires a change in grid step, and changes in the treatment of the inner boundary because of the movement of the dayside magnetopause towards the Earth. Another challenge is the development of appropriate metrics to test models in a relevant way. For example, this may mean the use of operational metrics designed specifically for certain forecast types or user communities, and not simply science-based assessments. In the context of the work discussed here, such a metric may correspond to GIC threshold, rather than polar cap potential for example.

A key question in the medium term future is the issue of developing models that deliver forecasts of improved accuracy. In some sense this is already in progress, with the standard approach of coupling different codes that describe different regions (notably the ring current, the plasmasphere and the ionosphere/thermosphere system). One route is to include reconnection physics by modifying the treatment of resistivity, or to embed PIC simulation capability within the larger fluid simulation (Kuznetsova et al. 2007; Tóth et al. 2016). A global Vlasov code has been developed—Vlasiator (Sandroos et al. 2013; von Alfthan et al. 2014;

[2] http://www.swpc.noaa.gov/products/geospace-global-geomagnetic-activity-plot.

Kempf et al. 2015) and this may ultimately be the simulation tool of choice, as computing power becomes sufficiently powerful and accessible. This may enable a more detailed exploration of the Axford conjecture, which argues that large-scale magnetospheric dynamics are governed by global processes and boundary conditions rather than the precise nature of the local resistivity, etc. associated with reconnection (Gonzalez et al. 2016).

7 Conclusions

In this review paper we have attempted to summarise the scientific foundations of magnetospheric space weather forecasting. The basic nature of the solar wind magnetosphere interaction is, we argue, well understood. It is driven by magnetic reconnection, and therefore depends on the orientation of the solar wind magnetic field. Whilst the original Dungey picture correctly shows the pattern of magnetospheric circulation, it is important to appreciate that dayside and nightside reconnection do not typically occur in lockstep as simple cartoons such as the one shown in Fig. 1 imply. In fact there is a process of loading and unloading which can be monitored by the size of the polar cap. The unloading process drives strong, structured electrical currents in the ionosphere, and these represent the significant threat to power grids, which is the primary socio-economic impact. Therefore a key goal is to improve the forecasting of substorms. This requires knowledge of both the loading and unloading of energy in the magnetosphere at the magnetopause and in the magnetotail respectively. Recent research has examined the evolution of the solar wind from L1 to the magnetopause, the location, timing, and extent of reconnection on the magnetopause, the physics and temporal variability of reconnection, the nature of bursty bulk flows and their coupling to the inner magnetosphere, and the effects of magnetospheric preconditioning. Observational studies go hand-in-hand with modelling, and improved metrics and benchmarking are aiding the transition to operations. In the future, improved forecasts are likely to derive from three areas: ongoing scientific investigation using existing and future space missions; availability of multipoint operational real time data from hosted payloads and dedicated platforms; and improved modelling schemes. In each of these areas, prospects are bright and we anticipate that there should be a significant improvement in our ability to forecast magnetospheric space weather over the coming years.

Acknowledgements The authors thank the International Space Science Institute for its hospitality in hosting the workshop on the scientific foundations of space weather and for enabling the writing of this article. JPE acknowledges STFC grant ST/N000692/1. LT acknowledges Marie Sklodowska-Curie grant agreement No. 704681. RN is supported by Austrian Science Fund (FWF) I2016-20. The authors thank S.E. Milan, T.D. Phan, E. Gordeev and A. Pulkkinen for their kind assistance with the preparation of Figs. 2, 5, 7 and 8. The editors and authors thank an anonymous reviewer for very useful comments and suggestions to the original manuscript.

References

S.-I. Akasofu, The development of the auroral substorm. Planet. Space Sci. **12**, 273–282 (1964)

B.J. Anderson, K. Takahashi, B.A. Toth, Sensing global Birkeland currents with iridium® engineering magnetometer data. Geophys. Res. Lett. **27**(24), 4045–4048 (2000). doi:10.1029/2000GL000094

B.J. Anderson, K. Takahashi, T. Kamei, C.L. Waters, B.A. Toth, Birkeland current system key parameters derived from Iridium observations: method and initial validation results. J. Geophys. Res. Space Phys. **107**(A6), SMP 11-11 (2002). doi:10.1029/2001JA000080
K. Andréeová, T.I. Pulkkinen, T.V. Laitinen, L. Přech, Shock propagation in the magnetosphere: observations and MHD simulations compared. J. Geophys. Res. Space Phys. **113**(A9), A09224 (2008). doi:10.1029/2008JA013350
V. Angelopoulos, W. Baumjohann, C.F. Kennel, F.V. Coroniti, M.G. Kivelson, R. Pellat, R.J. Walker, H. Lühr, G. Paschmann, Bursty bulk flows in the inner central plasma sheet. J. Geophys. Res. **97**(A4), 4027–4039 (1992)
V. Angelopoulos et al., Magnetotail flow bursts: association to global magnetospheric circulation, relationship to ionospheric activity and direct evidence for localization. Geophys. Res. Lett. **24**(18), 2271–2274 (1997). doi:10.1029/97GL02355
V. Angelopoulos, F.S. Mozer, R.P. Lin, T. Mukai, K. Tsuruda, R. Lepping, W. Baumjohann, Comment on "Geotail survey of ion flow in the plasma sheet: observations between 10 and 50 RE" by W.R. Paterson et al.. J. Geophys. Res. Space Phys. **104**(A8), 17521–17525 (1999). doi:10.1029/1999JA900198
V. Angelopoulos et al., Tail reconnection triggering substorm onset. Science **321**, 931–935 (2008)
V. Angelopoulos, A. Runov, X.Z. Zhou, D.L. Turner, S.A. Kiehas, S.S. Li, I. Shinohara, Electromagnetic energy conversion at reconnection fronts. Science **341**(6153), 1478–1482 (2013). doi:10.1126/science.1236992
M. Ashour-Abdalla, M. El-Alaoui, M.L. Goldstein, M. Zhou, D. Schriver, R. Richard, R. Walker, M.G. Kivelson, K.-J. Hwang, Observations and simulations of non-local acceleration of electrons in magnetotail magnetic reconnection events. Nat. Phys. **7**(4), 360–365 (2011). http://www.nature.com/nphys/journal/v7/n4/abs/nphys1903.html#supplementary-information
U. Auster et al., Space weather magnetometer set with automated AC spacecraft field correction for Geo-Kompsat-2A, in *Proc. '2016 ESA Workshop on Aerospace EMC'*. ESA SP, vol. 738 (2016). doi:10.1109/AeroEMC.2016.7504585
D.N. Baker, T.I. Pulkkinen, V. Angelopoulos, W. Baumjohann, R.L. McPherron, Neutral line model of substorms: past results and present view. J. Geophys. Res. **101**, 12975–13010 (1996)
D.N. Baker et al., Timing of magnetic reconnection initiation during a global magnetospheric substorm onset. Geophys. Res. Lett. **29**(24), 43 (2002). doi:10.1029/2002GL015539
D.N. Baker et al., A telescopic and microscopic examination of acceleration in the June 2015 geomagnetic storm: magnetospheric multiscale and Van Allen probes study of substorm particle injection. Geophys. Res. Lett. **43**(12), 6051–6059 (2016). doi:10.1002/2016GL069643
M.A. Balikhin, A. Runov, S.N. Walker, M. Gedalin, I. Dandouras, Y. Hobara, A. Fazakerley, On the fine structure of dipolarization fronts. J. Geophys. Res. Space Phys. **119**(8), 6367–6385 (2014). doi:10.1002/2014JA019908
W. Baumjohann, R.A. Treumann, J. LaBelle, R.R. Anderson, Average electric wave spectra across the plasma sheet and their relation to ion bulk speed. J. Geophys. Res. Space Phys. **94**(A11), 15221–15230 (1989). doi:10.1029/JA094iA11p15221
W. Baumjohann, Y. Kamide, R. Nakamura, Substorms, storms, and the near-Earth tail. J. Geomagn. Geoelectr. **48**, 177 (1996)
J. Birn, R. Nakamura, E.V. Panov, M. Hesse, Bursty bulk flows and dipolarization in MHD simulations of magnetotail reconnection. J. Geophys. Res. Space Phys. **116**(12), A01210 (2011). doi:10.1029/2010ja016083
J. Birn, M. Hesse, R. Nakamura, S. Zaharia, Particle acceleration in dipolarization events. J. Geophys. Res. **118**(5), 1960–1971 (2013). doi:10.1002/jgra.50132
M. Bodeau, Review of better space weather proxies for spacecraft surface charging. IEEE Trans. Plasma Sci. **43**(9), 3075–3084 (2015). doi:10.1109/TPS.2015.2441038
L. Bolduc, GIC observations and studies in the Hydro-Quebec power system. J. Atmos. Sol.-Terr. Phys. **64**, 1793–1802 (2002)
J.E. Borovsky, M. Hesse, J. Birn, M.M. Kuznetsova, What determines the reconnection rate at the dayside magnetosphere? J. Geophys. Res. Space Phys. **113**(A7), A07210 (2008). doi:10.1029/2007JA012645
J.E. Borovsky et al., Estimating the effects of ionospheric plasma on solar wind/ magnetosphere coupling via mass loading of dayside reconnection: ion-plasma-sheet oxygen, plasmaspheric drainage plumes, and the plasma cloak. J. Geophys. Res. **118**, 5695–5719 (2013)
G. Branduardi-Raymont et al., AXIOM: advanced X-ray imaging of the magnetosphere. Exp. Astron. **33**(2–3), 403–443 (2012). doi:10.1007/s10686-011-9239-0
J.L. Burch, T.D. Phan, Magnetic reconnection at the dayside magnetopause: advances with MMS. Geophys. Res. Lett. **43**(16), 8327–8338 (2016). doi:10.1002/2016GL069787
J.L. Burch, T.E. Moore, R.B. Torbert, B.L. Giles, Magnetospheric multiscale overview and science objectives. Space Sci. Rev. **199**(1), 5–21 (2016a). doi:10.1007/s11214-015-0164-9

J.L. Burch et al., Electron-scale measurements of magnetic reconnection in space. Science **352**(6290), aaf2939 (2016b). doi:10.1126/science.aaf2939
P. Cannon et al., *Extreme Space Weather: Impacts on Engineered Systems and Infrastructure* (Royal Academy of Engineering, London, 2013)
N.A. Case, J.A. Wild, A statistical comparison of solar wind propagation delays derived from multispacecraft techniques. J. Geophys. Res. **117**, A02101 (2012). doi:10.1029/2011JA016946
M.D. Cash, S. Witters Hicks, D.A. Biesecker, A.A. Reinard, C.A. de Koning, D.R. Weimer, Validation of an operational product to determine L1 to Earth propagation time delays. Space Weather **14**(2), 93–112 (2016). doi:10.1002/2015SW001321
P.A. Cassak, M.A. Shay, Scaling of asymmetric magnetic reconnection: general theory and collisional simulations. Phys. Plasmas **14**(10), 102114 (2007). doi:10.1063/1.2795630
L.B.N. Clausen, S.E. Milan, J.B.H. Baker, J.M. Ruohoniemi, K.H. Glassmeier, J.C. Coxon, B.J. Anderson, On the influence of open magnetic flux on substorm intensity: ground- and space-based observations. J. Geophys. Res. Space Phys. **118**(6), 2958–2969 (2013). doi:10.1002/jgra.50308
E.W. Cliver, W.F. Dietrich, The 1859 space weather event revisited: limits of extreme activity. J. Space Weather Space Clim. **3**(15), A31 (2013). doi:10.1051/swsc/2013053
I.J. Coleman, A multi-spacecraft survey of magnetic field line draping in the dayside magnetosheath. Ann. Geophys. **23**(3), 885–900 (2005). doi:10.5194/angeo-23-885-2005
M.R. Collier, J.A. Slavin, R.P. Lepping, A. Szabo, K. Ogilvie, Timing accuracy for the simple planar propagation of magnetic field structures in the solar wind. Geophys. Res. Lett. **25**(14), 2509–2512 (1998)
S.W.H. Cowley, M. Lockwood, Excitation and decay of solar wind-driven flows in the magnetosphere-ionosphere system. Ann. Geophys. **10**, 103–115 (1992)
S.W.H. Cowley, C.J. Owen, A simple illustrative model of open flux tube motion over the dayside magnetopause. Planet. Space Sci. **37**, 1461–1475 (1989)
L. Dai et al., Near-Earth injection of MeV electrons associated with intense dipolarization electric fields: Van Allen Probes observations. Geophys. Res. Lett. **42**(15), 6170–6179 (2015). doi:10.1002/2015GL064955
A.P. Dimmock, K. Nykyri, The statistical mapping of magnetosheath plasma properties based on THEMIS measurements in the magnetosheath interplanetary medium reference frame. J. Geophys. Res. Space Phys. **118**(8), 4963–4976 (2013). doi:10.1002/jgra.50465
A.P. Dimmock, K. Nykyri, H. Karimabadi, A. Osmane, T.I. Pulkkinen, A statistical study into the spatial distribution and dawn-dusk asymmetry of dayside magnetosheath ion temperatures as a function of upstream solar wind conditions. J. Geophys. Res. Space Phys. **120**(4), 2767–2782 (2015). doi:10.1002/2014JA020734
A.P. Dimmock, T.I. Pulkkinen, A. Osmane, K. Nykyri, The dawn–dusk asymmetry of ion density in the dayside magnetosheath and its annual variability measured by THEMIS. Ann. Geophys. **34**(5), 511–528 (2016). doi:10.5194/angeo-34-511-2016
S. Dubyagin, V. Sergeev, S. Apatenkov, V. Angelopoulos, A. Runov, R. Nakamura, W. Baumjohann, J. McFadden, D. Larson, Can flow bursts penetrate into the inner magnetosphere? Geophys. Res. Lett. **38**(8), L08102 (2011). doi:10.1029/2011GL047016
J.W. Dungey, Interplanetary magnetic field and the auroral zones. Phys. Rev. Lett. **6**(2), 47–48 (1961)
M.W. Dunlop et al., Extended magnetic reconnection across the dayside magnetopause. Phys. Rev. Lett. **107**, 025004 (2011a)
M.W. Dunlop et al., Magnetopause reconnection across wide local time. Ann. Geophys. **29**, 1683–1697 (2011b)
J.P. Eastwood, Observing magnetic reconnection: the influence of Jim Dungey, in *Magnetospheric Plasma Physics: The Impact of Jim Dungey's Research*, ed. by S.W.H. Cowley Frs, D. Southwood, S. Mitton (Springer, Cham, 2015), pp. 181–197. doi:10.1007/978-3-319-18359-6_9
J.P. Eastwood, S.A. Kiehas, Origin and evolution of plasmoids and flux ropes in the magnetotails of Earth and Mars, in *Magnetotails in the Solar System* (Wiley, New York, 2015), pp. 269–287. doi:10.1002/9781118842324.ch16
J.P. Eastwood, T.D. Phan, R.C. Fear, D.G. Sibeck, V. Angelopoulos, M. Oieroset, M.A. Shay, Survival of flux transfer event (FTE) flux ropes far along the tail magnetopause. J. Geophys. Res. Space Phys. **117**(9), A08222 (2012). doi:10.1029/2012ja017722
J.P. Eastwood, T.D. Phan, M. Oieroset, M.A. Shay, K. Malakit, M. Swisdak, J.F. Drake, A. Masters, Influence of asymmetries and guide fields on the magnetic reconnection diffusion region in collisionless space plasmas. Plasma Phys. Control. Fusion **55**, 124001 (2013)
J.P. Eastwood, M.V. Goldman, H. Hietala, D.L. Newman, R. Mistry, G. Lapenta, Ion reflection and acceleration near magnetotail dipolarization fronts associated with magnetic reconnection. J. Geophys. Res. **120**(1), 511–525 (2015a). doi:10.1002/2014JA020516

J.P. Eastwood, H. Hietala, G. Toth, T.D. Phan, M. Fujimoto, What controls the structure and dynamics of Earth's magnetosphere? Space Sci. Rev. **188**(1), 251–286 (2015b). doi:10.1007/s11214-014-0050-x
J.P. Eastwood, D.O. Kataria, C.R. McInnes, N.C. Barnes, P. Mulligan, Sunjammer. Weather **70**(1), 27–30 (2015c). doi:10.1002/wea.2438
J.P. Eastwood, E. Biffis, M.A. Hapgood, L. Green, M.M. Bisi, R.D. Bentley, R. Wicks, L.A. McKinnell, M. Gibbs, C. Burnett, The economic impact of space weather: where do we stand? Risk Anal. **37**(2), 206–218 (2017). doi:10.1111/risa.12765
S. Eriksson et al., Magnetospheric multiscale observations of magnetic reconnection associated with Kelvin-Helmholtz waves. Geophys. Res. Lett. **43**(11), 5606–5615 (2016). doi:10.1002/2016GL068783
I.A. Erinmez, J.G. Kappenman, W.A. Radasky, Management of the geomagnetically induced current risks on the national grid company's electric power transmission system. J. Atmos. Sol.-Terr. Phys. **64**(5–6), 743–756 (2002). doi:10.1016/s1364-6826(02)00036-6
G. Facskó, I. Honkonen, T. Živković, L. Palin, E. Kallio, K. Ågren, H. Opgenoorth, E.I. Tanskanen, S. Milan, One year in the Earth's magnetosphere: a global MHD simulation and spacecraft measurements. Space Weather **14**(5), 351–367 (2016). doi:10.1002/2015SW001355
C.J. Farrugia, N.V. Erkaev, V.K. Jordanova, N. Lugaz, P.E. Sandholt, S. Mühlbachler, R.B. Torbert, Features of the interaction of interplanetary coronal mass ejections/magnetic clouds with the Earth's magnetosphere. J. Atmos. Sol.-Terr. Phys. **99**, 14–26 (2013). doi:10.1016/j.jastp.2012.11.014
R.C. Fear, S.E. Milan, A.N. Fazakerley, E.A. Lucek, S.W.H. Cowley, I. Dandouras, The azimuthal extent of three flux transfer events. Ann. Geophys. **26**, 2353–2369 (2008)
R.C. Fear, S.E. Milan, K. Oksavik, Determining the axial direction of high-shear flux transfer events: implications for models of FTE structure. J. Geophys. Res. Space Phys. **117**(19), A09220 (2012). doi:10.1029/2012ja017831
H.S. Fu, Y.V. Khotyaintsev, A. Vaivads, M. André, S. Huang, Electric structure of dipolarization front at sub-proton scale. Geophys. Res. Lett. **39**(6), L06105 (2012)
S.A. Fuselier, W.S. Lewis, Properties of near-Earth magnetic reconnection from in-situ observations. Space Sci. Rev. **160**, 95–121 (2011)
S.A. Fuselier, H.U. Frey, K.J. Trattner, S.B. Mende, J.L. Burch, Cusp aurora dependence on interplanetary magnetic field B_Z. J. Geophys. Res. **107**(A7), 1111 (2002). doi:10.1029/2001JA900165
C. Gabrielse, V. Angelopoulos, A. Runov, D.L. Turner, Statistical characteristics of particle injections throughout the equatorial magnetotail. J. Geophys. Res. Space Phys. **119**(4), 2512–2535 (2014). doi:10.1002/2013JA019638
M.V. Goldman, D.L. Newman, G. Lapenta, What can we learn about magnetotail reconnection from 2D PIC Harris-sheet simulations? Space Sci. Rev. **199(1)**, 651–688 (2016). doi:10.1007/s11214-015-0154-y
W.D. Gonzalez, J.A. Joselyn, Y. Kamide, H.W. Kroehl, G. Rostoker, B.T. Tsurutani, V.M. Vasyliunas, What is geomagnetic storm? J. Geophys. Res. **99**(A4), 5771–5792 (1994)
W.D. Gonzalez et al., Fundamental concepts associated with magnetic reconnection, in *Magnetic Reconnection: Concepts and Applications*, ed. by W. Gonzalez, E. Parker (Springer, Cham, 2016), pp. 1–32. doi:10.1007/978-3-319-26432-5_1
E. Gordeev, V. Sergeev, I. Honkonen, M. Kuznetsova, L. Rastätter, M. Palmroth, P. Janhunen, G. Tóth, J. Lyon, M. Wiltberger, Assessing the performance of community-available global MHD models using key system parameters and empirical relationships. Space Weather **13**(12), 868–884 (2015). doi:10.1002/2015SW001307
E. Gordeev, V. Sergeev, N. Tsyganenko, M. Kuznetsova, L. Rastäetter, J. Raeder, G. Tóth, J. Lyon, V. Merkin, M. Wiltberger, The substorm cycle as reproduced by global MHD models. Space Weather **15**(1), 131–149 (2017). doi:10.1002/2016SW001495
J.T. Gosling, D.J. McComas, J.L. Phillips, S.J. Bame, Geomagnetic activity associated with Earth passage of interplanetary shock disturbances and coronal mass ejections. J. Geophys. Res. **96**(5), 7831–7839 (1991)
M.A. Hapgood, Towards a scientific understanding of the risk from extreme space weather. Adv. Space Res. **47**(12), 2059–2072 (2011). doi:10.1016/j.asr.2010.02.007
M. Hapgood, Prepare for the coming space weather storm. Nature **484**(7394), 311–313 (2012)
H. Hasegawa, Structure and dynamics of the magnetopause and its boundary layers. Monogr. Environ. Earth Planets **1**(2), 71–119 (2012)
H. Hasegawa, M. Fujimoto, T.-D. Phan, H. Rème, A. Balogh, M.W. Dunlop, C. Hashimoto, R. TanDokoro, Transport of solar wind into Earth's magnetosphere through rolled-up Kelvin-Helmholtz vortices. Nature **430**, 755–758 (2004)
H. Hasegawa et al., Evidence for a flux transfer event generated by multiple X-line reconnection at the magnetopause. Geophys. Res. Lett. **37**, L16101 (2010). doi:10.1029/2010GL044219
M. Hesse, T. Neukirch, K. Schindler, M. Kuznetsova, S. Zenitani, The diffusion region in collisionless magnetic reconnection. Space Sci. Rev. **160**(1), 3–23 (2011). doi:10.1007/s11214-010-9740-1

M. Hesse, N. Aunai, D. Sibeck, J. Birn, On the electron diffusion region in planar, asymmetric, systems. Geophys. Res. Lett. **41**(24), 8673–8680 (2014). doi:10.1002/2014GL061586

E.W. Hones Jr., Substorm processes in the magnetotail: comments on 'On hot tenuous plasmas, fireballs, and boundary layers in the Earth's magnetotail' by L.A. Frank, K.L. Ackerson and R.P. Lepping. J. Geophys. Res. **82**(35), 5633–5640 (1977)

K.J. Hwang, M.M. Kuznetsova, F. Sahraoui, M.L. Goldstein, E. Lee, G.K. Parks, Kelvin-Helmholtz waves under southward IMF. J. Geophys. Res. **116**, A08210 (2011). doi:10.1029/2011JA016596

P. Janhunen, M. Palmroth, T. Laitinen, I. Honkonen, L. Juusola, G. Facsko, T.I. Pulkkinen, The GUMICS-4 global MHD magnetosphere-ionosphere coupling simulation. J. Atmos. Sol.-Terr. Phys. **80**, 48–59 (2012)

L. Juusola et al., Statistical comparison of seasonal variations in the GUMICS-4 global MHD model ionosphere and measurements. Space Weather **12**(10), 582–600 (2014). doi:10.1002/2014SW001082

Y. Kamide, What is an "Intense Geomagnetic Storm"? Space Weather **4**, S06008 (2006). doi:10.1029/2006SW000248

K. Kauristie, V.A. Sergeev, O. Amm, M.V. Kubyshkina, J. Jussila, E. Donovan, K. Liou, Bursty bulk flow intrusion to the inner plasma sheet as inferred from auroral observations. J. Geophys. Res. Space Phys. **108**(A1), 1040 (2003). doi:10.1029/2002JA009371

Y. Kempf, D. Pokhotelov, O. Gutynska, L.B. Wilson III, B.M. Walsh, S. v. Alfthan, O. Hannuksela, D.G. Sibeck, M. Palmroth, Ion distributions in the Earth's foreshock: hybrid-Vlasov simulation and THEMIS observations. J. Geophys. Res. Space Phys. **120**(5), 3684–3701 (2015). doi:10.1002/2014JA020519

L. Kepko, R.L. McPherron, O. Amm, S. Apatenkov, W. Baumjohann, J. Birn, M. Lester, R. Nakamura, T.I. Pulkkinen, V. Sergeev, Substorm current wedge revisited. Space Sci. Rev. **190(1)**, 1–46 (2015). doi:10.1007/s11214-014-0124-9

R.L. Kessel, E. Quintana, M. Peredo, Local variations of interplanetary magnetic field at Earth's bow shock. J. Geophys. Res. **104**(A11), 24869–24878 (1999)

J. Kissinger, R.L. McPherron, T.S. Hsu, V. Angelopoulos, Diversion of plasma due to high pressure in the inner magnetosphere during steady magnetospheric convection. J. Geophys. Res. Space Phys. **117**(A5), A05206 (2012). doi:10.1029/2012JA017579

L.M. Kistler et al., Ion composition and pressure changes in storm time and nonstorm substorms in the vicinity of the near-Earth neutral line. J. Geophys. Res. Space Phys. **111**(A11), A11222 (2006). doi:10.1029/2006JA011939

S. Knight, Parallel electric fields. Planet. Space Sci. **21**(5), 741–750 (1973). doi:10.1016/0032-0633(73)90093-7

D.J. Knipp, Advances in space weather ensemble forecasting. Space Weather **14**(2), 52–53 (2016). doi:10.1002/2016SW001366

E.A. Kronberg et al., Circulation of heavy ions and their dynamical effects in the magnetosphere: recent observations and models. Space Sci. Rev. **184**(1), 173–235 (2014). doi:10.1007/s11214-014-0104-0

M. Kubicka, C. Möstl, T. Amerstorfer, P.D. Boakes, L. Feng, J.P. Eastwood, O. Törmänen, Prediction of geomagnetic storm strength from inner heliospheric in situ observations. Astrophys. J. **833**(2), 255 (2016)

M.M. Kuznetsova, M. Hesse, L. Rastätter, A. Taktakishvili, G. Toth, D.L. De Zeeuw, A. Ridley, T.I. Gombosi, Multiscale modeling of magnetospheric reconnection. J. Geophys. Res. Space Phys. **112**(A10), A10210 (2007). doi:10.1029/2007JA012316

L.J. Lanzerotti, Comment on "Great magnetic storms" by Tsurutani et al. Geophys. Res. Lett. **19**, 1991–1992 (1992). doi:10.1029/92GL02238

B. Lavraud, J.E. Borovsky, Altered solar wind-magnetosphere interaction at low Mach numbers: coronal mass ejections. J. Geophys. Res. Space Phys. **113**(A9), A00B08 (2008). doi:10.1029/2008JA013192

B. Lavraud, J.E. Borovsky, A.J. Ridley, E.W. Pogue, M.F. Thomsen, H. Rème, A.N. Fazakerley, E.A. Lucek, Strong bulk plasma acceleration in Earth's magnetosheath: a magnetic slingshot effect? Geophys. Res. Lett. **34**(14), L14102 (2007). doi:10.1029/2007GL030024

B. Lavraud, C. Foullon, C.J. Farrugia, J.P. Eastwood, The magnetopause, its boundary layers and pathways to the magnetotail, in *The Dynamic Magnetosphere*, ed. by W. Liu, M. Fujimoto (Springer, Berlin, 2011)

L.C. Lee, Z.F. Fu, A theory of magnetic flux transfer at the Earth's magnetopause. Geophys. Res. Lett. **12**, 105–108 (1985)

R.P. Lepping, C.-C. Wu, K. McClernan, Two-dimensional curvature of large angle interplanetary MHD discontinuity surfaces: IMP-8 and WIND observations. J. Geophys. Res. **108**(A7), 1279 (2003). doi:10.1029/2002JA009640

X.L. Li, I. Roth, M. Temerin, J.R. Wygant, M.K. Hudson, J.B. Blake, Simulation of the prompt energization and transport of radiation belt particles during the March 24, 1991 SSC. Geophys. Res. Lett. **20**(22), 2423–2426 (1993). doi:10.1029/93gl02701

J. Liu, V. Angelopoulos, A. Runov, X.Z. Zhou, On the current sheets surrounding dipolarizing flux bundles in the magnetotail: the case for wedgelets. J. Geophys. Res. Space Phys. **118**, 2000–2010 (2013). doi:10.1002/jgra.50092

J. Liu, V. Angelopoulos, X.Z. Zhou, A. Runov, Magnetic flux transport by dipolarizing flux bundles. J. Geophys. Res. Space Phys. **119**, 909–926 (2014). doi:10.1002/2013JA019395

J. Liu, V. Angelopoulos, X.-J. Zhang, D.L. Turner, C. Gabrielse, A. Runov, J. Li, H.O. Funsten, H.E. Spence, Diolarizing flux bundles in the cisgeosynchronous magnetosphere: relationship between electric fields and energetic particle injections. J. Geophys. Res. Space Phys. **121**, 1362–1376 (2016). doi:10.1002/2015JA021691

M. Longmore, S.J. Schwartz, J. Geach, B.M.A. Cooling, I. Dandouras, E.A. Lucek, A.N. Fazakerley, Dawn-dusk asymmetries and sub-Alfvénic flow in the high and low latitude magnetosheath. Ann. Geophys. **23**(10), 3351–3364 (2005). doi:10.5194/angeo-23-3351-2005

M. Longmore, S.J. Schwartz, E.A. Lucek, Rotation of the magnetic field in Earth's magnetosheath by bulk magnetosheath plasma flow. Ann. Geophys. **24**(1), 339–354 (2006). doi:10.5194/angeo-24-339-2006

R.E. Lopez, M. Wiltberger, S. Hernandez, J.G. Lyon, Solar wind density control of energy transfer to the magnetosphere. Geophys. Res. Lett. **31**(8), L08804 (2004). doi:10.1029/2003GL018780

R.E. Lopez, S. Hernandez, M. Wiltberger, C.L. Huang, E.L. Kepko, H. Spence, C.C. Goodrich, J.G. Lyon, Predicting magnetopause crossings at geosynchronous orbit during the Halloween storms. Space Weather **5**(1), S01005 (2007). doi:10.1029/2006SW000222

R.E. Lopez, V.G. Merkin, J.G. Lyon, The role of the bow shock in solar wind-magnetosphere coupling. Ann. Geophys. **29**(6), 1129–1135 (2011). doi:10.5194/angeo-29-1129-2011

T.M. Loto'aniu, H.J. Singer, J.V. Rodriguez, J. Green, W. Denig, D. Biesecker, V. Angelopoulos, Space weather conditions during the Galaxy 15 spacecraft anomaly. Space Weather **13**(8), 484–502 (2015). doi:10.1002/2015sw001239

J.J. Love, Credible occurrence probabilities for extreme geophysical events: earthquakes, volcanic eruptions, magnetic storms. Geophys. Res. Lett. **39**(6), L10301 (2012). doi:10.1029/2012gl051431

J.G. Lyon, J.A. Fedder, C.M. Mobarry, The Lyon-Fedder-Mobarry (LFM) global MHD magnetospheric simulation code. J. Atmos. Sol.-Terr. Phys. **66**(15–16), 1333–1350 (2004). doi:10.1016/j.jastp.2004.03.020

P. MacNeice, M. Hesse, M. Kuznetsova, M. Maddox, L. Rastaetter, D. Berrios, A. Pulkkinen, Transforming community access to space science models. Eos **93**(15), 153–154 (2012). doi:10.1029/2012EO150002

R.L. McPherron, Growth phase of magnetospheric substorms. J. Geophys. Res. **75**(28), 5592–5599 (1970). doi:10.1029/JA075i028p05592

V.G. Merkin, J.G. Lyon, Effects of the low-latitude ionospheric boundary condition on the global magnetosphere. J. Geophys. Res. Space Phys. **115**(A10), A10202 (2010). doi:10.1029/2010JA015461

S.E. Milan, Both solar wind-magnetosphere coupling and ring current intensity control the size of the auroral oval. Geophys. Res. Lett. **36**, L18101 (2009). doi:10.1029/2009GL039997

S.E. Milan, Sun et lumiere: solar wind-magnetosphere coupling as deduced from ionospheric flows and polar auroras, in *Magnetospheric Plasma Physics: The Impact of Jim Dungey's Research*, ed. by S.W.H. Cowley Frs, D. Southwood, S. Mitton (Springer, Cham, 2015), pp. 33–64. doi:10.1007/978-3-319-18359-6_2

S.E. Milan, L.B.N. Clausen, J.C. Coxon, J.A. Carter, M.-T. Walach, K. Laundal, N. Østgaard, P. Tenfjord, J. Reistad, K. Snekvik, H. Korth, B.J. Anderson, Overview of solar wind-magnetosphere-ionosphere-atmosphere coupling and the generation of magnetospheric currents. Space Sci. Rev. **206**(1–4), 547–573 (2017). doi:10.1007/s11214-017-0333-0

Y. Miyashita et al., Geotail observations of signatures in the near-Earth magnetotail for the extremely intense substorms of the 30 October 2003 storm. J. Geophys. Res. Space Phys. **110**(A9), A09S25 (2005). doi:10.1029/2005JA011070

T. Nagai, M. Fujimoto, R. Nakamura, W. Baumjohann, A. Ieda, I. Shinohara, S. Machida, Y. Saito, T. Mukai, Solar wind control of the radial distance of the magnetic reconnection site in the magnetotail. J. Geophys. Res. Space Phys. **110**(A9), A09208 (2005). doi:10.1029/2005JA011207

R. Nakamura, W. Baumjohann, R. Schödel, M. Brittnacher, V.A. Sergeev, M. Kubyshkina, T. Mukai, K. Liou, Earthward flow bursts, auroral streamers, and small expansions. J. Geophys. Res. Space Phys. **106**(A6), 10791–10802 (2001). doi:10.1029/2000JA000306

R. Nakamura et al., Motion of the dipolarization front during a flow burst event observed by Cluster. Geophys. Res. Lett. **29**(20), 1942 (2002). doi:10.1029/2002GL015763

R. Nakamura et al., Spatial scale of high-speed flows in the plasma sheet observed by Cluster. Geophys. Res. Lett. **31**(9), L09804 (2004). doi:10.1029/2004gl019558

R. Nakamura, A. Retinò, W. Baumjohann, M. Volwerk, N. Erkaev, B. Klecker, E.A. Lucek, I. Dandouras, M. André, Y. Khotyaintsev, Evolution of dipolarization in the near-Earth current sheet induced by Earthward rapid flux transport. Ann. Geophys. **27**(4), 1743–1754 (2009). doi:10.5194/angeo-27-1743-2009

R. Nakamura et al., Flow bouncing and electron injection observed by Cluster. J. Geophys. Res. Space Phys. **118**(5), 2055–2072 (2013). doi:10.1002/jgra.50134

National Research Council, *Severe Space Weather Events—Understanding Societal and Economic Impacts: A Workshop Report* (The National Academies Press, Washington DC, 2008)

K. Nykyri, Impact of MHD shock physics on magnetosheath asymmetry and Kelvin-Helmholtz instability. J. Geophys. Res. Space Phys. **118**(8), 5068–5081 (2013). doi:10.1002/jgra.50499

M. Oieroset, D. Sundkvist, C.C. Chaston, T.D. Phan, F.S. Mozer, J.P. McFadden, V. Angelopoulos, L. Andersson, J.P. Eastwood, Observations of plasma waves in the colliding jet region of a magnetic flux rope flanked by two active X lines at the subsolar magnetopause. J. Geophys. Res. Space Phys. **119**(8), 6256–6272 (2014). doi:10.1002/2014ja020124

D.M. Oliveira, J. Raeder, Impact angle control of interplanetary shock geoeffectiveness. J. Geophys. Res. Space Phys. **119**(10), 8188–8201 (2014). doi:10.1002/2014JA020275

N. Omidi, D.G. Sibeck, Flux transfer events in the cusp. Geophys. Res. Lett. **34**, L04106 (2007). doi:10.1029/2006GL028698

M. Palmroth, T.I. Pulkkinen, P. Janhunen, C.C. Wu, Stormtime energy transfer in global MHD simulation. J. Geophys. Res. Space Phys. **108**(A1), 1048 (2003). doi:10.1029/2002ja009446

M. Palmroth, T.V. Laitinen, T.I. Pulkkinen, Magnetopause energy and mass transfer: results from a global MHD simulation. Ann. Geophys. **24**, 3467–3480 (2006)

E.V. Panov et al., Multiple overshoot and rebound of a bursty bulk flow. Geophys. Res. Lett. **37**(8), L08103 (2010). doi:10.1029/2009GL041971

E.V. Panov, W. Baumjohann, R. Nakamura, O. Amm, M.V. Kubyshkina, K.H. Glassmeier, J.M. Weygand, V. Angelopoulos, A.A. Petrukovich, V.A. Sergeev, Ionospheric response to oscillatory flow braking in the magnetotail. J. Geophys. Res. Space Phys. **118**(4), 1529–1544 (2013a). doi:10.1002/jgra.50190

E.V. Panov, M.V. Kubyshkina, R. Nakamura, W. Baumjohann, V. Angelopoulos, V.A. Sergeev, A.A. Petrukovich, Oscillatory flow braking in the magnetotail: THEMIS statistics. Geophys. Res. Lett. **40**(11), 2505–2510 (2013b). doi:10.1002/grl.50407

G. Paschmann, B.U.Ö. Sonnerup, I. Papamastorakis, N. Sckopke, G. Haerendel, S.J. Bame, J.R. Asbridge, J.T. Gosling, C.T. Russell, R.C. Elphic, Plasma acceleration at the Earth's magnetopause: evidence for reconnection. Nature **282**, 243–246 (1979)

G. Paschmann, I. Papamastorakis, W. Baumjohann, N. Sckopke, C.W. Carlson, B.U.Ö. Sonnerup, H. Lühr, The magnetopause for large magnetic shear: AMPTE/IRM observations. J. Geophys. Res. Space Phys. **91**(A10), 11099–11115 (1986). doi:10.1029/JA091iA10p11099

G. Paschmann, M. Øieroset, T. Phan, In-situ observations of reconnection in space. Space Sci. Rev. **178**, 385 (2013). doi:10.1007/s11214-012-9957-2

K.I. Paularena, J.D. Richardson, M.A. Kolpak, C.R. Jackson, G.L. Siscoe, A dawn-dusk density asymmetry in Earth's magnetosheath. J. Geophys. Res. Space Phys. **106**(A11), 25377–25394 (2001). doi:10.1029/2000JA000177

A. Petrukovich, A. Artemyev, R. Nakamura, Magnetotail reconnection, in *Magnetic Reconnection: Concepts and Applications*, ed. by W. Gonzalez, E. Parker (Springer, Cham, 2016), pp. 277–313. doi:10.1007/978-3-319-26432-5_7

Y. Pfau-Kempf, H. Hietala, S.E. Milan, L. Juusola, S. Hoilijoki, U. Ganse, S. von Alfthan, M. Palmroth, Evidence for transient, local ion foreshocks caused by dayside magnetopause reconnection. Ann. Geophys. **34**(11), 943–959 (2016). doi:10.5194/angeo-34-943-2016

T.D. Phan et al., Extended magnetic reconnection at the Earth's magnetopause from detection of bi-directional jets. Nature **404**, 848–850 (2000)

T.D. Phan, J.T. Gosling, G. Paschmann, C. Pasma, J.F. Drake, M. Oieroset, D. Larson, R.P. Lin, M.S. Davis, The dependence of magnetic reconnection on plasma beta and magnetic shear: evidence from solar wind observations. Astrophys. J. Lett. **719**(2), L199–L203 (2010). doi:10.1088/2041-8205/719/2/l199

T.D. Phan, G. Paschmann, J.T. Gosling, M. Oieroset, M. Fujimoto, J.F. Drake, V. Angelopoulos, The dependence of magnetic reconnection on plasma beta and magnetic shear: evidence from magnetopause observations. Geophys. Res. Lett. **40**(1), 11–16 (2013). doi:10.1029/2012gl054528

C. Pollock et al., Fast plasma investigation for magnetospheric multiscale. Space Sci. Rev. **199**(1), 331–406 (2016). doi:10.1007/s11214-016-0245-4

D.H. Pontius Jr., R.A. Wolf, Transient flux tubes in the terrestrial magnetosphere. Geophys. Res. Lett. **17**(1), 49–52 (1990)

K.G. Powell, P.L. Roe, T.J. Linde, T.I. Gombosi, D.L. De Zeeuw, A solution-adaptive upwind scheme for ideal magnetohydrodynamics. J. Comput. Phys. **154**(2), 284–309 (1999). doi:10.1006/jcph.1999.6299

Z.Y. Pu, J. Raeder, J. Zhong, Y.V. Bogdanova, M. Dunlop, C.J. Xiao, X.G. Wang, A. Fazakerley, Magnetic topologies of an in vivo FTE observed by Double Star/TC-1 at Earth's magnetopause. Geophys. Res. Lett. **40**(14), 3502–3506 (2013). doi:10.1002/grl.50714

T.I. Pulkkinen, M. Palmroth, R.L. McPherron, What drives magnetospheric activity under northward IMF conditions? Geophys. Res. Lett. **34**(18), L18104 (2007). doi:10.1029/2007GL030619

A. Pulkkinen, M. Hesse, S. Habib, L. Van der Zel, B. Damsky, F. Policelli, D. Fugate, W. Jacobs, E. Creamer, Solar shield: forecasting and mitigating space weather effects on high-voltage power transmission systems. Nat. Hazards **53**(2), 333–345 (2010). doi:10.1007/s11069-009-9432-x

A. Pulkkinen et al., Geospace environment modeling 2008–2009 challenge: ground magnetic field perturbations. Space Weather **9**(2), S02004 (2011). doi:10.1029/2010SW000600
A. Pulkkinen et al., Community-wide validation of geospace model ground magnetic field perturbation predictions to support model transition to operations. Space Weather **11**(6), 369–385 (2013). doi:10.1002/swe.20056
A. Pulkkinen, E. Bernabeu, J. Eichner, A. Viljanen, C. Ngwira, Earth Planets Space **67**, 93 (2015). doi:10.1186/s40623-015-0255-6
T.I. Pulkkinen, A.P. Dimmock, A. Lakka, A. Osmane, E. Kilpua, M. Myllys, E.I. Tanskanen, A. Viljanen, Magnetosheath control of solar wind-magnetosphere coupling efficiency. J. Geophys. Res. Space Phys. **121**(9), 8728–8739 (2016). doi:10.1002/2016JA023011
T. Pytte, R.L. McPherron, E.W. Hones, H.I. West, Multiple-satellite studies of magnetospheric substorms: distinction between polar magnetic substorms and convection-driven negative bays. J. Geophys. Res. Space Phys. **83**(A2), 663–679 (1978). doi:10.1029/JA083iA02p00663
J.M. Quinn, D.J. Southwood, Observations of parallel ion energization in the equatorial region. J. Geophys. Res. Space Phys. **87**(A12), 10536–10540 (1982). doi:10.1029/JA087iA12p10536
J. Raeder, Flux transfer events: 1. Generation mechanism for strong southward IMF. Ann. Geophys. **24**, 381–392 (2006)
J. Raeder, Y.L. Wang, T.J. Fuller-Rowell, H.J. Singer, Global simulation of magnetospheric space weather effects of the Bastille Day storm. Sol. Phys. **204**(1), 323–337 (2001). doi:10.1023/a:1014228230714
J. Raeder, D. Larson, W. Li, E.L. Kepko, T. Fuller-Rowell, OpenGGCM simulations for the THEMIS Mission. Space Sci. Rev. **141**(1), 535–555 (2008). doi:10.1007/s11214-008-9421-5
A. Retinò et al., Cluster multispacecraft observations at the high-latitude duskside magnetopause: implications for continuous and component magnetic reconnection. Ann. Geophys. **23**(2), 461–473 (2005). doi:10.5194/angeo-23-461-2005
I.G. Richardson, E.W. Cliver, H.V. Cane, Sources of geomagnetic storms for solar minimum and maximum conditions during 1972–2000. Geophys. Res. Lett. **28**(13), 2569–2572 (2001)
I.G. Richardson et al., Major geomagnetic storms ($Dst \leq -100$ nT) generated by corotating interaction regions. J. Geophys. Res. **111**, A07S09 (2006). doi:10.1029/2005JA011476
A.J. Ridley, T.I. Gombosi, D.L. DeZeeuw, Ionospheric control of the magnetosphere: conductance. Ann. Geophys. **22**(2), 567–584 (2004). doi:10.5194/angeo-22-567-2004
P. Riley, On the probability of occurrence of extreme space weather events. Space Weather **10**(12), S02012 (2012). doi:10.1029/2011sw000734
R.G. Roble, E.C. Ridley, A thermosphere-ionosphere-mesosphere-electrodynamics general circulation model (time-GCM): equinox solar cycle minimum simulations (30–500 km). Geophys. Res. Lett. **21**(6), 417–420 (1994). doi:10.1029/93GL03391
G. Rostoker, S.-I. Akasofu, J. Foster, R.A. Greenwald, Y. Kamide, A.T.Y. Lui, R.L. McPherron, C.T. Russell, Magnetospheric substorms—definition and signatures. J. Geophys. Res. **85**(A4), 1663–1668 (1980)
A. Runov, V. Angelopoulos, X.Z. Zhou, X.J. Zhang, S. Li, F. Plaschke, J. Bonnell, A THEMIS multicase study of dipolarization fronts in the magnetotail plasma sheet. J. Geophys. Res. Space Phys. **116**(15), A05216 (2011). doi:10.1029/2010ja016316
C.T. Russell, R.C. Elphic, Initial ISEE magnetometer results: magnetopause observations. Space Sci. Rev. **22**, 681–715 (1978)
J. Šafránková, M. Hayosh, O. Gutynska, Z. Němeček, L. Přech, Reliability of prediction of the magnetosheath BZ component from interplanetary magnetic field observations. J. Geophys. Res. Space Phys. **114**(A12), A12213 (2009). doi:10.1029/2009JA014552
A. Sandroos, I. Honkonen, S. von Alfthan, M. Palmroth, Multi-GPU simulations of Vlasov's equation using Vlasiator. Parallel Comput. **39**(8), 306–318 (2013). doi:10.1016/j.parco.2013.05.001
D. Schmid, M. Volwerk, R. Nakamura, W. Baumjohann, M. Heyn, A statistical and event study of magnetotail dipolarization fronts. Ann. Geophys. **29**(9), 1537–1547 (2011). doi:10.5194/angeo-29-1537-2011
D. Schmid et al., A comparative study of dipolarization fronts at MMS and Cluster. Geophys. Res. Lett. **43**(12), 6012–6019 (2016). doi:10.1002/2016GL069520
R. Schödel, W. Baumjohann, R. Nakamura, V.A. Sergeev, T. Mukai, Rapid flux transport in the central plasma sheet. J. Geophys. Res. Space Phys. **106**(A1), 301–313 (2001). doi:10.1029/2000JA900139
M. Scholer, Magnetic flux transfer at the magnetopause based on single X-line bursty reconnection. Geophys. Res. Lett. **15**, 291–294 (1988)
C.J. Schrijver et al., Understanding space weather to shield society: a global road map for 2015–2025 commissioned by COSPAR and ILWS. Adv. Space Res. **55**(12), 2745–2807 (2015). doi:10.1016/j.asr.2015.03.023
S.J. Schwartz et al., Cross-scale: multi-scale coupling in space plasmas. Exp. Astron. **23**, 1001–1015 (2009)
V.A. Sergeev, V. Angelopoulos, J.T. Gosling, C.A. Cattell, C.T. Russell, Detection of localized, plasma depleted flux tubes or bubbles in the midtail plasma sheet. J. Geophys. Res. **101**(A5), 10817–10826 (1996)

V.A. Sergeev, N.A. Tsyganenko, V. Angelopoulos, Dynamical response of the magnetotail to changes of the solar wind direction: an MHD modeling perspective. Ann. Geophys. **26**(8), 2395–2402 (2008). doi:10.5194/angeo-26-2395-2008

V.A. Sergeev, V. Angelopoulos, R. Nakamura, Recent advances in understanding substorm dynamics. Geophys. Res. Lett. **39**, L05101 (2012). doi:10.1029/2012gl050859

V.A. Sergeev, D.A. Sormakov, V. Angelopoulos, A missing variable in solar wind-magnetosphere-ionosphere coupling studies. Geophys. Res. Lett. **41**(23), 8215–8220 (2014). doi:10.1002/2014GL062271

J.H. Shue et al., Magnetopause location under extreme solar wind conditions. J. Geophys. Res. Space Phys. **103**(A8), 17691–17700 (1998). doi:10.1029/98JA01103

M.I. Sitnov, V.G. Merkin, J. Raeder, Great mysteries of the Earth's magnetotail. EOS Trans. AGU **97** (2016). doi:10.1029/2016EO048185

J.A. Slavin, E.J. Smith, B.T. Tsurutani, D.G. Sibeck, H.J. Singer, D.N. Baker, J.T. Gosling, E.W. Hones, F.L. Scarf, Substorm associated traveling compression regions in the distant tail: Isee-3 Geotail observations. Geophys. Res. Lett. **11**(7), 657–660 (1984). doi:10.1029/GL011i007p00657

P. Song, C.T. Russell, Model of the formation of the low-latitude boundary layer for strongly northward interplanetary magnetic field. J. Geophys. Res. **97**(A2), 1411–1420 (1992)

D.J. Southwood, C.J. Farrugia, M.A. Saunders, What are flux transfer events? Planet. Space Sci. **36**, 503–508 (1988)

H.E. Spence et al., The Magnetospheric Constellation Mission Dynamic Response and Coupling Observatory (DRACO) understanding the global dynamics of the structured magnetotail, Greenbelt, MD (2001)

H.E. Spence et al., The Magnetospheric Constellation (MC) global dynamics of the structured magnetotail, Washington, DC (2004)

J.E. Stawarz et al., Observations of turbulence in a Kelvin-Helmholtz event on 8 September 2015 by the Magnetospheric Multiscale mission. J. Geophys. Res. Space Phys. **121**(11), 11021 (2016). doi:10.1002/2016JA023458

E.C. Stone, A.M. Frandsen, R.A. Mewaldt, E.R. Christian, D. Margolies, J.F. Ormes, F. Snow, The advanced composition explorer. Space Sci. Rev. **86**, 1–22 (1998)

M. Swisdak, B.N. Rogers, J.F. Drake, M.A. Shay, Diamagnetic suppression of component magnetic reconnection at the magnetopause. J. Geophys. Res. Space Phys. **108**(A5), 1218 (2003). doi:10.1029/2002ja009726

M. Swisdak, M. Opher, J.F. Drake, F.A. Bibi, The vector direction of the interstellar magnetic field outside the heliosphere. Astrophys. J. **710**(2), 1769–1775 (2010). doi:10.1088/0004-637x/710/2/1769

K.G. Tanaka, M. Fujimoto, I. Shinohara, Physics of magnetopause reconnection: a study of the combined effects of density asymmetry, velocity shear, and guide field. Int. J. Geophys. **2010**, 202583 (2010). doi:10.1155/2010/202583

T. Terasawa et al., Solar wind control of density and temperature in the near-Earth plasma sheet: WIND/GEOTAIL collaboration. Geophys. Res. Lett. **24**(8), 935–938 (1997)

F. Toffoletto, S. Sazykin, R. Spiro, R. Wolf, Inner magnetospheric modeling with the rice convection model. Space Sci. Rev. **107**(1–2), 175–196 (2003). doi:10.1023/a:1025532008047

G. Toth et al., Space Weather Modeling Framework: a new tool for the space science community. J. Geophys. Res. Space Phys. **110**(A12), A12226 (2005). doi:10.1029/2005ja011126

G. Toth et al., Adaptive numerical algorithms in space weather modeling. J. Comput. Phys. **231**(3), 870–903 (2012). doi:10.1016/j.jcp.2011.02.006

G. Tóth et al., Extended magnetohydrodynamics with embedded particle-in-cell simulation of Ganymede's magnetosphere. J. Geophys. Res. Space Phys. **121**(2), 1273–1293 (2016). doi:10.1002/2015JA021997

K.J. Trattner, J.S. Mulcock, S.M. Petrinec, S.A. Fuselier, Location of the reconnection line at the magnetopause during southward IMF conditions. Geophys. Res. Lett. **34**(3), L03108 (2007a). doi:10.1029/2006gl028397

K.J. Trattner, J.S. Mulcock, S.M. Petrinec, S.A. Fuselier, Probing the boundary between antiparallel and component reconnection during southward interplanetary magnetic field conditions. J. Geophys. Res. Space Phys. **112**(A8), A08210 (2007b). doi:10.1029/2007ja012270

K.J. Trattner, S.M. Petrinec, S.A. Fuselier, T.D. Phan, The location of reconnection at the magnetopause: testing the maximum magnetic shear model with THEMIS observations. J. Geophys. Res. Space Phys. **117**, A01201 (2012). doi:10.1029/2011ja016959

L. Trenchi, M.F. Marcucci, G. Pallocchia, G. Consolini, M.B. Bavassano Cattaneo, A.M. Di Lellis, H. Rème, L. Kistler, C.M. Carr, J.B. Cao, Occurrence of reconnection jets at the dayside magnetopause: Double Star observations. J. Geophys. Res. Space Phys. **113**(A7), A07S10 (2008). doi:10.1029/2007JA012774

B.T. Tsurutani, F.L. Guarnieri, G.S. Lakhina, T. Hada, Rapid evolution of magnetic decreases (MDs) and discontinuities in the solar wind: ACE and Cluster. Geophys. Res. Lett. **32**, L10103 (2005). doi:10.1029/2004GL022151

L. Turc, D. Fontaine, P. Savoini, E.K.J. Kilpua, Magnetic clouds' structure in the magnetosheath as observed by Cluster and Geotail: four case studies. Ann. Geophys. **32**(10), 1247–1261 (2014a). doi:10.5194/angeo-32-1247-2014

L. Turc, D. Fontaine, P. Savoini, E.K.J. Kilpua, A model of the magnetosheath magnetic field during magnetic clouds. Ann. Geophys. **32**(2), 157–173 (2014b). doi:10.5194/angeo-32-157-2014

L. Turc, C.P. Escoubet, D. Fontaine, E.K.J. Kilpua, S. Enestam, Cone angle control of the interaction of magnetic clouds with the Earth's bow shock. Geophys. Res. Lett. **43**(10), 4781–4789 (2016). doi:10.1002/2016GL068818

L. Turc, D. Fontaine, C.P. Escoubet, E.K.J. Kilpua, A.P. Dimmock, Statistical study of the alteration of the magnetic structure of magnetic clouds in the Earth's magnetosheath. J. Geophys. Res. Space Phys. **122**, 2956–2972 (2017). doi:10.1002/2016JA023654

V.M. Vasyliunas, Theoretical models of magnetic field line merging. Rev. Geophys. **13**(1), 303–336 (1975)

V.M. Vasyliunas, Relation between magnetic fields and electric currents in plasmas. Ann. Geophys. **23**(2589), 2597 (2005)

V.M. Vasyliūnas, The largest imaginable magnetic storm. J. Atmos. Sol.-Terr. Phys. **73**(11–12), 1444–1446 (2011). doi:10.1016/j.jastp.2010.05.012

S. von Alfthan, D. Pokhotelov, Y. Kempf, S. Hoilijoki, I. Honkonen, A. Sandroos, M. Palmroth, Vlasiator: first global hybrid-Vlasov simulations of Earth's foreshock and magnetosheath. J. Atmos. Sol.-Terr. Phys. **120**, 24–35 (2014). doi:10.1016/j.jastp.2014.08.012

R. Walker et al., Source and loss processes in the magnetotail. Space Sci. Rev. **88**(1–2), 285–353 (1999). doi:10.1023/a:1005207918263

B.M. Walsh, D.G. Sibeck, Y. Wang, D.H. Fairfield, Dawn-dusk asymmetries in the Earth's magnetosheath. J. Geophys. Res. Space Phys. **117**(A12), A12211 (2012). doi:10.1029/2012JA018240

B.M. Walsh, J.C. Foster, P.J. Erickson, D.G. Sibeck, Simultaneous ground- and space-based observations of the plasmaspheric plume and reconnection. Science **343**(6175), 1122–1125 (2014). doi:10.1126/science.1247212

W. Wang, M. Wiltberger, A.G. Burns, S.C. Solomon, T.L. Killeen, N. Maruyama, J.G. Lyon, Initial results from the coupled magnetosphere–ionosphere–thermosphere model: thermosphere–ionosphere responses. J. Atmos. Sol.-Terr. Phys. **66**(15–16), 1425–1441 (2004). doi:10.1016/j.jastp.2004.04.008

D.R. Weimer, J.H. King, Improved calculations of interplanetary magnetic field phase front angles and propagation time delays. J. Geophys. Res. **113**, A01105 (2008). doi:10.1029/2007JA012452

M. Wiltberger, W. Wang, A.G. Burns, S.C. Solomon, J.G. Lyon, C.C. Goodrich, Initial results from the coupled magnetosphere ionosphere thermosphere model: magnetospheric and ionospheric responses. J. Atmos. Sol.-Terr. Phys. **66**(15–16), 1411–1423 (2004). doi:10.1016/j.jastp.2004.03.026

R.A. Wolf, M. Harel, R.W. Spiro, G.H. Voigt, P.H. Reiff, C.K. Chen, Computer-simulation of inner magnetospheric dynamics for the magnetic storm of July 29, 1977. J. Geophys. Res. Space Phys. **87**(A8), 5949–5962 (1982). doi:10.1029/JA087iA08p05949

Y. Yu, A.J. Ridley, Response of the magnetosphere-ionosphere system to a sudden southward turning of interplanetary magnetic field. J. Geophys. Res. Space Phys. **114**(A3), A03216 (2009). doi:10.1029/2008JA013292

X.Z. Zhou, V. Angelopoulos, J. Liu, A. Runov, S.S. Li, On the origin of pressure and magnetic perturbations ahead of dipolarization fronts. J. Geophys. Res. Space Phys. **119**(1), 211–220 (2014). doi:10.1002/2013ja019394

Space Sci Rev (2018) 214:17
DOI 10.1007/s11214-017-0452-7

Space Weather Effects in the Earth's Radiation Belts

D.N. Baker[1] · P.J. Erickson[2] · J.F. Fennell[3] · J.C. Foster[2] · A.N. Jaynes[1] · P.T. Verronen[4]

Received: 6 May 2017 / Accepted: 22 November 2017 / Published online: 12 December 2017

Abstract The first major scientific discovery of the Space Age was that the Earth is enshrouded in toroids, or belts, of very high-energy magnetically trapped charged particles. Early observations of the radiation environment clearly indicated that the Van Allen belts could be delineated into an inner zone dominated by high-energy protons and an outer zone dominated by high-energy electrons. The energy distribution, spatial extent and particle species makeup of the Van Allen belts has been subsequently explored by several space missions. Recent observations by the NASA dual-spacecraft Van Allen Probes mission have revealed many novel properties of the radiation belts, especially for electrons at highly relativistic and ultra-relativistic kinetic energies. In this review we summarize the space weather impacts of the radiation belts. We demonstrate that many remarkable features of energetic particle changes are driven by strong solar and solar wind forcings. Recent comprehensive data show broadly and in many ways how high energy particles are accelerated, transported, and lost in the magnetosphere due to interplanetary shock wave interactions, coronal mass ejection impacts, and high-speed solar wind streams. We also discuss how radiation belt particles are intimately tied to other parts of the geospace system through atmosphere, ionosphere, and plasmasphere coupling. The new data have in many ways rewritten the textbooks about the radiation belts as a key space weather threat to human technological systems.

Keywords Radiation belts · Space weather · Spacecraft anomalies · Particle acceleration · Magnetospheric dynamics · Wave–particle interactions

The Scientific Foundation of Space Weather
Edited by Rudolf von Steiger, Daniel Baker, André Balogh, Tamás Gombosi, Hannu Koskinen and Astrid Veronig

✉ D.N. Baker
daniel.baker@lasp.colorado.edu

1 Laboratory for Atmospheric and Space Physics, University of Colorado at Boulder, 3665 Discovery Drive, Boulder, CO 80309, USA

2 Massachusetts Institute of Technology, Haystack Observatory, Westford, MA 01886, USA

3 Aerospace Corp., Space Sciences Laboratory, Los Angeles, CA 90009, USA

4 Finnish Meteorological Institute, Helsinki, 00101, Finland

1 Introduction

Many of the space systems on which modern human society depends operate in the region from altitudes of a few hundred km to ~40,000 km above Earth's surface. This region of space is filled with the populations of energetic particles that we collectively call the *radiation belts*. The fact that the Earth is surrounded by toroids of very energetic protons and electrons was the first great discovery of the Space Age in 1958 (see Van Allen et al. 1958, 1959). From the first realization that the terrestrial magnetic field could "trap" high-energy particles, we have come today to a much more sophisticated understanding of what are appropriately often dubbed the "Van Allen Radiation Zones" (see Summers et al. 2012).

More or less from the first discovery of the radiation belts, it was recognized that intense populations of penetrating particles could be highly damaging to electronic systems in space as well as to spacecraft structural material and spacefaring human beings (Van Allen 1966). Thus, from the earliest days of the space era it was realized that the Earth's radiation belts were a challenge to be reckoned with when it came to flying satellites and human missions in near-Earth orbits. Today we recognize that radiation belt particles are one of the most pervasive and persistent threats that constitute what we comprehensively term *space weather* (Baker and Lanzerotti 2016).

This chapter is intended to provide a brief overview of radiation belt space weather impacts. A more thorough treatment of such impacts is provided in the companion chapter by Luntama et al. (2017, this issue). The further intention of this paper is to describe from a scientific perspective our present understanding of radiation belt structure, dynamics, and coupling to other parts of the geospace system. In providing such a scientific survey, the goal is to characterize as completely as possible the increasing importance of radiation belt effects on technological systems.

2 Overview of Radiation Belt Space Weather Impacts

Figure 1 shows a schematic cutaway diagram of the Earth's radiation belts. Closest to the Earth's surface is the inner Van Allen zone. This belt extends from just above the dense atmosphere out to an equatorial altitude of about 10,000 km above the Earth's surface. The inner Van Allen belt is comprised dominantly of very energetic protons (ranging up to multiple GeV energies). Recent work using the Relativistic Electron-Proton Telescope (REPT) instruments on NASA's Van Allen Probes Mission (see Baker et al. 2013b) has shown fascinating temporal and spatial behavior for the inner zone proton population (see Selesnick et al. 2014). These results demonstrate that protons with energies from ~10 MeV to ~100 MeV are quite stable in time near the geocentric radial distance of $r \sim 1.5 R_E$ (Earth radii = 6372 km) at which the inner zone proton fluxes peak. However, an outer "shoulder" of the radial distribution from $1.7 \leq r \leq 2.5 R_E$ shows tremendous temporal variability for protons with $E \lesssim 60$ MeV. These variable proton fluxes are probably due primarily to evolution of trapped solar energetic protons (Selesnick et al. 2014).

The inner Van Allen zone also has copious fluxes of low-and medium-energy electrons (Fennell et al. 2015; Li et al. 2015) as revealed by Van Allen Probes and other spacecraft data sets. However, the Van Allen Probes era (September 2012 → present) has provided many new discoveries about inner magnetospheric ultra-relativistic electrons with energies $E \gtrsim 2.5$ MeV. In particular, initial results after major storm intervals surprisingly showed essentially no detectable prompt ultra-relativistic electron fluxes in the region $L \lesssim 2.8$ (Baker et al. 2014a), but over longer time scales these electrons did eventually move inward from locations beyond this radial distance to $L \lesssim 2.5$. The paucity of

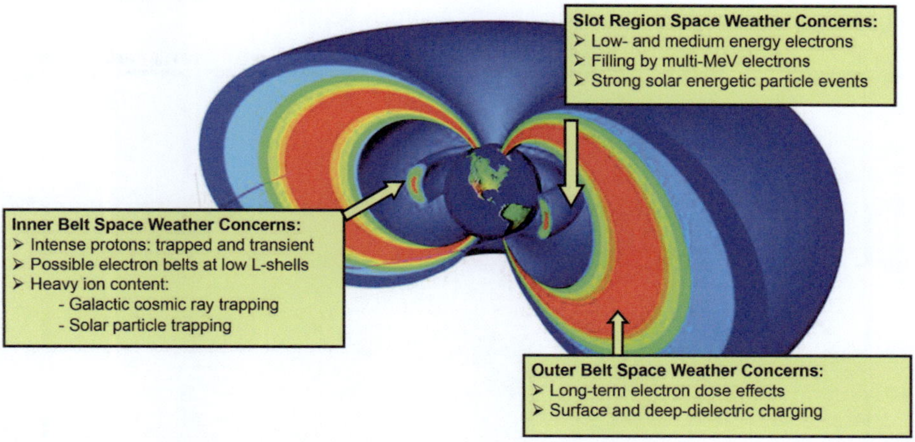

Fig. 1 Schematic diagram of Earth's radiation belt also designating some of the space weather concerns in each region

very energetic electrons in the inner magnetosphere immediately following major magnetic storms is quite striking (Foster et al. 2016; Baker et al. 2016) and the consequences for space weather effects will be discussed further in this chapter (see also Boyd et al. 2016; Baker et al. 2016; and Kim et al. 2016).

As shown by the leftmost label in Fig. 1, the space weather concerns for the inner radiation belt zone are several. The intense, high-energy trapped protons are extremely damaging to space systems (Vette 1966) and are quite hard to shield against. There are also the variable, trapped solar energetic particles in the inner zone that can cause dose and single-event effects (see Lorentzen et al. 2002; Baker 2002). The inner zone also has trapped galactic cosmic rays (see Klecker et al. 1995; Cummings et al. 1993). Finally, the trapped energetic electrons with $E \lesssim 1$ MeV (Fennell et al. 2015; Claudepierre et al. 2017) also represent a further significant source causing total dose effects.

The so-called "slot region" of the radiation belts extends from roughly $L \sim 2.0$ to $L \sim 3.0$ depending on energy and particle species. (L is the geocentric distance in Earth radii at which a dipole magnetic field line crosses the magnetic equatorial plane). The slot is a region often relatively devoid of energetic electrons. However, during strong geomagnetic storm periods, the gap between the inner and outer zone can be filled to a large degree by moderate (and even high) energy electrons (Fennell et al. 2005, 2012; Turner et al. 2016; see also Fig. 6). For example, in the intense "Halloween" storm period of late-October and November 2003, the slot region was filled with multi-MeV electrons for several weeks (Baker et al. 2004a). Thus, as indicated in Fig. 1, the slot region can present several space weather concerns including low- and medium-energy electron enhancements, multi-MeV electrons (on rare occasions), and strong solar energetic particle events (again on relatively rare occasions).

Finally, the outer Van Allen radiation belt represents in many ways the most pervasive space weather risks to operating spacecraft. As shown in Fig. 1, the outer zone is broad in spatial extent (from $r \sim 3R_E$ to $r \gtrsim 6.5R_E$). It is comprised of mildly to highly relativistic electrons ($\sim$100 keV to $\gtrsim$10 MeV) and varies widely in particle intensity. Commercial, military, and scientific satellites operating in medium-Earth orbit (MEO) and geostationary Earth orbit (GEO) number in the multiple hundreds worldwide. All of these operating spacecraft are subject to outer Van Allen belt space weather impacts.

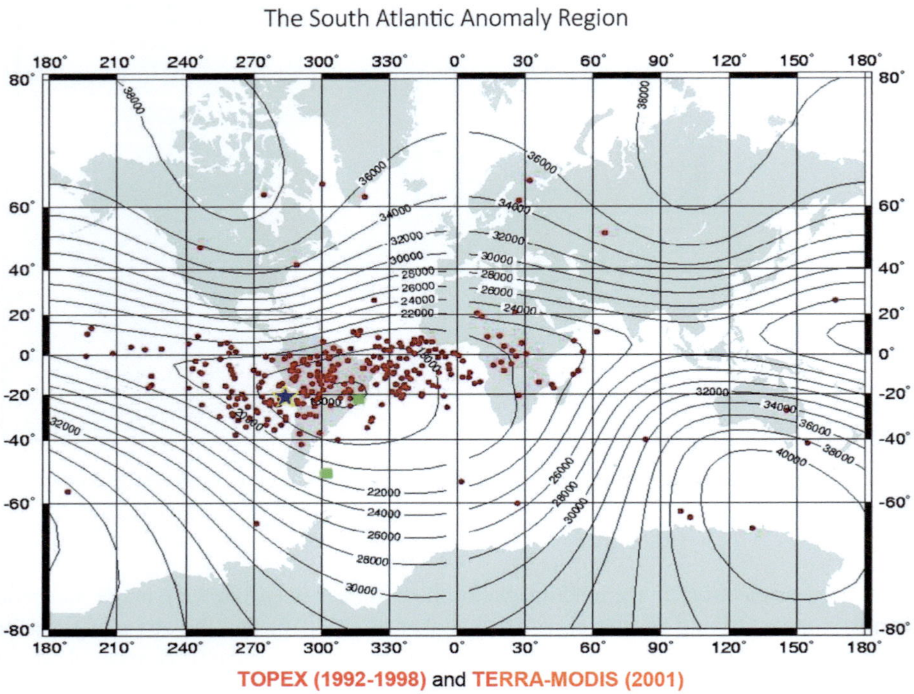

Fig. 2 A map of the Earth showing contours of constant surface magnetic field strength. The weakest field region over South America is called the South Atlantic Anomaly (SAA). The *red symbols* show where significant operational anomalies occurred for the TOPEX and TERRA spacecraft

The record of operating spacecraft over the Space Age is laden with examples of anomalies and outright satellite failures due to space weather effects (Allen 2002). Many operational problems occur in the so-called South Atlantic Anomaly (SAA) region as shown here in Fig. 2. As illustrated in the diagram, the Earth's surface magnetic field is weakest over Brazil and eastward into the South Atlantic Ocean region. This is due to the offset, tilted dipole nature of Earth's intrinsic magnetic field (Cain 1966). Trapped particles mirroring along magnetic field lines, especially high-energy protons in the inner zone, can approach closest to the Earth's surface in the SAA weak field region. Thus, this is the place where low-Earth orbit (LEO) spacecraft encounter the most intense particle fluxes. Figure 2 shows that the TOPEX (U.S./French Ocean Surface Topography mission) and TERRA (flagship Earth Observing System-1 satellite at 705 km altitude) spacecraft suffered a very large number of electronic anomalies and upsets in the SAA (Allen 2002), probably due to single-event effects from inner zone protons.

Figure 3 shows a schematic diagram (Baker 2002) of a spacecraft at some location within the Earth's magnetosphere. At relatively high altitudes, the spacecraft can be subject to a variety of effects. Both solar energetic particles (SEPs) and trapped energetic protons can cause significant energy deposition insensitive electronics on board a given spacecraft. Similarly, galactic cosmic ray nuclei can pass through electrical components and induce radiation degradation effects. As illustrated in Fig. 3, these ion interactions effects are termed "single event effects" (SEE).

Another effect of the space environment is called "deep-dielectric charging" and is indicated in Fig. 3b. Electrons with energies of hundred of keV (and up to multiple MeV

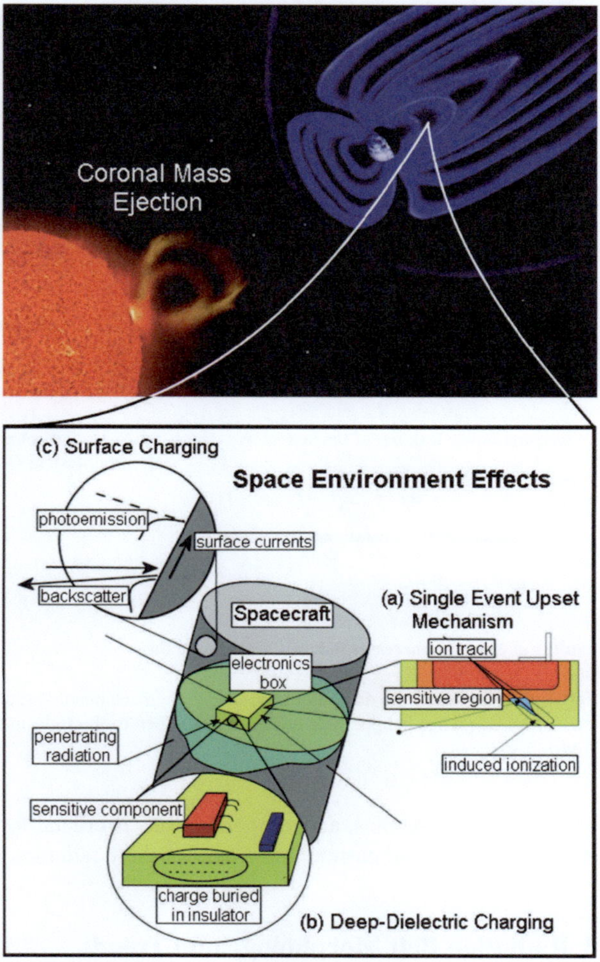

Fig. 3 A schematic diagram showing spacecraft anomaly mechanisms: (a) single-event ion effects; (b) Deep-dielectric charging due to high-energy electrons; and (c) Surface charging due to low-to-medium energy electrons (from Baker 2002)

energies) can penetrate shielding and bury themselves in dielectric materials inside spacecraft subsystems. As shown by Fig. 4, if the build up of buried charge is fast enough (and the leakage of charge out of the dielectric material is slow enough, then a powerful discharge can occur. This can damage materials and sensitive electronic components.

A third kind of space weather impact within the radiation belts is shown in Fig. 3c. This is called "surface charging". If the space system is subject to hot plasma in its vicinity and photo-electrons are not able to carry charge away promptly from surface insulator and dielectric material (as happens in solar eclipse conditions or on the dark side of a space vehicle) there can be a powerful build-up of charge on parts of the spacecraft surface. When the charge buildup is great enough, there can be a sudden discharge that can produce material damage and/or can produce a significant electromagnetic disturbance around the spacecraft. This can cause phantom signals or changes of state in memory systems, and even permanent damage to electronics.

Obviously, from the above brief description it is clear that the radiation belts can cause a wide variety of impacts on space systems. Thus, having a deeper understanding of radiation

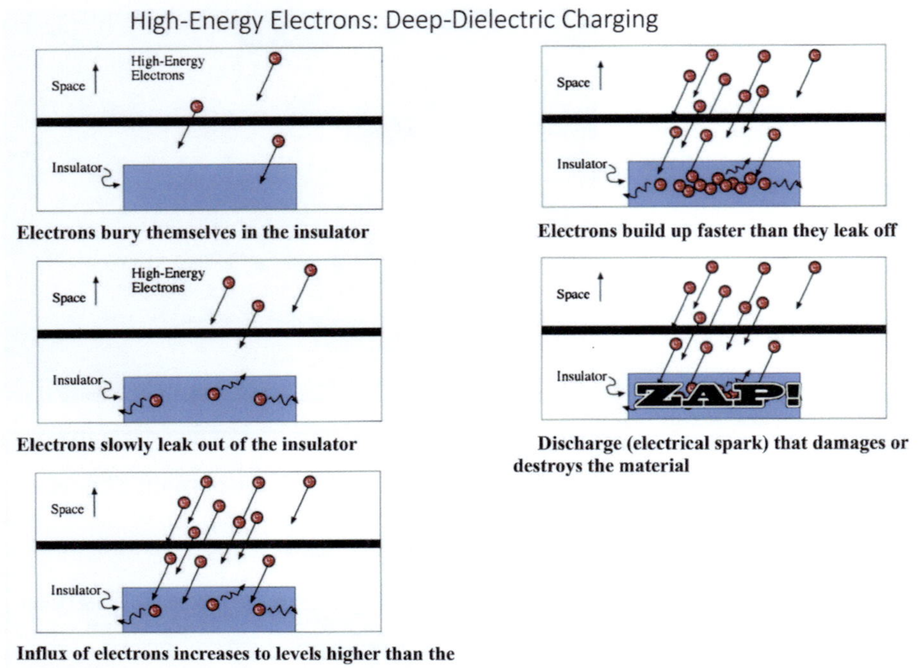

Fig. 4 Illustration of the deep-dielectric discharge mechanism that can cause severe spacecraft operational anomalies. Sequence is upper left downward and then to the right in the diagram (see Hastings and Garrett 1996)

belt structure, dynamics, and temporal trends is crucial for our technological society. This chapter describes our current knowledge of these radiation belt aspects.

3 Radiation Belt Morphology and Trends

This section discusses our present understanding of the temporal and spatial properties of the radiation belts as they relate to space weather effects. We begin with a description of the inner zone properties and impacts. This is followed by a description of the slot region and then outer zone structure and dynamics.

3.1 Inner Zone Electrons

The inner radiation belt was the first region that was accessible to early satellite measurements. Unfortunately, very early on, before comprehensive measurements of the natural radiation there had been completed, there was a succession of high-altitude nuclear tests that injected populations of electrons into these regions (Van Allen et al. 1959). Many of the injected electrons were lost fairly rapidly but some were still present many years later because they were injected into regions where the electron loss rates were quite low (Van Allen 1966). The STARFISH event was one of those that generated enduring electrons because of the L-position where its electron belt formed. This made it difficult to specify what the natural electron environment was and what the sources were during the early times in space science (see Gombosi et al. 2017, this issue). This was compounded by the fact

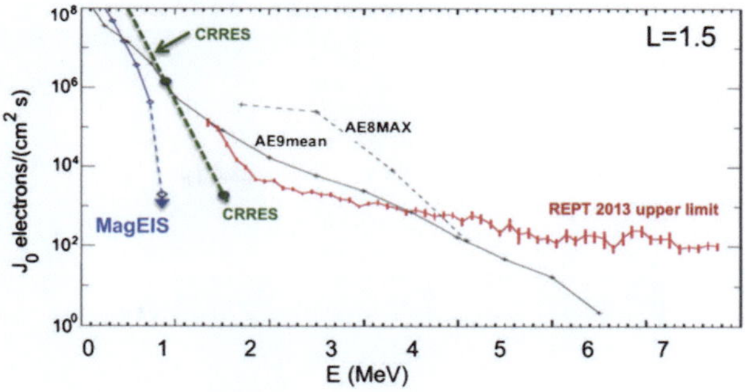

Fig. 5 Upper limit electron flux estimates for $L = 1.5$ from the Van Allen Probes mission (MagEIS and REPT) compared to CRRES observations and the electron models AE8MAX and AE9mean (after Fennell et al. 2015)

that the inner zone protons are of high energy and can penetrate electron sensors, contaminating many of the attempted electron measurements. The early electron radiation models (Vette 1966; Teague and Vette 1972, 1974; Teague et al. 1976; Sawyer and Vette 1976; Vette 1991) for the inner zone were built on a very limited set of observations and there was concern that their electron fluxes did not represent the real flux levels (Daly et al. 1996; Barth et al. 2003).

Recently, the NASA Van Allen probes mission made measurements in the inner zone and, subsequently, upper limits on the electron fluxes there were developed (Li et al. 2015; Fennell et al. 2015, 2016) as shown in Fig. 5. These new measurements corrected for much of the proton backgrounds using capabilities built into the sensors (Claudepierre et al. 2015; Fennell et al. 2016; Blake et al. 2013; Baker et al. 2013b). As a result, they showed that real electron fluxes >0.5 MeV are indeed significantly lower than the commonly used radiation model fluxes. They also showed that there were essentially no (or very few) $E > 1$ MeV electrons present, at least in the early Van Allen Probes mission time frame. Selesnick (2015) showed that the neutron albedo generated electron fluxes are too low to account for the observed $E < 1$ MeV inner zone electrons, indicating that a different source was responsible for the inner zone electron fluxes. However, his calculations also indicated that the process could generate multi-MeV electrons at low flux levels, below the upper limits set by the Van Allen Probe measurements.

As is noted below for Fig. 13, and is shown here in Fig. 6 from the pre-Van Allen Probe era, there are episodic incursions of $E \geq 1$ MeV electrons into the low-L regions and even into the inner zone, but their presence is not the norm. Figure 6 shows HEO3 satellite (Fennell et al. 2012) measurements for > 1.5 and >3 MeV electrons and >15 MeV protons in the $1.75 < L < 7$ region during the 1998 through 2011 period that includes solar sunspot cycle 23. The figure shows multiple instances of the electrons penetrating to $L < 3$ and a few to $L \leq 2$. The Halloween storm of 2003 (Day 304) accelerated and transported >1.5 and >3 MeV electrons through the slot region ($2 < L < 3$) and into the inner zone (see also Baker et al. 2004a; Baker et al. 2007) while other large storms during solar cycle 23 accelerated and transported electrons to $L \sim 2$. See also Fig. 7 from Van Allen Probes measurement.

Horne et al. (2005) and Shprits et al. (2006) have speculated that the erosion of the plasmasphere to low L could create temporary conditions that are conducive to local acceleration there. If MeV electrons can be accelerated, injected, or transported deep in the inner zone where the electron lifetimes typically exceed hundreds of days (Van Allen 1964; Roberts 1968) and they can persist for years at observable levels as happened with the electrons injected into the belts by the STARFISH event. These lifetimes are generally controlled by plasmaspheric hiss (Summers et al. 2007) however losses can be enhanced by chorus and EMIC waves during periods where the plasmasphere is strongly eroded.

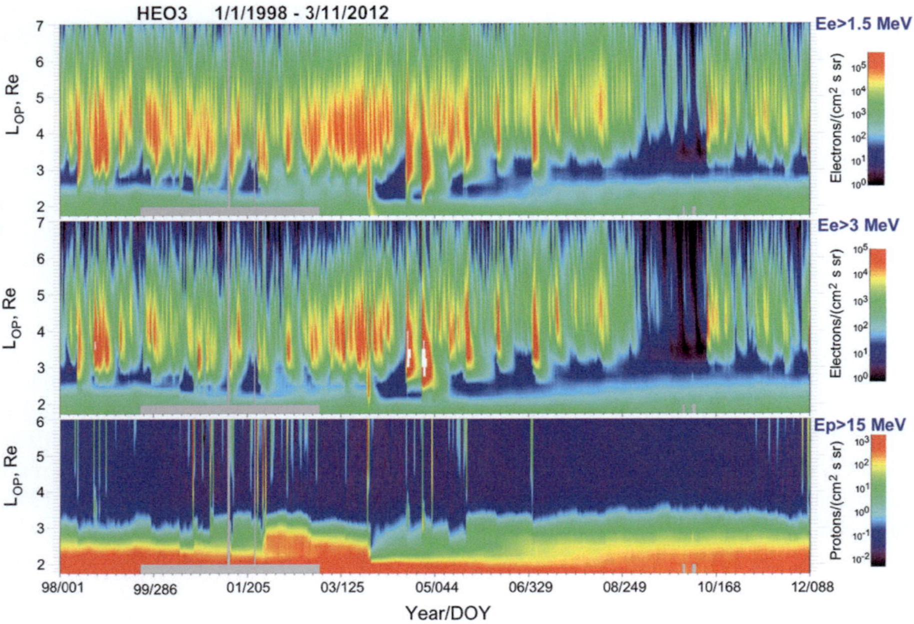

Fig. 6 Electron and Proton spectrograms covering the interval 1998–2012 for >1.5 and >3 MeV electrons and >15 MeV protons. The *gray bar* at $L \sim 1.75$ indicates no data taken at $L < 2$ during that interval

June 2015 was the most recent period where MeV electrons were injected into the inner zone (Claudepierre et al. 2017) as shown in Fig. 7. These electrons persisted for months at $L \sim 1.75$. As Fig. 8 shows, $E < 800$ keV electrons can be transported to low L-values and into the inner radiation zone during storms (see Baker et al. 2007). The lower the energy, the more common are such injections to the lowest L-values. Note, for example, that the 36 keV to 460 keV electrons in Fig. 8 were injected to $L < 2$ multiple times during the $\sim$2 years of data shown. In fact, many of these injections caused electron enhancements at <1.5 and, during the March 2015 storm, enhanced fluxes were observed down to $L \sim 1.2$ especially at energies <460 keV. Such penetrations of electrons into the inner zone where their lifetimes are long can be a hazard for satellites orbiting there from both the total dose perspective and from the subsurface/internal charging perspective. So, while the occurrence of >1 MeV electrons in the inner zone is relatively rare, they can become an enhanced hazard when they occur and their populations can persist for months (Shprits, 2009).

Turner et al. (2016) investigated "sudden particle enhancements at low L" (SPELLS) as slot filling events and as a source of inner zone electrons. The study found that the enhancements occurred much more often for electrons $E < 600$ keV than for those $E > 600$ keV as shown in Fig. 9, reproduced from that paper. Turner and colleagues argue that these electron enhancements or injections are most likely the dominant source for most of the electrons in the inner zone. These more common enhancements of the 100–800 keV electrons, as shown in Fig. 8, can penetrate thermal control material and wiring insulation to become a significant electrostatic discharge threat to space systems. For example, 300 keV electrons have sufficient range to penetrate typical thermal blanket material and deposit their charge on underlying insulators where it can build up to high-voltage breakdown levels as shown in Fig. 4.

A clear result of these studies found that the quiet-time phase space density radial profiles in the inner zone are characterized by positive slopes with increasing L (or L^*) for $L <$

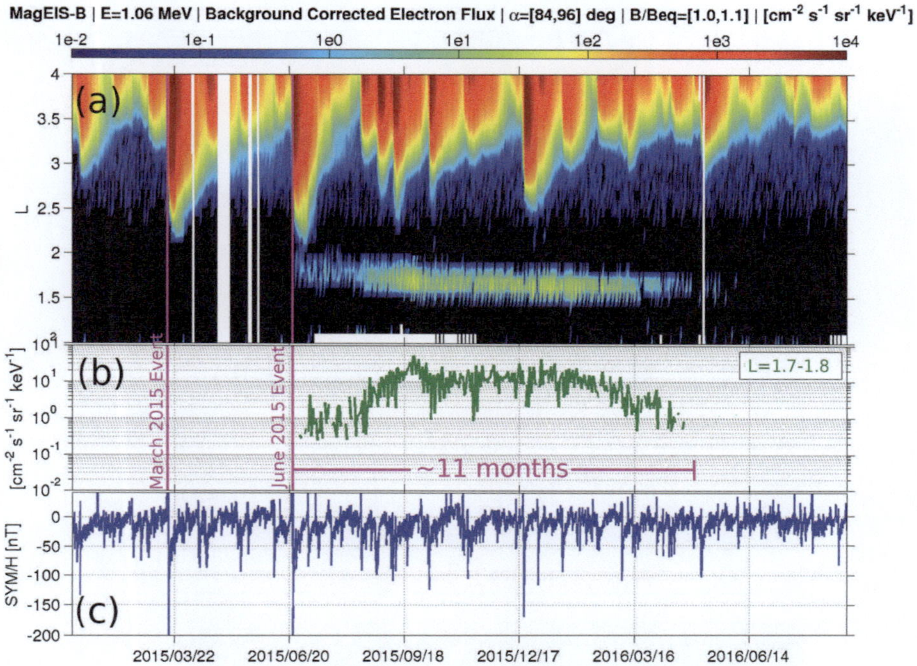

Fig. 7 MagEIS equatorially mirroring electron fluxes showing the injection/transport of 1.06 MeV electrons into the inner zone during the June 2015 storm period (after Claudepierre et al. 2017)

2 and beyond for the electrons with the lowest magnetic moments as shown in Fig. 10 (Fennell et al. 2015). Figure 10 also indicates that a peak in the radial profiles develops near $L \sim 2$ at the larger magnetic moment values. Turner et al. (2016) showed the peak in the phase space density radial profiles develops in this region as the electrons injected into and through the slot region began to decay away. Radial flux profiles also remained stable in the $L < 1.8$ inner radiation zone, as can be seen by comparing Figs. 10 and 11. Figure 11 is excerpted from a longer movie (O'Brien et al. 2016) and shows the radial profiles for equivalently constant magnetic moments, with clear dynamical variations in phase space density radial profiles at low L. The Fig. 11 frame is a snapshot of the PSD radial profiles observed following the 17 March 2013 storm time injection (see Baker et al. 2014b). This can be contrasted with the quiet time radial profiles in Fig. 10. The post-storm 20 March 2013 radial profiles do not show a peak near $L = 2$ and indicate there is a source of electrons external to the slot region that were transported inward to low L. This indicates that the electrons transported to low L are most likely the source of the inner zone electrons.

Features still in need of further examination are the PSD radial profiles at $L < 2$. First impressions, from examining many intervals of the PSD radial profiles (Fennell et al. 2015; Turner et al. 2016; O'Brien, private communication), are that for $L < 2$ the PSD profiles are relatively constant in shape and levels over time. This means that the electron losses there are being well matched by inward transport. Electron lifetime estimates indicate that at $L < 1.5$ they are consistent with losses from atmospheric scattering (Walt 1966). So, the processes of inward transport of electrons must balance the estimated loss rates. In any case, the transport, buildup and penetration of electron fluxes through the slot region and into the inner zone represent a time dependent threat from the subsurface/internal charging

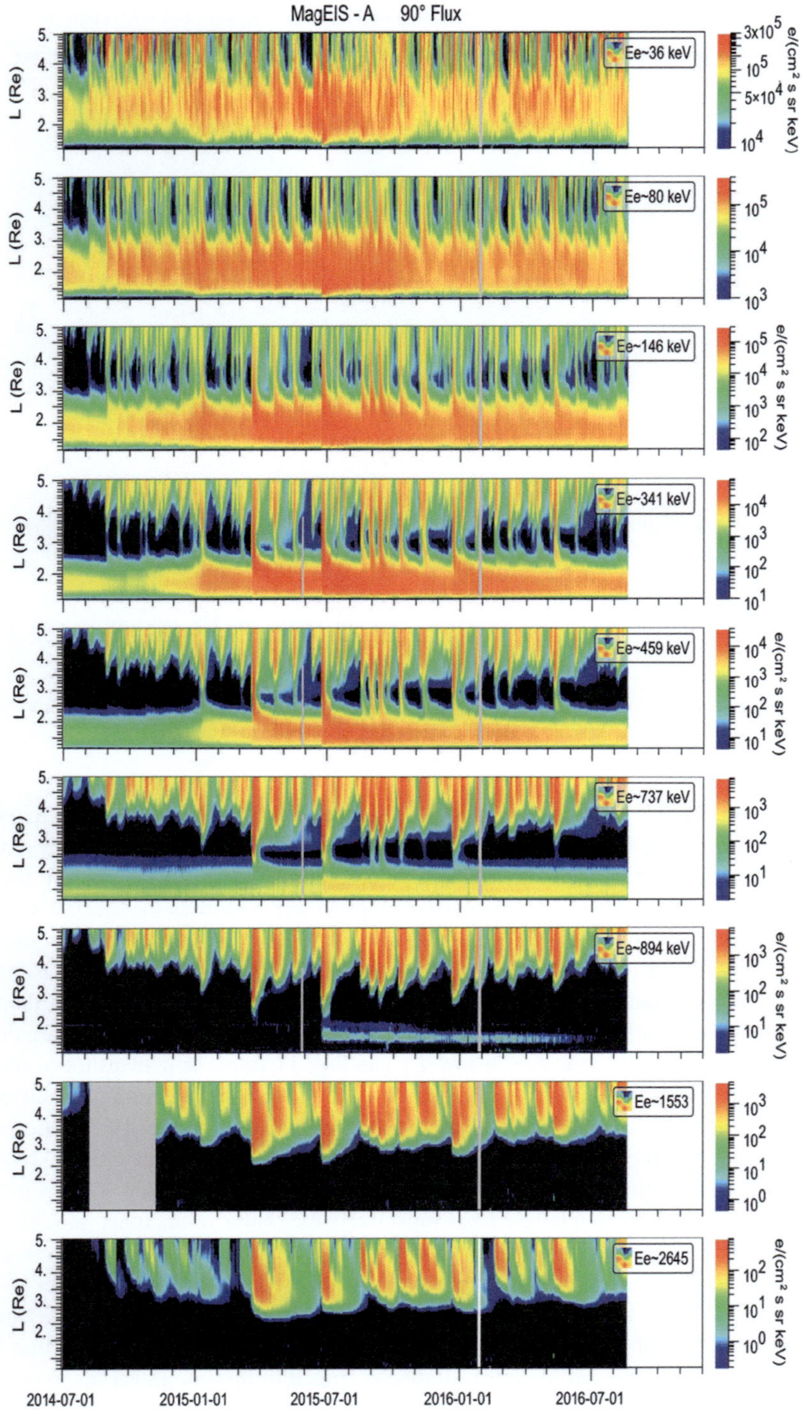

Fig. 8 Electron locally mirroring flux history for $1.2 < L < 5$ during the July 2014–August 2016 period. The electron energies range from 36 keV to 2.645 MeV as noted in the panel legends. The sudden flux enhancements or injections to low L values is easily seen at the lower energies (<900 keV) whereas at > MeV energies they do not penetrate deeply into the slot region

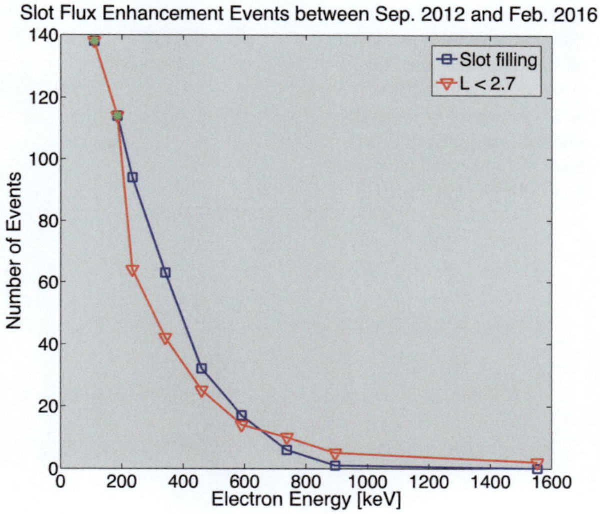

Fig. 9 Number of low L electron flux enhancements as a function of energy that penetrated to $L < 2.7$ and those which were slot filling during the Sep. 2012 to Feb. 2016 period (after Turner et al. 2016)

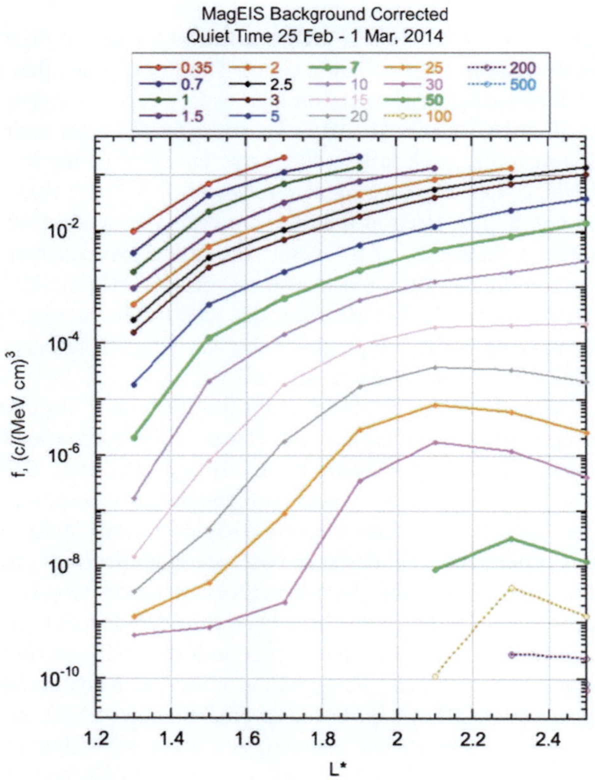

Fig. 10 Quiet time phase space density radial profile of quiet time inner zone electrons for a range of constant first adiabatic invariant (after Fennell et al. 2015)

perspective that cannot be ignored. Since inner zone is within the high-density plasmasphere, surface charging is not a threat there even though there are significant and variable fluxes of ≤80 keV electrons there as shown in Fig. 8.

Fig. 11 Phase space density (PSD) radial profiles at constant ξ in the inner zone and slot region for 20 March 2013. $\xi = M/\gamma^2$ where M is the first adiabatic invariant and γ is the sine of the equatorial pitch angle (after O'Brien et al. 2016)

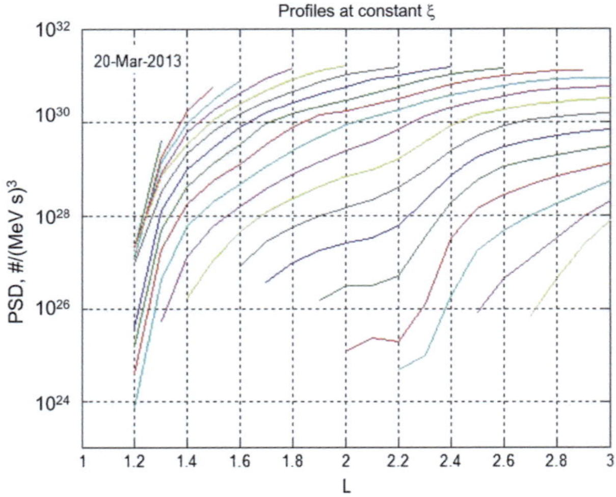

3.2 Slot Region Electrons

The region $2 \leq L \leq 3$ is known as the electron slot. It is a transition region between the outer and the inner radiation zones. The slot region's flux levels are highly time dependent and the spatial structure is strongly energy dependent (Turner et al. 2015 and 2016; Reeves et al. 2016; Zhao et al. 2016). During extended quiet periods the electron fluxes can decay to low levels, especially at the higher energies, giving rise to PSD radial profiles that peak in the $2 < L < 3$ range for electrons with $E \leq$ MeV. Both of these features are clearly seen in Fig. 8. The electron lifetimes are energy dependent with the dominant loss being from pitch angle scattering by waves. Inside the plasmasphere hiss scattering of the electrons is the dominant loss mechanism (Lyons and Thorne 1973; Lyons et al. 1972; Abel and Thorne 1998a, 1998b; Meredith et al. 2009). The energetic electrons (0.2–1.5 MeV) that are transported into the slot region decay fairly rapidly (Abel and Thorne 1998a, 1998b; Fennell et al. 2012; Claudepierre et al. 2017).

What has been notable is that ultra-relativistic electrons, $\geq$2.5 MeV, seldom penetrate deeply into the slot region at L values < 2.8 over prompt time scales following acceleration/injection events (Baker et al. 2014a; Foster et al. 2016) as is evident from Figs. 6, 9, and 12. (On longer $\sim$ weeks to months scales, however, gradual inward radial diffusion does seem to move these electron energies inwards of this radial distance to $L \sim 2.5$.) While this penetration limit has been observed multiple times preceding and during the Van Allen Probes mission, at this point the processes that are responsible for this condition are still under discussion. Three that have been considered are: (1) A balance between diffusive radial transport and scattering losses that limit the ability of highly relativistic electrons to penetrate to low L values during storms of the Van Allen Probes era (Lyons and Thorne (1973); Baker et al. 2014a); (2) localized precipitation of MeV electrons through wave mode coupling and wave–particle interactions induced by VLF transmitter signals at their maximum outward extent near $L = 2.8$ (Foster et al. 2016); and (3) the effect of VLF transmitter-induced waves on the coherency of the wave–particle interactions, creating an earthward limit to the local acceleration of electrons to MeV energies. (Regarding this latter possibility, see as well the discussion on possible anthropogenic VLF transmission effects within Gombosi et al., this issue.) Whatever the mechanism, the fact that high fluxes of multi-MeV electrons are seldom present deep in the slot means that the threat of internal charging

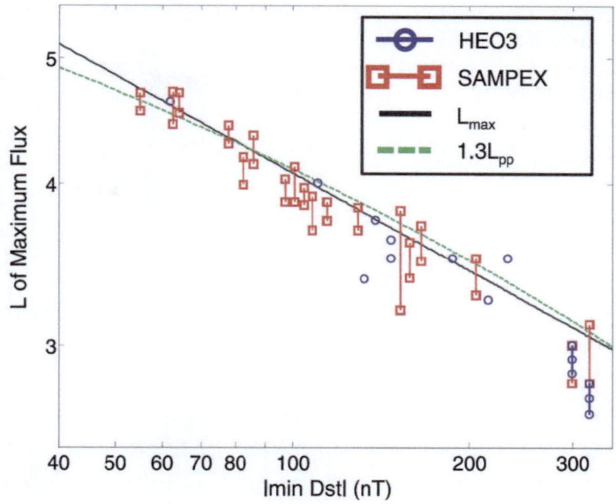

Fig. 12 Location of the peak in >1.5 MeV electron flux from HEO3 and >1 MeV electrons from SAMPEX as a function of minimum DST. Also plotted are the function LMAX from Tverskaya et al. (2002) and a scaled plasma pause model Lpp (after O'Brien et al. 2003)

and highly penetrating radiation dose is minimal throughout that region most of the time. In fact, as Fig. 12 shows, it takes large storms to position the peak of the MeV electron fluxes at $L < 3$. The figure also implies that there is a linkage between the plasmapause position and the L of maximum MeV fluxes. Thus, high fluxes of MeV electrons will populate the outer edge of the slot region for large storms, as measured by Dst or by the plasmapause erosion. When this occurs, deep dielectric charging will become a hazard for satellites orbiting there.

The mechanisms that cause injections of the lower energy electrons into and through the slot region (see Russell and Thorne 1970) have been investigated by Thorne et al. (2005), Zhao and Li (2013), and Turner et al. (2015, 2016). Zhao and Li modeled the transport into the slot region and inner zone as diffusive transport using a 1-D radial diffusion model. However, they used diffusion coefficients that were much larger than the usual Kp-dependent electromagnetic diffusion values and concluded that the usual Kp dependent coefficients cannot be simply extended to the slot region and inner zone. Turner et al. (2015, 2016) and others (Sergeev et al. 1998; Runov et al. 2014; Takahashi et al. 2003) considered how magnetosonic waves and compressional fast mode waves might be responsible for the sudden enhancements in the electron fluxes at low L-values in the plasmasphere. After examining the electric and magnetic field fluctuations during an injection of electrons to low L, Turner et al. (2015) proposed that Pi2 cavity mode waves could explain the electron flux enhancements inside the plasmasphere. They also speculated that the upper energy limit to the enhancements could result from the radial gradient of the PSD combined with the efficiency of the interaction with the waves. Thus, sudden enhancements of energetic electrons associated with substorm activity and injections at higher L shells can penetrate very deep into the inner magnetosphere and may ultimately play an activity-dependent role in the formation of the inner zone population as discussed above. In any case, as noted previously, the intense but variable hundreds of keV electron fluxes can be a hazard from the vehicle charging perspective which can include surface charging during those intervals when the plasmasphere is eroded to $L < 3$.

3.3 Inner Zone and Slot Region Protons

Only the more energetic ions/protons in space are of concern from a space weather hazard perspective. Very high energy protons and ions can penetrate shielding, depositing their

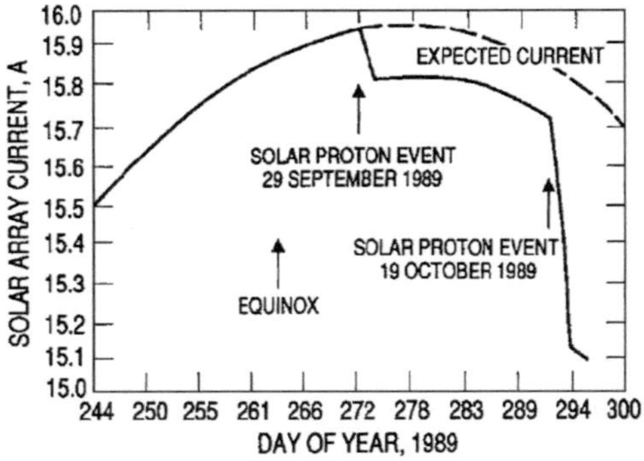

Fig. 13 Observed solar array current reduction on GOES-7 following solar energetic particle events in the fall 1989

energy in electronic devices. The hazardous proton populations of the inner zone and slot regions have two fundamental sources. In the inner zone the primary source is the neutron albedo process, which has been well defined (Selesnick et al. 2007 and references therein). The outer edges of the inner zone and the lower part of the slot region also host energetic protons and ions that consist of entrapped solar particles (Selesnick et al. 2007; Lorentzen et al. 2002; Mazur et al. 2005). Protons with $E > 3$ MeV are a primary contributor to solar array degradation while >15 MeV protons contribute to single event effects (e.g. Fig. 2) and non-ionizing radiation damage to electronic parts within satellites that orbit in or traverse these regions. In addition to the trapped population, there are also episodic solar particle events, with associated energetic solar ions. Those ions that eventually reach the Earth can penetrate deeply into the inner magnetosphere where they can cause single event upsets, as in Fig. 14, and enhance the radiation damage to satellite solar arrays and electronic parts, similar to that shown in Fig. 13 (Marvin and Gorney 1991. Figure 6 shows multiple incidences of >15 MeV solar protons penetrating to low L combined with the presence of the protons from cosmic ray albedo neutron decay (CRAND) at very low L and inherent variability of the proton fluxes in the $2.5 < L < 4$ region in response to magnetic storms (Lorentzen et al. 2002; Mazur et al. 2005).

Solar Energetic Particle (SEP) events are the mechanism by which very high-energy protons appear in the outer zone region of the radiation belts. However, these events are very short-lived (hours in duration), as the high-energy protons with such large gyro-radii are not able to be stably trapped at these locations. As mentioned earlier, they can often penetrate to the inner zone where stronger magnetic fields in a more rigid topology can entrain these particles for very long times. In general, highly energetic protons (multi-MeV energies) are not considered to be a regular feature in the outer zone.

3.4 Outer Zone Electrons

The characteristics and dynamics of ultra-relativistic and MeV electron fluxes in Earth's outer radiation belt are strongly influenced by geomagnetic storm time processes, leading both to significant electron loss and local acceleration deep within the inner magnetosphere (Ukhorskiy et al. 2014; Reeves et al. 2013). The outer Van Allen belt is highly variable on essentially all temporal scales and shows tremendous variability as well as significant variations in the outer belt spatial extent. This is well illustrated by Fig. 14 adapted from Li et al. (2006). The main (lower) panel of the figure shows integral fluxes of $2 \lesssim E \lesssim 6$ MeV electrons measured by sensors on board the LEO spacecraft called the Solar, Anomalous, and Magnetospheric Particle Explorer (SAMPEX) mission (see Baker et al. 1993). The data are color-coded representations of electron flux from $L = 2.0$ to $L = 7.0$ (vertical axis)

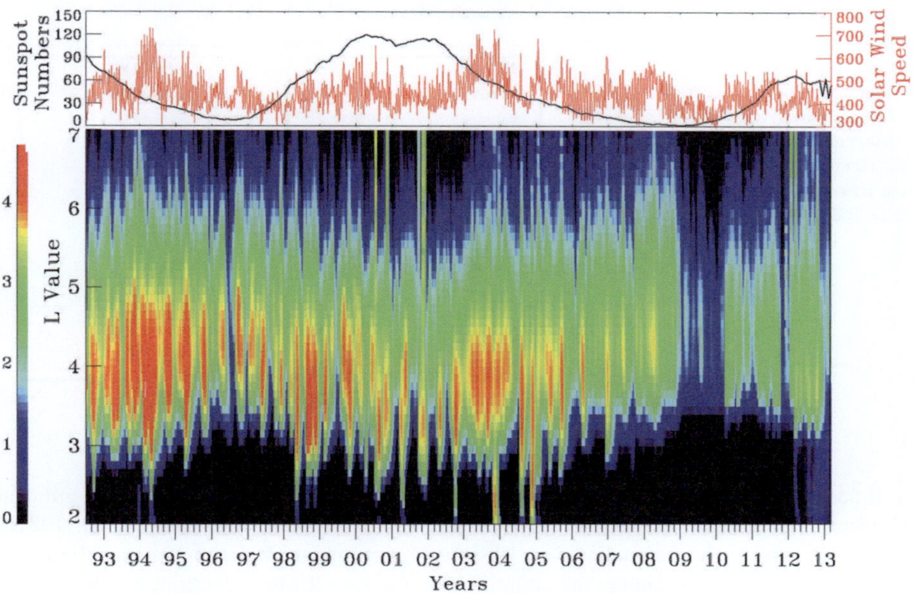

Fig. 14 Long-term (1992–2013) record of $2 \lesssim E \lesssim 6$ MeV electrons measured primarily by the SAMPEX spacecraft at low altitude. The *upper panel* shows smoothed sunspot number (*black*) and solar wind speed (*red*) for the same period of time. (Courtesy of X. Li)

versus time (horizontal axis) from 1992 to 2013. The color bar to the left is the log of electron flux (electrons/cm^2-s-sr). The upper panel of Fig. 14 shows in black the smoothed sunspot number over the period 1992–2013 while the red trace shows the solar wind speed (km/sec) measured upstream of the Earth's magnetosphere.

As is evident from the figure, the outer Van Allen belt (as monitored from the ~500 km altitude of SAMPEX) typically extends from $L \sim 3.0$ to roughly $L = 6.5$. However, the belt's spatial extent was highly variable over the ~ 20-year SAMPEX lifetime. Also, the peak electron flux levels varied widely as a function of time. For example, in 1994, the peak >2 MeV electron fluxes were over 10^4 electrons/cm^2-s-sr, while in 2008–2009 the electron flux was often barely above background levels (~ 10^0 electrons/cm^2-s-sr). As is obvious from inspection of the figure, a main controller of the electron fluxes in the outer belt is the solar wind speed impacting Earth's magnetosphere (Baker et al. 1979, 1986), Kellerman et al. (2013), Kellerman and Shprits (2012). Indeed, solar wind speeds with $V \gtrsim$ 500 km/sec generally are associated with high outer radiation belt electron "content" (Baker 2004). On the other hand, persistent solar wind speeds of $V \lesssim 300$ km/sec can lead to virtual disappearance of the outer belt (for electron energies $E \gtrsim 1$ MeV). This occurred quite prominently in the profound solar minimum of 2008–09.

The inner "edge" of the outer zone — as is obvious in Fig. 14 — shows considerable variability in spatial location. During particularly strong solar wind forcing conditions, the inward extent of the outer zone population can appear to reach down to $L \sim 2.5$. This is typically a rare and brief circumstance. During truly extreme forcing events such as the Halloween Storm events of late 2003 (see Baker et al. 2004a), the slot region can be filled with multi-MeV electrons for weeks of time. An event of great strength was also seen in late 2004 (see Fig. 13).

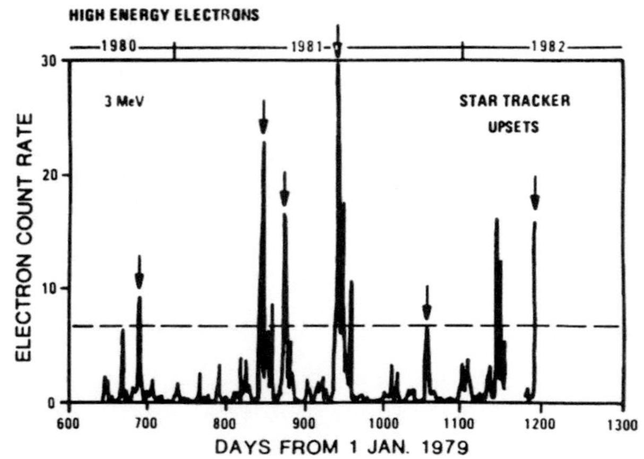

Fig. 15 Temporal correspondence between star tracker anomalies onboard a geostationary orbit spacecraft (*vertical arrows*) and enhancements in $E \sim 3$ MeV electron fluxes measured by sensors on the same spacecraft (from Baker et al. 1987a)

It has long been clear that large increases in multi-MeV electrons can cause significant spacecraft operational anomalies. For example, Fig. 15 (taken from Baker et al. 1987a) shows that star-tracker anomalies were observed on an operational geostationary Earth orbit (GEO) spacecraft whenever the counting rate for $E \sim 3$ MeV electrons reached or exceeded 7 counts/sec. The vertical arrows show the precise timing relationship between the star tracker problems and the enhanced electron flux episodes during this 1980–82 period.

Another example drawn from SAMPEX for the 2003 Halloween storm experience is shown here in Fig. 16. This figure from Baker et al. (2004b) shows that dozens of spacecraft operational anomalies occurred during the Halloween Storm (Oct.–Nov. 2003) period. Some of these anomalies were onboard spacecraft in distant locales (e.g., Genesis and Mars Odyssey), but most anomalies were seen on near-Earth spacecraft (GOES, RHESSI, ADEOS, NOAA-17, POLAR) and were due to the highly elevated radiation belt electron population (Baker et al. 2004b).

The extent and quality of observations of the radiation belts improved dramatically with the successful launch of the Radiation Belt Storm Probes (RBSP, renamed the Van Allen Probes mission in November 2012). The Van Allen Probes have a very comprehensive particle and fields detection payload (see Mauk et al. 2013). The higher energy electron measurements from the REPT instruments (Baker et al. 2013a) give unprecedented information throughout the inner and outer Van Allen belts. From the outer radiation belt standpoint, the REPT measurements show beautifully the spatial and temporal properties of electrons from $E \sim 1$ MeV to $E \gtrsim 10$ MeV. Figure 16, for example shows combined data from the Van Allen Probes A and B spacecraft for $E = 1.8$ MeV (1.6–2.0 MeV channel). Each panel shows color-coded flux levels for one-year periods. The vertical axis is L-value ($2 \leq L \leq 6$) in each case and each horizontal axis shows one year of time starting with 1 September 2012 (upper left) to 1 September 2016 (lower right). There are fascinating acceleration, transport, dropout, and recovery features seen in the data. Particularly strong storm events (such as October 2012, March 2015, and June 2015) are evident. See Reeves et al. (2013) and Baker et al. (2016) for details of these coronal mass ejection (CME)-driven events Thorne et al. (2013b). Also, note the very weak outer belt structures during much of 2014 as seen in panel (b) of the figure. The most recent year of data shown (1 September 2015 to 1 September 2016) is particularly notable for the frequency and intensity of the $E \sim 1.8$ MeV electron enhancements.

Examination of the inward extent of relativistic electrons in Fig. 17 and the earlier Fig. 14 demonstrate visually the striking conclusions of Baker et al. (2014a) and Baker

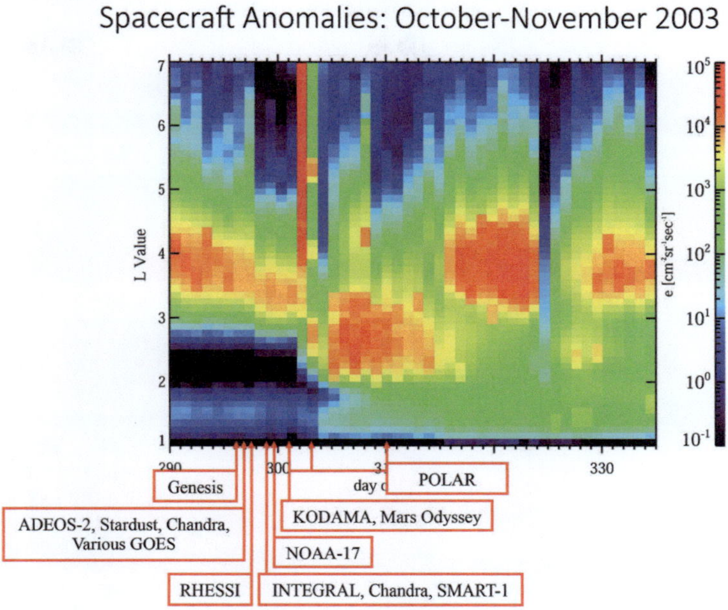

Fig. 16 Color-coded electron fluxes measured by the SAMPEX spacecraft (see Fig. 13) for Days 290 through 335 of 2003 in an L versus time format. Times of severe operational anomalies for various spacecraft are indicated by the *red arrows* and *boxes*

et al. (2016). Those studies found that in recent times since the launch of Van Allen Probes in August 2012, the Earthward extent of the outer belt of highly relativistic electrons appears to encounter an almost impenetrable barrier at an equatorial radial distance (L) near 2.8 R_E, such that almost no electrons with $E \geq 2.5$ MeV are seen inside this position in the hours and days following major disturbances and their associated inward injections and/or in-situ acceleration events. Slow radial diffusion to locations inwards of $L = 2.8$ does occur at longer time scales, but in no instance over four years of data do we see electrons with $E \gtrsim 1$ MeV penetrate inside $L \sim 2.5$. The greatest "intrusions" observed were for the March and June 2015 storms (see Baker et al. 2016; Foster et al. 2016). The nearly "impenetrable" barrier to inward injection or diffusive transport of relativistic electrons over prompt time intervals is one of the most unexpected findings of the Van Allen Probes era (see Baker et al. 2014a). The finding has important and nuanced space weather implications when considering the temporal response of the outer belt's inner edge, with the general conclusion that locations in the $L = 2.5$ to 2.8 range can be expected to have almost complete protection from prompt increases in ultra-relativistic electrons. However, over longer time scales of weeks to months typically associated with slower diffusion effects, these locations will gradually see the impacts of these extremely energetic electrons in the absence of further storm time or injection associated events.

The features for $E \sim 1.8$ MeV in Fig. 17 are amplified in the $E = 4.2$ MeV channels of REPT shown here as Fig. 18, and demonstrate once again that the "impenetrable" barrier to inward prompt injection holds across multiple energy levels. In particular, the 4.2 MeV electron fluxes (in identical format to Fig. 16) show a virtual absence of electrons for much of 2014 (see panel (b)). Numerous instances in 2013, 2015, and 2016, however show the fascinating "3-belt" structure indicative of the relativistic electron "storage ring" property reported by Baker et al. (2013a). Again – and even more strikingly at

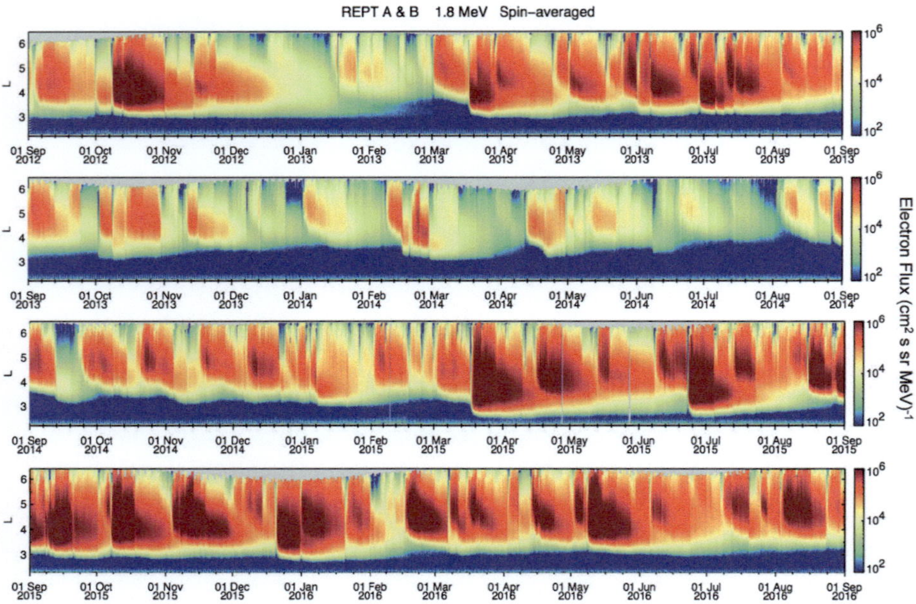

Fig. 17 Relativistic Electron–Proton Telescope (REPT) data ($E \sim 1.8$ MeV) from the Van Allen Probes mission from 1 September 2012 (*upper left*) to 1 September 2016 (*lower right*) as described in the text

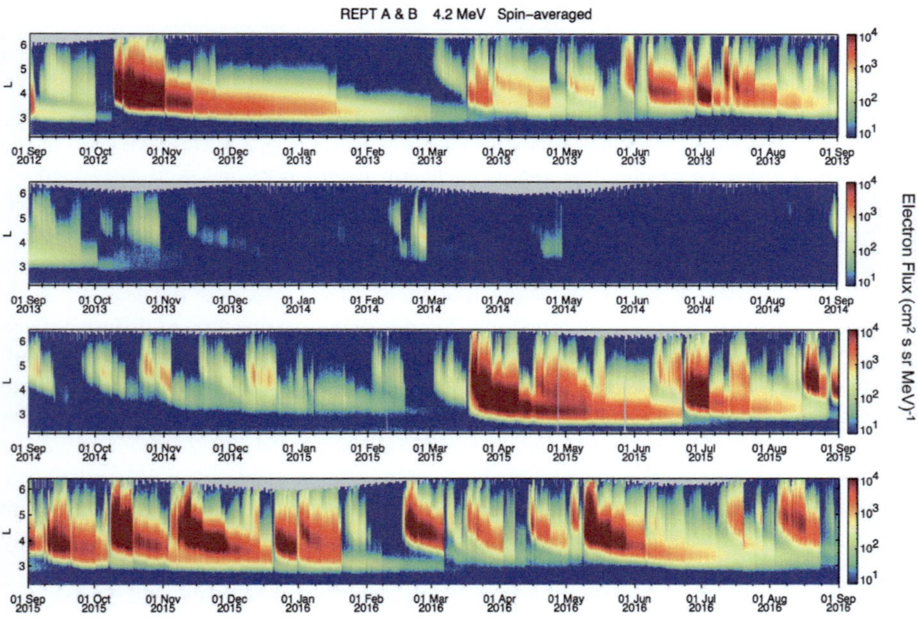

Fig. 18 Similar to Fig. 16 but for $E \sim 4.2$ MeV electrons from the REPT instruments

these higher energies — the outer zone electrons approach $L \sim 2.8$ (Baker et al. 2014a; Kellerman et al. 2014). However, in no cases do these highly energetic electrons pass through the impenetrable barrier either by direct injection or by radial diffusion, although on occasion these electrons later slowly penetrate inwards to $L = 2.5$ — e.g. by mid June 2015 (difficult to see on this plot scale) following the March 17, 2015 great storm. Clearly, the dynamic behavior of these electrons is far richer than previously thought, and more studies are under way to further explore these features and to quantify associated space weather effects across the range of natural injection and diffusion processes and characteristic scales.

4 Radiation Belt Acceleration, Transport, and Loss

4.1 Particle Diffusion Across Invariant Parameters

Around the time of the early Van Allen-led space missions in the late 1950s (as mentioned previously), theorists asserted that classical and relativistic physics laws have important and inherent consequences for the behavior of charged particles in the presence of a background magnetic field. In particular, for the approximate dipole field present in the terrestrial inner magnetosphere, particles under non-disturbed conditions undergo three quasi-periodic motions, each associated with an adiabatic invariant and a corresponding spatial and temporal characteristic scale. Further, particles that obey all three invariant motions define a stable drift shell encircling Earth, allowing trajectory predictions. The three invariants, arranged from strongest to weakest and from shortest time scale to longest time scale, are particle gyration in a homogeneous magnetic field, bounce gyrocenter motion along the field lines, and drift-bounce gyrocenter motion around Earth (Shulz and Lanzerotti 1974). As Ukhorskiy and Sitnov (2013) point out, for relativistic $\sim$1 MeV electrons in the center of the outer belt ($L \sim 4$–5), these motion invariants are well decoupled as they are separated by approximately six orders of magnitude: gyration frequency on $\sim$kHz scales, bounce frequency on $\sim$Hz scales, and gradient-curvature drift frequency on $\sim$mHz scales (Schulz and Lanzerotti 1974).

However, nature often provides non-adiabatic conditions e.g. during injection events, strong plasma wave generation at appropriate frequencies, or during strong non-dipole magnetic dipolarizations. The Van Allen-led early missions and a large number of subsequent observational experiments (e.g. CRRES, AMPTE, SAMPEX, Van Allen Probes) revealed that these highly dynamic conditions can often provide strong and variable particle forcing on time scales less than the characteristic invariant time scale, leading to direct violation of one (or often more) of the invariant particle motion laws. Intervals with these types of violations can therefore cause particle intensities across the radiation belts to vary significantly with time (see early review by Roederer 1968), with direct associated space weather impacts. In particular, large amplitude perturbations in electric and magnetic fields along particle orbits can violate invariants. Plasma waves, often (but not always) separately excited by differing energy pathways in the plasma, have associated electrostatic and electromagnetic fields that can resonantly interact with quasi-periodic particle motions to amplify their energy and alter their densities.

In the case where the radiation belt particle energy and density is small compared to other plasma populations, one can consider the forces mentioned as external, and quasi-linear theory then allows a modeling of the action of these forces as particle diffusion in the three adiabatic invariants, with various formulations to describe the effective diffusion coefficients caused by the external forcing. Historically, Ukhorskiy and Sitnov (2013) note that this quasi-linear particle diffusion theory was well established within the first decade after initial radiation belt discoveries, but non-linear modifications to this diffusion framework did not occur until much later in the 1980s and 1990s.

Excellent summaries of adiabatic invariant particle motion theory as well as violations of that theory are given by a number of previous works including Dungey (1965), Trakhtengerts (1966), Walt (1971), and the previously referenced Ukhorskiy and Sitnov (2013). As this article is devoted to a description of radiation belt space weather effects, however, we focus here on recent results with new insights into the general space weather consequences of dynamic, non-adiabatic particle motions. We also note that recent results have implied that multiple cycles of energy exchange between waves and particles may be involved in relativistic particle acceleration, since unstable plasmas use the energy of other particles for the ultimate acceleration energy source (Shklyar 2017). These situations make it complex to assess the full life cycle of particle acceleration, with associated implications and impacts on quasi-linear and non-linear particle invariant diffusion theories for radiation belt dynamics. Such topics remain at the heart of current theoretical research (e.g. Albert et al. 2016).

4.2 Radial Diffusion

A long recognized scenario for transport of energetic electrons from acceleration or injection zones is radial diffusion, caused when the first two adiabatic invariants are preserved in the outer zone whereas the third and slowest gradient curvature drift invariant is violated by particle–wave processes through the presence of ultralow-frequency (ULF) waves in the Pc-5 range between ~1.67 and ~6.67 mHz (Jacobs et al. 1964). Efficient interactions occur when the driving ULF wave frequency is an integer multiple of the azimuthal drift frequency (Schulz and Lanzerotti 1974). Extensions to the pure azimuthal drift based radial diffusion framework have been pursued more recently by Elkington et al. (1999, 2003) through inclusion of drift-resonant electron interactions with ULF toroidal and/or poloidal wave modes in a manner that inherently accommodates an asymmetric geomagnetic field. The general effect of third invariant violation is to cause a net change in the particle's radial coordinate, causing a net gain/loss in particle energy as the particle diffuses inwards/outwards. ULF waves in this frequency range can be produced by internal processes including both mirror (Hasegawa 1969) and drift (e.g. Chen and Hasegawa 1991) instabilities embedded in the inner magnetospheric plasma. ULF wave power can also be generated by external forcing events including magnetopause shear flow instabilities (e.g. Claudepierre et al. 2008), solar wind pressure variations (Kivelson and Southwood 1988), and transient ion foreshocks (Hartinger et al. 2013). It is important to note that ULF driven radial diffusion is inherently stochastic in nature rather than coherent, as typical ULF driving wave fields are rarely monochromatic and span a range of ~mHz frequencies.

A number of observational studies have attempted to calculate the effective total radial diffusion coefficient, itself a measure of the radial diffusion rate. This rate, D_{LL}, includes effects from perturbations in both the induced electric field and magnetic field, and various theoretical formulations (Fälthammar 1965; Fei et al. 2006) have been established to provide quantitative estimates. Many observational studies have been conducted using these frameworks with a notable focus on establishing when electric and/or magnetic field effects are dominant, and a recent comprehensive summary is provided by Ali et al. (2016). Notably, the work of Tu et al. (2012) using THEMIS and GOES data, along with the work of Ozeke et al. (2012, 2014) using AMPTE and GOES data, determined that, contrary to earlier expectations, electric field effects in radial diffusion dominate over magnetic field effects when setting radial diffusion coefficient amplitudes. In fact, Ali et al. (2015, 2016), using Van Allen Probes observations, finds that the radial diffusion electric component is on average larger than the magnetic component by one to two orders of magnitude for almost all L^* and Kp values. Thus, it appears that electric field perturbations in the ULF range are substantially more effective in driving radial diffusion of charged particles in the inner magnetosphere.

For space weather purposes, typical radial diffusion coefficients range from 10^{-2} to 10^{-4} days^{-1} for low Kp, setting a time scale of hundreds to thousands of days for these slow processes to effect the distribution of energetic particles. Thus, as referred to earlier, the slow movement of the inner edge of the outer belt in Fig. 17 seen in ultra-relativistic electrons between March and June 2015 is an example of observations compatible with inward radial diffusion of an accelerated population created earlier by stormtime injection or by local acceleration processes (cf. later discussion in this section). However, radial diffusion coefficients can be as high as 10^0 days^{-1} during times of heightened geomagnetic activity (Ali et al. 2016), allowing for events during which radial diffusion is arguably the dominant acceleration process in the heart of the outer radiation belt.

4.3 Role of Substorms

The role of geomagnetic storms versus substorms and high-energy radiation belt acceleration has been studied for decades. While the exact definitions of the two types of geomagnetic disturbance can be debated, they each represent fundamentally different forms of energy transfer into the Earth's magnetosphere as a result of solar wind driving. Long-term studies of correlations between radiation belt electron content and solar wind speed show a repeatable 'triangular distribution' (Baker et al. 2004b; Reeves et al. 2011), indicating that large enhancements of outer belt relativistic electron flux can occur at any solar wind speed — from the lowest to the highest velocities (Fig. 19). These distributions can be clarified by plotting in terms of occurrence rate contours (e.g. Fig. 1, Kellerman and Shprits 2012), yet still the outcome shows that the relationship between relativistic electron flux and solar wind drivers is not linear. Not only can enhancements occur over a wide range of solar wind speeds, the contents of the radiation belts can also change unpredictably from one geomagnetic storm to the next. Reeves et al. (2003) studied the response of high-energy electrons to almost 300 storms over an 11-year period. The result was a varied distribution, with the highest likely response being one of no change between pre- and post-storm flux levels. An almost equal likelihood existed of the total population increasing or decreasing with any given storm. These results were replicated even when very small storms or non-storm events ($Dst > -50$ nT) were taken into account (Anderson et al. 2015). Thus, geomagnetic storms and solar wind speed are not good predictors of energetic electron enhancement events, and electron response does not necessarily scale with the strength of a storm. Several recent studies during the Van Allen Probes era have in fact shown cases of strong outer belt relativistic electron enhancements during non-storm times (Schiller et al. 2014; Su et al. 2014). Both papers noted the increase in chorus wave power during the acceleration periods, but did not explicitly make the connection to substorm activity.

If strong enhancement events cannot be predicted by strength of a geomagnetic storm (measured by the Dst index) or solar wind speed alone, what then determines an efficient acceleration event? Several studies have attempted to bring light to this subject by examining both individual case studies and the statistics of many enhancements. Meredith et al. (2002) looked at several example periods of outer belt electron acceleration during both strong and weak storm conditions and determined that prolonged substorm activity is a requirement for effective enhancement of relativistic electrons. Kataoka and Miyoshi (2010) found that the main controller of outer belt electron fluxes was IMF orientation: weaker southward Bz resulted in weaker enhancement events. The authors note that northward or weak southward IMF orientation suppresses the amount of substorm activity generated. Comparing modeled

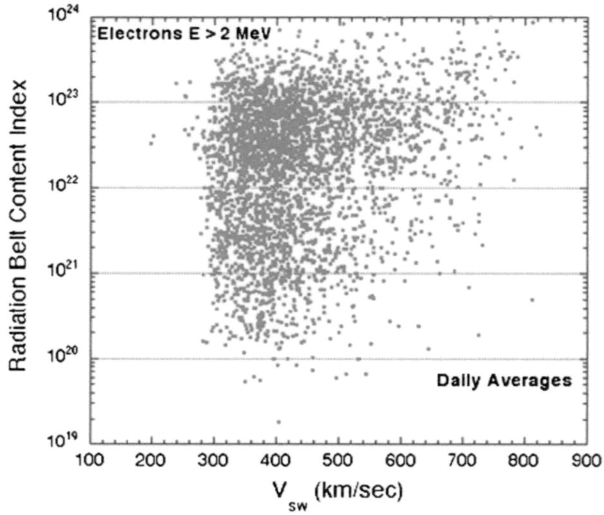

Fig. 19 Outer belt radiation belt content index vs. solar wind speed (after Baker et al. 2004b)

electron fluxes at GEO with measurements, Li et al. (2011) concluded that high-speed solar wind is not a necessary condition for MeV electron enhancements at geosynchronous orbit — only a southward orientation of the IMF is absolutely required. More recently, a superposed epoch analysis of both efficient and inefficient acceleration events from Van Allen Probes data has provided a statistical look at the underlying causes (Li et al. 2015). The comprehensive study found three main contributions to efficient acceleration events: (1) prolonged southward Bz, (2) high solar wind speed, and (3) low solar wind dynamic pressure. By examining global chorus wave distributions, the study also clearly shows that intense chorus waves are present for far longer periods during the efficient enhancements.

The common thread throughout this previous work is one of substorm-driven energy being introduced into the ring current region, thus driving relativistic electron enhancements. The acceleration can be achieved via resonant wave–particle interactions with VLF chorus waves, as discussed in a following section. In a quasi-linear interaction scenario, the enhancement progresses in a step-wise pattern, with lower energies appearing first, and high energies taking longer to be produced. This common picture has been noted as far back as Baker et al. (1979) where lower energy electron enhancements at GEO were seen to be closely related to solar wind speed variations, while >1 MeV electrons exhibited a similar profile but with a significant delay in time. The same delay pattern can often be observed, with the time lag increasing as energy does, in the Van Allen Probes data (e.g. Baker et al. 2016). In order for the process to be efficient at producing the very highest energy tail of the radiation belt population, VLF chorus (and thus, their source of energy in the form of plasmasheet particles) need to be continually energized over a long duration (many hours to days), as during High Intensity Long Duration Continuous AE Activity events (Gonzalez and Tsurutani 1987; Tsurutani et al. 1995). A recent case of where this mechanism was not put into place properly showed a failure of relativistic and ultra-relativistic electrons to appear, despite the high solar wind speeds and space weather models predicting enhancements of relativistic electron (Jaynes et al. 2015). The relativistic accelerator process, described in a schematic reprinted here as Fig. 20, requires both source (10's keV) and seed (100's keV) electrons to drive chorus waves and provide a base population for acceleration to much higher energies.

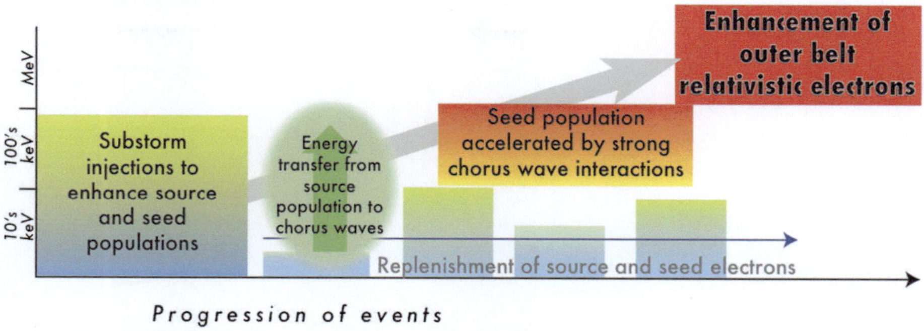

Fig. 20 Schematic of the relativistic accelerator mechanism in Earth's inner magnetosphere (reprinted from Jaynes et al. 2015)

The developments throughout the Van Allen Probes era have built on the foundational studies of this topic published over the preceding decades. The emerging picture is one of substorms (with or without geomagnetic storms) playing the dominant role in producing Earth's high-energy radiation belt population through the intermediary of chorus wave generation. This is a particularly important paradigm to consider as the need for more accurate space weather forecasting models becomes pressing.

4.4 Wave–Particle Interactions (Linear, Nonlinear)

4.4.1 Acceleration by ULF Waves

It has long been known that there is a direct correlation between ULF wave power in the inner magnetosphere and relativistic electron acceleration (Rostoker et al. 1998; Mathie and Mann 2000). ULF waves can accelerate radiation belt particles through two different mechanisms: inward radial diffusion and drift-resonance interaction. The first mechanism was extensively discussed in Sect. 4.2. and can accelerate outer belt electrons on timescales of hours or longer. The second mechanism has been continuously studied for the past 20 years, yet, despite these efforts, it is still unclear what extent of the overall acceleration is driven by drift-resonance. The resonance that can take place between a relativistic electron and the appropriate frequency ULF wave is fundamentally dependent upon the wave period matching the drift period (or multiples thereof) of the electron. This phase-coherent interaction is possible both with toroidal mode ULF waves, detailed nicely in Elkington et al. (1999), as well as poloidal mode (Claudepierre et al. 2013) as shown in Fig. 21. Section 4.2 already covered these processes thoroughly.

4.4.2 Local Acceleration by VLF Chorus

VLF waves in particular have a strong interaction potential with radiation belt electrons, and can cause significant acceleration characterized by a fundamentally local nature. During this local acceleration process, both a source of energy for wave growth and a population of energetic seed electrons are required. New observations are shedding considerable light on the interdependence and necessity of these requirements for local acceleration. In particular, recent Van Allen Probes observations indicate that rapid (<1 hour) energization of seed electrons to MeV energies is associated with (substorm) injections of 10s keV to 100s keV

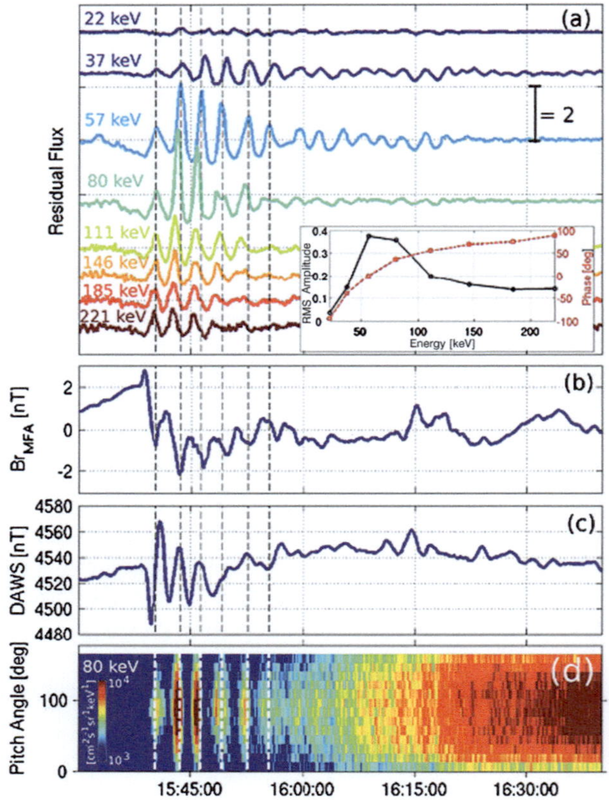

Fig. 21 Evidence of drift-resonance interaction between poloidal mode ULF wave (b) and 57–80 keV electrons (*blue and green trace* in (a)). From Claudepierre et al. (2013)

electrons into the inner magnetosphere and the subsequent enhancement of VLF chorus band waves in the inner magnetosphere (e.g. Foster et al. 2014a).

The acceleration of electrons to MeV energies in the core of the outer zone of the radiation belt involves the local acceleration (Horne and Thorne 1998; Reeves et al. 2013; Thorne et al. 2013a; Summers et al. 1998) of a seed population of energetic electrons (e.g. Jaynes et al. 2015) in interactions with (very low frequency) VLF chorus waves. Thorne et al. (2013a) demonstrated that quasi-linear modeling of this local acceleration process could reproduce the observed recovery of the inner radiation belt electrons. Li et al. (2014) modeled the recovery of the radiation belt over the course of ∼24 hours by quasi-linear interactions with VLF chorus. Foster et al. (2014a) described Van Allen Probes observations of substorm electron injection and chorus amplification associated with the rapid recovery of the MeV electron population following a storm time depletion of radiation belt electrons. Lower energy injected electrons in the ∼10 to 80 keV range provide a source of free energy for VLF waves growth and subsequently are driven into the loss cone, precipitating into the atmosphere (e.g. Foster and Rosenberg 1976). At the same time, the high-energy tail of the electron distribution gains perpendicular energy from the waves and is accelerated (Summers et al. 1998; Albert 2002; Omura et al. 2007).

A detailed example is given by Van Allen Probes dual spacecraft particle and wave observations spanning a substorm during the large March 17, 2013 storm with onset near 15:50 UT, as shown in Figs. 22 and 23. Lower energy injected electrons (50 keV, Fig. 23 panel a) arrive promptly at the Van Allen Probe A position near $L^* = 4.7$. The effects of their

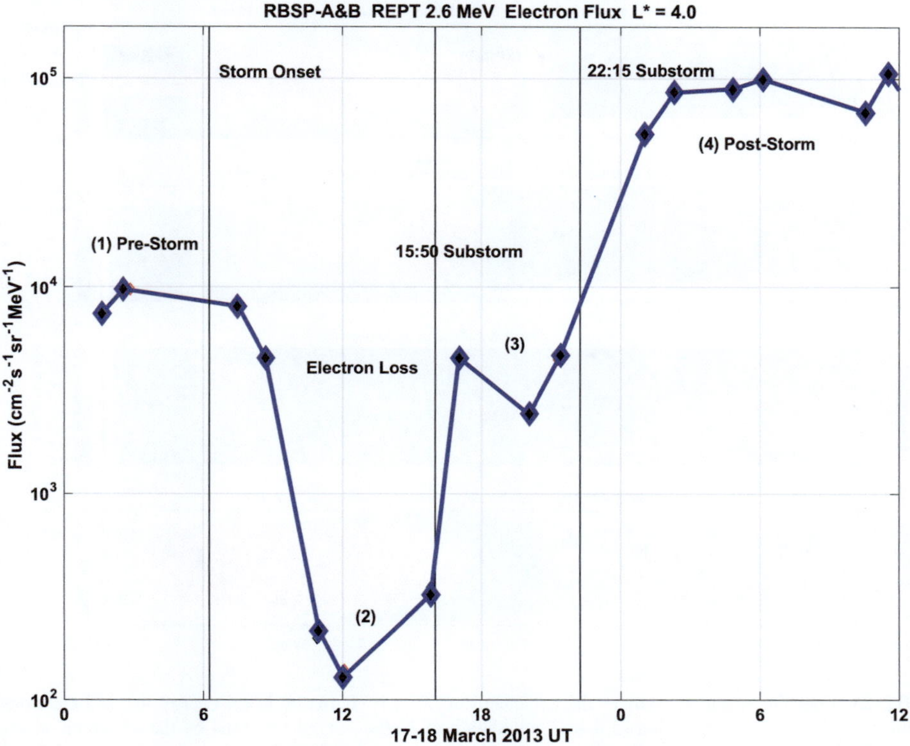

Fig. 22 Storm–time changes in the 2.6 MeV electron fluxes observed at $L^* = 4.0$ with the REPT instruments on both Van Allen Probes delineate the event time history of relativistic electron loss and recovery. Storm onset was at 06:15 UT on 17 March 2013 and prompt stepwise radiation belt energization accompanied substorm injections at 15:50 UT and 22:15 UT. Diamonds indicate the sequential observations by Probes A and B at $L^* = 4.0$ (after Foster et al. 2017)

involvement in chorus amplification are already seen at this time, as particle distribution is driven toward loss cone pitch angles. Panel c of Fig. 23 shows the development of strong chorus wave activity in step with the electron injections. The multi-MeV radiation belt electron population (panel d) that had been significantly depleted earlier in the storm increases by 10x during the substorm interval (with respect to that measured before the substorm 1 hour earlier by Van Allen Probe B).

It should be noted here that ultra-relativistic electron acceleration in the outer belt can also take place under the influence of fast magnetosonic waves which are thought to take their energy from enhanced/unstable proton ring current distributions at 10 s of keV energies, as demonstrated by several studies using GOES and CRRES observations among others (Perraut et al. 1982; Meredith et al. 2008; Chen et al. 2016; Balikhin et al. 2015). For example, the production of electron butterfly distributions has recently been observed in Van Allen Probes data during a 2015 storm and was linked to parallel acceleration caused by Landau resonance with magnetosonic waves (Li et al. 2016). Details on this rich and nonlinear process are still a matter of current research and are a good example of the importance of wave–particle interactions on the outer belt (e.g. Thorne 2010).

The rapid increase in MeV electron fluxes associated with substorm injections suggests that nonlinear processes, with associated rapid interaction time scales, play a significant role

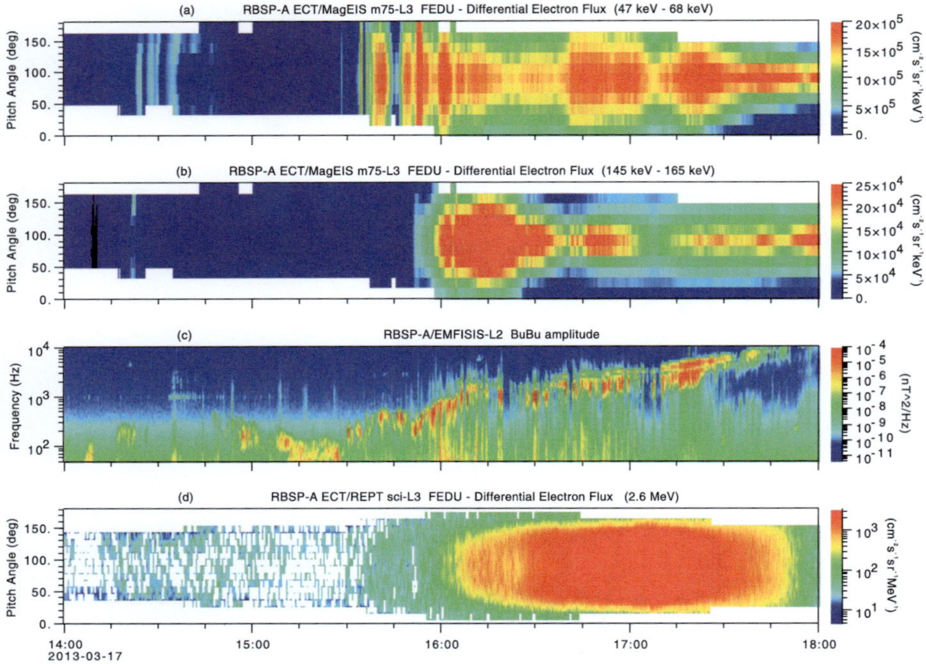

Fig. 23 Local acceleration involves the enhancement of VLF chorus by lower energy injected electrons, and the subsequent energization of a seed population of higher energy electrons by the amplified waves. (a) RBSP-A MagEIS observations of injected >50 keV and (b) >150 keV electrons and (c) associated chorus emissions observed by EMFISIS-A are shown for the 15:50 UT substorm interval on March 17, 2013. *The lower panel* (d) shows the growing enhancement of 2.6 MeV electrons resulting from the local acceleration of the near-relativistic seed population (after Foster et al. 2017)

in the local acceleration process (Foster et al. 2017). Wave-electron interactions during the generation of VLF rising tones are strongly non-linear, leading the formation of a highly coherent wave structure. As a result, a fraction of the relativistic electron seed population at resonant energies can be trapped by waves, leading to significant non-adiabatic energy exchange. Direct observations show that chorus elements consist of a sequence of wave subpackets (e.g. Santolík et al. 2014) each spanning multiple wave cycles and delineating a single cycle of continuous wave growth and damping (Fig. 24).

For phase trapped resonant electrons with near-relativistic initial energies, well-developed subpackets at and around $1/4\ f_{ce}$ result in strong (>100 keV) energy gain through nonlinear relativistic turning acceleration (RTA) (Omura et al. 2007) for electrons with relativistic initial energies (Foster et al. 2017). Electrons with initial energy 100s keV–3 MeV can be accelerated by 50 keV–200 keV in resonant interactions with a single VLF rising tone on a time scale of 10–100 msec (Foster et al. 2017).

To illustrate the effect of nonlinear RTA acceleration on the evolution of the relativistic electron flux spectrum, we have iteratively applied the calculated nonlinear energy gain profile (Fig. 25) to the observed initial electron flux profile for the 17 March 2015 event. The evolution of the ratio of the modeled electron flux profile at $L^* = 4$ over 30 cycles of nonlinear RTA modeling indicates a 10x flux enhancement confined to low MeV energies and centered near 2 MeV. For electrons with 0.5 sec bounce period, and assuming that strong chorus waves are encountered only 1 percent of the time, the 30 cycles modeled would occur

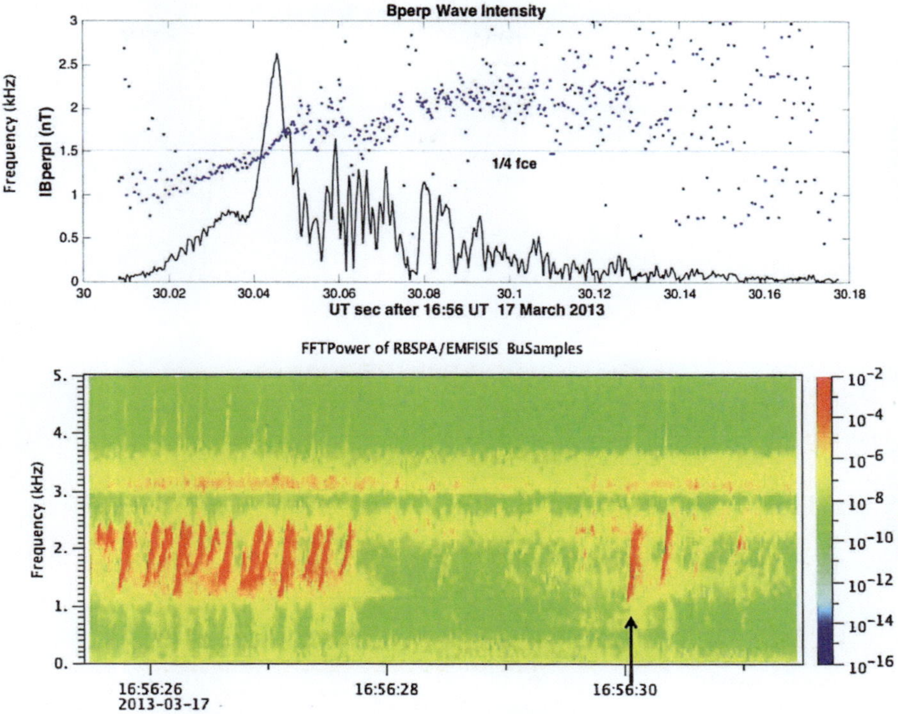

Fig. 24 For the strong VLF lower-band chorus rising tone seen at 16:56:30 in the lower panel, by RBSP-A during the March 17, 2013 storm, subpacket analysis of the single 180-msec chorus element is shown in the *upper panel*. The magnitude of the wave magnetic field in the plane perpendicular to B for individual wave cycles is shown in *black*. Clear modulation of the wave amplitude identifies a sequence of subpackets within the rising tone chorus element, each spanning multiple wave cycles. The instantaneous wave frequency determined from individual wave cycles is shown in *blue* (adapted from Foster et al. 2017)

over 1500 sec. or ~30 min the approximate characteristic time for the relativistic electron acceleration observed in this event. The good agreement between theory and observations indicates the efficiency of nonlinear interactions with chorus rising tones for accelerating a seed population of energetic electrons to MeV energies.

In aggregate, these observationally based calculations reinforce the conclusion that nonlinear acceleration processes are likely a fundamentally important part of local acceleration to relativistic levels, and that the very rapid time scales involved set a considerably faster minimum time scale for in-situ space weather effects at these energies than previously considered.

Energy gain for a single RTA interaction with the chorus rising tone shown in Fig. 24 is shown as a function of initial electron energy; B) the observed electron flux increase ratio at $L^* = 4$ is shown as the black curve. The nested blue curves indicate the iterative application of the RTA nonlinear energy gain to the observed initial electron flux profile over 30 cycles with 5% trapping ratio.

4.5 Shock Acceleration

Solar wind shocks are another process that produces significant space weather, through extremely fast acceleration of ultra-relativistic electrons deep within the radiation belts and

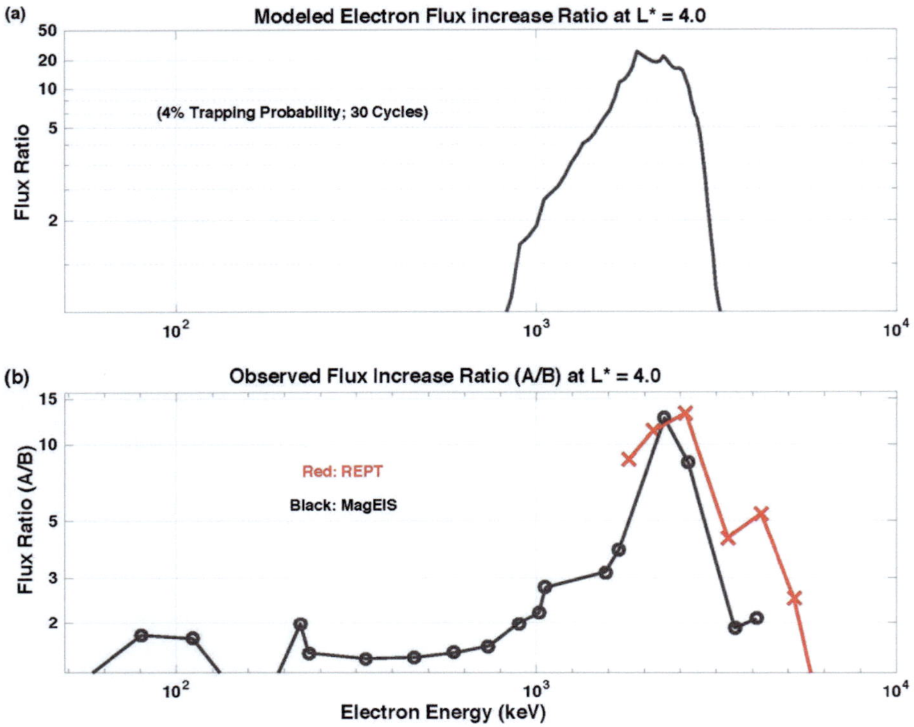

Fig. 25 a) The ratio of the modeled electron flux increase at $L^* = 4$ after 30 cycles of nonlinear RTA acceleration to the initial profile indicates a 10x flux enhancement confined to los-MeV energies and centered near 2 MeV. (b) The electron flux increase determined as the ratio of RBSP-A to RBSP-B observations shows good qualitative agreement with the RTA theoretical predictions (Foster et al. 2017)

inside the plasmasphere. Explosive eruptions on the Sun launch a supersonic blast wave of high-density particles that stream through the background solar wind. This solar wind shock front sweeps up the interplanetary magnetic field along its path, carrying significant particle and magnetic energy to Earth orbit in about one day's time. The high-speed solar wind shock strikes a sledgehammer blow to the protective barrier of Earth's magnetic field on the dayside, launching a powerful magnetosonic pulse that tears through the magnetosphere from front to back at speeds approaching 1000 km/sec. The strong electric field associated with this shock-induced pulse efficiently accelerates radiation belt electrons (Fig. 26). As the electrons circle the Earth, those with azimuthal velocity matching the magnetosonic pulse propagation speed can stay in the shock induced electric field for ~100 sec, and gain significant energy (up to millions of electron volts) (e.g. Foster et al. 2015).

Thus, prompt acceleration by impulsive shock-induced electric fields and subsequent ULF wave processes (Fig. 27) comprise a significant mechanism on space weather time scales of ~10s to ~100s of seconds for enhancing relativistic electron populations inside the plasmasphere and in the heart of the outer zone (e.g. Foster et al. 2015; Kanekal et al. 2016; Li et al. 1993; Blake et al. 1992). Of significance for space weather effects, shock acceleration can introduce nearly instantaneous, significant fluxes of ultra-relativistic electrons deep within the slot region where other electron acceleration processes are not operative (Blake et al. 1992).

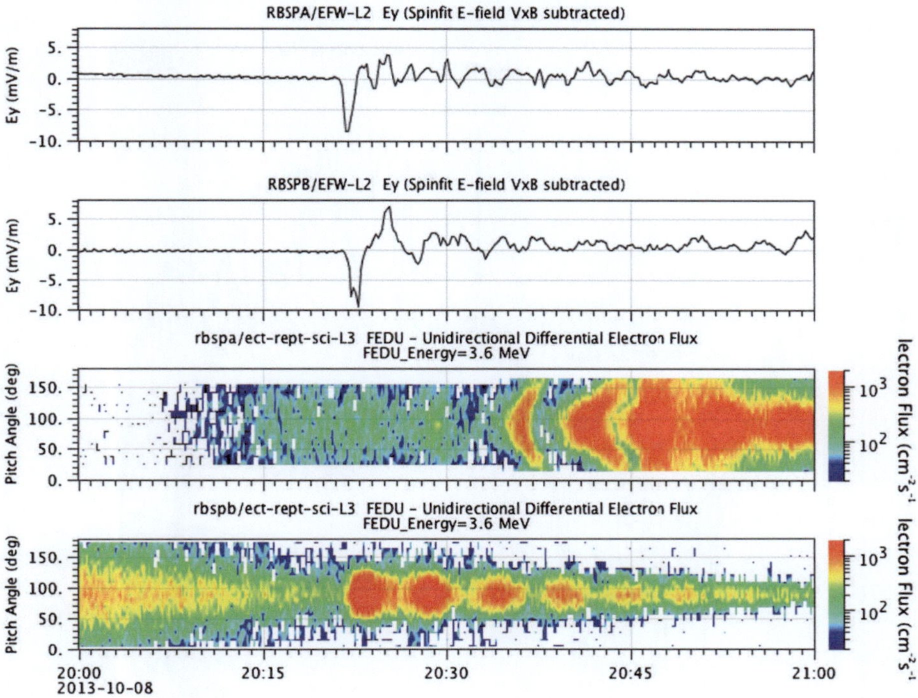

Fig. 26 Relativistic electron flux increases in the 3.6 MeV channel from the REPT instrument on the Van Allen Probes A and B is shown. In the *bottom panels*, measurements electron flux in the 3.6 MeV REPT channel are shown as functions of pitch angle and time from each of the two Probes. The shock-driven electric field at each spacecraft is included in the upper panels for reference (after Foster et al. 2015)

4.6 Solar Wind Drivers

Many studies have focused on the response of energetic electrons in the outer radiation belt to upstream solar wind conditions. It has been long recognized that energetic particles at geosynchronous Earth orbit (GEO) are modulated by solar wind forcing (e.g., Paulikas and Blake 1979; Baker et al. 1979, 1986, 1990; Blake et al. 1997; Reeves 1998; Mathie and Mann 2001; O'Brien et al. 2001; Li, 2001, 2004; Li et al. 2005, 2011; Lyatsky and Khazanov 2008a, 2008b; Borovsky and Denton 2009; Reeves et al. 2011; Boynton et al. 2013; Simms et al. 2016; Kellerman et al. 2015). Paulikas and Blake (1979), for example, analyzed energetic electron data from ATS-1, ATS-5 and ATS-6 spacecraft and found a pronounced positive correlation between the electron fluxes and solar wind speed. Concurrently and independently, Baker et al. (1979) found strong solar wind speed correlations with GEO electron fluxes as well. Baker et al. (1986), using data from the Spectrometer for Energetic Electrons (SEE) onboard two operational GEO satellites, showed a very good correlation between highly relativistic electrons and solar wind stream structures. Blake et al. (1997), based on measurements from a GEO satellite and the WIND spacecraft, presented a study of the correlation between relativistic electrons at GEO and properties of solar wind drivers during a solar minimum period. Their results indicated that relativistic electron enhancements are associated with a solar wind speed increase, but also with solar wind density increases and southward turnings of the *IMF*. Li (2001), using a model based on the radial diffusion equation, quantitatively predicted the electron fluxes at GEO based

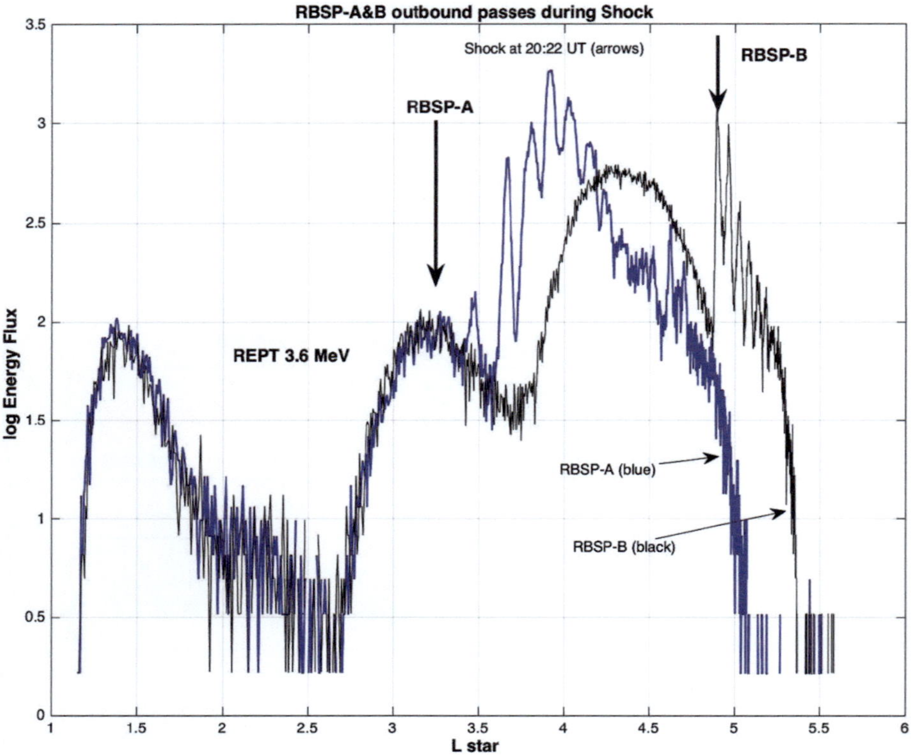

Fig. 27 Sequential Van Allen Probes A and B orbits during a shock event are shown. *Arrows* indicate the 20:22 UT time of shock observation at each spacecraft. At shock time, Probe B was at $L^* \sim 5$ and Probe A was further inward near $L^* \sim 3.3$. Rapid acceleration of 3.6 MeV electrons and subsequent periodic drift echoes were seen at each location (after Foster et al. 2015)

on solar wind velocity and *IMF* properties. They found a linear correlation of 0.9 between predicted and measured values for the period 1995–1996. The solar wind speed was found to be the most important factor in their prediction model.

Figure 28 is taken from Baker et al. (1979). It shows the temporal relationships between fluxes of moderate energy electrons ($E \sim 200$ keV) and highly relativistic electrons ($E \sim 1.5$ MeV) as compared to concurrent upstream solar wind speed. As is evident, the low-to-moderate energy electron fluxes at GEO track the solar wind speed quite closely in time while the very energetic electron fluxes are delayed by $\sim$2 days in time.

As mentioned in Sect. 4.2, Reeves et al. (2011) examined the relationship between relativistic electron fluxes at GEO and solar wind speed using LANL-GEO data from 1989 to 2010. Their results showed a triangle-shaped distribution of electron fluxes versus solar wind speed, indicating that the radiation belt electron fluxes can be high for essentially any concurrent solar wind speed. Li et al. (2011) emphasized that geomagnetic activity driven by a southward orientation of IMF is a necessary condition for MeV electron enhancements at GEO. Boynton et al. (2013), using the nonlinear autoregressive moving average approach, studied the relation between electron fluxes at GEO and various solar wind parameters. Their results suggested that for lower energy electrons, the solar wind speed is the most strongly controlling factor of electron fluxes at GEO. On the other hand, for MeV electrons, the solar wind speed 2 days preceding was found to be the most important factor in electron flux

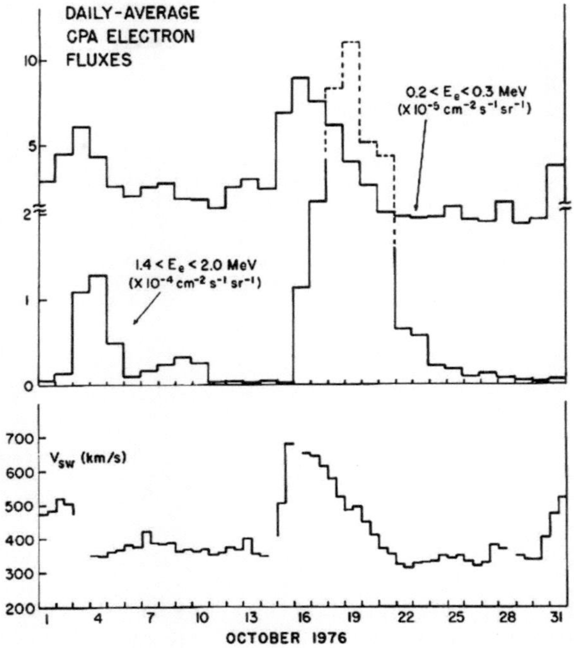

Fig. 28 *Top panel*: Daily average fluxes of $E \sim 200$ keV and $E \sim 1.4$ MeV electrons at GEO for October 1976. *Bottom panel*: Solar wind speeds for the same period of time. Data show that ~200 keV electrons track solar wind driver very closely in time while E 1 MeV electrons are delayed by 2–3 days (from Baker et al. 1979)

enhancements. This is consistent with the Baker et al. (1979) results shown in Fig. 28 here. Also, Wing et al. (2016), (Blake et al. 1992; Li et al. 1993) using information theory, investigated the solar wind driver for 1.8–3.5 MeV electrons at GEO. They showed that various solar wind parameters are causally related to 1.8–3.5 MeV electron fluxes at GEO while among all parameters the solar wind speed has shown to have the highest peak information transfer. These previous studies show the important role of solar wind on the dynamics of relativistic electrons at GEO.

There also have been statistical studies focusing on the electrons in the whole outer radiation belt and their relation to the solar wind parameters (e.g., Li et al. 1997a, 1997b; Baker et al. 1999; Iles et al. 2002; Vassiliadis et al. 2002; Mann et al. 2004; Rigler et al. 2007; Li et al. 2009). Iles et al. (2002), using >750 keV and >1 MeV electron data from two microsatellites, STRV-1a and 1b, which were in highly elliptical, near-equatorial orbits, studied the correlation between total relativistic electron content in the outer belt and solar wind and geomagnetic parameters during geomagnetic storms at the first six months of 1995. They found that the main requirements for relativistic electron enhancements were fast solar wind speed and fluctuating or southward IMF Bz. Baker et al. (2004b) constructed the radiation belt electron content index (RBC), which is a robust estimate of total relativistic electrons in the outer radiation belt, and showed that the RBC is well correlated with solar wind speed and the *Dst* index. The RBC index has proven very useful in its correlation with times of severe space weather anomalies as shown in Fig. 29.

The driving of magnetospheric processes by the solar wind has also been shown to be critical in radiation belt electron dynamics (e.g., Baker et al. 1990; Reeves 1998; Reeves et al. 2003; Li et al. 2009; Zhao and Li 2013; Jaynes et al. 2015). Geomagnetic storms and substorms play essential roles in transport and acceleration of radiation belt electrons. Reeves (1998) studied relativistic electron enhancement events at GEO from 1992 to 1995. He found correlation between maximum electron flux and minimum Dst, though geomagnetic storms were found to not necessarily cause flux enhancements. Li et al. (2009)

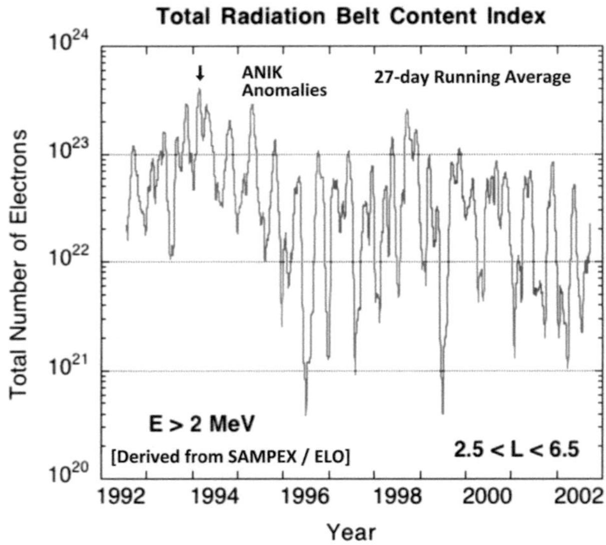

Fig. 29 The total Radiation Belt Content (RBC) index (see Baker et al. 1999, 2004b) for the period 1992–2000 (monthly smoothing). The ANIK spacecraft operational anomalies in January 1994 (Baker et al. 1994) occurred when the RBC was at an historic high level

analyzed total relativistic electron flux variations in the outer radiation belt during 18 storms from 1990 to 1991 and studied the statistical roles of storms and substorms on the radiation belt electron population. Substorm activity was well correlated with the net increases of total electron fluxes in the outer belt. Jaynes et al. (2015), using data from Van Allen Probes, studied the importance of source and seed populations in radiation belt electron variations and showed the crucial role of substorms in radiation belt electron acceleration. Most recently, Forsyth et al. (2016) examined the total radiation belt electron content variations following substorms and showed increases of total radiation belt electron content following substorms with a time lag of 1–3 days (see Baker et al. 1979, 2004b).

As the above review shows, the roles of solar wind and magnetospheric processes on the radiation belt electrons have been extensively studied. However, due to the limitations of observations, most previous studies only focused on relativistic electrons at GEO or on a very limited energy range. With the launch of Van Allen Probes, which provide clean radiation belt electron measurements in a wide L range with very fine energy resolution, researchers have been able to study the correlation between radiation belt electrons and solar wind parameters/geomagnetic indices and its dependence on the energy of electrons and L shell for the first time. Also, the majority of previous studies only examined the correlation between various parameters and radiation belt electron fluxes. However, these studies cannot rule out the influence of adiabatic effects in their statistical analyses, which can greatly affect the measured electron fluxes during geomagnetic storms especially at higher L shells.

In a very recent paper, Zhao et al. (2017), using data from the Magnetic Electron Ion Spectrometer (MagEIS) (Blake et al. 2013) and the Relativistic Electron–Proton Telescope (REPT) (Baker et al. 2013b) instruments of the Energetic Particle, Composition, and Thermal Plasma (ECT) suites (Spence et al. 2013) on the Van Allen Probes, have studied the relation between radiation belt electron phase space density (PSD) and solar wind parameters/geomagnetic indices while focusing on its dependence on the first adiabatic invariant μ and L^* for a wide μ and L^* range. Using the electron PSD instead of fluxes, they excluded the influence of the adiabatic effect and thus the actual effects of solar wind/magnetospheric processes on radiation belt electrons are appropriately highlighted. Besides the radiation belt electron PSD, Zhao et al. (2017) also show that the solar wind and magnetospheric processes

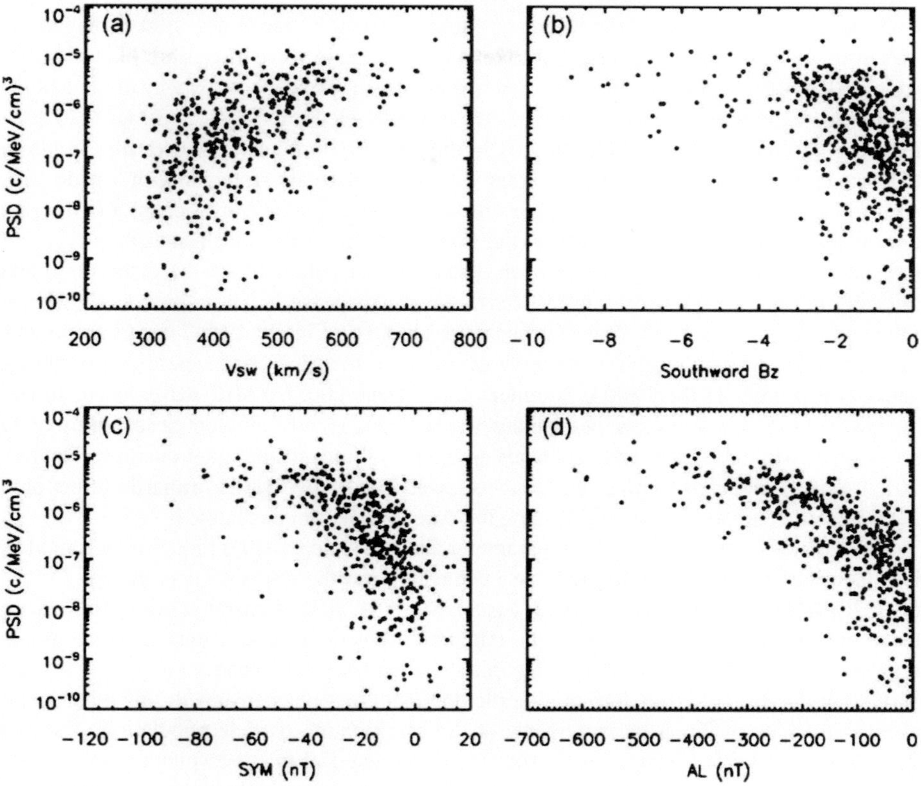

Fig. 30 A scatterplot of dependences on (a) solar wind speed, (b) southward IMF B_Z, (c) SYM-H index, and (d) AL index for electrons with the $\mu = 1000$ MeV/G. These are phase space density by (PSD) enhancements at $L^* = 4.5$ using REPT data from the Van Allen Probes (from Zhao et al. 2017)

play important roles in radiation belt electron acceleration by analyzing the relation between electron PSD enhancements and solar wind/geomagnetic parameters. Examples of such correlations are shown here in Fig. 30 from Zhao et al. (2017). Furthermore, the important role of substorms on radiation belt electron PSD enhancements and their prolonged effects are unveiled by this most recent work.

4.7 Loss Mechanisms

Despite decades of radiation belt research and monitoring, quantifying the total loss from the belts during both quiet and active times is a daunting task. Yet a part of this loss, in the form of precipitation to the atmosphere, contributes to the overall space weather effects on satellites and other technology in low to medium Earth orbit. There are many processes, usually acting in concert, which produce irreversible losses in the radiation belts. Several of these, such as the complimentary ring current expansion and inward magnetopause motion during active times, are discussed in the next section. Several different wave–particle interactions drive pitch angle scattering of relativistic and ultra-relativistic electrons and protons, which can result in loss from part or all of the full pitch angle distribution. We refer the readers to comprehensive reviews written previously on the subject (Summers et al. 2007; Thorne et al. 2010).

In addition, the Van Allen Probes era has produced a multitude of new studies on particle loss through wave–particle interactions. Notably, three main wave modes are likely responsible for the vast majority of particle losses in the inner magnetosphere by way of pitch angle scattering. VLF chorus can interact with lower energy electrons (up to ~30 keV) to scatter them into the loss cone, which results in the optical consequence called diffuse aurora (Thorne et al. 2010). The more incoherent VLF hiss waves are resonant with a wide variety of electron energies, but pitch angle diffusion timescales vary drastically, and hiss are only present inside the high density of the Earth's plasmasphere or plasmaspheric plumes (see next section). Recent work has been done to better parameterize the distribution and intensity of hiss wave power with respect to the plasmapause location, over varying local time and geomagnetic conditions (Malaspina et al. 2016). Finally, EMIC waves are known to interact strongly with highly energetic electrons in the MeV range, particularly during active storm times (Albert 2003; Summers and Thorne 2003). EMIC waves occur in frequency bands and new studies indicate that the H^+-band is most efficient at scattering up to a few MeV, while the He^+- and O^+-bands are more significant for higher energies (Ni et al. 2015). Interestingly, Usanova et al. (2014) showed that EMIC waves primarily affect only the lower pitch angles, essentially leaving the core distribution unchanged.

The task of implementing newly parameterized waves in existing radiation belt models is a daunting one, but much progress has been made over the last several years (e.g. Orlova et al. 2014, 2016; Agapitov et al. 2015; Spasojevic et al. 2015; Drozodov et al. 2017). Still, these models cannot always predict the outcome in event-specific instances, since less is known about which loss mechanism may play a larger role at a given time.

One of the difficulties in determining the dominant loss mechanisms during a particular event is the lack of direct measurements of the precipitating particle population. The loss can be observed as a decay or a dropout of near-equatorial magnetosphere particles, but the low-altitude measurement is rarely available; thus, the loss could be an apparent loss to other energies or equatorial locations, rather than a true loss from the system. Several platforms are ideal for filling in this measurement gap. LEO CubeSats, for example, can make substantial contributions to our understanding of precipitation. The Colorado Student Space Weather Experiment (CSSWE) CubeSat (Li et al. 2012) measured radiation belt particles in the loss cone at low altitudes simultaneously with the Van Allen Probes mission, and characterized the precipitating portion of several radiation belt loss events (Li et al. 2013; Blum et al. 2013; Schiller et al. 2014; Jaynes et al. 2014). The Balloon Array for RBSP Relativistic Electron Losses (BARREL) mission of opportunity (Millan et al. 2013) was designed to fly at very low (sub-orbital) altitudes to quantify and constrain precipitation timescales, amounts, and locations. To date, they have flown four separate campaigns in conjunction with NASA's Van Allen Probes. A major accomplishment of the mission was to find a distinct coherence between hiss wave modulation and variation in precipitation observed on the balloons (Breneman et al. 2015), leading to an enhanced understanding of overall radiation belt loss. The two generations of Focused Investigations of Relativistic Electron Burst Intensity, Range, and Dynamics (FIREBIRD) CubeSats (Spence et al. 2012) had the objective to spatially and temporally constrain microbursts, a type of magnetospheric wave thought to be responsible for sudden relativistic electron precipitation (Lorentzen et al. 2001). The second flight (FIREBIRD II) did indeed measure microburst precipitation and has provided the first in-situ confirmation of these spatial scales (Crew et al. 2016).

A sensor-fusion technique for studying loss from the radiation belts in detail incorporates in-situ wave measurements with conjunction observations from the ground using incoherent scatter radar instrumentation. Incoherent or Thomson scatter radars (Evans 1969) use radar scatter at VHF to UHF frequencies from thermal electron density fluctuations to directly probe the plasma state vector for the cold, dense plasma (∼0.1–0.4 eV) at the foot of inner magnetospheric field lines. Analysis of the very weak backscattered signal yields full ionospheric altitude profiles of ion and electron temperature, ion composition, electron density, and line-of-sight velocity. In particular, the electron density profile deep in the ionosphere at altitudes below 100 km is directly enhanced by precipitating medium energy particles >10 s of keV beyond those levels provided by solar EUV illuminations. With appropriate assumptions, inversion techniques can convert the radar observed ionospheric electron density profile to provide estimates of the incoming particle energy distribution (e.g. Semeter and Kamalabadi 2005). These can then be correlated with in-situ wave activity on orbiting platforms that are in magnetic conjunction with the ground based radar field of view to assess radiation belt loss and impacts on the dynamic atmosphere-ionosphere system. These impacts on such things as global circulation patterns and the production of molecules such as NOx are of intense interest in the middle atmosphere modeling community (cf. Sect. 5.3).

Another long-established technique for examining radiation belt losses to the atmosphere uses the relative ionospheric opacity meter (riometer), a broadband HF ground radio receiver system examining >25 keV electron precipitation levels through induced changes in received transionospheric signals originating from the relatively constant cosmic noise background (e.g. Hargreaves 1969). The mechanism behind the riometer principle involves enhanced electron density produced in the D region ionosphere, causing increased electron collision rates with sufficient frequency to attenuate HF radio waves transiting the atmosphere. Riometers can be used in event and statistical studies (e.g., Baker et al. 1981; Spanswick et al. 2007) and can also be used in infer temporal evolution of population fluxes from the overlying magnetosphere (Kellerman et al. 2015). Riometer technology has evolved from simpler wide beam systems to analog and now digital beam forming, wide frequency capture imaging systems, with the latter providing much better measurement precision and/or much higher spatial resolution for temporal and spatial precipitation loss studies (e.g., Honary et al. 2011).

5 Radiation Belt Coupling

5.1 Ring Current and Plasmasphere Coupling

The ring current is a westward-directed electric current carried by a toroidal population of ions surrounding Earth, often overlapping with both the outer radiation belt and the plasmasphere. The ring current varies in density and structure, but often exists between 2 and 9 R_{E}. Changes in composition can occur suddenly during times of geomagnetic storms and heightened geomagnetic activity; the source of ring current plasma can very during these times, with ionospheric outflow becoming dominant over protons of solar wind origin during storm times. For excellent reviews on the ring current, see Daglis et al. (1999) and Daglis and Kozyra (2002) along with Keika et al. (2013) summarizing work on inner magnetospheric O^+ energization and associated ring current injection/pressure effects.

The Dst index is a measure of the ring current strength as measured from a network of ground-based magnetometer stations, and is central in space weather investigations. The so-called "Dst effect" (Li et al. 1997a, 1997b; Kim and Chan 1997) appears as adiabatic changes in electron populations during times of dramatic magnetic field changes that manifest as variations in the Dst index. The result can be a sudden loss or enhancement in radiation belt fluxes that appears to be real, but is in fact an entirely adiabatic reaction that eventually reverses as the Dst index recovers. The real loss and acceleration during the main phase of storms can therefore be very difficult to separate from this Dst effect, as all mechanisms are likely taking place concurrently.

Enhancements in the stormtime ring current can have direct impacts on radiation belt losses in two ways: through wave–particle interactions or through increased magnetopause shadowing due to diamagnetic effects. Electromagnetic ion-cyclotron (EMIC) waves are driven by unstable ring current proton distributions (e.g. Anderson and Hamilton 1993) and are able to produce losses of the very highest-energy radiation belt electrons and protons (Meredith et al. 2003). The other mechanism involves a pumping of the partial ring current by enhanced convection and repeated small-scale injections (Gkioulidou et al. 2014), thereby diamagnetically expanding electron drift orbits to a point where magnetopause shadowing (West et al. 1972) can have a greater effect on the core population (Ukhorskiy et al. 2006) and lead to dramatic dropout features in the observations (Shprits et al. 2006). This process irreversibly alters the content of the outer radiation belt region, and can lead to almost total drainage of the preexisting populations (Turner et al. 2014).

Acceleration events of relativistic electrons can also be influenced by ring current dynamics. Sustained periods of electron injections can build up and maintain the source and seed populations, which are required to produce relativistic and ultra-relativistic electrons. Chorus wave growth, which plays a central role in relativistic electron enhancements, is driven by temperature anisotropies in ring current energy electrons (Sudan 1964; Kennel and Petschek 1966). See Sect. 4.4 for more details on this process.

The plasmasphere is a cold, dense toroidal plasma population that typically extends from the ionosphere out to an extremely variable outer boundary known as the plasmapause (e.g., Carpenter and Lemaire 2004). The plasmapause varies in local time, and can be as low as 2 R_E or as far out as >6 R_E, depending on geomagnetic activity (Chappell 1972). There is often much structure observed in the plasmapause/plasmasphere boundary layer, including persistent features such as the duskside bulge and plume region (Darrouzet et al. 2009). Plasmaspheric densities are often dramatically restructured during large geomagnetic storm events due to e.g. coronal interaction region or coronal mass ejection impacts. Afterwards, several processes act to refill the pre-storm population from the ionosphere through e.g. field-aligned diffusion, charge exchange, and cross-field transport. During recovery phases, the rate of field-aligned refilling is quite slow ($\sim$days) and therefore variability in local plasmaspheric densities can be accordingly quite high during dynamic conditions associated with repeated storms. Even after more than five decades of study (see review by Singh and Horwitz 1992), plasmasphere refilling dynamics and their impacts on overall plasmasphere structuring are still a subject of intense study with many unsolved problems (Gallagher and Comfort 2016).

The dynamic plasmasphere has long been recognized as a key agent in radiation belt configurations due to multiple interaction pathways between the cold plasma and more energetic particles. Early work showed a clear correlation between the plasmapause location and innermost extent of the multi-MeV electron outer radiation belt (Li et al. 2006). However, more recent detailed observations across a wide range of energies have revealed that the relationship between the plasmapause and relativistic electrons is complex and highly energy- and activity-dependent (Reeves et al. 2016). The plasmapause can provide both a protective effect as well as causing greater loss rates, depending on the energy of the electrons and their location within the magnetic field. The primary loss mechanism for radiation belt electrons within the plasmasphere is by cyclotron resonance interactions with VLF hiss waves. This wave mode is fairly efficient at pitch angle scattering relativistic electrons into the loss cone (Thorne et al. 1973). At lower L-shells, the resonance is more favorable for higher energies (Summers et al. 2007) and thus the outer radiation belt can be quickly carved out by the

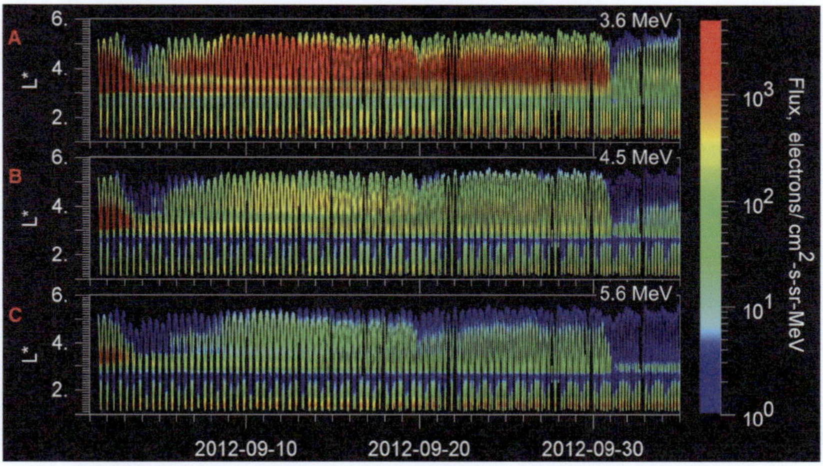

Fig. 31 A long-lived remnant belt of ultra-relativistic energies protected by a refilled plasmasphere. Reprinted from Baker et al. (2013a)

location of the plasmasphere and the hiss contained within. Several models have attempted to quantify this loss rate in a parameterized sense (e.g. Orlova et al. 2014), but event-specific studies have also shown faster loss rates for relativistic electrons than previously predicted (Jaynes et al. 2014), thus this is still an open question.

The lifetimes for relativistic electrons due to hiss scattering vary with magnetic field strength, magnetic local time, electron energy, and geomagnetic activity. In addition to this constant loss mechanism, radial diffusion is at work in this region, although it acts much more slowly here than in the outer belt region. Therefore, the dynamics inside the plasmasphere are a complicated balance of inward diffusion and slow scattering losses, highly dependent on the location and portion of the population being observed. There are times when the plasmasphere can retreat quickly, allowing for acceleration at very high energies to take place just outside the plasmapause boundary at low L-shells. After a time, the plasmasphere can refill to higher radial distances, effectively enveloping the newly accelerated electrons in a protective shell, as during the formation of the third belt in September 2012 (Baker et al. 2013a). Deep inside the plasmasphere, ultra-relativistic electrons can have lifetimes of 10's to 100's of days. Coupled with very slow diffusion rates, this means that a so-called remnant belt can persist for a long time untouched inside the plasmasphere (Fig. 31).

5.2 Ionosphere Coupling

5.2.1 Ionospheric Plasma in the Magnetosphere

In the inner magnetosphere at L shells $< \sim 4$, high density solar-produced ionospheric plasma is redistributed upward along closed magnetic field lines until a diffusive equilibrium (Angerami and Thomas 1964) is reached between ion species partial pressures and gravity. The underlying ionosphere is the source of the overlying plasmasphere and the plasmasphere can be considered as the high-altitude extension of the topside F region ionosphere. Disturbance effects observed in the ionosphere and magnetosphere during geomagnetic disturbances, and their space weather effects, have a direct relation to processes occurring in

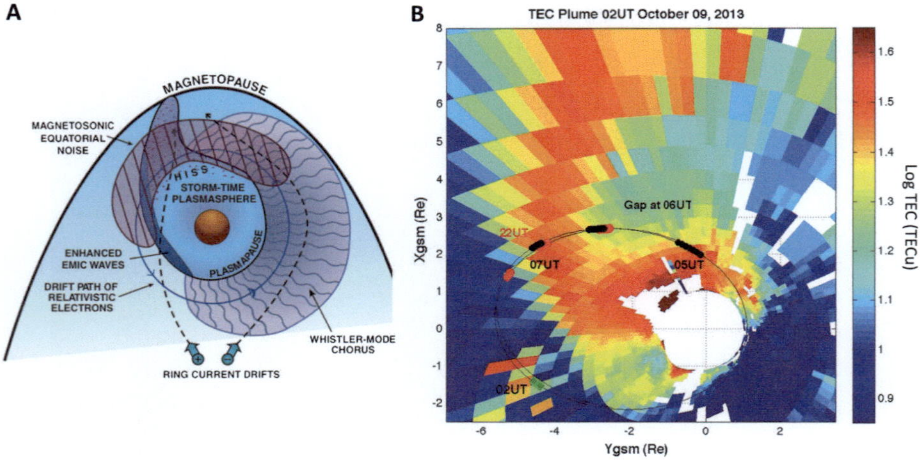

Fig. 32 (A) Schematic representation of cold plasma density structures in the magnetospheric equatorial plane (the plasmapause and plasmaspheric plume) indicates their relationship to the regions where electron interactions with different wave modes are important (after Summers et al. 1998). (B) The orbits of the Van Allen Probes spacecraft intersect the plume and the plasmapause (shown with mapped GPS TEC observations) facilitating the investigation of these significant space weather wave particle interactions

the plasmasphere boundary layer (PBL) (Carpenter and Lemaire 2004), the boundary between the corotating field lines of the plasmasphere and the outer magnetosphere. This section discusses space weather phenomena observed near this outer boundary of the coupled plasmasphere–mid-latitude ionosphere with an emphasis on related effects in the radiation belts. The dense low energy plasmas of the ionosphere and plasmasphere play a significant role in modulating the wave–particle interactions (WPI) that are very important for both radiation belt energization and loss. In general, an increase in density inhibits local acceleration, and an increase in heavy ions generates EMIC waves that produce loss of relativistic electrons. Interactions with waves of both natural (e.g. Thorne et al. 2013a) and anthropogenic origin (e.g. Foster et al. 2016) are important. The redistribution of low energy plasma of ionospheric origin into the high-latitude outer magnetosphere (cf. Foster 2016) creates density gradients and localized structure that influence radiation belt space weather effects.

5.2.2 Cold Plasma Redistribution

The theoretical study of Summers et al. (1998) pointed out the importance cold plasma density in modulating the interactions of magnetospheric waves and electrons. Figure 32A, based on that work, describes the relationship of cold plasma density structure in the magnetosphere to regions where RB interactions with VLF chorus, hiss, and EMIC waves would be important. The plasmaspheric plume, the plasmapause, and the characteristics of the plasmasphere boundary layer are important regions in this regard.

During disturbed conditions, magnetosphere–ionosphere coupling processes at the plasmasphere boundary layer drive a large-scale redistribution of cold ionospheric/plasmaspheric plasma through the geospace system. Foster (1993) investigated the disturbed mid-latitude ionosphere using scanning observations with the Millstone Hill incoherent scatter radar that provided a two-dimensional picture of the evolution of ionospheric F region structure. Spatially extended sunward convecting density enhancements seen immediately equatorward of the dusk-sector ionospheric trough were termed storm-enhanced density (SED). SED was

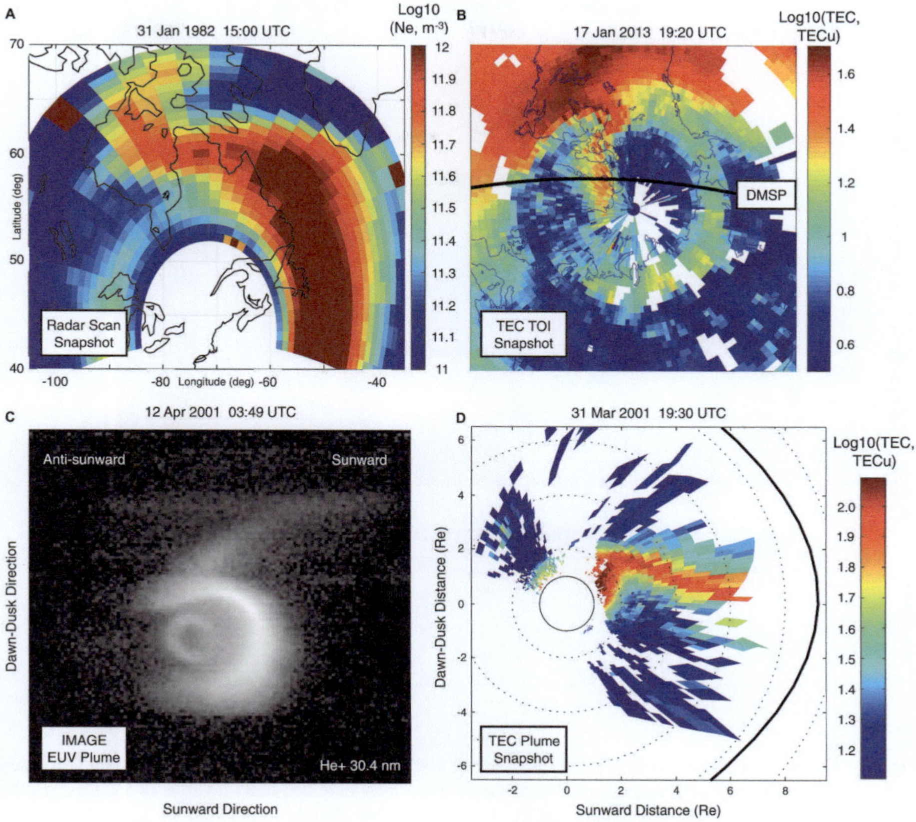

Fig. 33 (A) Azimuth scan observations with the Millstone Hill radar provide a spatial "snapshot" of a continuous stream of plasma carried from low latitudes in the evening sector to the cusp and polar cap. The observations of ionospheric electron density were made over a ~30 minute interval near the noon meridian and are plotted in latitude//longitude coordinates with north at the top. (B) Aggregation of 5 minutes of northern hemisphere GPS TEC observations provided global imagery of a nearly-identical feature some 30^+ years later during a March 2013 storm. Presented in polar local time geodetic coordinates with noon at the top, the plume extending poleward from North America is seen to stretch anti-sunward across polar latitudes and to merge into the auroral oval in the midnight sector. The orbital track of a DMSP satellite crossing polar latitudes is shown as a black line. (C) Space based imagery of the outer extent of the plasmasphere depicts the erosion and sunward extension of the cold plasma plume. An IMAGE EUV snapshot looking down on the plasmasphere and plume from above north polar latitudes is shown. (D) GPS TEC observations, when mapped along the magnetic field into the magnetospheric equatorial plane, reveal plasma boundaries closely reproducing the evolution of the geospace plume at magnetospheric heights, indicating its full sunward extension to the magnetopause

observed as a continuous plume of ionospheric plasma spanning a large portion of the post-noon to midnight sector (Fig. 33A). The study concluded that SED plasma was transported sunward (westward) in the sub-auroral polarization stream (SAPS, Foster and Vo 2002) at speeds of ~1 km/s. SED flows at ionospheric altitudes provide significant heavy ion fluxes into the high latitude F-region cusp ionosphere with magnitude $\geq 1.0 \times 10^{14}\ \mathrm{m^{-2}\,s^{-1}}$ (Foster et al. 2004, 2014b; Erickson et al. 2011; Makarevich et al. 2009, 2011). In this way a continuous stream of ionospheric plasma is carried in the convective flow channel from low latitudes in the evening sector to the cusp and into the polar cap at noon (Fig. 33B) creating a

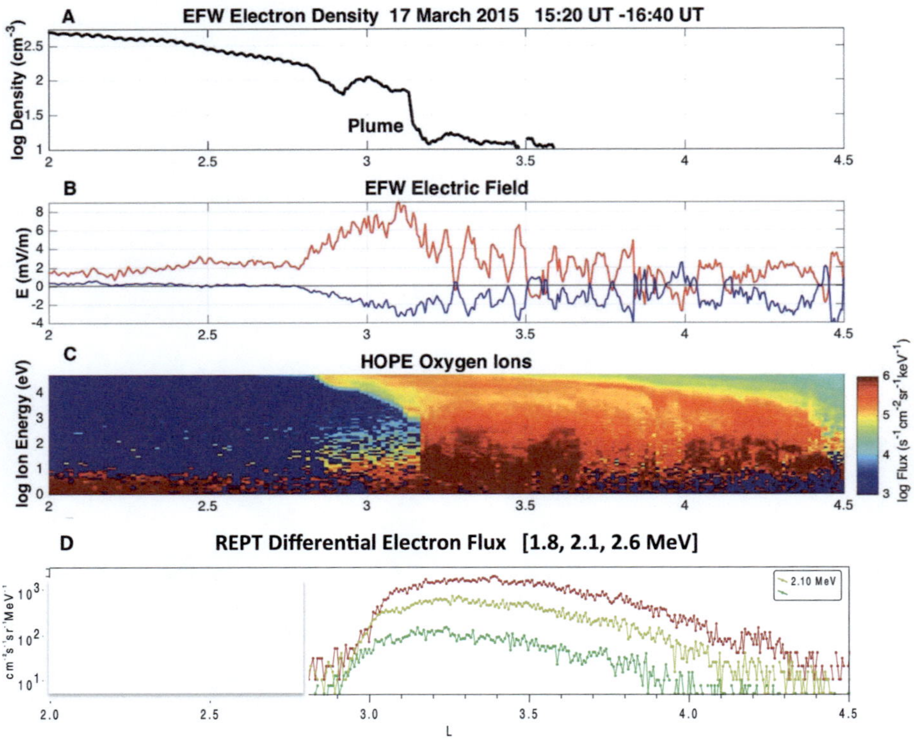

Fig. 34 The relationship of the SAPS, the ring current, and the geospace plume were observed by Van Allen Probe A as its near-equatorial outbound orbit crossed the plasmasphere boundary layer at 18 MLT during the 17 March 2015 storm. (A) In situ plasma density observations identified the plume near its point of formation near $L \sim 3$. (B) Strong radially outward electric fields (*red curve*; azimuthal component is shown in *blue*) were observed across the SAPS flow channel responsible for the sunward advection of outer plasmaspheric ions in the geospace plume. (C) A 'nose' of hot ring current O^+ ions extended across the region of SAPS electric field. Ring current pressure gradients associated with the hot ions drive the SAPS convection. (D) The Earthward extent of the recovering outer zone multi-MeV electrons falls within the region of SAPS, ring current, plume PBL interactions

tongue of ionization (TOI, Foster et al. 2005) stretching across the polar cap to the midnight sector. In general, an increase in density inhibits local acceleration, and an increase in heavy ions generates EMIC waves that produce loss of relativistic electrons.

A geospace system perspective of SED has been provided with distributed ground-based Global Positioning System (GPS) observations of TEC (Foster et al. 2002) and with space-based imagery of the outer plasmasphere by the IMAGE EUV instrument (Goldstein and Sandel 2005). Combining these ground and space-based plasma imaging techniques, Foster et al. (2002) demonstrated that the ionospheric SED plume was magnetically connected to the low altitude signature of a drainage plume (plasmaspheric tail) associated with the storm-time erosion of the outer plasmasphere (Figs. 33C and 33D). Outflowing cold ionospheric plasma can have large effects on the radiation belts due to mechanisms such as inhibition of local relativistic acceleration processes through detuning of cyclotron resonances which depend on cold plasma density (Abel and Thorne 1998a, 1998b). The O+ rich ion content of these flows also is a source of EMIC waves that are known to be highly efficient loss mechanisms for relativistic particles (e.g. Jordanova et al. 2008).

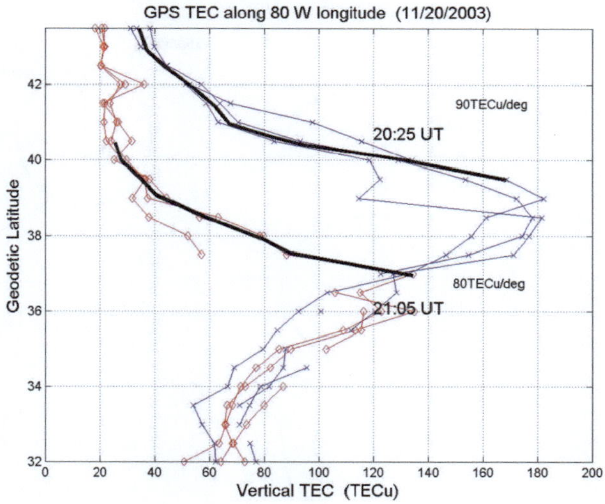

Fig. 35 The TEC gradients across the poleward edge of the SED plume observed over the central US at two times during the 20 November 2003 storm are shown. TEC in excess of 180 TECu is observed equatorward of the PBL gradients. The steep decreases in TEC with latitude across the PBL extend over ~2 deg of latitude with a characteristic gradient of -85 TECu per deg of latitude. Such gradients constitute a severe space weather hazard for positioning and navigation systems

The combination of ground based (e.g. radar or GPS TEC) observations with in situ Van Allen Probes or MMS observations of waves and electrons, as depicted in Fig. 33B, has addressed the radiation belt associated space weather effects of SAPS (fundamental magnetosphere–ionosphere electrodynamic coupling agent) and cold plasma plumes in the $L = 4 - 8+$ region in the post-noon sector (e.g. Foster et al. 2014a; Erickson et al. 2016). Substorms also influence SAPS occurrence (e.g., Makarevich et al. (2009, 2011) leading to plume-related space weather effects.

The characteristic interrelationship of outer plasmaspheric M–I coupling phenomena to the density structure at the PBL and their relationship to the inward extent of MeV RB electrons are shown in Fig. 34 for disturbed storm-recovery conditions.

Very sharp cold plasma density gradients can develop in the mid-latitude ionosphere as the result of strong M–I coupling effects during magnetic disturbances. Plasma transport in combination with flow-dependent collisional ion-neutral recombination can produce a deep low-density ionospheric trough associated with strong SAPS flows (Schunk and Walker 1969; Anderson et al. 1991). An example of the steep ionospheric density gradients observed during a major storm is shown in Fig. 35 using GPS TEC observations.

The magnetosphere–ionosphere coupling effects producing the sharp ionospheric density gradients do not only affect the ionosphere, but, owing to their electrodynamic nature, map along magnetic field lines into the outer plasmasphere. At these locations, they are associated with steep electron density gradients seen along the outer edge of the erosion plume in the dusk sector (e.g. Figure 34). The cold plasma density gradient at the plasmapause fundamentally controls the stability of injected electrons for wave–particle interactions generating VLF chorus rising tones. Importantly, in those nonlinear interactions, the ~50 keV electrons give up perpendicular energy to the VLF waves and are rapidly precipitated into the atmosphere. This process was illustrated in the study of Foster and Rosenberg (1976) as shown in Fig. 36.

Simultaneously, plasma wave effects in the form of discrete chorus rising tones within the inner magnetosphere undergo nonlinear interactions with the high-energy tail of the electron population, rapidly accelerating electrons to multi-MeV energies (Foster et al. 2017). These effects constitute important space weather features. Figure 23 (above) illustrates the rapid recovery of the MeV RB electron population observed simultaneously with the M–I

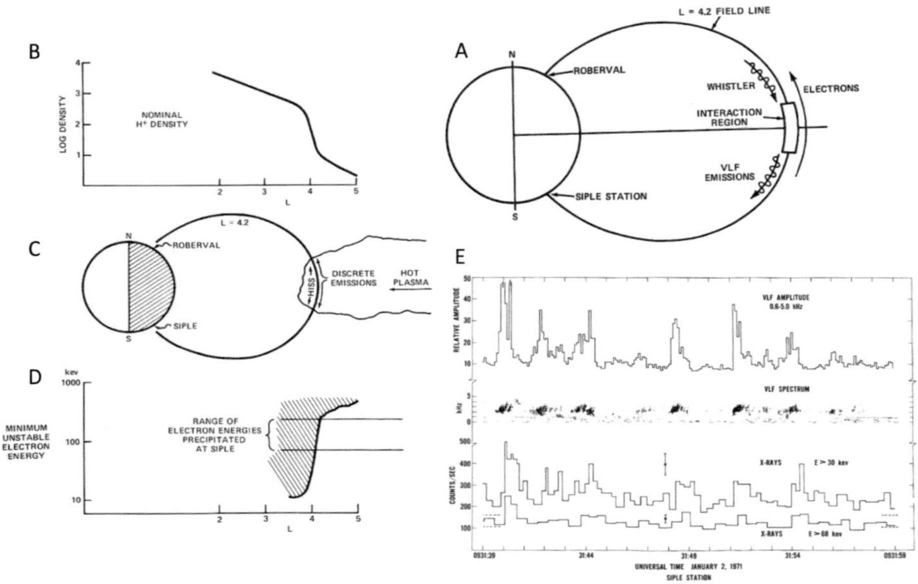

Fig. 36 (A) Cyclotron interactions between 20–80 keV electrons and VLF waves on field lines near the cold plasma density gradient at the PBL (B) lead to the generation of VLF chorus rising tones and scatter substorm-injected electron (C) into the atmospheric loss cone. (D) The cyclotron resonant energy drops sharply into the energy range of the injected electrons as the cold plasma density increases. (E) Bursts of precipitating electrons (seen with a balloon-borne X-ray detector) are correlated on a one-to-one basis with bursts of chorus rising tones (Foster and Rosenberg 1976)

coupling and density structure shown in Fig. 34. Multi-MeV electron fluxes increased by a factor of 10 over 1-hour interval in the low-density region immediately outside the PBL in association with $\sim$100 keV electron injection and the generation of strong VLF chorus rising tones.

The examples provided in this section have concentrated on the processes associated with the interaction of lower energy and relativistic electrons with VLF chorus and the effect of ambient cold density on those processes. A similar scenario could be presented for the density dependence of the electron interactions with VLF hiss and EMIC waves at other regions in the inner magnetosphere. More generally, results of these recent studies underline the fundamental conclusion that the dynamic variations of the highest energy radiation belt electrons are closely coupled to the characteristics of the lowest energy (ionospheric) particle populations.

5.3 Middle Atmospheric Coupling

Part of the radiation belt electron population is continuously lost to the Earth's atmosphere. When wave–particle interactions scatter electrons into the bounce loss cone, they penetrate to low-enough altitudes ($\sim$100 km) and deposit their energy in collisions with atmospheric gas molecules instead of mirroring back to higher altitudes. This process, termed "electron precipitation", takes place in the auroral and sub-auroral latitudes, is strongly driven by the solar wind and geomagnetic storms (as radiation belt processes are in general), and leads to ionization and dissociation of atmospheric molecules (see Baker et al. 1987b).

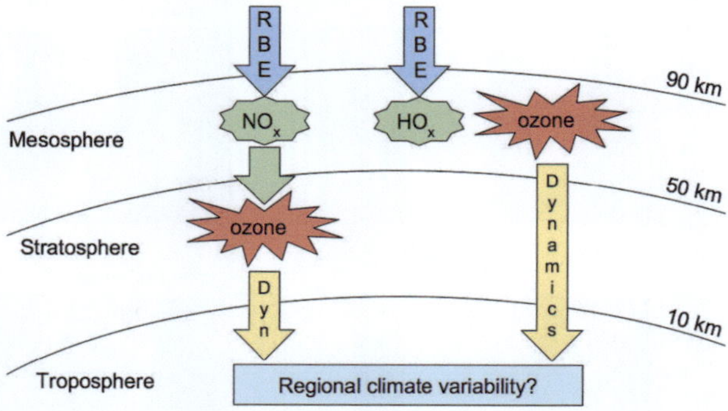

Fig. 37 Rough schematic view of RBE, mesospheric production of HO_X and NO_X, loss of ozone in the mesosphere and stratosphere, and connection to ground-level climate through atmospheric dynamics

Energetic electron precipitation from the outer radiation belt (from now on denoted RBE) is one of the several populations of energetic particles ionizing the atmosphere (e.g. Verronen and Rodger 2015). In addition to RBE, CME-driven solar proton events (SPE) and substorm-driven auroral electron (AE) precipitation are also important. At high-latitude ($L = 2$–7) mesospheric altitudes, RBE ionization dominates over AE and protons (except during strong SPEs which, however, are infrequent) while mesospheric ionization caused by solar ultraviolet and X-ray radiation and galactic cosmic rays is typically marginal. The energy of an electron determines its atmospheric penetration depth and the altitude region of ionization. Thus, RBE with energy between about 30 keV and 1 MeVs is most important for mesospheric altitudes (90–50 km) (see Fig. 37).

Energetic particle precipitation (EPP) can lead to important changes in the neutral composition of the atmosphere at high polar latitudes (Sinnhuber et al. 2012). Most studied have been the production of a) odd hydrogen HO_x ($H + OH + HO_2$) through chemistry associated with ion pair production, water cluster ion formation, and subsequent recombination; and b) odd nitrogen NO_x ($N + NO + NO_2$) through dissociation of molecular nitrogen via charged particle impact and ion chemistry. Production of HO_x and NO_x is important because these species participate in catalytic chemical reactions that play a key role in the ozone balance of the mesosphere and stratosphere, respectively. Both short-term (days) and longer-term (weeks) ozone depletion has been observed after strong precipitation events, e.g. SPEs (Reid et al. 1991; Jackman et al. 2001; Seppälä et al. 2004).

Most of the precipitating electrons deposit their energy in the mesosphere-lower thermosphere (MLT) region. Unlike HO_x, which can directly affect ozone at mesospheric altitudes, the produced NO_x needs to be transported from MLT down to the upper stratosphere in order to have an impact on ozone. The NO_x descent can only take place in the winter pole, inside the polar vortex, which restricts mixing with low-NO_x air in the lower latitudes. In the absence of solar radiation, NO_x has a long chemical lifetime and is thus subject to transport. Many observational studies have shown NO_X descent to take place during individual winters (Callis and Lambeth 1998; Siskind et al. 2000; Randall et al. 2009; Päivärinta et al. 2013), and a decadal analysis of satellite data has shown that NO_x descent is in fact occurring over every winter, in both hemispheres (Funke et al. 2014a, 2014b). The amount of NO_X, however, varies from year-to-year with respect to electron forcing and polar vortex dynamics, the South typically having more NO_x in the stratosphere than the North (Fig. 38).

Direct, mesospheric impacts of RBE have been recently quantified using satellite-based OH observations. At geomagnetic latitudes between 55 and 72 degrees, there is a clear and robust connection between satellite-based measurements of RBE fluxes and mesospheric

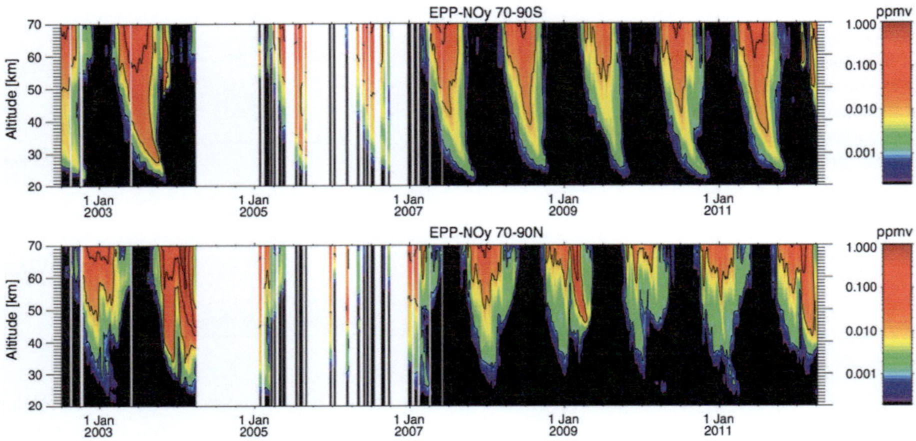

Fig. 38 EPP-produced, excess NOy (NO_x + NO_x converted to other nitrogen species, in ppmv mixing ratio) descent observed between 2002 and 2012 by the MIPAS instrument. *Upper panel*: latitudes between 70 and 90 degrees South. *Lower panel*: the same but for North. Adopted from Funke et al. (2014b)

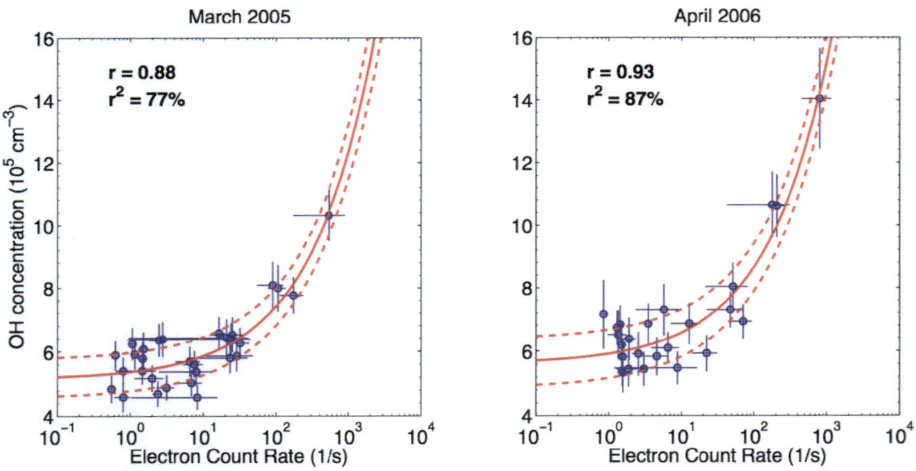

Fig. 39 Relation between OH concentration at 71–78 km and count rate of precipitating electrons in the outer radiation belt as observed by the MLS/Aura and MEPED/POES instruments, respectively. Shown are daily mean data in March 2005 and April 2006 (*blue dots*). *Horizontal and vertical blue lines* mark the standard error of the mean. The *red solid* and *dashed lines* mark a line fit and its estimated standard deviation r is the correlation coefficient. The counts/s measured by the MEPED instrument can be converted to "normal" electron flux units ($cm^{-2}\,s^{-1}\,sr^{-1}$) by multiplying the count rate by 100. Reproduced from Verronen et al. (2011)

OH concentration (Verronen et al. 2011; Andersson et al. 2012), with strong RBE significantly increasing night-time OH (Fig. 39). Impacts are observed down to about 50 km altitude but not below, which indicates that even during the strongest RBE periods the fluxes of MeV electrons are too low to enhance OH above its background concentration. The strongest effects are seen during the declining phase of the solar cycle while practically no effect was seen during the geomagnetically record-low year of 2009. The longitudinal distribution of the OH response follows the geomagnetic latitudes in both hemispheres and indicates a more

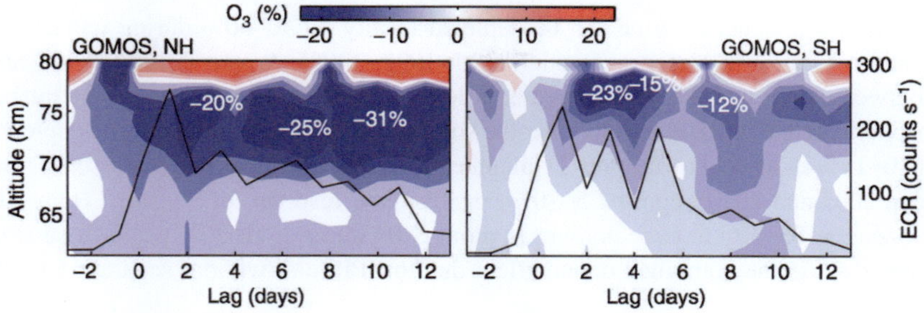

Fig. 40 Short-term RBE effects on mesospheric ozone at geomagnetic latitudes between 55 and 65 degrees, North and South. Results are from a superposed epoch analysis of the 60 largest RBE events between 2002 and 2012. Relative ozone anomalies from the GOMOS instrument are shown in *color scale*, and count rates of 100–300 keV electrons (ECR) from the MEPED/POES instrument with the *black line*. Day zero marks the beginning of RBE events. Reproduced from Andersson et al. (2014a)

inhomogeneity in the Southern Hemisphere polar region (Andersson et al. 2014b), similar to those seen in satellite-based RBE flux observations. While there is a stronger OH signature in the sector poleward of the South Atlantic Anomaly (SAA) region when compared to other longitudes, the SAA region itself shows no OH response. This indicates negligible fluxes of >100 keV RBE in the SAA region.

Connected to the OH response discussed above, there are also studies analyzing the response of mesospheric ozone from satellite-based measurements, considering both daily variability during RBE events (individual as well statistical) and long-term variability over (partial) solar cycle (Verronen et al. 2013; Andersson et al. 2014a). Unlike in the case of HO_x, the ozone response is not strictly limited to the magnetic latitudes of the radiation belts but is observed over a more extended area in the polar cap region. This is because, in addition to chemistry, ozone is also affected by atmospheric dynamics on daily time scales. During strong RBE events, mesospheric ozone can decrease by tens of percent due to RBE driven HO_x enhancements (Fig. 40). The magnitude and duration of the ozone effect depends on the RBE event but also on the atmospheric state, e.g., the effect is relatively larger in the winter pole. The higher frequency of the events during solar active years leads to ozone variability by up to tens of percent on solar cycle timescales as well.

It has been suggested for a long time that the RBE contributes to polar NO_x production in the MLT region (Callis et al. 1996, 1998). However, distinguishing the contribution of RBE to the mesospheric NO_x budget is not straightforward using observations only. This is because the RBE-NO_x produced in the mesosphere is mixing in with large amounts of NO_x produced by auroral electrons above the mesopause. Both descend from MLT to the upper stratosphere during winter months. For specific years, there are studies reporting increases in mesospheric NO_x ranging from 100% to orders of magnitude due to in-situ production by RBE (Clilverd et al. 2009; Newnham et al. 2011). Over decadal time scales, however, the majority of mesospheric NO_x enhancement in wintertime has been reported to be due to descent of auroral NO_x from the lower thermosphere (Sinnhuber et al. 2011), with considerable in-situ production occurring above 70 km only (Sinnhuber et al. 2016). Nevertheless, any NO_x produced by RBE below the mesopause is more likely to descend to the upper stratosphere and affect ozone, because at the mesopause the wintertime mean circulation changes vertical direction to from downward below to upward above (Smith et al. 2011).

Understanding the response of upper stratospheric ozone on the descending NO_x has been challenging because of its relatively large overall variability due to atmospheric dynamics (e.g. Päivärinta et al. 2013). Nevertheless, satellite observations has been shown that polar upper stratospheric ozone can be depleted locally by 40–60% during winters of exceptionally strong NO_x descent (Randall et al. 1998, 2005). A recent study using satellite data between 1979 and 2014 has revealed a long-term response of Southern Hemispheric stratospheric ozone to RBE activity over the solar cycle, with an average ozone depletion of about 10–15% at 30–45 km altitude (Damiani et al. 2016). Another study, using observations between 2005 and 2010, reported 5–10% negative EPP response of Southern Hemispheric polar ozone at 25–50 km over the winter months (Fytterer et al. 2015). All these studies, however, discuss the combined ozone effect due to EPP total without separating the RBE contribution.

In atmospheric simulations, RBE impact can be separated. However, there have been serious issues with the quality and characteristics of available satellite-based data. Most useful long-term data sets of RBE fluxes have been provided by the instruments aboard the POES satellites, but even these data have restrictions concerning their applicability to atmospheric modeling (Rodger et al. 2010a). The measurement setup allows only part of the bounce loss cone fluxes (i.e. fluxes of precipitating electrons) to be measured, with a limited energy resolution. Based on ground-based ionospheric observations, it has been suggested that precipitating fluxes might be underestimated by as much as a factor of 10 (Hendry et al. 2012; Clilverd et al. 2012). On the other hand, the underestimation seems to be restricted to low-to-medium geomagnetic activity periods because a comparison of modeled and observed concentrations of mesospheric OH indicates a reasonable agreement during geomagnetic storms (Verronen et al. 2013). Contamination by energetic protons is another issue that has to be carefully considered (Rodger et al. 2010a). Despite these difficulties, precipitation models have been created based on the POES (and other) data for 3-D, long-term atmospheric applications. For example, the AIMOS model uses statistical maps of precipitation, scaled with observed electron fluxes to provide global RBE (and other EPP) forcing data from year 2002 onwards (Wissing and Kallenrode 2009). Recently, a RBE model driven by geomagnetic indices has been created for multi-decadal climate simulations (Van de Kamp et al. 2016) as a part of the solar forcing recommendation for Coupled Model Intercomparison Project Phase 6 (CMIP6) (Matthes 2016). An example of the model output, in Fig. 41, shows RBE forcing patterns on a world map and demonstrates the variability from solar maximum to minimum. To date, there are a few atmospheric modeling studies addressing the separate impact of RBE on the upper stratospheric and mesospheric composition (Codrescu et al. 1997; Semeniuk et al. 2011; Arsenovic et al. 2016). All these studies, ranging from monthly to solar-cycle time scales, suggest a considerable RBE contribution to EPP-driven polar stratospheric NO_x and ozone variability. In the mesosphere, it is already evident that RBE is the main driver of HO_x and ozone response. Although the work is still on-going, especially in terms of simulating the climate effects over several solar-cycles, uncertainties in the RBE forcing remain such that the assessment of its importance would benefit from improved satellite-based flux observations and their representation in models.

Ozone is one of the most important minor species in the middle atmosphere. It is a strong absorber of solar UV radiation and long-wave radiation from the surface, and thus important to the temperature balance and dynamics. It has been suggested that the polar ozone variability caused by energetic particle precipitation (including RBE), could connect to the troposphere and cause regional variability of up to 5 K in ground-level wintertime average temperatures over the solar cycle, in regions of Europe and North America (Rozanov et al. 2005; Seppälä et al. 2009; Baumgaertner et al. 2011; Maliniemi et al. 2013). The most likely physical explanation is the so-called top-down mechanism (Gray et al. 2010; Maycock and Misios 2015). It connects ozone and dynamical changes in the stratosphere to modulation of the North Atlantic Oscillation, which is a large-scale mode (i.e., pattern)

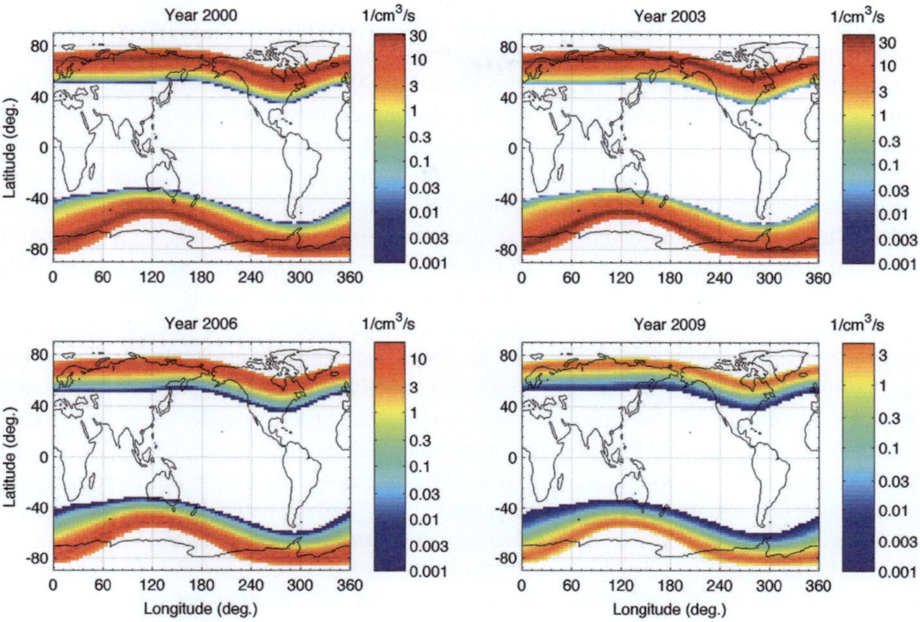

Fig. 41 Yearly average, zonal mean mesospheric ionization rates due to RBE, calculated using the model by Van de Kamp et al. (2016). The model provides data at L shells 2 to 10, *white areas* indicate no data

of climate variability around the North Atlantic. Although the RBE-climate connection has been reproduced in some climate simulations, so far the magnitude of the ground-level temperature variations have been underestimated compared to observations, and the details of the complex chemical-dynamical mechanism are not yet fully understood. A recent summary by Clilverd et al. (2016) gives a more detailed discussion on the proposed ground-level climate effects of RBE.

In summary, there is now a wealth of evidence of RBE effects in polar, wintertime atmosphere. In the mesosphere, substantial amounts of HO_x are produced, and this causes ozone response in time scales ranging from days to the solar cycle. NO_x production is probably an important addition to the indirect effect on upper stratospheric ozone through NO_x descent. The combination of ozone responses in both the mesosphere and stratosphere is likely needed to understand the suggested coupling to the troposphere and regional climate variability over the solar cycle.

6 Summary and Future Directions

As described in this chapter, the Earth's radiation belts were the first great discovery of the Space Age. With the realization that near-Earth space was filled with very energetic particles, there was recognition of the deleterious effects that could be produced in spacecraft and biological systems flying in space. The space weather impacts of the radiation belts have only increased over the past six decades as humans have used space more and more intensively for communication, navigation, reconnaissance, weather forecasting, and countless other applications. Deeper physical understanding of the radiation belts and their temporal variations has become a pressing requirement in the modern age.

In this chapter we have reviewed the structure and dynamics of the radiation belts from a scientific perspective. We have described our present understanding of the acceleration, transport, and loss of particles across the inner Van Allen zone, the slot region, and the outer Van Allen belt. We have shown how recent comprehensive observations have revealed new and unexpected aspects of the radiation belt properties. While many features of the radiation belts depend on external forcing of the system by the ambient solar wind near Earth, there still is a pressing requirement to make continual observations throughout the belts. The complexity and diversity of responses of both inner and outer zone properties are quite striking and, thus, we still do not have the knowledge to predict radiation belt properties at key locations based solely upon solar and solar wind forcing parameters.

As described in this chapter, the radiation belts are an integral part of the overall magnetospheric environment of Earth. As such, the Van Allen belts are intimately coupled to the ring current, plasmasphere, ionosphere, and magnetotail regions of our planet. The ebb and flow of cold plasmas and the electromagnetic waves these plasmas support play an essential role in the acceleration and loss of radiation belt particles. We also are finding new evidence that human activities such as powerful ground-based radio transmitters can directly affect the structure and dynamics of the radiation belts.

Quite importantly, there is ample evolving evidence that the loss of radiation belt particles into the Earth's atmosphere can have profound effects on the chemistry and electrodynamic state of our planet. This means that the coupling from the radiation belts into the mesosphere and stratosphere must be actively considered as contributing to the richly varying properties of the overall atmospheric system. This fact amplifies the importance of studying and monitoring the state of Earth's radiation belts now and into the future.

References

B. Abel, R.M. Thorne, Electron scattering loss in Earth's inner magnetosphere 1. Dominant physical processes. J. Geophys. Res. **103**(A2) (1998a). doi:10.1029/97JA02919

B. Abel, R.M. Thorne, Electron scattering loss in Earth's inner magnetosphere 2. Sensitivity to model parameters. J. Geophys. Res. **103**(A2) (1998b). doi:10.1029/97JA02920

O.V. Agapitov, A.V. Artemyev, D. Mourenas, F.S. Mozer, V. Krasnoselskikh, Empirical model of lower band chorus wave distribution in the outer radiation belt. J. Geophys. Res. **120**, 10425 (2015)

J.M. Albert, Nonlinear interaction of outer zone electrons with VLF waves. Geophys. Res. Lett. **29**(8) (2002). doi:10.1029/2001GL013941

J.M. Albert, Evaluation of quasi-linear diffusion coefficients for EMIC waves in a multispecies plasma. J. Geophys. Res. **108**(A6), 1249 (2003). doi:10.1029/2002JA009792

J.M. Albert, M.J. Starks, R.B. Horne, N.P. Meredith, S.A. Glauert, Quasi-linear simulations of inner radiation belt electron pitch angle and energy distributions. Geophys. Res. Lett. **43**, 2381–2388 (2016). doi:10.1002/2016GL067938

A.F. Ali, S.R. Elkington, W. Tu, L.G. Ozeke, A.A. Chan, R.H.W. Friedel, Magnetic field power spectra and magnetic radial diffusion coefficients using CRRES magnetometer data. Rev. Geophys. Space Phys. **120**, 973–995 (2015). doi:10.1002/2014JA020419

A.F. Ali, D.M. Malaspina, S.R. Elkington, A.N. Jaynes, A.A. Chan, J. Wygant, C.A. Kletzing, Electric and magnetic radial diffusion coefficients using the Van Allen probes data. J. Geophys. Res. Space Phys. **121**, 9586–9607 (2016). doi:10.1002/2016JA023002

J.H. Allen, Historical and recent solar activity and geomagnetic storms affecting spacecraft operations, in *Proc. GOMAC, Modern Space Syst. Issues*, Monterey, CA (2002)

B.J. Anderson, D.C. Hamilton, Electromagnetic ion cyclotron waves stimulated by modest magnetospheric compressions. J. Geophys. Res. **98**(A7) (1993). doi:10.1029/93JA00605

P.C. Anderson, R.A. Heelis, W.B. Hanson, The ionospheric signatures of rapid subauroral ion drifts. J. Geophys. Res. **96**(A4), 5785–5792 (1991). doi:10.1029/90JA02651

B.R. Anderson, R.M. Millan, G.D. Reeves, R.H.W. Friedel, Acceleration and loss of relativistic electrons during small geomagnetic storms. Geophys. Res. Lett. **42**, 10,113–10,119 (2015). doi:10.1002/2015GL066376

M.E. Andersson, P.T. Verronen, S. Wang, C.J. Rodger, M.A. Clilverd, B. Carson, Precipitating radiation belt electron and enhancements of mesospheric hydroxyl during 2004–2009. J. Geophys. Res. **117**(D09), 304 (2012). doi:10.1029/2011JD017246

M.E. Andersson, P.T. Verronen, C.J. Rodger, M.A. Clilverd, A. Seppälä, Missing driver in the Sun–Earth connection from energetic electron precipitation impacts mesospheric ozone. Nat. Commun. **5** (2014a). doi:10.1038/ncomms6197

M.E. Andersson, P.T. Verronen, C.J. Rodger, M.A. Clilverd, S. Wang, Longitudinal hotspots in the mesospheric OH variations due to energetic electron precipitation. Atmos. Chem. Phys. **14**, 1095–1105 (2014b). doi:10.5194/acp-14-1095-2014

J.J. Angerami, J.O. Thomas, Studies of planetary atmospheres: 1. The distribution of electrons and ions in the Earth's exosphere. J. Geophys. Res. **69**(21), 4537–4560 (1964). doi:10.1029/JZ069i021p04537

P. Arsenovic, E. Rozanov, A. Stenke, B. Funke, J.M. Wissing, K. Mursula, F. Tummon, T. Peter, The influence of Middle Range Energy Electrons on atmospheric chemistry and regional climate. J. Atmos. Sol.-Terr. Phys. **20** (2016). doi:10.1016/j.jastp.2016.04.008

D.N. Baker, How to cope with space weather. Science **297**, 1486–1487 (2002). doi:10.1126/science.1074956

D.N. Baker, Specifying and forecasting space weather threats to human technology, in *Effects of Space Weather on Technology Infrastructure*, ed. by I.A. Daglis (Kluwer, New York, 2004), pp. 1–25

D.N. Baker, L.J. Lanzerotti, Space weather. Am. J. Phys. **84**, doi.org/10.1119/1.4938403 (2016)

D.N. Baker, P.R. Higbie, R.D. Belian, E.W. Hones, Do Jovian electrons influence the terrestrial outer radiation zone? Geophys. Res. Lett. **6**, 531–534 (1979). doi:10.1029/GL006i006p00531

D.N. Baker, P. Stauning, E.W. Hones Jr., P.R. Higbie, R.D. Belian, Near equatorial, high-resolution measurements of electron precipitation and $L \cong 6.6$. J. Geophys. Res. **86**, 2295 (1981)

D.N. Baker, J.B. Blake, R.W. Klebesadel, P.R. Higbie, Highly relativistic electrons in the Earth's outer magnetosphere: 1. Lifetimes and temporal history 1979–1984. J. Geophys. Res. **91**(A4), 4265–4276 (1986)

D.N. Baker, R.D. Belian, P.R. Hibgie, J.B. Blake, Deep dielectric charging effects due to high energy electrons in Earth's outer magnetosphere. J. Electrost. **20**(3) (1987a). doi:10.1016/0304-3886(87)90082-9

D.N. Baker, J.B. Blake, D.J. Gorney, P.R. Higbie, R.W. Klebesadel, J.H. King, Highly relativistic magnetospheric electrons: a role in coupling to the middle atmosphere? Geophys. Res. Lett. **14**, 1027-10–1027-30 (1987b). doi:10.1029/GL014i010p01027

D.N. Baker, R.L. McPherson, T.E. Cayton, R.W. Klebesadel, Linear prediction filter analysis of relativistic electron properties at 6.6 RE. J. Geophys. Res. **95**(A9), 15133–15140 (1990). doi:10.1029/JA095iA09p15133

D.N. Baker, G.M. Mason, O. Figueroa, G. Colon, J.G. Watzin, R.M. Aleman, An overview of the Solar, Anomalous, and Magnetospheric Particle Explorer (SAMPEX) mission. IEEE Trans. Geosci. Remote Sens. **31**(3), 531–541 (1993). doi:10.1109/36.225519

D.N. Baker, S.G. Kanekal, J.B. Blake, B. Klecker, G. Rostoker, Satellite anomalies linked to electron increase in the magnetosphere. Eos **75**(35), 401–405 (1994). doi:10.1029/94EO01038

D.N. Baker, S.G. Kanekal, T.I. Pulkkinen, J.B. Blake, Equinoctial and solstitial averages of magnetospheric relativistic electron: a strong semiannual modulation. Geophys. Res. Lett. **26**(20), 3193–3196 (1999). doi:10.1029/1999GL003638

D.N. Baker, S.G. Kanekal, X. Li, S.P. Monk, J. Goldstein, J.L. Burch, An extreme distortion of the Van Allen belt arising from the 'Hallowe'en' solar storm in 2003. Nature **432**, 878–881 (2004a). doi:10.1038/nature03116

D.N. Baker, S.G. Kanekal, J.B. Blake, Characterizing the Earth's outer Van Allen zone using a radiation belt content index. Space Weather **2**, S02003 (2004b). doi:10.1029/2003SW000026

D.N. Baker, S.G. Kanekal, R.B. Horne, N.P. Meredith, S.A. Glauert, Low-altitude measurements of 2–6 MeV electron trapping lifetimes at $1.5 \leq L \leq 2.5$. Geophys. Res. Lett. **34**(20), L20110 (2007). doi:10.1029/2007GL031007

D.N. Baker, S.G. Kanekal, V.C. Hoxie, M.G. Henderson, X. Li, H.E. Spence, S.R. Elkington, R.H. Friedel, J. Goldstein, M.K. Hudson, G.D. Reeves, R.M. Thorne, C.A. Kletzing, S.G. Claudepierre, A long-lived relativistic electron storage ring embedded within the Earth's outer Van Allen Radiation Zone. Science **340**, 186–190 (2013a). doi:10.1126/science.1233518

D.N. Baker, S.G. Kanekal, V.C. Hoxie, S. Batiste, M. Bolton, X. Li, S.R. Elkington, S. Monk, R. Reukauf, S. Steg, J. Westfall, C. Belting, B. Bolton, D. Braun, B. Cervelli, The Relativistic Electron–Proton Telescope (REPT) instrument on Board the Radiation Belt Storm Probes (RBSP) spacecraft: characterization of Earth's radiation belt high-energy particle Populations. Space Sci. Rev. **179** (2013b). doi:10.1007/s11214-012-9950

D.N. Baker, A.N. Jaynes, V.C. Hoxie, R.M. Thorne, J.C. Foster, X. Li, J.F. Fennell, J.R. Wygant, S.G. Kanekal, P.J. Erickson, W. Kurth, W. Li, Q. Ma, Q. Schiller, L. Blum, D.M. Malaspina, A. Gerrard, L.J. Lanzerotti, An impenetrable barrier to ultra-relativistic electrons in the Van Allen Radiation Belt. Nature **515**, 531–534 (2014a). doi:10.1038/nature13956

D.N. Baker, A.N. Jaynes, Z. Li, M.G. Henderson, S.G. Kanekal, G.D. Reeves, H.E. Spence, S.G. Claudepierre, J.F. Fennell, M.K. Hudson, R.M. Thorne, J.C. Foster, P.J. Erickson, D.M. Malaspina, J.R. Wygant, A. Boyd, C.A. Kletzing, A. Drozdov, Y.Y. Shprits, Gradual diffusion and punctuated phase space density enhancements of highly relativistic electrons: Van Allen Probes observations. Geophys. Res. Lett. **41**(5), 1351–1358 (2014b)

D.N. Baker, A.N. Jaynes, S.G. Kanekal, J.C. Foster, P.J. Erickson, J.F. Fennell, J.B. Blake, H. Zhao, Z. Li, S.R. Elkington, M.G. Henderson, G.D. Reeves, H.E. Spence, C.A. Kletzing, J.R. Wygant, Highly relativistic radiation belt electron acceleration, transport, and loss: large solar storm events of March and June 2015. J. Geophys. Res. Space Phys. **121**(7), 6647–6660 (2016). doi:10.1002/2016JA022502

M.A. Balikhin, Y.Y. Shprits, S.N. Walker, L. Chen, N. Cornilleau-Wehrlin, I. Dandouras, O. Saantolik, C. Carr, K.H. Yearby, B. Weiss, Observations of discrete harmonics emerging from equatorial noise. Nat. Commun. **6**, 7703 (2015). doi:10.1038/ncomms8703

J.L. Barth, C.S. Dyer, E.G. Stassinopoulos, Space, atmospheric and terrestrial radiation environment. IEEE Trans. Nucl. Sci. **50**, 466–482 (2003)

A.J.G. Baumgaertner, A. Seppälä, P. Jöckel, M.A. Clilverd, Geomagnetic activity related NO_X enhancements and polar surface air temperature variability in a chemistry climate model: modulation of the NAM index. Atmos. Chem. Phys. **11**, 4521–4531 (2011). doi:10.5194/acp-11-4521-2011

J.B. Blake, W.A. Kolasinski, R.W. Fillius, E.G. Mullen, Injection of electrons and protons with energies of 10 s of MeV into $L < 3$ on March 24, 1991. Geophys. Res. Lett. **19**, 821–824 (1992). doi:10.1029/92GL00624

J.B. Blake, D.N. Baker, N. Turner, K.W. Ogilvie, R.P. Lepping, Correlation of changes in the outer-zone relativistic electron population with upstream solar wind and magnetic field measurements. Geophys. Res. Lett. **24**(8) (1997). doi:10.1029/97GL00859

J.B. Blake et al., The Magnetic Electron Ion Spectrometer (MagEIS) instruments aboard the Radiation Belt Storm Probes (RBSP) spacecraft. Space Sci. Rev. **179** (2013). doi:10.1007/s11214-013-9991-8

L.W. Blum, Q. Schiller, X. Li, R. Millan, A. Halford, L. Woodger, New conjunctive CubeSat and balloon measurements to quantify rapid energetic electron precipitation. Geophys. Res. Lett. **40**, 5833–5837 (2013). doi:10.1002/2013GL058546

J.E. Borovsky, M.H. Denton, Relativistic-electron dropouts and recovery: a superposed epoch study of the magnetosphere and the solar wind. J. Geophys. Res. **114**, A02201 (2009). doi:10.1029/2008JA013128

A.J. Boyd, H.E. Spence, C.-L. Huang, G.D. Reeves, D.N. Baker, D.L. Turner, S.G. Claudepierre, J.F. Fennel, J.B. Blake, Y.Y. Shprits, Statistical properties of the radiation belt seed population. J. Geophys. Res. **121**(8), 7636–7646 (2016)

R.J. Boynton et al., The analysis of electron fluxes at geosynchronous orbit employing a NARMAX approach. J. Geophys. Res. **118** (2013). doi:10.1002/jgra.50192

A.W. Breneman et al., Observations of a global coherence scale modulation of radiation belt electron loss resulting from plasmaspheric hiss. Nature **523**, 193–195 (2015). doi:10.1038/nature14515

J.C. Cain, Models of the Earth's magnetic field, in *Radiation Trapped in the Earth's Magnetic Field*, ed. by B. McCormac (Gordon and Breach, Newark NJ, 1966)

L.B. Callis, J.D. Lambeth, NO_Y formed by precipitating electron events in 1991 and 1992: descent into the stratosphere as observed by ISAMS. Geophys. Res. Lett. **25**, 1875–1878 (1998). doi:10.1029/98GL01219

L.B. Callis, R.E. Boughner, D.N. Baker, R.A. Mewaldt, J.B. Blake, R.S. Selesnick, J.R. Cummings, M. Natarajan, G.M. Mason, J.E. Mazur, Precipitation electrons: evidence for effects on mesospheric odd nitrogen. Geophys. Res. Lett. **23**, 1901–1904 (1996). doi:10.1029/96GL01787

L.B. Callis, M. Natarajan, D.S. Evans, J.D. Lambeth, Solar atmospheric coupling by electrons (SOLACE) 1. Effects of the May 12, 1997 solar event on the middle atmosphere. J. Geophys. Res. **103**, 28,405–28,419 (1998)

D.L. Carpenter, J. Lemaire, The plasmasphere boundary layer. Ann. Geophys. **22**, 4291–4298 (2004). doi:10.5194/angeo-22-4291

C.R. Chappell, Recent satellite measurements of the morphology and dynamics of the plasmasphere. Rev. Geophys. **10**(4), 951–979 (1972). doi:10.1029/RG010i004p00951

L. Chen, A. Hasegawa, Kinetic theory of geomagnetic pulsations: 1. Internal excitations by energetic particles. J. Geophys. Res. **96**(A2), 1503–1512 (1991). doi:10.1029/90JA02346

L. Chen, J. Sun, Q. Lu, X. Gao, Z. Xia, Z. Zhima, Generation of magnetosonic waves over a continuous spectrum. J. Geophys. Res. Space Phys. **121**, 1137–1147 (2016). doi:10.1002/2015JA022089

S.G. Claudepierre, S.R. Elkington, M. Wiltberger, Solar wind driving of magnetospheric ULF waves: pulsations driven by velocity shear at the magnetopause. J. Geophys. Res. **113**, A05218 (2008). doi:10.1029/2007JA012890

S.G. Claudepierre, I.R. Mann, K. Takahashi, J.F. Fennell, M.K. Hudson, J.B. Blake, J.L. Roeder, J.H. Clemmons, H.E. Spence, G.D. Reeves, D.N. Baker, H.O. Funsten, R.H.W. Friedel, M.G. Henderson, C.A. Kletzing, W.S. Kurth, R.J. MacDowall, C.W. Smith, J.R. Wygant, Van Allen, Probes observation of localized drift resonance between poloidal mode ultra-low frequency waves and 60 keV electrons. Geophys. Res. Lett. **40**, 4491–4497 (2013). doi:10.1002/grl.50901

S.G. Claudepierre, T.P. O'Brien, J.B. Blake, J.F. Fennell, J.L. Roeder, J.H. Clemmons, M.D. Looper, J.E. Mazur, T.M. Mulligan, H.E. Spence, G.D. Reeves, R.H.W. Friedel, M.G. Henderson, B.A. Larsen, A background correction algorithm for Van Allen Probes MagEIS electron flux measurements. J. Geophys. Res. Space Phys. **120**(7) (2015). doi:10.1002/2015JA021171

S.G. Claudepierre, T.P. O'Brien, J.F. Fennell, J.B. Blake, J.H. Clemmons, M.D. Looper, J.E. Mazur, J.L. Roeder, D.L. Turner, G.D. Reeves, H.E. Spence, The hidden dynamics of relativistic electrons (0.7–1.5 MeV) in the inner zone and slot region. J. Geophys. Res. Space Phys. **122**(3) (2017). doi:10.1002/2016JA023719R

M.A. Clilverd, A. Seppälä, C.J. Rodger, M.G. Mlynczak, J.U. Kozyra, Additional stratospheric NO_X production by relativistic electron precipitation during the 2004 spring NO_X descent event. J. Geophys. Res. **114** (2009). doi:10.1029/2008JA013472

M.A. Clilverd, C.J. Rodger, D. Danskin, M.E. Usanova, T. Raita, T. Ulich, E.L. Spanswick, Energetic particle injection, acceleration, and loss during the geomagnetic disturbances which upset Galaxy 15. J. Geophys. Res. **117**(A12), 213 (2012). doi:10.1029/2012JA018175

M.A. Clilverd, C.J. Rodger, M.E. Andersson, A. Seppälä, P.T. Verronen, Linkages between the radiation belts, polar atmosphere and climate, in *Electron Precipitation Through Wave Particle Interactions, Waves, Particles, and Storms in Geospace*, ed. by G. Balasis, I.A. Daglis, I.R. Mann (Oxford University Press, Oxford, 2016)

M.V. Codrescu, T.J. Fuller-Rowell, R.G. Roble, D.S. Evans, Medium energy particle precipitation influences on the mesosphere and lower thermosphere. J. Geophys. Res. **102**(A9) (1997). doi:10.1029/97JA01728

A.B. Crew, H.E. Spence, J.B. Blake, D.M. Klumpar, B.A. Larsen, T.P. O'Brien, S. Driscoll, M. Handley, J. Legere, S. Longworth, K. Mashburn, E. Mosleh, N. Ryhajlo, S. Smith, L. Springer, M. Widholm, First multipoint in situ observations of electron microbursts: initial results from the NSF FIREBIRD II mission. J. Geophys. Res. Space Phys. **121**(6), 5272–5283 (2016). doi:10.1002/2016JA022485

J.R. Cummings, A.C. Cummings, R.A. Mewaldt, R.S. Selesnick, E.C. Stone, T.T. Von Rosenvinge, J.B. Blake, SAMPEX measurements of heavy ions trapped in the magnetosphere. IEEE Trans. Nucl. Sci. **40**(6) (1993)

I.A. Daglis, J.U. Kozyra, Outstanding issues of ring current dynamics. J. Atmos. Sol.-Terr. Phys. **64**(2) (2002)

I.A. Daglis, R.M. Thorne, W. Baumjohann, S. Orsini, The terrestrial ring current: origin, formation, and decay. Rev. Geophys. **37**(4), 407–438 (1999). doi:10.1029/1999RG900009

E.J. Daly, J. Lamaire, D. Heynderickx, D.J. Rodgers, Problems with models of the radiation belts. IEEE Trans. Nucl. Sci. **43**(2) (1996). doi:10.1109/23.490889

A. Damiani, B. Funke, M.L. Santee, R.R. Cordero, S. Watanabe, Energetic particle precipitation: a major driver of the ozone budget in the Antarctic upper stratosphere. Geophys. Res. Lett. **43**, 3554–3562 (2016). doi:10.1002/2016GL068279

F. Darrouzet et al., Plasmaspheric density structures and dynamics: properties observed by the CLUSTER and IMAGE missions. Space Sci. Rev. **145**(55) (2009). doi:10.1007/s11214-008-9438-9

A.Y. Drozodov, Y.Y. Shprits, M.E. Usanova, N.A. Aseev, A.C. Kellerman, H. Zhu, EMIC wave parameterization in the long-term VERB code simulation. J. Geophys. Res. **122**, 8488 (2017)

J.W. Dungey, Effects of the electromagnetic perturbations on particles trapped in the radiation belts. Space Sci. Rev. **4**, 199 (1965)

S.R. Elkington, M.K. Hudson, A. Chan, Acceleration of relativistic electrons via drift-resonant interaction with toroidal-mode Pc-5 ULF oscillations. Geophys. Res. Lett. **26**(21), 3273–3276 (1999). doi:10.1029/1999GL003659

S.R. Elkington, M.K. Hudson, A.A. Chan, Resonant acceleration and diffusion of outer zone electrons in an asymmetric geomagnetic field. J. Geophys. Res. **108**(A3), 1116 (2003). doi:10.1029/2001JA009202

P.J. Erickson et al., Multipoint MMS observations of fine-scale SAPS structure in the inner magnetosphere. Geophys. Res. Lett. **43** (2016). doi:10.1002/2016GL069174

P.J. Erickson, F. Beroz, M.Z. Miskin, Statistical characterization of the American sector subauroral polarization stream using incoherent scatter radar. J. Geophys. Res. **116**, A00J21 (2011). doi:10.1029/2010JA015738

J.V. Evans, Theory and practice of ionosphere study by Thomson scatter radar. Proc. IEEE **57**(4), 496–530 (1969)

C.-G. Fälthammar, Effects of time-dependent electric fields on geomagnetically trapped radiation. J. Geophys. Res. **70**(11), 2503–2516 (1965). doi:10.1029/JZ070i011p02503

Y. Fei, A.A. Chan, S.R. Elkington, M.J. Wiltberger, Radial diffusion and MHD particle simulations of relativistic electron transport by ULF waves in the September 1998 storm. J. Geophys. Res. **111**, A12209 (2006). doi:10.1029/2005JA011211

J.F. Fennell, J.B. Blake, R. Friedel, S.G. Kanekal, The energetic electron response to magnetic storms: HEO satellite observations, in *The Inner Magnetosphere: Physics and Modeling*, ed. by T. Pulkkinen, Friedel. Geophysical Monograph, vol. 155 (2005). doi:10.1029/155GM10

J.F. Fennell, S.G. Kanekal, J.L. Roeder, Storm responses of radiation belts during solar cycle 23: HEO satellite observations, in *Dynamics of the Earth's Radiation Belts and Inner Magnetosphere*, ed. by D. Summers, I.R. Mann, D.N. Baker, M. Schulz. Geophysical Monograph Series, vol. 199 (American Geophysical Union, Washington, D.C., 2012). doi:10.1029/2012GM001356

J.F. Fennell, S.G. Claudepierre, J.B. Blake, T.P. O'Brien, J.H. Clemmons, D.N. Baker, H.E. Spence, G.D. Reeves, Van Allen Probes show that the inner radiation zone contains no MeV electrons: ECT/MagEIS data. Geophys. Res. Lett. **42**(5), 1283–1289 (2015). doi:10.1002/2014GL062874

J.F. Fennell, J.B. Blake, S. Claudepierre, J. Mazur, S.G. Kanekal, P. O'Brien, D.N. Baker, W. Crain, D. Mabry, J. Clemmons, Current energetic particle sensors. J. Geophys. Res. **121**(9) (2016). doi:10.1002/2016JA022588

C. Forsyth, I.J. Rad, K.R. Murphy, M.P. Freeman, C.-L. Huang, H.E. Spence, A.J. Boyd, J.C. Coxon, C.M. Jackman, N.M.E. Kalmoni, C.E.J. Watt, J. Geophys. Res. (2016). Virtual issue. doi:10.1002/2016JA0022620

J.C. Foster, Storm-time plasma transport at middle and high latitudes. J. Geophys. Res. **98**(A2), 1675–1689 (1993). doi:10.1029/92JA02032

J.C. Foster, Cold plasma redistribution throughout geospace. Sci. China, Technol. Sci. **59**(9), 1340–1345 (2016). doi:10.1007/s11431-016-6047-9

J.C. Foster, T.J. Rosenberg, Electron precipitation and VLF emissions associated with cyclotron resonance interactions near the plasmapause. J. Geophys. Res. **81**(13), 2183–2192 (1976). doi:10.1029/JA081i013p02183

J.C. Foster, H.B. Vo, Average characteristics and activity dependence of the subauroral polarization stream. J. Geophys. Res. **107**(A12), 1475 (2002). doi:10.1029/2002JA009409

J.C. Foster, P.J. Erickson, A.J. Coster, J. Goldstein, F.J. Rich, Ionospheric signatures of plasmaspheric tails. Geophys. Res. Lett. **29**(13) (2002). doi:10.1029/2002GL015067

J.C. Foster, A.J. Coster, P.J. Erickson, F.J. Rich, B.R. Sandel, Stormtime observations of the flux of plasmaspheric ions to the dayside cusp/magnetopause. Geophys. Res. Lett. **31**(8) (2004). doi:10.1029/2004GL020082

J.C. Foster, A.J. Coster, P.J. Erickson, J.M. Holt, F.D. Line, W. Rideout, M. McCready, A. van Eyken, R.J. Barnes, R.A. Greenwald, F.J. Rich, Multiradar observations of the polar tongue of ionization. J. Geophys. Res. **110**(A9), A09S31 (2005). doi:10.1029/2004JA010928

J.C. Foster, P.J. Erickson, D.N. Baker, S.G. Claudepierre, C.A. Kletzing, W. Kurth, G.D. Reeves, S.A. Thaller, H.E. Spence, Y.Y. Shprits, Prompt energization of relativistic and highly relativistic electrons during a substorm interval: Van Allen Probes observations. Geophys. Res. Lett. **41**(1), 20–25 (2014a). doi:10.1002/2013GL058438

J.C. Foster, P.J. Erickson, A.J. Coster, S. Thaller, J. Tao, J.R. Wygant, J.W. Bonnell, Storm time observations of plasmasphere erosion flux in the magnetosphere and ionosphere. Geophys. Res. Lett. **41**(3), 762–768 (2014b). doi:10.1002/2013GL059124

J.C. Foster, J.R. Wygant, M.K. Hudson, A.J. Boyd, D.N. Baker, P.J. Erickson, H.E. Spence, Shock-Induced Prompt Relativistic Electron Acceleration in the Inner Magnetosphere. J. Geophys. Res. **120**(3) (2015). doi:10.1002/2014JA020642

J.C. Foster, P.J. Erickson, D.N. Baker, A.N. Jaynes, E.V. Mishin, J.F. Fennell, Z. Li, M.G. Henderson, S.G. Kanekal, Observations of the impenetrable barrier, the plasmapause, and the VLF bubble during the 17 March 2015 storm. J. Geophys. Res. Space Phys. **121**(6), 5537–5548 (2016). doi:10.1002/2016JA022509

J.C. Foster, P.J. Erickson, Y. Omura, D.N. Baker, C.A. Kletzing, S.G. Claudepierre, Van Allen Probes observations of prompt MeV radiation belt electron acceleration in non-linear interactions with VLF Chorus. J. Geophys. Res. Space Phys. **122**(1) (2017). doi:10.1002/2016JA023429

B. Funke, M. López-Puertas, L. Holt, C.E. Randall, G.P. Stiller, T. von Clarmann, Hemispheric distributions and interannual variability of NO_y produced by energetic particle precipitation in 2002–2012. J. Geophys. Res. **119**, 13.565–13.582 (2014a). doi:10.1002/2014JD022423

B. Funke, M. López-Puertas, G.P. Stiller, T. von Clarmann, Mesospheric and stratospheric NO_y produced by energetic particle precipitation during 2002–2012. J. Geophys. Res. **119**, 4429–4446 (2014b). doi:10.1002/2013JD021404

T. Fytterer, M.G. Mlynczak, H. Nieder, K. Perot, M. Sinnhuber, G. Stiller, J. Urban, Energetic particle induced intra-seasonal variability of ozone inside the Antarctic polar vortex observed in satellite data. Atmos. Chem. Phys. **15**, 3327–3338 (2015). doi:10.5194/acp-15-3327-2015

D.L. Gallagher, R.H. Comfort, Unsolved problems in plasmasphere refilling. J. Geophys. Res. Space Phys. **121**, 1447–1451 (2016). doi:10.1002/2015JA022279

M. Gkioulidou, A.Y. Ukhorskiy, D.G. Mitchell, T. Sotirelis, B.H. Mauk, L.J. Lanzerotti, The role of small-scale ion injections in the buildup of Earth's ring current pressure: Van Allen Probes observations of the 17 March 2013 storm. J. Geophys. Res. Space Phys. **119**, 7327–7342 (2014). doi:10.1002/2014JA020096

J. Goldstein, B.R. Sandel, The global pattern of evolution of plasmaspheric drainage plumes, 1–22, in *Inner Magnetosphere Interactions: New Perspectives from Imaging*, ed. by J. Burch, M. Schultz (AGU Press, Washington, DC, 2005). doi:10.1029/159GM02

T.I. Gombosi, D.N. Baker, A. Balogh, P.J. Erickson, J.D. Huba, L.J. Lanzerotti, Anthropogenic space weather. Space Sci. Rev. **212**(3–4), 985–1039 (2017), this issue. doi:10.1007/s11214-017-0357-5

W.D. Gonzalez, B.T. Tsurutani, Criteria of interplanetary parameters causing intense magnetic storms ($Dst <$ -100 nT). Planet. Space Sci. **35**, 1101–1109 (1987). doi:10.1016/0032-0633(87)90015-8

L.J. Gray, J. Beer, M. Geller, J.D. Haigh, M. Lockwood, K. Matthes, U. Cubasch, D. Fleitmann, G. Harrison, L. Hood, J. Luterbacher, G.A. Meehl, D. Shindell, B. van Geel, W. White, Solar influences on climate. Rev. Geophys. **48**, 75 RG4001 (2010). doi:10.1029/2009RG000282

J.K. Hargreaves, Auroral absorption of HF radio waves in the ionosphere: a review of results from the First Decade of Riometry. Proc. IEEE **57**(8), 1348–1371 (1969)

M.D. Hartinger, D.L. Turner, F. Plaschke, V. Angelopoulos, H. Singer, The role of transient ion foreshock phenomena in driving Pc5 ULF wave activity. J. Geophys. Res. Space Phys. **118**, 299–312 (2013). doi:10.1029/2012JA018349

A. Hasegawa, Drift mirror instability in the magnetosphere. Phys. Fluids **12**, 2642 (1969)

D. Hastings, H. Garrett, *Spacecraft–Environment Interactions* (Cambridge University Press, Cambridge, 1996)

A.T. Hendry, C.J. Rodger, M.A. Clilverd, N.R. Thomson, S.K. Morley, T. Raita, Rapid radiation belt losses occurring during high speed solar wind stream driven storms: importance of energetic electron precipitation, in *Dynamics of the Earth's Radiation Belts and Inner Magnetosphere*, ed. by D. Summers, I.R. Mann, D.N. Baker, M. Schulz. American Geophysical Union Monograph, vol. 199 (American Geophysical, Union, 2012), pp. 213–223. doi:10.1029/2012GM001299

F. Honary, N. Borisov, M. Beharrell, A. Senior, Temporal development of the magnetic zenith effect, **116**(A6) (2011). doi:10.1029/2010JA0116029

R.B. Horne, R.M. Thorne, Potential waves for relativistic electron scattering and stochastic acceleration during magnetic storms. Geophys. Res. Lett. **15**, 3011–3014 (1998). doi:10.1029/98GL01002

R.B. Horne, R.M. Thorne, S.A. Glauert, J.M. Albert, N.P. Meredith, R.R. Anderson, Timescale for radiation belt electron acceleration by whistler mode chorus waves. J. Geophys. Res. **110**, A03225 (2005). doi:10.1029/2004HA010811

R.H.A. Iles et al., The relativistic electron response in the outer radiation belt during magnetic storms. Ann. Geophys. **20**, 957–965 (2002)

C.H. Jackman, R.D. McPeters, G.J. Labow, E.L. Fleming, C.J. Praderas, J.M. Russell, Northern hemisphere atmospheric effects due to the July 2000 solar proton events. Geophys. Res. Lett. **28**, 2883–2886 (2001)

J.A. Jacobs, Y. Kato, S. Matsushita, V.A. Troitskaya, Classification of geomagnetic micropulsations. J. Geophys. Res. **69**, 180–181 (1964). doi:10.1029/JZ069i001p00180

A.N. Jaynes, X. Li, Q.G. Schiller, L.W. Blum, W. Tu, B. Ni, J. Bortnik, D.N. Baker, S.G. Kanekal, J.B. Blake, J. Wygant, Evolution of relativistic outer belt electrons during an extended quiescent period. J. Geophys. Res. Space Phys. **119**, 9558–9566 (2014). doi:10.1002/2014JA020125

A.N. Jaynes, D.N. Baker, H.J. Singer, J.V. Rodriguez, T.M. Loto'aniu, A.F. Ali, S.R. Elkington, Z. Li, S.G. Kanekal, S.G. Claudepierre, J.F. Fennell, W. Li, R.M. Thorne, C.A. Kletzing, H.E. Spence, G.D. Reeves, Source and seed populations for relativistic electrons: Their roles in radiation belt changes. J. Geophys. Res. Space Phys. **120**(9) (2015). doi:10.1002/2015JA021234

V.K. Jordanova, J. Albert, Y. Miyoshi, Relativistic electron precipitation by EMIC waves from self-consistent global simulations. J. Geophys. Res. **113**, A00A10 (2008). doi:10.1029/2008JA013239

S.G. Kanekal, D.N. Baker, J.F. Fennell, A. Jones, Q. Schiller, I.G. Richardson, X. Li, D.L. Turner, S. Califf, S.G. Claudepierre, L.B. Wilson III., A. Jaynes, J.B. Blake, G.D. Reeves, H.E. Spence, C.A. Kletzing, J.R. Wygant, Prompt acceleration of magnetospheric electrons of ultrarelativistic energies by the 17 March 2015 interplanetary shock. J. Geophys. Res. (2016). virtual issue. doi:10.1002/2016JA022596

R. Kataoka, Y. Miyoshi, Why are relativistic electrons persistently quiet at geosynchronous orbit in 2009? Space Weather **8**, s08002 (2010). doi:10.1029/2010SW000571

K. Keika, L.M. Kistler, P.C. Brandt, Energization of O+ ions in the Earth's inner magnetosphere and the effects on ring current buildup: a review of previous observations and possible mechanisms. J. Geophys. Res. Space Phys. **118**, 4441–4464 (2013). doi:10.1002/jgra.50371

A.C. Kellerman, Y.Y. Shprits, On the influence of solar wind conditions on the outer-electron radiation belt. J. Geophys. Res. (2012). doi:10.1029/2011JA017253

A.C. Kellerman, Y.Y. Shprits, D.L. Turner, A Geosynchronous Radiation-belt Electron Empirical Prediction (GREEP) model. Space Weather **11**, 463–475 (2013). doi:10.1002/swe.29974

A.C. Kellerman, Y.Y. Shprits, D. Kondrashov, D. Subbotin, R.A. Makarevich, E. Donovan, T. Nagai, Three-dimensional data assimilation and reanalysis of radiation belt electrons: Observations of a four-zone structure using five spacecraft and the VERB code. J. Geophys. Res., Space Phys. **119** (2014). doi:10.1002/2014JA020171

A.C. Kellerman, Y.Y. Shprits, R.A. Makarevich, E. Spanswick, E. Donovan, G. Reeves, Characterization of the energy-dependent response of riometer absorption. J. Geophys. Res. Space Phys. **120**, 615–631 (2015). doi:10.1002/2014JA020027

C.F. Kennel, H.E. Petschek, Limit on stably trapped particle fluxes. J. Geophys. Res. **71**(1), 1–28 (1966). doi:10.1029/JZ071i001p00001

H.-J. Kim, A.A. Chan, Fully adiabatic changes in storm time relativistic electron fluxes. J. Geophys. Res. **102**(A10), 22107–22116 (1997). doi:10.1029/97JA01814

K.-C. Kim, Y.Y. Shprits, J.B. Blake, Fast injection of the relativistic electrons into the inner zone and the formation of the Split-zone structure during the Bastille Day storm in July 2000. J. Geophys. Res. **121**(9) (2016)

M.G. Kivelson, D.J. Southwood, Hydromagnetic waves and the ionosphere. Geophys. Res. Lett. **15**(11) (1988). doi:10.1029/GL015i011p01271

B. Klecker, M.C. McNab, J.B. Blake, D.C. Hamilton, D. Hovestadt, H. Kaestle, Charge state of anomalous cosmic ray nitrogen, oxygen, and neon: SAMPEX observations. Astrophys. J. **442** (1995)

X. Li, Quantitative prediction of radiation belt electrons at geostationary orbit based on solar wind measurements. Geophys. Res. Lett. **28**(9) (2001)

X. Li, Variations of 0.7–6.0 MeV electrons at geosynchronous orbit as a function of solar wind. Space Weather **2** (2004). doi:10.1029/2003W000017

X. Li, I. Roth, M. Temerin, J.R. Wygant, M.K. Hudson, J.B. Blake, Simulation of the prompt energization and transport of radiation belt particles during the March 14, 1991 SSC. Geophys. Res. Lett. **20**(22), 2423–2426 (1993). doi:10.1029/93GL02701

X. Li, D.N. Baker, M. Temerin, D. Larson, R.P. Lin, G.D. Reeves, M. Looper, S.G. Kanekal, R.A. Mewaldt, Are energetic electrons in the solar wind the source of the outer radiation belt? Geophys. Res. Lett. **24**(8) (1997a)

X. Li, D.N. Baker, M. Temerin, T.D. Cayton, E.G.D. Reeves, R.A. Christensen, J.B. Blake, M.D. Looper, R. Nakamura, S.G. Kanekal, Multisatellite observations of the outer zone electron variation during the November 3–4, 1993, magnetic storm. J. Geophys. Res. **102**(A7), 14123–14140 (1997b). doi:10.1029/97JA01101

X. Li, D.N. Baker, M. Temerin, G. Reeves, R. Friedel, C. Shen, Energetic electrons, 50 keV to 6 MeV, at geosynchonous orbit: their responses to solar wind variations. Space Weather **3**(4) (2005). doi:10.1029/2004SW000105

X. Li, D.N. Baker, T.P. O'Brien, L. Xie, Q.G. Zong, Correlation between the inner edge of outer radiation belt electrons and the innermost plasmapause location. Geophys. Res. Lett. **33**(14) (2006). doi:10.1029/2006GL026294

X. Li, A.B. Barker, D.N. Baker, W.C. Tu, T.E. Sarris, R.S. Selesnick, R. Friedel, C. Shen, Modeling the deep penetration of outer belt electrons during the "Halloween" magnetic storm in 2003. Space Weather **7**(2) (2009)

X. Li, M. Temerin, D.N. Baker, G.D. Reeves, Behavior of MeV electrons at geosynchronous orbit during last two solar cycles. J. Geophys. Res. **116**, A11207 (2011). doi:10.1029/2011JA016934

X. Li, S. Palo, R. Kohnert, D. Gerhardt, L. Blum, Q. Schiller, D. Turner, W. Tu, N. Sheiko, C.S. Cooper, Colorado student space weather experiment: differential flux measurements of energetic particles in a highly inclined low Earth orbit, in *Dynamics of the Earth's Radiation Belts and Inner Magnetosphere* (American Geophysical Union, Washington, D.C., 2012). doi:10.1029/2012GM001313

X. Li, Q. Schiller, L. Blum, S. Califf, H. Zhao, W. Tu, D.L. Turner, D. Gerhardt, S. Palo, S.G. Kanekal, D.N. Baker, J. Fennell, J.B. Blake, M. Looper, G.D. Reeves, H. Spence, First results from CSSWE CubeSat: characteristics of relativistic electrons in the near-Earth environment during the October 2012 magnetic storms. J. Geophys. Res. Space Phys. **118**(10), 6489–6499 (2013). doi:10.1002/2013JA019342

W. Li et al., Radiation belt electron acceleration by chorus waves during the 17 March 2013 storm. J. Geophys. Res. Space Phys. **119**, 6 (2014). doi:10.1002/2014ja019945

X. Li, R.S. Selesnick, D.N. Baker, A.N. Jaynes, S.G. Kanekal, Q. Schiller, L. Blum, J. Fennell, J.B. Blake, Upper limit on the inner radiation belt MeV electron intensity. J. Geophys. Res. **120**(2), 1215–1228 (2015). doi:10.1002/2014JA020777

J. Li, B. Ni, Q. Ma, L. Xie, Z. Pu, S. Fu, R.M. Thorne, J. Bortnik, L. Chen, W. Li, D.N. Baker, C.A. Kletzing, W.S. Kurth, G.B. Hospodarsky, J.F. Fennell, G.D. Reeves, H.E. Spence, H.O. Funsten, D. Summers, Formation of energetic electron butterfly distributions by magnetosonic waves via Landau resonance. Geophys. Res. Lett. **43**(7) (2016). doi:10.1002/2016gl067853

K.R. Lorentzen, M.D. Looper, J.B. Blake, Relativistic electron microbursts during the GEM storms. Geophys. Res. Lett. **28**(13), 2573–2576 (2001). doi:10.1029/2001GL012926

K.R. Lorentzen, J.E. Mazur, M.D. Looper, J.F. Fennell, J.B. Blake, Multisatellite observations of MeV ion injections during storms. J. Geophys. Res. **107**(A9), 1231 (2002). doi:10.1029/2001JA000276

J.P. Luntama et al. 2017, this issue

W. Lyatsky, G.V. Khazanov, A new polar magnetic index of geomagnetic activity. Space Weather **6**(6) (2008a). doi:10.1029/2007SW000382

W. Lyatsky, G.V. Khazanov, Effects of solar wind density on relativistic electrons at geosynchronous orbit. Geophys. Res. Lett. **35** (2008b)

L.R. Lyons, R.M. Thorne, Equilibrium structure of radiation belt electrons. J. Geophys. Res. **78**, 2142–2149 (1973). doi:10.1029/JA078i013p02142

L.R. Lyons, R.M. Thorne, C.F. Kennel, Pitch angle diffusion of radiation belt electrons within the plasmasphere. J. Geophys. Res. **77**(19), 3455–3474 (1972)

R.A. Makarevich, A.C. Kellerman, Y.V. Bogdanova, A.V. Koustov, Time evolution of the subauroral electric fields: a case study during a sequence of two substorms. J. Geophys. Res. **114**, A04312 (2009). doi:10.1029/2008JA013944

R.A. Makarevich, A.C. Kellerman, J.C. Devlin, H. Ye, L.R. Lyons, Yo. Nishimura, SAPS intensification during substorm recovery: a multi-instrument case study. J. Geophys. Res. **116**, A11311 (2011). doi:10.1029/2001JA016916

D.M. Malaspina, A.N. Jaynes, C. Boule, J. Bortnik, S.A. Thaller, R.E. Ergun, C.A. Kletzing, J.R. Wygant, The distribution of plasmaspheric hiss wave power with respect to plasmapause location. Geophys. Res. Lett. **43**, 7878–7886 (2016). doi:10.1002/2016GL069982

V. Maliniemi, T. Asikainen, K. Mursula, A. Seppälä, QBO dependent relation between electron precipitation and wintertime surface temperature. J. Geophys. Res., Atmos. **118**, 6302–6310 (2013). doi:10.1002/jgrd.50518

I.R. Mann, T.P. O'Brien, D.K. Milling, Correlations between ULF wave power, solar wind speed, and relativistic electron flux in the magnetosphere: solar cycle dependence. J. Atmos. Sol.-Terr. Phys. **66**(2) (2004). doi:10.1016/j.jastp.2003.10.002

D.C. Marvin, D.J. Gorney, Solar proton events of 1989: effects on spacecraft solar arrays. J. Spacecr. Rockets **28**, 713–719 (1991). doi:10.2514/3.26304

R.A. Mathie, I.R. Mann, A correlation between extended intervals of ULF wave power and storm-time geosynchronous relativistic electron flux enhancements. Geophys. Res. Lett. **27**, 3261 (2000). doi:10.1029/2000GL003822

R.A. Mathie, I.R. Mann, On the solar wind control of Pc5 ULF pulsation power at mid-latitudes: implications for MeV electron acceleration in the outer radiation belt. J. Geophys. Res. **106**, A12 (2001). doi:10.1029/2001JA000002

K. Matthes, Solar forcing for CMIP6 (v3.2). Geosci. Model Dev. Discuss. **10**, 2247–2302 (2016). doi:10.5194/gmd-2016-91

B.H. Mauk, N.J. Fox, S.G. Kanekal, R.L. Kessel, D.G. Sibeck, A. Ukhorskiy, Science objectives and rationale for the radiation belt storm probes mission. Space Sci. Rev. **179** (2013). doi:10.1007/s11214-012-9908-y

A.C. Maycock, S. Misios, Top-down versus bottom-up mechanisms for solar-climate coupling, in *Earth's Climate Response to a Changing Sun*, ed. by T. Dudok de Wit et al. (EDP Sciences, Les Ulis Cedex A, 2015), pp. 237–246. www.edpsciences.org

J.E. Mazur et al., The creation of new ion radiation belts associated with solar energetic particle events and interplanetary shocks, in *Solar Eruptions and Energetic Particles*, ed. by M. Gopalswamy, Torsti. Geophys. Monogr. Ser., vol. 165 (AGU, Washington, D.C., 2005), p. 345

N.P. Meredith, R.B. Horne, R.H.A. Iles, R.M. Thorne, D. Heynderickx, R.R. Anderson, Outer zone relativistic electron acceleration associated with substorm-enhanced whistler mode chorus. J. Geophys. Res. **107**(A7) (2002). doi:10.1029/2001JA900146

N.P. Meredith, R.M. Thorne, R.B. Horne, D. Summers, B.J. Fraser, R.R. Anderson, Statistical analysis of relativistic electron energies for cyclotron resonance with EMIC waves observed on CRRES. J. Geophys. Res. **108**, 1250 (2003). doi:10.1029/2002JA009700

N.P. Meredith, R.B. Horne, R.R. Anderson, Survey of magnetosonic waves and proton ring distributions in the Earth's inner magnetosphere. J. Geophys. Res. **113**, A06213 (2008). doi:10.1029/2007JA012975

N.P. Meredith, R.B. Horne, S.A. Glauert, D.N. Baker, S.G. Kaneka, J.M. Albert, Relativistic electron loss timescales in the slot region. J. Geophys. Res. **114**, A03222 (2009). doi:10.1029/2008JA013889

R.M. Millan, M.P. McCarthy, J.G. Sample, D.M. Smith, L.D. Thompson, D.G. McGaw, L.W. Woodger, J.G. Hewitt, M.D. Comess, K.B. Yando, A.X. Liang, B.A. Anderson, N.R. Knezek, W.Z. Rexroad, J.M. Scheiman, G.S. Bowers, A.J. Halford, A.B. Collier, M.A. Cliverd, R.P. Lin, M.K. Hudson, The balloon array for RBSP relativistic electron losses (BARREL). Space Sci. Rev. (2013). doi:10.1007/s11214-013-9971-z

D.A. Newnham, P.J. Espy, M.A. Clilverd, C.J. Rodger, A. Seppälä, D.J. Maxfield, P. Hartogh, K. Holmén, R.B. Horne, Direct observations of nitric oxide produced by energetic electron precipitation into the Antarctic middle atmosphere. Geophys. Res. Lett. **38**, L20104 (2011). doi:10.1029/2011GL048666

B. Ni, X. Cao, Z. Zou, C. Zhou, Z. Gu, J. Bortnik, J. Zhang, S. Fu, Z. Zhao, R. Shi, L. Xie, Resonant scattering of outer zone relativistic electrons by multiband EMIC waves and resultant electron loss tie scales. J. Geophys. Res. **120**, 7357–7373 (2015). doi:10.1002/2015JA021466

T.P. O'Brien et al., Which magnetic storms produce relativistic electrons at geosynchronous orbit. J. Geophys. Res. **106**, 15533–15544 (2001). doi:10.1029/2001JA000052

T.P. O'Brien, K.R. Lorentzen, I.R. Mann, N.P. Meredith, J.B. Blake, J.F. Fennell, M.D. Looper, D.K. Milling, R.R. Anderson, Energization of relativistic electrons in the presence of ULF power and MeV microbursts: evidence for dual ULF and VLF acceleration. J. Geophys. Res. **108**(A8), 1329 (2003). doi:10.1029/2002JA009784

T.P. O'Brien, S.G. Claudepierre, T.B. Guild, J.F. Fennell, D.L. Turner, J.B. Blake, J.H. Clemmons, J.L. Roeder, Inner zone and slot electron radial diffusion revisited. Geophys. Res. Lett. **43** (2016). doi:10.1002/2016GL069749

Y. Omura, N. Furuya, D. Summers, Relativistic turning acceleration of resonant electrons by coherent whistler mode waves in a dipole magnetic field. J. Geophys. Res. **112**, A06236 (2007). doi:10.1029/2006JA012243

K. Orlova, M. Spasojevic, Y. Shprits, Activity-dependent global model of electron loss inside the plasmasphere. Geophys. Res. Lett. **41**, 3744–3751 (2014). doi:10.1002/2014GL060100

K. Orlova, Y. Shprits, M. Spasojevic, New global loss model of energetic and relativistic electrons based on Van Allen Probes measurements. J. Geophys. Res. **121**, 1308 (2016)

L.G. Ozeke, I.R. Mann, K.R. Murphy, I.J. Rae, D.K. Milling, S.R. Elkington, A.A. Chan, H.J. Singer, ULF wave derived radiation belt radial diffusion coefficients. J. Geophys. Res. **117**, A04222 (2012). doi:10.1029/2011JA017463

L.G. Ozeke, I.R. Mann, K.R. Murphy, I.J. Rae, D.K. Milling, Analytic expressions for ULF wave radiation belt radial diffusion coefficients. J. Geophys. Res. Space Phys. **119**, 1587–1605 (2014). doi:10.1002/2013JA019204

S.-M. Päivärinta, A. Seppälä, M.E. Andersson, P.T. Verronen, L. Thölix, E. Kyrölä, Observed effects of solar proton events and sudden stratospheric warmings on odd nitrogen and ozone in the polar middle atmosphere. J. Geophys. Res., Atmos. **118**, 6837–6848 (2013). doi:10.1002/jgrd.50486

G.A. Paulikas, J.B. Blake, Effects of the solar wind on magnetospheric dynamics: energetic electrons at the synchronous orbit, in *Quantitative Modeling of Magnetospheric Processes*. Geophys. Monogr. Ser., vol. 21 (AGU, Washington, D.C., 1979)

S. Perraut, A. Roux, P. Robert, R. Gendrin, J.-A. Sauvaud, J.-M. Bosqued, G. Kremser, A. Korth, A systematic study of ULF waves above f_{H+} from GEOS 1 and 2 measurements and their relationships with proton ring distributions. J. Geophys. Res. **87**(A8), 6219–6236 (1982). doi:10.1029/JA087iA08p06219

C.E. Randall, D.W. Rusch, R.M. Bevilacqua, K.W. Hoppel, J.D. Lumpe, Polar Ozone and Aerosol Measurement (POAM) II stratospheric NO_2, 1993–1996. J. Geophys. Res. **103** (1998). doi:10.1029/98JD02092

C.E. Randall, V.L. Harvey, G.L. Manney, Y. Orsolini, M. Codrescu, C. Sioris, S. Brohede, C.S. Haley, L.L. Gordley, J.M. Zawodny, J.M. Russell, Stratospheric effects of energetic particle precipitation in 2003–2004. Geophys. Res. Lett. **32** (2005). doi:10.1029/2004GL022003

C.E. Randall, V.L. Harvey, D.E. Siskind, J. France, P.F. Bernath, C.D. Boone, K.A. Walker, NO_x descent in the Arctic middle atmosphere in early 2009. Geophys. Res. Lett. **36** (2009). doi:10.1029/2009GL039706

G.D. Reeves, Relativistic electrons and magnetic storms: 1992–1995. Geophys. Res. Lett. **25(11)** (1998)

G.D. Reeves, K.L. McAdams, R.H.W. Friedel, T.P. O'Brien, Acceleration and loss of relativistic electrons during geomagnetic storms. Geophys. Res. Lett. **30**, 1529 (2003). doi:10.1029/2002GL016513

G.D. Reeves, S.K. Morley, R.H.W. Friedel, M.G. Henderson, T.E. Cayton, G. Cunningham, J.B. Blake, R.A. Christensen, D. Thomsen, On the relationship between relativistic electron flux and solar wind velocity: Paulikas and Blake revisited. J. Geophys. Res. **116**, A02213 (2011). doi:10.1029/2010JA015735

G.D. Reeves, H.E. Spence, M.G. Henderson, S.K. Morley, R.H.W. Friedel, H.O. Funsten, D.N. Baker, Electron acceleration in the heart of the Van Allen radiation belts. Science **341**(6149) (2013). doi:10.1126/science.1237743

G.D. Reeves, R.H.W. Friedel, B.A. Larsen, R.M. Skoug, H.O. Funsten, S.G. Claudepierre, J.F. Fennell, D.L. Turner, M.H. Denton, H.E. Spence, J.B. Blake, D.N. Baker, Energy dependent dynamics of keV to MeV electrons in the inner zone, outer zone, and slot regions. J. Geophys. Res. **121**(1), 397–412 (2016). doi:10.1002/2015JA021569

G.C. Reid, S. Solomon, R.R. Garcia, Response of the middle atmosphere to the solar proton events of August–December, 1989. Geophys. Res. Lett. **18** (1991)

E.J. Rigler, M. Wiltberger, D.N. Baker, Radiation belt electrons respond to multiple solar wind inputs. J. Geophys. Res. **112**, A06208 (2007). doi:10.1029/2006JA012181

C.S. Roberts, Cyclotron-resonance and bounce resonance scattering of electrons trapped in the Earth's magnetic field, in *Earth's Particle and Fields*, ed. by B.M. McCormack (Reinhold, New York, 1968), pp. 317–336

C.J. Rodger, M.A. Clilverd, J.C. Green, M.M. Lam, Use of POES SEM-2 observations to examine radiation belt dynamics and energetic electron precipitation into the atmosphere. J. Geophys. Res. Space Phys. **115** (2010a). doi:10.1029/2008JA014023

J.G. Roederer, Experimental evidence on radial diffusion of geomagnetically trapped particles, in *Earth's Particles and Fields*, ed. by B.M. McCormac (1968)

G. Rostoker, S. Skone, D.N. Baker, On the origin of relativistic electrons in the magnetosphere associated with some geomagnetic storms. Geophys. Res. Lett. **25**, 3701–3704 (1998). doi:10.1029/98GL02801

E. Rozanov, L. Callis, M. Schlesinger, F. Yang, N. Andronova N, V. Zubov, Atmospheric response to NO_y source due to energetic electron precipitation. Geophys. Res. Lett. **32** (2005). doi:10.1029/2005GL023041

A. Runov et al., Diamagnetic oscillations ahead of stopped dipolarization fronts. J. Geophys. Res. Space Phys. **119** (2014). doi:10.1002/2013JA019384

C.T. Russell, R.M. Thorne, On the structure of the inner magnetosphere. Cosm. Electrodyn. **1**, 67–89 (1970)

O. Santolík, C.A. Kletzing, W.S. Kurth, G.B. Hospodarsky, S.R. Bounds, Fine structure of large-amplitude chorus wave packets. Geophys. Res. Lett. **41**, 293–299 (2014). doi:10.1002/2013GL058889

D.M. Sawyer, J.I. Vette, *AP-8 trapped proton environment for solar maximum and solar minimum* (Goddard Space Flight Center, Greenbelt, Md, 1976). NASA WDC-A-R&S 76-06, NASA-TM-X-72605

Q. Schiller, X. Li, L. Blum, W. Tu, D.L. Turner, J.B. Blake, A nonstorm time enhancement of relativistic electrons in the outer radiation belt. Geophys. Res. Lett. **41**, 7–12 (2014). doi:10.1002/2013GL058485

M. Schulz, L.J. Lanzerotti, Particle diffusion in the radiation belts, in *Physics and Chemistry in Space*, vol. 7 (Springer, New York, 1974)

R.W. Schunk, J.C.G. Walker, Thermal diffusion in the topside ionosphere for mixtures which include multiply-charged ions. Planet. Space Sci. **17** (1969)

R.S. Selesnick, High-energy radiation belt electrons from CRAND. J. Geophys. Res. **120**(4) (2015). doi:10.1002/2014JA010963

R.S. Selesnick, M.D. Looper, R.A. Mewaldt, A theoretical model of the inner proton radiation belt. Space Weather **5**(S04003) (2007). doi:10.1029/2006SW000275

R.S. Selesnick, D.N. Baker, A.N. Jaynes, X. Li, S.G. Kanekal, M.K. Hudson, B.T. Kress, Observations of the inner radiation belt: CRAND and trapped solar protons. J. Geophys. Res. **119** (2014). doi:10.1002/2014JA020188

K. Semeniuk, V.I. Fomichev, J.C. McConnell, C. Fu, S.M.L. Melo, I.G. Usoskin, Middle atmosphere response to the solar cycle in irradiance and ionizing particle precipitation. Atmos. Chem. Phys. **11** (2011). doi:10.5194/acp-11-5045-2011

J. Semeter, F. Kamalabadi, Determination of primary electron spectra from incoherent scatter radar measurements of the auroral **E** region. Radio Sci. **40**, RS2006 (2005). doi:10.1029/2004RS003042

A. Seppälä, P.T. Verronen, E. Kyrölä, S. Hassinen, L. Backman, A. Hauchecorne, J.L. Bertaux, D. Fussen, Solar proton events of October–November 2003: Ozone depletion in the Northern Hemisphere polar winter as seen by GOMOS/Envisat. Geophys. Res. Lett. **31** (2004). doi:10.1029/2004GL021042

A. Seppälä, C.E. Randall, M.A. Clilverd, E. Rozanov, C.J. Rodger, Geomagnetic activity and polar surface air temperature variability. J. Geophys. Res. **114** (2009). doi:10.1029/2008JA014029

V. Sergeev, M.A. Shukhtina, R. Rasinkangas, A. Korth, G.D. Reeves, H.J. Singer, M.F. Thomsen, L.I. Vagina, Event study of deep energetic particle injections during substorm. J. Geophys. Res. **103**(A5), 9217–9234 (1998)

D.R. Shklyar, Energy transfer from lower to higher energy electrons mediated by whistler waves in the radiation belts. J. Geophys. Res. **122** (2017). doi:10.1002/2016JA023263

Y.Y. Shprits, R.M. Thorne, R. Friedel, G.D. Reeves, J. Fennell, D.N. Baker, S.G. Kanekal, Outward radial diffusion driven by losses at magnetopause. J. Geophys. Res. **111**, A11214 (2006). doi:10.1029/2006JA011657

M. Shulz, L.J. Lanzerotti, *Particle Diffusion in the Radiation Belts* (Springer, Berlin, 1974)

L.E. Simms, M.J. Engebretson, V. Pilipenko, G.D. Reeves, M. Clilverd, Empirical predictive models of daily relativistic electron flux at geostationary orbit: Multiple regression analysis. J. Geophys. Res. **121**(4) (2016). doi:10.1002/2016JA022414

N. Singh, J.L. Horwitz, Plasmasphere refilling: recent observations and modeling. J. Geophys. Res. **97**, 1049–1079 (1992). doi:10.1029/91JA02602

M. Sinnhuber, S. Kazeminejad, J.M. Wissing, Interannual variation of NO_X from the lower thermosphere to the upper stratosphere in the years 1991–2005. J. Geophys. Res. **116** (2011). doi:10.1029/2010JA015825

M. Sinnhuber, H. Nieder, N. Wieters, Energetic particle precipitation and the chemistry of the mesosphere/lower thermosphere. Surv. Geophys. 33 (2012). doi:10.1007/s10712-012-9201-3

M. Sinnhuber et al., The response of mesospheric NO to geomagnetic forcing in 2002–2012 as seen by SCIAMACHY. J. Geophys. Res. Space Phys. **65**(121), 3603–3620 (2016). doi:10.1002/2015JA022284

D.E. Siskind, G.E. Nedoluha, C.E. Randall, M. Fromm, J.M. Russell, An assessment of Southern Hemisphere stratospheric NO_X enhancements due to transport from the upper atmosphere. Geophys. Res. Lett. **27** (2000)

A.K. Smith, R.R. Rolando, D.R. Marsh, J.H. Richter, WACCM simulations of the mean circulation and trace species transport in the winter mesosphere. J. Geophys. Res. **116**, D20115 (2011). doi:10.1029/2011JD016083

E. Spanswick, E. Donovan, R. Friedel, A. Korth, Ground based identification of dispersionless electron injections. Geophys. Res. Lett. **34**, L03101 (2007). doi:10.1029/2006GL02839

M. Spasojevic, Y.Y. Shprits, K. Orlova, Global empirical models of plasmaspheric hiss using Van Allen Probes. J. Geophys. Res. **120**, 10370 (2015)

H.E. Spence, J.B. Blake, A.B. Crew, S. Driscoll, D.M. Klumpar, B.A. Larsen, J. Legere, S. Longworth, E. Mosleh, T.P. O'Brien, S. Smith, L. Springer, M. Widholm, Focusing on size and energy dependence of electron microbursts from the Van Allen radiation belts. Space Weather (2012). doi:10.1029/2012SW000869

H.E. Spence, G.D. Reeves, D.N. Baker, J.B. Blake, M. Bolton, S. Bourdarie, A.A. Chan, S.G. Claudepierre, J.H. Clemmons, J.P. Cravens, S.R. Elkington, J.F. Fennell, R.H.W. Friedel, H.O. Funsten, J. Goldstein, Science goals and overview of the energetic particle, composition, and thermal plasma (ECT) Suite on NASA's Radiation Belt Storm Probes (RBSP) mission. Space Sci. Rev. **179** (2013). doi:10.1007/s11214-013-0007-5

Z. Su, F. Xiao, H. Zheng, Z. He, H. Zhu, M. Zhang, C. Shen, Y. Wang, S. Wang, C.A. Kletzing, W.S. Kurth, G.B. Hospodarsky, H.E. Spence, G.D. Reeves, H.O. Funsten, J.B. Blake, D.N. Baker, Nonstorm time dynamics of electron radiation belts observed by the Van Allen Probes. Geophys. Res. Lett. **41**, 229–235 (2014). doi:10.1002/2013GL058912

R.N. Sudan, Plasma electromagnetic instabilities. Phys. Fluids **6**(1) (1964). doi:10.1063/1.1724508

D. Summers, R.M. Thorne, Relativistic electron pitch-angle scattering by electromagnetic ion cyclotron waves during geomagnetic storms. J. Geophys. Res. **108**, 1143 (2003). doi:10.1029/2002JA009489

D. Summers, R.M. Thorne, F. Xiao, Relativistic theory of wave–particle resonant diffusion with application to electron acceleration in the magnetosphere. J. Geophys. Res. **103**(20), 487 (1998)

D. Summers, B. Ni, N.P. Meredith, Timescales for radiation belt electron acceleration and loss due to resonant wave–particle interactions: 2. Evaluation for VLF chorus, ELF hiss, and electromagnetic ion cyclotron waves. J. Geophys. Res. **112**, A04207 (2007). doi:10.1029/2006JA011993

D. Summers, I.R. Mann, D.N. Baker, M. Schulz, *Dynamics of the Earth's Radiation Belts and Inner Magnetosphere*. AGU Monograph, vol. 199 (2012)

K. Takahashi, D.-H. Lee, M. Nose, R.R. Anderson, W.J. Hughes, CRRES electric field study of the radial mode structure of Pi2 pulsations. J. Geophys. Res. **108**(A5), 1210 (2003). doi:10.1029/2002JA009761

M.J. Teague, J.I. Vette, The inner zone electron Model AE-5, NSSDC 72-10, NASA TMX-69987, Goddard Space Flight Center, Greenbelt MD (1972)

M.J. Teague, J.I. Vette, *A model of the trapped electron population for solar minimum* (Goddard Space Flight Center, Greenbelt, Md, 1974). NSSDC 74-03, NASA TMS-69909

M.J. Teague, K.W. Chan, J.I. Vette, *AE 6: A model environment of trapped electrons for solar maximum* (Goddard Space Flight Center, Greenbelt, Md, 1976). NSSDC/WDC-a-R&S 76-04, NASA TMX-72597

R.M. Thorne, Radiation belt dynamics: the importance of wave–particle interactions. Geophys. Res. Lett. **37**, L22107 (2010). doi:10.1029/2010GL044990

R.M. Thorne, E.J. Smith, R.K. Burton, R.E. Holzer, Plasmaspheric hiss. J. Geophys. Res. **78**(10), 1581–1596 (1973). doi:10.1029/JA078i010p01581

R.M. Thorne, T.P. O'Brien, Y.Y. Shprits, D. Summers, R.B. Horne, Timescale for MeV electron microburst loss during geomagnetic storms. J. Geophys. Res. **110**, A09292 (2005). doi:10.1029/2004JA010882

R.M. Thorne, B. Ni, X. Tao, R.G. Horne, N.P. Meredith, Scattering by chorus waves as the dominant cause of diffuse auroral precipitation. Nature **467** (2010). doi:10.1038/nature09467

R.M. Thorne, W. Li, B. Ni, Q. Ma, J. Bortnik, L. Chen, D.N. Baker, H.E. Spence, G.D. Reeves, M.G. Henderson, C.A. Kletzing, W.S. Kurth, G.B. Hospodarsky, J.B. Blake, J.F. Fennell, S.G. Claudepierre, S.G. Kanekal, Rapid local acceleration of relativistic radiation-belt electrons by magnetospheric chorus. Nature **504**, 411–414 (2013a). doi:10.1038/nature12889

R.M. Thorne et al., Evolution and slow decay of an unusual narrow ring of relativistic electrons near $L \sim 3.2$ following the September 2012 magnetic storm. Geophys. Res. Lett. **40**, 3507–3511 (2013b). doi:10.1002/grl.50627

V.Y. Trakhtengerts, Stationary states of the Earth's outer radiation zone. Geomagn. Aeron. **6**, 827 (1966)

B.T. Tsurutani, W.D. Gonzalez, A.L.C. Gonzalez, F. Tang, J.K. Arballo, M. Okada, Interplanetary origin of geomagnetic activity in the declining phase of the solar cycle. J. Geophys. Res. **100**(A11), 21717–21733 (1995). doi:10.1029/95JA01476

W. Tu, S.R. Elkington, X. Li, W. Liu, J. Bonnell, Quantifying radial diffusion coefficients of radiation belt electrons based on global MHD simulation and spacecraft measurements. J. Geophys. Res. **117**, A10210 (2012). doi:10.1029/2012JA017901

D.L. Turner, V. Angelopoulos, S.K. Morley, M.G. Henderson, G.D. Reeves, W. Li, D.N. Baker, C.-L. Huang, A. Boyd, H.E. Spence, S.G. Claudepierre, J.B. Blake, J.V. Rodriguez, On the cause and extent of outer radiation belt losses during the 30 September 2012 dropout event. J. Geophys. Res. **119**(3), 1530–1540 (2014). doi:10.1002/2013JA019446

D.L. Turner, S.G. Claudepierre, J.F. Fennell, T.P. O'Brien, J.B. Blake, C. Lemon, M. Gkioulidou, K. Takahashi, G.D. Reeves, S. Thaller, A. Breneman, J.R. Wygant, W. Li, A. Runov, V. Angelopoulos, Energetic electron injections deep into the inner magnetosphere associated with substorm activity. Geophys. Res. Lett. **42**(7), 2079–2087 (2015). doi:10.1002/2015GL063225

D.L. Turner, T.P. O'Brien, J.F. Fennell, S.G. Claudepierre, J.B. Blake, A.N. Jaynes, D.N. Baker, S.G. Kanekal, M. Gkioulidou, M.G. Henderson, G.D. Reeves, Investigating the source of near-relativistic and relativistic electrons in Earth's inner radiation belt. J. Geophys. Res. **122**(1) (2016). doi:10.1002/2016JA023600

L.V. Tverskaya, N.N. Pavlov, J.B. Blake, R.S. Selesnick, J.F. Fennell, Predicting the L-position of the storm-injected relativistic electron belt. Adv. Space Res. **31**(4), 1039–1044, doi.org/10.1016/S0273-1177(02)00785-8 (2002)

A.Y. Ukhorskiy, M.I. Sitnov, Dynamics of radiation belt particles. Space Sci. Rev. **179**(1), 545–578 (2013). doi:10.1007/s11214-012-9938-5

A.Y. Ukhorskiy, B.J. Anderson, P.C. Brandt, N.A. Tsyganenko, Storm time evolution of the outer radiation belt: transport and losses. J. Geophys. Res. Space Phys. **111**, A11S03 (2006). doi:10.1029/2006JA011690

A.Y. Ukhorskiy, M.I. Sitnov, R.M. Millan, B.T. Kress, D.C. Smith, Enhanced radial transport and energization of radiation belt electrons due to drift orbit bifurcations. J. Geophys. Res. Space Phys. **119**, 163–170 (2014)

M.E. Usanova, A. Drozdov, K. Orlova, I.R. Mann, Y. Shprits, M.T. Robertson, D.L. Turner, D.K. Milling, A. Kale, D.N. Baker, S.A. Thaller, G.D. Reeves, H.E. Spence, C. Kletzing, J. Wygant, Effect of EMIC waves on relativistic and ultrarelativistic electron populations: ground-based and Van Allen Probes observations. Geophys. Res. Lett. **41**(5), 1375–1381 (2014). doi:10.1002/2013GL059024

J.A. Van Allen, Lifetimes of geomagnetically-trapped electrons at several MeV energy. Nature **203**, 1006–1007 (1964)

J.A. Van Allen, Spatial distribution and time decay of the intensities of geomagnetically trapped electrons from the high altitude nuclear burst of July 1962, in *Radiation Trapped in the Earth's Magnetic Field*, ed. by B.M. McCormac (Dordrecht-Holland, Reidel, 1966), pp. 575–592

J.A. Van Allen, G.H. Ludwig, E.C. Ray, C.E. McIlwain, Observation of high intensity radiation by satellites 1958 alpha and gamma. Jet Propuls. **28**(9), 588–592 (1958)

J.A. Van Allen, C.E. McIlwain, G.H. Ludwig, Satellite observations of electrons artificially injected into the geomagnetic field. J. Geophys. Res. **64**, 877 (1959)

M. Van de Kamp, A. Seppälä, M.A. Clilverd, C.J. Rodger, P.T. Verronen, I.C. Whittaker, A model providing long-term datasets of energetic electron precipitation during geomagnetic storms. J. Geophys. Res., Atmos. **121**(20), 2015JD024212 (2016)

D. Vassiliadis, A.J. Klimas, S.G. Kanekal, D.N. Baker, R.S. Weigel, Long-term average, solar-cycle, and seasonal response of magnetospheric energetic electrons to the solar wind speed. J. Geophys. Res. **107** (2002). doi:10.1029/2001JA000506

P.T. Verronen, C.J. Rodger, M.A. Clilverd, S. Wang, First evidence of mesospheric hydroxyl response to electron precipitation from the radiation belts. J. Geophys. Res. **116**(D7) (2011). doi:10.1029/2010JD014965

P.T. Verronen, M.E. Andersson, C.J. Rodger, M.A. Clilverd, S. Wang, E. Turunen, Comparison of modeled and observed effects of radiation belt electron precipitation on mesospheric hydroxyl and ozone. J. Geophys. Res. **118** (2013). doi:10.1002/jgrd.50845

P.T. Verronen, C.J. Rodger, Atmospheric ionization by energetic particle precipitation, in *Earth's Climate Response to a Changing Sun*, ed. by T. Dudok de Wit et al. (EDP Sciences, Les Ulis Cedex, 2015), pp. 261–266. Chap. 4.5. www.edpsciences.org

J.I. Vette, *Models of the Trapped Radiation Environment, Vol. I: Inner Zone Protons and Electrons, NASA SP-3024* (Goddard Space Flight Center, Greenbelt, MD, 1966)

J.I. Vette, *The AE-8 Trapped Electron Model Environment* (Goddard Space Flight Center, Greenbelt, Md, 1991). NSSDC/WDC-A-R&S 91-24

M. Walt, Loss rates of trapped electrons by atmospheric collisions, in *Radiation Trapped in the Earth's Magnetic Field*, ed. by Reidel, B.M. McCormac. Astrophysics and Space Science Library, vol. 5, Dordrecht (1966), p. 337

M. Walt, Radial diffusion of trapped particles and some of its consequences. Rev. Geophys. **9**(1) (1971). doi:10.1029/RG009i001p00011

H.I. West, R.M. Buck, J.R. Walton, Shadowing of electron azimuthal-drift motions near the noon magnetopause. Nat. Phys. Sci. **240**, 6–7 (1972). doi:10.1038/physci240006a0

S. Wing, J.R. Johnson, E. Camporeale, G.D. Reeves, Information theoretical approach to discovering solar wind drivers of the outer radiation belt. J. Geophys. Res. **121**(10) (2016). doi:10.1002/2016JA022711

J.M. Wissing, M.-B. Kallenrode, Atmospheric Ionization Module Osnabruck (AIMOS): A 3-D model to determine atmospheric ionization by energetic charged particles from different populations. J. Geophys. Res. **114**(A6) (2009). doi:10.1029/2008JA013884

H. Zhao, X. Li, Modeling energetic electron penetration into the slot region and inner radiation belt. J. Geophys. Res. Space Phys. **118**, 6936–6945 (2013). doi:10.1002/2013JA019240

H. Zhao, X. Li, D.N. Baker, S.G. Claudepierre, J.F. Fennell, J.B. Blake, B.A. Larsen, R.M. Skoug, H.O. Funsten, R.H.W. Friedel, G.D. Reeves, H.E. Spence, D.G. Mitchell, L.J. Lanzerotti, Ring current electron dynamics during geomagnetic storms based on the Van Allen Probes measurements. J. Geophys. Res. **121**(4), 3333–3346 (2016). doi:10.1002/2016JA022358

H. Zhao, D.N. Baker, A.N. Jaynes, X. Li, S.R. Elkington, S.G. Kanekal, H.E. Spence, A.J. Boyd, C.-L. Huang, C. Forsyth, On the relation between radiation belt electrons and solar wind parameters/geomagnetic indices: dependence on the first adiabatic invariant and l*. J. Geophys. Res. **122**(2), 1624–1642 (2017). doi:10.1002/2016JA023658

Space Sci Rev (2017) 212:1315–1344
DOI 10.1007/s11214-017-0412-2

Space Weather Effects Produced by the Ring Current Particles

Natalia Ganushkina[1,2] · **Allison Jaynes**[3] ·
Michael Liemohn[2]

Received: 6 February 2017 / Accepted: 17 August 2017 / Published online: 2 October 2017

Abstract One of the definitions of space weather describes it as the time-varying space environment that may be hazardous to technological systems in space and/or on the ground and/or endanger human health or life. The ring current has its contributions to space weather effects, both in terms of particles, ions and electrons, which constitute it, and magnetic and electric fields produced and modified by it at the ground and in space. We address the main aspects of the space weather effects from the ring current starting with brief review of ring current discovery and physical processes and the Dst-index and predictions of the ring current and storm occurrence based on it. Special attention is paid to the effects on satellites produced by the ring current electrons. The ring current is responsible for several processes in the other inner magnetosphere populations, such as the plasmasphere and radiation belts which is also described. Finally, we discuss the ring current influence on the ionosphere and the generation of geomagnetically induced currents (GIC).

Keywords Inner Earth's magnetosphere · Ring current · Space weather

The Scientific Foundation of Space Weather
Edited by Rudolf von Steiger, Daniel Baker, André Balogh, Tamás Gombosi, Hannu Koskinen and Astrid Veronig

✉ N. Ganushkina
Natalia.Ganushkina@fmi.fi

A. Jaynes
Allison.Jaynes@lasp.colorado.edu

M. Liemohn
liemohn@umich.edu

1 Finnish Meteorological Institute, Helsinki, Finland

2 University of Michigan, Ann Arbor, MI, USA

3 Laboratory for Atmospheric and Space Physics, University of Colorado, Boulder, CO, USA

1 Introduction

The concept of space weather includes the effects imposed by the time-varying space environment which may be hazardous to technological systems in space and/or on the ground and/or endanger human health or life. One of the important regions where these variations of space environment matter in the Sun-Earth system is the inner Earth's magnetosphere. There a lot of satellites fly and there even rather quiet times are full of dynamical processes. The ring current is one of the populations in the inner magnetosphere consisting of ions and electrons with energies of a few to hundreds of keV. The ring current energies are rather far from the ultra-relativistic energies of "killer electrons" which are especially hazardous and can penetrate even the most protected satellites in space or from the MeV and GeV energies of solar energetic particles and galactic cosmic rays which can come very close to the Earth. At the same time, the ring current has rather significant influence in terms of space weather effects.

The ring current has several distinct contributions to the space weather effects, both in terms of particles, ions and electrons, which constitute it, and magnetic and electric fields produced and modified by it at the ground and in space. Ring current is a significant contributor to the Dst-index which, in its turn, can be used as an indicator, a measure and a predictor of geomagnetic storms. Ring current keV electrons are responsible for surface charging effects on satellites. The ring current dynamics including electric and magnetic field distortions due to this dynamics is tightly related to the radiation belts and plasmasphere populations in the inner magnetosphere. The storm-time ring current has several space weather effects on the ionosphere and thermosphere via influencing on the rearranging of the ionospheric density and heating the thermosphere. The ring current contributes to the Geomagnetically Induced Currents (GIC) effects via its role in the generation of Region 2 field-aligned currents.

There have been numerous reviews on the ring current formation, dynamics and decay and reviews on space weather. While reviews of space weather effects have examined the influence of the ring current, none of them specifically focused on the ring current as a source of space weather effects. The present review addresses the space weather effects from the ring current particles compiled into a single place and avoids repetitions of the general ring current physics. The goal of this review is to stress the importance of the ring current being an inner magnetosphere population with rather moderate energies as compared to the high energy electrons and protons present in the Earth's magnetosphere. The specifics of the ring current particles can result in the space weather effects which are rather extreme without the occurrence of the extreme geomagnetic conditions. Even a presence of a moderate storm is not necessary for significant surface charging event to happen, for example.

This review is an output of "The Scientific Foundation of Space Weather" Workshop which was held during 27 June–1 July 2016 at the International Space Science Institute, Bern, Switzerland. In the following Sect. 2, we start with a very brief review of ring current discovery and physical processes. The Dst-index, the role of the ring current in producing it and predictions of the ring current and storm occurrence based on the Dst-index are presented in Sect. 3. Section 4, the heart of the review, details the space weather effects of the ring current. This begins with the effects on satellites produced by the ring current electrons (Sect. 4.1). The processes in other inner magnetosphere populations, such as the plasmasphere and radiation belts controlled by the ring current, are described in Sect. 4.2. Then, in Sect. 4.3, we discuss the ring current influence on the ionosphere and generation of geomagnetically induced currents (GIC). Finally, the ring current space weather effects are summarized in Sect. 5.

2 Ring Current in the Earth's Magnetosphere: Brief Review

The ring current is one of the oldest concepts in magnetospheric physics (see the historical review by Egeland and Burke 2012). A current flowing around the Earth was first introduced by Störmer (1907) and supported by Schmidt (1917). Chapman and Ferraro (1931, 1941) used a ring current concept for the model of a geomagnetic storm. The ring current can be described in a simplified way as a toroidal shaped electric current, flowing westward around the Earth with variable density at geocentric distances between 2 and 9 R_E. Note that we use "ring current" here with a broad definition that includes any inner magnetospheric westward or eastward current that is either symmetric or not in local time. In fact, the symmetric ring current is often dominated by the partial ring current during magnetic storms (e.g., Takahashi et al. 1990; Liemohn et al. 2001b; Mitchell et al. 2001). Ring current consists of ions and electrons with energies covering the range from about 1 to 400 keV. The main ion species are H^+, O^+, and He^+ (see the recent review by Kronberg et al. 2014, where the observations and modeling of ion circulation and effects in the ring current dynamics were presented, and references therein). The intensity of quiet time ring current is about $\sim$ 1–5 nA/m^2 (Lui and Hamilton 1992), which can be as large as 50 nA/m^2 during storm times (e.g., Vallat et al. 2005), although there is some debate about this (Liemohn et al. 2016). Although, the early measurements of the ring current particles were made onboard OGO 3 (e.g., Frank 1967), Explorer 45 (e.g., Smith and Hoffman 1974) and CRRES missions (e.g., Grande et al. 1997; Friedel and Korth 1997), the first mission, which clarified the detailed picture of the ring current energy and composition was the AMPTE mission of the late 1980s (e.g., Lui et al. 1987; Lui and Hamilton 1992; De Michelis et al. 1997). Extensive observations of the ring current were done by the most recent Van Allen Probes mission (e.g., Yu et al. 2014; Zhao et al. 2015; Gkioulidou et al. 2016; Kistler et al. 2016; Shi et al. 2016; Menz et al. 2017).

Inner magnetosphere current density can be measured in several ways. They include (1) direct measurements of ion and electron current densities, (2) particle measurements which can provide the plasma pressure and the current itself can be estimated from it, (3) multi-point magnetic field measurements and computing the current density from the curl of magnetic field, (4) remote sensing of energetic neutral atoms (ENAs) emitted from the ring current (for more detailed description of these methods, see the recent reviews by Ganushkina et al. (2015b) and Lühr et al. (2017) and references therein and the commentary by Dunlop et al. 2016).

Two main sources for the ring current particles are the terrestrial ionosphere and the solar wind. Particles from both of them populates the plasma sheet, therefore, it is an important source of the ring current (see the reviews by Ebihara and Ejiri 2003; Daglis 2001, 2006; Kronberg et al. 2014; Ebihara 2016 and references therein). Plasma sheet particles move inward when acted upon by time-varying electromagnetic fields. There are still open questions regarding this process, especially, for storm times, such as, whether substorm-associated magnetic field variations and the associated (inductive) electric fields are necessary for the buildup of the ring current energy density. While there are arguments that large-scale convection alone can account for the ring current increase, other studies have concluded that energization of ring current particles to energies over 100 keV requires smaller-scale, time-varying electric fields (see one of the earliest studies on this subject by Kim et al. 1979 and more recent reviews by Daglis and Kozyra 2002; Ebihara and Ejiri 2003; Kozyra and Liemohn 2003; Ganushkina 2005; Ebihara and Miyoshi 2011; Bourdarie et al. 2016; Ebihara 2016 and references therein).

The concept of the partial ring current and its closure to the ionosphere was suggested by Alfvén in 1950's. One of the earliest observational evidences were given based on OGO

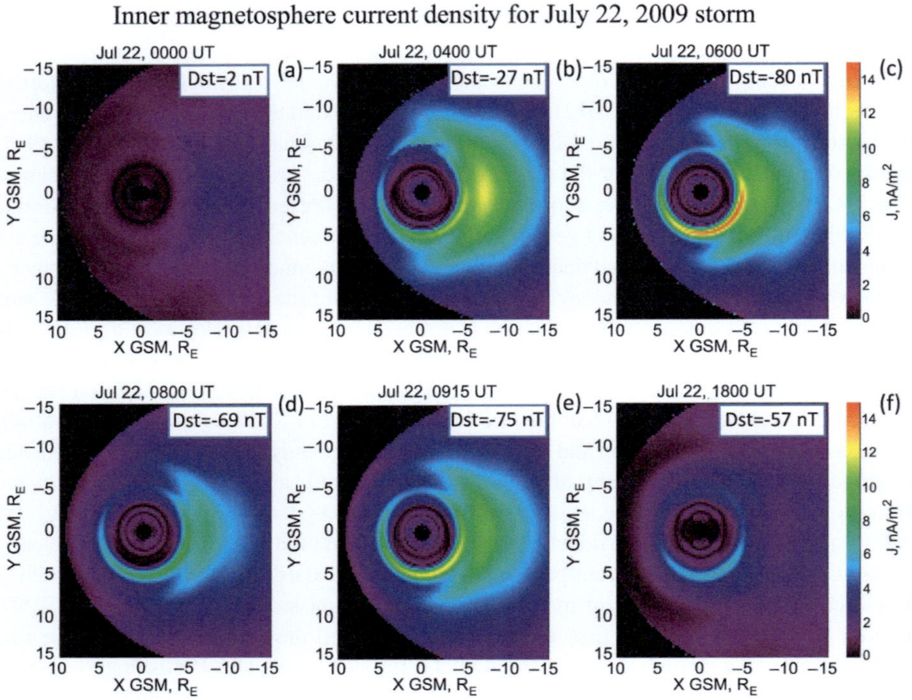

Fig. 1 Evolution of the inner magnetosphere current density in the equatorial plane computed from the curl of the magnetic field modeled with the Tsyganenko T02 global magnetospheric magnetic field model (Tsyganenko 2002a,b) during the storm event on July 22, 2009

3 measurements (Frank 1970). The magnetosphere is essentially asymmetric, compressed by the solar wind dynamic pressure on the dayside, and stretched by the tail current on the night-side. Plasma pressure distribution during disturbed times (storm times) becomes highly asymmetric due to plasma transport and injection from the night-side plasma sheet to the inner magnetosphere. The resulting plasma distribution presents a gradient in the azimuthal direction resulting in the spatial asymmetry of the ring current. The remnant of the perpendicular current must flow along a field line to complete a closure of the current (see the reviews by Ebihara and Ejiri 2003; Kozyra and Liemohn 2003; Ebihara and Miyoshi 2011; Ganushkina et al. 2015b).

Figure 1 presents an illustration on the storm time ring current dynamics. It demonstrates the distribution of the inner magnetosphere current density in the equatorial plane modeled with the Tsyganenko T02 global magnetospheric magnetic field model (Tsyganenko 2002a,b) during the storm event on July 22, 2009. The current densities of the inner magnetosphere current were computed from the curl of the model magnetic field using the Ampère's law with Maxwell's extension. Before the storm at 00 UT when Dst was 2 nT (Fig. 1a), a relatively symmetric ring current can be seen at distances inside 5 R_E and the near-Earth part of the tail current is present at distances of 10–15 R_E. The Dst becomes negative (−27 nT) at 04 UT and the ring current intensifies with the storm development (Fig. 1b) and becomes asymmetric with its peak shifted duskward at the storm maximum with Dst = −80 nT (Fig. 1c) at 06 UT. The July 22, 2009 event was a moderate storm. The increased tail current is also evident, although with lower current densities. During the storm recovery phase at 08 UT (Dst = −69 nT), the ring current density decreases (Fig. 1d), but remains asymmetric. There was a second Dst-index drop to about −75 nT at 0915 UT which can be clearly seen as a new intensification of the ring current (Fig. 1e). The July 22, 2009 storm had a long recovery phase, therefore, close to the end of the day at 18 UT when Dst was −57 nT (Fig. 1f), the ring current symmetrized but it was still not fully symmetric and its current density decreased but not to the non-disturbed level. This is rather typical storm-time evolution of the ring current.

Ring current ions are lost due to several loss mechanisms (for a recent comprehensive review, see Ebihara and Miyoshi 2011). Charge-exchange between energetic ring current ions and neutral exospheric hydrogen is thought to act as a dominant loss. Through this process, energetic ions colliding with neutral exospheric gas may acquire an electron from the cold neutral atom and become neutralized. Convection outflow occurs when a particle intersects the magnetopause, flows away along magnetosheath magnetic field lines and is lost from the ring current region. In the magnetosphere, the ring current particles experience Coulomb interaction with the thermal plasma in the plasmasphere. They transfer energy to the plasmaspheric plasma due to collisions (Coulomb drag) and some of them are also scattered into the loss cone (their pitch angle changes so that they can be lost in the atmosphere). Although classical collisional processes provide the dominant loss process for ring current ions, two classes of plasma waves are able to interact with ring current ions, namely, electromagnetic ion cyclotron (EMIC) waves and magnetosonic waves. This wave-particle interaction results in the pitch angle scattering which contributes to ion loss, especially during the main phase of a storm. For ring current electrons, charge-exchange is not a loss process. The main losses are Coulomb interactions, convection outflow and wave-particle interactions which are most important. Ring current electrons interact mainly with chorus waves observed outside the plasmasphere and hiss waves inside the plasmasphere.

This section gave a brief introduction to the ring current and the major physical processes controlling it. For additional details of ring current structure, dynamics and physics, please see the numerous reviews cited above.

3 Dst-Index as a Storm Indicator, Measure and Predictor

It has long been known that the horizontal component, H, of the geomagnetic field is depressed during periods of great magnetic disturbances and that the recovery to its average level is gradual. In 1741 Celsius observed a large magnetic disturbance in Uppsala, Sweden, and Graham observed the same in London, England, simultaneously. These large magnetic field disturbances were called "magnetic storms", and were shown to be non-local. Gauss and Weber founded a network of ground-based magnetic observatories expanded by the British and Russians. It was soon found that magnetic storms are worldwide phenomena. Much later, Akasofu and Chapman (1961), Kamide and Fukushima (1971), Kamide (1974) concluded that the near-equatorial ground disturbance of the magnetic field is a measure of the ring current. Sugiura (1964) and Iyemori (1990) used the averaged magnetic field depression observed at low latitudes to derive the Dst-index.

If the ring current and the Dst-index are directly related and the Dst-index serves as an indicator of a storm activity going on in the magnetosphere, the Dst-index can also serve as a predictor of a future storm occurrence. Therefore, it is useful to predict the Dst-index itself. The ability to predict storm occurrence by predicting of the Dst-index can greatly facilitate the prediction of many storm-related space weather effects. The Dst-index is computed from the observations of the ground-based magnetic field. To be able to achieve the goal of Dst-index prediction, models are necessary and these models need to be in the form of operational tools which can be further developed into forecasting tools.

There are several approaches for modeling the Dst-index. One of them is to use models which provide the magnetic field everywhere in the Earth's magnetosphere and on the ground. These models include the semi-empirical models based on the large amount of magnetic field data with specific representations of current systems in them, like the Tsyganenko models (see the review by Tsyganenko 2013 and http://geo.phys.spbu.ru/~tsyganenko/modeling.html). The most recent model version (Tsyganenko and Andreeva 2015) is intended to forecast the magnetospheric magnetic field using an optimal solar wind coupling function. A very different approach is to use the large volume of magnetic field data for reproducing the detailed structure of the magnetosphere as was done within the TS07 model (Tsyganenko and Sitnov 2007). This approach is somewhat similar to the modelling of the internal geomagnetic field with higher-order harmonics in the scalar potential expansion and it is based on extensible high-resolution expansions for the field of the equatorial current sheet. Early models of the magnetospheric magnetic field used such a method (for example, Mead and Fairfield 1975). Most recently, Andreeva and Tsyganenko (2016) and Tsyganenko and Andreeva (2016) also developed a model without a priori assumptions about the geometry of the magnetic field sources by splitting the model magnetic field into the toroidal and poloidal parts and then expanding each part into a weighted sum of radial basis functions. The Dst-index from these magnetic field models can be computed following the same procedure as for the real Dst-index using the magnetic field modeled at the locations of real Dst low-latitude stations at the Earth's surface as it is done at the World Data Center for Geomagnetism, Kyoto, Japan (http://wdc.kugi.kyoto-u.ac.jp/dstdir/dst2/onDstindex.html).

Global magnetohydrodynamic (MHD) models provide the magnetic field in the magnetosphere and on the ground and they are also used to compute the Dst-index. The detailed analysis of several MHD models' performance to calculate the Dst-index was presented in the study of Rastätter et al. (2013) which resulted from the 2008–2009 Geospace Environment Modeling (GEM) challenge defined at the 2008 GEM workshop. Two quiet and two disturbed events were selected for modeling. The three-dimensional MHD models of the magnetosphere coupled to an ionosphere electrodynamics solver evaluated were (1) the Space Weather Modeling Framework (SWMF) (Tóth et al. 2005, 2012), (2) the Open Geospace General Circulation Model (OpenGGCM) (Raeder et al. 2001), and (3) the Coupled Magnetosphere-Ionosphere-Thermosphere (CMIT) model, also referred to by the magnetospheric Lyon-Fedder-Mobarry (LFM) component (Lyon et al. 2004; Wiltberger et al. 2004). The contributions from the magnetospheric and ionospheric currents were taken into account in the Biot-Savart's integration for the approximation of the Dst-index by the North-South component of the perturbation magnetic field δB_Z (in SM coordinates) at the Earth's center location (for details, see Rastätter et al. 2013). Another challenge was focused on the reproducing the magnetic field variations at the ground-based stations (Pulkkinen et al. 2010, 2011, 2013; Rastätter et al. 2014). As a result of these challenges, the NOAA Space Weather Prediction Center (SWPC) added the SWMF to the operational forecasting tools. The real-time Dst-index together with the observed one is displayed at http://www.swpc.noaa.gov/products/geospace-global-geomagnetic-activity-plot. None of the other models are operational at present.

Another type of tools is particle, or kinetic, models (brief descriptions of the existing ring current models are give below in Sect. 4.1). From kinetic models, the Dst-index is very often obtained using the Dessler-Parker-Sckopke (DPS) relationship (Dessler and Parker 1959; Sckopke 1966). The DPS relation connects the total energy in the ring current and the change in the ground magnetic field strength. This perturbation can be considered as a rough equivalent to the Dst-index if it is assumed (Carovillano and Siscoe 1973) that it is close to the perturbation averaged around the equator of the Earth. The DPS relation was used in numerous studies for Dst-index estimates from modeling and observations and related ring current dynamics (e.g., Greenspan and Hamilton 2000; Ebihara and Ejiri 2000; Fok et al. 2001b; Jordanova et al. 2012; Ganushkina et al. 2012; Katus et al. 2015; Zhao et al. 2015). However, the Dst-index contains contributions from many other sources than the azimuthally symmetric ring current and their contributions can be significant or even largest during disturbed conditions. The suggested magnetospheric sources include (1) magnetopause currents, (2) cross-tail current, (3) partial ring current, and even (4) substorm current wedge (see the review by Maltsev 2004 and references therein) and also studies by Friedrich et al. (1999), Munsami (2000), Ohtani et al. (2001), Liemohn et al. (2001a,b), Liemohn (2003), Ganushkina et al. (2004, 2010), Kalegaev et al. (2005). The magnetic field measured on the ground contains contributions from all current systems and there is no other way to separate them but to use magnetospheric models. The 2008–2009 GEM Dst challenge (Rastätter et al. 2013), in addition to MHD models, included two kinetic ring current models, Ring Current-Atmosphere Interactions Model with Self-Consistent Magnetic Field (RAM-SCB) (Jordanova et al. 2010; Yu et al. 2011) and the Rice Convection Model (RCM) (Toffoletto et al. 2003). The Dst-index from these models was estimated by the DPS relation.

In addition to the DPS relation, the Biot-Savart's law is used to calculate the magnetic disturbance parallel to the Earth's dipole at the center of the earth induced by the azimuthal current component given by the modeled magnetospheric currents within the modeling region. Calculating the model Dst-index by the Biot-Savart's law instead of the DPS relation in nondipolar magnetic field gives larger and more realistic values, since, for example, the effect from the near-Earth tail current is included by the presence of the stretched magnetic field lines (e.g., Liemohn 2003; Ganushkina et al. 2012; Jordanova et al. 2012). None of the kinetic models with the Dst-index as an output is operational and run online in real-time. The main challenge for predicting of the Dst-index and, subsequently, storm occurrence and strength is the absence of reliable predictions of Interplanetary Magnetic Field (IMF) and solar wind parameters, days in advance. These parameters are the main drivers of global empirical and MHD magnetospheric models and kinetic magnetospheric models.

There is also a very different way to predict the Dst-index which does not include any magnetospheric modeling. Different kinds of empirical relations were obtained for the solar wind parameters and the Dst-index. One of the earliest is the relation between changes in Dst-index and the rate of ring current injection as a function of a delayed and filtered solar wind electric field and solar wind dynamic pressure derived by Burton et al. (1975). Following Burton et al. (1975), further corrections and improvements using general relation between the energy balance for the ring current given as the rate of energy input into the ring current and the decay time and DPS relation were done in later studies (e.g., Murayama 1986; Gonzalez et al. 1994; Thomsen et al. 1998; Klimas et al. 1998; Fenrich and Luhmann 1998; O'Brien and McPherron 2000; Wang et al. 2003). The model of this type which is run online in real time is the Dst forecast model by Temerin and Li (2006) at the Laboratory for Atmospheric and Space Physics (LASP, USA) (http://lasp.colorado.edu/space_weather/dsttemerin/dsttemerin.html). The Dst-index is calculated as a sum of several terms (Temerin and Li 2002, 2006). The first three terms are similar that their driver terms depend strongly on a negative IMF B_Z-component. They differ from each other by their decay terms. The next term is the pressure term which is proportional to the square root of the solar wind dynamic pressure and it is usually assumed to represent the magnetopause currents. The direct IMF B_Z term depends on the angle of the dipole with respect to the solar wind velocity. The last term in the offset term which does not depend on the IMF and solar wind parameters and can compensate for a portion of the secular and annual variation in the Dst index.

Other techniques for Dst-index predictions include the usage of neural networks, with delayed inputs, with feed-back connections (e.g., Lundstedt and Wintoft 1994; Lundstedt et al. 2001; Watanabe et al. 2002; Pallocchia et al. 2006; Bala and Reiff 2012; Dolenko et al. 2014; Lu et al. 2016). Bala and Reiff (2012) use the solar wind-magnetosphere coupling function, the Boyle index (Boyle et al. 1997), which is an empirical approximation for the polar cap potential dependent on solar wind parameters, as basis functions to train an artificial neural network to predict the Dst-index. These predictions are shown in real time at http://mms.rice.edu/realtime/forecast.html as 1, 3, and 6 hours ahead. Another operational model is the Lund Dst model (Lundstedt et al. 2001, 2002) run at the Swedish Institute of Space Physics (IRF-Lund, Sweden) at http://www.irfl.lu.se/rwc/dst/. This model is also based on a recurrent neural network that has been optimized to be as small as possible without degrading the accuracy and it is driven by the hourly averages of the IMF B_Z-component, solar wind number density and velocity. A similar approach is used for the Dst-index predictions (Dolenko et al. 2014) available at the Skobeltsyn Institute of Nuclear Physics Moscow State University (SINP MSU, Russia) at http://swx.sinp.msu.ru/models/dst.php?gcm=1.

Yet another method includes the models based on linear moving average (MA) filters or linear auto-regressive moving average (ARMA) filters, or similar (Vassiliadis et al. 1999) and the nonlinear auto-regressive moving average model with exogenous inputs (NARMAX) applied to Dst-index (Boynton et al. 2011a; Beharrell and Honary 2016). For example, the NARMAX algorithm (Leontaritis and Billings 1985a,b) is an advanced system identification technique, similar to neural networks, which can be applied for linear and non-linear systems with physically interpretable parameters. Boaghe et al. (2001) applied the NARMAX algorithm and derived a model for the Dst-index using vBs with v as the solar wind velocity and Bs as the southward IMF component as the input. Boynton et al. (2011a) developed a model for the Dst-index using NARMAX with a different coupling function according to Boynton et al. (2011b) as an input. The advantage of this type of models is that the output of the model at a specific time can be represented by rather simple polynomial function of the previous values of inputs, outputs, and error terms (Beharrell and Honary 2016). At present, these types of models are in great demand, since they can give exact (though perhaps not always accurate) values at some time moments in a future.

Ji et al. (2012) compared the outputs from six Dst-index forecast models during intense geomagnetic storms, namely, Burton et al. (1975), Fenrich and Luhmann (1998), O'Brien and McPherron (2000), Wang et al. (2003), Temerin and Li (2006), and Boynton et al. (2011a) models. They found that the Temerin and Li (2006) model gives most reasonable results. The 2008–2009 GEM Dst challenge (Rastätter et al. 2013) also included Burton-type representations (Burton et al. 1975; Murayama 1986), NARMAX model (Boynton et al. 2011a), and Rice Dst model (Bala and Reiff 2012). All of the models showed very reasonable results, NARMAX model (Boynton et al. 2011a) was specifically mentioned.

This brief over-review of Dst prediction methods is indented to give readers an understanding of the various types of models in existence and, in particular, those running online and in real time. These models can be used as an estimate of ring current intensity and, therefore, the severity of the possible space weather effects of the ring current, which will now be discussed.

4 Space Weather Effects Related to the Ring Current

4.1 Ring Current Electrons and Effects on Satellites

Electrons with energies less than 100 keV are one of the important constituents of the ring current. Their contribution has been studied starting from the early observations of the ring current. Based on the analysis of the low-energy (<50 keV) electron and proton fluxes onboard the OGO 3 satellite, Frank (1967) showed that the electron component may provide about 25% of the ring current energy during storm times. Liu et al. (2005) analyzed Explorer 45 electron data for energies from 1 to 200 keV and found the electron contribution of about 7.5% during quiet time and about 19% during storm time. Contribution of electrons to the ring current energy content based on the recent Van Allen Probes measurements was estimated to be about 10% of the ions during storm main phase (e.g., Zhao et al. 2016).

The electron fluxes at these keV energies vary significantly with the current activity on the scale of minutes or even shorter. They can be responsible for surface charging effects which is a serious risk for satellites (e.g., Garrett 1981; Lanzerotti et al. 1998; Koons et al. 1999; Whipple 1981; Davis et al. 2008). The electrons with energies of 10's of keV do not penetrate deep into the satellite materials but stay near the surface. Secondary electrons are generated due to the interactions between the surface materials and both the incident plasma and the solar UV. The satellite's surface materials will be charged in order to have the zero net current between the surfaces and the plasma. Therefore, the surface will have nonzero voltages. Usually, the sunlit areas of the satellite's surface are positive and the shadowed areas are negative. For the conducting surfaces, the potential of the surface is uniform for reaching the equilibrium for zero net current. For insulating materials, this equilibrium can be only on several points on the surface.

Electrons have higher speeds, so they constitute the main source of initial plasma current to a satellite. The currents of photo and secondary electrons from a surface are often higher than the plasma electron current to it. If the average electron energy in the surrounding plasma is larger than 1 keV, the regions of a satellite in shadows will be charged negative to significant potentials (several kilovolts). In the dense plasma also the sunlit regions of a satellite can charge to significant levels. Secondary and photoelectron currents are different for different materials. Therefore, there is a wide range of surface potentials. The differences in potential between adjacent materials can cause the local electrical stress resulting in vacuum arcs. Surface materials can also discharge into space or to structure ground. The resulting electrostatic discharge (ESD) currents can electromagnetically couple into electronic circuits and subsystems, causing damage.

Typically, most satellite anomalies due to surface charging at geostationary orbit occur at night and early dawn (e.g., Fennell et al. 2001; Gubby and Evans 2002; O'Brien 2009). During substorms, a hot plasma is injected from the magnetotail into the nightside inner regions. They perform the gradient and curvature drifts towards dawn. These newly injected electrons can cause significant changes in the satellite charging levels.

When an anomaly occurs, the radiation environment may be more extreme than that given by the specification models used for design. However, data may not be available at the location of the satellite to fully determine the cause of the anomaly. Therefore, there is a need for physical models with the correct dynamical behavior that can be used to reconstruct the radiation environment. To determine the impact of an extreme space weather event on satellites, we need to define the space radiation environment for different orbits. This can be done by using a combination of physical models and in-situ observations.

Significant surface charging events can happen during rather undisturbed conditions, like small, isolated substorms (e.g., Ganushkina et al. 2013; Matéo-Vélez et al. 2016). Occurrence of a storm, even a moderate one, can be not directly related to the occurrence of the surface charging. The electron flux at the keV energies is largely determined by convective (Korth et al. 1999; Friedel et al. 2001; Thomsen et al. 2002; Elkington et al. 2004; Miyoshi et al. 2006; Kurita et al. 2011; Jordanova et al. 2016) and substorm-associated (Vakulin et al. 1988; Grafodatskiy et al. 1987; Degtyarev et al. 1990; Fok et al. 2001a; Khazanov et al. 2004; Kozelova et al. 2006; Ganushkina et al. 2013; Yu et al. 2014; Turner et al. 2015) electric fields. Since it varies on short time scales, it is not possible to average the keV electron fluxes over an orbit/day/hour as it can be done for higher energies, like MeVs. This makes the modeling of keV electron flux variations very challenging.

When modeling using physics-based models (kinetic models for the ring current), it is necessary to follow keV electrons in large-scale, time-varying magnetic and electric fields. Even though there exist several models, the correct models for these fields are extremely hard to develop. Electrons in the inner regions come from the plasma sheet where specification of a realistic boundary conditions is very nontrivial. The question of how to introduce losses for low energy electrons is still open. Here, the issue about the importance of other waves, apart from chorus and hiss, in the pitch angle scattering resulting from wave-particle interactions is still unclear. Finally, the most important factor in the electron dynamics is substorms which play a significant role in keV electron transport and energy increase. Their proper modeling has not been done yet.

Fluxes of low-energy electrons have been modeled in several studies as a part of ring current simulations. One of the widely used inner magnetosphere models is the ring current-atmosphere interactions model (RAM) (Jordanova et al. 1996) mentioned above in Sect. 3. Jordanova and Miyoshi (2005), Miyoshi et al. (2006), and, more recently, Jordanova et al. (2014) extended the RAM model to relativistic energies and electrons investigating the effect of magnetospheric convection and radial diffusion during the October 2001 geomagnetic storm and later for October 2012 storm (Jordanova et al. 2016). This model is part of the SHIELDS project of Los Alamos National Laboratory (http://www.lanl.gov/projects/shields/index.php).

Chen et al. (2006) performed magnetically self-consistent ring current simulations during the 19 October 1998 storm calculating the ion and electron transport in the equatorial plane using the kinetic proton and electron drift-loss model (Chen et al. 1994) coupled to the force-balanced representation of the magnetic field intensity (Chen et al. 2005).

Another widely used inner magnetosphere model is the Rice Convection Model (RCM), also mentioned above in Sect. 3, which describes plasma electrodynamics in the inner and middle magnetosphere and its coupling to the ionosphere (see the review by Toffoletto et al. 2003 and references therein). The Rice Convection Model Equilibrium (RCM-E) (Lemon et al. 2004) combines the drift physics computing the bounce-averaged relativistic drifts of isotropic electrons and ions with a model of equilibrium magnetic field in a force balance with the RCM-computed pressures. In addition to magnetic self-consistency, the particle transport is also self-consistent with the electric field. This model was used for the assessing of the different electron loss models for the model's ability to reproduce the observed electron fluxes of 1.2 keV to 925 keV measured by LANL 1994-084 satellite during the large 10 August 2000 magnetic storm (Chen et al. 2015).

These studies were focused on the application to specific events. The models are not run online in real time. Taking into account the space weather effects requires models run in near real time, when no changes can be made and the models are supposed to perform reasonably well for wide range of geomagnetic conditions, not just for specific storm events.

The model which is run online at the Community Coordinated Modeling Center (CCMC) (http://ccmc.gsfc.nasa.gov/index.php) in near real time is Fok Ring Current Model (FRC). This is a bounce-averaged kinetic model which calculates the temporal and spatial variation of the phase space density of ring current species, including electrons in the energy range of 1–300 keV (Fok and Moore 1997; Fok et al. 1999, 2001b). The model inputs are the solar wind driven outputs of the global MHD modeling provided in near real time by the Space Weather Modeling Framework (Tóth et al. 2005, 2012). 3D magnetic fields, as well as equatorial plasma density and temperature are obtained from the global magnetosphere SWMF GM/BATSRUS model (Powell et al. 1999; De Zeeuw et al. 2000) and 2D ionospheric height-integrated potentials are given by the ionospheric electrodynamics SWMF IE/RIM model (Ridley et al. 2001, 2004). Equatorial plasma density and temperature are used to set the boundary conditions in the plasma sheet for electrons, which move inward in the provided magnetic and computed from the ionospheric potentials electric fields. The model output given as the electron fluxes as functions of time, energy and pitch angle for different keV energy ranges in the equatorial plane is presented. At the same time, no real time comparison with the observations is shown.

A variation of the FRC model is the Comprehensive Inner-Magnetosphere Ionosphere (CIMI) model which also calculates the temporal and spatial variation of the phase space density of electrons in the energy range of 1–4000 keV (Fok et al. 2011, 2014). Instead of MHD-related inputs, this model has different inputs based on empirical models. For 3D magnetic fields, it employs Tsyganenko magnetospheric magnetic field T96 (Tsyganenko 1995) and T04 (Tsyganenko and Sitnov 2005) models. Weimer (2005) representation for ionospheric electric potentials at the model's poleward boundary at 70.3° magnetic latitude is used to set up the electric field. The initial distribution for electrons is set according to the empirical model of AE8 (Vette 1991). Boundary conditions in the plasma sheet are specified by the empirical plasma sheet models (Ebihara and Ejiri 2000; Borovsky et al. 1998; Tsyganenko and Mukai 2003). This model also provides the equatorial electron fluxes as functions of time, energy and pitch angle but it is not run in real time with no comparison with the observations.

The model which runs online in real time and which outputs are directly compared to the real time observations for low energy (< 200 keV) electrons is Inner Magnetosphere Particle Transport and Acceleration model (IMPTAM) (Ganushkina et al. 2013, 2014, 2015a). IMPTAM operates online (imptam.fmi.fi) under the completed FP7 SPACECAST and SPACESTORM projects (http://fp7-spacecast.eu, http://www.spacestorm.eu/) funded by the European Union Seventh Framework Programme and under the on-going H2020 PROGRESS project (https://ssg.group.shef.ac.uk/progress2/html/) funded by the European Union's Horizon 2020 research and innovation programme. Its mirror site is also at the Center for Space Environment Modeling (CSEM) at the University of Michigan (http://csem.engin.umich.edu/tools/imptam). The driving parameters are determined by the models used in IMPTAM. The T96 model (Tsyganenko 1995) is used for the external magnetic field and the Boyle et al. (1997) polar cap potential mapped to the magnetosphere is used for the electric field. The model boundary is set at 10 R_E as a kappa electron distribution function with parameters given by the Tsyganenko and Mukai (2003) empirical model with necessary scaling for electrons. The solar wind and IMF parameters which drive the IMPTAM are solar wind density N_{SW}, dynamics pressure P_{SW}, velocity V_{SW}, IMF B_Y, B_Z, and total IMF strength. The model provides the low energy electron flux at all L-shells and at all satellite orbits, when necessary. Real time geostationary GOES 13 or GOES 15 (whenever which available) MAGED data on electron fluxes at three energies (40, 75, and 150 keV) are used for comparison and validation of IMPTAM running online (Ganushkina et al. 2015a). IMPTAM provides nowcast of keV electrons comparable to the observations.

A purely empirical model for ion flux and electron flux in the energy range ~1 eV to ~40 keV at geosynchronous orbit (GEO) was developed by Denton et al. (2015, 2016, 2017) based on a total of 82 satellite-years of observations from the Magnetospheric Plasma Analyzer instruments on Los Alamos National Laboratory satellites (http://gemelli.spacescience.org/mdenton/). The model provides the electron flux values at any energy and local time for given values of the 3-hour Kp-index or solar wind electric field ($-vB_z$). The Kp version of the model also provides flux values for given values of the daily F10.7 index. The limited number of driving parameters makes this model not suitable for fast variations of keV electrons. Even though the model is called empirical, no equations are provided which can be easily used to relate electron flux at a specific energy with the driving parameters.

A very different approach is used in the SNB3GEO models (Balikhin et al. 2011; Boynton et al. 2015, 2016) based on Multi-Input Single-Output (MISO) Nonlinear AutoRegressive Moving Average with eXogenous inputs (NARMAX) methodologies (Leontaritis and Billings 1985a,b). In the applied system identification approach, models are automatically deduced from input-output data by the system identification algorithms. NARMAX is capable of modeling nonlinear dynamics within the system and can provide a simple polynomial function with intuitive understanding of how the inputs change the output. The model is run online in real time at the University of Sheffield Space Weather Website (http://www.ssg.group.shef.ac.uk/USSW/UOSSW.html) and under the H2020 PROGRESS project (https://ssg.group.shef.ac.uk/progress2/html/index.phtml) funded by the European Union's Horizon 2020 research and innovation programme. The model inputs are the solar wind velocity, density and pressure, the fraction of time that the IMF was southward, the IMF contribution of a solar wind-magnetosphere coupling function following Boynton et al. (2011b) and the Dst-index. The model output is the fluxes for 30–50 keV, 50–100 keV, 100–200 keV, 200–350 keV and 350–600 keV directly compared to the GOES MAGED observations at geostationary orbit. This model also contains a forecast. For low energy electrons, it is not possible to forecast one day ahead because the same day variations in the solar wind affect the current low energy electron flux. Boynton et al. (2016) calculated the past 24 hour averages for each hour and, therefore, the input time lags in the algorithm were shifted hourly, not daily. It is stated that the model provides accurate forecasts with prediction efficiencies ranging between 66.9% and 82.3%. At the same time, smaller-scale variations are not forecasted but they constitute the most important factor when trying to estimate the surface charging effects from keV electron variations.

None of the models has explicit representation for substorms which are a crucial factor in the transport and acceleration of keV electrons. There have been several studies aimed at the modeling of electron injections as they are well-known signatures of a substorm in the near-Earth space (Li et al. 1998; Zaharia et al. 2000; Sarris et al. 2002; Gabrielse et al. 2012, 2014, 2016; Ganushkina et al. 2013, 2014). These models give relatively good agreement with the observed dispersionless electron injections at geostationary orbit during specific events (Ingraham et al. 2001; Fok et al. 2001a; Li et al. 2003; Mithaiwala and Horton 2005; Liu et al. 2009). The substorm-associated electromagnetic fields are very complex and their proper representation is still missing from the modeling efforts for keV electrons in the inner magnetosphere. No technique is implemented yet to include substorm influence on modeled fluxes in real time.

Ring current electrons in the keV energy range represent a real threat to equatorially orbiting spacecraft. In the midnight-to-dawn sector, fresh injections can intensify the low-energy electron fluxes and cause surface charging, sometimes with detrimental effects. Several models exist which nowcast or forecast these electrons, providing at least some level of awareness about this population. The critical influence of substorms on the near-Earth keV electrons, however, has not yet been fully implemented into a real time and online prediction tool.

4.2 Ring Current and Inner Magnetosphere Populations

4.2.1 Ring Current and Radiation Belts

The ring current is often thought of as a lower-energy part of the outer radiation belt that surrounds Earth in a toroidal shape. However, these are really two distinct populations with very different dynamics and responses to solar wind forcing (e.g. Schulz and Lanzerotti 1974; Liemohn 2006; Reeves et al. 2016). Naturally, there does exist a coupling between the two energy regimes, as they overlap in spatial extent most of the time. In fact, a very significant influence of the ring current on radiation belt losses during the main and recovery phases of geomagnetic storms (e.g. Ukhorskiy et al. 2015).

The first effect is related to the diamagnetic consequences of a partial ring current that develops during storms. Inflation of the near-Earth magnetic field due to ring current changes (e.g. Parker and Stewart 1967; Tsyganenko et al. 2003; Ganushkina et al. 2010) alters the drift paths of the relativistic electrons in the radiation belts, causing the adiabatic "Dst effect" (e.g. Kim and Chan 1997), which can become a real loss of particles at the magnetopause (e.g. West et al. 1972; Hudson et al. 1998; Ukhorskiy et al. 2006). This process has been identified as contributing rapid loss for up to 90% of the pre-existing population of both radiation belt electrons (100's keV to multi-MeV) energies and energetic ring current protons (100's keV to >1 MeV) (Turner et al. 2014).

The second effect is due to pitch angle scattering of relativistic electrons via wave-particle interactions. The notable mode here is electromagnetic ion-cyclotron (EMIC) waves, since these waves are able to very efficiently interact with high-energy radiation belt electrons (Summers et al. 1998; Meredith et al. 2003), and are driven by unstable ring current proton distributions (Anderson and Hamilton 1993). The anisotropy required for EMIC wave growth is fueled by ring current proton injections, which are convected inward to the inner magnetosphere during intervals of enhanced storm and substorm activity (e.g. Young et al. 1981; Jordanova et al. 2001). Recent simulations combined with the data from the Van Allen Probes (Mauk et al. 2013), the mission specifically designed for radiation belts, have demonstrated the dramatic effect on >1 MeV electron pitch angle distributions that EMIC waves can have (see, for example, Usanova et al. 2014). While this study did not find a clear way for EMIC waves to produce a full dropout of the core electron population, the flux levels were reduced substantially at lower pitch angles, indicating significant precipitation. Electron precipitation can have impacts on human society at the ground level, as well as terrestrial weather and climate systems (e.g. Rodger et al. 2010). Cutting edge models need to incorporate radiation belt losses from both of these effects: loss to the outer boundary of the magnetosphere via magnetopause shadowing and loss from atmospheric precipitation due to EMIC wave scattering.

Radiation belt enhancement and acceleration events are likewise tied to ring current dynamics. Local acceleration can provide significant acceleration of electron populations in the heart of the outer belt. The process is generated by VLF whistler-mode chorus waves, which grow as a result of a temperature anisotropy in few-10s keV electrons (Helliwell 1967) likely caused by substorm injections (Tsurutani and Smith 1974; Meredith et al. 2001). This anisotropic population of ring current energy electrons is often called the "seed" population, wherein the source of the free energy for chorus wave growth is available. Chorus waves interact with "seed" electrons at energies of 10's to 100's of keV through cyclotron resonance and can accelerate electrons to very high energies (Kennel and Petschek 1966; Kennel and Thorne 1967; Li et al. 2008, 2012; Thorne 2010). Whistler-mode chorus plays a critical role in accelerating ring current electrons to relativistic energies in the outer belt (Horne et al. 2005; Green 2006; Chen et al. 2007), including to "killer" electron energies that can have disastrous space weather consequences to sensitive systems (Baker 2001; Fennell et al. 2001).

4.2.2 Ring Current and Plasmasphere

The ring current overlaps the cold, dense plasmasphere surrounding Earth, with the center peak of the ring current typically co-located with the plasmapause boundary (Kozyra et al. 1997; Burch et al. 2001). The ring current can effect the shape of the plasmasphere (e.g. Goldstein et al. 2003; Liemohn et al. 2004; Gallagher et al. 2005), and the location of the plasmapause boundary, thereby sculpting the energy-dependent radiation belts and influencing electron precipitation to the atmosphere and locations of enhanced spacecraft charging.

Enhanced convection during geomagnetically active times leads to erosion of the plasmasphere such that the boundary contracts throughout local times (e.g. Nishida 1966), although a high-density plume can form on the duskside of Earth (e.g. Brice 1967; Carpenter 1970; Spasojević et al. 2003). Drainage plumes may also form, allowing direct transport of plasmaspheric particles to the outer magnetosphere (Grebowsky 1970; Chen and Wolf 1972; Rasmussen et al. 1993; Goldstein et al. 2004). Additionally, sub-auroral polarization streams (SAPS) can contribute to the erosion of the plasmasphere on very fast timescales (Gonzales et al. 1978; Burke et al. 1998; Foster and Vo 2002; Califf et al. 2016). During storm time, ring current injections influence the near-Earth electric field (Jaggi and Wolf 1973; Fok et al. 2001a) to the extent that the ionospheric convection pattern is affected which has space weather effects on human technologies such as GPS signal processing (Yeh et al. 1991; Foster et al. 2002).

The variation of plasmaspheric dynamics contribute to distinct consequences that effect the ring current population. First is the myriad roles of wave-particle interactions. When the plasmasphere begins to refill after a period of erosion to low radial distances, the outward expansion of the sharp density boundary once it overlaps the inner edge of the ring current can trigger lower hybrid wave growth (Roth and Hudson 1992). These waves can then accelerate low-energy ring current ions. Another effect is the growth of whistler-mode chorus waves much closer to Earth during an eroded plasmasphere, which will both scatter and accelerate a range of ring current electrons and ions. One more effect is the growth of EMIC waves in plasmaspheric plumes (Posch et al. 2010), which are triggered by ring current proton anisotropies. The existence of a high-density plume can encourage EMIC wave growth, which can contribute to losses of the "killer electrons" at ultra-relativistic energies (Horne and Thorne 1998). Finally, VLF hiss waves, which exist inside the dense plasmasphere, can produce a slow and steady loss of ring current electrons and protons (Thorne et al. 1973; Kozyra et al. 1994).

The position of many of these wave-particle interactions depends on the dynamics of the plasmasphere and plasmapause boundary. The location of the plasmasphere therefore effects where in space, and on the ground, many space weather consequences of the ring current are localized. Modeling of the plasmasphere, ring current, and associated plasma waves are essential for comprehensively defining the space weather environment.

4.3 Ring Current and Ionosphere and Below

The ring current not only influences the near-Earth magnetosphere with space weather consequences, but also alters the ionosphere and thermosphere. This is done through closure of the partial ring current and the generation of strong ionospheric flows. Inner magnetospheric plasma populations are often highly asymmetric in local time during active times, which creates localized pressure peaks and corresponding field aligned currents (FACs) at the two ends of the pressure crescent (Wolf 1970; Fok et al. 2001b; Liemohn et al. 2001a). These region 2 FACs become current source and sink regions in the ionosphere, where horizontal currents flow to close the circuit. The resulting horizontal electric fields in the ionosphere can be drastically different than the "standard" closure of the region 1 FACs, especially near the equatorward edge of the auroral zone and at subauroral latitudes. They are often poleward electric fields associated with a strong westward (sunward) flow in the midlatitude ionosphere. These flows have dramatic consequences for ionospheric density and thermospheric density, temperature, and composition.

4.3.1 Ring Current and Ionosphere and Thermosphere

The closure of the ring current FACs through the ionosphere create subauroral polarization streams (SAPS). The term SAPS was first coined by Foster and Burke (2002) to classify the evening-sector increase in electric field equatorward of the electron precipitation boundary. Because the plasma sheet ions drift closer to the Earth in this sector, region 2 FACs enter the ionosphere at a latitude without much ionization from the precipitation of keV-energy electrons. Foster and Vo (2002) and Erickson et al. (2011) presented statistical analyses of these electric fields and ionospheric flows, showing a clear dependence of SAPS intensity and spatial extent with geomagnetic activity. Ridley and Liemohn (2002) showed that they were directly linked to the partial ring current localized pressure peak, and other observations and numerical modeling has confirmed and further quantified this relationship (e.g. Garner et al. 2004; Mishin and Burke 2005; Liemohn et al. 2005; Yu et al. 2015; Yuan et al. 2016; Califf et al. 2016). SAPS are relatively stable during storms, being a persistent feature across the evening sector that can last for many hours (e.g. Huang and Foster 2007) as the hot ions in the inner magnetosphere slowly drift through near-Earth space. However, SAPS also occur during less active times, including isolated substorms (e.g. Karlsson et al. 1998; Wang et al. 2008).

These strong flows greatly perturb the storm-time ionospheric density configuration at midlatitudes. Pintér et al. (2006) showed that SAPS cause significant density depletion in the fast-flow channel. However, just equatorward of the peak of the flow channel, the higher densities from lower latitudes can become entrained in the westward flows, creating storm enhanced density (SED) plumes, which are anomalously high ionospheric electron densities on the dayside at mid-to-high latitudes during storms (e.g. Kelley et al. 2004; Heelis and Coley 2007). Yin et al. (2004) showed that SAPS created a SED over the USA during the 15 July 2000 storm, influencing Global Positioning System (GPS) communications. In fact, total electron content (TEC) maps created from GPS signal processing has been extensively used to quantify the timing and intensity of SEDs (e.g. Bust et al. 2007; Wen et al. 2007; Yizengaw et al. 2007; Basu et al. 2008; Coster and Skone 2009). SEDs have been shown to extend right through the dayside auroral zone into the polar cap, becoming tongues of ionization and, if breaking off, polar cap patches (e.g. Horvath and Lovell 2009, 2015). Interestingly, Sojka et al. (2004) and David et al. (2011) showed that SEDs might actually be a locally produced phenomenon, with the horizontal SAPS flows driving ExB drift in slant paths, suppressing recombination and allowing a new, higher equilibrium density at these locations. Schunk et al. (2012) assessed their online ionospheric prediction model, the Global Assimilation of Ionospheric Measurements (GAIM) model, especially for the low and midlatitude regions. They concluded that a significant part of the error of the results was due to uncertain magnetospheric inputs to the code. It is clear that the ionosphere is greatly perturbed by the electric fields related to the closure of the partial ring current with effects at all local times and across a range of latitudes.

The neutral upper atmosphere is also affected by these strong subauroral electric fields. Wang et al. (2011) conducted a statistical study of the influence of SAPS on the thermosphere, finding a 10% increase in density within the flow channel, presumably from heating. This was further explained with numerical modeling by Wang et al. (2012), who found that SAPS cause increases in global thermospheric temperature, but this effect takes days to fully develop. They also found that SAPS drive large thermospheric zonal winds due to ion-neutral drag. The heating also caused compositional changes to the thermosphere, with upwelling of molecular-rich atmosphere and downwelling of atomic-oxygen-rich gas elsewhere. Mishin et al. (2012) found that SAPS can create atmospheric gravity waves near 300 km altitude, presumably related to this localized thermospheric density and temperature enhancement.

A final example is Codrescu et al. (2012), who examined real time simulations of the Coupled Thermosphere-Ionosphere Plasmasphere-Electrodynamics model, CTIPe, who compared upper atmospheric neutral mass density results against CHAMP observations. They found that the total amplitude of the perturbations was well captured but the timing was not, indicating that nonlinear feedback in the system, such as that from the ring current, could be crucial in accurately predicting thermospheric space weather effects.

Note that SAPS have been related to the faster, narrower channels of flow known as polarization jets (Galperin et al. 1974) or sub-auroral ion drifts (SAIDs) (Anderson et al. 1993). SAPS, however, are broader in latitude and lower in intensity than these sharp spikes in electric field and ionospheric flow. One hypothesis for the relationship between these phenomena is the localized intensification of the SAPS field as increased temperature enhances the ionospheric recombination rate and lowers the conductivity in the SAPS channel (Wolf et al. 2007).

4.3.2 Ring Current and GICs

Geomagnetically induced currents (GIC) (Pirjola 2000; Boteler 2001) can disrupt normal transmission system operations triggering voltage collapse or damage transformers (e.g. Kappenman et al. 1997; Molinski 2002; Wik et al. 2009). The magnitude of GIC flowing in the high-voltage power transmission networks is determined by the horizontal geoelectric field (e.g. Pirjola 2002, 2013). The induced geoelectric field on the ground is mainly controlled by currents in the magnetosphere and ionosphere, and by the conductivity of the Earth (e.g. Pirjola 2000, 2002; Ngwira et al. 2013).

The large-scale electric currents in the ionosphere are coupled to the magnetosphere through field-aligned currents (FACs). The Region 1 currents that reside on open field lines are driven by the solar wind, which acts as a generator (Siscoe et al. 1991) by dayside reconnection. On closed field lines, their formation can be due to processes taking place in either the boundary layer (Lotko et al. 1987) or in the plasma sheet (e.g. Antonova and Ganushkina 1997; Toffoletto et al. 2001).

The Region 2 currents flow at lower latitudes and can be mapped to the ring current region. They are produced by the pressure gradient dynamics in the inner magnetosphere (Vasyliunas 1970; Birmingham 1992; Heinemann 1990; Heinemann and Pontius 1991). Thus, the ring current dynamics contributes to the GIC effects via its role in the generation of Region 2 FACs. Numerous other studies have examined the connection between conductance and the ring current (e.g. Fok et al. 2001b; Jordanova 2003; Sazykin et al. 2002; Khazanov et al. 2003; Ebihara et al. 2004; Liemohn et al. 2005), showing that the partial ring current closure can strongly perturb the inner magnetospheric electric fields.

GICs with large amplitudes are of primary concern for high-latitudes but, at the same time, the GICs were recently observed at the magnetically mid- and low-latitudes during storm times (e.g. Kappenman 2003; Trivedi et al. 2007; Marshall et al. 2012; Zhang et al. 2015; Barbosa et al. 2015; Watari 2015). The presence of the enhanced ring current during storms may be directly related to the GIC generation at these latitudes. The exact connection between the storm-time ring current and these subauroral GICs, however, has yet to be fully understood.

The geoelectric field can be computed in different ways but it is necessary to have the ground-based geomagnetic field obtained either from observations (Viljanen et al. 2006; Pulkkinen et al. 2012; Ngwira et al. 2013) or from the first-principles-based MHD modeling of the magnetosphere-ionosphere system as surface variations of magnetic field at the ground-based stations (Pulkkinen et al. 2013; Glocer et al. 2016) and a representative ground conductivity structure (Pirjola 2002). The ring current physics needs to be properly modeled for the geoelectric field calculations, especially during extreme space weather events.

5 Summary

The space weather effects from the ring current particles with keV energy range cannot be considered as highly obvious as those from the "killer" electrons or from the solar energetic protons with energies of tens of MeVs but they are nevertheless quite significant. Figure 2 summarized the space weather effects that can be produced by the ring current. One of the definitions of space weather includes the time-varying conditions in the space environment that may be hazardous to technological systems on the ground. Ring current has a direct influence on the Dst-index which is computed from the ground-based magnetic field observations and which is an indicator of a storm activity. Electrons with energies less than

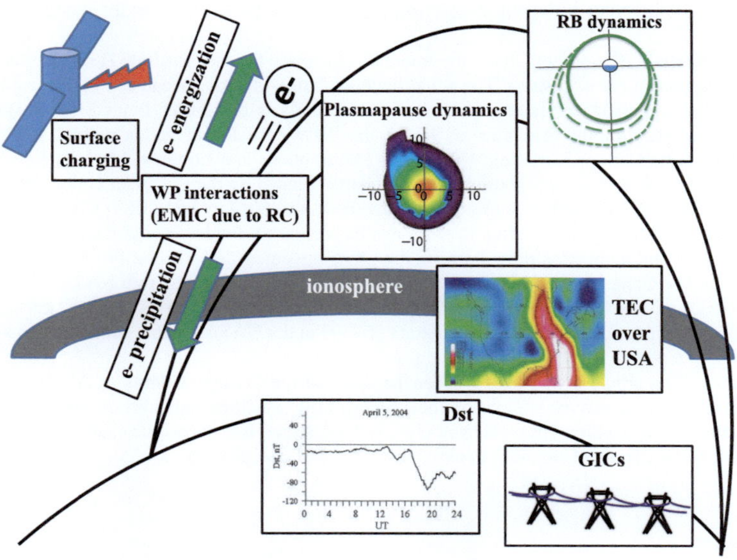

Fig. 2 Space weather effects due to ring current

100 keV are one of the important constituents of the ring current. The electron fluxes at these keV energies vary significantly with the current activity on the scale of minutes or even shorter. The keV electrons do not penetrate deep into the satellite materials but stay near the surface. They can be responsible for surface charging effects which is a serious risk for satellites.

Ring current dynamics are tied to both radiation belt losses and enhancements by affecting the efficiency of magnetopause shadowing and driving various wave-particle interactions. These consequences can be dramatic and impulsive, thereby altering the higher-energy space weather environment suddenly. With the addition of the overlapping plasmasphere, the picture is complicated further. Effective space weather models need to account for the myriad interacting density regimes and particle populations that are related to ring current changes.

The storm-time ring current has several detrimental space weather effects on the ionosphere and thermosphere. The partial ring current closes through the ionosphere, which leads to the SAPS phenomenon of strong westward flows at mid-latitudes. This rearranges the ionospheric density, creating SED plumes across the dayside middle and high latitude regions, extending even over the polar caps. These density enhancements adversely affect GPS signals, resulting in location errors of 50–100 meters during large events. Furthermore, the thermosphere is heated by the SAPS flows, leading to chemistry changes, and thermospheric winds ramp up to match the ionospheric flows during prolonged SAPS intervals.

The ring current contributes to the Geomagnetically Induced Currents effects via its role in the generation of Region 2 FACs. The magnitude of GIC is determined by the horizontal geoelectric field which is mainly controlled by currents in the magnetosphere and ionosphere, and by the conductivity of the Earth. The large-scale electric currents in the ionosphere are coupled to the magnetosphere through field-aligned currents (FACs). The Region 2 currents which can be mapped to the ring current region are generated by the pressure gradient dynamics in the inner magnetosphere.

The dynamics of the ring current is a preeminent factor in space weather forecasting, thereby of critical importance to the health and safety of our spacecraft systems. The ring current does not interact independently and alone, it is tied to the greater system.

Acknowledgements The authors thank the International Space Science Institute (ISSI) in Bern, Switzerland, for the support of the workshop "The Scientific Foundation of Space Weather". N. Ganushkina thanks the ISSI for the support of the international teams on "Analysis of Cluster Inner Magnetosphere Campaign Data, in Application the Dynamics of Waves and Wave-Particle Interaction within the Outer Radiation Belt" and "Ring Current Modeling: Uncommon Assumptions and Common Misconceptions." The research of N. Ganushkina leading to these results was partly funded by the European Union Seventh Framework Programme (FP7/2007-2013) under grant agreement No. 606716 SPACESTORM. The research of N. Ganushkina and M. Liemohn leading to these results was partly funded by the European Union's Horizon 2020 research and innovation programme under grant agreement No. 637302 PROGRESS. Support for M. Liemohn and N. Ganushkina at Michigan was provided by research grants from NASA (NNX14AF34G, NNX14AC02G, and NNX17AB87G) and NSF (AGS-1414517). The work of A. Jaynes was supported by RBSP-ECT funding provided by JHU/APL contract 967399 under NASA's Prime contract NAS5-01072.

References

S.-I. Akasofu, S. Chapman, The ring current, geomagnetic disturbance, and the Van Allen radiation belt. J. Geophys. Res. **66**, 1321–1350 (1961)

B.J. Anderson, D.C. Hamilton, Electromagnetic ion cyclotron waves stimulated by modest magnetospheric compressions. J. Geophys. Res. **98**, 11 (1993). doi:10.1029/93JA00605
P.C. Anderson, W.B. Hanson, R.A. Heelis, J.D. Craven, D.N. Baker, L.A. Frank, A proposed production model of rapid subauroral ion drifts and their relationship to substorm evolution. J. Geophys. Res. **98**, 6069–6078 (1993). doi:10.1029/92JA01975
V.A. Andreeva, N.A. Tsyganenko, Reconstructing the magnetosphere from data using radial basis functions. J. Geophys. Res. Space Phys. **121**, 2249–2263 (2016). doi:10.1002/2015JA022242
E.E. Antonova, N.Yu. Ganushkina, Azimuthal hot plasma pressure gradients and dawn-dusk electric field formation. J. Atmos. Terr. Phys. **59**, 1343–1354 (1997)
D.N. Baker, Satellite anomalies due to space storms, in *Space Storms and Space Weather Hazards*, ed. by I.A. Daglis (2001), p. 285
R. Bala, P. Reiff, Improvements in short-term forecasting of geomagnetic activity. Space Weather **10**, S06001 (2012). doi:10.1029/2012SW000779
M.A. Balikhin, R.J. Boynton, S.N. Walker, J.E. Borovsky, S.A. Billings, H.L. Wei, Using the NARMAX approach to model the evolution of energetic electrons fluxes at geostationary orbit. Geophys. Res. Lett. **38**(18), L18105 (2011)
C. Barbosa, L. Alves, R. Caraballo, G.A. Hartmann, A.R.R. Papa, R.J. Pirjola, Analysis of geomagnetically induced currents at a low-latitude region over the solar cycles 23 and 24: Comparison between measurements and calculations. J. Space Weather Space Clim. **5**, A35 (2015). doi:10.1051/swsc/2015036
Su. Basu, S. Basu, J.J. Makela, E. MacKenzie, P. Doherty, J.W. Wright, F. Rich, M.J. Keskinen, R.E. Sheehan, A.J. Coster, Large magnetic storm-induced nighttime ionospheric flows at midlatitudes and their impacts on GPS-based navigation systems. J. Geophys. Res. **113**, A00A06 (2008). doi:10.1029/2008JA013076
M.J. Beharrell, F. Honary, Decoding solar wind-magnetosphere coupling. Space Weather **14**, 724–741 (2016). doi:10.1002/2016SW001467
T.J. Birmingham, Birkeland currents in an anisotropic, magnetostatic plasma. J. Geophys. Res. **97**(A4), 3907–3917 (1992). doi:10.1029/91JA03029
O.M. Boaghe, M.A. Balikhin, S.A. Billings, H. Alleyne, Identification of nonlinear processes in the magnetospheric dynamics and forecasting of Dst index. J. Geophys. Res. **106**, 30047–30066 (2001). doi:10.1029/2000JA900162
J.E. Borovsky, M.F. Thomsen, R.C. Elphic, T.E. Cayton, D.J. McComas, The transport of plasma sheet material from the distant tail to geosynchronous orbit. J. Geophys. Res. **103**, 20297–20331 (1998)
D.H. Boteler, Space weather effects on power systems, in *Space Weather*, ed. by D. Song, H.J. Singer, G.L. Siscoe. AGU Geophysical Monograph, vol. 125 (2001), pp. 347–352
S. Bourdarie, V.K. Jordanova, M. Liemohn, T.P. O'Brien, Modeling the energetic particles of the inner magnetosphere, in *Waves, Particles, and Storms in Geospace: A Complex Interplay*, ed. by G. Balasis, I.A. Daglis, I.R. Mann (Oxford Scholarship, London, 2016). doi:10.1093/acprof:oso/9780198705246.003.0006
C. Boyle, P. Reiff, M. Hairston, Empirical polar cap potentials. J. Geophys. Res. **102**(A1), 111–125 (1997)
R.J. Boynton, M.A. Balikhin, S.A. Billings, A.S. Sharma, O.A. Amariutei, Data derived NARMAX Dst model. Ann. Geophys. **29**, 965–971 (2011a)
R.J. Boynton, M.A. Balikhin, S.A. Billings, H.L. Wei, N. Ganushkina, Using the NARMAX OLS-ERR algorithm to obtain the most influential coupling functions that affect the evolution of the magnetosphere. J. Geophys. Res. **116**(A5), A05218 (2011b)
R.J. Boynton, M.A. Balikhin, S.A. Billings, Online NARMAX model for electron fluxes at GEO. Ann. Geophys. **33**(3), 405–411 (2015)
R.J. Boynton, M.A. Balikhin, D.G. Sibeck, S.N. Walker, S.A. Billings, N. Ganushkina, Electron flux models for different energies at geostationary orbit. Space Weather **14**, 846–860 (2016). doi:10.1002/2016SW001506
N.M. Brice, Bulk motion of the magnetosphere. J. Geophys. Res. **72**, 5193 (1967)
J.L. Burch, D.G. Mitchell, B.R. Sandel, P.C. Brandt, M. Wuest, Global dynamics of the plasmasphere and ring current during magnetic storms. Geophys. Res. Lett. **28**, 1159–1162 (2001). doi:10.1029/2000GL012413
W.J. Burke, N.C. Maynard, M.P. Hagan, R.A. Wolf, G.R. Wilson, L.C. Gentile, M.S. Gussenhoven, C.Y. Huang, T.W. Garner, F.J. Rich, Electrodynamics of the inner magnetosphere observed in the dusk sector by CRESS and DMSP during the magnetic storm of June 4–6, 1991. J. Geophys. Res. **103**, 29399–29418 (1998)
R.K. Burton, R.L. McPherron, C.T. Russell, An empirical relationship between interplanetary conditions and Dst. J. Geophys. Res. **80**, 4204–4214 (1975)
G.S. Bust, G. Crowley, T.W. Garner, T.L. Gaussiran, R.W. Meggs, C.N. Mitchell, P.S.J. Spencer, P. Yin, B. Zapfe, Four-dimensional GPS imaging of space weather storms. Space Weather **5**, 2 (2007)

S. Califf, X. Li, R.A. Wolf, H. Zhao, A.N. Jaynes, F.D. Wilder, D.M. Malaspina, R. Redmon, Large-amplitude electric fields in the inner magnetosphere: Van Allen Probes observations of subauroral polarization streams. J. Geophys. Res. **121**, 5294–5306 (2016). doi:10.1002/2015JA022252

R.L. Carovillano, G.L. Siscoe, Energy and momentum theorems in magnetospheric processes. Rev. Geophys. Space Phys. **11**, 289–353 (1973)

D.L. Carpenter, Whistler evidence of dynamic behavior of the duskside bulge in the plasmasphere. J. Geophys. Res. **75**, 3837 (1970)

S. Chapman, V.C.A. Ferraro, A new theory of magnetic storms. Terr. Magn. **36**, 77–78 (1931)

S. Chapman, V.C.A. Ferraro, The geomagnetic ring-current: Its radial stability. Terr. Magn. **46**, 1–6 (1941)

A.J. Chen, R.A. Wolf, Effects on the plasmasphere of a time-varying convection electric field. Planet. Space Sci. **20**, 483 (1972)

M.W. Chen, L.R. Lyons, M. Schulz, Simulations of phase space distributions of storm time proton ring current. J. Geophys. Res. **99**, 5745 (1994)

M.W. Chen, M. Schulz, S. Liu et al., Simulated stormtime ring-current magnetic field produced by ions and electrons, in *Inner Magnetosphere Interactions: New Perspective from Imaging*, ed. by J. Burch, M. Schulz, H. Spence. Geophys. Monogr. Ser., vol. 159 (Am. Geophys. Union, Washington, 2005), pp. 237–250. doi:10.1029/GM159

M.W. Chen, S. Liu, M. Schulz, J.L. Roeder, L.R. Lyons, Magnetically self-consistent ring current simulations during the 19 October 1998 storm. J. Geophys. Res. **111**, A11S15 (2006). doi:10.1029/2006JA011620

Y. Chen, G.D. Reeves, R.H.W. Friedel, The energization of relativistic electrons in the outer Van Allen radiation belt. Nat. Phys. **3**, 614–617 (2007). doi:10.1038/nphys655

M.W. Chen, C.L. Lemon, K. Orlova, Y. Shprits, J. Hecht, R.L. Walterscheid, Comparison of simulated and observed trapped and precipitating electron fluxes during a magnetic storm. Geophys. Res. Lett. **42**, 8302–8311 (2015). doi:10.1002/2015GL065737

M.V. Codrescu, C. Negrea, M. Fedrizzi, T.J. Fuller-Rowell, A. Dobin, N. Jakowsky, H. Khalsa, T. Matsuo, N. Maruyama, A real-time run of the Coupled Thermosphere Ionosphere Plasmasphere Electrodynamics (CTIPe) model. Space Weather **10**, S02001 (2012). doi:10.1029/2011SW000736

A. Coster, S. Skone, Monitoring storm-enhanced density using IGS reference station data. J. Geod. **83**(3–4), 345 (2009)

I.A. Daglis, The storm-time ring current. Space Sci. Rev. **98**, 343–363 (2001)

I.A. Daglis, Ring current dynamics. Space Sci. Rev. **124**, 183–202 (2006). doi:10.1007/s11214-006-9104-z

I.A. Daglis, J.U. Kozyra, Outstanding issues of ring current dynamics. J. Atmos. Sol.-Terr. Phys. **64**, 253–264 (2002)

M. David, J.J. Sojka, R.W. Schunk, M.W. Liemohn, A.J. Coster, Dayside mid-latitude ionospheric response to storm-time electric fields. J. Geophys. Res. **116**, A12302 (2011). doi:10.1029/2011JA016988

V.A. Davis, M.J. Mandell, M.F. Thomsen, Representation of the measured geosynchronous plasma environment in spacecraft charging calculations. J. Geophys. Res. **113**, A10204 (2008). doi:10.1029/2008JA013116

P. De Michelis, I.A. Daglis, G. Consolini, Average terrestrial ring current derived from AMPTE/CCE-CHEM measurements. J. Geophys. Res. **102**, 14103–14111 (1997)

D.L. De Zeeuw, T.I. Gombosi, C.P.T. Groth, K.G. Powell, Q.F. Stout, An adaptive MHD method for global space weather simulations. IEEE Trans. Plasma Sci. **28**(6), 1956–1965 (2000). doi:10.1109/27.902224

V.I. Degtyarev, G.V. Popov, S.S. Sheshukov, Modelling the dynamics of fluxes of electrons with energies 30–300 keV in geostationary orbit. Geomagn. Aeron. **30**, 866–868 (1990)

M.H. Denton, M.F. Thomsen, V.K. Jordanova, M.G. Henderson, J.E. Borovsky, J.S. Denton, D. Pitchford, D.P. Hartley, An empirical model of electron and ion fluxes derived from observations at geosynchronous orbit. Space Weather **13**, 233–249 (2015). doi:10.1002/2015SW001168

M.H. Denton, M.G. Henderson, V.K. Jordanova, M.F. Thomsen, J.E. Borovsky, J. Woodroffe, D.P. Hartley, D. Pitchford, An improved empirical model of electron and ion fluxes at geosynchronous orbit based on upstream solar wind conditions. Space Weather **14**, 511–523 (2016). doi:10.1002/2016SW001409

M.H. Denton, G.D. Reeves, B.A. Larsen, R.H.W. Friedel, M.F. Thomsen, P.A. Fernandes, R.M. Skoug, H.O. Funsten, L.K. Sarno-Smith, On the origin of low-energy electrons in the inner magnetosphere: Fluxes and pitch-angle distributions. J. Geophys. Res. Space Phys. **122**, 1789–1802 (2017). doi:10.1002/2016JA023648

A.J. Dessler, E.N. Parker, Hydromagnetic theory of geomagnetic storms. J. Geophys. Res. **64**, 2239 (1959)

S.A. Dolenko, I.N. Myagkova, V.R. Shiroky, I.G. Persiantsev, Objective discrimination of geomagnetic disturbances and prediction of Dst index by artificial neural networks, in *Proceedings of the 10th Intl. Conf. Problems of Geocosmos* (2014), pp. 270–275

M.W. Dunlop, S. Haaland, P.C. Escoubet, X.-C. Dong, Commentary on accessing 3-D currents in space: Experiences from cluster. J. Geophys. Res. **121**, 7881–7886 (2016). doi:10.1002/2016JA022668

Y. Ebihara, Chapter 9: Ring current, in *Space Weather Fundamentals*, ed. by G.V. Khazanov (CRC Press, Boca Raton, 2016)

Y. Ebihara, M. Ejiri, Simulation study on fundamental properties of the storm-time ring current. J. Geophys. Res. **105**, 15843 (2000)

Y. Ebihara, M. Ejiri, Numerical simulation of the ring current: Review. Space Sci. Rev. **105**, 377–452 (2003). doi:10.1023/A:1023905607888

Y. Ebihara, Y. Miyoshi, Dynamic inner magnetosphere: A tutorial and recent advances, in *The Dynamic Magnetosphere*, ed. by W. Liu, M. Fujimoto (Springer, Berlin, 2011). doi:10.1007/978-94-007-0501-2

Y. Ebihara, M.-C. Fok, R.A. Wolf, T.J. Immel, T.E. Moore, Influence of ionospheric conductivity on the ring current. J. Geophys. Res. **109**, A08205 (2004). doi:10.1029/2003JA010351

A. Egeland, W.J. Burke, The ring current: A short biography. Hist. Geo- Space Sci. **3**, 131–142 (2012)

S.R. Elkington, M. Wiltberger, A.A. Chan, D.N. Baker, Physical models of the geospace radiation environment. J. Atmos. Sol.-Terr. Phys. **66**, 1371–1387 (2004)

P.J. Erickson, F. Beroz, M.Z. Miskin, Statistical characterization of the American sector subauroral polarization stream using incoherent scatter radar. J. Geophys. Res. **116**, A00J21 (2011). doi:10.1029/2010JA015738

J.F. Fennell, H.C. Koons, J.L. Roeder, J.B. Blake, Spacecraft charging: Observations and relationship to satellite anomalies, in *Proceedings of 7th Spacecraft Charging Technology Conference*, vol. 476 (European Space Agency, Noordwijk, Netherlands, 2001), pp. 279–285

F.R. Fenrich, J.G. Luhmann, Geomagnetic response to magnetic clouds of different polarity. Geophys. Res. Lett. **25**, 2999–3002 (1998)

M.-C. Fok, T.E. Moore, Ring current modeling in a realistic magnetic field configuration. Geophys. Res. Lett. **24**, 1775–1778 (1997)

M.-C. Fok, T.E. Moore, D.C. Delcourt, Modeling of inner plasma sheet and ring current during substorms. J. Geophys. Res. **104**, 14557–14569 (1999)

M.-C. Fok, T.E. Moore, W.N. Spjeldvik, Rapid enhancement of radiation belt electron fluxes due to substorm dipolarization of the geomagnetic field. J. Geophys. Res. **106**(A3), 3873–3882 (2001a). doi:10.1029/2000JA000150

M.-C. Fok, R.A. Wolf, R.W. Spiro, T.E. Moore, Comprehensive computational model of the Earth's ring current. J. Geophys. Res. **106**, 8417–8424 (2001b)

M.-C. Fok, A. Glocer, Q. Zheng, R.B. Horne, N.P. Meredith, J.M. Albert, T. Nagai, Recent developments in the radiation belt environment model. J. Atmos. Sol.-Terr. Phys. **73**, 1435–1443 (2011)

M.-C. Fok, N.Y. Buzulukova, S.-H. Chen, A. Glocer, T. Nagai, P. Valek, J.D. Perez, The comprehensive inner magnetosphere-ionosphere model. J. Geophys. Res. Space Phys. **119**, 7522–7540 (2014). doi:10.1002/2014JA020239

J.C. Foster, W.J. Burke, SAPS: A new categorization of subauroral electric fields. Eos Trans. AGU **83**(36), 393 (2002)

J.C. Foster, H.B. Vo, Average characteristics and activity dependence of the subauroral polarization stream. J. Geophys. Res. **107**(A12), 1475 (2002). doi:10.1029/2002JA009409

J.C. Foster, P.J. Erickson, A.J. Coster, J. Goldstein, F.J. Rich, Ionospheric signatures of plasmaspheric tails. Geophys. Res. Lett. **29**, 1623 (2002). doi:10.1029/2002GL015067

L.A. Frank, Several observations of low-energy protons and electrons in the Earth's magnetosphere with OGO 3. J. Geophys. Res. **72**, 1905–1916 (1967)

L.A. Frank, Direct detection of asymmetric increases of extraterrestrial 'Ring Current' proton intensities in the outer radiation zone. J. Geophys. Res. **75**, 1263–1268 (1970)

R.H.W. Friedel, A. Korth, Review of CRRES ring current observations. Adv. Space Res. **20**, 311–320 (1997). doi:10.1016/S0273-1177(97)00673-X

R.H.W. Friedel, H. Korth, M.G. Henderson, M.E. Thomsen, J.D. Scudder, Plasma sheet access to the inner magnetosphere. J. Geophys. Res. **106**, 5845–5858 (2001)

E. Friedrich, G. Rostoker, M.G. Connors, R.L. McPherron, Influence of the substorm current wedge on the Dst index. J. Geophys. Res. **104**(A3), 4567–4575 (1999). doi:10.1029/1998JA900096

C. Gabrielse, V. Angelopoulos, A. Runov, D.L. Turner, The effects of transient, localized electric fields on equatorial electron acceleration and transport toward the inner magnetosphere. J. Geophys. Res. **117**, A10213 (2012). doi:10.1029/2012JA017873

C. Gabrielse, V. Angelopoulos, A. Runov, D.L. Turner, Statistical characteristics of particle injections throughout the equatorial magnetotail. J. Geophys. Res. Space Phys. **119**, 2512–2535 (2014). doi:10.1002/2013JA019638

C. Gabrielse, C. Harris, V. Angelopoulos, A. Artemyev, A. Runov, The role of localized inductive electric fields in electron injections around dipolarizing flux bundles. J. Geophys. Res. Space Phys. **121**, 9560–9585 (2016). doi:10.1002/2016JA023061

D.L. Gallagher, M.L. Adrian, M.W. Liemohn, The origin and evolution of deep plasmaspheric notches. J. Geophys. Res. **110**, A09201 (2005). doi:10.1029/2004JA010906

Y. Galperin, V.N. Ponomarev, A.G. Zosimova, Plasma convection in the polar ionosphere. Ann. Geophys. **30**, 1–7 (1974)

N.Yu. Ganushkina, Drivers of the inner magnetosphere, in *The Inner Magnetosphere: Physics and Modeling*, ed. by T.I. Pulkkinen, N.A. Tsyganenko, R.H.W. Friedel. Geophysical Monograph Series, vol. 155 (Am. Geophys. Union, Washington, 2005), pp. 135–145. doi:10.1029/155GM16

N.Y. Ganushkina, T.I. Pulkkinen, M.V. Kubyshkina, H.J. Singer, C.T. Russell, Long-term evolution of magnetospheric current systems during storms. Ann. Geophys. **22**, 1317–1334 (2004)

N.Y. Ganushkina, M. Liemohn, M. Kubyshkina, R. Ilie, H. Singer, Distortions of the magnetic field by storm-time current systems in Earth's magnetosphere. Ann. Geophys. **28**, 123–140 (2010)

N.Yu. Ganushkina, M.W. Liemohn, T.I. Pulkkinen, Storm-time ring current: Model-dependent results. Ann. Geophys. **30**, 177–202 (2012)

N.Y. Ganushkina, O. Amariutei, Y.Y. Shpritz, M. Liemohn, Transport of the plasma sheet electrons to the geostationary distances. J. Geophys. Res. **118**, 82–98 (2013). doi:10.1029/2012JA017923

N.Y. Ganushkina, M. Liemohn, O. Amariutei, D. Pitchford, Low energy electrons (5–50 keV) in the inner magnetosphere. J. Geophys. Res. **119**, 246–259 (2014). doi:10.1002/2013JA019304

N.Y. Ganushkina, O.A. Amariutei, D. Welling, D. Heynderickx, Nowcast model for low-energy electrons in the inner magnetosphere. Space Weather **13**, 16–34 (2015a). doi:10.1002/2014SW001098

N.Y. Ganushkina, M.W. Liemohn, S. Dubyagin et al., Defining and resolving current systems in geospace. Ann. Geophys. **33**, 1369–1402 (2015b). doi:10.5194/angeo-33-1369-2015

T.W. Garner, R.A. Wolf, R.W. Spiro, W.J. Burke, B.G. Fejer, S. Sazykin, J.L. Roeder, M.R. Hairston, Magnetospheric electric fields and plasma sheet injection to low L-shells during the 4–5 June 1991 magnetic storm: Comparison between the rice convection model and observations. J. Geophys. Res. **109**, A02214 (2004). doi:10.1029/2003JA010208

H.B. Garrett, The charging of spacecraft surfaces. Rev. Geophys. **19**(4), 577 (1981). doi:10.1029/RG019i004p00577

M. Gkioulidou, A.Y. Ukhorskiy, D.G. Mitchell, L.J. Lanzerotti, Storm time dynamics of ring current protons: Implications for the long-term energy budget in the inner magnetosphere. Geophys. Res. Lett. **43**, 4736–4744 (2016). doi:10.1002/2016GL068013

A. Glocer et al., Community-wide validation of geospace model local K-index predictions to support model transition to operations. Space Weather **14**, 469–480 (2016). doi:10.1002/2016SW001387

J. Goldstein, B.R. Sandel, M.R. Hairston, P.H. Reiff, Control of plasmaspheric dynamics by both convection and sub-auroral polarization stream. Geophys. Res. Lett. **30**(24), 2243 (2003). doi:10.1029/2003GL018390

J. Goldstein, B.R. Sandel, M.F. Thomsen, M. Spasojevic, P.H. Reiff, Simultaneous remote sensing and in situ observations of plasmaspheric drainage plumes. J. Geophys. Res. Space Phys. **109**, A03202 (2004). doi:10.1029/2003JA010281

C.A. Gonzales, M.C. Kelley, L.A. Carpenter, R.H. Holzworth, Evidence for a magnetospheric effect on mid-latitude electric fields. J. Geophys. Res. **83**, 4397–4399 (1978)

W.D. Gonzalez, J.A. Joselyn, Y. Kamide, H.W. Kroehl, G. Rostoker, B.T. Tsurutani, V.M. Vasyliunas, What is a geomagnetic storm? J. Geophys. Res. **99**, 5771–5792 (1994)

O.S. Grafodatskiy, V.I. Degtyarev, A.G. Kozlov, V.I. Lazarev, O.I. Platonov, G.V. Popov, M.V. Teltsov, Relationship between characteristics of low-energy electrons and geomagnetic disturbance in geostationary orbit. Geomagn. Aeron. **27**, 494–496 (1987)

M. Grande, C.H. Perry, A. Hall, J. Fennell, B. Wilken, Survey of ring current composition during magnetic storms. Adv. Space Res. **20**, 321–326 (1997). doi:10.1016/S0273-1177(97)00675-3

J.M. Grebowsky, Model study of plasmapause motion. J. Geophys. Res. **75**, 4329 (1970)

J.C. Green, Using electron phase space density signatures to identify the electromagnetic waves responsible for accelerating relativistic electrons in Earth's magnetosphere, in *Magnetospheric ULF Waves: Synthesis and New Directions*, ed. by K. Takahashi, P.J. Chi, R.E. Denton, R.L. Lysak. Washington DC American Geophysical Union Geophysical Monograph Series, vol. 169 (2006), p. 225

M.E. Greenspan, D.C. Hamilton, A test of the Dessler-Parker-Sckopke relation during magnetic storms. J. Geophys. Res. **105**, 5419–5430 (2000)

R. Gubby, E. Evans, Space environment effects and satellite design. J. Atmos. Sol.-Terr. Phys. **64**(16), 1723–1733 (2002)

R.A. Heelis, W.R. Coley, Variations in the low- and middle-latitude topside ion concentration observed by DMSP during superstorm events. J. Geophys. Res. **112**, A08310 (2007). doi:10.1029/2007JA012326

M. Heinemann, Representations of currents and magnetic fields in anisotropic magnetohydrostatic plasma. J. Geophys. Res. **95**, 7789–7797 (1990)

M. Heinemann, D.H. Pontius, Representation of currents and magnetic fields in anisotropic magnetohydrostatic plasma. 2. General theory and examples. J. Geophys. Res. **96**, 17609–17626 (1991)
R.A. Helliwell, A theory of discrete VLF emissions from the magnetosphere. J. Geophys. Res. **72**, 4773–4790 (1967). doi:10.1029/JZ072i019p04773
R.B. Horne, R.M. Thorne, Potential waves for relativistic electron scattering and stochastic acceleration during magnetic storms. Geophys. Res. Lett. **25**, 3011–3014 (1998). doi:10.1029/98GL01002
R.B. Horne, R.M. Thorne, Y.Y. Shprits, N.P. Meredith, S.A. Glauert, A.J. Smith, S.G. Kanekal, D.N. Baker, M.J. Engebretson, J.L. Posch, M. Spasojevic, U.S. Inan, J.S. Pickett, P.M.E. Decreau, Wave acceleration of electrons in the Van Allen radiation belts. Nature **437**, 227–230 (2005). doi:10.1038/nature03939
I. Horvath, B.C. Lovell, Storm-enhanced plasma density features, daytime polar cap plasma enhancements, and their underlying plasma flows investigated during superstorms. J. Geophys. Res. **114**, A11304 (2009). doi:10.1029/2009JA014465
I. Horvath, B.C. Lovell, Investigating storm-enhanced density and polar tongue of ionization development during the 22 October 1999 great storm. J. Geophys. Res. Space Phys. **120**, 1428–1444 (2015). doi:10.1002/2014JA020598
C.-S. Huang, J.C. Foster, Correlation of the subauroral polarization streams (SAPS) with the Dst index during severe magnetic storms. J. Geophys. Res. **112**, A11302 (2007). doi:10.1029/2007JA012584
M.K. Hudson, V.A. Marchenko, I. Roth, M. Temerin, J.B. Blake, M.S. Gussenhoven, Radiation belt formation during storm sudden commencements and loss during main phase. Adv. Space Res. **21**(4), 597–607 (1998)
J.C. Ingraham, T.E. Cayton, R.D. Belian, R.A. Christensen, R.H.W. Friedel, M.M. Meier, G.D. Reeves, M. Tuszewski, Substorm injection of relativistic electrons to geosynchronous orbit during the great magnetic storm of March 24, 1991. J. Geophys. Res. **106**, 25759–25776 (2001)
T. Iyemori, Storm-time magnetospheric currents inferred from mid-latitude geomagnetic field variations. J. Geomagn. Geoelectr. **42**, 1249–1265 (1990)
R.K. Jaggi, R.A. Wolf, Self-consistent calculation of the motion of a sheet of ions in the magnetosphere. J. Geophys. Res. **78**, 2852 (1973). doi:10.1029/JA078i016p02852
E.-Y. Ji, Y.-J. Moon, N. Gopalswamy, D.-H. Lee, Comparison of Dst forecast models for intense geomagnetic storms. J. Geophys. Res. **117**, A03209 (2012). doi:10.1029/2011JA016872
V.K. Jordanova, New insights on geomagnetic storms from model simulations using multi-spacecraft data. Space Sci. Rev. **107**, 157–165 (2003)
V.K. Jordanova, Y.S. Miyoshi, Relativistic model of ring current and radiation belt ions and electrons: Initial results. Geophys. Res. Lett. **32**, L14104 (2005). doi:10.1029/2005GL023020
V.K. Jordanova, L.M. Kistler, J.U. Kozyra, G.V. Khazanov, A.F. Nagy, Collisional losses of ring current ions. J. Geophys. Res. **101**, 111 (1996)
V.K. Jordanova, L.M. Kistler, C.J. Farrugia, R.B. Torbert, Effects of inner magnetospheric convection on ring current dynamics: March 10–12, 1998. J. Geophys. Res. **106**, 29705–29720 (2001)
V.K. Jordanova, S. Zaharia, D.T. Welling, Comparative study of ring current development using empirical, dipolar, and self-consistent magnetic field simulations. J. Geophys. Res. **115**, A00J11 (2010). doi:10.1029/2010JA015671
V.K. Jordanova, D.T. Welling, S.G. Zaharia, L. Chen, R.M. Thorne, Modeling ring current ion and electron dynamics and plasma instabilities during a high-speed stream driven storm. J. Geophys. Res. **117**, A00L08 (2012). doi:10.1029/2011JA017433
V.K. Jordanova, Y. Yu, J.T. Niehof, R.M. Skoug, G.D. Reeves, C.A. Kletzing, J.F. Fennell, H.E. Spence, Simulations of inner magnetosphere dynamics with an expanded RAM-SCB model and comparisons with Van Allen Probes observations. Geophys. Res. Lett. **41**, 2687–2694 (2014). doi:10.1002/2014GL059533
V.K. Jordanova, W. Tu, Y. Chen, S.K. Morley, A.-D. Panaitescu, G.D. Reeves, C.A. Kletzing, RAM-SCB simulations of electron transport and plasma wave scattering during the October 2012 “double-dip” storm. J. Geophys. Res. Space Phys. **121**, 8712–8727 (2016). doi:10.1002/2016JA022470
V.V. Kalegaev, N.Y. Ganushkina, T.I. Pulkkinen, M.V. Kubyshkina, H.J. Singer, C.T. Russell, Relation between the ring current and the tail current during magnetic storms. Ann. Geophys. **23**, 523–533 (2005)
Y. Kamide, Association of DP and DR fields with the interplanetary magnetic field variation. J. Geophys. Res. **79**, 49–55 (1974). doi:10.1029/JA079i001p00049
Y. Kamide, N. Fukushima, Analysis of magnetic storms with DR indices for equatorial ring-current field. Radio Sci. **6**, 277–278 (1971). doi:10.1029/RS006i002p00277
J.G. Kappenman, Storm sudden commencement events and the associated geomagnetically induced current risks to ground-based systems at low-latitude and midlatitude locations. Space Weather **1**, 1016 (2003). doi:10.1029/2003SW000009
J.G. Kappenman, L.J. Zanetti, W.A. Radasky, Geomagnetic storms can threaten electric power grid. Earth Space **9**(7), 9–11 (1997)

T. Karlsson, G.T. Marklund, L.G. Blomberg, A. Mälkki, Subauroral electric fields observed by the Freja satellite: A statistical study. J. Geophys. Res. **103**, 4327–4341 (1998). doi:10.1029/97JA00333
R.M. Katus, M.W. Liemohn, E.L. Ionides, R. Ilie, D. Welling, L.K. Sarno-Smith, Statistical analysis of the geomagnetic response to different solar wind drivers and the dependence on storm intensity. J. Geophys. Res. Space Phys. **120**, 310–327 (2015). doi:10.1002/2014JA020712
M.C. Kelley, M.N. Vlasov, J.C. Foster, A.J. Coster, A quantitative explanation for the phenomenon known as storm-enhanced density. Geophys. Res. Lett. **31**, L19809 (2004). doi:10.1029/2004GL020875
C.F. Kennel, H.E. Petschek, Limit on stably trapped particle fluxes. J. Geophys. Res. **71**, 1 (1966)
C.F. Kennel, R.M. Thorne, Unstable growth of unducted whistlers propagating at an angle to the geomagnetic field. J. Geophys. Res. **72**(3), 871–878 (1967). doi:10.1029/JZ072i003p00871
G.V. Khazanov, T.S. Newman, M.W. Liemohn, M.-C. Fok, R.W. Spiro, Self-consistent magnetosphere-ionosphere coupling: Theoretical studies. J. Geophys. Res. **107**(A3), 1122 (2003). doi:10.1029/2002JA009624
G.V. Khazanov, M.W. Liemohn, T.S. Newman, M.-C. Fok, A.J. Ridley, Magnetospheric convection electric field dynamics and stormtime particle energization: Case study of the magnetic storm of 4 May 1998. Ann. Geophys. **22**, 497 (2004)
H.-J. Kim, A.A. Chan, Fully adiabatic changes in storm time relativistic electron fluxes. J. Geophys. Res. **102**, 22107 (1997)
J.S. Kim, D.A. Graham, C.S. Wang, Inductive electric fields in the ionosphere-magnetosphere system. Rev. Geophys. Space Phys. **17**, 2049–2056 (1979)
L.M. Kistler et al., The source of O^+ in the storm time ring current. J. Geophys. Res. Space Phys. **121**, 5333–5349 (2016). doi:10.1002/2015JA022204
A.J. Klimas, D. Vassiliadis, D.N. Baker, Dst index prediction using data-derived analogues of the magnetospheric dynamics. J. Geophys. Res. **103**, 20435–20447 (1998)
H.C. Koons, J.E. Mazur, R.S. Selesnick, J.B. Blake, J.F. Fennell, J.L. Roeder, P.C. Anderson, The impact of the space environment on space systems. Aerospace Rep. TR-99 (1670)-1, Aerospace Corp., El Segundo, CA (1999)
H. Korth, M.F. Thomsen, J.E. Borovsky, D.J. McComas, Plasma sheet access to geosynchronous orbit. J. Geophys. Res. **104**, 25047–25061 (1999)
T.V. Kozelova, L.L. Lazutin, B.V. Kozelov, N. Meredith, M.A. Danielides, Alternating bursts of low energy ions and electrons near the substorm onset. Ann. Geophys. **24**, 1957–1968 (2006)
J.U. Kozyra, M.W. Liemohn, Ring current energy input and decay. Space Sci. Rev. **109**, 105–131 (2003)
J.U. Kozyra, C.E. Rasmussen, R.H. Miller, L.R. Lyons, Interaction of ring current and radiation belt protons with ducted plasmaspheric hiss. 1: Diffusion coefficients and timescales. J. Geophys. Res. **99**, 4069–4084 (1994). doi:10.1029/93JA01532
J.U. Kozyra, A.F. Nagy, D.W. Slater, High-altitude energy source(s) for stable auroral red arcs. Rev. Geophys. **35**(2), 155–190 (1997). doi:10.1029/96RG03194
E.A. Kronberg, M. Ashour-Abdalla, I. Dandouras et al., Circulation of heavy ions and their dynamical effects in the magnetosphere: Recent observations and models. Space Sci. Rev. **184**, 173–235 (2014). doi:10.1007/s11214-014-0104-0
S. Kurita, Y. Miyoshi, F. Tsuchiya, Y. Nishimura, T. Hori, Y. Miyashita, T. Takada, A. Morioka, V. Angelopoulos, J.P. McFadden, H.U. Auster, J.M. Albert, V. Jordanova, H. Misawa, Transport and loss of the inner plasma sheet electrons: THEMIS observations. J. Geophys. Res. **116**, A03201 (2011). doi:10.1029/2010JA015975
L.J. Lanzerotti, K. LaFleur, C.G. Maclennan, D.W. Maurer, Geosynchronous spacecraft charging in January 1997. Geophys. Res. Lett. **25**(15), 2967–2970 (1998)
C. Lemon, R. Wolf, T.W. Hill, S. Sazykin, Magnetic storm ring current injection modeled with the rice convection model and a self-consistent magnetic field. Geophys. Res. Lett. **31**, L21801 (2004). doi:10.1029/2004GL020914
I.J. Leontaritis, S.A. Billings, Input-output parametric models for non-linear systems part i: Deterministic non-linear systems. Int. J. Control **41**(2), 303–328 (1985a)
I.J. Leontaritis, S.A. Billings, Input-output parametric models for non-linear systems part ii: Stochastic non-linear systems. Int. J. Control **41**(2), 329–344 (1985b)
X. Li, D.N. Baker, M. Temerin et al., Simulation of dispersionless injections and drift echoes of energetic electrons associated with substorms. Geophys. Res. Lett. **25**, 3763–3766 (1998)
X. Li, T.E. Sarris, D.N. Baker, W.K. Peterson, H.J. Singer, Simulation of energetic particle injections associated with a substorm on August 27, 2001. Geophys. Res. Lett. **30**, 1004 (2003). doi:10.1029/2002GL015967
W. Li, R.M. Thorne, N.P. Meredith, R.B. Horne, J. Bortnik, Y.Y. Shprits, B. Ni, Evaluation of whistler-mode chorus amplification during an injection event observed on CRRES. J. Geophys. Res. **113**, A09210 (2008). doi:10.1029/2008JA013129

W. Li, R. Thorne, J. Bortnik, R. McPherron, Y. Nishimura, V. Angelopoulos, I.G. Richardson, Evolution of chorus waves and their source electrons during storms driven by corotating interaction regions. J. Geophys. Res. **117**, A08209 (2012). doi:10.1029/2012JA017797

M.W. Liemohn, Yet another caveat to using the Dessler-Parker-Sckopke relation. J. Geophys. Res. **108**(A6), 1251 (2003). doi:10.1029/2003JA009839

M.W. Liemohn, Introduction to the special section on "Results of the national science foundation geospace environment modeling inner magnetosphere/storms assessment challenge". J. Geophys. Res. **111**, A11S01 (2006). doi:10.1029/2006JA011970

M.W. Liemohn, J.U. Kozyra, C.R. Clauer, A.J. Ridley, Computational analysis of the near-Earth magnetospheric current system. J. Geophys. Res. **106**, 29531 (2001a)

M.W. Liemohn, J.U. Kozyra, M.F. Thomsen et al., Dominant role of the asymmetric ring current in producing the stormtime Dst. J. Geophys. Res. **106**, 10883–10904 (2001b)

M.W. Liemohn, A.J. Ridley, D.L. Gallagher, D.M. Ober, J.U. Kozyra, Dependence of plasmaspheric morphology on the electric field description during the recovery phase of the April 17, 2002 magnetic storm. J. Geophys. Res. **109**(A3), A03209 (2004). doi:10.1029/2003JA010304

M.W. Liemohn, A.J. Ridley, P.C. Brandt, D.L. Gallagher, J.U. Kozyra, D.G. Mitchell, E.C. Roelof, R. DeMajistre, Parametric analysis of nightside conductance effects on inner magnetospheric dynamics for the 17 April 2002 storm. J. Geophys. Res. **110**, A12S22 (2005). doi:10.1029/2005JA011109

M.W. Liemohn, N.Y. Ganushkina, R. Ilie, D.T. Welling, Challenges associated with near-Earth nightside current. J. Geophys. Res. Space Phys. **121**, 6763–6768 (2016). doi:10.1002/2016JA022948

S. Liu, M.W. Chen, J.L. Roeder, L.R. Lyons, M. Schulz, Relative contribution of electrons to the stormtime total ring current energy content. Geophys. Res. Lett. **32**, L03110 (2005). doi:10.1029/2004GL021672

W.L. Liu, X. Li, T. Sarris, C. Cully, R. Ergun, V. Angelopoulos, D. Larson, A. Keiling, K.H. Glassmeier, H.U. Auster, Observation and modeling of the injection observed by THEMIS and LANL satellites during the 23 March 2007 substorm event. J. Geophys. Res. **114**, A00C18 (2009). doi:10.1029/2008JA013498

W. Lotko, B.U. Sonnerup, R.L. Lysak, Nonsteady boundary layer flow including ionospheric drag and parallel electric fields. J. Geophys. Res. **92**(A8), 8635–8648 (1987). doi:10.1029/JA092iA08p08635

J.Y. Lu, Y.X. Peng, M. Wang, S.J. Gu, M.X. Zhao, Support vector machine combined with distance correlation learning for Dst forecasting during intense geomagnetic storms. Planet. Space Sci. **120**, 48–55 (2016)

H. Lühr, C. Xiong, N. Olsen, G. Le, Near-Earth magnetic field effects of large-scale magnetospheric currents. Space Sci. Rev. **206**, 521–545 (2017). doi:10.1007/s11214-016-0267-y

A.T.Y. Lui, D.C. Hamilton, Radial profiles of quiet time magnetospheric parameters. J. Geophys. Res. **97**, 19325–19332 (1992)

A.T.Y. Lui, R.W. McEntire, S.M. Krimigis, Evolution of the ring current during two geomagnetic storms. J. Geophys. Res. **92**, 7459–7470 (1987)

H. Lundstedt, P. Wintoft, Prediction of geomagnetic storms from solar wind data with the use of a neural network. Ann. Geophys. **12**, 19–24 (1994)

H. Lundstedt, H. Gleisner, P. Wintoft, Operational forecasts of the geomagnetic Dst index. Geophys. Res. Lett. **106**(A6), 1–4 (2001)

H. Lundstedt, H. Gleisner, P. Wintoft, Operational forecasts of the geomagnetic Dst index. Geophys. Res. Lett. **29**, 2181 (2002). doi:10.1029/2002GL016151

J.G. Lyon, J.A. Fedder, C.M. Mobarry, The Lyon-Fedder-Mobarry (LFM) global MHD magnetospheric simulation code. J. Atmos. Sol.-Terr. Phys. **66**, 1333–1350 (2004). doi:10.1016/j.jastp.2004.03.020

Y.P. Maltsev, Points of controversy in the study of magnetic storms. Space Sci. Rev. **110**, 227–277 (2004)

R.A. Marshall, M. Dalzell, C.L. Waters, P. Goldthorpe, E.A. Smith, Geomagnetically induced currents in the New Zealand power network. Space Weather **10**, S08003 (2012). doi:10.1029/2012SW000806

J.-C. Matéo-Vélez, A. Sicard-Piet, D. Lazaro, V. Inguimbert, P. Sarrailh, S. Hess, V. Maget, D. Payan, Severe geostationary environments: Numerical estimation of spacecraft surface charging from flight data. J. Spacecr. Rockets **53**(2), 304–316 (2016). doi:10.2514/1.A33376

B.H. Mauk, N.J. Fox, S.G. Kanekal, R.L. Kessel, D.G. Sibeck, A. Ukhorskiy, Science objectives and rationale for the radiation belt storm probes mission. Space Sci. Rev. **179**, 3–27 (2013). doi:10.1007/s11214-012-9908-y

G.D. Mead, D.H. Fairfield, A quantitative magnetospheric model derived from spacecraft magnetometer data. J. Geophys. Res. **80**, 523–534 (1975)

A.M. Menz, L.M. Kistler, C.G. Mouikis, H.E. Spence, R.M. Skoug, H.O. Funsten, B.A. Larsen, D.G. Mitchell, M. Gkioulidou, The role of convection in the buildup of the ring current pressure during the 17 March 2013 storm. J. Geophys. Res. Space Phys. **122**, 475–492 (2017). doi:10.1002/2016JA023358

N.P. Meredith, R.B. Horne, R.R. Anderson, Substorm dependence of chorus amplitudes: Implications for the acceleration of electrons to relativistic energies. J. Geophys. Res. **106**, 13165–13178 (2001)

N.P. Meredith, R.M. Thorne, R.B. Horne, D. Summers, B.J. Fraser, R.R. Anderson, Statistical analysis of relativistic electron energies for cyclotron resonance with EMIC waves observed on CRRES. J. Geophys. Res. Space Phys. **108**, 1250 (2003). doi:10.1029/2002JA009700

E.V. Mishin, W.J. Burke, Stormtime coupling of the ring current, plasmasphere, and topside ionosphere: Electromagnetic and plasma disturbances. J. Geophys. Res. **110**, A07209 (2005). doi:10.1029/2005JA011021

E. Mishin, E. Sutton, G. Milikh, I. Galkin, C. Roth, M. Förster, F2-region atmospheric gravity waves due to high-power HF heating and subauroral polarization streams. Geophys. Res. Lett. **39**, L11101 (2012). doi:10.1029/2012GL052004

D.G. Mitchell, K.C. Hsieh, C.C. Curtis, D.C. Hamilton, H.D. Voss, E.C. Roelof, P. C:son Brandt, Imaging two geomagnetic storms in energetic neutral atoms. Geophys. Res. Lett. **28**, 1151–1154 (2001)

M.J. Mithaiwala, W. Horton, Substorm injections produce sufficient electron energization to account for MeV flux enhancements following some storms. J. Geophys. Res. **110**, A07224 (2005). doi:10.1029/2004JA010511

Y.S. Miyoshi, V.K. Jordanova, A. Morioka, M.F. Thomsen, G.D. Reeves, D.S. Evans, J.C. Green, Observations and modeling of energetic electron dynamics during the October 2001 storm. J. Geophys. Res. **111**, A11S02 (2006). doi:10.1029/2005JA011351

T.S. Molinski, Why utilities respect geomagnetically induced currents. J. Atmos. Sol.-Terr. Phys. **64**(16), 1765–1778 (2002)

V. Munsami, Determination of the effects of substorms on the storm-time ring current using neural networks. J. Geophys. Res. **105**(A12), 27833–27840 (2000). doi:10.1029/2000JA000041

T. Murayama, Coupling function between the solar wind and the Dst index, in *Solar Wind-Magnetosphere Coupling*, ed. by Y. Kamide, J.A. Slavin (Terra Scientific Publishing Co., Tokyo, 1986), p. 119

C.M. Ngwira, A. Pulkkinen, F.D. Wilder, G. Crowley, Extended study of extreme geoelectric field event scenarios for geomagnetically induced current applications. Space Weather **11**, 121–131 (2013). doi:10.1002/swe.20021

A. Nishida, Formation of plasmapause, or magnetospheric plasma knee, by the combined action of magnetospheric convection and plasma escape from the tail. J. Geophys. Res. **71**, 5669 (1966)

T.P. O'Brien, SEAES-GEO: A spacecraft environmental anomalies expert system for geosynchronous orbit. Space Weather **7**, S09003 (2009). doi:10.1029/2009SW000473

T.P. O'Brien, R.L. McPherron, Forecasting the ring current Dst in real time. J. Atmos. Sol.-Terr. Phys. **62**, 1295–1299 (2000)

S. Ohtani, M. Nos, G. Rostoker, H. Singer, A.T.Y. Lui, M. Nakamura, Storm-substorm relationship: Contribution of the tail current to Dst. J. Geophys. Res. **106**(A10), 21199–21209 (2001). doi:10.1029/2000JA000400

G. Pallocchia, E. Amata, G. Consolini, M.F. Marcucci, I. Bertello, Geomagnetic Dst index forecast based on IMF data only. Ann. Geophys. **24**, 989–999 (2006)

E.N. Parker, H.A. Stewart, Nonlinear inflation of a magnetic dipole. J. Geophys. Res. **72**, 5287 (1967)

B. Pintér, S.D. Thom, R. Balthazor, H. Vo, G.J. Bailey, Modeling subauroral polarization streams equatorward of the plasmapause footprints. J. Geophys. Res. **111**, A10306 (2006). doi:10.1029/2005JA011457

R. Pirjola, Geomagnetically induced currents during magnetic storms. IEEE Trans. Plasma Sci. **28**(6), 1867–1873 (2000)

R. Pirjola, Review on the calculation of the surface electric and magnetic fields and geomagnetically induced currents in ground based technological systems. Surv. Geophys. **23**, 71–90 (2002)

R. Pirjola, Practical model applicable to investigating the coast effect on the geoelectric field in connection with studies of geomagnetically induced currents. Adv. Appl. Phys. **1**(1), 9–28 (2013)

J.L. Posch, M.J. Engebretson, M.T. Murphy, M.H. Denton, M.R. Lessard, R.B. Horne, Probing the relationship between electromagnetic ion cyclotron waves and plasmaspheric plumes near geosynchronous orbit. J. Geophys. Res. Space Phys. **115**, A11205 (2010). doi:10.1029/2010JA015446

K. Powell, P. Roe, T. Linde, A solution-adaptive upwind scheme for ideal magnetohydrodynamics. J. Comput. Phys. **154**, 284–309 (1999). doi:10.1006/jcph.1999.6299

A. Pulkkinen, L. Rastätter, M. Kuznetsova, M. Hesse, A. Ridley, J. Raeder, H.J. Singer, A. Chulaki, Systematic evaluation of ground and geostationary magnetic field predictions generated by global magnetohydrodynamic models. J. Geophys. Res. **115**, A03206 (2010). doi:10.1029/2009JA014537

A. Pulkkinen, M. Kuznetsova, A. Ridley et al., Geospace environment modeling 2008–2009 challenge: Ground magnetic field perturbations. Space Weather **9**, S02004 (2011). doi:10.1029/2010SW000600

A. Pulkkinen, E. Bernabeu, J. Eichner, C. Beggan, A.W.P. Thomson, Generation of 100-year geomagnetically induced current scenarios. Space Weather **10**, S04003 (2012). doi:10.1029/2011SW000750

A. Pulkkinen, L. Rastätter, M. Kuznetsova et al., Community-wide validation of geospace model ground magnetic field perturbation predictions to support model transition to operations. Space Weather **11**, 369–385 (2013). doi:10.1002/swe.20056

J. Raeder, R.L. McPherron, L.A. Frank, S. Kokubun, G. Lu, T. Mukai, W.R. Paterson, J.B. Sigwarth, H.J. Singer, J.A. Slavin, Global simulation of the geospace environment modeling substorm challenge event. J. Geophys. Res. **106**, 381–395 (2001). doi:10.1029/2000JA000605

C.E. Rasmussen, S.M. Guiter, S.G. Thomas, A two-dimensional model of the plasmasphere: Refilling time constants. Planet. Space Sci. **41**, 35 (1993)

L. Rastätter, M.M. Kuznetsova, A. Glocer et al., Geospace environment modeling 2008–2009 challenge: Dst index. Space Weather **11**, 187–205 (2013). doi:10.1002/swe.20036

L. Rastätter, G. Tóth, M.M. Kuznetsova, A.A. Pulkkinen, CalcDeltaB: An efficient postprocessing tool to calculate ground-level magnetic perturbations from global magnetosphere simulations. Space Weather **12**, 553–565 (2014). doi:10.1002/2014SW001083

G.D. Reeves, R.H.W. Friedel, B.A. Larsen, R.M. Skoug, H.O. Funsten, S.G. Claudepierre, J.F. Fennell, D.L. Turner, M.H. Denton, H.E. Spence, J.B. Blake, D.N. Baker, Energy-dependent dynamics of keV to MeV electrons in the inner zone, outer zone, and slot regions. J. Geophys. Res. Space Phys. **121**, 397–412 (2016). doi:10.1002/2015JA021569

A.J. Ridley, M.W. Liemohn, A model-derived description of the penetration electric field. J. Geophys. Res. **107**(A8), 1151 (2002). doi:10.1029/2001JA00051

A.J. Ridley, D.L. De Zeeuw, T.I. Gombosi, K.G. Powell, Using steady state MHD results to predict the global state of the magnetosphere-ionosphere system. J. Geophys. Res. **106**(A12), 30067 (2001). doi:10.1029/2000JA002233

A.J. Ridley, T.I. Gombosi, D.L. De Zeeuw, Ionospheric control of the magnetosphere: Conductance. Ann. Geophys. **22**, 567 (2004)

C.J. Rodger, M.A. Clilverd, A. Seppala, N.R. Thomson, R.J. Gamble, M. Parrot, J.-A. Sauvaud, T. Ulich, Radiation belt electron precipitation due to geomagnetic storms: Significance to middle atmosphere ozone chemistry. J. Geophys. Res. Space Phys. **115**, A11320 (2010). doi:10.1029/2010JA015599

I. Roth, M.K. Hudson, Wave-particle interaction at the plasmasphere-ring current interface. IEEE Trans. Plasma Sci. **20**, 829–832 (1992). doi:10.1109/27.199535

T.E. Sarris, X. Li, N. Tsaggas, N. Paschalidis, Modeling energetic particle injections in dynamic pulse fields with varying propagation speeds. J. Geophys. Res. **107**, 1033 (2002). doi:10.1029/2001JA900166

S. Sazykin, R.A. Wolf, R.W. Spiro, T.I. Gombosi, D.L. De Zeeuw, M.F. Thomsen, Interchange instability in the inner magnetosphere associated with geosynchronous particle flux decreases. Geophys. Res. Lett. **29**(10), 88-1–88-4 (2002). doi:10.1029/2001GL014416

A. Schmidt, Erdmagnetismus, in *Enzyklopeadie der Mathem. Wiss., VI* (1917), pp. 265–396

M. Schulz, L. Lanzerotti, *Particle Diffusion in the Radiation Belts* (Springer, New York, 1974)

R.W. Schunk, L. Gardner, L. Scherliess, L. Zhu, Problems associated with uncertain parameters and missing physics for long-term ionosphere-thermosphere forecasting. Radio Sci. **47**, RS0L23 (2012). doi:10.1029/2011RS004911

N. Sckopke, A general relation between the energy of trapped particles and the disturbance field near the Earth. J. Geophys. Res. **71**, 3125–3130 (1966)

R. Shi, D. Summers, B. Ni, J.W. Manweiler, D.G. Mitchell, L.J. Lanzerotti, A statistical study of proton pitch angle distributions measured by the radiation belt storm probes ion composition experiment. J. Geophys. Res. Space Phys. **121**, 5233–5249 (2016). doi:10.1002/2015JA022140

G.L. Siscoe, W. Lotko, B.U. Sonnerup, A. High-Latitude, Low-latitude boundary layer model of the convection current system. J. Geophys. Res. **96**(A3), 3487–3495 (1991). doi:10.1029/90JA02362

P.H. Smith, R.A. Hoffman, Direct observations in the dusk hours of the characteristics of the storm time ring current particles during the beginning of magnetic storms. J. Geophys. Res. **79**, 966–971 (1974)

J.J. Sojka, M. David, R.W. Schunk, J.C. Foster, H.B. Vo, A modeling study of the F region response to SAPS. J. Atmos. Sol.-Terr. Phys. **66**, 415–423 (2004). doi:10.1016/j.jastp.2003.11.003

M. Spasojević, J. Goldstein, D.L. Carpenter, U.S. Inan, B.R. Sandel, M.B. Moldwin, B.W. Reinisch, Global response of the plasmasphere to a geomagnetic disturbance. J. Geophys. Res. Space Phys. **108**, 1340 (2003). doi:10.1029/2003JA009987

C. Störmer, Sur les trajectories des corpuscles electrises dans l'espace sous l'action du magnetisme terrestre avec application aux aurores boreales. Arch. Sci. Phys. Nat. **24**, 317–364 (1907)

M. Sugiura, Hourly values of the equatorial Dst for IGY, in *Annals of the International Geophysical Year*, vol. 35 (Pergamon, Oxford, Tarrytown, 1964), pp. 945–948

D. Summers, R.M. Thorne, F. Xiao, Relativistic theory of wave-particle resonant diffusion with application to electron acceleration in the magnetosphere. J. Geophys. Res. **103**(A9), 20487–20500 (1998). doi:10.1029/98JA01740

S. Takahashi, T. Iyemori, M. Takeda, A simulation of the storm-time ring current. Planet. Space Sci. **38**, 1133–1141 (1990)

M. Temerin, X. Li, A new model for the prediction of Dst on the basis of the solar wind. J. Geophys. Res. **107**, 1472 (2002). doi:10.1029/2001JA007532

M. Temerin, X. Li, Dst model for 1995–2002. J. Geophys. Res. **111**, A04221 (2006)
M.F. Thomsen, J.E. Borovsky, D.J. McComas, M.R. Collier, Variability of the ring current source population. Geophys. Res. Lett. **25**, 3481–3484 (1998)
M.F. Thomsen, H. Korth, R.C. Elphic, Upper cutoff energy of the electron plasma sheet as a measure of magnetospheric convection strength. J. Geophys. Res. **107**, 1331 (2002). doi:10.1029/2001JA000148
R.M. Thorne, Radiation belt dynamics: The importance of wave-particle interactions. Geophys. Res. Lett. **37**, L22107 (2010). doi:10.1029/2010GL044990
R.M. Thorne, E.J. Smith, R.K. Burton, R.E. Holzer, Plasmaspheric hiss. J. Geophys. Res. **78**, 1581–1596 (1973). doi:10.1029/JA078i010p01581
F.R. Toffoletto, R.W. Spiro, R.A. Wolf, J. Birn, M. Hesse, Modeling inner magnetospheric electrodynamics, in *Space Weather*, ed. by P. Song, H.J. Singer, G.L. Siscoe (Am. Geophys. Union, Washington, 2001). doi:10.1029/GM125p0265
F. Toffoletto, S. Sazykin, R. Spiro, R. Wolf, Inner magnetospheric modeling with the rice convection model. Space Sci. Rev. **107**, 175–196 (2003)
G. Tóth, I.V. Sokolov, T.I. Gombosi et al., Space weather modeling framework: A new tool for the space science community. J. Geophys. Res. **110**(A12), A12226 (2005). doi:10.1029/2005JA011126
G. Tóth, B. van der Holst, I.V. Sokolov et al., Adaptive numerical algorithms in space weather modeling. J. Comput. Phys. **231**(3), 870–903 (2012). doi:10.1016/j.jcp.2011.02.006
N.B. Trivedi et al., Geomagnetically induced currents in an electric power transmission system at low latitudes in Brazil: A case study. Space Weather **5**, S04004 (2007). doi:10.1029/2006SW000282
B.T. Tsurutani, E.J. Smith, Postmidnight chorus: A substorm phenomenon. J. Geophys. Res. **79**, 118–127 (1974). doi:10.1029/JA079i001p00118
N.A. Tsyganenko, Modeling the Earth's magnetospheric magnetic field confined within a realistic magnetopause. J. Geophys. Res. **100**, 5599–5612 (1995)
N.A. Tsyganenko, A model of the near magnetosphere with a dawn-dusk asymmetry 1. Mathematical structure. J. Geophys. Res. **107**, 12-1–12-15 (2002a). doi:10.1029/2001JA000219
N.A. Tsyganenko, A model of the near magnetosphere with a dawn-dusk asymmetry 2. Parameterization and fitting to observations. J. Geophys. Res. **107**, 10-1–10-17 (2002b). doi:10.1029/2001JA000220
N.A. Tsyganenko, Data-based modelling of the Earth's dynamic magnetosphere: A review. Ann. Geophys. **31**, 1745–1772 (2013). doi:10.5194/angeo-31-1745-2013
N.A. Tsyganenko, V.A. Andreeva, A forecasting model of the magnetosphere driven by an optimal solar wind coupling function. J. Geophys. Res. Space Phys. **120**, 8401–8425 (2015). doi:10.1002/2015JA021641
N.A. Tsyganenko, V.A. Andreeva, An empirical RBF model of the magnetosphere parameterized by interplanetary and ground-based drivers. J. Geophys. Res. Space Phys. **121**, 10786–10802 (2016). doi:10.1002/2016JA023217
N.A. Tsyganenko, T. Mukai, Tail plasma sheet models derived from Geotail particle data. J. Geophys. Res. **108**(A3), 1136 (2003). doi:10.1029/2002JA009707
N.A. Tsyganenko, M.I. Sitnov, Modeling the dynamics of the inner magnetosphere during strong geomagnetic storms. J. Geophys. Res. **110**, A03208 (2005). doi:10.1029/2004JA010798
N.A. Tsyganenko, M.I. Sitnov, Magnetospheric configurations from a high-resolution data-based magnetic field model. J. Geophys. Res. **112**, A06225 (2007). doi:10.1029/2007JA012260
N.A. Tsyganenko, H.J. Singer, J.C. Kasper, Storm-time distortion of the inner magnetosphere: How severe can it get? J. Geophys. Res. **108**(A5), 1209 (2003). doi:10.1029/2002JA009808
D.L. Turner, V. Angelopoulos, S.K. Morley, M.G. Henderson, G.D. Reeves, W. Li, D.N. Baker, C.-L. Huang, A. Boyd, H.E. Spence, S.G. Claudepierre, J.B. Blake, J.V. Rodriguez, On the cause and extent of outer radiation belt losses during the 30 September 2012 dropout event. J. Geophys. Res. Space Phys. **119**, 1530–1540 (2014). doi:10.1002/2013JA019446
D.L. Turner, S.G. Claudepierre, J.F. Fennell et al., Energetic electron injections deep into the inner magnetosphere associated with substorm activity. Geophys. Res. Lett. **42**, 2079–2087 (2015). doi:10.1002/2015GL063225
A.Y. Ukhorskiy, B.J. Anderson, P.C. Brandt, N.A. Tsyganenko, Storm time evolution of the outer radiation belt: Transport and losses. J. Geophys. Res. Space Phys. **111**, A11S03 (2006). doi:10.1029/2006JA011690
A.Y. Ukhorskiy, M.I. Sitnov, R.M. Millan, B.T. Kress, J.F. Fennell, S.G. Claudepierre, R.J. Barnes, Global storm time depletion of the outer electron belt. J. Geophys. Res. Space Phys. **120**, 2543–2556 (2015). doi:10.1002/2014JA020645
M.E. Usanova, A. Drozdov, K. Orlova, I.R. Mann, Y. Shprits, M.T. Robertson, D.L. Turner, D.K. Milling, A. Kale, D.N. Baker, S.A. Thaller, G.D. Reeves, H.E. Spence, C. Kletzing, J. Wygant, Effect of EMIC waves on relativistic and ultra-relativistic electron populations: Ground-based and Van Allen Probes observations. Geophys. Res. Lett. **41**, 1375–1381 (2014). doi:10.1002/2013GL059024

I.I. Vakulin, O.S. Grafodatskii, V.I. Degtiarev, V.I. Dovgii, G.A. Zherebtsov, The radiation environment of the geostationary orbit for magnetically quiet conditions according to data from the Raduga communication satellites. Space Res. **26**, 120–126 (1988)
C. Vallat, I. Dandouras, M. Dunlop, A. Balogh, E. Lucek, G.K. Parks, M. Wilber, E.C. Roelof, G. Chanteur, H. Reme, First current density measurements in the ring current region using simultaneous multi-spacecraft CLUSTER-FGM data. Ann. Geophys. **23**, 1849–1865 (2005). doi:10.5194/angeo-23-1849-2005
D. Vassiliadis, A.J. Klimas, J.A. Valdivia, D.N. Baker, The Dst geomagnetic response as function of storm phase and amplitude and the solar wind electric field. J. Geophys. Res. **104**(A11), 24957–24976 (1999)
V.M. Vasyliunas, Mathematical models of magnetospheric convection and its coupling to the ionosphere, in *Particles and Fields in the Magnetosphere*, ed. by B.M. McCormac (D. Reidel, New York, 1970)
J.I. Vette, The AE-8 trapped electron model environment, National Space Science Data Center (NSSDC), World Data Center A for Rockets and Satellites (WDC-A-RS), National Aeronautics and Space Administration, Goddard Space Flight Center (1991)
A. Viljanen, A. Pulkkinen, R. Pirjola, K. Pajunpää, P. Posio, A. Koistinen, Recordings of geomagnetically induced currents and a nowcasting service of the Finnish natural gas pipeline. Space Weather **4**, S10004 (2006). doi:10.1029/2006SW000234
C.B. Wang, J.K. Chao, C.H. Lin, Influence of the solar wind dynamic pressure on the decay and injection of the ring current. J. Geophys. Res. **108**, 1341 (2003). doi:10.1029/2003JA009851
H. Wang, A.J. Ridley, H. Lühr, M.W. Liemohn, S.Y. Ma, Statistical study of the subauroral polarization stream: Its dependence on the cross–polar cap potential and subauroral conductance. J. Geophys. Res. **113**, A12311 (2008). doi:10.1029/2008JA013529
H. Wang, H. Lühr, K. Häusler, P. Ritter, Effect of subauroral polarization streams on the thermosphere: A statistical study. J. Geophys. Res. **116**, A03312 (2011). doi:10.1029/2010JA016236
W. Wang, E.R. Talaat, A.G. Burns, B. Emery, S. Hsieh, J. Lei, J. Xu, Thermosphere and ionosphere response to subauroral polarization streams (SAPS): Model simulations. J. Geophys. Res. **117**, A07301 (2012). doi:10.1029/2012JA017656
S. Watanabe, E. Sagawa, K. Ohtaka, H. Shimazu, Prediction of the Dst index from solar wind parameters by a neural network method. Earth Planets Space **54**, 1263–1275 (2002)
S. Watari, Estimation of geomagnetically induced currents based on the measurement data of a transformer in a Japanese power network and geoelectric field observations. Earth Planets Space **67**, 77 (2015). doi:10.1186/s40623-015-0253-8
D.R. Weimer, Improved ionospheric electrodynamic models and application to calculating Joule heating rates. J. Geophys. Res. **110**(A5), A05306 (2005). doi:10.1029/2004JA010884
D. Wen, Y. Yuan, J. Ou, X. Huo, K. Zhang, Ionospheric temporal and spatial variations during the 18 August 2003 storm over China. Earth Planets Space **59**(4), 313 (2007)
H.I. West, R.M. Buck, J.R. Walton, Shadowing of electron azimuthal-drift motions near the noon magnetopause. Nat. Phys. Sci. **240**, 6–7 (1972). doi:10.1038/physci240006a0
E.C. Whipple, Potentials of surfaces in space. Rep. Prog. Phys. **44**, 1197 (1981). doi:10.1088/0034-4885/44/11/002
M. Wik, R. Pirjola, H. Lundstedt, A. Viljanen, P. Wintoft, A. Pulkkinen, Space weather events in July 1982 and October 2003 and the effects of geomagnetically induced currents on Swedish technical systems. Ann. Geophys. **27**, 1775–1787 (2009)
M. Wiltberger, W. Wang, A.G. Burns, S.C. Solomon, J.G. Lyon, C.C. Goodrich, Initial results from the coupled magnetosphere ionosphere thermosphere model: Magnetospheric and ionospheric responses. J. Atmos. Sol.-Terr. Phys. **66**, 1411–1423 (2004). doi:10.1016/j.jastp.2004.03.026
R.A. Wolf, Effects of ionospheric conductivity on convective flow of plasma in the magnetosphere. J. Geophys. Res. **75**(25), 4677–4698 (1970)
R.A. Wolf, R.W. Spiro, S. Sazykin, F.R. Toffoletto, How the Earth's inner magnetosphere works: An evolving picture. J. Atmos. Sol.-Terr. Phys. **69**, 288–302 (2007). doi:10.1016/j.jastp.2006.07.026
H.-C. Yeh, J.C. Foster, F.J. Rich, W. Swider, Storm time electric field penetration observed at mid-latitude. J. Geophys. Res. **96**, 5707–5721 (1991). doi:10.1029/90JA02751
P. Yin, C.N. Mitchell, P.S.J. Spencer, J.C. Foster, Ionospheric electron concentration imaging using GPS over the USA during the storm of July 2000. Geophys. Res. Lett. **31**, L12806 (2004). doi:10.1029/2004GL019899
E. Yizengaw, M.B. Moldwin, P.L. Dyson, E.A. Essex, Using tomography of GPS TEC to routinely determine ionospheric average electron density profiles. J. Atmos. Sol.-Terr. Phys. **69**(3), 314 (2007)
D.T. Young, S. Perraut, A. Roux, C. de Villedary, R. Gendrin, A. Korth, G. Kremser, D. Jones, Wave-particle interactions near VHe1 observed on GEOS 1 and 2, 1, propagation of ion cyclotron waves in He1-rich plasma. J. Geophys. Res. **86**, 6755 (1981)

Y. Yu, V. Jordanova, S. Zaharia, J. Koller, J. Zhang, L.M. Kistler, Validation study of the magnetically self-consistent inner magnetosphere model RAM-SCB. J. Geophys. Res. **117**, A03222 (2011). doi:10.1029/2011JA017321

Y. Yu, V. Jordanova, D. Welling, B. Larsen, S.G. Claudepierre, C. Kletzing, The role of ring current particle injections: Global simulations and Van Allen Probes observations during 17 March 2013 storm. Geophys. Res. Lett. **41**, 1126–1132 (2014). doi:10.1002/2014GL059322

Y. Yu, V. Jordanova, S. Zou, R. Heelis, M. Ruohoniemi, J. Wygant, Modeling subauroral polarization streams during the 17 March 2013 storm. J. Geophys. Res. Space Phys. **120**, 1738–1750 (2015). doi:10.1002/2014JA020371

Z. Yuan, Y. Xiong, Z. Qiao, H. Li, S. Huang, D. Wang, X. Deng, T. Raita, J. Wang, A subauroral polarization stream driven by field-aligned currents associated with precipitating energetic ions caused by EMIC waves: A case study. J. Geophys. Res. Space Phys. **121**, 1696–1705 (2016). doi:10.1002/2015JA021804

S. Zaharia, C.Z. Cheng, J.R. Johnson, Particle transport and energization associated with substorms. J. Geophys. Res. **105**(A8), 18741–18752 (2000). doi:10.1029/1999JA000407

J.J. Zhang, C. Wang, T.R. Sun, C.M. Liu, K.R. Wang, GIC due to storm sudden commencement in low-latitude high-voltage power network in China: Observation and simulation. Space Weather **13**, 643–655 (2015). doi:10.1002/2015SW001263

H. Zhao, X. Li, D.N. Baker et al., The evolution of ring current ion energy density and energy content during geomagnetic storms based on Van Allen Probes measurements. J. Geophys. Res. Space Phys. **120**, 7493–7511 (2015). doi:10.1002/2015JA021533

H. Zhao, X. Li, D.N. Baker et al., Ring current electron dynamics during geomagnetic storms based on the Van Allen Probes measurements. J. Geophys. Res. Space Phys. **121**, 3333–3346 (2016). doi:10.1002/2016JA022358

Space Sci Rev (2017) 212:1041–1067
DOI 10.1007/s11214-017-0379-z

Locations Where Space Weather Energy Impacts the Atmosphere

Jan J. Sojka[1]

Received: 23 November 2016 / Accepted: 10 May 2017 / Published online: 10 July 2017

Abstract In this review we consider aspects of space weather that can have a severe impact on the terrestrial atmosphere. We begin by identifying the pre-conditioning role of the Sun on the temperature and density of the upper atmosphere. This effect we define as "space climatology". Space weather effects are then defined as severe departures from this state of the atmospheric energy and density. Three specific forms of space weather are reviewed and we show that each generates severe space weather impacts. The three forms of space weather being considered are the solar photon flux (flares), particle precipitation (aurora), and electromagnetic Joule heating (magnetosphere–ionospheric (M-I) coupling).

We provide an overview of the physical processes associated with each of these space weather forms. In each case a very specific altitude range exists over which the processes can most effectively impact the atmosphere. Our argument is that a severe change in the local atmosphere's state leads to atmospheric heating and other dynamic changes at locations beyond the input heat source region. All three space weather forms have their greatest atmospheric impact between 100 and 130 km. This altitude region comprises the transition between the atmosphere's mesosphere and thermosphere and is the ionosphere's E-region. This region is commonly referred to as the Space Atmosphere Interaction Region (SAIR). The SAIR also acts to insulate the lower atmosphere from the space weather impact of energy deposition. A similar space weather zone would be present in atmospheres of other planets and exoplanets.

Keywords Space weather · Atmospheric temperature · Flares · Aurora · Joule heating · Energy deposition

The Scientific Foundation of Space Weather
Edited by Rudolf von Steiger, Daniel Baker, André Balogh, Tamás Gombosi, Hannu Koskinen and Astrid Veronig

✉ J.J. Sojka
jan.sojka@usu.edu

[1] Center for Atmospheric and Space Sciences, Utah State University, 4405 Old Main Hill, Logan, UT 84322-4405, USA

1 Introduction

The terrestrial atmosphere, especially its upper regions, is thermally controlled by the solar irradiance. Under quiescent solar conditions the neutral temperature dependence specifically the exospheric temperature is well understood. This dependence can be described by three distinct cycles, (i) the day-night 24-hour cycle; (ii) the terrestrial seasonal cycle, winter-summer 1-year cycle; and (iii) the solar minimum-maximum 11-year cycle. With 60 years of observation during the space era the upper atmosphere's dependence on all three cycles has been well established. In Sect. 2 we will review this dependence.

Our objectives for this review are to address the ways in which space weather processes can significantly impact the atmosphere and to locate the altitudes, latitudes, and longitudes where this could happen. A significant short-term local energy deposition comparable to the atmospheric local energy density would result in hydrodynamic processes that would in turn, generate density, composition, and wind changes. We also distinguish between impulsive processes and sustained long-term increases in the solar irradiance. A number of studies of the response to increases in solar irradiances in the 1 to 120 nm spectral band by up to a factor of 10 have been carried out (Smithtro and Sojka 2005a, 2005b; Tian et al. 2008a, 2008b). These results show how the upper atmosphere attains new equilibrium with potentially high exospheric temperatures leading to severe Jean's atmospheric escape conditions. In this review, we consider which space weather processes can in short time periods i.e., substorm time scale, lead to local energy deposition that will drive a hydrodynamic atmospheric response.

The space weather energy deposition rates for three mechanisms have been studied and global averages have been obtained. Table 1 is a representative example taken from Knipp (2011). These energy deposition rates are based on many different observations and hence represent regional averages or peak estimates. Table 1 lists the three main mechanisms as electromagnetic energy (Poynting flux), particle precipitation (auroral), and photon (solar flares). The upper value of space weather power for each mechanism is listed in the table. The electromagnetic mechanism is dominant and through Joule heating provides a local thermalization process that would raise atmospheric temperature among other dynamic responses. Burke (2008) considered the details of the solar wind-magnetosphere-ionospheric electromagnetic energy flow. Overall the information in Table 1 is representative of how space weather processes deliver energy into the atmosphere. Our review will attempt to identify the atmospheric altitude location where this energy deposition is locally significant enough to cause a local hydrodynamic atmospheric response. These atmospheric changes

Table 1 Space weather energy fluxes at top of the atmosphere*

Source	Energy flux (W/m^2)	Latitude/Longitude
Solar radiation		
x-ray–EUV (5 nm–120 nm)	0.0032	Dayside (solar zenith dependent)
Particles		
Magnetospheric electrons	0.03	High latitude (auroral region)
Magnetospheric protons	0.006	High latitudes
Magnetospheric Poynting Flux	0.15	Polar and auroral regions (mid latitudes during storms)

*Knipp (2011).

adversely impact mankind's space infrastructures, i.e., satellite orbit changes, degradation of HF radio communications as well as accuracy and availability of GPS geo-location information.

Independently the three processes are well understood as they have been studied for well over 50 years. Hence our citations are in two forms, the first a series of graduate level texts and the second specific papers that clearly present aspects of the impact of space weather on atmospheric heating despite being several decades old. The most directly related text is Delores Knipp's book "Understanding Space Weather and the Physics Behind it" (Knipp 2011). This book provides information and citations for these three space weather processes. Prior research has established the energy flow into the upper atmosphere as shown in Table 1. We will establish the altitudes over which this energy flow leads to an energy density contribution comparable to the atmosphere's thermal energy content. Additional texts used were by Bauer (1973), Jones (1974), Brasseur and Solomon (1984), Kivelson and Russell (1995), Kintner et al. (2008), Schunk and Nagy (2009), Bosinger et al. (2013), and Khazanov (2016). The first two chapters of Bosinger et al. (2013) written by Smith (2012) and Pfaff (2012) respectively provide extensive descriptions of the upper atmosphere and near Earth plasma environment that establish the background for our analysis. In Sect. 2, we describe the upper atmosphere's climatology dependence on the Sun. Section 3 has three subsections that describe the severe impacts of space weather flares, the aurora, and the impacts of electromagnetic heating on the upper atmosphere. Section 4 identifies present-day shortfalls in knowledge concerning these space weather processes. Section 5 describes a pioneering effort to determine the SAIR neutral temperature response to storms. Finally in Sect. 6 the science conclusions are listed along with suggested observations that would be needed to make significant progress in understanding the upper atmosphere's response to severe space weather. A brief statement on how this space atmosphere interaction region (SAIR) is expected to be present both in other planetary atmospheres and those of the exoplanets.

2 The Atmosphere's Response to Space Climate Conditions

Our knowledge of the quiescent state of the atmosphere with respect to solar quiet conditions is excellent. Both ground-based Incoherent Scatter Radar (ISR) and satellite in-situ measurements as well as remote sensing measurements have enabled the compilation of very complete data bases of the atmosphere's temperature, density, and composition. For over four decades, a series of ever more comprehensive empirical atmospheric models has been created from these data bases (Jacchia 1970; Hedin et al. 1977 (MSIS); Hedin 1987; Hedin 1991 (MSIS-90); Picone et al. 2002 (NRLMSISE-00)). Our theoretical understanding of the atmosphere under these solar quiescent conditions is also very comprehensive and is demonstrated by computational models that solve the first principles physics of the upper atmosphere (Roble et al. 1988 (TIGCM); Fuller-Rowell and Rees 1980; Fuller-Rowell et al. 1996 (CTIM); Roble 1995 (TIEGCM); Ma and Schunk 1995 (GTM); Fuller-Rowell et al. 2010 (WAM); Gardner and Schunk 2011 (GTM)). These models are all global in nature and couple to/with the embedded ionosphere. For solar cycle seasonal and local time dependences, these models reproduce the empirical observations well. Roble et al. (1987) and later Smithtro and Sojka (2005a) used a one-dimensional global average upper atmospheric model to study the atmospheric sensitivity to extremes in the quiescent solar climate such as under Maunder Minimum conditions. These studies are considered to be space climate studies and are not included in our review of space weather impacts that are comparatively short-lived events.

Fig. 1 Atmospheric temperature from 50 km to 450 km for four specific conditions defined in Table 2 obtained from NRLMSISE00

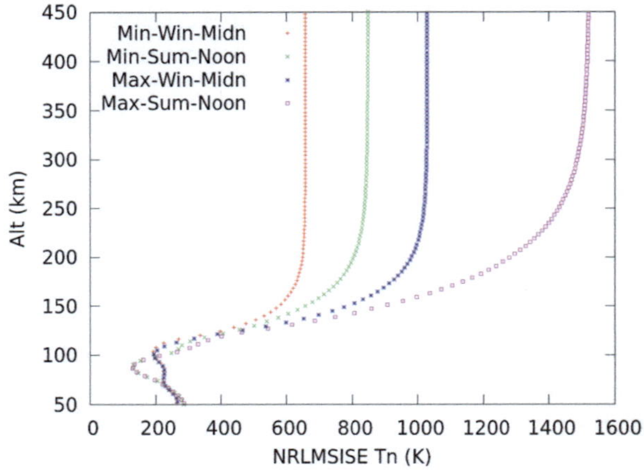

Table 2 Solar cycle, season, and LT parameters

Latitude (Deg)	Longitude (Deg)	Local time	F10.7	Season	Color key*	Symbols*
80	0	Midnight	70.0	Winter	Red	Plus
10	0	Noon	70.0	Summer	Green	Cross
80	0	Midnight	250.0	Winter	Blue	Star
10	0	Noon	250.0	Summer	Mauve	Square

*Plot colors and symbols used in Figs. 2, 3, 4, 8, 12, and 16

Fig. 2 Atmospheric $[N_2]$ density from 50 km to 450 km for the four conditions shown in Fig. 1 and defined in Table 2. The density was obtained from NRLMSISE-00 and plotted as a logarithm base10

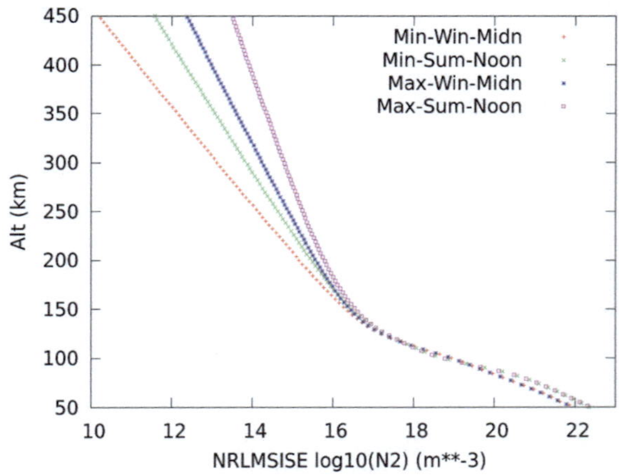

Figure 1 shows four examples of the atmospheric temperature from 50 to 450 km altitude. Each represents a different combination of solar cycle, seasonal, and local time conditions. Table 2 provides the parameters that specify these climate conditions. The solar cycle minimum and maximum are defined by the solar radio flux 10.7 cm index values of 70 and 250; the seasonal cycle contrasts summer and winter solstice, and noon and midnight local time represent the diurnal cycle. The Picone et al. (2002) NRLMSISE-00 empirical model has been used to generate the temperatures shown in Fig. 1 and also the data in Figs. 2 and 3. Figure 1 shows that the EUV space climate's control of the Earth's at-

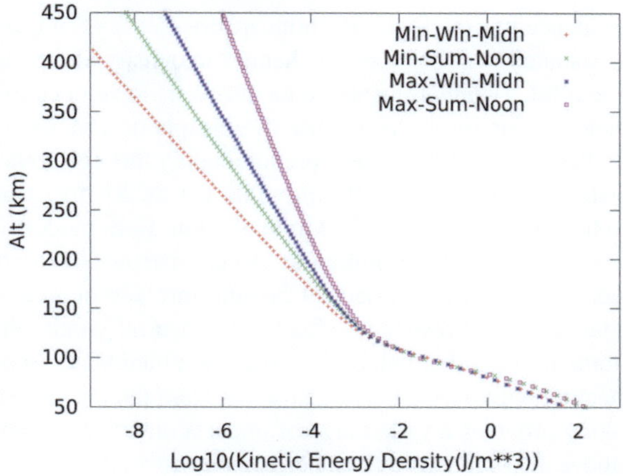

Fig. 3 Computed atmospheric kinetic energy density for the four specific conditions defined in Table 2. The energy density is the sum for the molecules $[N_2]$ and $[O_2]$ and given in logarithm base10 of J/m^3

mosphere's exospheric temperature ranges from as low as 600 K to almost 1600 K. The lower temperatures are found at solar minimum at midnight in winter in the polar cap while the higher values are found near the subsolar point at solar maximum around noon. The transition to exospheric temperature occurs above 200 km at solar minimum and above 300 km at solar maximum. Below 100 km these space climate variations have a negligible effect and a transition region exists between 100 and 200/300 km. This transition region also corresponds to the transition between a well-mixed lower atmosphere and a diffusive atmosphere above. The exception being around the mesopause, 80 to 100 km, where the summer is almost 50K colder than winter. This is due to a well understood global atmospheric circulation characterized by interhemispheric mesospheric flows (McLandress 1998; Fritts and Alexander 2003). Figure 2 shows the $[N_2]$ density corresponding to the four conditions shown in Fig. 1. Again above 150 km a significant solar cycle, seasonal and local time dependence is present. The simple expectation that as the atmosphere has become diffusive at higher altitudes, then a colder atmosphere (Fig. 1) leads to a corresponding lower density (Fig. 2) is confirmed. Below 100 km the summer density slightly exceeds the winter density.

For the objectives of this review, the important question is to identify under what conditions "space weather" deposits an energy density comparable to the local energy density and locate the altitudes at which this occurs. Figures 1 and 2 show that the quiescent solar cycle already generates a significant upper atmosphere (thermosphere) dependence which is associated with the solar cycle variability in the EUV irradiance. Considering space weather to be specifically solar flares, auroral precipitation, and electromagnetic Joule heating, our analysis reviews where in altitude these processes will deposit energy that is comparable to or larger than the quiescent atmospheric thermal energy density. Figure 3 shows results of a calculation of the thermal kinetic energy density for the same four conditions defined in Table 2. Using the temperatures shown in Fig. 1, a root mean thermal speed was calculated for both $[N_2]$ and $[O_2]$ and a kinetic energy for each of these molecules. Using densities for $[N_2]$ in Fig. 2 and from NRLMSISE-00 for $[O_2]$, an energy density was calculated and the sum of these two is plotted in Fig. 3. The solar cycle, seasonal, and LT trends in Fig. 3 are consistent with the temperature and density dependences. In subsequent sections we will compare the three space weather energy deposition mechanisms with these data (Fig. 3).

The space climate introduces systematic changes in the thermosphere above about 100 km, as shown in Figs. 1, 2, and 3. The magnitude of these changes increases with increasing altitude above 100 km. During severe geomagnetic storms, the thermospheric changes are evidenced primarily by the atmospheric density's impact on satellite orbits via drag effects associated with unexpected changes in density along the orbit or cross drifts orthogonal to the orbit. Two thermospheric models were developed to study how strong localized electric fields in the ionosphere cause thermospheric changes (St. Maurice and Schunk 1981 and Fuller-Rowell 1984, 1985) predominantly through Joule heating. The Fuller-Rowell model extended from 70 to 420 km while the St. Maurice and Schunk emphasized the thermosphere spanning from 120 km to 600 km. Both models were 2-D, latitude and altitude with sufficient latitude resolution to capture narrow electric field structures. Both techniques incorporated a wide variety of heating rate and cooling rate terms while emphasizing how hydrodynamic responses affected the neutral winds. Energy deposition was primarily via Joule heating although both studies identified how viscous processes led to the extension of heating both horizontal and upwards from the original source. The St. Maurice and Schunk study provides a wide range of impacts on neutral wind dynamics as a function of altitude above the source as well as horizontally beyond the source's horizontal width. Fuller-Rowell 1984 developed his model, tested the thermospheric response to the polar cap dark ionospheric conditions and showed a minor impact. The Fuller-Rowell 1985 study used auroral ionospheric conditions and produced wind responses similar to those of the St. Maurice and Schunk study. During the initial stages of the electric field event, his study showed gravity waves being launched. Though the auroral conditions were not of a severe space weather event, the results did show how thermospheric temperature, composition, and by inference neutral density were

changed above and beyond the source's horizontal width. With respect to Figs. 1, 2, and 3, these auroral-electric field simulations placed the source region around 120 km. This is the altitude above which the different climate cycles introduce solar control via changes in the EUV irradiance. Both thermospheric models were designed to study equilibrium conditions, i.e., run events for several hours and simulate how the thermosphere comes to steady-state with the imposed driver. Severe space weather events, flares, auroral, and Poynting flux events are all temporally dynamic or have moved in latitude significantly on the scale of an hour. Therefore the thermosphere would never reach equilibrium/steady state. These studies do however provide theoretical underpinning that localized energy deposition, in these examples, Joule heating, results in significant thermospheric dynamic responses in wind and to a lesser extent in temperature and density to altitudes well above the source region.

3 The Atmosphere's Response to Severe Space Weather

3.1 Solar Flare Energy Deposition

The quiescent solar irradiance described in the prior section is also the conveyor of severe space weather in the form of solar flares. This photon space weather impact is very wavelength dependent. Smith and Gottlieb (1974) presented example spectra of the solar irradiance from hard X-rays (0.3 nm) to wavelengths longer than 10^5 nm. Figure 4, taken from Smith and Gottlieb, shows that for wavelengths longward of Lyman α (121.6 nm) the irradiance is insensitive to space weather. However the historic Carrington white light flare observation (Carrington 1860) is the famous example that contradicts this generalization. The strongest flare dependence is at the shortest wavelengths, and almost all the flare energy enhancement occurs at wavelengths shorter than 10 nm. The example of a flare spectrum shown in Fig. 4 has an increase of 4 orders of magnitude at 1 nm and one order of magnitude at 10 nm. Figure 5 shows the atmospheric penetration depth (unitary optical depth) for

Fig. 4 Solar irradiance spectra and flare response over the 0.3 to 3×10^5 nm wavelength range (from Smith and Gottlieb 1974)

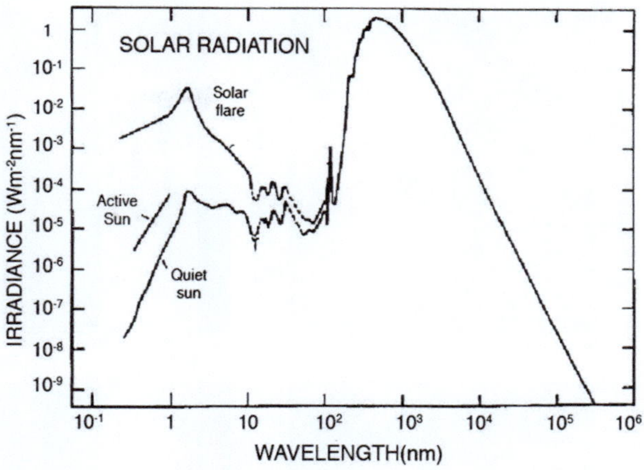

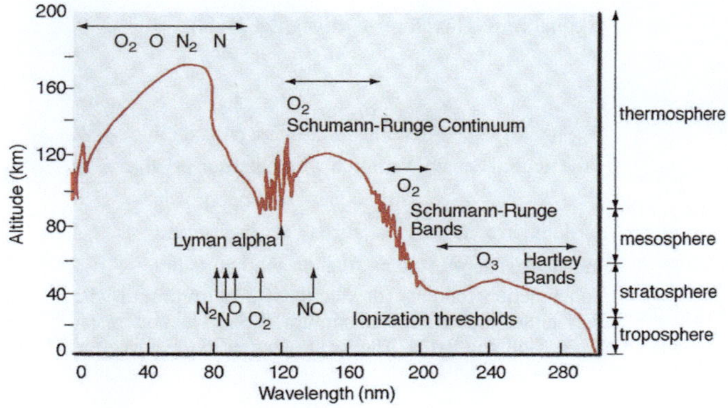

Fig. 5 Solar irradiance atmospheric penetration depth, unitary optical depth, for photons from hard X-rays to 300 nm (from Chamberlain 1978)

solar photons from hard X-rays to 300 nm. Figure 5 is taken from Chamberlain (1978) and is still valid as shown by consistency with that of Haigh (2004). The 1 to 10 nm photons penetrate to 100 km as do photons of wavelengths greater than 180 nm. For the purposes of identifying the solar flare impact on the atmospheric energy budget, the emphasis will lie with the less than 10 nm photons since their irradiances increase by orders of magnitude during strong flares. Short wavelength photons are also more energetic and are therefore capable of delivering more energy. The 1 to 10 nm photons have corresponding energies of 1240 to 124 eV. These produce energetic photoelectrons that in turn generate a cascade of secondary electrons during the thermalization process.

That the upper atmosphere, specifically the temperature and density, responds to strong solar flares was clearly shown by Sutton et al. (2006). Using a suite of solar irradiance instruments (SEE on TIMED, GOES-12) and atmospheric density accelerometers on GRACE and CHAMP satellites, the atmospheric response to the strong X17.2 and X17.3 Halloween Storm flares (at 400 and 490 km) (Lopez et al. 2004) was observed. Figure 6, taken from Sutton et al. (2006), shows the atmospheric density at 490 km (left panels observed from GRACE) and at 400 km (right panels observed from CHAMP). The upper panels represent the sunlit hemisphere while the lower panels represent the dark hemisphere. Neutral densi-

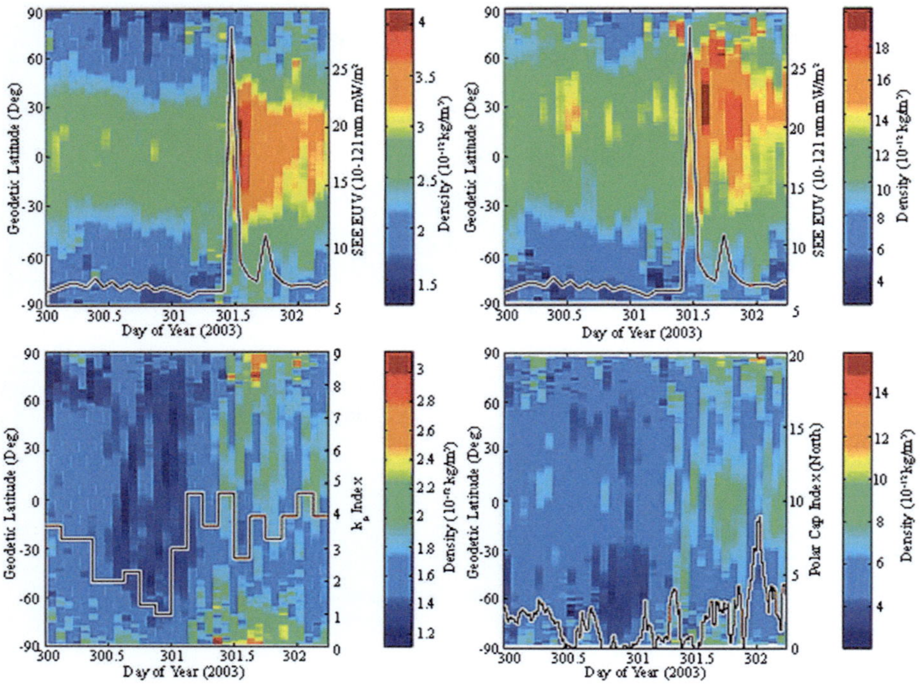

Fig. 6 Flare response of the atmosphere shown by increases in atmospheric density when a strong X17.2 flare occurred. *Upper panels* are dayside observations and *lower panels* are night side observations. The *left panels* are at 490 km altitude observed by the GRACE satellite while the *right panels* are at 400 km observed by the CHAMP satellite. The time history of the TIMED SEE EUV flare response is shown as a *black line* in the *upper two panels* (from Sutton et al. 2006)

ties are shown from Day 300 into Day 302 with the X17.2 flare occurring on 28 October (Day 301) at 11:10 UT. The upper two panels also show the integrated irradiance between 10 and 121 nm as the white/black line. This line shows that the wavelength band (EUV) increased by a factor of almost 5 at the peak of the storm. In contrast the GOES-12 hard X-ray (0.1 to 0.8 nm) increased by almost 3 orders of magnitude (Sutton et al. 2006). In Fig. 6 on the dayside at latitudes around the equator (top left panel), the neutral density at 490 km increases by about 25% for several orbits (hours) with a peak increase of almost 50% within about 30 to 120 minutes of the flare peak. At 400 km (right top panel), the density enhancements are similar or slightly larger by up to 65% but are observed at a somewhat higher latitude; an effect associated with the satellites having different orbital local times. Sutton et al. (2006) also show that the maximum change in the neutral atmosphere's exospheric temperature was 125K at 490 km and 175 km at 400 km. Note that in Fig. 6 both lower panels that are associated with density measurements in the dark hemisphere do not show a direct flare response.

We now consider the altitude where the flare energy deposition is most effective. The focus is on the soft X-rays 1 to 10 nm. In Fig. 5 this solar irradiance reaches below 120 km and even below 100 km. Very few spectrally resolved observations of the 1 to 10 nm (1.240 to 0.124 keV) solar irradiance have been made. Recently Caspi et al. (2015) described a new soft X-ray irradiance instrument whose spectral range is 0.5 to 5 keV. They point out the present day problem, "There is a crucial observational gap between ~0.2 and ~4 keV, outside the ranges of existing spectrometers." Although their new instrument covers part of this range, it only reaches a wavelength shortward of about 3 nm which is the key Auger cut-off wavelengths. Currently, integral measurements are available from the NASA SDO, EVE instrument that has a 0.1 to 7 nm channel. Unfortunately as Fig. 4 shows, the spectral response to large flares can differ by several orders of magnitude over this wavelength range. A further complication is that over the same wavelength range, the ionization cross section for the main atmospheric constituents N_2 and O_2 also varies by over an order of magnitude. Additionally, the Auger process energy cutoffs for oxygen (2.3 nm) and nitrogen (2.9 nm) lie in this spectral band (Fennelly and Torr 1992; Henke et al. 1993).

As modeling of the soft X-ray irradiance based on solar physics can be used as a proxy for the missing spectral information, a description of this procedure follows. Both the continuum and line emission processes of the Sun are reasonably well understood and models such as CHIANTI (Dere et al. 1997; Del Zanna et al. 2015) can simulate these emission processes given certain information about temperature and ionic content of the Sun's atmosphere. This information is available from line emission analysis at longer wavelengths where adequate spectral resolution is available. Warren (Private communication 2015) has used this information in the Differential Emission Measure (DEM) model (Cheung et al. 2015; Aschwanden et al. 2015) to reconstruct the irradiance from 0.1 to 7 nm with very high wavelength resolutions (0.01 nm) for four X5-class flares that were observed by SDO and other satellites in May 2013.

In order to determine where these soft X-rays deposit their energy and hence compare it with the local atmospheric energy density, we use the following modeling sensitivity approach. We have used these high-resolution spectra and the longer wavelength SDO EVE EUV observations as input to the Time Dependent Ionospheric Model (TDIM) to carry out a series of six sensitivity simulations to determine the ionosphere's altitude dependence on soft X-ray wavelengths. The TDIM has a heritage of over three decades of ionospheric simulation (Schunk 1988; Sojka 1989). Most recently TDIM has been used to study the soft X-ray impact on the ionosphere E-region (Sojka et al. 2013). SDO EVE has a ESP detector that integrates the irradiance between 0.1 and 7 nm, the soft X-ray band. However, no spectral information is available. To test the sensitivity of the E-region to this spectral range a set of 6 TDIM simulations were carried out. These six simulations differed in how the soft X-rays irradiance was described. Each simulation used a 1 nm wide spectra bin to represent the 0.1 to 7 nm spectrum whose mid-point wavelength was either 0.5, 1.5, 2.5, 3.5, 4.5, or 5.5 nm. Irradiance at wavelengths longer than 7 nm were obtained from SDO EVE channels and were the same for all 6 simulations. The total ESP irradiance was then placed in the selected 1 nm wide bin: bin 1, 0.1 to 1 nm; bin 2, 1.0 to 2.0 nm; bin 3, 2.0 to 3.0 nm; bin 4, 3.0 to 4.0 nm; bin 5, 4.0 to 5.0 nm and bin 6, 5.0 to 6.0 nm. Figure 7 shows the 6 TDIM electron density profiles at the peak of the flare event. We present the altitude dependence of the E-region electron density profile as a proxy for the altitude dependence of the X-ray energy deposition profile. In so doing we identify the altitudes over which the X-class solar flares deposit their energy into the atmosphere. The first three simulations for bin 1 (0–1 nm), bin 2 (1–2 nm), and bin 3 (2–3 nm) show a clear altitude dependence that reflects how as wavelength increases from 0 to 3 nm, the ionization cross sections also increase causing the ionization to occur at higher altitudes. For these three bands the peak height ranges from 95 to 118 km. The Auger process edges for oxygen are at 2.3 nm and for nitrogen are at 2.9 nm. The cross sections for each molecule drops by almost an order of magnitude at these edges. This is reflected in the simulations by the deeper penetration of the bin 4 (3–4 nm) wavelengths and subsequent altitude increase of the peaks in electron density for bins 5 and 6 as the cross sections begin to increase. The approximate peak height for each of the 6, 1 nm wide soft X-ray spectra are tabulated in Table 3. The altitude profiles in Fig. 7 are representative not only of the strong wavelength dependence of the optical depth of the respective

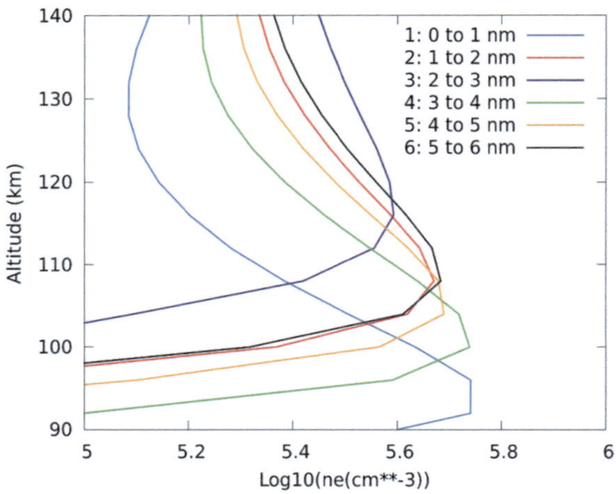

Fig. 7 6 TDIM simulations of the ionosphere electron density based upon 6 different ionizations rates associated with the 6 synthetic 1 nm wide pseudo flare spectra spanning 0 to 6 nm of the X-ray spectrums. The 6 separate simulations are color coded as well as labeled with their corresponding wavelength band

Table 3 Peak altitudes for the six soft X-ray ionospheric simulations

Bin	Lower wavelength (nm)	Upper wavelength (nm)	Altitude (km)	Mean energy (keV)
1	0.1	1.0	94 ± 2	2.26
2	1.0	2.0	108 ± 2	0.83
3	2.0	3.0	117 ± 2	0.50
4	3.0	4.0	101 ± 2	0.35
5	4.0	5.0	116 ± 2	0.28
5	5.0	6.0	118 ± 2	0.23

photons but also where their energy is deposited. In later sections this altitude dependence of the electron density will be indicative of how the Pedersen conductivity depends on altitude for Joule heating analysis. It is this spectral range, 1 to 6 nm, that exhibits the many orders of magnitude irradiance increase in response to space weather flare events. The 1 to 6 nm soft X-ray band deposits energy in the 100 to 120 km range. This energy deposition also includes that of the majority of the photoelectrons and that of the secondary electron cascade whose energies are less than a few hundred eV. These electrons have short mean free paths at these altitudes that ensure they are locally thermalized.

3.1.1 Flare Energy Calculations

Our objective is then to compare the solar flare energy density deposition with that of the local atmosphere's energy density. The assumptions are that the duration of sustained irradiance is 30 minutes and that 50% of this energy is deposited as heating of the neutral gas. The deposition is assumed to be uniform over the 100 to 120 km altitude range based on Fig. 7. Note the 50% efficiency argument is taken from Knipp (2011) and is qualitatively justified by considering that a fraction of the incoming photons will ionize above 120 km and that other energy loss channels exist, e.g. CO_2 and NO radiative cooling. Based upon these assumptions, duration and efficiency, the Fig. 4 flare spectrum between 1 and 7 nm would deliver an average energy density of $\sim 1 \times 10^{-2}$ J/m^3 and the DEM irradiance for the 13 May 2013 X5-class flare would deliver an average energy density of $\sim 2 \times 10^{-2}$ J/m^3 in the 100 to 120 km altitude range.

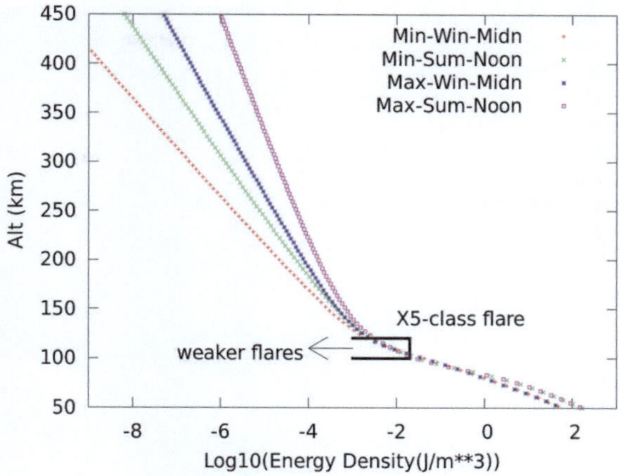

Fig. 8 Same as Fig. 3 with the X5-class flare energy density deposition estimates identified as black symbols between 100 and 120 km

Figure 8 shows the flare energy densities altitude range superimposed on the earlier Fig. 3. Clearly, these flare examples would have an impact on the equilibrium of the atmosphere in the 100–120 km range since they are similar to the thermal content of the atmosphere at these altitudes. This altitude range is similar to that used by Fuller-Rowell (1985) for the Joule Heating deposition for strong electric fields in aurorally produced enhanced Pedersen conductivity. Hence in a summer sunlit auroral-polar cap region the flare would also enhance Joule heating energy deposition by increasing the Pedersen conductivity.

3.1.2 Other Flare Wavelengths

Do the longer or shorter wavelengths also have a significant impact? For longer wavelengths, the energy of each photon is less and the flare multiplying factor decreases to tens of percent (Fig. 4). Also as shown in Fig. 5, the deposition altitudes are higher and, with the larger neutral scale heights, the deposition is spread over a much broader altitude range. For the Halloween Storm X-class flares reported by Sutton et al. (2006) the EUV irradiances increase by almost a factor of 5. This was still small compared to the three orders of magnitude increased in the hard X-ray irradiances. These flares were very strong X-class representing quite intense space weather flare activity. Hence, the longer wavelengths progressively contribute less significantly to the atmospheric dynamics. More energetic photons of the shorter wavelength penetrate much deeper as their ionization cross sections decrease. The total energy content of the irradiance at wavelengths short of 1 nm is also small compared to the atmospheric thermal energy content below 100 km. Even the 3 to 4 order of magnitude flare increase at these short wavelengths does not offset these trends. Therefore the less than 1 nm solar hard X-rays do not contribute significantly to the thermal structure of the atmosphere. High energy X-ray spectra are available from the NASA RHESSI instrument at energies above 10 keV (Holman et al. 2011). They provide photon fluxes both before and during solar flares that confirm the earlier space weather arguments that it is shortward of 10 nm where the irradiance spectrum can increase by one to several orders of magnitude. At their lowest energy, 10 keV (about 0.1 nm), their example shows a flare increase of about 2 orders of magnitude.

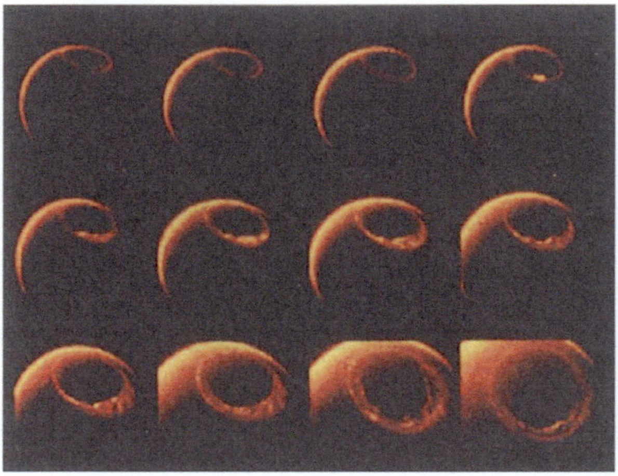

Fig. 9 Twelve consecutive false color images of the northern auroral oval that record the onset and evolution of a substorm as the spacecraft Dynamics Explorer 1 descends from apogee position during the period 0529 through 0755 UT on 2 April 1982 (from Pfaff 2012)

3.2 Auroral Energy Deposition

For those of us living in the northern hemisphere, the most endearing aspect of space weather is undoubtedly the observation of the northern light's display, the aurora borealis; a colorful visual phenomenon whose dynamics outpaces even our best computer-aided artistic products. The observer of the northern lights sees a very local display of this space weather phenomenon and appreciates that it is structured over many scales in both space and time. However, the casual observer does not appreciate that the power being unleashed is on a global scale. Our objective is to identify where in the atmosphere this energy disposition could result in a significant change to the atmosphere. The first step is to establish the specifics of the auroral global distribution. Figure 9 is a sequence of auroral images obtained by a camera on the NASA Dynamics Explorer satellite (DE-1) (Frank et al. 1981; Frank and Craven 1988). This sequence of 12 images is representative of both the global nature and structural forms found in the auroral "oval". Each image takes 12 minutes to acquire and hence the full 12 frame sequence spans 144 minutes (2 hrs. 24 min.) as the DE-1 satellite moves from near apogee towards perigee. Between the first image, top left, and the last image bottom right, an auroral substorm has occurred and the auroral oval has widened as well as expanded equatorward. For reference the sunlit dayside is identified by the crescent arc seen in the top left of each image and the pole lies near the center of the auroral oval. Clearly, the global nature of the aurora is predominantly restricted to an oval region and therefore this is where we expect to find the maximum energy deposition. This analysis is complementary to Table 1 in which global auroral electron energy deposition is reported as reaching up to 0.03 $\mathrm{W/m^2}$ locally or 35 GW integrated over a hemisphere (Knipp 2011). The auroral oval occupies between 5 and 20% of the global surface area depending on the severity of space weather. Paxton and Zhang (2016) provide an in-depth review of auroral imager missions with examples from each of the instruments. From high altitudes these instruments capture the entire oval while from lower altitudes a high-resolution swath along the orbit track is captured. By using imagery at multiple wavelengths, the images can be converted to information about the auroral source particle energy deposition (Strickland et al. 1989, 1999; Zhang and Paxton 2008). At the time of writing, there are no high-altitude global images in operation to capture the complete spatial and temporal evolution of the auroral energy deposition.

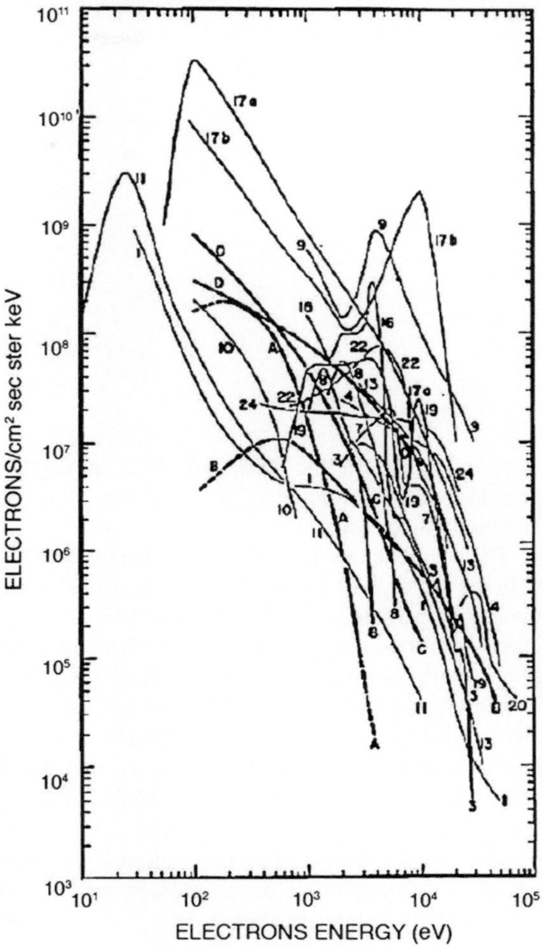

Fig. 10 Auroral electron spectra measured by 24 rocket flights (labeled 1 through 24) and satellites (A, B, C, D) (from Hones et al. 1971)

In the early days of rocket and satellite exploration of the space environment, particle detectors, both ion and electron, made extensive measurements of auroral spectra. For our purposes, we direct the reader to a few sources of these measurements. We have focused on the electron precipitation, but acknowledge that proton precipitation also provides energy but usually at a somewhat lower intensity level (see Table 1). Figure 10 (from Hones et al. 1971) shows a collage of observed electron spectra from 24 auroral rocket flights. These spectra are typical of a wide variety of the major auroral forms and hence are a good representation of the space weather spectra that will impact the neutral atmospheric temperature. From Fig. 10 the strongest events could well be represented by electrons of energy in the few to less than 10 keV range with fluxes reaching almost $10^9\ \text{cm}^{-1}\,\text{s}^{-1}\,\text{str}^{-1}\,\text{kev}^{-1}$. Subsequently, these ranges of energy and fluxes are used to determine where in altitude energy deposition occurs, and also how this energy deposition compares with the ambient energy content of the atmosphere. Today these auroral particle detectors are typically flown as part of an environmental sensor package on satellites. Both the USAF and NOAA have weather satellite programs with over three decades of such in-situ auroral particle monitoring. Although these data sets are more comprehensive with improved energy and angular resolutions as well as cadence, the net result is that Fig. 10 still represents the diversity

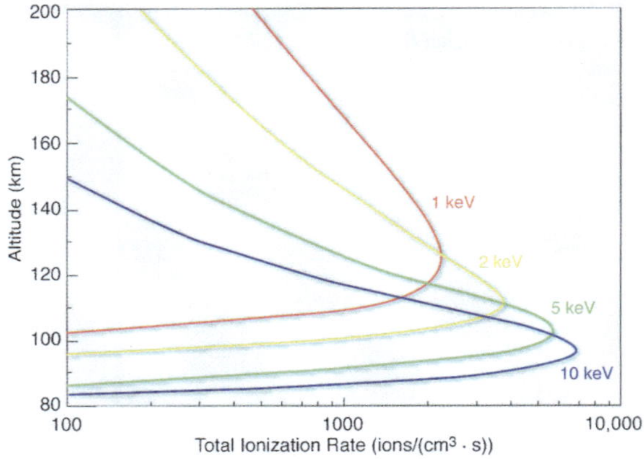

Fig. 11 Model ionization profiles for energetic electrons (from Knipp 2011)

of precipitating electron fluxes. However, although the in-situ technique does provide the necessary information about particle energy input, it fails to capture spatial and temporal variability time scales.

The energetic electron entering the atmosphere undergoes elastic and inelastic collisions with the atmosphere's molecules and atoms. The inelastic collisions lead to ionizations of the neutral particle and a secondary electron. These secondary electrons along with the primary electrons cause further ionization. This ionization cascade leads to an energy deposition scenario in which much of the energy is converted to neutral atmospheric heating and dynamics. Rees and Maeda (1973) summarized the knowledge of this complex process at a level of completeness consistent with that of today (Thayer and Semeter 2004). The depth into the atmosphere that a primary auroral electron can penetrate depends upon its energy and pitch angle. A graphic summary of this altitude deposition is shown in Fig. 11 (from Knipp 2011); see also Rees (1963) and Berger et al. (1970) for examples showing the effect of pitch angle distribution on penetration depth. These ionization rates are readily related to energy by considering that on average each ionization requires 35 eV (Rees and Maeda 1973). Many computational models have simulated the ionization rate profiles from auroral electrons: for example the Rees (1963) work is based on functions representing absorption of electrons in air while Berger et al. (1970) use the Monte Carlo technique to follow individual particles through the sequence of atmospheric collisions and secondary electron thermalization. The sophistication of first principles software to infer auroral ion production rates and subsequently energy deposition in the atmosphere has advanced, but the identification of the precipitating particle input distribution with sufficient spatial and temporal variability has not kept pace. The limitation from inadequate characterization of particle input is present in all present day global ionosphere-atmosphere models even though these models have a comprehensive first principle's description of the ionization/thermalization processes.

Figure 11 shows that the ionization rate peaks at deeper altitudes as the electron's energy increases. Further, the profile becomes narrower as the energy increases and penetration is deeper into the atmosphere where the atmospheric scale height is smaller. Figure 10 shows that auroral electron spectra span the few keV to 10 keV range. This corresponds to peak ionization rates between 100 and 120 km. At these incoming electron energies, the profile has an altitude extent of about 10 km. Hence, the incoming electrons will deposit the majority of their energy as they are thermalized in the atmosphere in this 10 km band in the 100 to 120 km range. Before computing the effective deposition energy density, it is noteworthy that this altitude range is the same as that found in Sect. 3.1 for the intense space weather flares.

3.2.1 Auroral Energy Calculation:

Knipp (2011) page 549 provides an example of "heating rate from particle energy deposition". In the example, the assumption was made that each ionization requires 35 eV and that 50% of the energy goes to heating. Thayer and Semeter (2004) provide an in-depth analysis showing that about 50% of the incoming particle energy is effective in heating the neutral gas. Then the heating rate is proportional to the ionization rate which is proportional to the square of the electron density. This is summarized by the energy deposition equation:

$$W = AB\alpha n_e^2$$

where A is the efficiency taken to be 0.5, B is the energy per ionization 35 eV, α is the recombination rate $9 \times 10^{-14}\ \text{m}^3/\text{s}$ and n_e is the electron density at 150 km. A heating rate of $10^{-8}\ \text{W/m}^3$ was obtained at 150 km. Thayer and Semeter (2004) used a polar cap auroral example that had the same heating rate at 150 km and a peak heating rate of about $2 \times 10^{-6}\ \text{W/m}^3$ at an altitude of 110 km. An alternative method of inferring this auroral energy density involves integrating the incoming energy flux spectrum to obtain total energy flux and then based on altitudes over which this is distributed, see Fig. 11, infer an average energy density. To make this estimate the auroral spectrum "9" in Fig. 10, which was measured by Evans (1967) is approximated as: peak energy 4 keV, width of spectrum 3 keV, and average number flux is $7 \times 10^8\ \text{cm}^2\,\text{s}\,\text{str}\,\text{keV}$. This approximated spectrum then yields an energy flux of $4.22 \times 10^{-2}\ \text{W/m}^2$. From Fig. 10, a 4 keV auroral electron spectrum will have an ionization rate peak at about 110 km with an approximate half-width of 15 km. If the auroral deposition is maintained locally for 20 minutes, this corresponds to an energy density of $3.378 \times 10^{-3}\ \text{J/m}^3$ and if 50% is assumed to be converted to atmospheric thermal energy, this would be $1.7 \times 10^{-3}\ \text{J/m}^3$.

Rosenqvist et al. (2005) analyzed the large geomagnetic storm of 30 October 2003. In their analysis they found that during the peak of the storm, a local particle energy flux reached $3.0 \times 10^{-1}\ \text{W/m}^2$ for about one minute and for about 20 minutes sustained a level of $5.0 \times 10^{-2}\ \text{W/m}^2$. This analysis was based on measurements made by the Tromso-UHF incoherent scatter radar (EISCAT) system with receivers at Kiruna and Sodankylä. The sustained 20-minute energy flux level yields similar deposition energy densities to that inferred from the Evans (1967) auroral spectrum. Furthermore, the EISCAT electron density profiles (Fig. 9 in Rosenqvist et al. 2005) show an intense ionization peak from 125 km extending down below 100 km. Figure 12 expands on the prior Fig. 3 by adding the effect of intense space weather electron precipitation. In this space weather case, the energy density is based on 20 minutes of continuous electron precipitation. These values reach a few 10's of percent of the local thermal energy density in the altitude region of 110 ± 7.5 km.

3.2.2 Impact of Other Auroral Energies

The less than 1 keV electrons are composed of both the incoming auroral population and secondary electrons generated from high energy primaries in the tail of the low energy distribution. Their interactions with the atmosphere occur at higher altitudes, 150 to 250 km. These electron fluxes generate a lower density plasma in the F-region and also deposit a significantly lower energy density. Hence their impact on the thermosphere would be at and above these altitudes only. For auroral electrons of energy greater than 10s of keV, they penetrate well below 100 km where the atmospheric energy density is much larger than theirs. Also the 10's of keV fluxes are typically short-lived events and occur over limited regions.

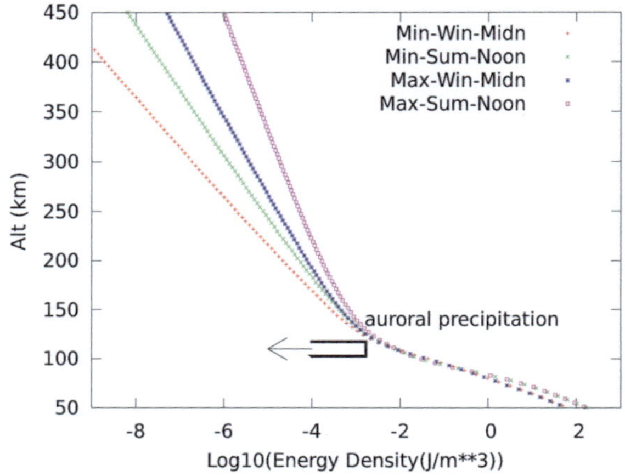

Fig. 12 Same as Fig. 3 with the auroral electron precipitation energy deposition estimates identified as black symbols between 100 and 125 km

Ions, particularly H^+, also contribute to auroral particle precipitation, but as Table 1 indicates, the energy fluxes are significantly smaller than those of the electrons. In our summary we raise the question of how unique is the fact that the solar flare and auroral electron space weather impacts both have energy depositions in the same altitude range.

3.3 Electromagnetic Energy Deposition

Severe space weather's dominant energy deposition mechanism is the Poynting flux (Richmond and Lu 2000). The electromagnetic energy propagates from the magnetosphere into the polar auroral ionosphere. Only on a few occasions does the neutral atmosphere flywheel transfer energy from the ionosphere to the magnetosphere (Kelley et al. 1991). The Poynting flux high-latitude distribution can be inferred from combinations of electric and magnetic field measurements (Thayer and Semeter 2004). However to define the altitude at which the energy deposition occurs, information on the altitude dependence of the Pedersen conductivity is needed to evaluate the local Joule heating rate. We review aspects of both the Poynting flux and the local Joule heating.

Magnetosphere-Ionosphere (M-I) Poynting flux associated with space weather geomagnetic storms extends over the polar auroral latitudes and is qualitatively well understood. Figure 13 shows schematically the key elements of the Poynting flux as energy flows into the upper atmosphere (blue arrows). In this schematic, the yellow contours simplify the convection electric field potential as a symmetric 2-cell pattern, and the electric field is represented by the red arrows. The upper atmosphere contains the ionosphere and the Hall and Pedersen conductivities provide closure to the generalized Ohms law current system (orange arrows). Although overly simplified, the schematic provides the reader with an indication of how anisotropic the Poynting flux distribution will be. During the severe space weather events of interest, the region defined as polar auroral latitudes in Fig. 13 extends to encompass the traditional mid-latitude regions. This is readily substantiated when auroral sightings are made at lower mid latitudes, e.g., in the Mediterranean and Caribbean. Adequate modeling of such storm time expansions of the M-I electrodynamics (severe weather) is however, still in its infancy. Figure 14 shows a statistical electric field, E_{conv}, pattern (Weimer 1995) for various geomagnetic/solar wind conditions. The altitude distributions of these electric fields have also been made from sounding rockets (Pfaff et al. 1992). Such electric field distributions can be combined with the magnetic field variability associated with the M-I currents as is shown in Fig. 13 to infer an incoming Poynting flux, and subsequently the

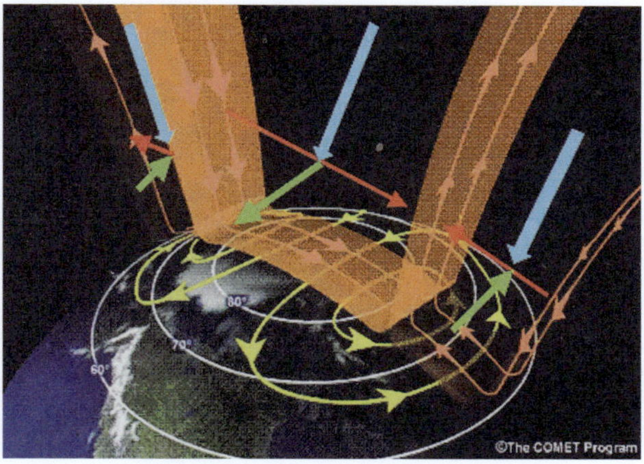

Fig. 13 Schematic of Poynting Flux (*blue arrows*) and the M-I electrodynamic parameters being imposed on a high-latitude polar region. The other parameters are convection electric field (*red arrows*), convection pattern (*yellow contours*), auroral currents (*orange*) and magnetic field perturbations (*green arrows*) (from Knipp 2011)

Joule heating. Knipp (2011) carried out sample calculations to show that the magnitude of the Poynting flux could be on the order of a few tens of mW/m^2 (Knipp 2011, page 555). For this calculation Knipp used a E_{conv} of 50 mV/m and a magnetic perturbation ΔB of 500 nT. These numbers represent disturbed geomagnetic conditions. The Poynting flux is not uniformly distributed across the high latitudes. The electric field distributions shown in Fig. 14 indicate that the largest electric fields are found across the polar cap (regions of closely spaced potential lines). However, the specific observations of when these large electric fields coincide with the large magnetic perturbations is not available. The Poynting flux deposits energy into the atmosphere via Joule heating-frictional heating (Strangeway 2012; Thayer 1998, 2000).

3.3.1 Electromagnetic Energy Density Calculations

To evaluate the energy density altitude distribution, it is necessary to use a Joule heating formulation for Q_{Joule} (Cole 1971) such as:

$$Q_{\text{Joule}} = \sigma_p E_{\text{conv}}^2$$

Where σ_p is the Pedersen conductivity and E_{conv} is the imposed magnetospheric convection electric field. In general the Joule heating needs to be referenced to the neutral gas frame. However since we are considering severe space weather conditions, the neutral velocity will be uncorrelated to the imposed $\vec{E}_{\text{conv}} \times \vec{B}$ flow or will be small compared to this imposed flow during these initial 10's of minutes. Figure 15 shows two examples of the auroral ionosphere's Hall and Pedersen conductivity profiles (Pfaff 2012). The left hand pair of panels shows a night-time auroral zone location devoid of electron precipitation. The very low E-region electron density leads to the conductivity remaining below 1×10^{-6} mho/m. At the same location with moderate auroral precipitation, the density and subsequently the conductivities increase by two orders of magnitude. This auroral zone conductivity is controlled by the auroral particle energy and energy fluxes as described in Sect. 3.2. Hence, the altitude of energy deposition via Joule heating is correlated to auroral particle precipitation during extreme space weather events. The altitude profiles in Fig. 15 (right panel) indicate that the majority of the Pedersen conductivity is located between 110 and 130 km. This altitude depends on the auroral ionization source. If the auroral electrons are lower in energy, ionization and hence Pedersen conductivity will be present at somewhat higher altitudes (Thayer and Semeter 2004).

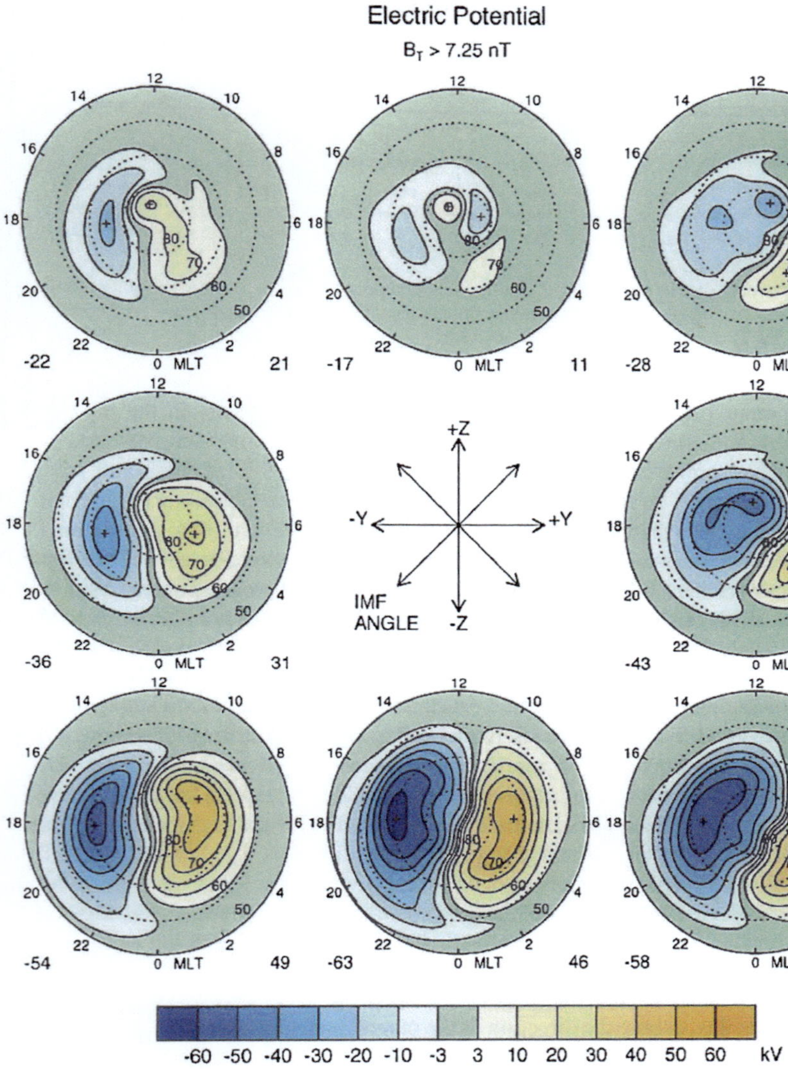

Fig. 14 A series of 9 high latitude electric potential distributions for IMF magnetic field strength greater than 7.25 nT. The 9 patterns correspond to different clock angles of the IMF $B_y - B_z$ orientation (from Wiemer 1995)

The Poynting flux varies over a wide range. We used the upper value of 0.15 W/m^2 given in Table 1 and then assumed the altitude dependence of the subsequent Joule heating as represented by the Pedersen conductivity profile. For example Fig. 15, right panel suggests the Pedersen conductivity could be approximated as a 20 km altitude band between 110 and 130 km. Then assuming that the Poynting flux remains constant at this location for one hour, a total energy would be 540 J/m^2. Distributing this uniformly over the 20 km results in a volume energy deposition of 27 mJ/m^3 and then based on a 50% efficiency argument as seen in Sect. 3.2 produces a volume energy available for heating of 13.5 mJ/m^3. Alternative estimates in the context of the Joule heating formulation could be based on a 50 mV/m electric

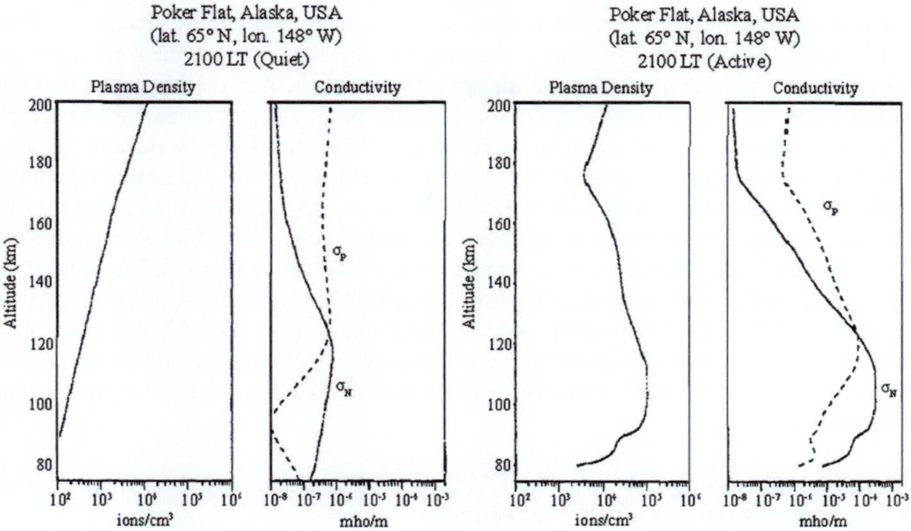

Fig. 15 Altitude profiles of the E-region electron density and corresponding Hall and Pedersen conductivities at Poker Flat, Alaska during nighttime quiet conditions (*left 2 panels*) and during moderate auroral conditions (*right 2 panels*) (from Pfaff 2012)

Fig. 16 Same as Fig. 3 with the Joule heating energy deposition estimates identified as black symbol between 110 and 130 km

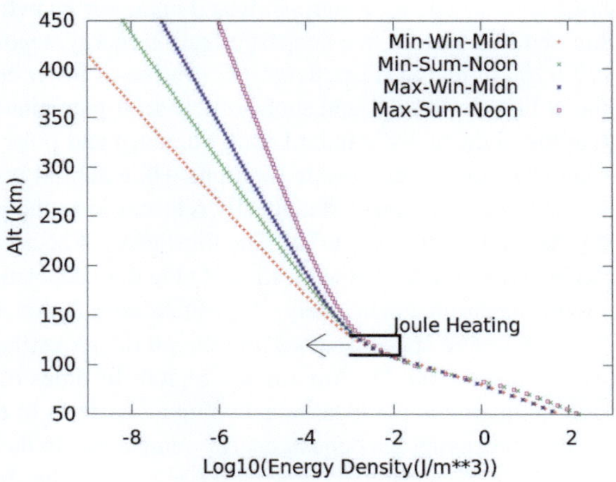

field imposed on a Pedersen conductivity of 4×10^{-3} mho/m for one hour with a heating efficiency of 50%. This results in a volume energy deposition of 18 mJ/m^3. The overall consistency of estimates, i.e., Joule heating and topside Poynting flux is reassuring, but they are still very crude estimates. An average of these two values, 16 mJ/m^3 has been plotted on Fig. 16 for the Poynting flux/Joule heating comparisons with the Fig. 3 atmospheric energy density.

3.3.2 Joule Heating Caveats

The Poynting flux at the top of the atmosphere is inferred from various satellite in situ measurements. Synchronization between the peaks in electric fields and magnetic perturbations whether on the same or separate satellites is still difficult. Even theoretically the expectation that Joule heating maximizes where both σ_p and E_{conv} have their largest values is unclear. In Fig. 14, the electric field is the largest in the polar cap while in Fig. 9 the auroral precipitation, hence σ_p, is largest in the auroral oval. Another refinement on this spatial concern was described by Codrescu et al. (1995) in that Joule heating is proportional to the square of the electric field. In the auroral and polar region, these electric fields are very variable and structured. Therefore average "smooth" electric fields would readily underestimate the heating by a factor of two. Codrescu et al. deduced this as they identified a missing geomagnetic heat source in their thermospheric modeling work. Another uncertainty concerning the Poynting flux is the degree to which it has a strong Alfvénic component. Waygant et al. (2000) studied the Alfvén wave nature of the magnetic field perturbations, and deduced that as much as 10% of the Poynting flux was associated and structured by Alfvén waves. This finding further impacts the implementation of Joule heating calculations. For these reasons Fig. 16 may well either under or overestimate the efficiency of converting Poynting flux into atmospheric heating.

4 Space Weather Atmospheric Impacts

The three processes highlighted in Sect. 3; flare, auroral, and electromagnetic energy deposition are all present during intense space weather events. Both flares and auroral precipitation lead to energy deposition via the thermalization of a primary photon or electron and their cascade of secondary electrons. Based upon our knowledge of severe space weather in our Sun and M-I system, the deepest effective energy deposition for flares and aurora is in the 100–120 km region (Figs. 8 and 12). Aurora of lower characteristic energy would be effective at higher altitudes and such soft electron precipitations would occur at locations other than the night sector standard oval, i.e., cusp and polar cap regions. Flares deposit energy homogeneously over dayside locations while auroral precipitation leads to inhomogeneous deposition over a region statistically referred to as the auroral oval. These two regions can at times and locations overlap. The third process because of its dependence upon Pedersen conductivity couples to the locations of the flare deposition and auroral precipitation. Therefore the electromagnetic energy deposition altitudes overlap that of the other two processes. All three severe space weather impacts on the terrestrial atmosphere are depositing energy over a narrow 100–130 km altitude region. In times of 10s of minutes to an hour, that is short compared to the local radiative time constants of days (Brasseur and Solomon 1984), all three processes can deposit energy comparable to the local thermal energy density of the atmosphere. As a result, there would be changes in the atmospheric energy budget, leading to hydrodynamic effects. Fuller-Rowell (1985) modeled the thermospheric response to energy deposited in a channel located about the altitude identified in Sect. 3. His model extended downward to 70 km and upwards to over 400 km. The effect of the energy deposition was to create waves, winds, temperature and composition changes upwards in altitude as well as horizontally out of the channel, but has very little impact below the lower boundary of the channel. *Hence, the atmosphere had a natural mechanism to insulate the lower atmosphere from the space weather energy deposition*. The major impacts would be towards higher altitudes where the neutral densities are much smaller. The Sutton et al. (2006) observations and correlations with the X17-class flares are evidence of such higher altitude impacts where the observations are well above the region of energy deposition.

In Figs. 8, 12, and 16 a very specific upper limitation on the energy density is shown. How variable are these large energy densities? What are the caveats on these altitudes and energy density calculations? Spatially, the flare impact is over the entire dayside with its maximum effect at the sub-solar point. This effect diminishes towards the dark hemisphere controlled by the solar zenith angle. In Fig. 8, the maximum energy density associated with the flare is 20 mJ/m^3 and this would be at the sub-solar location for a X5-class flare. Hence the X17-class flares of the Sutton et al. (2006) would have significantly higher energy density but in the same altitude range. A key caveat is that during solstice conditions one entire polar region is sunlit. This enables the flare process to have an impact on the auroral and electromagnetic processes through modification of the conductivity. Currently this is qualitatively known but as Fig. 4 indicated, the key flare wavelengths from about 1 to 10 nm have respectively a flare increase factor of from 10^3 at 1 nm to a few at 10 nm. These are the photons that control the 100 to 130 km altitude ionospheric conductivity response. Unfortunately at this time, no high-temporal, high-spectral resolution observations of this part of the solar irradiance are available.

The auroral electrons like the soft X-rays deposit energy via the thermalization of collision processes of the primary electrons and a cascade of secondary electrons. The specific energy range of the auroral electrons from a few keV to less than 10 keV is a characteristic of severe space weather that ensures the deposition region is from 100 to 130 km. The latitude longitude distribution is associated with the high-latitude locations of the auroral oval. However, a caveat is that severe space weather causes the oval's equatorward boundary to move to low latitudes; in fact, probably encompassing the region usually termed mid-latitudes. An outstanding question in this severe space weather scenario is identifying the locations within the expanded oval where the duration of the sustained electron precipitation is as long as the 20 minutes used in Sect. 3.2.

Energy deposition via the Poynting flux into the upper atmosphere is on a global scale and is the largest space weather source of energy to the atmosphere (see Table 1). The altitude deposition is defined by the Joule heating mechanism and this in turn, depends upon the Pedersen conductivity profile. This profile in both its magnitude and shape is dependent upon the aurora and potentially the flare processes. Hence, exact knowledge of this is rarely available even if the electric fields are observed. In Sect. 3.3, the Joule heating altitude dependence overlaps and is slightly higher than that of both the flare and auroral deposition. Figure 16 shows that the electromagnetic energy source can readily exceed the normal upper atmospheric thermal energy density between 110 and 130 km. The latitude and longitude regions of maximum Joule heating depend upon both the electric fields (Fig. 13) and the Pedersen conductivity, the auroral regions (Fig. 9). If a major X-class flare occurs during solstice, then a large hemispheric difference in energy deposition will also be present. Comparing the electric field and auroral distributions, both of which are averages or low-resolution representations, indicates the two are not well correlated. Indeed, local regions of large electric fields, auroral currents and conductivity enhancements are controlled by local M-I MHD physics, not global averages. Therefore, at this time Fig. 16 analysis needs to be considered as a rough estimate whose specific latitude-longitude locations is poorly determined. However the altitude is very well defined.

All three space weather processes deposit energy into a narrow altitude band that overlaps between about 100 to 130 km. All three can also provide enough energy in under an hour in this region to equal or exceed the local atmosphere's thermal energy content. The atmospheric response to this localized heating results in local changes in winds, temperature, density and composition which in turn are propagated horizontally and upward to the more rarefied atmosphere (St. Maurice and Schunk 1981; Fuller-Rowell 1984, 1985; Thayer and Semeter 2004). These authors have used atmospheric simulations to model the responses. All the simulations produce wind modifications on the order of 10's to 100's of m/s. Thayer and Semeter (2004) indicate that the local heating rate would be on the order of 0.1 K/s, and this rate integrated over 20 minutes leads to a neutral temperature change of 120 K. In the source region, this corresponds to many 10's of % of the local neutral temperature. However, this source region is quite confined and in the next section we will discuss how difficult it is to make atmospheric measurements in this source region. We point out that this narrow altitude region is colocated with the Space Atmosphere Interaction Region (SAIR) and forms a space weather boundary to the atmosphere below.

5 Measurements of Tn in the SAIR Region

The 100 to 130 km altitude region of the atmosphere is almost devoid of in-situ measurements. The exception would be probes mounted on sounding rockets that traverse this region. However, this technique is unable to amass sufficient measurements to address the response of the atmosphere to severe space weather depositions of energy. The study by Sutton et al. (2006) highlights the changes in satellite drag at 400 and 490 km altitude and hence infers how much the atmosphere's exospheric temperature has increased. Significant changes will also be found in both neutral composition and the wind patterns (St. Maurice and Schunk 1981; Fuller-Rowell 1985; Thayer 2000; Thayer and Semeter 2004).

Remote sensing spectroscopy can be used to infer the temperature in the 100–130 km altitude region. Since the atmospheric response to the space weather drivers is usually on local spatial scales and short lived, these spectroscopic measurements need to be made with spatial resolution and on time scales commensurate with the thermospheric response. The cited thermospheric storm response model studies used energy deposition channels that had latitude extents of 10s of km to a few 100 km and durations on the order of an hour. These early studies (St. Maurice and Schunk 1981; and Fuller-Rowell 1984) however considered the final steady state response which was typically several hours later. Present day modeling capability is approaching such a spatial scale, 10s of km. However, the limiting factors are the space weather energy inputs themselves rather than the modeling capability. Recently, Suresh (2016) successfully defended her dissertation research in which a significant start had been made to observe the atmospheric temperatures (100–125 km) in response to space weather. Suresh used over a solar cycle of observations from the NASA's TIMED satellite's SABER instrument to create a temperature data base. The temperature inversion techniques were described by Mertens et al. (2001) and Christensen et al. (2013). A calibrated data base was developed in which the solar cycle-seasonal quiet time atmosphere is modeled. This model varies with latitude and has a day and night sector that is constrained by the TIMED satellite local time precession. Then the storm period changes in atmospheric temperature are computed relative to this background model.

Suresh's dissertation research focused on the geomagnetic storm of May 5, 2003. This storm had two impulsive phases before reaching a Dst of -100 nT four days later. Such a storm would be expected to deposit energy at levels discussed in Sects. 3.2 and 3.3. But as already pointed out by Sutton et al. (2006), the time constants are up to an hour and the satellite orbit period is about 1 hour 40 minutes. Hence a single satellite's capability to capture the temporal evolution of the atmospheric storm response is quite limited. Figure 17 taken from Suresh (2016) provides a glimpse into how challenging the task of deriving the atmospheric T_n storm responses is, let alone correlating with the space weather drivers. The following general findings were consistent with the results Suresh (2016) showed as seen in Fig. 17.

(1) Even over the narrow 100 to 125 km range the lower altitude had small T_n storm responses.

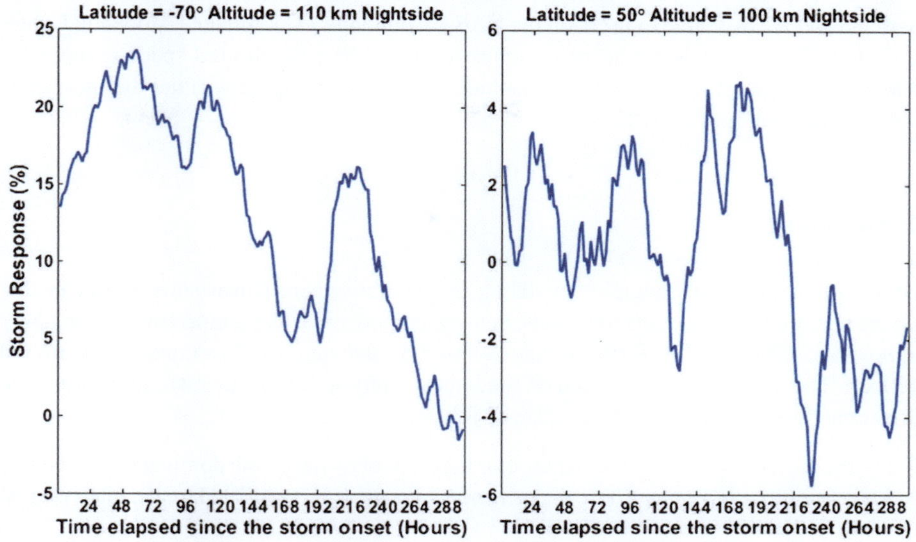

Fig. 17 Auroral high latitude T_n response to storm beginning on 5 May 2003 in the night side (*left panel*) and the T_n response at mid-latitudes in the night side (*right panel*). The *left panel* is at 110 km while the *right side panel* is at 100 km (from Suresh 2016)

(2) There was no evidence of a direct input and output time correlation between storm indices and T_n storm response temporal evolutions.
(3) Mid and low latitudes showed a weaker T_n storm response than that at high latitudes.

These findings for one geomagnetic storm do not provide statistically convincing information that other atmospheric wave disturbance mechanisms can be ruled out as the source of the Fig. 17 dynamics. Such analysis needs to be extended over many geomagnetic storms as well as quiet geomagnetic periods to determine if a causal relationship can be identified. This geomagnetic storm would be classed as a moderate storm since Dst only reached −100 nT while Kp just reached 7 for one 3-hour period. Note the fluctuations are as large as the trend. This complicated response is representative of the findings Suresh (2016) presented. Suresh's pioneering effort to separate the storm neutral temperature response from the ambient temperature in the 100–130 km altitude range, the SAIR, is significant. Her findings indicate that present day modeling is quite unable to reproduce the high degree of variability she found. Specifically, the overall T_n variability would range in the 10s of K and this could be consistent with the Thayer and Semeter (2004) estimates of about 0.1 K/s leading to 10s K over 10s of minutes. However, the latitudinal, longitudinal, and altitude structures in T_n which are associated with space weather structures in space and time are only known in a statistical sense and cannot be correlated with the drivers. Zesta and Huang (2016) carried out a review of how well satellite drag during space weather events could be specified and forecast. Apart from empirical models, their review pointed out that although first principles thermosphere-ionosphere models are always improving in their resolution, it is the driver that cannot currently be adequately defined. As they state, "There is no simple remedy to the driver problem. In the three GCMs discussed in this chapter, all dynamics are provided by the drivers. If the drivers are empirical models of energy input, the GCMs cannot provide faithful predictions of magnetic storm dynamics, which include accurate forecasts of thermospheric densities." All three GCMs (CTIPe, TIE-GCM, GITM) have extensive heritage and are available to users on the NASA Community Coordinated Modeling Center (CCMC). These models are all being improved to achieve the desired space weather capability, but without improved driver resolution obtaining the space weather thermospheric response accurately is impossible.

6 Summary

Observations and first principles modeling of the atmosphere consistently shows in that severe space weather impacts the winds, density, composition, and temperature of the upper atmosphere. The modeling further defines a lower altitude range for this impact to be around 100 km. This lower altitude is based on fundamental physics of the interactions between the driver energy source and the atmosphere as follows:

- For the most effective space weather solar flares, X-class flares, the dominant wavelengths are between 1 and 10 nm. These deposit energy between 100 and 120 km. i.e., their optical depth in the terrestrial atmosphere.
- For the most effective space weather auroral precipitation, electron fluxes in the few to 10 keV characteristic energy deposit their energy between 100 and 120 km. This is again based upon the optical depth of penetration into the terrestrial atmosphere of these electrons.
- For the most effective space weather, electromagnetic Poynting flux atmospheric heating is via Joule heating which in turns depends on the Pedersen conductivity. This Pedersen conductivity is a layer located within the altitude region where both severe flares and aurora are depositing energy and generating enhanced conductivity. The electromagnetic energy deposition lies over a 110 to 130 km range.

As the location of the most severe space weather energy is a layer from 100 to 130 km, it provides quantitative measures of the Space Atmosphere Interaction Region (SAIR) from a first principles physics viewpoint. However it is also the reason why our ability to specify and forecast the impact of severe space weather on the upper atmosphere is poor. The SAIR is a region where only sounding rockets can make in situ measurements. All other information must come from remote sensing which from the ground would be only one location per site and from a satellite would be a twice per day visit per terrestrial location although future missions plan to place imagers on geostationary satellites that will provide continuous observations of one hemisphere. Almost all present day knowledge of the severe space weather aurora or electromagnetic drivers is based on statistical models. The deficiency for X-flares is reconstructing the 1 to 10 nm spectral hardening during the flaring process. The outstanding observational requirements to address the severe space weather atmospheric needs are:

- X-class flares require the 1 to 10 nm irradiance to be observed with better than 1 nm resolution and cadence on the order of 10's of seconds. This needs to be in conjunction with instruments similar to the hard X-ray monitoring by NOAA GOES satellites and the EUV spectrum by EVE instrument on SDO.
- Multiple wavelength (UV) of the global auroral regions. The requirement is to be able to infer the auroral precipitation characteristic at both space (better than 100 km) and temporal (better than one minute) resolution to capture the dynamics of severe space weather.

- The electromagnetic energy deposition observation is significantly more challenging since no remote sensing instrument is available to image either electric fields and/or magnetic field perturbations. During severe space weather events present day modeling of the M-I system is unable to adequately reproduce the spatial and temporal dynamics of these Poynting flux fields. The use of low Earth orbit (LEO) satellite constellations such as the commercial iridium fleet has demonstrated the potential for capturing the magnetic field variability (Anderson et al. 2014). Hence one approach would be to use in situ measurement techniques on such constellations that infer the electric fields and magnetic field variability. The Pedersen conductivity, essential for the altitude distribution of energy deposition, needs to be generated by I-T models driven by the higher resolution X-class flare irradiance and auroral images.

The location of the SAIR also depends on the atmosphere which, in turn is in a climate state that also depends upon the solar irradiance, the solar constant. In the terrestrial atmospheric system, the SAIR's location is the result of collision processes between specific energies of photons and electrons and is physically an optical depth process. Therefore, the concept is readily applied to other planets with atmospheres. The relative role of the aurora and the electromagnetic Poynting flux would depend on the planet also having a magnetosphere. An important question to be addressed is can the SAIR altitude range be bifurcated with flares depositing their energy in one part of the SAIR altitude range and aurora, because of the precipitation energy, depositing its energy in a higher or lower altitude range? The electromagnetic deposition however is linked to both through their control of the Pedersen conductivity. The importance of the SAIR altitude is that the impact of severe space weather on atmospheric temperature and density occurs above this altitude thereby insulating the lower atmosphere from space weather heating. Another implication of this review is that all exoplanets with atmospheres with or without magnetospheres will have a SAIR whose stellar source can be dramatically different from that of our Sun. Under these different conditions, the question of whether the space weather events can deliver sustained energy fluxes comparable to the star's "solar constant" would be the key to establishing how successful the star's SAIR is in protecting the lower atmosphere from rapid atmospheric loss processes.

Acknowledgements This research was supported by National Science Foundation grant AGS1441774 to Utah State University and by a subcontract from the University of Colorado from National Aeronautics Space Agency NAS5-02140 to Utah State University. The author is grateful to the organizers of the ISSI-Scientific Foundation of Space Weather, 2016 Workshop in Bern, Switzerland for the invitation to present this work.

References

B.J. Anderson et al., Geophys. Res. Lett. **41**, 3017–3025 (2014)

M. Aschwanden, P. Boerner, A. Caspi, J.M. McTiernan, D. Ryan, H. Warren, Sol. Phys. **209**, 2735 (2015). doi:10.1007/s11207-015-0790-0

S.J. Bauer, *Physics of Planetary Ionospheres* (Springer, New York, 1973), pp. 1–230

M.J. Berger, S.M. Seltzer, K. Maeda, J. Atmos. Terr. Phys. **32**, 1015–1045 (1970)

T. Bosinger, J. LaBelle, H.J. Opgenoorth, J.-P. Pommereau, K. Shiokawa, S.C. Solomon, R.A. Treumann (eds.), in *Dynamic Coupling Between Earth's Atmospheric and Plasma Environments*. Space Sciences Series of ISSI, vol. 9 (Springer, New York, 2013), pp. 1–609

G. Brasseur, S. Solomon, *Aeronomy of the Middle Atmosphere* (Reidel, Dordrecht, 1984), pp. 1–441

W.J. Burke, in *Mid Latitude Ionospheric Dynamics and Disturbances*. AGU Monograph Series, vol. 181 (2008). doi:10.1029/181GM21

R.C. Carrington, *Monthly Notices of the Royal Astronomical Society*, vol. 20 (Royal Astronomical Society, London, 1860)

A. Caspi, T.N. Woods, H.P. Warren, Astrophys. J. Lett. **802**, L2 (2015)
J.W. Chamberlain, *Theory of Planetary Atmospheres: An Introduction to Their Physics and Chemistry*. International Geophysical Series, vol. 22 (Academic Press, New York, 1978)
M.C.M. Cheung, P. Boerner, C.J. Schrijver, P. Testa, F. Chen, H. Peter, A. Malanushenko, Astrophys. J. **807**, 143 (2015). doi:10.1007/s11207-15-0790-0
A.B. Christensen, R.L. Bishop, S.A. Budzien, J.H. Hecht, M.G. Mlynezak, J.M. Russell, A.W. Stephan, R.W. Walterscheid, J. Geophys. Res. **118**(6), 2146–2152 (2013)
M.V. Codrescu, T.J. Fuller-Rowell, J.C. Foster, Geophys. Res. Lett. **22**, 2399 (1995)
K.D. Cole, Planet. Space Sci. **19**, 59–75 (1971)
G. Del Zanna, K.P. Dere, P.R. Young, E. Landi, H.E. Mason, Astron. Astrophys. **582**, A56 (2015)
K.P. Dere, E. Landi, H.E. Mason, B.C. Monsignovi Fossi, A.R. Young, Astron. Astrophys. Suppl. Ser. **125**, 149 (1997)
D.S. Evans, J. Geophys. Res. **72**, 4281 (1967)
J.A. Fennelly, D.G. Torr, At. Data Nucl. Data Tables **51**, 321–363 (1992)
L.A. Frank, J.D. Craven, Rev. Geophys. **26**(2), 249–283 (1988)
L.A. Frank, J.D. Craven, L.L. Ackerson, M.R. English, R.H. Eather, R.L. Carovillano, Space Sci. Instrum. **5**, 369–393 (1981)
D.C. Fritts, M.J. Alexander, Rev. Geophys. **41**, 1003 (2003). doi:10.1029/2001RG000106
T.J. Fuller-Rowell, J. Geophys. Res. **89**, 2971 (1984)
T.J. Fuller-Rowell, J. Geophys. Res. **90**, 6567 (1985)
T.J. Fuller-Rowell, D. Rees, J. Atmos. Sci. **37**, 2545 (1980)
T.J. Fuller-Rowell, D. Rees, S. Quegan, R.J. Moffett, M.V. Codrescu, G.H. Millward, A coupled thermosphere ionosphere model (CTIM), in *Handbook of Ionospheric Models*, ed. by R.W. Schunk. STEP Report, vol. 217 (1996)
T.J. Fuller-Rowell, F. Wu, R. Akmaev, T.W. Fang, E. Araujo-Pradere, J. Geophys. **115**, A00G08 (2010)
L.C. Gardner, R.W. Schunk, J. Geophys. Res. **116**, A06303 (2011). doi:10.1029/2010JA015629
J.D. Haigh, *Fundamentals of the Earth's Atmosphere and Climate*. Geophysical Monograph, vol. 141 (Am. Geophys. Union, Washington, 2004)
A.E. Hedin, J. Geophys. Res. **92**, 4649 (1987)
A.E. Hedin, J. Geophys. Res. **96**(A2), 1159 (1991)
A.E. Hedin et al., J. Geophys. Res. **82**, 2139 (1977)
B.L. Henke, E.M. Gullikson, J.C. Davis, At. Data Nucl. Data Tables **54**(2), 181–342 (1993)
G.D. Holman, M.J. Aschwanden, H. Aurass, M. Battaglia, P.C. Grigis, E.P. Kontar, W. Liu, P. Saint-Hilaire, V.V. Zharkova, Space Sci. Rev. **159**, 107–166 (2011)
E.W. Hones Jr., J.R. Asbridge, S.J. Bame, S. Singer, J. Geophys. Res. **76**, 63–87 (1971)
L. Jacchia, Spec. Rep. **131**, Smithsoni. Astrophys. Observ., May 6 (1970)
A.V. Jones, *Aurora* (Reidel, Dordrecht, 1974), pp. 1–301
M.C. Kelley, D.J. Knudsen, J.F. Vickrey, J. Geophys. Res. **96**, 201–207 (1991)
G.V. Khazanov (ed.), *Space Weather Fundamentals* (CRC Press, Boca Raton, 2016)
P.M. Kintner Jr., A.J. Coster, T. Fuller-Rowell, A.J. Mannucci, M. Mendillo, R. Heelis (eds.), *Midlatitude Ionospheric Dynamics and Disturbances* (Am. Geophys. Union, Washington, 2008), pp. 1–327
M.G. Kivelson, C.T. Russell (eds.), *Introduction to Space Physics* (Cambridge University Press, Cambridge, 1995), pp. 1–568
D.J. Knipp, *Understanding Space Weather and the Physics Behind It* (McGraw-Hill, New York, 2011), pp. 1–727
R.E. Lopez, D.N. Baker, J. Allen, Eos **85**, 105 (2004)
T.-Z. Ma, R.W. Schunk, J. Geophys. Res. **100**, 19701–19713 (1995). doi:10.1029/95JA01122
C. McLandress, J. Atmos. Sol.-Terr. Phys. **60**, 1357–1383 (1998)
C.J. Mertens, M.G. Miynczak, M. Ldpez-Puertas, P.O. Wintersteiner, R. Picard, J.R. Winick, L.L. Gordley, J.M. Russell III, Geophys. Res. Lett. **28**, 1391–1394 (2001)
L.J. Paxton, Y. Zhang, in *Space Weather Fundamentals*, ed. by G.V. Khazanov (CRC Press, Boca Raton, 2016), Chap. 13
R.F. Pfaff Jr., Space Sci. Rev. **168**, 23–112 (2012). doi:10.1007/s11214-012-9872-6
R.F. Pfaff, J. Sahr, J.F. Providakes, W.E. Swartz, D.T. Farley, P.M. Kintner, I. Haggstrom, A. Hedberg, H. Opgenoorth, G. Holmgren, J. Atmos. Sol.-Terr. Phys. **54**, 779–808 (1992)
J.M. Picone, A.E. Hedin, D.P. Drob, A.C. Aikin, J. Geophys. Res. **107**(A12), 1464 (2002)
M.H. Rees, Planet. Space Sci. **11**, 1209–1218 (1963)
M.H. Rees, K. Maeda, J. Geophys. Res. **78**, 8391–8394 (1973)
A.D. Richmond, G. Lu, J. Atmos. Sol.-Terr. Phys. **62**, 1115–1127 (2000)
R.G. Roble, Energetics of the mesosphere and thermosphere, in *The Upper Mesosphere and Lower Thermosphere: A Review of Experiment and Theory*, ed. by R.M. Johnson, T.L. Killen. Geophysical Monograph Series, vol. 87 (Am. Geophys. Union, Washington, 1995), pp. 1–21

R.G. Roble, E.C. Ridley, R.E. Dickinson, J. Geophys. Res. **92**(A8), 8745 (1987)
R.G. Roble, E.C. Ridley, A.D. Richmond, R.E. Dickinson, Geophys. Res. Lett. **15**, 1325 (1988)
L. Rosenqvist, H. Opgenoorth, S. Buchaert, I. McCrea, O. Amm, C. Lathuillere, J. Geophys. Res. **110**, A09523 (2005). doi:10.1029/2004JA010927
R.W. Schunk, Pure Appl. Geophys. **127**, 225 (1988)
R.W. Schunk, A.F. Nagy, *Ionospheres*, 2nd edn. (Cambridge University Press, Cambridge, 2009), 628 pp.
A.K. Smith, Space Sci. Rev. **168**, 1–21 (2012). doi:10.1007/s1124-011-9791-y
E.V.P. Smith, D.M. Gottlieb, Space Sci. Rev. **16**, 771 (1974)
C.G. Smithtro, J.J. Sojka, J. Geophys. Res. **110**, A08305 (2005a)
C.G. Smithtro, J.J. Sojka, J. Geophys. Res. **110**, A08306 (2005b)
J.J. Sojka, Rev. Geophys. **27**, 371 (1989)
J.J. Sojka et al., J. Geophys. Res. **118**, 1 (2013)
J.-P. St. Maurice, R.W. Schunk, J. Geophys. Res. **86**, 11299 (1981)
R.J. Strangeway, J. Geophys. Res. **117**, A02310 (2012)
D.J. Strickland, R.R. Meier, J.H. Hecht, A.B. Christensen, J. Geophys. Res. **94**, 13537–13539 (1989)
D.J. Strickland, J. Bishop, J.S. Evans, T. Majeed, P.M. Shen, R.J. Cox, R. Link, R.E. Huffman, J. Quant. Spectrosc. Radiat. Transf. **62**, 689 (1999)
P. Suresh, Global Thermosphere Response to Geomagnetic Storms. Ph.D. Dissertation, Utah State University (2016)
E.K. Sutton, J.M. Forbes, R.S. Nerem, T.W. Woods, Geophys. Res. Lett. **33**, L22102 (2006). doi:10.1029/2006GL027737
J.P. Thayer, J. Geophys. Res. **103**, 471–487 (1998)
J.P. Thayer, J. Geophys. Res. **105**, 23015–23025 (2000)
J.P. Thayer, J. Semeter, J. Atmos. Sol.-Terr. Phys. **66**, 807–824 (2004)
F. Tian, J.F. Kasting, H-L. Liu, R.G. Roble, J. Geophys. Res. **113**, E05008 (2008a). doi:10.1029/2007JE002946
F. Tian, S.C. Solomon, L. Qian, J. Lei, R.G. Roble, J. Geophys. Res. **113**, E07005 (2008b). doi:10.1029/2007JE003043
J.R. Waygant et al., J. Geophys. Res. **105**, 18675 (2000)
D.R. Weimer, J. Geophys. Res. **99**, 19595 (1995)
E. Zesta, C.Y. Huang, in *Space Weather Fundamentals*, ed. by G.V. Khazanov (CRC Press, Boca Raton, 2016)
Y. Zhang, L.J. Paxton, J. Atmos. Sol.-Terr. Phys. **70**, 1231–1242 (2008)

Space Sci Rev (2018) 214:21
https://doi.org/10.1007/s11214-017-0456-3

Extreme Space Weather Events: From Cradle to Grave

**Pete Riley[1] · Dan Baker[2] · Ying D. Liu[3,4] ·
Pekka Verronen[5] · Howard Singer[6] · Manuel Güdel[7]**

Received: 20 July 2017 / Accepted: 5 December 2017 / Published online: 20 December 2017

Abstract Extreme space weather events, while rare, can have a substantial impact on our technologically-dependent society. And, although such events have only occasionally been observed, through careful analysis of a wealth of space-based and ground-based observations, historical records, and extrapolations from more moderate events, we have developed a basic picture of the components required to produce them. Several key issues, however, remain unresolved. For example, what limits are imposed on the maximum size of such events? What are the likely societal consequences of a so-called "100-year" solar storm? In this review, we summarize our current scientific understanding about extreme space weather events as we follow several examples from the Sun, through the solar corona and inner heliosphere, across the magnetospheric boundary, into the ionosphere and atmosphere, into the Earth's lithosphere, and, finally, its impact on man-made structures and activities, such as spacecraft, GPS signals, radio communication, and the electric power grid. We describe preliminary attempts to provide probabilistic forecasts of extreme space weather phenomena, and we conclude by identifying several key areas that must be addressed if we are better able to understand, and, ultimately, predict extreme space weather events.

The Scientific Foundation of Space Weather
Edited by Rudolf von Steiger, Daniel Baker, André Balogh, Tamás Gombosi, Hannu Koskinen and Astrid Veronig

✉ P. Riley
pete@predsci.com

1 Predictive Science Inc., 9990 Mesa Rim Rd., Suite 170, San Diego, CA 92121, USA

2 LASP, University of Colorado, Boulder, CO, USA

3 State Key Laboratory of Space Weather, National Space Science Center, Chinese Academy of Sciences, Beijing 100190, China

4 University of Chinese Academy of Sciences, Beijing 100049, China

5 Finnish Meteorological Institute, Erik Palménin aukio 1, 00560 Helsinki, Finland

6 NOAA, Space Weather Prediction Center, Boulder, CO, USA

7 Department of Astrophysics, University of Vienna, Türkenschanzstrasse 17., 1180 Vienna, Austria

Keywords Extreme space weather events · Coronal mass ejections · Geomagnetic storms · Carrington event

1 Introduction

Extreme space weather events include a range of solar, heliospheric, magnetospheric, ionospheric, atmospheric, and ground phenomena, all driven by coronal mass ejections (CMEs). While rare, such events can have a disproportionately large impact on our technology-reliant society. In this review, we summarize the current state of our knowledge with respect to these extreme phenomena. In particular, we: (1) provide a brief historical review; (2) highlight several specific events that have been well studied; (3) discuss extreme stellar phenomena and their possible relationship to the Sun; (4) summarize the societal consequences of extreme events; (5) describe forecasting techniques that can be used to estimate the occurrence rate of these events; and (6) speculate on the future path that research into extreme events may take.

1.1 What Is an Extreme Event?

An extreme event refers to some phenomenon that is out of the ordinary, i.e., it is rare, or unusual. We could create an arbitrary threshold for such an event, but there are no such values that are agreed upon by the scientific community (although there is a National Space Weather Strategy and Action Plan that calls for establishing benchmarks for space weather events that can have "detrimental effects on the Nation's economic and social well-being" Jonas et al. 2017). Instead, extreme events are identified subjectively, the threshold primarily driven by the specific study being undertaken. Moreover, extreme space weather events can be extreme with respect to one parameter, but only moderate or even weak with respect to another. Thus, any definition of an "extreme" event must include which parameter or parameters make it extreme. As we move from the Sun to the Earth, the quintessential example of an extreme space weather would manifest itself in terms of: (1) massive soft and hard X-rays; (2) a CME traveling extremely rapidly away from the Sun; (3) driving a strong fast-forward shock; (4) the creation of large fluxes of accelerated particles; (5) substantial variations in a range of geomagnetic parameters (e.g., Dst, Kp, and AE), sudden ionospheric disturbances; (6) strong ground-induced currents in the lithosphere; and (7) possibly, the deposition of nitrates in the polar ice. Figure 1 summarizes some of these phenomena.

From a societal perspective, the same event may be viewed in terms of adverse technological consequences: satellite failures, health problems for astronauts and long-haul air passengers, communication blackouts, power outages, and any range of cascading problems from these primary hazards, including food and water shortages, all the way to a breakdown in society (NRC 2008).

Given their rarity, it is reasonable to ask whether extreme solar events are merely "black swans" such as crashes in the stock market (Taleb 2007). Such phenomena are defined as highly improbable events with the following characteristics: (1) unpredictable; (2) massive impact; and (3) an inherently random event for which we retrospectively concoct an explanation to explain it, making it apparently more predictable. Extreme space weather events are not unpredictable, at least once their signatures have been observed at the Sun. They undoubtedly have a massive impact, ranging from failures in the power grid, navigation assets, and communication. However, they do not need us to invent reasons for their occurrence. Given the substantial knowledge we have about them, preparing for them is within our grasp; that is, this is a solvable policy and investment problem. Extreme solar events are not "black swans."

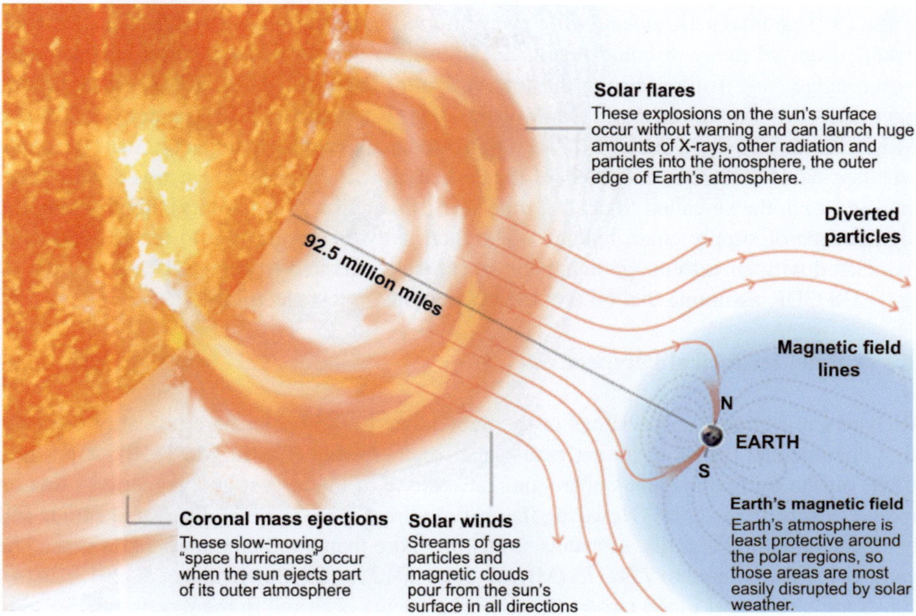

Fig. 1 Extreme space weather events include a broad range of phenomena all the way from the Sun's surface to the Earth's atmosphere. Note that features in this image are not to scale. Image courtesy of the Washington Post

1.2 Why Do We Care?

From a scientific perspective, extreme solar phenomena, leading to extreme space weather events, are vital for understanding the dynamic behavior and variability of the Sun. They represent the outliers of a wide spectrum of transient phenomena. And, while it could be argued that because these events are rare, they are more of a curiosity than fundamental process, it is only by understanding how these events are produced that we can gain a comprehensive understanding of their underlying physical mechanisms. But, more than this, extreme solar events are truly spectacular to contemplate. How is the Sun capable of producing them? Why are they so infrequent? How frequently do they occur? Are there unique physical processes required to produce them? How extreme can they be?

From a societal perspective, the reasons we care about extreme solar events are even clearer. Even moderately severe events have resulted in significant damage to a wide range of technologies. During the March 1989 storm, for example, power to many residents in Quebec was knocked out. Similarly, the $200M Lockheed-Martin built Telstar 401 satellite suffered a massive—and terminal—power failure, likely the result of the January 7, 1997 CME. In January 1994 the ANIK E-1 tumbled out of control, and, later, in May, 1998 the Galaxy IV satellite was lost. The latter event was attributed to a failure of the satellite's primary control processor; a result of a hole developing in the wax coating that allowed whiskers to develop, according to engineers. Solar physicists, however, pointed out that a more likely (but disputed) reason was the high levels of high-energy, solar storm produced electrons (the so-called "killer" electrons), in which the spacecraft was bathed for several weeks, and which slowly fried the electronics (Baker et al. 1994, 1998).

More extreme space weather events will have more extreme societal consequences. A report commissioned by the National Research Council, relying on earlier studies, stated that if the 1921 geomagnetic storm (which was inferred to be ten times stronger than the 1989 event) occurred today, it could result in 130 million people without power and damages in the range \$1–2 trillion during the first year after the event (NRC 2008). Other reports have arrived at similar estimates (McMorrow 2011; Kappenman 2012; Cannon et al. 2013). Most recently, Oughton et al. (2017) estimated the daily economic impact of extreme space weather, concluding that only 49% of the cost of an extreme solar event was due to direct causes within the so-called 'blackout zone" (i.e., disruption of electricity), highlighting the significance of supply chain linkages. The total daily economic loss to the US economy associated with an extreme geomagnetic storm was estimated to be $\sim$ \$40B, or $\sim$ 90% of daily US GDP, assuming that 66% of the US population lost electricity.

1.3 A Brief Historical Perspective

Richard Carrington's observations of the September 1, 1859 solar flare probably mark the first documented record of an extreme space weather event (Carrington 1859). A massive CME hit the Earth's magnetosphere and caused one of the largest geomagnetic storms on record (Siscoe et al. 2006). However, there are indirect records, both observer accounts and evidence in natural records, of events stretching more than a millennium back in time. The carbon-14 spikes dated to 774/775 (Miyake et al. 2012) and 993 (Miyake et al. 2013, 2014), for example, suggest much larger extreme solar events occurred in the recent past. In spite of these, as well as more recent events in the 19th century, the significance of extreme events did not become widely appreciated until much later (i.e., the space era). This significance has grown as society has become more dependent on technology that is sensitive to such events.

1.4 Extreme Events in the Corona

While it could be argued that extreme space weather events, and more specifically, the extreme CMEs that give rise to them, are 'born' below the Sun's surface, at least currently, our first observations come from disk and limb observations of the photosphere and corona. What separates extreme CMEs from the vast majority of otherwise spectacular solar activity? At the least, the following factors are important: (1) They are massive; (2) They are propelled with substantially higher speeds and drive strong fast-forward shocks ahead of them; (3) They carry large internal magnetic fields strengths within them; and (4) They are directed towards the Earth. This last factor is what distinguishes an extreme solar event from an extreme space weather event (at least from the Earth's perspective). Just how large, fast, and magnetic must the event be? Although arguable, an event with a mass in excess of 5×10^{13} Kg, a speed in excess of 3,000 km/s, and internal field strength (as measured at 1 AU) of greater than 100 nT would certainly qualify it as being extreme (Riley 2012). It does not need all three criteria to be met, and which are necessary is application dependent. However, generally, the strength and orientation of the embedded magnetic field will be a major factor in whether or not an event is geo-effective, as will its speed as it approaches 1 AU. Thus, a CME initially traveling at $> 3000\ \mathrm{km\,s^{-1}}$, but which slows down to 1200 km/s by the time it reaches 1 AU will have much less impact than the same event that through the presence of favorable conditions ahead of it (e.g., a high-speed stream or an earlier CME that swept up ambient plasma prior to the passage of the second CME, i.e., a preconditioning CME) maintains a high speed all the way to 1 AU.

To produce an extreme CME requires lots of magnetic flux. The magnetic field in the low corona provides both the energy to drive the eruption of the CME as well as the internal field that it carries away with it. There are several studies that have described empirical relationships between the properties of CMEs and the regions from which they erupt. Vršnak et al. (2007), for example, showed that CMEs from more spatially compact source regions tended to be faster and accelerated more impulsively, presumably due to the higher fields associated with the source regions. Thus, we anticipate that the most extreme CMEs will require high magnetic flux contained within a presumably large enough active region to fuel the eruption. The Carrington event was associated with a large and complex sunspot region, as drawn by Lord Carrington. It also resulted in a white-light flare lasting 5 minutes—the first such case, at least in recorded history—suggesting the conversion of significant amounts of magnetic energy through magnetic reconnection into kinetic motion.

The eruption of an extreme CME also produces solar proton events (SPEs), where particles (primarily protons) are accelerated—either close to the Sun during the flaring process or further away by the CME-driven shock. These particles can pose a significant radiation hazard to both astronauts and spacecraft. Additionally, the can penetrate the geomagnetic field, ionizing particles in the atmosphere (as discussed in more detail in Sect. 1.8).

1.5 Extreme Events in the Solar Wind

In some sense, extreme space weather events in the solar wind are just extensions of what occurred in the corona back at the Sun. However, it is becoming increasingly clear that the preexisting conditions in the solar wind, into which the CME propagates, can have a profound effect on the evolution of the ejecta, and, in turn dictate whether the event will remain "extreme" by the time it reaches the Earth's location (Liu et al. 2014). Consider the following scenarios. First, a CME that is launched into slow and dense ambient solar wind. The interaction will generate a strong magnetosonic shock ahead of the ejecta, which in turn will produce a sheath region between the shock and the leading edge of the ejecta. This will likely contain large fluctuations in the magnetic field that are compressed and amplified. On the other hand, the exchange of momentum from the CME to the wind ahead will cause the speed of the CME to decrease, which in turn will reduce the dawn-dusk electric field (a driver of magnetospheric activity, and estimated by $B_z \times v_x$, where B_z is z-component of the interplanetary magnetic field and v_x is the x-component of the solar wind velocity) that is imposed across the magnetosphere when the CME sweeps over it. Second, consider the same fast CME propagating into fast ambient solar wind. In this case, the shock will not likely be as strong, and the sheath region too will be more modest. In contrast, the CME will retain more of its momentum, and input more energy into the magnetosphere. Finally, consider an example involving the eruption of multiple CMEs from the same active region in relatively rapid succession, such as was inferred for the July 23, 2012 extreme event (discussed in more detail in Sect. 2.3), which missed Earth but was observed by a spacecraft (STEREO A) far from Earth but at a similar distance from the Sun (Liu et al. 2014). The initially fast CME propagates through predominantly slow solar wind, sweeping it up, and generating a shock and sheath in front of it. Behind, it outruns the slower ambient solar wind creating an expansion wave, or rarefaction region; a volume of space with low density and an accelerating velocity profile (Riley et al. 1997; Riley 1999). A second CME launched into this background has even less difficulty maintaining its high speed as it propagates through a tenuous medium. Such conditions, which are optimal for maintaining, or even increasing the ejecta's extreme properties, can lead to a perfect storm scenario that retains the CME's extreme properties far from the Sun. Moreover, these conditions, where multiple CMEs are launched from large, energetic and complex active regions are not that rare. Ruzmaikin et al. (2011), for example, showed that fast CMEs, in particular, tend to arrive (and hence are produced) in clusters.

1.6 Extreme Events in the Magnetosphere

Energy is transferred from the solar wind into the magnetosphere by magnetic reconnection, with the southward component of the interplanetary magnetic field and bulk solar wind flow speed being the prime properties of the solar wind driving the reconnection rate (Dungey 1961).

Extreme events in the magnetosphere are often known as extreme geomagnetic storms. The societal consequences of such events could be severe, with our increasing reliance on a large number of commercial and military satellites currently orbiting within magnetospheric boundaries. Scientifically, the effects of an extreme event have not been well established. With our quintessential space-era event occurring almost 30 years ago (March 1989), only limited data were available. Thus we must rely primarily on models to help us understand what the potential or probable effects of an extreme geomagnetic storm would be.

Current global, coupled MHD magnetospheric (and ionospheric) models are extremely sophisticated. An example, which is used operationally by NOAA's SWPC, is the University of Michigan's BatsRus model. However, they have, for the most part, only been tested against recently observed non-extreme events. Ngwira et al. (2014) performed a global MHD simulation that suggested that the ground electric fields produced by the Carrington event would be twice as large as those measured during the 1989 'extreme' storm. However, the magnetospheric system is incredibly complex, and driving it at such extremes could produce phenomena that are not captured in current models. Saturation effects, in particular, may become increasingly important. A related question concerns the most extreme value that *Dst* could attain during an extreme solar event. While, in principle, it could reach −32,000 nT (resulting in the complete cancellation of the Earth's magnetic field), it is quite possible that it is constrained to a much lower limit of a few thousand nT (Toffoletto et al. 2016).

As we have noted, 'extreme' space weather events may be extreme with respect to some parameter or property, but almost normal with respect to another. Based on an analysis of the *aa* index for all storms from 1868 through 2010, Vennerstrom et al. (2016) inferred that the May 1921 and March 1989 events were the most extreme. Moreover, they argued that neither *Kp* nor *Dst* were as good an indicator of the extremity of the storms. In addition to suggesting that extreme geomagnetic storms were often associated with interacting interplanetary CMEs (ICMEs), they also noted that a necessary precondition for extreme geomagnetic events was that geomagnetic activity was already high.

1.7 Extreme Events in the Ionosphere

At least indirectly, it could be argued that the first space weather effects to be systematically analyzed occurred in the Earth's ionosphere. In 1722, George Graham first reported deflections in a compass needle, shown later in 1822 by Balfour Stewart to be due to currents flowing in the ionosphere (Whitten and Poppoff 1965). And of course, the long-observed aurora point to the influence of the Sun's variability on the ionosphere.

During severe space weather events the ionosphere can be disturbed substantially. This can have a significant impact on the propagation of radio waves, which is of both scientific and societal interest. The scintillation (or distortion) of these waves may be so drastic as to render the signals unrecognizable. More specifically, Global Positioning System (GPS) signals can be impacted by space weather, to the extent that the GPS-based navigation tool for commercial aviation in the US (Wide Area Augmentation System, WAAS) can be disabled during very severe geomagnetic storms (Webb and Allen 2004). Additionally, transpolar flights are also impacted during severe space weather events. In addition to the poorly understood radiation effects that may be suffered by passengers, crew, and equipment, solar energetic protons affect ionospheric conditions in the polar regions that degrade reliable radio communication with the aircraft and can sometimes result in the diversion of these flights.

Space weather also affects shortwave (high-frequency, HF) communications. Today, their use is limited to ships and aircraft that do not have more modern capabilities (such as Iridium), but they also act as the backup for others. Shortwave transmissions, which are generally reflected by the ionosphere and ground (allowing beyond-the-line of-sight communication), may be scattered by the ionospheric irregularities produced by major solar events, causing disruptions to communications. At least three mechanisms contribute to this: (1) solar radio bursts; (2) X-rays produced from solar flares, which disturb the ionospheric D-layer; and (3) Total Electron Content (TEC) enhancements resulting from a geomagnetic storm.

How severe would these impacts be? Cannon et al. (2013) estimated that an extreme space weather event would result in the complete loss of GPS service for at least one day, and possibly up to three days, although they anticipated that it might be possible to receive signals from at least one spacecraft, potentially reducing the effects. The "spill-over" effects of a disruption to the GPS system would be significant too. In the US, in particular, where cellular service relies on GPS, mobile communications would likely be vulnerable to outages (Cannon et al. 2013).

1.8 Extreme Events in the Atmosphere

In the atmospheric context, extreme events are most closely related with SPEs, which can lead to substantial increase of high-energy (> 1 MeV) proton fluxes in the polar atmosphere for several days. Although SPEs are typically identified and sorted based on > 10 MeV proton observations by the GOES satellites in geosynchronous orbit (US Dept. of Commerce 2010), another way to rank them is to look directly at the magnitude of their atmospheric effect, e.g., the amount of odd nitrogen (NOx) produced in the middle atmosphere (Vitt and Jackman 1996; Jackman et al. 2008). In the space age, very intense SPEs have occurred in 1972 (August), 1989 (October), 1991 (March), 2000 (July), 2001 (November), and 2003 (October). Such strong events are infrequent and most likely occur during the maximum or declining phase of solar cycle.

Although SPE protons can have sufficient rigidity (momentum per charge) to pass through the Earth's magnetosphere and precipitate directly into the atmosphere (indeed for some events, the most energetic particles can penetrate to the ground—as so-called "ground level enhancements" Shea and Smart 2012), in practice this only takes place in the polar regions where the Earth's magnetic shielding is weakest. Typically, geomagnetic latitudes poleward of $\sim 60°$, in both hemispheres, experience a more-or-less homogeneous SPE forcing (Rodger et al. 2006; Verronen et al. 2007). The region between 90 and 30 km is significantly affected (e.g. Verronen and Lehmann 2013), corresponding to proton energies between 1 and 200 MeV, respectively. At higher and lower altitudes, the impact of SPEs is negligible compared to the continuous and dominant forcing from precipitating magnetospheric electrons and galactic cosmic rays, respectively.

Due the their strong impact over a rather extensive latitudinal region, SPEs have played an important role in understanding the atmospheric effects of energetic particle precipitation (Jackman and McPeters 2004; Verronen and Lehmann 2013) and references therein. In the 1970s, it was theoretically proposed that atmospheric ionization during SPEs leads to production of odd hydrogen HO_x (H, OH, HO_2) and odd nitrogen NO_x (N, NO, NO_2), and enhanced catalytic depletion of ozone (Swider and Keneshea 1973; Crutzen et al. 1975). Later, satellite observations confirmed that ozone concentration can decrease by tens of percent in the mesosphere and upper stratosphere during large SPEs (e.g. Jackman et al. 2001; Verronen et al. 2006). In the mesosphere, the ozone response to large SPEs may lead to a cooling by a few Kelvins, which modulates zonal, meridional, and vertical winds by up to tens of percent over a period of 4–6 weeks after a event (Jackman et al. 2007). Because NOx is chemically long-lived in the upper stratosphere, a longer-term ozone decrease lasting months beyond the end of an SPE has also been predicted and observed (Seppälä et al. 2004; Jackman et al. 2008, 2009). The effect on total ozone column (most of ozone lies at altitudes below 30 km in the so-called ozone layer) is estimated to be just a few percent even after very large SPEs (Jackman et al. 1996, 2011). Thus, the overall societal impact through modulation of solar UV radiation affecting flora and fauna can be expected to be moderate (although even small increases in UV-B radiation can be harmful for many life forms). However, if the structure and strength of the geomagnetic field changed significantly, e.g. during a polarity transition (which tends to occur at irregular intervals of $\sim 200{,}000$ years), SPEs could also affect mid- and low-latitudes. Thereby reducing the total ozone column by tens of percent (Sinnhuber et al. 2003).

1.9 Extreme Events in the Lithosphere

Space weather effects have been followed all the way to the Earth's conduction crust and upper mantle, known as the lithosphere. The March 1989 magnetic storm, for example, resulted in large electric fields being induced into the lithosphere, the generation of uncontrollable ground-induced currents (GICs) within the Quebec electric-power grid, and an ensuing blackout affecting nine million people (Allen et al. 1989). Additional effects may range from the disruption of telecommunication systems to the corrosion of buried pipelines (Pirjola 2000).

In spite of the potential for a massive storm to cause the failure of the electric power grid over an entire continent (Cannon et al. 2013), and all the consequences that such an event might spawn, the effects of extreme space weather in the lithosphere are the least well known in the entire chain from the Sun to the Earth. Recognizing this, Love et al. (2016) developed geoelectric hazard maps for the continental US showing the likely extreme values for electric field amplitudes of much of the United States. They found that the size of these fields depended sensitively on location, so-called 'once in a century' storms producing electric fields that varied by as much as two orders of magnitude.

An added complication for modeling the response of the lithosphere to extreme events is that GICs are driven not only by large-scale solar events, but also small-scale turbulent features in the solar wind (Huttunen et al. 2008; Pulkkinen 2015). Thus, a robust prediction scheme must account for a wide range of spatial and temporal scale drivers, which, at least currently, cannot be accurately forecasted (Riley et al. 2017). Moreover, small-scale ionospheric features may also be implicated in the generation of GICs (Pulkkinen et al. 2015). Since their origin has not yet been well established, current global magnetospheric/ionospheric models are unlikely to be able to produce them.

2 The Science of Extreme Space Weather Events: Case Studies

2.1 Introduction

Case studies allow us to investigate the complex processes that bring the CME from the solar surface to the Earth. And, for extreme phenomena, in particular, they are crucial. However,

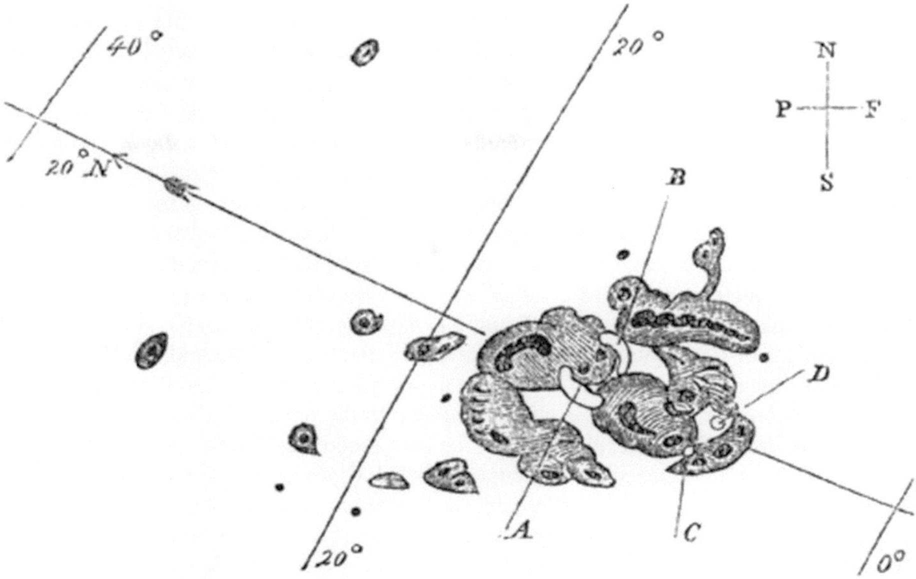

Fig. 2 Lord Carrington's drawing of the sunspots associated with the "Carrington" event illustrate the remarkable detail he captured by hand (Carrington 1859). The two points, A and B mark the initial positions of an intense, bright event, which, over approximately five minutes, moved to points C and D, before disappearing

two caveats must be made. First, as with all case study research, we must take care that our findings hold in general, or at least recognize any biases that might be present. Second, our supposed "case studies" for truly extreme events, may not be sufficiently good fiduciaries. If there are fundamental differences between the events that we can analyze in detail and strictly extreme events, our extrapolations may not be valid.

With that in mind, here, we review two undoubtedly extreme solar events, the so-called Carrington event of September 1, 1859, and the July 23, 2012 event. Additionally, we discuss two events much further back in time, one in 774/775 and the other in 993 AD. Finally, we summarize the limited information available on Sun-like stars, as determined by the remarkable observations by the Kepler spacecraft and their implications for understanding extreme space weather events.

2.2 The "Carrington" Event

Perhaps the most (in)famous extreme solar event is the "Carrington" event, so named because Richard C. Carrington witnessed and reported it. This event is particularly interesting—and frustrating—because it straddles a period in science where many advances were being made both in observational techniques and our understanding of the Sun. Figure 2 shows Lord Carrington's drawing of the Sunspot that produced the September 1st, 1859 solar eruption. Even though this was the first direct report of what would later become to be known as an extreme event, the richness and complexity of the drawing foretells some of the latest advances in understanding some of the key features that make such events "extreme".

Occurring when it did, there is a tantalizing collection of data associated with the Carrington event, but many crucial measurements are missing, relegating it to an event with considerable uncertainty. Inferences of Dst, for example, range from −850 nT to more than −1700 nT (Siscoe et al. 2006). In spite of this, at least until recently, it likely held the record as the largest space weather event observed in over 400 years (Shea et al. 2006). (Only the July 23, 2012 event, observed at the STEREO A spacecraft, and to be discussed below, might have been larger if it had encountered the Earth.) Additionally, the shock arrival time was estimated to be 16.5 hours (from the initial flare observation), making the CME one of, but not the fastest events (Siscoe et al. 2006). The event also occurred during the beginning of the electronic revolution. At the time, the telegraph had been invented and was in regular use (although communications from them during the storm were disrupted), but many other technologies, which we now appreciate are extremely sensitive to geomagnetic activity, remained to be discovered. Thus, it is difficult to extrapolate, with any degree of certainty, the specific effects the Carrington event would have had on our current technological society.

The assessment of atmospheric impacts of the Carrington event, as well as those of any other SPE before the space age, is challenging due to the obvious lack of reliable proton flux measurements. On longer time scales, ice core records have been used to determine the spectral shape and fluence of very large SPEs (McCracken et al. 2001), assuming that the nitrate deposition and sedimentation reflects the amount of stratospheric NOx production during SPEs. However, there are concerns about the proposed connection between SPE-NOx and nitrate deposition, such that ice core records have been deemed unusable for SPE studies (Duderstadt et al. 2016). In particular, many ice core data do not show a nitrate spike related to the Carrington event (Wolff et al. 2008).

Although atmospheric observations during the Carrington event are lacking, there are a few modeling studies that compare its implications to those of the space-age SPEs (Thomas et al. 2007; Calisto et al. 2012). Using SPE characteristics (fluence and spectrum) derived (at least partly) from the ice core records, the applied SPE forcing was set to be 4–6.5 times larger than observed for the strongest events of the space age. Reported effects on HOx, NOx and ozone are qualitatively very similar to smaller SPEs, while the magnitude of the response increases less than the SPE forcing. However, consistently in all studies, an unusually strong and long-lasting ozone depletion develops in the middle-to-upper stratosphere due to larger NOx production. This leads to an estimated 4% decrease in globally-averaged ozone column, maximizing two months after the event, followed by a full recovery after several years. In the polar upper stratosphere, ozone is depleted by 20–30% for months. These results seem to be weakly dependent on the assumed SPE spectral shape. Climate simulations indicate that ozone changes may lead to cooling in the polar mesosphere and upper stratosphere, which would impact zonal winds (polar jets) and surface air temperatures by up to 7 K. However, these effects seem to be most pronounced shortly after the event, and the sign of the response as well as its statistical significance seem to be dependent, e.g., on the SPE spectral shape and seasonality. In summary, although the atmospheric effects of the Carrington event are estimated to be larger than for a typical space-age SPE, they are not drastic in comparison.

2.3 The July 23, 2012 CME

For many years, the operational space weather community has sought a defensible, plausible "worst case" space weather scenario that was better documented than the Carrington event. The Sun provided an almost ideal active experiment on July 23, 2012, when it launched a powerful and extreme CME away from, but within sight of the Earth's many observing platforms that fortuitously encountered the STEREO A spacecraft. Thus, we were able to obtain phenomenal measurements of the event and conclude that, had it hit the Earth, it would likely have caused catastrophic damage (Baker et al. 2013).

The July 23, 2012 event demonstrated how the "perfect" storm" could arise through a combination of circumstances (Liu et al. 2014). In particular, interactions between consecutive CMEs led to a significantly more powerful event than it would have otherwise been. The event was discussed in detail by Russell et al. (2013), Baker et al. (2013), Liu et al. (2014), Riley and Love (2017), Zhu et al. (2016). Here, we provide a brief summary, focusing on the properties of the event within the context of understanding extreme events in general and potentially inferring what consequences such an event could have had, had it hit the Earth.

At 02:08 UT on July 23, 2012 a fast CME erupted from solar active region 11520 on the Sun. It was located at S15° and W133°, and, thus, 43° behind the west limb as viewed from the Earth. STEREO A, however, was 121° west of the Earth's location at this time. Hence, from this spacecraft's perspective, the eruption occurred 12° west of its central meridian, and, as best as *in situ* measurements and related numerical simulations suggest (see below), the spacecraft bore the full brunt of the ejecta and its associated disturbance (Baker et al. 2013). We do not, however, believe that the event was a single CME (Liu et al. 2014). Two prominence eruptions occurred from the same AR, separated in time by 10–15 minutes, which were subsequently seen as CMEs in COR1 observations by STEREO's sister spacecraft (see Fig. 3).

Figure 4 shows *in situ* measurements at STEREO A ($\sim$ 1 AU) during the arrival of the structures associated with the July 23 complex CME. The solar wind speed downstream of the shock is as high as 2246 $\mathrm{km\,s^{-1}}$. Comparison with the CME speed near the Sun indicates a very weak deceleration during the transit from the Sun to STEREO A: The speed of the CME near the Sun was estimated to be 3050 $\mathrm{km\,s^{-1}}$ while the average transit speed was inferred to be 2150 $\mathrm{km\,s^{-1}}$ (Liu et al. 2014). The cause of the weak deceleration is interpreted as the preconditioning of the upstream solar wind by earlier CMEs including the July 19 event, which swept through the corona with an initial speed of 1900 $\mathrm{km\,s^{-1}}$, propagated through the inner heliosphere, and created a tenuous medium into which the July 23 CME propagated (Liu et al. 2014). The shock is quasi-parallel with its normal predominantly along the radial direction from the Sun (Riley et al. 2016), which supports the "preconditioning" idea. Two interplanetary CMEs (ICMEs) can be identified from the magnetic field measurements behind the shock, which may have already merged. The maximum magnetic field inside the ejecta is 109 nT, and the ejecta magnetic fields larger than 60 nT lasted for over 6 hours. The southward magnetic field component was larger than 20 nT for more than 5 hours with a maximum value of 52 nT inside ICME2. Liu et al. (2014) suggest that the extremely enhanced ejecta magnetic field was generated by the in-transit interaction between the two closely launched eruptions. Those magnetic field values would have driven a massive geomagnetic storm, if the event had hit the Earth. Baker et al. (2013), for example, estimated that the July 2012 storm would have driven Dst to under -1200 nT, far exceeding the record set by the March 1989 storm (-589 nT). It should be noted, however, that the storm size is dependent on the empirical scheme used to model Dst (Fig. 4), with one predicting a truly extreme event (-1154 nT) and the other a very severe one (-598 nT).

The magnitude of the associated flare on the Sun was estimated (based on EUVI observations) to be at most X2.5 (Nitta et al. 2013), which is not particularly intense compared with the Carrington ($>$ X10) and 2003 November (X28) flares. This is surprising given how extreme the solar wind disturbance was. The July 23 event indicates a "perfect storm" scenario for the formation of extreme events, i.e., a combination of circumstances results in an event of unusual magnitude. A recent study of the 2015 March 17 intense geomagnetic storm (minimum $Dst = -223$ nT) shows a similar scenario (Liu et al. 2015), although the storm is not "super" in the same sense as the 2012 July 23 event. This may suggest that the "perfect storm" scenario is not as rare as the phrase implies.

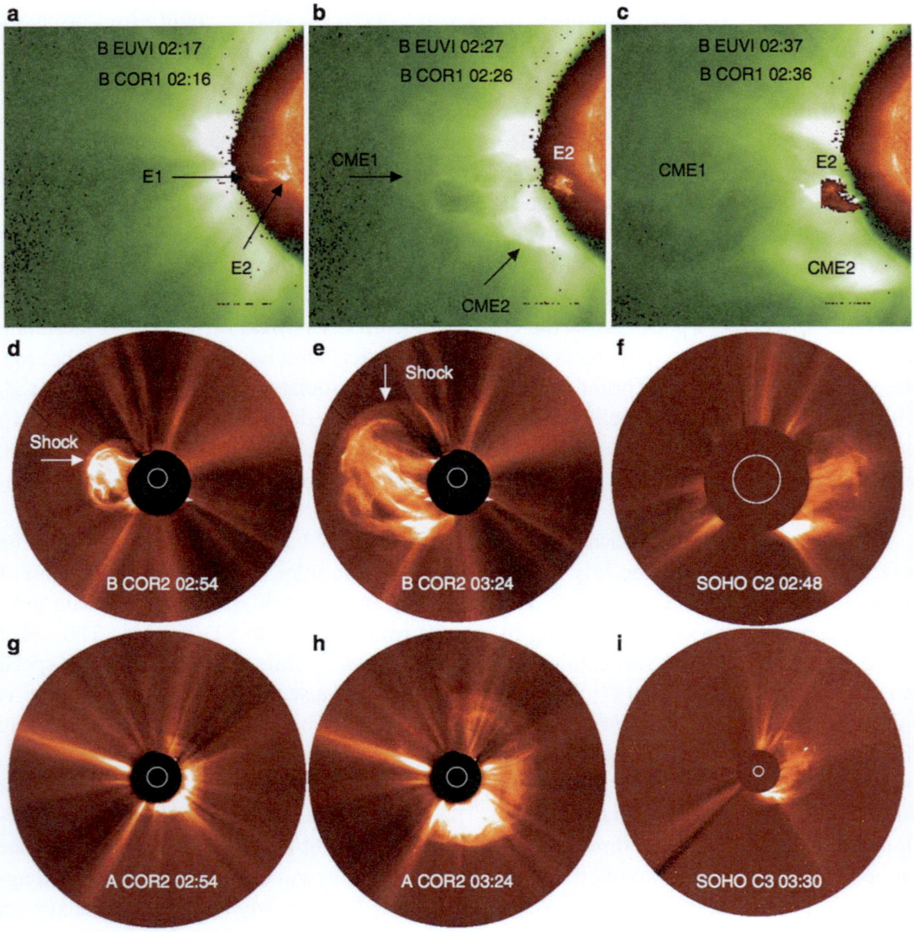

Fig. 3 Origin and initial evolution of the July 23, 2012 CME as viewed from STEREO B (**a**, **b**, **c**, **d**, and **e**), STEREO A (**g** and **h**) and SOHO (**f** and **i**). The *top row* show two prominence eruptions (E1 and E2) and resulting CMEs, while the *middle* and *bottom rows* show the CMEs' evolution through the corona. Adapted from Baker et al. (2013), Liu et al. (2014)

What would have been the consequences to society had the July 23, 2012 CME hit the Earth? Answering this question brings us squarely into the realm of speculation. Some predictions are on firmer ground than others. The X-class flare, for example, would have undoubtedly resulted in immediate radio blackouts. Similarly, it's likely that aurora would have been observed at least as far south as Central America, and the polar cap would have been pushed down to 20° magnetic latitude. The CME and its associated shock would have pushed the magnetopause far further toward the Earth than any other event in the space era; perhaps as close as 1–2 R_E altitude. At this distance, GEO spacecraft, which provide weather, communication, and surveillance, would have found themselves outside the Earth's bow shock, in the solar wind. Similarly, GPS spacecraft would no longer have been located within the magnetosphere. Finally, it is possible that such a storm would completely eradicate the outer radiation belt (with it reforming at low L), and totally deplete the inner radiation belt. Further down, it has been estimated that the July 23 storm would have resulted

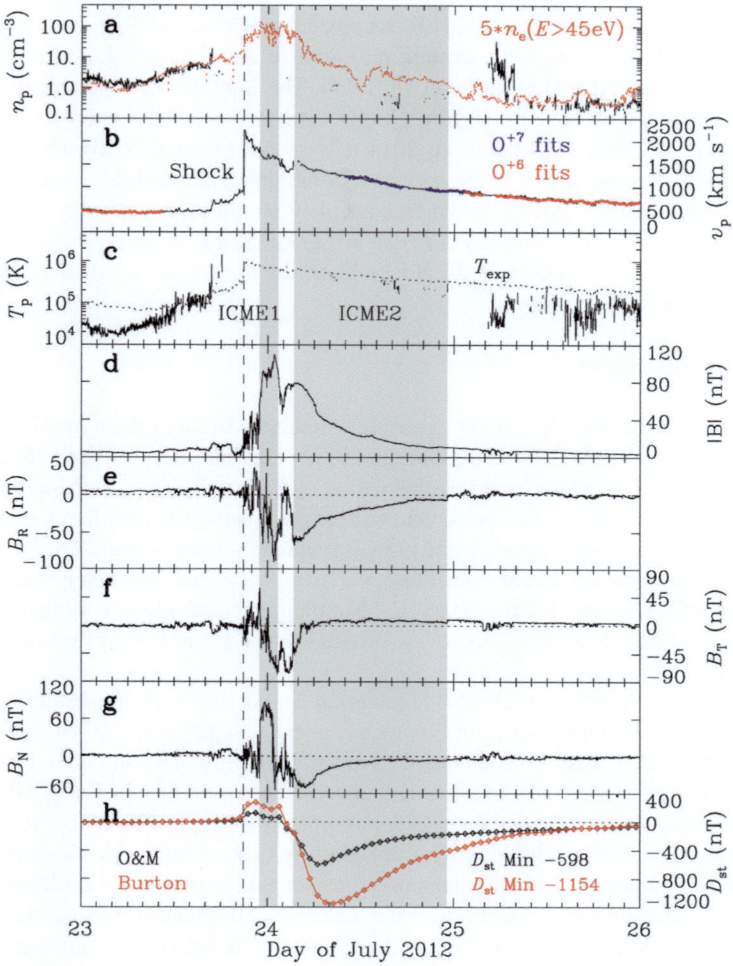

Fig. 4 In situ measurements of the 2012 July 23 complex CME at STEREO A (after Liu et al. 2014). The panels show: (**a**) the proton density—as directly computed from the plasma moments (*black*) and as inferred from electrons at energies greater than 45 eV (*red*); (**b**) bulk speed; proton temperature—as directly computed from the plasma moments (*black*) and as inferred from the "expectation temperature" based on solar wind speed; (**d–g**) magnetic field strength and components (B_R, B_T, and B_N, where B_N is approximately the same as the *z*-component of the IMF, B_z); and (**h**) modeled *Dst* index using two different empirical schemes. The *shaded regions* indicate the two ICME intervals, and the *vertical dashed line* marks the associated shock. For further details, please see Liu et al. (2014)

in only modest geomagnetically induced currents (GICs) in and low-latitude power grids (Zhang et al. 2016), although this work did not account for equatorial electrojet effects, which might substantially increase the risk for at least some power networks. More work remains to be done on the effects at mid- and high-latitudes.

2.4 The 774/775 and 993 AD Events

If the July 2012 provided a giant step forward in terms of data availability over the 1859 event, two recently discovered but nevertheless exciting events represent a commensurately large step backward. Miyake et al. (2012) described an increase in 14C concentrations from tree rings that she attributed to a cosmic ray event in AD 774/775. Later, another event was identified from AD993 (Miyake et al. 2014). The inferred flare size for these events was $\sim 10^{34}$ erg, which is some 20 times larger than the bolometric energy estimated for a Carrington-class flare (Miyake et al. 2014; Cliver and Dietrich 2013). Currently, there is no consensus on the origin of these events. Some potential candidates are: (1) the Sun (Usoskin and Kovaltsov 2012), in which case, they would represent "Super-Carrington" events; (2) a nearby supernova (Miyake et al. 2012); or (3) a gamma-ray burst (Hambaryan and Neuhäuser 2013). Thus, currently, at least some doubt exists as to whether these events are extreme solar events or not.

2.5 Kepler Superflares

The Kepler satellite was developed to detect exoplanets through their transit of the star around which they orbit. From 2009 through 2013 it continuously observed 160,000 stars, approximately half of which could be defined as "solar type" stars. Maehara et al. (2012) used the publicly available data (at 30 min cadence) to search for superflares on solar type stars. They identified 365 of them from 148 G-type main sequence stars. They also studied brightness variations associated with these stars, inferring that they likely resulted from the rotation of a star with a large "starspot." Simple model calculations suggest that these starspots would have a radius of $0.16\ R_{star}$, which is considerably larger than any spot seen on the Sun.

More recently, Maehara et al. (2015) analyzed 1-min Kepler data, identifying 187 superflares on 23 solar-type stars with bolometric energy ranges from 10^{32} to 10^{36} erg. For reference, the largest solar flares recorded have a bolometric luminosity of $< 10^{30}$ erg. Using these results, they were able to estimate that occurrence rates of superflares with energies of 10^{33} or larger (i.e., $\times 100$ solar flare) is approximately once per 500–600 years. They further estimated that the maximum energy released by these events was comparable to the amount of energy stored within the starspots. A reasonable criticism that could be leveled at these stellar extrapolations is that perhaps the underlying mechanism for solar flares and the solar-like stellar flares is different. For example, in star-disk systems the large-scale field of the star might be connected to a co-rotating disk, the interaction of which can lead to some form of disruption and the resulting superflare (Hayashi et al. 1996). Karoff et al. (2016) studied stars observed by the LAMOST telescope, 48 of which produced superflares. They argued that magnetic reconnection in the corona of these stars remained the best candidate mechanism based on the increased chromospheric emissions from these stars.

3 Societal Consequences of Extreme Space Weather Events

One of the advances—or realizations—over the last decade has been to recognize the importance of the interdependence of various components in our complex and technology reliant society. As illustrated in Fig. 5, disruptions in electric power, and in turn oil/gas supplies can have significant and cascading effect in a wide range of sectors, including, but not limited to: banking and finance, government services, and emergency response. In particular, it is worth pointing out that since 2008, when this figure was published, society now relies more heavily on other technologies, such as GPS and cell phones, suggesting that the impact may be even wider. Moreover, as society becomes evermore dependent on newer technologies, the impact will only become larger; possibly in quite unanticipated ways.

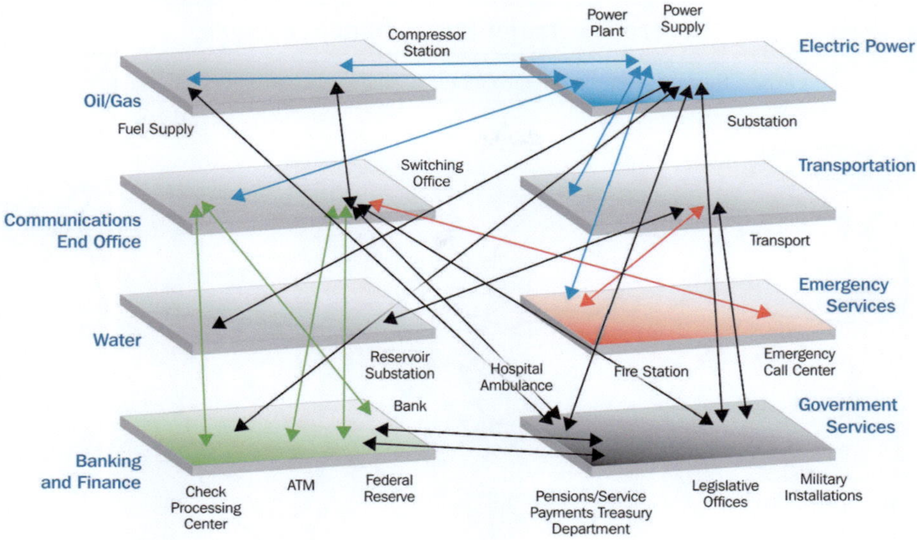

Fig. 5 The interdependency of Society and the cascading effects of an extreme space weather event. Image courtesy of NRC (2008)

One of the more controversial potential consequences of an extreme geomagnetic storm concerns the widespread outage of the US power system. Kappenman (2012) projected that the worst-case scenario for a one-in-a-hundred-year storm would include: (1) 1–2 Trillion dollars of economic loss; and (2) 130 million people without electrical power for several years, based on the destruction of several hundred transformers. Figure 6 illustrates such a scenario, where the red and green circles show the size of ground induced currents (GICs) at transformers. Two areas—the Eastern Interconnect and the Pacific Northwest—are highlighted as areas of probable power system collapse, which would result in more than 2/3 of the population suffering power loss. Because of long (15 month) delivery times for replacement transformers, estimates of four years were made for full service to be restored.

Inferring the probable consequences of an extreme space weather event, however, is fraught with uncertainty. Our experience is limited to a handful of driving events providing, at best, anecdotal evidence for what might happen. The Carrington event of 1859, for example, resulted in a several reports of fires starting in telegraph offices. But how can we translate this into effects within today's electronic society? As we have noted above, more recent storms, such as that of March 1989, provide a glimpse of how technology and the electric grid, in particular might be affected. The geomagnetic storm of October 2003 (the so-called "Halloween Storms"), for example, may have caused the destruction of five transformers and the damage of ten more in South Africa (McMorrow 2011). Nevertheless, extrapolation of these results to truly extreme events is challenging. First, current cases of transformer failure are relatively rare, making it difficult to predict how many more would be destroyed when GICs spike or due to longer-term degradation from overheating. Second, our ability to model and forecast the US grid remains in its infancy. Third, estimates for the replacement time for destroyed transformers may change/improve, as, more generally will happen as the grid is incrementally improved over time.

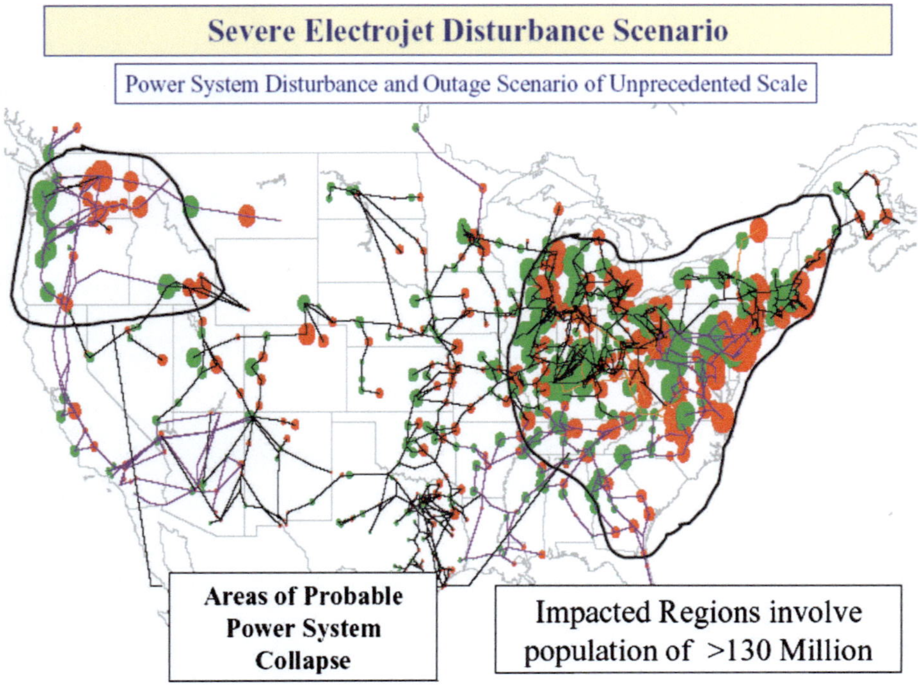

Fig. 6 Prediction of regional power grid disruptions from an extreme space weather event in the USA: 4800 nT/min geomagnetic field disturbance at 50° geomagnetic latitude scenario. *Red* and *green circles* indicate the magnitude of GICs into and out of transformers from the ground. The areas of probable power system collapse are enclosed in *thick black curves*. An estimated 130 million or more people would be affected by this event. Image courtesy of J. Kappenman, Metatech Corp.

4 Forecasting Extreme Space Weather Events

4.1 Event-Based Predictions

Ultimately, one of the primary goals of any operational space weather service is to predict the properties of specific CMEs long before they reach the Earth. Ideally, this prediction should be several days to more than a week in advance to give the relevant authorities sufficient time to implement their preventative strategies. In reality, such an advance warning would require a prediction of the event prior to the appearance of any reliable CME-related signatures at the Sun. Although as we discuss below, this may be a realizable goal in a probabilistic sense, no observational signatures have yet been discovered that would allow us to predict the occurrence of a specific event from a particular active region. Thus, at least for the foreseeable future, we must focus on forecasts that begin when the first signs of a CME eruption are present. This limits our prediction window to as little as 12 hours in the most extreme cases, limiting the actions that can be taken, for example, by power grid operators.

In order of importance, we would like to predict the z-component of the magnetic field, B_z, the x-component of the velocity, v_x, and the plasma density, ρ. These produce the dawn-dusk electric field ($B_z \times v_x$) and the solar wind momentum flux ($\rho x v_2$). For extreme events, in particular, knowledge even of the average value of these quantities, particularly the sign (or variation in sign) of B_z, as well as their duration, would provide vital information for estimating the probable impact of an extreme event.

If as Kepler data suggest, some solar-type stars are capable of producing flares with energies as high as 10^{36} erg, we might ask what would be the ramifications for Earth and civilization? Although considerably more speculative, we might forecast the damage and loss of all Earth-orbiting satellites, loss of power and potential long-term or permanent damage to power stations, including nuclear stations, a global blackout, loss of essentially all modern modes of communication. Aircraft would be exposed to fatal doses of radiation, and depletion of the ozone layer may even occur (NRC 2008).

4.2 Probabilistic Forecasts

Probabilistic forecasts avoid, or perhaps more accurately evade the difficult issue of predicting when a particular event is going to occur, by addressing the significantly easier question of the likelihood of an event occurring. While this provides little day-to-day guidance for space weather operations, in analogy with climatology, it can be useful for longer term policy decisions.

Riley et al. (2012) first addressed the likelihood of extreme space weather events occurring using a rigorous statistical model. Assuming that distributions of some space weather datasets could be approximated by a power-law distribution, they estimated the probability of another Carrington-like event occurring over the next decade. Depending on the parameter, that probability ranged from a few percent up to 12%. Love (2012) found similar estimates, but, more importantly estimated the errors associated with those forecasts, finding them to be substantial. More recently, Love (2012), Love et al. (2015) questioned the assumption that the data could be adequately represented by a power law, instead, suggesting that a log-normal distribution was more appropriate; at least for Dst. In spite of this, they concluded that a Carrington event was likely to occur approximately 1.13 times per century, with 95% CI [0.42, 2.41], consistent with the 10% per decade estimates obtained previously.

We can extend these studies in several important ways. First we can ask whether the data are distributed to one of several families (exponential, log-normal, or power-law). Second, we can remove the subjectivity associated with choosing the regions of the distribution that are fitted. Third, we can include non-parametric bootstrapping to estimate the confidence intervals (i.e., the uncertainties in the forecasts). And fourth, we can apply a likelihood test to discriminate between the various models.

Figure 7 summarizes geomagnetic storms where Dst exceeded −100 nT over the last 60 years. We note that only one event approached 600 nT (the March 1989 storm) and only five events exceeded 400 nT. Additionally, it is apparent that the storms tend to cluster on a timescale of 11 years, mimicking, but trailing the sunspot cycle. Bimodel peaks in these clusters are also present, matching those observed in CME rates (Riley et al. 2006). Finally, there is an apparent tendency for the strongest storms to become stronger from 1965 through 2007; while the largest five storms around 1970 were 200–300 nT, the five most intense storms around 2005 were 350–450 nT. In contrast the most recent decade shows a paucity of events, particularly severe ones. Taken together, these observations suggest that Dst displays both periodic and secular trends during this 60 year period, which attacks a basic assumption in this approach that the data are time stationary over the period of interest.

With this caveat in mind, following Riley and Love (2017) we can estimate the likelihood of an extreme space weather event (as large, or larger than the Carrington event) for a set of distributions of storm sizes. Figure 8 summarizes the procedure. The left panel shows the complementary cumulative distribution function (CCDF), which is the probability that an event as large, or larger than some critical value will occur during a unit time interval. The open circles show all of the events. One of the advantages in using the CCDF, as oppose to

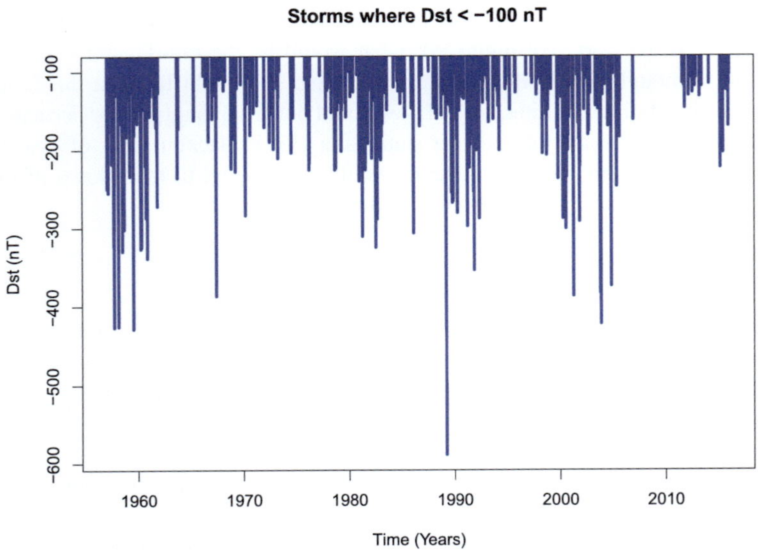

Fig. 7 Magnetic storms, defined by events where $|Dst|$ exceeds 100 nT, are shown as a function of time. Individual storms are identified as contiguous intervals where $|Dst|$ exceeded 100 nT. The data were obtained from NASA's COHOWEB, which in turn obtained the data from World Data Center for Geomagnetism at the Data Analysis Center for Geomagnetism and Space Magnetism, Kyoto University

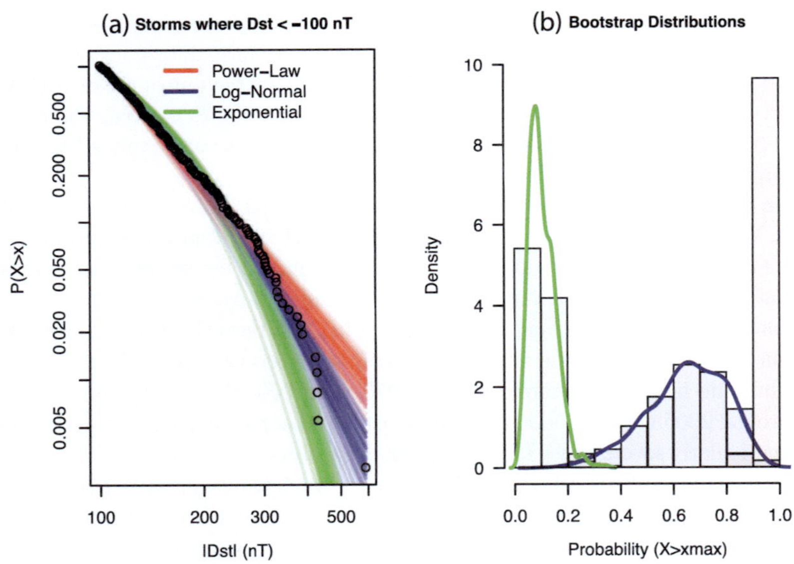

Fig. 8 (**a**) Complementary cumulative distribution function (CCDF) for the geomagnetic storms shown in Fig. 7. Bootstrap fits for the three distributions (power-law, log-normal, and exponential) are superimposed. (**b**) Histogram and density plots showing the probability of an event as large as, or larger than the largest event in the dataset (−589 nT) over the duration of the dataset. The density curve colors follow the convention given in the legend within panel (**a**)

the probability of an event of size x occurring is that the data are not binned, but summed from right to left. More practically, we are also not interested in events of a particular size, but events as large or larger than a particular size, since those more severe events will also cause at least as much damage. The data seem to follow a straight line (in log-log space), at least out to 280 nT. Beyond this, with the exception of the most severe storm (589 nT), they appear to fall off more rapidly. The colored curves summarize a selection of bootstrap fits for each of the three distributions considered. We infer that the power-law profiles capture the lower severity measurements well, but over-estimate the likelihood of the most severe events. In contrast, the log-normal profiles underestimate the low-severity events but capture the trends at the highest severities. Finally, the exponential profiles both over-estimate the low-severity events and underestimate the high-severity events. Figure 8(b) summarizes the likelihood of observing an event as severe as or more severe than the most severe event observed, that is, $Dst < -589$ nT, during the entire span of the data (~ 57 years). The probabilities were calculated for each bootstrap iteration and the distributions derived from the probabilities for each of them.

Based on these results, our best estimate for the probability of another extreme geomagnetic event comparable to the Carrington event occurring within the next 10 years is 10.3% with 95% confidence intervals (CI) in the range [0.9, 18.7] for a power-law distribution, but only 3.0% with 95% CI [0.6, 9.0] for a log-normal distribution (see also Riley and Love 2017). Our results, however, depend on: (1) how an extreme event is defined; (2) the statistical model used to describe how the events are distributed in intensity; (3) the techniques used to infer the model parameters; and (4) the data and duration used for the analysis. Riley and Love (2017) tested the assumption that the data are approximately represented by a time-stationary process, arguing that while not strictly true, it could be approximately factored into one's interpretation of the forecasts. One of the key results from these studies is that both the uncertainties and the assumptions used to derive the forecasts are at least as important as the forecasts themselves.

It has been argued that, extreme events may be associated with smaller solar cycles. This stems from the observation that the 1859 and 1921 storms, arguably the largest two geomagnetic storms on record, occurred during cycles where the peak sunspot number reached only 100. While suggestive, more analysis needs to be done. In fact, based on the work by Riley and Love (2017), it is more likely that extreme events will be associated with solar active conditions.

5 How Extreme Can Extreme Events Become?

Having witnessed events with speeds of greater than 3,000 km/s near the Sun, with embedded fields exceeding 100 nT at 1 AU, and producing Dst values in excess of 850 nT, it is clear that the Sun is capable of producing events that would have catastrophic effects in the Earth's environment. But, just how bad could the most extreme event be?

Considering the Earth's magnetosphere, there's a physical limit to the most extreme value of Dst: $-32{,}000$ nT. Why? Because this would be enough to completely nullify the Earth's geomagnetic field. What kind of CME could produce such a deviation? Something immeasurably larger than we have yet witnessed, and, possibly larger than anything inferred from Kepler observations. Instead, we might consider what the Sun is capable of producing. What would be the most extreme CME that could be generated by the largest active region we can imagine being supported by the Sun? Knowledge of the magnetic field, velocity, and mass of such an event could be used to estimate the maximum possible deviation in Dst.

The largest active region observed during the past 100+ years was the "great sunspot" of 1947, which had an approximate area of $\sim 7.5 \times 10^{-3}$ of the visible area of the Sun. This of course does not tell us about whether or not the magnetic fields embedded within it will erupt, but it does provide a rough limit on the available energy that could be used to fuel an eruption. Shibata et al. (2013) estimated that to generate a flare with energy of 10^{34}–10^{35} erg would require a massive sunspot housing a magnetic flux of $\sim 2 \times 10^{23}$–1×10^{24} Mx. Using simple ideas related to our understanding of the mechanisms that drive the solar dynamo, he estimated that it would take approximately 40 years to produce such a region on the Sun. Thus, even if the Sun were physically able to support such a structure (and it is not clear that the mechanisms that actually produce the stellar superflares have analogous processes on the Sun) we would be able to anticipate the potential likelihood of such an event well in advance.

6 Conclusions and Future Directions

In this review, we have discussed the basic features of extreme space weather phenomena, in part, through the use of two case studies: The Carrington event of 1859 and the STEREO A event of July 23, 2012. While infrequent, extreme space weather phenomena occurring in the future will likely produce catastrophic effects on Earth. Although we have developed a basic understand of how they are produced, evolve, and interact with the Earth's environment, many mysteries remain. What, for example, is the largest possible CME (in terms of mass, speed, and embedded magnetic field strength) that the Sun can produce? Can isolated extreme CMEs produce catastrophic effects, or must there be a "perfect storm" of conditions? What is the largest variation in the Earth's horizontal magnetic field during an extreme event? How large are the GICs that can be produced? These questions all state the more general question: Just how bad can things get?

From a scientific perspective, there are a number of avenues we could pursue, which would allow us to better understand, and, ultimately predict extreme space weather events. Observational-based studies of our limited set of events may identify further knowledge about what makes these events so extreme. The concept of "preconditioning" appears to be a necessary criterion. If further substantiated, this suggests that even rudimentary forecasts may be able to distinguish between "severe" and "extreme" based on whether a prior CME erupted from the same or nearby active region. Additionally, if the origin of the 774/775 and 993 events can be better constrained they, together with extreme events from solar-like stars, may provide further clues about just how extreme these events can become. Modeling-based studies will, at some point, be able to reproduce all the key features within extreme ICMEs as well as their likely effects in the Earth's magnetosphere. Current modeling efforts can capture the hydrodynamic effects of such events (Baker et al. 2013), and recent work is beginning to explore the complex evolution of the magnetic field within the erupting CME and the role it plays in the structure's evolution as it propagates to 1 AU (Linker et al. 2016).

From a societal point of view, low-frequency but high-impact extreme events are clearly a danger. With an estimated \$1–2 trillion in damages from the consequential harm caused by a widespread power grid blackout, and a projected recovery time of 4–10 years, some questions that naturally arise are: (1) How much will it cost to mitigate these damages? (2) How long will it take to implement hardening procedures? (3) How do the hazards from extreme space weather events compare with those from other natural disasters?

We propose that policy makers and space weather professionals should adopt the July 23, 2012 solar storm to "war game" emergency preparedness planning for extreme space weather events. It is more likely than not that the Earth will be hit by such an event within the next 50 years, and the consequences could be absolutely devastating.

To address these questions, however, requires that we assess what assets are most at risk, and under what types of extreme phenomena. What, for example, are the effects on the US power grid, and dependent interconnected systems, under the influence of severe space weather? What techniques or engineering solutions are required to keep GICs isolated from key infrastructure? These are not the types of questions that can be addressed by scientists. They require work by the power engineers and transformer experts.

While our emphasis here has centered on a scientific discussion, the obvious societal impacts that an extreme space weather event may have require that we consider more practical approaches for understanding and, ultimately, forecasting these events. For example, it appears that more extreme space weather events require more time to develop. If a truly extreme "superflare" event is possible, it is likely that it will take decades or more to develop (Shibata et al. 2013). If so, this suggests that we have time to prepare for it. More importantly, by studying more deeply the relationship between the dynamo and visible photosphere/corona, we may be able to better constrain the timescales for production of events of different sizes. Since the sunspot associated with the 1859 event developed rapidly during the month prior to its eruption, it appears that even for "modestly" extreme events, there would be little to no warning, since they could develop just past the west limb and have ~ 14 days to develop. Observations near the east limb, which would afford us, at best one week's advance warning serve as our long term forecast. But even then, even with a massive active region to observe, our models are currently not sophisticated enough to predict when (or, in some cases, if) an eruption will take place. Consider the July 23, 2012 event. There were multiple eruptions associated with that active region. The earlier July 12 event is the one that is currently the focus of several studies because it intercepted Earth (e.g., Hu et al. 2016). However, it was considerably more modest than the July 23 event that measured by STEREO A. Can we identify patterns in our complex active regions erupt, which will ultimately result in an estimate for the region's propensity to erupt?

In closing, these are but a few of the scientific challenges that must be addressed if we are to successfully protect our technologically-dependent society from the impacts of extreme space weather.

Acknowledgements The authors would like to thank the International Space Science Institute (ISSI) for their gracious hospitality while hosting our team for a workshop on space weather. PR would also like to acknowledge support from NASA's Living with a Star program. Finally, we would like to extend our sincerest thanks to a reviewer who provided extremely detailed and constructive comments on a draft of the paper.

References

J. Allen, H. Sauer, L. Frank, P. Reiff, Effects of the March 1989 solar activity. Eos **70**(46), 1479–1488 (1989)

D.N. Baker, S. Kanekal, J.B. Blake, B. Klecker, G. Rostoker, Satellite anomalies linked to electron increase in the magnetosphere. Eos **75**(35), 401–405 (1994)

D.N. Baker, J.H. Allen, S.G. Kanekal, G.D. Reeves, Disturbed space environment may have been related to pager satellite failure. Eos **79**(40), 477–483 (1998)

D.N. Baker, X. Li, A. Pulkkinen, C.M. Ngwira, M.L. Mays, A.B. Galvin, K.D.C. Simunac, A major solar eruptive event in July 2012: defining extreme space weather scenarios. Space Weather **11**(10), 585–591 (2013)

M. Calisto, P.T. Verronen, E. Rozanov, T. Peter, Influence of a Carrington-like event on the atmospheric chemistry, temperature and dynamics. Atmos. Chem. Phys. **12**(18), 8679–8686 (2012)

P. Cannon, M. Angling, L. Barclay, C. Curry, C. Dyer, R. Edwards, G. Greene, M. Hapgood, R.B. Horne, D. Jackson et al., Extreme space weather: impacts on engineered systems and infrastructure. Royal Academy of Engineering (2013)

R.C. Carrington, Description of a singular appearance seen in the Sun on September 1, 1859. Mon. Not. R. Astron. Soc. **20**, 13–15 (1859)
E.W. Cliver, W.F. Dietrich, The 1859 space weather event revisited: limits of extreme activity. J. Space Weather Space Clim. **3**, 31 (2013)
P.J. Crutzen, I.S.A. Isaksen, G.C. Reid, Solar proton events: stratospheric sources of nitric oxide. Science **189**(4201), 457–459 (1975)
K.A. Duderstadt, J.E. Dibb, N.A. Schwadron, H.E. Spence, S.C. Solomon, V.A. Yudin, C.H. Jackman, C.E. Randall, Nitrate ion spikes in ice cores not suitable as proxies for solar proton events. J. Geophys. Res. Atmos. **121**(6), 2994–3016 (2016)
J.W. Dungey, Interplanetary magnetic field and the auroral zones. Phys. Rev. Lett. **6**(2), 47 (1961)
V.V. Hambaryan, R. Neuhäuser, A galactic short gamma-ray burst as cause for the 14c peak in ad 774/5. Mon. Not. R. Astron. Soc. **430**(1), 32–36 (2013)
M.R. Hayashi, K. Shibata, R. Matsumoto, Astrophys. J. **468**, L37 (1996)
H. Hu, Y.D. Liu, R. Wang, C. Möstl, Z. Yang, Sun-to-Earth characteristics of the 2012 July 12 coronal mass ejection and associated geo-effectiveness. Astrophys. J. **829**(2), 97 (2016)
K.E.J. Huttunen, S.P. Kilpua, A. Pulkkinen, A. Viljanen, E. Tanskanen, Solar wind drivers of large geomagnetically induced currents during the solar cycle 23. Space Weather **6**(10), S10002 (2008)
C.H. Jackman, R.D. McPeters, The effect of solar proton events on ozone and other constituents, in *Solar Variability and Its Effects on Climate* (2004)
C.H. Jackman, E.L. Fleming, S. Chandra, D.B. Considine, J.E. Rosenfield, Past, present, and future modeled ozone trends with comparisons to observed trends. J. Geophys. Res., Atmos. **101**(D22), 28753–28767 (1996)
C.H. Jackman, R.D. McPeters, G.J. Labow, E.L. Fleming, C.J. Praderas, J.M. Russell, Northern Hemisphere atmospheric effects due to the July 2000 solar proton event. Geophys. Res. Lett. **28**(15), 2883–2886 (2001)
C.H. Jackman, R.G. Roble, E.L. Fleming, Mesospheric dynamical changes induced by the solar proton events in October–November 2003. Geophys. Res. Lett. **34**(4), L04812 (2007)
C.H. Jackman, D.R. Marsh, F.M. Vitt, R.R. Garcia, E.L. Fleming, G.J. Labow, C.E. Randall, M. López-Puertas, B. Funke, T. von Clarmann et al., Short-and medium-term atmospheric constituent effects of very large solar proton events. Atmos. Chem. Phys. **8**(3), 765–785 (2008)
C.H. Jackman, D.R. Marsh, F.M. Vitt, R.R. Garcia, C.E. Randall, E.L. Fleming, S.M. Frith, Long-term middle atmospheric influence of very large solar proton events. J. Geophys. Res., Atmos. **114**(D11), D11304 (2009)
C.H. Jackman, D.R. Marsh, F.M. Vitt, R.G. Roble, C.E. Randall, P.F. Bernath, B. Funke, M. López-Puertas, S. Versick, G.P. Stiller et al., Northern Hemisphere atmospheric influence of the solar proton events and ground level enhancement in January 2005. Atmos. Chem. Phys. **11**(13), 6153–6166 (2011)
S. Jonas, W. Murtagh, M. Bonadonna, Released for public comment: space weather benchmarks and operations-to-research plan. Space Weather **15**(2), 282 (2017)
J.G. Kappenman, Geomagnetic disturbances and impacts upon power system operation, in *Electric Power Generation, Transmission, and Distribution* 3rd edn. (CRC Press, Boca Raton, 2012), pp. 1–22
C. Karoff, M.F. Knudsen, P. De Cat, A. Bonanno, A. Fogtmann-Schulz, J. Fu, A. Frasca, F. Inceoglu, J. Olsen, Y. Zhang et al., Observational evidence for enhanced magnetic activity of superflare stars. Nat. Commun. **7**, 11058 (2016)
J. Linker, T. Torok, C. Downs, R. Lionello, V. Titov, R.M. Caplan, Z. Mikić, P. Riley, MHD simulation of the Bastille day event, in *AIP Conference Proceedings*, vol. 1720 (AIP Publishing, New York, 2016)
Y.D. Liu, J.G. Luhmann, P. Kajdič, E.K.J. Kilpua, N. Lugaz, N.V. Nitta, C. Möstl, B. Lavraud, S.D. Bale, C.J. Farrugia et al., Observations of an extreme storm in interplanetary space caused by successive coronal mass ejections. Nat. Commun. **5**, 3481 (2014)
Y.D. Liu, H. Hu, R. Wang, Z. Yang, B. Zhu, Y.A. Liu, J.G. Luhmann, J.D. Richardson, Plasma and magnetic field characteristics of solar coronal mass ejections in relation to geomagnetic storm intensity and variability. Astrophys. J. Lett. **809**(2), 34 (2015)
J.J. Love, Credible occurrence probabilities for extreme geophysical events: earthquakes, volcanic eruptions, magnetic storms. Geophys. Res. Lett. **39**, 10301 (2012). https://doi.org/10.1029/2012GL051431
J.J. Love, E.J. Rigler, A. Pulkkinen, P. Riley, On the lognormality of historical magnetic storm intensity statistics: implications for extreme-event probabilities. Geophys. Res. Lett. **42**(16), 6544–6553 (2015)
J.J. Love, A. Pulkkinen, P.A. Bedrosian, S. Jonas, A. Kelbert, E.J. Rigler, C.A. Finn, C.C. Balch, R. Rutledge, R.M. Waggel et al., Geoelectric hazard maps for the continental united states. Geophys. Res. Lett. **43**(18), 9415–9424 (2016)
H. Maehara, T. Shibayama, S. Notsu, Y. Notsu, T. Nagao, S. Kusaba, S. Honda, D. Nogami, K. Shibata, Superflares on solar-type stars. Nature **485**(7399), 478–481 (2012)

H. Maehara, T. Shibayama, Y. Notsu, S. Notsu, S. Honda, D. Nogami, K. Shibata, Statistical properties of superflares on solar-type stars based on 1-min cadence data. Earth Planets Space **67**(1), 1 (2015)

K.G. McCracken, G.A.M. Dreschhoff, E.J. Zeller, D.F. Smart, M.A. Shea, Solar cosmic ray events for the period 1561–1994: 1. Identification in polar ice, 1561–1950. J. Geophys. Res. **106**(15), 21585–21598 (2001). http://adsabs.harvard.edu/abs/2001JGR...10621585M

D. McMorrow, *Impacts of Severe Space Weather on the Electric Grid* (JASON, Virginia, 2011)

F. Miyake, K. Nagaya, K. Masuda, T. Nakamura, A signature of cosmic-ray increase in AD 774–775 from tree rings in Japan. Nature **486**(7402), 240–242 (2012)

F. Miyake, K. Masuda, T. Nakamura, Another rapid event in the carbon-14 content of tree rings. Nat. Commun. **4**, 1748 (2013)

F. Miyake, K. Masuda, M. Hakozaki, T. Nakamura, F. Tokanai, K. Kato, K. Kimura, T. Mitsutani, Verification of the cosmic-ray event in AD 993–994 by using a Japanese Hinoki tree. Radiocarbon **56**(3), 1189–1194 (2014)

C.M. Ngwira, A. Pulkkinen, M.M. Kuznetsova, A. Glocer, Modeling extreme Carrington-type space weather events using three-dimensional global MHD simulations. J. Geophys. Res. Space Phys. **119**(6), 4456–4474 (2014)

N.V. Nitta, M.J. Aschwanden, P.F. Boerner, S.L. Freeland, J.R. Lemen, J-P. Wuelser, Soft x-ray fluxes of major flares far behind the limb as estimated using STEREO EUV images. Sol. Phys. **288**(1), 241–254 (2013)

NRC, *Severe Space Weather Events—Understanding Societal and Economic Impacts: A Workshop Report* (National Academies Press, Washington, DC, 2008)

E.J. Oughton, A. Skelton, R.B. Horne, A.W. Thomson, C.T. Gaunt, Quantifying the daily economic impact of extreme space weather due to failure in electricity transmission infrastructure. Space Weather **15**(1), 65–83 (2017)

R. Pirjola, Geomagnetically induced currents during magnetic storms. IEEE Trans. Plasma Sci. **28**(6), 1867–1873 (2000)

A. Pulkkinen, Geomagnetically induced currents modeling and forecasting. Space Weather **13**(11), 734–736 (2015)

A. Pulkkinen, E. Bernabeu, J. Eichner, A. Viljanen, C. Ngwira, Regional-scale high-latitude extreme geoelectric fields pertaining to geomagnetically induced currents. Earth Planets Space **67**(1), 1–8 (2015)

P. Riley, CME dynamics in a structured solar wind, in *Solar Wind Nine*, Am. Inst. Phys. Conf. Proc., vol. 471, ed. by S.R. Habbal, R. Esser, V. Hollweg, P.A. Isenberg (1999), p. 131

P. Riley, On the probability of occurrence of extreme space weather events. Space Weather **10**(2), S02012 (2012)

P. Riley, J.J. Love, Extreme geomagnetic storms: probabilistic forecasts and their uncertainties. Space Weather **15**(1), 53–64 (2017)

P. Riley, J.T. Gosling, V.J. Pizzo, A two-dimensional simulation of the radial and latitudinal evolution of a solar wind disturbance driven by a fast, high-pressure coronal mass ejection. J. Geophys. Res. **102**(A7), 14677 (1997)

P. Riley, C. Schatzman, H.V. Cane, I.G. Richardson, N. Gopalswamy, On the rates of coronal mass ejections: remote solar and in situ observations. Astrophys. J. **647**, 648–653 (2006). https://doi.org/10.1086/505383

P. Riley, R. Lionello, J.A. Linker, Z. Mikic, J. Luhmann, J. Wijaya, Global MHD modeling of the solar corona and inner heliosphere for the whole heliosphere interval. Sol. Phys. **274**, 361–3775 (2012). https://doi.org/10.1007/s11207-010-9698-x

P. Riley, R.M. Caplan, J. Giacalone, D. Lario, Y. Liu, Properties of the fast forward shock driven by the July 23 2012 extreme coronal mass ejection. Astrophys. J. **819**, 57 (2016). https://doi.org/10.3847/0004-637X/819/1/57

P. Riley, M. Ben-Nun, J.A. Linker, M.J. Owens, T.S. Horbury, Forecasting the properties of the solar wind using simple pattern recognition. Space Weather **15**(3), 526–540 (2017)

C.J. Rodger, M.A. Clilverd, P.T. Verronen, T. Ulich, M.J. Jarvis, E. Turunen, Dynamic geomagnetic rigidity cutoff variations during a solar proton event. J. Geophys. Res. Space Phys. **111**(A4), A04222 (2006)

C.T. Russell, R.A. Mewaldt, J.G. Luhmann, G.M. Mason, T.T. von Rosenvinge, C.M.S. Cohen, R.A. Leske, R. Gomez-Herrero, A. Klassen, A.B. Galvin, K.D.C. Simunac, The very unusual interplanetary coronal mass ejection of 2012 July 23: a blast wave mediated by solar energetic particles. Astrophys. J. **770**, 38 (2013). https://doi.org/10.1088/0004-637X/770/1/38

A. Ruzmaikin, J. Feynman, I. Jun, Distribution of extreme solar energetic proton fluxes. J. Atmos. Sol.-Terr. Phys. **73**, 300–307 (2011). https://doi.org/10.1016/j.jastp.2009.12.016

A. Seppälä, P.T. Verronen, E. Kyrölä, S. Hassinen, L. Backman, A. Hauchecorne, J.L. Bertaux, D. Fussen, Solar proton events of October–November 2003: ozone depletion in the Northern Hemisphere polar winter as seen by GOMOS/Envisat. Geophys. Res. Lett. **31**(19), L19107 (2004)

M.A. Shea, D.F. Smart, Space weather and the ground-level solar proton events of the 23rd solar cycle. Space Sci. Rev. **171**(1–4), 161–188 (2012)

M.A. Shea, D.F. Smart, K.G. McCracken, G.A.M. Dreschhoff, H.E. Spence, Solar proton events for 450 years: the Carrington event in perspective. Adv. Space Res. **38**, 232–238 (2006). https://doi.org/10.1016/j.asr.2005.02.100

K. Shibata, H. Isobe, A. Hillier, A.R. Choudhuri, H. Maehara, T.T. Ishii, T. Shibayama, S. Notsu, Y. Notsu, T. Nagao et al., Can superflares occur on our Sun? Publ. Astron. Soc. Jpn. **65**(3), 49 (2013)

B.-M. Sinnhuber, M. Weber, A. Amankwah, J.P. Burrows, Total ozone during the unusual Antarctic winter of 2002. Geophys. Res. Lett. **30**(11), 1580 (2003)

G. Siscoe, N.U. Crooker, C.R. Clauer, Dst of the Carrington storm of 1859. Adv. Space Res. **38**, 173–179 (2006). https://doi.org/10.1016/j.asr.2005.02.102

W. Swider, T.J. Keneshea, Decrease of ozone and atomic oxygen in the lower mesosphere during a PCA event. Planet. Space Sci. **21**(11), 1969–1973 (1973)

N.N. Taleb, *The Black Swan: The Impact of the Highly Improbable* (Random House, New York, 2007)

B.C. Thomas, C.H. Jackman, A. Melott, L Modeling atmospheric effects of the September 1859 solar flare. Geophys. Res. Lett. **34**(6), L06810 (2007)

F. Toffoletto, R. Wolf, J. Yang, Dst during extreme events—how low can it get? in *AGU Fall Meeting Abstracts* (2016)

US Dept. of Commerce NOAA, Space Weather Prediction Center, Solar proton events affecting the earth environment (2010) (accessed July 11, 2017). ftp://ftp.swpc.noaa.gov/pub/indices/SPE.txt

I.G. Usoskin, G.A. Kovaltsov, Occurrence of extreme solar particle events: assessment from historical proxy data. Astrophys. J. **757**(1), 92 (2012)

S. Vennerstrom, L. Lefevre, M. Dumbović, N. Crosby, O. Malandraki, I. Patsou, F. Clette, A. Veronig, B. Vršnak, K. Leer, T. Moretto, Extreme geomagnetic storms—1868–2010. Sol. Phys. **291**, 1447–1481 (2016). https://doi.org/10.1007/s11207-016-0897-y

P.T. Verronen, R. Lehmann, Analysis and parameterisation of ionic reactions affecting middle atmospheric HOx and NOy during solar proton events. Ann. Geophys. **31**, 909–956 (2013)

P.T. Verronen, A. Seppälä, E. Kyrölä, J. Tamminen, H.M. Pickett, E. Turunen, Production of odd hydrogen in the mesosphere during the January 2005 solar proton event. Geophys. Res. Lett. **33**(24), L24811 (2006)

P.T. Verronen, C.J. Rodger, M.A. Clilverd, H.M. Pickett, E. Turunen, Latitudinal extent of the January 2005 solar proton event in the Northern Hemisphere from satellite observations of hydroxyl. Ann. Geophys. **25**, 2203–2215 (2007)

F.M. Vitt, C.H. Jackman, A comparison of sources of odd nitrogen production from 1974 through 1993 in the Earth's middle atmosphere as calculated using a two-dimensional model. J. Geophys. Res., Atmos. **101**(D3), 6729–6739 (1996)

B. Vršnak, D. Maričić, A.L. Stanger, A.M. Veronig, M. Temmer, D. Roša, Acceleration phase of coronal mass ejections: I. Temporal and spatial scales. Sol. Phys. **241**(1), 85–98 (2007)

D.F. Webb, J.H. Allen, Spacecraft and ground anomalies related to the October-November 2003 solar activity. Space Weather **2**(3) (2004)

E.W. Wolff, A.E. Jones, S.J.-B. Bauguitte, R.A. Salmon, The interpretation of spikes and trends in concentration of nitrate in polar ice cores, based on evidence from snow and atmospheric measurements. Atmos. Chem. Phys. **8**(18), 5627–5634 (2008). https://doi.org/10.5194/acp-8-5627-2008. http://www.atmos-chem-phys.net/8/5627/2008/

R.C. Whitten, I.G. Poppoff, *Physics of the Lower Ionosphere* (Prentice Hall, New York, 1965)

J.J. Zhang, C. Wang, T.R. Sun, Y.D. Liu, Risk assessment of the extreme interplanetary shock of 23 July 2012 on low-latitude power networks. Space Weather **14**(3), 259–270 (2016)

B. Zhu, Y.D. Liu, J.G. Luhmann, H. Hu, R. Wang, Z. Yang, Solar energetic particle event associated with the 2012 July 23 extreme solar storm. Astrophys. J. **827**(2), 146 (2016)

Space Sci Rev (2018) 214:11
DOI 10.1007/s11214-017-0443-8

The Varying Core Magnetic Field from a Space Weather Perspective

Mioara Mandea[1] · **Michael Purucker**[2]

Received: 2 April 2017 / Accepted: 13 November 2017 / Published online: 30 November 2017

Abstract This paper summarizes recent advances in our understanding of geomagnetism, and its relevance to terrestrial space weather. It also discusses specific core magnetic field features such as the dipole moment decay, the evolution of the South Atlantic anomaly, and the location of the magnetic poles that are of importance for the practice of space weather.

Keywords Core field · Secular variation · Secular acceleration · Dipole decay · South Atlantic Anomaly

1 Introduction

The geometry of the geomagnetic field, and its evolution in space and time contribute to the complexities of space weather observed at the surface of the Earth, and in the near-Earth space. This paper is devoted to a review of our knowledge of the geomagnetic field as it pertains to a better understanding of space weather effects. Space weather is defined as the collection of physical processes, beginning at the Sun, and ultimately affecting human activities on Earth and in space (Natural Resources Canada 2017). Radiation and particles emitted from the Sun, with variable delays, interact with the geomagnetic field, and the atmosphere, in ways that cause electrical currents to flow in regions of the ionosphere and magnetosphere. The radiation and particles also contribute to energetic particle populations in the near-Earth space. The electrical currents, and the changing energetic particle populations result in geomagnetic variations, aurora, and can profoundly affect a number of technologies.

The Scientific Foundation of Space Weather
Edited by Rudolf von Steiger, Daniel Baker, André Balogh, Tamás Gombosi, Hannu Koskinen and Astrid Veronig

✉ M. Mandea
mioara.mandea@cnes.fr

[1] CNES—Centre National d'Etudes Spatiales, 2 Place Maurice Quentin, 75039 Paris, France

[2] NASA—Goddard Space Flight Center, Greenbelt, MD 20771, USA

The geomagnetic field can be described in terms of magnetic poles. These poles are either deduced from the dipole term exclusively, and dominate within the plasma source regions in the magnetic tail that feed the auroral oval, or they include higher-order terms, and can describe the near-Earth fields. The northern and southern magnetic poles change positions with time, and these changes affect the location of the auroral ovals, the cusp, the location of open and closed field lines, and where those field lines connect. On a global scale, the decay of the geomagnetic field over the past few centuries has engendered a significant change in the near-Earth radiation environment. For example, the expansion of the well-known South Atlantic anomaly (SAA) creates more single-event upsets (SEU) in the electronics systems of near-Earth satellites.

To understand the Earth's complicated magnetic environment, a tour is needed through the origins of the geomagnetic field. The Earth's magnetic field comprises contributions from sources inside the Earth (internal contributions) and outside it (external contributions). Internal sources include the core field (also called the main field) originating in the external fluid core of the Earth) and the lithospheric field (caused by magnetic minerals in the crust and subordinately the upper mantle). The external sources have their origins in the ionosphere, the magnetosphere and also from electrical currents coupling the ionosphere and magnetosphere (field aligned currents, or FAC). These sources external to the solid Earth induce secondary fields in the Earth. The core magnetic field acts like a shield to the solar wind that the Sun continually emits. The need to understand the variations of the core field is then essential when space weather is considered. However, to analyze the core field and its variations, we also need to understand the interactions and separations of different geomagnetic field sources. The data and tools to obtain the best possible description of the magnetic field constituents are multi-faceted, so their review can bring us the capability to better describe and understand core field variations.

To take into consideration these issues, the remainder of the paper is organized as follows. Section 2 looks at the Earth's magnetic field contributions and how to measure them on the ground and in the near- Earth space. In Sect. 3 some information on modeling the magnetic field is given, mainly to show the possibilities to describe the core field. Section 4 discusses the core field and its variations, and finally Sect. 5 draws some conclusions and maps out future directions.

2 The Magnetic Field: Sources and Measurements

2.1 Geomagnetic Field Sources

The sources of internal magnetic fields include those originating in the core and lithosphere of the Earth. These natural magnetic fields largely originate in electric currents and magnetized media, respectively. The core field (Hulot et al. 2015) is dominated by a field generated from a self-sustaining dynamo in the Earth's molten outer core. The lithospheric fields (Purucker and Whaler 2015) are generated in rocks from the upper portion of the Earth's lithosphere that remain below the Curie temperature and containing minerals carrying the magnetization. The largest scale, and magnitude of these formally inseparable core and lithospheric fields can be approximated by an offset magnetic dipole at some distance from the Earth's rotation axis. The strongest scalar magnetic fields originating from the Earth's core are found in the Earth's polar regions. The area just offshore Antarctica, in the ocean between Australia and the Antarctic, possesses the largest scalar magnetic fields, averaging 66000 nT.

In contrast, the magnetic fields originating from the Earth's lithosphere have a bimodal character, lineated above the ocean lithosphere and weaker than the fields associated with continental lithospheres, which are characterized by much more varied patterns. At the surface of the Earth, the weak oceanic lithospheric fields are dominantly a function of the difference in thickness of the oceanic vs. continental crust (7 vs. 35 km, respectively). Two kinds of magnetization can exist within the cool lithosphere, permanent (remanent) magnetization and induced magnetization. The direction and intensity of permanent magnetization is dependent on the origin and history of the rock. Induced magnetization is proportional in magnitude, and generally parallel to the ambient core field. Another first-order observation about the strength of the lithospheric field is that the field, while variable in detail, varies in magnitude from the magnetic poles to the equator by a factor of two, a reflection of a strong contribution from induced magnetization.

The sources of external magnetic fields include those originating in the ionosphere, those which couple the magnetosphere and ionosphere, and those associated with more distant regions of the magnetosphere, for example the ring current, the magnetotail, and the magnetopause. These fields are produced by electric current systems in the ionosphere and magnetosphere. These current systems in the ionosphere originate in currents driven by dynamo processes in the ionospheric plasma. In the magnetosphere, currents are often produced by the movement of charged particles. The magnitude of these fields at the surface varies in local time (LT), in space, and in response to solar forcing. At ground level, a large-scale magnetospheric field of some tens of nT is found at all times. The ionospheric field at ground level in magnetically quiet times is of the order of a few nT or so, and varies with LT, with a maximum of several tens of nT associated with a commonly known Solar quiet (Sq) variation, a consequence of electric currents flowing in the region of ionosphere between some 90–150 km (Campbell 2003). In disturbed magnetic conditions, ionospheric and magnetospheric fields become much larger, and more dynamic, and can reach magnitudes of hundreds, or even thousands, of nT, at the ground. The Carrington solar event in 1859, the largest recorded solar magnetic event, has been associated with external field changes of more than 1500 nT at the Colaba Indian magnetic observatory (Cliver and Dietrich 2013).

2.2 Measuring the Geomagnetic Field

Measurements of the geomagnetic field include archeomagnetic and paleomagnetic time series for magnetic intensity, marine compass records for magnetic declination, magnetic observatories for vector measurements, and since the 1960s, near-Earth satellites measuring the scalar magnetic field, and only since the end of the 1970s the vector magnetic field. Human observations of the geomagnetic field began with measurements of declination (D) less than 500 years ago. One of the longest time series is Paris one. The declination and its first time derivative (secular variation) for the Paris series (Alexandrescu et al. 1996; Mandea and Le Mouël 2016) is plotted in Fig. 1. This figure shows that the declination varies over tens of degrees over a couple of centuries. Of course the interest is to be able to characterize these variations at a global scale and to understand them. The declination measurements expanded to measurements of inclination (I) some 400 years ago.

It has been less than 200 years, since 1832, that Gauss demonstrated the ability to measure the absolute intensity of the geomagnetic field, and the first magnetic observatories were installed. Magnetic observatories require a full description of the geomagnetic field so two types of measurements are needed: vector and scalar. The vector measurements are generally made with a fluxgate magnetometer, and they are subject to instrument drift. To minimize this drift a scalar magnetometer is used in a calibration procedure. A type of proton

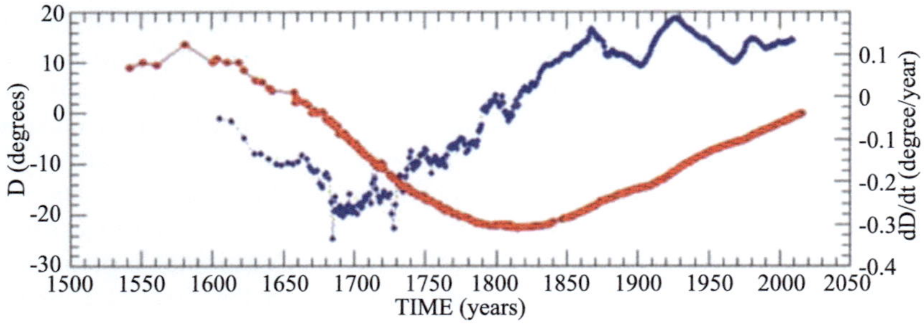

Fig. 1 Paris series: Annual mean values of declination for the period 1541–2015 (D, *red curve, left vertical axis*), adjusted for the Chambon la Forêt observatory location. The secular variation is computed by annual differences starting with 1600 and applying an 11-yr smoothing (dD/dt, *blue curve, right vertical axis*)

precession or resonance magnetometer, such as the Overhauser magnetometer, is typically installed at magnetic observatories to measure the field intensity. Such a device, however, makes measurements of the strength of the magnetic field only, and provides no information about its direction. To complete a set of measurements information about the direction of the geomagnetic field is needed, obtained by measuring the declination and inclination, commonly using a fluxgate-theodolite (named Declination Inclination fluxgate, or DI-flux).

Modern magnetic observatories use similar instrumentation to produce similar data products (Intermagnet 2017). For many observatories the fundamental measurements recorded are one-second values, used thereafter to compute one-minute values of the vector components and of scalar intensity. From the one-minute data, hourly, daily, monthly and annual mean values are produced. The magnetic field variation over a day gives an important indication about the status of the day, i.e. a magnetically "disturbed" or "quiet" day. An example is given in Fig. 2, the choice being to show the "Halloween" geomagnetic storm in October 2003, and a quiet day, in October 2016.

Future observatory developments might include remote sensing of the scalar magnetic field (Kane et al. 2016) in the mesosphere at some 90 km altitude based on sodium atoms, the residue of meteor ablation. These laser-based instruments might be included in selected observatories to capture the time variation of the scalar magnetic field close to the ionosphere.

High accuracy and precision measurements of the near-Earth geomagnetic field from space began with NASA's MAGSAT satellite that flew for six months from 1979 to 1980. This was followed in the first decade of the 21st century by the Ørsted (2017), CHAMP (2017), and SAC-C magnetic field satellites.

Swarm, the 3-satellite ESA constellation (ESA 2017), is the most current, and most advanced geomagnetic observatory in space. One of the novelties of the mission is not only the constellation concept, but also the new absolute scalar magnetometer, an optically pumped Helium magnetometer, which provides absolute scalar measurements of the magnetic field with high accuracy and stability for the calibration of the vector field magnetometer (Leger et al. 2009). The mission was designed to derive the first global representation of the geomagnetic field variations on times scales from an hour to several years, addressing the crucial problem of source separation. The three Swarm satellites are identical, and the constellation consists of two satellites (A and C) flying almost side-by-side at an altitude close to 460 km (April 2016), longitude separation of 1.4 degrees and on circular and almost polar orbits with an inclination of 87.4 degrees. The third satellite (B) flies higher, close to

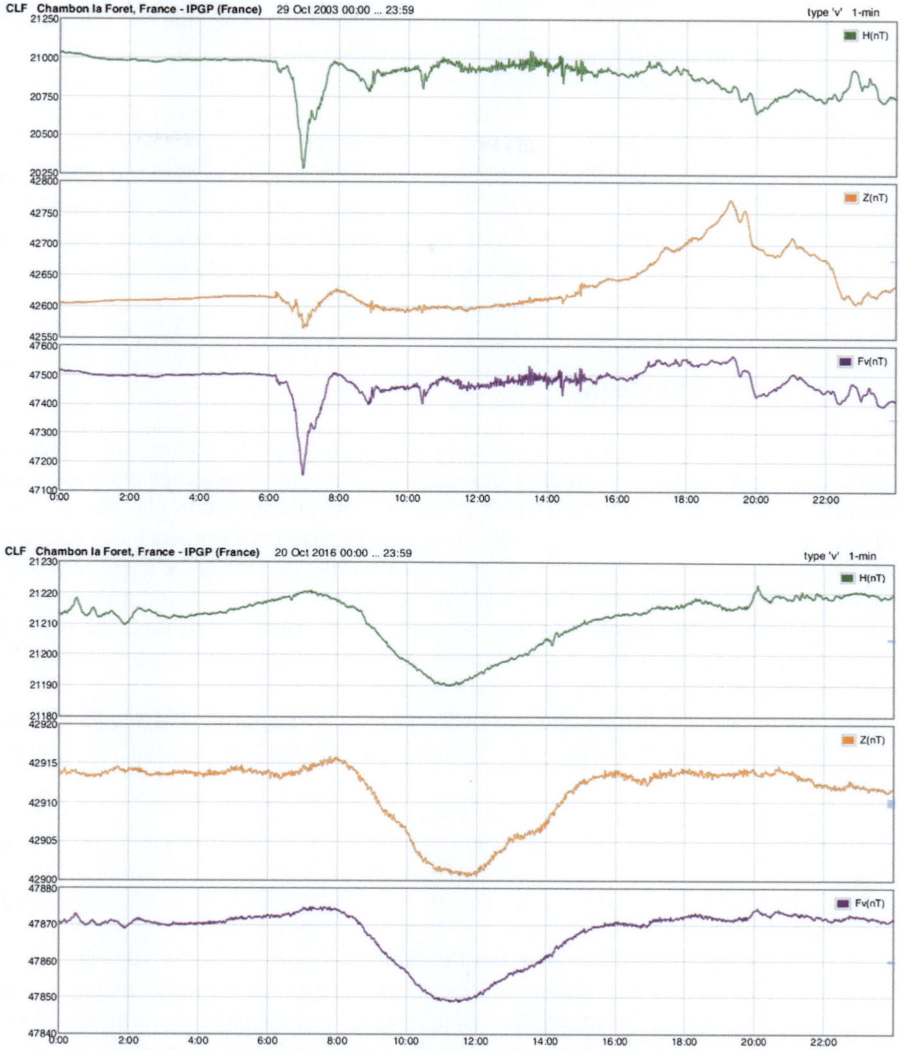

Fig. 2 Magnetograms for a disturbed day (29 October 2003) and a quiet day (20 October 2016), as recorded at Chambon la Forêt observatory (*H*—horizontal component, *Z*—vertical component and *F*—field intensity). Note the differences for the vertical axis

510 km (April 2016), on a more polar orbit (inclination of 87.8 degrees) and is progressively separated in LT with respect to A and C.

As mentioned above, the Swarm magnetometers measure a superposition of magnetic fields produced by a number of different sources: core, lithosphere, ionosphere, and magnetosphere. Note that contributions due to external sources depend largely on latitude, LT, and solar activity, and that other satellites have the capability of making highly precise and accurate magnetic field measurements with fluxgate magnetometers within the magnetosphere or ionosphere. Examples include Cluster (2017), MMS (2017), and the Van Allen Probes (RBSP 2017). These can augment the results of the Swarm mission in the distant magnetosphere, and illuminate the process of magnetic reconnection for example.

The Swarm mission may continue to operate until the 2030s, when the lower two satellites will reenter the atmosphere, leaving only the higher altitude satellite. In order to be assured of high-quality geomagnetic field data in the future, it is important to develop concepts for future missions now. One possibility is the use of one or more CubeSats with scalar Helium magnetometers, a concept under development with support from CNES (Hulot 2017). This implies new instrument developments, including the need to miniaturize the optically pumped Helium magnetometer. Diamond vector magnetometry, using nitrogen-vacancy centers in diamond (Wolf et al. 2015) is another promising class of absolute scalar and vector magnetometers that may see use in observatories and satellite missions of the future.

3 Magnetic Field Modeling

Geomagnetic field modeling in the satellite era has taken many approaches. Those that use large quantities of satellite data directly in inversions or parameter searches often use some combination of selected quiet-time satellite observations and magnetic observatory data, and often use spherical harmonic basis functions, or more localized functions like wavelets. Here, we provide details of the modeling approaches in order to provide the interested reader with an overview of the process. This is useful in space weather practice and we provide regions of validity of the models, and their accuracy and resolution. The time-varying core magnetic field, and models that describe it, are relevant to space weather because they determine where space weather effects are most severe.

Spherical harmonic functions allow us to describe the geomagnetic field as a global phenomenon. The magnetic field, following Maxwell's equations, is source-free, so it can be written:

$$\nabla \times \boldsymbol{H} = \boldsymbol{J} + \frac{\partial \boldsymbol{D}}{\partial t} \tag{1}$$

$$\nabla \cdot \boldsymbol{B} = 0 \tag{2}$$

where $\boldsymbol{H}$ is the magnetic field, $\boldsymbol{B}$ is the magnetic induction, $\boldsymbol{J}$ is the current density, and $\partial \boldsymbol{D}/\partial t$ is the electric displacement current density. In a region without magnetic field sources, say from the Earth's surface up to about 50 km, it is reasonable to assume that $\boldsymbol{J} = 0$ and $\partial \boldsymbol{D}/\partial t = 0$, so $\nabla \times \boldsymbol{H} = 0$, meaning that the vector field is conservative in the region of interest, and the magnetic field $\boldsymbol{H}$ can be expressed as $\boldsymbol{H} = -\nabla V$, where V is a scalar potential. Because $\boldsymbol{B} = \mu_0 \boldsymbol{H}$ above the Earth's surface (where $\mu_0 = 4\pi \times 10^{-7}$ Hm^{-1}), it follows that $\nabla \cdot \boldsymbol{H} = 0$ and that V has to satisfy Laplace's equation:

$$\nabla^2 V(\theta, \varphi, r) = 0 \tag{3}$$

In geomagnetism the magnetic induction $\boldsymbol{B}$ is traditionally called "magnetic field". We use the same notation in the following. In spherical coordinates (θ, φ, r), which are the geocentric co-latitude, longitude, and radial distance, the solution to Laplace's equation takes the form:

$$\begin{aligned} V &= V^{\text{int}} + V^{\text{ext}} \\ &= a \sum_{n=1}^{N_{\text{int}}} \sum_{m=0}^{n} (g_n^m \cos m\varphi + h_n^m \sin m\varphi) \left(\frac{a}{r}\right)^{n+1} P_n^m(\cos\theta) \end{aligned}$$

$$+a\sum_{n=1}^{N_{\text{ext}}}\sum_{m=0}^{n}(q_n^m\cos m\varphi + s_n^m\sin m\varphi)\left(\frac{r}{a}\right)^n P_n^m(\cos\theta) \tag{4}$$

where $a = 6371.2$ km is a reference radius, (θ, φ, r), are geocentric coordinates, $P_n^m(\cos\theta)$ are the associated Schmidt-normalized Legendre functions, N_{int} is the maximum degree and order of the internal potential coefficients g_n^m, h_n^m, and N_{ext} is that of the external potential coefficients q_n^m, s_n^m. Note that this equation is expressed in such a form so as to formally separate two important contributions in the potential field: a part, V^{int} describing the internal (core and lithospheric) sources, and a part, V^{ext} describing the external (mainly magnetospheric) sources.

The Gauss coefficients can be interpreted in terms of "sources". The first term, g_1^0, is associated with the geocentric dipole oriented along the vertical axis, i.e. axis of the Earth's rotation, with dipole moment $g_1^0 4\pi/\mu_0$. The potential for a dipole falls off as r^{-2} and the strength of the field components as r^{-3}. The next two terms, g_1^1, h_1^1, characterize the geocentric dipole oriented along the two last orthogonal axis. Similarly, for $n = 2$ the terms represent the geocentric quadrupole, for $n = 3$ the terms represent the geocentric octopole, and so forth.

To describe only the part of the geomagnetic field with sources beneath the Earth's surface, only the part V^{int} of Eq. (4) is used. However, this first term of Eq. (4) describes both the core and lithospheric fields. It is possible to see the transition from magnetic fields originating dominantly in the core from those originating in the lithosphere by computing the mean square value of the field over the Earth's surface produced by harmonics of a given degree n (Lowes 1966):

$$W_n = (n+1)\sum_m\left((g_n^m)^2 + (h_n^m)^2\right) \tag{5}$$

Variation of the power W_n depends on degree n and is known as the power or Lowes-Mauersberger spectrum. The steep part of the spectrum ($n \leq 13$) indicates the signature of the long wavelengths of the core field, and the much weaker signatures of the long wavelengths of the lithospheric field. The transition degrees ($n = 14$–15) can be attributed to subequal signals from both the core and lithospheric sources, and the higher degrees ($n \geq 16$) are dominantly lithospheric in origin. Examples of models based on spherical harmonics include the IGRF series, the CHAOS series (Olsen et al. 2014; Finlay et al. 2016a, 2016b), the Comprehensive models (Sabaka et al. 2015), and the GRIMM series (Lesur et al. 2015). The attributes of these models are described in Table 1 (Appendix). The newest release of CHAOS and GRIMM models are built along the same lines as their previous versions.

The most widely used magnetic field models are those of the IGRF series (International Geomagnetic Reference Field). Released every five years, they describe the Earth's core field and its secular variation. The core field is described to spherical harmonic degree 13, and the secular variation is described up to spherical harmonic degree 8 (Table 1). The purpose of the IGRF, and their relative, the World Magnetic Model (WMM) is as a practical way to predict the core field. The uses are myriad but they are focused on navigation, and as a supplement to GPS. The IGRF models are derived from candidate models submitted by independent worldwide teams. The candidates are assessed, and the final model is generally a weighted mean of the candidate models. The virtue of the IGRF is its simplicity, and the candidate models are often simplified and truncated models of more advanced models (cf. Gillet et al. 2015a, 2015b) used to describe the geomagnetic field. We discuss those next.

The more advanced models (Olsen et al. 2014; Finlay et al. 2016a, 2016b; Sabaka et al. 2015; Lesur et al. 2015) use the process of coestimation, and robust least-squares estimation, to separate the sources of the field. They differ in whether they describe quiet and disturbed magnetic conditions (the Comprehensive model series), or only quiet magnetic conditions (CHAOS and GRIMM). The Comprehensive model is the only model to use data from all local times, and all latitudes. All three of these models use vector and scalar magnetic observations from the CHAMP, Ørsted and SAC-C satellites over the first decade of the 21st century, and hourly or monthly mean observations from the 150 or so magnetic observatories. However, the latest variant of the GRIMM model (Lesur et al. 2015) does not use observatory data. Instead, it uses the observatory data to provide an independent set of measurements with which to test their models. Indeed, the secular variation values estimated from the observatories show a good fit to the model predictions. We illustrate many of our points in subsequent section using the latest model published by Lesur et al. (2015). This model has been specifically designed with the goal of improving the time-varying part of the core magnetic field.

Spherical harmonic analyses are widely used in representing the magnetic field and its variations at the Earth's surface. Here, because of the interest in the manner the radial magnetic field can be represented at the core-mantle boundary (see Sect. 4), these quantities are plotted in Fig. 3.

Another prominent category of models are those which use data assimilation techniques adapted from weather forecasting in an attempt to forecast the state of the core magnetic field years into the future, or into the past using historical databases (Fournier et al. 2010). These inverse methods rely on a numerical model of core dynamics that can be used to predict secular variation at the Earth's surface from the initial conditions. The goal of geomagnetic data assimilation is to combine the geomagnetic data sets with a dynamic numerical model by adjusting the trajectory of the model to adequately fit the data. Many of these models take as their starting point not the measurements of the fields themselves, but rather the models described in the previous paragraphs. The term "data assimilation" in meteorology describes the inverse techniques used to improve the forecast of atmospheric flow, and the state of the atmosphere. In geomagnetism, the observations directly access a much smaller fraction of the state vector than they do in meteorology. To illustrate this statement, the only "observations" that geomagnetism can bring to bear on the state of the core magnetic field are an estimate of the radial magnetic field, and its time derivative, at the core-mantle boundary. Those observations are generated by downward continuing magnetic observations from the surface, using regularization. We also lack knowledge of the background state of the core, the equivalent of a "climatological mean", about which the dynamics is likely to take place. In spite of these limitations, several promising approaches are evident. In particular, the assimilation of laboratory observations of liquid metal analogs of the Earth's core (Adams et al. 2015) offer the promise of both characterizing an appropriate background state, and perhaps offering access to more of the state vector of the core. The approach of Whaler and Beggan (2015) is another promising, and less computationally demanding, approach for short period (< 5 years) prediction, and generates core surface flow models that can be used for forecasting secular variation. The work of Aubert (2015) is the longest-term geomagnetic forecast using data assimilation, extending out a century. This model predicts that the focal (or minimum intensity) point of the South Atlantic Anomaly, discussed later in Sect. 4.3, is predicted to enter the South Pacific region in the next century, with the feature deepening and widening.

Models based on data from the pre-satellite era include the gufm model series (Jackson et al. 2000) and the CALSK-x model series (Korte et al. 2011). Using historic data from the last 400 years, largely land observatory records and ship navigation data, the gufm model is a spherical harmonic model extending to degree 14. The gufm-sat model (Finlay et al. 2012)

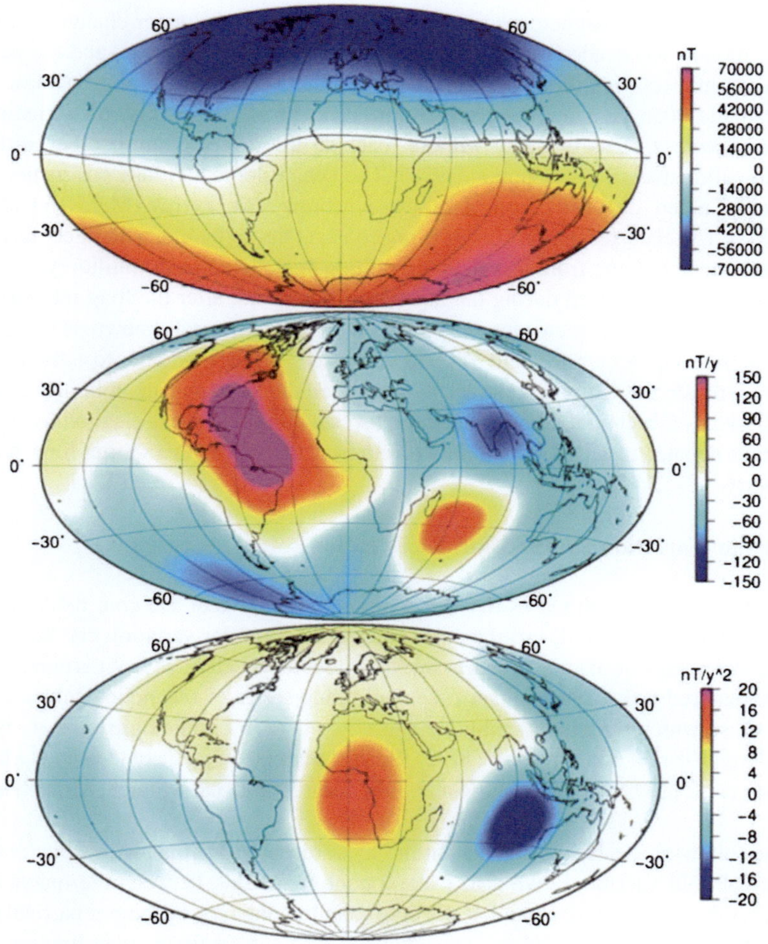

Fig. 3 Radial component of the magnetic field and its variations (secular variation and secular acceleration) at the Earth's surface as obtained from the model by Lesur et al. (2015)

is a potential successor to gufm, and incorporates satellite data from the modern era, with a similar parameterization to that of gufm. The CALSK-x model series is also a spherical harmonic model but uses paleo- and archeomagnetic data, and covers the last few thousand years, with some recent models extending back to a hundred thousand years. The model extends to degrees 8 to 10, but it is based on a very geographically restricted set of data that covers mainly the northern hemisphere.

4 The Core Field and Its Variations

After summarizing the geomagnetic field sources and modeling techniques, we turn to a discussion of core field variations and some of its specific features.

The geomagnetic field is in a permanent state of change on time scales ranging from milliseconds to millions of years. The shortest time scales are linked to processes arising in the ionosphere and magnetosphere, and their coupling. The longer changes, over time-scales of months to decades, mostly reflect changes in the Earth's core, and the very long ones, from centuries to millions of years, reveal different aspects of core processes.

A dramatic change in the core field is the reversal of its polarity, or, sometimes, an important decrease of its intensity without reaching the reversed, antiparallel, directional state—so-called excursions (Gubbins 1999). The ocean floors and lake sediments are one of the only, albeit imperfect, record of these geomagnetic events (Merrill et al. 1996; Valet 2003). The occurrence of reversals is not constant through time, with intervals between such phenomena ranging from less than 0.1 million years to some 50 million years. Reversal frequency has increased during the last 85 million years, after the long interval of normal polarity during the Cretaceous. The last geomagnetic reversal, known as the Brunhes-Matuyama reversal occurred about 780000 years ago. It is remarkable to note that recent estimates indicate the actual directional shift between the two opposite polarities lasted less than 1000 years (Valet et al. 2012). Recently, investigating this event, Sagnotti et al. (2014) advance the debated idea that this polarity reversal occurred even in a shorter time, a fraction of a century.

4.1 Secular Variation and Secular Acceleration

The secular variation, defined as the first temporal derivative of the core field, carries information about the dynamics of the core. Directly, the secular variation can be computed from measurements at magnetic observatories, some of which have been around for more than one hundred years. Over longer times only long series of magnetic declination exist, such as those which have been compiled and given here in chronological order, by Malin and Bullard (1981), Cafarella et al. (1992), Barraclough (1995), Alexandrescu et al. (1996) and Korte et al. (2009) for London, Rome, Edinburgh, Paris and Munich, respectively (see Fig. 1).

In the following we however focus on the most recent variations, with some references to historical field variations. We also make a continuous link between the magnetic field behavior at the Earth's surface and at the core-mantle boundary, without a particular focus on the topic of core flow modeling. This choice is twofold: (i) firstly, over the last decades the magnetic field variations have been better and better described, at the global scale, due to availability of satellite data; (ii) secondly, for the purpose of this review, the rapid variations of the core field are the most interesting and appropriate to be described, considering the space weather/climatology time-scales.

A geomagnetic field model defining the core field and its secular variation can be downward continued to the core-mantle boundary, considering the mantle as an insulator from an electromagnetic point of view. We can then express the Gauss coefficients describing the core field and its secular variation at the Earth's surface into descriptions of the field at the core-mantle boundary and produce maps describing their spatial variations. The radial (or vertical) component is usually considered, the one which is continuous through this limit and which is used in computing flows at the top of the core. Figure 4 shows a map of the radial component of the main field and its secular variation at the top of the core for epoch 2005.0, as obtained from the model by Lesur et al. (2015). One of the most recent models, CHAOS-6 (Finlay et al. 2016a, 2016b) captures the secular variation up to at least spherical harmonic degree 16 (though the smaller length-scales mainly consist in time-averaged values).

At the core-mantle boundary, the secular variation map displays a different behavior over the Atlantic and Pacific hemispheres. In the Atlantic hemisphere the variations are very

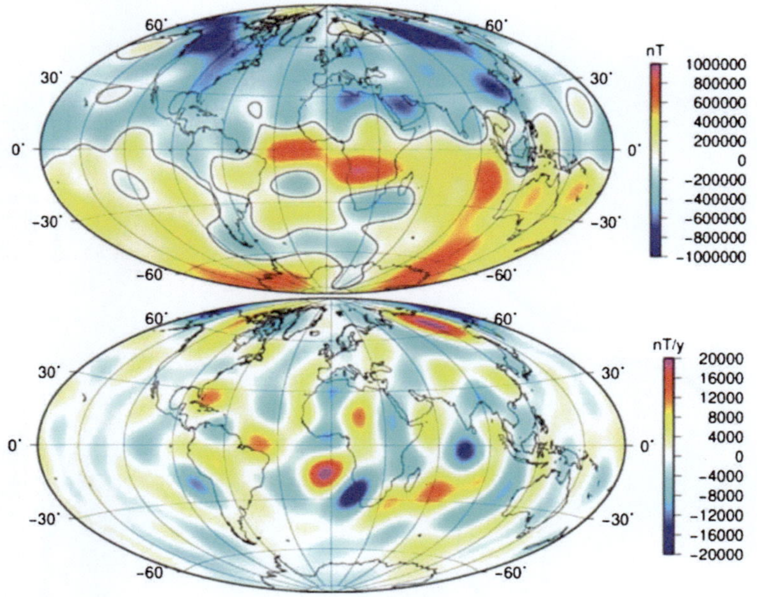

Fig. 4 Radial component of the core magnetic field and its secular variation at the core-mantle boundary as obtained from the model by Lesur et al. (2015)

important, up to some 10000 nT/yr, mainly in a region centered on the southern African continent, indicating very active core dynamics. The Pacific hemisphere remains, generally at the same low rate of secular variation, as well as the southern polar region. A combined explanation of the westward drift and hemispherical localisation of the secular variation is provided by Aubert et al. (2013).

It is interesting to compare the completely different behavior of the secular variation over the polar regions, with strong pairs of secular variation patches in the northern polar region. Some analysis of the regional secular variation inferred core upwelling in the northern polar region where reversed flux patches have been detected (see Fig. 5). Olson and Aurnou (1999) find a persistent westward polar vortex and a fluid upwelling inside the tangent cylinder, and Chulliat et al. (2010a, 2010b) note the existence of a bipolar secular variation structure situated at an emerging reversed flux patch and consider it as a harbinger of an upwelling and magnetic flux expulsion by radial diffusion. Amit (2014) shows that the geomagnetic secular variation morphology, in particular at some specific core-mantle boundary regions, may reflect the kinematics of magnetic field stretching by fluid upwelling or downwelling.

The origin of the reversed flux patches has been an outstanding question; previously they had been attributed to westward motion of a large flux patch under Siberia (Bloxham and Gubbins 1985). Recently, Livermore et al. (2017) considering the dynamical regime of the core, suggest that this behavior can be linked to a localized westward cylindrical jet of about 400 km width centered on the tangent cylinder. Since the epoch 2000, they argue that its strength has increased by a factor of three. Let us note here that the whole core flow geometry, including the local jet has been revealed by Pais and Jault (2008). Computing core flows from high resolution secular variation models, these authors show the presence of a large jet encircling the inner core, which carries an important part of the core angular momentum.

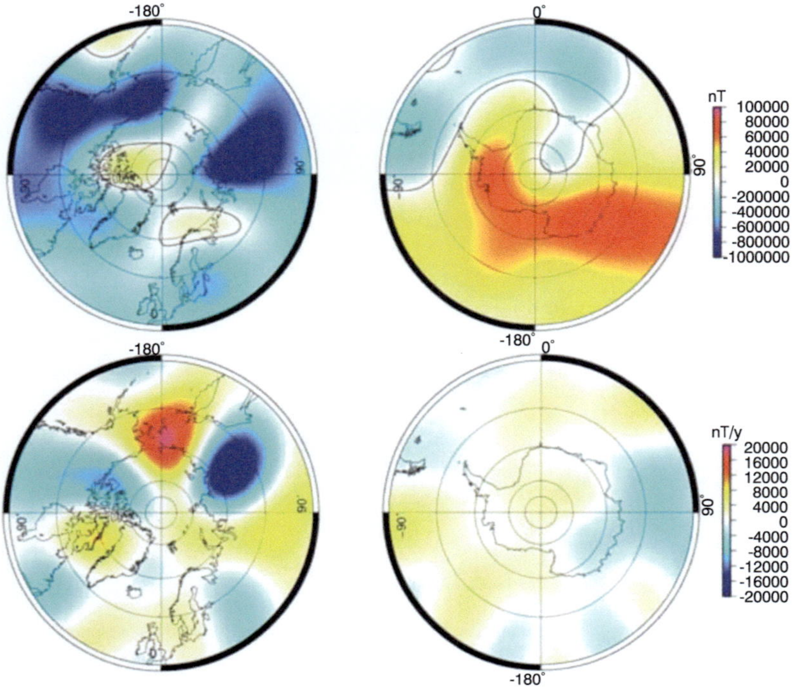

Fig. 5 Radial component of the magnetic field and its secular variation at the core-mantle boundary as obtained from the model by Lesur et al. (2015). Polar projection

For the secular variation the characteristic time-scale is considered of the order of a century and may reflect the convective overturn time of the core (Christensen et al. 2012). The shortest observable time-scales over a decade or less could be related to fast torsional waves (Gillet et al. 2010; Asari and Lesur 2011). Christensen et al. (2012) suggest from the analysis of geodynamo simulations that core dynamics at short time-scales (decadal and subdecadal) is conducted by processes similar to those at centennial timescales. The short timescale features presently observed in the satellite-based models have been connected to the possible mechanism through which geomagnetic jerks occur (Olsen and Mandea 2007; Wardinski et al. 2008; Chulliat et al. 2010a, 2010b; Finlay et al. 2016a, 2016b).

4.2 Rapid Fluctuations and Geomagnetic Jerks

Geomagnetic jerks are sharp changes in the trend of the first-time derivative of the magnetic field corresponding to abrupt jumps in the second-time derivative (e.g. Mandea et al. 2010). The secular variation and geomagnetic jerks then appear as a series of nearly straight lines separated by these events, as seen in Fig. 6 for the East (Y) component at two European observatories, Chambon la Forêt and Niemegk, located respectively in France and Germany. It is interesting to note the different behavior after 1940, when the rate of secular variation is different between the two locations.

During the past nearly two decades when high quality satellite data has been available the core secular variation and secular acceleration have been described to high spherical harmonic degree, reliable up to degree-order 16. This allows a focus on explaining rapid core field variations on a global scale. At Earth's surface, large-scale patterns of secular

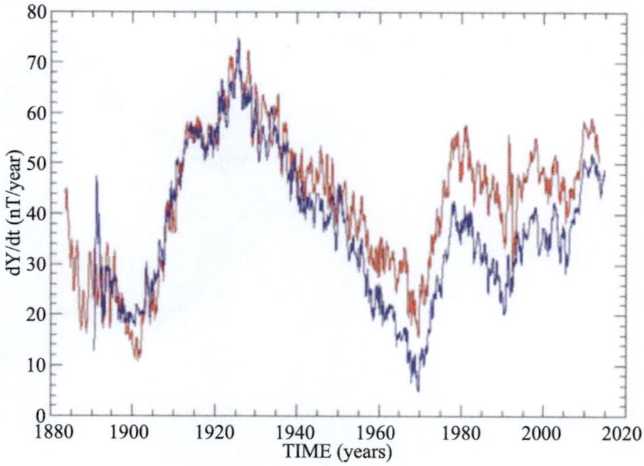

Fig. 6 Secular variation as obtained from monthly means at Chambon la Forêt (*red curve*) and Niemegk (*blue curve*) observatories

acceleration change on short, sub-decadal timescales. To better describe these rapid changes, time series of monthly point estimates at satellite altitude have been proposed and exploited, as "virtual observatories" (Mandea and Olsen 2006; Whaler and Beggan 2015; Saturnino et al. 2017). This new concept allows us to observe geomagnetic jerks directly in satellite data. Satellite data also reveal secular acceleration pulses in the Earth's geomagnetic field (Chulliat et al. 2010a, 2010b). The most recent events in 2007 and 2012 appear to have their origin in a succession of core field acceleration pulses occurring predominantly in West Africa and the South Atlantic region (Chulliat and Maus 2014; Finlay et al. 2015). The last reported jerk is in 2014 (Torta et al. 2015) occurred as a strong secular acceleration in the Africa–South Atlantic region, extending into Europe.

In spite of many attempts, the physical origin of these events is not yet fully established. Torsional waves have often been considered as the main source of the intense and rapidly changing patterns evident in the secular variation. Recently, however, Gillet et al. (2015a, 2015b) suggest that enhanced non-zonal flow at low latitudes, particularly around 90°W, as well as 60 and 120°E appear to be responsible. Whether the secular variation is due to wave or advection is a controversial issue, and likely it is a combination of both. How to separate the signal due to waves from the signal due to advection by the flow can take advantage of the new magnetic observations provided by satellite missions as well as the new advanced numerical simulations.

Finally, let us recall that geomagnetic jerk occurrences may be linked to other geophysical phenomena, such as sudden changes in the length of the day (Olsen and Mandea 2008; Mandea et al. 2010; Holme and de Viron 2013), or to changes in the phase of the free core nutation (Malkin 2013). We can notice here the complexity of these phenomena, and the importance of understanding their possible multiple origins, especially when considering topics such as space weather.

4.3 Specific Core Features and Their Temporal Variations

Some specific features linked to the core field and its variations are of signal importance to space weather, such as the location of the magnetic poles, the dipole moment decay, and the SAA. The locations of the magnetic poles controls the locations of the northern and southern cusps of the magnetosphere. The dipole moment decay, and the associated increase in size and deepening of the SAA, have important consequences with respect to radiation damage of satellites, and the electric power grid.

Fig. 7 Magnetic dip (*blue* and *red*) and dipole (*green*) positions as obtained from IGRF-12 model

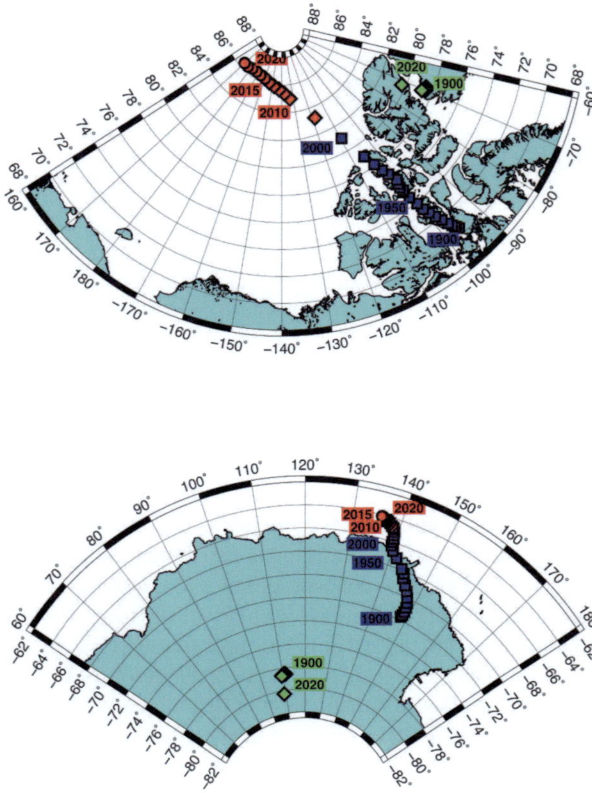

Magnetic Poles The positions of the north and south magnetic poles gradually change. Using the gufm model (Jackson et al. 2000) or recent IGRF models (Thébault et al. 2015), the locations and velocities for the two magnetic poles can be computed. Such computations indicate that over more than 400 years both magnetic poles have changed their positions. However, the rate for each pole is very different, and the movement depends on whether the pole is calculated solely from the dipole term, or from the entire spherical harmonic series (termed the dip pole). The symmetry of the auroral oval, at the boundary of open and closed magnetic field lines, is dominated by the dipole term because the proximal source of the aurora components is in the Earth's magnetic tail. Higher order magnetic components decay faster than the dipole term, and in the magnetic tail, the dipole term is dominant. For longer time-scales, the geomagnetic and magnetic pole locations from 7000 BC to A.D. 2006.6 are discussed in Korte and Mandea (2008).

Over the last century, the magnetic dip poles have been moving in a northwest direction. During the first half of the 20th century their velocity was comparable, around some 5–10 km/yr. Since 1970, the time of a well-documented geomagnetic jerk, the north magnetic pole has moved from Canada towards Siberia, with velocities reaching some 60 km/yr (Newitt et al. 2002; Olsen and Mandea 2007) and the south magnetic pole towards Australia, with a much lower velocity of around 5 km/yr.

Figure 7 shows the magnetic poles positions as obtained from IGRF models. A striking feature appears for the north magnetic pole velocity, which stopped its advance, entering into a deceleration phase, its velocity decreasing from about 53 km/yr in 2015 to an estimated

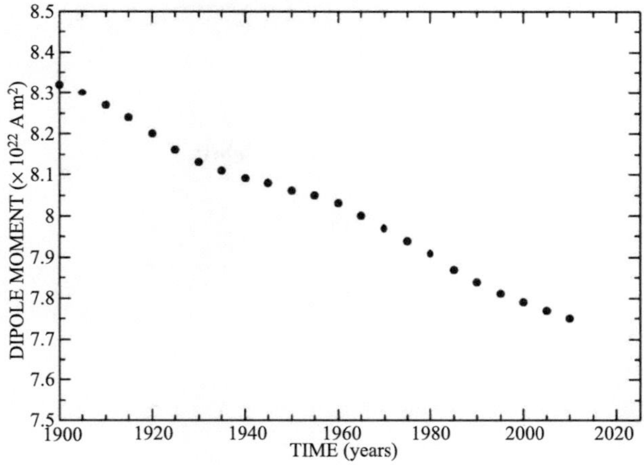

Fig. 8 Geomagnetic dipole moment as computed from the IGRF-12 model

value of 43 km/yr in 2020 (but note that the latter value relies on the predictive secular variation).

A glance over the last century of the rapid change in the north magnetic pole positions can be thought to be a signature of a sudden change in the behavior of the field, which has some signatures of an early phase of a field reversal. However, in the other hemisphere the south magnetic pole has kept nearly the same velocity rate (Mandea and Dormy 2003).

Dipole Moment Decay As discussed previously, direct intensity measurements are available since Gauss elaborated a method for determining intensity in 1832. These measurements allow us to compute the dipolar field, and its dominant axial part, parallel to the Earth's rotation axis. The dipole moment has decreased by nearly 6% per century since intensity magnetic field measurements exist. Figure 8 shows the axial component of the dipole field as computed from the IGRF field model. The general decreasing rate is of some 16 nT/yr. However, at some epochs the field has very different rate fluctuations. For example around 1980, it diminished twice as fast as nowadays. If the same rate of decay is extrapolated in time, the axial dipole would reach zero in less than 2000 yrs.

A question arises: what may be the future evolution and the implications of this significant dipole moment decay? To offer possible answers it is crucial to understand the physical process responsible for the dipole moment decay. Fluid motion in the outer core transport magnetic field lines, converting kinetic energy into magnetic energy. This ensures the maintenance of the magnetic field against Ohmic dissipation. Considering the free decay of the dipole, if the fluid motions ceased, it is estimated that in some 55000 years the axial dipole would reach a zero level. This time interval is more than one order of magnitude larger that the value of 2000 yrs., based on recent observations. An explanation could be found in proposed scenarios by Finlay et al. (2016a, 2016b) for the fluid motions within the core.

South Atlantic Anomaly One of the striking characteristics of the geomagnetic field is named the South Atlantic Anomaly, the region where the total field intensity is unusually low and the particle flux unusually high (Fig. 9), with a minimum value of about 22,000 nT for the epoch 2015. That minimum is situated close to the Paraguayan city of Asuncion. This central minimum has moved from southern Africa to South America (Mandea et al. 2007) in the recent past. Comparing MAGSAT and CHAMP satellite measurements, as well as considering only Swarm measurements, it has been possible to estimate the core field

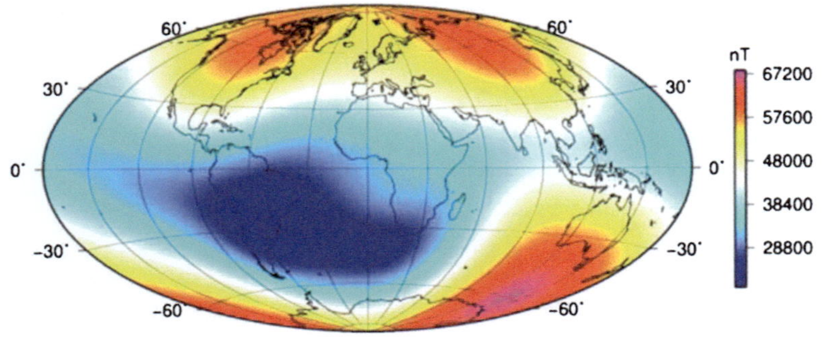

Fig. 9 Map of the total field intensity at the Earth's surface. The South Atlantic Anomaly is defined as the region of the lowest magnetic field values

intensity changes. The result indicates that the field strength has dramatically changed over the last decades, in a region situated again on the southern Atlantic Ocean–African subcontinent, by nearly 1%/yr.

The SAA is not only a characteristic of the present geo-magnetic field but it has been observed during the last four centuries (Jackson et al. 2000). Over longer time-scales the model resolution does not allow a well-determined description of its spatial variations. However, it is also present in the CALSxK series of models (based on paleo- and archeomagnetic data) with different locations and morphologies (Korte et al. 2011).

The SAA behavior seems to indicate that it might be connected to the general decrease of the dipole moment and to the significant increase of the non-dipolar field in this region. The weakness of the field intensity is caused by an increasing patch of opposite magnetic flux compared to the dipole direction at the core-mantle boundary. Pavón-Carrasco and De Santis (2016) analyze in detail the behavior of the SAA over the last 200 yrs. Their results indicate that one of the reversed polarity patches located under the South Atlantic Ocean grew with a rate of 2.54×10^5 nT per century. At the same time the quadrupole field seems to control this decreasing field strength. This behavior is similar to that obtained by numerical modeling, i.e. a reversal is preceded by the development of reverse flux patches at low-mid latitude, which migrates poleward and reduce the axial dipolar field (Wicht et al. 2010). However, the current value of the dipole moment is not as low as the mean moment in the Holocene.

The importance of understanding the evolution of the South Atlantic Anomaly for space weather is evident. This regional weakening in field intensity allows energetic particles and cosmic rays to penetrate much deeper into the magnetosphere than in other regions, resulting in significant space weather effects, such as satellite outages, even affecting ground-based systems. As a space weather phenomena, Geomagnetically Induced Currents (GIC) impact ground-based technological structures at all latitudes. Recently, Barbosa et al. (2015), investigating GIC occurrences in the power networks at low latitudes in the central Brazilian region, show significant effects in these regions during solar cycles 23 and 24.

5 Conclusions and Future Directions

A comprehensive separation and understanding of the internal and external processes contributing to the Earth's magnetic fields is crucial for space weather research. Continuous monitoring of the magnetic field, combining ground and space platforms, aims to address such needs.

What can the practice of space weather learn from current research in geomagnetism? Two first order features of the geomagnetic field, the rapid decay of the dipole field, and the expansion of the South Atlantic Anomaly, are of prime importance to space weather because they accentuate the impact of space weather events. Both of these features point to the importance of combining data assimilation approaches (Fournier et al. 2010) to geomagnetic weather prediction with experimental approaches to determine more of the state vector at the core mantle boundary, and a better understanding of its mean state. That means investing in experimental dynamos that are good analogs to the Earth's outer core, and in forecasting and hindcasting of the geomagnetic field. Understanding the puzzling processes that produce geomagnetic jerks also hold the promise of a deeper understanding of the mean state of the core.

Acknowledgements We thank N. Gillet for valuable comments and V. Lesur for providing GRIMM model inputs. Data collected at magnetic observatories Chambon la Foret and Niemegk were used, and we thank the observatory staff for providing them (BCMT: doi:10.18715/BCMT.MAG.DEF). Part of this was supported by a grant of the Ministry of National Education and Scientific Research, RDI Programme for Space Technology and Avanced Research—STAR, project number 513. We gratefully acknowledge constructive suggestions from two anonymous reviewers.

Appendix

Table 1 Attributes of recent geomagnetic field models

	Version	Comp.[a]	S.H. degree[b]	Data[c]	Selec.[d]	Ref.	Web[e]	Usable[f]	Uses[g]
CM	5	In, Io, T, M, O	107/20/60/1	S, Sg, O	Q, D	Sabaka et al. (2015)	1	≤1000 km	R., Sw V
CHAOS	4	In, M	85/16/–/2	S, Om	Q	Olsen et al. (2014)	2	≤1000 km	R.
GRIMM	2	In, M	30/16/–/2	S, O	Q	Lesur et al. (2010)	3	CMB	R., CMB
IGRF	12	In	13/8/–/–	S, Om	Q	Thébault et al. (2015)	4	≤1000 km	Navigation

[a]Components explanation: In = Internal (core and lithosphere) components; Io = Ionosphere and associated induced components; T = Toroidal components; M = Magnetosphere and associated induced components; O = Oceanic M2 tide

[b]S. H. (Spherical Harmonic) degree expansion: 1st number is the degree of the spatial expansion for core and lithospheric fields; 2nd number is the maximum degree for the time varying core field; 3rd number is the maximum degree of the ionospheric field expansion; 4th number is the maximum degree of the magnetospheric field expansion

[c]Data explanation: S = Satellite vector and scalar data (CHAMP, Ørsted, SAC-C or some subset); Sg = Along-track satellite gradient data; O = Observatory hourly mean values; Om = Observatory monthly mean values

[d]Selection explanation: Selection indicates whether quiet (Q) and/or Disturbed (D) data have been used in the model

[e]Webpages: 1 = www.planetary-mag.net/planets.html (for the previous CM model, CM4); 2 = www.spacecenter.dk/files/magnetic-models/CHAOS-6; 3 = www.gfz-potsdam.de/magmodels; 4 = www.ngdc.noaa.gov/IAGA/vmod/igrf.html

[f]Usable: Region where the model is pertinent; CMB = Core mantle boundary

[g]Uses: Specific use of the models; R = Research; Sw V = Swarm validation; CMB = Core mantle boundary

References

M.M. Adams, D.R. Stone, D.S. Zimmerman, D.P. Lathrop, Liquid sodium models of the Earth's core. Prog. Earth Planet. Sci. (2015). doi:10.1186/s40645-015-0058-1

M. Alexandrescu, V. Courtillot, J.L. Le Mouël, Geomagnetic field direction in Paris since the mid-sixteenth century. Phys. Earth Planet. Inter. **98**, 321–360 (1996)

H. Amit, Can downwelling at the top of the Earth's core be detected in the geomagnetic secular variation? Phys. Earth Planet. Inter. **229**, 110–121 (2014). doi:10.1016/j.pepi.2014.01.012

S. Asari, V. Lesur, Radial vorticity constraint in core flow modeling. J. Geophys. Res. **116**, B11101 (2011). doi:10.1029/2011JB008267

J. Aubert, Geomagnetic forecasts driven by thermal wind dynamics in the Earth's core. Geophys. J. Int. **203**(3), 1738–1751 (2015). doi:10.1093/gji/ggv394

J. Aubert, C.C. Finlay, A. Fournier, Bottom-up control of geomagnetic secular variation by the Earth's inner core. Nature **502**, 219–223 (2013). doi:10.1038/nature12574

C. Barbosa, L. Alves, R. Carballo, G.A. Hartmann, A.R.R. Papa, R.J. Pirjola, Analysis of geomagnetically induced currents at a low-latitude region over the solar cycles 23 and 24: comparison between measurements and calculations. J. Space Weather Space Clim. **5**, A35 (2015). doi:10.1051/swsc/2015036

D. Barraclough, Observations of th Earth's magnetic field in Edinburgh, from 1670 to the present day. Trans. R. Soc. Edinb. Earth Sci. **85**, 239–252 (1995)

J. Bloxham, D. Gubbins, The secular variation of Earth's magnetic field. Nature **317**, 777–781 (1985)

L. Cafarella, A. De Santis, A. Meloni, Secular variation in Italy from historical geomagnetic field measurements. Phys. Earth Planet. Inter. **73**, 206–221 (1992)

W.H. Campbell, *Introduction to Geomagnetic Fields* (Cambridge Univ. Press, New York, 2003), 304 pp.

CHAMP, http://www.gfz-potsdam.de/champ/. Accessed Sept. 2017

U.R. Christensen, I. Wardinski, V. Lesur, Timescales of geomagnetic secular acceleration in satellite field models and geodynamo models. Geophys. J. Int. **190**(1), 243–254 (2012). doi:10.1111/j.1365-246X.2012.05508.x

A. Chulliat, S. Maus, Geomagnetic secular acceleration, jerks, and a localized standing wave at the core surface from 2000 to 2010. J. Geophys. Res. (2014). doi:10.1002/2013JB010604

A. Chulliat, G. Hulot, L. Newitt, Magnetic flux expulsion from the core as a possible cause of the unusually large acceleration of the north magnetic pole during the 1990s. J. Geophys. Res. **115** (2010a). doi:10.1029/2009JB007143

A. Chulliat, E. Thébault, G. Hulot, Core field acceleration pulse as a common cause of the 2003 and 2007 geomagnetic jerks. Geophys. Res. Lett. (2010b). doi:10.1029/2009GL042019

E.W. Cliver, W. Dietrich, The 1859 space weather event revisited: limits of extreme activity. J. Space Weather Space Clim. **3**, 15 (2013)

Cluster, http://sci.esa.int/cluster/. Accessed Nov. 2017

ESA, http://www.esa.int/Our_Activities/Observing_the_Earth/Swarm. Accessed Sept. 2017

C.C. Finlay, A. Jackson, N. Gillet, N. Olsen, Core surface magnetic field evolution 2000–2010. Geophys. J. Int. **189**, 761–781 (2012). doi:10.1111/j.1365-246X.2012.05395.x

C.C. Finlay, N. Olsen, L. Tøffner-Clausen, DTU candidate field models for IGRF-12 and the CHAOS-5 geomagnetic field model. Earth Planets Space **67**, 114 (2015). doi:10.1186/s40623-015-0274-3

C.C. Finlay, J. Aubert, N. Gillet, Gyre-driven decay of the Earth's magnetic dipole. Nat. Commun. (2016a). doi:10.1038/ncomms10422

C.C. Finlay, N. Olsen, S. Kotsiaros, N. Gillet et al., Recent geomagnetic secular variation from swarm and ground observatories as estimated in the CHAOS-6 geomagnetic field model. Earth Planets Space **68**, 112 (2016b). doi:10.1186/s40623-016-0486-1

A. Fournier, G. Hulot, D. Gault, W. Kuang, A. Tangborn, N. Gillet, E. Canet, J. Aubert, F. Lhuillier, An introduction to data assimilation and predictability in geomagnetism. Space Sci. Rev. **155**, 247–291 (2010)

N. Gillet, D. Jault, E. Canet, A. Fournier, Fast torsional waves and strong magnetic field within the Earth's core. Nature **465**, 74–77 (2010). doi:10.1038/nature09010

N. Gillet, D. Jault, C.C. Finlay, Planetary gyre, time dependent eddies, torsional waves, and equatorial jets at the Earth's core surface. J. Geophys. Res., Solid Earth **120**(6), 3991–4013 (2015a)

N. Gillet, O. Barrois, C.C. Finlay, Stochastic forecasting of the geomagnetic field from the COV-OBS.x1 geomagnetic field model, and candidate models for IGRF-12. Earth Planets Space **67**, 71 (2015b). doi:10.1186/s40623-015-0225-z

D. Gubbins, The distinction between geomagnetic excursions and reversals. Geophys. J. Int. **137**, 1–3 (1999)

R. Holme, O. de Viron, Characterization and implications of intradecadal variations in length of day. Nature **4999**, 202–204 (2013). doi:10.1038/nature12282

G. Hulot, *The Swarm Delta NanoMagSat Project, Latest News, IAPSO-IAMAS-IAGA 2017 Scientific Assembly* (2017)

G. Hulot, T. Sabaka, N. Olsen, A. Fournier, The present and future geomagnetic field, in *Treatise on Geophysics, Geomagnetism*, 2nd edn., ed. by M. Kono (Elsevier, Amsterdam, 2015), pp. 33–78

Intermagnet, www.intermagnet.org. Accessed Sept 2017

A. Jackson, A.R.T. Jonkers, M.R. Walker, Four centuries of geomagnetic secular variation from historical records. Philos. Trans. R. Soc. Lond. A **358**, 957–990 (2000)

T. Kane, P. Hillman, C. Denman, M. Hart, R. Scott, M. Purucker, S. Potashnik, Laser remote magnetometry using mesospheric sodium (2016), arXiv:1610.05385v2

M. Korte, M. Mandea, Magnetic poles and dipole tilt variation over the past decades to millennia. Earth Planets Space **60**, 937–948 (2008)

M. Korte, M. Mandea, J. Matzka, A historical declination curve for Munich from different data sources. Phys. Earth Planet. Inter. **174**, 161–172 (2009)

M. Korte, C. Constable, F. Donadini, R. Holme, Reconstructing the Holocene geomagnetic field. Earth Planet. Sci. Lett. **312**, 497–505 (2011). doi:10.1016/j.epsl.2011.10.031

J.M. Leger, F. Bertrand, T. Jager, M. Le Prado, I. Fratter, J.C. Lalaurie, Swarm absolute scalar and vector magnetometer based on helium 4 optical pumping, in *Proceedings of the Eurosensors XXIII Conference 1:1* (2009), pp. 634–637. doi:10.1016/j.proche.2009.07.158

V. Lesur, I. Wardinski, S. Asari, B. Minchev, M. Mandea, Modelling the earth's core magnetic field under flow constraints. Earth Planets Space **62** (2010)

V. Lesur, K. Whaler, I. Wardinski, Are geomagnetic data consistent with stably stratified flow at the core-mantle boundary? Geophys. J. Int. 929–946 (2015). doi:10.1093/gji/ggv031

P.W. Livermore, R. Hollerbach, C.C. Finlay, An accelerating high-latitude jet in Earth's core. Nat. Geosci. **10**, 62–68 (2017). doi:10.1038/ngeo2859

F.J. Lowes, Mean-square values on sphere of spherical harmonic vector fields. J. Geophys. Res. **71**, 2179 (1966)

S.R.C. Malin, E. Bullard, The direction of the Earth's magnetic field at London, 1570–1975. Philos. Trans. R. Soc. Lond. **299**, 357–423 (1981)

Z. Malkin, Free core nutation and geomagnetic jerks. J. Geodyn. **72**, 53–58 (2013). doi:10.1016/j.jog.2013.06.001

M. Mandea, E. Dormy, Asymmetric behaviour of magnetic dip poles. Earth Planets Space **55**, 153–157 (2003)

M. Mandea, J.L. Le Mouël, After some 350 years—zero declination again in Paris. Hist. Geo- Space Sci. (2016)

M. Mandea, N. Olsen, A new approach to directly determine the secular variation from magnetic satellite observations. Geophys. Res. Lett. **33**, L15306 (2006). doi:10.1029/2006GL026616

M. Mandea, M. Korte, D. Mozzoni, P. Kotzé, The magnetic field changing over the Southern African continent—a unique behaviour. S. Afr. J. Geol. **110**, 193–202 (2007)

M. Mandea, R. Holme, A. Pais, K. Pinheiro, A. Jackson, G. Verbanac, Geomagnetic jerks: rapid core field variations and core dynamics. Space Sci. Rev. **155**, 147–175 (2010). doi:10.1007/s11214-010-9663-x

R.T. Merrill, M.W. McElhinny, P.L. McFadden, *The Magnetic Field of the Earth* (Academic Press, London, 1996)

MMS, http://mms.gsfc.nasa.gov. Accessed Nov. 2017

Natural Resources Canada, Space Weather, http://spaceweather.gc.ca/sbg-en.php. Accessed 7 Sep. 2017

L. Newitt, M. Mandea, L. McKee, J.J. Orgeval, Recent acceleration of the North magnetic pole linked to magnetic jerks. Eos Trans. AGU **83**(35), 381–389 (2002)

N. Olsen, M. Mandea, Will the magnetic north pole wind up in Siberia? EOS (2007)

N. Olsen, M. Mandea, Rapidly changing flows in the Earth's core. Nat. Geosci. **1**, 390–394 (2008). doi:10.1038/ngeo203

N. Olsen, H. Lühr, C. Finlay, T. Sabaka, I. Michaelis, L. Tøffner-Clausen, The CHAOS-4 geomagnetic field model. Geophys. J. Int. **197**, 815–827 (2014). doi:10.1093/gji/ggu033

P. Olson, J. Aurnou, A polar vortex in the Earth's core. Nature **402**, 170–173 (1999)

Ørsted, https://en.wikipedia.org/wiki/Ørsted_(satellite). Accessed Sept. 2017

M.A. Pais, D. Jault, Quasi-geostrophic flows responsible for the secular variation of the Earth's magnetic field. Geophys. J. Int. **173**, 421–443 (2008)

F.J. Pavón-Carrasco, A. De Santis, The South Atlantic Anomaly: the key for a possible geomagnetic reversal. Front. Earth Sci. **4** (2016). doi:10.3389/feart.2016.00040

M. Purucker, K. Whaler, Crustal magnetism, in *Treatise on Geophysics, Geomagnetism*, 2nd edn., ed. by M. Kono (Elsevier, Amsterdam, 2015), pp. 185–218

RBSP, http://www.nasa.gov/mission_pages/rbsp/mission/. Accessed Nov. 2017

T. Sabaka, N. Olsen, R. Tyler, A. Kuvshinov, CM5, a pre-Swarm comprehensive geomagnetic field model derived from over 12 yr of CHAMP, Ørsted and SAC-C data. Geophys. J. Int. **200**, 1596–1626 (2015). doi:10.1093/gji/ggu493

L. Sagnotti, G. Scardia, B. Giaccio, J.C. Liddicoat, S. Nomade, P.R. Renne, C.J. Sprain, Extremely rapid directional change during Matuyama–Brunhes geomagnetic polarity reversal. Geophys. J. Int. **199**(2), 1110–1124 (2014). doi:10.1093/gji/ggu287

D. Saturnino, B. Langlais, H. Amit, F. Civet, M. Mandea, E. Beucler, Combining virtual observatory and equivalent source dipole approaches to describe the geomagnetic field with Swarm measurements. Phys. Earth Planet. Inter. (2017). doi:10.1016/j.pepi.2017.06.004

E. Thébault, C.C. Finlay, C.D. Beggan, P. Alken, J. Aubert, O. Barrois, F. Bertrand, T. Bondar, A. Boness, L. Brocco, E. Canet, A. Chambodut, A. Chulliat, P. Coisson, F. Civet, A. Du, A. Fournier, I. Fratter, N. Gillet, B. Hamilton, M. Hamoudi, G. Hulot, T. Jager, M. Korte, W. Kuang, X. Lalanne, B. Langlais, J.M. Leger, V. Lesur, F.J. Lowes, S. Macmillan, M. Mandea, C. Manoj, S. Maus, N. Olsen, V. Petrov, V. Ridley, M. Rother, T.J. Sabaka, D. Saturnino, R. Schachtschneider, O. Sirol, A. Tangborn, A. Thomson, L. Tøffner-Clausen, P. Vigneron, I. Wardinski, T. Zvereva, International geomagnetic reference field: the 12th generation. Earth Planets Space **67**(1), 1–19 (2015). doi:10.1186/s40623-015-0228-9

J.M. Torta, F.J. Pavón-Carrasco, S. Marsal, C.C. Finlay, Evidence for a new geomagnetic jerk in 2014. Geophys. Res. Lett. (2015). doi:10.1002/2015GL065501

J.P. Valet, Time variations in geomagnetic intensity. Rev. Geophys. (2003). doi:10.1029/2001/RG000104

J.P. Valet, A. Fournier, V. Courtillot, E. Herrero-Bervera, Dynamical similarity of geomagnetic field reversals. Nature **490**(7418), 89–93 (2012)

I. Wardinski, R. Holme, S. Asari, M. Mandea, The 2003 geomagnetic jerk and its relation to the core surface flows. Earth Planet. Sci. Lett. **267**, 468–481 (2008). doi:10.1016/j.epsl2007.12.008

K. Whaler, C. Beggan, Core Flow modelling using acceleration. J. Geophys. Res., Solid Earth **120**, 1400–1414 (2015). doi:10.1002/2014JB011697

J. Wicht, H. Harder, S. Stellmach, Numerical dynamo simulations—from basic concepts to realistic models, in *Handbook of Geomathematics*, ed. by W. Freeden, Z. Nashed, T. Sonar (Springer, Heidelberg, 2010)

T. Wolf, P. Neumann, K. Nakamura, H. Sumiya, T. Ohshima, J. Isoya, J. Wrachtrup, Subpicotesla diamond magnetometry. Phys. Rev. X **5**, 041001 (2015). doi:10.1103/PhysRevX.5.041001

Space Sci Rev (2017) 212:985–1039
DOI 10.1007/s11214-017-0357-5

Anthropogenic Space Weather

T.I. Gombosi[1] · D.N. Baker[2] · A. Balogh[3] · P.J. Erickson[4] · J.D. Huba[5] · L.J. Lanzerotti[6,7]

Received: 15 December 2016 / Accepted: 24 March 2017 / Published online: 13 April 2017

Abstract Anthropogenic effects on the space environment started in the late 19th century and reached their peak in the 1960s when high-altitude nuclear explosions were carried out by the USA and the Soviet Union. These explosions created artificial radiation belts near Earth that resulted in major damages to several satellites. Another, unexpected impact of the high-altitude nuclear tests was the electromagnetic pulse (EMP) that can have devastating effects over a large geographic area (as large as the continental United States). Other anthropogenic impacts on the space environment include chemical release experiments, high-frequency wave heating of the ionosphere and the interaction of VLF waves with the radiation belts. This paper reviews the fundamental physical process behind these phenomena and discusses the observations of their impacts.

Keywords High-altitude nuclear explosions · Artificial radiation belts · Electromagnetic pulse (EMP) · Damage to satellites · Space Debris · Chemical releases · HF heating · VLF waves and radiation belts

The Scientific Foundation of Space Weather
Edited by Rudolf von Steiger, Daniel Baker, André Balogh, Tamás Gombosi, Hannu Koskinen and Astrid Veronig

✉ T.I. Gombosi
tamas@umich.edu

1 University of Michigan, Ann Arbor, MI 48109, USA

2 University of Colorado, Boulder, CO 80303, USA

3 Imperial College, London, UK

4 MIT Haystack Observatory, Westford, MA 01886, USA

5 Naval Research Laboratory, Washington, DC 20375, USA

6 New Jersey Institute of Technology, Newark, NJ 07102, USA

7 Alcatel Lucent Bell Laboratories, Murray Hill, NJ 07974, USA

1 Introduction

The "victory" of Nikola Tesla's alternating current (AC) as implemented by George Westinghouse (to whom Tesla sold most of his patents) over Thomas Edison's direct current (DC) as the means to power and to light up the US and the world beginning in the late 19th century can be considered the initiation point of potential anthropogenic modifications of Earth's space environment. However, it was not until nearly a century later when the very quiet electromagnetic environment of Siple Station, Antarctica, came online that the evidence of power line harmonic radiation was discovered in Earth's magnetosphere (Helliwell et al. 1975).

Since this report by the Stanford group, a considerable number of ground- and space-based studies have been published that have examined the potential effects of these harmonics in the VLF range on the electron population of the magnetosphere. The published discussions have ranged from "control" of the magnetosphere through actions on, or production of, chorus emissions; (e.g. Park and Helliwell 1977; Luette et al. 1979; Bullough 1983; Parrot 1994) to considerable skepticism (e.g. Tsurutani et al. 1979; Tsurutani and Thorne 1981). The possible effects on the space environment of human activities via use of electrical power sources in industrial activities have also been reported from statistical analyses of weekly variations in geomagnetic activity (e.g. Park and Miller 1979; Fraser-Smith 1979), and refuted from other analyses (e.g. Karinen et al. 2002). While it is agreed that anthropogenic power line harmonic radiation does exist in the magnetosphere, the magnitude of any effects of this radiation on the trapped electron populations remains uncertain.

Another persistent anthropogenic radiation in the magnetosphere is that produced by the widespread distribution of VLF and RF transmitters around the world. The radiation from these transmitters, used for navigation and communications, is known to disturb the trapped electron population of the magnetosphere. In addition, over several decades, purposeful VLF transmissions in the form of controlled experiments have been conducted from spacecraft and from the ground (one of the more notable and long-lasting set of ground experiments was from Siple Station, beginning in the early 1970s and extending to the late 1980s). These topics are addressed in Sect. 8.

In addition to purposeful short-term VLF radiation experiments to understand wave growth and trapped particle interactions, a number of other short-term experimental techniques have been developed and exploited over time. The most dramatic of these were the explosion of nuclear bombs in the near-space environment, discussed in Sect. 2. While these nuclear experiments were short term, their effects on the space environment in terms of enhanced radiation belt electrons persisted for weeks and months, forming artificial radiation belts (Sect. 3), with damaging radiation effects on flying spacecraft, as described in Sect. 4. The geophysical and geomagnetic effects of these explosions are reviewed in Sect. 5.

Other short-term experiments directed toward specific understanding goals and with limited lasting effects on the magnetosphere include the injection of clouds of barium into the ionosphere and lower altitude magnetosphere, and the heating of the ionosphere by high power RF transmitters. Ionosphere RF heating experiments have had several objectives, including the generation of ULF waves with $f \lesssim 1$ Hz. Such waves, if propagated into the magnetosphere, could potentially interact with the gyro frequency of magnetosphere ions, as VLF waves do with electrons. These topics are discussed in Sect. 9.

The exhaust from launches of large rockets and the controlled firings of the engines of the space shuttles were used specifically for modifications of the ionosphere—the production of ionosphere "holes" (e.g. Mendillo 1981; Mendillo et al. 1987; Meier et al. 2011). Such holes can change the propagation conditions for ground-to-satellite (and the reverse) signals, and can in some instances enable the measurement of galactic cosmic radio noise at frequencies lower than normally possible under the conducting ionosphere.

At times, other possibilities of exciting ULF waves from the ground into the magnetosphere have been discussed. Generation possibilities have included large current loops on Earth's surface, vertical or horizontal dipoles, and driving a current in seawater around a peninsula (Fraser-Smith 1981). Fraser-Smith and Coates (1978) reported ULF noise in ground-based measuring systems that is produced by the San Francisco BART (Bay Area Rapid Transit) system. No evidence of such frequency noise propagating into the magnetosphere has been reported. This BART-effect is similar to the long-recognized magnetic noise that is produced by trains passing in the vicinity of a magnetic field measuring site.

The permanent existence, and growth, of power grids and of VLF transmitters around the globe means that it is unlikely that Earth's present-day space environment is entirely "natural"—that is, that the environment today is the environment that existed at the onset of the 19th century. This can be concluded even though there continue to exist major uncertainties as to the nature of the physical processes that operate under the influence of both the natural environment and the anthropogenically-produced waves. As new techniques are considered for human modification of elements of Earth's space environment, it is important to carefully assess the short-term and long-term implications of anthropogenic modifications in order to arrive at final experiment design, and even decision to proceed.

2 High-Altitude Nuclear Explosions

Missiles that were able to launch the Soviet Sputnik satellite could also deliver weapons across the world. At the Livermore branch of the Lawrence Radiation Laboratory (now the Lawrence Livermore National Laboratory) where he was employed, Nicholas Christofilos proposed in October 1957 (just after the launch of Sputnik and months before the discovery of the Van Allen Radiation belts) a defensive way to intercept and destroy intercontinental missiles (see pp. 55–56 in Finkbeiner 2006). He suggested that by exploding a nuclear bomb in the upper atmosphere, the electrons from the fission process would be trapped in Earth's magnetic field. This huge cloud of trapped electrons would destroy an incoming missile.

Christofilos's idea was followed up and supported by the Director of the Advanced Research project Agency (ARPA), Herbert York, as the Argus experiments in 1958 (Christofilos 1959a,b). The Argus shots demonstrated the feasibility of the concept and also showed that such injections of electrons could damage spacecraft flying through such a cloud: anthropogenic space weather. Thus, high-altitude explosions of nuclear devices could be offensive as well as Christofilos's original proposal of defense against missiles. On the other hand, the devastating electromagnetic pulse (EMP) that could be produced by the high altitude explosion of a nuclear device was only observed years later following the Starfish Prime event of July 9, 1962. A list of high-altitude nuclear explosions is given in Table 1.

The EMP generated by a high altitude nuclear explosion is one of a small number of threats that can hold our society at risk of catastrophic consequences. The increasingly pervasive use of electronics of all forms represents the greatest source of vulnerability to attack by EMP. When a nuclear explosion occurs at high altitude, the EMP signal it produces will cover the wide geographic region within the line of sight of the detonation. This broad band, high amplitude EMP, when coupled into sensitive electronics, has the capability to produce widespread and long lasting disruption and damage to the critical infrastructures that underpin the fabric of US society.

Table 1 List of high-altitude nuclear explosions (Wikipedia Contributors 2016)

Designation	Country	Date	Altitude	Yield (kt)
Yucca	USA	Apr 28, 1958	26 km	1.7
Teak	USA	Aug 1, 1958	77 km	3.8×10^3
Orange	USA	Aug 12, 1958	43 km	3.8×10^3
Argus I	USA	Aug 27, 1958	200 km	1.7
Argus II	USA	Aug 30, 1958	240 km	1.7
Argus III	USA	Sep 6, 1958	540 km	1.7
Test#88	USSR	Sep 6, 1961	23 km	10.5
Test#115	USSR	Oct 6, 1961	41 km	40
Test#127	USSR	Oct 27, 1961	150 km	1.2
Test#128	USSR	Oct 27, 1961	300 km	1.2
Starfish Prime	USA	Jul 9, 1962	400 km	1.4×10^3
Checkmate	USA	Oct 20, 1962	147 km	7
Test#184	USSR	Oct 22, 1962	290 km	300
Bluegill Triple Prime	USA	Oct 26, 1962	50 km	410
Test#187	USSR	Oct 28, 1962	150 km	300
Kingfish	USA	Nov 1, 1962	97 km	410
Test#195	USSR	Nov 1, 1962	59 km	300

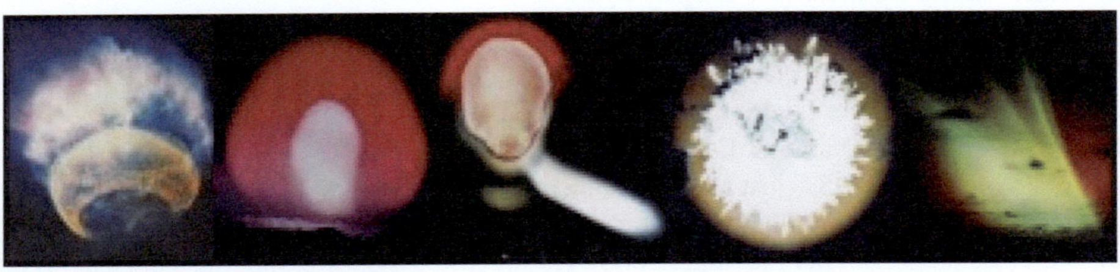

Fig. 1 From *left to right*, the Orange, Teak, Kingfish, Checkmate, and Starfish high-altitude nuclear tests conducted in 1958 and 1962 by the United States near Johnston Island in the mid-Pacific (from Foster et al. 2008)

High-altitude nuclear explosions have vastly different EMP effects depending on the geomagnetic location and burst altitude. Figure 1 shows images of atmospheric nuclear bursts carried out by the US over Johnston Island (geomagnetic latitude 10.5°) during the 1958–1962 time period (Foster et al. 2008).

From the very first nuclear test in July 1945 on, electromagnetic effects were always a threat for experimenters trying to record weapon performance data electronically. With bunkered and electrically shielded apparatus, data could be recorded successfully, but the power supplies used to convert AC current to direct current for the vacuum tubes would be burned out after the event.

Around 1960, questions were raised about possible damaging effects of EMP on the Minuteman missile system, which was in development at that time. This system was supposed to be able to launch retaliatory missiles after being subjected to nuclear attack. It was suggested that a large nuclear burst in the missile farms could expel the geomagnetic field from a large volume of air and ground, rapidly changing the magnetic flux. This would induce currents in cables that might be large enough to burn them out, preventing the retaliatory launches.

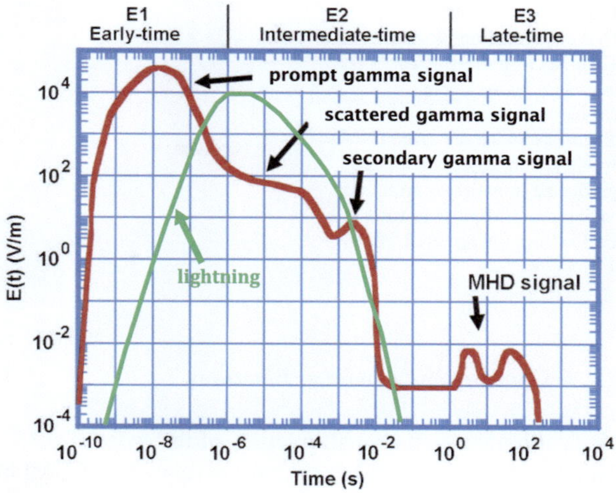

Fig. 2 The various phases of a generic HEMP signal (from Savage et al. 2010). For comparison, we also show a typical electromagnetic signal from lightning

The Argus nuclear tests were the first experiment to explode nuclear weapons above the dense atmosphere. While these tests produced some interesting effects, the yield of the fission device (1.7 kt) was not large enough to produce electromagnetic effects over a wide area.

High-altitude nuclear EMP (HEMP) is a complex multi-pulse phenomenon, usually described in terms of three components, "E1" (early phase), "E2" (intermediate phase) and "E3" (late phase, or MHD phase). A concise summary of these phases is shown in Fig. 2. Since the E2 phase is often compared to lightning the figure also shows a typical electromagnetic signal from lightning.

2.1 E1 Phase

The true importance of the early phase of HEMP was revealed by the three high-altitude devices exploded at 400 km (Starfish), 97 km (Kingfish) and 50 km (Bluegill) altitudes that showed early electric field pulses of tens of kV/m peaking around 10 ns and lasting about a μs (see Fig. 2). The extremely short duration of the initial signal presented a puzzle that was solved by Conrad Longmire in 1964. Two decades later Longmire (1986) published an unclassified report that explains the fundamental physics of the early phase EMP (the E1 phase).

Figure 3 shows a general diagram of the E1 HEMP process. A nuclear burst puts out a fast pulse of gamma rays. This thin shell of photons streams outward, including downward toward the Earth and its exponentially increasing air density. Once low enough in altitude, the gammas start striking air molecules, knocking electrons off (which mostly move outward from the explosion). The Earth's magnetic field causes the electrons to turn coherently looping around the magnetic field, and this constitutes an electric current, which generates an EM signal, much like the currents on a transmitting loop antenna (magnetic dipole). This EM field propagates downward as an EM wave—the E1 signal. This process was first explained by Longmire (1986).

Longmire (1986) explains the basic physics of the E1 pulse in terms of an oversimplified "flat Earth" scenario. While this scenario is clearly unphysical, it clarifies the basic physics and explains the extremely short duration of the initial pulse. Here we briefly outline his derivation.

Fig. 3 General basis of the E1 HEMP generation process. Gammas from the nuclear burst interact with the upper atmosphere generating Compton electrons, which are turned in the Earth's geomagnetic field, and produce a transverse current that radiates an EM pulse towards the Earth (from Savage et al. 2010)

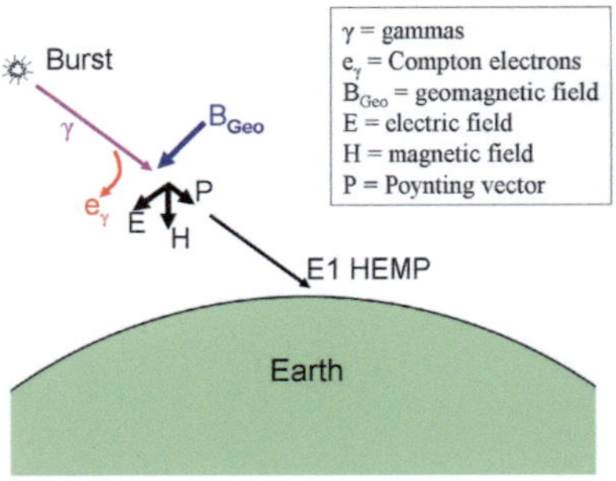

2.1.1 Gammas and Compton Electrons

Nuclear bombs emit a small fraction, of the order of 0.3%, of their energy in gamma rays. Thus a 1 Mt bomb (that has the explosive power of 1 Mt of TNT), produces a total energy of about 4.2×10^{15} J and emits ~3 kt or about 1.2×10^{13} J in gamma rays with a typical average energy of 2 MeV (3.2×10^{-13} J). Thus the total number of gammas emitted by a 1 Mt bomb is of the order $N_\gamma \approx 4 \times 10^{25}$ gammas.

The principal interaction of gamma rays with air molecules is Compton scattering. In this process, the gamma collides with an electron in the air molecule and knocks it out. In so doing, the gamma transfers part of its energy (on the average about half) to the electron, and is scattered into a new direction. The Compton recoil electron goes generally near the forward direction of the original gamma, thus a directed flux of gammas produces a directed electric current of Compton recoil electrons (see Fig. 4).

A ~2 MeV gamma penetrates an air column of ~220 kg/m^2 before it is scattered. Assuming that the mass of the air column above the burst altitude is negligible (this was certainly true for the Starfish experiment) the gammas will have their first collision at an altitude above which the column density of air is ~220 kg/m^2. This altitude is around $z_0 = 30$ km. Most gammas will suffer their first collision in a layer around z_0 with a thickness of about an atmospheric scale-height (at 30 km altitude the typical scale-height is $H_n \approx 7$ km).

Assuming that gammas are uniformly emitted by the burst in every direction the total number of gammas crossing a unit cross sectional area at a distance r from the burst is $N_\mathrm{A} = N_\gamma/4\pi r^2$. Assuming a 100 km burst altitude and 1 Mt bomb we get $N_\mathrm{A} \approx 2.3 \times 10^{13}$ gammas/m^2. These gammas are scattered within an atmospheric scale-height around 30 km altitude, so the number of Compton electrons produced in a unit volume is $n_\gamma \approx 10^{11}$ electrons/m^3. The density of Compton electrons (n_{Ce}) would be the same as n_γ if they did not move. Because they move in the same direction as the gammas with an average speed of about $0.94c$ (where c is the speed of light), their actual density is about an order of magnitude larger than n_γ (due to relativistic effects). In his estimates Longmire (1986) used $n_{Ce} \approx 10^{12}$ electrons/m^3. We note that if the Compton electrons all moved together at nearly the speed of light, they would make a current density of tens of A/m^2, that is a very substantial current density (about ten orders of magnitude larger than typical magnetospheric currents).

2.1.2 Motion of Compton Electrons

For ~2 MeV gammas the Compton recoil electrons have an average kinetic energy of about 1 MeV and their angular distribution strongly peaks in the forward direction. The velocity of a 1 MeV electron is about $0.94c$, while the relativistic mass is about three times larger than

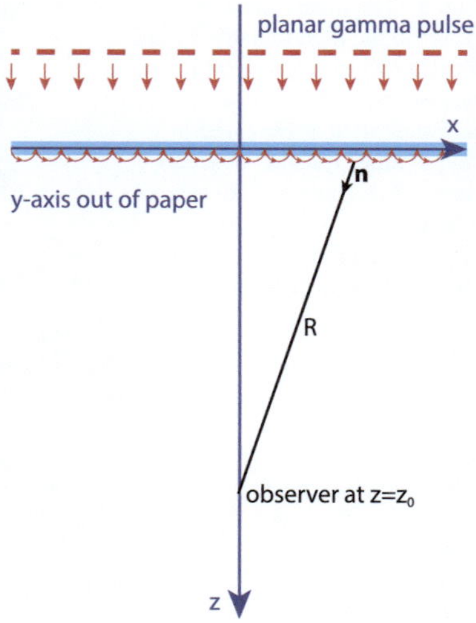

Fig. 4 Schematic diagram of plane parallel E1 generation (Longmire 1986)

the rest mass. In his estimates Longmire (1986) used a magnetic field value of $B_0 = 5.6 \times 10^{-5}$ T that results in a gyroradius for 1 MeV electrons of $r_c = 85$ m. The mean stopping range of the 1 MeV electrons due to collisions with air at 30 km altitude is $R_m \approx 170$ m, taking into account the fact the MeV electron gradually loses energy as it undergoes multiple collisions. In the process of stopping the MeV electron about 3×10^4 low energy electron-ion pairs (typical energies are $\sim$10 eV) are created. Because the velocities of secondary electrons are randomly distributed they generate no significant current. However, they create an electrically conducting layer that plays a role in the generation of the EMP pulse.

During its first gyration around the local magnetic field an average Compton electron would travel about $2\pi r_c = 530$ m. However the stopping distance is only 170 m, so a Compton electron makes only a third of gyration before it is stopped. Substituting the numerical values used by Longmire (1986) we obtain a Compton current density of about $j_C = -ecn_{Ce} \approx 50\ \mathrm{A/m^2}$.

2.1.3 Radiation

Let us consider a planar, impulse function of gammas propagating through vacuum and arriving perpendicularly to a thin insulating sheet, as indicated in Fig. 4. We assume that electrons are knocked out of the sheet in the forward direction of the gammas, and that there is a magnetic field B_0 parallel to the sheet, in the y-direction. This field deflects the electrons in semicircles, until they return to the sheet and are stopped. An observer is located at a distance z_0 below the sheet, which is very large compared to the gyroradius of the electrons. These simplifications are introduced for ease of calculation and they still preserve the underlying physics.

At $t = 0$ the gamma pulse arrives at the thin sheet. The observer, however, will only get a start signal when the gammas arrive at his location, therefore the "observing time" (or delayed time) T is related to the "event time" (t) at any value of z as $T = t - z/c$. Longmire (1986) found the relation between event time and observing time for a radiating electron to be

$$T = t - \frac{r_c}{c} \sin \omega_e t \tag{1}$$

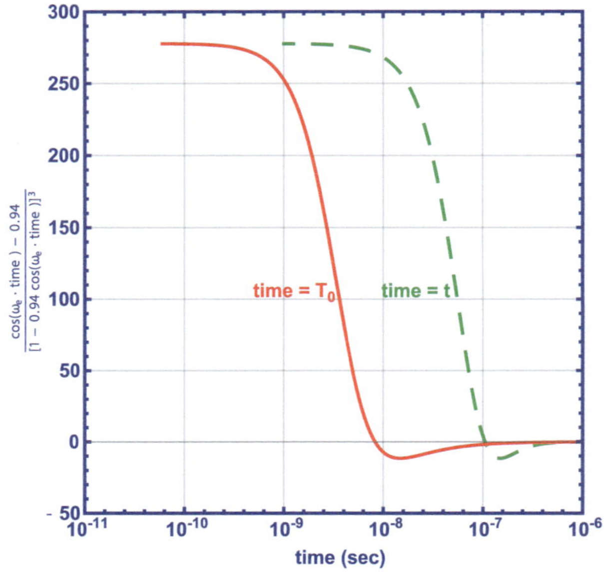

Fig. 5 The time-dependent factor in the field radiated by an electron with $v_e/c = 0.94$ that makes one-half turn in a magnetic field as observed in "event time" (t) and "observer time" (T_0). Due to relativistic effects the observer detects a <10 ns signal of positive E_x (from Longmire 1986)

where ω_e is the electron gyrofrequency. For short times ($\omega_e t \ll 1$) this expression becomes

$$T = \left(1 - \frac{v_e}{c}\right)t. \tag{2}$$

This means that for MeV electrons the observing time initially advances nearly twenty times slower than the event time, resulting in an extremely short observed pulse.

Longmire (1986) also derived the radiated electric field associated with the motion of an electron that was born at $t = 0$ at the origin of the coordinate system. The signal is detected by an observer located at a distance z_0 below the origin (see Fig. 4). In our simple model the horizontal electric field is

$$E_x = E_0 \frac{\cos \omega_e t - \frac{v_e}{c}}{(1 - \frac{v_e}{c} \cos \omega_e t)^3} \tag{3}$$

where E_0 is a time-independent constant. The time-dependent factor in E_x is shown in Fig. 5 as a function of both T_0 (i.e., as it would appear to the observer) and t (as the electron experiences it). We note the very large amplitude at early times. The end of the electron trajectory after one-half turn in the magnetic field corresponds to $\omega_e T_0/\pi = 1$ or $T_0 \approx 1$ µs. No more field is radiated to our observer after this time (the radiation emitted in the starting and stopping of the electron, which may be called bremsstrahlung, vanishes in the z-direction) and the observed E_x signal becomes small and negative after less than 10 ns. Integrating the electric field signal coming from electrons within a column of radius ρ gives the total electric field seen by the observer (Longmire 1986):

$$\mathcal{E}_x = \frac{1}{2} Z_0 e v_e N_A \frac{\sin \omega_e t}{1 - \frac{v_e}{c} \cos \omega_e t} \tag{4}$$

where N_A is the column density of electrons in the thin layer where the gammas are absorbed. Since in this simple model every gamma creates an electron, N_A is the column density of 2 MeV gammas hitting the atmosphere shortly after the burst. As we discussed

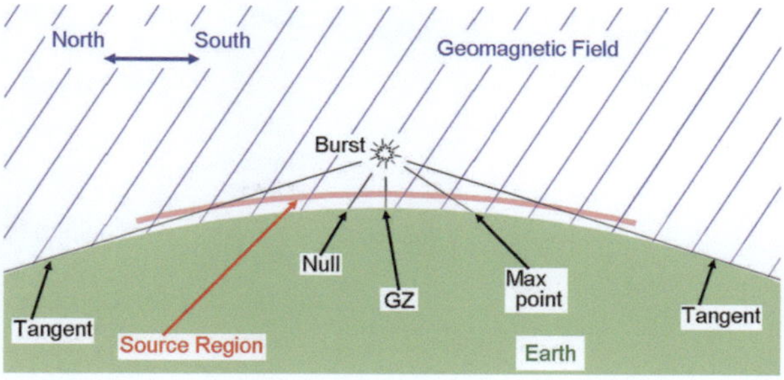

Fig. 6 Typical geometry of E1 phase. There are three special points on the ground: *Ground Zero* (*GZ*)—is the point directly below the burst, *Null Point*—where the observer ray and geomagnetic field lines are parallel, and *Max Point*—where the E1 HEMP has its maximum peak level (from Savage et al. 2010)

earlier $N_A \sim 2.3 \times 10^{13}$ m^{-2} and thus the amplitude of the total electric field at the observer is

$$\mathcal{E} = \frac{1}{2} Z_0 e v_e N_A \approx 2 \times 10^5 \text{ A/m}. \tag{5}$$

Below the source region, where the gammas have been mostly scattered, the HEMP propagates as a free wave without further buildup or attenuation. This is the signal that reaches the surface and can cause major disruptions in technological systems.

2.1.4 Exposed Area

The area impacted by HEMP E1 phase is controlled by the burst altitude and the direction of the local geomagnetic field. Figure 6 shows three special points. The rays originating at the burst and that are tangent to the Earth's surface define the maximum extent of the HEMP E1 exposure region. Ground Zero is the point directly below the burst where the observer ray goes straight down. The red region in the atmosphere is the "source region" where most of the burst gammas interact with the atmosphere. The other two special positions in Fig. 6 are related to the geomagnetic field. The "null point" is where the observer ray and geomagnetic field lines are parallel. For this point the E1 HEMP is very low (ideally it would be zero). On the opposite side of Ground Zero is the "max field point". This is the location where the HEMP E1 has its maximum peak level (where the observer ray is perpendicular to the geomagnetic field lines). This point could be called the "geomagnetic max point".

Figure 7 shows the area coverage for a HEMP over the US, for several burst altitudes. One can see that a nuclear explosion at an altitude of ~400 km above Oklahoma would impact the entire continental United States.

2.2 E2 Phase

As we discussed above the E1 phase of the HEMP is associated with unscattered gammas directly coming from the nuclear explosion. However, as we discussed in Sect. 2.1 the direct gammas travel through ~220 kg/m^2 of air before they hit the E1 source layer. About a fraction of $1/e \approx 37\%$ of the 2 MeV gammas interact with air molecules before reaching the E1 source layer resulting in a short time delay in the arrival time of these scattered gammas to the source layer. These variously-produced gammas lead to an impulsive Compton electron current (due to the separation of the electrons from their parent molecules) that depends on the polar angle because of the atmospheric density gradient. There is also a non-compensated vertical current directly below the explosion.

The end result of these secondary gammas is that at the end they also produce an electromagnetic pulse similar to the E1 phase. While the basic physics of the generation of this

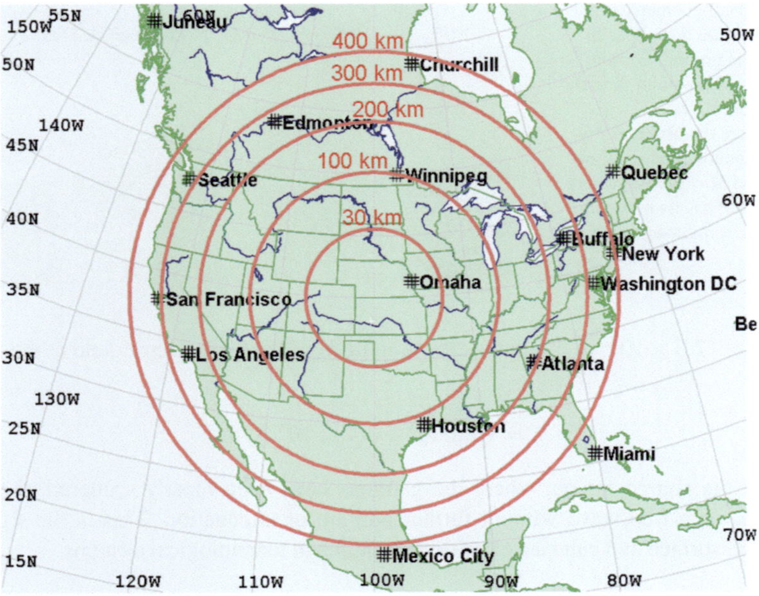

Fig. 7 Samples of E1 HEMP exposed regions for several burst heights. The *red circles* show the exposed regions for the given burst heights, for a nuclear burst over the central US (from Savage et al. 2010)

secondary signal is basically the same as we discussed in Sect. 2.1, there is one big difference: the secondary E2 signal lasts much longer (actually several thousand time longer) than the E1 signal. Consequently the peak of the E2 signal is several hundred times weaker than the peak of the E1 signal (see Fig. 2). Even though most of the E2 signal takes place within about 10 ms after the burst, the E2 phase only ends at about 1 s after the burst.

The E2 phase covers roughly the same geographic area as the E1 component and is similar to lightning in its time-dependence, but is far more geographically widespread in its character and somewhat lower in amplitude. In general, it would not be an issue for critical infrastructure systems since they have existing protective measures for defense against occasional lightning strikes. The most significant risk is synergistic, because the E2 component follows a small fraction of a second after the devastating E1 impact, which has the ability to impair or destroy many protective and control features. Thus the energy associated with the second component thus may be allowed to pass into and damage systems.

2.3 E3 Phase

The E3 component is very different from E1 and E2. E3 is a very slow pulse, lasting tens to hundreds of seconds (see Fig. 2). It is produced by two different physical mechanisms, both of which are associated with the continuum behavior of the medium. This is the reason why the E3 phase is also called the MHD (magnetohydrodynamic) phase of the HEMP.

The first phase that typically lasts about 1 to 10 seconds is called the E3A or "Blast Wave" phase. This is characterized by the explosive expansion of the fireball containing a large mass of ionized material. The expanding plasma cloud expels the geomagnetic field and creates a diamagnetic bubble. The distortion of the geomagnetic field generates a transient current system that creates the first MHD peak shown in Fig. 2 (cf. Gilbert et al. 2010).

The second phase of E3, called E3B, takes place between about 10 and 300 seconds. It is created when the hot, ionized, diamagnetic debris bubble buoyantly rises in the upper atmosphere. As this conducting patch crosses geomagnetic field lines it generates currents that flow in the patch and distorts the ground magnetic fields on the surface. This late phase is also called "Heave" and it creates the second MHD peak shown in Fig. 2 (cf. Gilbert et al. 2010).

2.3.1 *Blast Wave*

About 75% of the energy of a high-altitude nuclear explosion is emitted in the form of X-rays with an average energy of a few keV. About half of these X-rays are emitted downward and are absorbed in the atmosphere in the altitude range of 80 and 110 km. The primary X-rays interact with K-shell electrons in neutral molecules and knock them out producing primary photoelectrons. The missing K-shell electron causes secondary X-ray emission when an electron jumps from a high-shell orbit to fill the gap on the K-shell. The primary photoelectrons create an additional ionization cascade resulting in a high level of ionization and heating in the 80–110 km layer. The horizontal extent of this layer is similar to the E1 phase, since both the gamma rays and X-rays expand radially from the blast location. This hot, ionized layer acts to shield the ground from direct electromagnetic signals generated in the burst region and "anchors" the geomagnetic field lines.

About a quarter of the blast energy is kinetic energy of the weapon debris, which has become highly ionized and therefore has a high electrical conductivity. The debris expands outward, pushing the geomagnetic field out of the conducting region, and this forms a diamagnetic cavity or "magnetic bubble." The initial expansion of the bubble is determined by the velocity of the debris, which is greater than 1000 km/s. Later expansion of the bubble depends on the blast altitude. For altitudes below about 300 km, the dominant effect slowing the expansion is the outside atmospheric pressure, and the bubble becomes asymmetric as it expands more easily upward into more rarefied air than downward into denser air. For higher blast altitudes, the atmospheric pressure is negligible and the expansion is slowed by the anisotropic magnetic pressure gradient force. At the end the kinetic energy of the debris is converted into magnetic energy of distorted field lines. Due to the anisotropy of the magnetic pressure the bubble expands more rapidly along the geomagnetic field lines and less rapidly in the perpendicular direction (Karzas and Latter 1962). This situation is depicted in Fig. 8. For observers at large distances from the burst, the perturbation of the magnetic field looks like a field-aligned dipole at the burst point. The resulting electric and magnetic fields are low, but they can exist over long time periods and large areas.

2.3.2 *Heave Phase*

For bursts above $\sim$150 km altitude the second part of the EMP E3 phase is the Heave effect (also called the E3B phase). The bomb debris and shock heated air ions stream downward along geomagnetic field lines until they deposit their energy at altitudes near 130 km. These two processes contribute to both the heating and additional ionization of the ionospheric E-layer. There is also heating of this region by UV radiation from the burst, and this heating is centered beneath the burst. After an initial brief period of expansion of the heated air, it buoyantly rises. The basic phenomenology of the heave generation is shown in Fig. 9. The region with enhanced conductivity is depicted as a reddish "hat" with the heated center rising more rapidly than the brim. As this conducting layer rises across the geomagnetic field, a current is induced by the dynamo effect, as indicated by the solid arrow on the hat in Fig. 9.

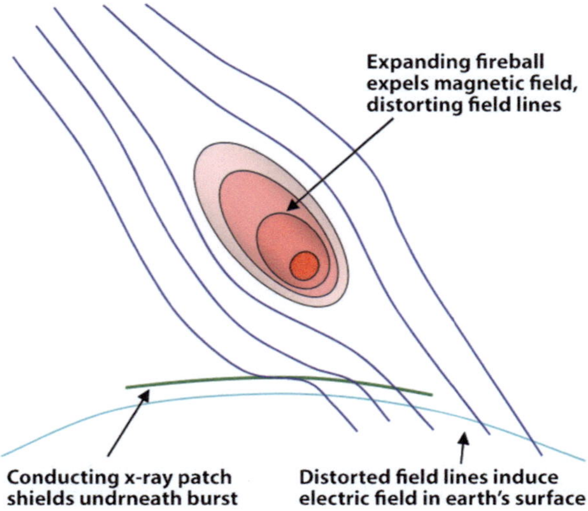

Fig. 8 Schematic of E3A Blast Wave phenomenology. The X-ray patch is shown at less that its actual radius for illustrative purposes (from Gilbert et al. 2010)

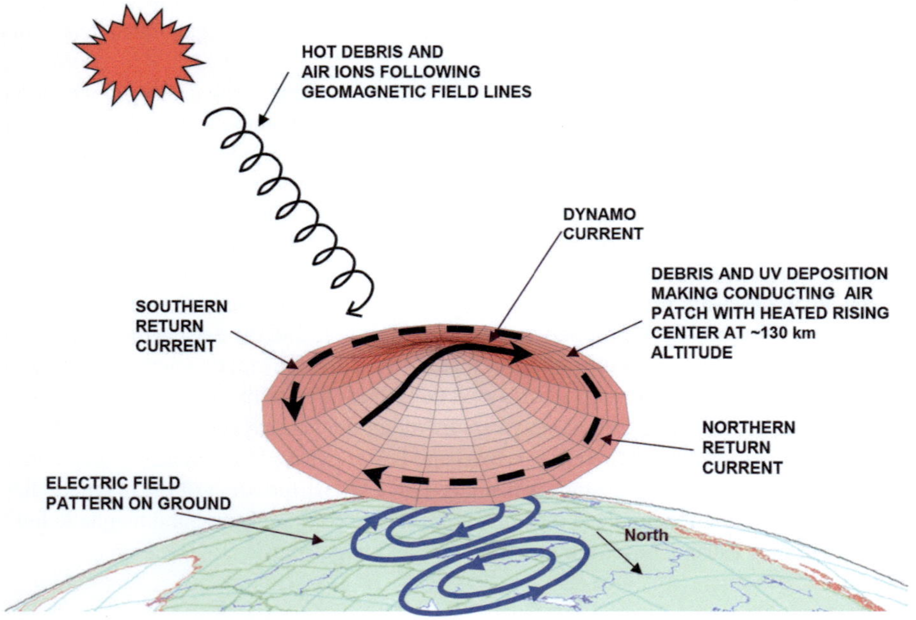

Fig. 9 Schematic of E3B Heave phenomenology (from Gilbert et al. 2010)

The dynamo current flowing to the west is accompanied by northern and southern return currents in regions where the heating and buoyant rise is smaller. This current system induces an oppositely oriented "two-cell" current system in the ground, and the finite conductivity of the ground means that the current in the ground is accompanied by an electric field in the same direction, as shown by the blue pattern (Gilbert et al. 2010).

The maximum electric field strength can be found in the region beneath the most highly heated portion of the atmosphere. Because the heating of the atmosphere is relatively localized (several hundred kilometers), the heave phase impacts a much smaller area than the blast wave. The electric field pattern mirrors the current flow in the conducting heave region. The centers of the "two-cell' patterns themselves are null field regions, where there is large cancellation between dynamo current flow and the return current flow.

3 Artificial Radiation Belts

The first major scientific discovery of the Space Age was that the Earth is enshrouded in toroids, or "belts," of very high-energy magnetically trapped charged particles (Van Allen et al. 1958; Van Allen 1959). Early observations of the radiation environment suggested that the Van Allen belts could be delineated into an inner zone dominated by high-energy protons and an outer zone dominated by high-energy electrons (Van Allen and Frank 1959). Subsequent studies showed that electrons in the energy range 100 keV $< E <$ 1 MeV often populated both the inner and outer zones with a pronounced "slot" region relatively devoid of energetic electrons existing between them. The energy distribution, spatial extent and particle species makeup of the Van Allen belts has been subsequently explored by several space missions.

Experience has shown that the near-Earth space environment can cause significant operational anomalies and even spacecraft failures under certain circumstances (e.g. Fennell et al. 2001). The primary sources of spacecraft operational problems include energetic ions causing so-called single-event upsets (SEUs), moderate-energy electrons producing surface differential charging effects, and high-energy radiation belt electrons inducing deep-dielectric charging conditions (see Baker 2002). SEU effects can be due to galactic cosmic rays, solar energetic particles, or trapped ions in Earth's inner Van Allen radiation belt. Surface charging is associated most closely with $\sim$10 to $\sim$100 keV energetic electrons during geomagnetically active times when spacecraft surfaces are in shadowed regions (or the entire satellite is in solar eclipse). Deep dielectric charging occurs most prominently when electrons of hundreds of keV to multiple-MeV energy are enhanced within the outer Van Allen radiation belt.

The natural space environment of the Earth has been extensively explored through the past six decades. The Radiation Belt Storm Probes (Van Allen Probes) mission of NASA (Mauk et al. 2013), launched on August 30, 2012, are currently bringing exceptionally detailed observations of the near-Earth space environment and its dynamic properties. The salient properties of the multi-MeV electron environment of the Earth, as reported in a recent study (Baker et al. 2016), include: (1) The outer Van Allen zone during strong geomagnetic storms can vary in absolute intensity by some six orders of magnitude on time scales of less than one day; (2) The magnetosphere is fully capable of accelerating relativistic electrons up to energies approaching 10 MeV on time scales of a few hours; (3) The radiation belt boundaries are determined both by natural and human-induced wave effects on the electron spatial and spectral properties (see below); and (4) The inner Van Allen radiation zone ($1.1 \lesssim L \lesssim 2.5$) has been essentially devoid of multi-MeV electrons for the entire Van Allen Probes era (2012-present).

As introduced in Sect. 2, secret activities were underway in 1957–1958 to explode nuclear weapons in space to create artificial radiation regions in Earth's magnetic field (Christofilos 1959a,b). Christofilos envisioned two physical effects: (1) A strong, prompt enhancement of ionospheric ionization in the vicinity of the nuclear explosion, thereby producing strong disruption of radio communication at VHF frequencies; and (2) Strong injection and trapping of high-energy ($E >$ 1 MeV) electrons in Earth's dipole-like magnetic field. A key question in Christofilos' work was what fraction of emitted fission electrons

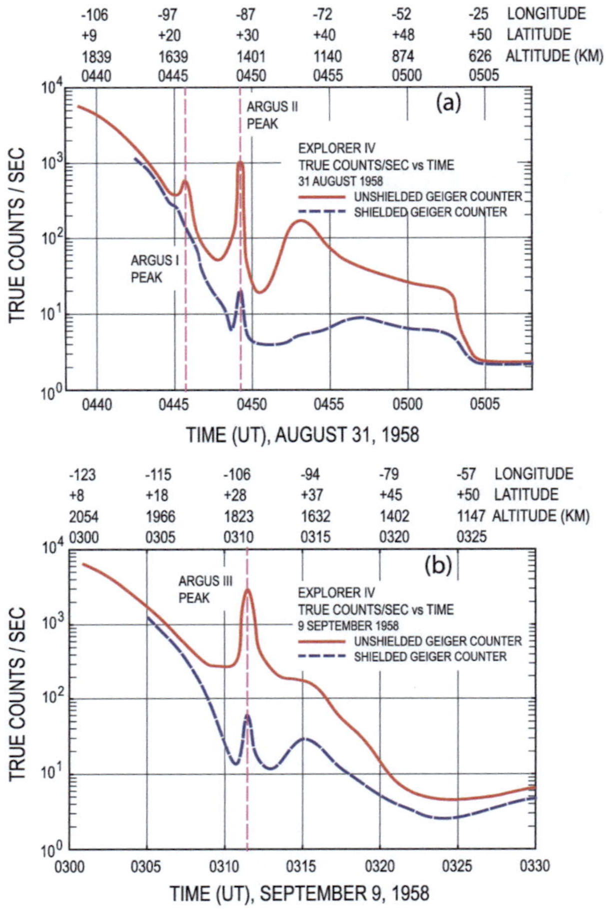

Fig. 10 Plots of passes by the Explorer IV satellite through the enhanced energetic electron fluxes in the artificial radiation belts created by the Argus nuclear explosions. (**a**) A pass through the Argus II belt, with the decaying Argus I belt still visible on 31 August 1958. (**b**) A pass through the Argus III artificial belt on September 6, 1958 (from Van Allen et al. 1959c)

and positrons from the nuclear explosion would become trapped in the Earth's field (see Van Allen et al. 1959c; Walt 1997).

After the successful deployment and initial operation of the Explorer I and III missions, Van Allen and coworkers were brought into the secret program to test the Christofilos effect (Van Allen 1997). In anticipation of high-altitude nuclear tests, starting in May 1958, the Explorer IV spacecraft was instrumented by Van Allen and his team in record time and was launched on July 26, 1958 (Van Allen et al. 1959b). As seen in Table 1 above, there had already been by that time, one relatively small (1.7 kt) atmospheric test (codenamed Yucca) on April 28, 1958. In August 1958 there were large weapon tests of about 4 Mt yields (Teak and Orange), but these were at relatively low heights (some tens of km altitude) and there were only small effects seen in the radiation belts (Van Allen 1997). The truly high-altitude weapons tests were in Operation Argus (Jones et al. 1982) and, as shown in Table 1, these were small weapons at several hundred kilometers altitude. As reported (Van Allen et al. 1959c; Brown 1966; Van Allen 1997; Walt 1997) the Argus tests produced discernible, but not major, perturbations of the radiation environment. All three artificial radiation shells were clearly observable by the Explorer IV instruments designed for the purpose (see Fig. 10). Van Allen (1997) summarized that Argus I and II each produced shells of artificially injected electrons that had a lifetime of about 3 weeks. The explosions, on August

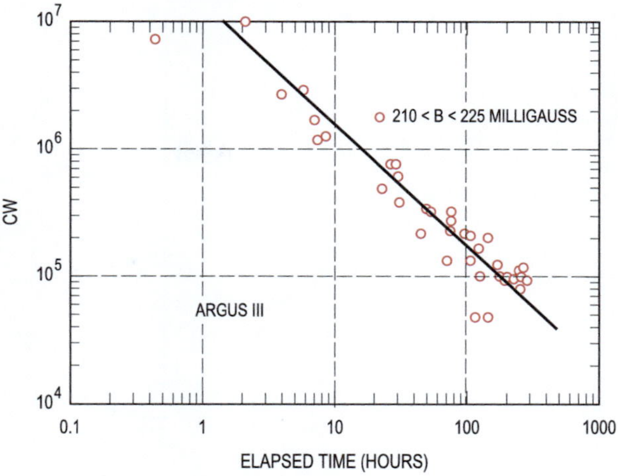

Fig. 11 The decay of energetic electron fluxes in the artificial radiation belt created by the Argus III nuclear explosion, observed by the Explorer IV satellite over several weeks. The plot shows the product of the maximum true counting rate at the center of the Argus III shell with the geometric width of the shell at half maximum, to provide a measure of the electron content of the belt and its decrease with time (from Van Allen et al. 1959c)

27 and 30, 1962 were followed by major—natural—geomagnetic storms. It is likely that these shortened the lifetime of the artificial radiation belts, as the loss of energetic trapped electron due to geomagnetic activity is well documented (e.g. Horne et al. 2009). The Argus III shot also produced a shell of electrons. Argus III had a lifetime of about a month, see the decay of the observed peak intensities in Fig. 11. All of these nuclear weapon-injected electrons had—as measured by Explorer IV sensors—a very different energy spectrum than the naturally occurring radiation belt population. Thus, observations could readily distinguish fission electrons from "normal" Van Allen belt particles (Van Allen et al. 1959c).

Neither the Explorer IV satellite, nor the Argus experiments were a part of the contribution of the United States to the International Geophysical Year (IGY) that ran from July 1957 to December 1958 during which over 60 countries collaborated on investigations of a range of geophysical phenomena. But Van Allen, in a letter dated February 21, 1959, requested the declassification of the geophysically relevant results of both the Argus experiments and the observations made by Explorer IV. In the letter, Van Allen wrote that the Argus experiments "undoubtedly constitute the greatest geophysical experiment ever conducted by man." The earliest published record, describing both geomagnetic effects and an artificial aurora, identified the nuclear tests as the cause of these geophysical phenomena (Cullington 1958). Following Van Allen's letter and the declassification of the Argus tests, the Explorer IV observations were presented at a Symposium of the US National Academy of Sciences by Van Allen et al. (1959a). Observations of the trapped electrons from the Argus tests were also made by a series of 19 dedicated sounding rockets (Allen et al. 1959) who noted that the energy spectrum of the electrons was a modified form of the fission spectrum, with a lower than expected high energy contribution, confirming the observations mentioned above by Explorer IV.

The series of high-altitude nuclear detonations in 1962 by both the United States and the Soviet Union (see Table 1) provided a greater opportunity to observe and analyze the formation of artificial radiation belts. Their effects on satellites and on geophysical phenomena are discussed below in Sects. 4 and 5. Here we describe the formation and evolution of the radiation belts formed from energetic electrons originating from the decay of nuclear fission products.

The best documented explosion, with the longest lasting effects, was the Starfish burst on July 9, 1962 at 400 km altitude over Johnston Island in the Pacific, with a yield of about 1.4 Mt. From classified observations at the time that became public only much later (Dyal

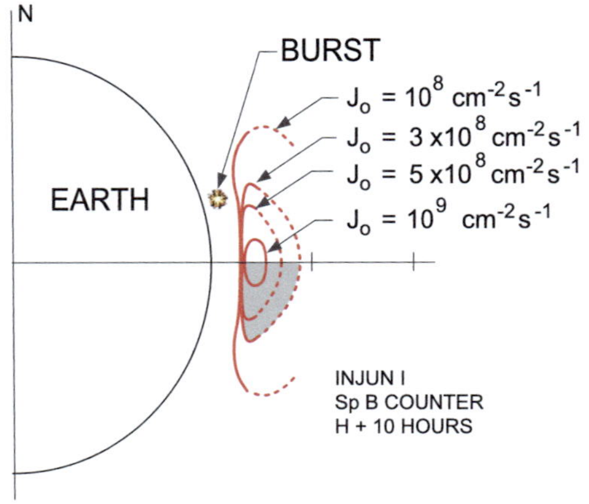

Fig. 12 The spatial distribution of high energy electron fluxes measured by the Injun I satellite in the newly formed artificial radiation belt in the hours following the explosion on July 9, 1962 (from Van Allen et al. 1963; Van Allen 1966)

2006), it is now known that in the diamagnetic cavity that was formed at the time of the explosion the flux of energetic beta particles (electrons and positrons) was approximately uniform through the volume, with intensities of about 3×10^{11} particles $cm^{-2}\,s^{-1}$, for a duration of at least 7 seconds. At 34 s after the explosion, the flux injected into the geomagnetic field from the diamagnetic cavity was 2.5×10^{10} beta $cm^{-2}\,s^{-1}$. Observations by two of the rocket flights above the detonation site that were reported by Dyal (2006) were made at a location (in McIlwain's B–L coordinate system) at $B = 0.153$ gauss and $L = 1.269$. The fluxes measured were 4×10^{10} beta $cm^{-2}\,s^{-1}$ at 25 seconds after the explosion and 6×10^9 beta $cm^{-2}\,s^{-1}$ at 10 minutes after the explosion.

It is this population of energetic electrons that were trapped and entrained by the Earth's magnetic field and formed the artificial radiation belt. The first observations of the Starfish explosion were made by the UK-US Ariel-1 satellite (Durney et al. 1962), the Soviet Kosmos-5 satellite (Galperin and Boliunova 1964) and by the University of Iowa's Injun I satellite (O'Brien et al. 1962a; Van Allen et al. 1963), as described in more detail in Sect. 4. In the hours that followed the detonation, Injun I successfully mapped the core of the newly created artificial belt of high energy electrons, as shown in Fig. 12 (taken from Van Allen 1966). The peak flux in the central volume of the belt, shortly after the explosion, was $>10^9$ electrons $cm^{-2}\,s^{-1}$, well matched to the estimated injected population calculated by Dyal (2006).

The launch of Telstar-1, on the day following the Starfish explosion, brought a very powerful observation platform into near-earth space, extremely well instrumented (for the time) to monitor the energetic particle population of the natural as well as of the artificial radiation belts (Brown et al. 1963). Helped by both its orbit (perigee = 952 km, apogee = 5632 km, inclination = 45°) and its instrumentation, it was able to map the Starfish radiation belt and follow its evolution. Figure 13 shows the distribution of energetic electron fluxes within the first week following the explosion. Two aspects need to be noted. First, the contours do not show the apparently much higher intensity core of the artificial belt that the Injun I observations implied. This led at the time to a possible ambiguity between the two sets of measurements. However, the Telstar-1 data do not contradict the existence of a higher intensity core, but represent simply different results due to the different energy response of the instruments and the different spatial sampling. The second point to note is that the artificial radiation belt, at least initially, had a much greater extent in volume than implied by the explosion site. The extent of the diamagnetic cavity, already described and the implication

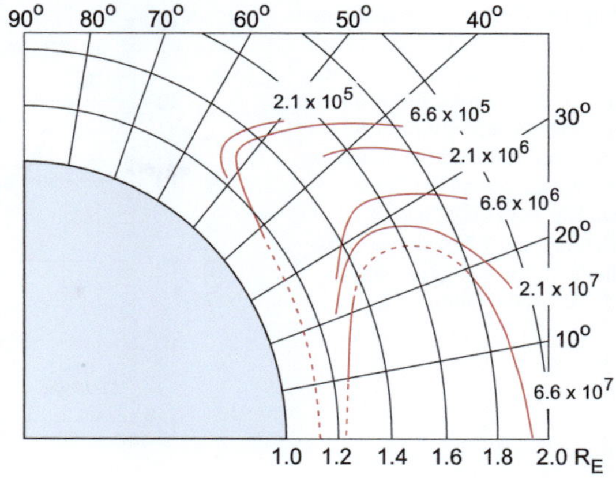

Fig. 13 Contours of omnidirectional counting rates (OCRs), electron channel 3 by Telstar-1, on Days 193–197, 1962, starting three days following the Starfish explosion. OCR is measured in counts/sec; the electron channel 3 responds to electrons of energy greater than about 400 keV, with maximum efficiency at about 500 keV. The plot is in the *R*–λ coordinate system, related to the McIlwain's *B*–*L* coordinate system (see Roberts 1964). (From Brown et al. 1963)

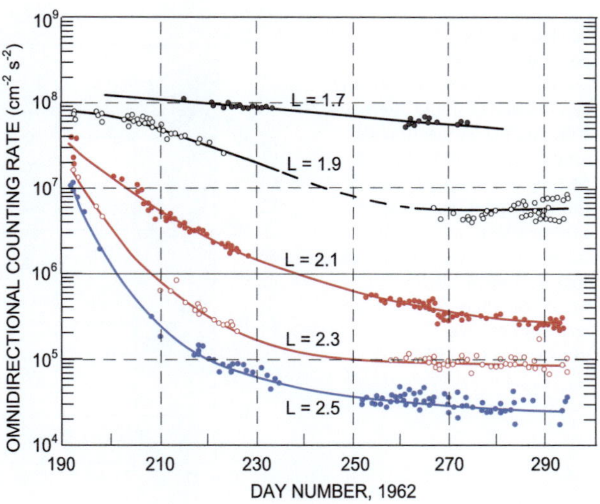

Fig. 14 The omnidirectional counting rate between the launch of Telstar-1 and the first of the Soviet high altitude blasts on day 295. The curves show the counting rates in a narrow λ ranges on individual *L* shells, close to the equator for $L = 1.7$ and increasingly larger *l* values for the higher *L* shells. The clustering of observational points is due to the precession of the Telstar-1 orbit (from Brown 1966)

of the vertical expansion, driven by jets and driving a shock wave to a height of 1000 km or higher (Colgate 1965), can explain the population of *L* shells up to and beyond $L = 2$.

Given the quasi-instantaneous injection of the population at the time of the explosion, the decay of electron fluxes in the artificial belts provided an opportunity to study the possible loss mechanism of these particles. Figure 14 follows the decay, as a function of *L*, of energetic electrons observed by Telstar-1 (from Brown 1966). The decay is faster as *L* increases towards the radiation belt slot region where electrons are lost mainly to pitch-angle scattering due to wave-particle interactions. However, the artificial belt is quite stable at $L = 1.7$. At lower *L*-values ($L < 1.2$), atmospheric interactions caused the decay of the fluxes (Walt 1964). But at intermediate *L*-values, $1.3 < L < 2$, the loss mechanism is not fully understood, and the lifetimes in the artificial belt were found to be several years.

The three Soviet nuclear explosions in late October and on November 1, 1962 replenished the artificial radiation belt, although apparently at somewhat higher *L*-values (likely to be due to the more northerly location of the explosions). The new population, superimposed on the decaying Starfish belt, is well illustrated in Fig. 15 (from Brown 1966). The peak intensities appeared close to that from Starfish; the decay constants appeared to be quite comparable to those experienced by the Starfish fluxes.

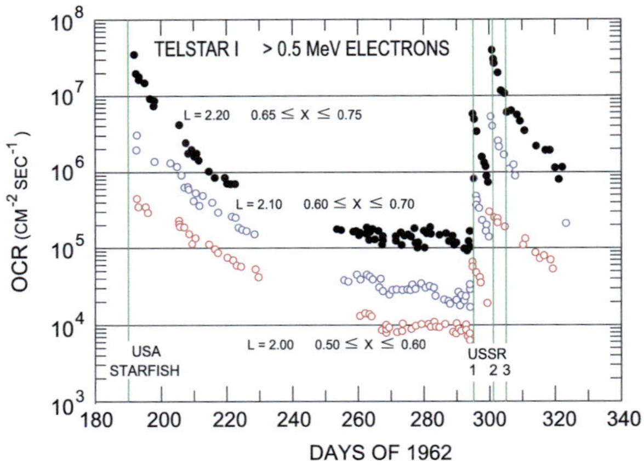

Fig. 15 Observations of the time evolution of energetic electron fluxes by Telstar-1, showing the decay of the fluxes following the Starfish nuclear test on 9 July 1962 and showing the temporary replenishment by the Soviet nuclear tests on 22 and 28 October and 1 November 1962. In the figure $x = \sqrt{1 - B_0/B}$ (from Brown 1966)

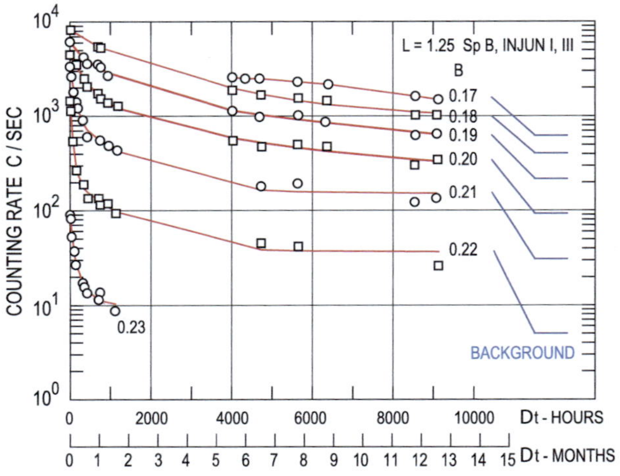

Fig. 16 The long-lasting energetic electron fluxes at $L = 1.25$, measured first by the Injun-I satellite in the aftermath of the Starfish detonation and, from its launch on December 13, 1962, by the Injun-III satellite. After the initial fast decay, fluxes in the artificial radiation belt stabilized and decayed over many months. Similar instruments on the two satellites responded to electrons between 1.5 and 5 MeV (from Van Allen 1964)

The longer term stability at $L = 1.25$ was measured by comparing the fluxes observed by Injun I in the weeks after Starfish and those measured by the Injun III satellite (launched on December 13, 1962) by similar instruments. The results are shown in Fig. 16. It can be seen that following a fast initial decay, the near-equatorial Starfish artificial belt stabilized and the fluxes remained up to an order of magnitude higher for well over a year than the pre-Starfish natural fluxes. It is unclear, but likely, that the fluxes observed by Injun III were exclusively due to Starfish and were not influenced by any replenishment from the Soviet explosions that occurred about six weeks before The Injun III launch, because the core high intensities of the Soviet artificial belts were at higher L-values than the Starfish belt.

The third Soviet high-altitude explosion on November 1, 1962 created a narrow band of high energy electrons around $L = 1.76$ (Brown 1966). Figure 17 shows the peak in this belt at four epochs after the explosion, at one, four eight and eighteen days after the blast, as a function of L. As noted by Brown (1966), the position of the peak in L remains constant while the intensity decreases by about a factor four. At the same time, the full-width half-maximum of the peak increases monotonically from 0.037 to 0.052, implying the diffusion across L-shells of the electrons in the belt. While this is a good illustration of the process well observed in the natural radiation belts (e.g. Williams et al. 1968), it does not completely

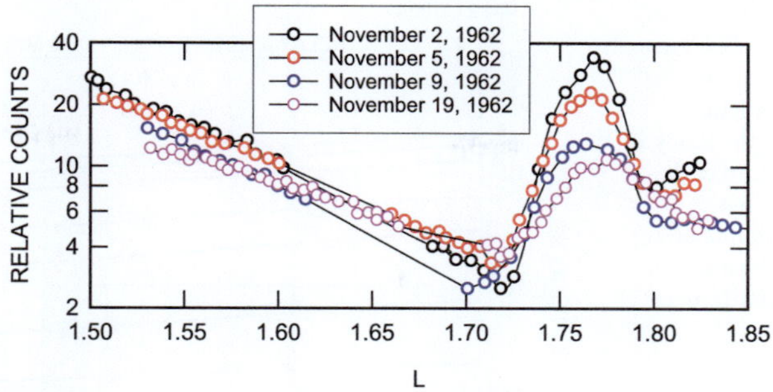

Fig. 17 The peak of energetic electron fluxes ($E > 1.9$ MeV) due to the third USSR test on November 1, 1962 were distributed in a narrow range of L-shells around $L = 1.765$ at four epochs following the nuclear blast. The height of the peak is decreasing with a time constant of days. It is also broadening, due to cross-L diffusion. The fluxes at lower L-values are due to the slowly decaying electrons from the Starfish explosion (from Brown 1966)

Fig. 18 A schematic representation of the evolution of the Starfish radiation belt over a year. The decay continued for more than two years, until observations became difficult (from Van Allen 1964). (The quantity plotted is according to the original labelling by James Van Allen)

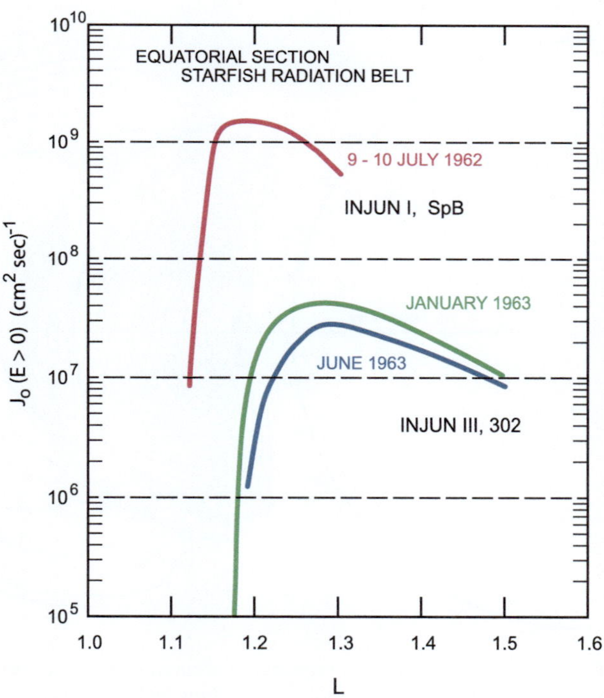

explain the reduction in flux in the Soviet artificial radiation belt, as the broadening of the peak accounts for only 30% of the losses.

A summary sketch of the Injun I and Injun III observation is shown in Fig. 18 (from Van Allen 1964). The peak fluxes decreased by close to two orders of magnitude, by the erosion of the artificial belt at the low L values where atmospheric losses operated. Removing that part of the artificial belt left the more stable belt at about $L = 1.3$ to persist for at least a year. The complementary longer term observations by the Relay II satellite, launched on January 21, 1964 and fully equipped to measure the trapped radiation belts particles, in-

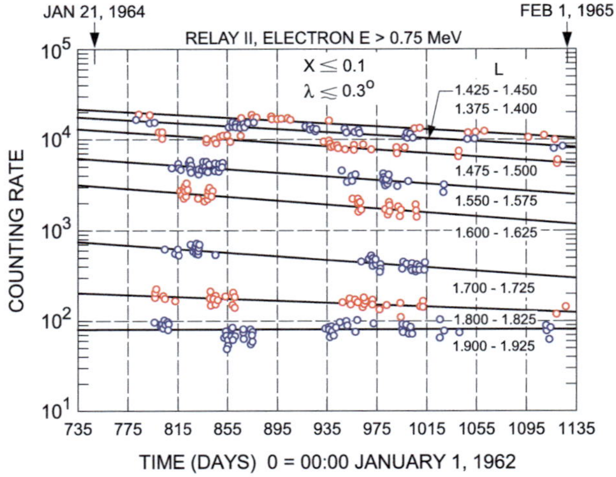

Fig. 19 Long term survival of the artificial radiation belt created by the Starfish detonation, observed by the energetic electron detector on the Relay II satellite. The clustering of the data points is due to the precession of the satellite orbit. The *solid lines* are *least square fits* of exponentials to the data (from Brown 1966)

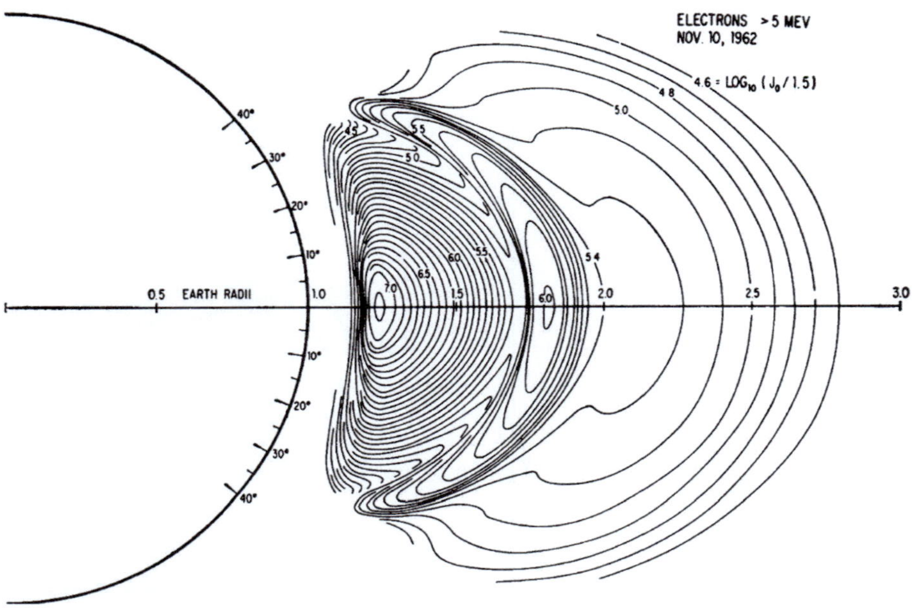

Fig. 20 Spatial distribution of high-energy electrons on 10 November 1962. The intensities represented by adjacent contours differ by a factor of 1.259 (from McIlwain 1963)

cluding the remnants of the artificial belt populations, are shown in Fig. 19 (from Brown 1966). The interesting aspect of these observations is that where the earlier decay observations showed decay constants of months at most, the still decreasing fluxes through to the end of 1964, more than two years after Starfish, now show that the time constants are in fact longer, as much as two years. It is clear that the artificial belts created by the high altitude nuclear explosions endured at detectable levels for several years in a range of L-values about 1.5 to 1.8. Both below and above these values loss mechanisms actively removed the excess population. In the more stable range, however, it is likely that cross-L diffusion

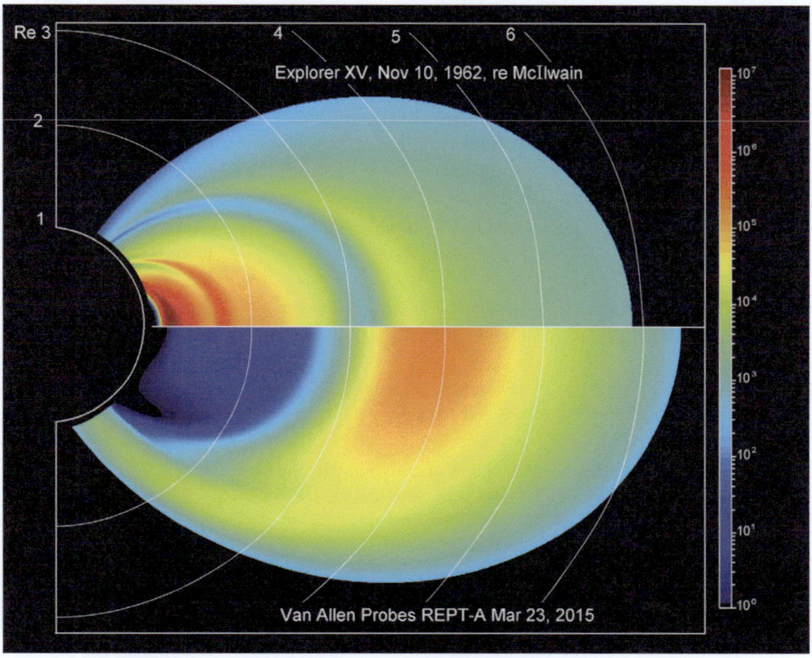

Fig. 21 Comparison of Explorer XV data from 10 November 1962 with Van Allen Probes data taken with the Relativistic Electron-Proton Telescope (REPT) on 23 March 2015. The Explorer XV data are adapted from McIlwain (1963) (see Fig. 20) and the Van Allen Probes data are adapted from measurements described in Baker et al. (2016). It is noted that the March 2015 outer zone fluxes of $E > 5$ MeV electrons were more intense and broader in spatial extent than were such electrons in the November 1962 case. On the other hand, the inner zone electron fluxes in the 1962 post-Starfish era were at least 10 million times more intense than fluxes seen in the present-day magnetosphere. Such high fluxes were deadly to orbiting spacecraft of the early 1960s (as described in the text)

slowly moved the particles towards regions where more effective loss mechanisms finally removed them.

Finally, it is instructive to compare the exceedingly high energetic electron fluxes injected by the high-altitude nuclear explosions to today's state of the locations of the erstwhile artificial belts.

Figure 20 (taken from McIlwain 1963) shows that in mid-November 1962 (i.e., following the Starfish and major Soviet exoatmospheric tests), the omnidirectional flux of electrons with $E > 5$ MeV exceeded 10^7 cm^{-2} s^{-1} at $L \sim 1.2$. The study by (Brown 1966) similarly showed (see his Fig. 9) that electrons with $E > 1.9$ MeV were above 10^7 cm^{-2} s^{-1} for $1.2 \lesssim L \lesssim 2.5$ and had a peak intensity of $>10^8$ cm^{-2} s^{-1} at $L \sim 1.4$ in late October 1962.

Figure 21 compares the post Starfish observations with the modern equivalent measurements from the NASA Van Allen Probes mission for a period (March 20–23, 2015) just following the most powerful geomagnetic storm of the last decade (see Baker et al. 2016). The Van Allen Probes data show quite graphically that the "natural" radiation environment at present has no significant fluxes of $E > 5$ MeV electrons at $L \lesssim 2.5$ (see, also Fennell et al. 2015; Li et al. 2015). The inner zone electron fluxes in the 1962 post-Starfish era were at least 10^7 times more intense than even the highest fluxes seen in the present-day magnetosphere.

Table 2 Satellites in orbit and damaged by the high-altitude nuclear explosions

Satellite	Launch	End	Damage history	Comments
Injun I	Jun 29, 1961	Mar 6, 1963	Likely damage to solar cells and transmitter electronics, but not documented as associated with the Starfish belt.	Low power drain probably limited the damage due to reduced power from the solar cells
TRAAC	Nov 15, 1961	Aug 12, 1962	Failed due to radiation damage to solar cells	Solar cell test circuits enabled monitoring and diagnosing the failure of the power system
Transit-4B	Nov 15, 1961	Aug 2, 1962	Failed due to radiation damage to solar cells	Solar cell test circuits enabled monitoring and diagnosing the failure of the power system
OSO-1	Mar 7, 1962	May 1964	Eventual failure was due to solar cell damage	Operated only intermittently after the tape recorder failure in May 1962, prior to Swordfish.
Ariel-1	Apr 26, 1962	Aug 5, 1962	Failed due to radiation damage to solar cells. Command system electronics damaged; intermittent failures.	
Kosmos-5	May 28, 1962	Decay date: May 2, 1963	Last reported data in the literature is from November 1962. Reportedly damaged by Starfish	
Telstar-1	Jul 10, 1962	Feb 21, 1963	Well documented damage to the solar cells, but eventual failure was caused by radiation damage to transistors and other electronic devices	This spacecraft was extensively instrumented to measure radiation damage, both to solar cells and to active devices (transistors) in the electronics.
Alouette-1	Sep 29, 1962	Jan 28, 1968	Reportedly no adverse effects from the Starfish radiation	Very conservative power supply design, allowing for 40% degradation of solar cell performance
Explorer XIV	Oct 2, 1962	Aug 11, 1963	Intermittent electronic failures in the spacecraft encoder	
Explorer XV	Oct 27, 1962	Jan 30, 1963	Intermittent electronic failures in the spacecraft encoder, failure due to under-voltage caused by radiation damage	
ANNA-1B	Oct 31, 1962	?	Deterioration of the solar cells noted, but not fatal	Flew the first GaAs solar cells in space, making the solar array power source more resistant to radiation

4 Damage to Satellites

Satellites in orbit at the time of the nuclear tests experienced energetic particle fluxes that were intense not only in the short term, but—as can be seen in Fig. 15—remained significant long enough for the accumulated radiation dose to cause damage. Starfish Prime, having been the largest explosion, with the best documented consequences in terms of the artificial radiation belt created is also the one that had the best documented consequences in terms of satellites destroyed or damaged. A first comprehensive review of the effects on satellites was given by Hess (1963). The near-immediate satellite losses due to Starfish: Ariel-1, TRAAC

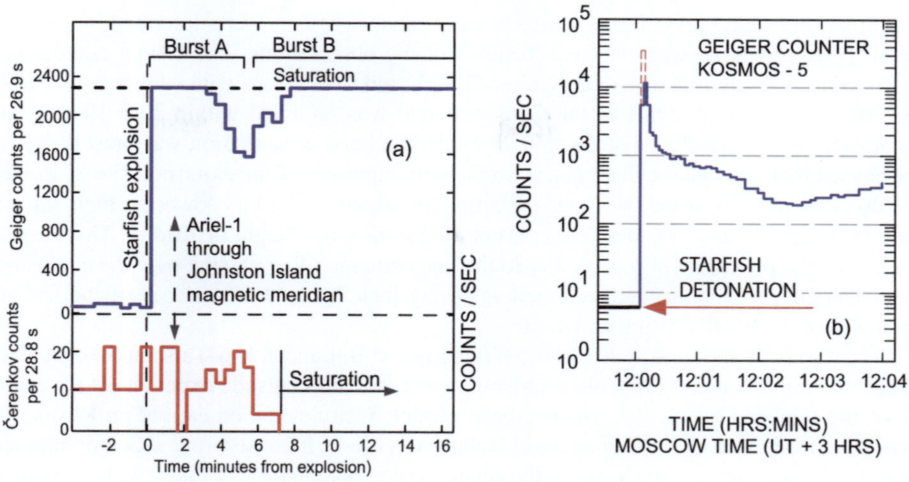

Fig. 22 High time resolution observations on the immediate radiation detected following the Starfish explosion. (**a**) Observations by Ariel-1, the first spacecraft to be fatally damaged by the Starfish explosion. In the first minutes after the explosion, both the Geiger counter and the Čerenkov detector saturated, as Ariel crossed the magnetic meridian of the explosion site within two minutes. (Durney et al. 1964). (**b**) Observations by the Geiger counter on the Kosmos-5 satellite showing the near-instantaneous increase of three orders of magnitude or more. The instrument may have entered a near-saturation mode, but its counting rate was corrected using the cross-calibration with the non-saturating fluorescent electron detector as shown by the *red line* (Galperin and Boliunova 1964)

and Transit-4B, had been documented earlier (Durney et al. 1962; Hess and Nakada 1962; Fischell 1962a).

Hess (1963) lists 6 satellites that sustained damage from the energetic particles generated by Starfish. These are Ariel-1 TRAAC, Transit-4B, Tiros 5 and OSO-1. More extended lists were published later by Wenaas (1978) and Conrad et al. (2010). The most relevant contemporary missions that were affected (or not) by the increased radiation due to the artificial radiation belt are listed in Table 2, together with their failure modes and any reason for mitigation for those missions that had their useful life shortened but did not immediately fail after Starfish.

The nuclear explosion on July 9, 1962 at 09:00:09 GMT was very quickly sensed by the radiation detectors on all five spacecraft: Kosmos-5, Ariel-1, TRAAC, Transit B, and Injun I then operating.

The two satellites closest to the explosion were Ariel-1 and Kosmos-5 (Durney et al. 1962; Galperin and Boliunova 1964). Kosmos-5 was on a meridian $\sim 75°$ to the west and Ariel-1 at $\sim 27.5°$ also to the west of Johnston Island. Their latitudes and height were 44.9°N and 1442 km (Kosmos-5) and 52°S and 819 km (Ariel-1). Their distances to Johnston Island were very comparable: 7400 km for Ariel-1 and 7500 km for Kosmos-5. While Ariel-1 was closer to the magnetic meridian of the detonation site, on shell $L \sim 5$, the greater height of Kosmos-5 allowed viewing above the site over the horizon at a height of $\sim$1200 km, while for Ariel-1 direct visibility over the site was above 2000 km. These differences probably explain the differences in the observations as well as their interpretation, although both provide evidence of the geographically very broad reach of the immediate effects.

A plot showing the observations by Ariel-1 immediately following the detonations are shown in Fig. 22a. Ariel-1 (Baumann 1963) was a joint US/UK mission carrying, among other instruments, two energetic particle detectors with the original objective of measuring the fluxes of cosmic rays. The characteristics of the instruments were described by Durney et al. (1962) together with the main features of the observations. The Geiger counter responded to both protons with energy $E > 43$ MeV and to electrons with energy >4.7 MeV. In Fig. 22a, it can be seen that the Geiger counter was saturated within 20 ± 10 s of the explosion. It was concluded at the time that this first burst of saturation was most likely to be caused by the magnetic disturbance propagating upwards of the explosion site at speeds $\sim$100 km/s that disturbed the outer radiation belt electrons and by lowering their mirror points effectively dumped these electrons at the location and height of Ariel-1. This is supported by the proximity of the satellite to the magnetic meridian passing over the explosion site. The Geiger counter then saturated again, by then from the beta decay of the fission products of the bomb drifting eastward.

The observations made by Kosmos-5 (Galperin and Boliunova 1964) also make clear that there was an effectively immediate and very large radiation episode experienced far away from the Johnston Island site. The energetic particle instrumentation carried on Kosmos-5 was described in detail by Galperin and Boliunova (1964). It consisted of a Geiger counter apparently very similar to those on the other, contemporaneous satellites, with a comparable shielding (3.4 g/cm^2 Pb and 0.8 g/cm^2 Al) and therefore had a comparable energy response to high energy electrons and protons. The observations made immediately after the detonation are shown in Fig. 22b. A first clear difference with the Ariel-1 observations is the apparent absence of saturation that allowed the initial peak intensity, about four orders of magnitude greater than the pre-starfish level, to be recorded. Although, as explained by Galperin and Boliunova (1964), the counter departed from the linear regime, it had been possible to calibrate the counting rate for high particle fluxes by the use of an additional instrument, an "electron indicator", effectively an analogue electron detector using a fluorescent screen and a photomultiplier monitoring the input electron flux. As the fluorescent detector did not saturate even in the maximum flux intensity regions of the artificial radiation belt as it evolved, the two detectors could be cross-calibrated.

Even if the first explanation of the origin of the first burst of particles is the dumping of particles from the disturbance to the natural radiation belt, the sustained high intensities after the initial saturation are best explained by the decay of fission products. This was the conclusion reached by O'Brien et al. (1962a), on the basis of pre- and post-burst observations of the high energy particles on the Injun I satellite. Their instrument, also a shielded Geiger counter, had a closely similar response (protons with $E > 45$ MeV and electrons with $E > 6$ MeV) to the Ariel-1 detector. The first observations on Injun I were made at the first pass over a ground station in Southern Rhodesia (now Zimbabwe) at 09:42 to 09:50 GMT, about 45 minutes after the explosion. By that time, electrons from fission products had time to form the eastward drifting shell of fission product electrons.

Another explanation of the very fast, first burst of saturation seen by Ariel-1 could be associated with the ionization and very fast motion of the fission products along the magnetic field lines, as reported later by D'Arcy and Colgate (1965), based on the interpretation of the high energy gamma radiation measurements near the magnetically conjugate point of the explosion site. Their argument is that a large population of quickly trapped electrons was generated by the decay of the fission products during the flight along the magnetic field line. The rise time of the excess gamma rays was very short, reaching a peak at $\sim$20 seconds from the explosion, showing that the decay-product electrons reached far from the explosion site to generate the saturation seen in the Ariel-1 detector. These arguments also apply to the results reported by Galperin and Boliunova (1964) about the peak increase of radiation immediately following the detonation.

D'Arcy and Colgate (1965) also suggest that the interaction of the plasma associated with the explosion with the geomagnetic field could have caused a mixing on neighboring flux

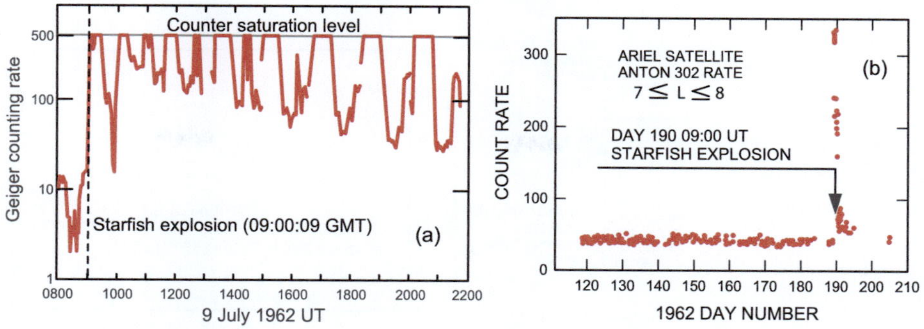

Fig. 23 (**a**) The Geiger counting rate observed on the Ariel-1 satellite. In the hours following the explosion the instrument regularly saturated on each orbit as the spacecraft crossed the forming artificial radiation belt. (**b**) In the long term observations of the stable high energy cosmic rays by Ariel-1, the high energy electrons injected by the explosion stand out even at high value L-shells (Figs. from Durney et al. 1962 and Elliot 1966)

surfaces, so that the original Ariel-1 explanation for the first burst of saturation, involving the disturbance in the magnetic field lines, was a factor in the minutes after the explosion. But given the very intense gamma ray radiation described by D'Arcy and Colgate (1965), it is also possible that these contributed (maybe significantly) to the initial burst of saturation observed by Ariel-1 (Durney et al. 1964).

In the left hand panel (a) of Fig. 23, the periodic passage through the then forming and evolving artificial electron belt, when the detector saturated, can clearly be seen. The minima occur when the satellite moved to the outer radiation belt that had large fluxes of the transient energetic electrons from the explosion out to $L = 4$ and beyond. However, the decrease in the intensity of the minima shows that at these high L-shells the fluxes decayed quite rapidly. The right hand panel (b) shows the average counting rates at the highest L values ($L \sim 7$ to 8), normally measuring the cosmic ray intensity. The peak at the time of the Starfish explosion provides another proof that the high intensity radiation reached far from the explosion site, but that at these heights the radiation was only short-lived.

The complexity of the early phase of the artificially formed radiation belt is illustrated by the possibility that a short-lived second "belt" was formed (Pieper 1963) below the longer lasting artificial belt which, however, may not have survived the eastward drift of the trapping shells past the South-Atlantic anomaly. This population of electrons had also been found by Ariel-1 (Durney et al. 1962) and Kosmos-5 (Galperin and Boliunova 1964). Further analysis of the Ariel-1 observations (Durney et al. 1964) also found evidence for a short-lived (about one day) shell of energetic electrons, probably due to fission products remaining in the vicinity of Johnston Island after the explosion.

Thus the early radiation exposure of the satellites in orbit on July 9, 1962 may have been markedly different, depending on their actual orbits and the timing of their orbits with respect the explosion site and their magnetic connectivity. This, together with their somewhat different design and components (in particular the solar cells) has led to the differences in the timing of their failures.

All evidence points to Ariel-1 and Kosmos-5 being the most exposed, particularly early after the explosion. Ariel-1 was the first spacecraft to show functional disturbances in the electronics of both the spacecraft system and in the scientific instruments. Intermittent electronic failures were noted within a few days, but 104 hours after the explosion the power supply showed the first undervoltage condition (Wenaas 1978), triggering a protective switch off of the payload, resulting from a 25% drop in power from the solar cells. Although the

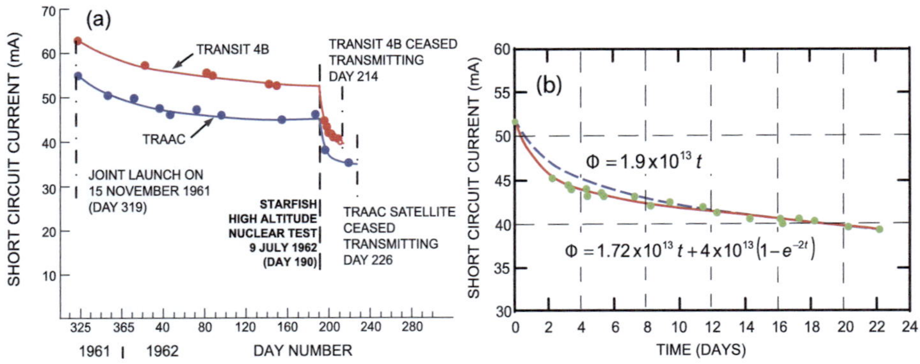

Fig. 24 Observation of the decreasing power from the solar cells on the TRAAC and Transit-4B satellites. (**a**) Time history from launch, showing the gradual damage from the natural radiation environment, followed by the much accelerated loss of power after the Starfish explosion (from Fig. 5 in Fischell 1962a). (**b**) Transit-4B solar-panel performance after the Starfish event, showing that the performance loss is best modeled by an initial period of higher intensity radiation, followed by a constant radiation flux (from Fig. 7 in Fischell 1962a)

satellite partially recovered, this condition dominated the next three weeks, finally leading to the end of the mission on August 5, 1962. There is no concrete report on damage to Kosmos-5, although it is generally considered to have had its operational life shortened by the Starfish radiation exposure (Wenaas 1978; Conrad et al. 2010). It continued to provide data at least until late October 1962, as evidenced by the published observations. It re-entered the atmosphere on 2 May 1963.

The TRAAC and Transit-4 satellites were launched simultaneously into the same orbit (1100 × 950 km, 32° inclination) on November 15, 1961. TRAAC was a satellite to test the concept of gravity gradient attitude control, but also carried energetic proton and electron detectors (Fischell 1962b). Transit-4-B was a US Navy navigation satellite, an early, experimental member of the long-running Transit series (Danchik 1998). Transit-4B carried no scientific instruments and was powered by both solar cells and one of the first Radioactive Thermoelectric Generators, the SNAP-3. The two satellites carried different circuits to measure the output of the solar cells. This monitoring function showed that from launch to July 9, Transit-4B and TRAAC solar cells lost, respectively, 17% and 18% of their initial power output. However, following the Starfish explosion, the deterioration in the Transit-4B solar cell output was 22% in 20 days. The equivalent figure for TRAAC was 22% in 28 days. Transit-4B ceased transmitting on August 2, 1962, and TRAAC also stopped transmitting on August 14, 1962 (Fischell 1962a). It should be noted that the nuclear power source on Transit-4b that could have continued to power the satellite, itself failed a month before the Starfish explosion, on June 12, 1962.

The time history of the output of the solar cells on TRAAC and Transit-4B is shown in Fig. 24(a) (Fig. 5 in Fischell 1962a). The difference between the rates of deterioration in the natural environment and in the artificial radiation belt after Starfish is clearly shown. In Fig. 24(b), the decrease in short circuit current (a measure of the power output) is shown in detail for TRAAC (from Fig. 7 in Fischell 1962a). Two empirical curves were fitted to the observed measurements. One (dashed line) assumed that there was a constant high energy electron flux incident on the solar panels, about 1.9×10^{13} 1 MeV electrons cm^{-2} per day. This is to be compared with the average pre-Starfish value of 8.9×10^{10} 1 MeV electrons cm^{-2} per day, a factor 225 less than after Starfish. The second empirical curve (solid line) assumes a constant flux of 1.72×10^{13} 1 MeV electrons cm^{-2} per day, with the addition of a flux component 4×10^{13} 1 MeV electrons cm^{-2} per day that decayed with a time constant of 12 hours. This curve is a better fit, confirming that there was a large initial flux that decayed over the first two days. It is this large initial component that was probably both space and time dependent that led, at least in part, to the different spacecraft being affected in different ways. However, the very close history of TRAAC and Transit-4B is due to their proximity in orbit, as well to the similar components used in their construction.

The fourth spacecraft also in orbit at the time of Starfish was Injun I (Pieper 1961), extensively instrumented with radiation detectors. Injun I was launched on 29 June 1961 into an orbit 1020 km $\times$820 km that was similar to Transit-4B and TRAAC, but with a higher, 67° inclination. It was able to measure both the pre-Starfish (O'Brien et al. 1962b) and post-Starfish energetic particle populations in the radiation belts (O'Brien et al. 1962a; Pieper 1962; Van Allen et al. 1963). The observations, possibly limited in spatial coverage by the orbit, identified the maximum intensities of the trapped electrons from fission products in a thin shell centered at $L = 1.2$, with omnidirectional intensities up to 10^9 energetic electrons $cm^{-2}\,s^{-1}$ shortly after the detonation on 9 July 1962. This interpretation of the data was challenged by the Telstar-1 observations that showed a much broader artificial radiation belt out to $L > 2$ (Brown et al. 1963). The high values of the energetic electron fluxes initially, of order 10^9 electrons/cm^2/sec, were confirmed by Galperin and Boliunova (1964) who, however, also argue that these high fluxes had a greater spatial extent (up to $L \sim 2$) than seen by Injun I which was presumably limited by its orbital coverage. It should be noted that the Kosmos-5 Geiger counter was very similar in shielding and performance characteristics to the one onboard Injun I, so the differences in finding clearly represent the differences in orbital coverage, all the more so, that where they were at similar positions in orbit, the results are quite comparable.

Injun I operated until December 1962, and again, for a brief period, in early 1963. However, Van Allen (1966) stated that "useful data" from Injun I was only received until late in August 1962. The failure mode has not been reported, nor the design margin built into the power system. However, the power used by Injun I was only 2 W (Pieper 1961) and this very low power drain may have allowed the satellite to operate longer than TRAAC or Transit-4B, despite flying the same solar cells as the other two.

As Kosmos-5 continued to operate to at least October 1962, it was able to monitor the decay of the intensities observed with its instruments. Figure 25 shows the decrease in intensity by the two instruments onboard Kosmos-5 (from Galperin and Boliunova 1964). Figure 25a shows the decrease of 1 MeV electron intensity measured by the "electron indicator" fluorescent detector. Figure 25b is the decrease in intensity of the high energy electrons observed by the Geiger counter. The very early highest intensity points are caused at least in part by delayed gamma ray fluxes from fission products.

Telstar-1 was launched on July 10, 1962, the day after the Starfish explosion. It provided, thanks to its instrumentation and orbit, some of the most important and unique observations made in the aftermath of Starfish. The satellite was considered an experiment in active satellite communications by the AT&T company, who paid for the spacecraft, and paid NASA for its launch (Dickieson 1963). As outlined by Dickieson, the experiment had six objectives, the fifth of which was "To gain a better numerical knowledge of the character and intensity of radiation in the Van Allen belt." Dickieson further wrote "knowledge of the ... radiation in the Van Allen belt was not adequate as a basis for design of a long-lived satellite." Hence, Telstar was the first commercial, non-scientific spacecraft that was instrumented to understand for operations and design what are now called effects of space weather on commercial technologies.

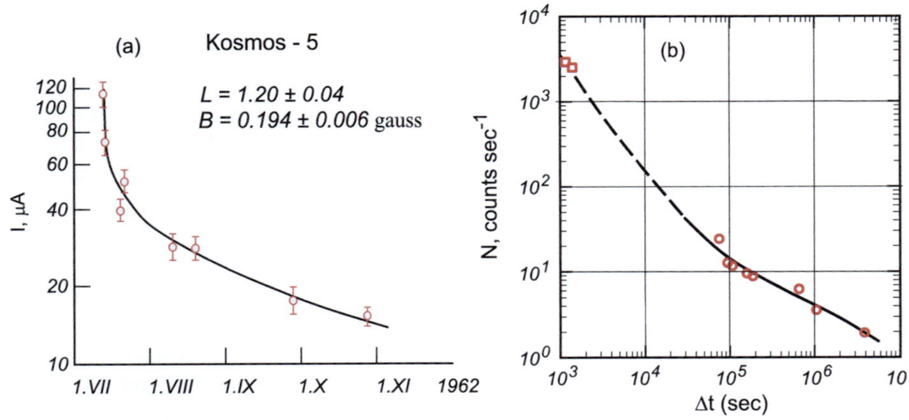

Fig. 25 The decay of excess radiation observed by the Kosmos-5 satellite following the Starfish explosion. (**a**) Excess electrons of energy ≤ 1 MeV measured by the electron indicator, an analogue, fluorescent detector near the magnetic equator, at $L \sim 1.2$. (**b**) The decay of the Geiger counter rate from the early, very high values at $L \sim 1.1$, near the magnetic equator. Note the log scale along the time axis. The initial measurements (*red squares*) can be in part attributed to gamma rays from fission products above the detonation site. (Both figures from Galperin and Boliunova 1964)

Telstar was instrumented specifically to understand radiation effects on solar cells and on transistors, and to specifically measure the electron and proton radiation in the Telstar orbit (Brown et al. 1963). The design and use of the four specific semiconductor particle detectors (three for protons, 2.4 to >50 MeV, and one for electrons, 250–1000 keV) was considered a spin-off from the active nuclear physics program headed by Walter Brown at Bell Laboratories, and that had been initiated by Bell Labs management in the late 1950s (see Chap. 8 in Millman 1983). Semiconductor particle detectors had been invented at Bell Laboratories shortly after the invention of the transistor (McKay 1949), and the nuclear physics program extensively developed and employed them.

The satellite's semiconductor detector array made crucial measurements of the fission-product electrons produced by the Starfish Prime high altitude nuclear explosion (Brown and Gabbe 1963), see Sect. 2. In addition, Telstar-1 and the team behind it were extremely well equipped to assess the space radiation effects on solar cells and electronic devices (primarily transistors). Solar cells had been developed in the first instance by Bell Labs scientists (Chapin et al. 1954), so the solar cells used by Telstar-1 used new technology that also helped the radiation resistance, as is illustrated in Fig. 26(a) that compares the performance of solar cells used by contemporary satellites [based on Fig. 3 in Rosenzweig et al. (1963)]. Figure 26(b) summarizes the energetic electron fluxes as a function of energy; the figure shows clearly the significantly higher flux rates that were experienced by Telstar-1 when compared, for instance, to Ariel-1.

The solar cell performance monitoring equipment on Telstar-1 logged the performance of the solar array, as shown in Fig. 27 (Brown and Gabbe 1963). The rate of deterioration of the performance cannot be compared between pre- and post-Starfish epochs for Telstar-1, as was the case for TRAAC and Transit-4B as illustrated in Fig. 26. However, the average energetic electron flux, referenced to 1 MeV electrons, 6×10^{12} electrons cm^{-2} per day is very high for the approximately 200 day lifetime of the mission. This can be compared to the equivalent fit for the degradation of the TRAAC solar cells (see above) given as 1.9×10^{13} 1 MeV electrons cm^{-2} per day in the days following the Starfish detonation.

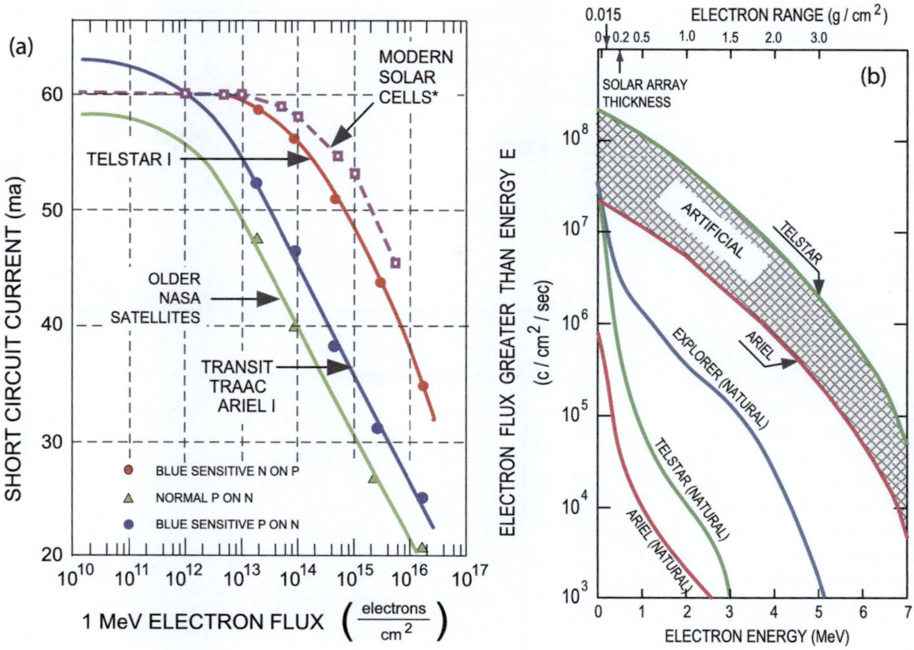

Fig. 26 (**a**) Solar cell damage curves. Based on Fig. 3 in Rosenzweig et al. (1963) and Fig. 8 in Hess (1963). The performance of the modern solar cells is taken from Xin et al. (2014). (**b**) Natural and artificial energetic electron fluxes showing cumulative spectra above 40 keV (from Fig. 4 in Wenaas 1978)

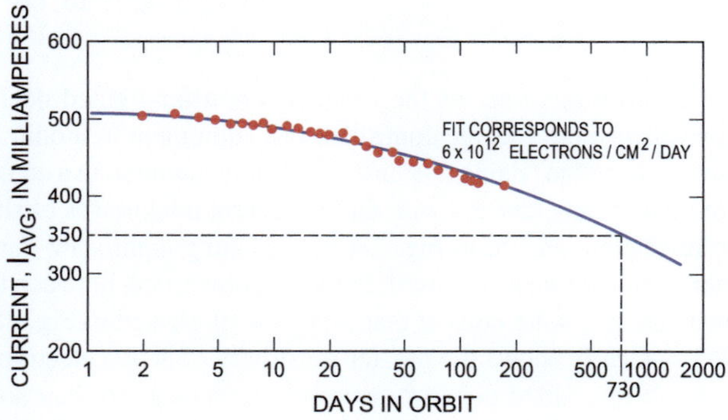

Fig. 27 The evolution of the average output of the current supplied by the solar cells on Telstar from launch to February 1963, corrected for mean solar distance. The fit corresponds to the damage calculated for a mission-averaged 1 MeV equivalent energetic electron flux rate of 6×10^{12} electrons cm^{-2} per day. The fit has been extrapolated to two years in orbit at the same energetic particle flux rate; the power output would then be reduced to 68% of its initial performance. (Taken from Fig. 28, Brown and Gabbe 1963)

Despite the damage to the solar cells by the greatly increased radiation levels after Starfish, Telstar-1 in fact failed because of failures of electronic components in its command system. Early indications were noted already a month after launch, but unexplained but fortunate partial recovery occurred so the mission could continue for several months more. There was very little known about radiation damage effects in the active electronic

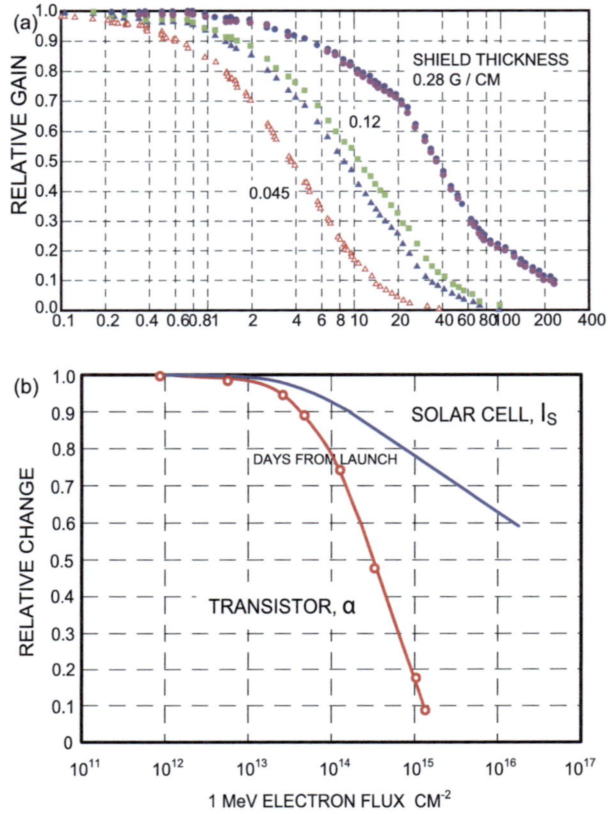

Fig. 28 (**a**) Deterioration of transistor performance measured on Telstar-1. These were specially instrumented "damage-transistors" for the purpose of observing the nature and magnitude of radiation damage (from Fig. 30, Brown and Gabbe 1963).
(**b**) Comparison of the radiation damage to transistors and solar cells. Solar cells are more rugged, but spacecraft are linearly dependent on the power supplied by the cells, whereas circuits can be designed to be radiation resistant by taking into account the degradation in transistor parameters. For today's spacecraft, solar cells remain critical, but the design and fabrication of electronic components have developed very significant radiation resistance (from Brown et al. 1963)

components, diodes and transistors, of the time. It was a far-sighted decision to include in the Telstar experiment some specific controlled test equipment to monitor the damage to transistors. The seven so-called "damage-transistors" were mounted in a carefully controlled circuit to measure their parameters and with three different thicknesses of shielding (Brown et al. 1963). The results are shown in Fig. 28(a), indicating significant damage to critical functional parameters of the transistors used. In fact, a comparison between the damage susceptibility of transistors and solar cells at that time—as illustrated in Fig. 28(b) taken from Brown et al. (1963)—clearly shows the greater sensitivity of active electronic devices. The fact that the spacecraft that failed did so because of the damage to their solar cells can be explained that most functional electronic components were usually well shielded within the spacecraft, even if this was not in the end the case for Telstar-1.

The radiation data from the Telstar sensors, together with extensive laboratory investigations of electron radiation effects with exposures as high as those measured in-situ, identified the malfunction and ultimate total failure of the satellite's command system in February 1963, some six months after launch (Mayo et al. 1963). Although the investigations pointed to surface damage in some of the transistors used from the enhanced radiation in the Van Allen belts, a highly ingenious procedure (foreshadowing many such efforts in later missions) had been worked out, using the detailed functionality of the command system that allowed circumventing to some extent the failure mode. However, this recovery procedure also proved ineffective after repeated use. It remains unclear whether the radiation damage process was identified with complete certainty [see doubts expressed by Wenaas (1978)],

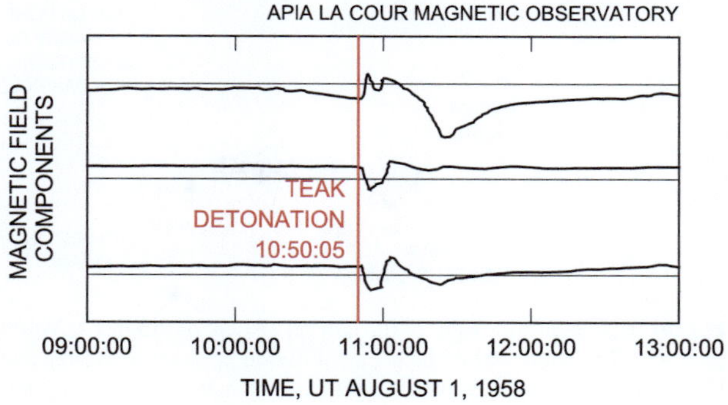

Fig. 29 The first geomagnetic signature of a high-altitude nuclear detonation. Observations by the Apia La Cour magnetic observatory, Western Samoa, on August 1, 1958, following the TEAK nuclear test. The event was likened to a magnetic storm, preceded by a Storm Sudden Commencement (from Lawrie et al. 1961)

but as the design and fabrication of semiconductor components was evolving very fast even then, new devices came later to replace the transistors used in Telstar.

It is notable that early 1960s-vintage spacecraft were rather robustly built using the technology of the time. However, hazards in space operations and the effects of the space environment were only beginning to be quantitatively assessed at the time of Starfish. The fact that several operational scientific, DoD and commercial spacecraft of that era failed due to damage from the artificial radiation belts is very sobering. Were nuclear explosions of the Starfish or Soviet high-altitude sort to be carried out today, the number of spacecraft that could be affected would be staggering. Given the much greater complexity of modern spacecraft electronics and the susceptibility of many such spacecraft to radiation dose effects (Baker 2002), the impact on modern infrastructure might be almost incalculable.

5 Geomagnetic and Geophysical Signatures of High-Altitude Nuclear Explosions

Although the longest-lasting effect of the high-altitude nuclear blasts between 1958 and 1962 was the artificial radiation belt created by Starfish (and to a much lesser extent some of the others), the largest of them did generate other geophysical effects, including ionospheric and geomagnetic effects. These were remarkable phenomena by both their propagation speed and (for the largest) their global reach. However, in all cases these remained transients, lasting from minutes to hours, but nevertheless of sufficient magnitude that they qualify as "anthropogenic" space weather events.

The series of high-altitude nuclear explosions in 1958 (see Table 1) led to a series of studies based on the observation on magnetic and geophysical observatories where the detonations generated sudden, unusual signatures that were quickly associated with the nuclear tests. The first account of such signatures was given by Cullington (1958) who described a highly unusual auroral display on August 1, 1962 at the tropical Apia Observatory, Western Samoa (latitude 14° South), together with geomagnetic signatures characterized as a Storm Sudden Commencement (SSC) at the same location.

The magnetic observations are shown in Fig. 29 (from Lawrie et al. 1961). The Western Samoan aurora was explained by Kellogg et al. (1959) and Elliot and Quenby (1959) as having been caused by the nuclear explosion on August 1, 1958 over Johnston Island (Teak, see Table 1). The aurora was all the more remarkable, because of the tropical location of the Apia Observatory. Kellogg et al. (1959) argued that the aurora was caused by energetic electrons traveling along the magnetic field lines from above the explosion site to its magnetically conjugate region, in this case Western Samoa. Elliot and Quenby (1959) recalculated

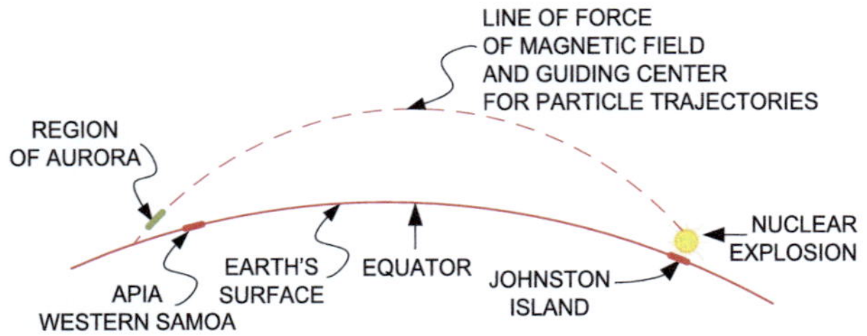

Fig. 30 Sketch of the origin of the auroral display on August 1, Western Samoa (from Kellogg et al. 1959). Electrons that originated in the decay of fission products above the nuclear detonation site are guided by the Earth's magnetic field lines to the region magnetically conjugate where they cause an aurora when interacting with the atmosphere

the expected magnetically conjugate location of Johnstone Island using a different, surface-field-based magnetic field line model to get a better agreement with the auroral observations. The sketch, in Fig. 30, reproduced from Kellogg et al. (1959), illustrates the path of the particles that cause the aurora. Note that these calculations predate the B–L coordinate system devised by McIlwain (1961) and later refinements that have provided much greater accuracy in magnetic conjugacy calculations.

Geomagnetic signatures following the August 1, 1962 detonation were noted, away from the conjugate region in Honolulu, by Maeda (1959) who pointed out the similarities of the signatures (and their timing) with the observations in Western Samoa. However, the worldwide scale of the geomagnetic signatures could only be observed after Argus III on September 6, 1958 and catalogued by Berthold et al. (1960). They examined the geomagnetic signatures from six ground stations, well separated in longitude and covering a distance range of 9800 to 13,700 km from the detonation site. Plotting on the same scale allowed a comparison of the timing of the signatures and to deduce apparent velocities. The results are shown in Fig. 31. Two signals were identified, a prompt one, with approximate velocity 3000 km/s and a delayed one, with velocity between 760 and 430 km/s. Their conclusion was that these velocities are comparable with theoretical values calculated for hydromagnetic waves propagating in the ionosphere. The propagation was approximately at constant speed; the magnitudes of the signals decreased with propagation direction away from the magnetic meridian of the detonation site. The analysis for the onset was challenged by Caner (1964) who, in the light of the Starfish experience, proposed that the signal front was effectively synchronous worldwide. That interpretation also fits with the (noisy) data—the accuracy of the Azores data signal onset having been questioned by Berthold et al. (1960). This is also illustrated in Fig. 31.

These conclusions following the 1958 series of explosions came to be significantly revised after the analysis of the Starfish and other 1962 explosions. The reason was the considerably increased height of Starfish, as well as its much greater yield than the Argus series, in addition to the greater readiness of ground-based instrumentation available for observations. It is arguable that of these two parameters the height, at 440 km, was the more important. While many of the associated geophysical phenomena were confirmed, the better coverage as well as the stronger signatures allowed more detailed conclusions to be drawn from the event.

A well-prepared and well-instrumented ground magnetometer network prepared by the US Army Electronics Research and Development Laboratories was ready to record the geomagnetic signatures of the Starfish explosion on July 9, 1962. As described by Bomke et al. (1964), all observatories made high-speed and detailed recording of the Starfish ef-

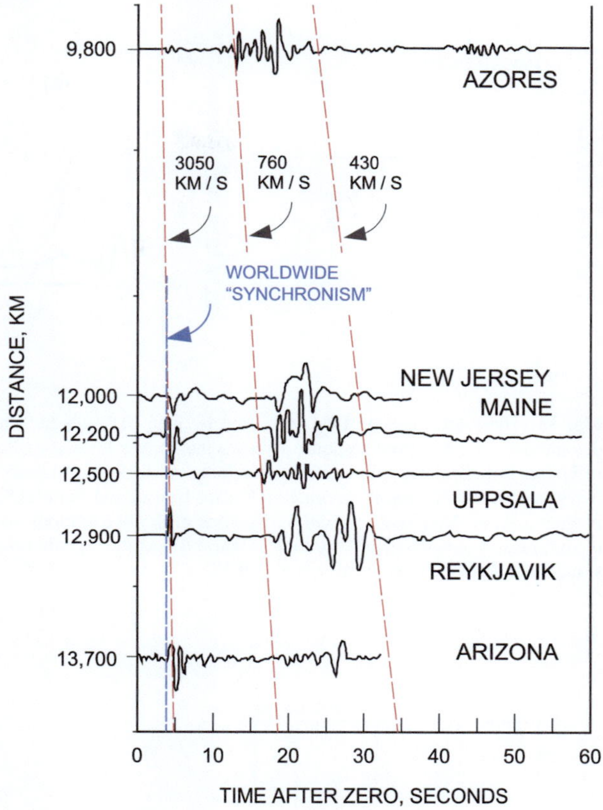

Fig. 31 Magnetograms from six stations distributed in distance from Johnston Island following the Argus III nuclear detonation on September 6, 1958. It shows two groups of signatures; the first very prompt, the second showing reasonably clear velocity dispersion. The fitted velocities are by Berthold et al. (1960); for the prompt signal, Caner (1964) argued that it was arguably synchronous worldwide, as was later found for the prompt signature from Starfish

fects. The onset of the disturbance at the Hawaii observatory is shown in Fig. 32a; it is seen to be effectively instantaneous on the time resolution of the magnetometer, better than 0.1 s. The more intense signal arrives, with a very fast rise time, about 2 s after the detonation. The full amplitude of the signal can be seen in Fig. 32b; it peaked at about 4 s, but then, following another large-amplitude interval, decayed relatively fast on this scale. The same figure also shows the corresponding signal recorded at the Western Samoa observatory. The amplitude of the signal is significantly smaller, but the early time profile matches that observed in Hawaii: near-instantaneous onset, followed by a second, larger amplitude signal. The observations of the signature in the total field observed at La Hambra, California are shown in Fig. 33a, illustrating again the very prompt onset, then the arrival of the higher magnitude signal and the concluding phase of the relatively short duration event. The micropulsations observed in Tasmania, together with the simultaneously observed energetic electron counting rate onboard a stratospheric balloon are shown in Fig. 33b.

The geomagnetic disturbance caused by Starfish was reported to have been observed worldwide: in the Pacific and the continental United States (Unterberger and Byerly 1962; Breiner 1963; Bomke et al. 1964; Miles and Lepping 1964), in Canada (Baker and Strome 1962), in Tasmania, Australia (Edwards and Reid 1964), in Peru (Casaverde et al. 1963), in India (Pisharoty 1962), in France and in the Kerguelen Islands, South Indian Ocean (both these by Roquet et al. 1963). A noted characteristic was the near-instantaneous arrival of the prompt onset, implying the propagation of the initial disturbance by electromagnetic means, at or near the speed of light. Roquet et al. (1963) report that at the two locations,

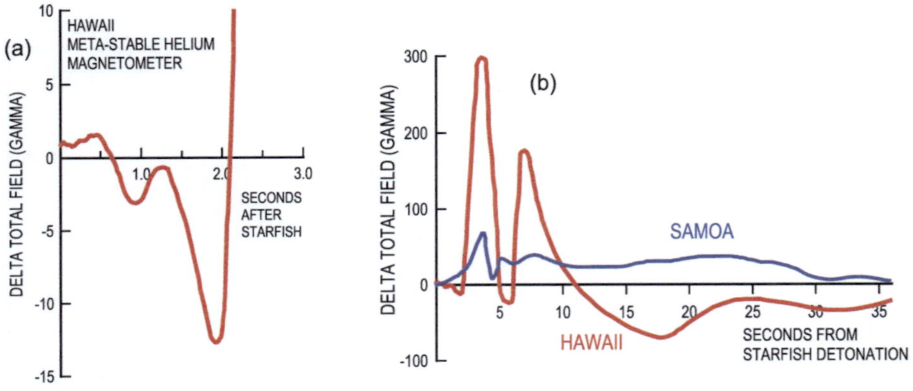

Fig. 32 The total magnetic field measured by meta-stable helium magnetometers following the Starfish detonation. (**a**) The initial response of a magnetometer located in Puako, Hawaii, showing an immediate response, less than one second after the detonation, followed by the second, high amplitude response after two seconds. (**b**) The traces recorded in Puako, Hawaii and Tutuila, Samoa by identical meta-stable helium magnetometers. Note that the initial responses at the two stations are in phase, although the Samoa peak is significantly smaller than that seen in Hawaii and the signals do not remain in phase after the initial, simultaneous peaks (from Bomke et al. 1964)

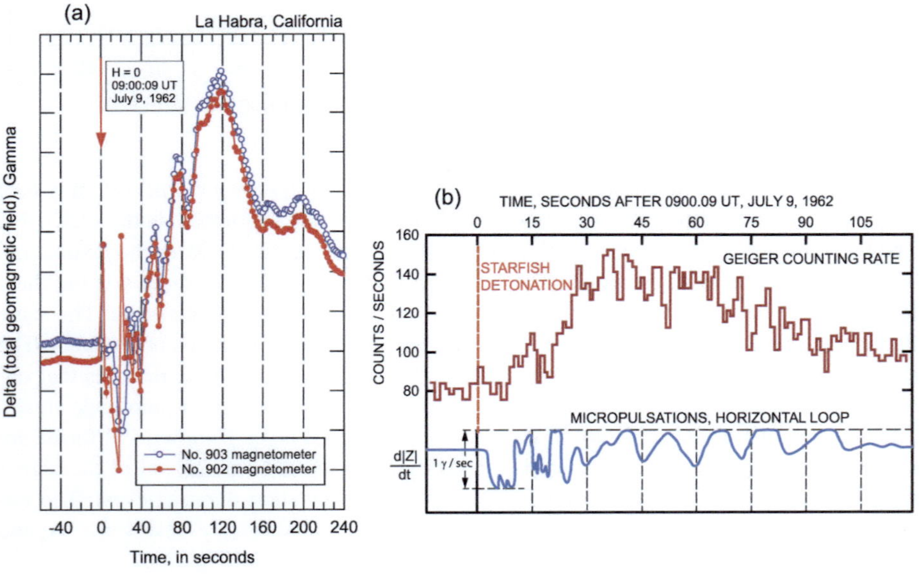

Fig. 33 (**a**) The magnetic disturbance observed with two rubidium-87 optically pumped vapour magnetometers at La Habra, California. The *onset* of the disturbance is quasi-instantaneous and it lasts ~7 minutes. The rate of change at the onset was too fast for the sampling averages, hence the two instruments only track each other after about 20 seconds. (from Unterberger and Byerly 1962). (**b**) The *upper panel* shows the counting rate of a balloon-borne Geiger counter at 80 g/cm^2 atmospheric depth (~18,200 m height) near Hobart, Tasmania, following the Starfish explosion. In the lower panel, the record of micropulsations that followed the detonation is shown. After the initial transient, irregular pulsation, a quasi-periodic oscillation was observed. The 16-second period corresponds to the fundamental V-mode oscillation of the geomagnetic field line through Hobart (from Edwards and Reid 1964)

in France and in the South Indian Ocean, the onset was timed at 09:00:08.8 $\pm$ 0.2 UT and 09:08:07.7 $\pm$ 0.5 UT, respectively, to be compared with the most accurate available timing of the Starfish detonation at 09:00:09 UT.

Following the global synchronism of the onset of the magnetic signal to within measurement accuracies, generally within a fraction of a second, a second, much higher magnitude signal was observed, as already indicated in connection with Fig. 32a and 32b. The onset of this second signal was similarly global and synchronous and contained a significant oscillatory component with period about 3 to 4 seconds—difficult to measure more precisely, due to the diversity of instrumental non-linearities that affected the observations. The oscillations decayed rapidly, after five to eight periods.

In reviews of the collected observations, Caner (1964) and Kahalas and Newman (1965) listed the characteristics of the two components of the prompt signal and offered explanations for them. The first signal, at the time of the explosion, may be due to the absorption of the gamma rays in the atmosphere that generate an electromagnetic pulse which propagates in the earth-ionosphere cavity to generate a global signal. The second component, detected globally after a $\sim$2 second delay, proved much more difficult to explain. Caner (1964) considered all the alternatives and came to the conclusion that hydromagnetic standing waves along the magnetic field line from the explosion site are at the origin of waves that are converted to electromagnetic waves in the lower ionosphere. These can then propagate in the ionosphere-Earth surface cavity at close to the speed of light. The $\sim$2 second delay is then accounted for by the initial propagation delay along the magnetic field line.

A very comprehensive modeling of the geomagnetic observations in the form of iso-intensity contours of the magnetic disturbances by Bomke et al. (1966) have yielded the unexpected result that the maxima were in magnetically conjugate regions about the Johnston Island magnetic meridian plane, but at 45° magnetic latitude, rather than closer to the Johnston Island magnetic latitude of 14°N (see Fig. 34). Their tentative explanation for this result is the coupling of magnetoacoustic waves with guided Alfvén waves that is most effective at the conditions about 1000 km height, corresponding to the magnetic field lines at the higher latitude. Auroral observations following Starfish (Gabites and Rowles 1962) support a very prompt activity at higher magnetic latitudes, in the mid-40° range, than the conjugate area of explosion site. The conclusions of Bomke et al. (1966), together with the observation of auroral activity confirm that the magnetic field lines in the vicinity of the meridian plane of the explosion site were strongly disturbed to much greater heights and magnetic latitudes than was the case for the earlier, 1958 series of nuclear detonations. In particular the sketch in Fig. 30 would need to be complemented by the inclusion of a significant volume of space reaching out to at least $L = 5$. This volume is now known to have corresponded to the diamagnetic cavity generated by the explosion, described below. However, at the time, no information was available on the properties and dynamics of the diamagnetic cavity.

These considerations have received a late support, more than 40 years after the event, when Dyal (2006) published a detailed analysis of observations made at the time of the Starfish explosion by five fully instrumented sounding rockets near the Johnston Island site. The data had not been in the public domain until this analysis was published and the observations throw an interesting light on the "initial conditions" following the explosion that could not be included in the contemporaneous studies in the 1960s. The details relate to the dynamics and evolution of the diamagnetic cavity formed immediately after the explosion. The cavity reached a length of 1840 km along the magnetic field lines and an extent in height of 680 km, reaching beyond 1000 km above the explosion site. This extent of the cavity was reached in 1.2 seconds, and the bubble took about 16 seconds to collapse. The intensity of electrons in the bubble was extremely high, of order 10^{13} cm^{-2} s^{-1} initially. The magnetic

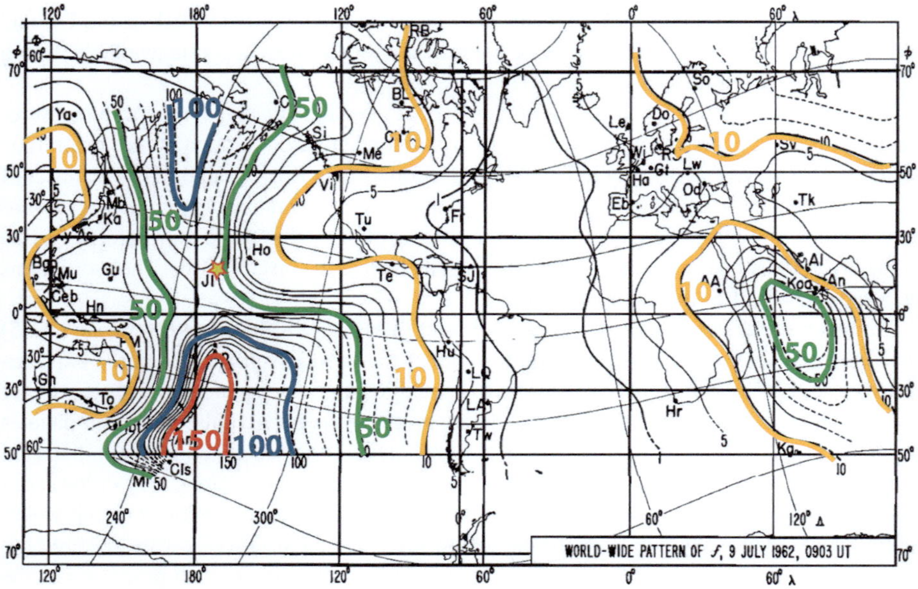

Fig. 34 Isopleths of the amplitude of the total-disturbance magnetic vector at 0903 UT on July 9, 1962 (2 min 51 s after Starfish), plotted on a Mercator projection. The detonation took place 400 km above Johnston Island (JI). Values are in gammas. A contour interval of 10 gammas is used from the 10-gamma isopleth upward; the color lines are 10, 50, 100, and 150 gammas (from Bomke et al. 1966)

field lines were clearly significantly distorted to relatively high L values above the explosion site, and relaxed once the cavity collapsed, but by that time high fluxes of electrons became trapped and formed the injection population for the artificial radiation belts as well as contributing to the generation of auroral precipitation in a broad range of magnetic latitudes, as described above.

In summary, the high-altitude nuclear detonations in 1958, but even more obviously those in 1962, and in particular the Starfish event on July 9, 1962, have generated widely-observed, worldwide geophysical signatures. However, the geomagnetic effects were not long-lasting, unlike the artificial radiation belt. The signatures were complex and despite the abundant contemporaneous literature, the physical phenomenology was at best tentatively identified. It is not clear whether this was because the near-Earth space environment (including the ionosphere and exosphere) and its processes were not well understood at the time or because the observations, although plentiful, were neither sufficiently comprehensive nor sufficiently accurate. In the absence of such events since then, it could be worthwhile to revisit the unique set of observations of the time to investigate if more could be learnt from them, particularly in the light of the observation made public by Dyal (2006).

6 Impact on the Power Grid

Since the advent of the electrical telegraph in the mid-19th century, it has been recognized that long metallic conductors grounded at both ends—e.g., power grids, communication cables, pipelines—can efficiently carry electrical currents that may exist in the Earth. That is because the conductors fashioned by human technology have far less resistivity than does the Earth, and so the conductors provide an easy path for electrical currents flowing in the Earth. Geomagnetic disturbances in Earth's magnetosphere produce variations in the current systems that flow in the ionosphere. These ionosphere current variations produce in turn variations in the magnetic field at Earth's surface. These magnetic variations generate electrical currents flowing in the solid Earth (illustrated in Fig. 9 under a high altitude nuclear explosion). Significant electrical potential differences can occur over both short and long distances depending upon the scale sizes of the changing electrical currents in the ionosphere and the spatial conductivity of the underlying Earth, both horizontally and vertically.

This geophysical effect was evident during the Starfish Prime event, when some 300 streetlights in Honolulu, about 1500 km from the detonation site, were extinguished. An analysis by Vittitoe (1989) indicated that the orientation of the magnetic fields from the bomb-produced EMP was consistent with the production of the streetlight outage (Vittitoe 1989). The reported outage at the same time of a microwave telecommunications link between Kauai and other Hawaiian islands could have been caused by damage or outage of the electrical power to the microwave station by the EMP event. The damaging effects of EMP that can be expected on the electrical power grid of the nation (as well as on other infrastructures such as communications, banking and finance, transportation) are outlined in Foster et al. (2004).

While the streetlight failures in Oahu were apparently due to fuses in the system (Vittitoe 1989), the outages of electrical power systems due to geomagnetic activity can be produced by a number of different factors depending upon the grid system, the overlying ionosphere, and Earth's conductivity. The Quebec outage of 1989 was caused by induced current-produced voltage depressions that could not be compensated by the system's automatic compensation equipment. This event also saw generators tripping out of service and other effects in other systems (Vittitoe 1989). The picture of the damage of a transformer at the Salem nuclear power plan in New Jersey in the 1989 event remains an iconic symbol of natural space weather effects on technical systems.

The 2008 Workshop report from the National Academies (Baker et al. 2008) indicates that the least understood aspect of power system failure during a geomagnetic event is the permanent damage and loss of grid assets, especially large and costly transformers. These transformers can experience excessive levels of internal heating that is caused by the induced Earth currents flowing through the devices. The heating can produce melting and burn-through of copper windings. Such damaged transformers cannot be repaired in the field, and have to be replaced. The Report discusses the implications of the long lead times for producing new transformers, and therefore the implications of long outages if many are damaged in a geomagnetic event. The same considerations apply to the societal implications of many damaged transformers that will occur in an EMP event, as outlined in Foster et al. (2004).

7 Other HEMP Impacts

Much of the literature on HEMP is either classified or not easily accessible. Savage et al. (2010) and Foster et al. (2008) discuss some "real life" effects. Here we briefly outline some of these phenomena.

7.1 Cars

Some say that all vehicles traveling will come to a halt, with all modern vehicles damaged because of their use of modern electronics. A car does not have very long cabling to act as antennas, and there is some protection from metallic construction. As non-metallic materials are used more and more in the future to decrease weight and increase fuel efficiency, this advantage may disappear. More recently, the number of microprocessors in cars and the reliance on microprocessors in all motor vehicles has increased greatly. Also, the sensitivity of the electronic circuitry to EMP has increased due to the use of smaller electronic components designed to operate on lower voltages. The fact is that not all cars need to be damaged to make the traffic stop. It is enough to permanently damage less than half (some sources say even 15%) of cars to block cities, highways and supply lines.

7.2 Mobile Devices

About 95% of these devices already have an internal electromagnetic interference shielding (which is there to protect the components from affecting each other). Plus, these devices are very small (compared to the power grid) so there is a very good chance that mobile devices can be used immediately after an EMP. But not to make calls or to search the internet. Telecommunication antennas will be unusable so there will be no signal of any kind. But if the phone is connected to the power grid—charging—at the moment when the EMP hits it will be damaged beyond repair.

7.3 Pacemakers

The hermetically sealed can is a good Faraday cage so the pacemaker will suffer no direct damage from an EMP. A bipolar system with, say, 5 mm electrode separation will, with a major high-altitude EMP generating ~25 kV/m in the northern US, deliver a voltage pulse to the circuitry of about 900 V—this will not damage a pacemaker, they are designed to cope with external defibrillator voltages. A unipolar system, where the electrodes may be 15 cm apart, will generate a bigger voltage, maybe 4–5 kV, to the generator, and this could be damaging.

7.4 Airplanes

Older planes use hydraulics and cables attached to the pilot controls (with manual valve actuation and direct pressurization from the ram air turbine) which means they will still have semi-functional flight controls. Newer planes (almost all airlines) will be extremely difficult to control after all their electronic parts will be damaged. But even so, all Airbus and Boeing planes are demonstrated to be controllable with complete electrical failure. They would be extremely difficult to land, but it would still be possible. All planes will be turned into semi-controllable gliders… with newer planes having almost no control and older planes having almost full control over the plane. And pilots will face a very hard task: to land the plane, with highways being full of broken cars, no emergency services and if it is night then add no ground lights to this. But an EMP will not cause planes to enter in a spin and pancake into the ground.

7.5 Maximum Conductor Length

There is a suggestion that equipment will not be damaged if all connected conductors are less than a specific length. Certainly shorter lengths are generally better, but there is no magic length value, with shorter always being better and longer not. Coupling is much too complex for such a blanket statement—instead it should be "the shorter the better, in general". (There can be exceptions, such as resonance effects, which depend on line lengths.)

7.6 Turn Equipment off

There is truth to this recommendation (if there were a way to know that a burst was about to happen). Equipment is more vulnerable if it is operating, because some failure modes involving HEMP E1 phase trigger the system's energy to damage itself. However, damage can also happen, but not as easily, to systems that are turned off.

8 Space Weather Effects of Anthropogenic VLF Transmissions

8.1 Brief History of VLF Transmitters

By the end of World War 1, the United States military began use of very low frequency radio transmissions (VLF; 3–30 kHz) for long-distance shore to surface ship communications (Gebhard 1979). Since very high power can be radiated from large shore-based antenna complexes, worldwide VLF communication coverage was feasible, and along with LF and HF systems (300–30 MHz) these bands carried the major portion of naval communications traffic before later higher frequency systems came online. Early experiments also showed that VLF could penetrate seawater to a limited depth, a fact realized by the British Royal Navy during World War I (Wait 1977). Given this realization, when the modern Polaris nuclear submarine era began in the 1950s, the US Naval Research Laboratory conducted a series of thorough radio propagation programs at VLF frequencies to refine underwater communications practices (Gebhard 1979). Subsequent upgrades in transmission facilities led to the current operational US Navy VLF communications network, and other countries followed suit at various times. For example, Soviet naval communication systems were likely brought online in the late 1920s and 1930s during the interwar expansion period, and high power VLF transmitters were later established in the late 1940s and 1950s for submarine communications and time signals. These included Goliath, a rebuilt 1000 kW station first online in 1952 which partly used materials from a captured German 1940s era megawatt class VLF station operating at 16.55 kHz (Klawitter et al. 2000).

Table 2 of Clilverd et al. (2009) lists a variety of active modern VLF transmitter stations at distributed locations with power levels ranging from 25 to 1000 kW. These transmissions typically have narrow bandwidths (<50 Hz) and employ minimum shift keying (Koons et al. 1981). Along with these communications signals, a separate VLF navigation network (named Omega in the US and Alpha in the USSR) uses transmissions in the 10 kW range or higher (Inan et al. 1984, e.g. Table 1 of) with longer key-down modulation envelopes of up to 1 second duration.

8.2 VLF Transmitters as Probing Signals

Beginning in the first half of the 20th century, a vigorous research field emerged to study the properties of VLF natural emissions such as whistlers, with attention paid as well to information these emissions could yield on ionospheric and magnetospheric dynamics. Due to the high power and worldwide propagation of VLF transmissions, the geophysical research field was well poised to use these signals as convenient fixed frequency transmissions for monitoring of VLF propagation dynamics into the ionosphere and beyond into the magnetosphere (e.g. Chap. 2 of Helliwell 1965; Carpenter 1966). This was especially true since VLF transmissions had controllable characteristics as opposed to unpredictable characteristics of natural lightning, another ubiquitous VLF source. Beginning in the 1960s and continuing to the present, a vast amount of work was undertaken by the Stanford radio wave group and others (e.g. Yu. Alpert in the former USSR) on VLF wave properties, including transmitter reception using both ground-based and orbiting satellite receivers. These latter experiments occurred both with high power communications and/or navigation signals and with lower power (∼100 W), controllable, research grade transmitter signals.

The transmitter at Siple Station in Antarctica (Helliwell 1988) is worthy of particular mention, as the installation lasted over a decade (1973–1988) and is arguably the largest and widest ranging active and anthropogenic origin VLF experiment series. Two different VLF transmitter setups were employed at Siple covering 1 to ∼6 kHz frequency, with reception occurring both in-situ on satellites and on the ground in the conjugate northern hemisphere within the province of Quebec. Of particular note, the second Siple "Jupiter" transmitter, placed in service in 1979, had the unique property of having flexible high power modulation on two independent frequencies. This allowed targeted investigations of VLF propagation, stimulated emissions, and energetic particle precipitation with a large experimental program employing a vast number of different signal characteristics not available from Navy transmitter operations. These included varying transmission lengths, different modulation patterns (e.g. AM, SSB), polarization diversity, and unique beat frequency experiments employing two closely tuned VLF transmissions. Furthermore, the ability to repeat these experiments at will, dependent on ambient conditions, allowed assembly of statistics on propagation and triggered effects. These led to significant insights that were not possible for studies that relied on stimulation from natural waves (e.g. chorus) that are inherently quite variable.

Several excellent summaries of the literature on VLF transmission related subjects are available with extensive references, including the landmark work of Helliwell (1965) as well as the recent Stanford VLF group history by Carpenter (2015). As it is another effect of anthropogenic cause, we mention briefly here that a number of studies in the 1960s also examined impulsive large amplitude VLF wave events in the ionosphere and magnetosphere caused by above-ground nuclear explosions (e.g. Zmuda et al. 1963; Helliwell 1965).

Observations of VLF transmissions included as a subset those VLF signals that propagated through the Earth-ionosphere waveguide, sometimes continuing into the magnetosphere and beyond to the conjugate hemisphere along ducted paths (Helliwell and Gehrels 1958; Smith 1961). Ground based VLF observations (Helliwell 1965) and in-situ satellite observations of trans-ionospheric and magnetospheric propagating VLF transmissions were extensively used as diagnostics. For example, VLF signals of human origin were observed and characterized in the topside ionosphere and magnetosphere for a variety of scientific and technical investigations with LOFTI-1 (Leiphart et al. 1962), OGO-2 and OGO-4 (Heyborne et al. 1969; Scarabucci 1969), ISIS 1, ISIS 2, and ISEE 1 (Bell et al. 1983), Explorer VI and Imp 6 (Inan et al. 1977), DE-1 (Inan and Helliwell 1982; Inan et al. 1984; Sonwalkar and Inan 1986; Rastani et al. 1985), DEMETER (Molchanov et al. 2006; Sauvaud et al. 2008), IMAGE (Green et al. 2005), and COSMOS 1809 (Sonwalkar et al. 1994). VLF low Earth orbital reception of ground transmissions have been used also to produce worldwide VLF maps in order to gauge the strength of transionospheric signals (Parrot 1990).

8.3 VLF Transmitter Induced Effects in the Ionosphere and Magnetosphere

Given ambient terrestrial ionosphere and magnetosphere magnetic field and electron density values, VLF fixed frequency transmissions can not only pass through the medium but can under certain conditions trigger a variety of natural stimulated wave emissions, often interacting with existing particle populations in these regions. Observations exploring these physical mechanisms have been and continue to be conducted both from the ground and in-situ within the ionosphere and magnetosphere. Sect. 3 of Parrot and Zaslavski (1996) provides a mid 1990s snapshot review of observations and mechanisms.

Work in the 1960s and 1970s found that during sufficiently long key-down transmitter periods, triggered electron precipitation and stimulated wave events could occur, known as Trimpi effects (Helliwell et al. 1973; see also the observational and theoretical reviews by Helliwell 1988 and Omura et al. 1991). These findings, other electron precipitation observations (Inan et al. 1984, 1982, 1985; Imhof et al. 1986; Vampola 1987, 1990; Sauvaud et al. 2006) and later theoretical work (e.g. Abel and Thorne 1998) led to the realization that VLF waves of sufficient amplitude might be useful in modifying the radiation belt electron populations through so-called Radiation Belt Remediation (Inan et al. 2003; Rodger et al. 2006), consisting of precipitation by alteration of particle pitch angle and therefore lowering of the minimum bounce altitude (2nd adiabatic invariant). For example, the US Air Force DSX satellite scheduled for a 2017 launch will carry an onboard VLF transmitter for in-situ attempts to initiate radiation belt remediation mechanisms.

Unmodulated (i.e. carrier only) monochromatic waves from high power ground VLF transmitters have also been observed to quench natural magnetospheric hiss emissions (Raghuram et al. 1977), and VLF transmitted pulses were thought to be amplified in the magnetosphere during storms through wave-electron interactions (Smith and Clilverd 1991). Ionospheric heating and perturbations in electron and ion densities have been associated with VLF transmitters as well (Inan et al. 1992; Parrot et al. 2007).

8.4 VLF Transmitter Interactions Within the Radiation Belts

The theoretical work of Abel and Thorne (1998) predicted that the lifetimes of relativistic radiation belt electrons from $L = 1.3$ to 2.8 should be strongly influenced by VLF transmitter signals, and perhaps that this mechanism was responsible for the 'slot' region between the inner and outer radiation belts, although updated VLF wave propagation modeling seemed to cast some doubt on these findings by suggesting a potential overestimation of wave amplitudes (Starks et al. 2008, 2009; Cohen et al. 2012). However, Abel and Thorne (1998) and Kulkarni et al. (2008) asserted that typical VLF transmitter frequencies could not resonate beyond $L = 2.2$ to 2.4 at the equator due to propagation characteristics and an increase in wave absorption as the wave frequency approaches the electron gyrofrequency. Significantly, these results were obtained using an electron density typical of conditions inside the plasmasphere (10^3 cm^{-3}).

The recent launch of the equatorial plane Van Allen Probes dual satellite mission in late 2012 (Mauk et al. 2013) has provided a large amount of new observational information on radiation belt electron populations and their dynamics. The satellites carry a high fidelity electromagnetic wave package (Kletzing et al. 2013), and data from this sensor has led to renewed study of the role of VLF transmission effects on relativistic electron dynamics within the magnetosphere. It has been well established that electrons in the variable outer belt are affected by substorm injections and other inner magnetospheric dynamic processes, causing associated changes in populations at general ring current energies in the 10 to 100 s of keV. Through a multi-step process (see Fig. 1 of Jaynes et al. 2015), recent research has found that these seed particles can provide significant energy reservoirs for highly efficient and dynamic wave-particle interactions on very short time scales, through mechanisms such as cyclotron resonance that directly affect relativistic electrons. Resonant interactions can remove energy from pre-existing inner magnetospheric particle populations and can subsequently amplify coherent waves (e.g. rising tone chorus) that are highly efficient at affecting relativistic electrons through multiple mechanisms. These factors result in large overall changes both in particle loss and/or acceleration for electrons from ring current energies out to ultra-relativistic energies.

New observations from Van Allen Probes wave and particle instruments have provided high fidelity equatorial plane in-situ measurements of ground based VLF transmissions out to >24,000 km radial distance ($L \sim 2.8$), at the edge of what Foster et al. (2016) term the VLF bubble. VLF transmissions that propagate to these locations are strong and nearly omnipresent due to naval military operational considerations for submarine communications as mentioned previously. Initial investigations point to these strong VLF fixed frequency transmissions of anthropogenic origin as playing a potentially important role in shaping dynamic responses of the radiation belts, primarily through interactions with wave amplification processes for those coherent plasma wave modes known to be highly resonant with relativistic particles outside the plasmasphere.

In particular, the plasmapause and associated plasmasphere boundary layer (Carpenter and Lemaire 2004) overlaps the region where the equatorial outward extent of the VLF bubble lies. During disturbed intervals, the plasmapause can be retracted at or inward of the nominal VLF bubble edge, causing peak resonance energies at typical VLF transmission bands to shift dramatically away from the lower energies calculated by Abel and Thorne (1998) and towards relativistic particle populations (Foster and Rosenberg 1976; Foster et al. 2016). Significant spatial variation in resonance effects can therefore occur under retracted plasmasphere situations, with associated variations in wave-particle interactions possessing high degrees of coupling to relativistic electrons.

Foster et al. (2016) presented Van Allen Probes wave data from the great 17 March 2015 storm showing a dramatic increase in wave intensity around a ~21 kHz VLF transmitter frequency band, in the region where the transmitter signal extended beyond the contracted plasmapause (cf. Fig. 35). These intensifications occurred at the radial location where the ground based VLF transmitter frequency bands were near half the local electron cyclotron frequency, under conditions appropriate for the generation of whistler-mode stimulated emissions. The resultant electric field intensity in the transmitter frequency band was observed up to 100,000 times background. Sharp gradients in highly-relativistic outer zone electron fluxes were also observed to be spatially coincident with the region of VLF enhancement, suggestive of the significant space weather effects of VLF transmitters in defining the earthward extent of outer radiation belt "killer electrons" (e.g. Baker et al. 2014; Foster et al. 2016). In the region at the outer extent of the VLF bubble, beyond the contracted plasmapause, interactions involving signals from ground based VLF transmitters have a high cyclotron resonant potential to cause energetic electron precipitation, wave amplification, and/or other associated effects.

We note in passing that these effects may potentially be congruent with previously reported observations of whistler mode amplification (>10×) in ground based VLF transmitter signals propagating through the magnetosphere along $L = 2.5$ field lines during large storms, with such observations interpreted by the authors as due to wave-particle interactions (Smith and Clilverd 1991).

Investigations of these inner magnetospheric VLF transmitter stimulated effects near the VLF bubble edge are in early stages, and work is underway to examine the detailed relationship between VLF transmitter signals, stimulated emissions, and resonant particle populations. Relevant for this discussion, wave-particle interactions in this category depend by their nature on the characteristics of pre-existing inner magnetospheric particle populations placed there by e.g. substorm injections or earlier disturbances with subsequent radial diffusion. Further understanding of the potential anthropogenic space weather effects of ground based VLF transmitters therefore will depend on elucidation of the multi-step pathways involved as well as detailed examinations of the time history and characteristics of the required

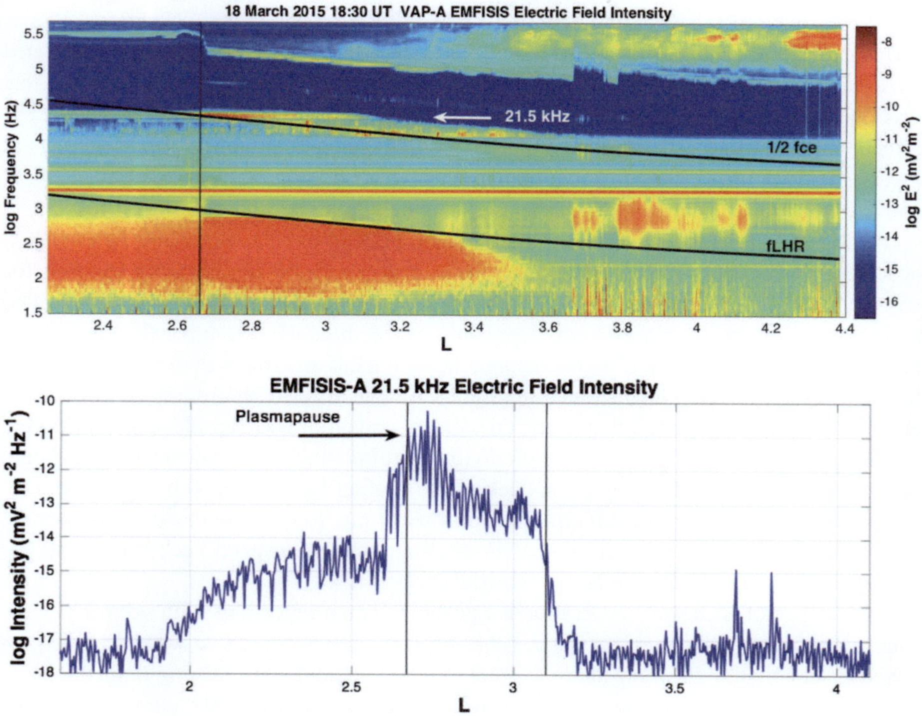

Fig. 35 (*Top*) Van Allen Probes A EMFISIS electric field intensity spectra from 30 Hz to 500 kHz are shown during a period following the 17 March 2015 great storm on 18 March 2015 at 18:30 UT, with 0.5 f_{ce} and f_{LHR} denoted by sloping black lines. The plasmapause boundary is indicated by the *vertical black line* at $L = 2.65$. In the region where a ~21 kHz transmitter signal extended beyond the plasmapause, enhanced emissions were observed associated with frequencies close to a VLF transmitter band (*white arrow*). (*Bottom*) Electric field wave intensity near ~21 kHz increased by $>10^5$ as L decreased from 3.2 to 2.7. The plasmapause position at $L = 2.65$ determined from the EMFISIS upper hybrid frequency and the outer extent of enhanced emissions near the VLF transmitter frequency are indicated with *black vertical lines*. Adapted from Foster et al. (2016)

resonant particles. Nevertheless, VLF transmissions of anthropogenic origin may constitute a key space weather influence on pathways that fundamentally alter the storm-time radiation belt. Under these assumptions, it is interesting for the reader to consider what the terrestrial radiation belt environment might have been in the pre-transmitter, and pre-observation, era.

9 High Frequency Radiowave Heating

Modification of the ionosphere using high power radio waves has been an important tool for understanding the complex physical processes associated with high-power wave interactions with plasmas. There are a number of ionospheric heating facilities around the world today that operate in the frequency range ~2–12 MHz. The most prominent is the High Frequency Active Auroral Research Program (HAARP) facility in Gakona, Alaska. HAARP is the most powerful radio wave heater in the world; it consists of 180 cross dipole antennas with a total radiated power of up to 3.6 MW and a maximum effective radiated power (EFR) of ~4 GW. The other major heating facilities are EISCAT, SURA, and Arecibo. EISCAT is near Tromso, Norway and has an EFR of ~1 GW. SURA is near Nizhniy Novgorod, Russia and is capable of transmitting ~190 MW ERP. A new heater has recently been completed at Arecibo, Puerto Rico with ~100 MW ERP. There was a heating facility at Arecibo that was operational in the 1980s and 1990s but it was destroyed by a hurricane in 1999. The science investigations carried out at heating facilities span a broad range of plasma physics topics involving ionospheric heating, nonlinear wave generation, ducted wave propagation, and ELF/VLF wave generation to name a few.

During experiments using the original Arecibo heating facility, Bernhardt et al. (1988) observed a dynamic interaction between the heater wave and the heated plasma in the 630 nm airglow: the location of HF heating region changed as a function of time. The heated region drifted eastward or westward, depending on the direction of the zonal neutral wind, but eventually "snapped back" to the original heating location. This was independently validated using the Arecibo incoherent scatter radar for plasma drift measurements (Bernhardt et al. 1989). They suggested that when the density depletion was significantly transported in longitude, the density gradients would no longer refract the heater ray and the ray would snap back, thereby resulting in a snapback of the heating location as well. However, a recent simulation study using a self-consistent first principles ionosphere model found that the heater ray did not snap back but rather the heating location snapped back because of the evolution of the heated density cavity (Zawdie et al. 2015).

The subject of ELF wave generation is relevant to communications with submarines because these waves penetrate sea water. It has been suggested that these waves can be produced by modulating the ionospheric current system via radio wave heating (Papadopoulos and Chang 1989). Experiments carried out at HAARP (Moore et al. 2007) demonstrated this by sinusoidal modulation of the auroral electrojet under nighttime conditions. ELF waves were detected in the Earth's ionosphere waveguide over 4000 km away from the HAARP facility.

VLF whistler wave generation and propagation have also been studied with the HAARP facility. This is important because whistler waves can interact with high-energy radiation belt electrons. Specifically, they can pitch-angle scatter energetic electrons into the loss cone and precipitate them into the ionosphere (Inan et al. 2003). One interesting finding is that the whistler waves generated in the ionosphere by the heater can be amplified by specifying the frequency-time format of the heater, as opposed to using a constant frequency (Streltsov et al. 2010).

New observations were made at HAARP when it began operating at its maximum radiated power 3.6 MW. Specifically, impact ionization of the neutral atmosphere by heater-generated suprathermal electrons can generate artificial aurora observable to the naked eye (Pedersen and Gerken 2005) and a long-lasting, secondary ionization layer below the F peak (Pedersen et al. 2009). The artificial aurora is reported to have a "bulls-eye" pattern which is a refraction effect and is consistent with ionization inside the heater beam. This phenomenon was never observed at other heating facilities with lower power (e.g., EISCAT, SURA).

10 Chemical Releases

Chemical releases in space have been carried out since the early 1960s. The first release experiments were performed by scientists at the Max Planck Institute for Extraterrestrial Physics. Barium was used because it rapidly ionizes in sunlight. One of the first successful experiments was at an altitude of 2000 km (Lüst 2001) but most of the subsequent experiments were in the altitude range 150–250 km. The original purpose of these experiments was to create an artificial cometary tail in order to understand the interaction of the solar wind with cometary ions Biermann (1951), however, it was soon realized that these releases could be used as a diagnostic of the ionosphere (e.g., the neutral wind and electric field), as well as a method to study basic plasma physics processes (e.g., ion-neutral interactions, plasma wave generation). Recognizing the scientific value of chemical release experiments, two chemical release missions were carried out by NASA: AMPTE (Active Magnetospheric Particle Tracer Explorers) (Haerendel et al. 1985) and CRRES (Combined Release and Radiation Effects Satellite) (Reasoner 1992).

AMPTE was launched in August, 1984. Lithium and barium were released in the solar wind and in the geomagnetic tail. A key objective of the solar wind releases echoed the original motivation of chemical releases: the generation and evolution of an artificial cometary tail. An interesting and surprising result was that the barium ion cloud did not immediately move in the direction of the solar wind, rather, the cloud initially moved perpendicular to the IMF and solar wind direction. Subsequent analysis indicated that barium ion inertial length was sufficiently large that the Hall term dominated the early evolution of the cloud (Harold and Hassam 1994). A second objective was to trace lithium ions from the solar wind into the magnetosphere. However, analysis of data from the Charge Composition Explorer (CCE) inside the magnetosphere following two lithium releases detected no lithium in the magnetosphere (Krimigis et al. 1986). AMPTE barium releases in the earth's magnetotail revealed complex motions (Mende et al. 1989) and rapid plasma structuring (Bernhardt et al. 1987). The structuring was explained by a new Rayleigh-Taylor instability mediated by the Hall term (Hassam and Huba 1987).

The CRRES satellite was launched in July, 1990 and the chemical releases were from the CRRES satellite itself as well as from a number of sounding rockets. The original mission plan was to launch the satellite from the space shuttle. It would spend 90 days in low Earth orbit (LEO) and perform chemical release experiments. It would then transfer to a geosynchronous orbit and perform more releases. However, because of the 1986 Challenger disaster the mission plan changed and the satellite was inserted into a geosynchronous transfer orbit (GTO). To accommodate the loss of the LEO chemical releases, sounding rockets were used. A number of diverse science objectives were addressed by the mission: critical ion velocity ionization, generation and evolution of diamagnetic cavities, magnetosphere-ionosphere coupling, modification of energetic electron distribution functions, stimulation of magnetospheric waves, and possible inducement of enhanced auroral activity (Reasoner 1992). The bulk of the chemical releases used lithium and barium. However, one rocket experiment used sulfur hexafluoride (SF_6). SF_6 reacts with cold electrons and becomes a negative ion. The idea was to release SF_6 in the post-sunset, bottomside, equatorial ionosphere to create a large electron 'hole' which would act as a trigger for equatorial spread F. Equatorial spread F did occur during this experiment and believed to be triggered by the chemical release (Mendillo, private communication, 2016) but the cause was not unambiguous.

Another chemical that has been used extensively to detect and measure thermospheric winds in the altitude range 80–140 km is trimethylaluminum (TMA). TMA reacts with oxygen to produce chemiluminescence that is easily observable at night. A large database of wind profiles has been developed (Larsen 2002) that show variability in thermospheric winds, as well as strong wind shears in the E-region between 100 and 110 km that are often associated with sporadic E.

Recently the United States Air Force conducted chemical release experiments at the Kwajalein Missile Range to assess their impact on radio wave propagation in the ionosphere. The Metal Oxide Space Cloud (MOSC) experiment was carried out in 2013 in which samarium vapor was released at 170 km and 180 km. The samarium vapor created an ionized plasma cloud through both photoionization and chemical ionization; this cloud created observable signatures in optical sensors, radar backscatter, and HF propagation diagnostics. A major goal of this experiment is to determine the viability of using chemical releases to maintain communications during disturbed ionospheric conditions by suppressing ionospheric instabilities. In fact the Air Force is sponsoring research to develop small satellites known as CubeSats to release chemicals to improve radio frequency communications (Hambling 2016).

In summary, chemical releases have been used for over 50 years to study plasma processes in the space environment and still continue to be used. The two overarching objectives of these experiments are to diagnose the environment (e.g., measure physical quantities such as thermospheric winds and electric fields) and to artificially generate plasma processes to better understand them (e.g., plasma waves, diamagnetic cavities, ion-neutral interactions).

11 Space Debris

Space debris consists of human-made waste, junk, and defunct objects orbiting in space (Klinkrad 2010). The debris consists of objects as large as spent rocket stages, and as small as flakes of paint. Additional debris can be produced by collisions between debris objects. Some debris has been produced by intentional tests of anti-satellite systems. As the use of space has increased dramatically since Sputnik, the amount of human space debris has also increased substantially. The space debris literature often discusses whether the amount of debris, and its interactions with itself, has reached the "Kessler syndrome" (Kessler and Cour-Palais 1978), where the density of debris in some orbital locations becomes high enough that debris collisions will cascade. If so, some orbits could become unusable.

Space debris can be considered an anthropogenic space hazard. But debris is not a space weather phenomenon. Nevertheless, space debris can damage and even destroy space assets, just as can naturally-occurring micrometeoroids. Debris can "sand-blast" and pockmark surfaces in space, from thermal blankets to solar cells and optical telescopes and star trackers. Large debris objects can potentially render an entire mission inoperable through a collision. Debris in the lowest orbits will be de-orbited and burned up in the atmosphere during times of solar maximum when the atmosphere increases at these altitudes, causing increased drag on the debris. The removal of debris from higher orbits, including at geosynchronous where many communications and national security space assets fly, remains an important engineering challenge.

12 Summary Discussion

While only recognized in recent times, human intervention in the Earth's space environment has occurred ever since the decision was made for alternating current as the electrical grid power source. These interventions, both inadvertent (such as the emission of power line radiation into the magnetosphere) and intentional (such as high altitude nuclear explosions) can produce a range of effects that not only resemble, but in many cases are significantly more severe, than what can be expected from naturally occurring space weather events. As described in this paper, some anthropogenic effects have been demonstrated to have the potential for globally destructive results.

The sections in this paper that describe the phenomenology of high altitude nuclear explosions, the actual formation and phenomenology of long-lasting artificial radiation belts, their damaging effects on contemporary satellites and their observable geophysical effects present a picture from the past, but with a magnified contemporary relevance in the context of space weather. These early, unrepeatable experiments remain an important reference point in understanding the phenomena that a disturbed space environment can present—nowadays expected to be the result of solar activity, rather than human intervention.

It is remarkable how the early space pioneers, within the first decade of the space age, were able to make progress in understanding the Earth's space environment, thanks to increasingly sophisticated and targeted instrumentation on a range of Earth satellites, as well as taking advantage of the man-made space weather events. The situation currently is well exemplified by the successes of the detailed observations of the dual Van Allen Probes in the radiation belts, as well as by the measurements of numerous other satellite missions all contributing to building an ever more detailed picture of phenomena in the near-Earth space environment. At the same time, modeling and simulations have progressed to an incomparable extent since the early means at the disposal of the observers and theoreticians of the Starfish events in 1962.

Current human intervention that can affect the space environment is rather more benign, such as using VLF signals as probes of near-Earth phenomena and ionospheric heating for clarifying the complex interactions of the ionosphere with many different wave fields. These experiments are carried out with a reasonable understanding of their potential effects and present only well-contained risks. Other experiments, such as the chemical releases that have been performed in the ionosphere and magnetosphere, have—generally speaking—negligible and evanescent effect on the environment, while contributing in a more limited way, to understanding relatively small-scale dynamic processes.

James Van Allen could rightly claim that the Argus nuclear explosions in 1958 "... undoubtedly constitute the greatest geophysical experiment ever conducted by man. The observations have great significance in understanding the nature of the geomagnetic field, the mechanisms of trapping of particles in the geomagnetic field, the origin of aurora andmagnetic storms, and the density of the very high atmosphere." (Letter from James Van Allen to James R. Killian, Office of the President of the United States, dated February 21, 1959.) It is to be hoped that no high altitude nuclear explosions will be carried out in the future, as the reliance of civilization in this first part of the 21st century on a complex technology infrastructure in the near-Earth space environment has increased enormously when compared to the situation half a century ago. In fact the "experiments" in 1958, originally designed for mitigation purposes, together with the rather more destructive ones in 1962, did bring a better early understanding of the Earth's natural radiation belts. Since that era, near-Earth space research has progressed to a far greater level of sophistication so that such experiments are not only unthinkable, but they would destroy much of the infrastructure of our current civilization.

The most dramatic—and unexpected—anthropogenic space weather phenomenon is the high-altitude nuclear EMP (HEMP). As it can be seen in Fig. 2 a megaton-class nuclear explosion at an altitude of $\sim$400 km produces a very fast electromagnetic pulse. The rise-time of this pulse is about 100 times faster than the rise-time of lightning while the generated electromagnetic fields are comparable ($\sim$50 kV/m). In addition, this pulse covers a continent-size area at the same time. The effect can be devastating for modern society: it can disrupt transportation, energy production and distribution, global and local computer networks and many other aspects of modern life we take for granted. The Hiroshima and Nagasaki nuclear devices had explosive powers of 15 kt and 21 kt, respectively. A relatively crude, but factor of 20 larger device can cause devastating effects in the US or Western Europe. Such a device in the hands of terrorists or a rogue nation would pose a major threat to the world.

While the study of space weather, as a scientific discipline, has deep roots from the time of installation of the first telegraph lines, to the early years of the space age, the subject remains a lively and timely topic that has become essential as part of the global technological infrastructure in space and on the ground. Anthropogenic intervention forms a portion of the past, and a modest component of the present, but it is now rightly dominated by the natural phenomena from the Sun to the Earth.

Acknowledgements The authors thank the International Space Science Institute, Bern, Switzerland and its staff for organizing and supporting the Workshop on the Scientific Foundations of Space Weather that motivated the work in this paper. The work performed at the University of Michigan was supported by National Science Foundation grant AGS-1322543. JDH was supported by NRL Base Funds. Work at the Massachusetts Institute of Technology was sponsored by US National Science Foundation grant AGS-1242204. Work at the University of Colorado/LASP was supported by funding from NASA and the National Science Foundation. The authors thank Vaughn Hoxie, Scot Elkington, Hong Zhao, and Tom Mason for extraordinary efforts in adapting and portraying data from the Explorer XV and Van Allen Probes missions.

References

B. Abel, R.M. Thorne, Electron scattering loss in Earth's inner magnetosphere: 1. Dominant physical processes. J. Geophys. Res. **103**(A2), 2385–2396 (1998). doi:10.1029/97JA02919

L. Allen, J.L. Beavers, W.A. Whitaker, J.A. Welch, R.B. Walton, Project Jason measurement of trapped electrons from a nuclear device by sounding rockets. Proc. Natl. Acad. Sci. USA **45**(8), 1171–1190 (1959)

D.N. Baker, How to cope with space weather. Science **297**(5586), 1486–1487 (2002). doi:10.1126/science.1074956

R.C. Baker, W.M. Strome, Magnetic disturbance from a high-altitude nuclear explosion. J. Geophys. Res. **67**(12), 4927–4928 (1962)

D.N. Baker, R. Balstad, J.M. Bodeau, E. Cameron, J.F. Fennell, G.M. Fisher, K.F. Forbes, P.M. Kintner, L.G. Leffler, W.S. Lewis, J.B. Reagan, A.A. Small III, T.A. Stansell, L. Strachan Jr., Severe Space Weather Events-Understanding Societal and Economic Impacts Workshop Report. Technical report ISBN: 0-309-12770-X, Committee on the Societal and Economic Impacts of Severe Space Weather Events, National Research Council (2008)

D.N. Baker, A.N. Jaynes, V.C. Hoxie, R.M. Thorne, J.C. Foster, X. Li, J.F. Fennell, J.R. Wygant, S.G. Kanekal, P.J. Erickson, W. Kurth, W. Li, Q. Ma, Q. Schiller, L. Blum, D.M. Malaspina, A. Gerrard, L.J. Lanzerotti, An impenetrable barrier to ultrarelativistic electrons in the Van Allen radiation belts. Nature **515**(7528), 531–534 (2014). doi:10.1038/nature13956

D.N. Baker, A.N. Jaynes, S.G. Kanekal, J.C. Foster, P.J. Erickson, J.F. Fennell, J.B. Blake, H. Zhao, X. Li, S.R. Elkington, M.G. Henderson, G.D. Reeves, H.E. Spence, C.A. Kletzing, J.R. Wygant, Highly relativistic radiation belt electron acceleration, transport, and loss: large solar storm events of March and June 2015. J. Geophys. Res. **121**(7), 6647–6660 (2016). doi:10.1002/2016JA022502

R.C. Baumann, Ariel I: The First International Satellite. Technical report NASA SP-43, NASA (1963)

T.F. Bell, H.G. James, U.S. Inan, J.P. Katsufrakis, The apparent spectral broadening of VLF transmitter signals during transionospheric propagation. J. Geophys. Res. **88**(A6), 4813 (1983). doi:10.1029/JA088iA06p04813

P.A. Bernhardt, R.A. Roussel-Dupre, M.B. Pongratz, G. Haerendel, A. Valenzuela, D.A. Gurnett, R.R. Anderson, Observations and theory of the AMPTE magnetotail barium releases. J. Geophys. Res. **92**(A6), 5777–5794 (1987). doi:10.1029/JA092iA06p05777

P.A. Bernhardt, L.M. Duncan, C.A. Tepley, Artificial airglow excited by high-power radio waves. Science **242**(4881), 1022–1027 (1988). doi:10.1126/science.242.4881.1022

P.A. Bernhardt, L.M. Duncan, C.A. Tepley, Heater-induced cavities as optical tracers of plasma drifts. J. Geophys. Res. **94**(A6), 7003–7010 (1989). doi:10.1029/JA094iA06p07003

W.K. Berthold, A.K. Harris, H.J. Hope, World-wide effects of hydromagnetic waves due to Argus. J. Geophys. Res. **65**(8), 2233–2239 (1960)

L. Biermann, Kometenschweife und Solare Korpuskularstrahlung. Z. Astrophys. **29**, 274–286 (1951)

H.A. Bomke, I.A. Balton, H.H. Grote, A.K. Harris, Near and distant observations of the 1962 Johnston Island high-altitude nuclear tests. J. Geophys. Res. **69**(15), 3125–3136 (1964)
H.A. Bomke, A.K. Harris, J.W. Walker, W.J. Ramm, The nature of worldwide geomagnetic disturbances generated by the Starfish explosion of July 9, 1962. J. Geophys. Res. **71**(11), 2777–2789 (1966). doi:10.1029/JZ071i011p02777
S. Breiner, Effect of nuclear detonation on the geomagnetic field at Palo Alto, California. J. Geophys. Res. **68**(1), 335–337 (1963). doi:10.1029/JZ068i001p00335
W.L. Brown, in *Observations of the Transient Behavior of Electrons in the Artificial Radiation Belts*, ed. by B.M. McCormac (Springer, Dordrecht, 1966), pp. 610–633. doi:10.1007/978-94-010-3553-8_44
W.L. Brown, J.D. Gabbe, The electron distribution in the Earth's radiation belts during July 1962 as measured by Telstar. J. Geophys. Res. **68**(3), 607–618 (1963). doi:10.1029/JZ068i003p00607
W.L. Brown, J.D. Gabbe, W. Rosenzweig, Results of the Telstar radiation experiments. Bell Syst. Tech. J. **42**(4), 1505–1560 (1963)
K. Bullough, Satellite observations of power line harmonic radiation. Space Sci. Rev. **35**(2), 175–183 (1983). doi:10.1007/BF00242242
B. Caner, Prompt world-wide geomagnetic effects of high-latitude nuclear explosions, Master's thesis, The University of British Columbia, Vancouver, BC Canada, 1964. https://open.library.ubc.ca/cIRcle/collections/ubctheses/831/items/1.0053563
D.L. Carpenter, Whistler studies of the plasmapause in the magnetosphere: 1. Temporal variations in the position of the knee and some evidence on plasma motions near the knee. J. Geophys. Res. **71**(3), 693–709 (1966). doi:10.1029/JZ071i003p00693
D.L. Carpenter, *Very Low Frequency Space Radio Research at Stanford 1950–1990*, 1st edn. (Lulu.com, Stanford, 2015). ISBN 9781329884106
D. Carpenter, J. Lemaire, The plasmasphere boundary layer. Ann. Geophys. **22**, 4291–4298 (2004)
M. Casaverde, A. Giesecke, R. Cohen, Effects of the nuclear explosion over Johnston Island observed in Peru on July 9, 1962. J. Geophys. Res. **68**(9), 2603–2611 (1963). doi:10.1029/JZ068i009p02603
D.M. Chapin, C.S. Fuller, G.L. Pearson, A new silicon *p-n* junction photocell for converting solar radiation into electrical power. J. Appl. Phys. **25**(5), 676–677 (1954). doi:10.1063/1.1721711
N. Christofilos, The Argus experiment. Proc. Natl. Acad. Sci. **45**, 1144–1152 (1959a)
N.C. Christofilos, The Argus experiment. J. Geophys. Res. **64**(8), 869–875 (1959b). doi:10.1029/JZ064i008p00869
M.A. Clilverd, C.J. Rodger, N.R. Thomson, J.B. Brundell, T. Ulich, J. Lichtenberger, N. Cobbett, A.B. Collier, F.W. Menk, A. Seppälä, P.T. Verronen, E. Turunen, Remote sensing space weather events: Antarctic-Arctic Radiation-Belt (Dynamic) Deposition-VLF Atmospheric Research Konsortium network. Space Weather **7**(4), S04001 (2009). doi:10.1029/2008SW000412
M.B. Cohen, N.G. Lehtinen, U.S. Inan, Models of ionospheric VLF absorption of powerful ground based transmitters. Geophys. Res. Lett. **39**(24), L24101 (2012). doi:10.1029/2012GL054437
S.A. Colgate, The phenomenology of the mass motion of a high altitude nuclear explosion. J. Geophys. Res. **70**(13), 3161–3173 (1965). doi:10.1029/jz070i013p03161
E.E. Conrad, G.A. Gurtman, G. Kweder, M.J. Mandell, W.W. White, Collateral Damage to Satellites from an EMP Attack. Technical report DTRA-IR-10-22, Defense Threat Reduction Agency, Fort Belvoir, Virginia (2010)
A.L. Cullington, A man-made or artifical aurora. Nature **182**(4646), 1365–1366 (1958). doi:10.1038/1821365a0
R.J. Danchik, An overview of transit development. APL Tech. Dig. **1**(1), 18–26 (1998)
R.G. D'Arcy, S.A. Colgate, Measurements at the southern magnetic conjugate region of the fission debris from the Starfish nuclear detonation. J. Geophys. Res. **70**(13), 3147–3159 (1965). doi:10.1029/JZ070i013p03147
A.C. Dickieson, The Telstar experiment. Bell Syst. Tech. J. **42**, 739–746 (1963)
A.C. Durney, H. Elliot, R.J. Hynds, J.J. Quenby, Satellite observations of the energetic particle flux produced by the high-altitude nuclear explosion of July 9, 1962. Nature **195**, 1245–1248 (1962). doi:10.1038/1951245a0
A.C. Durney, H. Elliot, R.J. Hynds, J.J. Quenby, The artificial radiation belt produced by the Starfish nuclear explosion. Proc. R. Soc. Lond. Ser. A, Math. Phys. Sci. **281**(1384), 565–583 (1964)
P. Dyal, Particle and field measurements of the starfish diamagnetic cavity. J. Geophys. Res. **111**(A12), 12211 (2006). doi:10.1029/2006JA011827
P.J. Edwards, J.S. Reid, Effects of nuclear explosion starfish prime observed at Hobart, Tasmania, July 9, 1962. J. Geophys. Res. **69**(17), 3607–3612 (1964). doi:10.1029/JZ069i017p03607
H. Elliot, in *Some Cosmic Ray and Radiation Belt Observations Based on Data from the Anton 302 G-M Counter in Ariel I*, ed. by B.M. McCormac (Springer, Dordrecht, 1966), pp. 76–99. doi:10.1007/978-94-010-3553-8_7

H. Elliot, J.J. Quenby, The Samoan artificial aurora. Nature **83**, 810 (1959). doi:10.1038/183810a0

J.F. Fennell, H.C. Koons, J.L. Roeder, J.B. Blake, Spacecraft charging: observations and relationship to satellite anomalies, in *Spacecraft Charging Technology, Proceedings of the Seventh International Conference*, ed. by R.A. Harris (European Space Agency ESTEC, Noordwijk, 2001), pp. 279–285

J.F. Fennell, S.G. Claudepierre, J.B. Blake, T.P. O'Brien, J.H. Clemmons, D.N. Baker, H.E. Spence, G.D. Reeves, Van Allen probes show that the inner radiation zone contains no MeV electrons: ECT/MagEIS data. Geophys. Res. Lett. **42**(5), 1283–1289 (2015). doi:10.1002/2014GL062874

A. Finkbeiner, *The Jasons: The Secret History of Science's Postwar Elite* (Viking, New York, 2006)

R.E. Fischell, Effect of the artificial radiation belt on solar power systems. APL Tech. Dig. **2**(2), 8–13 (1962a)

R.E. Fischell, The TRAAC satellite. APL Tech. Dig. **1**(3), 2–9 (1962b)

J.C. Foster, T.J. Rosenberg, Electron precipitation and VLF emissions associated with cyclotron resonance interactions near the plasmapause. J. Geophys. Res. **81**(13), 2183–2192 (1976). doi:10.1029/JA081i013p02183

J.S. Foster, E. Gjelde, W.R. Graham, R.J. Hermann, H.M. Kluepfel, R.L. Lawson, G.K. Soper, L.L. Wood, J.B. Woodard, Report of the Commission to Assess the Threat to the United States from Electromagnetic Pulse (EMP) Attack: Executive Report. Technical report, United States Congress, Washington, DC (2004)

J.S. Foster, E. Gjelde, W.R. Graham, R.J. Hermann, H.M. Kluepfel, R.L. Lawson, G.K. Soper, L.L. Wood, J.B. Woodard, Report of the Commission to Assess the Threat to the United States from Electromagnetic Pulse (EMP) Attack: Critical National Infrastructures. Technical report A2473, United States Congress, Washington, DC (2008)

J.C. Foster, P.J. Erickson, D.N. Baker, A.N. Jaynes, E.V. Mishin, J.F. Fennel, X. Li, M.G. Henderson, S.G. Kanekal, Observations of the impenetrable barrier, the plasmapause, and the VLF bubble during the 17 March 2015 storm. J. Geophys. Res. Space Phys. **121**(6), 5537–5548 (2016). doi:10.1002/2016JA022509

A.C. Fraser-Smith, A weekend increase in geomagnetic activity. J. Geophys. Res. **84**(A5), 2089–2096 (1979). doi:10.1029/JA084iA05p02089

A.C. Fraser-Smith, Effects of man on geomagnetic activity and pulsations. Adv. Space Res. **1**(2), 455–466 (1981). doi:10.1016/0273-1177(81)90321-5

A.C. Fraser-Smith, D.B. Coates, Large-amplitude ULF electromagnetic fields from bart. Radio Sci. **13**(4), 661–668 (1978). doi:10.1029/RS013i004p00661

J.F. Gabites, D.S. Rowles, Summary of visual observations of the aurora following the nuclear explosion above Johnston island on 9 July 1962. N.Z. J. Geol. Geophys. **5**(6), 920–924 (1962). doi:10.1080/00288306.1962.10420041

Y.I. Galperin, A.D. Boliunova, Recording of effects of high-altitude thermonuclear explosion of July 9, 1962, on the Cosmos 5 satellite. Kosm. Issled. (Cosm. Res.) **2**(5), 763–772 (1964)

L.A. Gebhard, Evolution of Naval Radio-Electronics and Contributions of the Naval Research Laboratory. Technical report, Naval Research Laboratory, Washington, DC (1979)

J. Gilbert, J. Kapperman, W. Radasky, E. Savage, The Late Time (E3) High-Altitude Electromagnetic Pulse (HEMP) and Its Impact on the US Power Grid. Technical report Meta-R-321, Metatech Corporation, Goleta, California (2010)

J.L. Green, S. Boardsen, L. Garcia, W.W.L. Taylor, S.F. Fung, B.W. Reinisch, On the origin of whistler mode radiation in the plasmasphere. J. Geophys. Res. **110**(A3), 03201 (2005). doi:10.1029/2004JA010495

G. Haerendel, A. Valenzuela, O.H. Bauer, M. Ertl, H. Foppl, K.-H. Kaiser, W. Lieb, J. Loidl, F. Melzner, B. Merz, H. Neuss, P. Parigger, E. Rieger, R. Schoning, J. Stocker, E. Wiezorrek, E. Molona, The Li/Ba release experiments of the ion release module. IEEE Trans. Geosci. Remote Sens. **GE-23**(3), 253–258 (1985). doi:10.1109/TGRS.1985.289523

D. Hambling, US Air Force wants to plasma bomb the sky using tiny satellites. New Sci. (August 20, 2016)

J.B. Harold, A.B. Hassam, Two ion fluid numerical investigations of solar wind gas releases. J. Geophys. Res. **99**(A10), 19325–19340 (1994). doi:10.1029/94JA00790

A.B. Hassam, J.D. Huba, Structuring of the AMPTE magnetotail barium releases. Geophys. Res. Lett. **14**(1), 60–63 (1987). doi:10.1029/GL014i001p00060

R.A. Helliwell, *Whistlers and Related Ionospheric Phenomena*, 1st edn. (Stanford University Press, Stanford, 1965). ISBN 0486445720

R.A. Helliwell, VLF wave stimulation experiments in the magnetosphere from Siple Station, Antarctica. Rev. Geophys. **26**(3), 551 (1988). doi:10.1029/RG026i003p00551

R. Helliwell, E. Gehrels, Observations of magneto-ionic duct propagation using man-made signals of very low frequency. Proc. Inst. Radio Eng. **46**(4), 785–787 (1958)

R.A. Helliwell, J.P. Katsufrakis, M.L. Trimpi, Whistler-induced amplitude perturbation in VLF propagation. J. Geophys. Res. **78**(22), 4679–4688 (1973). doi:10.1029/JA078i022p04679

R.A. Helliwell, J.P. Katsufrakis, T.F. Bell, R. Raghuram, VLF line radiation in the Earth's magnetosphere and its association with power system radiation. J. Geophys. Res. **80**(31), 4249–4258 (1975). doi:10.1029/JA080i031p04249

W.N. Hess, The artificial radiation belt made on July 9, 1962. J. Geophys. Res. **68**(3), 667–683 (1963). doi:10.1029/JZ068i003p00667

W.N. Hess, P. Nakada, Artificial radiation belt discussed in symposium at Goddard Space Center. Science **138**(3536), 53–54 (1962)

R.L. Heyborne, R.L. Smith, R.A. Helliwell, Latitudinal cutoff of VLF signals in the ionosphere. J. Geophys. Res. **74**(9), 2393–2397 (1969). doi:10.1029/JA074i009p02393

R.B. Horne, M. Lam, J.C. Green, Energetic electron precipitation from the outer radiation belt during geomagnetic storms. Geophys. Res. Lett. **36**(19), L19104 (2009). doi:10.1029/2009gl040236

W.L. Imhof, H.D. Voss, M. Walt, E.E. Gaines, J. Mobilia, D.W. Datlowe, J.B. Reagan, Slot region electron precipitation by lightning, VLF chorus, and plasmaspheric hiss. J. Geophys. Res. **91**(A8), 8883 (1986). doi:10.1029/JA091iA08p08883

U.S. Inan, R.A. Helliwell, DE-1 observations of VLF transmitter signals and wave-particle interactions in the magnetosphere. Geophys. Res. Lett. **9**(9), 917–920 (1982). doi:10.1029/GL009i009p00917

U.S. Inan, T.F. Bell, D.L. Carpenter, R.R. Anderson, Explorer 45 and Imp 6 observations in the magnetosphere of injected waves from the Siple Station VLF transmitter. J. Geophys. Res. **82**(7), 1177–1187 (1977). doi:10.1029/JA082i007p01177

U.S. Inan, T.F. Bell, H.C. Chang, Particle precipitation induced by short-duration VLF waves in the magnetosphere. J. Geophys. Res. **87**(A8), 6243 (1982). doi:10.1029/JA087iA08p06243

U.S. Inan, H.C. Chang, R.A. Helliwell, Electron precipitation zones around major ground-based VLF signal sources. J. Geophys. Res. **89**(A5), 2891 (1984). doi:10.1029/JA089iA05p02891

U.S. Inan, H.C. Chang, R.A. Helliwell, W.L. Imhof, J.B. Reagan, M. Walt, Precipitation of radiation belt electrons by man-made waves: a comparison between theory and measurement. J. Geophys. Res. **90**(A1), 359–369 (1985). doi:10.1029/JA090iA01p00359

U.S. Inan, J.V. Rodriguez, S. Lev-Tov, J. Oh, ionospheric modification with a VLF transmitter. Geophys. Res. Lett. **19**(20), 2071–2074 (1992). doi:10.1029/92GL02378

U.S. Inan, T.F. Bell, J. Bortnik, J.M. Albert, Controlled precipitation of radiation belt electrons. J. Geophys. Res. **108**(A5), 051186 (2003). doi:10.1029/2002JA009580

A.N. Jaynes, D.N. Baker, H.J. Singer, J.V. Rodriguez, T.M. Loto'aniu, A.F. Ali, S.R. Elkington, X. Li, S.G. Kanekal, S.G. Claudepierre, J.F. Fennell, W. Li, R.M. Thorne, C.A. Kletzing, H.E. Spence, G.D. Reeves, Source and seed populations for relativistic electrons: their roles in radiation belt changes. J. Geophys. Res. **120**(9), 7240–7254 (2015). doi:10.1002/2015JA021234

C.B. Jones, M.K. Doyle, L.H. Berkhouse, F.S. Calhoun, E.J. Martin, Operation ARGUS 1958, Technical report DNA 6039F, Defense Nuclear Agency, Washington, DC (1982)

S.L. Kahalas, P. Newman, Interpretation of early magnetic transients caused by high-altitude nuclear detonations. J. Res. Natl. Bur. Stand. D **69**, 1179–1183 (1965)

A. Karinen, K. Mursula, T. Ulich, J. Manninen, Does the magnetosphere behave differently on weekends? Ann. Geophys. **20**(8), 1137–1142 (2002). doi:10.5194/angeo-20-1137-2002

W.J. Karzas, R. Latter, Electromagnetic radiation from a nuclear explosion in space. Phys. Rev. **126**, 1919–1926 (1962). doi:10.1103/PhysRev.126.1919

P.J. Kellogg, E.P. Ney, J.R. Winckler, Geophysical effects associated with high-altitude explosions. Nature **183**(4658), 358–361 (1959). doi:10.1038/183358a0

D.J. Kessler, B.G. Cour-Palais, Collision frequency of artificial satellites: the creation of a debris belt. J. Geophys. Res. **83**(A6), 2637–2646 (1978). doi:10.1029/JA083iA06p02637

G. Klawitter, K. Herold, M. Oexner, *Langwellen- und Längstwellenfunk*, 3rd edn. (Siebel: Verlag für Technik und Handwerk, Amazon.com, 2000). ISBN 3896320432

C.A. Kletzing, W.S. Kurth, M. Acuna, R.J. MacDowall, R.B. Torbert, T. Averkamp, D. Bodet, S.R. Bounds, M. Chutter, J. Connerney, D. Crawford, J.S. Dolan, R. Dvorsky, G.B. Hospodarsky, J. Howard, V. Jordanova, R.A. Johnson, D.L. Kirchner, B. Mokrzycki, G. Needell, J. Odom, D. Mark, R. Pfaff, J.R. Phillips, C.W. Piker, S.L. Remington, D. Rowland, O. Santolik, R. Schnurr, D. Sheppard, C.W. Smith, R.M. Thorne, J. Tyler, The Electric and Magnetic Field Instrument Suite and Integrated Science (EMFISIS) on RBSP, in *The Van Allen Probes Mission* (Springer, Boston, 2013), pp. 127–181. doi:10.1007/978-1-4899-7433-4_5

H. Klinkrad, *Space Debris* (Wiley, New York, 2010). doi:10.1002/9780470686652.eae325

H.C. Koons, B.C. Edgar, A.L. Vampola, Precipitation of inner zone electrons by whistler mode waves from the VLF transmitters UMS and NWC. J. Geophys. Res. **86**(A2), 640 (1981). doi:10.1029/JA086iA02p00640

S.M. Krimigis, G. Haerendel, G. Gloeckler, R.W. Mcentire, E.G. Shelley, R.B. Decker, G. Paschmann, A. Valenzuela, T.A. Potemra, F.L. Scarf, A.L. Brinca, H. Lühr, AMPTE lithium tracer releases in the solar wind: observations inside the magnetosphere. J. Geophys. Res. **91**(A2), 1339–1353 (1986). doi:10.1029/JA091iA02p01339

P. Kulkarni, U.S. Inan, T.F. Bell, J. Bortnik, Precipitation signatures of ground-based VLF transmitters. J. Geophys. Res. Space Phys. **113**(A7), A07214 (2008). doi:10.1029/2007JA012569

M.F. Larsen, Winds and shears in the mesosphere and lower thermosphere: results from four decades of chemical release wind measurements. J. Geophys. Res. **107**(A8), 28–12814 (2002). doi:10.1029/2001JA000218

J.A. Lawrie, V.B. Gerard, P.J. Gill, Magnetic effects resulting from the Johnston island high altitude nuclear explosions. N.Z. J. Geol. Geophys. **4**(2), 109–124 (1961). doi:10.1080/00288306.1961.10423131

J. Leiphart, R. Zeek, L. Bearce, E. Toth, Penetration of the ionosphere by very-low-frequency radio signals-interim results of the LOFTI I experiment. Proc. IRE **50**(1), 6–17 (1962). doi:10.1109/JRPROC.1962.288269

X. Li, R.S. Selesnick, D.N. Baker, A.N. Jaynes, S.G. Kanekal, Q. Schiller, L. Blum, J. Fennell, J.B. Blake, Upper limit on the inner radiation belt MeV electron intensity. J. Geophys. Res. **120**(2), 1215–1228 (2015). doi:10.1002/2014JA020777

C.L. Longmire, Justification and Verification of High-Altitude EMP Theory: Part I. Technical report Technical Note 368, Mission Research Corporation, Santa Barbara, California (1986)

J.P. Luette, C.G. Park, R.A. Helliwell, The control of the magnetosphere by power line radiation. J. Geophys. Res. **84**(A6), 2657–2660 (1979). doi:10.1029/JA084iA06p02657

R. Lüst, in *Barium Cloud Experiments in the Upper Atmosphere*, ed. by J.A.M. Bleeker, J. Geiss, M.C.E. Huber (Springer, Dordrecht, 2001), pp. 179–187. doi:10.1007/978-94-010-0320-9_6

H. Maeda, Geomagnetic disturbances due to nuclear explosion. J. Geophys. Res. **64**(7), 863–864 (1959). doi:10.1029/JZ064i007p00863

B.H. Mauk, N.J. Fox, S.G. Kanekal, R.L. Kessel, D.G. Sibeck, A. Ukhorskiy, Science objectives and rationale for the Radiation Belt Storm Probes mission. Space Sci. Rev. **179**(1–4), 3–27 (2013). doi:10.1007/s11214-012-9908-y

J.S. Mayo, H. Mann, F.J. Witt, D.S. Peck, H.K. Gummel, W.L. Brown, The command system malfunction. Bell Syst. Tech. J. **42**, 1631–1657 (1963)

C.E. McIlwain, Coordinates for mapping the distribution of magnetically trapped particles. J. Geophys. Res. **66**(11), 3681–3691 (1961). doi:10.1029/JZ066i011p03681

C.E. McIlwain, The radiation belts, natural and artificial. Science **142**(3590), 355–361 (1963). doi:10.1126/science.142.3590.355

K.G. McKay, A germanium counter. Phys. Rev. **76**, 1537 (1949). doi:10.1103/PhysRev.76.1537

R.R. Meier, M.H. Stevens, J.M.C. Plane, J.T. Emmert, G. Crowley, I. Azeem, L.J. Paxton, A.B. Christensen, A study of space shuttle plumes in the lower thermosphere. J. Geophys. Res. **116**(A12), 12322 (2011). doi:10.1029/2011JA016987

S.B. Mende, G.R. Swenson, S.P. Geller, J.H. Doolittle, G. Haerendel, A. Valenzuela, O.H. Bauer, Dynamics of a barium release in the magnetospheric tail. J. Geophys. Res. **94**(A12), 17063–17083 (1989). doi:10.1029/JA094iA12p17063

M. Mendillo, The effect of rocket launches on the ionosphere. Adv. Space Res. **1**(2), 275–290 (1981). doi:10.1016/0273-1177(81)90302-1

M. Mendillo, J. Baumgardner, D.P. Allen, J. Foster, J. Holt, G.R.A. Ellis, A. Klekociuk, G. Reber, Spacelab-2 plasma depletion experiments for ionospheric and radio astronomical studies. Science **238**(4831), 1260–1264 (1987). doi:10.1126/science.238.4831.1260

D.P. Miles, R.P. Lepping, Magnetic disturbances due to the high-altitude nuclear explosion of July 9, 1962. J. Geophys. Res. **69**(3), 547–548 (1964). doi:10.1029/JZ069i003p00547

S. Millman (ed.), *A History of Engineering and Science in the Bell System: Physical Sciences (1925–1980)* (Bell Telephone Laboratories, New Jersey, 1983)

O. Molchanov, A. Rozhnoi, M. Solovieva, O. Akentieva, J.J. Berthelier, M. Parrot, F. Lefeuvre, P.F. Biagi, L. Castellana, M. Hayakawa, Global diagnostics of the ionospheric perturbations related to the seismic activity using the VLF radio signals collected on the DEMETER satellite. Nat. Hazards Earth Syst. Sci. **6**(5), 745–753 (2006)

R.C. Moore, U.S. Inan, T.F. Bell, E.J. Kennedy, ELF waves generated by modulated HF heating of the auroral electrojet and observed at a ground distance of ~4400 km. J. Geophys. Res. **112**(A5), 05309 (2007). doi:10.1029/2006JA012063

B.J. O'Brien, C.D. Laughlin, J.A. Van Allen, Geomagnetically trapped radiation produced by a high-altitude nuclear explosion on July 9, 1962. Nature **195**(4845), 939–943 (1962a). doi:10.1038/195939a0

B.J. O'Brien, C.D. Laughlin, J.A. Van Allen, L.A. Frank, Measurements of the intensity and spectrum of electrons at 1000-kilometer altitude and high latitudes. J. Geophys. Res. **67**(4), 1209–1225 (1962b). doi:10.1029/JZ067i004p01209

Y. Omura, D. Nunn, H. Matsumoto, M.J. Rycroft, A review of observational, theoretical, and numerical studies of VLF triggered emissions. J. Atmos. Terr. Phys. **53**(5), 351–368 (1991)
K. Papadopoulos, A.S. Sharma, C.L. Chang, On the efficient operation of a plasma ELF antenna driven by modulation of ionospheric currents. Comments Plasma Phys. Control. Fusion **13**, 1 (1989)
C.G. Park, R.A. Helliwell, Whistler precursors: a possible catalytic role of power line radiation. J. Geophys. Res. **82**(25), 3634–3642 (1977). doi:10.1029/JA082i025p03634
C.G. Park, T.R. Miller, Sunday decreases in magnetospheric VLF wave activity. J. Geophys. Res. **84**(A3), 943–950 (1979). doi:10.1029/JA084iA03p00943
M. Parrot, World map of ELF/VLF emissions as observed by a low-orbiting satellite. Ann. Geophys., Atmos. Hydrospheres Space Sci. **8**(2), 135–146 (1990)
M. Parrot, Observations of power line harmonic radiation by the low-altitude AUREOL 3 satellite. J. Geophys. Res. **99**(A3), 3961–3969 (1994). doi:10.1029/93JA02544
M. Parrot, Y. Zaslavski, Physical mechanisms of man-made influences on the magnetosphere. Surv. Geophys. **17**(1), 67–100 (1996). doi:10.1007/BF01904475
M. Parrot, J.A. Sauvaud, J.J. Berthelier, J.P. Lebreton, First in-situ observations of strong ionospheric perturbations generated by a powerful VLF ground-based transmitter. Geophys. Res. Lett. **34**(11), 11111 (2007). doi:10.1029/2007GL029368
T.R. Pedersen, E.A. Gerken, Creation of visible artificial optical emissions in the aurora by high-power radio waves. Nature **433**(7025), 498–500 (2005). doi:10.1038/nature03243
T. Pedersen, B. Gustavsson, E. Mishin, E. MacKenzie, H.C. Carlson, M. Starks, T. Mills, Optical ring formation and ionization production in high-power HF heating experiments at HAARP. Geophys. Res. Lett. **36**(18), 18107 (2009). doi:10.1029/2009GL040047
G.F. Pieper, Injun: a radiation research satellite. APL Tech. Dig. **1**(1), 3–7 (1961)
G.F. Pieper, The artificial radiation belt. APL Tech. Dig. **2**(2), 3–7 (1962)
G.F. Pieper, A second radiation belt from the July 9, 1962, nuclear detonation. J. Geophys. Res. **68**(3), 651–655 (1963). doi:10.1029/JZ068i003p00651
P.R. Pisharoty, Geomagnetic disturbances associated with the nuclear explosion of July 9. Nature **196**, 822–824 (1962). doi:10.1038/196822b0
R. Raghuram, T.F. Bell, R.A. Helliwell, J.P. Katsufrakis, A quiet band produced by VLF transmitter signals in the magnetosphere. Geophys. Res. Lett. **4**(5), 199–202 (1977). doi:10.1029/GL004i005p00199
K. Rastani, U.S. Inan, R.A. Helliwell, DE 1 observations of siple transmitter signals and associated sidebands. J. Geophys. Res. **90**(A5), 4128 (1985). doi:10.1029/JA090iA05p04128
D.L. Reasoner, Chemical-release mission of CRRES. J. Spacecr. Rockets **29**(4), 580–584 (1992). doi:10.2514/3.25502
C.S. Roberts, Coordinates for the study of particles trapped in the Earth's magnetic field: a method of converting from B, L to R, λ coordinates. J. Geophys. Res. **69**(23), 5089–5090 (1964). doi:10.1029/JZ069i023p05089
C.J. Rodger, M.A. Clilverd, T. Ulich, P.T. Verronen, E. Turunen, N.R. Thomson, The atmospheric implications of radiation belt remediation. Ann. Geophys. **24**(7), 2025–2041 (2006). doi:10.5194/angeo-24-2025-2006
J. Roquet, R. Schlich, E. Selzer, Evidence of two distinct synchronous world impetuses for the magnetic effects of the nuclear high-altitude detonation of July 9, 1962. J. Geophys. Res. **68**(12), 3731–3732 (1963). doi:10.1029/JZ068i012p03731
W. Rosenzweig, H.K. Gummel, F.M. Smits, Solar cell degradation under 1 MeV electron bombardment. Bell Syst. Tech. J. **42**(2), 399–414 (1963)
J.A. Sauvaud, T. Moreau, R. Maggiolo, J.-P. Treilhou, C. Jacquey, A. Cros, J. Coutelier, J. Rouzaud, E. Penou, M. Gangloff, High-energy electron detection onboard DEMETER: the IDP spectrometer, description and first results on the inner belt. Planet. Space Sci. **54**(5), 502–511 (2006). doi:10.1016/j.pss.2005.10.019
J.-A. Sauvaud, R. Maggiolo, C. Jacquey, M. Parrot, J.-J. Berthelier, R.J. Gamble, C.J. Rodger, Radiation belt electron precipitation due to VLF transmitters: satellite observations. Geophys. Res. Lett. **35**(9), 09101 (2008). doi:10.1029/2008GL033194
E. Savage, J. Gilbert, W. Radasky, The Early Time (E1) High-Altitude Electromagnetic Pulse (HEMP) and Its Impact on the US Power Grid. Technical report Meta-R-320, Metatech Corporation, Goleta, California (2010)
R.R. Scarabucci, *Interpretation of VLF Signals Observed on the OGO-4 Satellite* (Stanford University, Stanford, 1969)
R.L. Smith, Propagation characteristics of whistlers trapped in field-aligned columns of enhanced ionization. J. Geophys. Res. **66**(11), 3699–3707 (1961). doi:10.1029/JZ066i011p03699
A.J. Smith, M.A. Clilverd, Magnetic storm effects on the mid-latitude plasmasphere. Planet. Space Sci. **39**(7), 1069–1079 (1991). doi:10.1016/0032-0633(91)90114-P

V.S. Sonwalkar, U.S. Inan, Measurements of siple transmitter signals on the DE 1 satellite: wave normal direction and antenna effective length. J. Geophys. Res. **91**(A1), 154 (1986). doi:10.1029/JA091iA01p00154
V.S. Sonwalkar, U.S. Inan, T.F. Bell, R.A. Helliwell, V.M. Chmyrev, Y.P. Sobolev, O.Y. Ovcharenko, V. Selegej, Simultaneous observations of VLF ground transmitter signals on the DE 1 and COSMOS 1809 satellites: detection of a magnetospheric caustic and a duct. J. Geophys. Res. **99**(A9), 17511 (1994). doi:10.1029/94JA00866
M.J. Starks, R.A. Quinn, G.P. Ginet, J.M. Albert, G.S. Sales, B.W. Reinisch, P. Song, Illumination of the plasmasphere by terrestrial very low frequency transmitters: model validation. J. Geophys. Res. Space Phys. **113**(A9), A09320 (2008). doi:10.1029/2008JA013112
M.J. Starks, T.F. Bell, R.A. Quinn, U.S. Inan, D. Piddyachiy, M. Parrot, Modeling of Doppler-shifted terrestrial VLF transmitter signals observed by DEMETER. Geophys. Res. Lett. **36**(12), 12103 (2009). doi:10.1029/2009GL038511
A.V. Streltsov, M. Gołkowski, U.S. Inan, K.D. Papadopoulos, Propagation of whistler mode waves with a modulated frequency in the magnetosphere. J. Geophys. Res. **115**(A9), 09209 (2010). doi:10.1029/2009JA015155
B.T. Tsurutani, R.M. Thorne, A skeptic's view of PLR effects in the magnetosphere. Adv. Space Res. **1**(2), 439–444 (1981). doi:10.1016/0273-1177(81)90318-5
B.T. Tsurutani, E.J. Smith, S.R. Church, R.M. Thorne, R.E. Holzer, in *Does ELF Chorus Show Evidence of Power Line Stimulation?* ed. by P.J. Palmadesso, K. Papadopoulos (Springer, Dordrecht, 1979), pp. 51–54. doi:10.1007/978-94-009-9500-0_5
R.R. Unterberger, P.E. Byerly, Magnetic effects of a high-altitude nuclear explosion. J. Geophys. Res. **67**(12), 4929–4932 (1962). doi:10.1029/JZ067i012p04929
A.L. Vampola, Electron precipitation in the vicinity of a VLF transmitter. J. Geophys. Res. **92**(A5), 4525 (1987). doi:10.1029/JA092iA05p04525
A.L. Vampola, In-situ observations of magnetospheric electron scattering by a VLF transmitter. J. Atmos. Terr. Phys. **52**(5), 377–384 (1990). doi:10.1016/0021-9169(90)90106-W
J.A. Van Allen, The geomagnetically trapped corpuscular radiation. J. Geophys. Res. **64**(11), 1683–1689 (1959). doi:10.1029/JZ064i011p01683
J.A. Van Allen, Lifetimes of geomagnetically trapped electrons of several MeV energy. Nature **203**(4949), 1006–1007 (1964). doi:10.1038/2031006a0.
J.A. Van Allen, in *Spatial Distribution and Time Decay of the Intensities of Geomagnetically Trapped Electrons from the High Altitude Nuclear Burst of July 1962*, ed. by B.M. McCormac (Springer, Dordrecht, 1966), pp. 575–592. doi:10.1007/978-94-010-3553-8_42
J.A. Van Allen, in *Energetic Particles in the Earth's External Magnetic Field*, ed. by C.S. Gillmor, J.R. Spreiter (American Geophysical Union, Washington, 1997), pp. 235–251. doi:10.1029/HG007p0235
J.A. Van Allen, L.A. Frank, Radiation around the Erth to a radial distance of 107,400 km. Nature **183**(4659), 430–434 (1959). doi:10.1038/183430a0
J.A. Van Allen, G.H. Ludwig, E.C. Ray, C.E. McIlwain, Observation of high intensity radiation by satellites 1958 alpha and gamma (Explorers I and III). Jet Propuls. **28**(9), 588–592 (1958). doi:10.2514/8.7396
J.A. Van Allen, C.E. McIlwain, G.H. Ludwig, Satellite observations of electrons artificially injected into the geomagnetic field. Proc. Natl. Acad. Sci. USA **45**(8), 1152–1171 (1959a)
J.A. Van Allen, C.E. McIlwain, G.H. Ludwig, Radiation observations with satellite 1958 ε. J. Geophys. Res. **64**(3), 271–286 (1959b). doi:10.1029/JZ064i003p00271
J.A. Van Allen, C.E. McIlwain, G.H. Ludwig, Satellite observations of electrons artificially injected into the geomagnetic field. J. Geophys. Res. **64**(8), 877–891 (1959c). doi:10.1029/JZ064i008p00877
J.A. Van Allen, L.A. Frank, B.J. O'Brien, Satellite observations of the artificial radiation belt of July 1962. J. Geophys. Res. **68**(3), 619–627 (1963). doi:10.1029/JZ068i003p00619
C.N. Vittitoe, Did high-altitude EMP cause the Hawaiian streetlight incident? System Design and Assessment Notes (1989)
J. Wait, Propagation of ELF electromagnetic waves and project Sanguine/Seafarer. IEEE J. Ocean. Eng. **2**(2), 161–172 (1977). doi:10.1109/JOE.1977.1145337
M. Walt, The effects of atmospheric collisions on geomagnetically trapped electrons. J. Geophys. Res. **69**(19), 3947–3958 (1964). doi:10.1029/jz069i019p03947.
M. Walt, in *From Nuclear Physics to Space Physics by Way of High Altitude Nuclear Tests*, ed. by C.S. Gillmor, J.R. Spreiter (American Geophysical Union, Washington 1997), pp. 253–263. doi:10.1029/HG007p0253
E.P. Wenaas, Spacecraft Charging Effects on Satellites Following Starfish Events. Technical report RE-78-2044-057, JAYCOR, Alexandria, Virginia (1978)
Wikipedia Contributors, *High-altitude nuclear explosion* (Wikipedia, The Free Encyclopedia, 2016)
D.J. Williams, J.F. Arens, L.J. Lanzerotti, Observations of trapped electrons at low and high altitudes. J. Geophys. Res. **73**(17), 5673–5696 (1968). doi:10.1029/ja073i017p05673

G. Xin, F. Zhan-zu, C. Xin-yu, Y. Sheng-sheng, Z. Lei, Performance evaluation and prediction of single-junction and triple-junction GaAs solar cells induced by electron and proton irradiations. IEEE Trans. Nucl. Sci. **61**(4), 1838–1842 (2014). doi:10.1109/TNS.2014.2306991

K.A. Zawdie, J.D. Huba, D.P. Drob, P.A. Bernhardt, A coupled ionosphere-raytrace model for high-power HF heating. Geophys. Res. Lett. **42**(22), 9650–9656 (2015). doi:10.1002/2015GL066673

A.J. Zmuda, B.W. Shaw, C.R. Haave, VLF disturbances caused by the nuclear detonation of October 26, 1962. J. Geophys. Res. **68**(13), 4105–4114 (1963). doi:10.1029/JZ068i013p04105

MIX
Papier aus verantwortungsvollen Quellen
Paper from responsible sources
FSC® C105338

If you have any concerns about our products,
you can contact us on
ProductSafety@springernature.com

In case Publisher is established outside the EU,
the EU authorized representative is:
Springer Nature Customer Service Center GmbH
Europaplatz 3, 69115 Heidelberg, Germany

Printed by Libri Plureos GmbH
in Hamburg, Germany